Physiology

Physiology

Edited by

Nicholas Sperelakis, Ph.D.

Professor and Chairman
Department of Physiology and Biophysics
University of Cincinnati College of Medicine
Cincinnati, Ohio

Robert O. Banks, Ph.D.

Professor
Department of Physiology and Biophysics
University of Cincinnati College of Medicine
Cincinnati, Ohio

LITTLE, BROWN AND COMPANY   Boston/Toronto/London

Library of Congress Cataloging-in-Publication Data
Physiology : essentials of basic science / edited by
  Nicholas Sperelakis, Robert O. Banks.—1st ed.
      p.   cm.
  Includes bibliographical references and index.
  ISBN 0-316-80629-3
  1. Human physiology.   I. Sperelakis, Nicholas,
1930– . II. Banks, Robert O., 1939– .
  [DNLM: 1. Physiology.   QT 104 P5775]
QP34.5.P4967   1993
612—dc20
DNLM/DLC
for Library of Congress   92-23669
                                               CIP

Printed in the United States of America

MV-NY

Contents

Contents

Contents

Contributing Authors

Douglas K. Anderson, Ph.D.
Professor, Department of Neurology, University of Cincinnati College of Medicine; Research Physiologist, Medical Research Service, Veterans Administration Hospital, Cincinnati, Ohio

Gary L. Anderson, Ph.D.
Assistant Professor, Department of Physiology and Biophysics, University of Louisville School of Medicine, Louisville, Kentucky

Robert O. Banks, Ph.D.
Professor, Department of Physiology and Biophysics, University of Cincinnati College of Medicine, Cincinnati, Ohio

Frank C. Barone, Ph.D.
Associate Fellow, Department of Pharmacology, SmithKline Beecham Pharmaceuticals, King of Prussia, Pennsylvania

Michael M. Behbehani, Ph.D.
Professor, Department of Physiology and Biophysics, University of Cincinnati College of Medicine, Cincinnati, Ohio

Michael A. Berk, M.D.
Clinical Instructor, Department of Internal Medicine, Washington University School of Medicine; Attending Physician, Barnes Hospital, St. Louis, Missouri

Alvin S. Blaustein, M.D.
Associate Professor of Medicine, University of Cincinnati College of Medicine; Chief, Cardiac Non-Invasive Diagnostic Laboratory, Veterans Administration Hospital, Cincinnati, Ohio

Barbara M. Campaigne, Ph.D.
Adjunct Assistant Professor, Department of Physiology and Biophysics, University of Cincinnati College of Medicine; Research Scientist, Division of Cardiology, Children's Hospital Medical Center, Cincinnati, Ohio

John Cuppoletti, Ph.D.
Associate Professor, Department of Physiology and Biophysics, University of Cincinnati College of Medicine, Cincinnati, Ohio

Lawrence M. Dolan, M.D.
Associate Professor, Department of Pediatrics, University of Cincinnati College of Medicine; Attending Physician, Children's Hospital Medical Center, Cincinnati, Ohio

William C. Farr, M.D., Ph.D.
Professor of Family and Community Medicine, University of Arizona College of Medicine, Tucson, Arizona

Donald G. Ferguson
Research Associate Professor, Department of Physiology and Biophysics, University of Cincinnati College of Medicine, Cincinnati, Ohio

Joseph D. Fondacaro, Ph.D.
Senior Director, Department of Sponsored Research, Marion Merrell Dow, Kansas City, Missouri

Ernest C. Foulkes, Ph.D.
Professor, Department of Environmental Health, University of Cincinnati College of Medicine, Cincinnati, Ohio

Lawrence A. Frohman, M.D.
Professor of Medicine, and Director, Division of Metabolism and Endocrinology, University of Cincinnati College of Medicine; Director, Division of Endocrinology and Metabolism, University Hospital, Cincinnati, Ohio

John H. Galla, M.D.
Professor of Medicine, University of Cincinnati College of Medicine; Director, Division of Nephrology and Hypertension, University of Cincinnati Medical Center, Cincinnati, Ohio

Gunter Grupp, M.D., Ph.D.
Professor of Experimental Medicine, Department of Physiology and Biophysics, University of Cincinnati College of Medicine, Cincinnati, Ohio

Ingrid L. Grupp, M.D.
Associate Professor, Department of Pharmacology and Cell Biophysics, University of Cincinnati College of Medicine, Cincinnati, Ohio

Patrick D. Harris, Ph.D.
Professor and Chairman, Department of Physiology and Biophysics, University of Louisville School of Medicine, Louisville, Kentucky

Judith A. Heiny, Ph.D.
Associate Professor, Department of Physiology and Biophysics, University of Cincinnati College of Medicine, Cincinnati, Ohio

James E. Heubi, M.D.
Professor, Department of Pediatrics, University of Cincinnati College of Medicine; Director, Clinical Research Center, and Staff Physician, Division of Gastroenterology and Nutrition, Children's Hospital Medical Center, Cincinnati, Ohio

Robert F. Highsmith, Ph.D.
Professor, Department of Physiology and Biophysics, University of Cincinnati College of Medicine, Cincinnati, Ohio

Brian D. Hoit, M.D.
Assistant Professor of Medicine, University of Cincinnati College of Medicine; Staff Cardiologist, University Hospital, Cincinnati, Ohio

Harriet S. Iwamoto
Associate Professor of Pediatrics, and Physiology and Biophysics, University of Cincinnati College of Medicine; Division of Neonatology, Children's Hospital Medical Center, Cincinnati, Ohio

Shahrokh Javaheri, M.D.
Professor of Medicine, University of Cincinnati College of Medicine; Attending Physician, Veterans Administration Hospital, Cincinnati, Ohio

Ira R. Josephson, Ph.D.
Assistant Professor, Department of Physiology and Biophysics, University of Cincinnati College of Medicine, Cincinnati, Ohio

Andrew R. LaBarbera, Ph.D.
Associate Professor, Departments of Obstetrics and Gynecology, and Physiology and Biophysics, University of Cincinnati College of Medicine; Director, Andrology Laboratory, University Hospital, Cincinnati, Ohio

Danuta H. Malinowska, Ph.D.
Research Assistant Professor, Department of Physiology and Biophysics, University of Cincinnati College of Medicine, Cincinnati, Ohio

Richard J. Paul, Ph.D.
Professor, Department of Physiology and Biophysics, University of Cincinnati College of Medicine, Cincinnati, Ohio

Edward S. Redgate, Ph.D.
Associate Professor, Department of Physiology, University of Pittsburgh School of Medicine, Pittsburgh, Pennsylvania

R. John Solaro, Ph.D.
Professor and Head, Department of Physiology and Biophysics, University of Illinois College of Medicine, Chicago, Illinois

Nicholas Sperelakis, Ph.D.
Professor and Chairman, Department of Physiology and Biophysics, University of Cincinnati College of Medicine, Cincinnati, Ohio

Janusz B. Suszkiw, Ph.D.
Associate Professor, Department of Physiology and Biophysics, University of Cincinnati College of Medicine, Cincinnati, Ohio

Richard A. Walsh, M.D.
Professor of Medicine, and Pharmacology and Cell Biophysics, and Director of Cardiology, University of Cincinnati College of Medicine, Cincinnati, Ohio

Laura F. Wexler, M.D.
Associate Professor of Medicine, University of Cincinnati College of Medicine; Chief, Cardiology Section, Veterans Administration Medical Center, Cincinnati, Ohio

Timothy C. Williams, M.D.
Assistant Professor of Medicine and Endocrinology, University of Cincinnati College of Medicine; Attending Physician, University Hospital, Cincinnati, Ohio

Preface

Our goal in undertaking *Physiology: Essentials of Basic Science* was to provide students of medical physiology with an intermediate-level textbook. That is, we wanted to fill what we perceived to be a gap between the numerous overly simplistic "survey-of-physiology" texts and the ultrasophisticated, encyclopedic texts that contain an overwhelming amount of material. We also wanted to cover key information and principles at the cellular and subcellular levels, as well as at the tissue, organ, and whole-animal levels, because providing a proper balance among these various levels is necessary for explaining the mechanisms that underlie physiologic phenomena. Thus, the student will find herein adequate coverage of membrane biophysics, electrogenesis of excitability, and up-to-date organ function.

The chapters are organized with the beginning student and the introductory course in mind. Each chapter is written in a clear, concise, and didactic manner and includes numerous illustrations. Chapter objectives, summaries, and questions with annotated answers facilitate review. The length and complexity of each topic discussed in our textbook closely match the amount of curriculum time allocated to the various physiology topics in many medical schools. The book was also written and organized with the instructors in mind. Instructors should not find it necessary to provide supplemental reading assignments or to omit parts of the book because they lack relevance or go into too great a level of detail. Our textbook should provide the student with the key information and principles required for a sound understanding of human physiology.

In addition to medical and dental students taking introductory physiology courses, graduate students in the basic medical sciences will find this textbook very helpful, as will students in allied health professions such as nursing, pharmacy, and physical therapy.

Physiology offers complete coverage of all the physiologic systems except the nervous system. In most institutions, physiology of the central nervous system is taught in a separate neuroscience course, and it is covered in a number of texts designed for use in that course. Therefore, for the nervous system, we cover only the peripheral somatic nervous system and autonomic nervous system, including synaptic transmission, basis of excitability, and nerve conduction. That is, most of our discussion of the nervous system is restricted to the cellular level.

Physiology and biophysics is the key discipline for imparting an understanding of the fourth state of matter: the living state. The physics and chemistry of life—which is one way of describing or defining physiology and biophysics—are understandably complex. A good foundation in physics, chemistry, physical chemistry, and mathematics is of immense help in understanding physiology, especially at the cellular level. One of the strengths of this textbook is that it provides the student with a review of the relevant principles of the physical sciences in conjunction with physiologic principles and phenomena.

The contributing authors were carefully chosen for their ability to explain complex phenomena and principles clearly, and their chapter topics correlate closely with the subdisciplines of physiology in which they have extensive teaching experience and expertise.

The nucleus of our book grew out of an extensive physiology syllabus prepared for the medical students and graduate students at the University of Cincinnati College of Medicine—a syllabus we have been compiling during the past nine years.

We wish to thank the many students and faculty members who provided constructive criticism of many sections of this textbook. Their feedback was of great help to us.

N.S.
R.O.B.

Cellular Physiology

Notice

The indications and dosages of all drugs in this book have been recommended in the medical literature and conform to the practices of the general medical community. The medications described do not necessarily have specific approval by the Food and Drug Administration for use in the diseases and dosages for which they are recommended. The package insert for each drug should be consulted for use and dosage as approved by the FDA. Because standards for usage change, it is advisable to keep abreast of revised recommendations, particularly those concerning new drugs.

1 Basic Principles

ROBERT O. BANKS AND RICHARD J. PAUL

Objectives

After reading this chapter, you should be able to:

List the various fluid compartments of the body and their approximate relative sizes

Describe several methods for measuring each compartment and explain why operational definitions are used

State the major cations and anions of each compartment and their approximate concentrations

Define an *osmole* and *osmotic pressure* as well as how much pressure one osmole generates

Explain the difference between total osmotic pressure and effective osmotic pressure; explain the term *reflection coefficient*

Contrast hydrostatic pressure and osmotic (or effective osmotic) pressure

Explain why osmotic pressure is expressed in terms of solute rather than solvent (H_2O) concentration.

Explain why only the number, not the kind, of particles is important in calculating osmotic pressure

Describe what the effective osmotic pressure is between the vascular compartment and the interstitial fluid space, and between the interstitial fluid space and the intracellular fluid space

Define *tonicity* and describe how it is related to osmolarity

Summarize the forces that result in movement of water and electrolytes across biologic membranes

Explain the movements of fluid between the major compartments in conditions such as hemorrhage, sweating, and starvation, as well as other disturbances leading to hypoalbuminemia

---

Physiology is the study of organ systems and their role in the homeostatic mechanisms that attempt to maintain the internal environment of an organism in a steady-state. This chapter focuses on the fluid compartments of the body, the composition of these compartments, the methods used to measure these fluid spaces, and the osmotic and hydrostatic forces that maintain a stable distribution of fluid among these fluid compartments.

BODY FLUID COMPARTMENTS

There are two major fluid compartments in most multicellular organisms: the **intracellular fluid** (ICF) **space** and the **extracellular fluid** (ECF) **space.** The ECF space can be divided functionally into the **plasma space** (the volume occupied by plasma) and the **interstitial fluid** (ISF) **space.** Normally the fluid in the ISF space is an ultrafiltrate of plasma entrapped in minute spaces of a gel-like matrix composed of proteoglycan filaments. However, a number of pathologic conditions, particularly those characterized by low plasma protein concentrations, increased venous pressure, increased lymphatic pressure, or increased capillary permeability to plasma proteins, are characterized by marked expansion of the ISF space that results in a state known as *edema.* Some fluid spaces (e.g., ocular fluid, cerebrospinal fluid, intestinal fluid) are relatively small but unique ECF compartments and are referred to as *transcellular fluids.*

The total water content of the body, the ECF plus the ICF, constitutes approximately 45 to 60 percent of the body weight. As shown in Table 1-1, the fraction of the body weight that is water varies with both the age and gender of the individual. The difference in the fractional water composition between males and females is due to the fact that fat, which contains less water than other body tissues, represents a greater portion of the body weight in females.

In healthy individuals, the plasma volume, ISF volume, and ICF volume are approximately 5 percent, 15 to 25 percent, and 30 to 40 percent of the total body weight, respectively. Because the normal hematocrit, defined as the ratio of the red blood cell volume to the whole blood volume, is about 40 percent; the whole blood volume in a healthy individual is approximately 7 to 9 percent of the body weight. The balance of body weight (i.e., 40 percent) is roughly composed of 18 percent proteins, 7 percent minerals, and 15 percent fat. Thus, a healthy 20-year-old, 70-kg male contains about 42 liters of body water, of which 3.5 liters are plasma, 14 liters are ISF, and 24 liters are ICF. Whole blood volume is about 6 liters.

Table 1-1. Total Body Water (percentage of body weight)

Age (yr)	Male	Female
17–34	60%	55%
50–86	54%	46%

VOLUME MEASUREMENTS OF BIOLOGIC COMPARTMENTS
Theory

Estimates of the size of body fluid spaces are based on the **dye** (indicator) **dilution principle,** which is based on conservation of mass and the relationship between concentration, volume, and mass, or:

(Concentration)(Volume) = Mass
(milligrams/milliliter or moles/liter, etc.)
(milliliters or liter, etc.) = (milligrams or moles, etc.)

Following are two examples to illustrate the application of this principle to the measurement of fluid volumes.

Example 1

In the first example, consider one compartment with an unknown volume (V_2) to which samples are added or removed. If a relatively small volume (V_1) containing a substance (not initially present in the unknown volume) at a concentration C_1 is added to the unknown volume, then, following establishment of a **steady-state** (a condition in which there is uniform mixing of the substance within V_2, and therefore the concentration of the substance does not change with time), V_2 can be calculated. Since conservation of mass must apply, it follows that:

$$(C_1)(V_1) = (C_2)(V_2)$$

Since C_1 and V_1 are known and C_2 can be measured, then:

$$V_2 = (C_1)(V_1)/C_2$$

Two requirements apply: (1) uniform mixing following addition of substance to V_2; and (2) no loss or gain of the substance in the unknown volume. If there has been loss or gain, then $(C_1)(V_1) = (C_2)(V_2) +$ amount lost or $-$ amount gained.

Example 2

Add 1 ml of a solution containing 20 mg/ml of an indicator substance to an unknown volume (V_2). After mixing, the concentration of the indicator in V_2 is 0.2 mg/ml. V_2 is calculated as follows:

$V_2 = (C_1)(V_1)/C_2$ = Amount added/
Final concentration
$V_2 = (20 \text{ mg/ml})(1 \text{ml})/0.2 \text{ mg/ml}$
$V_2 = 100$ ml or 0.1 liters

Three-Compartment Model

The body, of course, actually represents a complex communicating system of compartments with intraorgan subdivisions. Nonetheless, a three-compartment analysis based on an ICF, ISF, and plasma space can be used to approximate the distribution of total body water (Fig. 1-1). These three spaces are separated by two membranes (or membrane groups), the **capillary unit** and the **cell membrane.** For this analysis, samples of fluid can only be added to or extracted from compartment *1* (i.e., the plasma space).

Idealized Analysis of a Three-Compartment System

Assume that three substances (*X, Y,* and *Z*) are injected into compartment *1* of the model illustrated in Figure 1-1, that none of these substances was present in the system before the injection, that membrane *a* is permeable to *Y* and *Z* but not to *X*, and that membrane *b* is permeable only to *Z.*

If equal amounts of *X, Y,* and *Z* are injected into compartment *1*, and the concentration of these substances in compartment *1* is analyzed as a function of time, a concentration profile can be obtained (Fig 1-2). For this analysis, the initial portions of the curves are not drawn because there would be mixing artifacts occurring during this time interval.

Thus, based on the information provided, the volume (V) of distribution of *X* would be within compartment *1* or V_1, of *Y* would be $V_1 + V_2$, and of *Z* would be $V_1 + V_2 + V_3$. Because substance *Z* is distributed in the largest volume, the concentration of *Z* would be the lowest of the three solutes.

SPECIFIC SUBSTANCES FOR MEASUREMENT OF BIOLOGIC COMPARTMENTS

Plasma Volume

Proteins or substances that bind to plasma proteins and thus do not readily pass through the capillary membrane can be used to measure the **plasma volume.** For example, albumin, a plasma protein that can be labeled with radioactive iodine (i.e., radioiodinated serum albumin), or agents such as Evans blue, which bind to plasma proteins with high affinity, can be used to estimate the plasma space. Red blood cells can be labeled with isotopes of chromium or iron and used to measure the vascular space. By determining the volume of distribution of these agents and the hematocrit (hct), where hct = red blood cell volume/whole blood volume, one can measure the plasma volume and whole blood volume. The red blood cell volume can also be measured if one neglects the small volume of non–red blood cell elements. Thus, 1 − hct = plasma volume/whole blood volume.

ECF (ISF + Plasma Volume)

As noted above, the volume of the ECF space can be measured with substances that cross the capillary unit but do not enter the cells. Examples of such solutes are: isotopes of sodium ions (Na^+), isotopes of chloride ions (Cl^-), sulfate, thiocyanate, and molecules such as inulin (a polymer of fructose). Because some of these molecules do enter cells to a limited extent (e.g., Na^+ and Cl^-) or can slowly penetrate areas such as the transcellular spaces, each molecule yields a slightly different value for ECF volume estimates. Therefore, as is the case with most estimates of biologic var-

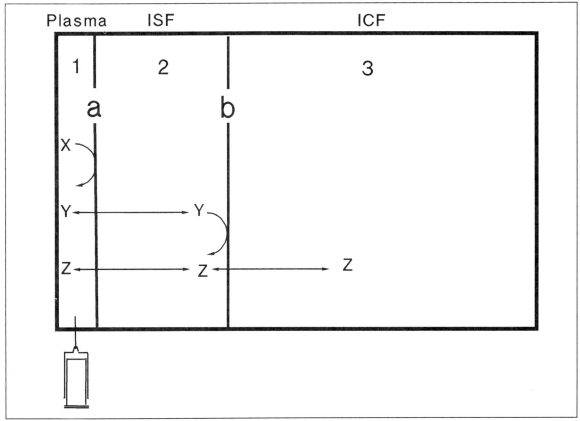

Fig. 1-1. An illustration of a three-compartment model in which sampling and injections are limited to compartment 1. Membrane *a*, which represents the capillary membrane unit, separates compartments *1* and *2* (the plasma and interstitial fluid [*ISF*] compartments, respectively), and membrane *b* separates compartments 2 and 3 (the ISF and the intracellular fluid [*ICF*] spaces respectively). Membrane *a* is permeable to substances *Y* and *Z* but not to *X* and membrane *b* is permeable only to substance *Z*.

iables, an **operational definition** is usually employed, meaning that a Na^+ space or an inulin space is reported rather than a value for the "ECF space" per se. Thus, an operational definition is one in which the resulting value depends on the method used. Estimates of most biologic variables are, in fact, operationally defined because these values depend on the method selected.

Total Body Water

Any substance that is uniformly distributed between the ECF and ICF spaces has a volume of distribution equivalent to the **total body water** content. A number of substances, including labeled water (^3H_2O or D_2O), urea, and various lipid-soluble (membrane permeable) molecules such as antipyrine, can be used.

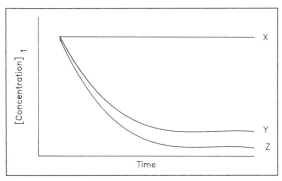

Fig. 1-2. Equal amounts of substances X, Y, and Z are injected into compartment *1* at time zero, and the concentration of each solute in compartment *1* is measured as a function of time.

COMPOSITION OF BODY FLUIDS

Mass, Amount, or Quantity

Quantities of solutes are usually expressed as grams (also as kilograms = 10^3 gm, milligrams = 10^{-3} gram, micrograms = 10^{-6} gram, nanograms = 10^{-9} gram, picograms = 10^{-12} gram, or femtograms = 10^{-15} gram), moles (also as mmoles, etc.), equivalents (also as mequivalents, etc.), or osmoles (also as mosmoles, etc.). Similarly, volumes are generally expressed in liters, milliliters (10^{-3}), microliters (10^{-6}), etc.

One gram **molecular weight** of a substance represents 1 mole of the substance and consists of approximately 6×10^{23} molecules. For example, the atomic weight of sodium chloride (NaCl) is 23 + 35.5 = 58.5; hence, 1 mole of NaCl represents 58.5 gm (1 mmole is 58.5 mg).

The concept of **equivalents** accounts for the presence of an electrical charge on the molecule. Thus, one equivalent (eq) represents 1 mole of a substance divided by its valence. Since NaCl, for example, dissociates into two particles in solution, 1 mole of NaCl contains 1 eq (23 gm) of Na^+ and 1 eq of Cl^- (35.5 gm). By contrast, 1 eq of calcium (Ca^{2+}) is 40/2, or 20 gm.

The concept of **osmoles** accounts for the numbers of particles released into solution when the solute is dissolved. Thus, 1 osmole is the gram molecular weight of the substance divided by the number of particles generated when the substance is dissolved in solution.

Mass Per Unit Volume

Concentration represents an amount (mass) per unit volume. Each of the units described in the previous section (grams, moles, equivalents, or osmoles) can be used to express a concentration. The notation "-ar," such as in a 1-mol*ar* or 1-osmol*ar* solution, indicates a concentration. Thus, 1 mole or 1 osmole dissolved in sufficient water to yield 1 liter represents a 1-molar and 1-osmolar solution, respectively. Alternatively, the notation "-al" is used when, for example, 1 mole of a substance is dissolved in 1 kg of water, resulting in a 1-mol*al* solution. Since physiologic solutions are relatively dilute, the difference between their molarity and molality is small.

A number of units can be used to express concentration, though moles or grams per liter and milliliter are often employed. In addition, many biologic concentrations are expressed as mass% (e.g., mg%), which indicates a mass per 100 ml, or deciliter (another expression for 100 ml). One example of this notation is the plasma glucose concentration that is often expressed as mg% or as mg/dl.

Concentrations of Major Cations and Anions in Biologic Fluids

Most of the major **anions** and **cations** present in the ICF and ECF spaces are illustrated in Figure 1-3 and Table 1-2 (a more detailed list is provided

Table 1-2. Plasma Concentrations of Some Major Cations and Anions

Cations	Anions
Sodium = 135–145 mEq/L	Chlorine = 100–106 mEq/L
Potassium = 3.5–5.0 mEq/L	Bicarbonate = 24–30 mEq/L
Calcium = 4.3–5.3 mEq/L	Protein^{-15} = 6–8.4 gm/dl

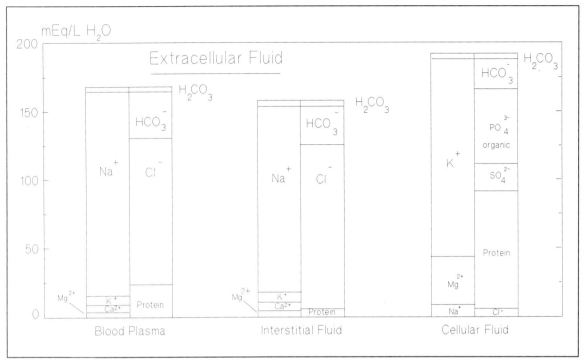

Fig. 1-3. The ionic composition of body fluid spaces.

in Appendix 1). The bar graph in Figure 1-3 is referred to as a *Gamblegram*, named after one of the first investigators to organize the concentration of anions and cations in the major body fluid spaces in this fashion. Some important facts to note are that: Na^+ is the major ECF cation, the potassium ion (K^+) is the major ICF cation, Cl^- is the major ECF anion, and inorganic PO_4^{3-} is the major ICF anion. Another important cation is Ca^{2+}.

A large difference in the protein concentration exists between the plasma space and ISF space. The reason for this is that the **net charge** on most proteins is negative (for example, the net negative charge on albumin is about 15) and thus some elements are bound to plasma proteins. Therefore the total concentration of Ca^{2+} in plasma is about 2.5 mM/liter (5.0 mEq/liter) but 40 to 50% of the total plasma Ca^{2+} is bound to albumin and globulins.

WATER MOVEMENT ACROSS BIOLOGIC MEMBRANES

Forces Involved

Two forces lead to the net movement of water across biologic membranes: (1) **hydrostatic and hydraulic pressure differences,** and (2) **osmotic pressure differences.** Both forces are physiologically important.

Hydrostatic and Hydraulic Pressure

Pressure within a fluid system has both a hydrostatic component, resulting from the gravitational force exerted on a column of fluid, and a hydraulic component, caused by fluid compression, resulting, for example, from the action of a pump. Both forces are important when evaluating fluid movement in the cardiovascular system, both in the vascular space and between the plasma compartment and the ISF space. Conceptually the importance of hydrostatic/hydraulic pressure as a contributing factor to fluid movement is easy to appreciate: water moves down a hydrostatic/hydraulic pressure gradient (i.e., from higher to lower pressure).

Osmotic Pressure

Osmotic pressure is one of the so-called colligative properties of solutions. To appreciate the contribution of osmotic pressure to net water movement across biologic membranes, one must also understand the distinction between the total or **calculated osmotic pressure** (sometimes referred to as the *potential osmotic pressure*) and the physiologically important component, known as the **effective osmotic pressure.** The effective osmotic pressure is determined by the permeability properties of the membrane separating two solutions.

To illustrate the importance of the membrane as the determinant of the effective osmotic pressure of a solution, two relatively simple experiments can be performed. In one, a **semipermeable membrane** (a membrane that is not permeable to solute but is permeable to water) is attached to the end of a thistle tube and suspended in a beaker of water; in the second experiment, a **permeable membrane** (one that is freely permeable to both solute and water) is placed at the end of the tube. The experiments are illustrated in Figure 1-4. If a solution of water plus solute is added to the inside of the tube with the semipermeable membrane (see Fig. 1-4A), a net flow of water will travel into the tube until the column of water (i.e., the hydrostatic pressure) offsets the osmotic pressure of the solution in the tube. Alternatively, if hydraulic pressure is applied to the solution inside the thistle tube, the amount of pressure necessary to prevent water movement would be equal to the osmotic pressure. The reason why water enters the tube is that water, like solute (see Chapter 2), diffuses from regions of high concentration to regions of low concentration. The concentration of pure water is 55.5 M (the molecular weight of water is

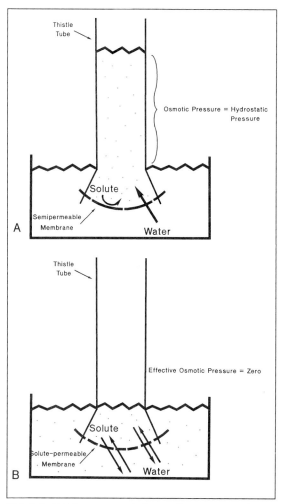

Fig. 1-4. (A) Measurement of osmotic pressure using a semipermeable membrane (i.e., a membrane that is only permeable to water). The addition of solute to the thistle tube results in a concentration gradient for water between the outside and inside of the tube; consequently, water diffuses into the tube until the height of the fluid column offsets this movement (osmotic pressure = τgh, where τ is the density of the solution, g is the gravitational force, and h is the height of the solution in the tube). (B) The permeability properties of the membrane determine the physiologically important component of the calculated osmotic pressure (i.e., the effective osmotic pressure of the solution). When the membrane is permeable to both water and solute, the solute equilibrates between the two solutions and no effective osmotic pressure is generated.

18.02 and 1 liter of water weights 1 kg; therefore, the concentration is 1,000 gm/18.02, or 55.5 M), whereas the concentration of water following addition of solute is lowered in proportion to the amount of solute added. Thus, a higher concentration of an impermeant solute on one side of a membrane will result in net diffusion of water into the solution. By contrast, if the same solution is added to the thistle tube with a solute/water–permeable membrane (see Fig 1-4B), the solute will equilibrate across the membrane and the effective osmotic pressure of the solution will be zero. Thus, if the membrane is as permeable to the solute as it is to water, no net water flow will occur even though the theoretical osmotic pressure of the solution (calculated by the van't Hoff equation) is the same as in the experiment illustrated in Figure 1-4A. Effective osmotic pressure is therefore dependent on the relative permeability of the membrane to the solute (a discussion of the reflection coefficient follows in the next section).

Osmotic pressure (π) can be calculated by the **van't Hoff equation:** $\pi = C_sRT$, where C_s is the osmolar concentration of the solute, R is the universal gas constant, and T is the absolute temperature; at 0°C, RT has a value of 22.4 liter atm per osmole (at 37°C, RT = 25.4).

Osmolarity (i.e., the osmolar concentration) is defined as the number of moles/liter multiplied by the number of dissociating ions. Osmolarity must be used when considering osmotic pressure, as osmotic pressure is dependent on the number of nonpenetrating molecules. Thus:

1 mole urea = 1 osmole	or	1 mole urea/liter = 1-osmolar solution
1 mole NaCl = 2 osmoles	or	1 mole NaCl/liter = 2-osmolar solution
1 mole calcium chloride (CaCl$_2$) = 3 osmoles	or	1 mole CaCl$_2$/liter = 3-osmolar solution

Since the NaCl concentration in the ECF space is about 150 mM/liter, the osmolar concentration is about 300 mOsm/liter.

To better appreciate the **equilibrium** aspect of the van't Hoff equation and its limitations, as well as to understand why osmotic pressure is expressed in terms of solute concentration when it is a property of water, it is useful to consider a nonrigorous derivation of this equation using principles of physical chemistry. One criterion for equilibrium is that Gibbs energy change (ΔG) is equal to zero. Therefore, for water at equilibrium:

$\Delta G = 0 = \pi + RT \ln(x_w)$,
where x_w = mole fraction of H_2O

$$x_w = \frac{n_w}{n_w + n_s}$$

$$= \frac{(\text{\# moles of } H_2O)}{(\text{\# moles of } H_2O + \text{\# moles of solute})}$$

Substituting for x_w in the equation, we get:
$\Delta G = \pi + RT \ln [n_w/(n_w + n_s)]$.
Since $\ln(z) = -\ln(1/z)$, x_w can be rewritten to equal:

$$\ln\left[\frac{n_w}{n_w + n_s}\right] = -\ln\left[\frac{n_w + n_s}{n_w}\right] = -\ln\left[1 + \frac{n_s}{n_w}\right]$$

Thus, with this substitution, the equation is:

$$\Delta G = \pi - RT \ln\left[1 + \frac{n_s}{n_w}\right]$$

However, under physiological conditions, "ln" can be approximated because the concentration of water is very high (55.5 M), whereas that of the solute is very small (on the order of millimolars). Thus:

$$\text{for } n_s/n_w << 1, \ln\left[1 + \frac{n_s}{n_w}\right] \cong \frac{n_s}{n_w}$$

$$\therefore G \cong \pi - RT\left[\frac{n_s}{n_w}\right] = 0$$

But the volume is proportional to n_w: $V \propto n_w$, so the van't Hoff equation is:

$$\pi \cong RT\frac{n_s}{V} \cong RTC_s$$

Thus, it is the total number of moles or particles in solution that is significant in determining the osmotic pressure, not the kind of solute. The presence of solute lowers the mole fraction or concentration of water and creates the imbalance that drives water movement.

This is true for colligative properties in general, in that the number of particles is the critical factor in freezing point depression or boiling point elevation of water as well as for the flow of water. However, when considering water flows in situations where solute flows also occur—the more general case for biologic membranes separating compartments—the kind of solute which determines its permeability across biologic membranes becomes a major factor in the osmotic flow of water.

In general, biologic membranes show varying degrees of permeability to many substances. If both water and solute flows are taking place, an effective osmotic pressure is defined as: $\pi \cong \sigma RTC$, where σ is the reflection coefficient. The reflection coefficient depends on the properties of both the membrane and solute and has a value that ranges from 0 to 1.

Osmotic Behavior of Cells

Water diffuses freely across cell membranes, whereas sodium ions, glucose, and large organic molecules do not. Thus, **osmotic pressure** plays an important role in water movements across the cell membrane. This is particularly critical for animal cells, which, unlike plant cells, do not have a cell wall and cannot withstand large osmotic forces.

Osmotic pressure is a very powerful driving force for water movement, and a small change in solute concentration can result in a large change in pressure. This can be illustrated, as follows. If the concentration of osmotically active particles inside a cell is 0.3 Osm/liter, then a cell placed in pure water would experience an osmotic pressure:

$$\pi = (22.4 \text{ liter/atm/Osm}) \times 0.3 \text{ Osm/liter}$$
$$= 6.7 \text{ atm}$$

Since 1 atm equals 760 mm Hg, the osmotic pressure is more than 5,000 mm Hg. This is sufficient to support a column of water approximately 68 meters high (mercury is about 13 times heavier than water).

Oxygen, carbon dioxide, urea, and certain amino and fatty acids are relatively permeable to the plasma membrane, and do not develop an osmotic pressure equal to their number of particles. Thus it is possible for cells, placed in an isosmotic solution (one containing the same number of particles per volume as inside the cell) to gain water, depending on the effective osmotic pressure of the solution (which is dependent on the relative permeability of the solutes).

The **tonicity** of a solution is defined in terms of the water movement. Cells placed in an isotonic solution neither gain nor lose volume. Cells placed in a hypertonic solution shrink because of an osmotic loss of water. Conversely, cells placed in a hypotonic solution swell because water flows into the cell.

Osmotic equilibrium is achieved very rapidly through the flow of water across the cell membrane. Therefore, cells are always in osmotic equilibrium with plasma. If a red blood cell is placed in a hypotonic solution, its volume will increase until osmotic equilibrium is reached (osmotic uptake of water decreases the intracellular tonicity until it is isotonic with the originally hypotonic solution). Once osmotic equilibrium is reached, the volume of the cell, unless it ruptures, will remain constant. The volume changes of red blood cells in hypotonic and hypertonic solutions can thus be used to measure the tonicity of plasma. First the volume of red blood cells that are in osmotic equilibrium with plasma is measured, and then the volume of the same number of red blood cells in osmotic equilibrium with solutions that are slightly hypotonic and hypertonic is measured. Plasma tonicity will be equal to that of a solution in which red blood cells have the same volume as they did in plasma (Fig. 1-5).

Utilizing the relationship between the relative

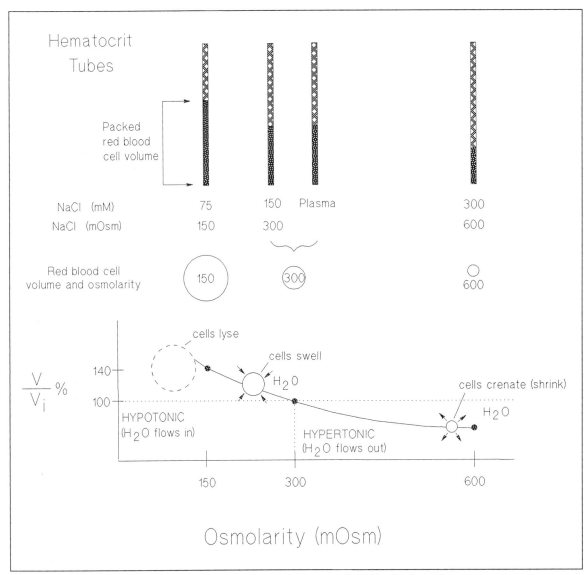

Fig. 1-5. Osmotic behavior of red blood cells. The osmotic behavior of red blood cells can be measured easily, as shown. Red blood cells placed in various NaCl solutions shrink (or swell) due to water loss (or gain). This can be measured in terms of the volume of a fixed number of cells in a hematocrit tube. The relationship between red blood cell volume and the solution osmolarity is shown in the graph. As described in the text, the relative red blood cell volume is inversely proportional to the osmolarity.

red blood cell (RBC) volume and the osmolarity (of impermeant solute), as depicted in Figure 1-5, at equilibrium, one gets:

$$Osm_{RBC} = Osm_{out} \text{ or } \pi_{RBC} = \pi_{soln}$$

Thus, using the van't Hoff equation: $RTC_{RBC} = RTC_{soln}$, where:

$$C_{RBC} = \frac{n_{RBC}}{V_{RBC}}$$

and n_{RBC} is the number of moles of osmotically active particles inside the red blood cell, which is constant, and V_{RBC} is the red blood cell volume. So,

$$C_{RBC} = \frac{n_{RBC}}{V_{RBC}} = C_{soln}$$

Thus, $V_{RBC} = 1/C_{soln}$, as shown in Figure 1-5.

As already mentioned, a permeant solute represents a more complex situation. A solution containing only a penetrating solute, whether or not it is isosmotic with the cell interior, cannot be isotonic. In fact, from the definition of tonicity described previously, all solutions containing only penetrating solutes must be regarded as hypotonic, whatever their actual osmolarity may be. The rate of swelling (or the time required for hemolysis in red blood cells) in an isosmotic solution of a penetrating solute can be used as an index of its permeability.

As an example, consider the case of a red blood cell placed in a 300-mM urea solution. This solution is isosmotic but hardly isotonic, as the rapid hemolysis of the cell indicates a large inward flow of water (Fig. 1-6).

Therefore, in order to predict the tonicity of solutions, one has to have some idea of the permeability of the solutes in the solution, in addition to its concentration, and, for electrolytes, the number of particles upon dissociation.

The **reflection coefficient** (σ) is used to modify the van't Hoff equation, to take into account the effects of permeant solutes. It is most simply defined as:

$$\sigma_i = \pi_{meas}/\pi_{expected} = \pi_{meas}/RTC_i$$

In the example using urea, because hemolysis occurs rapidly, the effective osmotic pressure elicited by the urea solution is very small, and hence the reflection coefficient for urea must be near zero. To calculate the effective osmotic pressure of a solution containing many different solutes, the following equation is used:

$$\pi = RT \Sigma_i \sigma_i C_i = RT (\sigma_1 C_1 + \sigma_2 C_2 + \ldots)$$

Thus, to understand water flows across the cell membrane into or out of the ICF space, one must additionally understand the factors involved in solute transport across this barrier. Although this will be reviewed in Chapter 2, at this point it is instructive to consider the osmotic behavior of red blood cells with respect to various solutes. This is because the term *permeable* is used in various contexts in cellular physiology. For example, K^+ and Cl^- are considered "permeable" in an electrophysiologic context, and glucose is clearly "permeable" in terms of cellular metabolism. However, in terms of the consequential water flows, neither is very permeable, as indicated by the time required for hemolysis to occur, as shown in Table 1-3. Cl^- is readily permeable, however, if its counterion, such as Na^+, is not, then Cl^- cannot move alone without violating electroneutrality. While clearly permeable in an electrophysiologic context, K^+ is not very permeable compared to water. Considering the larger size of thiourea compared to urea, the longer time for the hemolysis of thiourea is reasonable. It is, however, somewhat surprising that ammonium chloride (NH_4Cl) appears to be so permeable, in light of the fact that NaCl is impermeable. The insight here is that NH_4Cl is always in equilibrium with ammonia, which is very lipid soluble

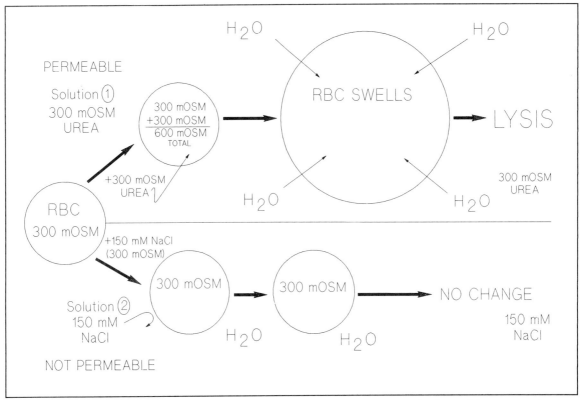

Fig. 1-6. Demonstration of the difference between osmolarity and tonicity. Red blood cells (*RBCs*) are exposed to solutions of 300 mM urea or 150 mM NaCl; both solutions are 300 mOsm. Cells in NaCl remain intact whereas those in the urea solution lyse, due to an inward flow of water. Water flowing into the red blood cells in the urea solution is attributable to a larger number of particles inside the red blood cell than outside, due to the penetration of urea as shown. Thus, the urea solution is hypotonic, though isosmotic.

and rapidly permeates biologic membranes. Because cells metabolize glucose, it is also surprising that it too is quite impermeable when compared to water. This is explained by the fact that only relatively small amounts of glucose can penetrate cells via a saturable, carrier-mediated diffusion mechanism (see Chapter 2). Finally, sucrose, a disaccharide, is virtually impermeable and often used as a marker for the ECF space. When studying membrane transport in subsequent chapters, bear in mind the effects on the duration of time before red blood cell hemolysis of these various solutions in terms of their relative permeabilities and consequent water flows.

Fluid Movement Into and Out of the Capillary

If a membrane is permeable to both water and solute, no effective osmotic pressure is realized. A good example of this relates to osmotic forces across the capillary wall. Na^+ and Cl^-, the major ECF ions, move rapidly between capillary plasma and ISF, and therefore are not effective osmotic agents in that system. Conversely, plasma proteins are restricted primarily to the plasma compartment and are therefore the only effective osmotic agents mediating osmotic water flow between plasma and ISF. Thus, the effective osmotic pressure difference between plasma and

Table 1-3. Properties of Various Solutes

Solute	Concentration (mM)	Osmolarity (mOsm/liter)	Tonicity	Time to Hemolysis (approximate)
Urea	300	300	Hypo	<1 sec
Urea	600	600	Hypo	<1 sec
Thiourea	300	300	Hypo	90 sec
Sodium chloride	150	300	Iso	∞
Ammonium chloride	150	300	Hypo	12 min
Potassium chloride	150	300	Hypo	24 hr
Glucose	300	300	Hypo	8 hr
Sucrose	300	300	Iso	∞

ISF is only a small fraction of the theoretical (potential) osmotic pressure. As noted previously, the potential osmotic pressure, as calculated from the total solute osmolar concentration, is more than 5,000 mm Hg. By contrast, the effective osmotic pressure (due to plasma proteins) is about 25 mm Hg. This latter pressure is usually referred to as the **colloid osmotic pressure** or **plasma oncotic pressure.**

In the early 1900s, Starling was one of the first investigators to recognize the forces involved in fluid movement across the capillary and formulated what is generally referred to as **Starling's law of the capillary** (see Chapter 22).

Finally, it is important to understand the direction of fluid and water shifts that occur between the three major fluid compartments (the plasma, ISF, and ICF spaces) during various physiologic and pathophysiologic conditions. For example, during a hemorrhage, among the many events that occur is a decrease in blood pressure and a reflex increase in the activity of the sympathetic nervous system. Consequently, there is a decrease in capillary pressure and a shift in the balance of the Starling forces that favor net reabsorption of fluid from the ISF space into the plasma compartment. Because the reabsorbate does not contain protein, the plasma oncotic pressure falls after a hemorrhage. On the other hand, because the initial volume of blood lost during a hemorrhage is isosmotic, neither the total osmolarity of the ECF nor the volume of the ICF space changes as a consequence of bleeding. By contrast, during conditions such as sweating, ICF is mobilized because sweat is hypotonic (Na^+ = 65, K^+ = 8, Cl^- = 39, and bicarbonate = 16 mEq/liter, i.e., the osmolarity is about 150 mOsm/liter). Therefore, during sweating there is increased osmolarity of the ECF, bringing about extraction of water from the ICF space into the ECF space. Net fluid shifts out of the plasma space and into the ISF space (the formation of edema) occur whenever there are decreases in the plasma concentration of proteins (e.g., liver disease), during reduced protein intake, and during starvation.

SUMMARY

Water constitutes 45 to 60 percent of body weight and is distributed among the three major fluid compartments: the plasma space (about 5 percent), the ISF space (about 15 percent), and the ICF space (about 40 percent). The non-water components are fat, minerals, and proteins. Na^+ and Cl^- represent the primary osmotic components in ECF (the plasma space plus the ISF space), whereas K^+ and proteins account for the major elements in ICF. The size of these fluid spaces can be estimated using agents that, when injected into the vascular space, remain in plasma, distribute within the ECF space, or diffuse throughout the entire body fluid. Hydraulic and hydrostatic as well as osmotic pressures produce net water flow. Hydraulic and hydrostatic pressures are created through compression of fluid by a pump and by a column of fluid, respectively. Osmotic pressure is a property of solutions and can be calculated using the van't Hoff equation. The physiologically important component of the total osmotic pressure (i.e., the effective osmotic pressure) depends on the solute permeability of the membrane separating two solutions. If the membrane is not permeable to any of the solute, the effective osmotic pressure of the solution is equal to the calculated osmotic pressure. By contrast, as the solute permeability of the membrane increases, the effective osmotic pressure decreases.

NATIONAL BOARD TYPE QUESTIONS

Select the one most correct answer for each of the following

1. Which of the following statements concerning the ISF is correct?
   A. Its volume can be measured with Evans blue and heavy water
   B. Its volume increases as the colloid osmotic pressure of plasma decreases
   C. It is not in osmotic equilibrium with the ICF
   D. Its volume is slightly less than that of the plasma compartment
   E. Its volume decreases when a hypertonic NaCl solution is infused intravenously

2. A patient has 14 liters of ECF, about 28 liters of ICF, and a plasma osmolarity of 270 mOsm/liter. If 100 ml of a 3,000-mM/liter urea solution were infused intravenously, which of the following is likely to occur (assume no loss through kidneys)?

   A. About 600 ml of water will move from the ICF to the ECF
   B. The new steady-state plasma osmolarity will be 320 mOsm/liter
   C. The new steady-state ICF volume will be about 28 liters
   D. The plasma Na^+ concentration will decrease to 134 mEq/liter
   E. The plasma Na^+ concentration will increase to 146 mEq/liter

3. Which one of the following solutions of identical osmolarity is isotonic with respect to the red blood cell?
   A. 150-mM sucrose + 75-mM NaCl
   B. 150-mM urea + 75-mM NaCl
   C. 150-mM NH_4Cl
   D. 300-mM urea
   E. 150-mM urea + 75-mM NH_4Cl

4. Which of the following answers would be consistent with the data given: 100 mg of substance X (dissolved in 1 ml of saline) is injected intravenously into a 100-kg person. At steady-state, the plasma concentration of substance X is 5×10^{-3} mg/ml (assume no loss of substance X from the body during the experiment).
   A. Substance X could be urea or a substance similar to urea
   B. Substance X would not be contained in red blood cells
   C. If 100 mg of Evans blue were injected at the same time as substance X, the concentration of Evans blue in plasma at steady-state would be lower than the concentration of substance X
   D. Substance X would be a good plasma expander (i.e., it is a protein-like substance)
   E. The volume of distribution of substance X cannot be calculated from these data

5. The osmotic pressure of a 0.9-gm/dl NaCl solution calculated from the van't Hoff equation is 5,776 mm Hg. What is the calculated osmotic pressure of a solution containing 100-mg/dl of glucose (the atomic weight of NaCl is 58.5 and glucose, 180)?
   A. Approximately 20 mm Hg
   B. Approximately 100 mm Hg

C. Approximately 200 mm Hg
D. Approximately 1,000 mm Hg
E. More than 2,000 mm Hg

ANNOTATED ANSWERS

1. B. Edema (an increase in the ISF volume) develops as the plasma colloid pressure decreases.
2. C. Cell membranes are permeable to urea. Therefore, the infused urea, and consequently the infused water, will distribute throughout the ECF and ICF.
3. A. The cell membrane is impermeant to both sucrose and NaCl. Because the total osmolarity of the solution is 300 mOsm, the solution is isotonic (i.e., no net water flow into or out of cells will occur with this solution).
4. B. The calculated volume of distribution of substance X is 20 liters, a value that would be expected for the ECF in a 100-kg person. Therefore, the substance could be inulin.
5. B. The calculation is based on the osmolarity of the two solutions. The NaCl solution is 150 mм (300 mOsm/liter) and the glucose solution is 5.6 mм (5.6 mOsm/liter). Therefore, the osmotic pressure of the glucose solution would be about 0.02 times that of the NaCl solution.

BIBLIOGRAPHY

Rose, B. D. *Clinical Physiology of Acid-Base and Electrolyte Disorders*. New York: McGraw-Hill, 1989.

2 Membrane Transport of Solute

JOHN CUPPOLETTI

Objectives

After reading this chapter, you should be able to:

Describe the organization of biologic membranes

Describe the barrier properties of biologic membranes

List the pathways for flow of solutes across biologic membranes

List the transport mechanisms operative in biologic membranes

Explain the regulation of membrane transport

The plasma membrane marks the outer boundary of the cell. Within cells are the organellar membranes that create compartments where otherwise competing processes such as synthesis and degradation of macromolecules occur.

Membranes provide a barrier function and mediate the selective flow of nutrients, specialized secretions, and signals that underlie the ability of the cell to respond to its environment. Numerous therapeutic agents must either cross the membrane or interact with membrane proteins in order to exert their effects. Defects in membrane transport proteins or regulation of the proteins are responsible for several diseases, including cystic fibrosis and diabetes. This chapter deals with the structures, molecules, and mechanisms involved in membrane transport.

STRUCTURE OF MEMBRANES

Membranes are between 4- and 9-mm thick and are composed of lipids and proteins in about equal amounts. Membrane lipids are **amphipathic,** with hydrophylic (water-loving) head groups and hydrophobic tails (which avoid water). The lipids are organized into a bilayer: their head groups are oriented toward aqueous solutions on the inner (cytosolic) and outer (extracytosolic) faces of the membrane, and the hydrophobic tails are located together in the center of the membrane. In living cells, membrane lipids are asymmetrically distributed among the **bimolecular leaflet,** with positively charged lipids enriched in the outer leaflet and negatively charged phospholipids enriched in the inner leaflet. The maintenance of this asymmetric distribution of phospholipids requires metabolic energy and may involve specialized proteins that flip phospholipids from one leaflet to another. The rate of exchange of lipids from one side of the membrane to the other is extremely slow in the absence of such proteins and is on the order of 10 to 20 hours. In contrast, membrane lipids are free to move about their own axis or within the plane of the membrane.

Membrane proteins are either deeply embedded in the bilayer (integral membrane proteins) or are attached to the membrane (peripheral membrane proteins). Transport proteins and hormone receptors are **integral membrane proteins** and the cytoskeletal proteins are **peripheral membrane proteins.** Membrane proteins do not flip across the membrane, but are generally free to rotate and diffuse within the plane of the membrane unless tethered to the underlying cytoskeletal network. The membrane is a dynamic structure, where insertion and uptake of membrane components occur as a consequence of cell growth, uptake of materials from the extracellular milieu, and regulation of the complement of membrane proteins.

Membrane transport has been studied in living cells and in a variety of model membrane systems. Among the most useful of these model systems are **unilamellar liposomes** (proteoliposomes) and **black lipid films.** Liposomes are formed by the sonication of lipids into aqueous solution. Membrane proteins such as transport proteins can be incorporated into these liposomes by removal of detergent from detergent-solubilized mixtures of lipids and proteins. Black lipid films are bilayers of lipids into which membrane proteins can be inserted by fusion with proteoliposomes or using membrane patches from biologic membranes. Such model systems are widely employed to study purified membrane transport proteins.

PASSIVE PERMEABILITY

Membranes provide a barrier to the flow of solutes. However, the barrier function is not absolute, and passive leakage of molecules across membranes occurs. The net flow of solute across biologic membranes can be described by the following relationship:

$$P = K_{part} - D_{mem}/\lambda$$

where P is the permeability coefficient in centimeters per second, K_{part} is the lipid–water partition coefficient of solute, D_{mem} is the diffusion coefficient of solute in the membrane in square centimeters per second, and λ is the thickness of the membrane in centimeters.

Permeability coefficients range from approximately 10^{-3} for water to 10^{-9} or higher for simple sugars. Thus, flow across the membrane should be linearly related to the oil–water partition coefficient, which is dependent on the relative hydrophobicity of the solute. The importance of hydrophylic groups in determining the rate of flow across the membrane can be appreciated by comparing a hydrophobic compound, ethyl alcohol, with a more hydrophylic substance, urea. These compounds are similar in size, but ethyl alcohol is 30,000 times more permeable than urea. Permeability is also dependent on D_{mem}, which in turn is highly dependent on the size of the solute. Larger solutes are less permeant than smaller solutes, as the passage of solutes across the membrane must occur through openings produced by random motions of the membrane lipids. Low-molecular-weight substances such as oxygen and small hydrophobic (polar) substances such as alcohols are more permeant than higher-molecular-weight and more hydrophylic (polar) substances such as sugars.

Although all membranes show some leak or passive pathways for the flow of some gases and solute through the bilayer, the movement of most physiologically relevant substances occurs through membrane transport proteins that specifically transport solutes across biologic membranes. Transport processes mediated by proteins can be classified into several categories: **Ion channels** for Ca^{2+}, K^+, and Na^+ carry ions down an electrochemical gradient and **facilitated diffusion systems** carry solutes down a chemical concentration gradient. Eventually, ion channels, such as the cardiac Ca^{2+} channel, and facilitated diffusion systems, such as the red blood cell (RBC) D-glucose carrier, would equilibrate solutes on opposite sides of the membrane. Countertransporters, such as the Cl^-–bicarbonate (HCO_3^-) exchanger or the Na^+–H^+ exchanger, carry out the exchange of ions on opposite sides of the membrane. Net transport continues until the concentration of both solutes is equal on both sides of the membrane, or the countertransporter stops operating. In all of these systems, net transport continues until the concentration of solute is equal on both sides of the membrane. In contrast, some transport proteins can catalyze the net concentrative uptake of solutes by secondary or primary active transport. Cotransporters such as the intestinal Na$^+$–D-glucose transporter move two solutes in the same direction. In this case, the concentration gradient of Na$^+$ can be used to drive the concentrative uptake (**secondary active transport**) of D-glucose. The ion pumps, such as the sodium-potassium adenosinetriphosphatase [(Na,K)-ATPase], can directly use metabolic energy, usually in the form of adenosine triphosphate (ATP), to generate ion and charge gradients. The ion pumps carry out **primary active transport.** Examples of each of these systems will be discussed in more detail in the following sections.

ION CHANNELS

Ion channel proteins facilitate the movement of ions across biologic membranes. These proteins are intimately involved in regulating the movement of ions and thus become charged in the electrically excitable tissues such as muscle and nerve. A variety of related ion channel proteins have been characterized at the molecular level. These ion channels show specificity for Na$^+$, Ca^{2+}, or K$^+$, and are important targets for therapeutic agents. The function and regulation of these ion channels will be discussed in detail in the sections on "Synaptic Transmission," "Muscle Physiology," and "Cardiovascular Physiology." The channels in these issues are largely involved in the movement of small amounts of cations for signaling and electric events. Other channels in secretory epithelia, such as the Cl$^-$ channels of airway epithelia, are responsible for the movement of chemical amounts of salts and water. In cystic fibrosis, there is a genetic defect affecting salt and water transport due to Cl$^-$ channel regulation in the airway epithelia, which causes death in affected individuals.

FACILITATED DIFFUSION: D-GLUCOSE TRANSPORT

Neonatal mammalian RBCs transport D-glucose to support glycolysis. When most mammals (ex-

cept humans) mature, they lose the ability to transport D-glucose across their RBCs. Shown in Figure 2-1 is a typical glucose transport experiment for non-human adult RBCs, neonatal RBCs, and neonatal RBCs that have been poisoned with inhibitory mercuric salts. Neonatal RBC D-glucose transport is fast, while the rate of transport in adult cells is much slower. In the curve for neonatal cells, the transport mediated by passive diffusion is superimposed on the transport mediated by facilitated diffusion. Mercuric salts inhibit the D-glucose carrier by binding to essential sulfhydryl groups. Treatment of the neonatal RBCs with mercuric salts inhibits facilitated diffusion, providing a means of obtaining the passive diffusion rate. When the passive diffusion rate is subtracted from the overall rate, the rate of transport due to facilitated diffusion is obtained. Transport by the neonatal cells is saturable at high concentrations of glucose due to saturation of the carrier. In contrast, transport by adult cells occurs largely by passive leak, and is essentially linear. The slower rate and nonsaturability of uptake of D-glucose by adult cells is expected for a passive uptake of the sugar. The increased rate of the transport process in the neonatal RBC suggests that D-glucose uptake is **facilitated.** Saturability implies that there are a finite number of glucose carriers in the RBC membrane which can carry D-glucose. The plot of the rate of facilitated uptake versus concentration approaches a maximum velocity (V_{max}), and the concentration of substrate that yields the half-maximal rate is the apparent K_m (Michaelis constant).

The amino acid sequence of a D-glucose carrier from non-insulin-sensitive tissues such as brain, as well as that of the insulin-sensitive tissues such as fat and muscle, have now been deduced by molecular biologic techniques. Models of the folding pattern of the D-glucose carrier have been prepared from the amino acid sequence through computer-aided determination of the relative hydrophobicity of regions of the transporter. Figure 2-2 contains a diagram of the RBC D-glucose transporter, showing the proposed transmembrane folding pattern. The most hydrophobic re-

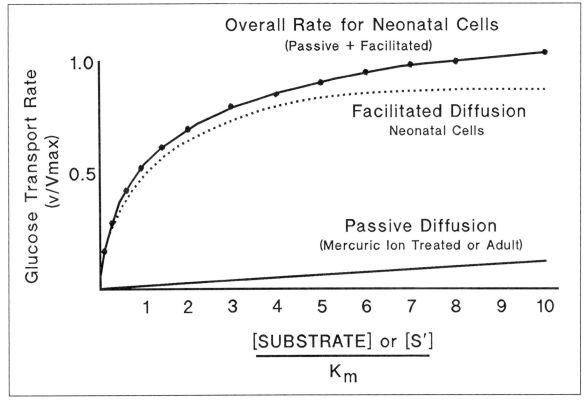

Fig. 2-1. Facilitated diffusion and passive diffusion. The highest curve represents the concentration dependence for the transport of D-glucose into the red blood cells of neonatal pigs, and the lowest curve is the rate expected in adult pigs or neonatal pigs treated with the glucose transport inhibitor, mercuric chloride. In the dotted curve (*Facilitated Diffusion*), the rate of passive diffusion has been subtracted from the overall rate. Neonatal D-glucose transport is fast, saturable, and inhibitable, indicating that D-glucose transport by the neonatal cell occurs largely by facilitated diffusion. The concentration dependence of facilitated diffusion is used to calculate the maximal rate (V_m) and the amount of substrate required to achieve half of the maximal rate (K_m), using the Henri-Michaelis-Menten equation: $v/V_m = [S']/K_m + [S']$, where S' = the substrate concentration equal to K_m (the Michaelis constant).

gions are assumed to be within the membrane, and most hydrophylic regions are thought to be exposed. The folding pattern has been adjusted to reflect the fact that glycosylation sites are known to be exposed to the extracellular fluid on the extracytoplasmic surface. The amino and carboxyl terminal regions and inhibitor binding sites, which are known to be on the inside of the cell, are used as markers. D-glucose presumably passes through the membrane aided by a structure formed by the folds on the D-glucose carrier.

PHYSIOLOGICALLY RELEVANT, INSULIN-REGULATED D-GLUCOSE TRANSPORT

In diabetes, D-glucose uptake, which is rate-limiting to D-glucose metabolism, is defective in the insulin-sensitive tissues (adipocytes, heart, and other muscle tissue). Because the bulk of the body mass is in these tissues, equilibration of D-glucose within these tissues is an important determinant of serum D-glucose levels. These insulin-sensitive tissues contain an additional

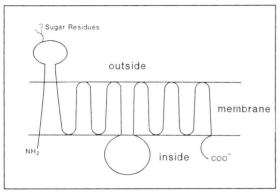

Fig. 2-2. Proposed structure of the insulin-regulated D-glucose carrier. The complete amino acid sequence of the D-glucose carrier from the insulin-sensitive tissues is known. Based on what is also known about charged amino acids and the relative hydrophobicity of short stretches of amino acids, a folding pattern has been suggested. The folding of the D-glucose carrier into the membrane or the interaction of more than one copy of the D-glucose carrier may provide a stereospecific pathway for the facilitated diffusion of D-glucose across the membrane (coo^- = carboxyl terminal; NH_2 = amino terminal.)

form of the D-glucose carrier, called the *insulin-regulated D-glucose carrier*. This form of the D-glucose carrier is similar in size, sequence, folding pattern, and sensitivity to inhibitors. When messenger RNA for the insulin-sensitive D-glucose carrier was injected into frog oocytes, the oocyte synthesized the D-glucose carrier. When glucose transport across the oocyte plasma membrane was measured, the glucose carrier was shown to have a lower V_{max} and a lower K_m (higher affinity) than did the D-glucose carrier from brain and RBCs (Fig. 2-3A). The high affinity of the insulin-regulated D-glucose carrier is consistent with its role in maintaining the physiologically low serum D-glucose concentration. The insulin-sensitive form constitutes the bulk of the D-glucose carrier in insulin-sensitive tissues. In the absence of insulin, the D-glucose carrier exists in intracellular vesicles and is unavailable to serum glucose. In normal individuals, insulin binding to the insulin receptor of cells that express this form of the D-glucose carrier results in movement of the D-glucose carrier to the plasma membrane, whereupon the vesicles containing the D-glucose carrier fuse and are thus available to carry out D-glucose transport (see Fig. 2-3B). Upon removal of insulin from the insulin receptor, vesicles containing the insulin-sensitive D-glucose carriers apparently "pinch off" from the plasma membrane, forming an intracellular storage pool (see Fig. 2-3C). Long-term insulin starvation, as might occur in pancreatic disease, can lead to a reduced number of insulin-sensitive D-glucose carriers in this intracellular pool, producing a reduction in the insulin sensitivity of glucose transport in these tissues. In animals, long-term insulin supplementation can increase the size of the pool, with a concomitant increase in the insulin sensitivity of these tissues. Whether similar long-term effects occur in humans routinely receiving insulin injections is not known.

Cl^-–HCO_3^- EXCHANGE

HCO_3^- is formed by the action of carbonic anhydrase on metabolically produced equivalents of base (hydroxide ion $[OH^-]$) and carbon dioxide. The RBC Cl^--HCO_3^- exchanger plays an important role in maintaining intracellular pH and eliminating CO_2 from the tissues by taking up this HCO_3^- in exchange for Cl^-. As the RBC passes through the pulmonary circulation with a transit time of approximately $1/20^{th}$ second, the process is reversed and the trapped HCO_3^- is released in exchange for Cl^-. The RBC is well equipped for this purpose. The Cl^-–HCO_3^- exchanger of the RBC is the most highly characterized, but this exchanger is also present in most cells. By labeling the RBC exchanger with radioactive distilbenes (which are potent inhibitors of the anion carrier), the carrier was shown to be a 95,000-Da protein present at approximately 10^6 copies per cell and comprising approximately 25% of the cell protein content. The anion exchanger is glycosylated at an extracytosolic domain and passes through the membrane several times. A continuous transmembrane domain present on the cytosolic surface of the anion carrier, which

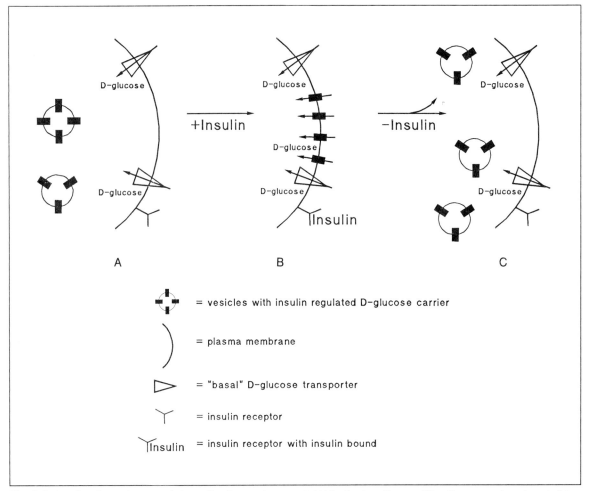

Fig. 2-3. Insulin-dependent regulation of D-glucose transport. (A) In the insulin-sensitive tissues such as heart, fat, and muscle, there is a low basal complement of D-glucose carriers which provide a low basal rate of D-glucose transport. (B) Insulin binding to an insulin receptor at the cell surface sends signals that produce the movement and fusion of vesicles containing D-glucose carrier proteins to the plasma membrane, and these proteins carry out the rapid D-glucose transport characteristic of the insulin-treated cell. (C) Upon removal of insulin from the insulin receptor, patches of membrane containing D-glucose carriers are apparently removed from the membrane into intracellular stores, and insulin-stimulated D-glucose transport ceases due to lack of access of the D-glucose carriers to the extracellular space.

constitutes approximately 40% of the mass of the anion carrier, can be proteolytically removed without affecting membrane transport. This intracellular domain of the exchanger binds cytoskeletal elements that form a link in anchoring membrane components to the cytoskeleton.

As an obligate exchanger, the protein must be fully occupied by anions before transport can occur. Accordingly, during a single cycle of transport, the Cl^-–HCO_3^- exchanger moves an equal number of charges in opposite directions, and the exchange is therefore electroneutral. However, Cl^-—Cl^- exchange, chloride-sulfate (Cl^-—SO_4^{2-}) exchange, and SO_4^{2-}—SO_4^{2-} exchange will also occur at different maximal rates. In the case of Cl^-—SO_4^{2-} exchange, neutrality is maintained by protonation of the carrier.

Na^+—H^+ EXCHANGER

Another ion exchanger that plays a role in the maintenance of intracellular pH is the Na^+—hydrogen ion (H^+) exchanger. This protein exchanges intracellular protons for extracellular Na^+. Most animal cells maintain a low intracellular concentration of Na^+ of between 5 and 10 mM, and the concentration of Na^+ in the extracellular fluid is in the range of 140 mM, thus providing a large driving force for the removal of intracellular H^+. Under most physiologic circumstances, the Na^+ gradient is sufficiently large to cause over-alkalinization of the cell. This is prevented by an allosteric site on the Na^+—H^+ exchanger, which requires occupation for operation of the exchanger. The Na^+—H^+ exchanger plays a role in regulating intracellular pH as cells undergo development and as the intracellular pH is altered in response to the changing needs of the cell. This presumably occurs through alteration of the set-point pH of the cell, accomplished through covalent modification of the Na^+—H^+ exchanger by intracellular protein kinases.

SECONDARY ACTIVE TRANSPORT

In all of the examples described, the transport systems serve to equilibrate substrates across cell membranes. With such systems, the concentration of solute within cells cannot exceed the extracellular concentration. Trapping of solutes by chemical modification of the substrate (for example, by phosphorylation of D-glucose to form the charged, membrane-impermeant glucose-6-phosphate) allows the D-glucose transport system to operate in a unidirectional manner. However, some transport systems can serve to concentrate solutes within cells. An example is the Na^+-dependent D-glucose transport system of the intestine, which accomplishes the concentrative uptake of D-glucose from the intestine into the cells lining the intestine. Glucose is then moved into the circulation by facilitated diffusion. Na^+ is present in the intestine at high levels, and is kept at low levels in the cells lining the intestines by the (Na,K)-ATPase. This Na^+-dependent D-glucose carrier allows the concentrative (active) uptake of glucose from the intestines by coupling the downhill movement of Na^+ to the uphill movement of D-glucose into the cell. This is an example of secondary active transport, driven by the Na^+ gradient produced eventually by (Na,K)-ATPase.

PRIMARY ACTIVE TRANSPORT: ION PUMPS

The (Na,K)-ATPase is found in virtually all animal cells. It consists of an approximately 100,000-Da alpha subunit and an approximately 50,000-Da beta subunit. The beta subunit is essential for complete catalytic function, stability, and movement to the plasma membrane, but most of the ligand binding sites (discussed below) have been located on the alpha subunit. Na^+ and K^+-dependent ATP hydrolysis by this enzyme results in the outward movement of three Na^+ in exchange for the inward movement of two K^+. Since the movement of ions is unequal, the (Na,K)-ATPase produces a membrane potential (**charge gradient**) in addition to **ion gradients** of Na^+ and K^+. The (Na,K)-ATPase is the receptor for cardiac glycosides, which act through inhibition of the catalytic and transport functions of the (Na,K)-ATPase. A close relative of the (Na,K)-ATPase is the gastric (H^+ + K^+)-ATPase, which is found in the gastric mucosa and which accomplishes the ATP- and K^+-dependent production of gastric hydrochloric acid. Ca^{2+}-Stimulated ATPases of the plasma membrane of most animal cells act to export Ca^{2+} from the cell, and similar Ca^{2+}-ATPases are found in the sarcoplasmic reticulum of muscles and related organelles of other cells. The sarcoplasmic reticulum contains Ca^{2+} stores. Hormones and other agents cause a release of Ca^{2+} from these intracellular storage organelles. Re-uptake of Ca^{2+} is accomplished by Ca^{2+}-ATPase in the sarcoplasmic reticulum. This class of Ca^{2+}-ATPase is under regulation by cyclic adenosine 5'-monophosphate (cAMP). In the heart, cAMP-dependent protein kinase phosphorylates the regulator of the Ca^{2+} pump, phospholamban. When nonphosphorylated, phospholamban binds to the Ca^{2+}-ATPase, and the binding of phospholamban inhibits the Ca^{2+}-ATPase, thereby maintaining high levels of intracellular free Ca^{2+}.

All of these ion pumps contain an essential aspartyl group that can be phosphorylated. Vanadate, which prevents phosphoenzyme formation, is an inhibitor of these pumps. Phosphoenzyme formation and vanadate sensitivity distinguish this class of ion pump from others such as the mitochondrial ATPase and the multicomponent vacuolar H^+-ATPases, which are found in a variety of intracellular organelles (discussed below).

The bulk of the energy (ATP) of many cells is used by these pumps for the generation of ion gradients and membrane potential. The precise mechanism by which the ion pumps utilize the energy of ATP hydrolysis remains largely unknown, although ligand binding and ATP hydrolysis are associated with changes in the conformation of the ion pumps which are intimately involved in the transport process. The magnitude of ion and potential gradients which are theoretically possible is dictated only by the energy released by hydrolysis of ATP. The rate of production of ion gradients is dictated by the rates of change of conformation of the ion pumps as they progress through the catalytic cycle. A simplified diagram of the reaction cycle of the (Na,K)-ATPase is shown in Figure 2-4. Na^+ binds to the (Na,K)-ATPase at a site on the cytosolic face of the enzyme, providing a unique conformation (E_1), with which ATP reacts, creating the phosphorylated form (E_1P) of the (Na,K)-ATPase. As Na^+ leaves the pump at the extracytosolic face, K^+ binds. The binding of K^+ is followed by loss of phosphate from the aspartyl group, and a change in the conformation of the pump in which K^+ is exposed to neither the extracellular nor intracellular face of the pump, and is therefore trapped, or occluded, within the pump. When ATP noncovalently binds to the (Na,K)-ATPase, the rate of release of occluded K^+ is increased and

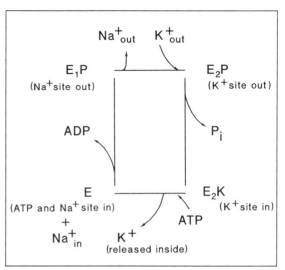

Fig. 2-4. Reaction cycle of the Na^+- and K^+-stimulated ATPase ([Na,K]-ATPase). The (Na,K)-ATPase catalyzes the transport of three Na^+ out of the cell in return for two K^+ transported into the cell for each ATP hydrolyzed: $3 Na^+_{out} + 2 K^+_{in} + 1 ATP \longrightarrow 3 Na^+_{in} + 2 K^+_{out} + ADP + P_i$. Substrate binding to the (Na,K)-ATPase causes a conformational change in the enzyme. Na^+ binding to the cytosolic face of the enzyme causes a conformational change in the enzyme to the E_1 form, to which ATP also binds. Phosphorylation of the enzyme, forming E_1NaP and trapping Na^+ (occlusion), follows. After Na^+ is released to the outside of the cell (E_2P), K^+ binds. Phosphate is released inside the cell, and if ATP is present and binds, K^+ is released, completing the catalytic cycle. For a given enzyme molecule, this reaction occurs approximately 120 times each second. The rate of change of enzyme conformation from one state to another determines the physiologically relevant rate of ion pump activity. (E = enzyme; P_i = inorganic phosphate.)

the pump returns to the E_1 form with bound Na^+.

Other ion pumps are known. These do not form phosphoenzyme intermediates and are not inhibited by vanadate. For example, the **mitochondrial proton pump** is responsible for producing ATP, and the **vacuolar ATPases** are present in the membrane of intracellular organelles. The vacuolar ATPases serve to generate membrane potentials (charge gradients) and to acidify the intracellular organelles. The pH and potential gradients formed by these pumps are utilized for secondary active transport processes. In the case of the vacuolar ATPases of the lysosome and the clathrin-coated vesicles, organelle acidification is directly related to the physiologic function of the organelle. For example, the lysosome must be acidified for the lysosomal hydrolase to be active, and the vacuolar ATPase is required for receptor-mediated uptake of some nutrients and hormones. The uptake of iron occurs through receptor-mediated endocytosis of iron bound to transferrin. Pump-dependent acidification of the vesicle containing transferrin with bound iron results in the release of iron. Transferrin bound to its receptor then returns to the cell surface.

SUMMARY

Biologic membranes are composed of lipids and proteins. Membranes pose a barrier to the passage of solutes. Smaller and more hydrophobic solutes are more readily permeant by passive diffusion across biologic membranes than are larger and more hydrophylic agents. The passage of some solutes requires specific proteins that catalyze the movement of solutes across the membrane. The presence of such a transport protein is indicated by a saturable uptake process, and, in many cases, specific inhibitors of the transport proteins have been identified that prevent the catalysis of transport. Facilitated diffusion results in equilibration of solutes across the membrane, driven by a concentration gradient, whereas primary and secondary active transport systems catalyze the concentrative uptake of solutes and require the input of metabolic energy.

NATIONAL BOARD TYPE QUESTIONS

Select the one most correct answer for each of the following

1. Which of the following statements regarding membranes is correct? Membranes are composed of:
   A. One layer of lipid with hydrophobic tails immersed in water
   B. One layer of lipid sandwiched between two layers of protein
   C. A rigid matrix of lipid with mobile protein
   D. A dynamic bimolecular leaflet of lipids with integral and peripheral proteins
   E. A lipid bilayer that is more permeant to hydrophylic compounds than to hydrophobic compounds

2. Arrange the following compounds in order according to their expected permeability across a lipid bilayer: ethane, methanol, hemoglobin, K^+, oxygen. (The molecular weight of O_2, ethane, methanol, and K^+ are approximately equal.)
   A. O_2 > ethane > hemoglobin > K^+ > methanol
   B. Methanol > ethane > K^+ > O_2 > hemoglobin
   C. K^+ > methanol > ethane > O_2 > hemoglobin
   D. K^+ > O_2 > methanol > ethane > hemoglobin
   E. Ethane > O_2 > methanol > K^+ > hemoglobin

3. Which one of the following statements is correct?
   A. D-glucose transport is regulated by insulin in all tissues
   B. The rate of utilization of D-glucose by the brain is the primary determinant of serum D-glucose concentration
   C. D-glucose transporters from all tissues are identical
   D. Diabetes results only from defective insulin receptors
   E. The glucose carriers of heart and skeletal muscles as well as adipocytes are important determinants of serum D-glucose concentration

4. Which of the following statements regarding the (Na,K)-ATPase is correct?
   A. Drives the movement of Na^+ from cells
   B. Transports unequal numbers of Na^+ and K^+
   C. Accomplishes primary active transport
   D. Generates a charge and ion gradient
   E. All of the above

5. Cell pH can be regulated by:
   A. The $Cl^- - HCO_3^-$ exchanger alone
   B. The $Na^+ - H^+$ exchanger alone
   C. Passive diffusion of H^+
   D. Vacuolar H^+ pumps
   E. The $Cl^- - HCO_3^-$ exchanger *and* the $Na^+ - H^+$ exchanger

---

ANNOTATED ANSWERS

1. D. The lipids are arranged in a bimolecular leaflet with the hydrophobic tails buried in the interior and the charged head groups on the surface. Lipids are mobile within the plane of the membrane, proteins often rotate about their own axis, and changes in membrane constituents occur through normal cellular processes.

2. E. Charge or polarity and size are important determinants of membrane permeability. Ethane is less polar than O_2, which is less polar than methanol, and all are approximately the same molecular weight. K^+ is charged. It will not easily partition into the membrane lipids, and therefore would be expected to move slower than the other compounds of similar molecular weight. Hemoglobin will not move across an intact membrane at a measurable rate.

3. E. Insulin stimulates the transport of D-glucose in the heart and skeletal muscle and adipocytes. These tissues constitute the mass of body tissue, and the equilibration of glucose across these tissues is responsible for most D-glucose transport in the tissues. In diabetes, D-glucose transport is defective. In a small number of diabetics, the insulin receptor is defective, but, in most diabetics, the insulin receptors are intact; it is other steps in the regulation of D-glucose transport that are defective.

4. D. The (Na,K)-ATPase transports three Na^+ out of the cell and two K^+ into the cell for each ATP molecule that is hydrolyzed. Charge and ion gradients are established at the expense of cellular ATP. The (Na,K)-ATPase is thus an example of a primary active transport system.

5. E. The $Cl^- - HCO_3^-$ exchanger and the $Na^+ - H^+$ exchanger both operate in the regulation of intracellular pH.

BIBLIOGRAPHY

Cabantchik, Z. I., and Rothstein, A. The nature of the membrane sites controlling anion permeability of human red blood cells as determined by studies with disulfonic stilbene derivatives. *J. Membr. Biol.* 10:311–330.

Hediger, M. A., Coady, M. J., Ikeda, T. S., and Wright, E. M. Expression cloning and cDNA sequencing of the Na^+/glucose co-transporter. *Nature* 330:379–381, 1987.

Simpson, I. A., and Cushman, S. W. Hormonal regulation of mammalian glucose transport. *Annu. Rev. Biochem.* 55:1059–1089, 1986.

Singer, S. J., and Nicholson, G. L. The fluid mosaic model of the structure of cell membranes. *Science* 175:720–731, 1972.

Stein, W. D. *Transport and Diffusion across Cell Membranes.* New York: Academic, 1986.

3 Origin of the Resting Membrane Potential

NICHOLAS SPERELAKIS

Objectives

After reading this chapter, you should be able to:

Describe the molecular structure and function of the membrane of cells at rest (i.e., not electrically excited)

Explain the maintenance of the cell's internal ion concentrations through the action of ion pumps and exchangers

Explain the production of voltages by separation of ionic charges

Describe the equilibrium potentials for ions distributed across the cell membrane

Identify the driving forces for ionic currents

Explain the meaning of permeability and ionic conductance

List the factors that determine the value of the resting potential

Explain why the resting potential decreases when the potassium ion concentration in the blood plasma is elevated above normal (life threatening)

Describe the influence of electrogenic pump potentials on the resting potential

Explain how spontaneous pacemaker potentials develop in some cell types

Explain the importance of the resting potential in the operation of action potentials and their velocity of propagation

---

The cell membrane exerts tight control over the contractile machinery of muscle cells during the process of **excitation—contraction** (electromechanical) **coupling.** Some drugs and toxins exert primary or secondary effects on the electric properties of the cell membrane, and thereby affect automaticity, arrhythmias, conduction, and the force of contraction. Therefore, to understand the mode of action of toxins, therapeutic agents, neurotransmitters, hormones, and plasma electrolytes in terms of the electric activity, it is necessary to comprehend the electric properties and the behavior of the cell membrane at rest and during excitation. The first step is to examine the electric properties of nerve and muscle cells at rest, including the origin of the resting membrane potential (E_m). The resting potential and action potential result from properties of the cell membrane and the ion distributions across the cell membrane.

PASSIVE ELECTRIC PROPERTIES AND CABLE PROPERTIES

Membrane Structure and Composition

The cell membrane is composed of a bimolecular leaflet containing phospholipid molecules (e.g., phosphatidylcholine and phosphatidylethanolamine) sandwiched between two layers of adsorbed protein. The membrane is about 70 Å thick. The nonpolar hydrophobic tails of the phospholipid molecules project toward the middle of the membrane, and the polar hydrophilic heads project toward the edges of the membrane bordering on the water phases (Fig. 3-1). This orientation is thermodynamically favorable. The lipid bilayer membrane is about 50 to 70 Å thick, and the phospholipid molecules are the right length (30–40 Å) to stretch across half the membrane thickness. Cholesterol molecules are highly concentrated in the cell membrane (of animal cells), yielding a phospholipid–cholesterol ratio of about 1.0; these molecules are inserted between the heads of the phospholipid molecules. Large protein molecules are also inserted in the lipid bilayer matrix. Some proteins protrude through the entire membrane thickness (e.g., the sodium-potassium adenosinetriphosphatase [Na^+K^+]-ATPase and the various ion channel proteins). These proteins "float" in the lipid bilayer matrix, and the membrane has fluidity (reciprocal of microviscosity), such that the protein molecules can move around laterally in the plane of the membrane. Thus, these floating proteins do not appear to be anchored in one place, in most cases.

The outer surface of the cell membrane is lined with strands of mucopolysaccharides (the cell coat or glycocalyx) that endow the cell with immunochemical properties. The cell coat is highly negatively charged, and therefore can bind cat-

Fig. 3-1. Cell membrane substructure showing the lipid bilayer. Nonpolar hydrophobic tails of the phospholipid molecules project toward the middle of the membrane, and polar hydrophilic heads border on the water phase at each side of the membrane. The lipid bilayer is about 50 to 70 Å thick. For simplicity, the cholesterol molecules are not shown. Large protein molecules protrude through the entire membrane thickness or are inserted into one leaflet only. These proteins include various enzymes associated with the cell membrane as well as membrane ionic channels. The membrane has fluidity, so that the protein and lipid molecules can move around in the plane of the membrane and fluorescent probe molecules inserted into the hydrophobic region of the membrane have freedom to rotate.

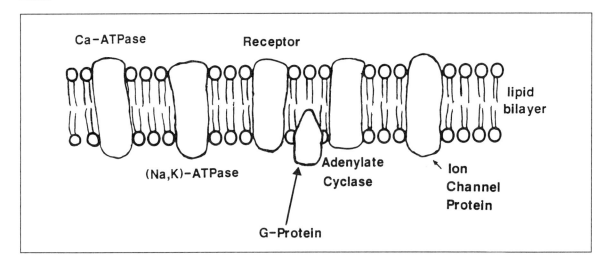

ions, such as Ca^{2+}. Treatment with neuraminidase, to remove sialic acid residues, destroys much of the cell coat.

Membrane Capacitance and Resistivity

Artificially made lipid bilayer membranes have a specific **membrane capacitance** of 0.4 to 1.0 $\mu F/cm^2$, which is close to the value for biologic membranes. The capacitance of cell membranes is due to the lipid bilayer matrix. A capacitor consists of two parallel plate conductors separated by a high-resistance dielectric material, and has the ability to store electric charges across its two plates. The following equation gives the parameters that determine membrane capacitance (C_m):

$$C_m = \left[\frac{\epsilon A_m}{\delta}\right]\left[\frac{1}{4\pi k}\right]$$

where A_m is the membrane area (in square centimeters) and k is a constant (9.0×10^{11} cm/F). Most oils have dielectric constants of 3 to 5. The more dipolar the material, the greater the dielectric constant (e.g., water has a value of 81). Calculation of membrane thickness (δ) from the above equation, assuming a measured membrane capacitance of 0.7 $\mu F/cm^2$ and a dielectric constant (ϵ) of 5, gives 63 Å.

The artificial lipid bilayer membrane, on the other hand, has an exceedingly high specific resistance of 10^6 to 10^9 Ω-cm^2, which is several orders of magnitude higher than that of the biologic cell membrane (about 10^3 Ω-cm^2). The membrane resistance is greatly lowered, however, if the bilayer is treated with certain proteins or substances, such as macrocyclic polypeptide antibiotics (ionophores). (The added ionophores may be of the ion-carrier type, such as valinomycin, or of the channel-former type, such as gramicidin.) Therefore, the presence of proteins that span the thickness of the cell membrane must account for the relatively low resistance (high conductance) of the cell membrane. These proteins include those associated with the voltage-dependent, gated ion channels of the cell membrane.

Membrane Fluidity

The ion transport and electric properties of the cell membrane are determined by the molecular composition of the membrane. The lipid bilayer matrix even influences the function of the membrane proteins (e.g., the [Na, K]-ATPase activity is affected by the surrounding lipid). A high cholesterol content lowers the fluidity of the membrane and a high degree of unsaturation and branching of the tails of the phospholipid molecules raises the fluidity; chain length also affects fluidity. The polar portion of cholesterol lodges in the hydrophilic part of the membrane, and the nonpolar part of the planar cholesterol molecule wedges between the fatty acid tails, thus restricting their motion and lowering fluidity. Phospholipids with unsaturated and branched-chain fatty acids cannot be packed tightly because of steric hindrance due to their greater rigidity; hence such phospholipids increase membrane fluidity. Low temperature decreases membrane fluidity, as expected. Ca^{2+} and Mg^{2+} may diminish the charge repulsion between the phospholipid head groups; this allows the bilayer molecules to pack more tightly, thereby constraining the motion of the tails and reducing fluidity. Each phospholipid tail occupies about 20 to 30 $Å^2$, and each head group, about 60 $Å^2$. Membrane fluidity changes occur in muscle development and in certain disease states, such as cancer, muscular dystrophy (Duchenne type), and myotonic dystrophy.

The hydrophobic portion of local anesthetic molecules may interpose between the lipid molecules. This further separates the acyl chain tails of the phospholipid molecules, reducing the Van der Waals forces of interaction between adjacent tails, and so increasing membrane fluidity. Local anesthetics depress the resting conductance of the membrane for K^+ and Na^+ (g_K and g_{Na}) and depress the voltage-dependent changes in g_{Na}, g_K, and g_{Ca}; that is, the local anesthetics produce a nonselective depression of most conductances of the resting and excited membrane. The estimated concentration of local anesthetic in the lipid bilayer required to completely block excitability is more than $100,000/\mu m^2$. Part of the depression of ion conductances and (Na,K)-ATPase activity could arise indirectly from the anesthetics' effect on the fluidity of the lipid matrix.

Potential Profile Across Membrane

The cell membrane has **fixed negative charges** at its outer and inner surfaces. The charges are presumably due to acidic phospholipids in the bilayer and to protein molecules either embedded in the membrane (islands floating in the lipid bilayer matrix) or tightly adsorbed to the surface of the membrane. Most proteins have an acidic isoelectric point, so that they possess a net negative charge at a pH near 7.4. The charge at the outer surface of the cell membrane, with respect to the solution bathing the cell, is known as the **zeta potential.** This charge is responsible for the electrophoresis of cells in an electric field (i.e., the cells moving toward the anode [positive electrode]) because unlike charges attract. The **surface charge** affects the true potential difference across the membrane, as illustrated in Figure 3-2A. At each surface, the fixed charge produces an electric field that extends a short distance into the solution and causes each surface of the membrane to be slightly more negative (by a few millivolts) than the extracellular and intracellular solutions. The potential theoretically recorded by an ideal electrode as it is driven through the solution perpendicular to the membrane surface should become negative as the electrode approaches within a few Angstroms of the surface. The potential difference between the membrane surface and the solution declines exponentially as a function of distance from the surface, the length constant being larger at low ionic strength (or high resistivity) of the solution. The magnitude of the potential difference depends on the density of the charge sites (number per unit area); the number of charges is also affected by the ionic strength and pH.

The membrane potential (E_m) measured by an intracellular microelectrode is the potential of the outer solution (ψ_o; the reference electrode) minus the potential of the inner solution (ψ_i; the active microelectrode), thus: $E_m = \psi_o - \psi_i$. The true potential difference across the membrane (E'_m), however, is that which is directly across the membrane (see Fig. 3-2A). If the surface charges at each surface of the membrane are equal, then $E'_m = E_m$. If the outer surface charge is decreased to zero by extra binding of protons or cations, such as Ca^{2+}, then the membrane becomes slightly hyperpolarized ($E'_m > E_m$) (see Fig. 3-2B). Conversely, if the inner surface charge is neutralized, then the membrane becomes slightly depolarized ($E'_m < E_m$). These effects are not measurable by an intracellular microelectrode.

Because the membrane ionic conductances are controlled by the potential difference directly across the membrane (i.e., by E'_m and not E_m), changes in the surface charges (e.g., by drugs, ionic strength, or pH) can lead to apparent shifts in threshold potential. For example, an elevated extracellular Ca^{2+} concentration ($[Ca]_o$) is known to raise the threshold potential (i.e., the critical depolarization required to reach the electric threshold), as expected from the small increase in E'_m.

ION DISTRIBUTIONS AND THEIR MAINTENANCE

Ion Distributions

The transmembrane potential in resting nerve and muscle cells varies with cell type. In nerve cells, the **resting potential** is about -70 mV, whereas in skeletal muscle fibers and myocardial cells, the resting potential is close to -80 mV. The resting potential or maximum diastolic potential in cardiac Purkinje fibers is somewhat greater (about -90 mV). In smooth muscle cells and nodal cells of the heart, the resting potential is lower (about -55 mV). The values of the resting potentials of some nerve and muscle cells are summarized in Table 3-1.

The ionic composition of the extracellular fluid

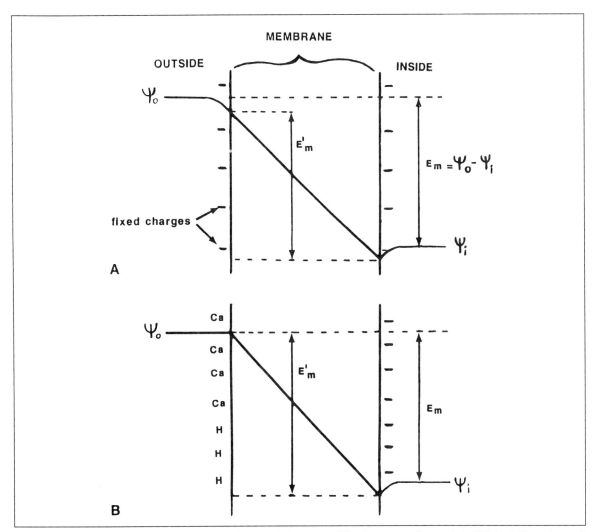

Fig. 3-2. Potential profile across the cell membrane. (A) Because of fixed negative charges (at pH 7.4) at the outer and inner surfaces of the membrane, there is a negative potential that extends from the edge of the membrane into the bathing solution on both sides of the membrane. This surface potential falls off exponentially with added distance into the solution. The magnitude of the surface potential is a function of the charge density. ψ_p is the electric potential of the outside solution, ψ_i is that of the inside solution, and membrane potential (E_m) is the difference between the two ($\psi_o - \psi_i$). E_m is determined by the equilibrium potentials and relative conductances. The profile of the potential through the membrane is shown as linear (the constant-field assumption), although this need not be true for the present purpose. If the outer surface potential is exactly equal to that on the inner surface, then the true transmembrane potential (E'_m) is exactly equal to the microelectrode-measured E_m. (B) If the outer surface potential is different from the inner potential, for example by elevating the extracellular Ca^{2+} concentration or lowering the pH to bind Ca^{2+} or H^+ to more of the negative charges, then the E'_m is greater than the measured E_m. Diminution of the inner surface charge decreases E'_m. The membrane ion changes are controlled by E'_m.

Table 3-1. Comparison of the Resting Potentials in Different Types of Cells

Cell Type	Resting Potential (mV)
Neuron	− 70
Skeletal muscle (mammalian)	− 80
Skeletal muscle (frog)	− 90
Cardiac muscle (atrial and ventricular)	− 80
Cardiac Purkinje fiber	− 90
Atrioventricular nodal cell	− 65
Sinoatrial nodal cell	− 55
Smooth muscle cell	− 55

Table 3-2. Summary of the Ion Distributions in Most Types of Cells, and the Equilibrium Potential Calculated From the Nernst Equation

Ion	Extracellular Distribution (mM)	Intracellular Distribution (mM)	Equilibrium Potential (mV)
Na^+	145	15	+ 60
Cl^-	125	7*	− 80
K^+	4.5	160	− 94
Ca^{2+}	1.8	0.0001	+ 129
H^+	0.0001	0.0002	− 18

*Assuming Cl^- is passively distributed.

bathing the cells is similar to that of blood plasma. It is high in Na^+ (about 145 mM) and Cl^- (about 120–130 mM), but low in K^+ (about 4.5 mM). The Ca^{2+} concentration is about 2 mM. In contrast, the intracellular fluid has a low concentration of Na^+ (about 15 mM or less) and Cl^- (about 6–8 mM), but a high concentration of K^+ (about 150–170 mM). The free intracellular Ca^{2+} concentration ($[Ca]_i$) is about 10^{-7} M, but during contraction it may rise as high as 10^{-5} M. The ion distributions are listed in Table 3-2. The total intracellular Ca^{2+} is much higher (about 2 mmol/kg), but most of this is bound to molecules such as proteins or is sequestered in compartments such as mitochondria and the sarcoplasmic reticulum. Most of the intracellular K^+ is free, and it has a diffusion coefficient only slightly less than K^+ in free solution. Thus, under normal conditions, the cell maintains an internal ion concentration markedly different from that in the medium bathing the cells, and it is these ion concentration differences that create a resting potential. The existence of the resting potential enables action potentials to be produced in those types of cells that are excitable. The ion distributions and related pumps and exchange reactions are depicted in Figure 3-3.

Inhibition of the Na^+-K^+ pump (e.g., by cardiac glycosides such as digitalis) gradually runs down (depletes) the ion concentration gradients. The cells loose K^+ and gain Na^+, and therefore the K^+ (E_K) and Na^+ (E_{Na}) equilibrium potentials become smaller. The cells thus become depolarized (even if the relative permeabilities are unaffected), which causes them to gain Cl^- (because the intracellular Cl^- concentration $[Cl]_i$ was kept low by the large resting potential) and therefore also water (cells swell) because of the resultant gain in intracellular osmotic strength.

Na^+—K^+ Pump

The intracellular ion concentrations are maintained differently from those in the extracellular fluid by active ion transport mechanisms that expend metabolic energy to propel specific ions against their concentration or electrochemical gradients. These ion pumps are located in the cell membrane at the cell surface and probably also in the transverse tubular membrane in striated muscle cells. The major ion pump is the **Na^+-K^+—linked pump,** which transports Na^+ out of the cell against its electrochemical gradient while simultaneously transporting K^+ in against its electrochemical gradient (see Fig. 3-3). The coupling of Na^+ and K^+ pumping is obligatory, because if the extracellular K^+ concentration ($[K]_o$) is zero,

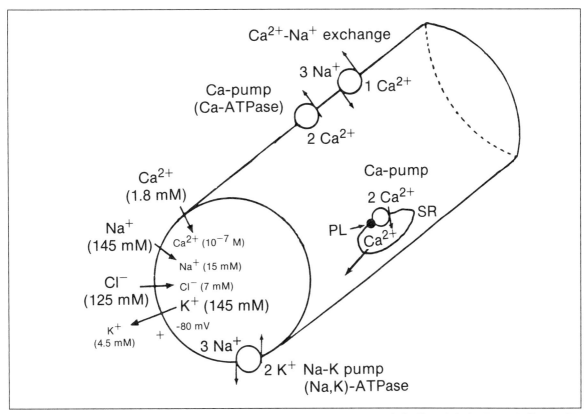

Fig. 3-3. Intracellular and extracellular ion distributions in vertebrate myocardial cells and skeletal muscle fibers. Also shown are the polarity and magnitude of the resting potential. Arrows indicate direction of the net electrochemical gradient. The $Na^+ - K^+$ pump and $Ca^{2+} - Na^+$ exchange carrier are located in the cell surface membrane. A Ca-ATPase and Ca^{2+} pump, similar to that in the sarcoplasmic reticulum (*SR*), is located in the cell membrane. Phospholamban (*PL*) is depicted close to the CA^{2+} pump of the sarcoplasmic reticulum and serves to regulate pump activity.

Na^+ can no longer be pumped out; that is, a coupling ratio of 3 Na^+:0 K^+ is not possible. The coupling ratio of Na^+ pumped out to K^+ pumped in is generally 3:2. The Na^+-K^+ pump is half-inhibited (Michaelis constant [K_m] value) when [K]$_o$ is lowered to 2 mM.

If the ratio were 3:3, the pump would be electrically neutral or nonelectrogenic: a potential difference across the membrane would not be directly produced because the pump would pull in three positive charges (K^+) for every three positive charges (Na^+) it pushed out. When the ratio is 3:2, the pump is electrogenic and directly produces a potential difference that causes the E_m to

be greater (more negative) than it would be otherwise, solely on the basis of the ion concentration gradients and relative permeabilities or net diffusion potential (E_{diff}). A coupling ratio of 3 Na^+:1 K^+ would produce a greater electrogenic pump potential (V_{ep}). Under normal steady-state conditions, the contribution of the Na^+ V_{ep} to E_m (ΔV_{ep}) is only a few millivolts in myocardial cells; the contribution is greater in smooth muscle (e.g., 8 mV) and even greater in mammalian skeletal muscle (e.g., up to 16 mV).

The driving mechanism for the Na^+-K^+ pump is a membrane ATPase, the (Na,K)-ATPase that requires both Na^+ and K^+ for activation. This enzyme requires Mg^{2+} for activity. ATP, Mg^{2+}, and Na^+ are thus required at the inner surface of the membrane, and K^+ is required at the outer surface. A phosphorylated intermediate of the (Na,K)-ATPase occurs in the transport cycle, its phosphorylation being Na^+ dependent and its dephosphorylation being K^+ dependent. The pump enzyme usually drives three Na^+ in and two K^+ out for each ATP molecule hydrolyzed. The (Na,K)-ATPase is specifically inhibited by the cardiac glycosides (digitalis) acting on the outer surface, by competing with K^+ for the K^+-binding site. The pump enzyme is also inhibited by sulfhydryl reagents (such as N-ethylmaleimide and mercurial diuretics), thus indicating that the sulfhydryl groups are crucial for activity.

Blockade of the Na^+-K^+ pump produces a small, but important, immediate effect on the resting E_m: a depolarization of about 2 to 16 mV, depending on cell type, representing the contribution of V_{ep} to E_m (ΔV_{ep}). Since the generation of action potentials is almost unaffected for short intervals, excitability is independent of active ion transport. However, over a period of many minutes, depending on the ratio of surface area to volume of the cell, the resting E_m slowly declines because of gradual dissipation of the ionic gradients. The progressive depolarization depresses the rate of rise of the action potential and hence the propagation velocity, and eventually all excitability is lost. Thus, a large resting potential and excitability, although not immediately dependent on the Na^+-K^+ pump, are ultimately so.

The rate of Na^+-K^+ pumping in excitable cells must change with cell activity in order to maintain relatively constant intracellular ion concentrations. A higher frequency of action potentials results in a greater overall movement of ions down their electrochemical gradients, and these ions must be repumped to maintain the ion distributions. For example, the cells tend to gain Na^+, Cl^-, and Ca^{2+} and to lose K^+. The factors that control the rate of Na^+-K^+ pumping include the intracellular Na^+ concentration ($[Na]_i$) and $[K]_o$. In cells with a large surface area to volume ratio (such as small-diameter nonmyelinated axons), $[Na]_i$ may increase by a relatively large per-centage during a train of action potentials, and this would stimulate the pumping rate. Likewise, an extracellular accumulation of K^+ occurs, and also stimulates the pump (the K_m for K^+ is about 2 mM).

Chloride Ion Distribution

In many invertebrate and vertebrate nerve and muscle cells, Cl^- does not appear to be actively transported: that is, there is no Cl^- pump (i.e., no Cl^--ATPase). In such cases, Cl^- distributes itself passively (no energy used) in accordance with E_m: the Cl^- equilibrium potential (E_{Cl}) is equal to E_m in a resting cell. For example, in mammalian myocardial cells, Cl^- also seems to be close to passive distribution, because $[Cl]_i$ is at, or only slightly above, the value predicted from the resting E_m. When passively distributed, $[Cl]_i$ is low because the negative potential inside the cell (the resting potential) pushes out the negatively charged Cl^- (like charges repel) until the Cl^- distribution is at equilibrium with the resting E_m. Hence, for a resting E_m of -80 mV, and an extracellular Cl^- concentration ($[Cl]_o$) of 120 mM, $[Cl]_i$ would be 5.9 mM, that is: $[E_m = +61$ mV log $([Cl]_i/[Cl]_o)]$. However, during the action potential, the inside of the cell changes in a positive direction, and a net Cl^- influx (outward Cl^- current) occurs, and thus increases $[Cl]_i$. The magnitude of the Cl^- influx (I_{Cl}) depends on the Cl^- conductance (g_{Cl}) of the membrane, that is: $[I_{Cl} = g_{Cl} (E_m - E_{Cl})]$. Thus, the average $[Cl]_i$ in excitable cells should depend on the frequency and duration of the action potential, that is, on the mean E_m averaged over many action potential cycles. In smooth muscle, $[Cl]_i$ is much higher than the value predicted from passive distribution; this elevated $[Cl]_i$ may be due to an exchange carrier (e.g., Cl^-—bicarbonate exchange) or a cotransporter (e.g., Na^+,K^+-Cl_2).

Calcium Ion Distribution

Need for Calcium Pumps

For the positively charged Ca^{2+}, there must be some mechanism for removing Ca^{2+} from the cytoplasm. Otherwise the cell would continue to gain Ca^{2+} until there was no electrochemical gradient for the net influx of Ca^{2+}. Theoretically, Ca^{2+} loading would occur until the free $[Ca]_i$ in the cytoplasm was even greater than that outside (approximately 2 mM) because of the negative potential inside the cell. Therefore, one or more **Ca^{2+} pumps** must be in operation. The sarcoplasmic reticulum membrane contains a Ca^{2+}-activated ATPase (which also requires Mg^{2+}) that actively pumps two Ca^{2+} from the myoplasm into the lumen of the sarcoplasmic reticulum at the expense of one ATP, and is capable of reducing the $[Ca]_i$ to less than 10^{-7} M. The Ca-ATPase of the sarcoplasmic reticulum is regulated by an associated low-molecular-weight protein, phospholamban. Phospholamban is phosphorylated by cyclic adenosine 5'-monophosphate–dependent protein kinase, and when phosphorylated, stimulates the Ca-ATPase and Ca^{2+} pumping. The sequestration of Ca^{2+} by the sarcoplasmic reticulum is essential for muscle relaxation. (The mitochondria can also actively take up Ca^{2+} but this Ca^{2+} pool probably does not play an important role in normal excitation–contraction coupling processes.) However, the resting Ca^{2+} influx and the extra Ca^{2+} influx that occurs with each action potential must be returned to the interstitial fluid. Mechanisms proposed for this include the Ca-ATPase and $Ca^{+}–Na^{+}$ exchanger present in the sarcolemma. The Ca-ATPase in the sarcolemma actively transports two Ca^{2+} outward against an electrochemical gradient, utilizing one ATP in the process. Phospholamban is not associated with the sarcolemmal Ca-ATPase.

$Ca^{2+}–Na^{+}$ Exchange Reaction

The **$Ca^{2+}–Na^{+}$ exchange reaction** exchanges one internal Ca^{2+} for three external Na^{+} via a membrane carrier molecule (see Fig. 3-3). This reaction is facilitated by ATP, but ATP is not hydrolyzed (consumed) in this reaction. Instead, the energy for the pumping of Ca^{2+} against its large electrochemical gradient is derived from the Na^{+} electrochemical gradient. That is, the uphill transport of Ca^{2+} is coupled with the downhill

movement of Na^{+}. Therefore, the energy required for this Ca^{2+} movement is effectively derived from the (Na,K)-ATPase. Thus, the $Na^{+}–K^{+}$ pump, which uses ATP to maintain the Na^{+} electrochemical gradient, indirectly helps to maintain the Ca^{2+} electrochemical gradient. Hence, the inward Na^{+} leak is greater than it would be otherwise.

The energy cost (in joules per mole) for pumping Ca^{2+} out of the cytoplasm is directly proportional to its electrochemical gradient, namely: $\Delta G_{Ca} = zF(E_m - E_{Ca})$, where ΔG_{Ca} is the change in free energy for Ca^{2+} and zF is coulombs/mol. The energy available from the Na^{+} distribution is directly proportional to its electrochemical gradient: $\Delta G_{Na} = zF(E_m - E_{Na})$. Depending on the exact values of $[Na]_i$ and $[Ca]_i$ at rest, the energetics would be about adequate for an exchange ratio of 3 Na^{+}:1 Ca^{2+}. An exchange ratio of 3:1 would produce a small depolarization (unless the three Na^{+} were accompanied by a univalent counter-ion such as Cl^{-}), because there is one net positive charge moving inward for every cycle of the carrier, i.e., a net inward current is produced.

The exchange reaction depends on relative concentrations of Ca^{2+} and Na^{+} on each side of the membrane and on relative affinities of the binding sites to Ca^{2+} and Na^{+}. Because of this $Ca^{2+}–Na^{+}$ exchange reaction, whenever the cell gains Na^{+} it also gains Ca^{2+} because the Na^{+} electrochemical gradient is reduced and the exchange reaction slows down. In addition, when the membrane is depolarized during the action potential plateau, the exchange carriers exchange the ions in reverse (internal Na^{+} for external Ca^{2+}), and thus increase Ca^{2+} influx. The net effect of both mechanisms is to elevate $[Ca]_i$. The $Ca^{2+}–Na^{+}$ exchange process has been proposed as the mechanism that stimulates contraction of the heart resulting from cardiac glycoside inhibition of the $Na^{+}-K^{+}$ pump.

EQUILIBRIUM POTENTIALS

For each ionic species distributed unequally across the cell membrane, an **equilibrium potential** (E_i) or battery can be calculated for that ion from the Nernst equation (for 37°C): $E_i = -61$ mV/z log (C_i/C_o), where C_i is the intracellular concentration of the ion, C_o is the extracellular concentration, and z is the valence of the ion (with sign). The -61 mV constant (2.303 RT/F) becomes -59 mV at 22°C. (R is the gas constant [8.3 joules/mol·deg K], T is the absolute temperature [deg K = 273 + degrees Celsius], F is the Faraday constant [96,500 coulombs/eq; zF = coulombs/mol], and 2.303 is the conversion factor for the natural log to $\log_{10}$.) The Nernst equation defines the potential difference (electric force) that would exactly oppose the concentration gradient (diffiusion force). Only a very small charge separation ([Q] in coulombs) is required to build up a very large potential difference: $E_m = Q/C_m$, where C_m is the membrane capacitance. For the ion distributions given in Table 3-2, the approximate equilibrium potentials are: $E_{Na} = +60$ mV; $E_{Ca} = +129$ mV; $E_K = -94$ mV; and $E_{Cl} = -80$ mV (Fig. 3-4). The sign of the equilib-

rium potential represents the inside of the cell with reference to the outside.

In a two-compartment system separated by a membrane, the *side of higher concentration becomes negative* for positive ions (cations) and positive for negative ions (anions). Any ion with an E_i different from the resting potential (e.g., -80 mV for a myocardial cell or skeletal muscle fiber) is off-equilibrium, and therefore must effectively be pumped at the expense of energy. In a myocardial cell, only Cl$^-$ appears to be at or near equilibrium, whereas Na$^+$, K$^+$, and Ca^{2+} are actively transported. Even H$^+$ is off-equilibrium, the H$^+$ equilibrium potential (E_H) being closer to zero potential (see Table 3-2). If H$^+$ were passively distributed, the negative intracellular potential would cause intracellular H$^+$ concentration [H]$_i$ to be much greater, and therefore the cell interior more acidotic. The mechanism for development of the equilibrium potential is depicted in Figure 3-5.

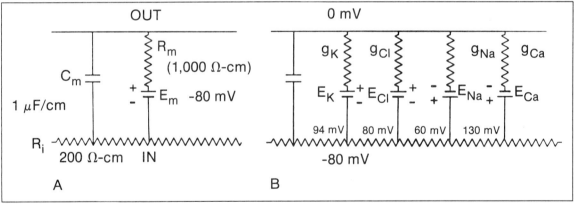

Fig. 3-4. Electric equivalent circuit for a cell membrane at rest. (A) Simplified circuit of resting membrane. The membrane serves as a parallel resistance–capacitance circuit, the membrane resistance (R_m) being in parallel with the membrane capacitance (C_m). Resting membrane potential (E_m) is represented by an 80-mV battery in series with the membrane resistance, the negative pole facing inward. (R_i = internal longitudinal resistivity.) (B) Expanded circuit. Membrane resistance is divided into four parts, one representing each of the four major ions of importance: K$^+$, Cl$^-$, Na$^+$, and Ca^{2+}. Resistances for these ions are parallel to one another, and represent totally separate and independent pathways for permeation of each ion through the resting membrane. These ion resistances are depicted as their reciprocals, namely, ion conductances (g_K, g_{Cl}, g_{Na}, and g_{Ca}). The equilibrium potential for each respective ion (e.g., E_K), determined solely by the ion distribution in the steady-state and calculated from the Nernst equation, is shown in series with the conductance path for that ion. The resting potential of -80 mV is determined by the equilibrium potentials and by the relative conductances.

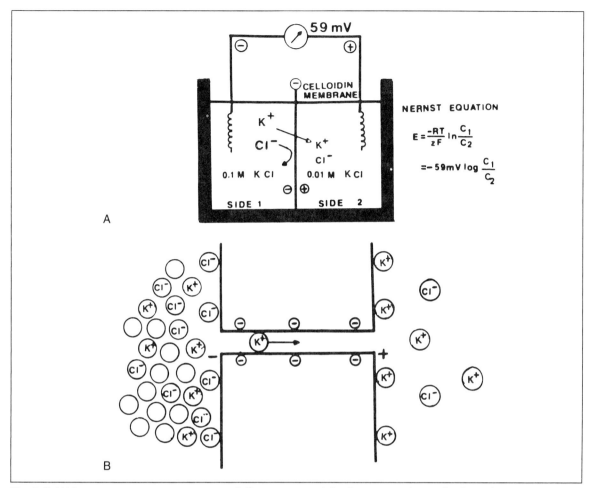

Fig. 3-5. (A) In a two-compartment system, a diffusion potential develops across the artificial membrane containing negatively charged pores. The membrane is impermeable to Cl⁻, but permeable to cations such as K⁺. The concentration gradient for K⁺ causes a potential to be generated, the side of higher K⁺ concentration becoming negative. (See text for elaboration of the Nernst equation.) (B) Expanded diagram of a water-filled pore in the membrane, showing the permeability to K⁺, but lack of penetration of Cl⁻. The potential difference is generated by the charge separation, a slight excess of K⁺ being held close to the righthand surface of the membrane; a slight excess of Cl⁻ is clustered close to the lefthand surface.

ELECTROCHEMICAL DRIVING FORCES AND MEMBRANE IONIC CURRENTS

The electrochemical driving force for each species of ion is the algebraic difference between its equilibrium potential (E_i) and E_m. The total driving force is the sum of two forces: an electric force (the negative potential in a cell at rest tends to pull in positively charged ions because unlike charges attract) and a diffusion force (based on the concentration gradient) (Fig. 3-6). Thus, in a resting myocardial or skeletal muscle cell, the driving force for Na⁺ is: $(E_m - E_{Na}) = -80$ mV $-$

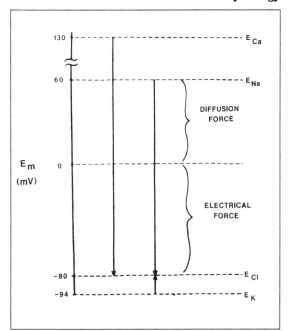

Fig. 3-6. Representation of the electrochemical driving forces for Na^+, Ca^{2+}, K^+, and Cl^-. The equilibrium potentials for each ion (e.g., E_{Na}) are positioned vertically according to their magnitude and sign; they were calculated from the Nernst equation for a given set of extracellular and intracellular ion concentrations. The measured resting potential is assumed to be -80 mV. The electrochemical driving force for an ion is the difference between its equilibrium potential and the membrane potential (E_m): $E_i - E_m$. Thus, at rest, the driving force for Na^+ is the difference between E_{Na} and the resting E_m: if E_{Na} is $+60$ mV and resting E_m is -80 mV, the driving force is 140 mV; that is, the driving force is the algebraic sum of the diffusion force and the electric force (represented by the length of the arrows in the diagram). The driving force for Ca^{2+} (209 mV) is even greater than that for Na^+, whereas that for K^+ is much less (14 mV). The direction of the arrows indicates the direction of the net electrochemical driving force; namely the direction for K^+ is outward, whereas that for Na^+ and Ca^{2+} is inward. If Cl^- is passively distributed, its distribution across the cell membrane can only be determined by the net E_m; for a cell sitting a long time at rest, $E_{Cl} = E_m$ and there is no net driving force.

$(-60$ mV$) = -140$ mV. The negative sign means that the driving force is in the direction that brings about the net movement of Na^+ inward. The driving force for Ca^{2+} is: $(E_m - E_{Ca}) = -80$ mV $- (+129$ mV$) = -209$ mV. The driving force for K^+ is: $(E_m - E_K) = -80$ mV $- (-94$ mV$) = +14$ mV; hence, the driving force for K^+ is small and directed outward. For a cell at rest, and if Cl^- is passively distributed, the driving force for Cl^- is nearly zero: $(E_m - E_{Cl}) = -80$ mV $- (-80$ mV$) = 0$. However, during the action potential, when E_m is changing, the driving force for Cl^- becomes large and there is a net driving force for inward Cl^- movement (Cl^- influx is an outward Cl^- current). Similarly, the driving force for the K^+ outward movement increases during the action potential, whereas those for Na^+ and Ca^{2+} decrease.

The net current (I) for each ionic species is equal to its driving force times its conductance (g; reciprocal of the resistance) through the membrane. This is essentially Ohm's law, $I = V/R = g \cdot V$, modified to reflect the fact that, in an electrolytic system, the total force tending to drive net movement of a charged particle must take into account both the electric and the concentration (or chemical) force. Thus, for the four ions, the net current can be expressed as:

$$I_{Na} = g_{Na} (E_m - E_{Na})$$
$$I_{Ca} = g_{Ca} (E_m - E_{Ca})$$
$$I_K = g_K (E_m - E_K)$$
$$I_{Cl} = g_{Cl} (E_m - E_{Cl})$$

In a resting cell, Cl^- and Ca^{2+} can be neglected, and the Na^+ current (inward) must be equal and opposite to the K^+ current (outward) to maintain a steady resting potential: $I_K = -I_{Na}$. Thus, although the resting membrane has a driving force for Na^+ much greater than that for K^+, g_K is much larger than g_{Na}, so the currents are equal. Hence, there is a continual leak of Na^+ inward and K^+ outward, and the system would run down even in a resting cell if active pumping were blocked. Because the ratio of the Na^+/K^+ driving forces $(-140$ mV$/-14$ mV$)$ is about 10, the ratio of conductances (g_{Na}/g_K) is about 1:10. The fact that g_K

is much greater than g_{Na} accounts for the resting potential being closer to the E_K than the E_{Na}.

The cell membrane of some cells (e.g., myocardial) has at least two separate voltage-dependent K^+ channels. One channel allows K^+ to pass more readily inward (against the usual net electrochemical gradient for K^+) than outward, the so-called **inward-going rectifier** or **anomalous rectification.** This gated channel is responsible for the rapid decrease in K^+ conductance upon depolarization (and increase in conductance with repolarization). The second type of voltage-dependent K^+ channel, which is commonly found in excitable membranes, slowly opens (increasing total g_K) upon depolarization, the so-called **delayed rectifier.** This channel allows K^+ to pass more readily outward (down the usual electrochemical gradient for K^+) than inward, and so is also known as the **outward-going rectifier.** In cardiac muscle, this delayed rectifier channel turns on much more slowly than it does in nerve or skeletal muscle, and the activation of this channel produces the increase in total g_K that terminates the cardiac action potential plateau.

DETERMINATION OF RESTING POTENTIAL AND NET DIFFUSION POTENTIAL

For given ion distributions, which normally remain nearly constant under usual steady-state conditions, the **resting potential** is determined by the relative membrane conductances (g) or permeabilities (P) for Na^+ and K^+. That is, the resting potential (about -80 mV in cardiac muscle and skeletal muscle) is close to the E_K (about -94 mV) because $g_K >> g_{Na}$ or $P_K >> P_{Na}$. (There is a direct proportionality between permeabilities and conductances at constant E_m and concentrations.) From simple circuit analysis (using Ohm's law and Kirchhoff's laws), this can be proven true. Therefore, the membrane potential is always closest to the battery (E_i) with the lowest resistance (highest conductance) in series with it (see Figs. 3-4 and 3-6). In the resting membrane, this battery is E_K, whereas in the excited membrane it is

E_{Na} (or E_{Ca}) because there is a large increase in g_{Na} (and/or g_{Ca}) during the action potential.

Any ion that is passively distributed across the membrane cannot determine the resting potential; instead, the resting potential determines the distribution of that ion. Therefore, Cl^- is essentially not a factor in the resting potential for neurons, myocardial cells, and skeletal muscle fibers, because it seems to be nearly passively distributed. However, transient net movements of Cl^- across the membrane do influence E_m (e.g., washout of Cl^- [in Cl^--free solution] produces a transient depolarization and reintroduction of Cl^- produces hyperpolarization), and Cl^- movement is involved in the production of some inhibitory postsynaptic potentials. Because of its relatively low concentration along with its relatively low resting conductance, the Ca^{2+} distribution has only a relatively small effect on the resting E_m.

Therefore, a simplified version of the Goldman-Hodgkin-Katz **constant-field equation** can be given (for 37°C):

$$E_m = -61 \text{ mV} \log \frac{[K]_i + P_{Na}/P_K [Na]_i}{[K]_o + P_{Na}/P_K [Na]_o}$$

This equation shows that, for a given ion distribution, the resting E_m is determined by the P_{Na}/P_K ratio, or the relative permeability of the membrane to Na^+ and K^+. For myocardial cells and skeletal muscle fibers, the P_{Na}/P_K ratio is about 0.04, whereas for nodal cells of the heart and smooth muscle cells, this ratio is between 0.10 and 0.20. Inspection of the constant-field equation shows that the numerator of the log term is dominated by the $[K]_i$ term—since the (P_{Na}/P_K) $[Na]_i$ term is very small—whereas the denominator is affected by both the $[K]_o$ and (P_{Na}/P_K) $[Na]_o$ terms. This relationship thus accounts for the deviation of the E_m versus log $[K]_o$ curve from a straight line (having a slope of 61 mV/decade) in normal Ringer's solution (Fig. 3-7). When $[K]_o$ is elevated ($[Na]_o$ reduced by an equimolar amount), the denominator becomes increasingly dominated by the $[K]_o$ term and decreasingly by the (P_{Na}/P_K) $[Na]_o$ term. Therefore, in bathing solutions containing high K^+ concentrations, the

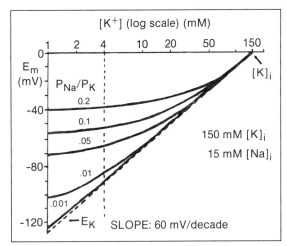

Fig. 3-7. Theoretical curves calculated from the Goldman-Hodgkin-Katz constant-field equation for the resting potential (E_m) as a function of the extracellular K+ concentration ($[K]_o$). The family of curves is given for the various relative membrane permeabilities to Na+ and K+ (P_{Na}/P_K) (0.001, 0.01, 0.05, 0.1, and 0.2). The K+ equilibrium potential (E_K) was calculated from the Nernst equation (*broken straight line*). The curves were calculated for an intracellular K+ concentration ($[K]_i$) of 150 mM and an intracellular Na+ concentration ($[Na]_i$) of 15 mM. Calculations were made holding $[K]_o$ and the extracellular Na+ concentration ($[Na]_o$) constant at 154 mM (i.e., as $[K]_o$ was elevated, $[Na]_o$ was lowered by an equimolar amount). A change in the P_K as a function of $[K]_o$ was not taken into account for these calculations. The point at which E_m is zero gives $[K]_i$. The potential reverses in sign when $[K]_o$ exceeds $[K]_i$.

constant-field equation approaches the simple Nernst equation for K+, and E_m approaches E_K. As $[K]_o$ is raised stepwise, E_K becomes correspondingly reduced because $[K]_i$ stays relatively constant; therefore, the membrane becomes increasingly depolarized (see Fig. 3-7).

An alternative method of approximating the resting potential (E_m) is by the **chord-conductance equation:**

$$E_m = \left[\frac{g_K}{g_K + g_{Na}}\right] E_K + \left[\frac{g_{Na}}{g_K + g_{Na}}\right] E_{Na}$$

This equation can be derived simply from Ohm's law and circuit analysis for the condition when net current is zero ($I_{Na} + I_K = 0$). The chord-conductance equation again illustrates the important fact that the g_K/g_{Na} ratio determines the resting potential. When $g_K \gg g_{Na}$, E_m is close to E_K; conversely, when $g_{Na} \gg g_K$ (as during the action potential spike), E_m shifts close to E_{Na} (or to E_{Ca} in the case of many smooth muscle cells).

When $[K]_o$ is elevated (e.g., to 10 mM) in some cells, a hyperpolarization of up to about 10 mV may be produced. Such behavior is often observed in cells with a high P_{Na}/P_K ratio (due to low P_K) and therefore a low resting E_m, such as in young embryonic hearts. This hyperpolarization could be explained by several factors: (1) stimulation of the Na+ V_{ep}, (2) an increase in P_K (and therefore g_K) due to the effect of $[K]_o$ on P_K, or (3) a direct increase in g_K due to the concentration effect. A similar explanation may apply to the fall-over in the E_m versus log $[K]_o$ curve, hence depolarizing the cells, when $[K]_o$ is lowered to 1 mM or less. This effect is prominent in rat skeletal muscle, for example. That is, lowering $[K]_o$ inhibits the Na+ V_{ep} and lowers P_K and g_k. The K+ concentration at which the (Na,K)-ATPase activity is half maximal is about 2 mM.

Electrogenic Sodium Pump Potentials

In the presence of ouabain (short-term exposure only), a drug that inhibits the Na+-K+ pump and V_{ep}, the resting potential is equal to the **net diffusion potential** (E_{diff}), as determined by the ion concentration gradients for K+ and Na+ and by the relative P_K and P_{Na} (Fig. 3-8A). If there were no v_{ep} contribution to the resting potential, E_m would equal E_{diff}. However, a direct contribution of the pump to the resting E_m can be demonstrated. For example, if the Na+-K+ pump is blocked by ouabain, there usually is an immediate depolarization of between 2 and 16 mV, depending on the type of cell. Thus, the direct contribution of the Na+-K+ V_{ep} to the measured resting E_m is relatively small under physiologic conditions. Because two K+ are usually carried in for every three Na+ moved out, the pump is

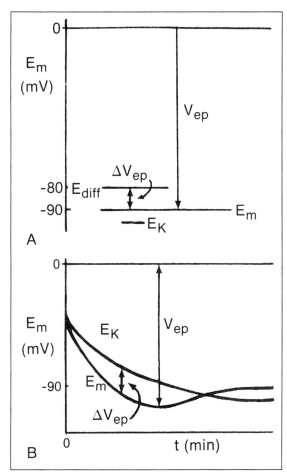

Fig. 3-8. An electrogenic Na$^+$ pump potential (V$_{ep}$). (A) Muscle cell with a net ionic diffusion potential ($[E_{diff}]$, function of ion equilibrium potentials and relative conductances) of -80 mV, but a measured membrane resting potential (E_m) that is greater. The difference between E_m and E_{diff} represents the contribution of the electrogenic pump to the resting E_m. Usually the direct contribution of the pump is only a few millivolts, and can be measured by the amount of depolarization produced immediately after complete inhibition of the (Na,K)-ATPase by cardiac glycosides. Because the pump pathway is separate from and parallel to the ionic conductance pathways, the electrogenic pump potential must be equal to V$_{ep}$. The contribution of the V$_{ep}$ to the resting potential ($E_m - E_{diff}$) is equal to ΔV_{ep}. (B) Cell that was run down (Na$^+$ loaded, K$^+$ depleted) over several hours by inhibition of Na$^+$-K$^+$ pumping, resulting in a low resting potential. Returning the muscle cell to a pumping solution allows the resting E_m

electrogenic; that is, it produces a net current (and hence potential) across the membrane. The coupling ratio cannot be 3:0 because external K$^+$ must be present for the pump to operate.

The fact that the pump is electrogenic can be demonstrated by injection of Na$^+$ into the cell through a microelectrode; this rapidly produces a small transient hyperpolarization, which is immediately abolished by ouabain. Under conditions in which the pump is stimulated to operate at a high rate (e.g., when [Na]$_i$ or [K]$_o$ is abnormally high), the direct electrogenic contribution of the pump to the resting potential can be much greater, and E_m can actually exceed E_K by as much as 20 mV or more. For example, if the ionic concentration gradients are allowed to run down, then after allowing the tissues to restart pumping, the measured E_m can exceed the calculated E_K (e.g., by 10 to 20 mV) for a period of time (see Fig. 3-8B). The Na$^+$ loading of the cells is facilitated by placing them in low [K]$_o$ solutions for several hours at reduced temperature, since external K$^+$ is necessary for the Na$^+$-K$^+$ pump to operate. The internal concentrations of Na$^+$, K$^+$, and Cl$^-$ approach the concentrations in the bathing Ringer's solution, and the resting potential is low (e.g., -20 to -50 mV). When the tissue is transferred to Ringer's solution (containing normal K$^+$ and at normal temperature), the pump turns over at a maximal rate, because the major factors that regulate the pump rate are [Na]$_i$ and [K]$_o$. The measured E_m of such Na$^+$ preloaded cells increases more rapidly than E_K (see Fig. 3-8B).

Rewarming previously cooled cardiac muscles leads to the rapid restoration of the normal resting potential (within 10 minutes), whereas recovery of the [Na]$_i$ and [K]$_i$ is slower. During prolonged hypoxia, the resting potential of cardiac muscle decreases much less than the E_K

to build back up as a function of time. The buildup in E_m occurs faster than that for E_K. Whenever E_m is greater (more negative) than the K$^+$ equilibrium potential (E_K), the difference (ΔV_{ep}) must reflect the contribution of the Na$^+$ pump potential (t = time.)

decreases. The electrogenic pump attempts to hold the resting potential constant despite dissipating ionic gradients. Thus, the V_{ep} delays depolarization under adverse conditions (e.g., ischemia and hypoxia) and speeds repolarization to the normal resting potential during recovery. It is crucial that the excitable cell maintains its normal resting potential as much as possible, because of the effect of small depolarizations on: action potential rate of rise, conduction velocity, and rate of firing of pacemaker cells. Therefore, the V_{ep} has physiologic importance in cells.

The ΔV_{ep} is in series with the net cationic E_{diff}:

$$E_m = E_{diff} + R_m I_p = E_{diff} + \Delta V_{ep}$$

where I_p is the net (electrogenic component) pump current, and E_{diff} is the E_m that would exist solely on the basis of the ionic gradients and relative permeabilities in the absence of a V_{ep} (as calculated from the constant-field equation). This equation states that E_m is the sum of E_{diff} and a voltage drop produced by the electrogenic pump. V_{ep} is the full pump potential with respect to ground (zero) (see Fig. 3-8A).

One possible equivalent circuit for an electrogenic Na^+ pump is given in Figure 3-9. The V_{ep} parallels the E_{diff}. If the pump is stopped by ouabain, V_{ep} goes to zero. The pump resistance may be about tenfold higher than membrane resistance. The pump potential contribution to E_m (ΔV_{ep}) increases when membrane resistance is higher. Using circuit analysis and the parameter values given in Figure 3-9, E_m would be -81.8 mV, which is moderately close to E_{diff} (-80 mV). If membrane resistance is raised twofold (to 2,000 Ω), E_m would be -83.3 mV.

The density of Na^+-K^+ pump sites, estimated by specific binding of [^{3}H] ouabain, is usually about 700 to 1,000/μm^2 (about a hundredfold greater than that of Na^+ and K^+ channels). The turnover rate of the pump is estimated to be 20 to 100 sec^{-1}. The net pump current (I_p) may be estimated by: $I_p = \Delta V_{ep}/R_m$, where R_m is the membrane resistance. Values of about 20 pmol/cm²-sec were obtained. A density of 1,000 sites/μm^2 (10^{11} sites/cm²) times a turnover rate of 40 per

second gives 4×10^{12} turnovers/cm²-sec. Since three Na^+ are pumped with each turnover, this yields 12×10^{12} Na^+/cm²-sec. Dividing by Avogadro's number (6.02×10^{23} ions/mol) yields 20×10^{-12} mol/cm²-sec, which is the same value as the 20 pmol/cm²-sec measured. Ion flux (J) can be converted to current (I) by the following: $I = J \times zF$ (amp/cm² = mol/sec-cm² $\times$ coulombs/mol). Thus, a flux of 20 pmol/cm²-sec is approximately equal to 2 μamp/cm² (20×10^{-12} mol/sec-cm² $\times$ 0.965 $\times 10^5$ coulombs/mol). Since $\Delta V_{ep} = I_p \times R_m$, if R_m were 1,000 ohm-cm² and I_p were 2 μamp/cm², the ΔV_{ep} contribution to E_m would be 2 mV.

In cells that have lower resting potentials (e.g., smooth muscle cells and cardiac nodal cells) (see Table 3-1), the ΔV_{ep} can be considerably larger. Sinusoidal oscillations in the Na^+-K^+ pumping rate could produce oscillations in E_m, which could exert important control over the spontaneous firing of the cell. Oscillation of the pump may be triggered by changes in $[Na]_i$. For example, the firing of several action potentials should raise $[Na]_i$ (some cells have a small volume/surface area ratio) and stimulate the electrogenic pump. The increased pumping rate, in turn, hyperpolarizes and suppresses firing, thus allowing $[Na]_i$ to become lower again and stopping stimulation of the pump; the latter depolarizes and triggers spikes and the cycle is repeated. When stimulated at a high rate, cardiac Purkinje fibers undergo a transient period of inhibition of automaticity after cessation of the stimulation, known as **overdrive suppression of automaticity,** which is caused by stimulation of the electrogenic pump due to elevation in $[Na]_i$.

PACEMAKER POTENTIALS AND AUTOMATICITY

To maintain a steady resting potential, the outward K^+ current (I_K) must be equal and opposite to the inward current, primarily Na^+ (I_{Na}) (but also Ca^{2+}), assuming Cl^- is passively distributed: $I_K = -I_{Na}$. If the inward current exceeds the outward current, then the membrane will depolarize along a certain time course (i.e., slope of the pacemaker

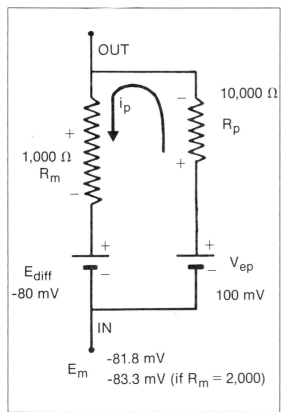

OUT

i_p

10,000 Ω

R_p

1,000 Ω

R_m

E_{diff}
-80 mV

V_{ep}

100 mV

IN

-81.8 mV

E_m -83.3 mV (if R_m = 2,000)

Fig. 3-9. Hypothetical electric equivalent circuit for the electrogenic Na⁺ pump. Model consists of a pump pathway in parallel with the membrane resistance (R_m) pathway and the membrane capacitance (C_m) pathway. This model conforms to the evidence that the pump is independent of short-range membrane excitability and that the pump and channel proteins are embedded in the lipid bilayer as parallel elements. The net diffusion potential ($[E_{diff}]$, determined by the ion equilibrium potentials and relative permeabilities) of − 80 mV is depicted in series with R_m. The pump leg is assumed to consist of a battery in series with a fixed resistor (pump resistance $[R_p]$) that does not change with changes in R_m and whose value is tenfold higher than R_m. The pump battery is charged up to some voltage (electrogenic pump potential $[(V_{ep})]$ of − 100 mV, for example) by a pump current generator. The net electrogenic pump current is developed by pumping in only two K⁺ for every three Na⁺ pumped out. For the values given in the figure (R_m = 1,000 Ω; E_{diff} = − 80 mV; R_p = 10,000 Ω; V_{ep} = − 100 mV), the measured membrane potential (E_m) can be calculated by circuit analysis to be − 81.8 mV; that is, the direct electrogenic

potential), depending on the excess (or net) inward current. The inward leak of Na⁺ and Ca²⁺ currents is called the **background inward current.** For the inward current to exceed the outward current (i.e., to achieve a net inward current), either the inward current can be increased or the outward K⁺ current can be decreased. Both of these mechanisms are used for genesis of **pacemaker potentials** (automaticity). For example, if a time-dependent decrease (decay) in g_K occurs following an action potential and hyperpolarizing afterpotential, then the outward K⁺ current decreases and the membrane depolarizes (g_{Na}/g_K progressively increases). Conversely, if an agent such as acetylcholine increases the resting g_K in cells such as heart nodal cells, then the outward K⁺ current is increased, the membrane hyperpolarizes and the slope of the pacemaker potential decreases, thus reducing the frequency of firing. Some agents, such as norepinephrine, increase the background inward current, thereby increasing the slope of the pacemaker potential.

A prerequisite for **automaticity** is that the cells must have a relatively low g_{Cl}. Most types of heart cells and neurons have a low g_{Cl}. A high g_{Cl} clamps E_m, making it difficult for a pacemaker potential to develop.

During the time course of the pacemaker potential, membrane resistance increases progressively due to a decrease in g_K. The progressive turnoff of the inwardly rectifying K⁺ channels causes a progressive increase in membrane resistance. In addition, there is a progressive turnoff of the g_K increase (delayed rectification) responsible for the rapid repolarizing phase of the action potential and the subsequent hyperpolarizing afterpotential. The decreasing g_K helps to produce the depolarization. The pacemaker depolarization is linear (or a ramp). Accommodation does not occur in a pacemaker cell; the cell

---

pump potential contribution to the resting potential is − 1.8 mV. If R_m were raised to 2,000 Ω (e.g., by placing the membrane in Cl⁻-free solution, or by adding Ba²⁺ to decrease the permeability to K⁺, or both), the calculated E_m would be − 83.3 mV.

fires, no matter how slowly E_m is brought to the threshold potential.

In any pacemaker cell, if the membrane is hyperpolarized by a current pulse, the frequency of spontaneous firing is slowed and stopped; that is, automaticity is suppressed at high resting potentials. Conversely, application of depolarizing current increases the frequency of discharge. Thus, the slope of the pacemaker potential is exquisitely sensitive to small changes in E_m. Elevations of $[K]_o$, which increase g_K, suppress automaticity despite depolarization.

Pacemaker cells of the heart also possess a unique type of ion channel that is turned on by hyperpolarization (the repolarization occurring during termination of the action potential). Therefore, this current is called **hyperpolarization-activated current.** This current channel is nonselective, allowing both Na^+ and K^+ to pass through, and thereby depolarizing.

Myocardial cells can exhibit abnormal automaticity of another type under certain pathophysiologic conditions. A large **depolarizing afterpotential** arises from the **hyperpolarizing afterpotential** and triggers a subsequent spike once the threshold potential is reached. In this type of automaticity, each action potential is triggered by the preceding one, and this type of abnormal automaticity is known as **triggered automaticity.**

EFFECT OF RESTING POTENTIAL ON THE ACTION POTENTIAL

Any agent that affects the resting potential (e.g., depolarizes) has important repercussions on the action potential. Depolarization reduces the rate of rise of the action potential, and thereby also slows its velocity of propagation. This effect is progressive as a function of the degree of depolarization. Such slowed propagation affects the function of nerves and muscles. For example, a slow spread of excitation throughout the heart interferes with the heart's ability to act as an efficient blood pump. If the myocardial cells, Purkinje fibers, skeletal muscle fibers, and neurons are depolarized to about -50 mV, then the rate

of rise goes to zero and all excitability (and contraction) is lost.

Hyperpolarizations usually produce only a small increase in the rate of rise, and large hyperpolarizations may actually slow propagation velocity (because the critical depolarization required to bring the membrane to threshold is increased) or block propagation. The explanation for the effect of resting E_m (or takeoff potential) on the maximum rate of rise of the action potential is based on the sigmoidal (h value at infinite time) h_∞ versus E_m curve (see Chapter 4).

The resting potential also affects the duration of the action potential. With polarizing current, depolarization lengthens the action potential, whereas hyperpolarization shortens it, due to anomalous rectification (i.e., a decrease in g_K with depolarization). In contrast, when an elevated $[K]_o$ is used to depolarize the cells, the action potential is usually shortened. One important determinant of action potential duration is the g_K. Agents that increase g_K tend to shorten the duration, whereas agents that decrease g_K or slow its activation (such as barium ion or tetraethylammonium ion) lengthen the action potential duration.

SUMMARY

Most of the factors that determine or influence the resting potential of cells have been discussed in this chapter. The structural and chemical compositions of the cell membrane were correlated with the membrane's resistive and capacitive properties. The factors that determine the intracellular ion concentrations in cells were examined, including the Na^+-K^+ pump, the Ca^{2+}–Na^+ exchange reaction, and the sarcolemmal Ca^{2+} pump. The (Na,K)-ATPase requires both Na^+ and K^+ for activity, and transports three Na^+ outward and usually two K^+ inward per ATP hydrolyzed. Cardiac glycosides are specific blockers of this transport ATPase. The Na^+-K^+ pump is only indirectly related to excitability, through its role in maintaining the Na^+ and K^+ concentration gradients. The carrier-mediated Ca^{2+}-Na^+ exchange reaction, which exchanges one internal Ca^{2+} for three external Na^+, is driven by the Na^+ electrochemical gradient (i.e., the energy for removing internal Ca^{2+} by this mechanism ultimately comes from the [Na,K]-ATPase).

The mechanism whereby the ionic distributions give rise to diffusion potentials was discussed, as were the factors that determine the magnitude and polarity of each equilibrium potential (E_i). The E_i for any ion and the transmembrane potential determine the total electrochemical driving force for that ion, and the product of this driving force and the membrane conductance for that ion determine the net ionic current or flux. The net ionic movement can be inward or outward across the membrane, depending on the direction of the electrochemical gradient. The key factor that determines the resting potential is the relative permeability of the various ions, particularly of K^+ and Na^+ (i.e., the P_{Na}/P_K ratio [or g_{Na}/g_K ratio], as calculated from the Goldman-Hodgkin-Katz constant-field equation). In some cells, Cl^- is distributed passively in accordance with the E_m; in other cells, the $[Cl]_i$ is higher than that predicted from the E_m, thus giving an E_{Cl} value less than the resting potential. If Cl^- is passively distributed, it cannot determine the resting potential, but transient net movements of Cl^- (e.g., during the action potential) can affect E_m.

Elevation of $[K]_o$ above the normal concentration of about 4.5 mM decreases the E_K, as predicted from the Nernst equation, and depolarization is produced. The resting potential is the potential energy storehouse that is drawn upon for propagation of the action potentials. Because the voltage-dependent cationic channels are inactivated by sustained depolarization, the rate of rise of the action potential, and hence propagation velocity, is critically dependent on the level of the resting potential. For example, an elevation of K^+ concentration in the blood by 3 to 5 mM has dire consequences for functioning of the heart, nerves, and skeletal muscles.

The degree of contribution of the Na^+-K^+ pump to the resting potential depends on: the coupling ratio of Na^+ pumped out to K^+ pumped in, the turnover rate of the pump, and the magnitude of membrane resistance. The rate of Na^+-K^+ pumping is controlled by $[Na]_i$ and by $[K]_o$. The V_{ep} parallels the E_{diff}, determined by the equilibrium potentials and relative permeabilities. Because the effect of ΔV_{ep} on the measured resting potential of myocardial cells and neurons is generally small, the immediate depolarization produced by complete Na^+-K^+ pump stoppage with cardiac glycosides is only a few millivolts. Long-term pump inhibition produces an increasingly larger depolarization as the ionic gradients are progressively dissipated. The V_{ep} might be physiologically important under certain conditions that tend to depolarize the cells, such as transient ischemia or hypoxia, but the actual depolarization produced may be less because of a relatively constant pump potential.

NATIONAL BOARD TYPE QUESTIONS

Select the one most correct answer for each of the following

1. The resting potential (E_m) of a skeletal muscle fiber diminishes (membrane depolarizes) progressively as the $[K]_o$ in the bathing Ringer's solution is progressively elevated (K^+ substituted for Na^+ on an equimolar basis). When E_m passes (or extrapolates) to zero, this point indicates which one of the following?
   A. $[Na]_o$
   B. $[K]_i$
   C. $[Cl]_i$
   D. $[H]_i$
   E. E_{epp}

2. In the E_m versus $\log [K]_o$ curve for mammalian skeletal muscle fibers, at very low $[K]_o$ (e.g., 1.0 mM to 0.1 mM) the curve "falls over"; that is, the membrane depolarizes rather than hyperpolarizes as $[K]_o$ is progressively lowered below about 1.0 mM. Assuming Cl^- to be passively distributed, which one of the following statements is incorrect?
   A. The fiber loses substantial intracellular K^+ (lower $[K]_i$)
   B. The electrogenic Na^+-K^+ pump is inhibited

C. The P_K is decreased when $[K]_o$ is lowered

D. The cell should gain Cl^-

E. There is a greater deviation between E_K and resting E_m

3. If Cl^- is passively distributed across the cell membrane of a resting nerve fiber, which one of the following is not true?

A. E_{Cl} equals the resting E_m

B. There is no Cl^- pump

C. Cl^- influx would occur during an action potential

D. Cl^- movement could not affect the E_m

E. Cl^- efflux that occurs transiently during bathing in Cl^--free solution would transiently depolarize the fiber

4. In a two-compartment system, with 100-mM potassium chloride on side No. 1 and 1.0-mM potassium chloride on side No. 2, the two solutions being separated by a thin membrane permeable to Cl^- but impermeable to K^+, the potential difference built up across the membrane (side No. 1 with respect to side No. 2) would be closest to (at 30°C)?

A. -120 mV

B. -60 mV

C. 0 mV

D. $+60$ mV

E. $+120$ mV

5. In a nerve cell, if E_K is -90 mV and the resting potential is -70 mV, what is the net electrochemical driving force for K^+ efflux (ignore directionality sign)?

A. 90 mV

B. 70 mV

C. 20 mV

D. 160 mV

E. None of the above

6. If a drug were to cause the fractional conductances for K^+ and Na^+ in a neuron at rest (resting potential, -70 mV) to become 0.40 and 0.60, respectively (ignoring Ca^{2+} and assuming: Cl^- is passively distributed, $E_K = -90$ mV, and $E_{Na} = +60$ mV), the chord-conductance equation predicts that the resting potential would be closest to?

A. $+60$ mV

B. 0 mV

C. -70 mV

D. -80 mV

E. -90 mV

ANNOTATED ANSWERS

1. B. When $[K]_o$ is equal to $[K]_i$, E_K is zero and the resting potential is zero.

2. A. Answers B, C, D, and E are all correct. The $[K]_i$ value remains almost unchanged.

3. D. Answers A, B, C, and E are all correct. Cl^- influx or efflux does affect E_m.

4. E. $E_m = +60$ mV log $\dfrac{[Cl]_1}{[Cl]_2}$

$$= +60 \text{ mV log } \frac{100 \text{ mM}}{1 \text{ mM}}$$

$$= +60 \text{ mV log } 100$$

$$= +60 \text{ mV } (2.0)$$

$$= +120 \text{ mV.}$$

5. C. Electrochemical driving force $= E_m - E_K$
$= -70$ mV $- (-90$ mV$) = -70 + 90 = +20$ mV.

6. B. $E_m = \left[\dfrac{g_K}{g_K + g_{Na}}\right] E_K + \left[\dfrac{g_{Na}}{g_K + g_{Na}}\right] E_{Na} =$

$(0.40)(-90$ mV$) + (0.60)(+60$ mV$) = -36$ mV $+ 36$ mV $= 0$ mV.

BIBLIOGRAPHY

Sperelakis, N. Origin of the cardiac resting potential. In R. M. Berne and N. Sperelakis (Eds.), *Handbook of Physiology, The Cardiovascular System. Vol. 1: The Heart.* Bethesda, MD: American Physiological Society, 1979. Pp. 187–267.

Sperelakis, N. Basis of the resting potential. In N. Sperelakis (Ed.), *Physiology and Pathophysiology of the Heart,* 2nd ed. Norwell, MA: Kluwer, 1989. Pp. 59–80.

Sperelakis, N. and Fabiato, A. Electrophysiology and excitation-contraction coupling in skeletal muscle. In C. Roussos and P. Macklem (Eds.), *The Thorax: Vital Pump.* New York: Dekker, 1985. Pp. 45–113.

4 Basis of Membrane Excitability

NICHOLAS SPERELAKIS AND IRA R. JOSEPHSON

Objectives

After reading this chapter, you should be able to:

Describe the molecular structure and function of ion channels in the cell membrane

Explain how action potentials are generated

List the properties of the action potentials

Explain the importance of the resting potential on the action potentials and their velocity of propagation

Explain why elevation of the potassium ion concentration above normal in the blood plasma has life-threatening consequences

---

Excitability is an intrinsic membrane property that allows a cell to generate an electric signal or action potential in response to environmental stimuli of sufficient magnitude. The highly elongated nerve axon transmits information over long distances in the form of action potentials. The action potential mechanism is required in order to propagate a uniform depolarization in a nondecremental manner. In muscle cells, the action potential spreads excitation over the entire cell surface and is involved in triggering cell contraction.

The energy source for the generation of the action potential is stored in the excitable cell itself. The initial depolarization, caused by a given stimulus, merely triggers the intrinsic action potential mechanism. The immediate source of energy (or battery) for the action potential comes from the transmembrane ionic gradients for K^+ and Na^+. The K^+ concentration gradient is mainly responsible for generating the resting potential, which causes an excess of negative charges to be built up on the inner surface of the membrane.

Upon depolarization to threshold, the Na^+ electric and chemical driving forces, which are directed inwardly, cause a large and rapid inward Na^+ current that generates the action potential upstroke. Over a longer time frame, the Na^+-K^+ pump is responsible for generation of the Na^+ and K^+ gradients, and for their restoration following repetitive action potential activity. The Na^+-K^+ pump derives chemical energy from the hydrolysis of adenosine triphosphate.

As in other scientific fields, important technical improvements have led to significant advances in the understanding of the basis of membrane excitability. Prior to the late 1930s, several theories had been proposed to explain the mechanism that caused the action potential. Julius Bernstein (1902, 1912) proposed that, at rest, the excitable cell membrane was selectively permeable to K^+ (hence, the resting potential), and that during excitation the membrane became permeable to all ions. Several years earlier, Overton had demonstrated that Na^+ was essential for excitability. However, without a means to directly measure

49

the transmembrane potential during excitation, Bernstein's membrane hypothesis and Overton's observation could not be tested.

INTRACELLULAR RECORDING

By the 1940s, improvements had been achieved in electronic instrumentation, especially in the high-input impedance amplifiers necessary to record bioelectric phenomena. In addition, biophysicists began to study the squid giant axon (500–1,000 μm in diameter), which permitted insertion of intracellular electrodes, yielding the first measurements of true transmembrane potential. The transmembrane potential is recorded as the difference between an intracellular and an extracellular electrode (Fig. 4-1).

Much of the information gained during 1930 to 1950 concerning the ionic basis of the action potential was derived from experiments using preparations of squid giant axons. These findings were successfully applied to neurons from the vertebrate nervous system, whose much smaller diameter (1–20 μm) made direct experimentation difficult at that time (Fig. 4-2).

EXTERNAL RECORDING

During propagation of an action potential down a nerve fiber, current flow accompanies the propagating change in the membrane voltage. This current is called the **local-circuit current,** which has both longitudinal and radial (transverse) components and makes a complete circuit (see also Chapter 5). For our purpose here, we will focus on the longitudinal component. The intracellular and extracellular (external) longitudinal current are exactly equal in amplitude, but flow in opposite directions (Fig. 4-3A). The intracellular current is confined to the cross-sectional area of the nerve fiber (neuroplasm), whereas the extracellular current can take the path of least resistance. Thus, if the fiber is bathed in a large volume of fluid (so-called volume conductor), the extracellular current will use the entire bath volume (more resistors in parallel yield a smaller total resistance). It is this fact that enables the

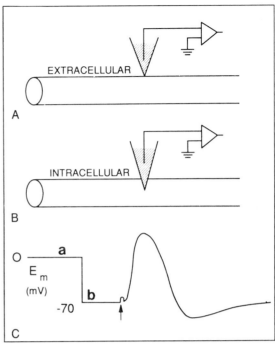

Fig. 4-1. Recording the transmembrane potential of a squid giant nerve axon. The transmembrane potential is measured as the difference between the intracellular and extracellular electrodes. In A, the electrode is outside the axon, and measures 0 mV (segment *a* in C). As the electrode is advanced and crosses the membrane (B), the resting potential of − 70 mV is measured (segment *b* in C). The cell membrane makes a tight seal around the glass electrode. In C, the membrane potential (ϵ_m) is depicted. Stimulation (*arrow*) elicits an action potential.

electrocardiogram to be recorded from the surface (skin) of the body; the same is true for the electromyogram recording of electric activity from skeletal muscles and for the electroencephalogram recording of electric activity from the brain. For example, the electrocardiogram records the extracellular action currents from the heart that spread throughout the torso.

If the single nerve fiber or entire nerve bundle is mounted in air, the extracellular action current is confined to the surface film of fluid adhering to the single fiber; in the case of a nerve bundle, the external current can also flow through the inter-

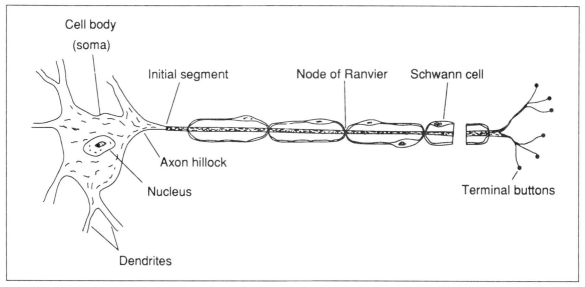

Fig. 4-2. A motor neuron with myelinated axon. The major structural elements are identified.

stitial fluid space between fibers in the bundle.

To illustrate the principles involved, we will consider what happens when we record externally with a pair of electrodes from a single nerve fiber (e.g., squid giant axon) bathed in room air. This is a so-called **bipolar recording.** In the example depicted in Figure 4-3C, as the action potential moves from left to right past the pair of recording electrodes, the first electrode becomes negative with respect to the second electrode. Therefore, the voltmeter swings in one direction (an upward deflection is defined as negative in extracellular recording) due to the potential difference. When the wavefront reaches electrode B, both electrodes record almost zero voltage. When the active region of the action potential passes beyond electrode A, but has not yet gone beyond electrode B, the voltmeter swings in the opposite direction, because B is now negative with respect to A. Thus, the bipolar electrodes record a biphasic change in voltage as the action potential sweeps past. If the distance between A and B is short relative to the wavelength of the action potential, the biphasic voltage change will approximate the first time derivative of the action potential (as recorded with an intracellular microelectrode).

The wavelength of the action potential is a product of the velocity of propagation and the action potential duration; for example, the wavelength in a motor axon or large sensory axon is about 12 cm (120 m/sec × 1.0 msec).

The action potential is often described as a wave of negativity sweeping down the fiber, when recording externally. When recording internally, the action potential is a wave of positivity.

The reason for the wave of negativity in external recording is as follows. Consider that the fluid between the two electrodes constitutes a resistance (R), and that longitudinal current (I) flow through this fluid resistance produces an IR drop (Ohm's law: V = IR). As the wavefront approaches A, the external local-circuit current goes from right to left and produces an IR drop: positive (B) to negative (A). Therefore, electrode A becomes negative with respect to B. Conversely, when the action potential tail passes beyond A, the action current (smaller amount) is now reversed in direction and produces a smaller IR drop: positive (A) to negative (B). This produces the biphasic voltage recording.

When recording externally from a nerve bundle in room air using bipolar electrodes, the greater

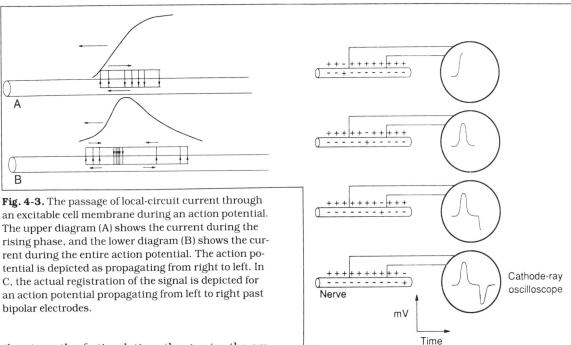

Fig. 4-3. The passage of local-circuit current through an excitable cell membrane during an action potential. The upper diagram (A) shows the current during the rising phase, and the lower diagram (B) shows the current during the entire action potential. The action potential is depicted as propagating from right to left. In C, the actual registration of the signal is depicted for an action potential propagating from left to right past bipolar electrodes.

the strength of stimulation, the greater the amplitude of the external signal recorded, up to a maximum. The external signal is called the **compound action potential,** because it is composed of the summated activities of the hundreds of individual fibers in the nerve bundle. The compound action potential is **graded** in that it becomes larger and larger with increasing strength of stimulation, up to a maximum amplitude. The maximum amplitude occurs when all of the individual fibers capable of responding are responding. The nerve bundle is composed of hundreds of axons of differing diameters, differing degrees of myelination, differing excitabilties, and differing propagation velocities. Therefore, increasing the strength of stimulation (e.g., the applied voltage) recruits a larger and larger fraction of the nerve fibers. The action potential in each fiber is all-or-none (not graded), but the compound action potential is graded. As each fiber is recruited, its local-circuit current is added to the total current passing through the fluid between the pair of electrodes, thus producing a greater IR voltage drop and greater signal.

The stimulus strength that just recruits all the axons in the bundle is known as the **maximal stimulus;** strength beyond that is known as **supramaximal stimulus.** In the graded region of the compound action potential, the stimulus strength is **submaximal,** and when the strength is insufficient to excite any of the fibers, it is **subthreshold.**

The compound action potential recorded often has a second smaller peak after the initial large peak; sometimes even a third and fourth tiny peak can be detected. The initial large peak reflects activation of the many large-diameter myelinated motor and sensory axons, which have a lower threshold (more excitable). The second smaller peak reflects activation of the fewer smaller-diameter myelinated fibers, which have a higher threshold (less excitable). The third peak, when present, represents activation of the small-diameter, non-myelinated axons. Propagation velocity is a function of axon diameter and myelination (see Chapter 5, on propagation), such that, the larger the diameter, the greater the velocity. Myelinated axons conduct much faster than nonmyelinated axons (see Chapter 5). Therefore, the first peak in the compound action potential consists of the fast-propagating, low-threshold fibers. Like runners in a race, the slower runners cross the finish line late. Therefore, the slower-propagating, higher-threshold fibers compose the second peak.

ACTION POTENTIAL CHARACTERISTICS

Threshold and All-or-None Property

Nerve membrane responses near the site of application of brief current pulses vary depending on the magnitude and direction of the pulses (Fig. 4-4). Inward currents produce hyperpolarizing responses and outward currents produce depolarizing responses. Hyperpolarizing and subthreshold depolarizing responses are graded in magnitude according to the stimulus current. However, a somewhat greater-intensity outward current produces a depolarizing response with a different waveform and a longer-lasting duration. This condition is referred to as the **local excitatory state** (see Fig. 4-4A and B). It occurs when a small area of the membrane near the stimulus electrode becomes excited, but it does not have sufficient area to generate an action potential. This local membrane activity is not propagated and decays with distance along the axon. A slightly greater stimulus is, however, sufficient to bring the membrane potential of a large enough membrane area to the **threshold potential,** thereby initiating an **all-or-none** action potential. At the threshold potential, there is a greater amount of inward (depolarizing) than outward (repolarizing) current, and the membrane will continue to depolarize. It is important to remember that, in regard to the applied stimulus, outward current depolarizes the membrane, whereas, for the membrane itself, inward currents are depolarizing.

A **subthreshold** depolarization is defined as one that does not reach the threshold and therefore does not elicit an action potential. It is pro-portional to the applied stimulus, and it decrements with distance along the nerve axon "cable." A subthreshold voltage change occurs because the stimulus current first changes the charge on the membrane capicitance, which in turn changes the transmembrane voltage. As the voltage changes, the electrical driving force increases and the current then leaks out across the membrane resistance pathway. In the steady-state, the stimulus current flowing into the cell equals the resistive current flowing out. The product of the membrane capacitance and resistance determines the time course of the voltage change. If a stimulus current has sufficient magnitude (i.e., enough positive charges are transferred into the cell), the resulting membrane depolarization reaches a critical value, called the **threshold potential,** at which an action potential is initiated (see Fig. 4-4). The action potential parameters, including **overshoot, duration,** and **rate of rise,** are characteristic for each type of excitable cell. For example, the duration of the action potential of the squid giant axon is about 1 msec, whereas the cardiac action potential lasts for several hundred milliseconds. These differences in the action potentials subserve the functions performed by the different excitable tissues. The overshoot and a hyperpolarizing potential following the spike are illustrated in Figure 4-4C.

Strength–Duration Curve

Whether the threshold potential is reached depends on the amount of charge transferred across the membrane. Figure 4-5 shows that the total charge transfer across the membrane necessary to produce excitation is approximately constant, since charge moved equals current times time ($Q = I \cdot t$). The **strength–duration curve** can be derived from the equation for the exponential charge of the membrane capacitance.

The strength–duration curve deals only with the stimulus parameters (e.g., strength and duration of the applied current pulses) necessary to bring the membrane to threshold. It shows that, the greater the duration of the applied pulse, the smaller the current intensity required to just

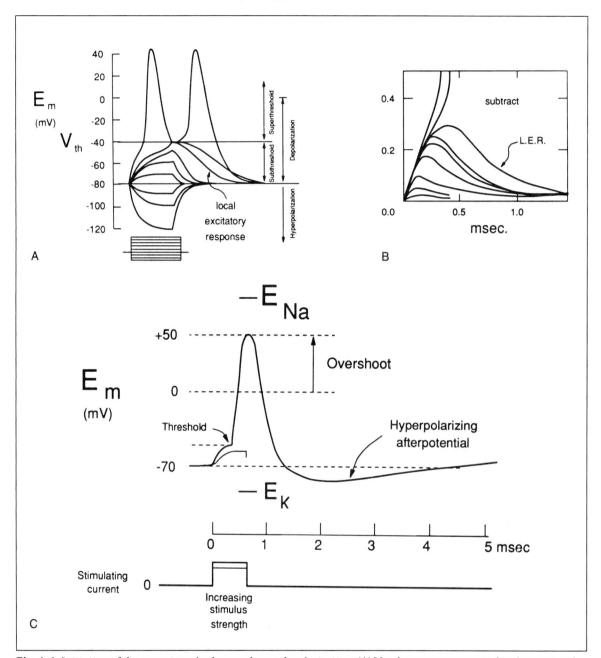

Fig. 4-4. Initiation of the nerve impulse by membrane depolarization. (A) Membrane responses to depolarizing and hyperpolarizing current pulses are shown. The nonlinear local excitatory response occurs just below the threshold (V_{th}) for the all-or-none action potential. (B) Illustration of the local excitatory response (*LER*) obtained by subtraction of the membrane responses to hyperpolarizing current pulses from the responses to depolarizing pulses. (C) A nerve action potential, illustrating the sharp threshold, overshoot, and hyperpolarizing afterpotential. (ϵ_m = membrane potential; ϵ_{Na}, ϵ_K = equilibrium potential to Na^+ and K^+, respectively.)

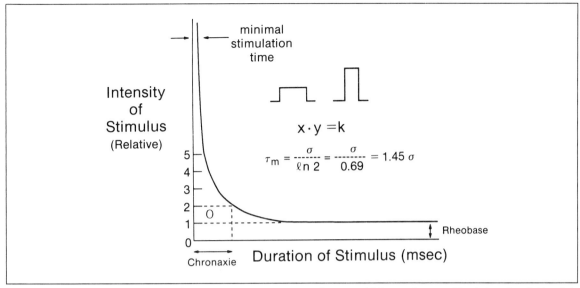

Fig. 4-5. Strength–duration curve for initiating action potentials in excitable membranes. The intensity of stimulating pulses is plotted against their duration for stimuli which are just sufficient to elicit an action potential. The rheobase current and chronaxie (σ) are indicated. (ℓn/ = natural logarithm.)

excite the fiber. The asymptote parallel to the X-axis is the **rheobase,** which is the lowest intensity of current capable of producing excitation, even when the current is applied for an infinite time (greater than 10 msec for myelinated nerve fibers). The asymptote parallel to the Y-axis is the **minimal stimulation time** and is the shortest duration of stimulation capable of producing excitation, even when huge currents are applied.

The rheobase is useless when comparing the excitability of one nerve with another, because only the relative current intensity is meaningful. Furthermore, it is difficult to measure the stimulation time of a current of rheobasic strength because it is an asymptote. Thus, a graphic measurement is made of the *time* during which a stimulus of double the rheobasic strength must act in order to reach threshold. This time is the **chronaxie.** Chronaxie values tend to remain constant regardless of the geometry of the stimulating electrodes. The shorter the chronaxie, the more excitable the fiber. Some nerve disorders in humans can be detected early by changes observed in the chronaxies.

The strength–duration curve shows that current pulses of very short duration (e.g., <0.1 msec) are less effective for stimulation. Thus, a sinusoidal a.c. at frequencies above 10,000 Hz is less capable of stimulation. Another way to view this is that, because the membrane impedance decreases greatly at high frequencies (since the cell membrane is a parallel RC network*), the potential difference that can be produced across the membrane by current flow across it (IR drops) is very small. Hence, very-high-frequency a.c. has less tendency to electrocute, but the energy of such currents can be dissipated as heat in body tissues and thus may be used in diathermy for therapeutic warming of injured tissues.

Refractoriness

Once an action is initiated, a finite and characteristic time must elapse before a second action potential can be generated. This time interval is

*RC = resistance times capacitance.

called the **refractory period,** and its value depends on the type of excitable cell. Cells with long-duration action potentials (e.g., myocardial cells) have long refractory periods; cells with brief action potentials (e.g., neurons) have short refractory periods. That is, the refractory periods are proportional to the duration of the action potential. Two types of refractory periods are usually defined: an **absolute refractory period** and a **relative refractory period** (Fig. 4-6). The absolute refractory period denotes the interval following an action potential during which a second action potential cannot be elicited, regardless of the intensity of the applied stimulus. During the **relative refractory period,** a second action potential may be elicited, provided that a greater-than-usual stimulus is applied. The second action potential often is subnormal in amplitude

and in rate of rise. Therefore, the physiologically important refractory period is the **functional refractory period,** or effective refractory period. This is defined by the highest frequency of action potentials that the excitable cells (e.g., neurons) can propagate. For example, if a myelinated nerve axon can propagate impulses up to 1,000/sec, the functional refractory period is 1.0 msec. The triggering of a second propagating impulse at a given point is limited by the amount of action current available (not like an electronic stimulator). Therefore, the functional refractory period encompasses all of the absolute and part of the relative refractory period.

The **absolute refractory period** extends from when threshold is reached at the initial portion of the rising phase of the action potential to when repolarization has reached about −50 mV. Dur-

Fig. 4-6. Refractory periods of nerve action potentials, both without (A) and with (B) a hyperpolarizing afterpotential. The absolute (*ARP*), relative (*RRP*), and functional (*FRP*) refractory periods are labeled.

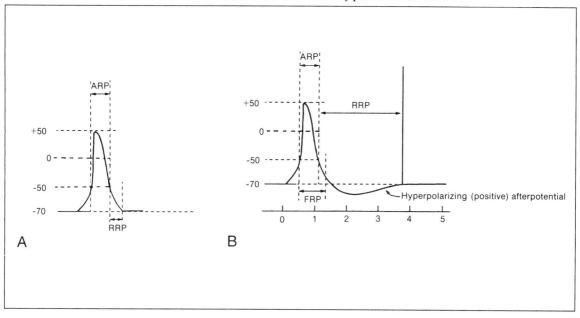

ing further repolarization beyond -50 mV (e.g., to -70 mV), a larger and larger fraction of the fast Na^+ channels has recovered from inactivation, and so is again available to be reactivated to produce another action potential. This is the period that inscribes the relative refractory period. The greater the degree of repolarization (toward the resting potential), the larger the subsequent action potential. Both voltage and time are factors in the recovery of the ion channels. Therefore, the absolute and relative refractory periods persist briefly beyond the theoretic voltages. Therefore, the relative refractory period actually exceeds the action potential duration.

Membrane excitability is greatly altered during refractory periods. Excitability is zero during the absolute refractory period and is depressed during the relative refractory period, becoming less and less depressed as the membrane repolarizes back to the resting potential (Fig. 4-7).

The afterpotentials that many cells exhibit also affect membrane excitability: hyperpolarizing afterpotentials depress excitability (greater critical depolarization required to reach threshold) and depolarizing afterpotentials enhance excitability. The latter produces a **supernormal period of excitability** and the former blends into and extends the relative refractory period. Thus, for example, neurons usually exhibit hyperpolarizing

afterpotentials, and so the functional refractory period actually includes part of the afterpotential.

Propagation velocity is slowed during the relative refractory period, achieving the normal value at the end of the relative refractory period. Propagation velocity is faster than normal during the supernormal period of excitability.

Accommodation

Accommodation, in physiological terms, refers to the loss of sensitivity of a cell or tissue to an applied stimulus. Sensory organs exhibit the property of accommodation, as do many neurons and other excitable membranes (e.g., skeletal muscle fibers). Automatic (spontaneously discharging) cells do not exhibit accommodation of their membranes to stimuli capable of evoking action potential responses (e.g., nodal [pacemaker] cells of the heart and some sensory neurons).

When a rectangular (square-wave) current pulse is used to depolarize a quiescent motor neuron ("step depolarization") from the resting potential (e.g., -70 mV) to the threshold potential or beyond, the neuron quickly responds with an all-or-none action potential. However, if the applied pulse is ramp shaped (triangular), the neuron may or may not respond, even if the normal threshold is exceeded, depending on the slope of the ramp. If the slope of the ramp is steep, the neuron will respond, but at a higher threshold (more critical depolarization is required). If the slope is shallow, the neuron will fail to fire an action potential, regardless of what level it is depolarized to. This is accommodation. That is, when the membrane is depolarized gradually, the stimulus is ineffective in producing an action potential response (Fig. 4-8A).

This is not true of pacemaker cells (see Fig. 4-8B). Such cells will discharge an action potential no matter how gradually the membrane is brought to the threshold point. In fact, the pacemaker potential in cardiac nodal cells is of the ramp type, producing depolarization to threshold over a period of about 200 to 800 msec.

The explanation for this phenomenon of

Fig. 4-7. The time course of recovery of nerve excitability during the relative refractory period. As shown, during the absolute refractory period, an action potential cannot be elicited, and there is a progressive increase in excitability during the relative refractory period.

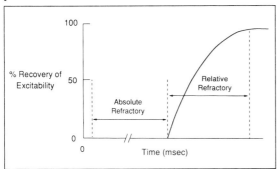

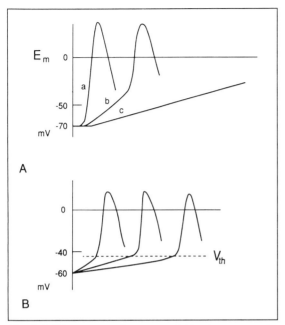

Fig. 4-8. The process of accommodation in response to a ramp stimulus in a motor neuron (A), and contrasted with the lack of accommodation in a nodal pacemaker cell from the heart (B). In A, as the slope of the ramp stimulus is decreased ($a{\rightarrow}c$), the action potential is delayed (b), and then fails completely (c). (E_m = membrane potential; V_{th} = threshold potential.)

accommodation is as follows: As the membrane is slowly depolarized toward threshold, the positive-feedback cycle between membrane potential (E_m), conductance for Na^+ (g_{Na}), and Na^+ current (I_{Na}) begins to operate, beginning at about 80 percent of the critical depolarization. Therefore, some of the fast Na^+ channels are turned on (activated), but they spontaneously inactivate (inactivation [I]–gates close) within 1 to 2 msec. If a critical number (critical mass) of fast Na^+ channels are not activated simultaneously, the positive-feedback cycle does not become explosive and a regenerative action potential is not produced. That is, the slow depolarization does not allow a critical number of fast Na^+ channels to be simultaneously in the open conducting state; the channels that opened and spontaneously inactivated cannot be reactivated until they return to the resting state, which requires re-

polarization to about the resting potential. Hence, they are lost from the pool of available channels.

Added to this is the fact that K^+ channels (delayed rectifier type) open during the slow depolarization, thus increasing the total K^+ conductance (g_K). Whenever g_K is increased, excitability becomes depressed because it tends to hyperpolarize and keep the membrane from depolarizing (to produce an action potential).

Therefore, accommodation to low-slope stimuli occurs for two reasons: (1) spontaneous inactivation of fast Na^+ channels that have been inactivated, and therefore lack of a critical mass to open simultaneously, and (2) increase in g_K, which depresses excitability.

The lack of accommodation in automatic cells (e.g., sinoatrial nodal cells of the heart), then, may be due to: less spontaneous ion channel inactivation and less g_K increase during the applied ramp stimulus or natural ramp pacemaker potential. Both of these conditions appear to apply. In cardiac nodal cells, the inward current responsible for the rising phase of the action potential is not a fast Na^+ current but a slow Ca^{2+} current; the Ca^{2+} slow channels inactivate very slowly at more positive potentials. In cardiac nodal cells, the kinetics of turn-on of the delayed rectifier K^+ current are also very slow.

As stated, accommodation also occurs in sensory organs. For example, some stretch receptors accommodate to a sustained stretch. When the stretch is first applied, there is a burst of action potentials, but the bursting frequency of discharge gradually slows down and then stops, even though the stretch is maintained. Other types of sense organs also possess the property of accommodation.

Anodal–Break Excitation

Excitation occurs on the "make" (the beginning) of a square-wave depolarizing stimulus. If the applied stimulus duration is very long (relative to the action potential duration), repetitive firing of action potentials will occur if the membrane is the nonaccommodating type. If the membrane is the accommodating type, then only the initial action potential is produced, because accommodation sets in.

If the cathode (negative) and anode (positive) electrodes are placed directly on an isolated single-nerve axon (e.g., squid giant axon), an action potential will be triggered at the cathode region on the make of the square-wave stimulus. This happens because, as illustrated in Figure 4-9A, depolarization occurs under the cathode, whereas hyperpolarization occurs under the anode. However, something unexpected occurs under the anode on the "break" of the stimulating pulse, namely an action potential is triggered from this hyperpolarized region of the axon (Fig. 4-9B). The explanation for this is that ion channel changes occur during the hyperpolarization, such that the excitability of that membrane is transiently increased (lower threshold point) immediately following cessation of the applied pulse. The increase in excitability is due to two factors: (1) increase in h value (at infinite time, h_∞) during the hyperpolarization, reflecting that almost 100 percent of the fast Na^+ channels have their I-gates open and hence are capable of conducting (open state) when the membrane is depolarized (removal of hyperpolarizing pulse) ($g_{Na} = \bar{g}_{Na}m^3h = \max g_{Na}m^3h$); and (2) decrease in n value (at infinite time, n_∞) during hyperpolarization, hence decreasing K^+ conductance ($g_K = \bar{g}_K n^4 = \max g_K n^4$). These changes in the h and n parameters persist for a short period after termination of the applied pulse, and hence increase membrane excitability during this brief period and trigger an action potential. This is called **anodal–break excitation** and **postanodal enhancement of excitability** and (postanodal depolarization).

In contrast, under the cathode, after termination of the applied pulse, an opposite change occurs in excitability and E_m. The membrane is hyperpolarized transiently and excitability is depressed. This is known as **postcathodal depression of excitability** and **postcathodal hyperpolarization.** In Hodgkin-Huxley terms, this phenomenon is also due to two factors: (1) decrease in h_∞ during depolarization, reflecting that a smaller fraction of fast Na^+ channels have their I-gates open, and hence are incapable of conducting, thus decreasing g_{Na}; and (2) increase in n_∞, hence increasing g_K. The increased g_K and decreased g_{Na} produce hyperpolarization (see Chapter 3, on the resting potential) and thereby depress excitability.

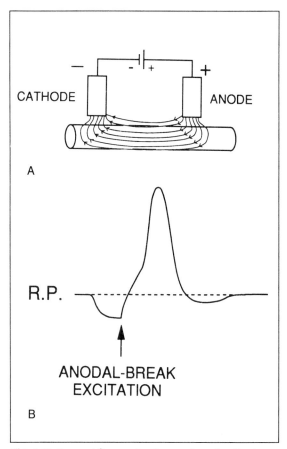

A

R.P.

ANODAL-BREAK EXCITATION

B

Fig. 4-9. Current flow under the anode and cathode during extracellular stimulation (A) and anodal-break stimulation of an action potential (B). Excitation occurs under the cathode on the "make" of a current pulse and under the anode at the "break" of the pulse. An intracellularly applied hyperpolarizing current pulse produces an anodal-break potential (B). (*RP = resting potential.*)

ELECTROGENESIS OF THE ACTION POTENTIAL

Early experimentation in electrophysiology was focused on determining the mechanism for the generation of the action potential. One important finding was that, during the action potential, the membrane resistance (but not the capacitance) changed dramatically (Fig. 4-10). The large **reduction in membrane resistance** that occurred during the action potential supported the hypothesis that the action potential resulted from a large increase in the ionic permeabilty of the membrane.

To determine which ionic species might be involved in generating the action potential, subsequent experimentation was directed toward varying the concentrations of the different ions bathing the axon. Figure 4-11 depicts a classic experiment in which the concentration of Na$^+$ bathing the squid axon was altered. It was found that the **over-**

shoot and the **rate of rise** of the action potential were proportional to the extracellular Na$^+$ concentration. This represented the first indirect demonstration that the action potential resulted from an increase in the membrane permeability to Na$^+$. Several years later, this idea was confirmed directly by the voltage-clamp method.

Fig 4-11. Na$^+$-dependence of the nerve action potential. (A) The method used to record the transmembrane action potential (E_m and E_M) from a squid axon. (B) The voltage traces show the reduction in action potential overshoot as the Na$^+$ concentration of the solution bathing the axon was reduced to 33 percent, or 50 percent of normal. Traces *1* were recorded with 100 percent of the normal Na$^+$ concentration; traces *2*, after reduction of the Na$^+$ concentration; and traces *3*, after return to the normal Na$^+$ concentration. (*ms* = milliseconds.) (Adapted from Hodgkin, A.L., *The Conduction of the Nervous Impulse*, Springfield, IL: Thomas, 1964; and Katz, B., *Nerve, Muscle and Synapse*, New York: McGraw-Hill, 1966.)

Fig. 4-10. Time course for the decrease in membrane resistance to the passage of ions (R_m) measured during the action potential (V) in a nerve axon. R_m falls from about 1000 Ω-cm^2 at rest to about 20 Ω-cm^2 near the peak of the action potential. (Adapted from K. C. Cole, *Membranes, Ions and Impulses: A Chapter of Classical Biophysics.* Berkeley: University of California Press, 1968.)

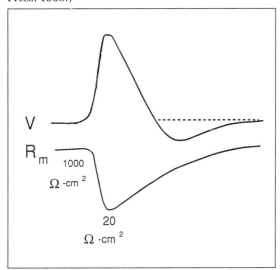

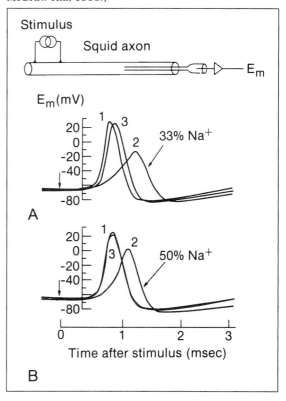

The Voltage-Clamp Method

The membrane current (I_m) that generates the action potential is composed of the **ionic current** (I_i) and **capacitive current** (I_c): $I_m = I_i + I_c$. The flow of ionic currents across their respective resistive membrane pathways (or channels) causes a change in the membrane potential (from Ohm's law: $V = IR$). The change in membrane voltage, in turn, causes a capacitive current to flow: $I_c = C_m \, dV/dt$, where dV/dt is the rate of change of the action potential.

Because the membrane potential during an action potential is constantly changing, it is difficult to separate the contributions of these interacting ionic and capacitive components. In addition, the total ionic current may be composed of multiple individual currents, carried by specific ions.

To analyze and separate the membrane currents into their capacitive and ionic components, a new and revolutionary method, called **voltage clamping,** was introduced in the early 1950s by Cole and Curtis and Hodgkin and Huxley. A diagram of the voltage-clamp method is shown in Figure 4-12. During a voltage-clamp experiment, the membrane potential or voltage is held constant ("clamped") by a **negative-feedback amplifier,** and the amount of current necessary to perform this task is recorded. The key to the success of the voltage-clamp experiment is that, if membrane potential (V_m) is held constant, the capacitive current is equal to zero. This can be rendered as follows:

$$I_m = I_i + C_m \, dV/dt$$

if v_m is constant, then:

$$dV/dt = 0$$
$$I_c = 0$$

and

$$I_m = I_i$$

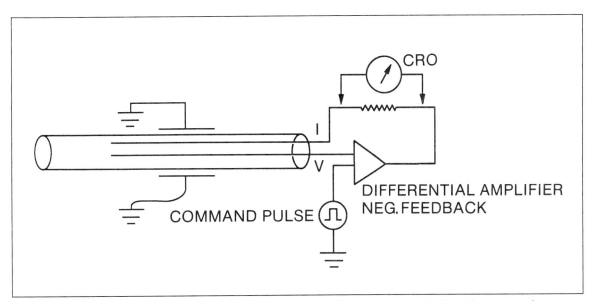

Fig. 4-12. The voltage-clamp method used in a squid giant axon. The two wires inserted into the axon are for measuring membrane potential (*V*) and passing current (*I*). The high-gain negative-feedback amplifier compares the command pulse with the membrane potential, and outputs the amount of current necessary to hold the potential constant (or "clamped"). The magnitude of the feedback current can be measured as the IR voltage drop across a resistor and displayed on an oscilloscope (*CRO*).

The voltage-clamp experiment identifies the magnitude and time course of the ionic currents at a given clamp potential. By clamping the membrane to many different potentials, information about the flow of ionic currents and the underlying conductance changes during the action potential is obtained.

Another advantage of the voltage-clamp experiment is that individual ion currents (such as Na^+, Ca^{2+} or K^+ currents) can be isolated from the total ionic current and analyzed separately. For example, in the squid axon experiments, the total ionic current consists of an early inward current, followed by a delayed outward current (Fig. 4-13). By varying the external Na^+ concentration, it has been shown that the early inward current is carried by Na^+. Similarly, by changing the K^+ concentration bathing the axon, one finds that the delayed outward current is carried by K^+. The Na^+ and K^+ currents can also be separated by blocking their respective pathways through the membrane. Na^+ channels can be blocked with **tetrodotoxin** (TTX), derived from the ovaries of the Japanese puffer fish, and K^+ channels can be blocked by several organic compounds, including tetraethylammonium ions (TEA^+). The remaining current can then be subtracted from the total ionic current to reveal the time course for the current that was blocked.

The **current–voltage relationship** is obtained by measuring the peak inward Na^+ and peak outward K^+ currents during a series of voltage-clamp steps (Fig. 4-14). Depolarizing voltage steps above the resting potential produce a small outward current. In this region, the membrane behaves in an ohmic fashion. With greater depolari-

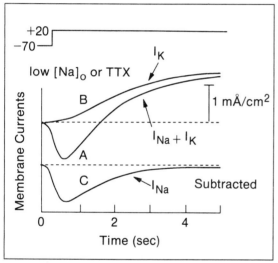

Fig. 4-13. The ionic currents that flow when a squid giant axon in sea water is clamped from its resting potential (-70 mV) to a transmembrane potential of $+20$ mV. Trace *A* shows the net inward (Na^+) current and outward (K^+) current in normal medium. Trace *B* shows the net ionic current when the axon is placed in artificial sea water with most of the Na^+ replaced by choline (an impermeant cation), so that the intracellular and extracellular Na^+ ($[Na]_0$) concentrations are equal. This current is due to K^+ only. Tetrodotoxin (*TTX*) can also be used to block the Na^+ current. Trace *C* shows the difference between *A* and *B*, which represents the Na^+ current (I_K, I_{Na} = K^+ and Na^+ currents, respectively.) (Redrawn from Hodgkin, A. L., and Huxley, A. F. *J. Physiol.* 116:449, 1952.)

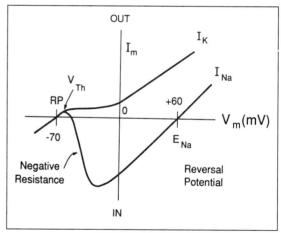

Fig. 4-14. Current–voltage relationship for the peak early inward current and delayed outward current obtained from a squid axon under voltage clamp. The inward current is carried by Na^+ (I_{Na}), the outward current by K^+ (I_K). The reversal potential for I_{Na} is the voltage at which the current changes from inward to outward. The region of the I_{Na} curve that has a negative slope is known as *negative resistance*. (RP = resting potential; V_{Th} = threshold; E_{Na} = Na^+ equilibrium potential; I_m = total membrane current; V_m = membrane potential.)

zation, the inward Na$^+$ current is activated and the current–voltage relationship displays a negative slope, or **negative resistance** region. A positive slope is seen at potentials above the peak of the current–voltage curve, and the current magnitude decreases as the Na$^+$ equilibrium potential (E$_{Na}$) is approached, actually becoming outward at voltages above E$_{Na}$. The voltage at which the current reverses in direction is the **reversal potential.** The reason that the current reverses at potentials above E$_{Na}$ is that the net electrochemical driving force for Na$^+$ becomes outwardly directed, whereas the conductance for Na$^+$ remains constant over this entire voltage range: I$_{Na}$ = g$_{Na}$ (E$_m$ − E$_{Na}$). The outward K$^+$ current activates above − 20 mV and increases with depolarization.

The voltage-clamp experiments have revealed the most fundamental property of the ionic conductances of excitable membrane—that these conductances are both **voltage** and **time dependent** (Fig. 4-15). Both the g$_{Na}$ and g$_K$ activate with depolarization, but with different time courses (Fig. 4-16). The g$_{Na}$ spontaneously turns off, or inactivates, with time. That is, the inward Na$^+$ current shuts off within 1 to 2 msec and causes the regenerative depolarization of the action potential. The depolarization is limited by the approach of the membrane potential toward E$_{Na}$, and by the Na$^+$ inactivation process. As the membrane is depolarized, both the g$_K$ and driving force for K$^+$ current increase, and the outward K$^+$ current repolarizes the membrane. The increase in g$_K$ is self-limited, namely the increase in g$_K$ produces repolarization, which, in turn, shuts off the increase in g$_K$.

The slow kinetics of the turnoff of the g$_K$ results in a transient hyperpolarization (the hyperpolarizing afterpotential) and the membrane potential is brought closer to the K$^+$ equilibrium potential (E$_K$) than at rest. The membrane conductance changes during the action potential are shown in Figure 4-17.

The time course for the ionic currents during the nerve action potential is shown in Figure 4-18. The total ionic current (I$_i$) is separated into its two major components, I$_{Na}$ and the net K$^+$ current (I$_K$). Since I$_K$ is slower to activate than I$_{Na}$, the

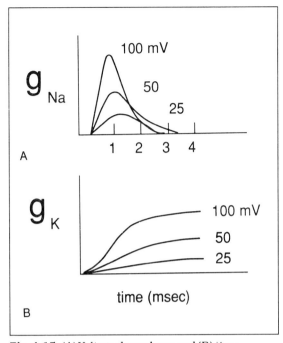

Fig. 4-15. (A) Voltage-dependence and (B) time-dependence of the changes in Na$^+$ (g$_{Na}$) and K$^+$ (g$_K$) conductances during voltage clamp of the squid giant axon. The numbers refer to the magnitude of depolarization (in millivolts) from the resting potential. The g$_{Na}$ turns on rapidly and then spontaneously declines (time is in milliseconds); g$_K$ turns on more slowly and is maintained during the long depolarizing clamp.

inward I$_{Na}$ predominates initially, giving rise to the upstroke of the action potential. Later, I$_K$ dominates, causing a net outward current that repolarizes the membrane.

The specific ionic currents are a product of the membrane conductance for the ionic species and the **electrochemical driving force** exerted on the ion. Thus:

$$I_{Na} = g_{Na}(E_m - E_{Na})$$
$$I_K = g_K (E_m - E_K)$$

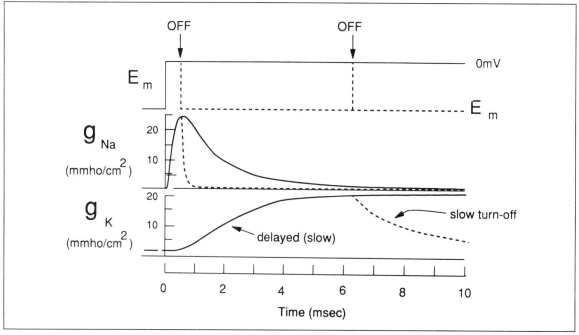

Fig. 4-16. Na$^+$ (g_{Na}) and K$^+$ (g_K) conductance changes are shown in response to voltage clamp of the squid giant axon from the resting potential to a membrane potential (E_m) of zero. The g_{Na} inactivates after a short time, even when the voltage clamp is maintained; in contrast, the g_K remains elevated until the clamp is released. When the clamp pulse is terminated early, the g_{Na} increase is quickly turned off; the turnoff of g_K is considerably slower. (mmho = milli-mho.) (Redrawn from Hodgkin, A. L. *Proc. R. Soc.* B148:1, 1958.)

The driving forces on Na$^+$ and K$^+$ continually change during the time course of the action potential, as diagrammed in Figure 4-19. At the resting potential, there is a large driving force for Na$^+$ to flow into the cell because ($E_m - E_{Na}$) is large. Conversely, at the peak of the action potential, ($E_M - E_{Na}$) is at its lowest value and the driving force for Na$^+$ entry is small. The driving force for K$^+$ efflux is, however, large at the peak of the action potential, when ($E_m - E_K$) is maximal. It is important to remember that ionic current flow depends on both an available conductance and a driving force. There is no net current if only one, but not the other, is present.

The biologic elements of the excitable membrane may be represented in terms of an **electric equivalent circuit model,** as shown in Figure 4-20. In the circuit model, current flow through the individual conductance pathways for each type of ion is represented by electron flow. The conductances for Na$^+$ and K$^+$ are variable, and depend on the transmembrane potential and time. Batteries, which are directed inwardly for Na$^+$ and outwardly for K$^+$, provide the driving force for ion flow. A passive leak conductance for Cl$^-$ is also included in the model. If the correct values for the elements are incorporated, the model circuit will generate an action potential.

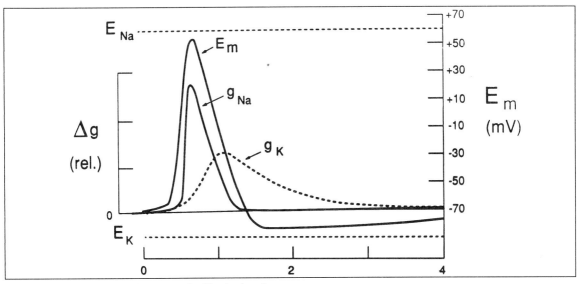

Fig. 4-17. The relative conductances for Na$^+$ (g_{Na}) and K$^+$ (g_K) during an action potential in nerve fibers. The rising phase of the action potential is caused by an increase in g_{Na}. The falling phase of the action potential is due to the rise of g_K (delayed rectification) and to the decrease in g_{Na}. The hyperpolarizing afterpotential is explained by the fact that g_K remains elevated for a short time following repolarization, tending to hold the membrane potential (E_m) near the K$^+$ equilibrium potential (E_K). (E_{Na} = Na$^+$ equilibrium potential; Δg = change in conductance.) (Redrawn from Hodgkin, A. L., *The Conduction of the Nerve Impulse, 1964.*)

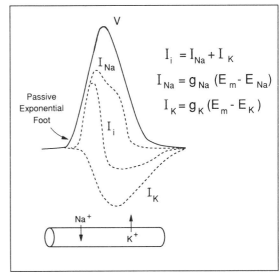

Fig. 4-18. Ionic currents that flow during the nerve action potential. The total current (I_i) is separated into an inward Na$^+$ current, (I_{Na}) and an outward K$^+$ current (I_K). I_i is the algebraic sum of I_{Na} and I_K. The appropriate equations for I_i, I_{Na}, and I_K are given. Also depicted is the fact that a net inward Na$^+$ flux occurs during the rising phase of the action potential and a net K$^+$ efflux occurs during the repolarizing phase. (E_m = membrane potential; E_{Na}, E_K = equilibrium potentials for Na$^+$ and K$^+$, respectively; V = voltage.)

Fast Na$^+$ Channel Activation

During an action potential, the increase in g_{Na} is related to the membrane potential in a positive-feedback fashion ("vicious cycle") (Fig. 4-21A). That is, a small depolarization leads to a increase in g_{Na}, which allows a larger inward I_{Na} that causes further depolarization. This greater depolarization produces a greater increase in g_{Na}. This positive-feedback process is an explosive one with a sharp trigger point (threshold), resulting from the exponential (positive) relationship between g_{Na} and E_m (see Fig. 4-21B). It is this

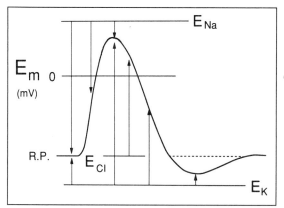

Fig. 4-19. Driving forces for Na⁺ and K⁺ currents during the action potential. The total electrochemical driving force is equal to the membrane potential (E_m) minus the equilibrium potential for the ion (E_i): ($E_m - E_i$). As depicted, the driving force for the Na⁺ current decreases during the action potential, whereas that for K⁺ increases. Even if Cl⁻ is passively distributed, a net driving force for Cl⁻ influx (outward Cl⁻ current) occurs during the action potential. (*RP* = resting potential; E_{Cl}; E_{Na}, E_K = equilibrium potential for Cl⁻, Na⁺, K⁺, respectively.)

positive-feedback relationship that accounts for the negative resistance (slope) in the current–voltage curve (see Fig. 4-14). As Figure 4-21B shows, the g_{Na} reaches a maximum (saturates) at positive potentials that produces maximal activation of the population of fast Na⁺ channels.

The fast Na⁺ channels (and the slow Ca²⁺ channels) have a **double gating mechanism,** consisting of an **inactivation gate** (I-gate) and an **activation gate** (A-gate) (Fig. 4-22). For a channel to be conducting, both the A-gate and I-gate must be open; if either is closed, the channel is nonconducting. The A-gate is located somewhere near the middle of the channel; it is not at the outer surface because even TTX does not prevent the movement of this gate, and it is not at the inner surface because proteases do not affect it. The A-gate is closed at the resting E_m and opens rapidly upon depolarization (whereas the I-gate is open at the resting E_m and closes slowly upon depolarization). In the Hodgkin-Huxley (1952) analysis, the opening of the A-gate requires simultaneous occupation of three negatively

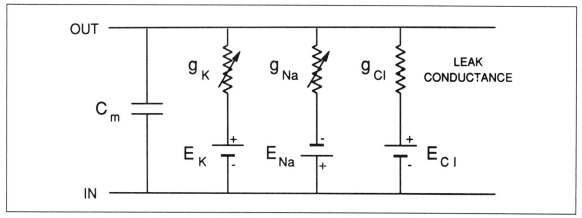

Fig. 4-20. Hodgkin-Huxley electric equivalent circuit for the squid giant nerve axon. The K⁺ conductance (g_K) is in series with the K⁺ equilibrium potential (E_K), and Na⁺ conductance (g_{Na}) is in series with the Na⁺ equilibrium potential E_{Na}. The arrows indicate that g_{Na} and g_K vary with voltage and time. The low conductance for Cl⁻ (g_{Cl}) was termed the *leak conductance*. (C_m = membrane capacitance; E_{Cl} = equilibrium potential for Cl⁻.) (Redrawn from Hodgkin, A. L. *The Conduction of the Nervous Impulse.* Springfield, IL: Thomas, 1964.)

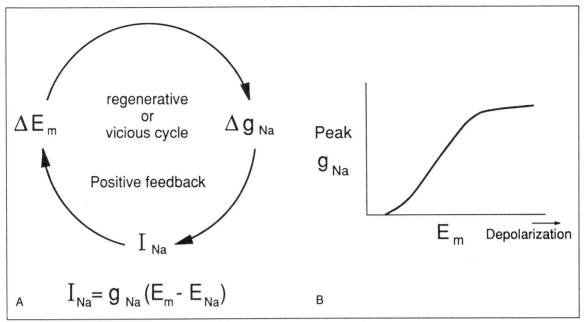

Fig. 4-21. The positive feedback relationship between Na^+ conductance (g_{Na}) and membrane potential (E_m), leading to the all-or-none action potential. (A) The increase in g_{Na} allows an increase in the inward Na^+ current: [$I_{Na} = g_{Na}(E_m - E_{Na})$], which is depolarizing and so triggers a further increase in g_{Na}. This explosive feedback cycle is caused by the voltage dependency of the gated fast Na^+ channels. ($I_{Na} = Na^+$ current; $E_{Na} = Na^+$ equilibrium potential.) (B) Plot of g_{Na} versus depolarization, showing the initial exponential (positive) increase in g_{Na} as a function of voltage. (Adapted from Hodgkin, A. L. *The Conduction of the Nervous Impulse.* Springfield, IL: Thomas, 1964.)

charged sites by three positively charged (m^+) particles. If m, the activation variable, is the probability of one site being occupied, then m^3 is the probability that all three sites are occupied; therefore: $g_{Na} = max\ g_{Na}\ m^3\ h$, where h is the inactivation variable and max g_{Na} is the maximum conductance (Fig. 4-23).

A **gating current** has been measured that corresponds to the movement of the m^+ particles (or rotation of an equivalent dipole). The gating current is very small in intensity and is measured by subtracting the linear capacitive current (from a hyperpolarizing clamp step) from the total capacitive current (linear plus nonlinear) that occurs with a depolarizing clamp step beyond threshold. The outward gating current precedes the inward I_{Na}. TTX does not block the gating current, although it does block I_{Na}. Thus, the gating current is a nonlinear, outward capacitive current (not an ionic current) obtained during depolarizing clamps that reflects movement of the A-gates from the closed to the open configuration. The linear capacitive current results mainly from the lipid bilayer matrix, whereas the nonlinear capacitive current arises from the charge movement within the protein channels.

Na^+ Inactivation

The fast I_{Na} lasts only for 1 to 2 msec because of the spontaneous voltage inactivation of the fast Na^+ channels. That is, the fast Na^+ channels inactivate quickly, even if the membrane remains depolarized. (In contrast, the slow Ca^{2+} channels inactivate very slowly.) Inactivation is produced in the fast Na^+ channels (and in the slow channels)

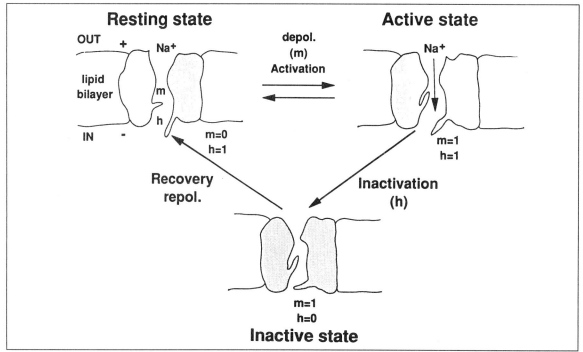

Fig. 4-22. Model of the three hypothetical states of the fast Na^+ channel. In the resting state, the activation gate (A or m) is closed and the inactivation gate (I or h) is open: m = 0; h = 1. Depolarization to the threshold activates the channel to the active state, the A-gate opening rapidly and the I-gate still being open: m = 1; h = 1. The activated channel spontaneously inactivates to the inactive state due to delayed closure of the I-gate: m = 1; h = 0. The recovery process upon repolarization returns the channel from the inactive state back to the resting state, making the channel available again for reactivation. Na^+ is depicted as being bound to the outer mouth of the channel and poised for entry down its electrochemical gradient when both gates are open (active state of channel). The reaction between the resting and active states is reversible, whereas the other reactions may not be. The Ca^{2+} slow channels pass through similar states. (Adapted from Hodgkin, A. L. *The Conduction of the Nervous Impulse.* Springfield, IL: Thomas, 1964.)

by the voltage-sensitive, slow closing of the I-gate (see Fig. 4-22). The I-gate is located near the inner surface of the membrane; this is shown by the fact that the addition of proteolytic enzyme to the inside of a perfused giant axon chops off the I-gate and eliminates inactivation, whereas its presence outside does not have this effect. The I-gate is presumably charged positively to allow it to move with changes in the membrane potential. During depolarization, the inside of the membrane becomes positive, and this causes the I-gate to close. At the normal resting potential, the I-gate is open but the A-gate is closed.

The voltage dependency of inactivation is depicted by the h_∞ versus E_m curve (Fig. 4-24A). The inactivation variable (h; probability function) varies between 0 and 1.0, perhaps reflecting occupation of a negatively charged site by a positively charged inactivation particle, and h_∞ is the value of the inactivation variable at infinite time (> 10 msec), or at steady-state. When the inactivation variable is 1.0, the I-gates of all the fast Na^+ channels are open; conversely, when the inactivation variable is zero, all the I-gates are closed. Since the g_{Na} at any time is equal to the maximal value (max g_{Na}) times m^3h, when h = 0, and $g_{Na} = 0$;

Fig. 4-23. Time course of the changes in the Hodgkin-Huxley m, n, and h factors during a depolarizing voltage-clamp step. The lower traces give the resulting time course for the changes in Na^+ (Δg_{Na}) and K^+ (Δg_K) conductories during the step. Since g_{Na} is proportional to the product of the m and h variables, it returns to about the original level within approximately 2 msec after the clamp step is applied. (V = membrane voltage.)

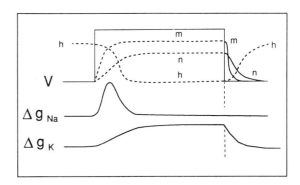

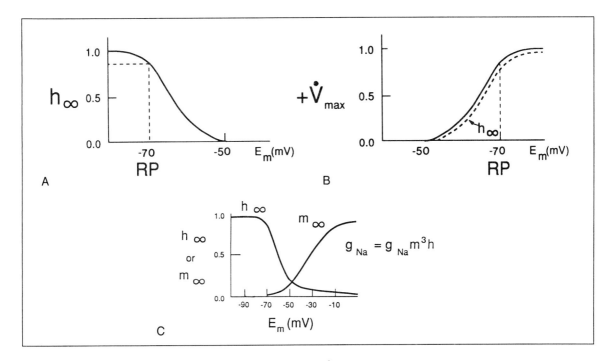

Fig. 4-24. Voltage inactivation of the fast Na^+ channels as a function of the resting membrane potential (E_m). (A) h_∞ plotted against E_m, where h_∞ is the inactivation factor of Hodgkin-Huxley. g_{Na} = max g_{Na} m^3h, where g_{Na} is the Na^+ conductance, m and h are variables, and max g_{Na} is the maximal conductance. The h_∞ represents h at time = infinity or steady-state (practically, after 10 msec). This graph illustrates that fast Na^+ channels begin to inactivate at about -75 mV, and nearly complete inactivation occurs at about -50 mV. In contrast, slow channels inactivate between about -45 and -5 mV. (B) Maximal rate of rise of the action potential $\dot{V}$ max, or (max dV/dt) as a function of resting E_m is also depicted for the normal action potential (dependent on inward current through the fast Na^+ channels). Max dV/dt is a measure of the inward current intensity, which is dependent on the number of channels available for activation. Therefore, max dV/dt decreases as h_∞ decreases. (Note that the abscissa is opposite in B compared to A.) (C) Plot of both steady-state inactivation (h_∞) and activation (m_∞) against E_m to illustrate the overlap of the two curves, depicting the "window" current region. [RP = resting potential.]

and when h = 1.0, g_{Na} = max g_{Na} (if m = 1.0). At the normal resting potential, h_∞ is nearly 1.0 and diminishes with depolarization, becoming zero at about −50 mV. Because the maximal rate of rise of the action potential is directly proportional to the net inward I_{Na}, which is directly proportional to g_{Na}, the decrease in h_∞ is the cause of decrease in max g_{Na}. At about −50 mV, h_∞ is zero and there is complete inactivation of the fast Na^+ channels. Therefore, depolarization by any means—e.g., elevated extracellular K^+ concentration or applied depolarizing current pulses—decreases max g_{Na} and excitability disappears at about −50 mV (see Fig. 4-24B). A plot of the inactivation curve along with the activation curve is given in Figure 4-24C to illustrate the overlap in these curves, which gives rise to a so-called "window" current (steady-state) over a certain voltage region.

The slow Ca^{2+} channels (Fig. 4-25) behave much the same way as the fast Na^+ channels with respect to inactivation, with one main difference being the voltage range over which the slow channels inactivate: −60 to 0 mV for the slow channels compared with −120 to −50 mV for the fast Na^+ channels. (The slow action potential disappears at a lower voltage [e.g., −20 mV] perhaps because of a minimum density requirement for a regenerative response.) Another major difference is that slow channels inactivate much more slowly than do fast channels, that is, they have a long inactivation time constant. (In myocardial cells, the inactivation variable for the slow channel is referred to as the *f variable,* and the m activation variable as the *d variable,* but the exponents are 1.0.)

Recovery

Any Na^+ or Ca^{2+} channel that has been activated and then spontaneously inactivated must go through a **recovery process** before it can return to the resting state from which it can be **reactivated** (see Fig. 4-22). This recovery process is dependent on voltage and time. The membrane must be repolarized beyond −50 mV before the recovery process can begin (i.e., traveling up the

h_∞ versus E_m curve). At any given E_m, it takes time for the recovery process to occur, namely, the time required for the charged A- and I-gates to be restored to their resting configuration (A-gate closed, I-gate open) with the electric field. The recovery process is fast for fast Na^+ channels (e.g., 1–10 msec) and slow for the slow channels. This recovery process of the slow channels is slowed by the effect of organic calcium antagonists.

K^+ Activation

The K^+ channel (outward-going delayed rectifer) is generally believed to have only an A gate, because it does not inactivate. This gate is thought to be located near the inner surface of the membrane, because TEA^+ blocks the K^+ channel more readily from the inner surface. The block is use dependent, the proposed mechanism being that, as the A-gate opens, the TEA^+ molecule can bind in the channel behind the gate. The A-gate is believed to be positively charged, and depolarization (inside becoming positive) opens the gate. In the Hodgkin-Huxley analysis of squid giant axon, the A-gate opens when four positively charged (n^+) particles simultaneously occupy four favorable positions (negatively charged sites). If n is the probability that one site is occupied, then n^4 is the probability that all four sites are occupied; therefore g_K = max $g_K n^4$. The power to which n is raised varies in different tissues from 2 to about 25.

Mechanisms of Repolarization

The action potential is terminated primarily by the turn-on of the g_K, the **delayed rectification.** It is called this because its turn-on is slower and delayed with respect to the turn-on of the fast Na^+ channels. Because this channel allows K^+ to pass more readily outward than inward, and more readily down the electrochemical gradient, it is also known as *outward-going* rectification. The increase in g_K acts to bring E_m toward the K^+ equilibrium potential (about −90 mV), because the membrane potential at any time is deter-

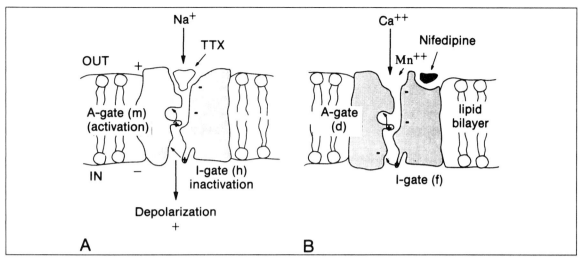

Fig. 4-25. Model for a fast Na^+ channel (A) and for a Ca^{2+} slow channel (B). The ionic channels are proteins that float in the lipid bilayer matrix of the cell membrane. These voltage-dependent channels have two gates, as depicted (conformational changes in the protein could serve as the gates): the activation gate (A or m) and the inactivation gate (I or h). The I-gate is located near the inner surface of the membrane because proteolytic enzymes added internally destroy this gate; the A-gate is located somewhere in the middle of the channel—tetrodotoxin *(TTX)* does not prevent movement of this gate. TTX binds to the outer mouth of the channel and plugs it physically, as depicted. The gates are presumably charged positively, so that upon depolarization (inside going less negative [i.e., in a positive direction]), the gates are electrostatically repelled outward. The A-gate opens relatively quickly (0.25 msec), whereas the I-gate closes relatively slowly (inactivation time constant of 1–2 msec). Therefore, for a short period, both gates are open and the channel is in the conducting mode with Na^+ entering the cell down its electrochemical gradient. However, when the I-gate closes (after about 2 msec), the channel again becomes nonconducting. During recovery (reactivation) upon repolarization, the gates return to their original resting positions.

The slow Ca^{2+} channels (B) behave similarly to the fast Na^+ channels (A), except that their gates appear to move more slowly on a population basis. In other words, the slow-channel conductance activates, inactivates, and recovers more slowly. The slow-channel gates operate over a different voltage range from the fast channels (i.e., less negative, more depolarized). TTX does not block the slow channels. Drugs such as nifedipine block the slow channels (but not the fast Na^+ channels) by binding to the channel and somehow inactivating it. In addition, the voltage inactivation curve of the Ca^{2+} channels is shifted to the right, so that inactivation begins at about -45 mV and is not complete until about -5 mV. The slow channels also have a lower activation (threshold) potential of about -35 mV (compared to about -60 mV for the fast Na^+ channel). The activation gate variable is known as d, and the inactivation gate variable as f, in the case of slow channels.

mined mainly by the ratio of g_{Na} to g_K (see Chapter 3). This type of g_K channel is activated by depolarization and turned off by repolarization. Therefore, this g_K channel is self-limiting, in that it turns itself off as the membrane is repolarized by its action.

In addition to the g_K turn-on, there is also some turnoff of g_{Na}, which would contribute to repolarization for two reasons: (1) spontaneous inactivation of fast Na^+ channels that had been acti-

vated (i.e., closing of their I-gate with an inactivation time constant of 1–3 msec), and (2) reversible shifting of activated channels directly back to the resting state (deactivation) because of the rapid repolarization occurring due to the g_K mechanism. Theoretically, it is possible to have an action potential that would repolarize (but slowly) even if there were no g_K mechanism, because the g_{Na} channels would spontaneously inactivate and so the g_{Na}/g_K ratio and E_m would slowly be restored to their original resting values.

In skeletal muscle, there is an important third factor involved in repolarization of the action potential: the Cl⁻ current. The Cl⁻ permeability and conductance (g_{Cl}) are very high in skeletal muscle. In fact, the Cl⁻ permeability of the surface membrane is much higher than that of K⁺, the ratio being about 3. However, as discussed in Chapter 3, Cl⁻ is passively distributed, or nearly so, and thus cannot determine the resting potential under steady-state conditions, although net Cl⁻ movements inward (hyperpolarizing) or outward (depolarizing) can and do affect E_m transiently until re-equilibration occurs. There is no net electrochemical driving force for the net Cl⁻ current at the resting potential, since $E_m = E_{Cl}$. However, during action potential depolarization, there is a larger and larger driving force for outward Cl⁻ current (I_{Cl}) (i.e., Cl⁻ influx), since $I_{Cl} = g_{Cl}$ ($E_m - E_{Cl}$), where E_{Cl} is the Cl⁻ equilibrium potential (see Fig. 4-19). In other words, the large electric field that was keeping Cl⁻ out (i.e., intracellular Cl⁻ concentration << extracellular Cl⁻ concentration), such that $E_{Cl} = E_m$ diminishes during the action potential, and so Cl⁻ enters the fiber. This Cl⁻ entry is hyperpolarizing, and so tends to repolarize the membrane more quickly than would otherwise occur. That is, repolarization of the action potential is "sharpened" by the Cl⁻ mechanism. For this mechanism to occur, there need not be any voltage-dependent, gated g_{Cl} channels. Only a high g_{Cl} is needed; the higher the g_{Cl}, the greater this effect.

To illustrate some of the above points, if skeletal muscle fibers are placed in Cl⁻-free Ringer's solution (e.g., methanesulfonate substitution), depolarization and spontaneous action potentials and twitches occur for a few minutes until most or all of the intracellular Cl⁻ is washed out. After equilibration, the resting E_m returns to its original value (approximately −90 mV for frog muscle), clearly indicating that Cl⁻ does not determine the resting potential and that net Cl⁻ efflux produces depolarization. Re-addition of Cl⁻ to the bath produces a rapid, large hyperpolarization (e.g., to −120 mV), due to net Cl⁻ influx. The E_m then slowly returns to its orig-inal value (−90 mV) as Cl⁻ re-equilibrates (i.e., redistributes itself passively). These same effects occur in cardiac muscle, smooth muscle, and nerve, but to a lesser extent because Cl⁻ permeability is much lower in these tissues (e.g., the Cl⁻ to K⁺ permeability ratio is about 0.5 in vascular smooth muscle).

Effect of Resting Potential on Action Potential

Any agent that affects the resting potential has important repercussions on the action potential. Depolarization reduces the rate of rise of the action potential, and thereby also slows its velocity of propagation. A slow spread of excitation throughout the nerve or muscle interferes with its ability to act efficiently. This effect is progressive as a function of the degree of depolarization. If nerve or muscle fibers are depolarized to about −50 mV, by any means, the rate of rise goes to zero and all excitability is lost.

Hyperpolarization usually produces only a small increase in the rate of rise. Larger hyperpolarization may actually slow the velocity of propagation, because the critical depolarization required to bring the membrane to its threshold potential is increased, or it may cause propagation block.

The explanation for the effect of resting E_m (or takeoff potential) on the maximal rate of rise of the action potential is based on the sigmoidal h_∞ versus E_m curve (see Fig. 4-24B). In Hodgkin-Huxley notation, h is the inactivation varible for the fast Na⁺ conductance; it is a probability factor that deals with the open (h = 1.0) versus closed (h = 0) position of the I-gate of the channel (see Fig. 4-22). The h varies as a function of E_m and time (t), and h_∞ is the h value at steady-state or infinite time (practically, t > 20 msec). At the resting potential of −80 mV, h_∞ is 0.9 to 1.0 and diminishes with depolarization, becoming zero at about −50 mV. The I-gates are open in a resting membrane and close slowly.

The resting potential also affects the duration of the action potential. With polarizing current, depolarization lengthens the action potential, whereas hyperpolarization shortens it. In contrast, when elevated extracellular K^+ levels are used to depolarize the cells, the action potential is usually shortened. One important determinant of the action potential duration is the g_K. Agents or conditions that increase g_K, such as elevation of extracellular K^+ content, tend to shorten the duration. In contrast, agents that decrease g_K or slow its activation, such as the barium ion or TEA^+, tend to lengthen the action potential duration. Due to **anomalous rectification** (i.e., a decrease in g_K with depolarization and an increase with hyperpolarization), depolarization by current prolongs the action potential and hyperpolarization shortens it. The effects of anomalous rectification are seen in muscle, but not in some nerve cells, which may lack these channels.

Other factors are also important in determining the action potential duration. For example, agents that slow closing of the I-gates of the fast Na^+ channels, such as veratridine, prolong the action potential. At high rates of activity, two factors could contribute toward changes in the membrane potentials: (1) the increase in intracellular Ca^{2+} (resulting from an increase in intracellular Na^+) produces an increase in g_K (the $g_{K[Ca]}$ channel) and (2) K^+ accumulates outside the cell membrane, thereby increasing g_K and decreasing E_K.

Electrogenesis of Afterpotentials

The action potentials of nerve and muscle cells usually consist of two components: an initial spike followed by an early afterpotential (Fig. 4-26). The afterpotentials may be of two types: depolarizing or hyperpolarizing. In addition, late afterpotentials may appear following a brief train of spikes.

Early Depolarizing Afterpotentials

The action potential spike in skeletal muscle fibers is followed by a prominent depolarizing afterpotential (also called a "negative" afterpotential, based on the old terminology used

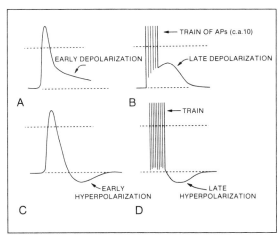

Fig. 4-26. Examples of the different types of afterpotentials. (A) Early depolarizing ("negative") afterpotential recorded after a single action potential in a skeletal muscle fiber. (B) Late depolarization afterpotential recorded after a train (e.g., 10) of action potentials (*APs*) in a skeletal muscle fiber. (C) Early hyperpolarizing ("positive") afterpotential recorded after a single action potential in a nerve fiber. (D) Late hyperpolarizing afterpotential recorded following a train of action potentials in a nerve terminal.

when recording was done externally) (see Fig. 4-26A). In addition to this "early" afterpotential (i.e., emerging from the spike), there is a late depolarizing afterpotential that follows a tetanic train of spikes (e.g., 10 spikes) (see Fig. 4-26B). The electrogeneses of the early and the late afterpotentials are different. The early afterpotential is due to a conductance change, whereas the late afterpotential may be due to K^+ accumulation in the transverse (T) tubules (periodic invaginations of the surface membrane into the fiber interior, involved in the excitation–contraction coupling mechanism).

The early depolarizing afterpotential of frog skeletal fibers is about 25 mV in amplitude, immediately after the spike component, and gradually decays to the resting potential over a period of 10 to 20 msec. It results from the fact that the delayed rectifier K^+ channel that opens during depolarization to terminate the spike is less selective for K^+ (the ratio of K^+ to Na^+ permeability is about 30:1) than is the K^+ "channel" in the resting membrane: P_K/P_{Na}, about 100:1. Therefore, from the constant-field equation (see Chapter 3), one can predict that the membrane should be partly depolarized when the membrane potential is dominated by this delayed K^+ conductance. Thus, the early depolarizing afterpotential seems to be due to the persistence of, and slow decay of, this less-selective K^+ conductance.

Late Depolarizing Afterpotentials

The late afterpotential results from accumulation of K^+ in the T-tubules (see Fig. 4-26B). During the action potential depolarization and turn-on of the g_K (delayed rectifier), there is a large driving force for K^+ efflux from the myoplasm coupled with a large K^+ conductance, resulting in a large outward K^+ current [$I_K = g_K (E_m - E_K)$] across all surfaces of the fiber, namely the surface sarcolemma and T-tubule walls. The K^+ efflux at the fiber surface membrane can rapidly diffuse away and mix the relatively large interstitial fluid volume, whereas the K^+ efflux into the T-tubules is trapped in this restricted diffusion space. The resulting high concentration of K^+ in the T-tubules decreases E_K across the T-tubule membrane, and thereby depolarizes this membrane. Due to cable properties, part of this depolarization is transmitted to the surface sarcolemma and is recorded by an intracellular microelectrode. The K^+ accumulation in the T-tubules can only be dissipated relatively slowly by diffusion out of the mouth of the T-tubules and by active pumping back into the myoplasm across the T-tubule wall. Thus, the decay of the late afterpotential is a function of these two processes.

The amplitude and duration of the late depolarizing afterpotential of frog skeletal fibers is a function of the number of spikes in the train and their frequency. That is, the greater the spike activity, the greater the amplitude and duration of the late afterpotential. If the train consists of 20 spikes at a frequency of 50 per second, a typical value for the amplitude of the late afterpotential is about 20 mV. When the diameter of the T-tubules is increased by placing the fibers in hypertonic solutions, the amplitude of the late afterpotential decreases, as expected, because of the great dilution of K^+ accumulating in the T-tubule lumen. When the T-tubule system is disrupted and disconnected from the surface membrane by the glycerol osmotic shock method, the late afterpotential disappears (whereas the early afterpotentials persist).

An alternative explanation for the late afterpotential is that it may be due to a slow g_K change. Hence, the late afterpotential may arise from the slow relaxation of a component of the K^+ conductance increase.

Hyperpolarizing Afterpotentials

Neurons, pacemaker heart cells, and vascular smooth muscle cells often exhibit **early hyperpolarizing** ("positive") **afterpotentials** (see Fig. 4-26C). These are due to the delayed rectifier K^+ conductance increase (which terminates the spike) persisting after the spike, thereby bringing E_m closer to E_K. The maximum amplitude of the positive afterpotential is the difference between E_K and the normal resting potential. The time course of this afterpotential is determined by the decay of the K^+ conductance increase (which is a function of the kinetics of the K^+ channel gates).

Some cells, such as nonmyelinated neurons, exhibit **late hyperpolarizing afterpotentials** following a train of spikes (see Fig. 4-26D). These hyperpolarizing afterpotentials are due to the Na^+-K^+ pump, because inhibition of the pump by any means (such as ouabain, cold, or lithium ions) abolishes the late hyperpolarizing afterpotential. Two mechanisms have been proposed for this phenomenon: (1) an increased electrogenic Na^+ pump potential stimulated by an increase in the intracellular Na^+ content (since these neurons are small in diameter, and hence have a large surface area to volume ratio) or by an increase in the extracellular K^+ concentration (since these neurons are surrounded by Schwann cells and hence a narrow intercellular cleft and restricted diffusion space, or both actions; and (2) an increased E_K caused by K^+ depletion in the intercellular cleft due to the Na^+-K^+ pump overpumping the

K+ back in. It is generally believed that the first mechanism, namely a larger pump potential, is the most probable, although it is difficult to distinguish these two possibilities.

Importance of Afterpotentials

All afterpotentials have physiologic importance because they alter excitability and the propagation velocity of the cell. A depolarizing afterpotential should enhance excitability (lower threshold) and a hyperpolarizing afterpotential should depress excitability to a subsequent action potential. This is because the critical depolarization required to reach the threshold potential would be decreased or increased, respectively. A large, late depolarizing afterpotential, such as that due to K+ accumulation in the T-tubules, can, under certain pathologic conditions, trigger repetitive action potentials.

The effect of afterpotentials on the velocity of propagations is more difficult to predict because there are two opposing factors: (1) the change in critical depolarization required and (2) the change in the maximal rate of rise of the action potential, which is a function of the takeoff potential (h_∞ versus E_m curve). For example, during a depolarizing afterpotential in skeletal muscle fibers, the critical depolarization required is decreased, but the maximal rate of rise of the action potential is decreased.

Molecular Basis of Excitability: Properties of Individual Ion Channels

Recently, a new electrophysiologic method, the **patch-clamp technique,** has enabled researchers to examine the basis of excitability at the molecular level. The method allows recording of the currents that flow through individual ion channels in the cell membrane. To record the channel currents, a small-tipped glass pipette is pressed against the cell membrane and negative pressure is created in the interior of the pipette. A small "patch" of membrane is drawn into the tip of the glass pipette, and a high-resistance seal [giga (10^{12})-ohm] forms spontaneously (Fig. 4-27). The

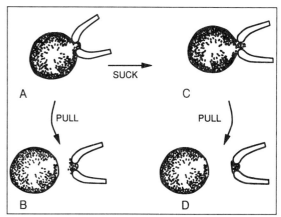

Fig. 4-27. The patch-clamp technique in isolated single cells. (A) Cell-attached patch mode, produced by applying light suction to the patch pipette to produce the gigaseal (>10 GΩ). (B) Isolated inside-out patch produced by pulling on the patch pipette to tear off the membrane patch from the cell. (C) Whole-cell clamp mode produced by applying strong suction to the patch pipette to blow out the membrane patch and allow the lumen of the pipette to be continuous with the lumen of the cell. This mode is used to record the macroscopic currents from the entire complement of ion channels in the cell membrane. (D) Isolated outside-out patch produced by pulling on the patch pipette to tear off a small patch of membrane from the cell, which then seals, as depicted. In all cases, a gigaseal is produced between the fire-polished patch pipette and the membrane. The modes depicted in A, B, and D are used to record the microscopic currents from single channels in the patch (single-channel currents). (Modified from Hille, B. *Ionic Channels of Excitable Membrane.* Boston: Sinauer Associates, 1984).

patch pipette is connected to a low-noise, high-gain amplifier. If one or more channels reside in the patch of membrane, then their open (conducting) and closed times can be recorded.

Single-channel opening and closing behavior appears to occur randomly in time (Fig. 4-28). If, however, the average behavior over time of only one channel (or a small group of channels) in a given patch is considered, the random behavior becomes deterministic. For example, the opening-and-closing sequence of fast Na+ channels, shown in Figure 4-29, appears to be random in each trial, yet the sum of the individual currents

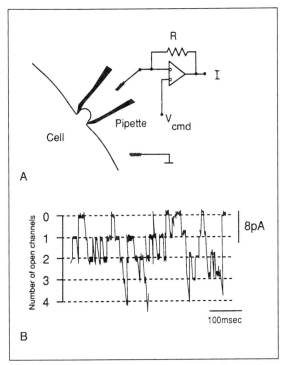

Fig. 4-28. Single-channel recording using the patch-clamp method. (A) The single-channel currents from the patch of membrane isolated by the pipette tip are amplified and recorded. (R = feedback resistor; V_{cmd} = command voltage; I = current.) (B) Single-channel recordings appear as small square steps of current as a channel opens, then closes, then reopens. Several channels were present in this patch, so several levels of superimposed channel currents are seen. (A modified from Sigworth, F. J., and Neher, E. *Nature* 287:447–449, 1980. B modified from Hammill, O. P., et al. *Pflugers Arch.* 391:85–100, 1981.)

is similar to the time course for the whole-cell Na^+ current. The distribution of the individual open and closed times is characteristic of each type of channel, and is used to make kinetic models of channel function.

The other information that the single-channel recordings contain is the single-channel conductance. This value (given in pS [picosiemens] or 10^{-12}S) tells us the maximal rate of ion flux through the channel, and it is also characteristic of each channel type.

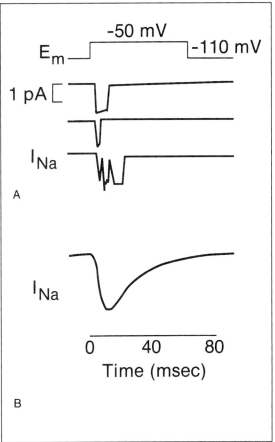

Fig. 4-29. Single Na^+ channel currents (I_{Na}) (A) Successively applied voltage steps (top trace, E_m) activate a Na^+ channel that opens and spontaneously closes. Three current traces are shown from a patch-clamp experiment using a rat skeletal myotube. (B) The averaged current from 144 such traces displays the characteristic rise and fall (activation and inactivation), as recorded for the Na^+ current in a whole-cell voltage clamp. (Modified from Hille, B. *Ionic Channels of Excitable Membrane.* Boston: Sinauer Associates, 1984.)

The gating of ionic channels involved in generating the action potential is voltage dependent. This means that part of the channel protein contains a charged group, or dipole, that can sense the electric field across the membrane and respond to a change in the transmembrane voltage.

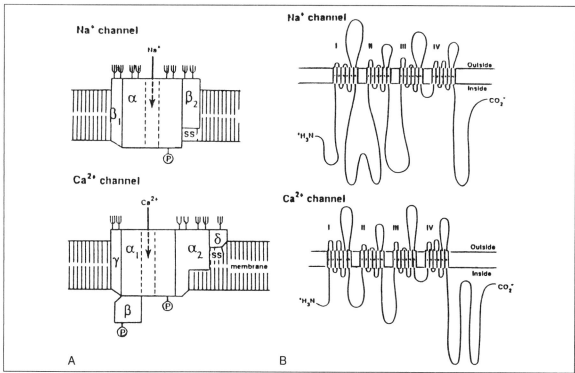

Fig. 4-30. Models of the fast Na^+ channel and the Ca^{2+} channel proteins based on recently acquired structural information. (A) Both Na^+ and Ca^{2+} channels have multiple protein subunits, labelled α, β, γ, and δ. The Ca^{2+} channel has two α_1 subunits: α_1 and α_2. The α_1 subunit is the one that contains the water-filled pore through which Ca^{2+} passes. Two sites that can be phosphorylated by cyclic adenosine 5'-monophosphate–dependent protein kinase are present: one on the α_1 and one on the β subunits. Only one α subunit exists in the Na^+ channel, but there are two β subunits: β_1 and β_2. The α subunit has a phosphorylatable site, as depicted. (SS = disulfide bonds.) (B) The structure of the central pore-forming α subunit for the Na^+ and Ca^{2+} channels. In each case, there are four homologous intramembrane-polypeptide repeat domains, connected by intracellular polypeptide segments and arranged in a circular structure within the plane of the membrane to form a channel. Each domain consists of six units that span the membrane, as depicted. (From Catterall, W. A. Structure and function of voltage-sensitive ion channels. *Science* 242: 50–61, 1988.)

When the gating region moves, it causes a shift in the overall conformation of the channel, which allows it to conduct ions. Most channels continue to open, close, and reopen many times during depolarization, forming "bursts." Some channels, such as the fast Na^+ channels, spontaneously inactivate (close) during depolarization, without reopening. It is thought that inactivation is related to a specific domain of the Na^+ channel polypeptide, which resides on the inner (cytoplasmic) surface of the protein.

The amino acid sequences of Na^+, Ca^{2+}, and K^+ channels are known, and their putative tertiary structures have been suggested (Fig. 4-30). Na^+ and Ca^{2+} channels have several subunits; the ionophore itself, and one or more regulatory subunits. Both Na^+ and Ca^{2+} channels have multiple sites for phosphorylation. Phosphorylation has been shown to alter the behavior of Na^+, Ca^{2+}, and K^+ channels. For example, in heart muscle, the Ca^{2+} channel activity is increased by adrenaline, a neurotransmitter that increases cyclic adenosine 5'-monophosphate levels and leads to Ca^{2+} channel phosphorylation.

As mentioned previously, ionic channels in excitable membrane are both voltage and time dependent. How are these properties incorporated into the structure of the ion channel? Located along the primary sequence of amino acids, which comprise the channel polypeptide, are specific charged residues. These charged groups can "sense" the transmembrane electric field and actually move in response to changes in the electrical field. It is thought that the movement of these voltage-sensing, gating charges initiates a conformational change in the channel structure, which then permits ions to flow through a central pore region of the molecule. The movement of the gating charge produces a small, but measurable, gating current. Gating currents have been recorded from Na^+, Ca^{2+}, and K^+ channels, and give us information concerning the steps leading to channel opening.

A number of biologic toxins have been discovered, which act at the level of specific ion channels. For example, TTX, which is extracted from the ovary of the Japanese puffer fish, has a high affinity for the fast Na^+ channel of nerves. Upon binding to Na^+ channel, TTX blocks the passage of Na^+ through the channel. Another type of toxin, batrachotoxin, counters inactivation of the Na^+ channel, so that the Na^+ currents are maintained during prolonged depolarization. Agents, such as these toxins, have proved to be valuable tools in understanding ion channel function because of their specific actions.

The information gained from patch-clamp experiments, performed on a variety of excitable cell types, has greatly increased our knowledge of the properties and regulation of many ionic currents. The macroscopic, or whole-cell, current (I) is the product of the number of functional channels (N), the probability that the "average" channel is open (p_o), and the single-channel current (i): $I = i N p_o$.

Single-channel recording allows direct measurement of these parameters (i.e., N, p_o, i), which were previously difficult or impossible to estimate indirectly from whole-cell recordings. In addition, the mechanism of action of channel regulatory agents can be examined at the level of the single channel. For example, adrenaline increases the Ca^{2+} current in heart cells by increasing the probability of opening.

Another discovery made using the patch-clamp method is that most ionic currents possess multiple types of channel behavior. For example, the Ca^{2+} currents in nerve cells result from three distinct Ca^{2+} channel populations with different conductances, kinetics, and voltage dependencies. In addition, both Na^+ and Ca^{2+} channel currents have been shown to exhibit multiple modes of gating behavior, which can be activated by certain agents.

The single-channel current results from the behavior of a single, albeit complicated, molecule. The open and closed periods of the channel give us information about possible conformational structures. By no other method available can the behavior of a single protein molecule, on a millisecond time scale, be observed.

SUMMARY

Membrane excitability is a fundamental property of nerve and muscle cells (skeletal, cardiac, and smooth), as well as certain other types, such as some endocrine cells. An excitable cell is one that, in response to certain environmental stimuli (electrical, chemical, or mechanical), generates an **all-or-none electrical signal** or **action potential.** The action potential is sometimes called an "impulse," for example, a **nerve impulse.** The action potential is triggered by depolarization of the membrane, which is triggered by the applied stimulus. The depolarization initiates an increase in the membrane permeability to Na^+, which then flows into the cell, causing a transient reversal in the membrane potential. A slower increase in the permeability of the membrane to K^+ contributes to repolarization of the membrane, in addition to the spontaneous **inactivation** of the Na^+ channel. Some cells display **automaticity,** in that they produce action potentials spontaneously without any externally applied stimulus.

In some excitable cell types (such as cardiac), a **slow inward Ca²⁺ current** contributes to the long plateau of the action potential, and in others (such as smooth muscle) the Ca^{2+} current generates the upstroke itself. The membrane currents (ionic and capacitive currents) that contribute to the action potential can be studied by the voltage-clamp method, which allows isolation and characterization of each membrane current as a function of membrane potential and time. Ionic currents flow across the membrane by means of numerous ion-specific protein pores, or **channels.** Each channel molecule has a region that senses the transmembrane potential, and acts as a gate to open or close the channel to ion passage through its central pore.

Some types of channels (e.g., Na⁺ channels) have an additional gating system that closes (or inactivates) the channel during a maintained depolarization. The pattern of minute currents that flow through individual voltage-dependent ion channels can be studied using the **patch-clamp method.** Single-channel current measurement, and recent structural information, have provided a greater understanding of the molecular basis of membrane excitability.

The action potentials in vertebrate nerve fibers consist of a spike followed by a hyperpolarizing afterpotential. A large fast inward Na⁺ current, passing through fast Na⁺ channels, is responsible for electrogenesis of the spike, which rises rapidly (approximately 1,000 V/sec). Subsequently, a small, slow inward Ca²⁺ current, passing through kinetically slow channels, may be involved in excitation–secretion coupling at the nerve terminals. The nerve cell membrane has voltage-dependent K⁺ channels that allow K⁺ to pass more readily outward (down the net electrochemical gradient for K⁺). This channel population is responsible for repolarization. The voltage-dependent K⁺ channel, which opens more slowly than g_{Na} upon depolarization, is called the **delayed rectifier.** This channel allows K⁺ to pass more readily outward (down the electrochemical

gradient for K⁺) than inward, and so is also known as the **outward-going rectifier.** The inactivation of this channel produces the large increases in total g_K that terminate the action potential.

The nerve action potential amplitude is about 110 mV, from a resting potential of -70 mV to a peak overshoot potential of about 40 mV. The duration of the action potential (at 50% repolarization) ranges between 0.5 and 1 msec, depending on the species and temperature. The threshold potential for triggering the fast Na⁺ channels is about -55 mV; a critical depolarization of about 15 mV is required to reach threshold. The turn-on of the fast g_{Na} (fast I_{Na}) is very rapid (within 0.2 msec), and E_m is brought rapidly toward E_{Na}. There is an explosive (positive exponential initially) increase in g_{Na} caused by a positive-feedback relationship between g_{Na} and E_m.

In certain nerve cells as well as muscle cells, as E_m depolarizes it crosses the threshold potential (about -35 mV) for the slow Ca²⁺ channels. Turn-on of the slow Ca²⁺ conductance and the slow inward current (I_{si}) is slow and tends to bring E_m toward the Ca²⁺ equilibrium potential. The peak I_{si} is considerably smaller than the peak fast I_{Na}. From the voltage-clamp currents/voltage curves, the maximum inward fast current and slow current occur at an E_m of about -20 mV and 0 mV, respectively. The currents decrease at more depolarized E_m levels because of diminution of the electrochemical driving force as the membrane is depolarized, even though the conductance remains high. At the reversal potential for the current, the current goes to zero and reverses direction with greater depolarization.

In some nerve and muscle cells, the slow Ca²⁺ current, itself, is sufficient to depolarize the membrane and generate a regenerative, slowly rising action potential in the absence of the fast Na⁺ current. The fast Na⁺ current may be voltage-inactivated (by depolarization with high extracellular K⁺ levels or current) or blocked by TTX. The resulting slow Ca²⁺-dependent action potential, or "slow response," also propagates slowly, due to the relatively low Ca^{2+} current density. In cardiac muscle, slow response formation during certain pathologic states (e.g., infarction) may give rise to arrhythmias (see Chapter 17).

NATIONAL BOARD TYPE QUESTIONS

Select the one most appropriate answer for each of the following

1. The voltage-clamp method is used to:
   A. Study the voltage dependence of membrane currents
   B. Study the time dependence of membrane currents
   C. Study individual ionic currents
   D. Study the channel gating currents
   E. All of the above
2. *Threshold* for firing an action potential occurs when:
   A. $I_{Na} < I_K$
   B. I_{Na} inactivates
   C. g_K increases
   D. $I_{Na} > I_K$
   E. None of the above
3. The strength–duration curve describes:
   A. The firing rate of a neuron
   B. The fatigue rate of action potential activity
   C. The amount of charge necessary to reach threshold
   D. The charge generated by I_{Na}
   E. None of the above
4. The fast Na^+ current:
   A. Activates after I_K
   B. Causes repolarization
   C. Turns on slowly
   D. Spontaneously inactivates
   E. All of the above
5. At the peak of the action potential:
   A. $g_K = g_{Na}$
   B. $I_K = I_{Na}$
   C. $I_c = I_i$

D. V_m is zero
   E. None of the above
6. Extracellular recording is the basis for the:
   A. Electroencephalogram
   B. Electromyogram
   C. Electrocardiogram
   D. Biphasic nerve action potential
   E. All of the above
7. The compound action potential (recorded extracellularly) is
   A. All-or-none
   B. The activity of a single fiber
   C. Reflects activity of many fibers of varying propagation velocity
   D. Reflects activity of unmyelinated axons only
   E. None of the above
8. Accommodation involves:
   A. A sustained increased g_{Na} during a ramp stimulus
   B. A decrease in g_K during a ramp stimulus
   C. A gradual inactivation of I_{Na} during a ramp stimulus
   D. A faster recovery of excitability
   E. None of the above

ANNOTATED ANSWERS

1. E. The voltage clamp is used to measure the voltage-dependent and time-dependent ionic currents flowing through the ion channels as well as the channel gating currents, which are special capacitance-like currents reflecting the gating charge movement.
2. D. At the threshold voltage and above (more positive or greater depolarization), there is a net inward current. Therefore, the inward Na^+ current must exceed the outward K^+ current, and there is further depolarization (rising phase of action potential).
3. C. Because charge (coulombs) is equal to current (amp) times time (sec), the mid-region strength–duration curve reflects the amount of charge necessary to reach threshold. That is, if the current is doubled, the

duration need be only half (to give the same charge). Each point plotted in the strength–duration curve gives the current (strength) and time (duration) parameters when threshold is just reached.

4. D. The fast Na^+ current spontaneously inactivates because the I-gate (or h-gate) of the fast Na^+ channel closes after a delay of 1 to 2 msec. The fast Na^+ current turns on quickly and is responsible for the rising (depolarizing) phase of the action potential.

5. B. At the peak of the action potential, the inward current (I_{Na}) and outward current (I_K) must be equal, so that there is no net current. The g_{Na} and g_K are not quite equal, because the two driving forces, $(E_M - E_{Na})$ and $(E_M - E_K)$, are not equal.

6. E. Extracellular recording registers the extracellular current that flows during the action potential in excitable cells. The extracellular current takes the path of least resistance, and so uses the entire body tissue (the "volume" conductor). Thus, the electric activity of the heart (the electrocardiogram) can be recorded from the skin of the arms, legs, and torso. Likewise, the electromyogram can be recorded from the skin overlying a skeletal muscle and the electroencephalogram can be recorded from the scalp.

7. C. The compound action potential represents the external recording of the action potential produced in a nerve bundle or skeletal muscle bundle, and reflects the activity of a varying number of the fibers possessing varying propagation velocity and threshold. The more fibers that are simultaneously activated (by electrical stimulation), the greater the recorded signal. Therefore, the compound action potential is not all-or-none, but graded (although each fiber responds in an all-or-none manner).

8. C. In accommodation to electrical stimulation using ramplike depolarizing pulses, the motor nerve fiber never reaches threshold because the relatively slow depolarization allows the function of fast Na^+ channels that have been activated to spontaneously inactivate. Therefore, a critical mass of simultaneously activated channels is never attained, and so a regenerative propagating response is not produced. In addition, the increase in g_K depresses excitability.

BIBLIOGRAPHY

Hille, B. *Ionic Channels of Excitable Membrane.* Boston: Sinauer Associates, 1984.

Hodgkin, A. L. *The Conduction of the Nervous Impulse.* Springfield, IL: Thomas, 1964.

Katz, B. *Nerve, Muscle and Synapse.* New York: McGraw-Hill, 1966.

Sperelakis, N., and Fabiato, A. Electrophysiology and excitation-contraction coupling in skeletal muscle. In: C. Roussos and P. Macklem (Eds.), *The Thorax: Vital Pump.* New York: Marcel Dekker, 1985. Pp. 45–113.

5 Cable Properties and Propagation Mechanisms

NICHOLAS SPERELAKIS

Objectives

After reading this chapter, you should be able to:

Explain why a fast propagating electrical signal, the all-or-none action potential, is required for rapid communication in the body

Describe the mechanism of propagation of the action potential

Describe the nature of saltatory conduction

List the factors that affect propagation velocity

Identify the type of records that are obtained upon external recording of the action potentials from single fibers or a bundle of fibers

---

Now that we have considered the electrogenesis of the resting potential of cells (Chapter 3), which enables the electrogenesis of action potentials (APs) and excitability (Chapter 4), we will examine the mechanism for the propagation of APs and excitability from one part of a neuron or muscle to a distal part. It is imperative that the body be able to transmit a very rapid signal from one point to another. The only way that this can be accomplished is through an electrical mechanism. Blood flow and diffusion of signaling molecules are too slow to allow rapid signaling. In contrast, electricity flows very quickly, at the speed of light (300,000,000 m/sec) in a copper wire or about one-ninth the speed of light in a water solution (such as the body is made up of). Therefore, the body uses electricity for rapid signaling in the nervous system, skeletal muscle, heart, and smooth muscles. Propagation velocity in our fastest nerves is about 120 m/sec; it is

about 6 m/sec in skeletal muscle, about 0.5 m/sec in cardiac muscle, and about 5 cm/sec in smooth muscle (Table 5-1).

As one example of the need for very fast communication or signaling, we will consider the process of walking. Very rapid signals must travel from the motor cortex of the brain, down to the lower spinal cord region, and out the motor axons to the skeletal muscles of the lower extremities (Fig. 5-1). In this process, the signal crosses one or more synapses, which are regions in which one neuron ends and the next one begins and in which a special chemical neurotransmitter signal is involved (see Chapter 8) (Fig. 5-2). At the termination of each branch of a motor nerve axon on the skeletal muscle fiber, there is another synapse, known as the **neuromuscular junction** or **motor end-plate** (see Fig. 5-1). The signal crosses the neuromuscular junction and gives rise to an AP in the muscle fiber, which

Table 5-1. Conduction Velocity as a Function of Fiber Diameter in Various Nerve Axons and Muscle Fibers

Fiber Type	Fiber Diameter (μm)	Propagation Velocity (m/sec)	Velocity/ Diameter (m/sec/μm)
Myelinated axons	20	120	6.0
	12	75	6.2
	5	30	6.0
Nonmyelinated axons	1.5	2.0	1.3
	1.0	1.3	1.3
Squid giant axons (20°C)	500	25	0.05
Skeletal muscle fibers	50	6.0	0.12
Cardiac muscle fibers	15	0.5	0.03
Smooth muscle fibers	5	0.05	0.01

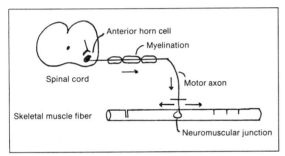

Fig. 5-1. A motor axon, with the cell body (soma) in the anterior horn of the spinal cord and its terminal branches ending on skeletal muscle fibers (only one muscle fiber depicted) to form the neuromuscular junctions (motor end-plates). Each motor axon with its attached skeletal muscle fibers is known as the *motor unit.* The motor axons have a large diameter (e.g., 20μm) and are myelinated; therefore they propagate at fast velocities (e.g., 120 m/sec). As a consequence of the chemical synaptic transmission process at the neuromuscular junction, an action potential is initiated in the muscle fiber and propagates in both directions, bringing about contraction.

propagates in both directions from the motor end-plate. The muscle AP elicits contraction. Receptors in the muscles (e.g., stretch receptors) transmit information (in the form of propagating APs) back into the central nervous system (CNS). Thus, in walking, there is a continual rapid flow

of information and instructions to the muscles in both directions: out of the CNS and into the CNS. Therefore, even a relatively simple skeletal activity such as walking is not possible without a very rapid signaling system. In various demyelinating diseases (e.g., caused by some viruses, heavy metals, and autoimmune reactions), loss of the myelin sheath (around the myelinated nerve fibers) causes propagation to become slowed and impaired in the affected nerve fibers, with associated incoordination and partial paralysis.

Since propagating all-or-none APs are all very similar to each other (in shape, duration, amplitude, rate of rise, and propagation velocity), in order to make the signal stronger or weaker, the body increases or decreases the frequency of the APs accordingly. That is, the body uses a **frequency-modulated system,** rather than an amplitude-modulated system (see Fig. 5-2). It is a **digital system** consisting of yes/no identical signals. At each synapse, the signal becomes **graded** in amplitude rather than all-or-none: the greater the amplitude and duration of the local postsynaptic potential, the higher the frequency of APs triggered. The same is true of the local graded receptor potential generated at some sensory organs and receptors. Stronger signals translate into higher frequency of impulses, and weaker

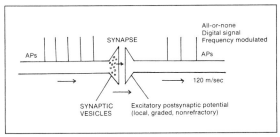

Fig. 5-2. A chemical excitatory synapse between two nerve fibers. An action potential *(AP)* in the presynaptic fiber on the left-hand side brings about release of the neurotransmitter at its nerve terminal. The transmitter molecules rapidly diffuse the short distance (e.g., 0.1 μm) across the synaptic cleft, bind to receptor sites on the postsynaptic membrane, and open the associated nonselective ion channels (Na^+, K^+ mixed conductance) complexed to the receptor (i.e., these are ligand-gated ion channels). The associated synaptic current depolarizes the postsynaptic membrane, producing the excitatory postsynaptic potential (EPSP), and this depolarization spreads passively into the adjacent conductile membrane (excitable) because of cable properties, thereby triggering one or more action potentials in the postsynaptic axon. The EPSPs are local, graded in amplitude, and nonrefractory, whereas the action potentials are all-or-none (maximal), refractory, and propagated actively. Thus, the amplitude-modulated synaptic process gives rise to a frequency-modulated or digital signal. The strength of a biologic response (e.g., contraction, secretion, or sensation) is a function of the frequency of the signal. That is, higher action potential frequency corresponds to stronger contraction or stronger sensation (in the sensory nervous system).

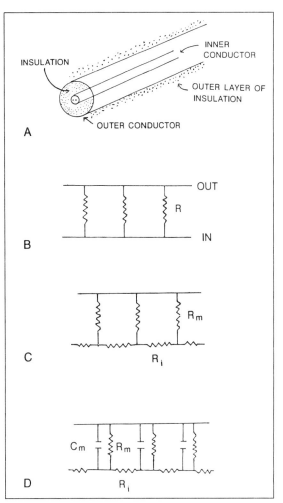

Fig. 5-3. (A) A co-axial cable and (B) associated electric equivalent circuit, the latter then being modified (C) to reflect the biologic cable (D). (A) The co-axial cable consists of an inner conductor (e.g., copper wire) and an outer concentric conductor separated by a layer of insulation (e.g., rubber). (B) The equivalent circuit for a co-axial cable. The inner and outer conductors depicted here have nearly zero resistances, and the transmembrane insulation resistance (R) is distributed along the length of the cable. (C) In the biologic cable, the inner conductor is the cytoplasm (axoplasm or myoplasm), which is not of negligible resistivity, and therefore is depicted as R_i distributed along the length of the fiber. The transverse insulation resistance is the cell membrane (R_m). (D) Addition of the capacitance elements to the biologic cable, which arise due to the lipid bilayer matrix of the cell membrane (C_m).

signals correspond to a lower frequency of impulses.

CABLE PROPERTIES

Biological Fiber as a Cable

An **electrical cable** consists of two parallel conductors separated by insulation material (e.g., two copper wires separated by rubber). Usually one of the conductors is arranged as a tubular sleeve surrounding a central solid rod (wire) (Fig. 5-3A). The equivalent electric circuit for a cable is shown in Figure 5-3B: two parallel conductors (wires) separated by a transverse resistance are distributed along the length of the cable. The resistance of the conductors is so small compared to the transverse insulation resistance that they are assumed to be zero. In the case of the **biologic cable** (a long narrow nerve fiber or skeletal muscle fiber), one parallel conductor is the inside fluid (cytoplasm) and the other parallel conductor is the outside fluid surrounding (bathing) the cell (the interstitial fluid). Because the conductivity of biologic fluid is much less (i.e., much higher resistance) than that of copper wire, and because the cross-sectional area of the cell cytoplasm is so small, the inside longitudinal resistance is high and cannot be ignored (see Fig. 5-3C). The outside longitudinal resistance is relatively small, as compared to the inside, because of the larger volume (cross-sectional area) of fluid available to carry the outside current; therefore it is assumed to be negligible.

In addition, there is a **stray capacitance** distributed along the length of the cable (see Fig. 5-3D), because a capacitance occurs when two parallel conductors ("plates") are separated by a high-resistance dielectric material. The **dielectric constant** of materials is related to vacuum, which is assigned a value of 1.000. Air has a value very close to vacuum; oils have a dielectric constant of 3 to 6 and that of pure water is 81. The biologic membrane, which has a matrix of phospholipid molecules, has a dielectric constant of about 5, typical of oils. The higher the dielectric constant, the higher the capacitance; the closer the parallel plates, the higher the capacitance. Because the biologic membrane is so thin (approximately 70 Å or 7 nm), its capacitance is relatively high: all cell membranes have a membrane capacitance of about 1.0 μF/cm^2, where F stands for farad.

Length Constant

In the electric cable depicted in Figures 5-3A and B, a voltage (or signal) applied at one end is transmitted to a distant end with little or no decrement (diminution or attenuation), and the so-called **length constant** is very long or infinite. In the

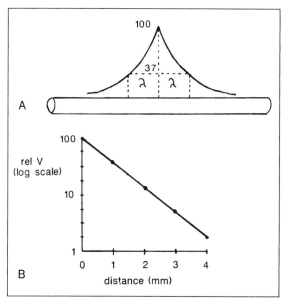

Fig. 5-4. Length constant (λ) of the biologic cable (fiber). (A) The exponential decay of voltage on both sides of an applied current (voltage) as a function of distance. The distance at which the voltage falls to 1/e (or 36.7 percent of voltage at site of current injection [at x = o]) gives the length constant. (B) When the voltage is plotted on a log scale against distance, a straight line is obtained.

biologic cable (Figs. 5-3C and D), however, a signal applied at one end rapidly falls off (decays) in amplitude as a function of distance, with a relatively short length constant (λ). This decay in voltage is exponential (Fig. 5-4A). An exponential process produces a straight line on a semilogarithmic plot (log V versus distance) (see Fig. 5-4B). In a cable, the relationship between the voltage at any distance (x) from the applied voltage (V_o) is:

$$V_x = V_o e^{-x/\lambda} \qquad\qquad \text{Eq. 5-1*}$$

Thus, when $x = \lambda$, $V_x = V_o \, 1/e^1 = V_o \, 1/2.718 = 0.37 \, V_o$. Hence, the distance at which the voltage

*The mathematical solution to Equation 5-1 is:

$$V_x = \text{antilog} \left[\frac{2.303 \log V_o - x/\lambda}{2.303} \right]$$

decays to 37% of the initial value gives the length constant. In nerve fibers and skeletal muscle fibers, the length constant has a value of only about 1 to 3 mm.

Therefore, the relatively short length constant, compared with the length of the neuron (e.g., 1.0 m for a lumbar anterior horn cell, motor neuron) or skeletal muscle fiber, means that a signal applied at one end (or midpoint) falls off very quickly over an increasing distance along the fiber (Fig. 5-5). If the length constant were 1.0 mm, at 4 mm the signal would become negligible. Hence, the electric signal cannot be conducted passively in the biologic cable, because it would decrement and disappear over relatively short distances. The AP (signal) is amplified to a constant value at each point (or each node) in the membrane, as discussed in Chapter 4. That is, conduction is active, not passive, with energy being put into the signal at each point to prevent any decay of the signal.

The parameters of a cable that determine its length constant are the square root of the ratio of the transverse resistance (r_m) to the sum of the inside (r_i) and outside (r_o) longitudinal resistances:

$$\lambda = \sqrt{\frac{r_m}{r_i + r_o}} = \sqrt{\frac{\Omega - cm}{\dfrac{\Omega}{cm} + \dfrac{\Omega}{cm}}} \qquad \text{Eq. 5-2}$$

where r_m, r_i, and r_o are the normalized resistances for a unit length (1 cm) of fiber. For surface fibers of a nerve or muscle bundle bathed in a large volume conductor, r_o is negligibly small, and Equation 5-2 reduces to:

$$\lambda = \sqrt{\frac{r_m}{r_i}} = \sqrt{\frac{R_m}{R_i} \frac{a}{2}} = \sqrt{\frac{\Omega\text{-cm}^2}{\Omega\text{-cm}} \frac{cm}{}} \qquad \text{Eq. 5-3}$$

where a is the fiber radius and R_m and R_i are the membrane resistance and longitudinal cytoplasmic resistance, respectively, normalized for

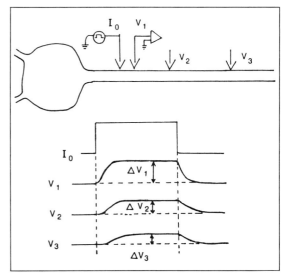

Fig. 5-5. A motor axon and the manner in which a voltage signal would decay with distance if the neuron were only a passive cable (nonexcitable). The voltage traces illustrate the voltage signals that would be simultaneously recorded at three different points along the axon (V_1, V_2, and V_3) from the site of injection of a rectangular current pulse (I_O). As depicted, the amplitude of the steady-state voltage pulse rapidly falls off with distance, because the length constant of the biologic cable is short (e.g., 1 mm) compared with the length of the axon (e.g., 1,000 mm). Therefore, an active response of the cell membrane at each point is required to faithfully propagate the signal. The voltage recorded at point V_1 (ΔV_1) also illustrates that the membrane potential changes in an exponential manner, both at the beginning of an applied rectangular current pulse and at the end. This exponential charge results from the capacitance of the cell membrane, the time constant being a product of the resistance and capacitance. The time it takes for the voltage to decay at the end of the pulse to 37 percent (1/e), or at the beginning of the pulse to build up to 63 percent (1 − 1/e) of the initial (maximal) value, gives the time constant.

both length (1 cm) and cell diameter. Thus, the greater the membrane resistance and the smaller the internal longitudinal resistance (larger cell diameter), the greater the length constant value. We will see below that the propagation velocity is a function of the length constant (e.g., larger-diameter fibers propagate faster). We will also see that myelination increases the effective membrane resistance and lowers the effective capacitance thereby increasing propagation velocity.

Time Constant

Because of the large capacitance of the cell membrane, the membrane potential cannot change instantaneously upon application of a step current pulse. Instead, the membrane potential changes in an exponential (negative) manner (see Fig. 5-5), both on the charge and discharge. The membrane **time constant** (τ_m) is a product of the resistance (R_m) and capacitance (C_m) of the membrane.

$$\tau_m = r_m \, c_m = R_m \, C_m \qquad \text{Eq. 5-4}$$
$$\text{sec} = \Omega \times F$$

where Ω is the resistance in ohms, and F is the capacitance in farads.

The **discharge** of the membrane (parallel RC [resistance $\times$ capacitance] network) is given by:

$$V_t = V_{max} \, e^{-t/\tau} \qquad \text{Eq. 5-5a}$$

where V_t is the voltage at any time, t (at the site of current injection), and V_{max} is the final maximum voltage attained during the pulse. When $t = \tau$, $V_t = V_{max} \, 1/e^1 = V_{max} \, 1/2.718 = 0.37 \, V_{max}$. Hence, the time at which the voltage decays to 37 percent of the initial (maximal) value gives the time constant (see Fig. 5-5). In nerve fibers and skeletal muscle fibers, the time constant has a value of about 1.0 msec.

When the membrane is **charging**, there is a similar exponential (negative) process, with the identical time constant (see Fig. 5-5). The corresponding relationship is given by:

$$V_t = V_{max} \, (1 - e^{-t/\tau}) \qquad \text{Eq. 5-5b}$$

The time it takes for buildup of the voltage to 63 percent ($1 - 1/e = 1 - 0.37 = 0.63$) of the final voltage gives the time constant. Thus, the time constant can be measured on the buildup (time to reach 63 percent of the final voltage) or on the decay (time to reach 37 percent of the initial voltage) of the pulse.

Local Potentials

In contrast to the active propagation of APs, synaptic potentials and sensory receptor potentials are not actively propagated. Such potentials decay exponentially (from their source of initiation) along the cell cable, as already described. Therefore, synaptic and receptor potentials are **local potentials**. When local potentials are depolarizing, they can give rise to APs, which are propagated; when hyperpolarizing, they inhibit production of APs. These local potentials are similar to the **local excitatory response** (Chapter 4) in that both are confined to a local region, but the electrogenesis of the two is different. As stated previously, the neuromuscular junction is an excitatory type of chemical synapse and produces excitatory postsynaptic potentials, known here as end-plate potentials. Most **synaptic potentials** are **graded,** in that they can add on one another, both in time and space (temporal and spatial), to produce larger responses; larger synaptic potentials exert a greater stimulatory or inhibitory effect on the production of APs.

CONDUCTION OF ACTION POTENTIALS

Local-Circuit Currents

The generation of APs was described in Chapter 4 and this section examines the mechanism for their rapid propagation (conduction). Propagation occurs by means of the **local-circuit currents** that accompany the propagating APs (Fig. 5-6). Such currents exist because, when two points are at a different potential (voltage) in a conducting medium, current (I) will flow between the two points, as governed by Ohm's law ($I = V/R$). We saw that, at the peak of the AP in one region of the fiber, the inside of the membrane at that region becomes positive with respect to

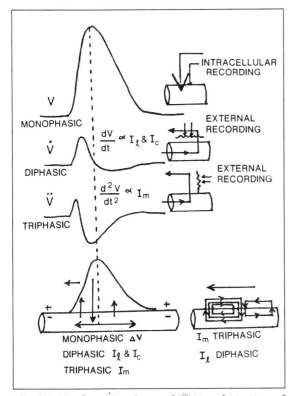

Fig. 5-6. The first ($\dot{V}$) and second ($\ddot{V}$) time derivatives of the action potential spike and the longitudinal I_l and radial currents associated with the propagating spike. The maximal rate of rise (dV/dt) ($\dot{V}$) is proportional to the capacitive current ($I_c = C_m \cdot$ dV/dt, where C_m represents the membrane capacitance) and the longitudinal (axial) current and is biphasic, having an intense forward phase and a less intense backward phase, as depicted in the diagram at the lower right. d²V/dt² ($\ddot{V}$) is proportional to the radial transmembrane current (I_m) and is triphasic, having a moderately intense initial outward phase, then a very intense inward phase (the current "sink"), followed by a less intense second outward phase, as depicted in the lower portion of the diagram. The *arrows* at the lower left depict the three phases of the membrane current and the two phases of the axial current. dV/dt can be recorded externally by a pair of closely spaced (relative to the spike wavelength) electrodes arranged parallel to the fiber axis, as illustrated. d²V/dt² can be recorded by a pair of electrodes arranged perpendicular to the fiber axis, as depicted. The *vertical dashed line* indicates that, when the slope of the spike goes to zero at the peak of the spike, dV/dt is zero; dV/dt is maximum at about the middle of the rising phase of the spike. (ΔV = change in voltage.)

the outside. The inside is also positive with respect to the inside cytoplasm at a region downstream from the active region. Therefore, current flows through the cytoplasm from the active region (current source) to the adjacent inactive region, then out of the fiber across the cell membrane, then through the interstitial fluid back to the active region (current "sink"), and finally through the membrane of the active region. This completes the closed loop for the current. The outward current through the membrane of the inactive region produces an IR voltage drop (Ohm's law)—positive inside to negative outside—that depolarizes this region, because the polarity of the voltage drop is opposite to that of the resting potential (negative inside, positive outside). When the depolarization exceeds the threshold potential, an AP is triggered. Thus, the inactive region is converted to an active region. This process is repeated in each segment of fiber, resulting in movement (propagation) of the impulse sequentially down the fiber.

If we examine a propagating AP in the middle region of a fiber (see Fig. 5-6), we see that there is also a small backflow of current internally, coupled with a corresponding small forward flow externally, associated with the repolarizing phase of the AP. Thus, as the AP propagates down the fiber, from right to left, there is a simultaneous double flow of local-circuit current: clockwise flow associated with the rising phase of the AP and counterclockwise flow associated with the repolarizing phase of the AP. The internal longitudinal current, sweeping past a transverse plane of the fiber, has two phases: first forward (right to left) and then reverse (left to right). The external longitudinal current also has two phases: left to right and then right to left. The transverse membrane current, sweeping through a point in the membrane, has three phases: first outward (still passive membrane), then inward (active membrane), and finally outward again (still active membrane).

It is the local-circuit current flow that is used to record the electrocardiogram (ECG) from the surface of the body, as well as the electromyogram

(EMG), electroencephalogram, and electroretinogram from the body surface over the tissue of interest (skeletal muscle, brain, eye). The internal longitudinal current is confined to the cytoplasm of the fiber, but the external current can use whatever conducting fluid is available (e.g., the entire torso volume conductor), because of the principle of resistors in parallel causing a lower total resistance (i.e., current takes the path of least resistance). Thus, this external local-circuit current causes the skin to be at different potentials, and these differences can be recorded by electrographs.

Propagation Velocity Determinants

The factors that determine the active velocity of propagation (θ) include: fiber diameter, length constant (λ), time constant (τ_m), local-circuit current intensity, threshold potential, and temperature. Some of these factors are interrelated, such as fiber diameter and length constant (since the length constant is proportional to the square root of radius), and length constant with time constant (since both depend on membrane resistance, R_m). Propagation velocity is directly proportional to the length constant and inversely proportional to the time constant.

$$\theta \propto \lambda / \tau_m \qquad \text{Eq. 5-6}$$

By substituting Equation 5-3 for the length constant and Equation 5-4 for the time constant, we arrive at the following relationship:

$$\theta \propto \frac{\sqrt{a}}{C_m \sqrt{R_i}\,\sqrt{R_m}}$$

$$\text{Eq. 5-6a}$$

Thus, propagation velocity is directly proportional to the square root of fiber diameter or radius (a) and inversely proportional to membrane capacitance (C_m). The larger the fiber diameter, the lower the absolute longitudinal resistance of the intracellular cytoplasm (R_i) (law of resistors in parallel), and therefore the greater the amount of local-circuit current flowing longitudinally and the greater the length constant. For example, it is well known that the larger the diameter of nerve fibers, the faster they propagate. Equation 5-6a shows that if membrane capacitance can be reduced (by myelination), then propagation velocity should increase in proportion. (This is discussed in the section "Saltatory Conduction.")

In addition, propagation velocity depends on the intensity of the local-circuit current, and hence on the rate of rise of the AP. The greater the AP rate of rise (max dV/dt), the greater is the longitudinal current and the transmembrane capacitive current. Therefore, all other factors being constant, faster-rising APs propagate faster. As discussed in Chapter 4, the AP rate of rise depends on the density of the inward current-carrying fast Na^+ channels that are available, on membrane capacitance, and on temperature. The max dV/dt decreases with increased membrane capacitance, with cooling, and with partial depolarization (due to the h_∞ versus membrane potential relationship). Cooling slows the rate of all chemical reactions, especially those with a high activation energy, such as the ion conductance changes in activated membrane.

Finally, the threshold potential affects propagation velocity. If the threshold is shifted to a more positive voltage (more depolarized), it takes longer for a given point in the membrane to reach threshold (and explode) during propagation of an AP from upstream. A greater **critical depolarization** (difference between resting potential and threshold potential) is required to bring the membrane to threshold. Therefore, propagation velocity would be slowed.

As stated previously, some of these factors are interrelated and some actually exert opposing effects, but these are not relevant to the introductory description presented here, and will not be discussed.

The preceding discussion applies to nonmyelinated nerve axons and skeletal muscle fibers. In myelinated nerve fibers, propagation velocity is greatly increased by the myelin sheath.

Saltatory Conduction

The nerve cable has been vastly improved by the evolutionary development of myelination in vertebrates. The myelin sheath improves the cable by increasing the effective membrane resistance by about a hundredfold and conversely decreasing the effective membrane capacitance by about a hundredfold. This increases the length constant and tends to decrease the time constant. Since the time constant tends to increase with the increase in effective membrane resistance (due to the myelin sheath), the decrease in membrane capacitance counteracts this effect, thus maintaining the time constant almost constant. Thus, Equations 6 and 6a predict that propagation velocity should increase with myelination.

One consequence of myelination is that **propagation velocity** is greatly increased. A second consequence is that the **energy cost** of signaling is greatly decreased because passive ion leaks are limited and active current losses are restricted to the small nodes of Ranvier, which are spaced relatively far apart: 0.5 to 2.0 mm internodal length. At each node, the length of exposed (naked) cell membrane is only a few micrometers. Therefore, the degree of energy-requiring active ion transport (Na^+-K^+ and Ca^{2+}) required to maintain the steady-state ion distributions and to keep the system in a state of high potential energy is greatly reduced. For example, the amount of Na^+ gained and K^+ lost per impulse is reduced as a result of myelination. The amount of oxidative metabolism in myelinated fibers reflects this lowered energy requirement.

The myelin sheath is produced by the Schwann cell, which is wrapped repeatedly in a spiral around the nerve fiber, forming 50 to 300 wrappings. For the purpose of the present discussion, we will assume an average of 100 wrappings. The myelin sheath covers the nerve axon like a coat sleeve, and is interrupted at each node. The cytoplasm of the Schwann cell is nearly completely extruded during formation of the myelin sheath, so the sheath essentially consists of 100 cell membranes in a series. Because of the law of resistors in series, the effective transmembrane resistance is increased a hundredfold. Because of the law of capacitors in series, the effective capacitance is reduced a hundredfold. Since the length constant is directly proportional to the square root of the membrane resistance, the length constant is increased accordingly. As described in the section "Propagation Velocity Determinants," increasing the length constant and lowering membrane capacitance increase the propagation velocity.

However, the myelin sheath is interrupted at every node of Ranvier. That is, the axon is bare (nude) at each node. The internodal distance is about 0.5 to 2.0 mm, depending on fiber diameter, and the width (length) of each node is only about 0.5 to 3 μm. The node is an annulus around the entire perimeter of the fiber.

Myelinated nerves usually have an optimal amount of myelin, such that the ratio of the diameter of axis cylinder (naked axon) to total fiber (including myelin sheath) is about 0.6 : 0.7. Assuming a maximal total diameter feasible (the body must pack many circuits within a limited space, e.g., the sciatic nerve bundle), a greater fraction of myelin, by infringing on the diameter of the axis cylinder, would raise the longitudinal cytoplasmic resistance too high, causing the active velocity of propagation to diminish. Thus, there are two opposing factors in determining the degree of optimal thickness of the myelin sheath: (1) the more the myelin, the greater the increase in the length constant and decrease in the membrane capacitance, and (2) the more the myelin, the lesser the diameter of the axis cylinder and therefore the higher the longitudinal cytoplasmic resistance.

In **saltatory** (L. *saltere*, to jump) **conduction**, the impulse jumps from one node to the next. The internodal membrane does not fire an AP for two reasons: (1) the internodal membrane is much less excitable (e.g., much fewer fast Na^+ channels), and (2) the depolarization of the neuron cell membrane at the internodal region is only about 1/100th of that at the node. The latter occurs because the IR voltage drop across the internodal cell membrane is only 1/100th of that across the entire series resistance network—neuron cell membrane plus many layers of Schwann cell membrane (Kirchhoff's laws dealing with voltage drops across resistors in series). Even though the internal potential in the internodal region swings positive (e.g., to +30 mV) when the adjacent nodes fire, the potential at the outer surface of the neuronal membrane swings nearly as positive (e.g., to +29 mV). Therefore, the depolarization of the neuronal membrane at the internode is only about 1 mV, which is well below threshold, and so the neuronal membrane does not fire. The potential that controls the membrane conductances (i.e., activates the voltage-dependent ion channels) is the potential difference directly across the membrane and not the absolute potential on either side.

As stated previously, the internodal membrane has only a few fast Na^+ channels. Because the cell membrane is fluid and proteins can diffuse (float) laterally in the lipid bilayer matrix, what then keeps the fast Na^+ channel proteins confined, at high density, in the nodal region? It appears that there are special anchoring proteins (e.g., anchorin) that tether the ion channel proteins to the cytoskeletal framework, thus preventing their lateral movement into the internodal membrane.

The combined effect of myelin and saltatory propagation makes propagation much faster. For example, a 20-μm-diameter myelinated nerve fiber conducts even faster than a 1,000-μm (1mm)-diameter nonmyelinated nerve fiber (e.g., the giant axon in squid, lobster, and earthworm): 120 m/sec versus approximately 25 to 50 m/sec. Thus, for invertebrates to achieve fast conduction in some essential circuits, they have to resort to giant neurons, resulting in a lower longitudinal cytoplasmic resistance and hence fast conductance. Because of space and size limitations, only a few critical neurons can have a giant diameter. In vertebrates, on the other hand, a large fraction of the nerve fibers in the peripheral nerves are myelinated for the purpose of fast propagation.

As was described previously, in nonmyelinated axons and skeletal muscle fibers, propagation velocity should vary with the square root of the cell diameter or radius ($a^{0.5}$). In myelinated axons,

propagation velocity varies with the first power of the cell radius (a^1), because propagation velocity (θ) varies with the length constant squared:

$$\theta \propto \frac{\lambda^2}{\tau_m} \propto \frac{a}{2R_iC_m} \qquad \text{Eq. 5-7}$$

The dependence of conduction velocity on the diameter of myelinated and nonmyelinated fibers is summarized in Table 5-1.

Wavelength of the Impulse

We can calculate the **wavelength** of the AP, which is the length of the axon simultaneously undergoing some portion of the AP. The wavelength is equal to the propagation velocity (θ) times the duration of the AP (APD_{100}):*

$$\text{Wavelength} = \theta \times APD_{100}$$

$$\text{cm} = \frac{\text{cm}}{\text{sec}} \times \text{sec}$$

$$\text{distance} = \text{velocity} \times \text{time} \qquad \text{Eq. 5-8}$$

The wavelength in a large myelinated nerve axon is about 12 cm: 120 m/sec × 1.0 msec. In a skeletal muscle fiber, it is about 1.8 cm (6 m/sec × 3.0 msec), and in a fine smooth muscle bundle, it is only about 1.5 mm (5 cm/sec × 30 msec).

EXTERNAL RECORDING OF ACTION POTENTIALS
Monophasic, Diphasic, and Triphasic Recording

As discussed in the section "Local-Circuit Currents," local-circuit currents accompany the propagating AP in each fiber. The intracellular

---

*Note the similarity of this relationship to that for the wavelength of an electromagnetic radiation: wavelength = velocity of light/frequency of the radiation. The reciprocal of frequency is the period (duration of one cycle).

and extracellular longitudinal currents are **biphasic**: initially they travel forward intracellularly and then they reverse direction. The forward-direction current is intense (high current density) and the reverse-direction current is weak (low current density). The transmembrane radial currents are **triphasic**: the first phase is outward (moderate intensity), the second phase is inward (high intensity), and the third phase is outward (low intensity). The first phase (outward) gives rise to the passive exponential foot of the AP, and is due to the passive cable spread of voltage and current. The second phase (inward) corresponds to the large inward fast Na^+ current, which occurs during the later portion of the rising phase and peak of the AP. The third phase (outward) corresponds to the net outward current (K^+), which occurs during the repolarizing phase.

These longitudinal and radial currents can be recorded by suitably placed external electrodes. The extracellular **longitudinal currents** can be recorded by two electrodes (bipolar) placed close together along the length of the fiber. If the interelectrode distance is short (relative to the wavelength), an approximate first (time) derivative of the AP is obtained (see Fig. 5-6 and Chap. 4).

The extracellular **radial currents** can be recorded by two electrodes placed close together in a plane perpendicular to the fiber axis. This gives an approximation of the second (time) derivative of the AP (see Fig. 5-6).

The internal axial currents are confined to the cytoplasm, whereas the external longitudinal currents can use the entire interstitial fluid space of the nerve bundle or muscle, or even the entire torso (so-called volume conductor), as current takes the path of least resistance (law of resistors in parallel). This allows recording of the ECG from the body surface and the EMG from the skin overlying an activated skeletal muscle. The ECG and EMG consist essentially of diphasic potentials, reflecting the external longitudinal currents during propagation of APs.

For biphasic recording in which the two phases are about equal, the two external electrodes can

be placed far apart (with respect to the wavelength) along a nerve or muscle fiber. First the proximal electrode records the wave of negativity (associated with the propagating AP), and then returns to isopotential. When the wave reaches the second electrode, it is recorded in the reversed polarity (because current flow through the voltmeter is reversed).

If the AP is prevented from reaching the second (distal) electrode, by crushing this region of the fiber or elevating the extracellular K^+ concentration to depolarize it, a **monophasic recording** is obtained (Fig. 5-7). This recording most resembles the true AP recorded by a microelectrode

Fig. 5-7. The waveforms that would be recorded externally during propagation of an action potential in a single fiber. (A) The two electrodes are far apart (relative to the action potential wavelength), and so a biphasic recording is obtained, with the two phases symmetrical and separated by an isopotential segment. The two phases are due to the current flow through the voltmeter recorder traveling first in one direction and then in the opposite direction. If the fiber between these two electrodes is damaged (e.g., by crushing) or depolarized (by elevating extracellular K^+ concentration), so the action potential cannot sweep past the second electrode (2), this second phase is prevented and the recording is monophasic. (B) If the two electrodes are brought progressively closer, the isopotential segment depicted in A would shorten and disappear. (C) If electrode 2 is brought very close to electrode 1, so that the interelectrode distance is short relative to the wavelength, then the second phase is smaller than the first phase and the recording resembles the first derivative of the true action potential.

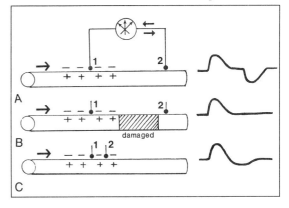

inserted into a fiber to record the transmembrane potential (see Fig. 5-7), but of course has a much smaller amplitude.

Compound Action Potential

When one records the APs externally, the records are graded and not all-or-none, as in the case of true APs recorded intracellularly from single fibers (see Chapter 4). That is, the recorded AP gets larger and larger, up to a maximum amplitude, as the intensity of stimulation is increased. This is the so-called **compound action potential**. It is graded because, as an increasingly greater number of the fibers are activated, the external longitudinal currents associated with the all-or-none AP in each fiber cut across the recording electrodes, and thereby produce a larger signal. The AP in each fiber is always all-or-none. The amplitude of the signal is determined by the resistance between the electrodes multiplied by the amount of current flowing through this resistance (V = IR). The recording of compound action potentials is diagrammed in Figure 5-8.

The compound action potentials can be demonstrated by recording the EMG in a human subject with one electrode placed on the skin of the ventral forearm and the other (reference) electrode on the wrist of the same arm. As the subject voluntarily produces stronger and stronger contractions to flex the hand, the electrical signals picked up get increasingly greater in amplitude and frequency. The amplitude gets larger because more muscle fibers are activated simultaneously. This is known as **fiber recruitment**. The frequency increases because the motor nerves fire at a higher frequency, causing the muscle fibers to fire at a higher frequency and thus producing a more powerful tetanic contraction.

SUMMARY

Although the biologic cable (i.e., nerve fiber or skeletal muscle fiber) is the best possible given the biologic circumstances, it is relatively poor compared with electrical wires and cables. This is because: (1) the resistivity of the cytoplasm of

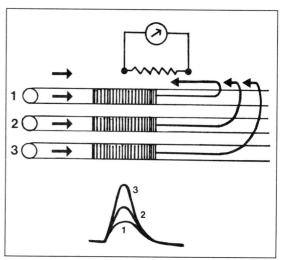

Fig. 5-8. A compound action potential in an isolated nerve trunk, such as the frog sciatic nerve, recorded externally by a pair of longitudinal electrodes. The voltmeter (oscilloscope) recorder records the voltage (IR) drop across the resistance (fluid) between the two electrodes. If only fiber *1* is activated, the current passing between the electrodes is small, and the voltage recorded is small. If fiber *2* is simultaneously activated with fiber *1*, the amount of current is doubled and the voltage is doubled. When all three fibers are activated simultaneously, the current is tripled and the voltage is tripled. Therefore, the externally recorded compound action potential is graded because it reflects the electrical activity of numerous fibers, each of which produces an all-or-none (nongraded) action potential.

nerve and skeletal muscle fibers is about 10^7 times that for copper wire; (2) the transverse membrane resistance is about 10^{-6} times that of a good insulator (like rubber) because it is so thin (approximately 70 Å or 7nm); and (3) the membrane has a high capacitance, and therefore a relatively long time constant. Because the biologic cable is poor, if the cable remained passive (no active impulse), there would be much signal lost in the transmission of information. Thus, the signal would become greatly attenuated and distorted after traveling only a short distance. Therefore, for faithful and rapid signal transmission over long distances, energy must be introduced into the system at each point along the way. The system that has evolved consists of AP generation; this involves all-or-none signals of constant amplitude and constant propagation velocity with refractory periods and sharp thresholds. It is a frequency-modulated system, whereby an increase in frequency of the AP signals produces increasing strength of sensation or motor response.

AP propagation occurs by means of the local-circuit currents. The transmembrane current has three phases: outward, inward, and outward. The internal and external longitudinal currents have two phases: forward and backward (for internal) or backward and forward (for external). The external currents use the path of least resistance, enabling electrograms (e.g., ECG, EMG) to be recorded from the body surface. The compound action potential is graded in amplitude, reflecting the summation of the external currents generated from each activated fiber: the more fibers activated simultaneously, the greater the amplitude of the electrogram signal.

Propagation velocity is faster, the larger the diameter of the fiber, the longer the length constant, and the lower its time constant and capacitance. The myelin sheath in vertebrates enables much faster propagation velocity and at a lower energy cost. The resulting myelination increases the effective membrane resistance and lowers the effective capacitance, with excitability occurring only at the short nodes of Ranvier that uniformly interrupt the myelin sheath. Therefore, the AP signal jumps from node to node in a saltatory pattern of conduction.

NATIONAL BOARD TYPE QUESTIONS

Select the one most correct answer for each of the following

1. Propagating APs are required in the peripheral nervous system
    A. Because the biologic cable is relatively poor and there would be resulting cable decrement
    B. To use up the excess nervous energy
    C. To allow collision of APs
    D. To increase membrane capacitance
    E. To allow the signal to "jump tracks," or cross over to neighboring fiber tracts

2. Which one of the following statements is not true of the length constant?
    A. Reflects an exponential voltage decay over distance
    B. Is directly proportional to the square root of the internal longitudinal resistance
    C. Is increased when fiber diameter is greater
    D. Is affected by membrane resistance
    E. Has a value of about 0.7 to 1.4 mm in skeletal muscle fibers

3. Which one of the following statements is not true of the membrane time constant?
    A. Reflects an exponential voltage decay (or buildup) with time
    B. Is a product of membrane resistance and capacitance
    C. The larger its value, the faster the propagation velocity
    D. Its value is equal to the time it takes for the voltage to decay to 37 percent of the initial (maximal) value
    E. Its value is greatly increased by myelination

4. Which one of the following statements is not true of the local-circuit currents?
    A. They are associated with propagating APs
    B. They are responsible for the ECG and EMG
    C. They have two internal longitudinal components: forward and backward
    D. They have two transmembrane radial components: inward and outward
    E. They are greater in intensity with fast-rising APs

5. Propagation velocity increases with which one of the following?
    A. Greater membrane capacitance
    B. Greater membrane time constant
    C. Cooling
    D. Greater fiber diameter
    E. Shorter length constant
    F. More positive (higher) threshold potential

6. Which one of the following statements is true with respect to the myelin sheath?
   A. Increases the effective membrane capacitance
   B. Increases the effective membrane resistance
   C. Decreases the length constant
   D. Allows for smooth continuous conduction
   E. Has great effect on the effective time constant

7. Which one of the following is true of saltatory conduction?
   A. Requires more energy for signaling
   B. Excitation does not occur at the nodes of Ranvier
   C. Propagation velocity is slower
   D. The internodal membrane segments fire APs
   E. There is an optimal ratio of the radius of the axis cylinder to the myelin thickness

8. Which one of the following is not true of the wavelength of the AP?
   A. It is increased if propagation velocitiy is increased
   B. It is increased if AP duration is decreased
   C. It is the length of axon (fiber) simultaneously undergoing some portion of the AP
   D. It is about 12 cm in myelinated axons, which are about 20 μm in diameter
   E. It is shorter in skeletal muscle fibers than in myelinated nerve fibers

9. Which one of the following is not true of the compound action potential?
   A. It is composed of all-or-none APs from individual fibers
   B. It is graded, reaching a maximum amplitude when all neurons are activated
   C. It is due to fiber recruitment
   D. It is due to summation of the action currents of individual fibers at the recording electrodes
   E. The ECG is a good example of a compound action potential

10. If two external electrodes are placed close together (with respect to the wavelength) along the longitudinal axis of an isolated giant nerve fiber, the recorded electric activity during a propagating AP would be
    A. Monophasic
    B. Diphasic
    C. Triphasic
    D. Quadriphasic
    E. None of the above

ANNOTATED ANSWERS

1. A. Since the nerve fiber is a relatively poor cable with a short length constant, graded signals cannot be used for long-distance communication because they would decrement over short distances, causing a very poor signal-to-noise ratio. Therefore, a system evolved that allowed an active signal to be propagated rapidly, in an all-or-none manner, over a long distance. For such a signaling system, energy has to be introduced into the signal at each point along the way.

2. B. The length constant (λ) is directly proportional to the square root of the ratio of membrane resistance (R_m) to the internal longitudinal resistance (R_l)

$$\lambda = \sqrt{\frac{r_m}{r_l}} = \sqrt{\frac{R_m}{R_l} \frac{a}{2}}$$

3. C. The larger the membrane time constant, the *slower* the propagation velocity. This is because the longer the time constant, the longer it takes for the membrane potential to change when a signal is applied, and hence the slower the AP propagation.

4. D. The local-circuit current has *three* transmembrane radial current components: outward (foot of AP), inward (rising phase of AP), and outward (repolarizing phase of AP).

5. D. Propagation velocity increases with smaller membrane capacitance, shorter time constant, longer length constant, and lower (more negative) threshold potential. Cooling slows most chemical and physical reactions, including propagation of APs.

6. B. The myelin sheath *decreases* the effective capacitance (because of capacitors in series), *increases* the length constant (because of the effective increase in membrane resistance due to resistors in series), and causes the saltatory (jumping) conduction. Myelination has relatively little effect on the time constant (τ_m) because the increase in membrane resistance (R_m) is offset by the decrease in membrane capacitance (C_m): $\tau_m = R_m \times C_m$.

7. E. Saltatory conduction requires *less* energy for signaling because the AP jumps from one node of Ranvier to the next. Since excitability is confined to the short nodal membrane, the amount of Na^+ gained and K^+ lost by each nerve fiber per AP is considerably less than that for nonmyelinated axons. Propagation velocity is also faster with myelination.

8. B. The wavelength of the AP is *decreased* if the AP duration is decreased, since: wavelength = velocity × duration; thus, in large-diameter myelinated axons: wavelength = 120 m/sec × 0.001 sec = 0.12 m.

9. E. The ECG is not a compound action potential because the entire atrial or ventricular myocardium normally responds in an all-or-none manner, and therefore does not exhibit the graded behavior of a compound action potential.

10. B. The recorded signal would be diphasic (or biphasic), reflecting the forward and backward components of the external longitudinal current (part of the local-circuit current) accompanying the propagated AP.

BIBLIOGRAPHY

Sperelakis, N. Origin of the cardiac resting potential. In: R.M. Berne and N. Sperelakis (Eds.), *Handbook of Physiology, The Cardiovascular System. Vol. 1: The Heart.* Bethesda, MD: American Physiological Society, 1979. Pp. 187–267.

Sperelakis, N., and Fabiato, A. Electrophysiology and excitation-contraction coupling in skeletal muscle. In: C. Roussos and P. Macklem (Eds.), *The Thorax: Vital Pump.* New York: Dekker, 1985. Pp. 45-113.

Cole, K. C. *Membranes, Ions and Impulses: A Chapter of Classical Biophysics.* Berkeley: University of California Press, 1968.

6 Hormonal and Chemical Transduction of Information

LAWRENCE A. FROHMAN

Objectives

After reading this chapter, you should be able to:

Distinguish different modes of communication between cellular elements in individual tissues and in different tissues

List the various types of hormones and understand the differences in their modes of biosynthesis

Distinguish how the different classes of hormones are transported through the circulation and the differences in the mechanisms by which they are metabolized

Explain the concept of feedback regulation

Describe the principles of hormone action and distinguish between the action of the various types of hormones

Explain the principle of radioimmunoassay and understand the differences between this assay method and the bioassay and radioreceptor assay, in terms of what is actually being measured

---

The endocrine system, together with the nervous system, has evolved to provide a means of communication between cells and organs. As single-cell organisms have developed into multicellular organisms, the endocrine system has assumed a critical role in the regulation of many processes crucial to life, including growth and development, metabolic homeostasis, reproduction, and responses to environmental perturbations such as stress. The evolutionary increases in size and complexity of organisms made it impractical for all cells to have direct contact with each other, such as occurs within the nervous system, and thus a hormone signaling system was required.

Endocrine glands are known as the **glands of internal secretion** or the **ductless glands.** The term *internal secretion* was first used by Claude Bernard in 1855, while the term *hormone*, referring to the product of the endocrine glands, was first proposed by Starling in 1905. The word is derived from the Greek verb *hormao*, which means to arouse, excite, or set in motion. Thus, a hormone is defined, in the strict sense, as a substance that is secreted by one cell and transported in the circulation to act on other cells.

The classic definition of a hormone has required modification in recent years as substances defined as hormones have been shown to exert actions in multiple settings. Thus, the same chemical compound may also function as a neurotransmitter in synaptic clefts (*neurocrine*), by exerting its effects on neighboring cells (*paracrine*), or on itself (*autocrine*). This multiplicity of function underscores an important concept in endocrinology: the same hormone may cause multiple effects by means of separate modes of action which are coordinated with one another, subserving a single homeostatic (or reproductive) function. For example, norepinephrine, acting as a hormone, inhibits insulin secretion by the pancreatic beta cells and, as a neurotransmitter, enhances hepatic glucose output, both of which result in hyperglycemia. In a similar manner, gonadotropin hormone–releasing hormone, acting as a hypothalamic hormone, stimulates the release of luteinizing hormone and follicle-stimulating hormone from the pituitary and, as a neurotransmitter acting in the brain, stimulates sexual behavior. Selective chemical compounds present in simple organisms and serving as cell-to-cell communicators have been co-opted as hormones to reach distant parts of the body via the circulatory system to exert coordinated effects necessary for life.

HORMONE BIOSYNTHESIS AND SECRETION

Hormones can be divided into three chemical types: peptides and proteins, steroids, and modified amino acids. The first group includes primarily neuropeptides, pituitary hormones, and gastroenteropancreatic hormones; the second consists of adrenal and gonadal steroids and vitamin D; and the third comprises thyroid hormones and catecholamines.

The biosynthesis of hormones is specific to certain cell types, though numerous hormones are synthesized in more than one type of cell. The **protein hormones** are synthesized through classic mechanisms involving stimulation of gene expression, transcription of messenger RNA (mRNA), and ribosomal translation of the mRNA to generate a hormone precursor. The hormone precursor is characterized by an amino–terminal sequence (signal peptide) that contains processing information to ensure that the protein enters the rough endoplasmic reticulum and is enveloped in a secretory granule to facilitate exocytosis (secretion). This multiplicity of events provides many opportunities for variations to occur. For example, a single mRNA precursor may be differentially spliced in individual tissues to provide separate mature mRNA species (e.g., calcitonin mRNA is generated in the C cells of the thyroid, and calcitonin gene–related product mRNA is formed in the brain from the same mRNA percursor). A single protein hormone precursor may also generate one or more hormones (e.g., ACTH and beta-lipotropin, vasoactive intestinal peptide and peptide histidine-methionine) and may also give rise to multiple forms of the hormone (e.g., somatostatin[1-28] and somatostatin[1-14]). Finally, **peptide hormones** undergo post-translational processing in which they are glycosylated (thyroid-stimulating hormone and the gonadotropins) or complexed with metal ions (insulin with the zinc ion). The hormones are packed in secretory granules that migrate to the cell surface while final processing occurs. Thus, under conditions of stimulated secretion, during which intracellular transit time is shortened, hormone precursors are frequently secreted and can be detected in the circulation (Fig. 6-1).

Steroid hormone biosynthesis is initiated by cholesterol side-chain cleavage, the rate-limiting step in steroidogenesis, followed by a cytochrome c P_{450}-linked dehydrogenase–isomerase complex and several hydroxylases. The intermediary compounds are shuttled from the mitochondria to the endoplasmic reticulum and back again to the mitochondria for the final steps in biosynthesis. The cell specificity of certain enzymes explains why various steroid hormones are synthesized only in selected cell types. Vitamin D synthesis, in contrast, requires transport of the precursor between several different organs (skin, liver, and kidney), each of which contains specific enzymes required for biosynthesis. While most steroids are secreted in the biologically active form, further processing to more active forms occurs in some target tissues (e.g., testosterone is converted to 5-α-dihydrotestosterone in skin and male reproductive tissues). In contrast to peptide hormones, steroid hormones are not stored in great quantities in their

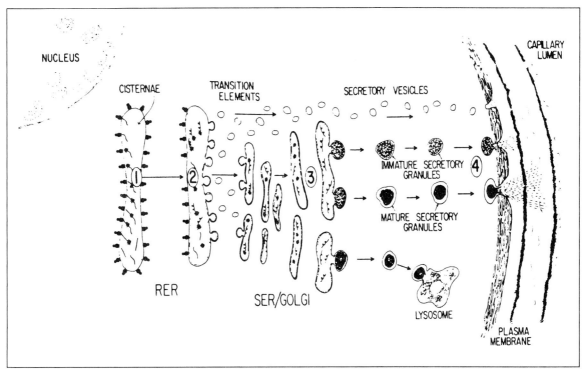

Fig. 6-1. Intracellular transport and secretory pathways in a protein-secreting endocrine cell. (*RER* = rough endoplasmic reticulum; *SER* = smooth endoplasmic reticulum; *Golgi* = Golgi complex.) Proteins are synthesized on polyribosomes attached to the endoplasmic reticulum and directed through the membrane to the cisternal space. Shuttling vesicles are formed from the endoplasmic reticulum and transported to and incorporated by the Golgi complex, where secretory granules are formed. Granules are then transported to the plasma membrane where they fuse with the membrane, leading to exocytosis and resulting in discharge of the granule contents into the extracellular space. Secretion may also occur by transport of secretory vesicles and both immature and mature granules. Some granules are taken up by lysosomes and degraded (crinophagy) without ever being secreted (From Gill, G. N. In: Felig, P. F., Baxter, J. D., Broadus, A. E., and Frohman, L. A., eds. *Endocrinology and Metabolism*, (1987).

cells of origin. Thus, a stimulus to hormone secretion is intimately linked to new hormone synthesis.

The two distinct categories of **modified amino acid hormones** (monoamines) also have different biosynthetic mechanisms. The catecholamines, consisting of dopamine, norepinephrine, and epinephrine, are derived from the amino acid tyrosine by a series of enzymatic conversions involving a hydroxylase, a decarboxylase, and a methyl transferase. The specific cell types containing these enzymes are known as **amine precursor uptake and decar-**

boxylation cells, which are derived primarily from neural crest tissues, but probably also from endoderm. A second set of monoamines, serotonin and melatonin, are derived from the amino acid tryptophan by a similar series of reactions. These hormones are stored in neurosecretory granules in nerve terminals and specialized cell types (pinealocytes and adrenal medulla), from which they are released upon stimulation.

Thyroid hormones are iodinated derivatives of tyrosine. They are formed in only one tissue, the thyroid gland, by the unique mechanism of iodination of tyrosine residues on a large protein, thyroglobulin. A large storage pool of thyroglobulin, known as "colloid," exists within the follicular lumen of the thyroid cells. The process of secretion thus requires reabsorption of thyroglobulin, its enzymatic degradation, and release of the two forms of the hormone, thyroxine (T_4) and triiodothyronine (T_3). T_4 may be considered as a hormone precursor (*prohormone*), since it is much less bioactive than is T_3. An enzymatic mechanism exists for its conversion to T_3 in peripheral tissues.

HORMONE TRANSPORT AND METABOLISM

Separate mechanisms exist for transport of the different hormone types in circulation. Many peptide and protein hormones, which are water soluble, do not require transport proteins and exist in their natural state. Others, such as insulin-like growth factors, are almost completely bound to a carrier protein. These proteins may be present uniquely in circulation or may be similar to tissue hormone receptors (the plasma growth hormone–binding protein is actually identical to the extracellular domain of the hepatic growth hormone receptor).

Steroid and thyroid hormones, in contrast, are tightly bound to specific binding proteins that provide a pool of circulating hormone that, particularly for the thyroid hormones, has a relatively prolonged half-life (several days). Their presence serves to increase the period that a hormone remains in the circulation, thereby providing another readily available reservoir of hormone for biologic action. Protein-bound hormone is in equilibrium with free hormone in the circulation, and it is the frcc hormone that is the most important determinant of the hormone's biologic activity. Thus, individuals with a genetic deficiency of thyroid-binding globulin (TBG) have decreased total T_4 levels in plasma, but their free T_4 levels are normal and they show no evidence of thyroid hormone deficiency. Binding proteins are synthesized in the liver, are under their own regulatory control, and may bind multiple hormones (e.g., cortisol-binding globulin also serves as the binding protein for progesterone; testosterone and estrogen both bind to the sex hormone–binding globulin; and both T_3 and T_4 bind to TBG).

Hormones are also metabolized by specific mechanisms. Peptide hormones and some protein hormones are degraded by circulating and tissue peptidases that destroy the hormone's biologic activity by cleavage at critical locations in the molecule. The specific enzymes involved are numerous, though by no means unique. The half-life of most peptide hormones is on the order of 1 to 20 minutes, and, although small amounts of some hormones are recovered intact in the urine, indicating renal clearance, most of the degradation occurs intravascularly. Some of the bigger protein hormones, particularly those that are glycosylated (e.g., the gonadotropins) are cleared intact by the kidneys and found in large quantities in the urine. Peptide hormones are also metabolized at the site of their action. They are internalized and transported to intracellular lysosomes where they are degraded.

Steroid hormones are metabolized primarily by reduction or hydroxylation, mainly in the liver, to biologically inactive metabolites, which are then often sulfated or glucuronidated before being excreted in the urine. Less than one percent of cortisol, for example, is excreted intact in the urine.

Catecholamines and indolamines are metabolized in the synaptic cleft and in circulation primarily by monoamine oxidation and transmethylation. However, termination of action of the monoamines at the synaptic cleft occurs primarily by reuptake into nerve terminals. Thyroid hormones are metabolized in target organ tissues by means of two separate deiodinases, one of which is the same enzyme that converts T_4 to T_3.

REGULATION OF HORMONE SECRETION: THE FEEDBACK CONCEPT

The control of hormone secretion is, of necessity, complex because it involves multiple factors, including external neural inputs, endogenous neural rhythmicity, circulating substrate concentrations and hormones, and other products of the target glands on which the hormones act (Fig. 6-2). Each of the various hormone systems (e.g., those mediating growth, metabolic homeostasis, stress, reproduction, and water and electrolyte balance) have their unique regulatory mechanisms. There are, however, certain general principles that apply. First, the cascade of responses is usually initiated by hypothalamic releasing and inhibiting hormones that are secreted in a pulsatile manner which reflects neu-

ronal activity. Their half-lives are extremely short and their concentrations vary considerably within short periods. Second, their effects on the pituitary are greatly influenced by the ambient concentrations of target gland hormones—primarily but not exclusively those in the same hormonal system. Third, levels of circulating target gland hormones change in a smoother manner, with less short-term fluctuation, than do those of hypothalamic or pituitary hormones. Their effects "feed back" on both hypothalamic and pituitary hormone secretion. The acute effects are usually inhibitory and relate to hormone release, while the longer-term effects, which influence hormone synthesis, may be either inhibitory or stimulatory. Thus, removal of a target gland will cause elevated levels of the hormones upstream in the particular axis (e.g., adrenalectomy increases the secretion of both corticotropin-releasing hormone and ACTH). These feedback effects are unique for each hormone system and their disturbances constitute the basis for many endocrine diseases. Knowledge of **feedback concepts** has also formed the basis for diagnostic tests to assess hormone secretory status.

Fig. 6-2. Endocrine gland organization. The signals flow from the central nervous system, through the hypothalamus, the pituitary, and the target glands in order to achieve the desired effect: hormone action on peripheral tissue. A complex feedback system exists that is mediated by any or all of the hormones in the cascade or by the physiologic action mediated by the hormones. Feedback effects may occur at one or more loci, though not all of the possible sites are effective for any particular hormonal system.

HORMONE RECEPTORS AND HORMONE ACTION

In order for hormones to act only on their target cells, a specific mechanism of interaction was required and this led to the development of **hormone receptors**. These protein substances, to which hormones bind, are present in multiple cell compartments (surface membranes, cytoplasm, and nucleus) and serve two functions. First, they are required for selectivity: they have three-dimensional shapes or conformations that allow them to distinguish the particular hormone that they bind from among the many other substances in the bloodstream. Second, they are connected to an effector mechanism: they must transmit a signal to the interior of the cell to activate the processes that are stimulated by hormones. In many ways, hormone receptors are similar to enzyme systems. Their conformation is changed in response to hormone binding, and this leads to

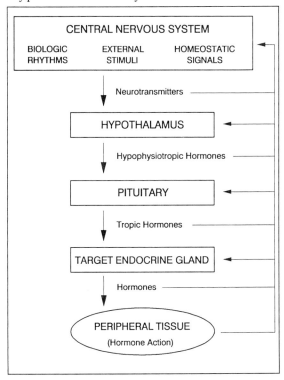

activation of a tightly coupled enzyme system that serves as an amplifier. Several peptide hormone receptors function as enzymes or as substrates for enzymes. In the cytoplasm, multiple "second-messenger" systems have evolved to serve these purposes, while in the nucleus, the hormone-receptor complex binds to DNA and has an important role in the regulation of gene expression (Fig. 6-3).

The hormone receptor, therefore, has two important domains: that which recognizes and binds to the hormone and that which couples to the **effector mechanism**. Peptide and mono-amine hormone receptors are located on the plasma membrane with the hormone-binding domain located extracellularly and the effector domain located intracellularly. The receptor-binding affinity must be high, as there is frequently competition for hormone binding with plasma-binding proteins. Binding is rapid and reversible. Although there is evidence for internalization of the hormone-receptor complex, this process is not required for the initiation of hormone action after binding and may represent pri-

Fig. 6-3. Target cell activation by hormones acting at membrane receptors (monoamine transmitters and peptide hormones) or cytoplasmic and nuclear receptors (thyroid and steroid hormones). The membrane receptors are coupled by guanine nucleotide regulatory proteins to one of several second-messenger systems for the stimulation of phosphorylation, through which their actions are mediated. The intracellular receptors exhibit a change in conformation after binding to their respective hormones and then bind to specific sites (response elements) on DNA for regulation of gene transcription. (*mRNA* = messenger RNA; *cAMP* = cyclic adenosine 5'-monophosphate; IP$_3$ = inositol trisphosphate.)

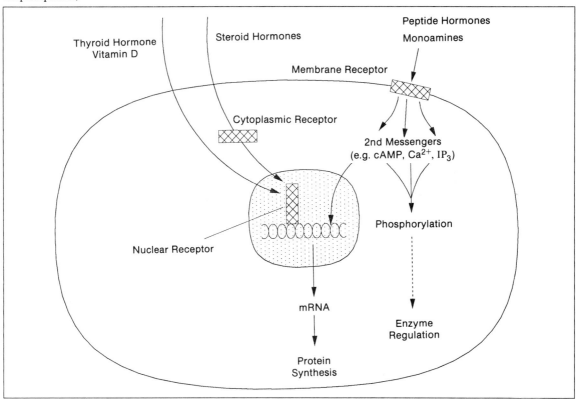

marily a means of terminating hormone action. The concentration of most receptors is greater than that required for the full biologic effects of the hormone. Thus, maximal effects of hormones are generally observed at a receptor occupancy of only 30 to 50 percent.

The effector domain of the receptor is tightly coupled to the regulatory portion of the effector enzymes, such as adenylate cyclase, which are located on the inner surface of the plasma membrane. These effector systems, in turn, control the production of cyclic nucleotides, the breakdown of phospholipids, membrane transport systems, and ion fluxes. They may also be coupled to high-energy phosphorylation of proteins (including the receptor itself), which mediate other cell processes such as the synthesis, transport, and metabolism of molecules critical for cell viability.

There are numerous **second-messenger systems** associated with plasma membrane receptors, including adenylate cyclase, guanylate cyclase, phospholipase C, protein kinase C, calcium mobilization, sodium–chloride channel activation, and tyrosine phosphorylation. Many hormone receptors use multiple mechanisms with a single cell, whereas others appear to be more specific. Extensive studies directed at the nature of the interaction of the hormone-receptor complex and the effector mechanism have provided evidence for guanyl nucleotide regulatory proteins that mediate this interaction. These proteins, which may exhibit either a stimulatory or an inhibitory action, perform a major role in transmembrane signaling that is common to all peptide and monoamine hormone receptors.

Nearly all hormone-regulated cells have the ability to respond to changes in ambient hormone concentrations by altering the number or affinity, or both, of their cell surface receptors. This phenomenon is most evident when the cell is exposed to high concentrations of the hormone and results primarily in a decrease in receptor-binding sites, which is associated with a corresponding decrease in the cellular response to the hormone. This decrease is caused both by internalization (endocytosis) of the hormone-receptor complex and by alterations in the dissociation of the complex, making the receptor unavailable for subsequent binding to fresh hormone. Thus, exposure of target cells to hormones initiates a local feedback response that limits the cellular effects of the hormone by limiting its ability to initiate the signaling process.

Steroid hormones are believed to be taken up by cells by passive diffusion through cell membranes rather than by a transport-mediated process. They were originally believed to bind to specific intracellular receptors that were then translocated into the nucleus. More recent information, however, suggests that the receptors may exist in the nucleus, even when unbound. Receptor occupation results in a conformational change in the molecule, which increases its affinity to bind to DNA. Specific binding sites have been identified on regions of genes upstream from the initiation site of transcription. The binding of the steroid-receptor complex to the specific genomic response element enhances the rate of initiation of mRNA transcription. In addition, effects of the hormone-receptor complex on other genes can affect the stability of specific mRNAs, thereby enhancing gene expression by an additional mechanism.

Although T_3-binding proteins have been identified in both cytoplasm and mitochondria, the most important actions of T_3 are mediated by nuclear-binding proteins. These receptors, when bound to T_3, attach to a thyroid response element in the gene in a manner similar to that of steroid receptors, and result in increased RNA polymerase activity and mRNA formation. Many genes have response elements for both T_3 and steroid receptors, and the two systems frequently work in conjunction with one another. Consequently, the amplification of gene expression seen in the presence of both steroid and thyroid hormones can be considerably greater than the sum of the effects of either hormone alone.

HORMONE MEASUREMENT

The measurement of hormones in tissues and circulating fluids has undergone many changes during the past several decades. Initial assays were based on the biologic properties of the hormones using *in vivo* tests that were often not only imprecise, but also very insensitive, requiring large amounts of tissue or biologic fluids. With the availability of pure hormones, the development of methods for radiolabeling, and the generation of high-affinity antibodies to the hormones, the technique of **radioimmunoassay** (RIA) was developed and has become the mainstay of hormone measurements. Concomitantly, more sensitive and precise bioassays evolved, which are also currently in use, though not for routine clinical diagnosis and treatment because of their complexity and expense. Newer techniques based on nonisotopic measurements have been introduced recently. Their concepts, advantages, and pitfalls are described in the next section.

Bioassays

Initial bioassays were based on *in vivo* animal models, in which graded doses of the hormone were injected and the biologic response quantified. In addition to the large variability created by the imprecision of the techniques, these assay systems were frequently incapable of distinguishing between multiple hormones that had overlapping effects. For example, growth hormone and prolactin from many species (including humans) both have lactogenic properties, and the quantitation of prolactin in a sample containing both hormones was quite unreliable. Furthermore, the sensitivity of most *in vivo* bioassays precluded measurement of circulating hormone levels.

With the development of pure hormones and, more importantly, *in vitro* culture systems that permitted detection of specific biochemical effects of hormone action at the cellular level, such as cyclic adenosine 5′-monophosphate (cAMP) accumulation, sensitivity comparable to or even greater than that of RIA has been achieved. This method is generally regarded as the "gold standard" because it measures the hormone's biologic activity rather than its chemical or immunologic reactivity. This is of considerable importance because it is possible to have subtle modifications in the chemical composition of a hormone that are undetectable by any method other than bioassay. One caveat that must be remembered with regard to any bioassay relates to its specificity. While the effects may appear to be specific to all known substances, a yet unidentified compound (hormone) may exhibit similar biologic properties in the particular system and its presence erroneously attributed to the hormone for which the bioassay was designed. One such example is the parathyroid hormone–like peptide seen in patients with malignancies associated with hypercalcemia. This peptide binds to the parathyroid hormone receptor on renal cortical cells and stimulates cAMP formation. A second caveat relates to the nature of the bioassay system. Some substances will exhibit biologic activity in a particular *in vitro* bioassay system, though not in others or *in vivo*. Thus, extrapolation of *in vitro* results must be done with caution. One of the current major applications for *in vitro* bioassays is in screening synthetic hormone analogs for agonist and antagonist properties.

An elegant variation of the bioassay is the cytochemical bioassay, which uses densitometric changes in tissue sections based on changes in the redox state induced by hormone stimulation. The sensitivity of this technique is several-fold greater than that of RIA, though the highly specialized instrumentation that it requires has limited its widespread use.

Radioimmunoassays

The concept of RIA was developed nearly thirty years ago by Berson and Yalow, and virtually revolutionized the field of hormone measurements. The technique spread rapidly to other disciplines and is currently the most commonly used system for measuring peptide, protein, steroid, and thyroid hormones concentrations, as well as numerous other natural and synthetic compounds. The principle of this technique is based on the competition of a labeled ligand (hormone) with an unlabeled ligand for a fixed number of binding sites on a specific antibody (Fig. 6-4). The quantity of radiolabeled ligand bound is inversely related to the quantity of competing non-radiolabeled hormone, and, following mathematical transformation of the results, unknown samples can be directly read off of the standard curve. The most commonly used label is iodine 125, though tritium and carbon 14 have also been employed. Currently, with the great interest in decreasing the use of radioisotopes, fluorescent and chemiluminescent labels are being rapidly developed.

The antibodies developed for RIA require a high affinity as well as capacity, and have been successfully produced in a variety of species. The binding reaction is reversible and can be performed under equilibrium or nonequilibrium conditions. The nonequilibrium conditions actually provide greater sensitivity. A technique is required to separate bound antibody from free hormone, in order to determine the fraction of total labeled ligand in either the bound or free state. A number of physicochemical separation methods have been used, though the most commonly used technique at present is a co-precipitation reaction using an antibody generated in a second species against immunoglobulins of the species of the primary antibody.

Although the precision and sensitivity of RIAs are generally very high and repeated measurements over short time intervals can be performed readily using small quantities of serum, there are many limitations to the use of RIA, some of which were not initially apparent. The radioiodinated

Fig. 6-4. The principle of radioimmunoassay. Labeled antigen (hormone) competes with unlabeled antigen (endogenous hormone) for binding to a limited number of sites on specific antihormone (*anti-IGG*) antibodies. A second antibody directed against the primary antibody is then used for precipitation of the bound complex. Other physicochemical methods of separation of free from bound hormone may also be used. Increasing amounts of unlabeled hormone result in progressive displacement of labeled hormone from antibody. With careful selection of antibodies, labeling techniques, and incubation conditions, sensitivities in the low picomolar range can be readily achieved.

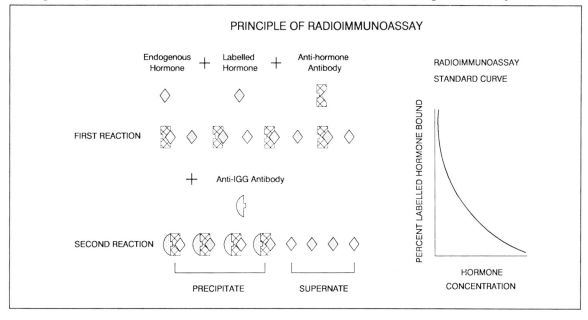

ligand is the most vulnerable component of the assay. If addition of the iodine atom alters the chemical stability or immunologic properties of the hormone, competition between the radioligand and the unlabeled hormone in serum may be altered, yielding incorrect results. Certain sera contain protein components that may alter binding of the ligand to the antibody, artificially elevating or reducing the calculated value. Plasma peptidases may degrade the radioligand or the endogenous hormone during the incubation period, leading to unpredictable results. Circulating hormone fragments that are biologically inactive or other hormones with common sequences may compete for binding with the radioligand. Endogenous antibodies to the hormone (e.g., diabetics receiving insulin) or to the primary antibody species (e.g., animal handlers exposed to rabbits) may markedly alter the ratio of bound to free hormone. Therefore, careful attention must be given to the assay conditions, so that the results can be interpreted with confidence. Despite these limitations, however, this technique, more than any other, has contributed to our knowledge of endocrine physiology in experimental animals and in humans.

Other Immunoassays

A variation of the RIA technique that has recently appeared is the **immunoradiometric assay,** which is performed as a solid-phase assay that uses two separate antibodies (one or both of which is frequently monoclonal) each directed against separate binding sites on the hormone. One antibody is bound to the solid-phase support and the other is radioiodinated. In this assay, the problems associated with radioiodinated hormones are eliminated and the amount of radioactivity bound to the solid phase is directly proportional to the amount of hormone in the sample. A further variation of this technique, the **chemiluminescent assay,** substitutes a chemical that emits photons upon light activation for the radioactive label. Such assays are not only more sensitive than RIAs, but eliminate the prob-

lems of radioactive waste disposal. Another nonisotopic assay is the **enzyme-linked immunosorbent assay.** In this assay, an enzyme is linked covalently to the antibody or the hormone and the reaction product is quantified on a spectrophotometer. This technique has been less successful because of nonspecific interfering substances present in biologic samples.

Radioreceptor Assays

The **radioreceptor assay** differs from the RIA in that a hormone receptor is substituted for the antihormone antibody. This assay has been used primarily for peptide and protein hormone measurement and has the advantage that it measures biologically active hormone. The results of such assays are not always comparable to those of RIAs, as the antibody and receptor may bind different portions of the hormone. The technique is generally less sensitive than RIA, and receptor preparation and storage are technically more difficult. In addition, radioiodination of the peptide may modify its receptor-binding site. However, use of this assay in conjunction with RIA has led to important new concepts related to alterations in the biologic activity of secreted hormones (e.g., the degree of glycosylation of gonadotropins secreted at different times of the menstrual cycle varies, influencing its radioreceptor binding, biologic activity but not its RIA activity).

SUMMARY

Hormones have evolved as the mediators of communication between cells located at great distances in multicellular organisms. They are identical to or derived from chemicals that are used for local cell communication. They function in groups, affecting the major physiologic processes necessary for the life of the organism, including growth, metabolism, reproduction, and adaptation to the external environment. Their action is initiated by binding to specific cellular receptors that are coupled to second-messenger effector mechanisms, which amplify important cellular cytoplasmic processes, or to nuclear receptors that interact with specific DNA regulatory elements to initiate RNA transcription. Numerous sensitive techniques have been developed for hormone measurement that allow assessment of their storage, secretion, and metabolism.

NATIONAL BOARD TYPE QUESTIONS

Answer each of the following

1. Match the hormone in Column A with its family type in Column B:

Column A	Column B
I. Epinephrine	A. Peptide
II. Somatostatin	B. Steroid
III. Thyroxine	C. Substituted
IV. Estradoil	amino acid
	D. Catecholamine

2. Receptors for which of the following hormones are present on the plasma membrane?
   A. Peptide hormones and catecholamines
   B. Steroid hormones and peptide hormones
   C. Catecholamines and steroid hormones
   D. Thyroid hormones and catecholamines

3. Which one of the following statements concerning RIAs of hormones is true?
   A. The assays are capable of measuring circulating levels of most hormones
   B. The limiting reagent in the assay is the radioactive hormone
   C. Assay values indicate biologically active hormone
   D. The assay is uninfluenced by endogenous antihormone antibodies
   E. Assays must be performed under equilibrium conditions to be accurate

4. Second-messenger systems are used for signal transduction by which of the following hormones?
   A. Cortisol
   B. Triiodothyronine

C. Parathyroid hormone
D. Vitamin D
E. Testosterone

ANNOTATED ANSWERS

1. I D. Epinephrine, along with norepinephrine and dopamine, is a member of the catecholamine family, all of which are derived from the amino acid tyrosine.

   II A. Somatostatin is a peptide of 14 amino acids which is synthesized from a larger precursor molecule (prohormone) by selective-processing peptidases.

   III C. Thyroxine is an iodinated thyronine; thyronine is composed of a phenyl ring attached by an ether linkage to the amino acid tyrosine. Thyroxine contains four iodine atoms while triiodothyronine contains three.

   IV B. Estradiol, the primary estrogenic hormone in humans, is derived from cholesterol and is, along with glucocorticoids, mineralocorticoids, androgens, and progesterone, a member of the steroid family.

2. A. Peptide and catecholamine hormones are bound to plasma membrane receptors. In contrast, thyroid and steroid hormones are bound to soluble cytosolic receptors, which are then translocated into the nucleus.

3. A. The RIA measures the presence of the epitope on the hormone against which the antibody is directed. This may be present on both biologically active hormones as well as inactive precursors or metabolites. RIAs can also be performed under nonequilibrium conditions, which actually increase the sensitivity of the assay. The rate-limiting reagent is the antihormone antibody. Endogenous antihormone antibodies can bind with native and labeled hormone and give spurious results.

4. C. Only peptide and catecholamine hormones utilize second-messenger systems.

Steroid and thyroid hormones as well as vitamin D bind directly to nuclear or cytosolic receptors that are then transported to the nucleus.

BIBLIOGRAPHY

Baxter, J. D., Frohman, L. A., Broadus, A. E., and Felig, P. A. Introduction to the endocrine system. In P. A. Felig, J. Baxter, A. E. Broadus, L. A. Frohman, (Eds.), *Endocrinology and Metabolism,* 2nd ed. New York: McGraw-Hill, 1987. Pp. 3–22.

Catt, K. J. Molecular mechanisms of hormone action: control of target cell function by peptide, steroid, and thyroid hormones. In P. A. Felig, J. Baxter, A. E. Broadus, and L. A. Frohman, (Eds.), *Endocrinology and Metabolism,* 2nd ed. New York: McGraw-Hill, 1987. Pp. 82–165.

Clark, J. H., Schrader, W. T., and O'Malley, B. W. Mechanisms of steroid hormone action. In: G. D. Wilson and D. W. Foster (Eds.), *Textbook of Endocrinology,* 7th ed. Philadelphia: Saunders, 1985. Pp. 33–75.

Gill, G. N. Biosynthesis, secretion and metabolism of hormones. In P. A. Felig, J. Baxter, A. E. Broadus, and L. A. Frohman, (Eds.), *Endocrinology and Metabolism,* New York: McGraw-Hill, 1987. Pp. 59–82.

Roth, J., and Grunfeld, C. Mechanism of action of peptide hormones and catecholamines. In G. D. Wilson and D. W. Foster, (Eds.), *Textbook of Endocrinology,* 7th ed. Philadelphia: Saunders, 1985. Pp. 76–122.

7 Sensory Transduction

MICHAEL M. BEHBEHANI

Objectives

After reading this chapter, you should be able to:

Define a sensory receptor

Describe sensory receptor types

Give a few examples of a mechanoreceptor

Define the term *threshold*

Describe how the threshold of a sensory receptor can be measured

Define the term *adequate stimulus*

Describe the law of specific nerve energy

Define *adaptation*

Describe the function of adaptive processes in information transmission

Define the components of the transduction process

Explain the generator potential

Describe how the generator potential differs from the action potential

Give an example of a rapidly adapting and slowly adapting cell

Describe the types of signals that are transmitted by the slowly adapting receptor

Describe the types of signals that are transmitted by the rapidly adapting receptor

Describe the cellular mechanisms involved in the generation of receptor potentials

List the types of second messengers involved in the generation of receptor potentials

Before sensory perception occurs, the external energy associated with that sensation must be changed to a series of action potentials, the language of the nervous system. Sensory receptors are the first component of this translation system and change, or transduce, the information in the external world to action potentials.

SENSORY RECEPTORS

There are, in general, five categories of sensory receptors and these differ in their morphology and functional characteristics.

Mechanoreceptors

Mechanoreceptors are cells that respond to stretching of their cell membrane, thereby transforming mechanical energy into nerve impulses. Some important mechanoreceptors are: (1) **joint receptors** (including the **transient receptors**) which respond to movement of the joint regardless of direction of movement, **velocity detectors** which discharge continuously during changes of joint angle, and **position detectors** which signal the position of the joint in the absence of movement; (2) **free nerve endings** such as those in the nasopharyngeal epithelium and the epithelial lining of the airways extending from the trachea to the respiratory bronchioles, and those of the larynx which are involved in the cough reflex; (3) **hair cells** of the organ of Corti which are involved in hearing and hair cells of the semicircular canal which are involved in corrective reflexes of the body during movement; (4) **stretch receptors** that signal the length and tension of the muscle; and (5) **pacinian corpuscles** which respond to deep pressure and vibration.

Photoreceptors

There are two types of photoreceptors, rods and cones. The **rod system** is involved in scotopic vision (low-level light detection), and the **cone system** is involved in photopic vision (high-level light detection) and color vision.

Chemoreceptors

Chemoreceptors respond to different chemicals. The important receptors in this category are: (1) receptors of the tongue which are involved in the sensation of taste; (2) receptors in the nasal mucosa which are involved in smell; and (3) visceral chemoreceptors in the aortic arch and carotid body that respond to changes in CO_2 and O_2 tension technique as well as pH (these receptors produce no conscious sensation).

Thermoreceptors

There are two types of thermoreceptors; **warm receptors** which respond to an increase in skin temperature and **cold receptors** which respond to a decrease in the temperature of the skin.

Nociceptors

Nociceptors are involved in encoding the sensation of pain. There are two types of nociceptors: mechanical and thermal. **Mechanical nociceptors** respond to high-intensity mechanical stimulation and the **thermal nociceptors** respond to temperatures above 44°C.

SENSORY MODALITY

There are several distinct sensations that are perceived by human observers. These sensations include light, sound, vibration, heat, cold, and pain, and are called **sensory modalities**. In general there are specific receptors for each modality. Nonetheless, there is considerable interaction between them and our overall perception of external stimuli may involve the simultaneous activation of several types of receptors.

Although there are significant morphologic differences among sensory receptors, there are several physiologic properties that are common to all. These properties are:

1. **Threshold**. Threshold of a sensory receptor is the intensity of the stimulus that will produce an action potential in the sensory unit 50 percent of the time when it is applied. There are many methods for measuring the threshold of a receptor. The simplest is to select a stimulus that is barely detectable by the receptor and present it repeatedly. After many trials, the frequency of detection is measured. If the detection rate is lower than 50 percent, the stimulus strength is increased and the process repeated until a stimulus strength that produces a 50 percent detection rate is found. A more accurate method for measuring threshold is by the signal detection technique. In this method, the occurrence of false positives (the number of times a signal is detected with no stimulus) and false negatives (the number of times no signal is detected despite application of a stimulus) is used to determine threshold. A complete description of this technique is beyond the scope of this chapter, however.

2. **Adequate stimulus**. An adequate stimulus is that stimulus toward which a given receptor has the lowest threshold. For example, photoreceptors have a low threshold for electromagnetic waves of between 400 and 700 nm (i.e., in the range of visible light); therefore, the adequate stimulus for these receptors is visible light.

3. **Law of specific nerve energy**. This law states that stimulation of a sensory neuron produces the same sensation regardless of how the receptor is excited. For example, whether or not

a photoreceptor is stimulated by mechanical or electric stimulation, it leads to the sensation of sight.

4. **Adaptation**. Adaptation plays an important role in perception. For example, in the visual system adaptation allows one to perceive light intensity that varies between 1 and 10^{10} photons per cm^2. It is this adaptive process that allows one to see in very low-level indoor light as well as in bright sunlight. The adaptation process takes time. For example, if one moves from very bright sunlight into a dark room, it can take several minutes before one can see.

GENERATOR POTENTIAL

The **generator potential** or **receptor potential** is defined as the change in the membrane potential of a sensory receptor induced by application of a stimulus. The generator potential has two important properties: (1) it is a graded response, that is, its amplitude increases as the strength of the stimulus increases; and (2) it exhibits temporal and spatial summations. When two stimuli are applied to the receptor, the generator potentials produced by the stimuli combine to produce a larger receptor potential. Similarly, if a stimulus is applied repeatedly to the same site, the generator potential that arises from each application summates. (More detailed explanations of these properties are discussed in the next section where specific sensory receptors are considered.)

Examples of a Receptor

An example of the **stretch receptor** is the crustacean stretch receptor (Fig. 7-1A). This receptor is found in the tail muscle of crustaceans and consists of specialized muscle cells innervated by the dendrites of a sensory neuron. There are two types of muscle bundles: slow and fast. These differ considerably in their response to stretch and in their anatomy. The sensory neurons supplying both bundles are large and can be impaled with two microelectrodes, which makes this preparation useful in studying the basic properties of transduction.

Generator Potential

When the slow bundle is stretched moderately for a short time (see Fig. 7-1B, part c), the sensory neuron depolarizes and fires action potentials (see Fig. 7-1B, part b). At the beginning of the response, the firing frequency is high, but, over time, the firing frequency decreases to a steady-state that is maintained for as long as the muscle is stretched. In contrast to the slow bundle, the fast bundle responds to stretch by producing several action potentials, then stops firing even though the muscle is still stretched (see Fig. 7-1B, part a).

If impulse production is prevented by adding a small quantity of an agent that blocks the voltage-gated Na^+ channel (e.g., tetrodotoxin [TTX]) to the bathing media, the effect of stretch on the membrane potential can be investigated. In the presence of TTX, stretching either the fast or slow bundles causes depolarization of the sensory neuron; this depolarization, which exhibits an identical waveform for both slow and fast bundles, is the **generator potential**. When the magnitude of the generator potential reaches the threshold of the action potential, the cell produces an impulse (but, as noted above, no action potential is produced in a TTX-poisoned preparation).

As shown in Figure 7-2, the magnitude of depolarization is graded, in that it is proportional to the magnitude of stimulation. At a low stimulus, depolarization is small. When the stimulus is increased, a larger generator potential is produced. Thus, unlike the action potential, the greater the stimulus strength, the larger the magnitude of the generator potential.

Ionic Basis for the Generator Potential

It is possible to measure the reversal potential of a stretch receptor by inserting two microelectrodes into the sensory neuron: one electrode to transmit current and the other to record the generator potential. The reversal potential of the generator potential is usually near zero millivolt, which implies that the membrane becomes permeable to Na^+ during stretch. This can be verified by substituting Na^+ with choline (Ch^+) in

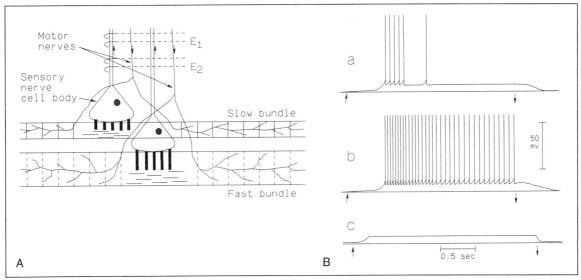

Fig. 7-1. (A) The anatomy of the crustacean stretch receptor. There are two sensory nerves, one innervating the slow and the other fast bundle. E_1 and E_2 are recording electrodes. (B) The physiologic properties of the two receptor types. (*a*) Stretching the fast bundle causes depolarization and firing. However, the firing does not last for as long as the fiber is stretched. (*b*) When the slow bundle is stretched, it fires action potentials for as long as the receptor is stretched. (*c*) Time duration of the stretching pulse.

the bathing medium. Because Na+ channels are far less permeable to Ch+ than to Na+, the stretch-induced movement of positive charges into the cell is prevented, and consequently no generator potential will occur.

Adaptation

A slow bundle fires as long as stimulus is applied, but with a decrease in frequency. By contrast, a fast bundle only fires a short burst of action potentials when stimulus is applied and ceases to fire even if the stimulus persists. This response of the fast bundle to a constant stimulus is an example of adaptation. The slow bundle is a slowly adapting receptor, whereas the fast bundle is a rapidly adapting cell. As already noted, the generator potential is the same in both fast and slow bundles. Therefore, the adaptive properties of the fast bundle are attributed to that section of the membrane that produces an action potential. This component of adaptation is called *accommodation*.

The mechanical properties of the receptor constitute another factor in adaptation. The amplitude of the generator potential is greater soon after the stimulus is applied, compared with that during prolonged stimulation. This phenomenon is attributed to mechanical factors such as slippage of the muscle fiber and a resultant decrease in the tension perceived by the stretch receptor.

PACINIAN CORPUSCLE

The pacinian corpuscle is an ellipsoid body composed of several concentric lamellae (Fig. 7-3A). A myelinated nerve fiber enters at one end and the final node of Ranvier is contained within the corpuscle.

Generator Potential

In an intact corpuscle, application of pressure leads to a depolarizing generator potential that diminishes rapidly; a second generator potential

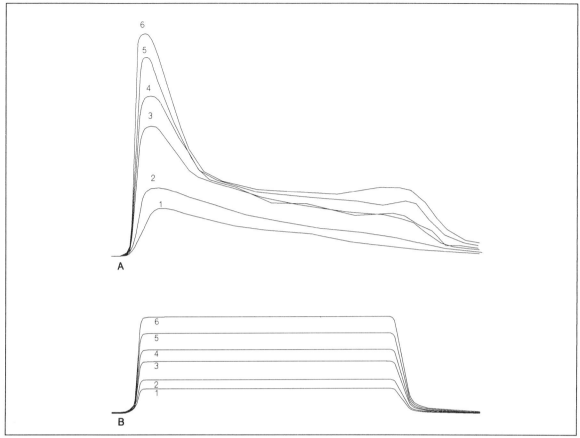

Fig. 7-2. In presence of tetrodotoxin, stretching of the slow or fast fibers produces a generator potential that lasts as long as the stretch pulse. The responses of the stretch receptor are shown in A and the stimulus strength and duration are shown in B. Note that the generator potential is maintained as long as the stimulus is applied and that, as the intensity of the stimulus is increased, the amplitude of the generator is also increased.

is also recorded when the pressure is released. Similar to the generator potential in the stretch receptor, the amplitude of the generator potential produced by the corpuscle is graded. As shown in Figure 7-3C, the amplitude of the generator potential increases as the magnitude of pressure increases. When the membrane depolarizes to approximately -40 mV, the cell fires an action potential (see Fig. 7-3C, e). If the amplitude of the generator potential exceeds the firing threshold, an action potential is produced only at the onset and offset of the pressure pulse (see Fig. 7-3B). Because the pacinian corpuscle only produces an action potential at the onset and offset of the pressure pulse even when the membrane is depolarized above the firing threshold, it is considered a rapidly adapting receptor.

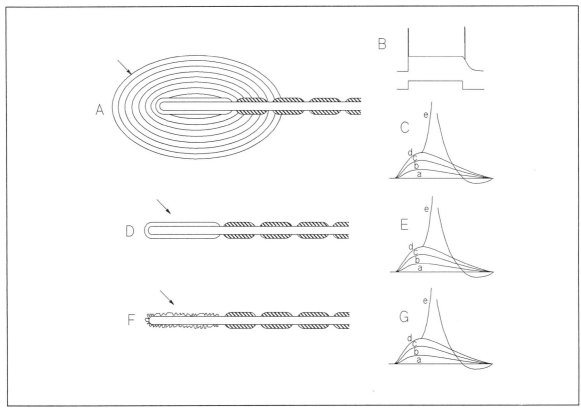

Fig. 7-3. (A) The laminated structure of the pacinian corpuscle. (B) If a pressure pulse of sufficient intensity is applied to this receptor, it produces an action potential at onset and offset of the pulse. (C) The response of an intact pacinian corpuscle to increasing pressure intensities (a–e). Note that the magnitude of the generator potential is proportional to the stimulus strength and, if the generator potential is large enough (e), it leads to generation of an action potential. (D) A receptor in which all outer laminae have been removed. (F) A receptor in which bits of the inner core have been removed. (E, G) Responses of receptors depicted in D and F, respectively. Note that the generator potential and generation of the action potential are independent of the outer and inner laminae.

Ionic Basis for the Generator Potential

As noted previously, by using two microelectrodes (one to pass current and one to record the receptor potential), the reversal potential of the generator potential can be determined for the pacinian corpuscle. Again, this value is near zero, which indicates that pressure causes an increased permeability of the membrane to Na^+. In a Na^+-free solution, the generator potential is abolished.

Adaptation

In an intact pacinian corpuscle, the laminated structure forms a mechanical filter and is responsible for its adaptive property. However, if the action potential is recorded in a corpuscle from which the laminae have been removed (see Fig. 7-3D), the application of pressure produces a generator potential that lasts as long as the stimulus. Nonetheless, in the absence of laminae, still only one or two action potentials are produced at the offset of the pressure, indicating that the axon also has accommodative properties.

Like other sensory receptors, pacinian corpuscles show spatial and temporal summation. If two weak stimuli are delivered sequentially at one point (Fig. 7-4, *probe 1*), the generator potentials produced by each stimulus summates to produce a larger generator potential (see Fig. 7-4, *temporal*). Similarly, if two stimuli are applied at two different locations on the receptor (see Fig. 7-4, *probes 1* and *2*), the magnitude of the generator potentials produced by these stimuli will be the sum of the generator potentials produced by each stimulus (see Fig. 7-4, *spatial*).

TRANSDUCTION

Application of the adequate stimulus to a receptor changes the permeability of the receptor to an ion (usually Na^+). This change produces a change in the membrane potential which, when large enough, generates action potentials. The exact mechanisms involved in sensory transduction are not known for all receptor types, but, recently, such information for rod photoreceptors and olfactory cilia has become available.

Transduction of light takes place at the outer-segment prime region of photoreceptors. These regions contain ion channels that are coupled to cyclic guanosine 5′-monophosphate (cGMP). In the dark, the concentration of cGMP inside the receptor is high, the ion channels are open, and the membrane potential is relatively low (around -30 mV). Exposure of the photoreceptor to light generates cGMP phosphodiesterase (an enzyme that hydrolyzes cGMP) and subsequently decreases the cGMP level. When the cGMP concentration drops below 10mM, the ion gates close and the membrane hyperpolarizes.

Recently, this hypothesis has been tested directly. Recordings have been made from the outer segment of frog rods using an excised patch-clamp procedure. Investigators have been able to expose the internal surface of the tissue to a variety of compounds and to measure their effect on membrane channels. Application of cGMP has been shown to increase the conductance of the membrane. The effect of cGMP can be reversed when the substance is removed.

Similar experiments on the cilia of olfactory receptors involved in the transduction of smell have indicated that cGMP and cyclic adenosine 5′-monophosphate (cAMP) are involved in generating receptor potentials in these receptors. In these sensory organs, the conductance of a single cilium increases when it is exposed to 10-μM cAMP. The time course of this conductance is a few hundred milliseconds, which is within the range of smell perception.

SUMMARY

Humans can perceive a variety of stimuli, such as light, sound, heat, cold, and pain. Such perceptions are mediated through the activation of receptors specific to each sensation. The receptor for each modality has a very low threshold to the energy that it encodes. For example, photoreceptors have a very low threshold to light. Activation of a sensory receptor produces a change in its membrane potential and generation of a receptor

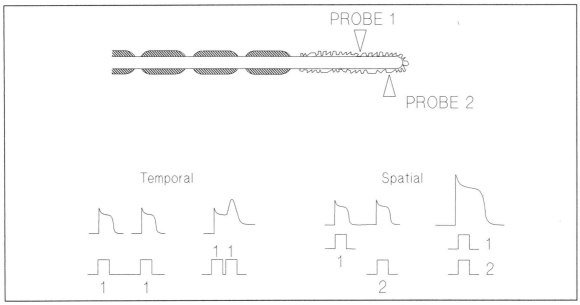

Fig. 7-4. The temporal and spatial summation properties of the pacinian corpuscle. The experimental setup in which two probes are used to stimulate a pacinian corpuscle in which all laminae have been removed. The spatial and temporal summation properties of the pacinian corpuscle. The top recordings are the receptor potential and the bottom recordings are the pressure pulses. (*1* and *2* represent the probe number.)

potential. The generator potential is a graded response and its magnitude increases as the stimulus strength increases. Activation of the generator potential, the sensory transduction, involves opening or closing the membrane channels in the sensory receptors. These channels are coupled to second messengers, such as cGMP and cAMP.

All sensory receptors adapt and their response to stimulation decreases with time. Some receptors adapt quickly and some very slowly. The rapidly adapting receptors encode stimuli that change very fast, such as vibration. The slowly adapting receptors encode sensations that change very slowly, such as muscle loading.

NATIONAL BOARD TYPE QUESTIONS

Select the one most correct answer for each of the following

1. A pacinian corpuscle is bathed in a solution containing a high concentration of TTX. You would expect that:
   A. The generator potential will be abolished
   B. The action potential in the nerve will be abolished
   C. The electronic spread of the generator will stop
   D. The amplitude of the generator will decrease
2. The major difference between the generator potential and the action potential is:
   A. The receptor has an all-or-none response but the action potential response is graded
   B. The action potential has no refractory period but the receptor potential does
   C. The receptor potential exhibits temporal and spatial summations but the action potential does not
   D. The action potential amplitude decreases

with distance, but the magnitude of the generator potential is independent of the distance from the stimulation site

3. A receptor that adapts rapidly is best suited to encode:
   A. A stimulus that has a very high strength
   B. A very rapidly changing stimulus
   C. A stimulus that does not change with time
   D. Changes in joint position

4. In the photoreceptors, transduction of light is initiated by:
   A. An increase in cGMP
   B. A decrease in cGMP
   C. Formation of cAMP
   D. Decrease in cGMP phosphodiesterase
   E. Opening of Na^+ channels

5. If the reversal potential of a generator potential is -10 mV, the ionic channels that can be involved in this process are:
   A. Opening of Na^+ channels only
   B. Opening of K^+ channels only
   C. Opening of both Na^+ and K^+ channels
   D. Closing of Ca^{2+} channels
   E. Blockade of K^+ channels

ANNOTATED ANSWERS

1. B. This is the only correct choice because Na^+ channels involved in the production of the generator potential are not blocked by TTX. Therefore, in this solution, no change in the generator potential is observed. TTX blocks Na^+ channels in the nerve, however, and pre-

vents the generation and propagation of the action potential.

2. C. Sensory receptors do not have Na^+ inactivation channels and stimulus-induced depolarization will summate.

3. B. Rapidly adapting receptors only signal the onset and offset of the stimulus. Therefore, if the frequency of stimulation is increased, these receptors can encode the change.

4. C. Light causes formation of cGMP phosphodiesterase, which breaks down cGMP and decreases its level.

5. C. Because the reversal potential for Na^+ is $+50$ mV and for K^+ is -90 mV, a reversal potential at -10 mV indicates opening of both Na^+ and K^+ channels.

BIBLIOGRAPHY

Ezyagurre, C., and Kuffler, S. W. Processes of excitation in the dendrites and in the soma of single isolated sensory nerves cells of the lobster and crayfish. *J. Gen. Physiol.* 39:87, 1955.

Fesenko, E. E., Kolesnikov, S. S., and Lyubarsky, A. L. Induction by cyclic GMP of cationic conductance in plasma membrane of retinal rod outer segment. *Nature* 313:310, 1985.

Lowenstein, W. R. Biological transducers. *Sci. Am.* 203:98, 1960.

Lowenstein, W. R., and Mendelson, M. Components of adaptation in a pacinian corpuscle. *J. Physiol.* (London), 177:377, 1965.

Nakamura, T., and Gold, G. H. A cyclic nucleotide-gated conductance in olfactory receptor cilia. *Nature* 325:442, 1987.

Stern, J. H., Kaupp, U. B., and MacLeish, P. R. Control of the light-regulated current in rod photoreceptors by cyclic GMP, calcium, and 1-*cis*-diliazem. *Proc. Natl. Acad. Sci. USA* 83:1163, 1986.

II Synaptic Transmission

8 Neuromuscular Junction of Skeletal Muscle

JANUSZ B. SUSZKIW AND NICHOLAS SPERELAKIS

Objectives

After reading this chapter, you should be able to:

Describe the microanatomy of the neuromuscular junction

Explain the factors that regulate acetylcholine release from motor nerve terminals

Explain the mechanism of end-plate potential

and its function in the genesis of the muscle action potential

Describe the effects and sites of action of drugs and toxins that affect neuromuscular transmission

Explain the biologic basis of the diseases of neuromuscular transmission and the effects of muscle denervation

---

Synaptic transmission refers to the transfer of signals between neurons or from neurons to effector cells (e.g., muscles, glands). The site of specialized contact between the cells where this transfer of signals occurs is called the *synaptic junction* or simply the **synapse.** The presynaptic element of the junction is the axon terminal of the transmitting nerve fiber. The postsynaptic element is the receiving structure.

Neurotransmission is mediated chemically at the vast majority of synapses in vertebrates. At chemical synapses, the presynaptic and postsynaptic membranes are separated by a 20- to 100-nm-wide synaptic cleft which is filled with extracellular fluid and prevents direct electrical excitation of the postsynaptic cell. Instead, excitation of the presynaptic neuron causes release of a chemical neurotransmitter substance that acts on specific receptors in the postsynaptic cell. Binding of the transmitter to its receptor opens

the receptor-gated ion channels and, depending on the ionic conductances that are activated, generates local, depolarizing, or hyperpolarizing **postsynaptic potentials**. The depolarizing potentials are called **excitatory postsynaptic potentials** because they increase the likelihood of an action potential. Conversely, the hyperpolarizing potentials are called **inhibitory postsynaptic potentials** because they decrease the likelihood of action potentials in the postsynaptic cell.

An alternate form of synaptic transmission involves direct transfer of current from the presynaptic to the postsynaptic elements, without mediation by a chemical transmitter. This can occur only if the presynaptic and postsynaptic membranes form a close contact, known as a **gap junction**. In contrast to chemical synapses which exhibit plasticity (i.e., the efficacy of synaptic transmission can be modulated by a variety of metabolic factors and chemical agents), the electrical synapses are designed for rapid all-or-none signal transmission. The electrical synapses are relatively rare in higher vertebrates but are fairly common in lower animal species where they mediate simple, stereotypic behaviors.

STRUCTURAL ELEMENTS OF THE SKELETAL NEUROMUSCULAR JUNCTION

The neuromuscular junction is formed by a motor neuron axon terminal in contact with the specialized postsynaptic muscle fiber membrane, forming the so-called **end-plate**. The presynaptic and postsynaptic elements are separated by an approximately 50-nm-wide synaptic cleft, and the synapse is insulated by a layer of Schwann cells. The nerve terminals contain spherical **synaptic vesicles** that store the transmitter **acetylcholine** (ACh). Synaptic vesicles tend to cluster at the presynaptic **active zones**, which are specialized membrane sites where transmitter is released into the synaptic cleft through the process of exocytosis. Receptors for ACh are concentrated in the crests of the postjunctional folds. These are localized opposite the presynaptic active zones and thus assure efficient interaction between the released transmitter and its postsynaptic receptors (Fig. 8-1).

OVERVIEW OF NEUROMUSCULAR TRANSMISSION

The skeletal neuromuscular junction is an example of an excitatory synapse that mediates the transfer of action potentials from motor neurons to skeletal muscle fibers. Depolarization of the nerve terminal membrane by the nerve action potential initiates a series of events collectively referred to as the **depolarization–release coupling** process. This consists of the activation of voltage-operated Ca^{2+} channels in the nerve terminal membrane and transient influx of extracellular Ca^{2+}, which triggers fusion of synaptic vesicles with the presynaptic membrane and the release (exocytosis) of intravesicular ACh into the synaptic cleft. The released molecules of ACh diffuse quickly (less than 0.1 msec) to the postsynaptic (muscle) membrane and combine with ACh receptors, causing the opening of the receptor-associated cationic channels, and lasting for 1 to 2 msec. The resultant transient increase in the permeability of the end-plate to Na^+ and K^+ results in local membrane depolarization, called the **end-plate potential** (EPP). In normal healthy muscle, the size of the EPP is always sufficient to trigger a muscle action potential that then propagates along the muscle fiber and elicits a muscle twitch (Fig. 8-2). The generation of a muscle action potential is not instantaneous, but instead registers about 0.3 to 0.5 msec after the arrival of the presynaptic action potential. This **synaptic delay** is a characteristic of all chemically transmitting synapses, and represents the time required for completion of the chemical processes associated with depolarization–release coupling and activation of postsynaptic receptor–gated permeability changes. The action of ACh at the neuromuscular junction is terminated in part by diffusion, but primarily through hydrolysis of ACh to acetate and choline through the action of the enzyme acetylcholinesterase (AChE).

PRESYNAPTIC PROCESSES
Synthesis and Storage of ACh

ACh is synthesized in the nerve terminal cytoplasm from acetyl coenzyme A, and choline is synthesized by the enzyme choline acetyltransferase. Acetyl coenzyme A most likely derives from intra-terminal mitochondria; however, choline is not produced by neuronal cells and must be transported from the extracellular space by means of a high-affinity choline carrier that resides in the membrane of cholinergic neuron terminals. A considerable proportion of choline (more than 50 percent) is recaptured from the synaptic cleft after the released ACh is degraded by AChE. Once synthesized in the cytoplasm, ACh is taken up and stored in the synaptic vesicles for subsequent release (Fig. 8-3).

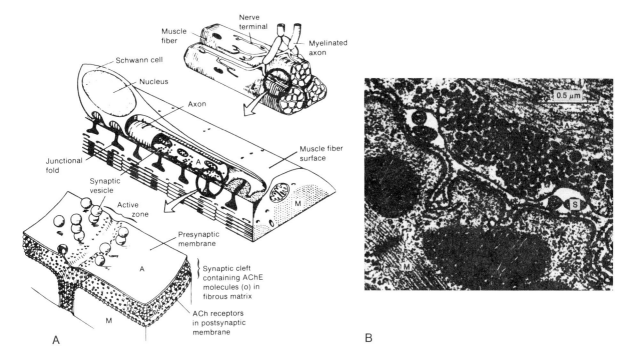

Fig. 8-1. (A) The neuromuscular junction of the frog. (Top) Individual muscle fibers are innervated by myelinated motor axons whose unmyelinated terminal branches contact the muscle membrane, forming a depressed groove several micrometers long. The mammalian neuromuscular junction is smaller and usually has the appearance of a plate-like depression about 1 to 2 μm in diameter, hence the name *end-plate*. The space between the presynaptic nerve and postsynaptic muscle membranes is the synaptic cleft. (Middle) A section through the portion of the end-plate reveals details of the junction. Within the terminal, the spherical, 500-Å synaptic vesicles congregate near the presynaptic active zones, opposite the postjunctional folds. The nerve terminal and the synaptic cleft are covered and insulated by a thin sheath of Schwann cell membrane, which also tends to separate individual active zone regions. (Bottom) The active zone area consists of parallel rows (bars) of particles that may represent presynaptic Ca^{2+} channels or structures where fusion of the synaptic vesicles with plasma membrane occurs, or both of these. The synaptic vesicles are frequently aligned along the presynaptic parallel bars and are presumably positioned for exocytosis. Acetylcholine (ACh) receptors span the postsynaptic membrane and are concentrated in the crests of postjunctional folds directly opposite the prejunctional active zones. The fibrous material (basal lamina) within the synaptic cleft contains molecules of the enzyme acetylcholinesterase (*AChE*). (B) Transverse view of the neuromuscular junction as seen in an actual electron micrograph from a portion of frog end-plate. (Reproduced from: Hille, B. *Ionic Channels of Excitable Membranes.* Sunderland, Mass.: Sinauer Associates, 1984. Micrograph was produced by Dr. John E. Heuser, Washington University School of Medicine, St. Louis.)

Release of ACh

ACh is released through the process of exocytosis, whereby there is a Ca^{2+}-dependent fusion of synaptic vesicles with the presynaptic plasma membrane and liberation of vesicle contents into the synaptic cleft.

Spontaneous Release

In the absence of presynaptic action potentials, the ambient concentration of ionized Ca^{2+} in the terminal cytoplasm is maintained at about 0.1 μM. Under these conditions, synaptic vesicle–plasmalemma fusion occurs relatively

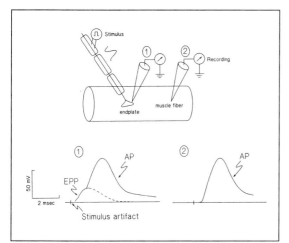

Fig. 8-2. Schematic representation of intracellular recording obtained from muscle in response to presynaptic nerve stimulation. (Top) A presynaptic action potential is evoked by an electrical pulse applied to the nerve. Postsynaptic membrane potentials are recorded with intracellular glass microelectrodes inserted at the muscle end-plate region (*1*) and at some distance (2 mm) from the end-plate (*2*).

(Bottom, 1) The resulting postsynaptic records show a stimulus artifact that represents approximately the time of the action potential (*AP*) in the presynaptic nerve terminal and the postsynaptic response (end-plate potential [*EPP*] recorded after about a 0.5-msec delay (synaptic delay). When recording intracellularly near the end-plate region, the EPP is largely masked by the AP that it triggers, but is evident as a step on the rising phase of the AP. If the AP is blocked (e.g., by application of tetrodotoxin), the EPP would be revealed in its entirety (*dashed trace*). In practice EPP properties are studied in the presence of micromolar concentrations of curare, which is used to depress the EPP to below the threshold for generation of muscle APs. (Bottom, 2) The APs recorded 2 mm away from the end-plate are smoothly rising without any evidence of the EPP. This is because the EPP is a local, nonregenerative, electrotonic potential that decrements with distance from the site of origin (i.e., the end-plate). The decrement of EPP with distance is determined by the muscle length constant, which for skeletal muscle is about 1 mm. (As explained in Chapter 4, the muscle length constant is the distance at which an electrotonic potential decays to 1/e or 37 percent of its initial amplitude.)

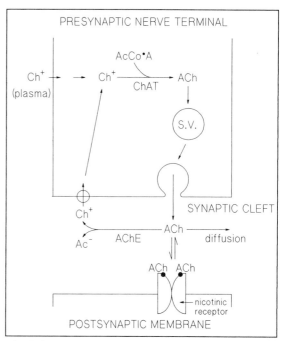

Fig. 8-3. The acetylcholine cycle at the neuromuscular junction. Acetylcholine (*ACh*) is synthesized in nerve terminal cytoplasm by the enzyme cholineacetyltransferase (*ChAT*) and is packaged and stored for subsequent release in the synaptic vesicles (*SV*). Once released by presynaptic action potentials, ACh in the synaptic cleft binds to the nicotinic receptors in the postsynaptic membrane. Its synaptic action is terminated by diffusion and by acetylcholinesterase (*AChE*)–catalyzed hydrolysis to inactive choline (Ch^+) and acetate (Ac^-). More than 50 percent of the choline produced in this way is taken up by the high-affinity choline transport system in the presynaptic plasma membrane and used in the resynthesis of ACh.

infrequently, at about one to three events per second. Such spontaneous, random release of ACh packets in each vesicle gives rise to small depolarizations (0.1–1mV) of the postsynaptic membrane, the so-called **miniature end-plate potentials**.

Evoked Release
The arrival of an action potential and consequent depolarization of the axon terminal opens voltage-

gated Ca^{2+} channels near the plasmalemmal release sites in the active zone. Transient influx of Ca^{2+} from the extracellular space and the consequent elevation of Ca^{2+} inside the terminal trigger near-synchronous exocytosis of from 100 to 300 vesicles and a release of ACh sufficient to generate a large depolarization of the subsynaptic membrane (i.e., the EPP). The release process terminates upon repolarization of the presynaptic membrane, closure of the Ca^{2+} channels, and rapid lowering of the cytosolic Ca^{2+} level, through the influence of intraterminal Ca^{2+} sequestration mechanisms, back to its ambient level of about 0.1 μM. The essential feature of the evoked release is its dependence on the extracellular Ca^{2+} level. Any reduction in the availability of Ca^{2+} to enter the terminals in turn reduces the amount of ACh released. The dependence of transmitter release on the entry of Ca^{2+} into the terminals is illustrated in Figure 8-4.

Quantal Nature of Transmitter Release

Research by Katz and co-workers in the 1950s demonstrated that spontaneous miniature end-plate potentials are the basic units of the postsynaptic response brought about by the random release of discrete packets, or **quanta**, of ACh. A quantum is thought to comprise between 5,000 and 10,000 molecules of ACh, and in all likelihood represents the transmitter content of a single synaptic vesicle. EPPs are generated by many quanta of transmitter released by a presynaptic action potential; they essentially represent a summation of multiple, synchronously generated miniature end-plate potentials (Fig. 8-5). The process of release is probabilistic and can be described by the expression:

$$m = n \cdot P \qquad \qquad \text{Eq. 8-1}$$

where *m* is the average amplitude of EPPs recorded in many trials and is called the *quantal content of the end-plate;* n is the number of quanta available for release; and *P* is the probability of any quantum being released.

It is evident that the amplitude of EPPs depends

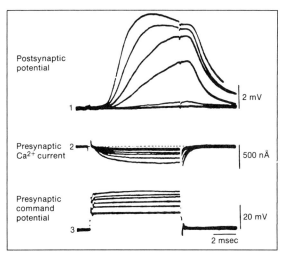

Fig. 8-4. Relationship between transmitter release and Ca^{2+} influx into the presynaptic terminal. This experiment was conducted at the squid giant synapse, which permits simultaneous microelectrode recordings from the presynaptic terminal and the postsynaptic cell. The presynaptic terminal was depolarized to various levels, as indicated by the bottom traces, giving rise to presynaptic Ca^{2+} currents recorded in the middle traces and the Ca^{2+}-dependent transmitter release monitored by the measurements of postsynaptic potentials in the top traces. The transmitter release, as reflected by the amplitude of postsynaptic potentials, increases as a function of the size of the presynaptic Ca^{2+} current (that is, Ca^{2+} influx into the terminals). (Reproduced from: Kandel, E. R., and Schwartz, J. M. *Principles of Neural Science* [Part III, Ch. 11]. New York: Elsevier Science Publishing Co., 1985. Original data from: Llinas R. R., and Heuser, J. E., *Neurosci Res. Program Bull.* 15:555–687, 1977.)

on the product of availability of quanta (i.e., presumably the number of vesicles available for discharge at the active zone) and the probability of discharge, which is a function of the ionized Ca^{2+} concentration at the release sites. Any factor that reduces the availability of quanta for release or reduces the entry of Ca^{2+} into the terminals will result in reduced transmitter release and a smaller EPP. For example, reduction of ACh synthesis, its uptake into the synaptic vesicles, or both, will produce less transmitter released and a lower EPP. Similarly, an increase in the extracellular concentration of Mg^{2+}, which

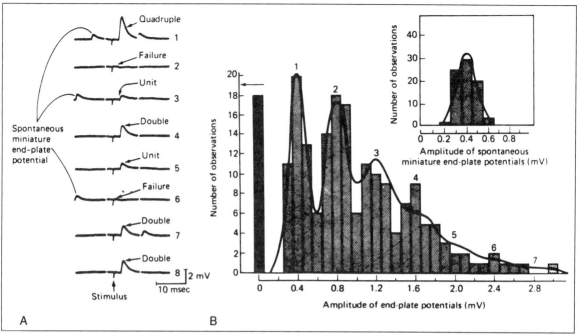

Fig. 8-5. Quantal nature of transmitter release. (A) Intracellular recordings from a rat phrenic nerve–diaphragm preparation in Ca^{+2}-deficient and Mg^{2+}-rich solutions. Under these conditions, the nerve-evoked acetylcholine release is greatly reduced and reveals fluctuation in the amplitude of end-plate potentials (EPPs), including occasional failures (*arrows* in traces *1–8*). Note that the evoked EPPs follow the stimulus artifact, whereas the miniature end-plate potentials (MEPPs) observed in the recordings occur spontaneously and randomly in the absence of nerve stimulation. (B) Histogram of MEPP distribution reveals that they have an average amplitude of about 0.4 mV. The histogram of EPPs indicates that they consist of integral n multiples of MEPPs, where n = 0, 1, 2, and so on. The solid trace describing the distribution of EPP amplitudes was calculated using the Poisson statistics. (Reproduced from: Kandel, E. R., and Schwartz, J. M. *Principles of Neural Science* [Part III, Ch. 11]. New York: Elsevier Science Publishing Co., 1985. Original data from: Liley, A. W. *J. Physiol.* [Lond.] 133:571–587, 1956 [A]; and Boyd, I. A., and Martin, A. R. *J. Physiol.* [Lond.] 132:75–91, 1956 [B].

antagonizes Ca^{2+} entry through the Ca^{2+} channels, or a reduction in the extracellular Ca^{2+} concentration, or both, will lead to reduced transmitter release and hence reduced EPPs.

POSTSYNAPTIC PROCESSES

The postsynaptic response to ACh is local depolarization of the postsynaptic membrane—the EPP. The EPP is generated when ACh binds to its receptors and opens the associated channels for a short, 1- to 2-msec duration (Fig. 8-6). The density of the receptors in the postsynaptic membrane is very high and probably excludes any significant number of voltage-gated channels. Thus, the end-plate region is chemosensitive but electrically inexcitable. Conversely, the density of ACh receptors decreases steeply with increasing distance from the end-plate, and the muscle membrane outside the junction is electrically excitable but unresponsive to ACh.

The Mechanism of EPP

At rest, most of the ACh-gated channels are closed and the conductance of the postsynaptic mem-

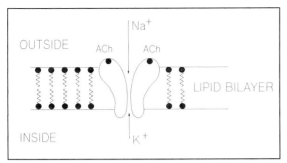

Fig. 8-6. Model of the acetylcholine (nicotinic) receptor/channel complex in the postjunctional membrane. The receptor/channel complex is formed by five membrane-spanning subunits. When activated by ACh, the complex undergoes a conformational change from a closed to an open state. In the open state, the channel permits passage of cations (Na$^+$ and K$^+$) but excludes anions.

brane to ions is very low. When opened, the ACh-receptor–gated channels conduct Na$^+$ and K$^+$ with about equal ease, but exclude anions. Thus, when ACh activates the receptors, this allows a Na$^+$ influx (and K$^+$ efflux) through the postsynaptic membrane, giving rise to the synaptic (end-plate) current (I_s):

$$I_s = I_{Na} + I_K = g_{Na}(V_m - E_{Na}) + g_K(V_m - E_K)$$
$$\text{Eq. 8-2}$$

where g_{Na} and g_K are Na$^+$ and K$^+$ conductances, E_{Na} and E_K are the Na$^+$ and K$^+$ equilibrium potentials, and I_{Na} and I_K are the Na$^+$ and K$^+$ currents, respectively. The direction and magnitude of ion flux is determined by the electrochemical driving force (V_{m-ion}). Because the Na$^+$ concentration is high on the outside and low on the inside, but the converse is true for K$^+$, Na$^+$ will flow into the cell and K$^+$ will travel out through the synaptic channels. Because $g_{Na}(V_m - E_{Na}) > g_K(V_m - E_K)$, the net synaptic current will be always inward current, carried by Na$^+$ and resulting in membrane depolarization.

Synaptic (End-Plate) Equilibrium Potential

It is evident from Equation 8-2 that, as the membrane potential (V_m) becomes depolarized, the

driving force acting on Na$^+$ to enter the cell decreases and that on K$^+$ to exit the cell increases. At a certain membrane potential, the driving forces acting on Na$^+$ to flow in and K$^+$ to flow out of the end-plate will be equal, so that the inward Na$^+$ and outward K$^+$ currents will be exactly balanced, yielding a zero net end-plate current. The membrane potential at which this occurs is termed the **end-plate equilibrium potential** (E_{EPP}):

$$I_s = 0 = g_{Na}(V_m - E_{Na}) + g_K(V_m - E_K) \qquad \text{Eq. 8-3}$$

$$V_m = E_{EPP} = [g_{Na}/(g_{Na} + g_K)] \cdot E_{Na} + [g_K/(g_K + g_{Na})] E_K$$
$$\text{Eq. 8-4}$$

Equation 8-4 shows that the E_{EPP} is a sum of Na$^+$ and K$^+$ equilibrium potentials weighted by the relative conductances of Na$^+$ and K$^+$. Because $g_{Na} \cong g_K$, $E_{Na} \cong +60$ mV, and $E_K \cong -90$ mV, the E_{EPP} can be calculated to be about -15 mV. This is the maximum theoretical potential to which the end-plate can be depolarized. Any further depolarization beyond this point would result in the K$^+$ current exceeding the Na$^+$ current, and consequently a reversal of the net end-plate current from inward to outward. Therefore, the E_{EPP} is also called a **reversal potential**. The relationship of the E_{EPP} to the equilibrium potentials for the major ions and the resting and threshold membrane potentials are illustrated in Figure 8-7.

Equivalent Circuit Analysis of EPP

The relationship between the end-plate current and muscle membrane depolarization can be analyzed in terms of an equivalent electrical circuit consisting of parallel synaptic and nonsynaptic branches (Fig. 8-8). The synaptic branch represents the synaptic receptor–gated channels in series with the synaptic battery (E_{EPP}). The nonsynaptic branch consists of membrane capacitance and leak channels in series with the battery representing the resting membrane potential E_m. During activation of the end-plate by ACh, the end-plate current (I_s) flows inward through the synaptic branch and outward

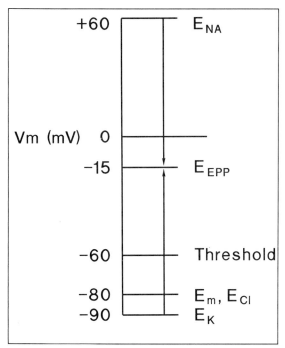

Fig. 8-7. Relationship of the end-plate equilibrium potential (E_{EPP}) to the resting membrane potential (E_m) and membrane threshold. Ionic equilibrium potentials for principal ions (E_{Cl}, E_K) are also shown for comparison. At the E_{EPP}, the electromotive forces (*arrows*) acting on Na^+ and K^+ are equal and opposite in direction. (*V_m = membrane potential in mV.*)

Fig. 8-8. Equivalent circuit description of the end-plate potential (*EPP*). (A) Relationship between the EPP and the underlying membrane currents, where I_s = synaptic, I_c = capacitive, and I_m = resistive membrane current. (B) Equivalent circuit of the postsynaptic membrane. The synaptic branch consists of a synaptic battery (E_{EPP}) in series with a synaptic conductance pathway (g_s), that is, the receptor-gated channels. The nonsynaptic branch consists of the battery of the resting membrane potential (E_m) in series with the conductance pathway (g_m), that is, the leak channels. Membrane capacitance (C_m) is in parallel with both the synaptic and nonsynaptic conductance pathways. In the *active phase of the EPP*, when the ACh-receptor–gated channels are activated (i.e., the switch in the electric circuit diagram is closed), the synaptic current (I_s) flows into the end-plate through the receptor-gated channels (g_s) and out through the capacitance (C_m) and leak channels (g_m) of the adjacent nonsynaptic membrane. The capacitive current (I_c) is due to the deposition of positive charges on the cytosolic aspect of the membrane and removal of an equal number of positive charges from the extracellular side, causing a discharge of capacitance and membrane depolarization. Once the membrane capacitance has become discharged to its final value, all outward current is through the nonsynaptic (g_m) pathway and $I_s = -I_m$. This condition corresponds to the peak of the EPP. In the *declining phase of the EPP*, as the receptor-gated channels are reclosed and I_s diminishes to zero, the outward I_m returns as I_c recharges the membrane capacitance and the membrane repolarizes to its resting value. The declining phase of EPP is a passive process, the rate of which is a function of the membrane time constant ($R_m \cdot C_m$). (*V_m = membrane potential.*) (Modified from: Kandel, E. R., and Schwartz, J. M. *Principles of Neural Science* [Part III, Ch. 9]. New York: Elsevier Science Publishing Co., 1985.)

through the parallel resistive and capacitive elements of the nonsynaptic branch (see Fig. 8-8A):

$$I_s = -(I_c + I_m) \qquad \text{Eq. 8-5}$$

where I_c is the capacitive current flowing into and out of membrane capacitance and I_m is the current flowing through the parallel leak channels in the nonsynaptic membrane. At the onset of synaptic action, most of the synaptic current flows through the capacitive branch because the outward driving force ($V_m - E_m$) on current flow through the nonsynaptic channels (g_m) is small. The deposition of positive charges on the interior of the lipid bilayer and removal of an equal number of positive charges from the outside results

in discharge of membrane capacitance and depolarization. Once the membrane capacitance is discharged to its final value, all the synaptic current exits through the leak channels (g_m), so that:

$$I_s = -I_m \qquad \text{Eq. 8-6}$$

This steady-state equation permits calculation of the peak value of the EPP. Noting that synaptic (end-plate) Na^+ and K^+ conductances can be expressed as a combined synaptic conductance (g_s), the currents flowing through the active synaptic

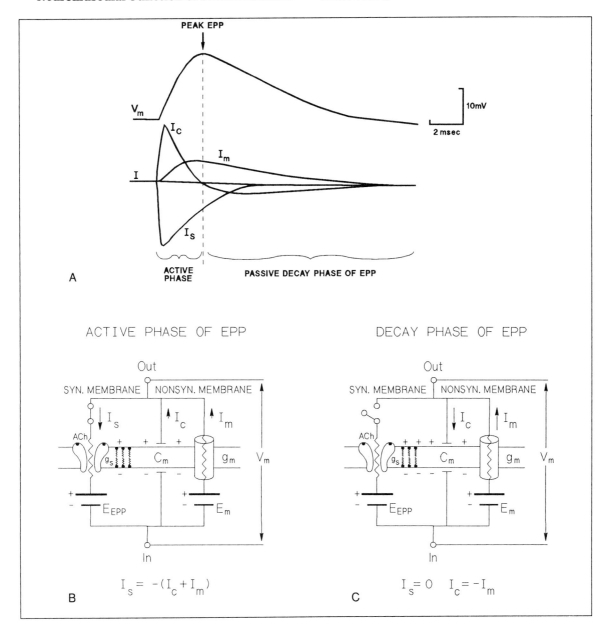

channels (I_s) and the leak nonsynaptic channels (I_m) can be written according to Ohm's law as:

$$I_s = g_s(V_m - E_{EPP}) \qquad \text{Eq. 8-7}$$

and

$$I_m = g_m (V_m - E_m) \qquad \text{Eq. 8-8}$$

By substituting Equations 8-7 and 8-8 into Equation 8-6 and solving for the membrane potential (V_m), one obtains the expression:

$$V_m = [g_s/(g_s + g_m)]E_{EPP} + [g_m/(g_s + g_m)]E_m$$
$$\text{Eq. 8-9}$$

Equation 8-9 shows that the value of the membrane potential at the peak of synaptic activation is a weighted average of E_{EPP} and E_m. The weighting factors are the relative magnitudes of synaptic and nonsynaptic conductances. During peak activation of the synaptic ACh channels, the synaptic conductance far exceeds the nonsynaptic conductance and the membrane potential will tend toward the E_{EPP} of about -15 mV. The amplitude of the EEP will however, be "loaded down" by the nonsynaptic conductance of the nonsynaptic membrane, so that the actual membrane potential will be somewhat less than the theoretical value of E_{EPP}. Nevertheless, as already noted, the amplitude of the EPP is usually considerably in excess of the depolarization needed to bring the membrane to the threshold, and thus the neuromuscular transmission is said to have a **high safety factor.**

Termination of the Synaptic Action of ACh and Decay of EPP

The active phase of synaptic ACh action lasts about 1 to 2 msec. As ACh in the synaptic cleft is degraded by AChE, the ACh-receptor complexes dissociate, the ACh-activated channels reclose, and the synaptic current declines to zero. The EPP decays passively as the membrane is repolarized over the next 5 to 10 msec (see Fig. 8-8).

Characteristics of the EPP and Generation of the Muscle Action Potential

Unlike a propagated action potential, the EPP is a locally generated electrotonic potential that passively decays with time and distance, according to the time and length constants of the muscle membrane. Because of the cable properties of the muscle fiber, with a length constant of about 1 mm (the distance at which the EPP decays to 37 percent of its peak value at the end-plate), the EPP essentially disappears by about 2 mm (see Fig. 8-2). However, the local depolarization of the electrically excitable membrane in the immediate vicinity of the end-plate (Fig. 8-9) is more than sufficient to reach the threshold potential (i.e., about -60 mV), so that a muscle potential is triggered. Since the neuromuscular junction is usually located somewhere near the middle of a muscle fiber, the action potentials triggered by EPPs are propagated toward each end of the fiber.

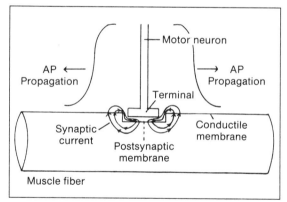

Fig. 8-9. The synaptic and adjacent nonsynaptic regions of the neuromuscular junction of vertebrate skeletal muscle, illustrating the direction of current flow. Depolarization of the electrically excitable nonsynaptic membrane by the local current flow to threshold triggers a muscle action potential (*AP*). The muscle potential propagates in both directions from the end-plate toward each end of the muscle fiber. (Reproduced from: Sperelakis, N., Suszkiw, J. B., and Hackett, J. T. *The Neuromuscular Junction.* In Roussos, C., and Macklem, P. T., eds. The Thorax. New York: Marcel Dekker, 1985, [Chapter 4, part A].

This propagating wave of depolarization elicits muscle twitch (depolarization–contraction coupling).

MODULATION OF SYNAPTIC TRANSMISSION

Although the the amount of transmitter released by each presynaptic action potential is usually constant, it can be modified as a function of the previous history of nerve activity. This gives rise to the phenomena of synaptic facilitation, post-tetanic potentiation, and depression.

Facilitation and Post-Tetanic Potentiation

When presynaptic action potentials are sufficiently close, the amount of transmitter released per each succeeding nerve impulse tends to increase. This is reflected in the progressively increased amplitude of postsynaptic potentials, and is termed **facilitation**.

When the nerve is activated with relatively prolonged or high-frequency (tetanic) bursts of impulses, or both, the quantity of transmitter is increased upon subsequent stimulation, even after a relatively long intervening rest period. This phenomenon is known as **post-tetanic potentiation**.

Both facilitation and potentiation are thought to be associated with the buildup of Ca^{2+} within the terminals, which increases the probability of transmitter release. However, the two processes are kinetically distinct. Facilitation is a relatively short-lived process, whereas post-tetanic potentiation may last for minutes and sometimes hours, possibly indicating a long-term modification of presynaptic function secondary to an increase in the cytosolic Ca^{2+} levels.

Depression or Fatigue of Neuromuscular Transmission

During prolonged, relatively high-frequency activation of the motor neurons, the quantity of transmitter released per nerve impulse tends to gradually diminish. This phenomenon, referred to as **depression** or **fatigue**, is probably due to progressive depletion of releasable transmitter stores. In other words, ACh probably is released at a faster rate than it is resynthesized and repackaged into synaptic vesicles. Recovery from fatigue may take from minutes to hours.

The phenomena of facilitation, post-tetanic potentiation, and depression are most readily demonstrated when neuromuscular transmission is impaired, either experimentally or by certain diseases, but these processes probably play a minor role, if any, in normal healthy individuals.

DRUGS AND TOXINS

A number of drugs and toxins interfere with synaptic transmission by acting either on the nerve terminal or the postsynaptic process.

Presynaptic Actions

Inhibition of Precursor Uptake and Deficiency of Transmitter Synthesis

Hemicholinium-3 is a potent competitive inhibitor of the high-affinity choline transport system, which provides choline for the synthesis of ACh in nerve terminals. In the presence of the drug, the supply of choline and hence the synthesis of ACh are depressed. During repetitive activation, the transmitter stores become depleted and synaptic fatigue ensues.

Disruption of Transmitter Packaging

D,L 2-(4-phenylpiperidino)cyclohexanol (Vesamicol) penetrates into presynaptic nerve terminals and blocks ACh uptake into synaptic vesicles. In the presence of the drug, synaptic vesicles cannot be refilled with the transmitter even though ACh synthesis itself is not impaired. Untimately, exhaustion of the releasable vesicular transmitter store leads to failure of neuromuscular transmission. α-**Latrotoxin,** a component of black widow spider venom, causes massive transmitter exocytosis and depletion of synaptic vesicles from terminals. Synaptic transmission fails in the presence of the toxin. This toxin is not specific to cholinergic nerve terminals.

Uncoupling the Depolarization—Secretion Process

Botulinus toxins, produced by *Clostridium botulinum,* are responsible for the symptoms of food poisoning, or botulism. The *C. botulinum* toxin is one of the most potent neuroparalytic agents known, with the median letal dose in humans estimated in the submicrogram range. The toxin's action is essentially irreversible and is exerted following internalization of the toxin or its subunit into nerve terminals, where it blocks synaptic vesicle exocytosis by an unknown mechanism. This toxin is relatively specific for cholinergic nerve endings. **Heavy metals** such as lead and cadmium are potent antagonists of Ca^{2+} entry through the presynaptic voltage-gated Ca^{2+} channels and, in submicromolar concentrations, inhibit the evoked transmitter release.

Postsynaptic Actions

Neuromuscular Blockers

Nicotine, succinylcholine, and **decamethonium** are examples of drugs that bind reversibly to the receptor active site and act as agonists. Because, unlike ACh, these drugs are not inactivated by AChE, they cause persistent receptor activation that initially results in repetitive discharge of muscle action potentials, followed by block of neuromuscular transmission, probably due to inactivation of the muscle spiking mechanism, receptor desensitization, or both. The result is flaccid paralysis of the muscle.

Curare is a classic nicotinic-receptor antagonist, which prevents binding of ACh to the receptor but by itself does not activate the receptor. At the neuromuscular junction curare causes a progressive decrease in the amplitude and shortening of the EPPs. In severe curare poisoning, EPPs are depressed below the threshold membrane potential, and thus are insufficient to initiate the muscle action potential. **α-Bungarotoxin**, a polypeptide component of the venom of the Formosan krait, essentially combines irreversibly with the receptor and, like curare, prevents its activation by ACh. As with curare poisoning, the result is flaccid muscle paralysis.

Reversible neuromuscular blockers, such as succinylcholine are sometimes used clinically for skeletal muscle relaxants during surgery.

Anticholinesterases

As discussed previously, the action of ACh is normally terminated through its hydrolysis by AChE to inactive acetate and choline. When AChE activity is inhibited, ACh which is released into the synaptic cleft builds up in concentration and persists for a longer time. This results in increased and multiple activation of the receptors by ACh, and hence increased amplitude and prolongation of the EPP. As with depolarizing blockers, after initial repetitive activation which may result in muscle spasm, the long-lasting depolarization inactivates the muscle spiking mechanism and ultimately leads to flaccid paralysis. There are two types of AChE inhibitors: (1) slowly reversible ones (e.g., eserine ([physostigmine]), which are generally compounds that carbamylate the active site of the enzyme; decarbamylation is slow (half-life is in minutes), but, once the inhibitor is removed, the enzyme eventually recovers its activity; and (2) irreversible ones, comprising organophosphorous compounds that phosphorylate the active site of the enzyme; since dephosphorylation is extremely slow (half-life of from days to weeks), the enzyme is essentially irreversibly inactivated. The classic example of organophosphorus anticholinesterase is diisopropylphosphofluoridate, originally developed as nerve gas. Anticholinesterases are used industrially as the principal component of various insecticides.

DISEASES OF THE NEUROMUSCULAR JUNCTION

Myasthenia Gravis

The characteristic feature of myasthenia gravis is muscle weakness, especially affecting cranial and upper limb muscles. The disorder is caused by an autoimmune response to ACh receptors, which leads to a reduced number of functional receptors and destruction of the postsynaptic membrane. The mechanism which causes the reduction of ACh receptors appears to involve an increase in the rate of receptor turnover, due to cross-linking of receptors by antibody and consequent stimulation of receptor endocytosis and degradation. Additionally, characteristic disorganization of the postsynaptic morphologic appearance is brought about by complement-mediated focal lysis of the membrane after the binding of antireceptor antibodies activates the complement.

In some patients, the myasthenic symptoms can be temporarily ameliorated by administration of anticholinesterases (e.g., physostigmine), which, as already discussed, elevate ACh concentration and prolong its action at the synaptic cleft. In such cases, the dosage of anticholinesterase must be carefully controlled in individual patients, because an excess may precipitate "cholinergic crisis" brought about by block of neuromuscular transmission. Thymectomy, steroids, immunosuppression drugs, and plasmapheresis have been employed as more prolonged treatment.

Lambert-Eaton Syndrome

The Lambert-Eaton syndrome is characterized by muscle weakness, which, in contrast to myasthenia gravis, tends to improve with vigorous exercise.

The syndrome is associated with the production of antibodies directed against the component, or components, of the presynaptic active zone, most likely the voltage-gated Ca^{2+} channels. The binding of antibodies leads to reduced Ca^{2+} entry and hence reduced probability of synaptic vesicle fusion and transmitter release. This, in turn, leads to a reduced EPP and a lowered probability of a muscle action potential. In Lambert-Eaton syndrome, the neuromuscular junction exhibits facilitation and potentiation when the nerve is stimulated at 20 to 100 Hz, that is, transmitter output improves, presumably because of buildup of Ca^{2+} in the terminals. This is probably why vigorous exercise reduces the weakness.

Slow-Channel Syndrome

The slow-channel syndrome is a hereditary (autosomal dominant) trait and is characterized by weakness, fatigability, and atrophy of the muscles.

The syndrome is associated with an abnormally long open time of the ACh-gated channels. This results in prolongation of the EPP, which then leads initially to discharge of multiple action potentials, followed by block of transmission.

Congenital End-Plate AChE Deficiency

The symptoms of congenital end-plate AChE deficiency, which is an extremely rare condition, are similar to those associated with slow-channel syndrome and anticholinesterase poisoning. As would be expected, cholinesterase deficiency results in an accumulation of ACh in the synaptic cleft and prolongation of ACh action on the receptors. This produces prolongation of EPPs, which in turn causes multiple discharges of muscle action potentials and then a depolarizing block of neuromuscular transmission, leading to muscle weakness. As the disease progresses, there is considerable atrophy of the muscles.

EFFECTS OF DENERVATION

Sectioning of a nerve produces immediate complete flaccid paralysis of the muscle, as action potentials cannot propagate past the site of section. The physiologic properties of the denervated muscle change over a period of days to weeks following the section, and these are: (1) the resting potential declines, (2) tetrodotoxin-insensitive Na$^+$ channels appear, (3) the muscle develops **denervation supersensitivity**, which is associated with the appearance and spread of receptors in the extrajunctional muscle membrane, (4) during early stages, fibrillation (spontaneous twitching of individual muscle fibers) develops, and (5) the muscle atrophies. The effects of denervation on muscle are partially due to interruption in the supply of trophic factors by the nerve. Muscle inactivity appears to be an important factor in the establishment of denervation changes, because direct stimulation of denervated muscles can slow or even prevent the changes caused by severance of a nerve.

SUMMARY

The transmission of signals from motor neurons to skeletal muscle fibers is mediated by ACh. This transmitter is released from presynaptic nerve endings and generates an EPP, which is a local depolarization of the muscle fiber end-plate. Skeletal neuromuscular transmission is very secure, such that normally every nerve action potential always elicits an EPP of sufficient amplitude and duration to trigger one muscle action potential. In abnormal states, when EPPs are reduced to below the threshold potential required for firing the muscle action potentials, neuromuscular transmission fails, causing muscle weakness and flaccid paralysis. These abnormalities may be brought about by drugs, neurotoxins, or diseases that reduce ACh release or interfere with ACh-receptor binding. Anticholinesterases, which prolong the life of synaptically released ACh, or drugs that act as nonhydrolyzable ACh-receptor agonists, generate abnormally prolonged EPPs, resulting in a depolarizing block of neuromuscular transmission and muscle paralysis.

NATIONAL BOARD TYPE QUESTIONS

Select the one most correct response for each of the following

1. The muscle end-plate potential:
   A. Is an all-or-none depolarizing potential
   B. Is mediated by the voltage-gated Na$^+$ and K$^+$ channels at the end-plate
   C. Is a response to the release of one quantum of ACh
   D. Can be elicited by direct electrical stimulation of the muscle
   E. In normal, healthy muscle is always of sufficient amplitude to trigger a muscle action potential
2. The E_{EPP} is approximately:
   A. -90 mV
   B. $+55$ mV
   C. -55 mV
   D. -15 mV
   E. $+15$ mV
3. Irreversible inhibition of AChE:
   A. Facilitates neuromuscular transmission
   B. Prevents binding of ACh to the postsynaptic receptors
   C. Causes permanent muscle contraction
   D. Can cause neuromuscular block and flaccid paralysis
   E. Results in long-term potentiation of neuromuscular transmission
4. The release of ACh evoked by presynaptic action potentials in motor neurons:
   A. Is independent of the extracellular Ca^{2+} concentration
   B. Is independent of the intracellular Ca^{2+} concentration
   C. Can be blocked by botulinus toxins
   D. Is not affected by the frequency of presynaptic action potentials
   E. Is not affected by inhibitors of choline uptake

5. Which of the following underlies facilitation?
   A. Increased sensitivity of the postsynaptic membrane
   B. Increased postsynaptic action potentials
   C. Decreased degradation of ACh
   D. Increased release of transmitter per each presynaptic action potential
   E. Upregulation of the presynaptic Ca^{2+} channels

ANNOTATED ANSWERS

1. E. The neuromuscular junction has a high safety factor. The EPP normally exceeds the threshold for a muscle action potential.
2. D. The E_{EPP} is a weighted average of Na^+ and K^+ equlibrium potentials (see Equation 8-4.)
3. D. Inhibition of AChE results in the accumulation of ACh in the synaptic cleft. This initially results in prolongation of EPPs and multiple muscle action potentials. The end result, however, is the block of neuromuscular transmission because of receptor desensitization, inactivation of the muscle spiking mechanisms (due to persistent local depolarization), or both. Flaccid paralysis of the muscle ensues.
4. C. Botulinus toxins prevent the release of ACh by irreversibly blocking synaptic vesicle exocytosis.
5. D. Facilitation, manifested as an increase in the amplitude of EPPs in the course of repetitive synaptic activation, is thought to reflect an increase in the transmitter quanta released due to accumulation of residual Ca^{2+} in the presynaptic nerve terminals.

BIBLIOGRAPHY

Kandel, E. R., and Schwartz, J. M. *Principles of Neural Science.* New York: Elsevier Science Publishing Co., 1985.

Katz, B. *Nerve, Muscle, and Synapse.* New York: McGraw-Hill, 1966.

9 Central Synapses

JANUSZ B. SUSZKIW AND MICHAEL M. BEHBEHANI

Objectives

After reading this chapter, you should be able to:

Explain the mechanisms of excitatory and inhibitory postsynaptic potentials

Explain the role of spatial and temporal summation in neuronal excitability

Distinguish between postsynaptic and presynaptic inhibition

List the major putative transmitters and describe their synaptic actions

---

Much of the present knowledge about synaptic mechanisms in the mammalian central nervous system is based on studies on spinal motor neurons, conducted by John Eccles and his colleagues. These studies have shown that the general principles of chemical synaptic transmission at the neuromuscular junction, elaborated on by Katz and his co-workers, also apply to central synapses. However, in contrast to powerful, single excitatory inputs received by the mammalian muscle fibers, central neurons usually receive several dozen to thousands of excitatory and inhibitory synaptic inputs. These synaptic inputs converge on the target neuron (e.g., a spinal alpha motor neuron) from a variety of other neurons in the brain. Because activation of any synapse alone produces a relatively small change in the membrane potential of the postsynaptic cell, several excitatory synaptic inputs must be active in order to bring the neuron to its threshold for firing an action potential. Furthermore, as will be discussed, the effectiveness of

the excitatory synaptic inputs can be diminished or nullified by activation of inhibitory synaptic inputs. Thus, whereas the sole function of the skeletal neuromuscular junction is secure transmission of neural command signals to the muscle to effect force generation, the function of central synapses is the integration of many inputs converging onto the postsynaptic cell.

Activation of excitatory synapses causes the release of transmitters that generate depolarizing excitatory postsynaptic potentials (EPSPs) in the postsynaptic neuron and bring it closer to the threshold for firing propagated action potentials. Activation of inhibitory synapses causes the release of transmitters that impede firing an action potential by hyperpolarizing the neuronal membrane (inhibitory postsynaptic potentials [IPSPs]) and diminishing the effectiveness of EPSPs. The postsynaptic cell integrates synaptic activity by summing the EPSPs and IPSPs, and the action potential is fired only if and when the net depolarization reaches threshold.

139

CENTRAL EXCITATORY AND INHIBITORY TRANSMITTERS

To qualify as neurotransmitter, a substance must satisfy several criteria: (1) enzymatic machinery for synthesizing the neurotransmitter must be present in identified presynaptic neurons and their nerve terminals; (2) the substance must be released by stimulation of presynaptic fibers in a Ca^{2+}-dependent manner; (3) exogenous application of the substance must produce effects on the postsynaptic cell that mimic those produced by neurally released transmitter; and (4) a mechanism for inactivation of the substance (i.e., specific uptake system and degrading enzymes) should be demonstrable. It is frequently technically difficult to satisfy all of these criteria at central synapses. For this reason, one speaks of transmitter "candidates" or **putative neurotransmitters.**

Well over a dozen established or putative transmitters have been identified in brain tissue, and some of the more common ones are listed in Table 9-1. The amino acid **glutamate** is the major excitatory transmitter in brain. **Gamma-aminobutyric acid** (GABA) and **glycine** serve as the major inhibitory transmitters.

Inhibition is particularly important in controlling neuronal activity in brain. The antagonists of inhibitory transmitters, such as picrotoxin and bicuculline which block GABA synapses and strychnine which blocks glycinergic synapses, are well-known convulsants. Tetanus toxin, a protein produced by the anaerobic bacterium *Clostridium tetani*, is taken up by spinal motor neuron terminals and transported to the spinal cord where it blocks the release of glycine. Spread of the toxin throughout the spinal cord and brain can lead to generalized convulsions and death.

Acetylcholine and the monoamines, adrenaline, noradrenaline, dopamine, and serotonin, all serve as neurotransmitters in the central nervous system and may have either excitatory or inhibitory actions, depending on the receptors with which they interact. This brings up an important point; namely, the excitatory versus inhibitory action of transmitter substances is not necessarily vested in the transmitter itself but rather is a function of the receptors and the associated ionic channels that the transmitter activates.

Finally, many small peptide hormones have now been identified that serve as neurotransmitters or neuromodulators. Examples are substance P, somatostatin, vasoactive intestinal polypeptide, and enkephalins. These **neuroactive peptides** frequently coexist with the classic small-molecular-weight transmitters in the same nerve endings and are co-released upon nerve stimulation. Conventional excitatory or inhibitory transmitter actions, such as those produced by glutamate or GABA, have a fast onset and short duration (a few milliseconds). In contrast to the relatively "fast" actions of conventional transmitters, the postsynaptic actions of peptides frequently are slower in onset and longer in

Table 9-1. Major Neurotransmitters and Neuroactive Peptides

Amino acidergic	Gamma-aminobutyric acid
	Aspartic acid
	Glutamic acid
	Glycine
	Histamine
Cholinergic	Acetylcholine
Monoaminergic	Dopamine
	Epinephrine (adrenaline)
	Norepinephrine (noradrenaline)
	Serotonin
Purinergic	Adenosine
	ATP
Neuroactive peptides	Angiotensin
	Cholecystokinin
	Enkephalin
	Neuropeptide Y
	Neurotensin
	Oxytocin
	Somatostatin
	Substance P
	Thyrotropin-releasing hormone
	Vasoactive intestinal peptide
	Vasopressin

duration, thus providing "background" for modulating the actions of fast-acting transmitters.

In the following sections we will consider the mechanisms involved in the generation and integration of excitatory and inhibitory postsynaptic potentials.

GENERATION OF POSTSYNAPTIC POTENTIALS

Excitatory Postsynaptic Potentials

The mechanism of EPSPs (Fig. 9-1) is analogous to that described for the end-plate potential at the neuromuscular junction (see Chapter 8). EPSPs are generated when a transmitter released from a presynaptic axon terminal acts on specific postsynaptic receptors to transiently increase membrane permeability to Na⁺ and K⁺, giving rise to a net inward synaptic current (I_{EPSP}) carried by Na⁺:

$$I_{EPSP} = I_{Na} + I_K = g_{Na}(V_m - E_{Na}) + g_K(V_m - E_K)$$
Eq. 9-1

where I_{Na} and I_K are the Na⁺ and K⁺ currents, g_{Na} and g_K are the Na⁺ and K⁺ conductances, and E_{Na} and E_K are the Na⁺ and K⁺ equilibrium potentials, respectively. In this equation, $g_{Na}(V_m - E_{Na}) > g_K(V_m - E_K)$, until the equilibrium potential (E_{EPSP}) is reached. The E_{EPSP}, that is, the membrane potential (V_m) at which $I_{Na} = -I_K$ and hence net synaptic current (I_{EPSP}) is zero, can be obtained readily by rearranging terms in Equation 9-1 to give an expression analogous to that for E_{EPP}.

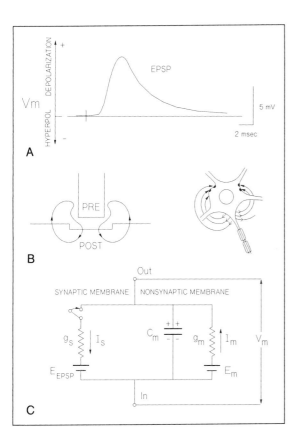

Fig. 9-1. Excitatory postsynaptic potentials (*EPSPs*). (A) Tracing of a single EPSP. EPSPs are depolarizing potentials that are a few millivolts in amplitude and a few milliseconds in duration. The notch preceding the EPSP is the trace of the presynaptic stimulus that initiated transmitter release. (V_m = membrane potential.) (B) Current flow that underlies the EPSP at excitatory synapses. The direction of current is inward through the synaptic channels and outward through the nonsynaptic membrane. Distribution of current flow over the entire neuron is illustrated at the right. (*pre* = presynaptic nerve terminal; *post* = postsynaptic cell.) (C) Electrical equivalent circuit of an excitatory synapse. Activation of receptor-gated channels in the postsynaptic membrane is equivalent to closing the synaptic switch. Current flows inward through the synaptic channels and outward through the nonsynaptic membrane, which is represented by capacitive (lipid bilayer) and conductive (non-gated channels) elements in parallel. g_s and g_m are the synaptic and nonsynaptic conductances, respectively; C_m is the membrane capacitance; E_{EPSP} is the synaptic battery (the equilibrium potential of EPSP); and E_m is the battery of the resting membrane potential. Note that $I_s = -(I_c + I_m)$, where I_s is the synaptic current, I_c is the capacitive current, and I_m is the membrane current. Once the membrane capacitance (C_m) has been fully charged, $I_c = 0$, and $I_s = -I_m$. (Adapted from Eccles, J. C. *The Physiology of Synapses*. Vienna: Springer-Verlag, 1964.)

$$V_m = E_{EPSP} = [g_{Na}/(g_K + g_{Na})] \, E_{Na} + [g_{K'}/(g_K + g_{Na})] \times E_K$$
$$\text{Eq. 9-2}$$

By substituting appropriate numerical values in Equation 9-2, the E_{EPSP} at the neuronal synapse can be shown to be somewhere between -15 and 0 mV.

Since the Na^+ and K^+ fluxes occur through the same synaptic channels, it is convenient to combine Na^+ and K^+ conductances into a single synaptic conductance (g_s) and express the synaptic current as:

$$I_{EPSP} = g_s(V_m - E_{EPSP})$$
$$\text{Eq. 9-3}$$

As discussed in detail in Chapter 8, the inward current through the synaptic channels exits through resistive and capacitive elements of the nonsynaptic membrane, that is:

$$I_{EPSP} = -(I_c + I_m)$$
$$\text{Eq. 9-4}$$

where I_c is the capacitative current and I_m is the membrane current.

Once the membrane capacitance (lipid bilayer) has been charged to its new value (i.e., $I_c = 0$) at the peak of synaptic potential, all inward synaptic current exits as membrane current through the nonsynaptic channels, that is:

$$I_{EPSP} = -I_m$$
$$\text{Eq. 9-5}$$

Starting with Equation 9-5 and the electrical equivalent circuit depicted in Figure 9-1C, one can derive the expression for calculating the amplitude of EPSP and the magnitude of current flow during peak synaptic action, as has been done for the end-plate potential in Chapter 8. More simply, the amplitude of EPSP (V_{EPSP}) can be calculated from I_{EPSP} and the cell's input resistance. From Ohm's law, the nonsynaptic current (I_m) can be written as:

$$I_m = \Delta V_m / R_m$$
$$\text{Eq. 9-6}$$

where R_m is the cell's input resistance and ΔV_m is the change in the voltage across the nonsynaptic

membrane. Combining Equations 9-5 and 9-6, one obtains a simplified expression for V_{EPSP}:

$$V_{EPSP} = \Delta V_m = I_{EPSP} \, R_m$$
$$\text{Eq. 9-7}$$

Equations 9-3 and 9-7 show that the size of the EPSP is determined by the magnitude of the synaptic current and the cell's input resistance. The higher the cell's input resistance, the greater the depolarizing efficacy of the I_{EPSP}. The magnitude of I_{EPSP} itself is determined by the number of synaptic channels opened (i.e., the synaptic conductance) and by the membrane potential (V_m). Membrane potential determines the size of the I_{EPSP} and thus of the V_{EPSP}, through its effect on the driving force ($V_m - E_{EPSP}$) in Equation 9-3.

The amplitudes of EPSPs generated at any single synaptic input are small, usually in the range of 0.5 to 2 mV. This is because very few transmitter quanta are released for each presynaptic action potential at the central synapses. Consequently, several EPSPs must summate in order to depolarize the membrane to the threshold potential for discharging the propagated action potential.

Inhibitory Synaptic Potentials

IPSPs are generated when a transmitter increases the conductance of the postsynaptic membrane to Cl^- or K^+, resulting in the influx of Cl^- or efflux of K^+, respectively. It is intuitively obvious that synaptic currents which consist of Cl^- influx or K^+ efflux produce a deficit of positive charges on the interior leaflet of the membrane, and hence there is membrane hyperpolarization (Fig. 9-2A). Note that the direction of currents associated with IPSPs is the reverse of that for EPSPs, in that the current flows out of the cell through the synaptic channels and into the cell through the nonsynaptic channels (Fig. 9-2B and C).

The inhibitory postsynaptic currents (I_{IPSP}) can be rendered in a fashion analogous to that for I_{EPSP}:

$$I_{IPSP} = g_s(V_m - E_{IPSP})$$
$$\text{Eq. 9-8}$$

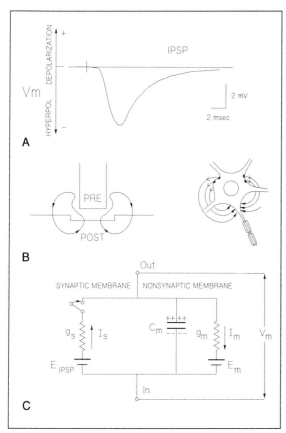

Fig. 9-2. Inhibitory synaptic potentials (*IPSPs*). (A) Tracing of an IPSP. Note that IPSPs are hyperpolarizing potentials, a few milliseconds in duration. The notch preceding the IPSP trace represents a stimulus artifact that provides the timing of the presynaptic stimulus which initiates transmitter release. (B) Direction of current flow during activation of an inhibitory synapse. The current flows outward through the receptor-gated synaptic channels and inward through the nonsynaptic membrane. At right is shown the distribution of currents over the entire area of the postsynaptic neuron. Compare the direction of the current flow with that for activation of the excitatory synapse (see Fig. 9-1B). (C) Electrical equivalent circuit of an inhibitory synapse. Activation of receptor-gated channels in the postsynaptic membrane (closing of the switch) results in outward current flow through the synaptic channels and inward flow through the capacitive and conductive elements of the nonsynaptic branch. Compare this with the equivalent circuit description of an excitatory postsynaptic potential in Figure 9-1C (g_s, g_m = synaptic and nonsynaptic conductances, respectively; C_m =

where the g_s refers to either Cl⁻ or K⁺ synaptic conductance and E_{IPSP} refers to the inhibitory synaptic equilibrium potentials, which are essentially the Cl⁻ or K⁺ equilibrium potentials. Thus, the membrane potential (V_m) during activation of Cl⁻ conductance tends toward the Cl⁻ equilibrium potential of −80 mV. In the case of the K⁺ permeability change, the membrane potential tends toward the K⁺ equilibrium potential of −90 mV.

The inhibitory effect of IPSPs on the cell electrical excitability (e.g., spike generation), however, is due not only to membrane hyperpolarization but also to the decrease in the cell's overall input resistance. This can be understood by examining Equation 9-6. As already discussed, this equation shows that the effectiveness of I_{EPSP} in depolarizing the membrane is a function of membrane input resistance. The depolarizing effectiveness of I_{EPSP} is reduced as the activation of inhibitory synapses and opening of the synaptic K⁺ or Cl⁻ channels, or both, reduces membrane resistance. This, in turn, provides low-resistance conductance pathways for current flow out of the cell, thus, in effect, short-circuiting the I_{EPSP}.

INTEGRATION OF SYNAPTIC INPUTS AND GENERATION OF THE ACTION POTENTIAL

Whether or not a neuron discharges a propagated action potential is determined by the integration of all synaptic inputs active at any given instant in time. Neurons integrate postsynaptic potentials by adding the EPSPs and subtracting the IPSPs from the membrane potential at any instant of time. When and if the resultant potential reaches the threshold value, a propagated action potential is triggered (Fig. 9-3). This action potential is usually triggered at the **axon hillock,** which is the region of the neuron with the lowest

---

membrane capacitance; I_s = synaptic current; I_m = membrane current; E_m = battery of the resting membrane potential; E_{IPSP} = equilibrium potential of the IPSPs.). (Adapted from Eccles, J. C. *The Physiology of Synapses.* Vienna: Springer-Verlag, 1964.)

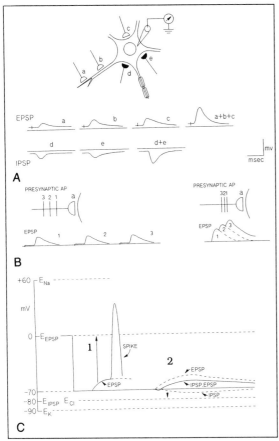

threshold. Because summation of synaptic currents at the axon hillock is the principal determinant of whether an action potential is fired or not, the initial segment is referred to as the **integrative zone of the neuron.**

Spatial and Temporal Summation

The synaptic inputs are integrated by means of spatial and temporal summation. **Spatial summation** occurs when two or more separate presynaptic inputs are activated nearly simultaneously, so that their postsynaptic potentials add together (see Fig. 9-3A). The effectiveness of spatial summation is critically dependent on the space constant of the neuronal membrane. As covered in Chapter 4, the membrane space constant is the distance at which electrotonic potentials decay to $1/e$ or 37% of their amplitude at the point of origin. Therefore, synaptic inputs separated by a distance smaller than the space constant can summate together much more effectively than those separated by a distance exceeding the space constant. In other words, the larger the membrane space constant, the greater the cell's ability to summate the postsynaptic potentials generated at various locations on the neuron.

Fig. 9-3. Integration of postsynaptic potentials. A postsynaptic neuron receiving excitatory (E-synapses a, b, and c) and inhibitory (I-synapses e and f) inputs. Activation of these synapses by presynaptic action potentials causes release of excitatory or inhibitory transmitters, respectively. Postsynaptic potentials are recorded with an intracellular electrode in the postsynaptic neuron. (A) Spatial summation. Individual activation of E-synapses (a, b, and c) results in excitatory postsynaptic potentials (EPSPs) a, b, and c. Simultaneous activation of all three excitatory inputs results in algebraic summation of the EPSPs produced at each synapse, giving a summated $EPSP_{a+b+c}$. Similarly, spatial summation is illustrated for the activation of I-synapses located at different regions of the neuron. (B) Temporal summation occurs when a presynaptic neuron fires at such a frequency that succeeding postsynaptic potentials overlap in time. For example, as shown, the application of sufficiently spaced (separated in time) stimuli to the presynaptic neurons produces correspondingly spaced EPSPs in the

postsynaptic neuron. When the frequency of stimulation is increased so that presynaptic action potentials (*AP*) occur before the EPSPs fully decay to the baseline potential, each succeeding EPSP adds to the preceding depolarization, thus giving a summated $EPSP_{1+2+3}$. (C) Integration of excitatory (*EPSPs*) and inhibitory postsynaptic potentials (*IPSPs*). A cat motor neuron showing the equilibrium potentials for Na^+ (E_{Na}), K^+ (E_K), and Cl^- (E_{Cl}). The resting membrane potential is at -70 mV, and the equilibrium potentials for IPSP (E_{IPSP}) and EPSP (E_{EPSP}) are at -80 mV and 0 mV, respectively. The threshold for neuronal spike generation is at about -50 mV. When the summated EPSP reaches the threshold membrane potential, an action potential is fired (*1*). If E- and I-synapses are activated approximately simultaneously, the IPSPs subtract from the EPSPs so that net membrane depolarization due to EPSP is reduced to below the threshold and action potentials cannot be fired (*2*).

Temporal summation occurs when action potentials in a single presynaptic neuron fire in rapid succession, such that the interval between successive presynaptic action potentials is less than the duration of the postsynaptic potential (e.g., 2–10 msec). In this event, the succeeding postsynaptic potentials add to the preceding ones, with a resultant increase in the overall amplitude of the postsynaptic potentials (see Fig. 9-3B). The effectiveness of temporal summation critically depends on the membrane time constant. The membrane time constant determines the time required for an electrotonic potential to decay to 1/e or 37% of its peak value. The longer the time constant, the longer the duration of the postsynaptic potential, and thus the greater opportunity for summation of successive postsynaptic potentials.

PRESYNAPTIC INHIBITION

Besides postsynaptic inhibition, the probability of a postsynaptic neuron firing an action potential may be decreased by **presynaptic inhibition** of an excitatory synaptic input.

Presynaptic inhibition is exerted at axo–axonic synapses, formed by the axon terminal of an inhibitory neuron onto the axon terminal of excitatory neuron. Activation of this type of synapse results in reduced excitatory transmitter released by the presynaptic action potential in the excitatory neuron (Fig. 9-4). This then leads to a smaller EPSP on the postsynaptic cell, and thus reduced effectiveness of the excitatory transmission.

Presynaptic inhibition tends to be found at sensory inflow points, where it presumably permits the selective "switching off" of some inputs without affecting the responsiveness of the postsynaptic neuron to other synaptic inputs. In the spinal cord, presynaptic inhibition has been documented in the primary afferent fibers from the muscle stretch receptors, and it is exerted by other primary afferents or afferents from mechanoreceptors by means of spinal inhibitory interneurons. In this case, stimulation of the neurons mediating presynaptic inhibition in the mammalian spinal cord causes depolarization of the terminals of the excitatory primary afferent fibers, known as **primary afferent depolarization.** Depolarization of presynaptic terminals reduces the quantity of transmitter released, probably because the amplitude of the presynaptic spike in a partially depolarized terminal is smaller, fewer Ca^{2+} channels are activated, and less Ca^{2+} enters the terminal. Lasting depolarization of the terminal arborizations may also block action potential invasion into the terminals. Primary afferent depolarization appears to be mediated by GABA and is blocked by the GABA-receptor antagonists, picrotoxin and bicuculline. Unlike the fast IPSP, presynaptic inhibition has a relatively long duration and may last up to several hundred milliseconds (see Fig. 9-4).

MODULATION OF SYNAPTIC TRANSMISSION

Facilitation, post-tetanic potentiation, and **synaptic fatigue,** described in Chapter 8, can all occur at central synapses. For example, monosynaptic EPSPs evoked in spinal motor neurons can change according to the frequency of the action potentials in the afferent fibers, and can show frequency-dependent facilitation at moderate frequencies, or depression of transmitter release at high frequencies of firing. Also, when the presynaptic neuron is activated at a high rate and then a monosynaptic EPSP is evoked at a slower rate, the size of the EPSP may increase above that of the control size (i.e., post-tetanic potentiation). The presynaptic modulation of transmitter release at central synapses is of particular interest because alterations in the synaptic efficacy as a function of previous activity are, in effect, examples of short-term synaptic memory.

In addition to the presynaptic modulation, synaptic efficacy can also be altered by postsynaptic mechanisms that lead to short- or long-term modification of the postsynaptic cell excitability. An important example is long-term potentiation. Long-term potentiation is thought to be an important mechanism involved in learning and

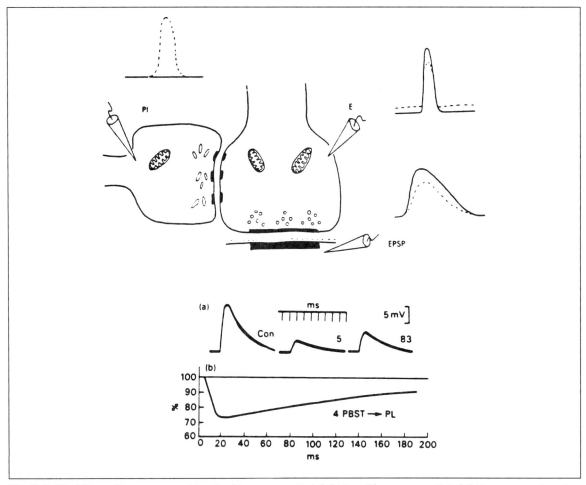

Fig. 9-4. (Top) Synaptic arrangement involved in presynaptic inhibition. The presynaptic inhibitory neuron terminal (*PI*) forms an axo–axonal synapse with the excitatory (*E*) terminal onto the postsynaptic neuron. In the absence of activity in the PI, the presynaptic potential (continuous trace) in the E-terminal is normal in size and generates a maximal excitatory postsynaptic potential (*EPSP*) in the postsynaptic cell. However, when PI input to the E-terminal is active, the presynaptic spike in the E-terminal is reduced (*dashed trace*) so that less transmitter is released from the terminal and consequently the EPSP in the postsynaptic cell is smaller (*dashed line*). (Reproduced from: Willis, W. D., and Grossman, R. G. *Medical Neurobiology.* St. Louis: C. V. Mosby, 1981.) (Bottom) Reduction of the EPSP in motor neurons by presynaptic inhibition. (*a*) EPSPs were recorded from a cat plantaris (*PL*) motor neuron in response to sensory I_a afferent nerve stimulation. The control EPSP (*Con*) is reduced following conditioning tetanus applied to group I afferents from the posterior biceps–semitendinosus (*PBST*) muscle nerve. The records of depressed EPSP were taken at 5 and 83 msec after the conditioning stimulus, and illustrate the relatively long duration of the effect. The time course of presynaptic inhibition of EPSP in the PL produced by application of four conditioning volleys applied to the PBST is illustrated in (*b*). (Reproduced from: Keele, C. A., Neil, E., and Joels, N., eds. *Samson Wright's Applied Physiology.* 13th ed. New York: Oxford University Press, 1982. Original data from Eccles, J. C., et. al, *J. Physiol.* 159:147, 1961).

memory processes in the hippocampus. Finally, as has already been briefly discussed, the responsiveness of the postsynaptic neuron to classic transmitters (those that generate fast, short-lasting postsynaptic potentials) can be altered by neuroactive peptides that produce more slowly developing but long-lasting changes in the membrane potential. These actions are thought to be exerted by second-messenger (e.g., cyclic adenosine 5′-monophosphate)–dependent modifications of ionic channels in the neuronal membrane.

SUMMARY

Postsynaptic potentials are electrotonic potentials that are elicited by a transmitter-induced, 1- to 2-msec change in the postsynaptic membrane permeability to specific ions. These postsynaptic potentials decay passively within 2 to 10 msec, depending on the membrane time constant, and decrement with increasing distance, depending on the membrane length constant. Synaptic signals converging onto a neuron are integrated through summation of the EPSPs and IPSPs, and the action potential is triggered if and when the summated membrane potential reaches the threshold at the initial segment of the axon.

NATIONAL BOARD TYPE QUESTIONS

Select the one most correct answer for each of the following

1. The EPSP:
   A. Is a hyperpolarizing potential
   B. Is always of sufficient amplitude to trigger an action potential in the postsynaptic neuron
   C. Has an equilibrium potential of about −80 mV
   D. Causes depolarization of the postsynaptic membrane
   E. Is associated with an outward current through the synaptic channels
2. The duration of the EPSP of a motor neuron is approximately:

A. 500 msec
B. 100 msec
C. 10 msec
D. 1 msec
E. 0.1 msec

3. An IPSP inhibits a neuron because:
   A. It leads to a reduction in the amount of transmitter released at excitatory synapses
   B. It hyperpolarizes the membrane potential
   C. It changes the threshold of the neuron
   D. It shortens the action potential
   E. It increases the resistance of the membrane
4. Presynaptic inhibition of an alpha motor neuron:
   A. Increases the amplitude of the IPSP of the motor neuron
   B. Has no effect on the amplitude of the EPSP of the motor neuron
   C. Is associated with hyperpolarization of the primary afferent nerve terminals
   D. Is associated with depolarization of the primary afferent nerve terminals
   E. Is due to an increase in transmitter release at inhibitory synapses
5. The integration of synaptic inputs:
   A. Refers to algebraic summation of miniature postsynaptic potentials
   B. Involves summation of EPSPs and IPSPs
   C. Determines the membrane threshold for neuronal action potential
   D. Is not affected by the membrane space constant
   E. Is not affected by the membrane time constant

ANNOTATED ANSWERS

1. D. EPSP is associated with the net inward synaptic Na^+ current, which depolarizes the neuronal membrane.
2. C. The postsynaptic potentials consist of a 1- to 2-msec active phase, followed by a 2- to 10-msec passive decay phase, for a total duration of 5 to 10 msec.

3. B. IPSPs are associated with an influx of Cl⁻ or efflux of K⁺, resulting in hyperpolarization of the membrane. This moves the membrane potential away from the threshold and makes firing of an action potential less likely.

4. D. Depolarization of primary afferent terminals associated with the activation of axo–axonal synapses reduces the amplitude of presynaptic action potentials invading the terminals, or even blocks the presynaptic action potentials. The end result is reduction in the amount of excitatory transmitter released from the primary afferent nerve terminals and reduced EPSP.

5. B. Summation of depolarizing (EPSP) and hyperpolarizing (IPSP) potentials occurring nearly simultaneously determines the net level of the membrane potential at any instant of time. Integration of EPSPs and IPSPs at the axon hillock determines whether an action potential is fired.

BIBLIOGRAPHY

Eccles, J. C. *The Physiology of Nerve Cells*. Baltimore: Johns Hopkins Press, 1957.

Kandel, E. R., and Schwartz, J. M. *Principles of Neural Science*. New York: Elsevier Science Publishing Co., 1985.

III Autonomic Nervous System

10 Peripheral Autonomic Nervous System

EDWARD S. REDGATE

Objectives

After reading this chapter you should be able to:

Compare the structural features distinguishing the somatic and autonomic nervous systems.

Describe how an autonomic reflex pathway is organized at cranial levels and at the thoracolumbar and sacral levels of the spinal cord

Explain how integration of signals may occur at synapses in the ganglia and at neuroeffector junctions

Describe the distinguishing characteristics of the enteric nervous system

Identify the neurosecretory functions of the hypothalamic output to the pituitary gland

Identify the different forms of synergism and antagonism that occur in interactions between the sympathetic and parasympathetic outflows

Compare the classic and modern views of neurotransmission and receptor activation

---

When an action potential is generated in a motor nerve to skeletal muscle, the ensuing events are very predictable. Depolarization of the nerve terminals of a skeletal muscle motor unit results in a release of acetylcholine, which binds to nicotinic cholinergic receptors in the motor end-plate region. The resulting motor end-plate potentials of the skeletal muscle fibers induce action potentials in the skeletal muscle fibers of the motor unit, and these muscle cells shorten. When one considers a comparable sequence of events in the peripheral autonomic nervous system (ANS), although the action potential conduction in the nerve fibers is similar, there are significant differences in each of the other events in the motor pathway that render the outcome less certain. Inasmuch as the ANS plays an important role in

the control of body functions, it is important to understand the differences in the output pathways of the somatic and ANS (Table 10-1).

The peripheral ANS possesses several distinctive features:

1. It consists of craniosacral (hypothalamus, the brainstem, and sacral spinal cord) and thoracolumbar outputs.
2. It innervates smooth muscle, the heart, and glands, as well as the enteric nervous system (ENS) which innervates the gastrointestinal tract.
3. The efferent pathway consists of two neurons in series, so that there is a neuron cell body located along the efferent pathway to the visceral target organs. The efferent pathway to

151

Table 10-1. Comparison of Somatic and Autonomic Nervous System

Somatic Nervous System	Autonomic Nervous System
OUTPUT NEURONS	
Brainstem Spinal cord	Hypothalamus Brainstem Thoracolumbar spinal cord Sacral spinal cord Enteric system
TYPE OF OUTPUT	
Cranial somatic or ventral horn cells of spinal cord to slow or fast skeletal muscle fibers	Hypothalamic secretory neuron Postganglionic neuron Enteric system motor neuron Adrenal medullary chromaffin cell
SECRETIONS FOR TARGET ORGANS	
Acetylcholine	Hypothalamic releasing factors Pituitary hormones Acetylcholine Norepinephrine Nonadrenergic, noncholinergic
POSTSYNAPTIC RECEPTORS	
Skeletal muscle nicotinic cholinergic	Ganglionic nicotinic cholinergic Muscarinic cholinergic Alpha and beta adrenergic Nonadrenergic, noncholinergic
TARGET ORGAN RESPONSE	
Skeletal muscle shortens	Smooth muscle contracts or relaxes Cardiac performance increases or decreases Gland secretion increases or decreases

the gastrointestinal system includes additional neurons in the ENS.

4. The nerve cell bodies in these efferent pathways are located in ganglionated chains lying alongside the vertebral column or in ganglia and plexuses in the abdomen, or close to and even in the walls of the target organ innervated.

5. It plays a major role in homeostasis but also controls nonhomeostatic organs, such as the reproductive system.

6. Unlike the somatic nervous system whose output to skeletal muscle is phasic and undergoes periods of repose, the ANS output to many visceral organs is continuous (tonic).

7. ANS activity changes in anticipation of demands.

8. The output of the ANS is coordinated with somatic activity to provide supportive visceral function (e.g., blood flow).

BRIEF HISTORY OF OUR KNOWLEDGE OF VISCERAL INNERVATION

In dissections performed by early anatomists, the highly visible white fibers linking the visceral organs were named *nerves* (from *nervus,* meaning "sinew"). In 200 A.D., Galen proposed that these sinewy fibers contained channels through which animal spirits flowed from one organ to another, thereby establishing a "sympathy" of interaction among visceral organs. Winslow, in 1732, thought that these visceral "sympathies" were regulated by the chains of ganglia located anterolateral to the bodies of the vertebrae, which he called "the great sympathetics."

In 1916 Gaskell demonstrated three separate outflows to the viscera: (1) a **cranial outflow** that included cranial nerves III, VII, IX, and X; (2) a **thoracolumbar outflow** that connected the spinal cord with the paravertebral ganglionic chain; and (3) a **sacral outflow** that innervated the pelvic viscera and was separate from the thoracolumbar outflow.

Langley, in 1905, gave the name *autonomic* to the visceral innervation, and divided it into: thoracolumbar or sympathetic, craniosacral or parasympathetic, and the enteric system of the gastrointestinal tract. The true status of the ENS is only now becoming widely accepted. It is now recognized that the size of the population of neurons in the ENS rivals that of the spinal cord. Immunohistochemical techniques have revealed an ENS extremely rich in transmitters and modulators, and, from these investigations, it has been reaffirmed that many gastrointestinal reflex pathways lie entirely within the ENS and may function independently of the central nervous system (CNS).

THE PERIPHERAL AUTONOMIC NERVOUS SYSTEM

In the most restricted point of view, the ANS is a system of peripheral nerves that extends from the CNS to the viscera and consists entirely of **efferent neurons** (Fig. 10-1). There are separate outflows from the CNS at the cranial, thoracolumbar, and sacral levels. The outflows consist of preganglionic and postganglionic neurons in series, and the various outflows innervate not only smooth muscles, glands, and the heart but also the enteric system.

In a broader context, the ANS is comparable to the somatic nervous system and includes afferent pathways and integrating mechanisms at all levels of the CNS. For example, visceral reflex pathways are similar to somatic reflex pathways in that they include visceral receptors, afferent neurons, integrating centers in the CNS, and efferent neurons to effectors.

VISCERAL AFFERENT NERVES

Afferent fibers from visceral structures are of great practical significance, as they are responsible for evoking appropriate reflex responses to changes in the internal environment, so it is unfortunate that our knowledge of visceral afferent nerves is less complete than our knowledge of somatic afferent nerves (Table 10-2). Little is known about **visceral receptors,** but many of them appear to be mechanoreceptors, chemoreceptors, or osmoreceptors. The reflex pathways appear to be **multisynaptic,** traveling by way of interneurons in the dorsal horn and connections to preganglionic neurons in the intermediolateral column of the thoracolumbar spinal cord or by ascending spinal pathways to the brainstem and connections via interneurons with preganglionic vagal neurons. Many of these afferent pathways are concerned with the mediation of visceral sensation, such as pain, or with vasomotor, respiratory, gastrointestinal, or other visceral reflexes (see Table 10-2). Some of these visceral afferents play a role in the various reflexes involved in **homeostasis,** but visceral reflexes are not initiated only by visceral afferent fibers, since somatic inputs also may elicit visceral responses, such as vomiting after vestibular stimulation. The **neurotransmitters** released at the synapses of these visceral afferent fibers are not clearly established, but a leading candidate for the mediation of pain is substance P. Somatostatin, vasoactive intestinal peptide, and cholecystokinin are present but are not known to be associated with a particular sensory modality.

Visceral afferent pathways differ from somatic afferent pathways (Fig. 10-2). Afferent nerve fibers from the viscera and blood vessels may be myelinated or unmyelinated and may enter the cerebrospinal axis by different pathways:

1. Afferent fibers from **inner organs** or arising from **large blood vessels,** such as the aorta and its major branches, travel along visceral nerves into the sympathetic trunk and then reach the dorsal root ganglia by way of white or gray rami communicantes.
2. Other visceral afferent neurons from **inner organs** may join somatic spinal nerves and travel to the dorsal root ganglia without traversing sympathetic nerves.
3. Afferent nerves originating from **blood vessels in the extremities** can join peripheral nerves, pass through the rami communicantes into the sympathetic trunks, and enter the dorsal root ganglia.
4. Visceral afferent fibers in the **pelvic nerves**

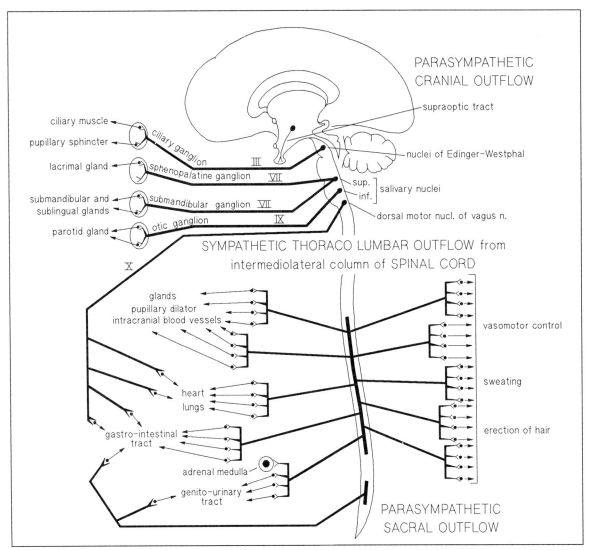

Fig. 10-1. The principles of the origins and distributions of the sympathetic and parasympathetic divisions of the autonomic nervous system. Only one side of the cranial (cranial nerves III, VII, IX, and X) and spinal outflows is shown.

also have their cell bodies in the dorsal root ganglia.

5. Visceral afferents from **taste receptors** (special visceral afferent) and general impulses from the **viscera** (general visceral afferent) enter the brainstem in cranial nerves VII, IX, and X. Visceral afferent fibers in the vagus

nerves (cranial nerve X) have their cell bodies in the ganglion nodosum and enter the brainstem.

The sensory column receiving these visceral afferent fibers in the cranial nerves is the **tractus solitarius,** and this is concerned chiefly with vis-

Table 10-2. Comparison of Inputs to Somatic and Autonomic Nervous System

Somatic	Visceral
Receptors	
Mechanoreceptors of skin	Mechanoreceptors
Thermoreceptors of skin	Chemoreceptors
Cutaneous pain	Osmoreceptors
Proprioceptors	Pain
Perception	
Usually conscious	Usually not conscious (exceptions: taste, pain)

ceral afferent input from the mouth, pharynx, lungs, heart, esophagus, and upper part of the gastrointestinal tract. Unlike the afferent projections in the sympathetic system, these afferents do not seem to be involved in pain sensation. Pain sensation from the viscera of the head project via the trigeminal nerve (cranial nerve V) to the sensory nuclei of the trigeminal nerve, such as pain from tooth pulp or the nasal sinuses.

AUTONOMIC GANGLIA

In the past, synaptic transmission in the autonomic ganglia was regarded as largely a relay process, with obligatory transmission occurring as at the neuromuscular junction. However, this was an oversimplification. The ganglion cells are **multipolar** with long dendrites. Preganglionic axons branch extensively and synapse in several ganglia with a number of ganglion cells. Fiber counts in preganglionic versus postganglionic nerves has shown that the number of postganglionic nerves exceeds that of the preganglionic nerves, so that stimulation of a single preganglionic neuron activates many postganglionic nerves. This is termed **divergence.** The degree of divergence varies and may be as large as 1:200.

On the other hand, there is **convergence** of preganglionic fibers on postganglionic neurons, which is about 50:1 in the human superior cervical ganglia. The preganglionic inputs elicit a variety of fast and slow synaptic potentials in the

postganglionic neuron. Summation of both these potentials, especially the fast excitatory synaptic potentials resulting from the action of ACh on nicotinic cholinergic receptors, is usually sufficient to produce transmission.

ANATOMY OF THE SYMPATHETIC SYSTEM

The sympathetic system (thoracolumbar outflow) originates in the preganglionic neurons lying in the **intermediolateral cell column** of the spinal cord, which extends from the first thoracic segment to the second or third lumbar segment. In these segments, this cell column is prominent and the total number of preganglionic neurons in it exceeds the total number of somatic motor neurons in the ventral horn. The preganglionic nerves are thin, myelinated fibers. They leave the spinal ventral roots in the white rami communicantes to terminate on postganglionic cells. There are four patterns exhibited by the sympathetic preganglionic fibers (Fig. 10-3):

1. The preganglionic fibers exit from the spinal cord in the ventral roots, traverse the white rami communicantes, and synapse in the first sympathetic chain ganglion they enter.
2. The preganglionic fibers pass through the first chain ganglion they enter and travel upward or downward in the sympathetic trunk before synapsing in other sympathetic chain ganglia.
3. The preganglionic fibers pass through the sympathetic chain without synapsing and travel in a visceral nerve, such as one of the splanchnic nerves, to synapse in a prevertebral ganglion.
4. The preganglionic neurons travel as in the third pattern but, instead of synapsing in a prevertebral ganglion, innervate adrenal medullary chromaffin cells, which secrete catecholamines.

Sympathetic fibers that innervate **peripheral structures,** such as sweat glands and blood vessels, exhibit the first pattern. Postganglionic

Afferent Nerves

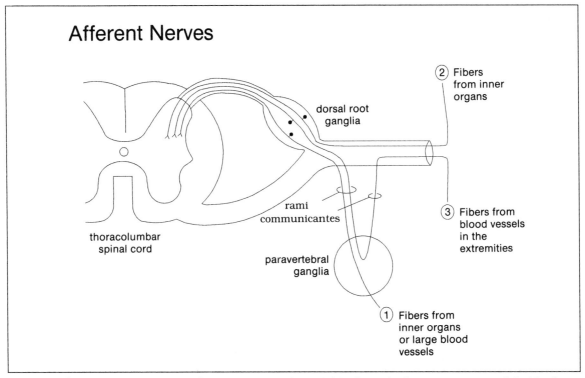

Fig. 10-2. The pathway of the afferent fibers innervating the large blood vessels, the inner organs, and the blood vessels in the extremities.

fibers rejoin the same spinal nerves from which the preganglionic fibers arose and travel with spinal nerves to the periphery. Sympathetic fibers that innervate structures in the **head** or **pelvic region** demonstrate the second pattern. Preganglionic fibers pass up the sympathetic chain from the thoracic region to the cervical ganglia where they synapse with postganglionic neurons to supply structures in the head, or the preganglionic fibers descend in the sympathetic chain to innervate the pelvic ganglia. Sympathetic fibers that innervate the **abdominal viscera** display the third pattern. The three prevertebral ganglia (celiac, superior, and inferior mesenteric) are supplied with preganglionic fibers, primarily through the splanchnic nerves, and postganglionic fibers arise from them to innervate the abdominal viscera and the ENS.

ANATOMY OF THE PARASYMPATHETIC SYSTEM

The **parasympathetic system** differs from the sympathetic system not only in its site of origin (craniosacral) but also in the location of its ganglia. Parasympathetic ganglia tend to be near the organ innervated. For example, the cell bodies of cardiac postganglionic parasympathetic fibers are beneath the epicardium in the wall of the heart itself. Similarly, postganglionic fibers innervating the urinary bladder muscle lie within the detrusor muscle of the bladder.

Even when the postganglionic cells are not actually within the walls of the innervated structure, they are in nearby ganglia. For example, the ciliary ganglion contains cell bodies of parasympathetic postganglionic fibers innervating the cil-

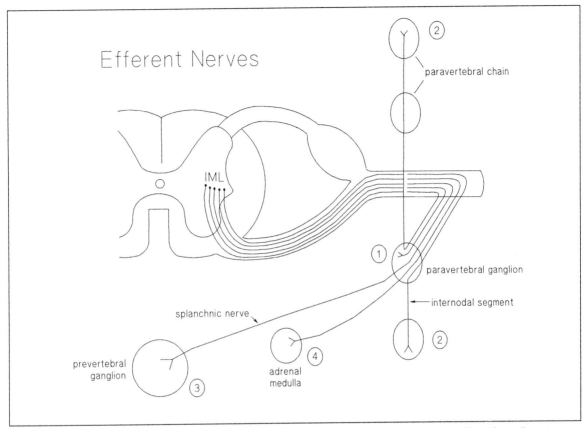

Fig. 10-3. The principles of the outflow and distribution of sympathetic preganglionic nerve fibers from the intermediolateral column of the spinal cord.

iary muscle and circular muscle of the iris. Other examples are the submandibular, otic, and sphenopalatine ganglia, which innervate the salivary glands and the nasal and pharyngeal mucosa. This arrangement requires that the parasympathetic preganglionic fibers are much longer than the parasympathetic postganglionic fibers, which is the opposite of the arrangement generally found in the sympathetic system.

The **parasympathetic** (craniosacral) **outflow** originates in neurons located in cranial nerves III, VII, IX, and X of the brainstem, and in the lateral columns of the spinal cord at sacral segments 2 and 3 (see Fig. 10-1). The divergence along the parasympathetic output is much less

than that of the sympathetic output (1:2), so the response to stimulation in this system is much more localized than the response in the sympathetic system.

THE ENTERIC NERVOUS SYSTEM

The ENS is that system of neurons lying within the walls of the gastrointestinal tract, including neurons in the pancreas and gallbladder. It is an intricate system of neuronal circuits that controls gastrointestinal motility, secretion, and blood flow. There are both internal and external connections of enteric neurons. **Internal connections** are among sensory, associative, and interneurons within the ENS. The **external con-nections** extend from the ENS to the sympathetic ganglia, the dorsal roots, and the vagal sensory ganglia.

THE NEUROENDOCRINE HYPOTHALAMUS

The **hypothalamus** exerts its influence on both the autonomic nervous and endocrine systems, and is the major crossroads of these two regulatory systems. The neural and endocrine systems are intimately interconnected in the hypothalamus. In this section, the ways in which the hypothalamus contributes to endocrine secretion will be considered.

There are numerous examples in the body of how ANS pathways control **exocrine** and **endocrine secretions.** One example is the cranial output to the parasympathetic system that travels over cranial nerves III, VII (intermediate nerve), IX, and X. Many of the nerve fibers included in these cranial nerves synapse with ganglion cells that innervate the glandular epithelium of the salivary, lacrimal, and gastric glands, and control their secretion by the release of ACh and a neuropeptide (e.g., vasoactive intestinal peptide) at their nerve terminals. Another example is the major endocrine output of the sympathetic nervous system. Secretion from the chromaffin cells of the adrenal medulla is controlled by preganglionic sympathetic fibers. In addition to the above examples, certain neurons that originate in the hypothalamic nuclei secrete hormones directly into the blood, and for this reason are called **neurosecretory neurons.**

One group of these hypothalamic neurons, referred to as **magnocellular neurons,** have their nerve terminals in the posterior pituitary. These hypothalamic neurons secrete either vasopressin (antidiuretic hormone) or oxytocin. The nerve terminals of these neurons end at capillaries in the posterior pituitary. The hormones are then released by exocytosis into the blood and act on target organs located at some distance from the pituitary.

Another group of hypothalamic neurons, called **parvocellular neurons,** originates in various nuclei of the hypothalamus, and these neurons terminate on the capillaries of a highly vascular structure at the base of the hypothalamus called the **median eminence.** Most of the blood supply of the anterior pituitary originates in these capillaries and flows through the **hypothalamohypophysial portal system** to supply the various types of secretory epithelium in this structure. This arrangement enables hypothalamic neurons to integrate input from a number of different sources, and then send an endocrine signal, called a **releasing** or **inhibiting hormone,** to the anterior pituitary which stimulates or inhibits secretion of a specific pituitary hormone. The names of several of these hypothalamic hormones are listed in Table 10-3. In each case, the primary action of the hormone is included in its name, but most of them have more than one effect.

ACTIONS AND INTERACTIONS OF THE SYMPATHETIC AND PARASYMPATHETIC SYSTEMS

Some visceral organs receive innervation from both the sympathetic and parasympathetic systems, and, in some of these organs, the impulse discharge from this dual innervation exerts **antagonistic effects,** so that the amount of activity depends on the balance between the discharges over the two autonomic outflows. The innervation of the **pacemaker of the heart** exhibits this antagonistic pattern. The heart receives excitatory innervation from the sympathetic outflow of the upper four or five thoracic segments, which acts on the sinoatrial pacemaker node, the atrioventricular conduction system, the atrial and ventricular myocardium, and the coronary vessels. Excitation of this outflow accelerates the heart rate and increases the force of contraction. The parasympathetic innervation originates in the **medulla oblongata** in the vicinity of the dorsal motor nucleus of the vagus and the nucleus ambiguus, and this influence is exerted on the sinoatrial and atrioventricular nodes and the atrial myocardium. Excitation of this output deceler-

Table 10-3. Hypothalamic Releasing and Inhibiting Hormones

Hypothalamic Hormone	Origin*	Pituitary Anterior Hormones
Corticotropin-releasing factor (CRF)	PVN	ACTH
Growth hormone–releasing hormone (GHRH)	ARC	GH
Growth hormone–inhibiting hormone (or somatostatin) (GHIH)	AHA	GH, TSH
Thyrotropin-releasing hormone (TRH)	PVN	TSH, PRL
Gonadotropin-releasing hormone (GnRH)	POA	LH, FSH
Prolactin-inhibiting hormone (or dopamine) (PIH)	ARC	PRL

*Hypothalamic nuclei.
ACTH = adrenocorticotropic hormone; AHA = anterior hypothalamic arcuate; ARC = arcuate; FSH = follicle-stimulating hormone; GH = growth hormone; LH = luteinizing hormone; POA = preoptic area; PRL = prolactin; PVN = paraventricular; TSH = thyroid-stimulating hormone.

ates the heart rate. When the impulse discharge of the sympathetic system dominates, the heart rate accelerates, but when the parasympathetic system is dominant, the heart rate slows. The impulse discharge of these two systems to the heart pacemaker is tonic. In the absence of innervation, the intrinsic heart rate is about 100 beats per minute, but in the innervated heart in a resting individual, vagal tone is dominant and it decreases heart rate to about 70 beats per minute. When a reflex adjustment of heart rate occurs in response to a decrease in arterial blood pressure, there is a reciprocal change in the discharge frequency in the sympathetic and parasympathetic systems: the sympathetic discharge frequency increases, while the parasympathetic discharge frequency decreases. In view of the reciprocal changes in discharge frequencies, the functional characteristics of the two subdivisions of the ANS are regarded as **synergistic** and not antagonistic.

A second example is found in the innervation of the smooth muscle of the **gastrointestinal tract,** but it is unlike the innervation of the pacemaker of the heart in that the pathways of the two subsystems terminate on different target tissues. The circular layer of smooth muscle in the nonsphincter regions of the gastrointestinal tract is innervated by a parasympathetic outflow that

consists of vagal preganglionic fibers synapsing with enteric neurons. Excitation of this pathway increases the force of contraction of circular smooth muscle during peristalsis. Sympathetic postganglionic fibers from the prevertebral ganglia mainly terminate on the excitatory parasympathetic pathway in the enteric system, instead of circular smooth muscle. Excitation of this pathway inhibits the excitatory parasympathetic outflow to the circular layer of smooth muscle and decreases the force of its contraction. In this instance, the opposing actions of the parasympathetic and sympathetic systems are exerted at different points along the pathway to the smooth muscle target organ, and the roles of the sympathetic and parasympathetic systems are reversed, in that the parasympathetic system is excitatory and the sympathetic system is inhibitory.

In the **pupil of the eye,** another form of interaction occurs between the sympathetic and parasympathetic innervation. In this instance, separate but antagonistic muscles are innervated by the autonomic outflows and both outflows are excitatory. Contraction of the constrictor muscle of the iris is controlled by parasympathetic preganglionic nerve cells in the oculomotor nucleus of the midbrain. These axons synapse with postganglionic neurons in the ciliary ganglion. Excitation of this pathway contracts the constrictor muscle, constricting the pupil and restricting the amount of light admitted. Contraction of the radial muscles of the iris is controlled by sympathetic preganglionic nerve cells in the upper thoracic intermediolateral cell column. These axons synapse with postganglionic neurons in the superior cervical ganglion of the sympathetic trunk. Excitation of this pathway contracts the radial muscle of the iris, dilating the pupil and increasing the amount of light admitted. When the impulse frequency in the sympathetic pathway increases, the pupil dilates, but, when the discharge frequency of the parasympathetic pathway increases, the pupil constricts.

The **salivary glands** are innervated by both sympathetic and parasympathetic fibers. Both cause secretion but activate different subcellular processes. However, the sympathetic innervation plays only a minor functional role.

These examples illustrate that the antagonism between the sympathetic and parasympathetic systems is far less absolute than originally proposed, and they show that the structures which receive autonomic innervation exhibit different forms of antagonistic and synergistic actions. Neither antagonism nor synergism can occur in other structures, as the autonomic effector receives innervation from only one division of the ANS. The effectors that receive only excitatory sympathetic innervation are sweat glands, vascular smooth muscle and the pilomotor muscles of hair follicles. The absence of a common pattern of autonomic innervation of target organs suggests different adaptive arrangements exist in the various tissues and organs of the body.

Probably no example of synergism between the sympathetic and parasympathetic systems has been so often portrayed to illustrate this concept as the **fight-or-flight response** so well described by Cannon. This response is elicited in an emergency situation. In the fight-or-flight response, the sympathetic innervation of many visceral structures becomes dominant, so that there is:

1. Increase in the rate and force of heart contraction
2. Redistribution of blood from the viscera to active skeletal muscles brought about by selective vasoconstriction in the visceral vascular bed and preferential vasodilation in active skeletal muscle
3. Inhibition of gastrointestinal activity
4. Glycogenolysis and lipolysis
5. Adrenal medullary and adrenocorticotropin (ACTH) secretion
6. Dilation of respiratory airways

A CNS neuropeptide may play a role in integrating this assortment of responses because they are elicited by administration of a single peptide (corticotropin-releasing factor) in the brain. This suggests that ACTH release and sympathetic activity in an emergency may be promoted by release of corticotropin-releasing factor into the pituitary portal systems and at various synapses in the brain.

NEUROTRANSMITTERS

Our understanding of the events that occur during transmission of a signal at a synapse between the preganglionic and postganglionic neurons and at the neuroeffector junction is undergoing rapid development, so that we will refer to a **classic** as well as a **modern** view of neurotransmitters of the ANS. In the classic view, **acetylcholine** (ACh) is the transmitter released onto the ganglionic neurons at the synapse between preganglionic and postganglionic neurons, and is responsible for evoking action potentials in these neurons in both the parasympathetic and sympathetic systems. According to the classical view, ACh is the transmitter at neuroeffector junctions in the parasympathetic system, and its action at these sites may be excitatory (e.g., constrictor muscle of the iris, salivary gland secretion) or inhibitory (e.g., cardiac pacemaker). The transmitter at neuroeffector junctions in the sympathetic system is **norepinephrine** (NE) and its action on effectors may be either excitatory (e.g., vascular smooth muscle, cardiac pacemaker) or inhibitory (e.g., bronchial smooth muscle, detrusor muscle). The visceral target organs are integrative and may summate excitatory and inhibitory signals.

A prominent feature of the classic view is **Dale's law,** which states that each neuron contains only one neurotransmitter. In the modern view, Dale's law, as stated above, has been found to be invalid, as numerous instances of the coexistence of two or more chemical messengers at ganglionic synapses and neuroeffector junctions have been discovered. These chemical messengers may modulate synaptic excitability in either an excitatory or inhibitory way and may elicit action potentials. Consequently, a number of these substances are regarded as putative neurotransmitters. Many of these putative neurotransmitters are peptides—enkephalin, neuropeptide Y, vasoactive intestinal peptide, luteinizing hormone–releasing hormone, neurotensin—and they may coexist with the classic neurotransmitters—ACh, NE, serotonin, dopamine—in nerve terminals.

The neurotransmitters involved in the innervation of blood vessels and sweat glands include both the classic neurotransmitters and the neuropeptides. In the classic view, control of vascular smooth muscle regulating blood flow to tissues is regarded as arising through alterations in vasoconstrictor neural activity mediated by NE and the local concentration of vasodilator metabolites. In the modern view, besides the classic adrenergic nerves, vasodilator nerves containing neither NE nor ACh (called *nonadrenergic, noncholinergic nerves*) supply a wide variety of vascular beds and can produce local hyperemia. According to this modern view, vasodilator neurotransmitters include dopamine, ACh, serotonin, purines (adenosine and purine nucleotides), vasoactive intestinal peptides, substance P, neuropeptide Y, calcitonin gene–related peptide, neurotensin, and dynorphin.

For most large terrestrial mammals indigenous to tropical regions, sweating constitutes their major heat loss mechanism. There are two types of sweat glands, eccrine and apocrine. Thermoregulation involves the eccrine type of gland. Eccrine glands are distributed generally over the body surface. They secrete a dilute salt solution and are controlled by sympathetic postganglionic fibers, which unlike other sympathetic innervations release Ach. As a result, if someone is treated wtih a muscarinic cholinergic blocker, like atropine, that individual may lose the ability to regulate heat loss by sweating.

The other type of sweat gland, apocrine, is distributed mainly in the axillary and pubic regions. This type secretes odoriferous substances which may have a social or sexual purpose rather thermoregulation. These glands have adrenergic receptors and appear to be influenced humorally by adrenaline rather than by innervation.

RECEPTORS FOR AUTONOMIC TRANSMITTERS

The concept of **cellular receptors for neurotransmitters** in the ANS was originally suggested by Langley in 1913, and states that a specific "receptive substance" mediates the actions of neurohumors or related drugs. As this concept evolved, a receptor came to be regarded as a specific component of a cell whose interaction with a neurohumor, a hormone, or a drug (agonists) produces a biologic response. The specificity of receptors is remarkable, as even minor differences in configuration (e.g., stereoisomers) of an agonist can lead to marked quantitative differences in the biologic response. Other molecules (antagonists) may interact with the receptor and block the biologic response to an agonist. The action of an agonist and of many antagonists may be completely reversible.

Two of the best understood responses to receptor activation are activation of a **second-messenger system** and change in **ion channel permeability.** In the second-messenger type of system, an extracellular chemical (first messenger) interacts with a receptor in the plasma membrane and, by changing the intracellular levels of second messengers (e.g., cyclic adenosine and guanine 5'-monophosphate, Ca^{2+}, diacylglycerol), causes phosphorylation of specific proteins that may be effectors of the biologic response. The nerve impulse may also stimulate an influx of Ca^{2+} which may regulate phosphorylation of specific proteins. In the channel-forming type of system, the neurotransmitter binds with receptors directly coupled to ion channels and this interaction produces a change in ion permeability. The second-messenger system includes the receptors to NE (alpha and beta) as well as the muscarinic receptor. The channel-forming system includes the nicotinic cholinergic receptors present in ANS ganglia. Receptors differ in their response to drugs, chemical structure, molecular configuration, actions, and location in a cell (e.g., plasma membrane, nucleus). Most receptors were first identified on the basis of their different responses to certain drugs.

Cholinergic Receptors

The effect of ACh released by preganglionic fibers of either the sympathetic or parasympathetic systems can be simulated by **nicotine.** Conversely, the action of ACh released by postganglionic parasympathetic terminals at target organs can be simulated by **muscarine.** The selective action of nicotine and muscarine is convincing evidence for the presence of two receptors for ACh—the nicotinic receptor at ganglia and the muscarinic receptor at target organs. There are drugs that are specific antagonists of nicotinic receptors (hexamethonium) and muscarinic receptors (atropine). Activation of nicotinic receptors in the autonomic ganglia produces a short-lived excitatory response of the ganglionic neurons which is capable of eliciting a nerve impulse in the postganglionic neuron. Activation of muscarinic receptors at neuroeffector junctions may be either excitatory or inhibitory, depending on the effector. At the neuroeffector junctions in smooth muscle, subthreshold excitatory (excitatory junction potentials) or inhibitory (inhibitory junction potentials) effects may be observed that can summate to produce a greater or lesser effect. Summation of junction potentials of like sign increases the excitatory or inhibitory effects, while summation of junction potentials of unlike sign decreases the effect.

The nicotinic cholinergic receptor is a large protein consisting of five subunits (alpha$_1$, alpha$_2$, beta, gamma, delta) that span the cell membrane. The two alpha subunits are the primary recognition sites for ACh. When each alpha subunit is occupied by an ACh molecule, a rapid conformational change of the receptor occurs, so that an aqueous channel is formed which permits fluxes of Na^+ and K^+ (inward Na^+, outward K^+). The duration of the channel opening is about 1 msec. In the presence of an agonist, the frequency of opening of the channels increases while the frequency decreases in the presence of a competitive antagonist. With each channel opening, a square-wave pulse of current carried by the cations crosses the cell membrane.

The sites of action and effects of the cholinergic receptors are summarized in Table 10-4.

Adrenergic Receptors

As with the innervation of target organs by the parasympathetic system, the innervation of target organs by the sympathetic system may exert either excitatory or inhibitory actions. An early explanation of this dual response to adrenergic stimulation, popular in the 1930s and 1940s, was that there were two substances—sympathin E and sympathin I—that mediated these excitatory or inhibitory actions through a single receptor. Contradicting this explanation were findings from experiments that showed the pressor response to intravenous epinephrine could be reversed by certain types of adrenergic antagonists. In 1948, Ahlquist proposed that adrenergic receptors could be divided into two types, alpha and beta, which could be distinguished by their relative sensitivity to adrenergic agents and by the actions of pharmacologic blocking agents. Each of these receptor types has subsequently been divided into two subtypes: alpha$_1$, alpha$_2$, beta$_1$, and beta$_2$.

Alpha receptors are responsible for the contractile responses of the spleen, vascular smooth muscle, sphincter muscles of the gastrointestinal tract and bladder, and radial iris muscle. Antagonists for alpha receptors are: phentolamine

Table 10-4. Cholinergic Stimulation

Tissue	Receptors	Action	Response
Sinoatrial node of the heart	Muscarinic	↓ Depolarization	↓ Heart rate
Postganglionic parasympathetic neurons of the enteric system	Nicotinic	↑ Discharge	↑ Motility ↑ Secretion
Smooth muscle of bronchi	Muscarinic	↑ Constriction	↑ Air flow resistance

(alpha nonselective), prazosin (alpha$_1$ selective), and rauwolscine (alpha$_2$ selective).

Beta receptors are responsible for adrenergic mediation of increased heart rate and force (beta$_1$), relaxation of bronchiolar smooth muscle (beta$_2$), and inhibition of the parasympathetic excitatory input to the ENS innervating gastrointestinal smooth muscle (beta$_2$). Examples of beta-receptor blockers are: propranolol (beta nonselective), metoprolol (beta$_1$ selective), and butoxamine (beta$_2$ selective).

The sites of action and effects of the adrenergic receptors are summarized in Table 10-5.

SUMMARY

The activity of smooth muscle, cardiac muscle, and glands is controlled by a distinctive outflow from the CNS called the *autonomic nervous system* (ANS). This outflow to viscera arises from several levels of the CNS: the hypothalamus, the brainstem, and the thoracolumbar and sacral spinal cord, and is divided into two subdivisions, the sympathetic and the parasympathetic. A distinctive characteristic of these autonomic outflows to the viscera is that there are synaptic connections to neurons that lie outside the CNS in various types of ganglia. Major formations of these ganglia, collectively called the *enteric nervous system* (ENS), reside in the walls of the gastrointestinal tract where they exert control over the vasomotor and secretomotor neurons and external muscle layer motility. Sympathetic and parasympathetic outflows innervate the ENS, but the system is to a large degree independent of the CNS. Within the CNS, there is a multitude of reflex pathways controlling other visceral functions. These reflex pathways consist of afferent neurons from visceral receptors that make synaptic contact through interneurons with autonomic outflows from the craniosacral (parasympathetic) and thoracolumbar (sympathetic) levels of the CNS. The terminations of these outflows overlap in many of the target organs and, as a result, interact in a variety of ways but usually synergistically. Each part of the afferent and efferent pathways is characterized by the presence of specific neurotransmitters. While the major classic neurotransmitters of the ANS are ACh and NE, modern view includes a number of monoamines (dopamine, serotonin, histamine, and ATP) and a growing list of neuropeptides, referred to as the nonadrenergic, noncholinergic system. This variety of neurotransmitters is coupled with different receptors, each of which is characterized by the presence of distinct subtypes, such as alpha$_1$, alpha$_2$, beta$_1$, beta$_2$ adrenergic or nicotinic, and muscarinic cholinergic. As a result, the ability to selectively influence autonomic outflows by treating subjects with selective agonists or antagonists of the various receptors has become a highly specialized science.

Table 10-5. Adrenergic Stimulation

Tissue	Receptors	Action	Response
Heart	Beta₁	↑ Heart rate ↑ Force	↑ Cardiac output
Arteries	Alpha₁	↑ Constriction in skin, kidney, mesentery	↑ Resistance
	Beta₂	↑ Dilation	↓ Resistance
Veins	Alpha₁	↓ Compliance	↑ Venous return
Bronchi	Beta₂	↑ Dilation	↓ Air flow resistance
Liver	Beta	↑ Glycogenolysis	↑ Blood sugar
Fat	Beta	↑ Lipolysis	↑ Blood free fatty acids

NATIONAL BOARD TYPE QUESTIONS

Select the one most correct answer for each of the following

1. Nicotine stimulates receptors on:
   A. The sinoatrial node of the heart
   B. Preganglionic neurons
   C. Skeletal muscle
   D. Bronchial muscle
   E. Sweat glands
2. Muscarine stimulates receptors on:
   A. The atrioventricular node of the heart
   B. Preganglionic neurons
   C. Skeletal muscle
   D. Adrenal medullary cells
   E. Coronary blood vessels
3. Stimulation of the vagus nerve would result in the following:
   A. Increase in stomach motility
   B. Increase in heart rate
   C. Decrease in small intestine motility
   D. Relaxation of bronchial smooth muscle
   E. Salivation
4. Which of the following functions is *not* induced by increased activity of the parasympathetic system?
   A. Constriction of the pupil
   B. Erection of the penis
   C. Reduction of heart rate
   D. Stimulation of gastric secretion
   E. Vasoconstriction of cutaneous vessels
5. The sympathetic division of the ANS is characterized by:
   A. Inhibition at neuroeffector junctions by atropine
   B. Thoracolumbar outflow from the spinal cord
   C. Short postganglionic fibers
   D. Adrenergic preganglionic fibers
   E. The vagus nerve, which is its major component
6. The sympathetic division of the ANS:
   A. Has its cells of origin in the lateral portion of the thoracic and sacral segments of the spinal cord
   B. Generally has short postganglionic fibers
   C. Is concerned primarily with energy conservation and secretion
   D. Utilizes ACh for transmission from postganglionic neurons to sweat glands
   E. Has its cells of origin in the brainstem

ANNOTATED ANSWERS

1. C. The nicotinic type of cholinergic receptor is found on skeletal muscle end-plates.

2. A. The muscarinic type of cholinergic receptor is found on viscera innervated by parasympathetic fibers, such as the sinoatrial and atrioventricular nodes of the heart.

3. A. The vagal output from the enteric system excites myenteric neurons activating circular smooth muscle.

4. E. Cutaneous vasoconstriction is mediated by sympathetic outflow to skin vascular smooth muscle.

5. B. The sympathetic outflow from the spinal cord to peripheral ganglia arises in the thoracolumbar segments.

6. D. ACh neurotransmitter at sympathetic terminals on sweat glands is cholinergic.

BIBLIOGRAPHY

Burnstock, G. The changing face of autonomic neurotransmission. *Acta Physiol. Scand.* 126:67, 1986.

Cooper, J. A., Bloom, F. E., and Roth, R. H. *The Biochemical Basis of Neuropharmacology*, 5th ed., New York: Oxford University Press, 1986.

Gabella, G. *Structure of the Autonomic Nervous System*. London: Chapman and Hall, 1976.

Gershon, M. D. The enteric nervous system. *Annu. Rev. Neurosci.* 4:227, 1981.

Janig, W. The autonomic nervous system. In: Schmidt, R. F., ed., *Fundamentals of Neurophysiology*, 2nd ed. New York: Springer-Verlag, 1978.

Pick, J. *The Autonomic Nervous System: Morphological, Comparative, Clinical and Surgical Aspects*. Philadelphia: J. B. Lippincott, 1970.

11 Central Autonomic Nervous System

EDWARD S. REDGATE

Objectives

After reading this chapter you should be able to:

Describe the operation of the baroreceptor reflex pathways controlling the activity of the sinoatrial and atrioventricular nodes and the atria and ventricles of the heart

Describe the autonomic outflow to vascular smooth muscle

Describe the innervation of pelvic organs by somatic, thoracolumbar, and sacral nerves

Explain how visceral innervation plays a role in controlling genital functions, micturition, and defecation and compare their salient features

NEURAL CONTROL OF CARDIAC FUNCTION AND VASOMOTION

Autonomic Pathways to the Heart and Neural Control of Cardiac Function

The heart rate is regulated by tonic discharge of both divisions of the autonomic nervous system. When this tonic discharge of the vagal innervation is removed, the heart rate increases to a rate greater than the resting rate. On the other hand, when the tonic discharge of the sympathetic innervation is removed, the heart rate slows to less than the resting rate. The heart itself has an **intrinsic rhythm** that is greater than the resting rate. Under resting conditions, there is a dominant vagal tone that causes the heart to beat slower than its intrinsic rhythm.

Sympathetic Innervation

The preganglionic sympathetic outflow to the heart originates in the upper eight thoracic segments of the spinal cord (Fig. 11-1A). The preganglionic fibers emerge in the **white rami communicantes** of the upper thoracic segments and enter the sympathetic trunk. Synaptic terminations of preganglionic cardiac sympathetic fibers may exist in the upper thoracic ganglia, including the stellate ganglia (a fusion of the first thoracic ganglion and the inferior cervical ganglion; rarely the second thoracic ganglion is also fused) and the superior, middle, intermediate (also called *vertebral*), and inferior cervical ganglia. The postganglionic fibers enter a complicated **cardiac plexus** at the base of the heart beneath the arch of the aorta.

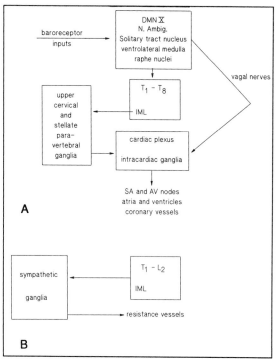

Fig. 11-1. (A) The major anatomic structures of the carotid sinus baroreceptor reflex pathway involved in innervation of the heart. The output from the integrating centers in the brainstem is divided into a parasympathetic pathway of vagal preganglionic fibers and a descending spinal pathway to the thoracic sympathetic preganglionic cells in the intermediolateral column (*IML*) of spinal segments T-1 to T-8. The postganglionic fibers of the sympathetic and parasympathetic outputs innervate the sinoatrial (*SA*) and atrioventricular (*AV*) nodes of the heart. (*DMN X* = dorsal motor nucleus; *N. ambig.* = nucleus ambiguus.) (B) The sympathetic fiber innervation of the vascular smooth muscle of the resistance vessels. This outflow originates in the intermediolateral column of spinal segments T-1 to L-2 and projects to ganglion cells in paravertebral and prevertebral ganglia.

Parasympathetic Innervation

The preganglionic parasympathetic fibers are located in the **dorsal motor nucleus** of the vagus or the **nucleus ambiguus,** or both. The preganglionic parasympathetic fibers travel in the vagus nerves and join the cardiac plexus containing the

synapses between vagal preganglionic and postganglionic neurons.

Innervation of the Sinoatrial and Atrioventricular Nodes

The sinoatrial and atrioventricular nodes are innervated by both sympathetic and vagal fibers from the cardiac plexus, but, although the left and right sympathetic innervations of these structures are relatively uniform, the left and right vagal fibers are not distributed equally to these two nodes. The **sinoatrial node** is primarily innervated by fibers in the right vagus and the **atrioventricular node** is primarily innervated by fibers in the left vagus, so that the dominant effect of right vagal stimulation is cardiac deceleration and the dominant effect of left vagal stimulation retards atrioventricular conduction.

Innervation of the Cardiac Atria and Ventricles

The atria and ventricles are innervated by both sympathetic and vagal fibers from the cardiac plexus, but the vagal innervation is distributed unevenly. Sympathetic activity strongly affects the atria and ventricles. It increases atrial and ventricular contractility, coronary blood flow, and cardiac output. The vagal fibers are much more densely distributed in the atria than in the ventricles. Vagal stimulation causes a small reduction in ventricular contractility and a large reduction in atrial contractility. Because the postganglionic sympathetic and vagal nerve terminals lie very close to one another within the walls of the heart, complex interactions occur between them. There is cholinergically mediated inhibition of norepinephrine (NE) release from the sympathetic nerve terminals and interactions between second-messenger systems of the myocardial cells.

NEURAL CONTROL OF SYSTEMIC BLOOD VESSELS

Perfusion of individual tissues and organs as they perform their diverse tasks requires continuous adjustment of both the **perfusion pressure** and the **resistance to flow** of each vascular bed. To maintain a constant systemic arterial pressure in the face of alterations in resistance of various vascular regions, the resistance in other vascular beds must be changed and cardiac output adjusted in a compensatory direction. These changes in resistance are primarily due to adjustments in the caliber of the arterioles. This is accomplished by the central nervous system (CNS), which integrates information from both the **baroreceptors** and the CNS so that the autonomic output to the cardiovascular system may be apprised of current and anticipated requirements. While noradrenergic nerves play a key role in the reflex regulation of the resistance vessels, certain vasomotor nerves may release transmitters other than NE; these include acetylcholine (ACh), ATP (purinergic), indolamines, and peptides.

CENTRAL BARORECEPTOR INPUT TO THE INTERMEDIOLATERAL CELL COLUMN

Baroreceptor input from the carotid and aortic regions projects to neurons in the vicinity of the **solitary tract nucleus** (see Fig. 11-1A). Solitary tract nucleus outflow is involved in controlling several descending bulbospinal pathways that converge on preganglionic sympathetic neurons in the **intermediolateral cell column.** There is evidence that the activity of one major excitatory bulbospinal path to this cell column arising in the **ventrolateral medulla** is restrained by gamma-aminobutyric acid interneurons located between the solitary tract nucleus and the ventrolateral medulla. Other descending bulbospinal pathways appear to be serotonergic; they arise from the raphe nuclei and either excite the sympathetic preganglionic outflow or inhibit it at the spinal cord by means of interneurons. In addi-

tion, there is evidence of suprabulbar components in the baroreceptor reflex pathway and these include structures in the midbrain and hypothalamus.

AUTONOMIC NERVES TO BLOOD VESSELS

Adrenergic Innervation

The sympathetic nervous system innervates all blood vessels, including arterioles and veins (see Fig. 11-1B). Postganglionic sympathetic nerve terminals form a network of **unmyelinated varicose fibers** that are in close apposition to the smooth muscle of the blood vessels. NE is released from the varicosities as they are depolarized during the passage of action potentials. The NE binds to vascular smooth muscle cell alpha-adrenergic receptors which initiate contraction. In many vascular beds, the arterioles are endowed with both alpha and beta adrenergic receptors. Stimulation of the beta adrenergic receptors causes vasodilation, but these receptors are not close to sympathetic nerve endings and thus the vasoconstrictor effect of the NE released from sympathetic nerve fibers predominates.

Cholinergic Innervation

The vascular smooth muscle in skeletal muscle appears to be innervated by postganglionic sympathetic fibers that release ACh instead of NE. This innervation does not possess the tonic activity of the adrenergic innervation but is activated during emotional states such as rage or fear (defense reaction or fight-or-flight reaction). In certain organs, there is parasympathetic cholinergic innervation of resistance vessels (salivary and sweat glands) and release of ACh is accompanied by vasodilation, but it is difficult to distinguish between a direct vasodilator effect from ACh and vasodilation secondary to increased tissue activity. However, a parasympathetic cholinergic arteriolar dilation is responsible for the expansion of venous sinuses in the erectile tissue of the penis.

Nonadrenergic, Noncholinergic Innervation

The view that vasodilation which is not mediated by adrenergic nerves is mediated by cholinergic nerves is now being challenged by evidence indicating that many of these responses are mediated by substances such as peptides, purines, and indolamines (i.e., the **nonadrenergic, noncholinergic system**). These substances are found in nerves supplying blood vessels, including the cerebral arteries, salivary gland vessels, cutaneous blood vessels, and penile vessels. It appears that innervation by classic transmitters cannot account for all the vascular responses that follow stimulation or denervation of sympathetic innervation. In addition, there is evidence that the role of ACh in vasodilation may be indirect, in that it acts prejunctionally on adrenergic neurons to inhibit NE release or on endothelial cells to produce a relaxing factor called *endothelium-dependent relaxing factor.*

INNERVATION OF THE PELVIC ORGANS

Motor Innervation

Pelvic organ function depends on innervation by the sympathetic (T-10–L-2), parasympathetic (S-1–S-4), and somatic (ventral horn cells) outflows (Fig. 11-2). This innervation controls smooth muscle, striated muscle, and glandular secretions involved in genital function, micturition, and defecation. The **sympathetic preganglionic outflow** emerges from the lumbar spinal cord, passes through the sympathetic chain ganglia, and then synapses with ganglion cells in the pelvic plexuses. The outputs from the ganglion cells travel in the hypogastric, pelvic, vesical, and enteric plexuses to innervate the smooth muscle and glands of the urogenital organs, the rectum, and the internal anal sphincters. **Parasympathetic preganglionic axons** from the intermediolateral region of the sacral spinal cord pass through the pelvic nerves to ganglion cells in the pelvic plexuses, which innervate the urogenital and anorectal structures. **Sacral somatic path-**

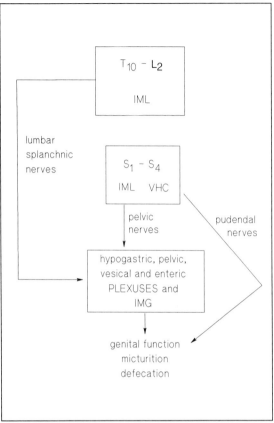

Fig. 11-2. The autonomic innervation of the pelvic organs. The sympathetic preganglionic outflow originates in the intermediolateral column (*IML*) cells of spinal segments T-10 to L-2. The parasympathetic preganglionic outflow originates in the lateral column cells of spinal segments S-1 to S-4 and passes in the pelvic nerves to ganglion cells in the pelvic organs. The pudendal nerves carry the somatic innervation. (*VHC* = ventral horn cell [somatic motor]; *IMG* = inferior mesenteric ganglion.)

ways from sacral ventral horn cells travel in the pudendal nerves to innervate the bulbocavernosus, ischiocavernosus, and the external urethral and external anal sphincter muscles.

Afferent Innervation

Afferent axons from the pelvic organs travel in the **pelvic, hypogastric,** and **pudendal nerves.**

These afferent fibers pass through the dorsal roots and enter the dorsal horn where they terminate primarily on the lateral side in the intermediolateral region. A number of neuropeptides, including vasoactive intestinal peptide, substance P, cholecystokinin, calcitonin gene–related peptide, and dynorphins, are present in these afferent fibers.

GENITAL FUNCTION

The sexual response of the male consists of several successive phases: erection of the penis, emission of semen from the cauda epididymidis into the posterior urethra, and ejaculation of semen from the penile urethra. Similar vascular and secretory responses are present in both sexes. The genital reflex pathways mediating these responses include the sympathetic, parasympathetic, and pudendal innervations.

Erection

Inputs evoking erection arise at both spinal and forebrain levels. At the **forebrain level,** erections induced by psychogenic stimuli involve descending pathways from an integrating center in the anterior hypothalamus. At the **spinal level,** afferent pathways from the penis and other genital regions which excite erection travel in the pudendal nerves to make reflex connections in the sacral spinal cord. Excitation of the parasympathetic output to the erectile tissue induces vasodilation and engorgement of the genital erectile tissue. Two putative neurotransmitter mechanisms, cholinergic and vasoactive intestinal peptide, are regarded as cotransmitters of this response.

Emission and Ejaculation

Reflex activity in the sympathetic outflow from an integrating center in the thoracolumbar spinal cord: (1) elicits contractions of the smooth muscles of the epididymis, vas deferens, seminal vesicles, and prostate; (2) propels semen into the posterior urethra, and at the same time (3)

causes contraction of the internal sphincter of the bladder to prevent reflux of secretions into the bladder. After emission, rhythmic contractions of the bulbocavernosus and ischiocavernosus striated muscles which compress the urethra provide the propulsive force of ejaculation. The afferent and efferent pathways controlling ejaculation travel in the **pudendal nerves.**

MICTURITION

Bladder function is divided into two relatively discrete phases: bladder filling and urine storage, and bladder voiding. **Bladder filling** and **urine storage** require: accommodation of increasing volumes of urine at low pressure and closure of bladder sphincters during the increased filling. **Bladder voiding** requires: contraction of bladder smooth muscle (detrusor muscle) and opening of the internal smooth and external striated muscle sphincters.

The neural mechanisms regulating micturition are operating in either the storage or voiding mode. In each mode there is reciprocal activity of the bladder and sphincters. During storage, the bladder is relaxed and the sphincters are contracting tonically. During voiding, the bladder contracts and the sphincters relax. Tension receptors in the bladder wall signal filling of the bladder. At low filling, there is low-frequency afferent activity in the pelvic nerves and bladder reflexes are in the storage mode.

During storage (Table 11-1), sacral ventral horn cells (**Onuf's nucleus**) are tonically active and cause contraction of the external striated muscle sphincter by means of axons in the pudendal nerve. The sympathetic output from intermediolateral column cells in the lumbosacral spinal cord travel in the **hypogastric nerve** and tonically excite the smooth muscle of the internal sphincter (alpha receptors) and tonically inhibit detrusor smooth muscle (beta receptors). The micturition center in the **dorsomedial pons** permits passive distention of the bladder during filling. When urine accumulation is 250 to 500 ml, afferent activity in the pelvic nerves increases to a high level, and bladder reflexes, if unopposed by

Table 11-1. Micturition

Operating Mode	Level of Integration	Efferent Nerve	Effector Organ	Muscle Receptor	Response of Muscle
Storing	Sacral VHC	Pudendal	External sphincter	N	Tonically excited
	IML T-10–L-2	Hypogastric (symp.)	Internal sphincter	α	Tonically excited
	IML T-10–L-2	Hypogastric (symp.)	Detrusor	β	Tonically inhibited
	Sacral IML	Pelvic (para-symp.)	Detrusor	M	Inactive
Voiding	Sacral VHC	Pudendal	External sphincter	N	Relaxes
	IML T-10–L-2	Hypogastric (symp.)	Internal sphincter	α	Relaxes
	IML T-10–L-2	Hypogastric (symp.)	Detrusor	β	Inactive
	Sacral IML	Pelvic (para-symp.)	Detrusor	M	Phasically excited

VHC = ventral horn cells; IML = intermediolateral column cells; N = nicotinic cholinergic receptor; M = muscarinic; α = alpha adrenergic receptor; β = beta adrenergic receptor.

forebrain influence, operate in the voiding mode. In the voiding mode (see Table 11-1), the high-frequency afferent activity in the pelvic nerves excites a **spinobulbospinal reflex** pathway to the micturition center in the dorsomedial pons. The dorsomedial pontine neurons project directly to the sacral spinal cord. Excitation of this pathway elicits detrusor muscle contraction and relaxation of the sphincter muscles.

DEFECATION

The external anal sphincter is under voluntary control and consists of striated muscle that is tonically contracted. The rectum is usually empty, but distention of the rectum initially increases the tone of the external anal sphincter and produces subjective rectal sensations. The internal anal sphincter is made up of smooth muscle, which is part of the circular layer of muscle of the rectum. This internal sphincter is not under voluntary control and relaxes when the rectum is distended. The mechanism of reflex relaxation is believed to depend on the enteric system of the rectum.

Distention of the rectum initiates reflex activity in the sacral spinal cord and sends impulses in **ascending sensory pathways** to the forebrain which reach consciousness and permit voluntary control over the spinal reflexes. If inhibitory signals are not sent by the forebrain mechanisms, distention of the rectum is followed by relaxation of the internal and external anal sphincters.

The **pattern of reflex activity** controlling defecation appears to be similar to that controlling micturition. The reflex pathway responsible for tonic innervation of the external anal sphincter consists of (1) afferent fibers to receptors in the anal and perianal skin region; (2) integration in Onuf's nucleus in the ventral horn of the sacral spinal cord of sensory input and descending signals from the forebrain; and (3) somatic output in the pudendal nerves to the striated muscle of the external sphincter. The tonic reflex activity in this pathway may be overruled by a descending signal from brain structures regulating voluntary defecation.

The **enteric neurons** controlling the internal anal sphincter are influenced by sympathetic and parasympathetic outputs. The sympathetic outflow is from the lumbar spinal cord. After synapsing in the inferior mesenteric ganglion, it travels in the hypogastric nerves to the enteric plexus. This sympathetic pathway activates noradrenergic neurons in the enteric plexus that excite tonic contraction of the internal anal sphincter by acting on alpha receptors of the sphincter smooth muscle. The parasympathetic outflow from the intermediolateral region of the sacral spinal cord synapses in the enteric plexus with the nonadrenergic, noncholinergic neurons that inhibit internal anal sphincter tone but are not active tonically.

Defecation involves: (1) the coordinated inhibition of the somatic motor output in the pudendal nerves to the external sphincter; (2) excitation of parasympathetic output to the enteric plexus neurons activating rectal and colonic smooth muscle; (3) inhibition of tonic sympathetic tone to the internal sphincter; and (4) excitation of the parasympathetic outflow to the nonadrenergic, noncholinergic neurons which relax internal sphincter tone.

SUMMARY

The autonomic outflows to the cardiovascular system, the gastrointestinal tract (Chapter 42), and the pelvic organs are among the most significant in the body. Autonomic nerve discharges control heart rate and contractility as well as vasoconstriction of arteriolar smooth muscle. The sympathetic and parasympathetic outflows control heart rate by releasing NE and ACh at the sinoatrial and atrioventricular nodes of the heart, and they control myocardial contractility by releasing these neurotransmitters at the myocardial cells. In contrast, the sympathetic outflow alone innervates the vascular smooth muscle of resistance vessels controlling systemic arterial pressure. There are additional autonomic innervations controlling blood flow in specialized regions where diverse neurotransmitters are released by sympathetic or parasympathetic nerve terminals. While many parts of the CNS play a role in regulating systemic blood pressure as well as blood flow in specialized regions, a major integrating center for reflex control lies in the medulla and receives multiple inputs, including inputs from baroreceptors. The synaptic connections of this integrating center are found in the solitary tract and ambiguus nuclei and in the ventrolateral medulla. These structures give rise to the vagal outflow to the heart and to the descending bulbospinal pathways controlling sympathetic innervation of the heart and blood vessels. In the lower abdomen, the smooth muscle and glandular tissues of the pelvic organs are innervated by sympathetic and parasympathetic outflows, while the striated muscle of the anal and urethral sphincters and the striated muscles involved in ejaculation are innervated by somatic nerves. There are spinal and spinobulbospinal reflex pathways involving both subdivisions of the autonomic nervous system, and these control genital function, micturition, and defecation. These reflex pathways are highly regulated by descending pathways from the forebrain.

NATIONAL BOARD TYPE QUESTIONS

Select the one most correct answer for each of the following

1. Sympathetic and parasympathetic innervations of the sinoatrial node of the heart differ from one another in that:

   A. Preganglionic sympathetic fibers innervate the cardiac plexus

   B. Parasympathetic postganglionic fibers are relatively long

   C. Parasympathetic preganglionic neuron cell bodies are clustered together in the paravertebral ganglia

D. The sinoatrial node is primarily innervated by fibers of the right vagus nerve

E. Sympathetic fibers are more densely distributed to the atria than to the ventricles

2. Adjustments in the caliber of the arterioles for the reflex control of blood pressure:

A. Is controlled by preganglionic neurons originating in the dorsal motor nucleus of cranial nerve X

B. Is mediated by postganglionic sympathetic fibers that release ACh

C. Is mediated by alpha$_1$ receptors on vascular smooth muscle

D. Is mediated by beta$_2$ receptors on vascular smooth muscle

E. Is limited to reflex pathways that lie in the lower brainstem

3. The most significant effect of the parasympathetic system on systemic circulation is its effect on:

A. Arteriolar resistance

B. Venous blood volume

C. Heart rate

D. Ventricular contractility

E. Capillary filtration

4. The transmitter released by sympathetic nerve terminals at the vascular smooth muscle of arterioles is:

A. Epinephrine

B. NE

C. Histamine

D. Serotonin

E. Gamma-aminobutyric acid

5. The pattern of reflex activity controlling defecation appears to be similar to that controlling micturition in that:

A. The external urethral and anal sphincters are phasically excited

B. The afferent fibers enter the spinal cord at T-2–T-4

C. Afferent activity from the rectum and bladder excites spinobulbospinal reflex pathways

D. The somatic output to the striated muscle sphincters travels in fibers of the pudendal nerves

E. The enteric nervous system controls detrusor muscle and rectal contraction

---

ANNOTATED ANSWERS

1. D. The sinoatrial node is primarily innervated by the right vagus nerve while the atrioventricular node is primarily innervated by the left vagus nerve.

2. C. Resistance vessel constriction is mediated by stimulation of the alpha$_1$ adrenergic receptors by NE.

3. C. The right vagus nerve exerts a powerful effect on heart rate.

4. B. The major transmitter at sympathetic nerve terminals in vascular smooth muscle is NE. Currently, the role of nonadrenergic, noncholinergic transmitters is being explored.

5. D. The striated muscle sphincters are innervated by ventral horn cells in the sacral spinal cord (Onuf's nucleus), and the fibers exit in the pudendal nerves.

BIBLIOGRAPHY

Brodal, A. *Neurological Anatomy*, 3rd ed. New York: Oxford University Press, 1981.

de Groat, W. C., and Booth, A. M. Autonomic systems to bladder and sex organs. In: Dyck, P. J., Thomas, P. K., Lambert, E., and Bunge, R., eds., *Peripheral Neuropathy*, 2nd ed. Philadelphia: W. B. Saunders, 1984.

Johnson, R. H., and Spalding, J. M. K. *Disorders of the Autonomic Nervous System*. Oxford, London, Edinburgh, Melbourne: Blackwell Scientific Publications, 1974.

Nishi, S. Cellular pharmacology of ganglionic transmission. In: Narahashi, T., and Bianchi, C. P., eds., *Advances in General and Cellular Pharmacology*. New York, London: Plenum Press, 1976, Pp. 179–245.

Schuster, M. M., and Mendeloff, A. I. Motor action of rectum and anal sphincters in continence and defecation. In: Code, C. F., and Heidel, W., eds., *Handbook of Physiology* (Section 6: Alimentary Canal. Volume IV: Motility). Washington, D.C.: American Physiological Society, 1968, Pp. 2121–2145.

IV Muscle Physiology

12 Muscle: Overview of Structure and Function at the Cellular Level

RICHARD J. PAUL AND JUDITH A. HEINY

Objectives

After reading this chapter, you should be able to:

Describe the sequence of events that occur at the cellular level and the relevant cellular structures involved in muscle contraction, from excitation through excitation–contraction coupling, the production of force by filament interaction, and the mobilization of energy metabolism

Draw a diagram of a skeletal muscle, showing the repeating subunits

Explain the difference between a muscle fiber and a myofibril

Define and draw a sarcomere, showing the arrangement of the thick and thin filaments

Describe the relationship between muscle length, sarcomere length, and isometric force

Cite the evidence for constant filament length and relative sliding of interdigitating filaments in contraction, known as the *sliding-filament model*

Describe the relationship of the force versus muscle length curve to the sliding-filament model

Explain how a muscle is stimulated to contract

Identify the initial membrane stimulus that triggers contraction in striated muscle

Describe how the membrane stimulus is conveyed to the filament system and converted to mechanical force

Describe the intermediate steps and the chemical second messengers involved in the process of excitation–contraction coupling

Explain the role of the sarcolemma, the transverse tubule system, the sarcoplasmic reticulum, and the regulatory proteins of the filament system in excitation–contraction coupling

List the features common to all muscle systems and explain the role of Ca^{2+} and ATP in these systems

---

Muscle physiology has long fascinated both the scientist and the lay person for a variety of reasons, the most obvious of which is that we are mostly muscle. Approximately 40 percent of our body weight is skeletal muscle; if one includes cardiac and smooth muscle, the total is closer to 50 percent. Consequently, muscle as an organ system plays many roles in physiological homeostasis in addition to its major function in motility. For example, muscle plays an important role in the metabolism of the whole organism: at rest, approximately 30 percent of basal metabolism is devoted to muscle, and during strenuous exercise up to 90 percent of metabolism can be directed to muscle function. Muscle also plays a major role in temperature regulation. Because approximately 50 percent of the energy mobilized for muscle contraction is degraded to heat, muscle heat production can be a significant load on the temperature regulation system. Finally, but not exclusively, muscle is a major site for storage and mobilization of glucose and such important body electrolytes as H^+, K^+, and Mg^{2+}.

While muscle's roles as an organ system should be kept in mind, the major focus of this section will be on the primary function of muscle, that of serving as the engine for the direct conversion of chemical energy to mechanical work. As a chemicomechanical energy converter, muscle is one of the most efficient systems known and one that has been the subject of extensive research. Due to its highly organized and repeating substructure, muscle is relatively easily studied. It has provided one of the best biological systems for understanding the relation between structure and function at the molecular level.

In this section we will study the relationship between muscle structure and function, focusing on the following key questions: (1) What are the mechanisms that underlie the generation of force and the ability to shorten? (2) What are the mechanisms that underlie the control of contractile activity? (3) How is chemical energy, in the form of ATP, provided by metabolism and transduced to mechanical work, and how is metabolism coordinated with the energetic requirements of muscle activity?

This chapter will examine muscle structure and function at the cellular level and serve as a general overview. In Chapter 13, there will be further discussion of structure–function relationships that exist at the molecular level. Chapter 14 will focus on the variability and flexibility of different muscle types to meet a variety of necessary mechanical tasks. Chapter 15 will go into further depth on the special features of excitation–contraction coupling in cardiac and smooth muscle.

MUSCLE STRUCTURE

The description throughout this chapter will pertain to generalized striated muscle. There are many aspects shared by different skeletal muscles as well as cardiac and smooth muscle. The differences in muscles with regard to their physiologic function will be considered in Chapter 14.

Muscle structure can be divided into two important systems: (1) a **filament system,** which underlies the mechanochemical energy conversion; and (2) an interrelated **membrane control system,** consisting of the sarcolemma (cell membrane), transverse-tubule network, and the sarcoplasmic reticulum, which together control contractile activity. The filament system gives muscle its most obvious structural characteristic, that of **banding** or **striations.** Figure 12-1 ilustrates muscle structure at various levels of resolution. The structure consists of both longitudinal and transverse repeating units, which constitute the organization of the filament system. Though most of the nomenclature arose from observations made by optical microscopy, the structure is probably most easily understood at the electron micrograph level. At this level, the longitudinal repeating unit of muscle, called the **sarcomere,** can be clearly seen to consist of two intersecting filament lattices. The thick (14 nm in diameter) filaments contain **myosin** and the thin (7 nm in diameter) filaments contain **actin.** The sarcomere is defined as the region between repeating structures, called **z-bands,** to which the thin filament lattices are anchored. At the rest length of the muscle, the sarcomere is about 2.2 μm long.

The terms **A-band** and **I-band** also arose from studies at the light microscope level. One of the bands is optically **anisotropic,** or nonhomogeneous with respect to its light-refracting properties, and the other is **isotropic,** or homogeneous. These optical properties arise because the I-band contains only thin filaments, while the A-band contains the overlap region of both thin and thick filaments. The A-band is 1.6 μm long and, as seen in Figure 12-1, is equal to the length of the thick filament. A less dense cen-

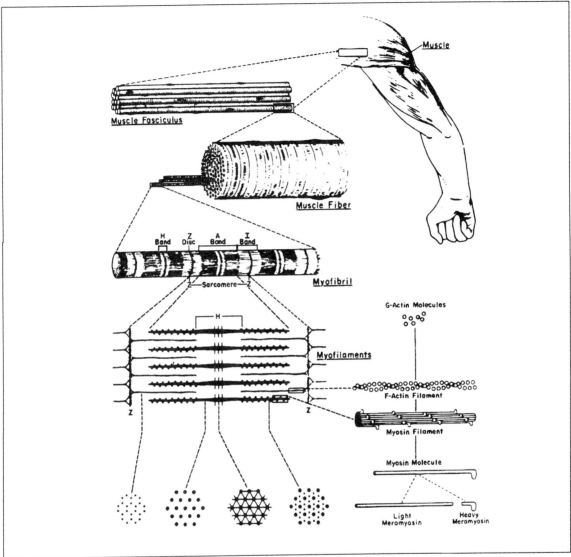

Fig. 12-1. Skeletal muscle structure. The whole muscle is constructed of fundamental repeating units. A muscle cell or fiber is composed of myofibrils, and its fundamental repeating unit is the sarcomere. All views are longitudinal except the cross-sections of the sarcomere shown in the bottom left drawings. (From Bloom, W., and Fawcett, D. W. *A Textbook of Histology.* Philadelphia: W. B. Saunders, 1968. Drawn by Sylvia Colard Keene.)

tral region of the A-band can be distinguished at higher resolution. This is known as the **H-zone** and corresponds to a region in which the thick filaments are not overlapped by thin filaments. The I-band varies in length because of interdi-gitation of the filaments. The thin filaments are 1 μm long on each side of the z-band.

Sarcomeres are organized into **myofibrils** (1 μm in cross-sectional diameter), which in turn form the repeating unit of the muscle cell, sometimes known as a **muscle fiber.** Muscle fiber diameters range from 20 to 100 μm and fiber length can be as long as the macroscopic length of the entire muscle. Single muscle cells can thus be centimeters long, and this likely accounts for their characteristic multinucleate structure. Muscle cells are organized into fiber bundles. At this level of organization, a **motor unit** consists of a motor neuron and all its innervated muscle fibers.

MUSCLE MECHANICS

Our next step toward understanding muscle function in terms of structure will be to describe the mechanical characteristics of muscle fibers, studies collectively known as muscle mechanics. The mechanical constraints under which muscles perform determine their function. The two most common are **isometric,** whereby the total length of the muscle is held constant, and **isotonic,** in which the load on the muscle is constant. Although most in vivo muscle contractions occur under mixed conditions, these imposed conditions, even if somewhat artificial, are important to our understanding of the muscle structure and function relationship.

The time course of the force generated by a muscle fiber under isometric conditions following stimulation is depicted in Figure 12-2.

Fig. 12-2. Effects of stimulus frequency on isometric force in a single muscle fiber cell. The additive nature of the force response is called *temporal summation.*

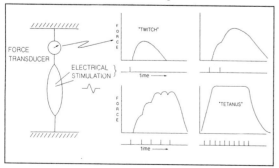

A single electric stimulation, sufficient to elicit a propagated action potential, provides a response known as a **twitch.** As seen in Figure 12-2, increasing the stimulus frequency leads to summation of the force responses until, at a certain frequency characteristic of the muscle type, a smooth fused response is obtained. This response is known as a **tetanus,** and, because in this case isometric conditions exist, this is an isometric tetanus. From our own experience of muscle contraction, we know that smooth contractions occur at submaximal levels of force. As shown in Figure 12-2, this does not occur at the single fiber level. We will return to this topic in Chapter 14, where behavior of muscle at the tissue level is discussed.

The next stage in correlating muscle function to structure involves understanding the link between the force generated under isometric tetanic conditions and the length of the muscle fiber. The experimental design for determining the isometric force–length relationship shown in Figure 12-3 is similar to that in Figure 12-2, but the apparatus permits the length of the muscle to be adjusted. In the absence of stimulation, increasing the length of the muscle causes an exponential increase in force. This behavior is similar to that of a rubber band, and the relationship between isometric force and muscle under these conditions is known as the **passive force–length curve.** This protocol can be repeated so that the muscle is stimulated tetanically at each length. Activation of the muscle produces a force that adds to the passive force at each length. This relationship between the total force and length is known as the **total force–length curve.** Subtracting the passive force from the total isometric force at each length yields the **active force–length relationship,** or that force attributable to the stimulation or activation of the muscle. This relationship is central to understanding muscle structure and function, and is unusual in that active force decreases at both long and short lengths. For most materials, such as rubber bands or springs, force simply increases with lengthening.

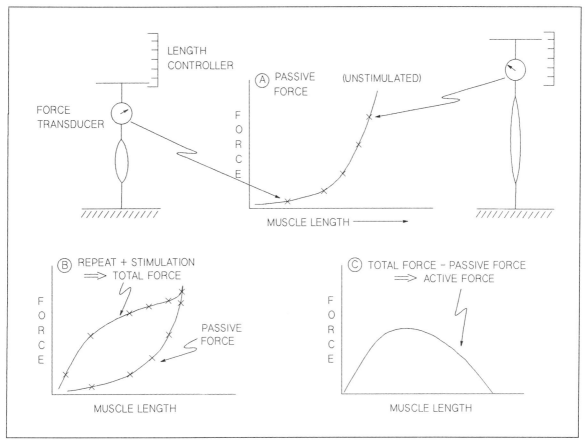

Fig. 12-3. The relationship between isometric force and muscle length. (For details of the experimental methodology, see text.)

CELLULAR MECHANISMS OF CONTRACTILITY: THE SLIDING-FILAMENT THEORY

The most widely accepted theory of muscle contraction is the **sliding-filament model.** This theory states that muscle force is generated by the interaction between thick and thin filaments of fixed length, which are free to interdigitate or slide past one another. The isometric force is proportional to an active region that corresponds to the overlap area between the thick and thin filaments. Figure 12-4 summarizes the structural and functional evidence supporting this mechanism. The proportionality between force and muscle length is best illustrated by the descending limb of the force–length relationship. For example, no overlap and hence no force would be expected to occur at a muscle length corresponding to a sarcomere length equal to the length of the thick filament plus twice that of two thin filaments, or 3.6 μm. The decline in force at short muscle lengths is less well understood, but the reduced interaction between actin and myosin may be partly attributable to double-filament overlap or filament compression at the z-bands.

The constancy of filament length is a critical component of this theory, and is supported by x-ray diffraction and electron microscopy findings. In addition, long before the advent of electron

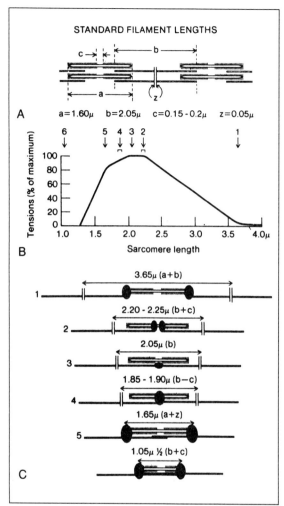

Fig. 12-4. Structural basis for the active isometric force–length relationship. (From Gordon, A. M., Huxley, A. F., and Julian F. J. The variation in isometric tension with sarcomere length in vertebrate muscle fibers. *J. Physiol.* 184: 170-192, 1966.)

microscopy, optical measurements of contraction in living muscle indicated that the A-band (and hence thick-filament length) remained constant over a wide range of total muscle lengths. Figure 12-5 shows how the interdigitating filaments of constant length are consistent with this observation.

MUSCLE ENERGETICS

In parallel with muscle mechanics, the study of the energy requirements for contraction has a long history. Collectively these studies are known as **muscle energetics.** The immediate source of energy for contraction is the free energy furnished by the hydrolysis of ATP. The protein myosin, the major component of the thick filament, is an ATPase, in that it acts as a catalyst for the hydrolysis of ATP. The ATPase activity of myosin is enhanced by interaction with actin, the major protein component of the thin filament. Production of heat by muscle is a consequence of this hydrolysis. Thus, in studies of muscle energetics, muscle heat production is often used to characterize the mechanochemical energy conversion transduced by the actin–myosin interaction. The nature of myosin ATPase and its role in muscle energetics will be considered in detail in Chapter 13. However, for our present purposes, the critical fact is that both ATP hydrolysis and its concomitant heat production exhibit a similar dependence on the length of muscle, as does active force. Thus, both force generation and the required energy transduction are dependent on the degree of interaction between thick and thin filaments, which in turn is governed by the geometry of the interdigitating filament lattices. This is the heart of the sliding-filament theory. How this interaction is regulated in muscle will be considered in subsequent sections.

EXCITATION–CONTRACTION COUPLING IN STRIATED MUSCLE

Skeletal muscle is optimized to contract rapidly and forcefully in response to neural input, in keeping with its primary physiologic role in movement. For example, to twitch the little toe, a signal is sent from the brain, via the spinal cord, to the motor neurons that innervate the toe muscles. In response, the toe muscle contracts once, then relaxes and remains relaxed until the next signal from the neuron arrives. At the muscle cell level, a sequence of control processes, collectively re-

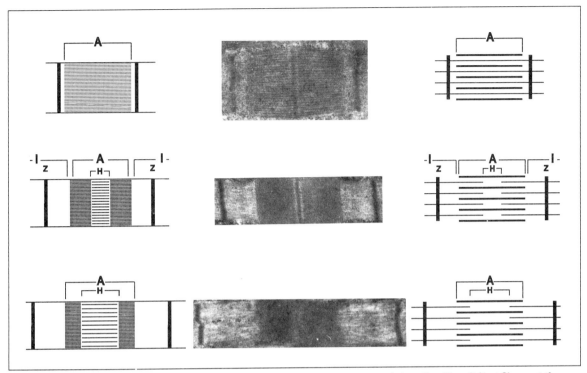

Fig. 12-5. Electron micrographs and diagrams of sarcomeres at various muscle lengths. The sliding-filament theory is based on the constancy of the thick and thin filaments. (A = A-bands; I = I-bands; Z = Z-bands; H = H-zone.)

ferred to as **excitation–contraction coupling,** receives the neural signal and converts it into mechanical force.

To understand these processes, remember that muscle consists of two major structural systems—a filament system, described in the preceding sections, and a **membrane control system,** which surrounds the filament system and triggers activation. This control system is ilustrated in Figure 12-6 and consists of three interrelated membrane compartments—the **sarcolemma,** the **transverse** (T) **tubules,** and the **sarcoplasmic reticulum** (SR). The sarcolemma, or outer cell membrane, is an electrically excitable membrane that propagates action potentials in a manner analogous to that of nerve fibers. At regular intervals, where the A-bands and I-bands intersect (two intersections per sarcomere), the

outer cell membrane invaginates and forms narrow, transversely oriented tubules that remain open to the extracellular fluid. These branch and interconnect with other tubules, so that the entire fiber cross-section at each junction of A-bands and I-bands, is uniformly crisscrossed with T-tubules. This network of T-tubules is referred to as the **T-system.** The T-tubules are also electrically excitable and serve as the electric pathway for propagation of the action potential into the fiber. Thus, both structurally and functionally, the T-tubules are an extension of the outer cell membrane. In addition, the T-tubules serve as a bridge between the external cell membranes and the internal SR.

Along most of their length, the T-tubules associate with the SR at specialized junctions called **triads,** so named because each T-tubule is

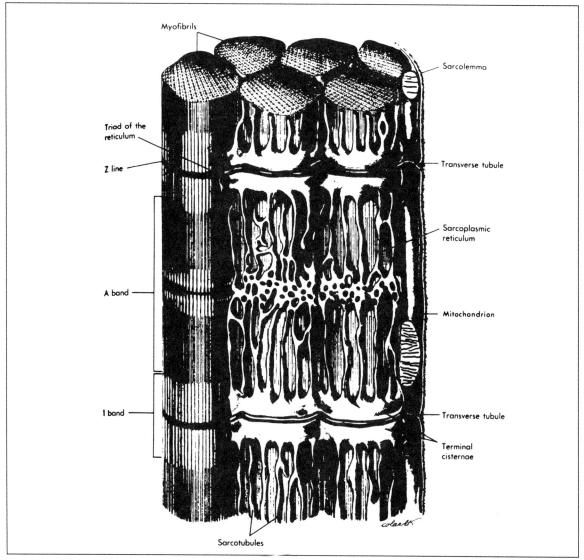

Myofibrils

Sarcolemma

Triad of the reticulum

Z line

Transverse tubule

A band

Sarcoplasmic reticulum

Mitochondrion

I band

Transverse tubule

Terminal cisternae

Sarcotubules

Fig. 12-6. Membrane systems controlling contractile activity in striated muscle. (Modified after L. Peachey; from Fawcett, D. W., and McNutt, S. The ultrastructure of the cat myocardium. I: Ventricular papillary muscle. *J. Cell Biol.* 42: 1–45, 1969. Drawn by Sylvia Colard Keene.)

flanked on two sides by SR. A narrow 10-nm gap, spanned by connecting "foot" proteins, separates the T-tubule and the SR membranes at the triad junctions. The specialized junctional surfaces of both the T-tubule and the SR contain proteins that play a major role in excitation–contraction coupling (described in Chapter 13). The SR is an internal membrane system that is specialized to store and rapidly release Ca^{2+} during activity. It has three anatomically and functionally distinct regions. Over most of the sarcomere, the SR consists of narrow, longitudinally oriented saclike structures with irregular shapes, called the **longitudinal SR.** The SR membrane in these regions contains a high density of Ca^{2+}-ATPase transport proteins, which actively accumulate Ca^{2+} in the SR and maintain resting cytosolic free Ca^{2+} levels in the low, 100-nM range. Near the regions of junctional contact with the T-tubules, the SR widens into larger saclike regions, termed **terminal cisternae.** Most of the Ca^{2+} that is accumulated by the SR is stored in these cisternae in bound form, complexed to the Ca^{2+}-binding protein **calsequestrin.** Those regions of the terminal cisternae membrane that lie immediately across from the T-system flatten out and are called the **junctional SR.**

Figure 12-7 summarizes the control sequence

that initiates contraction. It begins when the motor neuron releases the neurotransmitter acetylcholine (ACh) into the neuromuscular junction. The ACh diffuses and binds to ACh-activated channels concentrated at the muscle end-plate. The ACh channels then open and increase their permeability to Na^+ and K^+, producing local depolarization of the end-plate region. In turn, this local depolarization causes nearby voltage-activated Na^+ channels to open and initiate a propagating action potential. The action potential propagates longitudinally along the outer membrane and radially inward along the T-tubule membranes. This signal is transmitted, by a still incompletely understood mechanism, across the triad junctions to the SR. The SR responds to the T-tubule depolarization by releasing stored Ca^{2+} into the cytosol. Within a few milliseconds, the intracellular Ca^{2+} concentration increases more than a hundredfold, to micromolar levels.

The increased Ca^{2+} concentration is translated into increased actin–myosin interaction by the regulatory proteins, troponin and tropomyosin, which are located on the thin filament. A subunit of troponin, troponin-C, is the intracellular receptor protein for Ca^{2+}. The intracellular receptor for Ca^{2+} in smooth muscles differs from that in striated muscle, and this mechanism is described in Chapter 14. The Ca^{2+} released from the SR rapidly diffuses to and binds to troponin-C. The conformational change elicited by this binding is propagated along the thin filament by tropomyosin and releases the inhibition on the interaction between actin and myosin. When Ca^{2+} is bound to troponin-C, actin and myosin interact, slide past one another, and generate force. Contraction stops when Ca^{2+} is reaccumulated in the SR by the Ca^{2+}-ATPase transport proteins on the SR membrane, and the intracellular free Ca^{2+} concentration returns to resting levels.

Figure 12-8 shows the relative timing of these events in a fast limb skeletal muscle. Contraction is extremely rapid in skeletal muscle. The entire sequence of events takes a fraction of a second, and occurs only once in response to each input

Fig. 12-7. Schematic showing the membrane systems important for excitation–contraction coupling in skeletal muscle and the control sequence that initiates contraction. A T-tubule is shown invaginating from the sarcolemma. It forms a triad junction with the terminal cisternae of the sarcoplasmic reticulum (one-half triad drawn). "Foot" proteins span the gap between the T-tubule and terminal cisternae. The major steps are: nerve impulse, action potential propagation along the muscle sarcolemma membrane and into the transverse tubular membranes, signal coupling to the sarcoplasmic reticulum (SR) at the triad junctions, Ca^{2+} release from the SR, Ca^{2+} binding to the regulatory protein troponin-C, and resulting activation of the contractile proteins. Contraction stops when Ca^{2+} reaccumulates in the SR and intracellular Ca^{2+} concentration returns to resting levels.

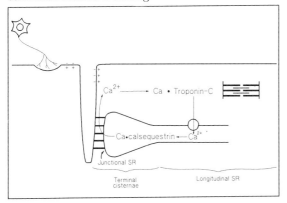

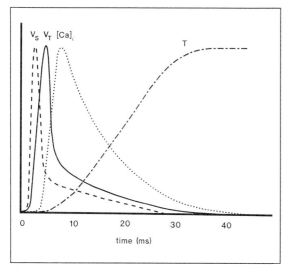

Fig. 12-8. Timing of the major triggering events in excitation–contraction coupling for a frog fast-twitch skeletal muscle fiber at 18°C. Traces V_s and V_T show the time course of the action potential on the sarcolemma and transverse-tubule membranes, respectively. Trace $[Ca]_i$ shows the time course of the myoplasmic Ca^{2+} concentration change. The last trace (T) shows the onset of tension development, which is complete in about 100 msec from the onset of the sarcolemmal action potential. All traces represent average data and have been normalized to (i.e., divided by) the maximum value for each response.

signal from the motor neuron. These features are in line with the major role of skeletal muscle in the control of body movement, most of which is under voluntary or central nervous system control. This is in contrast to cardiac and smooth muscle in which contraction is significantly slower and normally occurs repetitively in a cyclic pattern, in keeping with the major maintenance function of these tissues, which are under autonomic or hormonal control.

It can be seen that Ca^{2+} is the central intracellular messenger in this control sequence. The transient rise in intracellular Ca^{2+} concentration is the link between electrical excitation of the cell membrane and contraction.

SUMMARY

This chapter focuses on the mechanisms responsible for the generation of mechanical force and shortening, and the regulation of contractile activity at the cellular level. The *sliding-filament theory* of muscle contraction and its structural and functional bases are presented. The systems underlying regulation of contractility, termed *excitation–contraction coupling*, are also described. A flow chart of the events associated with muscle contraction is given in Table 12-1.

NATIONAL BOARD TYPE QUESTIONS

Select the one most correct answer for each of the following

1. An increase in the total isometric force with increases in length seen at long muscle lengths (sarcomere lengths greater than 3.6 μm) can best be attributed to:
   A. The passive elastic behavior of muscle
   B. An increase in overlap between thick and thin filaments
   C. An increase in thin-filament length
   D. Stimulation of ATP synthesis
   E. Increased release of Ca^{2+} from the SR
2. Relaxation of skeletal muscle is associated with:
   A. Rapid dissociation of thick filaments into myosin dimers
   B. Uncoupling of the T-tubules from the surface membrane
   C. Reduction of intracellular Ca^{2+} by uptake into the SR
   D. Inhibition of creatine kinase
   E. Formation of "rigor" links
3. Phosphorylation of certain proteins in the SR can enhance its rate of Ca^{2+}-uptake. The most likely effect of substances that promote SR phosphorylation (for example, β-adrenergic agonists) would be to:
   A. Increase the velocity of the spread of the action potential

Table 12-1. Summary of Muscle Contraction at the Cellular Level

Excitation
1. Propagation of an action potential and depolarization of the muscle cell membrane.
2. Inward spread of the action potential along the transverse tubule system.
3. Interaction of the transverse tubule system and sarcoplasmic reticulum at the triad junction, leading to release of Ca^{2+}.
4. Diffusion of Ca^{2+} to the contractile filaments.

Contraction
5. Binding of Ca^{2+} to troponin on the thin filament.
6. Alteration of the Ca^{2+}-troponin-tropomyosin-actin complex to remove its inhibition of myosin interaction with actin.
7. Myosin heads on the thick filaments interact in a cyclic manner with actin sites on the thin filaments, producing force. Force production is proportional to the overlap between thick and thin filaments, and hence the number of potential myosin head–actin interaction sites.
8. ATP provides the energy for this process, with one ATP molecule used for each crossbridge cycle.

Relaxation
9. Uptake of Ca^{2+} by the sarcoplasmic reticulum, lowering the intracellular concentration.
10. Release of Ca^{2+} from the contractile protein complex and restoration of the inhibition by troponin–tropomyosin of the actin–myosin interaction.
11. Decay of the tension-developing state as actin–myosin crossbridge links are broken.

B. Decrease muscle ATP content and metabolism

C. Shorten the duration of a single isometric twitch

D. Slow the rate of aggregation of thick filaments

E. Decrease the rate of A-band shortening

4. The activation of skeletal muscle by Ca^{2+} can best be ascribed to:

A. Shortening of the thick filament in the presence of Ca^{2+}

B. Ca^{2+} enhancement of SR formation

C. Stimulation of metabolism by Ca^{2+}

D. Binding of Ca^{2+} to troponin-C and subsequent removal of the inhibition of tropomyosin on actin–myosin interaction

E. Its enhancement of the formation of myosin filaments from myosin subunits

5. The plateau region found in the active isometric force versus length relationship in skeletal muscle can best be attributed to:

A. The decrease in passive force with lengthening

B. A decrease in excitation–contraction coupling, which offsets the inherent increase in active isometric force with lengthening

C. The equilibrium between filaments and protein monomers

D. The constant number of crossbridges available for interaction, due to the structure of the thick filament

E. A loss of T-tubules at this length

ANNOTATED ANSWERS

1. A. At long muscle lengths (sarcomere lengths greater than 3.6 μm), there is no overlap between thin and thick filaments, and hence no possibility for active force development. Therefore, the increases in total force are ascribable to the passive elastic behavior of muscle.

2. C. Relaxation is associated with lowering the intracellular Ca^{2+} concentration by the SR to the point where Ca^{2+} dissociates from troponin and there is subsequent restoration of the inhibition of tropomyosin on actin–myosin interaction.

3. C. A more rapid uptake of Ca^{2+} by the SR more rapidly decreases the intracellular concentration of Ca^{2+}, and consequently the dissociation of Ca^{2+} from troponin-C is more rapid, making relaxation faster. The net effect is to shorten twitch duration.

4. D. The troponin subunit, troponin-C, is the Ca^{2+} receptor for the thin-filament regulatory system in skeletal muscle. Activation is associated with the binding of Ca^{2+} to troponin-C and removal of the inhibition of the actin–myosin interaction by tropomyosin.

5. D. The assembly of myosin into bipolar thick filaments yields a "bare zone" in the middle of the thick filament that is devoid of myosin heads. In the sarcomere this structural feature leads to a constant number of cross-bridges in the plateau region of the force–length relationship. Figure 12-4 should be reviewed for visualization of this structure and its functional importance.

BIBLIOGRAPHY

Rüegg, J. C. *Calcium in Muscle Activation.* Heidelberg: Springer-Verlag, 1986.

Squire, J. *The Structural Basis of Muscular Contraction.* London: Plenum, 1981.

Woledge, R. C., Curtin, N. A., and Homsher, E. *Energetic Aspects of Muscle Contraction.* London: Academic Press, 1985.

13 Muscle Physiology: Molecular Mechanisms

RICHARD J. PAUL, DONALD G. FERGUSON, AND JUDITH A. HEINY

Objectives

After reading this chapter, you should be able to:

Describe the sequence of events that occur at the molecular level and the relevant subcellular structures involved in muscle contraction, from excitation through excitation–contraction coupling, filament interaction, and mobilization of energy metabolism

Describe the relationships that exist between muscle length, tension, and velocity of contraction and relate these to molecular structure

Discuss the molecular mechanisms involved in the conversion of chemical free energy into force and shortening

Describe the structure and protein composition of the thick and thin filaments

Describe the initial membrane stimulus that triggers contraction in striated muscle

Describe how this membrane event is conveyed to the filament system and transduced to mechanical force

Describe the intermediate steps and the chemical second messengers involved in excitation–contraction coupling

Describe the role of the sarcolemma, the transverse tubule system, the sarcoplasmic reticulum, and the regulatory proteins of the filament system in the regulation of contractility

Describe the biochemical interactions between the isolated muscle proteins actin and myosin

Explain the effects of adding ATP to a mixture of purified actin and myosin

Describe the interactions between tropomyosin, troponin, actin, and calcium

Describe the effects of ATP added to myosin plus "regulated" actin filaments (i.e., actin filaments with troponin and tropomyosin)

Explain the effects of adding 10^{-6}M Ca^{2+} to the components described in the preceding objective

Describe the structural considerations that lead to the concept of crossbridge cycling

Describe the molecular structure of thick and thin filaments and integrate this with the biochemical data on isolated proteins to explain potential mechanisms for sliding filaments

Give an order of magnitude for the number of crossbridge cycles per second and the consequent ATP use by muscle

Discuss the energy sources and stores in skeletal muscle

Describe how contractile ATP requirements are coordinated with metabolic energy sources

This chapter continues our study of the mechanisms underlying force production and shortening, their mechanical and energetic consequences, and the regulation of contractility. This will be described from the molecular perspective, and the approach will be similar to that used in Chapter 12 where the cellular aspect was considered. We will investigate structure and function, and then reconcile the two in terms of the mechanisms involved.

MUSCLE PROTEINS

Approximately 12 percent of muscle by weight is protein, not counting collagen and other connective tissue. The major proteins are listed in Table 13-1. Their assembly into the major filaments has been widely studied, for the highly organized structure of muscle makes them amenable to biophysical techniques such as x-ray diffraction, which are not generally applicable to other biologic tissues. As such, they serve as models for non-muscle motile systems.

PROTEINS OF THE CONTRACTILE MACHINE

The Thick Filament

Myosin, the major component of the thick filament, is a dimer of approximately 500 kDa. Each

Table 13-1. Relative Proportions of Myofibrillar Proteins in Rabbit Skeletal Muscle

Protein	Total Structural Protein (%)
Myosin	43
Actin	22
Titin	10
Nebulin	5
Tropomyosin	5
Troponin	5
C-protein	2
M-proteins	< 2
Alpha-actinin	2
Beta-actinin	2

monomer consists of a heavy chain (approximately 200 kDa) and two light chains (approximately 20 kDa). The role of the **light chains** is not completely understood, but they are thought to be involved in the modulation of contractility. The **heavy chain** consists of a long tail region (120 nm) and a globular head. The **globular head** contains the enzymic site for the catalysis of ATP hydrolysis and an actin-binding site. The long tail region is important in the self-assembly of myosin into thick filaments. Myosin has been studied biochemically using limited proteolysis, and the fragments obtained by this method have formed the basis of the nomenclature given in Figure 13-1.

The assembly of myosin molecules into a thick filament is depicted in Figure 13-2. Because the myosin molecules assemble with opposite polarity from the center of the filament, there is a central region, known as the **bare zone,** which is devoid of the **globular** or (**S₁**) **heads.** This structure underlies the plateau region in the **force—length relationship** (See Fig. 12-4).

Knowledge of the geometry of filament organization is crucial to an understanding of the molecular mechanisms of force generation. Based on x-ray diffraction evidence, sites for myosin head projections exist every 14.3 nm along the thick filament. In vertebrate striated muscle, the most reliable evidence suggests that the thick filament has three strands, yielding three myosin molecules arranged symmetrically at 120 degrees at each site (like a three-bladed propeller). Each consecutive site is rotated 40 degrees, so that an identical site occurs at 43-nm intervals. Because each myosin molecule contains two S₁ heads, there are six heads per site. Looking endwise at a thick filament, one would thus see nine projected myosin molecules, each containing two heads (try to work this out from the above discussion). Because the length of a thick filament is 1.6 μm, one filament has approximately a hundred sites (1.6 μm/0.014 μm), or about one site for each one percent of thick-filament length, except in the bare zone. There are thus approximately 300 myosin molecules

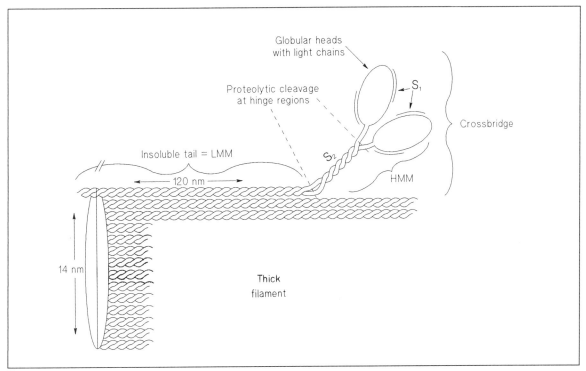

Fig. 13-1. The myosin molecule is a dimer containing two heavy chains and two pairs of light chains. Proteolytic fragments are shown as *S₁* (subfragment 1), *HMM* (heavy meromyosin), and *LMM* (light meromyosin).

and 600 S_l heads on each thick filament. Minor thick-filament proteins include C-protein. Its function is not certain, but it may be involved in the self-assembly of the thick filament.

The Thin Filament

Actin is the major protein component of the thin filament, shown schematically in Figure 13-3 and is a ubiquitous protein found in many cells. It has a globular form, known as **g-actin,** with a molecular weight of 44 kDa and a diameter of approximately 5.5 nm. Actin in muscle has a filamentous form known as **f-actin.** Actin filaments are often depicted as a double string of pearls, with the twist or repeat unit occurring every 7 g-actin units, or every 38.5 nm. Though represented as pearls, g-actin is not symmetrical. This is one of the characteristics that underlie the uni-

directional nature of muscle contraction. Muscle can only shorten; it cannot actively lengthen.

The native thin filament contains two accessory proteins, tropomyosin and troponin, which are involved in the regulation of contraction. Tropomyosin is a long molecule that is situated near the groove of the double strands of actin. Each tropomyosin molecule interacts with seven actin molecules and forms the regulated unit. Troponin is a more globular molecule and is composed of three subunits. The names of these subunits reflect their function. Troponin-C is the Ca^{2+} binding subunit, troponin-T is the subunit that interacts with tropomyosin, and troponin-I binds to actin and underlies the inhibition of the actin–myosin interaction.

Minor protein components include alpha- and beta-actinin, which are z-band components, and very large-molecular-weight proteins, titin

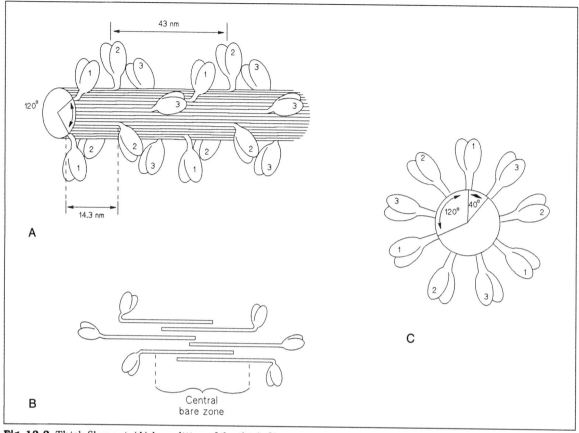

Fig. 13-2. Thick filament. (A) A rendition of the thick-filament structure based on current evidence. (B) The initiation of filament self-assembly in a tail-to-tail fashion. (C) The filament viewed on end. (Adapted from Murray, J. M., and Weber, A. The cooperative action of muscle proteins. *Sci. Am.* 230:58, 1974).

and nebulin, whose functions are unclear but which may serve as templates for the sarcomere superstructure. A high-resolution electron micrograph of a single sarcomere, which forms the basis for much of our structural knowledge, and a schematic summary are presented in Figure 13-4.

MOLECULAR MECHANICS AND BIOCHEMICAL FUNCTION

In parallel to our structure and function considerations at the cellular level (see Chapter 12), it

is important to consider the biochemical function of isolated actin and myosin filaments, as shown in the experiments depicted in Figure 13-5.

Using purified muscle proteins, let us consider the following test tube experiments. If we mix various combinations of these proteins, we can obtain an index of the actin–myosin interaction, based on the turbidity or cloudiness of the solution. This can be easily seen or measured with a spectrometer. In addition, we can measure the ATPase activity of the various combinations. This can be done by measuring the hydrolysis products, ADP or inorganic phosphate (P_i), or, alternatively, the disappearance of ATP. When ATP is

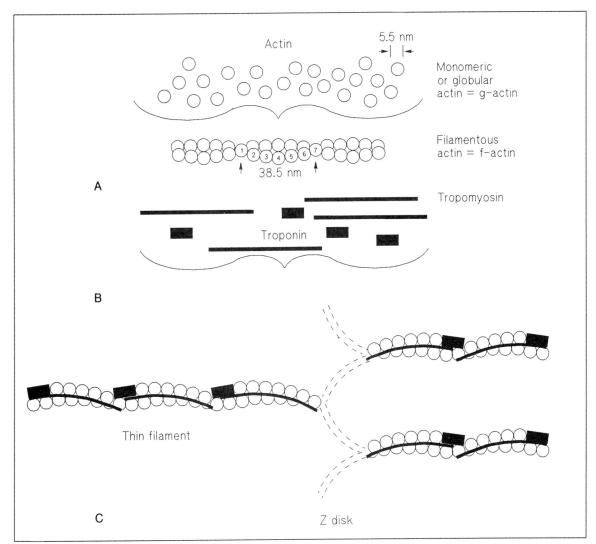

Fig. 13-3. Thin filament. (A) Globular or g-actin. (B) A thin filament and associated regulatory proteins, troponin, and tropomyosin. (C) The attachment of thin filaments to the z-disk (z-band) characteristic of striated muscle. (See text for further description.) (Adapted from Murray, J. M., and Weber, A. The cooperative action of muscle proteins. *Sci. Am.* 230:58, 1974.)

added to a solution of myosin, the solution remains relatively clear and a low level of ATPase activity is recorded. The addition of actin to a solution of myosin, on the other hand, causes a rapid increase of turbidity, indicating a strong interaction between actin and myosin. The actin-myosin complex formed in the absence of ATP is known as a **rigor complex,** because of the very stiff state of an intact muscle when ATP is depleted. The addition of ATP to these rigor complexes of actin and myosin initially leads to rapid clearing of the solution, followed by increased turbidity to the point where actin-myosin complexes precipitate out of the solution. This precipitation

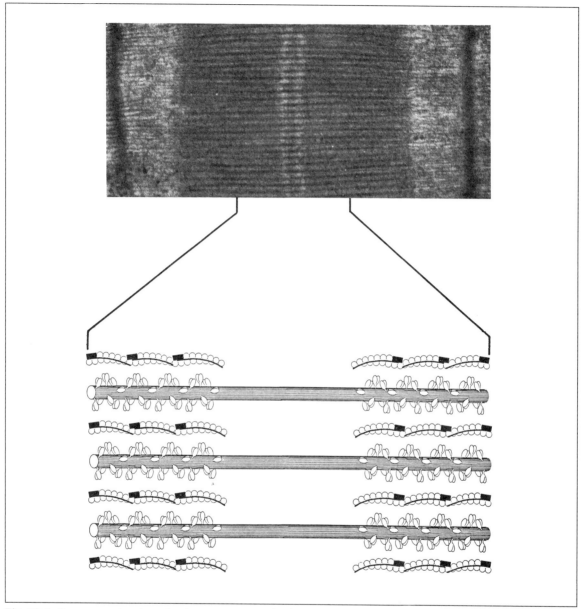

Fig. 13-4. Sarcomere structure. An electron micrograph of a sarcomere and an artistic rendition of the structure based on current knowledge. (Adapted from Murray, J. M., and Weber, A. The cooperative action of muscle proteins. *Sci. Am.* 230:58, 1974.)

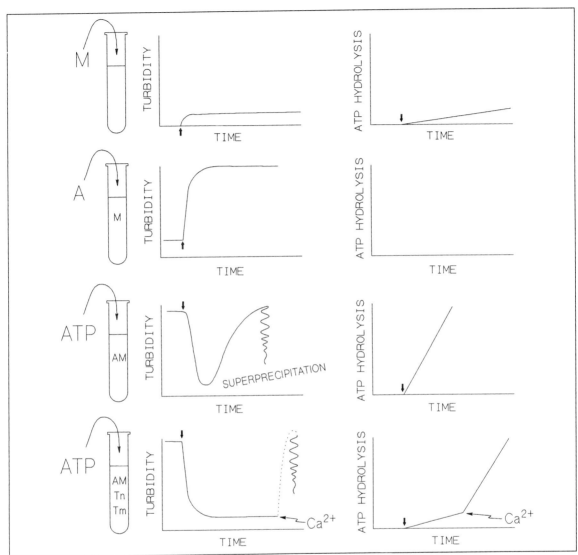

Fig. 13-5. Actin–myosin interactions. The effects of ATP and Ca^{2+} on solutions of actin (*A*) and myosin (*M*), and in combination with regulatory proteins troponin (*Tn*) and tropomyosin (*Tm*).

response results from contraction of the actin-myosin gel, and is known as **superprecipitation.** A large increase in ATPase activity is seen shortly following the clearing, after addition of ATP. The situation is quite different if the thin-filament regulatory proteins, troponin and tropomyosin, are present. In this case, the addition of ATP induces only the rapid-clearing response, and no increase in ATPase activity is seen. Finally, the addition of Ca^{2+} at micromolar levels initiates superprecipitation, with a concomitant increase in ATPase activity.

Several salient points can be made from these observations:

1. Myosin alone possesses only weak ATPase activity. Actin and myosin spontaneously form a complex, which can be dissociated by ATP.
2. The ATPase activity of myosin is greatly (200-fold) enhanced by actin.
3. Superprecipitation and the enhanced ATPase activity of actin-myosin is inhibited by the presence of troponin and tropomyosin.
4. Ca^{2+}, in micromolar quantities, can remove this inhibition.

THE CROSSBRIDGE CYCLE CONTRACTILE MECHANISM

Now we combine the structural and functional evidence to advance a molecular mechanism of muscle contraction. How does the interaction between actin and myosin, which is proportional to filament overlap, generate force and shortening? Figure 13-4 depicts the structural constraints in the overlap region. A very important consideration here is that muscle can operate over a wide range of lengths, roughly corresponding to sarcomere lengths of 1.6 to 3.6 μm. Translated into relative filament positions, a site for force generation on the thick filament can operate over a distance of about 1 μm, relative to a thin filament. Given that myosin heads exist every 14.3 nm (0.0143 μm) along the thick filament, it is difficult to imagine a contractile mechanism that alows a myosin head to remain attached to a thin filament actin site over the entire range of relative filament motion during shortening. It would appear that many attachment–detachment cycles of myosin to actin must occur during muscle contraction. This structural information, plus the biochemical data (discussed previously) indicating that ATP breaks the actin-myosin rigor complex and actin activates the myosin ATPase, has led to the **crossbridge cycling theory** of muscle contraction. This theory advances that there are multiple cycles of myosin-head attachment and detachment to actin during a contraction. Several biochemical steps in the crossbridge cycle have been proposed based on observations made during kinetic analysis of the actin-activated, myosin-catalyzed hydrolysis of ATP. In parallel, a

number of hypothetical crossbridge states have been postulated, based on findings from mechanical studies of muscle. Much of the current research in muscle physiology is directed at understanding how the steps delineated by biochemical analysis on isolated proteins can be related to the various mechanical states of the crossbridge cycle proposed for intact muscle. Though necessarily an oversimplification, some of the major features of the crossbridge cycle consist of the stages shown in Figure 13-6. For our purposes here, it is assumed that the muscle is activated, that is, Ca^{2+} is present. The regulatory system is considered in the section "Molecular Mechanisms of Excitation–Contraction Coupling" in this chapter.

In the resting state, in which the actin–myosin interaction is inhibited by the regulatory proteins and ATP is present, the myosin heads have ADP and P_i bound. The orientation of the heads is approximately perpendicular to the filament axes. Stimulation increases the intracellular Ca^{2+} concentration, which removes the inhibition, and myosin exhibits a high affinity for actin in this state. The S_2 region of the myosin molecule appears to serve as a flexible hinge, permitting movement of the head region (S_1) toward the thin filament (see Fig. 13-1). The binding of a myosin head to actin accelerates the removal of ADP and P_i, and the attached myosin head changes conformation by approximately 45 degrees with respect to the thin filament. This conformational change causes a relative filament movement of about 10 nm and underlies the generation of force. In the absence of ATP, the crossbridge cycle stops here, and when actin is bound to a myosin head in this configuration, it is known as a **rigor link,** analogous to the strong binding of actin to myosin seen in studies of isolated proteins (discussed previously). The next stage in the crossbridge cycle is the binding of ATP to the myosin head. This leads to a low affinity for actin and dissociation of the rigor complex. ATP is then hydrolyzed on the myosin head and the energy from this is stored in the myosin molecule by restoring the perpendicular conformation with a renewed high affinity for actin. A new cycle can

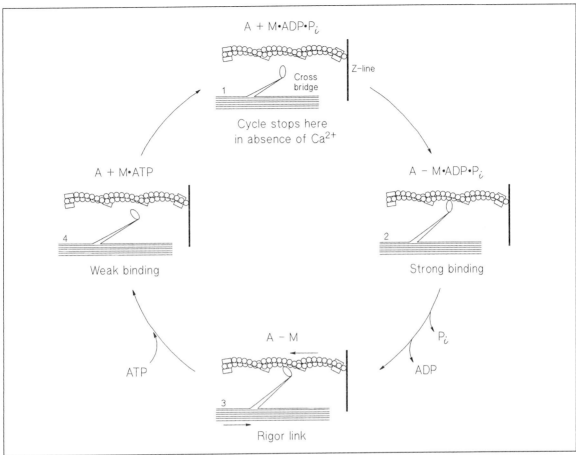

Fig. 13-6. The crossbridge cycle. This depiction represents the combined information gleaned from muscle mechanics and the biochemical kinetics of the actin-activated myosin ATPase. (P_i = inorganic phosphate; A = actin; M = myosin.)

be initiated at this point, or, if Ca^{2+} is absent and regulatory protein inhibition is restored, the cycle is terminated here and is equivalent to the state of the myosin head under resting conditions.

It is worth noting that only one ATP molecule is hydrolyzed per cycle. The free energy from this hydrolysis is transduced into mechanical energy by changing the conformation of the myosin head. One can get a rough estimate of the rate of crossbridge cycling by measuring ATP hydrolysis in active muscle. Though widely varying, a typical number for skeletal muscle ATP utilization is about 20 μmol/sec·gm (Table 13-2). The myosin

content of muscle is about 0.2 μmol/gm, so that the crossbridges cycle about a hundred times per second. This estimate depends on how many heads of each myosin molecule are interacting with actin. How the two S_1 myosin heads on each myosin molecule function during the cycle is unknown. Current evidence favors an independent action, but some form of cooperative interaction cannot be ruled out.

Though rough, this estimate of the crossbridge cycle rate can be useful when considering muscle function. For example, at 100 cycles per second and a potential movement of 10 nm per cycle,

Table 13-2. Skeletal Muscle Chemical Energy Utilization and Stores

Rate of ATP utilization ($\mu mol/sec \cdot gm$)	$ATP \xrightarrow{Ca^{2+}\text{-pump, myosin}} ADP + P_i = 5\text{–}30 \ \mu mol/gm$			
	Energy Store ($\mu mol/gm$)	Energy Source	ATP Equivalent ($\mu mol/gm$)	Reaction Time Scale
ATP	3–5		3–5	Immediate energy source
PCr	15–30	Lohmann reaction: $PCr + ADP \rightarrow Cr + ATP$ $ATP \rightarrow ADP + P_i$ Net: $PCr \rightarrow Cr + P_i$	15–30	Extremely rapid Only measurable chemical changes during contraction
Endogenous glycogen	75	Glycolysis	225	Rapid but limited energy store
		Oxidative phosphorylation	2,925	Slowest and most efficient source
Exogenous (plasma) free fatty acids, triglycerides, glucose, and amino acids (in apparent order of preference)		Oxidative phosphorylation	Very large	

PCr = phosphocreatine; Cr = creatine; P_i = inorganic phosphate.

each half of the sarcomere could shorten at a rate of 1 μm per second. Thus, maximal muscle velocities would be about one muscle length per second, such that, if a muscle is 3 cm long, the speed would be 3 cm per second. The ways in which the speed of muscle contraction and energy utilization depend on the mechanical constraints, such as the load on the muscle, will be considered in the next section.

CORRELATES OF THE CROSSBRIDGE CYCLE: THE FORCE–VELOCITY RELATIONSHIP AND ENERGETICS

With a clearer picture of the crossbridge cycle and its regulation, it is possible to consider other important mechanical and energetic characteristics of muscle. One of the more intriguing aspects is

that the mechanical constraints can alter the crossbridge cycle and the consequent energy turnover.

The relationship between muscle load (or force developed) and the speed of shortening had been studied long before the crossbridge mechanisms were understood. A number of the terms used to describe this process reflect the early experimental apparatus used, but are important to know because this terminology pervades the muscle literature, particularly that concerning the physiology of cardiac function (Chapters 16 and 19).

Force–Velocity Relationship

Figure 13-7 depicts the experimental apparatus used for measuring the **force–velocity relationship.** Using the lever system, a **preload** is placed on the unstimulated muscle. Because the muscle

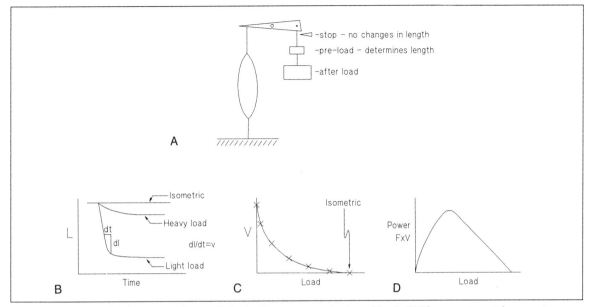

Fig. 13-7. Relationship between velocity of shortening, power output, and load. (A) The experimental apparatus used to simulate afterloaded, isotonic conditions. (B) The experimental results yielded by various loads. (C and D) The derived relationships between velocity and load and power and load, respectively. (L = length; V = velocity; dl = change in length; dt = change in time; $F \times V$ = force $\times$ velocity = power.)

is unconstrained, it stretches to the length at which the preload matches the force on the passive force–length relationship. A mechanical stop is then placed so that no further lengthening can occur. The important point here is that the preload determines the initial length of the muscle and consequently the **maximum force** possible, in keeping with the active force–length relationship. Additional loads, known as the **afterload,** can be placed on the apparatus. The name *afterload* derives from the fact that the muscle is not influenced by this load, as it is borne by the mechanical stop until "after" the muscle is stimulated. When the muscle is stimulated, and once the muscle generates an isometric force that just exceeds the afterload, the muscle shortens, as shown in Figure 13-7B. During the initial moments after shortening begins, there is a steady rate of shortening, which subsequently declines and ceases as the muscle reaches its final short-

ened length. This length corresponds to that point in the active force–length curve matching the afterload. Both the distance shortened and the velocity are dependent on the afterload. In Figure 13-7C, the relationship between shortening velocity and afterload is plotted. The interesting feature about this relation is that it is hyperbolic, indicating that the velocity decreases rapidly as the afterload is increased. The power output of muscle, which is the product of afterload times velocity, can be calculated from the relationship between afterload and velocity. Maximal power, shown in Figure 13-7D, occurs at a load of about one-third the maximal isometric force for skeletal muscle.

The molecular basis for the force–velocity relationship is not clear, but many hypotheses have been put forth. Most simply, it appears that the potential for actin–myosin interaction is reduced as the relative filament velocity increases.

Muscle Energetics

Knowledge about the energetics of muscle contraction has also influenced our understanding of the crossbridge cycle. One might anticipate that, for each ATP hydrolyzed, a fixed amount of energy is liberated. This energy could be recovered as work or dissipated as heat. However, it has been shown that the total energy released (heat plus work) during contraction is not constant but depends on the mechanical conditions. During an isotonic contraction, when work is produced by the muscle, the total energy liberated is up to threefold greater than that for a similar contraction occurring under isometric conditions. This phenomenon, known as the **Fenn effect,** indicates that there is feedback between the mechanical constraints and the rate of crossbridge cycling. Presumably some of the rate constants for the hydrolysis of ATP depend on the load transmitted by the filaments to the crossbridge. The molecular basis for the Fenn effect and the force–velocity relationship is a focus of much current research on muscle.

MUSCLE METABOLISM AND MECHANISMS FOR COORDINATION WITH CONTRACTILITY

Energy Utilization

Muscle contraction is accompanied by a large increase in energy transduction. The turnover of ATP during maximal contractile activity may be over one thousand times greater than the basal rate for an amphibian skeletal muscle at 0°C. A rough calculation can illustrate some of the consequences of muscle activity. As indicated previously, crossbridge cycling at a rate of 100 cycles per second leads to an ATP hydrolysis of about 20 μm/sec·gm. Other energy-dependent processes accompany muscle activity, predominantly Ca^{2+} translocation, which can use up to 25 percent of the energy attributed to crossbridge interaction. However, for our purposes here, 20 μmol/sec·gm is a reasonable estimate.

Muscle mass represents about 40 percent of total body weight in humans, which amounts to 28 kg in a 70-kg person. Thus an ATP hydrolysis of 0.56 mol per second would be necessary to achieve maximal levels of contractile activity. The heat production from this chemical reaction can be calculated from its **enthalpy** of 10 kcal/mol, and amounts to 5.6 kcal per second. Assuming that the **specific heat** of biologic materials is similar to that of water (their major component), that is, 1°C per cal/gm, 5.6 kcal per second would heat a 70-kg person at a rate of 0.08°C per second, or 1°C every 12.5 sec. The mechanisms available to the whole organism for dealing with this substantial heat load will be discussed in Chapter 59. However, the relevant point here is that muscle activity involves substantial metabolic activity, which has a number of consequences for the organism.

Energy Sources and Stores

Table 13-2 presents the major metabolic sources of ATP and the approximate size of these chemical energy stores. These sources and stores vary considerably with muscle type and function; however, some generalizations can be made. Skeletal muscle is often characterized by short bursts of intense activity. To meet these energy demands, skeletal muscle has adopted a "buy now, pay later" strategy. A high-energy compound, **phosphocreatine,** is used to meet the energy required for these short bursts of activity. The reaction involved in reforming ATP from ADP, driven by the conversion of phosphocreatine to creatine, is known as the **Lohmann** or **creatine kinase reaction.** The size of the phosphocreatine pool is relatively small, but, as only one reaction step is involved, ATP can be provided rapidly. This reaction is very efficient and, for most muscle activity, little change in ATP can be measured, with a breakdown of phosphocreatine the only consequence of muscle activity. The phosphocreatine broken down during muscle activity is restored following activity, a process sometimes referred to as **repayment of the oxygen debt.**

Skeletal muscle also stores energy in the form of **glycogen,** a glucose polymer. Glycolysis, with lactate as the end product, can also produce ATP relatively rapidly, but, compared with complete oxidation, it is relatively inefficient. Glycolysis is important in supporting moderate to heavy muscle activity, characterized by increasing levels of lactate. Lactic acid was so closely associated with muscle activity that one early theory held that the glycolytic production of lactate was the immediate chemical reaction underlying contraction.

The ultimate source of ATP for contraction is oxidative phosphorylation. As shown in Table 13-2, the stores of substrate for the oxidative production of ATP are large. This is the most efficient source but, as a multistep process, the slowest of the pathways for meeting contractile energy needs.

The optimal performance of muscle is limited by these energy sources and rate of ATP production. Figure 13-8 shows that the rate of running, as an index of muscle power output, and the time

over which a given level of power output can be maintained. The duration over which power output can be sustained appears to reflect these distinct metabolic pathways.

Coordination of ATP Production with Contractile ATP Utilization

Large and rapid changes in energy demand are characteristic of muscle, making the control mechanisms for metabolism critical for normal muscle function. There are a wide variety of regulatory mechanisms, but even though a complete discussion is beyond the scope of this text, the major systems will be reviewed.

The products of ATP hydrolysis, ADP and P_i, represent an important feedback mechanism. An increase in ADP levels resulting from contractile activity is a signal for increased mitochondrial

Fig. 13-8. The maximal power output of muscle, estimated from the rate of running derived from world track records, as a function of the duration of time for which it can be sustained. The data suggest a dependence on distinct energy sources, paralleling those listed in Table 13-2 and described in text. (*I* = Lohmann reaction; *II* = glycolysis; *III* = oxidative phosphorylation.)

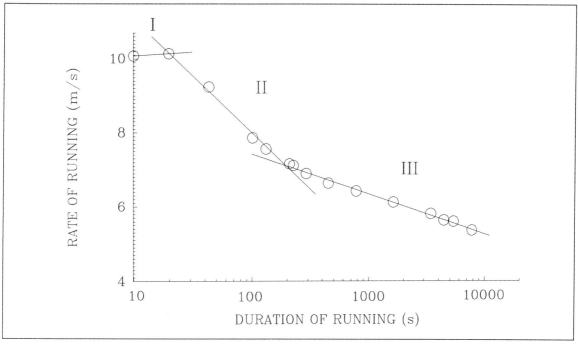

oxidative phosphorylation. This is a key mechanism for skeletal muscle, but the evidence for its existence in cardiac and smooth muscle is less certain. ADP production is also a key control for glycolysis, as an increase in ADP or decrease in ATP levels activates phosphofructokinase, an enzyme controlling a rate-limiting step in the glycolytic pathway. An increase in the P_i content also enhances the rate of glycolysis.

The mobilization of substrate also plays a role in the coordination of metabolism with contractility. Increases in the intracellular Ca^{2+} concentration, besides activating muscle contraction, are involved in activating the phosphorylase cascade that catalyzes the production of glucose units from glycogen. It is also postulated that increases in Ca^{2+} concentration heighten the permeability of the sarcolemma to glucose.

There are also several mechanisms involved in increasing blood flow to working muscles. This increased flow is important to tissue oxygenation and the removal of metabolic products, notably lactate.

MOLECULAR MECHANISMS OF EXCITATION–CONTRACTION COUPLING

Membrane Excitation and Intracellular Signaling at the Triad Junctions

As discussed in Chapter 12, the triggering event for muscle contraction is the electrical excitation of the sarcolemma and transverse (T)-tubule membranes during the action potential. At the triad junctions in the T-tubules, this depolarization is transduced into a signal for the sarcoplasmic reticulum (SR) to release Ca^{2+}.

The specialized junctional surfaces of both the **T-tubule** and the **terminal cisternae** contain proteins that play a major role in this signal transduction. The structural organization of the triad is depicted in Figure 13-9.

The junctional surface of the terminal cisternae contains rows of specialized proteins, spaced at periodic intervals, that bridge the gap between the terminal cisternae and the T-tubule. These

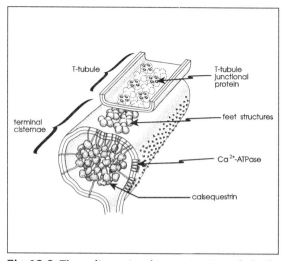

Fig. 13-9. Three-dimensional reconstruction of a half triad, showing the relative positions of foot proteins, transverse (T)-tubule system proteins, Ca^{2+}-ATPase, and calsequestrin. The junctional surface of the terminal cisternae of the sarcoplasmic reticulum bears two rows of feet. Each foot is composed of a cytoplasmic domain consisting of four identical globular subunits which occupy the junctional gap, and a transmembrane portion which spans the lipid bilayer. The junctional surface of the T-tubule also contains two rows of regularly disposed proteins which are arranged in groups of four. However, the only available structural evidence for the arrangement of these proteins indicates that the spacing is such that they line up with every second foot. In addition, the T-tubule junctional proteins from adjacent rows are not directly opposite each other, but are staggered such that they form a zig-zag pattern. One limitation of the information about the positioning of the T-tubule junctional proteins is that it was obtained from fish skeletal muscle. It has not yet been possible to observe mammalian skeletal muscle triads, where the spacing of the T-tubule proteins may differ from that depicted here. The nonjunctional sarcoplasmic reticulum contains densely packed Ca^{2+}-ATPase molecules. Although not shown here, the Ca^{2+}-ATPase molecules are abundant on the longitudinal or free sarcoplasmic reticulum, which is continuous with the terminal cisternae. The lumen of the terminal cisternae of the sarcoplasmic reticulum contains molecules of calsequestrin, a high-capacity Ca^{2+}-binding protein. (Modified from Block, B. A., et al. Structural evidence for direct interaction between molecular components of the transverse tubule/sarcoplasmic reticulum junction in skeletal muscle. *J. Cell Biol.* 107:2587, 1988.)

proteins, variously termed "feet," bridging, or spanning proteins, are aligned in two rows in a skewed pattern (see Fig. 13-9), with a center-to-center spacing of approximately 30 nm. A "foot" protein is composed of four identical subunits, each containing a membrane-spanning domain and a large cytosolic domain. The membrane-spanning domain inserts into the junctional SR bilayer, and the cytosolic domain extends across the junctional gap. Recent biochemical evidence has identified the "foot" protein as a large-molecular-weight protein, termed the **ryanodine receptor** because of its high-binding affinity to the plant alkaloid ryanodine. Electrophysiological studies have demonstrated that this ryanodine receptor functions as a type of Ca^{2+} channel, and can release Ca^{2+} from the SR at high rates.

The junctional face surfaces of the T-tubules have also been found to contain specialized proteins aligned with a regular periodicity. The T-tubule network of mammalian skeletal muscle cells is one of the richest sources of a voltage-activated Ca^{2+} channel, which is termed the **dihydropyridine** (DHP) **receptor** because of its high sensitivity to this class of Ca^{2+}-channel–blocking drugs. Biochemical and histochemical data suggest that the DHP receptors are localized in the junctional regions of the T-tubule membrane that directly face the junctional surfaces of the terminal cisternae of the SR. The available morphologic evidence from electron microscopy suggests that the DHP receptors are clustered in groups of four, termed *tetrads*, and these tetrads are aligned in parallel rows with a regular periodicity. Figure 13-9 illustrates this pattern for a fish skeletal muscle fiber, in which the tetrads line up opposite every *other* ryanodine receptor in a 1 to 2 ratio. This organization of tetrads has only been clearly worked out for skeletal muscle of lower vertebrates. In fast mammalian skeletal muscle, the tetrads may line up opposite every ryanodine receptor in a 1 to 1 ratio.

The complementary periodicity and symmetric fourfold molecular structure of these two major junctional proteins imply that they interact directly in the coupling mechanism between the T-tubules and SR. In effect, during excitation–contraction coupling, a DHP sensitive Ca^{2+} channel on the outer cell membrane controls the opening of a second type of Ca^{2+} channel, the ryanodine receptor, on an internal membrane. The triad junctions are the specialized membrane structures that bring these two kinds of Ca^{2+} channels together in a unique macromolecular complex which permits their interaction.

The exact mechanism by which the Ca^{2+} channels of the T-tubules communicate with the Ca^{2+}-release channels of the SR is not completely understood. However, it is known that junctional coupling in skeletal muscle is tightly controlled by the T-system membrane potential. If the ion channels mediating the action potential (i.e., the Na^+, K^+, and Cl^- channels of the surface and the T-system membranes) are blocked experimentally and the T-system membrane is depolarized to the same potential using electronic feedback, the SR will continue to release Ca^{2+}. In other words, it is the absolute value of the electrical potential difference across the T-system membrane, not the inflow of ions during the action potential, that controls junctional coupling. The Ca^{2+} channels of the T-tubules do not pass any significant Ca^{2+} current during the skeletal muscle action potential, because of their very slow opening kinetics. (This contrasts with cardiac muscle in which an influx of extracellular Ca^{2+} occurs during the slower cardiac action potential and is absolutely required for contraction [see Chapter 15].)

Therefore, the junctional DHP receptors of the T-system membranes are thought to function as voltage-sensor molecules that respond to the depolarization during the action potential, and somehow these receptors communicate the depolarization to the Ca^{2+}-release channels of the SR. The DHP receptor contains mobile, charged regions (either charge clusters or molecular dipoles termed *gating* or *voltage-sensing domains*) within its intramembrane domains. These charged regions move rapidly in response to a change in the transmembrane potential. This early fast movement of the gating domains controls a much slower conformational change that leads to channel opening and an influx of Ca^{2+}

current. However, this slower transition would not normally occur during a single twitch of skeletal muscle, which begins less than 10 msec after the start of the sarcolemmal action potential; it can only be seen experimentally when membrane depolarization is prolonged beyond the normal action potential duration. Thus, the early rapid conformational changes associated with the gating movement of the protein, and not the Ca^{2+} current that follows after a delay of hundreds of milliseconds, appear to control the opening of the SR Ca^{2+}-release channels during skeletal muscle excitation–contraction coupling. Figure 13-10 summarizes the essential steps in this junctional coupling. Current research into the molecular mechanisms underlying this junctional coupling is focused on whether the intermediate step is allosteric or chemical in nature, or some combination of both.

One theory, termed the **charge movement hypothesis,** postulates that the T-tubule–system (T-system) voltage sensors (DHP-sensitive, slow Ca^{2+} channels) are in physical contact with the "foot" proteins, either directly or through an intermediate binding protein which is not yet identified. This model postulates that the movement of charged intramembrane domains of the DHP receptor is transmitted allosterically along this macromolecular complex, thereby producing a conformational change on the SR ryanodine receptor which causes its opening. In support of this model, the kinetics and voltage dependence of the intracellular rise in free Ca^{2+} concentration, measured optically with Ca^{2+}-indicating dyes perfused into the myoplasm, closely match those of the charge movement, which can be de-

termined using standard electrophysiological techniques. Additionally, almost any pharmacologic or experimental manipulation that alters the charge movement produces a parallel effect on Ca^{2+} release. Conversely, any drug or manipulation that acts at the level of the ryanodine receptor to alter Ca^{2+} release also alters charge movement. This kind of "cross talk" between the two proteins is consistent with an allosteric coupling mechanism.

An alternative hypothesis for the triadic coupling step, termed the **junctional transmitter hypothesis,** postulates that movement of the T-system voltage sensor releases a chemical transmitter into the junctional gap. This postulated transmitter then diffuses and binds to a control site on the SR Ca^{2+}-release protein. This proposal is supported by the observation that the isolated SR Ca^{2+}-release channel behaves like a classic agonist-operated Ca^{2+} channel, and can open in response to a variety of agonist molecules, including the intracellular second messengers inositol-tris-phosphate (IP_3), adenine nucleotides, and calcium itself. Moreover, the ryanodine receptor of skeletal muscle is highly homologous with the IP_3 receptor, which, upon binding of the intracellular messenger IP_3, has been shown to release Ca^{2+} from internal membranes in smooth muscle and in a variety of other nonmuscle cells. It has also been demonstrated that the phosphoinositide lipids and enzymes involved in the IP_3 pathway are present in skeletal muscle T-tubule membranes. Normally IP_3 is produced from the breakdown of membrane lipids in response to stimulation by a variety of hormones and transmitters. However, the *de novo* production

Fig. 13-10. Intercellular communication at the triad junctions involves transducing an electric depolarization from the transverse tubules into a trigger for opening Ca^{2+}-release channels on the sarcoplasmic reticulum (*SR*). (*DHP* = dihydropyridine; *T-system* = transverse tubule system.)

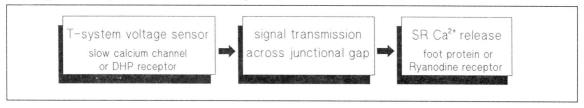

of IP_3 by this mechanism takes hundreds of milliseconds, which is too slow to be compatible with excitation–contraction coupling in skeletal muscle (see Fig. 12-8). Thus, there is clear evidence for the involvement of IP_3 in the activation of smooth muscle, but evidence for this in the fast twitch of a skeletal muscle is less compelling. Nevertheless, IP_3 may have some modulatory function.

Some combination of a charge-movement and transmitter hypothesis may exist at the triad junctions, and a variety of other models involving various combinations of these basic mechanisms have been proposed. For example, the intermediate step may involve a two-stage mechanism in which SR Ca^{2+} channels are opened initially by a conformational change communicated allosterically from the T-system DHP receptors, and then this release may be amplified by the binding of released Ca^{2+} itself to the SR Ca^{2+} channels. In this model, charge movement provides the initial trigger step and a Ca^{2+} induced Ca^{2+}-release mechanism provides the secondary amplification step. Another model, referred to here as the **voltage-modulated receptor hypothesis,** postulates that a conformational change on the voltage sensor may be transmitted mechanically to the SR Ca^{2+}-release protein, thereby increasing its affinity for an agonist such as IP_3. In this model, the agonist is not produced *de novo* upon stimulation, but rather endogenous levels of IP_3 suffice to modulate contraction once the receptor affinity is altered. Such a mechanism would allow for transmitter activation of the SR Ca^{2+}-release channels, consistent with the biochemical data pertaining to isolated ryanodine receptors, and could also work at speeds compatible with excitation–contraction coupling in fast skeletal muscle. The exact molecular mechanisms of this junctional coupling step are the subject of intense research. Failure at the level of excitation–contraction coupling underlies some skeletal muscle disease states, most notably malignant hyperthermia. This disorder is characterized by a hereditary predisposition to sustained Ca^{2+} release in response to certain general anesthetic agents. If not treated rapidly, the hyperthermia is lethal.

The genetic locus has been identified as an alteration of the gene coding for the ryanodine receptor. Many commonly used muscle relaxants, such as dantrolene sodium, also alter coupling at the triad junction.

Additional Mechanisms of Activation during Sustained Activity

During periods of sustained muscle contraction, action potentials may be elicited on the muscle membrane at a rate of about 50 per second. This rapid, repetitive electrical activity leads to a smooth contraction, referred to as a **tetanus.** In fact, tetanic contraction is the more common physiologic condition of skeletal muscle; a single twitch is not normally elicited except under experimental conditions. Each action potential triggers a bolus of Ca^{2+} released from the SR, following the mechanisms described. In addition, Ca^{2+} influx from the extracellular space begins to play a role in enhancing and maintaining the force of contraction. As noted previously, the influx of extracellular Ca^{2+} through voltage-dependent Ca^{2+} channels during a single action potential is not sufficient to elicit contraction. Nevertheless, if enough Ca^{2+} accumulates during a tetanus, it can help increase the force of contraction, and help maintain a steady level of cytosolic Ca^{2+} sufficient to ensure smooth, continuous contraction.

Transient Rise in Intracellular Ca^{2+} Concentration

In a resting skeletal muscle, cytoplasmic Ca^{2+} is maintained at a low level of about $0.1\ \mu M$. Most of the cell Ca^{2+}, which amounts to 2 to 3 mM, is sequestered in the SR, bound to calsequestrin. Upon stimulation, the SR can release Ca^{2+} at high rates, and thus the cytoplasmic Ca^{2+} can rise to micromolar levels within a few milliseconds.

Ca^{2+} diffuses rapidly to troponin-C, thereby removing the inhibition on interaction between actin and myosin. Consequently, the force of contraction is related to the amount of Ca^{2+} released, until all the Ca^{2+}-binding sites on troponin-C are saturated. Only a small fraction (about 350 μM) of the total Ca^{2+} stored in the SR is released in a single twitch. The large Ca^{2+} reserves allow the SR to respond repetitively during sustained muscle activity.

Ca^{2+} and Regulation of Actin–Myosin Filament Interation

Our knowledge of the molecular mechanisms by which Ca^{2+} removes the inhibition of the actin–myosin interaction is rapidly growing. In striated muscle, troponin-C is the intracellular receptor protein for Ca^{2+}. The binding of Ca^{2+} initiates a conformational change in the troponin protein complex, which is transmitted to tropomyosin. Tropomyosin lies in the groove of the double-stranded helix of the actin filament. X-ray diffraction indicates that the position of the tropomyosin molecule relative to the axis of the actin filament is altered by Ca^{2+}, as mediated by the Ca^{2+}–troponin interaction depicted in Figure 13-11. The details on how this interaction relieves the inhibition of the actin–myosin interaction are not yet completely resolved. In one theory, the **steric hindrance model,** this movement of tropomyosin elicited by Ca^{2+} removes a steric constraint or blockage of the site for the actin–myosin interaction. Recent findings, however, suggest that actin–myosin binding is not blocked in the absence of Ca^{2+}, but that the kinetics of the crossbridge cycle (the product-release step; see Fig. 13-6) are altered by tropomyosin. Thus the inhibition by tropomyosin may be governed by a combination of steric and kinetic mechanisms. Because the mode of actin–myosin interaction in striated muscle involves thin-filament proteins, regulation of striated muscle is sometimes termed *thin-filament regulated.* Regulation in some invertebrate muscle as well as in smooth muscle involves mechanisms on the myosin molecule, and this type of regulation, described further in Chapter 14, is called *thick-filament regulation.*

Relaxation

The rise in the intracellular free Ca^{2+} concentration is transient. The **Ca^{2+}-transport proteins of the SR** constitute the major Ca^{2+}-removal system in the muscle cell. These are highly concentrated in the longitudinal regions of the SR and use energy stored in ATP to transport Ca^{2+} back into the SR. There, Ca^{2+} is bound again to calsequestrin, a low-affinity high-capacity Ca^{2+}-binding protein, and reconcentrated in the terminal cisternae regions where it is available for the next twitch. Thus, Ca^{2+} is continuously recycled within the muscle cell, following the pathways outlined in Figure 12-7. Additional Ca^{2+}-removal systems operate during sustained muscle activity. Following a single twitch, the transport capacity of the SR Ca^{2+}-ATPase proteins is sufficient to rapidly restore the cytosolic Ca^{2+} concentration to resting levels. However, during sustained activity, as occurs in a tetanus, the cytosolic Ca^{2+} level may exceed the total capacity of the SR ATPase Ca^{2+}-transport system. In this event, a number of cytosolic proteins, such as parvalbumin, serve as Ca^{2+} buffers that temporarily hold then release Ca^{2+} to the SR Ca^{2+}-ATPase.

Besides the Ca^{2+} transport systems of the SR and cytosolic Ca^{2+}-buffering proteins, several other mechanisms operate in parallel to lower the cytosolic Ca^{2+} concentration. The coupling between the T-system and SR is turned off when the T-system transmembrane potential returns to the resting potential. There is also evidence that the Ca^{2+}-release channel of the SR–ryanodine receptor has a Ca^{2+}-dependent inactivating mechanism that automatically closes the channel. In other words, this channel has a bimodal response to Ca^{2+}. Micromolar levels of Ca^{2+} activate it and higher levels inactivate it.

SUMMARY

This chapter focuses on the mechanisms needed in the generation of mechanical force and shortening, and the regulation of contractile activity at the molecular level. Muscle contractile proteins

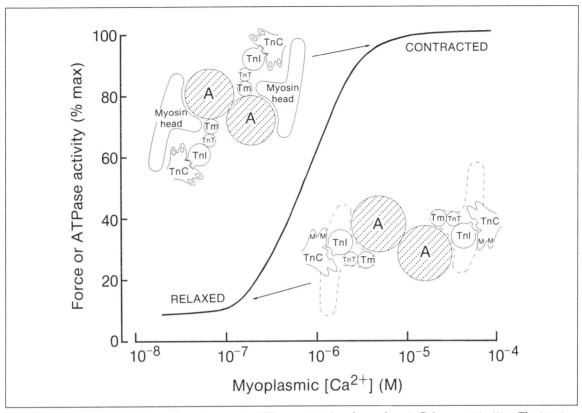

Fig. 13-11. Relationship between force (or myosin ATPase activity) and myoplasmic Ca^{2+} concentration. The insets indicate the relative positions of the myosin heads (*M*), actin (*A*), tropomyosin (*Tm*), and the troponin subunits (*TnI*, *TnC*, and *TnT*).

and their assembly into the sarcomeric structure are discussed. The crossbridge cycle is examined, and the mechanical and biochemical correlates presented. The behavior of intact muscle is then related to the crossbridge cycle and its regulation. This includes the force–velocity relationship, energetics, metabolism, and the coordination of metabolism with muscle function. Muscle regulatory proteins at the membrane system and filament levels are also considered. The mechanisms of excitation–contraction coupling at the molecular level are elaborated on, and extrapolated to the whole tissue level.

NATIONAL BOARD TYPE QUESTIONS

Select the one most correct answer for each of the following

1. During a single crossbridge cycle:
   A. Troponin is cleaved from tropomyosin
   B. Ca^{2+} dissociates from myosin, eliciting a change in conformation to that of a rigor link
   C. The hydrolysis of ATP on the myosin crossbridge is the force-generating step
   D. One ATP molecule is hydrolyzed
   E. One molecule of lactate is produced
2. The relationship between afterload and velocity of shortening in muscle:

A. Is the primary evidence supporting the sliding-filament theory

B. Indicates that the decrease in speed is linearly proportional to the increase in load

C. Indicates that Ca^{2+} uptake by the SR is proportional to the velocity of shortening

D. Is hyperbolic, in that shortening velocity decreases rapidly with increasing load

E. Differs substantially between skeletal and smooth muscle

3. A type of genetic myopathy involves a defect in the activity of phosphorylase, the initial enzyme in the glycogenolysis pathway. What symptoms might be characteristic of this syndrome?

A. Lack of expression of ryanodine

B. Inexcitability, due to failure of action potential propagation

C. Characteristic short sarcomeres

D. Though low levels of muscle activity are normal, moderate levels of activity cannot be sustained

E. Increases in shortening velocity for a given afterload, compared with normal muscle

4. The molecular mechanism by which Ca^{2+} activates mammalian skeletal muscle involves the:

A. Binding of Ca^{2+} to the thick filament

B. Binding of Ca^{2+} to troponin

C. Binding of Ca^{2+} to tropomyosin

D. Direct activation of C-protein

E. Enhancement of the SR permeability to Cl^-

5. If a muscular disorder is characterized by low-level activity of glycogen phosphorylase, an enzyme which is rate limiting for glycolysis, which of the following symptoms would be characteristic of this condition:

A. Moderate muscle activity would be normal, but intensive muscle activity would be disabling

B. Muscles would contract spontaneously and irregularly

C. Contraction would be normal but relaxation would be impaired

D. There would be pronounced atrophy in all muscle fibers

E. There would be insensitivity to normal Ca^{2+} levels

ANNOTATED ANSWERS

1. D. ATP plays two roles: it dissociates the actin–myosin complex, and its hydrolysis provides the energy for contraction; however, only one ATP per crossbridge cycle is hydrolyzed.

2. D. The relationship is hyperbolic. For relatively small increases in afterload, the velocity of shortening substantially decreases.

3. D. A lack of glycogenolysis, a metabolic pathway that can relatively rapidly produce ATP but of limited capacity, would eliminate ATP from this source. Levels of muscular activity that depend on the relatively rapid replenishment of ATP would be inhibited. Lower levels of muscular activity could be supported by the relatively slower oxidative synthesis of ATP.

4. B. The binding of Ca^{2+} to troponin-C of the troponin complex alters its conformation and that of tropomyosin, such that inhibition of the actin–myosin interaction is relieved.

5. A. Oxidative metabolism can sustain sufficient levels of ATP for moderate muscle activity. High levels of activity, however, would quickly exhaust the available phosphocreatine stores and normally depend on the glycolysis pathway to furnish ATP relatively rapidly. An increase in muscle lactate levels is a consequence of heavy activity.

BIBLIOGRAPHY

Rüegg, J. C. *Calcium in Muscle Activation*. Heidelberg: Springer-Verlag, 1986.

Squire, J. *The Structural Basis of Muscular Contraction*. London: Plenum, 1981.

Woledge, R. C., Curtin, N. A., and Homsher, E. *Energetic Aspects of Muscle Contraction*. London: Academic Press, 1985.

14 Diversity of Muscle

RICHARD J. PAUL, JUDITH A. HEINY, DONALD G. FERGUSON,
AND R. JOHN SOLARO

Objectives

After reading this chapter, you should be able to:

Describe the structural and functional differences at the cellular level of the major skeletal muscle fiber types.

Describe the potential physiologic significance of isoform differences among the major muscle proteins

List the major factors influencing the development of a smooth, graded contractile force in a whole muscle

Compare and contrast the major structural and functional differences between cardiac and skeletal muscle

Compare and contrast the major structural and functional differences between smooth and skeletal muscle

Compare and contrast the major modes of activation of striated and smooth muscle

Explain how the Ca^{2+}-sensitivity of smooth muscle may be altered

Motile systems in biology show great diversity which allows adaptation to specific functions. Two major and quite different protein systems underlie most motile cells. Muscle cells use **myosin** and **actin,** whereas the flagella and cilia of bacteria use **dynein** and **tubulin.** Within muscle cells, the diversity is equally as great; for example, contractile velocities for smooth and skeletal muscle differ by a factor of 500. Another striking characteristic of muscle is its **plasticity,** or ability to adapt to different conditions by changing its structure. How both the diversity and adaptation of muscle can arise using the same component proteins myosin and actin is a central question in the study of muscle physiology at the tissue level.

SKELETAL MUSCLE

Differences between skeletal muscle cells (fibers) are quite striking and obvious, as illustrated by the clear difference between red and white meat. Table 14-1 lists some of the characteristics of the different muscle fiber types and their nomenclature. The different systems of names have evolved from the various functional or histologic techniques used to differentiate the muscle cells. From both a histologic and functional perspective, there are basically three fiber types, as depicted in Figure 14-1. For our purposes here, we will use the fiber-type nomenclature that is based on function, namely **fast glycolytic** (FG), **fast oxidative–glycolytic** (FOG), and **slow oxidative** (SO). Most muscles are a mixture of these three types of fibers, though the proportions vary

Table 14-1. Morphologic and Histochemical Types of Twitch Fibers in Mammalian Skeletal Muscle

Classification	Fast Glycolytic (FG)	Slow Oxidative (SO)	Fast Oxidative–Glycolytic (FOG)
Other nomenclature	Fast, fatigable	Slow, fatigue-resistant	Fast, fatigue-resistant
	Fast-twitch white	Slow-twitch intermediate	Fast-twitch red
	White	Medium	Red
	A	B	C
	IIB	I	IIA
	I	III	II
	White	Intermediate	Red
Morphologic properties			
Mitochondrial content	Small	Intermediate	Large
Sarcoplasmic reticulum	Dense	Intermediate	Dense
Fiber diameter	Large	Intermediate	Small
Histochemical properties			
Oxidative enzyme activities	Low	High or intermediate	Intermediate or high
Mitochondrial ATPase	Low	Intermediate	High
Glycolytic activities	High	Low or variable	Intermediate or low
Myoglobin content	Low	High	High
Glycogen content	Intermediate	Low	High

considerably in different muscles. All such heterogeneous muscles appear pale compared with red muscle, which is mainly composed of SO fibers. The red color is imparted by the presence of myoglobin, a protein that facilitates the diffusion of oxygen through the cell. In primates, the prevalence of FOG fibers is relatively low.

The differences in fiber types are perhaps most easily understood in terms of their adaptation to the different power outputs required, and the corresponding difference in the fuels used. Many activities require large forces that are developed rapidly, but not necessarily sustained. The FG fibers are adapted to this task. FG fibers are large and their well-developed sarcoplasmic reticulum facilitates rapid control of contraction and relaxation. Their metabolism is highly glycolytic, and ATP synthesis can respond rapidly to demand, though with limited capacity.

Other tasks, such as those required of postural muscles, involve a more continuous level of activity. The fatigue-resistant SO fibers are designed for this. Highly oxidative, their metabolism can efficiently support moderate levels of

ATP use. Their relatively high capillary density and the presence of myoglobin are directed toward eliminating oxygen diffusion as a possible rate-limiting process. From the utilization aspect, the economical maintenance of force is facilitated by the lower actin-activated myosin ATPase of SO fibers relative to FG. The trade-off, however, is that the speed of shortening is also slower.

FOG fibers are a less common component, and it is difficult to identify a muscle composed of predominantly FOG fibers, unlike other fiber types. Some ocular muscles appear to pose the best example. Their contractile speed and twitch duration are intermediate between those of SO and FOG fibers, and they are characterized by a high level of both oxidative and glycolytic metabolism. They appear to be specialized for fine, fast movements and for near-continuous activity.

FO, FOG, and SO fiber types also differ significantly in their speed of activation, which is about threefold faster in FG than in SO fibers. In FG fibers, each step of the excitation–contraction coupling is enhanced, including the action potential, the amount of charge movement in the

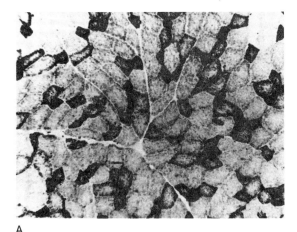

A

B

Fig. 14-1. Skeletal muscle fiber type: histology and function. (A) Cross-section of cat gastrocnemius muscle stained for mitochondrial ATPase activity. Large fibers are fast glycolytic; small dark fibers are fast oxidative–glycolytic; and fibers of intermediate size and density are slow oxidative. (From Henneman, E., and Olson, C. B., Relations between structure and function in the design of skeletal muscles. *J. Neurophysiol.* 28:581, 1965.) (B) Isometric force as a function of stimulus frequency for different fiber types. (*A* = fast glycolytic; *B* = slow oxidative, both from cat gastrocnemius; *C* = fast oxidative–gycolytic, from the soleus.) (From McPhedran, A. M., et al., Properties of motor units in a homogeneous red muscle (soleus) of the cat. *J. Neurophysiol.* 28:71, 1965.)

transverse tubule system (T-system), and the rate of Ca^{2+} release from the sarcoplasmic reticulum. These differences in the speed of excitation–contraction coupling correlate well with differences in the junctional surface area of the T-system and sarcoplasmic reticulum, and with the size of the junctional contact area. Some of the effects on the performance of the whole organism produced by the different fiber types and their respective metabolism are summarized in Table 13-2 and Figure 13-8.

Diversity of Contractile and Regulatory Proteins

An interesting feature of the structure–function relationship of mammalian striated muscle is that most of the myofilament proteins are now known to exist as populations of isoforms. This raises important questions concerning the functional significance of a particular isotype population as well as the regulation of the expression of the various isoforms.

Isoforms of Myosin Heavy and Light Chains

Myosin is the most well studied of the isoforms of myofilament proteins. Among various muscle types there is diversity in the expression of both the heavy and light chains. It is clear that one way nature has chosen to regulate muscle function for long-term adaptation to loading conditions is to alter the expression of myosin heavy- and light-chain **isoforms.**

There are three types of light chains in fast skeletal muscle: LC1, LC2, and LC3. There are two types of light chains in heart and slow skeletal muscle: LC1 and LC2. LC2 is the substrate for myosin light-chain kinase and is also known as the **regulatory light chain** or the **P-light chain.** The other forms of the light chain are essential to the activity of myosin. They can be extracted by alkali treatment, and are thus referred to as the **A1** and **A2 light chains** or the **essential light chains.** The composition of the particular light chain affects the rate of ATPase activity of myosin. In some cases, variants of the light chains are produced by alternative splicing of exons.

The heavy chain of myosin also exists in several isoforms. These isoforms appear to be products of a multigene family. In the case of heart muscle, which serves as a prototype, two heavy-chain isomers are known as **alpha** and **beta monomer.** Three isoforms are known; two are homodimers and one is a heterodimer. The ATPase activity and velocity of shortening are highly correlated with the relative proportions of isoforms in a given muscle fiber.

Isoforms of Actin and Tropomyosin

There are also several isoforms of actin that have subtle but potentially significant differences. Isoform populations vary with development and with long-term adaptation. As yet, though, there is no convincing evidence showing that shifts in the isoform population of actin have significant functional consequences.

Actin exists in many parts of the cell besides the myofilaments, and the diversity of actin may be related to activity unassociated with contraction. For example, both cardiac and skeletal alpha-actin are expressed in the fetal rat heart, whereas only the alpha-cardiac form is expressed in the adult. There is no strong evidence indicating that these two forms of actin substantially affect thin-filament Ca^{2+}-signaling.

Tropomyosin (Tm) also exists as alpha and beta isomers, the populations of which appear to be developmentally regulated. Alpha-Tm is predominant in the adult heart of small animal species, but, in the neonate, there is a relative abundance of striated muscle beta-Tm, and both smooth and non-muscle beta-Tm. As with actin, the structural changes are subtle and the functional significance is not clear.

Isoforms of Troponin-C, Troponin-T, and Troponin-I

The thin-filament Ca^{2+}-receptor protein **troponin** (Tn-C) exists as two tissue-specific isoforms—fast skeletal Tn-C and slow cardiac Tn-C. It appears that only one isotype exists within a particular mammalian muscle type. Fast Tn-C has four Ca^{2+}-binding sites, two of which are regulatory. Slow Tn-C, which is the same gene prod-

uct as cardiac Tn-C, has three Ca^{2+}-binding sites, one of which is regulatory. The nonregulatory metal-binding sites of Tn-C bind Mg^{2+} with high affinity, and appear to be important in thin-filament structure.

The presence of multiple isoforms of cardiac **troponin-T** (Tn-T) has been demonstrated in several species and muscle types. Alternative splicing of exons in the Tn-T gene suggests the existence of a wide variety of Tn-T isoforms. For example, five isoforms of Tn-T have been identified in rabbit heart. The proportion of each isoform varies with postnatal age, and there is some evidence that adaptational responses of muscle may involve shifts in the Tn-T isoform population. Variability occurs mainly in the N-terminal region of the Tn-T molecule, which lies adjacent to the region where Tm molecules overlap. Thus isoform populations of Tn-T and Tm have been suggested to be the major determinants of the different responses to Ca^{2+} observed in fast skeletal muscle fibers.

Three variants of **troponin-I** (Tn-I) are known: fast, slow, and cardiac. There is no evidence for alternative splicing in the processing of messenger RNAs for Tn-I. It is known that slow Tn-I is expressed as an embryonic isoform in heart muscle. Shifts in the Tn-I isotype population may be related to differential effects of acidic pH on Ca^{2+}-activation of myofilaments.

Post-transitional Modification of Myofilament Proteins

Activation of second-messenger cascades alters the state of phosphorylation of myofilament proteins. The Ca^{2+}-calmodulin–dependent pathway is associated with phosphorylation of myosin LC2 by activation of myosin light-chain kinase. In striated muscle, it is now clear that this phosphorylation is not responsible for activation. The state of phosphorylation of myosin does, however, modify the level of force achieved at submaximal levels of thin-filament Ca^{2+}-activation. The combined beta-adrenergic receptor–cyclic adenosine 3′: 5′ monophosphate (cAMP) pathway is associated with the phosphorylation of Tn-I, but only in heart muscle. Phosphorylation of Tn-I reduces the affinity of Tn-C for Ca^{2+}, and thus may act as a negative-feedback signal. An important topic that remains to be clearly addressed is the role of Tn-T phosphorylation. Tn-T, which is an elongated, tadpole-shaped molecule, is phorphorylated at one end by a cAMP- and Ca^{2+}-independent Tn-T kinase. Tn-T is also phosphorylated by phosphorylase kinase and protein kinase C at the other end of the molecule. These sites of phosphorylation are in regions of the molecule that might be involved in Ca^{2+} signaling, as well as in thin-filament cooperativity. There are also sites of phosphorylation in Tm that are strategically located at the region of overlap between adjacent molecules. The functional significance of Tn-T and Tm phosphorylation, however, remains unclear.

Plasticity of Muscle

While the physiologic significance of these isoforms is not yet completely understood, it is clear that muscle is a very plastic organ, and, depending on conditions, fiber type can be modified. For example, denervation of an SO fiber elicits a change toward the mechanical characteristics of an FG fiber. Cross-innervating an SO fiber with a nerve originally from an FG fiber also elicits a change in the SO fiber toward FG characteristics. Whether a trophic factor from the nerve or the stimulus frequency is the major stimulus for the transition is not certain. However, alteration of the muscle gene expression by a stimulus and by mechanical conditions is an area of considerable research interest.

Factors Influencing Total Force Developed during Contraction of Whole Muscle

Before leaving this discussion of skeletal muscle, it is useful to return to the question originally posed in Chapter 12. If submaximal·stimuli result in nonfused tetani (see Fig. 14-1), how do muscles produce smooth submaximal contractions?

The functional unit of muscle at the whole-tissue level is the **motor unit.** It consists of one motor neuron, its axon, and all the muscle cells innervated by that motor neuron. Individual motor units are composed of the same fiber types, but the individual fibers are not necessarily localized in the same area. Smooth submaximal contractions in whole muscle can result from the asynchronous firing of a large number of motor neurons. This type of force summation is different from the smooth contraction in a single fiber resulting from the temporal summation of depolarizations in a tetanus. Nontetanic, smooth contractions in whole muscle are sometimes referred to as **spatial summation** and involve the interplay of various factors, including: (1) twitch duration of individual fibers; (2) frequency of firing; (3) number of motor units recruited; and (4) size of the motor units, including the number of fibers per motor neuron and the fiber cross-section area.

The order of motor unit recruitment is also important in graded, smooth contractions of whole muscle. In the lower ranges, force is increased through the addition of motor units, which are recruited in order of their size. This is related to the fact that, as the size of the motor units increases, the size of the motor neuron also increases. Large neurons are less excited than are the smaller neurons of small motor units. Finally, at the highest forces, with most motor units involved, increases in force are achieved by increased frequency of firing. Interestingly, there is evidence that, with training, this so-called normal recruitment pattern can be bypassed.

CARDIAC MUSCLE

Cardiac Muscle Structure

Cardiac muscle cells are rod-shaped and intermediate in size between skeletal and smooth muscle. Typically, the working ventricular cell is

15 to 20 μm in width and 150 μm long. The basic elements in the cell are shown in Figure 14-2. These elements are similar to those in skeletal muscle, but the relative amounts differ. The sarcolemma contains transverse (T) tubules as in other striated muscle, but the openings are bigger in the heart. The T-tubules also snake along the length of the cell in ventricular heart muscle. The sarcoplasmic reticulum content of heart muscle cells is about half that of skeletal muscle. Overall, the role of the membrane systems, shown in Figure 14-2, in excitation–contraction coupling is similar to that described for skeletal muscle in Chapter 13. A major exception is that the regulated influx of extracellular Ca^{2+} is an important factor both in the mechanism of the action potential and for activation of the contractile apparatus. Details of these mechanisms are described in Chapter 15. Cardiac muscle has about one-half the number of myofilaments of skeletal muscle. Moreover, the relative amount of mitochondria is large in heart. These specialized features of cardiac muscle are related to its relatively slow but constant active contraction and relaxation cycles.

Cardiac Muscle Function

The force per unit cross-sectional area generated by cardiac muscle cells is about half that of skeletal muscle. The basic mechanism for force generation is the same as that described for skeletal muscle, but as just mentioned, there are about one-half as many myofilaments acting in parallel to generate force. The energy cost of force generation is about one-fifth that for fast skeletal

Fig. 14-2. Ultrastructure of the working myocardial cell. Contractile proteins are arranged in a regular array of thick and thin filaments. The A-band is that region of the sarcomere occupied by the thick filaments into which thin filaments extend from either side. The I-band is occupied only by thin filaments that extend toward the center of the sarcomere from the Z-lines; they bisect each I-band. The sarcomere is between each pair of Z-lines; it contains two half I-bands and one A-band. The sarcoplasmic reticulum consists of the sarcotubular network at the center of the sarcomere and the cisternae, which abut on the transverse (T) tubules and the sarcolemma. The T-tubular system is lined by a membrane that extends from the sarcolemma and carries the extracellular space into the myocardial cell. In contrast to the T-tubules of skeletal muscle, those of the myocardium can run in a longitudinal as well as a transverse direction. (From Katz, A. M., Congestive heart failure: Role of altered myocardial cellular control. *N. Engl. J. Med.* 293:1184, 1975.)

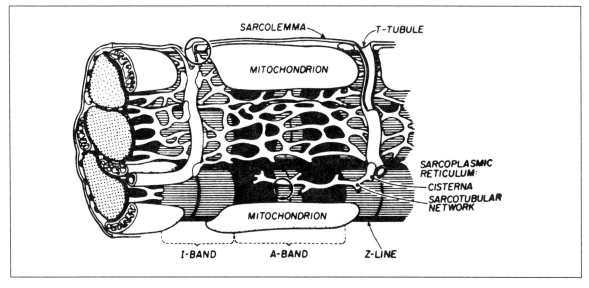

muscle, owing to a slower crossbridge cycling rate and unloaded velocity of shortening.

The force–velocity and length–tension relationships of cardiac muscle are essentially the same as those for fast and slow skeletal muscle. There is no strong indication that the fundamental operation of the sliding-filament mechanism differs for these muscle types. An interesting feature of the length–tension relationship in heart muscle is that force falls off much more steeply than it does in skeletal muscle, as the muscle shortens from lengths associated with optimal overlap. The shape of the length–tension relationship is important because it is the basis of the Frank-Starling relationship, which underlies the pump characteristics of the heart (see Chapter 19). The mechanism for this difference between heart and skeletal muscle appears to stem from the length dependence of excitation–contraction coupling, involving either Ca^{2+} release from the sarcoplasmic reticulum or the process by which Ca^{2+} activates the myofilaments.

Normally cardiac muscle contraction relies on aerobic metabolism, although there is a capability for anaerobic glycolysis. As expected for a system in which energy demand is normally matched by aerobic energy supply, heart muscle cells have a higher density of mitochondria and a lower amount of phosphocreatine than is found in fast skeletal muscle. When the oxygen supply is limited in heart muscle, force is inhibited by the accumulation of metabolites, especially protons and inorganic phosphate. This is important to the preservation of heart cells under pathophysiologic conditions, as this automatically reduces energy demand. This of course renders the myocardium dysfunctional as a pump, but permits preservation of the cells until oxygen can be restored. (Additional protective mechanisms are discussed in Chapters 15 and 17.)

SMOOTH MUSCLE

Smooth muscle, so-called because it lacks the sarcomeric banding characteristic of striated muscle, is an important tissue that typically lines the hollow organs of, for example, the vasculature and gastrointestinal tract. Historically it has taken a secondary role in studies of muscle, largely because its structure is less organized than that of skeletal muscle, making it less amenable to biophysical experimentation. However, for health professionals it is of major significance, as many of the diseases associated with industrial society, such as hypertension, coronary artery disease, stroke, asthma, and gastrointestinal disorders, are related to smooth muscle pathology. Many theories of smooth muscle contraction have been extrapolated from those of the more easily and extensively studied skeletal muscle. While there are many similarities, striking differences have recently been shown between striated and smooth muscle, particularly in terms of the regulation of contractility.

Perhaps the most impressive characteristic of smooth muscle is its ability to maintain large forces at relatively low energy cost. It is superbly adapted to this function. For example, if skeletal, rather than smooth, muscle lined the vasculature, the energy cost for just regulation of the circulation would be nearly twice that for the entire organism's basal metabolism. How this is accomplished using a basically similar actin-myosin–based contractile system is the focus of much research. Our presentation of smooth muscle will follow the structure–function analysis used for skeletal muscle.

Smooth Muscle Structure

Smooth muscle cells are spindle-shaped and very small compared with skeletal muscle. They are about 5 to 10 μm in diameter (one-tenth that of skeletal muscle) and up to several hundred micrometers long. Like striated muscle, smooth muscle cells contain thick, myosin-containing, and thin, actin-containing, filaments. However, in smooth muscle, the thick and thin filaments are not organized into myofibrils or regular sarcomeres. A cell cytoskeleton serves as attachment points for thin filaments and allows force transmission to the ends of cells. There are no z-bands per se, but specialized cytoskeleton regions, known as **dense bodies** or **patches,** appear to serve as comparable structures for the attachment of thin filaments. These areas contain alpha-actinin, a protein also found in the z-bands of skeletal muscle. Intermediate filaments with 10-nm diameters (intermediate between thin, 7-nm, and thick, 15 nm, filament diameters) link the dense bodies to form a cytoskeleton network. The ratio of thin to thick filaments in smooth muscle (approximately 15 : 1) is considerably higher than that of striated muscle (2 : 1), and the myosin content can be substantially lower, approximately one-fifth that of skeletal muscle. Electron micrographs of smooth muscle and a drawing of the filament structure are shown in Figure 14-3.

Smooth muscle also contains a reticular membrane system which, in terms of its ability to store and release Ca^{2+}, is analogous to the sarcoplasmic reticulum of skeletal muscle. The volume of the sarcoplasmic reticulum in smooth muscle ranges from about 2 percent in phasic, spike-generating smooth muscles, like the taeniae coli, to 5 percent to 7.5 percent in tonic smooth muscle, such as that found in the large elastic arteries. Thus, in extent, the sarcoplasmic reticulum of smooth muscles can be as abundant as that in striated muscle; however, its relationship to the contractile filaments is less clearly defined. This is not unexpected, as the time required for diffusion of intracellular Ca^{2+} would not be a limiting factor because of the small diameters of smooth muscle cells.

Smooth Muscle Function

The isometric force per cross-section area generated by smooth muscle can be as large as or larger than that of skeletal muscle, which is surprising in light of its smaller myosin content. As was previously indicated, the energy cost of tension maintenance (the rate of ATP hydrolysis per force per cross-section area) can be up to 500-fold lower than that of skeletal muscle. Again, the trade-off is in terms of contraction speed, with smooth muscle velocities being one

to two orders of magnitude lower.

The force–length relationships for smooth muscle are qualitatively similar to those for skeletal muscle. This can be taken as evidence for an analogous sliding-filament mechanism. How such a mechanism is visualized for the much less organized filament structure of smooth muscle is shown in Figure 14-3.

The force–velocity relationship is also qualitatively similar to that of skeletal muscle, albeit with much slower velocities. This lends credence to the existence of a similar crossbridge cycle for smooth muscle contractility. The actin-activated myosin ATPase activity of smooth muscle is substantially less than that of skeletal muscle. This and the lower crossbridge cycle rate can grossly account for the slower shortening speeds and lower tension cost.

In addition, smooth muscle appears to possess mechanisms for regulating the rate of crossbridge cycling as well as the number of activated bridges. Following stimulation, contraction velocities decrease, while the ability to maintain isometric force remains constant. This state of maintaining force despite reduced velocities is known as the **latch state.** As many smooth muscles are always tonically activated, for example, those in blood vessels, the latch state may be dominant under certain physiologic conditions. Thus the regulatory mechanisms involved are the focus of much current research, and will be discussed below.

Smooth muscle metabolism is primarily oxidative. The density of mitochondria is generally considered to be low, though it is of similar magnitude to that of FG skeletal fibers. Because the contractile energy requirements are low, oxidative ATP synthesis generally matches energy demand. Thus, in spite of low phosphocreatine pools compared to skeletal muscle, little change in either the phosphocreatine or ATP concentration can be measured during contraction, because increases in oxidative phosphorylation provide the ATP as needed. Thus, the oxygen debt phenomenon observed in skeletal muscle (oxidative resynthesis of phosphocreatine after ces-

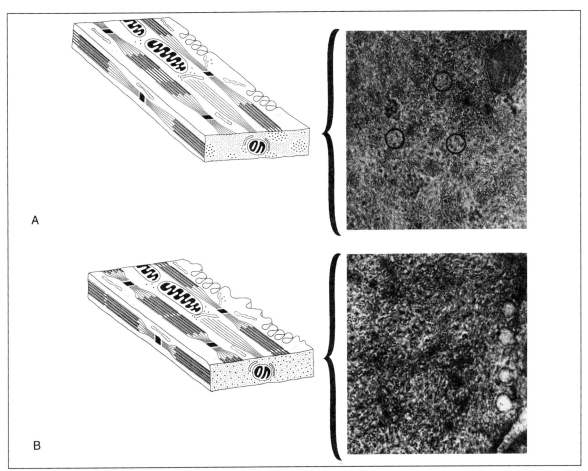

Fig. 14-3. Morphology of relaxed and contracted smooth muscle. (A) This diagram represents a portion of a relaxed smooth muscle cell, highlighting the arrangement of the actin-containing thin filaments and the myosin-containing thick filaments. This rendering is greatly simplified from the actual arrangement. From the longitudinal orientation, thin filaments arise at dense bodies and project to interact with thick filaments, forming the contractile apparatus. In the cross-section, the filaments are distributed in a nonuniform manner, with thick filaments forming clusters surrounded by groups of thin filaments. This arrangement is shown in the electron micrograph (*circled*), which is a cross-section of a relaxed visceral smooth muscle cell. (B) A contracted smooth muscle cell. The notable differences from the relaxed cell shown in A are that, in the longitudinal views, the thin filaments overlap considerably more of the thick filaments, drawing the dense bodies closer together and shortening the cell. In cross-section, the thin and thick filaments are randomly distributed to form a uniform pattern. The electron micrograph is a cross-section of a contracted visceral smooth muscle cell, which shows more uniform distribution of the thin and thick filaments. (Diagrams are modified from Heumann, H. G., Smooth muscle: Contraction hypothesis based on the arrangements of actin and myosin filaments in different states of contraction. *Philos. Trans. R. Soc. Lond.* [Biol.], 1973; 265:213.)

sation of contractile activity) in general does not exist for smooth muscle. An unusual aspect of smooth muscle metabolism is its production of substantial amounts of lactate when it is fully oxygenated. In some smooth muscles, notably vascular smooth muscle, this lactate production appears to be related to the energy requirements of membrane processes, such as ion pumps. Presumably the ATP-synthesizing, glycolytic enzyme cascades are co-localized in the plasma membrane. Thus the ATP-synthesizing machinery may be compartmentalized with the energy-dependent functions they subserve.

Excitation—Contraction Coupling in Smooth Muscle

Ca²⁺ Control at the Contractile Filament Level

Until the mid-1970s, the mechanism for the regulation of contractile activity in smooth muscle was postulated, by analogy to skeletal muscle, to be a thin-filament–linked mechanism involving Tn and Tm. Imagine both the consternation and excitement when further studies indicated that smooth muscle does not contain the Ca^{2+}-receptor protein Tn.

It is instructive to compare the biochemical behavior of purified actin and myosin from skeletal and smooth muscle. In skeletal muscle (see Fig. 13-5), when ATP is added to purified actin and myosin, the myosin ATPase is activated and actin and myosin interact. The presence of the thin-filament proteins, Tn and Tm, inhibit this interaction and Ca^{2+} removes this inhibition. In smooth muscle, on the other hand, purified actin and myosin are inactive and the myosin ATPase activity is low. Therefore, a Ca^{2+}-linked activator, rather than a de-inhibitor, is required for activity.

The activating factor for the smooth muscle actin—myosin interaction was found to be a covalent modification of the myosin itself, specifically the phosphorylation of the 20-kDa light chain. Ca^{2+} sensitivity for this phosphorylation resides in the Ca^{2+} dependence of the enzyme, **myosin light-chain kinase,** that catalyzes the phosphorylation of the myosin light chain. The Ca^{2+}-receptor protein involved is the ubiquitous Ca^{2+}-binding protein, **calmodulin.** The activation sequence involves the binding of Ca^{2+} to calmodulin. In turn this complex binds to myosin light-chain kinase and activates its ATP-dependent phosphorylation of myosin, leading to activation of the actin—myosin interaction.

In contrast to skeletal muscle, relaxation in smooth muscle does not simply involve reversal of the activation process. The dephosphorylation of myosin necessary to inactivate the actin—myosin interaction is catalyzed by a separate enzyme **myosin light-chain phosphatase.** Most current biochemical evidence suggests that this phosphatase is not itself regulated, but some functional studies suggest that this may also be a control point of contractility.

This basic scheme of Ca^{2+} control of smooth muscle contractility is depicted in Figure 14-4. It is generally accepted that phosphorylation of myosin plays an obligatory and major role in the regulation of smooth muscle contraction, though whether it is the only control mechanism remains uncertain. Other thin-filament–linked, Ca^{2+}-dependent modulators have been postulated, particularly for control of the latch state. What is clear is that even the basic scheme has a large number of potential regulatory sites, and, given the importance of smooth muscle in disease processes, this will be the active focus of much research.

Control of Intracellular Ca²⁺ Concentration

The control of intracelular Ca^{2+} concentration in smooth muscle is complex, involving both extracellular and intracellular sources, Ca^{2+} uptake by the sarcoplasmic reticulum, and extrusion by pumps and exchangers on the plasma membrane. These will be considered in detail in Chapter 15. It is important to note that, though alteration of the intracellular Ca^{2+} concentration is a major mechanism for regulating smooth muscle

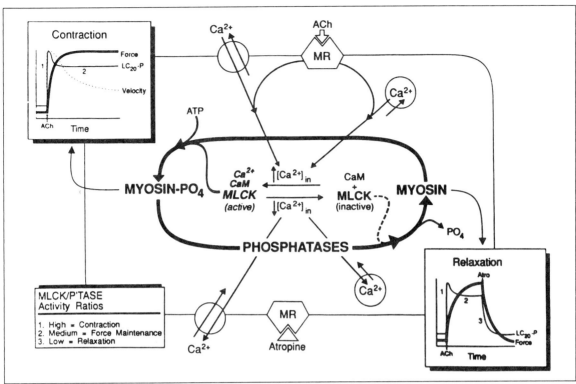

Fig. 14-4. The molecular basis of regulation of smooth muscle contraction. Stimulation of muscarinic receptors (*MR*) by acetylcholine (*ACh*) increases the intracellular Ca^{2+} concentration due to entry of external Ca^{2+} and release of Ca^{2+} from internal stores. Ca^{2+} binds to calmodulin (*CaM*); and the Ca^{2+}–CaM complex then binds to and activates myosin light-chain kinase (*MLCK*). Phosphorylation of myosin by MLCK stimulates actin-activated myosin-ATP hydrolysis, which produces contraction. The temporal relationships among the 20-kDa light-chain phosphorylation (LC_{20}-P), force, and velocity during contraction are shown in the top left. Relaxation begins with cessation of agonist stimulation, in this case due to atropine (*Atro*) binding to a muscarinic receptor. This results in decreased intracellular Ca^{2+} content, dissociation of Ca^{2+} from CaM, inactivation of MLCK due to dissociation of CaM from enzyme, dephosphorylation of myosin by phosphoprotein phosphatases (*P'tase*), and relaxation. The *broken line* from inactive MLCK signifies that it is passively involved in dephosphorylation. The temporal relationship between force and myosin dephosphorylation during relaxation is shown in the *bottom right*. Note that smooth muscles contract and relax because of changes in the relative rates of the enzymes phosphorylating (MLCK) and dephosphorylation myosin (phosphoprotein phosphatases). The activity ratios of these enzymes postulated for contraction and relaxation are summarized in the box at the bottom left. (From de Lanerolle, P., and Paul, R. J. Myosin phosphorylation/dephosphorylation and regulation of airway smooth muscle contractility. *Am. J. Physiol.* 261:L1–L14, 1991.)

tone, it is not the sole mechanism. Given the variety of control points, there are a number of cases in which the sensitivity of the contractile system to a given level of Ca^{2+} is the regulated parameter. For example, the phosphorylation of myosin light-chain kinase as a consequence of beta-adrenergic stimulation would inhibit kinase activity. Thus, for a similar level of intracellular Ca^{2+}, there would be a lower degree of myosin light-chain phosphorylation and corresponding force production.

Table 14-2. Summary of the Major Modes of Activation of Smooth, Cardiac, and Skeletal Muscle

Muscle Protein Type	Membrane Stimulus	Ca²⁺ Source	Ca²⁺-Binding Regulatory Protein
Skeletal	Depolarization, largely CNS controlled	Intracellular Ca^{2+} release from SR	Skeletal troponin-C
Cardiac	Depolarization, largely ANS-controlled or hormones	Ca^{2+} influx through voltage-activated Ca^{2+} channels; intracellular Ca^{2+} release mediated by Ca^{2+}; reverse Na^+–Ca^{2+} exchange	Cardiac troponin-C
Smooth	Depolarization, largely ANS-controlled or hormones	Intracellular Ca^{2+} release from SR mediated by IP_3; Ca^{2+} influx through voltage-activated and/or receptor-activated Ca^{2+} channels; reverse Na^+–Ca^{2+} exchange	Calmodulin

CNS = central nervous sytem; ANS = autonomic nervous system; SR = sarcoplasmic reticulum; IP_3 = inositol-tris-phosphate.

SUMMARY

The diversity of muscle is described at the cellular level for different types of skeletal muscle fibers. This is further developed at the molecular level in terms of the isoforms of the major muscle proteins. Our consideration of skeletal muscle concludes with a discussion of the factors influencing the generation of total force in a whole muscle. Next, cardiac and smooth muscle are considered in the context of the structure–function paradigm, paralleling that of skeletal muscle. The major differences in the contractile proteins and the regulation of contractility are also considered. Table 14-2 summarizes the major modes of activation of skeletal, cardiac, and smooth muscle.

NATIONAL BOARD TYPE QUESTIONS

Select the one most correct answer for each of the following

1. Under the light microscope, smooth muscle does not show the banding pattern characteristic of skeletal muscle. This is best attributed to:
   A. Activation of contraction in smooth muscle, which is not Ca^{2+}-dependent
   B. A high glycolytic capacity relative to skeletal muscle
   C. Lack of a T-system
   D. The less regular organization of thick and thin filaments in smooth muscle
   E. A contractile system based on tubulin and intermediate filaments
2. Relaxation of smooth muscle is associated with:
   A. Dephosphorylation of myosin light chains
   B. Activation of myosin light-chain kinase
   C. Increases in the concentration of inositol-tris-phosphate
   D. Increased oxidative metabolism
   E. Formation of the latch state of the cross-bridges

3. In comparison to skeletal muscle, cardiac muscle is characterized by:
   A. A qualitatively different force–length relationship
   B. The absence of Tn
   C. A less extensive sarcoplasmic reticulum
   D. A greater ratio of actin- to myosin-containing filaments
   E. The absence of striations at the light microscope level

4. In comparison to fast-twitch (type IIB, FG) fibers, slow-twitch (type I, SO) fibers have:
   A. A higher myoglobin content
   B. A higher glycolytic capacity
   C. More T-tubule–sarcoplasmic reticulum junctional surface area
   D. Similar myosin isoforms
   E. Different force–length relationship

5. A smooth, submaximal contraction in a whole muscle is dependent on:
   A. Synchronization of each crossbridge cycle
   B. Spatial summation of individual motor units
   C. Mixtures of fiber types within the whole muscle
   D. The presence of differing myosin isoforms
   E. The nearly identical twitch durations of all fibers

ANNOTATED ANSWERS

1. D. The banding pattern of skeletal muscle is ascribable to the regular spacing of the filaments within the sarcomere, and the repeating sarcomeres themselves. Although smooth muscle contains similar filaments, the order is considerably lower and no regular banding pattern is seen at the light microscope level.

2. A. Phosphorylation of the myosin light chain by the Ca^{2+}-dependent enzyme myosin light-chain kinase is an obligatory step in the activation of smooth muscle. Relaxation involves the dephosphorylation of the light chains, catalyzed by myosin light-chain phosphatase.

3. C. The less extensive sarcoplasmic reticulum reflects the slower speeds of activation and relaxation.

4. A. Slow-twitch fibers are more oxidative, with higher levels of myoglobin to facilitate diffusion of oxygen. They are slower, and this is paralleled by the decreased T-tubule–sarcoplasmic reticulum junctional surface area, which reflects slower speeds of activation and relaxation.

5. B.

BIBLIOGRAPHY

Paul, R. J. Smooth Muscle: Mechanochemical Energy Conversion, Relations between Metabolism and Contractility. In Johnson, L. R., et al., eds. *Physiology of the Gastrointestinal Tract*, 2nd ed. New York: Raven Press, 1987, vol. 1, pp. 483–506.

Rüegg, J. C. *Calcium in Muscle Activation*. Heidelberg: Springer-Verlag, 1986.

Solaro, R. J., *Protein Phosphorylation in Heart Muscle*. Boca Raton, FL: CRC Press, 1986.

15 Excitation–Contraction Coupling in Cardiac Muscle and Smooth Muscle

NICHOLAS SPERELAKIS

Objectives

After reading this chapter, you should be able to:

Describe the differences in excitation–contraction coupling between cardiac muscle and smooth muscle, and how these differ from that for skeletal muscle

List the various routes by which Ca^{2+} enters the cell to initiate contraction, and the various mechanisms by which free Ca^{2+} concentration in the myoplasm is reduced back to the resting level to bring about relaxation

Discuss the properties of the voltage-dependent Ca^{2+} channels and their regulation

CARDIAC MUSCLE

As discussed in previous chapters, Ca^{2+} is the key intracellular messenger for the regulation of contraction. In cardiac muscle, the force of contraction is dependent on the extracellular Ca^{2+} concentration ($[Ca]_o$) (Fig. 15-1). When $[Ca]_o$ goes below the normal level of about 1.8 mM, the contractile force is diminished, and ultimately abolished when the $[Ca]_o$ becomes zero. Conversely, when the $[Ca]_o$ is elevated, contractile force is augmented to a maximum level. This close dependence on $[Ca]_o$ in cardiac muscle is qualitatively and quantitatively different from that for skeletal muscle, which has much less immediate dependence on $[Ca]_o$. In isolated hearts perfused through the coronary arteries, the time constant for a change in contractile force following a step change in the $[Ca]_o$ of the perfusate is about 5 to 10 seconds, which is very short. Most of this time for equilibration is occupied by diffusion from the vascular compartment to the interstitial compartment. The effect of a change in the Ca^{2+} concentration at the outer surface of the membrane is virtually instantaneous.

The relationship between contractile force and intracellular Ca^{2+} concentration ($[Ca]_i$), on a logarithmic scale, is sigmoid. The $[Ca]_i$ of a myocardial cell at rest is about 1×10^{-7} M; elevation to 1×10^{-6} M produces about half-maximal force generation. The curve plateaus (saturates), so that elevation beyond 1×10^{-5} M does not further increase contraction. The **force versus Ca^{2+} sensitivity curve** can be shifted to the left or right under certain conditions, such as acidosis or alkalosis and the action of some drugs, thereby affecting how much force can be developed for a given rise in $[Ca]_i$.

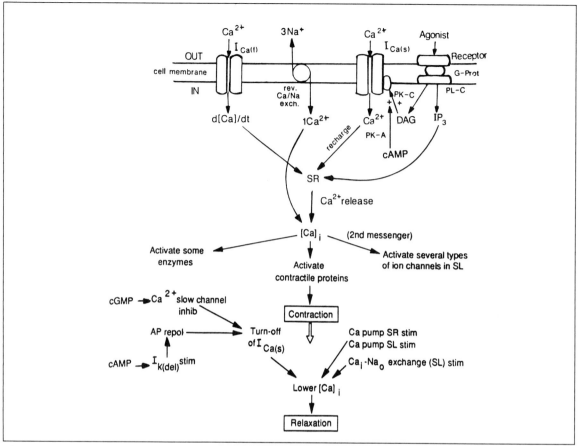

Fig. 15-1. Summary of excitation–contraction coupling in myocardial cells. There is Ca²⁺ influx through the sarcolemma (*SL*) via two types of voltage-dependent Ca²⁺ channels: a slow (or long-lasting) [L-type] channel and a fast (or transient) [T-type] channel. There is also some Ca²⁺ influx via the Ca²⁺–Na⁺ exchange reaction operating in the reverse mode because of the prolonged depolarization during the cardiac action potential (*AP*). The entering Ca²⁺ also triggers release of considerable additional Ca²⁺ stored in the sarcoplasmic reticulum (*SR*) by activating the Ca²⁺-release channels in the SR membrane. Thus, there are two pools of Ca²⁺ for contraction; the extracellular (interstitial) pool and the SR pool. Relaxation is produced when the intracellular Ca²⁺ concentration [Ca]ᵢ is lowered to the resting level by several mechanisms: turning off the inward slow Ca²⁺ current [$I_{Ca(s)}$] due to repolarization of the AP, Ca²⁺ pumping out of the myoplasm and into the SR lumen and the interstitial space, and exchange of intracellular Ca²⁺ for extracellular Na⁺ (Ca_i-Na_o) across the sarcolemma. ($I_{Ca(t)}$ = inward fast Ca²⁺ current; *G-Prot* = GTP-binding protein; *PK-C, PK-A* = protein kinase C and A; *DAG* = diacylglycerol; IP_3 = inositol-tris-phosphate; *PL-C* = phospholipase C; $d[Ca]/dt$ = rate of change of myoplasmic Ca²⁺; $I_{K(del)}$ = delayed-rectifier K⁺ current.)

Elevation of [Ca]ᵢ to Initiate Contraction

Ca²⁺ Slow Channels

The reason for the strong dependence of contraction on [Ca]ₒ is that Ca²⁺ influx into the myo-

cardial cell during excitation is greater when [Ca]ₒ is greater (see Fig. 15-1). This Ca²⁺ influx mainly passes through the voltage-gated slow Ca²⁺ channels (also known as the *L-type Ca²⁺ channels*) in the sarcolemma, and can be recorded as the

inward slow current or **Ca²⁺ slow current.**
Some of the major properties of the slow Ca^{2+}
channels are summarized in Table 15-1. As the
name *slow channel* implies, these channels (on
a population basis) are **kinetically slower** than
fast channels, in that they turn on (activate) more
slowly, turn off (inactive) much more slowly
(sometimes not at all), and recover (reactivate)
more slowly. In addition, slow channels operate
over a different voltage range, namely less negative (more depolarized); in other words, the slow
channels activate and inactivate at more depolarized voltages. Tetrodotoxin (TTX), a specific
blocker of fast Na^+ channels, has no effect on slow
channels. The conductance of each slow Ca^{2+}
channel is about 25 pS (25 × 10^{-12} siemens or
ohm^{-1}). These slow Ca^{2+} channels are opened
(made conducting) during the membrane depolarization produced during the cardiac action potential (AP). The slow Ca^{2+} channels stay open
during the entire period of the plateau compo-

Table 15-1. Summary of Major Differences between
the Slow (L-Type) and Fast (T-Type) of Ca^{2+} Channels

Properties	Ca^{2+} Channels	
	Slow (L-type)	*Fast (T-type)*
Duration of current	Long-lasting (sustained)	Transient
Inactivation kinetics	Slower	Faster
Activation kinetics	Slower	Faster
Threshold	High	Low
Inactivation potential	ca. −20 mV	ca. −50 mV
Single-channel conductance	High	Low
Regulated by cAMP and cGMP	Yes	No
Regulated by phos-phorylation	Yes	No
Blocked by Ca^{2+} antagonist drugs	Yes	No
Opened by Ca^{2+} agonist drugs	Yes	No

cAMP = cyclic adenosine 3′: 5′ monophosphate;
cGMP = cyclic guanosine 3′: 5′ monophosphate.

nent of the cardiac AP, and this inward Ca^{2+} slow
current actually contributes to the plateau component. When the inward Ca^{2+} slow current is
blocked by use of a calcium antagonist (such as
verapamil, diltiazem, or nifedipine), the plateau
overshoot is slightly less, the AP duration at 50%
repolarization (APD_{50}) is slightly less, and the AP
becomes a bit more "triangular" because of loss
of the Ca^{2+} slow current contribution to the plateau (see Chapter 17).

Ca²⁺ Release from the Sarcoplasmic Reticulum

The Ca^{2+} influx into the myocardial cell during
the cardiac AP plateau acts on the sarcoplasmic
reticulum (SR) to release additional Ca^{2+} into the
myoplasm. Thus, the Ca^{2+} required for contraction comes from two sources: (1) the extracellular
Ca^{2+} pool, as a Ca^{2+} influx through voltage-gated
slow Ca^{2+} channels; and (2) the Ca^{2+} pool in the
SR lumen (see Fig. 15-1). Ca^{2+} from the first extracellular pool triggers release of Ca^{2+} from the
second intracellular pool. It is estimated that, of
the Ca^{2+} required to elevate the myoplasmic Ca^{2+}
level ($[Ca]_i$) to that required for maximum force
development (about 10^{-5} M), about 10 percent is
derived from the Ca^{2+} influx across the sarcolemma and the remaining 90 percent is derived
from Ca^{2+} release from the SR. However, the Ca^{2+}
influx is the key controlling factor for determining the force of contraction, because it provokes
further Ca^{2+} release in a proportional manner.
Thus, the greater the Ca^{2+} influx (the inward slow
current), the greater the force of contraction.

The Ca^{2+} release from the SR, triggered by the
Ca^{2+} influx in a proportional manner, occurs
through the **Ca²⁺ release channels** embedded in
the SR membrane. These channels do not appear
to be voltage gated, but are gated by Ca^{2+} and
inositol-tris-phosphate (IP_3); ATP is also required
for the release. Hence, for example, Ca^{2+} acts on
the outer (myoplasmic) surface of the Ca^{2+}-
release channel to open the channel and allow
Ca^{2+} to pass into the myoplasm down a large electrochemical gradient. The Ca^{2+} concentration in
the SR lumen is much higher (e.g., >1.0 mM)

Fig. 15-2. Relationship between membrane receptors, guanosine-regulatory (*G*) proteins, membrane enzymes, second messengers, and protein kinase (*PK*) activation of cardiac and vascular smooth muscle. The following receptors are depicted: (A) atrial natriuretic peptide (*ANP*) receptor in vascular muscle, (B) beta-adrenergic receptor (for isoproterenol) in cardiac and vascular muscle as well as prostaglandin (*PGI*) receptor in vascular muscle, and (C) angiotensin-II (*Ang-II*) receptor in cardiac and vascular muscle. (*EDRF* = endothelium-derived relaxation factor; *NO* = nitric oxide; *PK-C* = protein kinase C; *PL-C* = phospholipase C; *PI* = phosphatidylinositol; IP_3 = inositol-trisphosphate; *DAG* = diacylglycerol.)

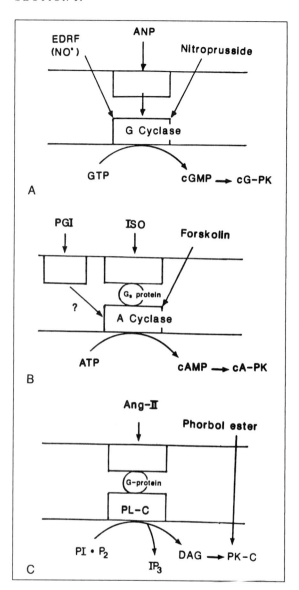

than that in the myoplasm (e.g., 1×10^{-7} M in a cell at rest), and so there is a large concentration gradient (diffusion force) for the transport of Ca^{2+} from the SR lumen into the myoplasm. It is not known whether a membrane potential normally exists across the SR membrane; if it does, it would contribute to the electrochemical driving force, depending on its magnitude and polarity (see Chapter 3).

The drug **ryanodine** has been shown to activate the Ca^{2+}-release channels in the SR of cardiac muscle and skeletal muscle. Therefore, as anticipated, this drug initially produces contracture and depletes the SR of its Ca^{2+} store.

For Ca^{2+} triggering of Ca^{2+} release from the SR, it appears that the most important factor is the **rate of change of myoplasmic Ca^{2+}** and not the absolute level of Ca^{2+} ($[Ca]_i$). That is, a rapid change in $[Ca]_i$ is a more effective activator of the Ca^{2+}-release channel. Therefore, the fast Ca^{2+} channels may be important in initiating contraction, as discussed in the next section.

It has been shown that IP_3 can activate the Ca^{2+} release channels of the SR to release Ca^{2+}. IP_3 is a second messenger produced by the activation of several types of sarcolemmal receptors for neurotransmitters and hormones, for example, angiotensin-II (Fig. 15-2). IP_3 is one end product of phosphatidylinositol metabolism in the cell membrane (along with diacylglycerol, which activates protein kinase C). The IP_3 so produced diffuses to the SR membrane, where it acts to bring about Ca^{2+} release. This mechanism is especially important in pharmacomechanical coupling in smooth muscles, but may also be involved in cardiac muscle and skeletal muscle excitation–contraction coupling.

Fast Ca^{2+} Channels

Myocardial cells possess a second type of voltage-gated Ca^{2+} channel, namely a **fast Ca^{2+} channel,** also known as the **T-type Ca^{2+} channel,** which is responsible for the fast Ca^{2+} current. As the name implies, this channel behaves much like a fast Na^+ channel, except that it is selective for Ca^{2+} (and not Na^+) and is not blocked by TTX. Table 15-1 summarizes the differences between the slow and fast Ca^{2+} channels. As shown, the Ca^{2+} fast channels have a low threshold, activate quickly, inactivate quickly (hence are transient [T-type]), are not blocked by the calcium antagonist drugs, are not regulated by cyclic nucleotides and phosphorylation, and have a lower single-channel conductance (about half that of the slow channels—about 10 to 12 pS). The number (per cell) of fast channels is usually much less than that of the slow channels. The function of the fast Ca^{2+} channels is not known, but some investigators believe that they provide the rapid change in $[Ca]_i$ required for the effective release of Ca^{2+} from the SR (see second to the last paragraph in the section "Ca^{2+} Release from the Sarcoplasmic Reticulum"), and that the slow Ca^{2+} channels provide the Ca^{2+} for recharging the SR stores for subsequent contraction.

Ca^{2+}–Na^+ Exchange Reversal

The Ca^{2+}–Na^+ exchange reaction contributes to the removal of Ca^{2+} from the cell, in that it helps the sarcolemmal Ca^{2+}-ATPase–Ca^{2+} pump in this job (see Fig. 15-1). In a cell at rest, the Ca^{2+}–Na^+ exchange helps maintain the ion distributions, namely a very low free $[Ca]_i$. The exchange in a resting cell is directed to trade one intracellular Ca^{2+} for three extracellular Na^+. The energy for the uphill transport (movement) of Ca^{2+} comes from the downhill transport of Na^+ (down its large electrochemical gradient), and the energetics are such that the outward movement of 1.0 mole of Ca^{2+} from the cell requires the inward movement of 3.0 moles of Na^+. Therefore, this exchange reaction contributes to the relaxation of cardiac muscle (immediately after the AP), which requires rapid lowering of the elevated $[Ca]_i$ (see the section "Lowering of the Elevated $[Ca]_i$ to Produce Relaxation").

In addition, the Ca^{2+}–Na^+ exchange probably contributes to excitation–contraction coupling by bringing extracellular Ca^{2+} into the cell via this pathway, by running in the reverse direction during excitation (see Fig. 15-1). In the reverse mode, the Ca^{2+}–Na^+ reaction exchanges three intracellular Na^+ for one extracellular Ca^{2+}. The exchange reverses because the cell membrane is depolarized during the long-duration cardiac AP, and the energetics are now more favorable for the reverse reaction. In some hearts, such as frog atrial myocardial cells, a significant fraction of the Ca^{2+} influx during the AP that initiates contraction occurs by means of this reversed-exchange pathway.

Because of this Ca^{2+}–Na^+ exchange, anything that causes $[Na]_i$ to rise will secondarily cause $[Ca]_i$ to rise, thereby leading to a more forceful contraction, known as the **positive inotropic effect** in the context of carciac muscle. For example, cardiac glycoside drugs, such as digitalis and ouabain, which inhibit the sarcolemmal (Na,K)-ATPase and Na^+-K^+ pump and thereby cause $[Na]_i$ to rise, have a potent positive inotropic effect. Such drugs are often used in patients with failing hearts, such as those with congestive heart failure.

Lowering of the Elevated $[Ca]_i$ to Produce Relaxation

For the heart to relax and refill after contraction, the elevated $[Ca]_i$ must be reduced back to the resting level of about 1×10^{-7} M. This is accomplished by the operation of three pathways stimulated by the higher $[Ca]_i$ during the AP: (1) the exchange of intracellular Ca^{2+} for extracellular Na^+ across the sarcolemma; (2) the sarcolemmal Ca^{2+}-ATPase–Ca^{2+} pump; and (3) the SR Ca^{2+}-ATPase–Ca^{2+} pump (see Fig. 15-1). As with (Na,K)-ATPase, in which intracellular Na^+ and extracellular K^+ stimulate this enzyme pump, a rise in $[Ca]_i$ stimulates the two Ca^{2+}-ATPases: the one in the sarcolemma and the one in the SR. These two pumps thereby act to lower $[Ca]_i$ to the resting level. The SR Ca^{2+} pump is probably the more important because it is present at a very high density in the SR membrane.

In addition, the high [Ca]$_i$ during the AP stimulates the Ca^{2+}–Na^+ exchange to operate in the *forward* direction: internal Ca^{2+} for external Na^+. When the membrane potential reverts to the original resting level following the AP, this makes the energetics favorable again for the exchanger to operate in the forward direction. A combination of these three factors (two pumps and one exchanger), coupled with the shutting off of the enhanced Ca^{2+} entry into the cell because the AP has terminated (and therefore the depolarization-gated slow Ca^{2+} channels have reclosed), lowers [Ca]$_i$ to the resting level and relaxes the muscle (see Fig. 15-1).

Regulation of Force of Contraction

The force of heart contraction can be quickly increased or decreased (Fig. 15-3). This fast regulation is primarily enabled by the special properties of the slow Ca^{2+} channels of the heart. Because there are more slow Ca^{2+} channels than

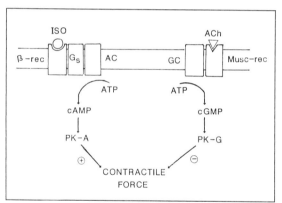

Fig. 15-3. Summary of the antagonism between cAMP and cGMP on the force of heart contraction. cAMP augments the contractile force and cGMP depresses contraction. Both cyclic nucleotides may activate their respective protein kinases and phosphorylate several membrane proteins, including the slow Ca^{2+} channel protein or associated regulatory proteins. (*ISO* = isoproterenol; β-*rec* = beta-adrenergic receptor; *AC* = adenylate cyclase; *GC* = guanylate cyclase; *ACh* = acetylcholine; *Musc rec* = muscarinic cholinergic receptors; *PK-A, PK-G* = protein kinase A and G; G_s = stimulatory GTP-binding protein).

fast channels, the conductance of the slow channels is about double that of the fast channels, and the slow-channel conductance remains activated for a much longer time; most of the Ca^{2+} influx into the myocardial cell during excitation is by means of slow Ca^{2+} channels. The activity of these slow Ca^{2+} channels is stimulated by phosphorylation with cyclic adenosine 3′: 5′ monophosphate (cAMP)–dependent protein kinase (cAMP-PK) and inhibited by phosphorylation with cyclic guanosine 3′: 5′ monophosphate (cGMP)–dependent protein kinase (cGMP-PK) (Fig. 15-4). Therefore, any intervention that elevates the cAMP level increases the force of heart contraction, and any intervention that elevates the cGMP level decreases the force of contraction. Thus, cAMP and cGMP act in an antagonistic manner (see Fig. 15-3).

The cAMP level is elevated by the action of a number of agents, including the sympathetic nerve neurotransmitter norepinephrine and circulating epinephrine, both of which bind to and activate the beta-adrenergic receptors. Other agents that elevate the cAMP level include histamine (via the H$_2$ receptor) and phosphodiesterase inhibitors, such as theophylline and caffeine. Drugs such as forskolin directly stimulate adenylate cyclase to elevate the cAMP level (see Fig. 15-2B). Therefore, all of these agents, as well as isoproterenol (a more selective beta-adrenergic activator), are positive inotropic agents, in that they make the heart contract more forcefully. Figure 15-2 (B and C) also illustrates that GTP-binding proteins (G-proteins) serve to couple membrane receptors to membrane enzymes (or to ion channels). G_s-protein stimulates adenylate cyclase, whereas G_i-protein inhibits this enzyme.

The cGMP level is elevated by the parasympathetic nerve neurotransmitter acetylcholine (ACh), which activates the muscarinic cholinergic receptors. Therefore, ACh is a negative inotropic agent, in that it makes the heart contract less forcefully (see Fig. 15-3).

Therefore, the sympathetic and parasympathetic nerves act in an antagonistic manner on the force of contraction of the heart. The sym-

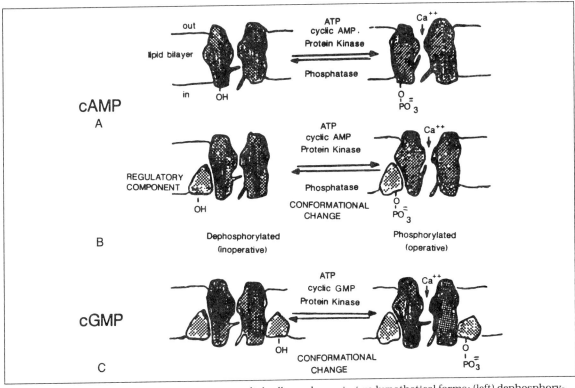

Fig. 15-4. Model for a slow channel in the myocardial cell membrane in two hypothetical forms: (left) dephosphorylated (or electrically silent) and (right) phosphorylated. The two gates associated with the channel, an activation gate and an inactivation gate, are kinetically much slower than those of the fast Na⁺ channel. The phosphorylation hypothesis states that (A) a protein constituent of the slow channel itself, or (B) a regulatory protein associated with the slow channel, must be phosphorylated for the channel to be available for voltage activation. Phosphorylation occurs through the action of a cAMP-dependent protein kinase in the presence of ATP. Presumably, a serine or threonine residue in the protein becomes phosphorylated. (C) The slow channel (or an associated regulatory protein) may also be phosphorylated by a cGMP-dependent phosphorylation, thus mediating the inhibitory effects of cGMP on the slow Ca^{2+} channel. (From Sperelakis, N., and Schneider, *J. Am. J. Cardiol.* 37:1079–1085, 1976.)

pathetic nerves, via norepinephrine release, stimulate the heart to contract more forcefully, whereas the parasympathetic nerves, via ACh release, inhibit the heart to contract less forcefully (see Fig. 15-3).

Figure 15-5 summarizes three mechanisms through which ACh inhibits contraction by activating the muscarinic receptor. In the first, the G_i coupling protein reverses the stimulation of adenylate cyclase produced by the beta-adrenergic receptor (via G_s coupling protein), thus lowering the elevated cAMP level. In the second

mechanism, the activation of guanylate cyclase raises the cGMP level. The third mechanism involves activation (gating) of a special type of K⁺ channel by means of a G_x coupling protein that increases the outward (repolarizing) K⁺ current, resulting in early termination of the cardiac AP that deactivates the slow Ca^{2+} channels. These G-protein-gated K⁺ channels, which are responsible for the ACh-activated K⁺ current, exist in atrial myocardial cells and nodal cells but not in ventricular myocardial cells.

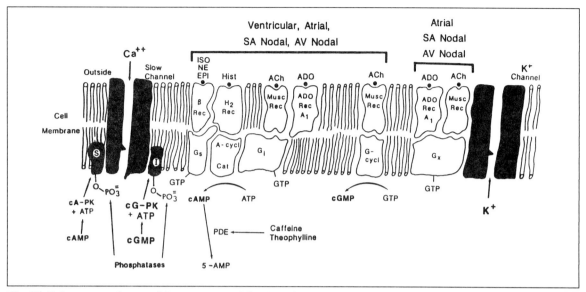

Fig. 15-5. Summary of some of the properties of the slow channels in the myocardial cell membrane. Included are the mechanisms of action of some positive inotropic agents such as beta-adrenergic agonists, histamine H_2 agonists, and methylxanthines (phosphodiesterase inhibitors). The beta and H_2 agonists act on a regulatory component (GTP-binding protein) of the adenylate cyclase complex to stimulate cAMP production. The voltage-dependent myocardial slow Ca^{2+} channels rely on cAMP and metabolism, presumably because a protein constituent (or regulatory component) of the slow channel must be phosphorylated for it to be in a form suitable for voltage activation. Other types of phosphorylation, such as cGMP-dependent, may also regulate the slow channel. (From Sperelakis, N. *Molec. Cell. Biochem.* 99: 97–109, 1990.) (*SA* = sinoatrial; *AV* = atrioventricular; *ISO* = isoproterenol; *NE* = norepinephrine; *EPI* = epinephrine; *Hist* = histamine; *ACh* = acetylcholine; *ADO* = adenosine; β *rec* = beta-adrenergic receptor; *Musc rec* = muscarinic cholinergic receptor; *A-cycl* = adenylate cyclase; *G-cycl* = guanylate cyclase; *PK* = protein kinase; *PDE* = phosphodiesterase.)

Special Properties of the Slow Ca^{2+} Channels

The slow Ca^{2+} channels in cardiac muscle possess some special properties that are different from those for other types of ion channels, and these features are summarized in Table 15-2. One special property is that the activity of the slow Ca^{2+} channels **depends on metabolic energy,** in the form of ATP. Thus, under conditions of hypoxia or ischemia, which are known to inhibit metabolism and lower ATP level, the slow Ca^{2+} channels are selectively inhibited. The other types of ion channels continue to function, at least initially, and so almost normal APs continue to be generated. Only the plateau of the AP becomes slightly more triangular because the Ca^{2+} current

contribution to the plateau is missing (see previous section, "Ca^{2+} Slow Channels," and Chapter 17). However, contraction is greatly depressed or abolished because of the loss of the Ca^{2+} influx through the slow Ca^{2+} channels, that is, there is uncoupling of contraction from excitation.

A second important property is that the slow Ca^{2+} channel activity is **selectively inhibited by acidosis** (see Table 15-2), for example, during ischemia. This allows almost normal APs to continue to be generated, although the contractions are greatly depressed. Therefore, ATP is conserved by the ischemic myocardial cells, and this protects them (to be discussed). The intracellular K_i value (for half-inhibition) has a pH of about 6.6, with almost complete inhibition occurring at pH 6.1. The effect of acidosis is quick in onset

Table 15-2. Special Properties of Ca^{2+} Slow Channels in Myocardial Cells

1. Activity dependent on metabolic energy (ATP)
2. Activity stimulated by cAMP and phosphorylation by cAMP-PK
3. Activity inhibited by cGMP and phosphorylation by cGMP-PK
4. Activity inhibited (quickly and reversibly) by acidosis (half-inhibition at pH 6.6)
5. Activity inhibited by calcium antagonist drugs and stimulated by calcium agonist drugs

cAMP = cyclic adenosine 3′: 5′ monophosphate; cGMP = cyclic guanosine 3′: 5′ monophosphate; PK = protein kinase.

and offset (i.e., rapidly reversible). Some of the rapid effects of ischemia and prolonged hypoxia are mediated by the accompanying acidosis.

A third important property is that Ca^{2+} slow-channel activity is **regulated by cyclic nucleotides** (see Table 15-2; Figs. 15-4, 15-5). As stated in the previous section, channel activity is **stimulated by cAMP** and **inhibited by cGMP,** so these two second messengers play **antagonistic roles** in regulating channel activity. This antagonistic relationship allows the two sets of autonomic nerve innervation to the heart to exert antagonistic effects on the force of contraction and heart rate by regulating Ca^{2+} entry into the cells, in accordance with the physiologic needs.

The two cyclic nucleotides exert their regulatory effects on the slow Ca^{2+} channels by **phosphorylation of the channel protein** itself or of **associated regulatory proteins** (see Fig. 15-4). Thus, cAMP activates cAMP-PK and cGMP activates cGMP-PK. Presumably, these two **protein kinases phosphorylate different sites** on the Ca^{2+} slow-channel protein or associated regulatory proteins. When phosphorylated, the cAMP-PK site stimulates channel activity, whereas the cGMP-PK site inhibits channel activity.

Phosphorylation by cAMP-PK increases the probability of the Ca^{2+} channels being opened at a given voltage, and the mean open time is increased. It is also likely that some silent (inactive) channels are recruited into activity by phosphor-

ylation of this site or sites (one to three sites have been proposed for phosphorylation by cAMP-PK). Phosphorylation by cGMP-PK decreases the probability of channel opening, and the mean open time is decreased.

A fourth special property of slow Ca^{2+} channels is that they are preferentially blocked by a class of drugs known as **calcium antagonists** or **slow-channel blockers** (see Table 15-2). Four major chemical types of these drugs are verapamil, bepridil, diltiazem, and nifedipine. The calcium antagonists also exert some blocking action on fast Na^+ channels and even on some K^+ channels. Nifedipine is one of these agents. A small chemical change in the dihydropyridine molecule causes the drug to act as a Ca^{2+} channel opener or so-called **calcium agonist** (e.g., Bay-K-8644). Thus, the activity of slow Ca^{2+} channels (in a variety of tissues) is inhibited by Ca^{2+} antagonist drugs and stimulated by Ca^{2+} agonist drugs. These drugs have many interesting properties, including an effect that is dependent on the frequency of stimulation.

The special properties of the slow Ca^{2+} channels of the heart, namely their dependence on metabolism for activity, rapid and reversible inhibition by acidosis, and regulation by cyclic nucleotides (cAMP and cGMP) and phosphorylation, protect the heart under ischemic conditions. For example, if vasospasm develops in one of the major coronary arteries supplying the heart muscle, the resultant acidosis quickly shuts off the slow Ca^{2+} channels but allows the other types of ion channels to function normally. Therefore, the myocardial cells in the ischemic zone stop contracting (or contract very weakly), because most of the Ca^{2+} influx into the cells has ceased. However, much of the ischemic area can still propagate more-or-less normal electrical activity through it, at least initially, and thus is less likely to cause the re-entrant type of arrhythmias. Therefore, these myocardial cells have become excitation–contraction *uncoupled*, in that contraction has been uncoupled from excitation.

In addition to direct regulation of slow Ca^{2+} channels, Ca^{2+} influx is indirectly controlled by regulating the activity of one type of K^+ channel.

This K^+ channel in the heart is inhibited by ATP. Therefore, in normal hearts, with a normal ATP level, the activity of this K^+ channel is continuously suppressed. This channel becomes unmasked (available to be voltage activated) when the ATP level is lowered in ischemic or hypoxic myocardial cells. Thus, during the AP, the outward repolarizing K^+ current is greatly increased, and increased earlier during the AP, thereby terminating the AP prematurely. This gives rise to progressively briefer APs. Hence, the Ca^{2+} channels, turned on by AP depolarization, are turned back off prematurely by AP repolarization, causing considerably less contraction.

Inhibition of contraction allows the ischemic cells to conserve their ATP content, because most of the ATP use in a myocardial cell is associated with contraction. Thus, when blood flow returns to normal (vasospasm relieved), the myocardial cells can fully recover and resume contracting. If the slow Ca^{2+} channels and K^+ channels (one type) did not have these special properties, the ischemic myocardial cells would not be able to recover when the ischemia ended.

SMOOTH MUSCLE

In smooth muscle, the force of contraction or state of tone is dependent on $[Ca]_o$, but some agents can trigger the release of intracellular Ca^{2+} from the SR (e.g., via IP_3 production), and so produce contraction even at very low $[Ca]_o$ (Figs. 15-6 and 15-7). As in cardiac muscle, the relationship between contractile force and $[Ca]_i$ is sigmoid on a logarithmic scale. In smooth muscle, Ca^{2+} acts as a **second messenger** to activate the myosin light-chain kinase for the phosphorylation of the myosin light chains and force production. (This has been described in Chapter 14.)

Two types of smooth muscles have been identified, based on their function: the **visceral,** or **unitary,** type and the **nonvisceral,** or **multiunit,** type. Examples of the visceral type are the smooth muscles of the gastrointestinal tract and uterus (i.e., the hollow viscera). An example of the nonvisceral type is the smooth muscle of the blood vessels. The visceral smooth muscles generally

fire APs that propagate at a velocity of about 5 cm per second. Some vascular smooth muscles also fire APs, in response to graded excitatory postsynaptic potentials—or excitatory junction potentials (EJPs)—that depolarize to a threshold potential. Other smooth muscles normally do not discharge APs, but contraction is controlled by graded changes in the membrane potential (graded depolarization) produced by neurotransmitters, hormones, and autacoids (local "hormones").

Resting Potential and APs

Resting Potential

The resting potential of smooth muscle cells is generally about -55 mV, although this may range from -40 to -70 mV, depending on the location of the smooth muscle. There is an electrogenic Na^+–K^+ pump contribution to the resting potential of about 8 mV. The resting potential in smooth muscle cells is considerably lower than that in cardiac muscle or skeletal muscle because of a higher ratio of Na^+ to K^+ permeability. (The origin of the resting potential is described in Chapter 3.)

AP Characteristics

In those smooth muscle cells that either normally fire APs or are induced to do so, the AP overshoots to about $+10$ mV. Thus, the AP amplitude is about 65 mV (from -55 to $+10$ mV). The maximum rate of rise of the AP is about 10 V per second, which is much slower than that in myocardial cells (approximately 200 V per second) and skeletal muscle fibers (about 600 V per second). The APD_{50} is about 30 msec, compared with about 3 msec in skeletal muscle fibers and 100 to 200 msec in myocardial cells. However, some vascular smooth muscles (e.g., aorta of rat) exhibit a prolonged plateau component that follows the spike, thus giving long APD_{50} values of about 100 msec. In some smooth muscles, the rate of repolarization of the AP is even faster than the rate of depolarization, suggesting a large, fast turn-on of the delayed-rectifier K^+ conductance and current. If activation of the K^+ conductance

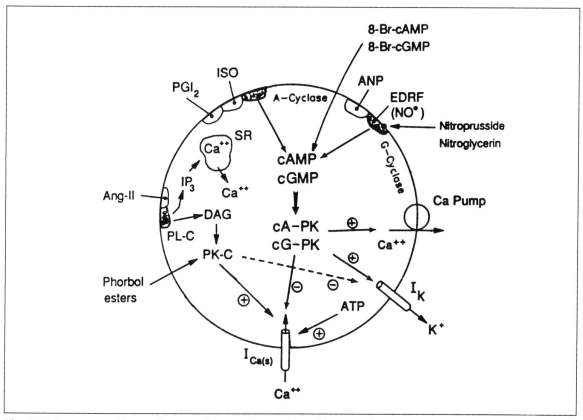

Fig. 15-6. Summary of the mechanism of action of several vasodilators and one vasoconstrictor (angiotensin-II [Ang-II]). The nitrovasodilators (nitroprusside and nitroglycerin) and endothelial-derived relaxing factor (*EDRF*) act directly on guanylate cyclase (*G-cyclase*) to stimulate its activity and elevate the cGMP level. Atrial natriuretic peptide (*ANP*) also stimulates G-cyclase, but indirectly via its membrane receptor. The beta-adrenergic agonist isoproterenol (*ISO*) and the prostaglandin (*PGI₂*; prostacyclin) act on their respective receptors, thus stimulating adenylate cyclase (*A-cyclase*) (via the G_s coupling protein) and elevating the cAMP level. cAMP and cGMP activate their respective protein kinases (*PK*), resulting in phosphorylation of the Ca^{2+} slow channel protein or an associated regulatory protein. This inhibits the activity of the slow Ca^{2+} channels. As depicted, cAMP and cGMP also stimulate the delayed-rectifier K^+ channels. This depresses excitability and action potential (AP) generation, and therefore indirectly inhibits Ca^{2+} influx. Thus, such vasodilators inhibit Ca^{2+} influx and stimulate K^+ efflux. In addition, cGMP (and perhaps cAMP) may stimulate Ca^{2+} efflux via the sarcolemmal Ca^{2+} pump, and thereby also act to lower the intracellular calcium concentration ($[Ca]_i$) and produce relaxation and vasodilation. Angiotensin acts on its membrane receptor to stimulate phosphatidylinositol turnover, and thereby inositol-tris-phosphate (*IP₃*) and diacylglycerol (*DAG*) production. IP_3 activates the Ca^{2+}-release channels of the sarcoplasmic reticulum (*SR*), resulting in Ca^{2+} release, $[Ca]_i$ increase, and hence contraction. DAG, like exogenously applied phorbol esters, activates protein kinase C (*PK-C*). PK-C presumably phosphorylates the slow Ca^{2+} channel protein (or an associated regulatory protein), stimulating channel activity and hence Ca^{2+} influx. Therefore, angiotensin produces contraction by at least these two mechanisms. Not depicted is the possibility that angiotensin may also inhibit the delayed-rectifier K^+ channel, and so prolong the AP, thereby prolonging the Ca^{2+} influx through the voltage-dependent slow Ca^{2+} channels (I_k = K^+ current; $I_{Ca(s)}$ = slow Ca^{2+} current; *NO* = nitric oxide.)

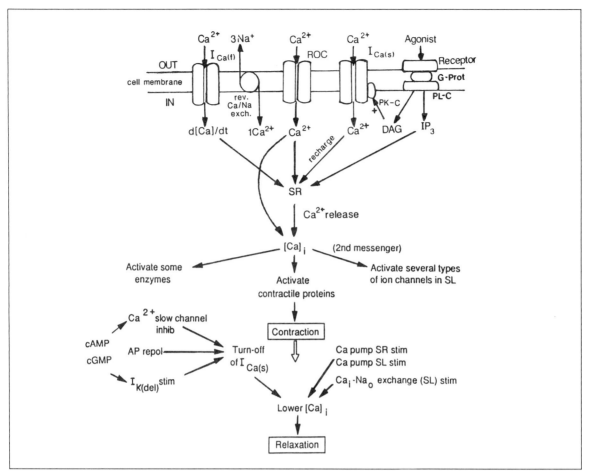

Fig. 15-7. Summary of excitation–contraction coupling in smooth muscle cells. There is Ca^{2+} influx through the sarcolemma (*SL*) via two types of voltage-dependent Ca^{2+} channels: a slow (long-lasting [L-type]) channel and a fast (transient [T-type]) channel. In addition, there is some Ca^{2+} influx via the Ca^{2+}–Na^+ exchange reaction operating in the reverse mode because of the depolarization during the action potential (*AP*). The entering Ca^{2+} also triggers release of additional Ca^{2+} stored in the sarcoplasmic reticulum (*SR*) by activating the Ca^{2+}-release channels in the SR membrane. Thus, there are two pools of Ca^{2+} for contraction: the extracellular (interstitial) and the SR pool. Also depicted is the fact that some chemical agents activate Ca^{2+}-selective receptor-operated channels (*ROC*) that allow Ca^{2+} influx across the sarcolemma; these channels are not voltage-gated. Also shown is the fact that some smooth muscle agonists act on their respective membrane receptors to stimulate phosphatidylinositol turnover, with resultant production of inositol-tris-phosphate (*IP₃*) and diacylglycerol (*DAG*). IP₃ acts on the SR to release Ca^{2+} via the Ca^{2+}-release channels. DAG activates protein kinase C, which stimulates the activity of the slow Ca^{2+} channels by phosphorylation. Relaxation is produced when the intracellular Ca^{2+} concentration ([*Ca*]ᵢ) is lowered back to the resting level by the several mechanisms depicted: turning off the inward slow Ca^{2+} current [$I_{Ca(s)}$] due to repolarization of the action potential, Ca^{2+} pumping from the myoplasm into the SR lumen and into the interstitial space, and intracellular Ca^{2+} for extracellular Na^+ exchange (Ca_i–Na_o) across the sarcolemma. The activity of the slow Ca^{2+} channels, K^+ channels, and sarcolemmal Ca^{2+} pump is regulated by phosphorylation (*PK-A, PK-G,* and *PK-C*); cAMP and cGMP inhibit the slow Ca^{2+} channels, stimulate the K^+ channels, and stimulate the Ca^{2+} pump. ($I_{Ca(f)}$) = inward fast Ca^{2+} current; $d[Ca]/dt$ = rate of change of myoplasmic Ca^{2+}; *PL-C* = phospholipase C; $I_{K(del)}$ = delayed-rectifier K^+ current; *G-prot* = GTP-binding protein.)

occurs earlier, this prevents the AP spike from attaining its normal overshoot potential ($+10$ mV), thus causing undershooting APs (e.g., a peak voltage of about -10 mV) and even inexcitability. The rising phase of the AP in most smooth muscles is produced by an inward Ca^{2+} current carried through slow Ca^{2+} channels. There are normally no Na^+ fast channels in smooth muscle cells.

Following the AP spike, there is often an **afterpotential** (see Chapter 4), which is usually of the hyperpolarizing type, produced by the increase in the delayed-rectifer K^+ conductance persisting beyond the spike and causing hyperpolarization toward the K^+ equilibrium potential. Membrane excitability is depressed during the **hyperpolarizing afterpotential.**

Spontaneous Activity

Many smooth muscles have spontaneous contractions and electric activity, that is, they possess **automaticity.** Automaticity is produced by pacemaker potentials that depolarize the cell to its threshold potential. (The ionic mechanism for the production of **pacemaker potentials** is discussed in Chapter 17.) Some smooth muscles, such as the longitudinal layer of small intestine, have a peculiar type of pacemaker potential known as **slow waves,** which are slow oscillations in the membrane potential with a periodicity of several seconds. During their depolarizing phase, a train of APs is elicited, the frequency of the burst gradually diminishes, and the APs stop when the repolarizing phase of the slow wave is under way. Thus, the APs occur in bursts during successive **slow-wave oscillations.** The peak-to-peak amplitude of the slow wave is about 15 to 30 mV.

The electrogenic mechanism underlying generation of the slow waves is not known. One hypothesis is based on **oscillations in the electrogenic Na_i–K_o pump (intracellular Na^+–extracellular K^+).** The pump is stimulated by the gain in $[Na]_i$ during the AP burst (a gain in $[Ca]_i$ leads to a gain in $[Na]_i$ via the Ca^{2+}–Na^+ exchange system), which hyperpolarizes, producing the hyperpolarizing phase of the slow wave. This shuts off the AP burst and gain in Ca^{2+} and Na^+, removing the stimulation of the electrogenic pump and causing depolarization (producing the depolarizing phase of the slow wave). When threshold is reached, a new burst is triggered. Another hypothesis is based on a K^+ channel whose activity is stimulated by Ca^{2+}, giving rise to a Ca^{2+}-activated K^+ conductance or current. The gain in $[Ca]_i$ during an AP burst would activate this current, which would then hyperpolarize and shut off the burst. When $[Ca]_i$ is lowered again, the cell would depolarize and initiate a new burst.

Some smooth muscles behave like **stretch receptors,** in that stretch (increased longitudinal applied tension) of the muscle leads to partial depolarization and initiation of AP bursts and contractions. Thus, a load applied to such a smooth muscle causes it to contract actively to counteract the deforming strain. This property allows smooth muscle to exercise local control over contraction, depending on the local environmental (mechanical) needs. Smooth muscles that behave like this include the longitudinal muscle layer of the gastrointestinal tract. One such widely studied muscle is the taenia coli of the guinea pig. The smooth muscles of the urinary bladder, uterus, and blood vessels may exhibit a similar property.

Nerve Control

As stated in the introductory paragraphs of this discussion on smooth muscles, some smooth muscles (for example, the nonunitary or multiunit muscles) are closely controlled by the autonomic nervous system. In these cases, most of the smooth muscle cells are not very far from a nerve terminal, and so can be influenced by the diffusion of a neurotransmitter (Fig. 15-8). Depending on the type of neuron (sympathetic or parasympathetic) and neurotransmitter (or neurotransmitters) released, the smooth muscle can either be stimulated to contract or be inhibited to relax. Because of cell-to-cell propagation of APs, the effect of the neurons can be quickly reflected

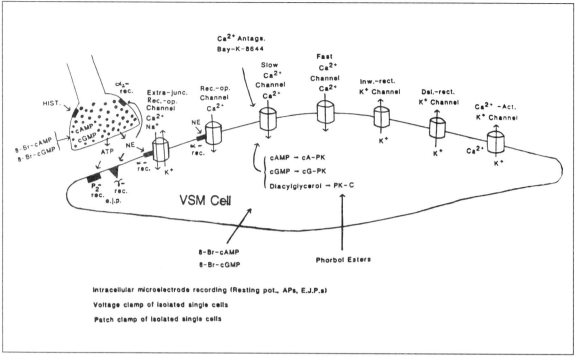

Fig. 15-8. Various ion channels in a vascular smooth muscle (*VSM*) cell and a number of agents that may either activate or inhibit these channels. Depicted are three different types of Ca^{2+} channels (fast, slow, and receptor-operated) and a nonselective ion channel (extrajunctional, receptor-operated), which allows Ca^{2+}, Na^+, and K^+ to pass through. The voltage-dependent slow Ca^{2+} channels are blocked by Ca^{2+} antagonists and enhanced by Ca^{2+} agonists (Bay-K-8644). The three different K^+ channels (inward-rectifier, delayed-rectifier, and Ca^{2+}-activated) are blocked by tetraethylammonium, barium, and cesium ions. Also shown is an adrenergic nerve terminal from which norepinephrine (*NE*) and ATP are released to activate alpha or gamma receptors and purinergic (P_2) receptors, respectively, on the postsynaptic membrane. Release of neurotransmitters may be modulated by substances such as NE, histamine, and cyclic nucleotides. Other vasoactive substances, such as angiotensin II, may also affect ion channels in the membrane. (*EJP* = excitatory junctional potential.)

over greater distances than the diffusion of the neurotransmitter alone would predict. In some smooth muscles (for example, the circular muscle layer of the small intestine and uterus), excitation can be propagated over relatively long distances with little decrement (fall-off or dying-out of activity).

Control of Contraction by $[Ca]_i$

Contraction

As discussed in the introductory paragraphs on cardiac muscle in this chapter, elevation of $[Ca]_i$

initiates contraction and lowered $[Ca]_i$ produces relaxation (see Fig. 15-7). Elevation of $[Ca]_i$ is produced by three events: (1) **Ca^{2+} influx** through the voltage-dependent slow and fast Ca^{2+} channels; (2) **Ca^{2+} release** from the SR stores through activation of the Ca^{2+}-release channels in the SR membrane by Ca^{2+} ("trigger" Ca^{2+}), or by IP_3, or both; and (3) reversed **Ca^{2+}—Na^+ exchange** (see Fig. 15-7). In addition, Ca^{2+} influx can occur by means of the **receptor-operated ion channels** (not voltage gated) that are selective for Ca^{2+}. Inhibition of the sarcolemmal **Ca^{2+}-ATPase—Ca^{2+} pump** would also lead to a rise in $[Ca]_i$ and contraction. Conversely, stimulation of the Ca^{2+} pump would lower $[Ca]_i$ and cause relaxation; there is some evidence that this mechanism operates in vascular smooth muscle.

In most smooth muscles examined, the number of **fast Ca^{2+} channels** is relatively small compared with the number of **slow Ca^{2+} channels,** as is true of myocardial cells (although in one vascular muscle, it has been reported that the Ca^{2+} fast current is as large as the Ca^{2+} slow current). Despite the relatively small Ca^{2+} fast current compared with the slow current, Ca^{2+} influx via this channel could be important in initiating contraction, because the **Ca^{2+} triggering of Ca^{2+} release** from the SR seems to depend on the rate of change of $[Ca]_i$, as well as on the absolute $[Ca]_i$ level (see introductory paragraphs on cardiac muscle in this chapter).

Thus, the Ca^{2+} current that is responsible for the **rising phase of the AP,** and hence underlies the spread of excitation over many cells by propagation, also serves as one **second messenger** involved in initiating contraction. The Ca^{2+} influx not only contributes to the elevation of $[Ca]_i$, but also **triggers the release** of more Ca^{2+} from the SR. As described in the first paragraph of this section, reversed Ca^{2+}–Na^+ exchange, activation of receptor-operated Ca^{2+}–selective channels, and inhibition of the sarcolemmal Ca^{2+} pump can also contribute to the elevation in $[Ca]_i$. As discussed in Chapter 14, the Ca^{2+}–calmodulin complex activates the myosin light-chain kinase, which phosphorylates the myosin light chains to bring about contraction.

Relaxation

Relaxation of smooth muscle, as in cardiac muscle, is produced by lowering $[Ca]_i$ (Fig. 15-7). This is accomplished by turning off Ca^{2+} influx and Ca^{2+} release, coupled with stimulation of the sarcolemmal Ca^{2+} pump. The Ca^{2+} pumps of both the sarcolemma and SR are stimulated by the elevated $[Ca]_i$, and perhaps by elevation of the cyclic nucleotides and phosphorylation. Stimulation of Ca^{2+} sequestering into the SR also speeds relaxation, as would an increase in the rate of the intracellular Ca^{2+}–extracellular Na^+ exchange (in the forward direction, that is, Ca^{2+} transported out). Inhibition of the Ca^{2+} APs and excitability lead to relaxation; reversal of membrane depolarization causes relaxation.

For relaxation, the elevated $[Ca]_i$ must revert to the resting level of about 1×10^{-7} M. This is accomplished through the operation of three mechanisms: (1) intracellular Ca^{2+}–extracellular Na^+ exchange across the sarcolemma; (2) stimulation of the sarcolemmal Ca^{2+}-ATPase–Ca^{2+} pump, because of the higher $[Ca]_i$ during the AP; and (3) stimulation of the SR Ca^{2+}-ATPase–Ca^{2+} pump, because of the higher $[Ca]_i$ during the AP. The rise in $[Ca]_i$ stimulates the two Ca^{2+}-ATPases: the one in the sarcolemma and the one in the SR. These two pumps then act to lower $[Ca]_i$ to the resting level.

In addition, the high $[Ca]_i$ during the AP stimulates the Ca^{2+}–Na^+ exchanger to operate in the *forward* direction: internal Ca^{2+} for external Na^+. When the membrane potential reverts to the original resting level following the AP, this makes the energetics again favorable for the exchanger to run in the forward direction. A combination of these factors (two pumps and one exchanger), coupled with the shutting off of the enhanced Ca^{2+} entry into the cell because the AP has terminated (and therefore the voltage-gated slow Ca^{2+} channels have reclosed), allows $[Ca]_i$ to be lowered to the resting level and the muscle to relax (see Fig. 15-7).

Regulation and Properties of the Ca^{2+} Slow Channels

In myocardial cells, we learned that cAMP stimulated the activity of the slow Ca^{2+} channels, whereas cGMP inhibited their activity. In contrast, in vascular smooth muscle, both cyclic nucleotides act in the same direction, namely to inhibit the activity of the slow Ca^{2+} channels. Therefore, any agent that elevates cAMP or cGMP levels in vascular muscle would tend to cause vasodilation. For example, atrial natriuretic peptide and endothelial cell nitric oxide, which stimulate guanylate cyclase and raise cGMP, and beta-adrenergic agonists, which stimulate adenylate cyclase and raise cAMP, are vasodilators. Drugs like nitroprusside, which directly activate guanylate cyclase to elevate cGMP, act as vasodilators by relaxing vascular smooth muscle. The effects of the cyclic nucleotides are presumably mediated by their respective protein kinases and by phosphorylation of the slow Ca^{2+} channel protein, or associated regulatory proteins.

ATP has been shown to be required for slow Ca^{2+} channels to function in vascular smooth muscle. Lowering the cellular ATP level (by metabolic inhibition) inhibits (half-inhibition [$K_{0.5}$] value of 0.3 mM) the slow Ca^{2+} channels and Ca^{2+} influx, and hence would tend to cause vasodilation. For example, if hypoxia were to lower [ATP]$_i$ sufficiently, this would tend to produce vasodilation by this mechanism.

One property of the slow Ca^{2+} channels is that they are preferentially blocked by the calcium antagonist drugs (see introductory discussion on cardiac muscle in this chapter). Because the binding of these drugs is sensitive to the membrane potential, and thus their binding and blocking effects are greater at lower (i.e., less negative) resting potentials, their effect is more pronounced on smooth muscle than on cardiac muscle. Therefore, they can exert substantial vasodilating effects (and thus be a treatment for such disorders as angina pectoris and hypertension) without significantly depressing the heart's pumping action.

SUMMARY

Cardiac Muscle

The electrophysiologic properties of the cell membrane of the myocardial cell, and its complement of ion channels, form the basis for excitability and the propagation of excitation over the heart, for arrhythmias of the heart, and for the ability to regulate the force and frequency of heart contraction, depending on the physiologic demands placed on the heart.

The activity of the slow Ca^{2+} channels is reg-

ulated by a number of mechanisms, including cyclic nucleotide levels, the ATP level, and pH. These Ca^{2+} channels can be opened or stimulated by some drugs (for example, dihydropyridine Ca^{2+} agonists) and blocked or inhibited by others (for example, dihydropyridine Ca^{2+} antagonists and local anesthetics). cAMP, presumably through phosphorylation of the slow Ca^{2+} channel protein or of an associated regulatory protein by cAMP-activated protein kinase (PK-A), stimulates the channels. cGMP, presumably through phosphorylation by cGMP-activated protein kinase (PK-G), inhibits the channels. In these instances, cAMP and cGMP have antagonistic effects.

The delayed-rectifer type of K^+ channel is stimulated by cAMP. Another type of K^+ channel, the ACh-activated one ($I_{K(ACh)}$), is directly gated by the alpha subunit of the G_k-coupling protein, which is activated when the muscarinic receptor is activated. This shortens the AP plateau and depresses contraction. Still another type of K^+ channel is constantly suppressed by the normal ATP level, but this is reversed when the ATP level falls (for example, during ischemia or hypoxia). This also shortens the AP plateau and depresses Ca^{2+} influx and contraction, and hence conserves ATP and protects the cells.

The heart has a number of intrinsic ways to regulate the Ca^{2+} influx under adverse conditions, and so can protect itself from irreversible damage. There are also a number of extrinsic ways for Ca^{2+} influx to be regulated, ensuring that the force and frequency of contraction can be controlled and adjusted to meet the physiologic demands.

Smooth Muscle

The process of excitation–contraction coupling in smooth muscle is similar in some ways to that in cardiac muscle. The ion channels and electrophysiologic properties of the membrane of the smooth muscle cell are the basis for the excitability and propagation of excitation over the smooth muscle. These ion channels and membrane receptors allow the force and frequency of contraction of the smooth muscle, and its state of tone (sustained contraction), to accommodate the physiologic demands. Smooth muscle cells generally do not possess Na^+ fast channels, and so the inward current during the rising phase of the AP is carried through the slow Ca^{2+} channels. Thus, Ca^{2+} influx during the AP also acts as a second messenger to alter the state of contraction.

In vascular smooth muscle, for example, the activity of the slow Ca^{2+} channels is regulated by a number of mechanisms, including the cyclic nucleotides (cAMP and cGMP), the ATP level, and the protein kinase C activity (see Fig. 15-6). Phosphorylation of the channel protein by PK-A and PK-G inhibits channel activity, whereas phosphorylation by protein kinase C stimulates channel activity (see Fig. 15-6). Therefore, agents that elevate the cAMP or cGMP level (such as atrial natriuretic peptide, nitroprusside, isoproterenol, or prostacyclin) act as vasodilators (inhibit contraction), whereas agents (like angiotensin-II) that stimulate diacylglycerol production, and thereby activate protein kinase C, act as vasoconstrictors (stimulate contraction) (see Figs. 15-2 and 15-6).

In addition, cAMP and cGMP stimulate the activity of K^+ channels (delayed-rectifier type), and thereby increase K^+ efflux, hyperpolarize, and depress excitability (see Fig. 15-6). This prevents the turn-on of the voltage-dependent slow Ca^{2+} channels, thereby inhibiting Ca^{2+} influx. Thus, contraction of the smooth muscle is inhibited both by direct inhibition of the Ca^{2+} slow channels and by indirect inhibition exerted through stimulation of the K^+ channels. Activation of the Ca^{2+}-sensitive K^+ channels or of the ATP-regulated K^+ channels would have the same effect, namely vasodilation. This latter type of K^+ channel, known to be present in vascular smooth muscle (and cardiac muscle), is normally inhibited or repressed ("masked") by ATP, and is de-repressed if ATP is lowered because of metabolic disturbance. A new class of antihypertensive vasodilator drugs (for example, pinacidil) has been developed recently; these act by opening K^+ channels, thereby hyperpolarizing and relaxing the vascular smooth muscle.

Another mechanism proposed for the inhibition of the tonic contraction of some smooth muscles by cyclic nucleotides is stimulation of the sarcolemmal Ca^{2+}-ATPase—Ca^{2+} pump (see Fig. 15-6). Stimulation of the Ca^{2+} pump lowers $[Ca]_i$, and so inhibits contraction. Thus, agents that elevate cAMP or cGMP levels in vascular smooth muscle produce vasodilation through the functioning of at least three mechanisms: (1) inhibition of the slow Ca^{2+} channels; (2) stimulation of the delayed-rectifier K^+ channels; and (3) stimulation of the sarcolemmal Ca^{2+} pump.

Some vasoconstrictor agents, such as angiotensin-II, stimulate phosphatidylinositol turnover and production of IP_3 and diacylglycerol (see Fig. 15-6). As just stated, diacylgylcerol activates protein kinase C, and IP_3 acts as a second messenger on the SR. IP_3 activates the Ca^{2+}-release channels in the SR membrane, leading to release of Ca^{2+} from the SR stores. Elevation of $[Ca]_i$ by this mechanism stimulates contraction. Thus, angiotensin-II stimulates contraction through the operation of at least three mechanisms: (1) production of diacylglycerol and activation of protein kinase C for phosphorylation of the slow Ca^{2+} channels; (2) production of IP_3 and release of Ca^{2+} from the SR; and (3) inhibition of the delayed-rectifier K^+ channel, thereby prolonging the APs and augmenting Ca^{2+} influx.

One major difference between smooth muscles and other excitable cells, such as cardiac muscle, is the presence of receptor-operated ion channels in smooth muscles (see Figs. 15-7 and 15-8). For example, there are ligand-gated ion channels selective for Ca^{2+}. The presence of such non-voltage-gated Ca^{2+} channels underlies the concept of **pharmacomechanical coupling,** such that some agents are able to produce contraction without producing a change in the resting potential or initiation of APs.

NATIONAL BOARD TYPE QUESTIONS

Select the one most correct answer for each of the following

1. The Ca^{2+} required to initiate contraction comes from two pools. One of the pools is the:
   A. Golgi apparatus
   B. Nerve terminals
   C. Nucleus
   D. Mitochondria
   E. Sarcoplasmic reticulum

2. Ca^{2+} release from the SR is brought about physiologically by at least two mechanisms, one of which is?
   A. Long-range forces
   B. IP_3 production
   C. Caffeine
   D. Ryanodine
   E. Mechanical squeezing

3. Which one of the following properties is *incorrect* for slow Ca^{2+} (L-type) channels, as compared with fast Ca^{2+} (T-type) channels?
   A. Long-lasting conductance or slow inactivation
   B. Sensitivity to organic Ca^{2+} antagonist drugs
   C. Smaller single-channel conductance
   D. Regulated by cyclic nucleotides and phosphorylation
   E. High threshold (activated at more depolarized potentials)

4. Which one of the following statements is *correct* for the Ca^{2+}–Na^+ exchanger working in the *reverse* direction?
   A. Exchanges intracellular Ca^{2+} with extracellular Na^+
   B. Exchange stoichiometry is one extracellular Ca^{2+} to three intracellular Na^+
   C. Exchange stoichiometry is three Ca^{2+} to one Na^+
   D. Exchange is nonelectrogenic
   E. None of the above

5. Which one of the following statements is *correct* for the APs of smooth muscle?
   A. Take-off potential (i.e., resting potential) is about -55 mV
   B. Maximum rate of rise of the AP is 200 V per second
   C. Overshoot potential is to $+30$ mV or higherD. APD_{50} is 3 msec
   E. Inward current is carried via fast Na^+ channels

ANNOTATED ANSWERS

1. E. The Ca^{2+} comes from: (1) the interstitial space via the Ca^{2+} channels and possibly the Ca^{2+}–Na^+ exchanger operating in reversed mode; and (2) Ca^{2+} release from the SR via the Ca^{2+}-release channels. The mitochondria have a Ca^{2+} pool, but it is not involved in excitation–contraction coupling.

2. B. The Ca^{2+}-release channels of the SR are activated by Ca^{2+}, known as *Ca^{2+}-triggered Ca^{2+} release*, and by IP_3, produced by phosphatidylinositol metabolism. The drugs caffeine and ryanodine bring about Ca^{2+} release from the SR, but they are not physiologic compounds.

3. C. The slow Ca^{2+} (L-type) channels have a *greater* single-channel conductance than do the fast T-type channels; about 24 pS versus 10 pS.

4. B. In the reverse mode of operation, the exchanger swaps one extracellular Ca^{2+} for three intracellular Na^+, and is therefore electrogenic. A significant amount of Ca^{2+} influx into the myocardial cell may occur by this mechanism for excitation–contraction coupling during the long-duration cardiac AP plateau.

5. A. The maximum rate of rise is only about 2 to 20 V per second, and the APD_{50} is about 20 msec or longer. Fast Na^+ channels (TTX sensitive) generally do not exist in smooth muscle. The inward current for the AP is usually a Ca^{2+} slow current.

BIBLIOGRAPHY

Sperelakis, N. Regulation of calcium slow channels of cardiac and smooth muscles by adenine nucleotides. In: Pelleg, A., Michelson, E. L., and Dreifus, L. S., eds. Proceedings of 1986 A. N. Richards Symposium on *Cardiac Electrophysiology and Pharmacology of Adenosine and ATP, Basic and Clinical Aspects.* New York: Alan R. Liss, 1987, pp. 135–193.

Sperelakis, N., and Ohya, Y. Electrophysiology of vascular smooth muscle. In: Sperelakis, N., ed. *Physiology and Pathophysiology of the Heart.* 2nd ed. Amsterdam: Kluwer Academic Press, 1988, pp. 773–811.

Sperelakis, N., and Wood, J. D. eds. *Frontiers in Smooth Muscle Research.* New York: Alan R. Liss, 1989.

V Cardiovascular Physiology

16 Basic Principles of Cardiovascular Physiology

RICHARD A. WALSH

Objectives

After reading this chapter, you should be able to:

Describe the functional anatomy of the heart and valves

Explain the normal sequence of blood flow through the heart and cardiovascular system

Identify the principal determinants of blood flow through the cardiovascular system

---

> A man is as old as his arteries.
> Thomas Sydenham

The principal function of the heart and circulatory system is to provide oxygen and nutrients and to remove metabolic waste products from tissues and organs of the body. This is accomplished by a closed-loop system in which large blood vessels provide the conduits for delivering and receiving these materials. The heart is a muscular pump that provides the energy for transporting the blood through this system to facilitate the exchange of oxygen, carbon dioxide, and other metabolites through the tiny thin-walled capillaries. An appreciation of normal cardiocirculatory physiology provides an essential foundation for understanding disease states that affect the heart and blood vessels and the actions of various cardioactive drugs.

Heart and vascular disease is the most common cause of morbidity and mortality in adults in the Western world. For example, over one-half million people die each year from heart attacks and over sixty million people in the United States have some form of cardiocirculatory disease. Cardiovascular disease is also the principal cause of all deaths in men between the ages of thirty-five and seventy years. These statistics emphasize the importance of medical students having a firm understanding of basic cardiocirculatory phenomena. The major elements of the heart and vascular system will be described in this chapter. More detailed considerations of the various aspects of normal cardiovascular physiology will be covered in subsequent chapters.

OVERVIEW OF THE HEART

The heart is a four-chambered pump that is largely composed of a special type of striated muscle, called **myocardium.** Two major pumps operate in the heart, and they are the **right ventricle,** which pumps blood into the pulmonary circulation, and the **left ventricle,** which pumps blood into the systemic circulation. Each of these pumps is connected to a special booster pump, called the **right** and **left atrium.** The heart possesses a specialized electrical system that

245

ensures the overall timing of ventricular and atrial pumping is optimal for producing the **cardiac output,** or the amount of blood pumped by the heart per minute.

Functional Anatomy of the Cardiac Conduction System

The normal pacemaker of the heart is a self-firing unit located in the right atrium and called the **sinoatrial node.** The electrical impulse generated by this tiny structure activates the two atria and stimulates atrial contraction. The electrical impulse then reaches the main specialized conduction system by means of conducting pathways within and between the atria. The impulse is delayed at the level of the **atrioventricular node** and is then transmitted down a rapid conduction system, composed of the **right** and **left bundle branches,** to stimulate the two ventricles and cause them to contract. The normal pacemaker and specialized conduction system are influenced by **intrinsic automatic activity** and by the **autonomic nervous system,** which modulates heart rate and the speed with which electrical impulses are conducted through the specialized system, as will be discussed in detail in Chapter 17.

There are many diseases that interfere with the specialized electrical system of the heart, and may result in abnormally fast, slow, or irregular heart rhythms.

Functional Anatomy of the Cardiac Chambers

The two atrial chambers are thin-walled structures possessing conduit reservoir and booster pump functions. They passively transmit blood during ventricular filling, store blood during ventricular contraction, and augment ventricular filling by contraction toward the end of ventricular relaxation. The **right atrium** receives unoxygenated venous blood from the **superior** and **inferior vena cava,** and the **left atrium** receives oxygenated blood from the **pulmonary veins** (Fig. 16-1). The right and left ventricles provide

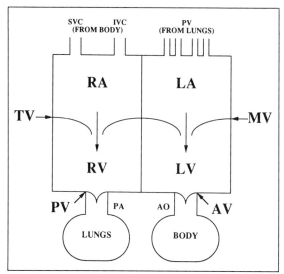

Fig. 16-1. The relationship among the four cardiac chambers, their valves, and the great vessels. Blood flows from the superior vena cava (*SVC*) and inferior vena cava (*IVC*) to the right atrium (*RA*) through the tricuspid valve (*TV*) and into the right ventricle (*RV*). Right ventricular contraction propels blood through the pulmonic valve (*PV*) into the pulmonary artery (*PA*). After oxygenation of blood and elimination of carbon dioxide at the level of the pulmonary capillaries, oxygenated blood returns to the left atrium (*LA*) through the pulmonary veins (*PV*). Left atrial blood traverses the mitral valve (*MV*) into the left ventricle (*LV*). During ventricular contraction, the aortic valve (*AV*) opens and blood is propelled into the aorta (*AO*). The cardiac phase is diastole or filling, hence the atrioventricular valves (*TV, MV*) are open while the semilunar valves (*PA, AO*) are closed.

the major energy for displacement of blood from the heart into the vascular system. The **right ventricle** pumps blood into the low-pressure pulmonary circulation. Consequently, the right ventricular wall is relatively thin (approximately 3 mm). By contrast, the **left ventricle** pumps blood into the high-pressure systemic circulation, and, although the cavity size of the right and left ventricle is similar, the left ventricle has much thicker walls (approximately 6 to 10 mm).

The force and rate of contraction of the heart is highly variable and dependent on many factors

(Chapter 19). These factors include heart rate, the amount of ventricular filling, the resistance to ejection of blood, and the intrinsic ability of the heart muscle to generate force and eject blood (Fig. 16-2). The autonomic nervous system may greatly modify the contractility of heart muscle in response to changing physiologic or pathologic conditions. For example, the sympathetic nerve terminals in the heart release norepinephrine, which is a powerful stimulant for heart muscle contraction during exercise or other stressful conditions (see Fig. 16-2).

A variety of diseases impair or destroy a variable amount of heart muscle and lead to **heart failure,** defined as a cardiac output that is inadequate for the metabolic needs of the peripheral tissues. These diseases include those caused by destruction of a region of the heart muscle by an inadequate blood supply, such as occurs during a heart attack, viral infections, and other pathologic processes.

Functional Anatomy of the Cardiac Valves

There are four **intracardiac valves** (see Fig. 16-1). Two of these valves are called the **atrioventricular valves** because they prevent leakage of blood backward from the ventricles into the atria when the right and left ventricles contract to eject blood into the pulmonary artery and aorta, respectively. The two-leaflet **mitral valve** serves this function between the left ventricle and left atrium, and the three-leaflet **tricuspid valve** is the barrier to retrograde flow of blood from the right ventricle to the right atrium. The other two valves are situated between the ventricles and the great arteries and are called **semilunar valves.** The **aortic valve** is interposed between the left ventricle and aorta and prevents retrograde flow from this great vessel into the left ventricle; the **pulmonic valve** serves a similar function between the pulmonary artery and the right

Fig. 16-2. The major determinants of blood flow in the cardiovascular system. Stroke volume (the amount of blood pumped per cardiac contraction [systole]) is determined by the preload (amount of cardiac filling during relaxation [diastole]), afterload (amount of resistance to ventricular ejection), and contractility (ability of heart muscle to develop tension and eject blood independent of preload and afterload). The product of stroke volume and heart rate is the cardiac output. The product of cardiac output and systemic vascular resistance is the systemic arterial pressure. These intrinsic circulatory factors are modulated by reflex, neurohormonal, and renal variables as well as intravascular volume.

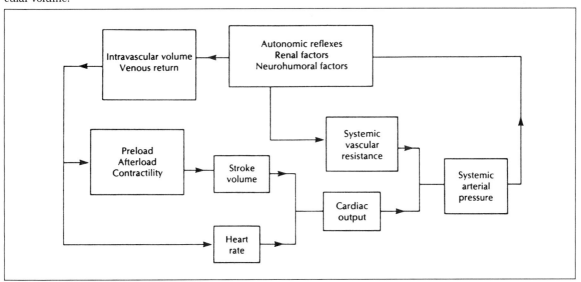

ventricle. During the contraction phase of the heart, called **systole,** when the ventricles develop pressure and eject blood, the mitral and tricuspid atrioventricular valves close, while the semilunar aortic and pulmonary valves open. During the relaxation phase of the cardiac cycle, called **diastole,** the atrioventricular tricuspid and mitral valves open and allow the ventricles to fill from the atria, while the semilunar valves are closed (see Fig. 16-1).

The timing of normal atrioventricular and semilunar valve closure is important in understanding the generation of normal and abnormal heart sounds as they are heard over the chest using a stethoscope (Chapter 19). The four cardiac valves may be affected by congenital or acquired disease, and this may result in leakage of blood through the valve or obstruction to blood flow.

Functional Anatomy of the Coronary Circulation

The **coronary arteries** are the first arterial branches to arise from the aorta just above the aortic valve. These vessels provide oxygen and nutrients to the heart muscle of all four cardiac chambers (Chapter 20). **Atherosclerosis** or variable obstruction of the coronary vessels by plaques composed of lipid fibrous tissue and calcium may variably obstruct or completely occlude one or more of the coronary arteries or their branches. Coronary atherosclerosis may produce electrical instability of the heart, death of the heart muscle (heart attack), or death of the patient.

Functional Anatomy of the Pericardium

The **pericardium** is an extremely thin-walled membranous sack that surrounds the heart. There is a small amount of fluid in the space between the surface of the heart and the pericardium that serves as lubrication. The pericardium anchors the heart within the chest cavity, prevents overfilling of the cardiac chambers, and protects the heart from infectious diseases that may affect the lungs.

OVERVIEW OF THE CIRCULATORY SYSTEM

Major Circulations

The two principal circulations of the body are linked in sequence, so that blood flows through the organs and tissues of the body, returns to the lungs, and recycles throughout the body after oxygen is acquired and carbon dioxide is eliminated through pulmonary respiration (see Fig. 16-1). Unoxygenated venous blood returns by means of the **superior and inferior vena cava** to the right atrium and subsequently to the right ventricle. The right ventricle pumps blood through the **pulmonary artery** to the lungs, where **pulmonary gas exchange,** defined as uptake of oxygen and release of carbon dioxide, occurs at the level of the capillaries. The oxygenated blood then flows through the pulmonary veins into the left atrium and subsequently to the left ventricle, where it is pumped to the arteries and capillaries of the rest of the body (Chapter 21).

The primary function of the **pulmonary circulation,** also called the **central** or **lesser circulation,** is to transport deoxygenated blood at low pressure through the lungs. The pulmonary circulation contains about 10 percent of the total blood volume at any given time.

The principal function of the **systemic circulation,** also called the **peripheral circulation,** is to provide a high-pressure source of oxygenated blood for tissue and organ perfusion, and to drain venous blood back to the right side of the heart. The phasicity and level of intravascular pressures progressively drop within the components of the systemic circulation, as blood flows from the arteries to the veins (Chapter 21). There is high pressure in the systemic arteries and low pressure in the veins. The major cause of reduced pulsatile pressure is the relatively high distensibility of the large arteries, as compared with the smaller arteries and arterioles. The progressive decline in pressure from the aorta to the capillary level is mainly brought about by the progressive increase in the total cross-sectional area of the circulation as it branches throughout the body.

The **lymphatic circulation** is composed of tiny

thin-walled channels that parallel the blood vessels of the body (Chapter 22). This circulation serves as an additional transport and drainage system. The branching network of these vessels throughout the tissues and the fluid contained within them, called **lymph,** is eventually channeled into the systemic venous circulation. Capillary filtration predominates slightly over capillary reabsorption and this produces tissue lymph flow. About 90 percent of net filtration is reabsorbed directly into the capillaries and 10 percent into the lymphatic system.

Hemodynamics and Regional Circulation

The circulation of blood in the pulmonary and systemic circulation largely depends on the interrelationships among **pressure** (force per unit area), **flow** (volume of blood transmitted per unit time), and **resistance** (the resistance to blood flow produced by structural or functional changes at the level of the arterioles). Multiple regional circulations exist in parallel off of the systemic circulation. Examples of specialized regional circulations include the **coronary,** the **cerebral, skeletal muscle,** and, during pregnancy, the **placental circulation** (Chapter 21). The determinants of blood flow and tissue perfusion in these specialized circulations may vary considerably, but involve a relative predominance of local, neural, or hormonal factors (Chapter 23). These factors ultimately affect resistance and hence perfusion of the relevant regional circulation. The major determinants of resistance can be understood by analogy to **Ohm's law of electrical circuits:** current flow = voltage/resistance. In circulatory physiology, this is rendered as:

$$flow = \frac{\text{the pressure difference across a given vascular bed}}{\text{vascular resistance}}$$

The systemic arterial pressure is in turn a product of flow and peripheral vascular resistance.

The level of systemic arterial pressure is held relatively constant by means of reflexes elicited from the autonomic nervous system. The prevailing level of systemic arterial pressure is ultimately determined by the interplay among intravascular volume, cardiac pump performance, and neural and endocrine factors (see Chapter 23 and Fig. 16-2). Alterations in any or all of these determinants of arterial pressure may provoke abnormally elevated blood pressure or systemic arterial hypertension.

NATIONAL BOARD TYPE QUESTIONS

Select the one most correct answer for each of the following

1. During ventricular systole:
   A. The semilunar valves are opened
   B. The atrioventricular valves are closed
   C. Neither A nor B
   D. Both A and B
2. Flow to a circulation or to an organ is:
   A. Directly related to the pressure difference across the system or organ
   B. Inversely related to the resistance of the system
   C. Neither A nor B
   D. Both A and B
3. Major functions of the pericardium include:
   A. Preventing overdistention of the ventricles
   B. Tethering the heart within the chest cavity
   C. Protecting the heart from inflammation and infection from the lungs
   D. All of the above
4. Normally the arteries carry oxygenated blood and the veins carry deoxygenated blood. The exception, or exceptions, to this rule is/are:
   A. The pulmonary artery
   B. The pulmonary vein
   C. The aorta
   D. Both A and B

ANNOTATED ANSWERS

1. D. During ventricular systole, the atrioventricular valves are closed to prevent backward flow of blood into the atria during ventricular contraction, while the semilunar valves are opened to permit forward flow into the great vessels.

2. D. Blood flow to an organ or circulation is a direct function of the pressure gradient across the system and an inverse function of the resistance in that system. This is analogous to Ohm's law of electrical current flow.

3. D. A mechanical function of the pericardium is to prevent the heart from overdistention and to anchor it in the chest cavity; it also protects the heart against the spread of inflammation and infection from the lungs.

4. D. The pulmonary artery transports deoxygenated blood from the right ventricle into the pulmonary capillaries, while the pulmonary veins transport oxygenated blood from the lungs into the left atrium.

17 Initiation and Propagation of the Cardiac Action Potential

IRA R. JOSEPHSON AND NICHOLAS SPERELAKIS

Objectives

After reading this chapter, you should be able to:

Describe the anatomic pathways for the spread of excitation through the heart

Compare the characteristics of the action potentials generated in different regions of the heart

Explain the ionic basis for the generation of cardiac action potentials

Describe the central role of the Ca^{2+} current in excitation–contraction coupling, and understand how it is regulated

Explain the electrophysiologic basis of cardiac arrhythmias

The normal heartbeat is initiated by a complex flow of electrical signals which provide the rhythm of the heart. The properties of the electrical signals, or **action potentials** (APs), that are generated in the different regions of the heart are diverse, but they all have a characteristic long duration (100 to 300 msec), as compared with nerve or skeletal muscle APs. The APs generated in cells residing in the **sinoatrial** (SA) **node** display **automaticity;** that is, they undergo spontaneous and rhythmic depolarization in the absence of external stimuli. The SA node is the primary pacemaker of the heart, and impulses originating in this region propagate over the heart and drive the quiescent tissues. The electrogenesis of the cardiac AP is similar to, but somewhat more complex than, that of nerve (Chapter 4). Besides contributions from Na^+ and K^+ currents, a third current, carried by Ca^{2+}, plays a central role in coupling cardiac excitation with

contraction. External influences, including the autonomic nervous system, hormones, and autacoids, modulate the force and rate of cardiac contraction through their action on the Ca^{2+} current and other currents.

In this chapter, we will first examine the pathways and mechanisms for the overall spread of electrical excitation in the heart. We will then focus on the properties of the APs in different regions of the heart, and the ionic currents which generate them.

FUNCTIONAL ANATOMY OF THE HEART AND SPREAD OF EXCITATION

For the heart to act as an efficient pump for the circulation of blood, it is essential that there is precise and sequential activation of the contraction of the atria and ventricles during each heartbeat. If the atria and ventricles were simultaneously depolarized, this would cause incomplete filling of the ventricles and reduced cardiac output. The pathway for the spread of excitation through the heart is diagrammed in Figure 17-1. The pattern of electrical activation of the heart can be recorded at the body surface, and is called the **electrocardiogram** (see Chapter 18).

The primary pacemaker for the normal heart beat is the SA node, which is located near the junction of the superior vena cava and the right atrium. It is composed of a small group of muscle cells that spontaneously and rhythmically fire APs. The SA nodal APs **propagate** (that is, they elicit APs) in neighboring right and left atrial muscle, leading to contraction of the atria. There are three bundles of muscle fibers—the internodal tracts of **Bachmann, Wenckebach,** and **Thorel**—which ensure that the atrial musculature is rapidly and nearly simultaneously activated, and that the depolarization reaches the **atrioventricular** (AV) **node.** The atria are electrically isolated from the ventricles by nonconductile connective tissue, and the spread of excitation in the ventricles can only occur by way of the AV node and the **bundle of His,** which are

fine-muscle fiber tracts linking the two regions. Once excited, the AV node conducts the impulse slowly (for example, about 3 cm per second), permitting ample time for ventricular filling prior to contraction.

The rate of firing of the SA (and AV) node is modulated by the **sympathetic** and **parasympathetic** divisions of the **autonomic nervous system.** Sympathetic stimulation speeds up the SA node (sinus tachycardia) and parasympathetic activation slows the SA node firing (sinus bradycardia). In addition to the effects on heart rate, the autonomic neurotransmitters increase (sympathetic) or decrease (parasympathetic) the velocity of propagation of excitation through the AV nodal region. Further, the force of contraction of the heart is augmented by sympathetic stimulation and diminished by parasympathetic stimulation. The autonomic nerves innervate the atrial and ventricular myocardium.

After passing through the AV node and the bundle of His, the excitation reaches a specialized conduction system, which serves to rapidly and nearly synchronously activate the ventricular myocardium. The propagation velocity in this **Purkinje system** is the fastest in the heart, or about 1.0 m per second. The Purkinje fibers, which comprise this system, are large-diameter muscle fibers that, at their terminals, ramify and form junctions with the ventricular myocardial cells. The Purkinje system terminates in the inner surface of the ventricular myocardium (the endocardium). Thus, the excitation is transferred from the Purkinje network to the myocardial cells, thus effecting rapid and nearly simultaneous activation of the two ventricles. Propagation velocity in the ventricular myocardium is about 0.4 m per second.

CONDUCTION OF THE CARDIAC AP

To understand the process by which excitation is propagated from cell to cell, it is necessary to appreciate the microscopic anatomy of the heart tissue. Cardiac muscle consists of a tight "brick-

Fig. 17-1. The pathway for electrical excitation in the heart. The cycle begins with depolarization of the sinoatrial node, which then propagates through the different regions of the heart until reaching the ventricular myocardium.

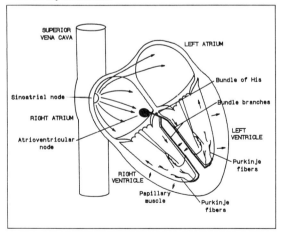

like" packing of cells, with dimensions of approximately 100 by 15 μm (Fig. 17-2). The ends of neighboring cells are joined at interdigitation regions called **intercalated disks.** Within the disks, the membranes of the two cells are occasionally in close apposition, and form a **nexus** or **gap junction.** Within the nexus there may be one or many **gap junction channels** that form a low-resistance pathway for the spread of current from one cell to another. The cardiac AP propagates along the myocardial fiber in a manner similar to that described for the propagation along a cable in elongated nerve and skeletal muscle cells (Chapter 5). **Local-circuit currents,** which

Fig. 17-2. Electrical communication between neighboring heart cells. (A) Arrangement of microelectrodes to measure electrical coupling between heart cells. Current pulses are injected into cell 1. The resulting voltage change is measured in the same cell and also in a distant cell (cell 4). The fact that a smaller, but measurable, voltage change occurs in the distant cell is interpreted to mean that the current flows between cells via an intercellular pathway. (B) The tight bricklike packing of individual myocardial cells and the flow of current (originating at *arrow*) through the tissue. Current passes from cell-to-cell through specialized gap junctions, located mainly at the ends of the myocytes. (S = stimulus current; V = voltage.) (Redrawn from: Woodbury, J.W. Cellular electrophysiology of the heart. In *Handbook of Physiology,* American Physiological Society, Washington, D.C., 1962.)

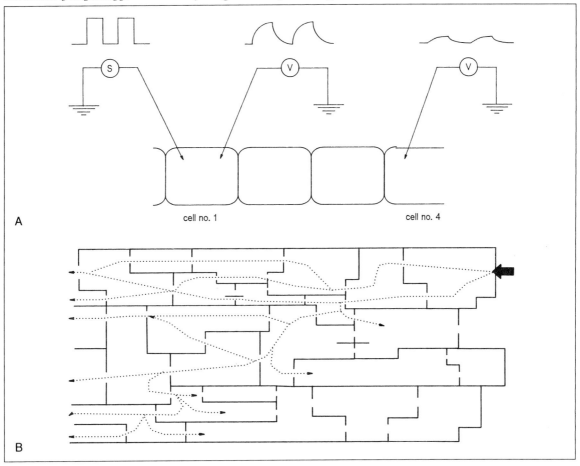

cell no. 1   cell no. 4

A

B

precede the AP wavefront, depolarize the adjacent membrane and bring it to the threshold potential. Above threshold, activation of the inward Na$^+$ currents causes further depolarization and further extension of the local-circuit currents. The process is, of course, continuous within one cell, until the AP reaches the end of the cell. When the depolarization reaches the disk region, the local-circuit currents can flow through the gap junction channels and thereby depolarize the neighboring cell membrane to generate an AP. It is also possible that the close apposition of the neighboring cell membranes and the associated electrical field generated by the AP causes transfer of excitation, in the absence of gap junction channels.

The **velocity of conduction** of the cardiac AP through each region of the heart depends on several factors. The **amplitude** and the **rate of rise of the AP** are the major determinants of its propagation velocity. The amplitude represents the potential difference between the polarized (resting) and depolarized regions of the cell. Because the magnitude of the local circuit currents is directly proportional to this potential difference, the larger the amplitude of the AP, the larger and more extensive the local-circuit currents. The result is faster conduction of the AP. The rate of rise of the AP is directly related to the density of the (net) inward current. The higher the density of the inward current, the more extensive the local-circuit currents, and hence faster excitation of adjacent membrane. For the normal, fast-rising AP, the inward current is supplied by the large, fast Na$^+$ current. For slowly rising APs in the SA and AV nodes, as well as the slow AP in the myocardium, the inward current is relatively small and is carried by the Ca^{2+} current. The difference in the density of the inward current produces a corresponding difference in the propagation velocity of these tissues. The **size** of the fiber is directly proportional to the velocity of propagation. For example, Purkinje fibers are among the largest cells in the heart, and have the largest conduction velocity. An increase, or decrease, in **temperature** also increases, or decrease, in **temperature** also increases, or de-

ceases, the conduction velocity, mainly by speeding, or slowing, the activation of the fast Na$^+$ currents.

GENERAL CHARACTERISTICS OF HEART ACTION POTENTIALS

The transmembrane potential can be recorded from individual myocardial cells by means of a microelectrode (a finely drawn glass capillary tube filled with a highly conductive solution of 3 M potassium chloride), which is connected to a high-input–impedance amplifier, and the output signal can be displayed on an oscilloscope. When the microelectrode tip is outside the cell membrane, there is a potential difference of zero millivolts between the microelectrode and the reference (ground) electrode. As the microelectrode tip is advanced and punctures the cell membrane, the potential difference abruptly shifts to a negative value. In the example shown for a ventricular cell (Fig. 17-3), the steady potential recorded when the cell is impaled is about −85 to −90 mV. This negative potential is referred to as the **resting potential,** because it represents the potential of the cell between successive APs and contractions.

If a myocardial cell is electrically stimulated to depolarize beyond a critical value (normally by ionic currents propagating from neighboring cells or, experimentally, by current applied through external or intracellular electrodes), called the **threshold potential,** then an **all-or-none, regenerative** electric response or AP is generated that is characteristic for that cell type. In the example shown in Figure 17-3, the cardiac AP can be divided into several phases: **Phase 0** represents the rapid upstroke or depolarization phase of the AP; **Phase 1** is the initial rapid repolarization; **Phase 2** is the long-duration plateau; **Phase 3** is the terminal repolarization; and **Phase 4** is the resting potential (or pacemaker potential). This general description applies to a ventricular AP, and should be slightly modified for the other cell types in the heart. The different properties of APs from each part of the heart will be reviewed in the following sections.

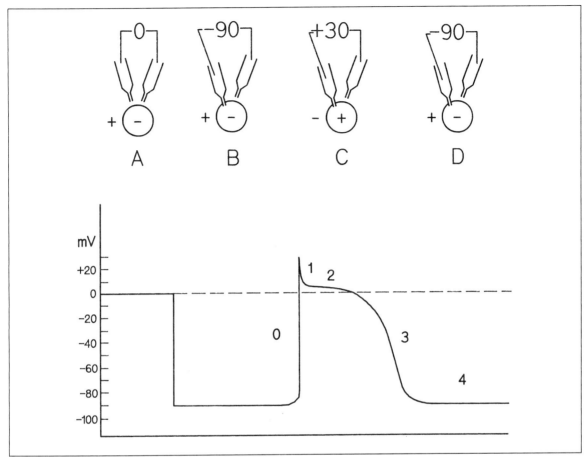

Fig. 17-3. The recording of the transmembrane resting and action potentials of a ventricular-like cell. The trace represents the potential difference (PD) between the recording microelectrode and a reference electrode in the bathing solution. At the beginning of the trace (A), the microelectrode tip is still outside the cell and there is no PD between the two electrodes. At B, the microelectrode is advanced and punctures the cell membrane; the cell interior then registers a PD or resting potential of approximately −90 mV, as compared with outside the cell. If the membrane is depolarized past a critical threshold potential, it generates an all-or-none action potential (C). The phases of the action potential are labeled: Phase 0, the upstroke; phase 1, rapid repolarization (notch); phase 2, the plateau; phase 3, terminal repolarization; and phase 4, the resting potential (D). (Redrawn from: Cranefield, P.F. *The Conduction of the Cardiac Impulse.* Mt. Kisco, NY: Futura, 1975.)

SA and AV Nodal APs

As mentioned previously, the SA node is the **primary pacemaker** of the heart, and APs recorded from these cells do not possess stable resting potentials but display a slow diastolic depolariza-

tion called the **pacemaker potential** (Fig. 17-4). The most negative potential reached (approximately −60 mV) is called the *maximum diastolic potential*. The membrane potential continues to depolarize until the threshold for firing an AP is reached (at about −40 mV). The upstroke of the

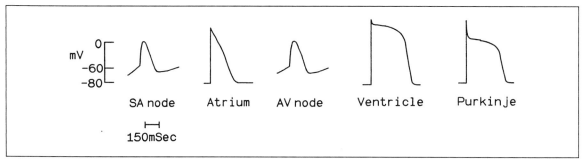

Fig. 17-4. Typical action potentials recorded from a sinoatrial (*SA*) node cell, atrial cell, atrioventricular (*AV*) node cell, ventricular cell, and a Purkinje fiber.

SA node AP is quite slow (about 1 to 5 V per second), as compared with the working myocardium (200 to 300 V per second). The maximum overshoot is usually 20 to 30 mV. The plateau phase may also be briefer than that in ventricular cells. Finally, Phase 3 repolarization causes the membrane potential to return to the maximum diastolic potential.

The AV node AP is similar in waveform to the SA node AP, as shown in Figure 17-4.

Atrial AP

The atrial AP (Fig. 17-4) displays a fast rate of rise (100 to 200 V per second) from a stable resting potential of -80 mV. The overshoot ranges from 20 to 30 mV. In most species there is a prominent Phase I; this may abbreviate the plateau and give the AP its characteristic triangular appearance.

Ventricular AP

The ventricular AP (Fig. 17-4) has a stable resting potential of -80 to -85 mV. The upstroke velocity is about 200 V per second, and the overshoot ranges from 20 to 30 mV. The duration of the ventricular AP is intermediate between the atrial and Purkinje APs, ranging from 200 to 250 msec.

Purkinje AP

Although normally large and stable (-95 mV), under pathologic conditions the resting potential

of the Purkinje fiber may depolarize spontaneously, giving rise to **automaticity.** Normally the SA node drives the heart at a rate which is faster than the intrinsic rate of discharge of the Purkinje system. The upstroke velocity of Purkinje fibers (200 to 500 V per second) is the greatest of all the different myocardial tissues, and accounts for the rapid conduction of these fibers. A long plateau and terminal repolarization are seen following a **prominent Phase 1** "notch" (see Fig. 17-4). The plateau phase of the Purkinje AP is longer than that for other cardiac tissues, and may help prevent re-excitation of the ventricles between beats.

Refractory Periods

In nerve cells, APs can be elicited at rates approaching 1,000 per second, but this is not the case for cardiac muscle. The long duration of the AP plateau in the heart limits its ability to be driven at a high frequency. This long plateau provides ample time for ventricular filling so that the heart can serve as an effective pump.

Based on their operation, it is possible to identify at least two types of refractory periods (Fig. 17-5). The **functional** (or effective) **refractory period** is the shortest interval after the initiation of the first AP when another normal AP can be initiated. The **absolute refractory period** is the interval during which no AP can be elicited, regardless of the stimulus intensity. The **relative refractory period** is the interval during which a

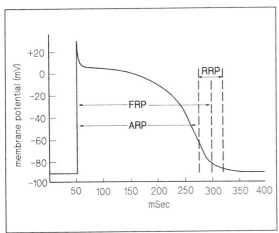

Fig. 17-5. Refractory periods during the cardiac action potential. The absolute refractory period (*ARP*) extends from the upstroke to approximately the middle of the repolarization phase; during this time, a second action potential cannot be elicited, regardless of the stimulus intensity. The interval between the ARP and the relative refractory period (*RRP*) is that time when a second action potential may be elicited, but usually with a slow upstroke velocity and slower propagation. The functional refractory period (*FRP*) is the shortest time interval after the first action potential when another action potential can be elicited.

second AP can be elicited, but at a higher stimulus intensity. The functional refractory period lasts for the entire absolute refractory period and part (about half) of the relative refractory period. As can be seen in Figure 17-5, the absolute refractory period lasts the entire upstroke and plateau of the AP and ends when the terminal repolarization of the AP reaches about −60 mV. The relative refractory period begins at the end of the absolute refractory period and continues until membrane excitability returns to normal (usually when the membrane potential returns to the resting potential). Because the current for stimulation of a not-yet-activated region is limited by the propagating AP, this defines the maximal current intensity available for stimulation, and thus defines the functional refractory period.

EFFECT OF RESTING POTENTIAL ON THE AP

Early investigators appreciated that the rate of rise and overshoot of the cardiac AP were sensitive to the magnitude of the resting membrane potential that preceded the AP. The resting membrane potential can be depolarized experimentally by two different methods. The first method involves injection of depolarizing (positive) current into the cell using a microelectrode; the second uses an elevated extracellular K+ concentration ($[K]_o$) to reduce the K+ equilibrium potential (E_K), and thereby depolarizes the membrane potential. In either case, membrane **depolarization reduces the rate of rise** of the AP, and eventually **decreases the overshoot.** At membrane potentials positive to −60 to −50 mV, APs (in ventricular, atrial, and Purkinje cells) may not be elicited, even by supranormal currents.

Conversely, if the membrane potential were **hyperpolarized** beyond the normal resting potential (for example, by lowering $[K]_o$ or by injecting negative current into the cell), this would **increase the rate of rise** and **overshoot.**

Basis of the Resting Potential

As described in Chapter 3, the resting potential that develops across the membrane of excitable cells results from the difference in concentration of certain ions (mainly K+ and Na+) inside and outside the cell, and from the selective permeability of the membrane to these ions. In cardiac muscle, as in nerve and skeletal muscle, K+ plays a major role in determining the magnitude of the resting potential.

If we were to assume that the cardiac cell membrane was permeable only to K+, then we could calculate the membrane potential (E_m) using the Nernst equation:

$$E_m = -\frac{RT}{zF} \ln \frac{[K]_i}{[K]_o}$$

Where $[K]_i$ is the K+ concentration inside the cell, R is the gas constant, T is the absolute temperature, z is the valence of the ion, F is Faraday's constant, and ln is the natural logarithm.

In this example, the membrane potential would be equal to E_K for all values of $[K]_o$, as plotted in Figure 17-6. Experimental microelectrode measurements of the membrane potential at varying $[K]_o$ show that the membrane potential does parallel E_K closely at high $[K]_o$ (exceeding 10 mM), but that the membrane potential deviates increasingly from E_K as $[K]_o$ goes below 10 mM to the physiologic level (see Fig. 17-6). This implies that, at lower $[K]_o$, the membrane potential is not determined solely by the membrane conductance to K^+. This occurs because, as $[K]_o$ is reduced, the membrane permeability for K^+ itself is also reduced. At low $[K]_o$, other ions, particularly Na^+, which have a small but finite permeability at rest, make a proportionately larger contribution to the resting potential. Because Na^+ has a positive equilibrium potential (+60 mV), it exerts a de-

polarizing influence on the membrane potential, which becomes more important at low $[K]_o$. It should be noted that the membrane potential is still negative, so it is still greatly influenced by, but does not exactly follow, E_K. This behavior at low $[K]_o$ can be accounted for by the **Goldman constant-field equation** (see Chapter 3).

The generation of the resting potential involves the transfer of only a very small amount of K^+ (less than a picomole per square centimeter) out of the cells, driven by the concentration gradient. K^+ moves out across the membrane, leaving a net negative charge inside the cell. Further net movement of the positively charged K^+ is counterbalanced by the resulting negative potential built up inside the membrane. The E_K reflects the electric force that must be built up to exactly counterbalance the chemical force (driven by the ionic concentration gradient); when the membrane potential is at E_K, there is no net flux of K^+ across the membrane.

Because the resting potential is generally more positive than E_K, there is a small efflux of K^+ from the cell, even at rest. As mentioned previously, the offset of the membrane potential from E_K is caused by the small influx of Na^+ into the cell at rest. Over time, the loss of K^+ and gain of Na^+ would lead to depolarization of the membrane. The rundown of the K^+ and Na^+ gradients is prevented by the **Na^+—K^+ pumps,** which are protein transport mechanisms residing in the cell membrane that translocate Na^+ out of and K^+ into the cell. In both cases, the replenishing action must work against the existing concentration and electric gradients (electrochemical) for Na^+ and K^+, and therefore requires chemical energy from the hydrolysis of ATP.

ION CHANNEL TYPES

The diversity of the waveforms of APs recorded from cells in different regions of the heart reflects the number and type of **voltage-dependent ionic channels** residing in their cell membranes. Channels are **transmembrane protein** (hydrophilic) **pores** that allow **ionic species,** such as Na^+, K^+, and Ca^{2+}, to cross the impermeable (lipid)

Fig. 17-6. Experimental measurements of the effect of changes in the external K^+ concentration ($[K^+]_o$) on the resting membrane potential (*RMP*) of myocardial cells. The data (*solid line*) represent values obtained by microelectrode measurements. The *dashed line* represents the value of K^+ equilibrium potential over the entire range of $[K^+]_o$ (plotted logarithmically), as calculated by the Nernst equation.

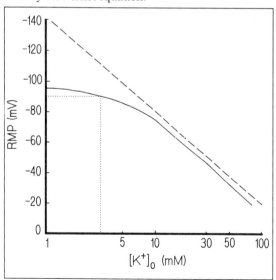

bilayer. Membrane channels are **gated** to open or close as the result of changes in the transmembrane potential. Following a voltage change, some channels open or close, as a function of time. The gating of other types of channels appears to be time-independent.

The major types of ionic (voltage-dependent) channels that are known to exist in certain heart cell membranes are listed in Table 17-1. Ionic channels can be identified on the basis of their **ionic specificity** (e.g., they pass Na^+, but not K^+), their **conductance** (i.e., how much current is allowed to pass through them), their **voltage dependence,** and their **kinetics** (i.e., how long it takes for them to turn on or off). A specific ionic current flowing into or out of a heart cell is the sum of the currents flowing through thousands of individual single channels.

IONIC BASIS OF CARDIAC AP

Rising Phase

The rapidly rising APs generated by ventricular, atrial, and Purkinje cells have in common that their **Phase 0** is caused by the rapid influx of Na^+ across the cell membrane. Depolarization leads to Na^+ influx, which leads to further depolarization, and so on. The upstroke is therefore a regenerative process, as in nerve. Early evidence that supported this conclusion was obtained from experiments in which the Na^+ concentration in the solutions bathing the heart cells was varied, and the effects on the rate of rise and overshoot of the AP were analyzed. It was found that both the overshoot and rate of rise of the AP were directly proportional to the external Na^+ concentration.

More recent experimentation, employing the voltage-clamp method to measure the membrane currents directly (see Chapter 4), has left little doubt that a **fast Na^+ current** is responsible for the upstroke of ventricular, atrial, and Purkinje APs. The term *fast* refers to the short time it takes for the current to turn on (activate) and turn off (inactivate). During a voltage-clamp step (from an initial, or holding, potential of -80 mV), the fast

Table 17-1. Types of Ion Channels in the Heart

Type	Heart Tissue
Na^+ fast channels	Atrial myocardial cells
	Ventricular myocardial cells
	Purkinje fibers
	AV nodal cells (low-density)
Ca^{2+} slow channels	Atrial myocardial cells
	Ventricular myocardial cells
	Purkinje fibers
	SA nodal cells
	AV nodal cells
K^+ channels	
Inwardly rectifying K^+ channels	Atrial myocardial cells
	Ventricular myocardial cells
	Purkinje fibers
Delayed K^+ channels	Atrial myocardial cells
	Ventricular myocardial cells
	Purkinje myocardial cells
	AV nodal cells
	SA nodal cells
Transient outward K^+ channels	Atrial myocardial cells
	Purkinje fibers
Pacemaker channels	
"Funny" currents	Purkinje fibers
Hyperpolarizing currents	SA nodal cells
	AV nodal cells
Ligand-operated channels	
Ca^{2+}-activated nonspecific channels	Ventricular myocardial cells
	Purkinje fibers
ATP-sensitive K^+ current	Ventricular and atrial myocardial cells
ACh-sensitive K^+ current	Atrial myocardial cells
	SA nodal cells
	AV nodal cells

AV = atrioventricular; SA = sinoatrial; ACh = acetylcholine.

Na^+ current **activates** and reaches a peak in 0.1 to 1.0 msec (depending on the potential at which it is measured). It then spontaneously declines, or inactivates, in 1 to 10 msec (Fig. 17-7). A rapid

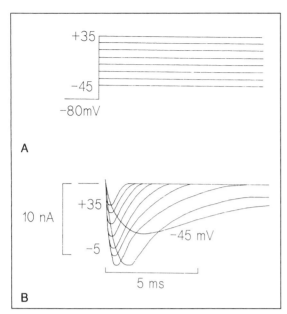

Fig. 17-7. The fast Na$^+$ current, recorded from a single heart cell under voltage-clamp conditions. (A) The voltage clamp steps from a holding potential of −80 mV to potentials of −45 to +35 mV. (B) The Na$^+$ currents elicited at those potentials.

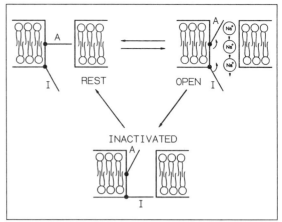

Fig. 17-8. The three states of a fast Na$^+$ channel. The channel spans the membrane and has an aqueous pore through which Na$^+$ can move down its concentration gradient into the cell. The channel has an activation (A) and an inactivation (I) gate. At rest, the A-gate is closed and the I-gate is open. Upon depolarization, the A-gate opens rapidly and the I-gate begins to close (the active, conducting channel). With time, the I-gate closes and the channel is inactivated (nonconducting). Following a period of depolarization, to reopen the I-gate, the channel must be at rest again.

activation of a fast Na$^+$ current of large magnitude (usually 10 to 100 times larger than the other ionic currents) is required to produce the rapid rates of depolarization measured for these cells.

We may think of the **fast Na$^+$ channel** (channels are the individual protein molecules that span the membrane and gate the movement of ions into or out of the cell) as existing in one of three **states** (Fig. 17-8). It may be in the **resting, active,** or **inactivated state.** What determines the particular state of the channel is the membrane potential, and the amount of time that the channel has "experienced" that potential. We say, therefore, that fast Na$^+$ channels are **voltage-dependent and time-dependent.** We can also imagine (and there is some experimental evidence to support the notion) that each Na$^+$ channel has two gates in series that govern the movement of Na$^+$ through the pore. One gate is called the **activation (A) gate** and the other is

the **inactivation (I) gate.** If both gates are open, Na$^+$ is permitted to flow through the channel and into the cell. If one or both gates are closed, the channel does not conduct Na$^+$.

If the membrane is depolarized (for example, by a voltage-clamp circuit) above a critical level from the resting potential, the A-gates of individual Na$^+$ channels open rapidly and Na$^+$ flows into the cell. At the same time, the I-gate, which was originally open at rest, begins to close. Na$^+$ flows through the channel until the I-gate is completely closed and the channel is inactivated. The channel must then experience repolarization of the membrane potential for a certain time, before it can conduct Na$^+$ again. This process is called **recovery.** As mentioned, all of the transitions between the three states of Na$^+$ channel are both voltage-dependent and time-dependent. Because activation is faster than inactivation, it is possible for the channels to activate briefly and then return to the resting state, without incurring inactivation. Such a procedure gives rise to **tail currents** that display the time course for channel deactivation, and these may be used to determine the voltage dependence of the conductance.

Direct measurements of single-channel currents in isolated heart cells can now be obtained experimentally using the **patch voltage-clamp method.** In this technique, a smooth-tipped pipette is pressed against the cell membrane and negative pressure is applied to the interior of the pipette. This procedure draws a small (several square micrometers) patch of membrane into the pipette tip which forms a **high-resistance seal** between the membrane and the glass pipette. The current signals from the isolated patch are amplified and displayed on an oscilloscope. For example, single Na^+-channel currents can be recorded from a membrane patch. The amplitude of the single Na^+-channel currents is 1 to 2 picoamp (10^{-12} amps), and their average duration is several milliseconds. Although the individual openings of the channel appear to occur **randomly** in time, an average of many samples shows the characteristic activation and inactivation time courses described for the total Na^+ current. During a voltage step, the channel may open and close more than once, before becoming inactivated. Individual cardiac cells may have hundreds of thousands of Na^+ channels.

Plateau

The long-duration plateau is a **unique characteristic** of the cardiac AP, and its electrogenesis has been the subject of continuing study for over three decades. The most obvious features of the plateau are that the membrane potential is maintained at a positive value, and that it is relatively constant for hundreds of milliseconds. Because the membrane potential is always the result of the sum of all the inward and outward membrane currents, the plateau must represent a fine balance of inward and outward currents. Experiments in which the membrane resistance was measured during the phases of the AP by the injection of current pulses have shown that the membrane resistance is highest during the pla-

teau. These results have led to the conclusion that both the inward and outward currents flowing during the plateau must be of small magnitude. The validity of this idea has been supported directly by the voltage-clamp experiments described in the next section, in which the inward and outward currents have been examined separately.

PROPERTIES OF THE CA^{2+} CURRENTS

In the absence of the fast Na^+ current (by voltage inactivation or by the addition of tetrodotoxin) a small, second inward current, carried by Ca^{2+}, is recorded when there is depolarization under voltage-clamp conditions (see Fig. 17-9). The **Ca^{2+} current** (sometimes called the **slow inward Ca^{2+} current** to differentiate it from the fast Na^+ current) activates at membrane potentials above -50 mV, is maximal around 0 to $+10$ mV, and

Fig. 17-9. The slow inward Ca^{2+} current. (A) The voltage-clamp protocol, in which the holding potential is -50 mV and the step potentials range from -35 to -5 mV. (B) The Ca^{2+} currents elicited in response to voltage steps from -45 to $+5$mV.

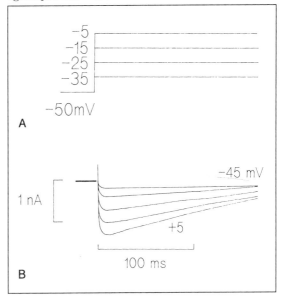

has a positive reversal potential (above +100 mV). The Ca^{2+} current peaks within a few milliseconds but inactivates very slowly over several hundred milliseconds.

There are **two types of Ca^{2+} channels** in certain cardiac cells. One type, called **low-threshold,** is activated over a more negative potential range, and these channels inactivate relatively rapidly. The other type, called **high-threshold,** is activated at less negative potentials, and the channels inactivate very slowly. At the single-channel level, these two types of Ca^{2+} channels may also be differentiated by their differing conductances and kinetics. (Some of their other properties will be discussed in the following sections on the regulation of Ca^{2+} currents.) Ca^{2+} channels are about ten times less numerous than Na^+ channels in the cell membrane, which partly accounts for the smaller magnitude of the total Ca^{2+} current.

Repolarization: Outward K^+ Currents

We have already learned that the plateau results from a balance of inward (Ca^{2+}) and outward (K^+) currents. The final repolarization of the AP must then be caused by inactivation of the Ca^{2+} current and an increase in the outward current, which then repolarizes the membrane back to the resting potential. Several **K^+ currents** contribute to the plateau and repolarization, and these will be discussed briefly.

The **inwardly rectifying K^+ current** (I_{K1}) is that K^+ conductance that is turned on at rest and generates the resting potential. The channels allow current to flow inward more easily than outward, hence inwardly rectifying. During depolarization of the AP upstroke, I_{K1} **turns off** and remains so until the membrane potential begins to repolarize. Late in the repolarization, I_{K1} generates a small outward current, which further aids in repolarization.

The **transient outward K^+ current** is mainly present in Purkinje and atrial cells, and is responsible for producing the **initial phase 1 repolarization** (or "notch") of the AP. It turns on

rapidly with depolarization and inactivates during the AP.

The **delayed outward K^+ current** (I_K) is the main current responsible for initiating the **final repolarization** of the AP. As the name suggests, I_K turns on very slowly, and its contribution occurs at the final phase of the AP. Recent evidence has shown that I_K is regulated by the autonomic nervous system, thereby controlling the AP duration. Norepinephrine (which increases cyclic adenosine $3':5'$ monophosphate [cAMP] levels) increases the magnitude of the outward K^+ current, thereby tending to shorten the AP duration.

A diagram of the ionic currents that contribute to the ventricular AP is given in Figure 17-10.

REGULATION OF CA^{2+} CURRENTS

The influx of Ca^{2+} serves as the **trigger** and **modulator** of the force of contraction of the myocardial cell (to be discussed in more detail in the section "Excitation—Contraction Coupling"). In light of this central role of the Ca^{2+} current, it is not surprising that several levels of regulation exist to control the activity of the voltage-dependent Ca^{2+} channels, and hence Ca^{2+} influx.

The first and most direct regulation of Ca^{2+} influx occurs as a result of Ca^{2+} flowing through the Ca^{2+} channel itself. Studies have demonstrated that inactivation of the Ca^{2+} channel is at least partially dependent on the **accumulation of Ca^{2+} ions at a binding site** at, or near, the inner surface of the channel. Thus, Ca^{2+} influx through the Ca^{2+} channel provides a negative-feedback signal that inhibits subsequent Ca^{2+} influx during depolarization. Such a mechanism serves a protective role by limiting Ca^{2+} influx and by helping to maintain $[Ca]_i$ at very low levels at rest.

Another level of regulation over the Ca^{2+} current is mediated by the **autonomic nervous system.** We are all familiar with the increased heart rate and force of cardiac contraction when fear or anger is provoked, or upon exertion. This primitive fight-or-flight response is triggered by the increased activity of the sympathetic nervous system. The neurotransmitter **norepinephrine**

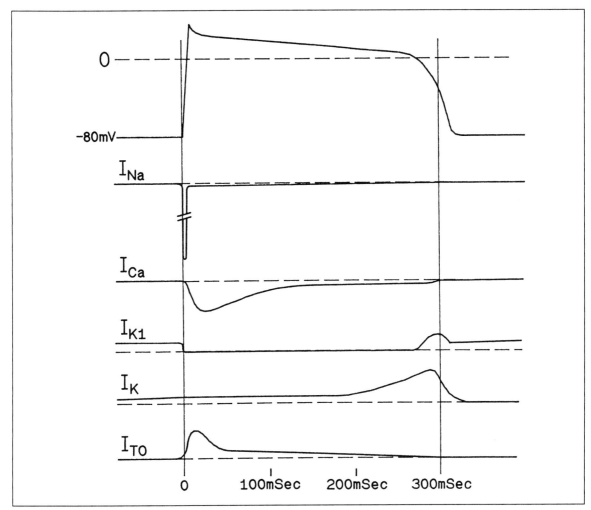

Fig. 17-10. A summary of the time course of the ionic currents that elicit a normal ventricular cardiac action potential. All of the currents, except the inwardly rectifying current (I_{K1}), are off at rest and turned on during depolarization, as indicated. I_{K1} is turned on at rest and turns off with depolarization. The fast Na$^+$ current (I_{Na}) is interrupted, as it is much larger than the other currents and would be off the scale in this diagram. (I_{Ca} = inward Ca^{2+} current; I_K = delayed outward K$^+$ current; I_{TO} = transient outward K$^+$ current.)

is released from nerve terminals located throughout the heart, in response to sympathetic activation. Norepinephrine molecules diffuse from the nerve terminals to the myocardial cells, where they bind to specific receptor proteins on the outer surface of the membrane. By binding to the receptor, a stimulatory signal is transmitted to an associated enzyme (located on the inner sur-

face of the membrane), called **adenylate cyclase.** The adenylate cyclase catalyzes the reaction that converts ATP to **cAMP.** cAMP is regarded as a second (intracellular) messenger, whereas norepinephrine is considered a first (extracellular) messenger. The importance of cAMP is that it activates a **protein kinase** that, among many other functions, **phosphorylates the Ca²⁺-**

channel protein, thereby increasing its activity. It is clear that norepinephrine elevates intracellular cAMP levels, and this is tightly correlated with an increase in the magnitude of the Ca^{2+} current and the force of contraction. At the single-channel level, norepinephrine increases the probability of the high-threshold Ca^{2+} channel being open during depolarization. Because the channel is more often open, more current can flow through it, thus increasing the total current.

The restorative branch of the autonomic nervous system, the **parasympathetics,** counteracts the sympathetics by slowing the heart rate and decreasing the force of contraction. Activation of parasympathetic nerve fibers leads to release of acetylcholine from nerve terminals, which can occupy and activate another set of specific receptors located on the heart cell membrane. **Acetylcholine** reduces the portion of the Ca^{2+} current that was increased by norepinephrine. Thus, acetylcholine restores the Ca^{2+} current to its original or basal magnitude. The reduction of the Ca^{2+} current by acetylcholine is thought to be related to a reduction in the cAMP levels and dephosphorylation of the Ca^{2+} channels. Alternatively, acetylcholine may increase in the **cyclic guanosine 3′: 5′ monophosphate level,** which itself acts to reduce the Ca^{2+} current.

Another level of regulation of the Ca^{2+} current is imposed by agents carried in the systemic circulation. An example of these agents is the peptide **angiotensin-II,** which has been shown to increase the Ca^{2+} current.

CA²⁺-DEPENDENT SLOW AP

Normally, the fast Na^+ current is responsible for the rapid depolarization of the upstroke of the cardiac AP. Experimentally, however, it is possible to block the fast Na^+ current by one of two means. The first method employs a biologic toxin, **tetrodotoxin,** which is extracted from the ovary of the Japanese pufferfish. Tetrodotoxin has been shown to bind specifically to and **block fast Na⁺ channels** at very low concentrations. The second method takes advantage of the fact that the Na^+ current is **voltage inactivated** (blocked) if the

membrane potential is held at a depolarized level, less than -50 mV. The membrane depolarization may be induced either by injection of a constant positive current into the cell, or by elevation of the external K concentration to about 25 mM, thereby decreasing E_K and consequently the membrane potential (Fig. 17-11).

At this stage, the membrane may be stimulated by applying depolarizing current pulses; however, it will not respond and will appear unexcitable. If the cardiac cell is then exposed to a positive inotropic agent (which increases the force of contraction) such as norepinephrine, electric stimulation causes a **regenerative, slowly rising** (10 to 20 V per second) **overshooting AP** (the slow response) and associated contractions (see Fig. 17-11). Studies in which $[Ca]_o$ is varied have demonstrated that the rising phase of the slow AP is caused by the influx of Ca^{2+}. The Ca^{2+} influx, and therefore the slow AP, is blocked by the addition of certain known Ca^{2+}-blocking (or antagonistic) agents. These include manganese, cobalt, cadmium, nickel, and lanthanum (inorganic blockers), as well as verapamil, nifedipine, and other **dihydropyridines** (organic blockers).

As discussed in the section dealing with the Ca^{2+} current, many positive inotropic agents increase the Ca^{2+} current, and thereby induce the slow AP. Elevation of the intracellular cAMP levels is a common mechanism for many of these agents. In addition, the slow AP may be induced simply by elevating the external Ca^{2+} concentration $[Ca]_o$ thereby increasing the current flowing through the "basal" Ca^{2+} channels. The slow AP may also be induced by reducing the amount of outward K^+ current by the addition of a K^+-channel blocker, such as tetraethylammonium ion.

The slow AP is important in that it may replace the normal fast AP in a region of the heart tissue during pathologic states, such as **ischemia.** During ischemic damage, there may be an elevated $[K]_o$, leading to depolarization and partial block of the fast Na^+ current, and increased concentrations of norepinephrine, brought about by the sympathetic discharge. These conditions will lead to slow AP formation. Because the slow AP

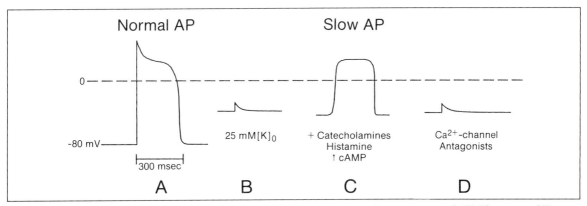

Fig. 17-11. The slow, Ca^{2+}-dependent action potential (*AP*). (A) The normal action potential. (B) The external K^+ concentration ($[K]_o$) is raised to 25 mM, causing partial depolarization of the membrane potential and loss of excitability. (C) The addition of certain positive inotropic agents (such as catecholamines, histamine, or cyclic adenosine 3′: 5′ monophosphate [cAMP]) induces a slowly rising, overshooting slow-response action potential. (D) The slow Ca^{2+}-dependent action potential is abolished by Ca^{2+}-channel antagonists, such as cobalt, cadmium, nickel, lanthanum, verapamil, and nifedipine.

propagates more slowly than does the fast AP (the propagation velocity is proportional to the density of inward current during the upstroke), propagation of excitation through the ischemic depolarized region may be disrupted, leading to **arrhythmia.**

EXCITATION–CONTRACTION COUPLING

The mechanism by which depolarization of the cardiac AP leads to contraction of the cardiac cell is referred to as **excitation–contraction coupling.** To understand the events leading to contraction, it is necessary to keep in mind the internal membrane structure of the myocardial cell and the regulation exerted over the free $[Ca]_i$. In myocardial cells, the surface membrane, known as the **sarcolemma,** exhibits numerous membrane invaginations, which form a system of narrow **transverse** (T) **tubules** that project into the cell interior. The purpose of these T-tubules is to conduct the AP radially into the interior of the cell. The T-tubules come into close contact with the **sarcoplasmic reticulum** (SR), an extensive intracellular membrane system that

sequesters Ca^{2+} by means of a **Ca^{2+} pump** driven by ATP.

At rest, $[Ca]_i$ is kept very low (10^{-8} to 10^{-7}M), but, during each contraction cycle, $[Ca]_i$ may rise to 10^{-5} M. Several mechanisms regulate the free $[Ca]_i$ concentration and restore it to the resting level between contractions. The **Na^+—Ca^{2+} exchange mechanism** uses the electrochemical gradient for Na^+ as the energy source to transport three Ca^{2+} out of the cell in exchange for one Na^+. Ca^{2+} is also pumped into the SR by the Ca^{2+}-ATPase. Without these restorative mechanisms, the myocardial cell would rapidly become overloaded with Ca^{2+} ions, undergo contracture, and subsequently die.

The shortening of the myofibrils is caused by a rise in the free Ca^{2+} concentration of from 10^{-6} to 10^{-5} M. In the heart, this elevation may be regulated by at least two routes. The first is the direct result of Ca^{2+} entering the cell during the AP plateau. Calculations suggest that at least part of the Ca^{2+} necessary for contraction comes from the Ca^{2+} current itself. The second source of Ca^{2+} for contraction originates from Ca^{2+} released from the SR store. It appears that the Ca^{2+} channels on the T-tubule membranes, in close physical association with the SR membrane, translate the excitation of the AP into a release of Ca^{2+} from the SR, perhaps by using the Ca^{2+} itself as the messenger.

IONIC BASIS OF AUTOMATICITY

The unceasing rhythmic contraction of the heart is normally driven by the spontaneous electrical activity of a small group of cells—the SA node. Other heart tissues (for example, the AV node and Purkinje fibers) are also capable of displaying **automaticity,** but are usually suppressed by the faster rate of firing of the SA node. The rate of firing is determined by several factors (Fig. 17-12), including the **slope of the pacemaker potential.** A more **positive slope** means that the membrane potential is depolarizing more rapidly, and will therefore more rapidly reach the threshold for firing an AP. The rate is also determined by the **maximum diastolic potential,** in that, given the same slope, the AP arising from a more negative potential will take longer to reach threshold. A third factor is the **threshold potential** itself, which may change and also affect the time it takes to reach threshold.

The rate of firing of the SA node AP is modulated by the **autonomic neurotransmitters.** Sympathetic stimulation (releasing norepinephrine) accelerates the rate of firing by increasing the slope of the pacemaker potential, and parasympathetic stimulation (releasing acetylcholine) reduces the rate of firing by decreasing the slope and hyperpolarizing the pacemaker potential (Fig. 17-13).

In contrast to the stable resting potential of ventricular and atrial cells (net inward and outward currents are balanced, that is, equal zero), cells of the SA and AV nodes as well as the Purkinje fibers have **imbalanced inward and outward currents,** and hence display automaticity. The slow depolarization of the pacemaker potential implies that there is either a time-dependent increase in the net inward current, or a time-dependent decrease in the net outward current, or both.

The ionic mechanisms that generate automaticity can be examined using the voltage-clamp

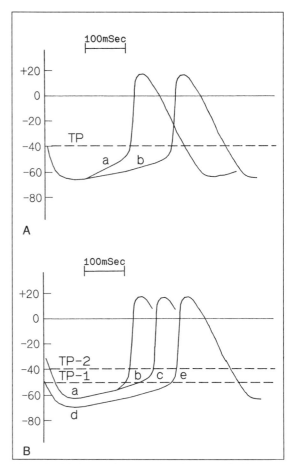

Fig. 17-12. The mechanisms for changing the firing rate of a pacemaking cell. (A) A decrease in the slope of the pacemaker potential (*b*) lengthens the time to reach the threshold potential (*TP*), and hence causes a greater cycle length. (B) A less-negative TP (compare *a* to *b* [*TP-1*] and *a* to *c* [*TP-2*]) lengthens the cycle. In addition, hyperpolarization of the maximum diastolic potential (*d* to *e*) also increases the time to TP and the cycle length.

method. Several ionic currents may be involved in generating the pacemaker potential, and the specific currents in the SA node may differ from those in the Purkinje fibers. In the SA node, a low-threshold Ca^{2+} current is responsible for contributing inward current to the pacemaker potential. This current is increased by norepi-

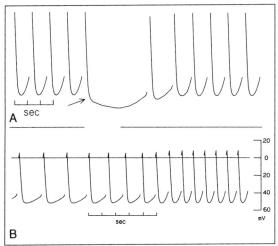

Fig. 17-13. The effects of acetylcholine (A) and norepinephrine (B) on the rate of firing of pacemaker cells from frog heart. In A, the vagus nerve was stimulated at the time indicated (*interruption in line*); this produced pronounced hyperpolarization (*arrow*) and slowing of the firing rate. In B, sympathetic stimulation (during the time indicated by the calibration) increased the rate of firing. (Redrawn from: Hutter, O., and Trautwein, W. *J. Gen. Physiol.* 39:715–733, 1956.)

nephrine and reduced by acetylcholine. An outward, delayed K+ current that is activated during the plateau and turned off (or deactivated) during the pacemaker potential also contributes to depolarization. Nodal cells do not have a large conductance for K+ at rest because they lack inwardly rectifying K+ channels. This condition also promotes automaticity, because the resulting high membrane resistance allows entry of only a small inward current to promote depolarization. In Purkinje fibers, a background Na+ current, which may result from the slow component of inactivation of the fast Na+ current, contributes inward current to the pacemaker potential, along with a low-threshold Ca^{2+} current. Deactivation of the delayed outward K+ current also promotes depolarization. The inwardly rectifying K+ current is present in Purkinje fibers, but tends to turn off with depolarization from the maximum diastolic potential (approximately −90 to −95 mV).

Another current, called the I$_f$ (**"funny" current**), is turned on by repolarization of the AP and contributes inward, depolarizing current. This "funny" current is carried by Na+ and K+; it is increased by norepinephrine and reduced by acetylcholine. A similar current exists in the SA node, where it is called the **hyperpolarizing current,** but it may not play a major role in pacemaking in those cells, since the membrane potential does not reach the negative values necessary for full activation.

CARDIAC ARRHYTHMIAS

Any disturbance in the normal rate or rhythm of cardiac electrical activation may lead to **cardiac arrhythmias.** Because, as already emphasized, the electrical behavior of cardiac muscle triggers and modulates the force of contraction, the inherent danger from arrhythmia is that the heart may not perform as an efficient pump of blood. Clinically, the electrocardiogram is a useful tool for detecting cardiac arrhythmia and discerning its origin within the heart (see Chapter 18).

Arrhythmia may arise from an alteration in the normal sequence of excitation. Normally, the SA node is the fastest pacemaker of the heart, and it determines the heart rate. Other cardiac tissues, including the AV node, Purkinje fibers, and even the atrial and ventricular muscle may, under pathologic conditions, display automaticity. If the rate of firing of these ectopic pacemakers is greater than that from the SA node, or if the signal from the SA node is blocked, they will supersede in the job of driving ventricular depolarization. The latter behavior is demonstrated when conduction between the atria and ventricles is slowed or blocked by an alteration in the properties of the AV node. Then, the AV node itself, or the Purkinje system, may take over as the pacemaker, usually driving the ventricles at a slower than normal rate.

Another type of arrhythmia, termed **re-entrant arrhythmia,** arises when cardiac excitation re-enters tissues of an **abnormal anatomic circuit** that would normally be refractory to excitation

(Fig. 17-14). The continued abnormal excitation is called **circus movement,** as it may travel around the path without cessation.

Re-entrant arrhythmia results from several causes. One of these was mentioned previously, and is the slow, Ca^{2+}-dependent AP that forms in a region of depolarization caused by ischemia or hypoxia, with consequential slow propagation through that region of the ventricle. Because the excitation travels slower through the damaged region, it may reach and re-excite normal tissue which, because of the delay, is no longer refractory. If the timing is "right," and a nonrefractory

Fig. 17-14. The role of unidirectional block in reentry. (Top left) Excitation travels down a single branch (S), bifurcates to left (L) and right (R) branches, and collides in a connecting branch (C). The collision annihilates the impulse in the C branch and the excitation continues normally along the L and R branches. (Top right) The L and R branches are both blocked, and propagation is blocked. (Bottom left) The R branch is blocked in both directions, so excitation may still travel down the L and lower R branches. (Bottom right) The impulse is blocked in the antegrade direction, but can propagate (abnormally) in the retrograde direction. This condition of unidirectional block can cause re-entry of the excitation back through S, leading to arrhythmia.

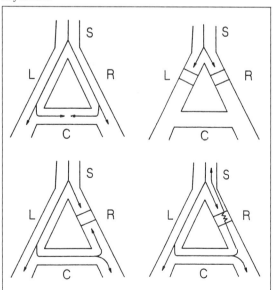

pathway is available, the ectopic impulse may be continuously reflected around the abnormal circuit, with dire effects on cardiac pumping.

Arrhythmias may result from **early** and **delayed afterdepolarizations.** An early afterdepolarization occurs when a second premature AP is generated during the plateau or repolarization phase of a preceding AP. A delayed afterdepolarization occurs after the repolarization of the preceding AP is completed, and may give rise to a premature AP. In either event, the premature AP may display abnormally slow conduction, leading to re-entry.

The most serious form of arrhythmia is **ventricular fibrillation,** in which multiple circus movements cause totally uncoordinated ventricular excitation. During fibrillation, the pumping ability of the heart is minimal, so this condition must be terminated within minutes for patient survival. Ventricular fibrillation can, in some cases, be converted to normal sinus rhythm by the application of a large electrical shock to the heart.

Alterations in the cellular electrophysiology of cardiac cells can be studied to provide insight into the mechanism of the generation of certain arrhythmias. For example, arrhythmia may be brought about by high doses of the cardiac glycoside **digitalis,** which is used to treat cardiac failure. This agent produces depolarizing APs, which may lead to the formation of premature APs and arrhythmia. Recently it has been shown that elevation of $[Ca_i]$, produced by digitalis, activates a **transient inward current,** which is responsible for the depolarization.

SUMMARY

The heart rate is determined by the firing of spontaneous APs in the SA node. Excitation is propagated throughout the myocardium by means of specialized conduction pathways. APs from the different heart tissues have different waveforms and properties. These different waveforms are a consequence of the type and amount of specific ionic currents that generate the APs. The upstroke of APs in rapidly conducting tissues and the myocardial cells is caused by the fast Na^+ current. The slowly rising APs of SA and AV nodes are caused by an inward Ca^{2+} current through voltage-dependent slow channels. The Ca^{2+} current is a major component of the AP plateau; it also initiates and modulates contraction of the myocardial cell. Outward K^+ currents repolarize the AP back to the resting potential. Cardiac arrhythmias are disturbances in the rate or rhythm, or both, of excitation, and arise from alterations in the generation or conduction of the AP.

NATIONAL BOARD TYPE QUESTIONS

Select the one most correct answer for each of the following

1. Which pathway does excitation follow through the heart?
   A. SA node → Purkinje → AV node → ventricle
   B. SA node → AV node → Purkinje → ventricle
   C. AV node → Purkinje → atrium → ventricle
   D. Purkinje → AV node → ventricle → atrium
2. Which ionic current helps generate the slow response?
   A. Na^+ current
   B. Hyperpolarizing current
   C. Ca^{2+} current
   D. Transient outward K^+ current
3. During Phase 2 of the ventricular AP, the inward current is carried by the:
   A. Na^+ current
   B. Transient outward K^+ current
   C. Ca^{2+} current
   D. "Funny" current
4. Automaticity in the SA node is primarily caused by which current?

A. "Funny" current
B. Transient outward K^+ current
C. Inwardly rectifying K^+ current
D. Ca^{2+} current
5. Which of the following may be important in the electrogenesis of cardiac arrythmia?
   A. Delayed afterdepolarization
   B. Re-entry
   C. Slow-response formation
   D. All of the above

ANNOTATED ANSWERS

1. B. The SA node is the primary pacemaker; excitation travels through the AV node and Purkinje system to excite the ventricles.
2. C. The Ca^{2+} current contributes the inward current necessary for depolarization of the slow-response AP.
3. C. The inward current during the AP plateau is carried by the Ca^{2+} current.
4. A. The "funny" (primary) current is a major contributor to pacemaker depolarization in the SA node.
5. D. All of these are electrophysiological mechanisms that may give rise to cardiac arrhythmias.

BIBLIOGRAPHY

Cranefield, P. F. *The Conduction of the Cardiac Impulse.* Mount Kisco, NY: Futura, 1975.
Hoffman, B. F., and Cranefield, P. F. *Electrophysiology of the Heart.* Mount Kisco, NY: Futura, 1976.
Noble, D. *Initiation of the Heartbeat.* Oxford: Clarendon, 1979.
Sperelakis, N. In: Berne, R. M., and Sperelakis, N., eds. *Handbook of Physiology, Vol 1: The Cardiovascular System.* New York: Oxford University Press, 1979. Pp. 187–267.
Sperelakis, N., ed. *Physiology and Pathophysiology of the Heart.* 2nd ed. Boston: Kluwer Academic Publishers, 1989.

18 The Physiologic Basis of the Electrocardiogram

GUNTER GRUPP, INGRID L. GRUPP, AND WILLIAM C. FARR

Objectives

After reading this chapter, you should be able to:

Identify the sequence of electrical events of the heart by obtaining a body surface electrogram—the electrocardiogram

Describe the normal conduction pathways from the sinoatrial node through the atria, atrioventricular conduction system, His–Purkinje system, to and through the ventricles

Explain the importance of the anatomic and electrophysiologic sequence of depolarization in an efficiently coordinated cardiac contraction

Describe the difference between the conduction pathways of depolarization and the process of repolarization

Name each component of the electrocardiographic complex and describe the underlying physiologic reasons for each component

Use the concept of a cardiac dipole to explain why the different electrocardiographic leads record different-appearing QRS complexes, although they are produced by the same electrical forces

Determine the mean electrical axis of the QRS complex of an electrocardiogram, using any two bipolar limb leads, and explain why any two limb leads yield the same finding

Suggest reasons for the different appearance of a QRS complex initiated by an ectopic ventricular pacemaker, as compared with a normally conducted impulse

The electrocardiogram (ECG) provides clinically useful information about the electrical orientation of the heart, the relative size of the heart chambers, conduction defects in the heart, consequences of changes in coronary blood flow (ischemia), and others. It does *not* reflect the state of the mechanical performance of the heart.

A great deal of our current knowledge of the equipment, recording techniques, and the theoretical basis of the ECG was gained from the work of Willem Einthoven. In 1903 he reported on the development of the string galvanometer and began a systematic study of the potential differences between different parts of the heart recorded on the body surface. The ECG is a composite picture of the changing potential differences during cardiac excitation, depolarization, and repolarization recorded between electrodes on the skin surface and plotted on moving paper (Fig. 18-1).

The ECG represented the first recording of its kind; years later, cardiac transmembrane and

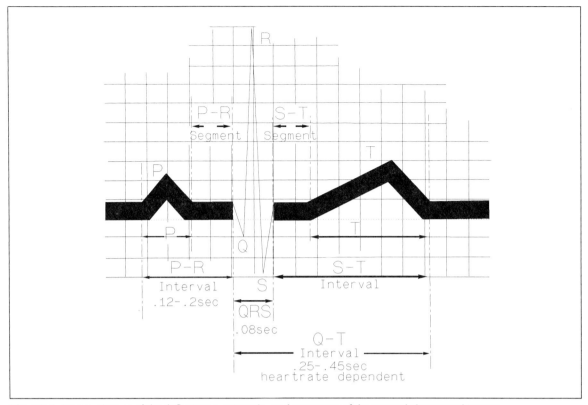

Fig. 18-1. Nomenclature of the deflections, intervals, and segments of the normal electrocardiogram.

action potentials and excitation patterns were discovered. Many theories have been proposed to explain the ECG tracing based on cellular events. Stated simply, depolarization of one sector of the heart causes a difference in the potential between the depolarized and polarized regions and produces an electric current that flows from the positive to the negative areas. Because electrolytes contained in the tissues and fluids of the body surrounding the heart are good conductors, these electrical currents reach the skin surface. Two ECG electrodes placed on the skin detect these changing potential differences. These continuously changing potentials are greatly amplified and recorded on calibrated moving paper.

The source of the electrical force, the heart, can theoretically be considered as a single electrochemical generator located in the center of a

sphere, and this "generator" is made up of positive and negative charges—**a dipole.** A dipole is best described mathematically using a vector. A **vector** can describe brief instants during the excitation process (instantaneous vector), or it can be used to summate a large number of instantaneous vectors (a mean vector). Over the 0.4 to 0.7 second (400 to 700 msec) duration of one heartbeat, a multitude of dipoles can be described by an equal number of vectors, giving insight into the electrical activity of the heart. The algebraic mean of all ventricular vectors is represented by the **mean electric axis.**

TERMINOLOGY OF THE ECG

As described in the previous chapter, the heartbeat is initiated by a spontaneous depolarization in the **sinoatrial** (SA) **node,** which is located at the junction of the superior vena cava and the right atrium. The impulses travel rapidly (1.0 to 1.5 m per second) through all right and left atrial fibers, depolarizing them. This atrial depolarization wave is recorded on the skin surface as the **P wave** of the ECG (see Fig. 18-1). This wave is normally a small, smooth wave without obvious notching, about 2 mm in height and 0.08 second in duration. The depolarization then traverses slowly (0.02 to 0.05 m per second) through the **atrioventricular** (AV) **node** to the bundle of His and its branches, and emerges again in the very fast-conducting Purkinje system (2.0 to 4.0 m per second) to enter both ventricles, thereby initiating the **QRS complex.** The conduction velocity through the ventricular myocardium is very fast (1.0 to 2.0 m per second). As a result, the duration of ventricular depolarization (QRS) is only 0.05 to 0.10 second. In general, the first downward deflection after the P wave is called the **Q wave** (see Fig. 18-1). The first upward component after the Q wave is the **R wave,** and the downward deflection after an R wave is the **S wave** (see Fig. 18-1). If there is a second upward deflection (caused by ventricular depolarization) after the S wave, it is labeled an **R′ wave,** and so on. Independent of the sequence of depolarization waves, the complex is always called the **QRS complex.** Ventricular repolarization occurs during the **T wave.** This wave is generally upright and about 0.15 to 0.25 second in duration.

The **PR interval,** which is the period between the beginning of the atrial depolarization and the beginning of the ventricular depolarization, is normally 0.12 to 0.2 second in duration. Shorter PR intervals may indicate AV conduction shortcuts, and longer intervals may indicate partial or complete AV block.

The **QT interval** encompasses the time from the beginning of the QRS complex to the end of the T wave, or that period from the beginning of ventricular depolarization to the end of repolarization. The duration of the QT interval therefore somewhat mimics the total of all ventricular action potential durations (see Chapter 17). Heart rate has a profound effect on the duration of the action potential, and consequently on QT interval duration; increases in the heart rate shorten the QT interval and decreases lengthen it. Table 18-1 lists the average QT intervals at different heart rates for men, women, and children.

The PR, ST, and TP segments are electrically quiet, and therefore **isoelectric. Elevation of the ST segment** often indicates disrupted conduction in the ventricular myocardium, frequently caused by **ischemia.**

THE BASIS OF THE ECG RECORDING

The following generalizations can be made about the recording of the ECG and the mechanisms that underlie it.

1. **Cell membranes** separate charges.
2. **Ion gradients** produce electrochemical gradients across semipermeable membranes.
3. A chemical force creates an electrical force that creates **equilibrium.**
4. Currents flow only during **depolarization** and **repolarization.**
5. Depolarization of the heart causes **potential differences** to form between polarized and depolarized regions. The same is true during repolarization.
6. Current flows from **polarized to depolarized areas,** from plus to minus, causing potential differences.
7. Arrows (force vectors) point in the direction of the **propagation of the wave of depolarization.**
8. The **polarity** of the recording voltmeter (ECG machine) is arranged so that, in bipolar recordings, the direction of the wave of propagation toward the positive pole produces an upright deflection; conversely, if the wave of propagation is toward the negative pole, it produces a downward deflection. In unipolar recordings, upright deflection indicates movement of the propagation wave toward the positive (exploring) electrode, and downward deflection, movement away from it.

Table 18-1. Heart Rate Dependence of QT Intervals

Heart Rate (bpm)	QT Interval (sec)			
	Men and Children		Women	
	Normal	Upper Limit	Normal	Upper Limit
50	0.41	0.45	0.43	0.46
60	0.39	0.42	0.40	0.43
80	0.34	0.37	0.35	0.38
100	0.31	0.34	0.32	0.35
120	0.28	0.31	0.29	0.32
150	0.25	0.28	0.26	0.28

9. The flow of electrical currents generates **electrical fields.**

10. The field always travels in the direction the charge wants to go in, from **positive to negative.**

11. Because **depolarization** runs from cell to cell until the whole heart is depolarized, current changes also run in this direction.

12. Current changes are picked up on the body surface in the form of the **changes in potentials surrounding the currents.**

13. The body is a nearly uniform **volume conductor** (a bag of saline), and the skin is a good conductor; therefore, potential differences in the heart, if and when they occur, can be measured and recorded from the surface of the body, that is, the skin.

In brief, there are currents flowing during depolarization and repolarization; there are potential differences when currents flow; there are no potential differences when there is *no* current flow; and there are no potential differences when the heart is completely polarized (at rest) or completely depolarized (after activation and before reactivation).

EQUIPMENT AND LEAD SYSTEM

The **electrocardiograph** is an elaborate **voltmeter** that measures the potential differences between the electrodes and records them. The arrangement of a pair of electrodes on the body surface constitutes a **lead.** In clinical practice, two lead systems are used: **bipolar** and **unipolar.**

There are **three bipolar leads,** and these are placed on the **frontal plane** of the body (Fig. 18-2). **Lead I** records the potential difference between the right and left arms (or shoulders); **lead II,** between the right arm and left leg; and **lead III,** between the left arm and left leg. The right leg is used to ground the patient. The three bipolar leads form the **Einthoven triangle,** with the heart at its center. An imaginary line between two electrodes of a lead is called the **lead axis** (denoted by *dashed lines* in Fig. 18-2).

Unipolar leads are used to add additional lead axes in the frontal plane and to obtain the V leads in the **horizontal plane.** In clinical practice, no truly unipolar lead system exists, but it can be approximated by use of **Wilson's central terminal** (denoted as *CT* in Figure 18-2). A unipolar system assumes that **one electrode** is an indifferent electrode, and the central terminal assumes this function. This "indifferent" electrode is considered to be close to zero potential (denoted as a *minus* in Figure 18-2). The **second,** or exploring, **electrode** constitutes the actual recording electrode, and this detects the potential changes relative to zero from its various locations on the body surface (denoted as *plus* in Figure 18-2). On the frontal plane, **Goldberger's augmented unipolar limb leads** are unipolar (shown as *aVF, aVL,* and *aVR* in Figure 18-2).

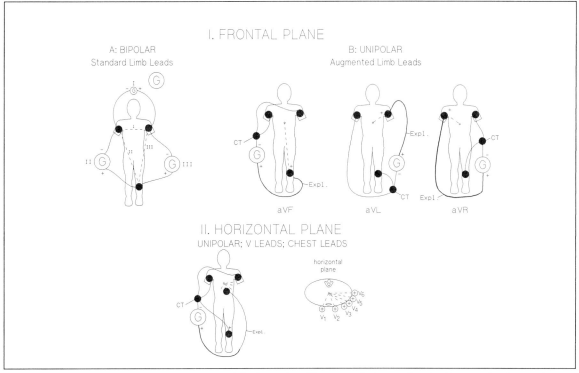

Fig. 18-2. Bipolar and unipolar lead systems of the frontal and horizontal planes. (*Dashed lines* = lead axes; *G* = galvanometer showing the potential differences on the electrodes; *CT* = Wilson's central terminal; *Expl.* = exploring electrode)

The exploring electrode of these three leads records potential changes manifested at the right arm, left arm, and left leg, respectively. The lead axes (*dashed lines* in Figure 18-2) fit reasonably well between the lead axes of the bipolar frontal leads I, II, and III (Fig. 18-3). These six frontal leads provide the so-called **hexaxial arrangement,** which constitutes a nearly all-encompassing frontal view of the electrical activity of the heart. The *arrowheads* on the *dashed lines* in Figure 18-3 indicate the positive poles of the respective leads. The term *augmented* has an historical origin, in that Wilson's earlier unipolar leads recorded 50 percent less voltage than did the bipolar leads. Goldberger's arrangement "augmented" the sensitivity of Wilson's bipolar leads by 50 to 87 percent. The six chest leads

(shown as *V leads* in Figure 18-2, part II) are also unipolar and record the local potentials at six points along the **horizontal axis** (plane) of the thorax (V_1 to V_6).

In clinical practice, one records all twelve leads. By doing so, the distribution of the potentials on the frontal and horizontal planes of the body can be studied. This provides a reasonable amount of information about the three-dimensional distribution of the heart's potential field. The potentials recorded from various areas of the body differ because of variations in the distance of the electrodes from the heart, the amount of intervening tissue, the tissue resistance, and the arrangement of the recording leads. Because of this, the shapes and directions of the P, QRS, and T waves *are different at the various leads,*

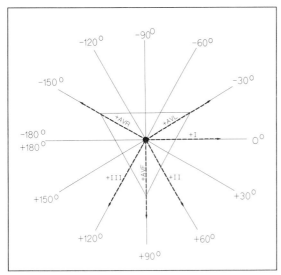

Fig. 18-3. Hexaxial arrangement of the six frontal leads from 0 to ± 180 degrees with relation to the heart. (*Broken lines* = lead axes; *arrowheads* = direction of positive half of the lead axis; *AVR, AVL,* and *AVF* = Goldberger's augmented unipolar limb leads; *I, II,* and *III* = bipolar frontal leads.)

although they all result from the same heartbeat.

THE HEART'S ELECTRICAL FIELD IN A VOLUME CONDUCTOR

When the cells of one area of the heart depolarize, their outside becomes electrically negative with respect to the quiescent positive areas (Fig. 18-4). *Current flows from the positive to the negative area; ions flow in the opposite direction.* In contrast, the direction of the propagation wave of the impulse goes from outside negative (depolarized) to outside positive (polarized) areas (Fig. 18-4), leaving depolarization in its wake. The pattern of current flow in different conductive media is shown in Figure 18-5. The greatest density of current flow is in a wire (see Fig. 18-5, *a*). It becomes more diffuse in a saline column (see Fig. 18-5, *b*) and very diffuse in a volume conductor (see Fig. 18-5, *c*), where the current is most dense in the immediate vicinity of the charges and de-

creases with distance from them (Fig. 18-6). The **body** is not a homogeneously conducting medium. It does, however, permit currents to reach the skin surface, and therefore it can be considered a **volume conductor.** A plot of the potential differences existing in an electrical field produced by currents flowing in a volume conductor is shown in Figure 18-6. There are areas of equal potential called **isopotential lines,** shown as *dotted lines* in Figure 18-6, and these are situated at right angles to the current flow lines. If two electrodes are placed in this electrical field, a potential difference is recorded, provided the electrodes are at different potentials. The potential recorded is dependent on how far the two electrodes are from the source, and on the resistance of the conducting medium. Different potentials are recorded when the electrodes are moved relative to the source. Likewise, if the electrode position is held in a fixed site (as is the case with a given ECG lead) and the electrical field changes over time (the fields in the heart change during electrical activity), a change in the potential difference over time is recorded, and *this is what the electrocardiograph does.*

THE ROTATING DIPOLE

The ever-changing depolarization pathways in the heart and the changing cardiac mass contributing to the potentials form a very complex relationship. At any instant, there are thousands of cardiac cells depolarizing in different locations in the heart and depolarizing in various directions at different distances from the electrodes. In addition, each cell has its own characteristic action potential and conduction velocity. Therefore, there is a **multiplicity of conductive resistances** between the cells of the heart and the skin surface. To better understand this complexity, the ECG theoreticians have represented the total excitation process as a single positive and negative charge—**a dipole** (Fig. 18-7). If one regards the cardiac dipole as the center of the chest, one can describe its magnitude and direction three dimensionally by using **vectors.** The complex excitation process of the heart can then be

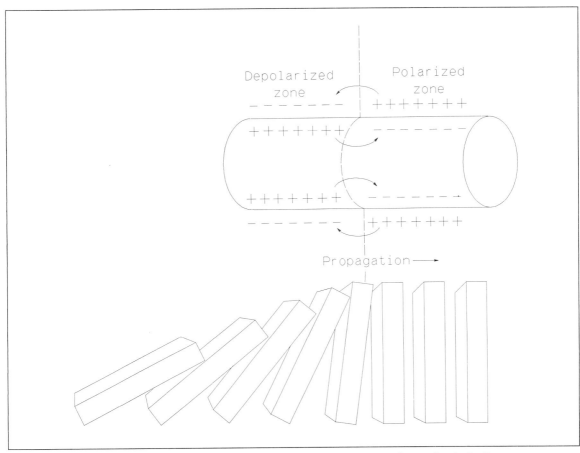

Fig. 18-4. Current flow and direction of the propagation of the excitation wave during depolarization.

described with an infinite number of vectors. In reality, only a few instants in time are used; these are called **instantaneous vectors.**

By convention, a vector has a positive and negative end (see Fig. 18-7A and B). The magnitude of the arrow representing the vector is determined by the magnitude of the potential difference recorded. The potential difference recorded by an ECG can be estimated using the simple rules of **vectorial projection.** To project a vector onto a lead axis (see Fig. 18-7B), perpendicular lines are dropped from the ends of the vector to the axis. If the vector projects onto the positive half of the lead axis, the voltage registered by the lead is positive and the ECG records an upward

deflection (see Fig. 18-7C). The magnitude of the deflection indicates the potential difference measured. The projected length of the vector in Figure 18-7B is equal to the length of the actual vector, because both vector direction and the lead axis are parallel. In most cases, the projected length is smaller because the in situ vectors diverge from the lead axis. Consequently, if the vector projects onto the negative portion of the lead axis, the lead voltage is negative and a downward deflection is recorded. If the vector parallels or coincides with the lead axis, it projects maximal voltage onto the lead. A vector directed perpendicular to a lead axis does not project onto the lead axis, and so the galvanometer shows zero

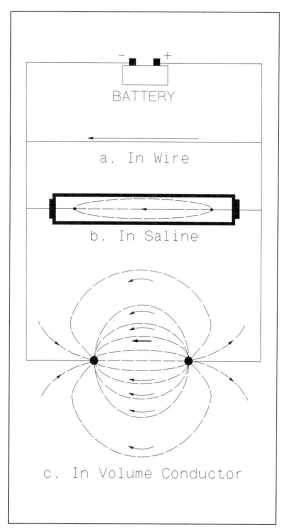

Fig. 18-5. Current flow in different conductors. The greatest current flow is in wire (*a*), becoming more diffuse in saline (*b*), and very diffuse in a volume conductor (*c*).

voltage. Rotation of the dipole vector between these two extremes yields a continuum of intermediate values. *Thus, the magnitude of the potential recorded is proportional to the length of a line projected onto the lead axis.* The polarity of the recorded event is upward when a positive-recording electrode is in the heart's positive field,

and downward or negative when in the negative field of the heart.

The **changing electrical forces of the heart** in three-dimensional space can be described with a rotating dipole and a series of instantaneous vectors.

MEAN INSTANTANEOUS VECTORS

The development of several mean instantaneous vectors is depicted in Figure 18-8. During the first part of ventricular depolarization, the left endocardial surface of the interventricular septum is activated (see Fig. 18-8A). This takes about 5 to 10 msec. If, at that instant, all these electrical forces are averaged, a **mean instantaneous vector** can be obtained (labeled *1* in Figure 18-8A). This vector is small and points to the right ventricle. During the subsequent phases of ventricular excitation, the forces occurring at other instants can also be averaged and represented by their respective vectors: **early apex activation** can be summarized by vector *2* (see Fig. 18-8B); **late apex activation** by vector *3* (see Fig. 18-8C); and the final **base activation** by vector *4* (see Fig. 18-8D). (Note that the direction and strength of the four instantaneous vectors shown in Figure 18-8, 18-9, and 18-10 are identical.) It follows from the example developed in Figure 18-8 that a whole continuum of vectors can describe ventricular depolarization. Figure 18-9A summarizes these events from beginning to end during sequential depolarization—from instantaneous vector *1* to vector *4*, showing the actual depolarization waves and their mean directions. These vectors can be used to produce a **continuous vectorial analysis** (see Fig. 18-9B). A loop can be inscribed within the Einthoven triangle, originating from the midpoint and going from there to the tips of vectors *1*, *2*, *3*, and *4* and back to the midpoint. This is a **vector cardiogram.** Such a loop can be constructed for P, QRS, and T wave activity for any of the ECG leads. Using the method of **vector summation** in the frontal plane (see Fig. 18-9D), a total **mean QRS vector,** also called the **mean electrical axis,** can be computed (see Fig. 18-9C).

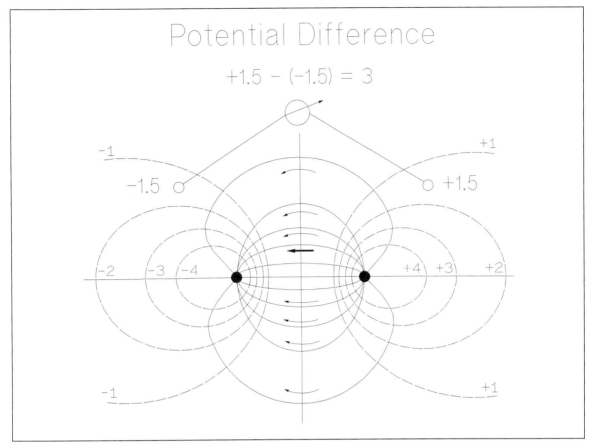

Fig. 18-6. Current flow (*solid lines*) and isopotential lines (*broken lines*) of a dipole.

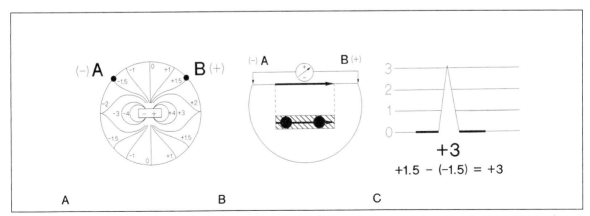

Fig. 18-7. (A) Dipole vector recorded in leads A(−) and B(+). (B) Vector projected onto lead axis. (C) Strength of vector recorded on moving paper.

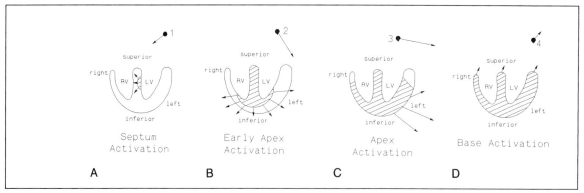

Fig. 18-8. Time sequence and instantaneous vectors 1 to 4 during ventricular depolarization. (*RV* = right ventricular; *LV* = left ventricular.)

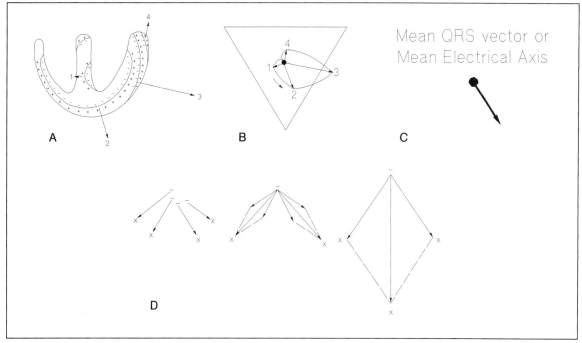

Fig. 18-9. Summation of instantaneous vectors 1 to 4 to produce the mean QRS vector or the mean electrical axis (A to C). (D) Summation of vectors.

QUANTITATION OF A QRS COMPLEX

Figure 18-10 shows two frontal plane projections, lead I and lead III, of the four instantaneous vectors elaborated in Figures 18-8 and 18-9.

These projections describe the electrical forces of the ventricles at four different instants during one ventricular depolarization. In the top part of Figure 18-10, the vectors are projected onto lead I. Using the method described previously (see Fig.

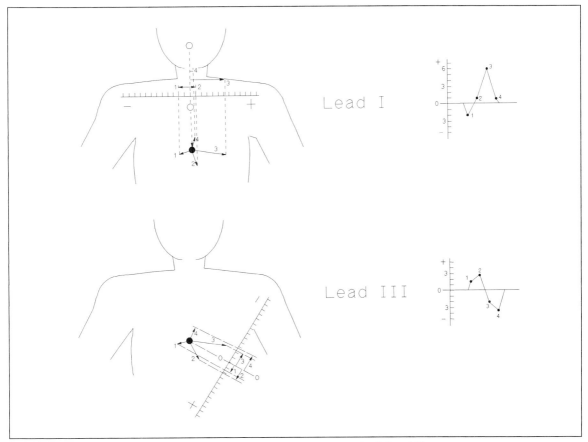

Fig. 18-10. Projection and recording of instantaneous vectors 1 to 4 in leads I and III.

18-7B), one can project these vectors onto the axis of lead I and quantitatively calculate the ECG deflections shown at the right of Figure 18-10. Vector *1* reflects onto the lead I axis with -2, vector *2* with $+1$, vector *3* with $+6$, and vector *4* with $+1$. These deflections can be plotted against time. The resulting QRS complex is shown in the upper right of Figure 18-10. If the exact same four electrical vectors are projected onto lead III (Fig. 18-10, lower part), a very different ECG complex is derived. Similarly, if lead II is used, or an augmented unipolar aVR, or even one of the V leads, the same electrical forces would produce *different* ECG complexes, depending on the spatial arrangement of the vectors in the re-

spective leads. These observations explain why it is important to record all twelve leads in clinical practice.

MEAN ELECTRICAL AXIS

Because vectors can be used to calculate precisely the type of the ECG deflection recorded (see Fig. 18-10), the reverse is also true. In clinical practice, an ECG tracing is obtained and then used to calculate mean vectors for the P, QRS, and T waves. Most often, the inscription of the QRS complex is used to provide general insight into the direction and distribution of the electrical forces of the ventricular myocardium. This is called the *mean QRS vector,* or *mean electrical axis.* In other instances, just the vector for the first part of the QRS complex (first 0.04-sec QRS vector) or the last part (last 0.04-sec QRS vector) is calculated.

The **mean electrical axis** is an important descriptor of the ECG, as it can be abnormal in a variety of cardiac disorders. It is determined by the anatomic position of the heart (vertical or horizontal, with the former more likely in young people and the latter more likely in obese individuals), and by the direction of electrical depolarization of the ventricles. The **mean QRS axis** is determined from the mean QRS vector and is reported in degrees, based on the **hexaxial reference system** (see Fig. 18-3). The normal mean QRS axis in the frontal plane lies between −30 and +100 degrees. If the mean QRS axis is −30 degrees or more negative, there is **left-axis deviation,** indicating left ventricular preponderance (for example, hypertrophy). If the mean QRS axis lies to the right of +100 degrees, there is **right-axis deviation,** indicating right ventricular preponderance (Fig. 18-11).

There are several ways to determine the **frontal plane mean electrical axis of the QRS.**

Three Standard Limb Leads

The mean electrical axis can be computed from any two of the three standard limb leads. In the example given in Figure 18-11, this was determined from leads I and III, the same leads described in Figure 18-10A and B. The sum of the downward deflections of the QRS complex is subtracted from the sum of the upward deflections. In this example, the R wave of lead I has a vertical height above baseline of +6 mm with a downward deflection of −2 mm. These values are added algebraically, giving a net value of +4. At a point 4 units toward the plus sign on the lead I axis of the triangle, a perpendicular line is erected. Similar measurement of Lead III yields a net amplitude of upward and downward deflections of −2 (+2 −4). A perpendicular line erected 2 units toward the minus sign on lead III is extended to intersect the perpendicular line from lead I. An

arrow drawn from the center of the triangle to the intersection of these two perpendicular lines indicates the **mean electrical axis,** pointing to 0 degrees. Any combination of any two of the three standard limb leads will give the same result.

Hexaxial Reference System

The frontal hexaxial reference system can also be used to determine the mean electrical axis (see Fig. 18-3). There are two applications of this approach.

The somewhat simpler of the two approaches uses the six limb leads to determine in which single lead the QRS complex is most nearly isoelectric (that is, the negative and positive areas are most equal), and in which lead the QRS deflection is largest. As an example, consider an ECG in which lead III shows a large upright deflection, whereas the aVR lead shows equal positive and negative deflections. Using Figure 18-3, the lead to which the QRS complex is most perpendicular (isoelectric) determines the axis and the lead to which it is most parallel (largest deflection area) aids in the identification. Because it is perpendicular to the aVR, the mean electrical axis is +120 degrees, indicating right-axis deviation (see Fig. 18-11). Lead III also points to +120 degrees, confirming the diagnosis.

The second method to determine the mean QRS axis using the hexaxial reference system involves finding two leads in which the QRS complex projects equally. The mean QRS axis then lies equidistant between these two leads. If, as an example, an ECG is used which shows a large negative deflection of lead III and an equally large positive deflection in the aVL lead, the mean electrical axis (according to Figure 18-3) lies between −60 degrees (minus lead III) and −30 degrees (plus lead aVL), or at about −45 degrees, indicating left-axis deviation (Fig. 18-11).

The normal mean electrical axis is somewhat different for each age group. The interpretation of normal requires judgment and experience. The mean electrical axis is changed from "normal" by changes in the heart's position in the thorax, ro-

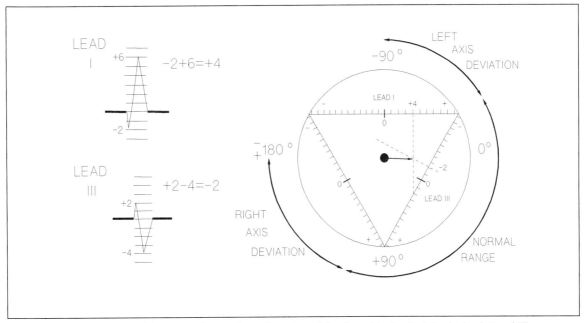

Fig. 18-11. Construction of the mean electrical axis from two of the three bipolar limb leads: leads I and III.

tation of the heart, thickness of the ventricular walls, disturbances of the conduction sequence, and pathologic processes. For instance, the electric forces generated by the heart are affected by the mass of muscle depolarized. When ventricular or atrial mass increases (hypertrophy), as occurs when there is sustained pressure (hypertension) or volume (high-output) work, the electrical axis changes. Also, if one side (or part) of the heart is destroyed by disease, as occurs in myocardial infarction, the electrical generating tissue is replaced by inactive fibrous scar tissue, which changes the vector direction. As a result, there is a change in the mean electrical axis.

VENTRICULAR REPOLARIZATION: THE T WAVE

Whereas the activation or **depolarization** of the ventricles occurs rapidly in the form of a progressive wave that passes from cell to cell, which is generally directed from the endocardium to the epicardium and from the apex to the base of the

heart, the recovery or **repolarization** is an independent, prolonged, and complicated process that seems to be an innate property of each cell. Consequently, repolarization *does not* occur as a propagated wave. Instead, there are **multiple areas of potential difference** that are oriented in many directions, and these relationships change frequently as the recovery is completed. Although the recovery process may be depicted by a single vector, recovery actually represents the summation of the effects of the very complicated process of repolarization. Another difference between depolarization and repolarization can be gleaned from the time courses of the two events. Based on the duration of the single action potential, it is obvious that depolarization, which is associated with Phase 0, is a rapid process. It requires less than 2 msec to depolarize a single cell and only 60 to 80 msec to depolarize both entire ventricles. On the other hand, repolarization, which occurs during Phases 2 and 3 of the action potential, is a much longer process, requiring 150 to 300 msec to repolarize each cell and 300 to 400 msec to repolarize both ventricles.

Intuitively one might expect the T wave to be of opposite polarity from the QRS complex, because repolarization is the reverse of depolarization at the cellular level. However, because of the considerations described, this assumption is not supported by the facts: QRS and T waves in leads I and II are both upright (positive) in most normal ECGs.

THE ISOELECTRIC LINE

Several segments of the ECG tracing are flat: the PR, ST, and TP segments. These segments, from which all deflections "take off," form the **isoelectric line.** The **PR segment** is not influenced by heart forces, as the mass of tissue activated during this period is too small (depolarization of the AV node, common-bundle branches). The **ST segment** probably represents that interval when most ventricular muscle cells are already completely depolarized and not yet repolarized. Because "all" cells are equally depolarized during the ST segment, no current flows and no potentials are recorded. During the **TP segment,** the ventricles are repolarized, and therefore no current flows until the next heartbeat.

SOME CLINICAL APPLICATIONS OF THE ECG

Heart rate and rhythm can be monitored with the ECG. **Tachycardia,** a heart rate of more than 100 beats per minute (bpm; Fig. 18-12A), and **bradycardia,** a heart rate of less than 60 bpm (see Fig. 18-12B), can be calculated from the R-R interval and the known paper speed of the ECG. The occurrence and frequency of extra beats originating from **premature atrial depolarization** (see Fig. 18-12C) or from **premature ventricular depolarization** (see Fig. 18-12D) can also be observed. When these events originate outside the normal pacemaker area, a compensatory pause ensues until the normal rate takes over again (see Fig. 18-12C and D).

The PR interval of the ECG is an important indicator of **AV conduction.** Its normal duration is 0.12 to 0.2 second. A PR interval shorter than 0.12 second may indicate a ventricular pre-excitation syndrome, showing a **Wolff-Parkinson-White pattern.** A PR interval longer than 0.2 second indicates **first-degree AV block** if each P wave is followed by an R wave. Intermittent failure of the conduction of the supraventricular impulse implies **second-degree AV block** with varying conduction ratios of 2:1 or 3:1, or other P waves per QRS complex. If the atrial and ventricular activities are completely independent, **third-degree** or **complete AV block** exists.

The ECG is very helpful in indicating the orientation of the heart within the body. The mean electrical axis (see also Fig. 18-11) gives an estimate of the orientation and preponderance of the heart. The normal mean electrical axis lies between about −30 and +100 degrees. Figure 18-13A is an example of an ECG in the normal range. In this, the QRS complexes of leads I and II are both positive overall. Figure 18-13B shows the recording from leads I, II, and III from a patient with left ventricular hypertrophy. In this instance, lead I shows an overall positive reading and lead III a negative value; this translates into a mean axis of −50 degrees, according to the text discussion for Figure 18-11. This indicates a left-axis deviation. Figure 18-13C shows a right-axis deviation of +120 degrees, representing right ventricular hypertrophy; in this example, lead I has an overall negative value and lead III, a positive one.

A **normal QRS duration** is between 0.06 and 0.1 second. About one-half of the adult population exhibits a QRS value of 0.08 second. Any ventricular conduction disturbance is reflected in the QRS complex. In particular, myocardial ischemia and infarction alter the QRS complex dramatically. Figure 18-13D shows the ECG recording from leads I, II, and III of a patient with acute anterior myocardial infarction (see Fig. 18-13D). Typical changes consist of the substantial elevation of the normally isoelectric ST segments in leads II and III. Similar radical changes in the QRS configuration can occur in a healthy ventricle that is depolarized outside its normal site (ectopic beat).

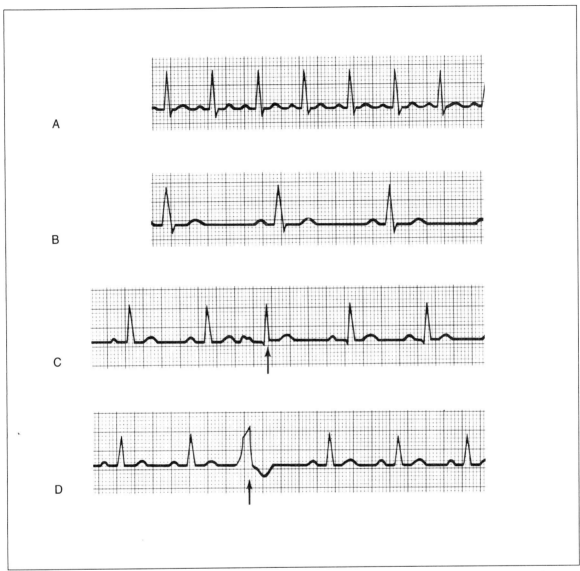

Fig. 18-12. Lead II tracings of (A) tachycardia (122 bpm), (B) bradycardia (47 bpm), (C) premature atrial depolarization with compensatory pause, and (D) premature ventricular depolarization with compensatory pause. (Adapted from S. Scheidt, *Basic Electrocardiography*. CIBA-GEIGY Pharmaceutical Co., 1986.)

SUMMARY

When depolarization proceeds in an orderly and predictable fashion over the atria and ventricles of the heart, this generates current flows. Because the human body behaves as a volume conductor, the potential differences created by these current flows can be measured and recorded on the surface of the body, producing the ECG. Depending on the placement of the recording electrodes, the sequence of the current flows

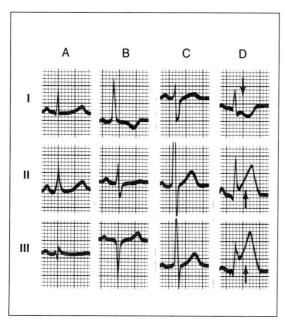

Fig. 18-13. (A–C) Mean electrical axis from leads I, II, and III. (A) Normal axis, +18 degrees; leads I and II positive. (B) Left-axis deviation, −50 degrees; lead I positive, lead II negative. (C) Right-axis deviation, +120 degrees; lead III negative, lead II positive. (D) Acute anterior myocardial infarction in which the ST segments are elevated in leads II and III. (Adapted from T. Winsor, The electrocardiogram in myocardial infarction. *Clin. Symp.* 29:16–17, 1977. CIBA-GEIGY Pharmaceutical Co.)

(depolarization) in the heart can be observed from many different aspects. Ordinarily, the frontal and horizontal planes of the heart are recorded, each in six different aspects; these consist of the hexaxial limb leads and the precordial V leads. The original three limb leads, introduced by Willem Einthoven in 1903, are bipolar; the augmented limb leads and V leads are unipolar. In contrast to the cell-to-cell conduction that occurs during depolarization, the process of repolarization is not conducted. Instead, it is produced by the innate properties of each cell, and is therefore prolonged and complicated. The ECG is a major tool in the diagnosis of conduction defects of the heart; these include first-, second-, and third-degree heart block, arrhythmias, ectopic

beats, and myocardial infarction. The recording from the different leads allows a quick assessment of the mean electrical axis of the heart, showing whether it is normal or has left- or right-axis deviation. The ECG does not, however, appraise the quality of the contractile function of the heart.

NATIONAL BOARD TYPE QUESTIONS

Select the one most correct answer for each of the following

1. Looking at Figure 18-1, which part of the normal ECG is most sensitive to the heart rate, that is, is prolonged at a slow rate and shortened at a fast rate?
   A. PR interval
   B. PR segment
   C. QRS duration
   D. QT interval
   E. T-wave duration
2. The AV conduction time is reflected in the:
   A. PR interval
   B. PR segment
   C. QRS duration
   D. QT interval
   E. T-wave duration
3. In heart failure, the ECG shows which of the following patterns?
   A. Absence of P waves
   B. Absence of QRS complexes
   C. Shortening of the PR interval
   D. Flattening of the ST segment
   E. None of the above
4. Using Figure 18-11, if the overall net deflection of lead I in a standard ECG is positive and that of lead III is negative, the mean electrical axis, calculated from leads I and III, should be between:
   A. −30 to −150 degrees
   B. +30 to −90 degrees
   C. −150 to +150 degrees
   D. −30 to −90 degrees
   E. +30 to +150 degrees

5. Using Figure 18-11, if the overall net deflection of lead I in a standard ECG is positive and that of lead II is negative, the mean electrical axis, calculated from leads I and II, should be between:
   A. −30 to −150 degrees
   B. +30 to −90 degrees
   C. −150 to +150 degrees
   D. −30 to −90 degrees
   E. +30 to +150 degrees

itive lead I must be between +90 and −90 degrees and the axis of a negative lead III must be between +30 and −150 degrees. The common overlap of the two leads (i.e., the mean electrical axis) must be between +30 and −90 degrees.

5. D. The axis of positive lead I is between +90 and −90 degrees; the axis of negative lead II is between −30 and +150 degrees; the common overlap is between −30 and −90 degrees.

ANNOTATED ANSWERS

1. D. Only the QT interval is strongly dependent on the heart rate. (See the third paragraph of the section "Terminology of the ECG" and Table 18-1.)
2. A. (See the second paragraph of the section "Terminology of the ECG.")
3. E. The ECG does not reflect the mechanical performance of the heart. (See the first paragraph in the introductory section.)
4. B. Figure 18-11 shows that the axis of a pos-

BIBLIOGRAPHY

Chou, T. *Electrocardiography in Clinical Practice*, 3rd ed. Philadelphia: W. B. Saunders, 1991.
Cooksey, J. D., Dunn, M., and Massie, E. *Clinical Vectorcardiography and Electrocardiography*, 2nd ed. Chicago: Year Book Medical Publishers, 1977, Part I.
Einthoven, W. *Pflugers Arch.* 99:472, 1903.
Pozzi, L. *Basic Principles in Vector Electrocardiography*. Springfield, IL: Charles C Thomas, 1961.
Scher, A. M., and Spach, M. S. *Handbook of Physiology*, Section 2, Vol. I. New York: Oxford University Press, 1979. Pp. 357–392.
Snellen, H. A. Contribution of Willem Einthoven to Physiology. *NIPS* 4:162–5, 1989.

19 Determinants of Left Ventricular Performance and Cardiac Output

BRIAN D. HOIT AND RICHARD A. WALSH

Objectives

After reading this chapter, you should be able to:

Describe the determinants of contraction in isolated cardiac muscle and in the intact left ventricle

Describe pressure, volume, and flow phenomena during the cardiac cycle and graphically depict these events using the pressure–volume loop

Explain the significance of ventricular–vascular coupling

Left ventricular performance may be assessed by evaluating the mechanics of constituent muscle fibers; these are: the **development of force** and **the velocity and extent of muscle shortening** (see Chapter 14). However, a more integrated analysis considers the left ventricle as a **muscle pump** coupled to the vascular (arterial and venous) system. The principal determinants of left ventricular function include the loading conditions (preload and afterload), the inotropic (contractile) state, and the heart rate. In this chapter, principles derived from studies in isolated cardiac muscle will be used to explain how performance is modulated in the intact left ventricle. The cardiac cycle will be described in terms of its temporal relationship with left ventricular pressure, volume, and flow and in the context of the ventricular pressure–volume relationship. Finally, coupling of the left ventricle to the vascular system will be discussed briefly.

DETERMINANTS OF MYOCARDIAL PERFORMANCE

Isometric Contraction

When a strip of heart muscle (such as cat papillary muscle) is attached at both ends, so that the length is fixed, and then electrically stimulated, the muscle develops force without shortening. A fundamental property of striated muscle is that the strength of this **isometric twitch** is dependent on the initial resting muscle length, or **preload.** As cardiac muscle is stretched passively (increased preload), the resting tension rapidly rises and prevents overstretching of the **sarcomeres.** If additional load is applied before contraction (preload), stimulation causes contraction with an increased **peak tension** and rate of **tension development** (dT/dt) (Fig. 19-1A and B). Thus, total tension includes both active and passive tension. The **length–tension relationship,** derived from variably preloaded fibers, is depicted in Figure 19-1C. Although the length–tension relationship is characteristic of all *isolated* striated muscles, skeletal muscle does not manifest this

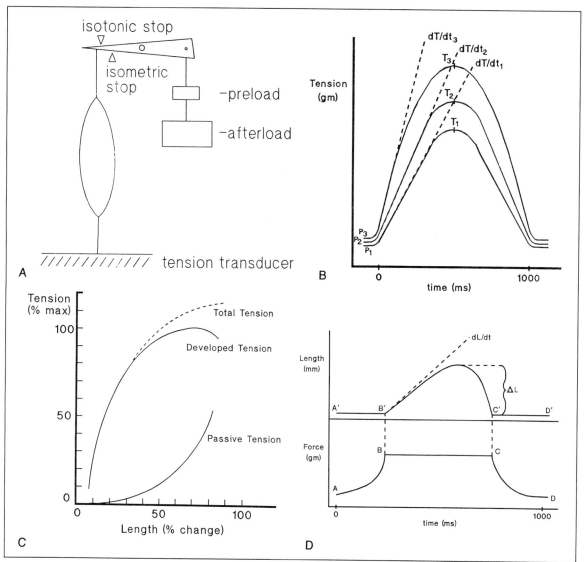

Fig. 19-1. Contractions in isolated muscle. (A) Isolated muscle preparation. Muscle is attached to a lever arm at one end and fixed to a tension transducer at the other. The muscle is stretched by applying a weight (preload) at one end of the lever arm. A stop prevents muscle shortening. (B) Tension–time curves of isometric twitches at three levels of preload. With increased preload, peak tension (T_1, T_2, and T_3) and the maximum rate of tension development (dT/dt_1, dT/dt_2, and dT/dt_3) are increased. The time to peak tension is unchanged. (C) Length–total tension relationship and its components, passive and active tension. As muscle is stretched, the absolute passive tension increases, along with its contribution to total tension. (D) Superimposed tension–time and length–time recordings from afterloaded isotonic contractions. After preload is applied, a stop is placed to prevent further stretching. Afterload is added and the muscle is stimulated. Muscle shortens when generated tension equals total load (preload and afterload). Measures of shortening in the isotonic contraction include total shortening (ΔL) and the initial velocity of contraction (dL/dt). (Redrawn with permission from: Ross, J. Jr. In *Best and Taylor's Physiological Basis of Medical Practice,* 12th ed. Baltimore: Williams & Wilkins, 1989. P. 213.)

property in vivo because of its fixed attachments to bone. The length–tension relationship forms the basis for the Frank-Starling relationship in the intact heart, which will be discussed later.

The **inotropic** (contractile) **state** is often defined operationally as a change in the rate or extent of force development that occurs independently of the loading conditions. The inotropic state is determined directly by subcellular processes that regulate myocyte cytosolic calcium and actin–myosin crossbridge cycling (Chapter 14). In isolated cardiac muscle, changes in the level of the inotropic state are measured by changes in the peak isometric tension and dT/dt at a fixed preload. Thus, positive inotropic agents, such as **catecholamines** or **digoxin,** increase the peak tension and dT/dt of an isometric contraction at a given preload. The opposite occurs when negative inotropic interventions, such as hypoxia and beta blockers, are used.

Isotonic Contraction

If isolated cardiac muscle is allowed to shorten, the contraction (Fig. 19-1D) is termed **isotonic.** Initial muscle length is determined by applying a preload; an additional load, known as the **afterload,** affects muscle behavior after stimulation. Muscle shortening occurs when tension development equals the total load (preload plus afterload). During shortening, tension remains constant (hence the term *isotonic*). With dissipation of the active state, the muscle returns to its initial preloaded length, and finally tension declines.

If preload is altered while the afterload is kept constant, **length–shortening** and **length–velocity curves** are obtained; these are analogous to the **length–tension curve** seen in isometric muscle (Fig. 19-2).

An inverse hyperbolic curve, relating afterload and the initial velocity of shortening, the **force–velocity curve,** can be obtained from a series of variably **afterloaded contractions.** As shown in Figure 19-3, when the afterload is so great that the muscle cannot shorten (the X intercept, P_o), the contraction becomes isometric. Because preload always exists, the velocity of an unloaded

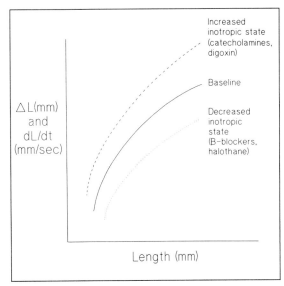

Fig. 19-2. Length–shortening and length–velocity relationships from isotonic contractions at a constant afterload. As muscle length (preload) increases, shortening (ΔL) and velocity of shortening (dL/dt) increase. An increased inotropic state shifts the curve upward and to the left; conversely, a decreased inotropic state shifts the curve downward and to the right.

contraction (the Y intercept, V_{max}) must be extrapolated from the force–velocity curve. V_{max} is determined by physiochemical properties unique to cardiac muscle, and is therefore considered a measure of the inotropic state. Although changes in preload shift P_o without changing V_{max}, a positive inotropic agent increases V_{max} and P_o by means of a parallel upward shift of the force–velocity curve; a negative inotropic agent causes the opposite (see Fig. 19-3). Similar operational definitions of the inotropic state may be applied to the **preloaded isotonic contraction,** in that a positive inotropic agent produces an upward shift of the length–shortening and length–velocity curves (see Fig. 19-2).

An important property of cardiac muscle is that the isometric passive length–tension curve establishes the limits of tension for an isotonic contraction. In other words, the tension at the end of an isotonic contraction is the same as the

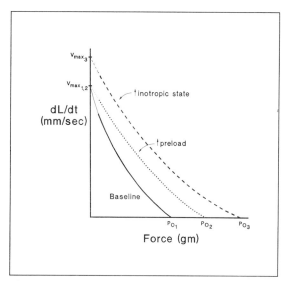

Fig. 19-3. Force–velocity relationship from variably afterloaded contractions. Increased preload causes an increase in the maximum isometric tension (P_{o_1} to P_{o_2}) without a change in the extrapolated velocity of an unloaded contraction (V_{max}). An increase in the inotropic state increases P_o and V_{max}. (dL/dt = velocity of contraction.)

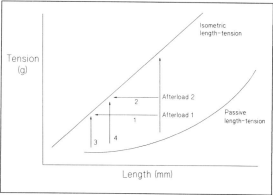

Fig. 19-4. Length–tension curves from variably loaded isotonic contractions. Beats *1* and *2* are isotonic and contract from the same preload but a variable afterload (beat 2 is greater than beat 1). Beats *3* and *4* are isometric and contract at a preload matched to the end-contraction length of beats *1* and *2*, respectively. Isotonic beats *1* and *2* contract to the same point on the isometric length–tension curve as isometric beats *3* and *4*.

tension developed for an isometric contraction *at the same resting muscle length* (Fig. 19-4).

Besides load and the contractile state, cardiac muscle performance is influenced by the frequency of stimulation (heart rate). An increase in stimulation frequently causes an increase in tension in isolated cardiac muscle, known as **Bowditch's phenomenon,** but is a relatively minor influence in the normal left ventricle.

Theoretical Models of Cardiac Muscle

A number of simplified models have been developed to explain the complex behavior of striated muscle as a function of **active contractile elements** and **passive viscoelastic properties** (Fig. 19-5). While these models have contributed significantly to our understanding of cardiac mechanics and thermodynamics, they have been unsatisfactory in simulating the behavior of myocardium. The hope that the velocity of the

Fig. 19-5. Theoretical models of cardiac muscle. Models contain either two or three elements. The two-element model (A) adequately predicts the behavior of skeletal muscle, but a parallel elastic element (B, C) is needed to account for the passive length–tension relationship in myocardium. (*CC* = contractile element; *SE* = series elastic element; *PE* = parallel elastic element.) (Reproduced with permission from: Milnor, W. R., *Hemodynamics,* 2nd ed. Baltimore: Williams & Wilkins, 1988. P. 270.)

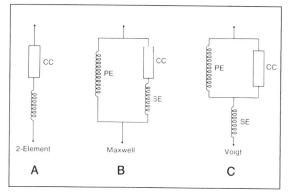

unloaded contractile element, V_{max} (deduced from experiments in which muscle strength was quickly released), could be a useful index of contractility has never materialized because the results are highly dependent on the type of model used.

CARDIAC CYCLE

The cardiac cycle describes **pressure, volume, and flow phenomena** in the ventricles as a function of time. This cycle is similar for both the left and right ventricles, although there are differences in timing stemming from differences in the depolarization sequence and the levels of pressure in the pulmonary and systemic circulations. For simplicity, the cardiac cycle for the left heart during one beat will be described (Fig. 19-6).

The **QRS complex** on the surface electrocardiogram reflects ventricular depolarization (Chapter 18). Contraction begins after a 50-msec delay and results in closure of the mitral valve. The left ventricle contracts isovolumetrically until the ventricular pressure exceeds the systemic pressure; the aortic valve opens and ventricular ejection occurs. Bulging of the mitral valve into the left atrium during isovolumetric systole causes a slight increase in left atrial pressure (C wave). Shortly after ejection begins, the active state declines and ventricular pressure begins to decrease. The aortic valve closes when left ventricular pressure falls below aortic pressure. In late systole, momentum briefly maintains forward flow despite greater aortic than left ventricular pressure. Ventricular pressure then declines exponentially during isovolumic relaxation, when both the aortic and mitral valves are closed. Left atrial pressure rises during ventricular systole (V wave) as blood returns to the left atrium by means of the pulmonary veins. When ventricular pressure declines below left atrial pressure,

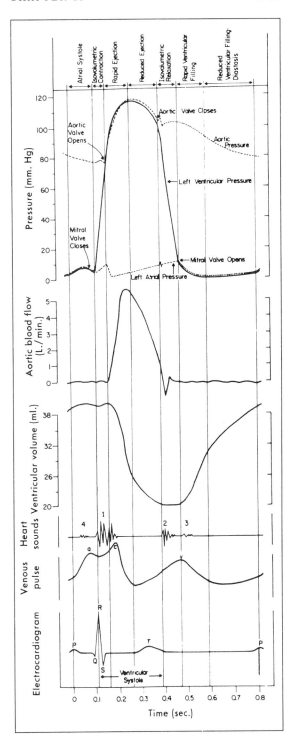

Fig. 19-6. Pressure, flow, volume, electrocardiographic and phonocardiographic events constituting the cardiac cycle. (See text for elaboration.) (Reproduced with permission from: Berne, R. M. and Levy, M. N. *Physiology*, 2nd ed. St Louis: Mosby, 1988, P. 444.)

the mitral valve opens and ventricular filling begins. Initially ventricular filling is very rapid, because of the relatively large pressure gradient between the atrium and ventricle. Ventricular pressure continues to fall after mitral valve opening because of continued ventricular relaxation; its subsequent rise (and the fall in atrial pressure) slows ventricular filling. Especially at low end-systolic volumes, ventricular **early rapid filling** may be facilitated by ventricular suction produced by elastic recoil. Ventricular filling slows during **diastasis,** or slow ventricular filling, when atrial and ventricular pressures and volumes rise very gradually. Atrial depolarization inscribes a P wave on the electrocardiogram and is followed by atrial contraction, increased atrial pressure, and a second, late rapid-filling phase. A subsequent ventricular depolarization completes the cycle.

Valve closure and **rapid-filling phases** are audible with a **stethoscope** placed on the chest, and may be recorded by a phonocardiograph after electronic amplification (see Fig. 19-6). The **first heart sound,** resulting from cardiohemic vibrations with closure of the atrioventricular (mitral, tricuspid) valves, heralds **ventricular systole.** The **second heart sound,** shorter and composed of higher frequencies than the first, is associated with **closure of the semilunar valves** (aorta and pulmonic) at the end of ventricular ejection. **Third** and **fourth heart sounds,** low-frequency vibrations caused by early, rapid filling and rapid filling due to atrial contraction, respectively, may be heard in normal children, but in adults usually indicate disease.

The Pressure—Volume Loop

An alternative, time-independent representation of the cardiac cycle is obtained by plotting instantaneous ventricular pressure and volume (Fig. 19-7A). During **ventricular filling,** pressure and volume increase nonlinearly (see Fig. 19-7A, *phase 1*). The instantaneous slope of the diastolic pressure—volume curve (dP/dV) is chamber stiffness and its inverse (dV/dP) is compliance. As chamber volume increases, the ventricle becomes stiffer (less compliant). In a normal ventricle, operative compliance is high, in that the ventricle operates on the flat portion of its diastolic pressure—volume curve.

During **isovolumetric contraction** (see Fig. 19-7A, *phase 2*), pressure increases and volume remains constant. During **ejection** (see Fig. 19-7A, *phase 3*), pressure rises and falls until the minimum ventricular size is attained. The maximum ratio of pressure to volume (maximal active chamber stiffness or elastance) usually occurs at the end of ejection. **Isovolumetric relaxation** follows (see Fig. 19-7A, *phase 4*), and, when left ventricular pressure falls below left atrial pressure, ventricular filling begins. Left ventricular pressure—volume diagrams can illustrate the effects of changing preload, afterload, and inotropic state in the intact left ventricle, and this will be discussed.

The Atrial Cycle

A pressure—volume loop can also be described for atrial events. During ventricular ejection, descent of the ventricular base lowers left atrial pressure and thus assists in atrial filling. Filling of the left atrium from the pulmonary veins results in a **V wave** on the atrial pressure tracing. (The **C wave** is inscribed on the left atrial pressure tracing during ventricular isovolumetric contraction.) When the **mitral valve** opens (see Fig. 19-8, *phase 1*), blood stored in the left atrium empties into the left ventricle. The atrium also acts as a conduit for blood flow from the pulmonary veins into the left atrium during passive atrial **emptying** (see Fig. 19-8, *phase 2*) and atrial **diastasis** (see Fig. 19-8, *phase 3*). Atrial **contraction,** denoted by an **A wave** on the left atrial pressure tracing, actively assists ventricular filling (see Fig. 19-8, *phase 4*). Thus, the atrium functions as a reservoir, conduit, and booster pump. In the normal ventricle, atrial systole contributes approximately 15 percent of the ventricular filling; when ventricular filling is impaired, such as occurs in

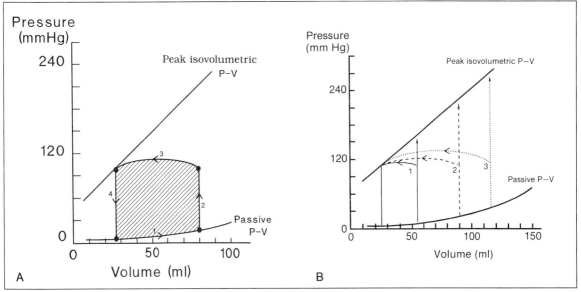

Fig. 19-7. Left ventricular pressure–volume (*P-V*) loops from an intact left ventricle. (A) Segments of the P-V loop correspond to events of the cardiac cycle: diastolic ventricular filling along the passive P-V curve (*phase 1*), isovolumetric contraction (*phase 2*), ventricular ejection (*phase 3*), and isovolumetric relaxation (*phase 4*). The ventricle ejects to an end-systolic volume determined by the peak isovolumetric P-V line. An isovolumetric contraction (*dashed line*) from the end-diastolic volume falls on the same isovolumetric P-V line. (B) A series of beats contracting from increasing preload to the same peak isovolumetric P-V line. Note the progressive increase in stroke volume with increasing preload.

hypertensive heart disease, the active atrial contribution may rise significantly.

DETERMINANTS OF PERFORMANCE AND CARDIAC OUTPUT IN THE INTACT LEFT VENTRICLE

Measures of Ventricular Performance

Measures of overall ventricular performance include cardiac output, stroke volume, and stroke work. **Cardiac output** is defined as the quantity of blood delivered to the circulation (usually expressed in liters per minute). Cardiac output is the product of stroke volume and heart rate. **Stroke volume** is the quantity of blood ejected by the heart in each beat and equals the ventricular end-diastolic volume minus the end-systolic

volume. **Stroke work** is the product of pressure and stroke volume and equals the area bounded by the ventricular pressure–volume loop. In the clinical setting, stroke work may be approximated as: [(LVSP − LVDP) × stroke volume × 0.0136], where LVSP and LVDP are the mean left ventricular systolic and diastolic pressures, respectively, and 0.0136 converts mm Hg·ml to gm−m.

Cardiac output responds to changes in the oxygen requirements of tissues, as provoked, for example, by exercise. The extraction of nutrients by tissue can be expressed as the **arteriovenous difference** across the tissue. According to the **Fick principle,** the consumption of a particular nutrient (for example, oxygen) by a tissue equals the rate of delivery of that nutrient, that is, the cardiac output times the arteriovenous difference (see Appendix 2). Changes in cardiac output

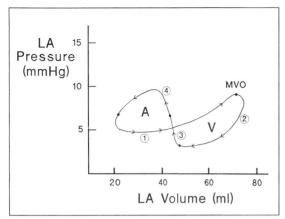

Fig. 19-8. Left atrial pressure–volume loop with its characteristic figure-of-eight configuration. Left atrial (*LA*) pressure and volume increase during ventricular systole (*phase 1*). The mitral valve opens (*MVO*), causing a decrease in atrial pressure and volume (*phase 2*). Atrial diastasis (*phase 3*) is followed by atrial contraction (*phase 4*). (*A* = atrium; *V* = ventricle.)

necessary to meet the metabolic needs of the tissues can be produced by changes in the stroke volume or heart rate, or both. Changes in stroke volume are mediated by altered loading conditions, inotropic state, and heart rate. Thus those factors that influence the strength of contraction in isolated muscle are the same factors that determine cardiac output.

Attempts to extrapolate experimental results from isolated muscle to the intact left ventricle have been complicated by the complexity of chamber geometry and myocardial fiber orientation, which make it difficult to estimate initial fiber length (preload) and the force opposing left ventricular ejection (afterload). In contrast to isolated cardiac muscle, contraction of the intact left ventricle is **auxotonic,** in that force rises and falls during ejection of viscous blood into a viscoelastic arterial system. As with isolated cardiac muscle, performance of the intact left ventricle depends on the interplay among preload, afterload, the inotropic state, and heart rate. However, unlike isolated cardiac muscle, ventricular performance is modulated by neurohumoral influences, central and local autonomic reflex pathways, right

and left ventricular interaction, the restraining effects of the pericardium, and atrial function (Chapter 23).

Preload and the Frank-Starling Relationship

The influence of preload on measures of ventricular performance defines the left ventricular function curve, known as the **Frank-Starling relationship:** increasing left ventricular end-diastolic volume increases stroke volume in ejecting beats and increases peak left-ventricular pressure in isovolumetric beats (see Fig. 19-7B). The modulation of ventricular performance by changes in preload, termed **heterometric regulation,** operates on a beat-by-beat basis and is responsible for matching outputs of the right and left ventricles (for example, after standing or with respiration). The Frank-Starling relationship also represents an important compensatory mechanism that maintains left ventricular stroke volume (by increasing left ventricular end-diastolic volume) when left ventricular shortening is impaired, owing either to myocardial contractile dysfunction or to excessive afterload. The atria also exhibit a Frank-Starling relationship which becomes clinically important during exercise and when there is resistance to early diastolic left ventricular filling (for example, caused by mitral stenosis, left ventricular hypertrophy, or impaired left ventricular relaxation).

Because a representative fiber length (preload) is difficult to determine in the left ventricle, changes in the myocardial fiber length are estimated from changes in the left ventricular end-diastolic volume. In the clinical setting, end-diastolic pressure or pulmonary capillary wedge pressure are used frequently as measures of preload. However, the passive pressure–volume relationship (analogous to the passive length–tension curve in isolated muscle) is not linear, but exponential (Fig. 19-9). Thus, the ratio of change in left ventricular pressure and volume is greater at higher than at lower left ventricular volumes. Not surprisingly, under certain circumstances, ventricular pressure may inac-

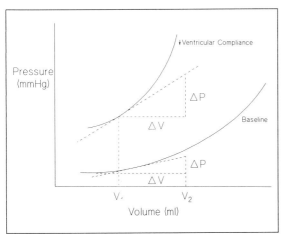

Fig. 19-9. Diastolic pressure–volume curves from a normal ventricle (*solid lines*) and a hypertrophied ventricle (*dashed lines*). The slope of the tangent of each curve at low (V_1) and high (V_2) volumes is drawn. Along each curve, the slope (dP/dV), or stiffness, increases with increased volume. The slope is greater at both volumes in the hypertrophied ventricle than in the normal ventricle. Compliance is the inverse of stiffness. (ΔP = change in pressure; ΔV = change in volume.) (Redrawn with permission from: Ross, J. Jr. In *Best and Taylor's Physiological Basis of Medical Practice*, 12th ed. Baltimore: Williams & Wilkins, 1989. P. 225.)

curately reflect the ventricular volume (preload). Moreover, changes in ventricular volume may erroneously be inferred from changes in cardiac pressures, which, in fact, result only from alterations in ventricular compliance. Compliance of the left ventricle is affected by pericardial pressure, right ventricular pressure and volume, and coronary artery perfusion (turgor), besides changes in the elastic properties of the left ventricle.

Afterload

Afterload in the intact heart may be considered as either the **tension** (stress) in the left ventricular wall during ejection, or as the **arterial input impedance** (a function of arterial pressures, elasticity, vessel dimension, and blood viscosity). Although force within the ventricular wall is

difficult to measure and varies throughout its thickness, initial estimates of systolic wall stress can be derived from application of the **Laplace relationship,** in which tension = P × r/2h, where P refers to the pressure, r to the ventricular radius, and h to the wall thickness. This relationship assumes spherical ventricular geometry, however. Complex derivations, based on more realistic geometric assumptions, are used in laboratory investigations. Measurement of aortic input impedance requires instantaneous measurement of pressure and flow, and is therefore impractical in a clinical setting. Accordingly, peak left ventricular pressure and systemic vascular resistance are used clinically as measures of afterload.

An increase in afterload (stress) causes a decrease in stroke volume (Fig. 19-10A) and the velocity of left ventricular shortening. The resulting stress–shortening and stress–velocity curves are analogous to those obtained from variably afterloaded, isotonic contractions in isolated muscle.

The Inotropic State

The ideal method of measuring the inotropic state in the intact left ventricle should incorporate the variables of **force, length, velocity, and time,** be **independent of external loading conditions,** and relate to **physicochemical processes** at the **sarcomeric level.** Because of these constraints, changes in the inotropic state are usually defined operationally by shifts of the various ventricular function curves. For example, a drug with positive inotropic activity (for example, digoxin) shifts the Frank-Starling relationship (analogous to the length–tension curve) upward and to the left, and changes the stress–shortening relationship (analogous to the force–velocity curve) upward and to the right (see Fig. 19-2 and 19-3).

The intact left ventricle can be made to contract isovolumetrically over a range of left ventricular end-diastolic volumes to produce an isovolumetric pressure–volume line, analogous to the isometric length–tension curves in isolated muscle. Moreover, end-systolic pressure-volume plot

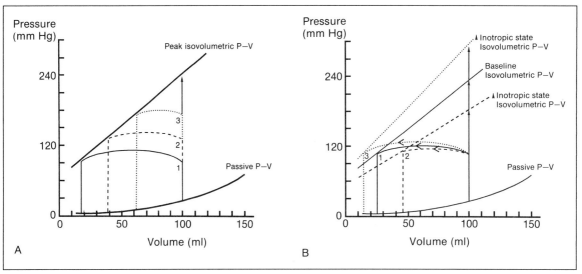

Fig. 19-10. Pressure–volume curves from variably afterloaded beats (A) and beats after different inotropic interventions (B). (A) As afterload increases (3 > 2 > 1), stroke volume decreases, although end-ejection points fall on the same isovolumetric pressure–volume (*P-V*) line. (B) End-ejection points from beats with different inotropic states (3 > 1 > 2) fall on different isovolumetric P-V lines. Beat *1* is from baseline, beat *2* is after administration of a negative inotropic drug, and beat *3* is after application of a positive inotropic drug.

points from ejecting beats (obtained from variably preloaded and afterloaded contractions) fall reasonably close to the isovolumetric pressure–volume line for a given inotropic state (see Fig. 19-7B). Thus, changes in the inotropic state, *independent of the loading conditions,* can be identified by changes in the slope of the end-systolic pressure–volume relationship (see Fig. 19-10B). **End systole** may be defined as end ejection or as that time of maximal elastance (the maximal pressure–volume ratio) during systole. In the normal heart, these two points are closely related in time.

The **rate of pressure development** in the left ventricle during isovolumic systole (dP/dt) has been used as an index of the inotropic state. However, because of the direct influence of preload on dP/dt, dP/dt at a common developed pressure and the slope of the dP/dt–end-diastolic volume curve have been proposed as preload-independent indices of the inotropic state.

Heart Rate

Increasing heart rate causes a small, but measurable, increase in the inotropic state through the force–frequency (Bowditch) relationship. This effect is more prominent in the depressed than in the normal left ventricle. Heart rate is an important determinant of left ventricular performance under certain circumstances, by virtue of the relationship between cardiac output and heart rate: **cardiac output = stroke volume × heart rate.** In a normal left ventricle, pacing between heart rates of 60 and 160 bpm has little effect on cardiac output, because the diminished diastolic filling time (and hence, the stroke volume determined by the Frank-Starling relationship) offsets the modest increase in the inotropic state. Heart rate is normally determined by the interplay between the intrinsic automaticity of the sinoatrial node and the activity of the autonomic nervous system. Sympathetic stimulation increases and parasympathetic stimulation decreases the heart rate (see Chapter 23).

VENTRICULAR VASCULAR COUPLING

In isolated muscle, loading conditions represent the force applied to muscle before (preload) and after (afterload) the onset of contraction. In the intact left ventricle, preload and afterload are also determined by the characteristics of the arterial and venous circulations. Thus, loading conditions are not only important direct determinants of left ventricular performance; they also function indirectly, by coupling the left ventricle to the vascular system.

The **venous return curve** describes the inverse relationship between venous pressure and cardiac output. This relationship is a function of arterial and venous capacitance (the change in volume per change in pressure [dV/dP]) and the peripheral resistance provided by the microcirculation (resistance = pressure gradient for flow divided by the cardiac output).

Ventricular contraction transfers blood from the venous to the arterial side of the circulation, and arterial and venous capacitances determine the respective pressures that result from the shift in blood volume. These pressures determine the **driving force** (pressure gradient) across the peripheral resistance and are primarily responsible for venous return to the heart. In contrast to convention, the venous return curve plots the independent variable (cardiac output) on the vertical axis and the dependent variable (venous pressure) on the horizontal axis. The X intercept is the mean circulatory pressure, or that pressure in the vascular system in the absence of cardiac pumping. The **mean circulatory pressure** is a function of the capacitance of the vascular system and the total blood volume. The plateau of the venous return curve and the Y intercept represent the **maximal obtainable cardiac output** as venous pressure is reduced. In the normal heart, cardiac output is limited by venous return, and the operating venous pressure is near the plateau of the venous return curve.

Coupling of the venous system to the heart is graphically represented in Figure 19-11. In this analysis, developed by Guyton and co-workers, the intersection of the ventricular function (Frank-Starling) curve and the venous return curve represents the steady-state operating values of cardiac output and venous pressure. At this **equilibrium point,** the ability of the venous system to provide venous return at a given pressure is matched with the ability of the ventricle to pump that venous return when distended to the same pressure. For example, an increase in venous pressure causes an increase in cardiac output (Frank-Starling), but a decrease in cardiac output, according to the venous return curve; the resultant cardiac output is determined by the dynamic equilibrium of these forces.

Increased blood volume and venoconstriction shift the venous function curve upward and to the right, increasing the mean circulatory pressure and the maximal cardiac output (see Fig. 19-11A). The venous system contains the major fraction of blood in the vascular system, because of the greater capacitance of veins than of arteries. As a result, venoconstriction shifts significant quantities of blood from the peripheral to central circulation. Because arteries contain only a small percentage of the total blood volume, their contractile state does not affect the mean circulatory pressure. Moreover, because venous pressure varies inversely with systemic vascular resistance, arteriolar constriction (increased afterload) shifts the curve downward and to the left without changing the mean circulatory pressure; conversely, arteriolar dilation (decreased afterload) shifts the curve upward and to the right (see Fig. 19-11B). An increased inotropic state (for example, that arising from sympathetic nervous stimulation [see Fig. 19-11C]) shifts the ventricular function curve to the left without significantly altering the venous return curve. Conversely, in chronic heart failure (see Fig. 19-11D) there is rightward shift of the ventricular function curve and, because of renal salt and water retention, a parallel rightward shift of the vascular function curve. In this way, cardiac output is initially maintained at the expense of increased venous pressure and congestion. If the compensatory mechanisms fail, venous pressure rises further and cardiac output falls.

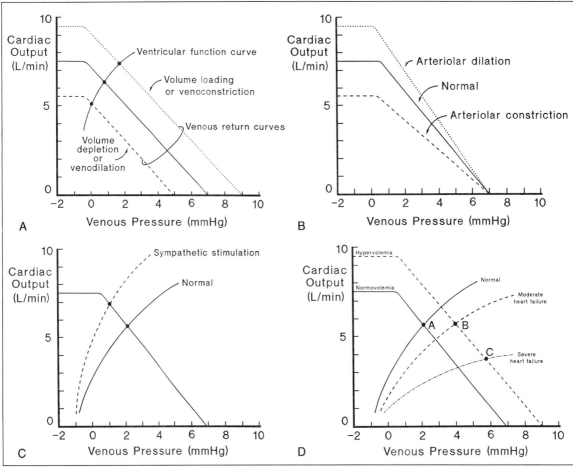

Fig. 19-11. Venous pressure–cardiac output curves. The equilibrium point is defined by the intersection of the ventricular function curve with the venous return curve. (A) Volume loading and venoconstriction shift the venous return curve to the right, resulting in an equilibrium point with a higher cardiac output and higher mean circulatory pressure (P_V). Volume depletion and venodilation shift the curve to the left, resulting in an equilibrium point with a lower cardiac output and a smaller P_V. (B) The effects of arteriolar constriction and dilation on the venous return curves are more complex. (C) Sympathetic nerve stimulation (a positive inotropic intervention) causes a leftward shift of the ventricular function curve, resulting in an equilibrium point with a lower venous pressure and higher cardiac output. (D) Chronic heart failure causes rightward shifts of both the ventricular function and venous return curves. At B (moderate heart failure), cardiac output is preserved at the expense of venous pressure. At C (severe heart failure), cardiac output is decreased and venous pressure further increased. (Redrawn with permission from: Berne, R. M., and Levy, M. N. *Physiology*, 2nd ed. St Louis: Mosby, 1988, Pp. 529, 530, 532, 533.)

SUMMARY

Although the complexities of the intact left ventricle are considerable, the Frank-Starling and force–shortening relationships, which are analogous to the length–tension and force–velocity curves, respectively, in isolated cardiac muscle, are useful for characterizing left ventricular performance. The Frank-Starling curve relates preload, and the force–shortening curve relates afterload, to measures of ventricular performance. The inotropic or contractile state is recognized by shifts in the ventricular function curves, such as the end-systolic pressure–volume relationship. The sequence of events in each cardiac cycle and changes in left ventricular performance can be represented graphically by the pressure–volume loop. Cardiac output is a measure of ventricular performance that, in turn, is linked to the metabolic needs of the tissues. The concept of ventricular–vascular coupling accounts for the interplay between the ventricular pump and the arterial and venous circulations that regulate cardiac output.

NATIONAL BOARD TYPE QUESTIONS

Select the one most correct answer for each of the following

1. Which of the following is *not* a determinant of left ventricular performance?
   A. Heart rate
   B. Preload
   C. Pressure–volume loop
   D. Afterload
   E. Inotropic state
2. A drug with a positive inotropic effect (increased inotropic state) will:
   A. Lower the blood pressure
   B. Shift the force–velocity relationship upward, increasing P_o and V_{max}
   C. Shift the length–tension relationship downward and to the right
   D. Increase the peak isometric tension, but will not change dT/dt
3. The Frank-Starling relationship:
   A. Relates the influence of changing preload on measures of ventricular performance
   B. Is an inverse hyperbolic curve, relating afterload (wall stress) and cardiac output
   C. Is determined during isovolumetric systole
   D. Is the principal force for ventricular filling

4. The period of early rapid filling:
   A. Follows atrial contraction
   B. Immediately precedes ventricular ejection
   C. Immediately follows isovolumetric relaxation
   D. Is represented by the second heart sound on the phonocardiogram
5. In the intact heart, the afterload:
   A. Is primarily responsible for matching outputs of the two ventricles
   B. May be estimated by the left ventricular end-diastolic volume or stress
   C. May be considered either as the tension in the left ventricular wall during ejection or as the arterial input impedance
   D. Is determined from the intersection of the left ventricular function and venous return curves

ANNOTATED ANSWERS

1. C. The pressure–volume loop is a time-independent graphic representation of the cardiac cycle. The determinants of ventricular performance are preload, afterload, the inotropic state, and heart rate.
2. B. The force–velocity relationship is an inverse hyperbolic curve that relates afterload to the initial velocity of shortening. The X intercept (P_o) is the load at which muscle cannot shorten; the extrapolated Y intercept (V_{max}) is the theoretical rate of shortening of an unloaded contraction. An increased inotropic state causes an increase in both P_o and V_{max}.
3. A. The Frank-Starling relationship states that increasing left ventricular end-diastolic volume (preload) increases stroke volume in ejecting beats and increases peak left ventricular pressure in isovolumic beats.
4. B. Early rapid filling represents the first phase of ventricular filling after mitral valve opening. Left ventricular pressure falls dur-

ing isovolumetric relaxation; when left ventricular pressure falls below left atrial pressure, the mitral valve opens.

5. C. Afterload is the force or tension borne by the ventricular wall during contraction; the aortic input impedance has also been used as a measure of afterload. Both measurements are impractical for clinical use, and therefore afterload is usually estimated from the peak systolic pressure of the systemic vascular resistance.

BIBLIOGRAPHY

Braunwald, E. Assessment of cardiac performance. In: *Heart Disease. A Textbook of Cardiovascular Medicine.* Philadelphia: W. B. Saunders, 1980, Pp. 472–492.

Guyton, A. C., Jones, C. E., and Coleman, T. G. *Circulatory Physiology: Cardiac Output and Its Regulation.* Philadelphia: W. B. Saunders, 1973, Pp. 146–233.

Ross, J., Jr., and Sobel, B. E. Regulation of cardiac contraction. *Annu. Rev. Physiol.* 34:47, 1972.

Sagawa, K. The ventricular pressure-volume diagram revisited. *Circ. Res.* 43:677, 1978.

Sonnenblick, E. H. Force-velocity relations in mammalian heart muscle. *Am. J. Physiol.* 202:931–939, 1962.

20 Coronary Circulation: Myocardial Oxygen Consumption and Energetics

LAURA F. WEXLER AND RICHARD A. WALSH

Objectives

After reading this chapter, you should be able to:

Describe the anatomy of the coronary circulation

Describe at least four factors that determine coronary blood flow

List four major determinants of myocardial oxygen consumption

Describe how coronary blood flow is regulated to match myocardial oxygen requirements

Characterize and compare the metabolic pathways used by the heart to obtain energy under normal conditions and under conditions of limited blood flow

Under normal conditions, cardiac muscle metabolism is almost exclusively aerobic, depending on oxidative phosphorylation to resynthesize the ATP continuously utilized for repetitive excitation–contraction–relaxation. Myocardial oxygen requirements are therefore high and unremitting, even under resting conditions. During stress or exercise, oxygen requirements may increase abruptly by three- to fourfold. Unlike skeletal muscle, cardiac muscle cannot obtain significantly more oxygen by extracting a greater percentage of the oxygen delivered to it, as myocardial oxygen extraction is near maximal at rest. The major mechanism by which oxygen delivery to the myocardium can be augmented is by an increase in the amount of coronary artery blood flow. The following section describes the anatomy of the coronary circulation and the means by which coronary flow is regulated to continuously meet the demands of a variable level of myocardial work.

ANATOMY OF THE CORONARY CIRCULATION

The Epicardial Vessels

Two main **coronary arteries** arise from the **aortic sinuses** just above the **aortic valve leaflets** and give rise to a series of branches that run along the outer (epicardial) surface of the heart (Fig. 20-1).

The **left main coronary artery** divides into two major branches just after it emerges from the left coronary sinus: the left anterior descending and the left circumflex arteries. The **left anterior**

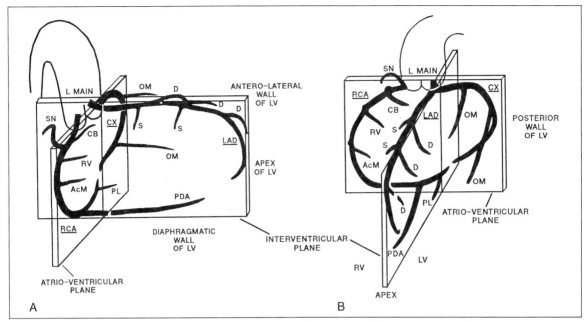

Fig. 20-1. The coronary anatomy relative to the atrioventricular groove and the interventricular groove, as seen in a right anterior oblique (A) and left anterior oblique (B) orientation. Coronary branches: *L main* = left main; *LAD* = left anterior descending; *D* = diagonal; *S* = septal perforator; *CX* = circumflex; *OM* = obtuse marginal; *RCA* = right coronary artery; *CB* = conus branch; *SN* = sinus node artery; *AcM* = acute marginal; *PDA* = posterior descending artery; *PL* = posterior left ventricular. (*LV* = left ventricle; *RV* = right ventricle.)

descending artery courses down the anterior surface of the heart in the interventricular groove to the apex, giving off large branches to the anterolateral wall of the left ventricle (diagonal branches) and small branches that penetrate and supply the upper two-thirds of the interventricular septum (septal perforators).

The **left circumflex artery** courses leftward and posteriorly in the atrioventricular groove, giving off obtuse marginal branches to the upper posterolateral wall of the left ventricle.

The **right coronary artery** emerges from the right aortic sinus and runs rightward and then posteriorly in the atrioventricular groove, giving off branches to the right atrium (including a branch to the sinus node) and the right ventricle (acute marginal branches). At the posterior aspect of the heart, the right coronary artery gives off the **posterior descending artery,** which runs

along the interventricular groove to the apex, giving off branches to the diaphragmatic and posterior walls of the left ventricle as well as small perforating branches that penetrate and supply the lower third of the interventricular septum. Beyond the origin of the posterior descending artery, the right coronary artery gives off the **posterior left ventricular branch,** which supplies the lower posterolateral wall of the left ventricle.

Anatomic Variations

There is considerable anatomic variability in the origin and distribution of specific coronary artery branches. The pattern described, in which the posterior descending artery and posterior left ventricular branches originate from the right coronary artery is termed **right dominant** and exists in about 50 percent of the population. A **left-**

dominant pattern, in which the right coronary artery is small and supplies only the right atrium and right ventricle, is found in approximately 20 percent of the population. In this variant, the posterior descending artery and posterior left ventricular branches originate from the **distal circumflex artery,** so that virtually all of the blood supply to the left ventricle originates from branches of the left main coronary artery. In another 30 percent of the population, the circulation is balanced; the posterior descending artery branches off from the right coronary artery and the posterior left ventricular branch originates from the circumflex. These variations are of considerable importance in patients with coronary artery disease, as will be discussed. The extent of myocardial damage resulting from occlusion of a coronary artery is determined by the extent of the territory supplied by branches of that artery.

Intramural Arteries and Capillary Network

The main epicardial vessels and their branches subdivide several times on the surface of the heart before giving off small penetrating branches that give rise to an extensively branching network of small **intramural arteries, arterioles,** and **capillaries,** which traverse the myocardial wall from the epicardium to endocardium. Capillary density is very high in heart muscle, commensurate with the high oxygen requirements of myocardial cells. Capillaries are situated close to each myocardial cell, providing a conduit for rapidly diffusible oxygen for ATP production in the cell and for removal of metabolic waste products. Considered collectively, the coronary vasculature and the blood it contains accounts for approximately 15 percent of the total mass of the heart (Fig. 20-2).

Venous Drainage

As in other parts of the systemic circulation, **myocardial capillaries** feed into a network of **intramural venules** that eventually drain into large

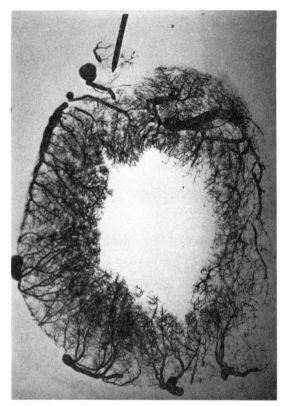

Fig. 20-2. Postmortem radiograph of the coronary vascular tree in a cross-section of the left ventricle. Note the extensive arborization of the intramural vessels.

epicardial collecting veins. Right ventricular venous blood drains into several anterior cardiac veins that empty into the right atrium. Most of the left ventricular venous blood drains into the **coronary sinus,** a large venous channel that runs along the atrioventricular groove and empties into the right atrium. The separate venous drainage of the right and left ventricles facilitates in vivo studies, in which catheter cannulation of the coronary sinus permits examination of isolated left ventricular venous effluent for metabolic studies. A small percentage of left ventricular venous blood (approximately 5 percent) drains directly into the left ventricular cavity from the venous ends of capillaries or from deep coronary veins (thebesian veins).

Coronary Artery Collaterals

Coronary collaterals constitute direct arterial connections between one coronary artery and another. If a coronary artery is occluded, collateral vessels may provide some degree of arterial blood supply beyond the obstruction, thus protecting the myocardium distal to the obstruction. Certain species (for example, dogs) normally possess an extensive network of large preformed arterial collaterals. Abrupt occlusion of a proximal coronary artery will cause only limited myocardial damage, as there is an alternative source of arterial flow provided by pre-existing collaterals. In contrast, other species (humans, nonhuman primates, pigs) have a minimal collateral network under normal circumstances. After acute coronary occlusion of a proximal coronary artery, small intramural collaterals can supply less than 10 percent of the normal flow, and a more extensive and predictable amount of myocardial damage results. However, in the event of gradual chronic coronary obstruction, as may occur with atherosclerotic coronary artery disease, these small collaterals may greatly enlarge over time (Fig. 20-3). Even after complete occlusion of the diseased coronary artery, well-developed coronary collaterals may provide normal or near-normal flow to the distal segment of the diseased artery. However, although the flow rate may be normal at rest, the capacity to augment myocardial blood flow during exercise or stress (coronary reserve) is usually limited in collateral vessels.

REGULATION OF CORONARY BLOOD FLOW

Resting coronary blood flow is normally between 60 and 90 ml/min/100 gm of myocardium and may rapidly increase by four- to fivefold during exercise or other conditions requiring augmented flow. The coronary **flow rate** is determined by the **coronary artery perfusion pressure** and by the **resistance to flow** exerted by forces generated within and without the coronary vascular bed.

The pattern of blood flow to the **left ventricle** (which receives the greatest proportion of coronary flow) is unique in that arterial flow is markedly decreased during **systole,** as the intramyocardial pressure generated by contracting myocardial fibers effectively shuts off flow from the epicardial to the intramural arteries. Most of the coronary flow to the left ventricle occurs during **diastole,** and coronary perfusion pressure is largely determined by aortic diastolic pressure. Blood flow to the right ventricular myocardium is also phasic, but, because the systolic pressure transmitted to the right ventricular myocardium is much lower, the difference between systolic and diastolic flow is less marked (Fig. 20-4).

Coronary Autoregulation

If there is a sudden change in aortic pressure (within certain limits), coronary vascular resistance will adjust itself proportionally within 8 to 12 seconds, so that a constant blood flow is maintained. This phenomenon is called **autoregulation** (Fig. 20-5), and it protects the myocardium from inadequate blood flow if there is a decline in coronary perfusion pressure. During periods of abnormally high aortic pressure, the role of autoregulation is less clear. It may attenuate endothelial wall stress, and result in vascular damage because of the elevated coronary distending pressures.

Autoregulation also occurs in **localized areas** of the coronary vasculature when partial obstruction of an artery causes a decrease in the coronary perfusion pressure. The vessel distal to the obstruction will dilate, thus normalizing flow by decreasing the coronary vascular resistance. The normal coronary vascular bed can autoregulate over a range of systemic arterial pressures, usually 60 to 140 mm Hg. Above or below these limits, autoregulation fails and coronary flow increases or decreases in a linear fashion, with corresponding increases or decreases in aortic pressure.

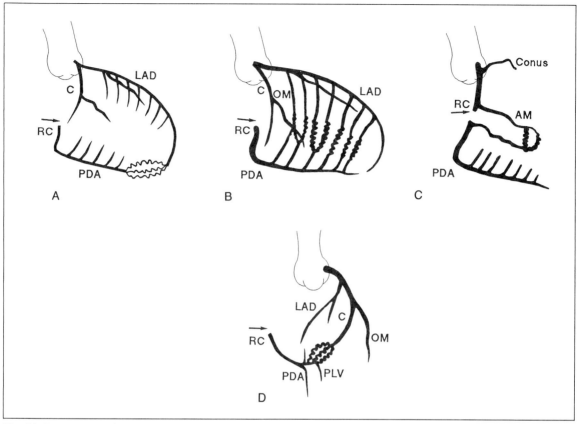

Fig. 20-3. Examples of some of the patterns of coronary collaterals development in response to gradual obstruction of a right coronary artery (*RC*). (A) The RC distal to the obstruction is supplied by collaterals from the left anterior descending (*LAD*) artery. (*C* = coronary artery; *PDA* = posterior descending artery.) (B) The supply to the distal RC is from septal-to-septal collaterals; (C) from RC marginal-to-marginal collaterals; and (D) from left circumflex to distal posterior descending artery (*PDA*) collaterals. (*OM* = obtuse marginal artery; *AM* = acute marginal artery; *PLV* = posterior left ventricular artery.)

Autoregulatory reserve refers to the maximal degree of vasodilation possible in the coronary vascular bed, and determines the range of perfusion pressures over which myocardial flow can be maintained. Autoregulatory reserve depends on the level of chronic vasodilatation in the coronary vasculature as a whole or in any specific region of the heart. If a region of the vascular bed is already vasodilated to compensate for a localized decrease in coronary perfusion pressure, the capacity to autoregulate during additional declines in aortic diastolic pressure will be impaired. In other words, autoregulatory reserve will be impaired and the affected area of myocardium will be more vulnerable to transient decreases in aortic pressure.

Several mechanisms have been proposed to explain autoregulation. Autoregulation may be attributable in part to the **myogenic response.** This is an intrinsic property of vascular smooth muscle, whereby an increase in passive stretch causes active smooth muscle contraction and

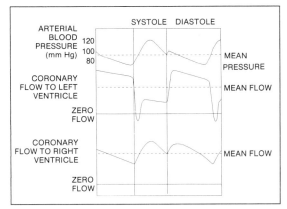

Fig. 20-4. Comparison of epicardial coronary blood flow to the left and right ventricles. Flow to the left ventricle is abolished in early systole because of high intraventricular pressures. Most flow takes place during early to mid-diastole. Coronary flow to the right ventricle is also cyclic, but is less affected during systole because of the lower pressure generated within the right ventricular myocardium.

diminished passive stretch leads to smooth muscle relaxation and vasodilation. Thus, mechanical stretching of the coronary artery as perfusion pressure rises is counteracted by vasoconstriction, and diminished wall stretch at low perfusion pressures triggers vasodilation, leading to increased flow.

However, most evidence points to the existence of a chemical mediator underlying autoregulation. **Adenosine,** a breakdown product of ATP, is the most likely candidate because it is a potent vasodilator that is continuously generated by the heart. In this mechanism, decreased perfusion pressure causes an initial decrease in coronary artery flow and this leads to a diminished rate of adenosine washout. This, in turn, would produce vasodilation and increased coronary flow. Increased perfusion pressure precipitates an initial increase in flow and more rapid washout of adenosine. This would result in diminished vasodilation and a subsequent decrease in the coronary flow rate (Fig. 20-6). It is also possible that the partial pressure of oxygen (Po_2) levels in tissue, by changing slightly as perfusion pressure rises and falls, may directly affect coronary artery tone.

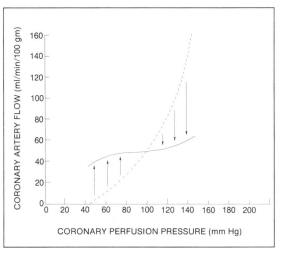

Fig. 20-5. Autoregulation of coronary artery flow during alterations in coronary perfusion pressure. The *dashed line* represents the instantaneous (nonautoregulated) coronary artery flow that would be observed immediately after altering coronary perfusion pressure between 20 and 160 mm Hg. The *solid line* depicts the effects of autoregulation: appropriate adjustments in coronary vascular resistance (*arrows*), occurring within 8 to 12 seconds, compensate for the acute changes in coronary perfusion pressure and return flow to a level that is maintained fairly constant between pressures of about 60 and 140 mm Hg. The level to which coronary flow is autoregulated at any given time is determined by the instantaneous oxygen requirements of the heart.

Metabolic Regulation of Coronary Artery Blood Flow

Coronary blood flow must be continuously modulated, so that a sufficient supply of oxygen is delivered to the myocardium to support **oxidative energy production** at a rate that matches energy utilization. The means by which coronary flow is so precisely regulated is still debated, but the mechanism clearly involves some signal that induces rapid changes in coronary vascular resistance. The most likely mediator again is **adenosine.** As the rate of ATP use increases with increased cardiac work, there is a relative accumulation of its breakdown products, including adenosine. This adenosine induces coronary

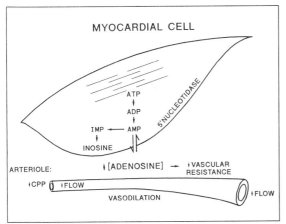

Fig. 20-6. Proposed mechanism of coronary autoregulation. As ATP is hydrolyzed during excitation, contraction, and relaxation, some is broken down to AMP and then to adenosine (which is freely diffusible across the cell membrane) by an enzyme (5'-nucleotidase) at the inner myocardial cell membrane, so that adenosine promptly diffuses out of the myocardial cell at a rate proportional to its rate of production. The degree of vasodilation it induces in surrounding arterioles would be determined by its rate of production and also by its rate of washout after diffusing into the intramural vessels. If the rate of washout decreases because coronary flow rate decreases, the local tissue concentration of adenosine would increase, leading to more vasodilation and increased flow rate. As the flow rate increases, washout would be accelerated, the local tissue concentration of adenosine would lower (assuming its rate of production has been stable), and vasodilation would diminish, thus readjusting flow. (*CPP* = coronary perfusion pressure; *IMP* = inosine 5'-monophosphate.)

vasodilation, and blood flow increases, which then furnishes the additional oxygen needed for accelerated ATP resynthesis (Fig. 20-7).

Several other factors have been hypothesized to be involved in the metabolic regulation of coronary blood flow, including Po_2, partial pressure of carbon dioxide (Pco_2), K^+ concentration, and pH. However, adenosine remains the most likely mediator in that its concentration is directly proportional to the rate of ATP turnover, it diffuses readily across the myocardial cell membrane, and it is a potent coronary artery vasodilator.

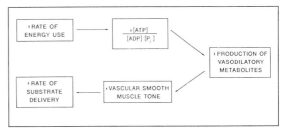

Fig. 20-7. Elements of a metabolic blood flow control system in which the cellular energy state (ATP potential) determines the rate of production of vasoregulatory metabolites, and thereby the rate of substrate delivery. The symbol (−) indicates that a change in one variable causes a reciprocal change in another. Thus, an increased rate of ATP hydrolysis leads to a decrease in the ratio of ATP to ADP + inorganic phosphate (P_i) and an increase in the rate of production of adenosine. This leads to a decrease in smooth muscle tone and an increase in the rate of oxygen delivery as coronary flow increases. (Adapted from: Olsson, R. A., and Bunger, R., Metabolic control of coronary blood flow. *Prog. Cardiovasc. Dis.* 29:369–387, 1987.)

Epicardial versus Endocardial Blood Flow

Several factors affecting blood flow are markedly different in the inner (subendocardial) and outer (subepicardial) layers of the myocardium. Systolic compression is much higher in the **subendocardial layers** than in the **subepicardial layers** of the heart, such that subendocardial blood flow is virtually absent in systole. There may also be some degree of mechanical interference with flow in the subendocardial vessels in late diastole as pressure is exerted on the inner layers of the heart by the blood filling and distending the ventricular chamber. This may account for the fact that flow in the subendocardial layers rises rapidly in early diastole but falls off in late diastole. In contrast, flow to the midwall of the heart is approximately equal during systole and diastole; in the outer layers of the subepicardium, flow is slightly higher during systole than during diastole.

The arrangement of intramural vessels partially compensates for the almost complete lack of blood flow to the subendocardium during systole, in that vascular density is increased in the subendocardium so that net flow is augmented. In addition, despite the greater external forces exerted on the subendocardial arteries, their intrinsic coronary vascular resistance is lower, such that the blood flow per gram of myocardium is normally higher in the subendocardium than in the subepicardium. The ratio of subendocardial to subepicardial flow is about 1.1 to 1. This is consistent with the fact that oxygen requirements are higher in the subendocardium because it develops higher wall stress and shortens more than the subepicardial layers. However, because of the lower resting coronary resistance of the subendocardial vessels, the capacity to further augment flow in response to increased metabolic demands (that is, the coronary reserve of the subendocardial vessels) is inherently less than that in the midwall and epicardial vessels. This makes the subendocardium more vulnerable to injury if coronary flow is impaired. Some oxygen does diffuse directly into the myocardium from the arterial blood which fills the left ventricular chamber, but it reaches only the most superficial layers of the subendocardium.

Neurohumoral Modulation of Coronary Blood Flow

The **autonomic nervous system** influences the smooth muscle tone of the coronary artery and coronary flow, although its role is overshadowed by metabolic and mechanical influences under normal conditions. The larger epicardial coronary arteries have both alpha-adrenergic receptors, which mediate vasoconstriction, and beta-adrenergic receptors, which mediate vasodilation. The smaller intramural arteries contain a larger percentage of beta receptors.

Release of **norepinephrine** during sympathetic stimulation can cause coronary artery vasoconstriction, but this response is normally overridden by metabolic factors because sympathetic stimulation also increases heart rate and contractility, thereby augmenting myocardial oxygen consumption and ATP turnover and re-

sulting in vasodilation from the input of autoregulatory and metabolic mechanisms. There does, however, appear to be a small degree of resting coronary vasoconstrictor "tone," as coronary denervation or pharmacologic alpha-adrenergic blockade will produce a small increase in coronary blood flow at rest and during exercise, suggesting release of a vasoconstrictor substance. The significance of sympathetic innervation of the normal coronary arteries is not clear. Abnormal increases in vasoconstrictor tone have been suggested as a mechanism underlying certain types of coronary artery disease states, but there is little evidence to support this.

Stimulation of **beta$_2$ receptors** in the smaller coronary arteries by endogenous circulating catecholamines or by pharmacologic beta agonists results in coronary vasodilation. The extent to which coronary beta receptors contribute to coronary blood flow regulation is difficult to assess, as beta stimulation on the myocardium increases oxygen consumption by increasing heart rate and contractility which leads to metabolically mediated vasodilation. However, autonomically mediated vasodilation can be demonstrated when sympathetic nerves are stimulated in a fibrillating (nonbeating) heart.

Endothelial Regulation of Coronary Blood Flow

Endothelium-derived relaxation factor (EDRF) is a potent vasodilator that is elaborated by vascular endothelial cells in response to a number of "stress" signals, including hypoxia, ADP accumulation, and serotonin secretion. EDRF release is also stimulated by distending forces in the vascular wall; that is, it is released when coronary perfusion pressure or flow increases. This may serve to amplify the coronary flow response to conditions such as exercise when it may be appropriate for both coronary perfusion pressure and flow to increase (in contrast to the autoregulatory response which keeps flow constant during inappropriate changes in coronary perfusion pressure).

DETERMINANTS OF MYOCARDIAL OXYGEN CONSUMPTION

Under normal conditions, the rate of coronary blood flow is regulated, so that oxygen delivery matches oxygen consumption. **Myocardial oxygen consumption** is relatively high, even under resting conditions in the beating heart (approximately 8.0 ml/min/100 gm of myocardium). The oxygen requirements of the heart are largely related to the energy requirements of cardiac contraction and relaxation, that is, the oxygen consumed in the resynthesis of ATP expended during crossbridge cycling and Ca^{2+} reuptake by the sarcoplasmic reticulum. Oxygen consumption markedly increases when the contractile work of the heart increases.

Left ventricular myocardium is responsible for the largest proportion of oxygen consumed by the heart, since it does the most contractile work. The **right ventricle** pumps the same volume of blood as the left ventricle during each cardiac cycle, but has to generate much less pressure because the resistance of the pulmonary arterial system is much lower than that of the systemic arterial system. The three most important determinants of left ventricular work (and therefore of myocardial oxygen consumption) are the **number of beats per minute** (heart rate), the **tension** or **pressure** developed during systole, and the **inotropic state** (the ability of the left ventricle to generate pressure and eject blood independent of loading conditions; see Chapter 19). To a lesser extent, the volume of **blood ejected** (that is, the extent of shortening of ventricular muscle) is also a factor.

Myocardial oxygen consumption increases with heart rate in an almost **linear fashion.** The energy cost of each heartbeat is actually slightly higher at rapid rates than at slow rates by virtue of a small increase in contractility that occurs at higher heart rates, a phenomenon termed the *positive staircase* or **Bowditch effect** (Chapter 19).

Myocardial oxygen consumption also increases in a near-linear fashion with the development of **systolic tension.** In experimental studies on isolated hearts in which the inotropic state can be held constant, there is a close positive correlation between oxygen consumption and the product of systolic pressure and heart rate. This product is usually derived as the **tension-time index,** which is calculated as the integrated area under the left ventricular pressure curve per beat or per minute.

Systolic tension, or wall stress, which is the force exerted per unit area of myocardium (Chapter 19), is determined not only by the level of the systolic pressure generated but also by the radius of curvature of the ventricle at the onset of contraction—the end-diastolic volume. For any degree of pressure development, the wall stress will be higher, and thus more oxygen consumed per beat, during contraction of a large dilated ventricle than during contraction of a small ventricle. Elevated wall stress is alleviated or normalized for any given pressure or chamber size when the myocardial walls thicken or hypertrophy. These components determining systolic wall stress or tension are expressed in the **Laplace relationship:** wall stress = $P \times r/2h$, where P is the systolic pressure, r is the radius of the ventricle, and h is the wall thickness.

The **energy cost** of developing tension (pressure work) can be considered separately from the cost of actual shortening (volume work) and is substantially greater. Ejection of blood only accounts for about 15 percent of the oxygen consumption. An increase in systemic vascular resistance requiring an increase in systolic tension is far more energy costly than an increase in the volume of blood ejected.

Contractility, or the inotropic state of the heart, is also a major determinant of oxygen consumption. In clinical situations, increased contractility enables the heart to do more work, that is, to pump a greater volume of blood or to develop more tension and pump against a greater resistance. However, the enhanced inotropic state in itself affects energy use. In an experimental preparation in which heart rate, stroke volume, and systolic pressure are held constant, a positive inotropic drug such as digitalis increases the velocity of contraction, or the rate at which pressure develops and oxygen consumption increases even though the actual mechanical work performed by the heart has not increased. It is not clear why myocardial oxygen consumption increases when there is an increase in the inotropic state. The intracellular event that appears to be a common factor in the mechanism of action of many positive inotropic drugs or interventions is that more calcium is made available to the contractile proteins. Thus, the increase in oxygen consumption is most likely related to the energy cost of the calcium pumps. Whether they cycle faster when presented with more calcium or whether additional pumps are recruited is not known.

In clinical situations, contractility and the other determinants of myocardial oxygen consumption are interrelated and usually cannot be analyzed separately. Drugs that increase contractility usually accelerate ventricular emptying (shortening), which in turn affects myocardial oxygen consumption. The resultant increase in cardiac output causes reflex changes in the heart rate and systemic resistance that also affect the myocardial oxygen consumption.

Energy is also required (but to a much lesser extent) for **noncontractile processes.** Even if there is no external work, the heart still consumes oxygen. A small component of oxygen consumption is related to **electrical excitation,** which utilizes membrane pumps and is consequently an energy-requiring process. There is also a **"basal" oxygen requirement** that is unrelated to contractile or electrical activity; it can be measured when the heart is arrested with potassium chloride. It accounts for approximately 15 to 20 percent of the total cardiac oxygen consumption and is associated with intracellular chemical reactions unrelated to excitation or contraction.

MYOCARDIAL ENERGY METABOLISM

ATP is the immediate source of energy for all energy-requiring processes in the heart; for electrical excitation, for contraction itself, and for relaxation and restoration of resting electrochemical gradients across the membranes of the heart. Compared with other tissues, the myocardium contains a low concentration of **high-energy phosphates,** given the continuous energy requirements of excitation, contraction, and relaxation. Furthermore, the heart can abruptly increase its work output at least sixfold, and thus requires a substantial energy reserve.

ATP levels are buffered in the heart by the much larger concentration of **phosphocreatine** (PCr), which regenerates ATP by the reaction: ADP + PCr = ATP + Cr, catalyzed by the enzyme creatine kinase. Regeneration of ATP from phosphocreatine can protect the heart from ATP depletion during a mild or brief increase in energy demand, but the heart is fundamentally dependent on continuous resynthesis of mitochondrial ATP.

The metabolic pathways by which ATP is synthesized in the myocardium utilize a variety of substrates but are almost exclusively **aerobic.** Unlike skeletal muscle, the heart cannot function anaerobically for extended periods of high demand, building up an "oxygen debt" to be repaid during subsequent periods of rest. Because the heart is continuously active, oxidative ATP synthesis must continuously match ATP utilization.

Under normal resting conditions, the heart generates 60 to 70 percent of its ATP from **beta oxidation** of free fatty acids and 30 percent from metabolism of **carbohydrates,** including exogenous glucose and lactate (Fig. 20-8). Amino acids and ketones are also used as substrates, but to a much lesser extent. During exercise, the large amounts of lactate produced by skeletal muscle become a major substrate for meeting the energy requirements of cardiac muscle, entering the Krebs' cycle after conversion to pyruvate. Oxidation of free fatty acids is inhibited and carbohydrates become the predominant substrate for energy metabolism (Fig. 20-9).

CORONARY ARTERY DISEASE AND IMPAIRED CORONARY ARTERY FLOW

The most common cardiovascular disease in the Western world is **atherosclerosis,** a process of

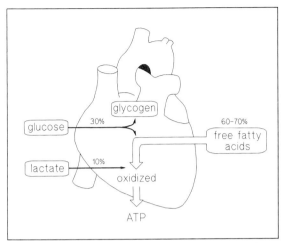

Fig. 20-8. Preferred substrate use by the heart under normal resting conditions. Oxidation of free fatty acids is the source of most of the ATP produced by the heart. (Adapted from: Opie, L. H., *The Heart: Physiology and Metabolism.* New York: Raven Press, 1991.)

Fig. 20-9. During acute exercise, the blood lactate level rises and lactate becomes the major fuel of the heart. Lactate inhibits the uptake of free fatty acids. Carbohydrate metabolism can account for 70 percent of the ATP generated. (Adapted from: Opie, L. H., *The Heart: Physiology and Metabolism.* New York: Raven Press, 1991.)

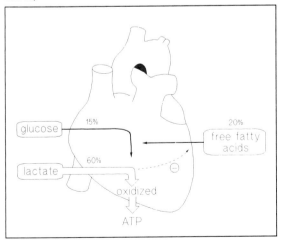

progressive thickening and calcification of the inner arterial walls that characteristically affects the lower abdominal aorta and the large arteries of the heart, brain, lower extremities, and kidneys. Atherosclerotic coronary artery disease remains a leading cause of death in the United States and in other industrialized nations. The disease may be diffuse, but most commonly is manifested as focal narrowings (plaques) in one or more of the **epicardial coronary arteries.** The diseased coronary artery segment may become gradually narrowed by progressive plaque enlargement, or it may be abruptly occluded by thrombus (clot) formation or vascular spasm, both of which can be triggered by vascular injury caused by the atherosclerotic process.

The functional and metabolic consequences of fixed focal narrowing of a coronary artery depend on the severity of the stenosis, whether collateral vessels have developed distal to the stenosis, and the extent of the myocardium supplied by the involved vessel. Resting flow does not decrease until there has been a very marked reduction in the arterial lumen, at least a 70 percent reduction in inner diameter (which corresponds to a 90 percent reduction in lumen size). Once the narrowing is severe enough to cause a drop in perfusion pressure across the stenosis, the vessel (and branches) distal to the stenosis vasodilates, so as to restore resting flow (Fig. 20-10). Although resting flow is not impaired until a very tight stenosis has developed, **coronary reserve,** which is the residual capacity of the coronary vascular bed to autoregulate and augment flow in response to an increase in myocardial oxygen consumption or further decrease in coronary perfusion pressure, becomes progressively impaired once the luminal area is reduced by more than 50 percent. Furthermore, neurogenic and endothelial modulation of coronary flow may become markedly abnormal in atherosclerotic vessels, leading to an impaired vasodilatory response or even inappropriate vasoconstriction (coronary spasm).

Once flow cannot adequately meet the oxygen requirements of the contracting muscle, the muscle is said to be ischemic. **Ischemia** can result from reduced coronary perfusion pressure

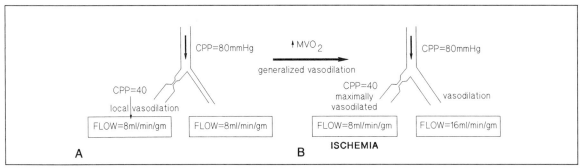

Fig. 20-10. Effect of coronary artery narrowing on blood flow at rest (A) and during stress or exercise (B) when myocardial oxygen demand (MVO_2) is increased. Despite the drop in coronary perfusion pressure (CPP) across the coronary artery narrowing in the left branch of the artery, flow through the diseased branch is normal at rest because the distal vessel has dilated to normalize flow. However, when oxygen demand increases such that flow in the normal artery doubles to meet the increased demand for oxygen, flow cannot increase further in the diseased segment because it is already maximally vasodilated. Because the rate of flow does not meet the demand for oxygen, the myocardium supplied by the diseased segment becomes ischemic.

below that to which autoregulation can restore flow (below the level of coronary reserve) or from an increase in myocardial oxygen demand beyond the capacity of the vascular bed to augment flow through metabolic regulation. Myocardial ischemia has almost immediate functional consequences, including electric instability leading to abnormal cardiac impulses, diminished force of contraction, diminished rate and extent of shortening, and impaired relaxation, presumably because of inadequate ATP stores for Ca^{2+} resequestration by the sarcoplasmic reticulum. In addition, myocardial ischemia often causes a very distressing type of chest pain, known as **angina pectoris.**

The combined influence of inadequate mitochondrial ATP synthesis due to lack of oxygen and the toxic metabolic waste products (particularly H^+) which accumulate at low flow rates because of impaired washout produce impaired cardiac function during myocardial ischemia.

When there is a lack of oxygen, myocardial cells can produce only limited amounts of ATP by anaerobic glycolysis of endogenous glycogen stores, and the heart is very limited in its ability to subsist on ATP generated anaerobically. Although the rate of glycolysis is accelerated in ischemia, the net yield of ATP from the initial nonoxidative

steps of the glycolytic pathway (two moles of ATP per mole of glucose; three moles of ATP per mole of glycogen) is meager compared with the yield from complete cycling of glucose or glycogen during oxidative phosphorylation (Fig. 20-11). Furthermore, myocardial glycogen stores are soon depleted and this combined with the accumulation of lactate and NADH, which cannot be metabolized in the absence of oxygen, have a negative-feedback effect on several key enzymes of the glycolytic pathway, thus effectively halting anaerobic glycolysis. Lactate accumulation produces intracellular acidosis. This further impairs cardiac function, as a high concentration of H^+ has a direct negative inotropic effect in that it diminishes the sensitivity of the contractile proteins to Ca^{2+}.

So long as there is a **small residual blood flow** through a narrowed coronary artery to provide some oxygen and substrate for oxidative phosphorylation and to wash out lactate produced by accelerated anaerobic glycolysis, the heart can function for extended periods during mild to moderate low-flow ischemia. However, the heart will be exceedingly vulnerable to any increase in myocardial oxygen demand (stress, exercise) or any further impairment of coronary flow.

The consequences of **total occlusion** of a cor-

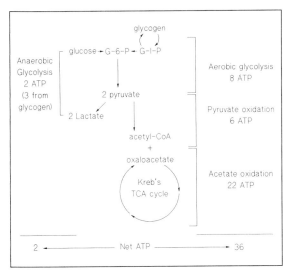

Fig. 20-11. ATP balance during anaerobic and aerobic glycolysis in the heart. Anaerobic glycolysis produces 4 moles of ATP during the conversion of 1 mole of glucose to 2 moles of lactate. However, 2 moles of ATP are consumed during the initial conversion of 1 mole of glucose to 2 moles of glyceraldehyde-3-phosphate and 1 mole of ATP is consumed during the initial metabolism of 1 mole of glycogen. Thus, the net yield of ATP from anaerobic glycolysis is 2 moles of ATP per mole of glucose and 3 moles of ATP per mole of glycogen. In contrast, complete oxidative metabolism of 1 mole of glucose, including aerobic glycolysis to 2 moles of pyruvate, oxidation of pyruvate to acetylcoenzyme A, and then oxidation of acetylcoenzyme A through the Kreb's tricarboxylic acid (*TCA*) cycle, yields 36 moles of ATP.

onary artery depend on whether the occlusion is permanent or transient (as might occur with spontaneous resolution of a clot or reversal of spasm), and whether there are collateral vessels supplying an alternative source of blood flow distal to the obstruction. **Complete cessation** of coronary flow within seconds causes profound malfunction of the myocardium supplied by that vessel and within 20 to 40 minutes leads to myocardial cell necrosis, or **myocardial infarction.**

The consequences of acute coronary occlusion include sudden death, which is estimated to occur in as many as 50 percent of patients before they reach a hospital. Another 10 to 15 percent

die within the first two weeks. The **mechanisms of death** in acute myocardial infarction include instability of electrical impulse formation, leading to abnormal cardiac rhythms (asystole or ventricular fibrillation); rupture of necrotic myocardial tissue, leading to pericardial tamponade or exsanguination; and cardiogenic shock, or progressive and irreversible cardiac failure related to a large area of injured tissue. If the patient survives the early stages of a myocardial infarction, the damaged myocardial tissue will eventually form a fibrotic scar, and the patient may or may not suffer overall impairment of cardiac function, depending on the location and extent of the scar.

Transient obstruction of a coronary artery, if relieved before the onset of irreversible cell damage, leads to a variable degree of cardiac dysfunction (diminished contractility and impaired relaxation). This may persist for hours, days, or weeks after flow has been restored, and is known as **myocardial stunning.** Patients with coronary artery disease often present with evidence of a "stuttering" coronary obstruction, in that the artery appears to be repetitively occluding and reperfusing, causing repeated episodes of chest pain at rest. This is called **unstable angina.** Many of these patients go on to suffer myocardial infarction when the artery occludes permanently or remains occluded long enough to cause irreversible damage.

The event that triggers this sequence of repetitive, transient coronary occlusions and unstable angina appears to be rupture of an atherosclerotic plaque, or some other injury to the endothelial surface of the coronary artery that leads to a sequence of platelet aggregation, clot formation, and coronary spasm. If stable flow can be restored promptly, myocardial infarction can be averted or at least limited in extent.

Once it was realized that coronary occlusion is often a dynamic process that can be interrupted, tremendous efforts were mounted to develop techniques that restore coronary vascular patency in patients with evidence of impending myocardial infarction. Drugs that prevent platelet aggregation, dissolve thrombi, and prevent coronary spasm are now used routinely in patients with unstable angina or in the early hours of my-ocardial infarction, in the hope that blood flow can be restored to areas of the heart not yet ir-reversibly damaged. Coronary artery segments that are severely narrowed by plaque formation can sometimes be opened by cannulating the dis-eased coronary artery with a very thin catheter and inflating a tiny balloon mounted on its end at the site of the stenosis. This procedure is known as **coronary angioplasty.** Patients with multiple severe narrowings of the coronary ar-teries may undergo **surgical bypass.** This in-volves channeling additional blood flow to the diseased coronary arteries using vein grafts har-vested from the patient's legs and anastomosed to the aorta and the coronary artery distal to the obstruction. Arteries dissected from the chest wall (internal mammary arteries) may also be an-astomosed to a coronary artery to bypass an ob-struction.

SUMMARY

Metabolism in cardiac muscle is predominantly **aerobic** and synthesizes the large amounts of ATP it requires for contraction and relaxation by oxidative phosphorylation in the mitochondria, with only a small contribution from anaerobic glycolysis. The **oxygen requirements** of the myo-cardium are therefore very high, even under rest-ing conditions, and markedly increase under conditions of stress or exercise. The major factor that determines the amount of oxygen supplied to the myocardium is **coronary blood flow,** as ox-ygen extraction by myocardial cells is near max-imal at rest and can be augmented only slightly in response to the oxygen requirements of in-creased cardiac work. The determinants of cor-onary flow are the **aortic perfusion pressure** (mainly diastolic pressure, as flow to the intra-myocardial vessels is largely shut off during sys-tole) and **coronary vascular resistance,** which is the key factor in the regulation of coronary flow. Three mechanisms regulate coronary vascular re-sistance: (1) metabolic regulation by **adenosine,** a vasodilatory breakdown product of ATP which

accumulates when ATP use is heightened (in-creased work) or ATP synthesis is decreased (is-chemia or hypoxia); (2) neurogenic regulation by means of the **autonomic nervous system;** and (3) regulation by **endothelial derived factors,** the best defined being EDRF, which is elaborated in response to a variety of stress signals.

Coronary artery disease, which is usually the result of **atherosclerosis,** produces not only me-chanical obstruction to blood flow but also func-tional alterations in the regulation of coronary flow. These changes prevent the coronary arteries from vasodilating appropriately in response to in-creased oxygen requirements. If the heart be-comes ischemic, such that coronary flow cannot adequately meet the oxygen requirements of the myocardium, cardiac contraction and relaxation become abnormal within seconds. These changes are reversible if normal flow is restored. However, if coronary blood flow is severely reduced for any significant length of time, myocardial necrosis and irreversible dysfunction will ensue.

NATIONAL BOARD TYPE QUESTIONS

Select the one most correct answer for each of the following

1. Which of the following factors regulate cor-onary artery flow?
   A. Systolic blood pressure
   B. Diastolic blood pressure
   C. Coronary artery diameter
   D. Arterial oxygen content
   E. All of the above
2. Myocardial oxygen consumption is deter-mined by all of the following factors except:
   A. Heart rate
   B. Contractility
   C. Diastolic blood pressure
   D. Systolic blood pressure
   E. Ventricular size
3. The coronary artery supply to the left ven-tricle:
   A. Always comes entirely from the left cor-

onary artery

B. Usually derives partially from the right coronary artery

C. Only comes from the right coronary artery if there is significant stenosis of the left coronary artery

D. Is always balanced between the left and right coronary arteries

E. Comes entirely from the left coronary artery except for the right side of the septum, which is supplied by the right coronary artery

4. Coronary venous effluent:

A. Drains entirely into the great coronary vein

B. Drains entirely into the coronary sinus

C. Drains entirely into the right side of the heart

D. Drains from the right coronary artery into the right ventricle and from the left coronary artery into the left ventricle

E. None of the above

5. When oxygen use by the myocardium increases, oxygen delivery is augmented to match this by:

A. Increased coronary perfusion pressure

B. A shift in the oxygen–hemoglobin dissociation curve, so that more oxygen is released per gram of hemoglobin

C. Increased extraction of oxygen from the coronary artery blood

D. Increased coronary artery blood flow

E. Increased oxygen uptake by the lungs

6. The most important mechanism by which coronary blood flow is regulated to match myocardial oxygen requirements is:

A. Adenosine-mediated vasodilation

B. Neurogenic reflex control of coronary artery tone

C. Autonomic reflex control of the heart rate and other determinants of myocardial work

D. Alterations in the coronary perfusion pressure

E. Augmentation of the intramyocardial blood flow during diastole by enhanced myocardial relaxation

ANNOTATED ANSWERS

1. E. Coronary flow is determined by the coronary perfusion pressure and coronary vascular resistance. Most coronary flow occurs during diastole because intramyocardial tension is high during systole, thus increasing coronary resistance in the intramural vessels, especially in the subendocardium where wall tension is highest. Some flow to the subepicardium occurs during systole. Coronary resistance is a function of coronary artery diameter. A decrease in the arterial oxygen content (hypoxia) produces an increase in coronary flow by means of adenosine-mediated vasodilation.

2. C. The major determinants of myocardial energy use, and thus oxygen consumption, are heart rate, wall stress (by the Laplace relationship, which is a function of systolic pressure generation and ventricular radius), and the contractile state.

3. B. The coronary artery blood supply to the diaphragmatic and posterior walls of the left ventricle usually originates entirely or partially from the right coronary artery.

3. E. Venous drainage from the left ventricle is primarily into the coronary sinus; a small percentage drains directly into the left ventricle through thebesian veins. The venous drainage from the right ventricle goes into the right atrium via several anterior cardiac veins.

5. D. Because myocardial oxygen extraction from coronary artery blood is near maximal at rest, the major mechanism by which oxygen supply is augmented to match increased demand is through increased coronary blood flow.

6. A. When energy (ATP) use increases, the relative concentration of ATP metabolites increases. Adenosine, a potent vasodilator, is the final product of the breakdown of ATP to ADP and then AMP. Adenosine-mediated vasodilation results in increased coronary blood flow, and thus more oxygen delivery

and augmented washout of adenosine so that flow is readjusted as the energy balance is restored with the synthesis of more ATP.

BIBLIOGRAPHY

Berne, R. M. The role of adenosine in the regulation of coronary blood flow. *Circ. Res.* 47:807–813, 1980.

Katz, A. M. *Physiology of the Heart.* New York: Raven Press, 1977.

Klocke, F. J. Measurement of coronary blood flow and degree of stenosis: current clinical implications and continuing uncertainties. *Am. J. Coll. Cardiol.* 1:31–41, 1983.

McAlpine, W. A. *Heart and Coronary Arteries: Anatomic Atlas for Radiologic Diagnosis and Surgical Therapy.* New York: Springer-Verlag, 1975.

Neely, J. R., Morgan, H. E. Myocardial utilization of carbohydrate and lipids. *Prog. Cardiovasc. Dis.* 15:289–329, 1972.

Olsson, R. A., and Bunger, R. Metabolic control of coronary blood flow. *Prog. Cardiovasc. Dis.* 29:369–387, 1987.

Opie, L. H. *The Heart: Physiology and Metabolism.* New York: Raven Press, 1991.

21 Hemodynamics and Regional Circulation

HARRIET S. IWAMOTO AND RICHARD A. WALSH

Objectives

After reading this chapter, you should be able to:

Describe in a general way how each of the following characteristics vary from one location in the peripheral circulation to another

Characteristic	Location
Compliance	Aorta
Resistance	Arterioles
Pressure	Capillaries
Capacitance	Venules
Volume	Veins
Vessel diameter	

Explain how total peripheral resistance is affected by amputation of a limb

List the major determinants of vascular resistance and identify which is the most important one under normal conditions

Describe the relationship between blood flow, resistance, and driving pressure

Define *compliance* and describe why it is important in the peripheral circulation

---

The adult cardiovascular system is a closed loop made up of two separate circulations arranged in series: the **systemic** and **pulmonary circulations** (Fig. 21-1). Blood ejected by the **left ventricle** perfuses the systemic circulation and returns to the right atrium and ventricle. Blood ejected by the **right ventricle** perfuses the pulmonary circulation and returns to the left atrium. The volume of blood ejected by the left or right ventricle per unit of time is the **cardiac output,** and normal output is 5 to 6 liters per minute. At rest, a major portion of the systemic blood flow is distributed to the renal and splanchnic circulations. During strenuous exercise, the relative distribution of the systemic blood flow changes

dramatically, such that perfusion of these two areas decreases and blood flow to the skin and skeletal muscle increases to constitute as much as 85 percent of the cardiac output. Regulation of blood flow to the different regions of the systemic circulation balances the needs of the body as a whole against the needs of the individual tissues for oxygen and nutrient uptake and waste removal. The factors that regulate the distribution of cardiac output under various conditions is the topic for this and the next two chapters.

The most important principle that governs the distribution of cardiac output is based on a physiologic analogy of **Ohm's law of electrical resistance** (Fig. 21-2). Ohm's law states that, for

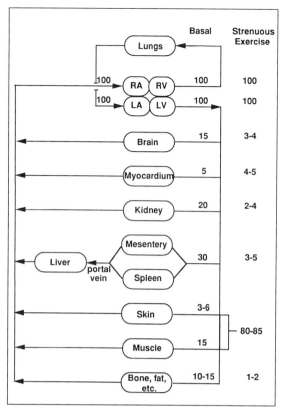

Fig. 21-1. The adult circulation in the human. The right ventricle (*RV*) ejects blood to perfuse the pulmonary circulation. Blood then returns to the left atrium (*LA*). Blood ejected by the left ventricle (*LV*) perfuses the systemic circulation. The cardiac output is the amount of blood (in liters) ejected by each ventricle per minute, and is indicated as 100 percent. Major components of the systemic circulation are depicted. The two columns of numbers represent the approximate percentage of cardiac output each organ receives under basal conditions (*left column*) and during strenuous exercise (*right column*). Cardiac output under basal conditions is 5 to 6 L/min; cardiac output during strenuous exercise can reach 25 L/min. (*RA* = right atrium.)

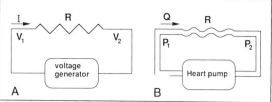

Fig. 21-2. Ohm's law as it applies to (A) an electrical circuit and (B) to the circulation. Flow of current through an electrical circuit and blood flow through the circulation are directly proportional to the gradient of voltage or pressure generated, and inversely proportional to the resistance. ([A] I = amount of current; R = resistance in the circuit; V_1, V_2 = voltages across system. [B] Q = blood flow; R = peripheral circulations; P_1, P_2 = pressures upstream and downstream, respectively.)

an electrical circuit composed of a voltage generator connected to a resistance element, the amount of current (I) that flows through the circuit is related directly to the voltage gradient across the system ($V_1 - V_2$) and inversely to the resistance in the circuit (R), according to the relationship: $I = (V_1 - V_2)/R$. Thus, the current flow increases when either the voltage gradient increases or resistance decreases. The heart and peripheral circulations (systemic and pulmonary) can be thought of as analogous circuits in which the voltage generator is the heart pump that generates a pressure gradient, the current is blood flow (Q), and the peripheral circulations are represented by the resistance element (R). By analogy with the electrical circuit, blood flow increases with an increase in the pressure gradient ($P_1 - P_2$) or a decrease in resistance, according to the relationship: $Q = (P_1 - P_2)/R$.

To understand why cardiac output is distributed as shown in Figure 21-1, one must realize that the peripheral circulation is not a single circuit but is composed of **many circuits**. An example of a peripheral circulation composed of three parallel circuits is shown in Figure 21-3. Blood pumped by the heart is propelled through the circulation in a clockwise direction. The extent to which blood enters each circuit depends on the relative magnitudes of the respective circuits. That is, most of the blood will flow through the circuit with the least resistance, and the least amount of blood will flow through the circuit with the greatest resistance. The rate at which blood flows through the circuit also depends on the pressure gradient ($P_1 - P_2$).

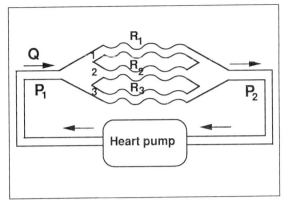

Fig. 21-3. Simplified example of the circulation, composed of a pump, conduit vessels, and three resistance circuits. The *arrows* indicate the direction of blood flow; Q is blood flow; P_1 and P_2 are pressures upstream and downstream, respectively, from the resistance units, R_1, R_2, and R_3.

PRESSURE

Definitions

Pressure is defined as a force applied per unit of area. Physiologic pressures, however, are usually expressed as centimeters of water (cm H_2O) or millimeters of mercury (mm Hg). These units are derived from measurements of pressure with a simple U-tube manometer filled with water or mercury, as illustrated in Figure 21-4A. When both sides of the tube are exposed to atmospheric pressure, gravity holds the fluid at the bottom, so that the level on both sides is even. When a higher pressure is applied to one side, the level of fluid on that side falls and that on the other side rises. The difference in the height of the two sides is the **absolute pressure** (P), or the difference between the applied and atmospheric pressures. Most vascular pressures are expressed in this manner, rather than as total pressure, which is the sum of the absolute and atmospheric pressures. **Transmural pressure** (P_{TM}) is illustrated in Figure 21-4B and constitutes the difference between the pressure inside and the pressure outside a structure. P_{TM} is the net pressure that distends structures, such as the cardiac chambers or blood vessels, and is used when consid-

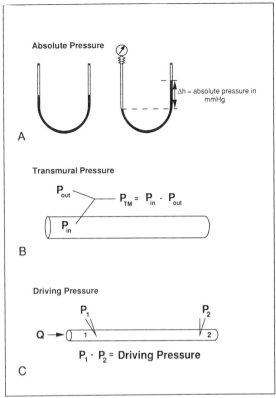

Fig. 21-4. Examples of pressure relationships important in cardiovascular physiology. ([B] P_{out}, P_{in} = pressures inside and outside the structure; P_{TM} = transmural pressure. [C] Q = blood flow; P_1, P_2 = pressure upstream and downstream, respectively.)

ering compliance. **Driving pressure** is also a concept central to the understanding of blood flow (see Fig. 21-4C). Driving pressure represents the difference between the absolute pressure upstream and the absolute pressure downstream. For blood to flow from one point to another, the pressure upstream must exceed that downstream.

Pressure Changes through the Circulation

The **absolute pressures** that exist in the circulation are depicted in Figure 21-5. The pumping

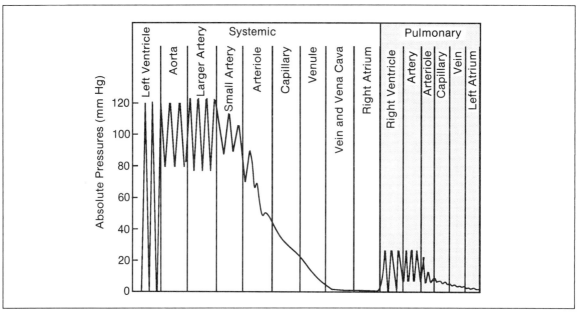

Fig. 21-5. Absolute pressures in major portions of the systemic and pulmonary (*shaded portion*) circulations. In the systemic circulation, blood flows from the left ventricle, through arteries, capillaries, venules, and veins to the right atrium. The driving pressure for the systemic circulation is the difference between aortic and right atrial pressures. In the pulmonary circulation, blood flows from the right ventricle, through pulmonary vessels to the left atrium. The driving pressure for the pulmonary circulation is the difference between right ventricular and left atrial pressures.

action of the left ventricle generates the energy that propels blood flow through the systemic circulation. Absolute pressure in the normal left ventricle oscillates between 3 and 120 mm Hg. In the aorta, which is the first vessel of the systemic circulation, blood pressure ranges between 80 and 120 mm Hg. The lowest pressure (80 mm Hg) is referred to as the **diastolic pressure** and the highest pressure (120 mm Hg) is referred to as the **systolic pressure**. The difference between these two values (systolic pressure − diastolic pressure = 40 mm Hg, in this instance) is the **pulse pressure**. The **mean arterial pressure** is the average arterial blood pressure during the entire cardiac cycle (see Chapter 19) and can be approximated from these two values. Mean arterial pressure is not simply the arithmetic mean of these values, because systole, or cardiac contraction, normally lasts only

half as long as diastole, that is, cardiac relaxation. A simple way to estimate the mean arterial pressure is to calculate a weighted average according to the relationship: mean arterial pressure = one-third the systolic pressure + two-thirds the diastolic pressure. Thus the mean arterial pressure in this example is 93 mm Hg.

As blood courses through the large arteries, systolic pressure increases and diastolic pressure decreases. The decrease in diastolic pressure becomes greater than the increase in systolic pressure, and, with this, pulse pressure increases gradually and the mean arterial blood pressure decreases in the systemic arteries as a direct function of the distance from the heart. The vascular resistance in the arterioles is greater than that anywhere else in the circulation. There is also a corresponding decrease in blood pressure in the arterioles. In addition, the oscillations of blood pressure are dampened and pressure waves are abolished in the arteriolar portion of the systemic circulation.

Blood enters the capillaries of the **systemic circulation** with pressures that range from 30 to 38 mm Hg, depending on the physiologic state and particular vascular bed. As blood flows through the capillaries, the blood pressure decreases to 25 mm Hg and, in the venules, decreases to 5 mm Hg. Blood pressure decreases further in the large veins and vena cava, so that blood returns to the right atrium with a blood pressure nearly equal to the atmospheric pressure.

The right ventricle is another pump; it has the same cavity size but smaller mass than that of the left ventricle. Consequently, the right ventricle generates pressure pulses that oscillate between 0 and 25 mm Hg. Pressure in the pulmonary artery oscillates between 8 and 25 mm Hg, and the mean pulmonary arterial blood pressure is about 15 mm Hg. As with the systemic circulation, the pulmonary arterioles exert the greatest resistance to blood flow in the **pulmonary circulation** and therefore the greatest pressure decrease occurs in the arteriolar portion. Unlike the systemic arterioles, the pulmonary arterioles do not completely dampen the pressure pulses. The mean pressure in the pulmonary capillaries is about 7 mm Hg. A further decrease in blood pressure occurs in the remainder of the pulmonary circulation, such that the mean pressure in the left atrium is ultimately about 2 to 3 mm Hg.

Relationship between Pressure and Volume

Changes in blood volume have very important effects on blood pressure and blood flow. The extent to which changes in blood volume affect blood pressure and blood flow depends on many factors; these include the physiologic state, cardiac pump performance, and degree of activation of neural, hormonal, and local mechanisms that alter vascular tone. The size of the blood volume

can change secondary to changes in red blood cell mass or extracellular fluid volume. One of the major variables affected by changes in blood volume is the **mean circulatory pressure,** which is defined as the equilibrium pressure of the circulation when blood flow is zero. The value is the same throughout the circulation and is dependent on the total blood volume and the physical properties of the circulation. Under normal circumstances, the mean circulatory pressure is 7 mm Hg. When blood volume is increased above normal, the mean circulatory pressure increases. To a certain extent, this increase in the mean circulatory pressure augments ventricular filling and cardiac output. When this happens, an increase in cardiac output produces an increase in the arterial blood pressure (remember Ohm's law, whereby a change in pressure is the product of the blood flow [current] and resistance). An increase in cardiac output, coupled with an increase in blood flow to the peripheral organs leads to a corresponding increase in resistance as peripheral organs decrease blood flow to baseline in accordance with autoregulatory mechanisms (see Chapter 23). This causes a further increase in the arterial blood pressure. **Compensatory responses** to an increase in the arterial blood pressure and the role of neural, hormonal, and local mechanisms in this process are discussed in detail in Chapter 23.

When a distensible structure such as a balloon is inflated, the volume it attains depends on the distending or transmural (P_{TM}) pressure and the elastic properties of the balloon material. A change in volume is accompanied by a corresponding change in the P_{TM}. The relationship between volume and P_{TM} for a distensible structure is **compliance**. Simply put, compliance (C) represents the ease with which a structure can be stretched. This is expressed mathematically as: $C = \Delta V / \Delta P_{TM}$. Note that differences (Δ) rather than absolute values are used in this expression, and that compliance is the *change* in pressure required to produce a *change* in volume. For a very compliant structure, a large change in volume can be achieved with a relatively small change in

P_{TM}. It is also important to note that the compliance of a given structure can change, depending on its initial volume (and wall tension).

Figure 21-6 depicts the **relationship between volume and P_{TM}** for the arterial and venous portions of the circulation. The slope of the relationship, as shown in the graph, is equal to compliance. As the graph also shows, arteries are much less compliant than veins. That is, a small amount of volume added to an artery produces a large increase in P_{TM}, whereas a large amount of volume added to a vein produces only a small increase in P_{TM}. Normal aging or an increase in sympathetic stimulation decreases the compliance of both arteries and veins. Arterial compliance decreases throughout development, from before birth well through the eighth decade of life. This is a gradual process that occurs primarily because, throughout development, there are gradual increases in arterial diameter, arterial blood pressure, and the relative abundance of relatively nondistensible structural proteins such as collagen in the arterial vessel wall. Sympathetic stimulation decreases vessel compliance by constricting vessels and increasing vascular tone. The slopes of the pressure–volume relationships in veins and arteries are reduced under these conditions (see Fig. 21-6).

Because of differences in compliance in the arterial and venous portions of the circulation, more blood volume resides in the venous than in the arterial component at any given time (Fig. 21-7). The greatest proportion of total blood volume exists in the accommodating, compliant venous aspect of the circulation, which is an important blood reservoir. When necessary, an increase in sympathetic stimulation contracts the large veins and increases venous pressure. Because only a small increase in venous pressure can markedly reduce venous blood volume, blood can be effectively returned to the heart to maintain cardiac output and arterial blood pressure. This is an

Fig. 21-6. Relationship between transmural pressure (P_{TM}) and volume in an artery and vein. Note that the slope of the relationship (Δ volume/ΔP_{TM}) is equal to compliance. Veins are more compliant than arteries. Aging and sympathetic stimulation render both arteries and veins less compliant.

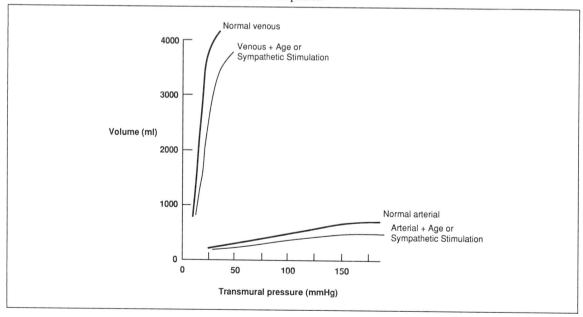

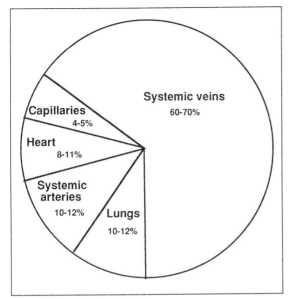

Fig. 21-7. Relative distribution of blood volume (instantaneous total blood volume) in different circulatory compartments.

important adaptive mechanism that operates not only in pathologic situations but also in normal day-to-day activities.

Arterial Blood Pressure

Transformation of the Ventricular Pressure Pulse

The arteries transport blood from the heart to the periphery, and ventricular pressure is the motive force that drives this blood flow. Although the output of blood from the ventricles is intermittent, blood flow in the capillaries is continuous. In the aorta and large arteries, the intermittent blood flow ejected from the ventricles is converted to a more continuous flow. The example shown in Figure 21-8 explains how this is accomplished. A rigid tube is connected to a piston pump on one end and a resistance unit on the other. A reservoir and a one-way valve are placed between the pump and the resistance unit. Pressure is

then measured between the valve and the resistance unit (Fig. 21-8A). Initially the reservoir is filled, the valve is closed, and the pressure measured is equal to the atmospheric pressure. When the pump is activated and the piston moves to the right, fluid flows out of the reservoir as the valve opens and a positive deflection of pressure is recorded. When the piston is pulled back, the pressure in the reservoir becomes less than that in the tube; the valve closes and the pressure returns to the atmospheric pressure. A pump connected to a rigid tube then delivers intermittent flow and pressure oscillates between zero and a positive value (Fig. 21-8B).

Such wide oscillations in the intraluminal pressure would damage those vessels of the systemic circulation that are thin-walled. By contrast, the system shown in Figure 21-8C is largely the same, except that a distensible tube has replaced the rigid tube. As fluid is ejected from the reservoir, the tube stretches and expands. Some of the energy imparted by the pump is converted into kinetic energy in the form of moving fluid and some of the energy is stored as potential energy in the tube wall. As the piston draws back, the one-way valve closes and pressure decreases in the reservoir and in the tube. However, the pressure in the tube decreases less slowly than that in the reservoir. During this phase, the stored potential energy in the distensible tube is converted to the kinetic energy of fluid flow as the wall of the tube recoils. If the next cycle of piston movement occurs relatively soon, the pressure does not fall to zero, but is maintained at a value greater than zero. In this way, driving pressure is maintained at a level greater than zero. Intermittent flow and large pressure changes in the reservoir are transformed into continuous flow and smaller pressure changes in the distensible tube.

Factors that Alter the Arterial Pressure Waveform

Normal Arterial Pressure Waveform Normally the circulation functions as a system in which large pressure changes in and

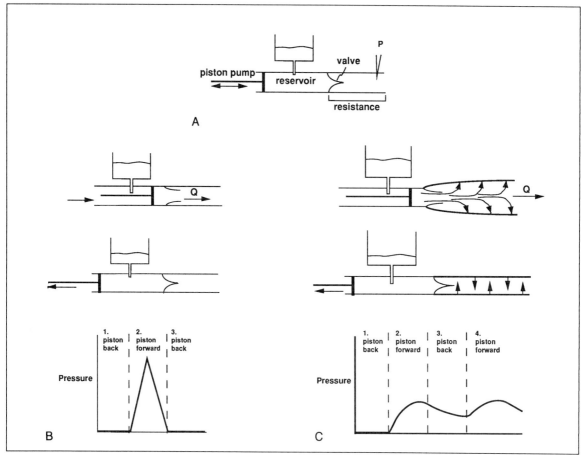

A

B

C

Fig. 21-8. Transformation of the left ventricular pressure pulse to an arterial pressure pulse (Q = fluid flow; P = pressure.)

intermittent ejection of blood from the cardiac ventricles are converted into smaller pressure oscillations and relatively continuous blood flow in the arteries (Fig. 21-9A). When the ventricles begin to relax at the beginning of ventricular diastole, pressure in the ventricles falls. When ventricular pressure becomes less than the pressure in the aorta and pulmonary artery, the aortic and pulmonic valves close. This causes the aortic valve leaflets to pull in slightly, bringing about a brief backflow of blood followed by cessation of backflow. A corresponding fall and rise in arterial pressure occurs, and this is inscribed as a **di-crotic notch** on the arterial pressure pulse. When the aortic and pulmonic valves close, there is a separation of pressures in the ventricles and arteries (see Chapter 19).

Effect of Heart Rate   Changes in heart rate largely affect the diastolic pressure. Figure 21-9B shows the effect of an increase in heart rate compared to the example shown in Figure 21-9A. Up to heart rates of approximately 120 beats per minute, an increase in heart rate shortens the period of ventricular diastole and has little influence on ventricular systole. As a result, aortic pressure

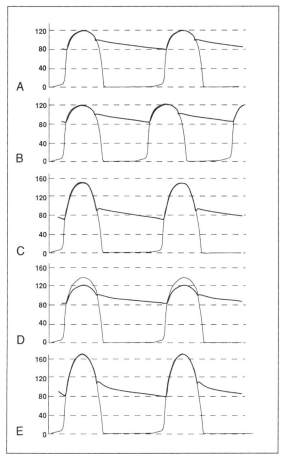

Fig. 21-9. Relationship between left ventricular (*thin line*) and aortic (*thick line*) pressures during two cardiac cycles. (A) The normal relationship in which left ventricular pressure oscillates between 0 and 120 mm Hg and aortic blood pressure oscillates between 80 and 120 mm Hg. Changes in the relationship are shown with (B) an increase in heart rate, (C) aortic regurgitation, (D) aortic stenosis, and (E) a decrease in aortic compliance. (See text for further description.)

does not have time to decrease to 80 mm Hg before ventricular systole increases ventricular pressure to a level sufficient to open the aortic valve. The net effect is an increase in diastolic pressure. The next heartbeat is initiated before aortic pressure falls to its previously low level. The effects on systolic pressure are normally minimal

if there is *only* a change in the heart rate. However, an increase in heart rate is often initiated by heightened sympathetic stimulation, and resultant increases in ventricular contractility would augment the forces (pressures) generated by the ventricular muscle. Thus, the general effect of increased heart rate on pulse pressure is increased diastolic pressure, but the net effect depends on the physiologic situation.

Effect of Altered Stroke Volume   Pulse pressure is related directly to stroke volume and inversely to arterial compliance. During ventricular systole, which lasts about 0.25 seconds, the stroke volume ejected by the ventricle, about 70 ml, enters the aorta. Most of the volume is ejected during the first 0.1 second of systole. This nearly instantaneous increase in aortic volume produces an increase in aortic pressure that is proportional to the increase in volume, as dictated by aortic compliance. Because the bulk flow of fluid is relatively slow (15 cm per second) during systole, most of the blood ejected remains in the aorta. An increase in stroke volume increases pulse pressure by (1) moving the pressure–volume compliance relationship up and to the right, which results in increased peak pressure; and (2) decreasing diastolic pressure because of more forceful recoiling of a stretched aorta.

Effect of Altered Aortic Valve Function **Aortic regurgitation** or **aortic valve insufficiency** occurs when the aortic valve does not close normally. In the extreme, the aortic valve does not close completely and there is no separation of ventricular and aortic pressures. However, in this syndrome, there is usually some degree of valve closure. The effects of abnormal aortic valve closure on ventricular and aortic pressures are illustrated in Figure 21-9C. During systole, ventricular and aortic pressures are nearly the same until the aortic valve closes. As the ventricular muscle relaxes during diastole, there is backflow of blood from the aorta to the ventricle, resulting in increased ventricular volume. Because of **Starling's law**, increased stretch of the myofibers results in more forceful contraction and increased stroke volume. The net effect of aortic valve insufficiency is an increase in the force of contraction, ultimately an increase in ventricular size and wall tension (the work of the heart), and a decrease in diastolic pressure. The end result of these effects is an increase in aortic pulse pressure.

Aortic stenosis is the inability of the aortic valve to open fully during systole. Aortic stenosis decreases the radius of the outflow tract and increases resistance to ventricular ejection (increased afterload) as shown in Fig. 21-9D. In turn, a pressure gradient forms between the left ventricle and aorta, in accordance with **Ohm's law**. To maintain normal aortic pressures, the left ventricle must contract more forcefully than normal. Thus, myocardial work is increased and left ventricular hypertrophy results. Stroke volume and, hence, pulse pressure may ultimately be reduced.

Decrease in Arterial Compliance Arterial compliance decreases with increased sympathetic stimulation and normal aging, and in certain diseases such as atherosclerosis (see Fig. 9E). Decreased compliance renders the arteries stiffer and less able to accommodate the stroke volume ejected by the ventricle with each heartbeat. The arteries become more like rigid tubes, as shown in Figure 21-8. As a result, pulse pressure increases and the arterial pressure wave becomes more peaked and angular because the stiffer arteries recoil more forcefully during diastole.

Pressure Wave Transmission throughout the Arterial Circulation As shown in Figure 21-5, the blood pressure contour in large arteries differs from that in the aorta. In large arteries, systolic pressure is greater, diastolic pressure is less, and pulse pressure is greater than these respective pressures in the aorta. The mean arterial blood pressure is slightly less than that in the aorta because the **decrease in diastolic pressure** is proportionately greater than the rise in systolic pressure. To understand this process, it is essential to consider the propagation of a pressure wave through fluid. An arterial pressure wave is composed of a number of different waves with various amplitudes and frequencies. Each frequency is transmitted at different rates, such that high-frequency pressure waves are transmitted faster than low-frequency pressure waves. At points some distance from the heart, these transmitted frequencies summate to form a different pressure profile. A second factor that alters the arterial pressure waveform is related to **arterial compliance**. As each segment of the arterial tree is stretched and recoils, the intraluminal pressure is altered, as previously described. Third, as the pressure wave hits major branch points, there is a **reflection** of the wave. These reflected waves summate with oncoming waves from subsequent beats. The net effect of these factors is that the arterial pressure waveform is continually altered as it is transmitted from the heart to the periphery.

Capillary Dynamics

For an in-depth discussion on capillary dynamics, the reader is referred to Chapter 22 on "Microcirculation and Lymphatic Circulation."

Venous Blood Pressure

Central venous blood pressure is that pressure measured at the level of the heart and normally is close to atmospheric pressure. Central venous blood pressure is the downstream pressure that influences **cardiac output**, according to the relationship: cardiac output = (aortic pressure − right atrial pressure)/peripheral vascular resistance. The pressure on the **venous side** of the circulation is expressed by the relationship: venous return = (capillary pressure − right atrial pressure)/venous resistance. Variations in venous pressure alter venous capacitance and the amount of blood stored in the venous reservoir; they also affect ventricular filling and cardiac function. In this section, the features of venous pressure are discussed.

Venous Pressure Waves

The pulmonary venous pulse is similar to the arterial pulse, in that the pulmonary arterioles and capillaries do not effectively dampen the pressure pulse generated by the right ventricle. In the systemic circulation, venous pressures are low and the pressure pulse waves generated by the left ventricle are dampened in the arterioles and capillaries. However, pulses can still be observed in the systemic veins near the heart. These pressure waves are quite unlike the arterial pressure pulse, though they are cyclic in nature. Instead they reflect the pressure changes in the right atrium.

There are **three pressure waves** in the right atrial pulse. The **A wave** occurs during atrial contraction. During late diastole the right ventricle is filled with nearly all of its end-diastolic volume. The right atrium contracts, right atrial pressure increases, and additional blood is transported to the right ventricle. The increase in right atrial pressure is reflected retrograde into the large veins near the heart. As the atrium relaxes, pressure in the right atrium and large veins falls. This rise and fall in pressure inscribes the A wave. The next wave is the **C wave** and this coincides with ventricular contraction. Right ventricular contraction increases ventricular pressure, which is transmitted retrograde into the right atrium. This produces a second increase in pressure in the atrium and large veins. When the pressure in the right ventricle exceeds that in the pulmonary artery, the pulmonic valve opens. Pressure in the ventricle, pulmonary artery, atrium, and large veins rises then falls. This pressure wave generated in the atrium and veins is the C wave. Throughout the cardiac cycle, blood returns to the heart from the periphery. During ventricular systole and isovolumetric relaxation, the tricuspid valve is closed (see Chapter 19). Toward the end of these phases, the systemic veins and right atrium are stretched and intraluminal pressure rises. This pressure buildup is relieved only when the tricuspid valve opens at the beginning of diastole. This rise and fall in pressure is called the **V wave**.

These three pressure waves can be seen in the Wigger's diagram in Chapter 19. In the normal person, these pressure waves are evident only when the person is quietly recumbent. They can become prominent in certain disease states such as atrioventricular nodal block. In such patients, the atrial and ventricular contractions are asynchronous. The atria often contract when the tricuspid and mitral valves are closed. This produces a prominent A wave that can be seen in the neck veins of an upright patient. In a patient with right heart failure caused by pulmonary hypertension, all the waves can become more prominent, as the right ventricle has to eject blood against a higher resistance. This produces higher diastolic pressures that are reflected in the venous system.

Normal Variations in Venous Pressure

Hydrostatic Effects   Pressure is not the same at all points in the circulation. In a recumbent person, the pressures as defined so far are fairly accurate. However, for a person standing upright (Fig. 21-10), blood pressure is much different. **Blood pressure** is a function of the energy imparted by the pumping action of the heart plus the hydrostatic pressure imposed by a column of fluid. In Figure 21-10, the pressure at the level of the heart is defined as being equal to atmospheric pressure. Blood in the hand, which is about two feet (60 cm) below the heart, has a two-foot column of blood resting on it. Gravity and the specific density of blood produces a **hydrostatic pressure** of about 35 mm Hg. The arterial blood pressure in the hand, relative to the atmospheric pressure at the level of the heart, is 95 + 35 mm Hg, or 130 mm Hg. Venous pressure in the hand is about 40 mm Hg. For this reason, arterial blood pressures are always measured at the same level as the heart. Similarly, in the foot, arterial blood pressure is 185 Hg and venous pressure is about 95 mm Hg. At each level, however, the relationship between the arterial and venous pressures, the **driving pressure,** is not altered.

Skeletal Muscle Pump   Gravitational effects on blood volume can reduce venous return. As

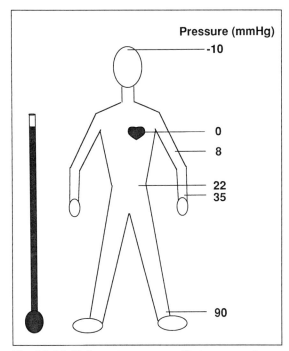

Fig. 21-10. Hydrostatic pressure effects on vascular blood pressures in the upright position. (See text for detailed explanation.)

much as 15 to 20 percent of the blood volume is retained in the lower extremities after only 15 minutes of quiet standing. The resultant increase in capillary hydrostatic and filtration pressures promotes capillary filtration and further loss of fluid from the vascular compartment. In this situation, **skeletal muscle activity** assists venous return. When a skeletal muscle contracts, the pressure surrounding the veins increases and this pressure increase is transmitted directly into the venous lumen. Peripheral venous pressure then increases and blood is squeezed from the area of muscle contraction. Blood is prevented from flowing backward by one-way valves in the lumen of peripheral veins. Regular cycles of muscle contraction and relaxation thus pump blood and greatly assist venous return. The importance of this phenomenon becomes obvious when, for example, a soldier stands at attention for a long time.

Skeletal muscle activity is not the only mechanism which enhances venous return in the peripheral veins. As will be discussed in detail in Chapter 23, stretching of the veins in the chest and right atrium is reduced when venous return to the right heart is decreased. This decreased stretch, perceived by sensory nerve endings located in these areas, activates the **sympathetic nervous system** and **endocrine mechanisms**. Vasoactive factors, thus stimulated, contract peripheral vessels. Venous pressure increases, venous compliance decreases, and additional blood is returned to the heart.

Thoracoabdominal Pump   The thoracoabdominal pump is another pumping mechanism that promotes venous return to the heart (Fig. 21-11). The inferior vena cava returns blood from the abdomen to the heart in the thoracic cavity. Intraabdominal pressure is a few millimeters mercury positive and intrathoracic pressure is negative relative to atmospheric pressure. During inspiration, the diaphragm contracts, the abdominal pressure increases, and intrathoracic pressure decreases. This presure gradient favors venous return. On expiration, abdominal pressure decreases and intrathoracic pressure increases yet remains negative. Rhythmic breathing thus produces a rhythmic oscillation of intraabdominal and intrathoracic pressures that alternately produces a more and less favorable pressure gradient for venous return.

BLOOD FLOW AND BLOOD FLOW VELOCITY

Blood flow refers to the bulk flow of fluid in the circulation. It is the volume of blood that passes all points equidistant from the heart per unit of time. This means that the flow of blood from the heart, through the aorta, through the total arteriolar system, through the capillaries, and through the venous system is equal to cardiac output and is 5 to 6 liters per minute (Fig. 21-12).

Blood flow velocity is the speed with which blood moves along the circulation in any partic-

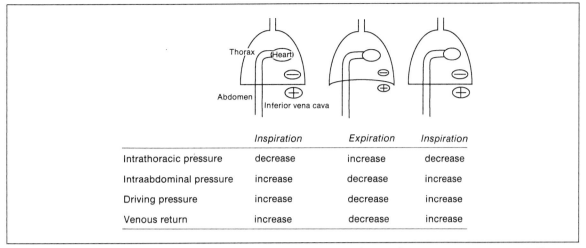

Fig. 21-11. The thoracoabdominal pump, produced by pulmonary ventilation, that assists venous return to the heart.

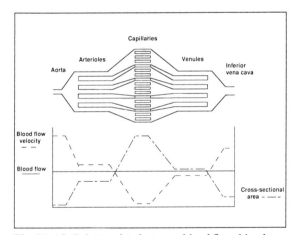

Fig. 21-12. Relationship between blood flow, blood flow velocity, and cross-sectional area of the peripheral circulation. These variables are related according to the relationship: average blood flow velocity = blood flow/cross-sectional area. Note that the total effective cross-sectional area increases with distance from the heart, going from the aorta to the capillaries, and decreases from the capillaries to the vena cava.

ular segment, and is expressed in units of distance per time. Blood flow velocity is related directly to blood flow and inversely to cross-sectional area, according to the relationship: average blood flow velocity = blood flow/cross-sectional area (see Fig. 21-12). Thus, blood flow velocity is greatest in the aorta and least in the capillary bed.

In the circulation blood flows predominantly in a streamline or laminar pattern. Blood flow velocity is zero next to a blood vessel wall and greatest in the center. Because of the friction between the blood vessel wall and that blood flowing directly against the vessel wall, the forward movement of blood is slowed. There are also shear stresses between adjacent layers of blood that cause blood to flow in a laminar pattern in much of the circulation. When blood flows in this fashion, it appears to move in concentric cylinders of ever-faster moving blood, with the plasma next to the blood vessel wall and the blood cells moving with their long axes parallel to the long axis of the vessel. The laminar pattern of blood flow is interrupted and converted to a turbulent flow pattern in the ventricles and, to a certain extent, in the atria. Turbulence in these portions of the circulation ensures mixing of the blood. Turbulence also occurs at branch points in the circulatory tree, when blood flow increases above a critical value, when there is an abrupt change in vessel diameter, such as one secondary to atherosclerotic plaque development, and in the aorta where turbulence is less than expected, probably because flow in the aorta is pulsatile. When blood travels in a laminar pattern, it makes little or no discernible noise. Turbulent blood flow in a region where wall resonance is prominent, such as the cardiac chambers and large vessels, produces a sound or murmur that can be discerned by auscultation with a stethoscope.

RESISTANCE

As stated previously, the relationship between blood flow and blood pressure is proportional and depends on vascular resistance. Several factors influence resistance to blood flow.

Viscosity refers to the "thickness" of a fluid. The more viscous a fluid is, the more energy is required to overcome frictional forces and initiate and maintain fluid movement. For a homogeneous fluid, such as water or plasma at a given temperature, viscosity is constant. A useful expression for viscosity is **relative viscosity,** which is the viscosity of a fluid relative to that of water. Thus, the viscosity of plasma at 37°C is about 1.7 times that of water. For suspension solutions such as blood, the relative viscosity is not constant. *In vitro,* the viscosity of blood is about three to four times greater than that of water. *In vivo,* however, the viscosity of blood is only one to two times greater than that of water, and several factors are responsible for this. In large blood vessels, **laminar blood flow** and the alignment of red blood cells parallel to the axis of motion greatly reduce viscosity. When blood flow falls below a critical level, the cells fall out of alignment and tumble end over end. This increases the relative viscosity. In small blood vessels, red blood cells flow in the center of the vessel, away from the no-flow zone next to the vessel wall. This reduces viscosity.

Temperature is another factor that affects viscosity. Viscosity increases approximately 2 percent for each 1°C decrease in temperature. This usually does not present a problem, but localized viscosity in a cold extremity can increase sufficiently to increase resistance to flow.

Another factor that affects viscosity is the **red blood cell mass** or **protein concentration**. As the hematocrit (red blood cell mass) increases, viscosity increases. The increase is not uniform throughout the circulation because of the friction-reducing alignment of red blood cells, and an increase in hematocrit largely affects viscosity in the resistance vessels that have a diameter of 200 μm. Viscosity increases with increased protein concentration (hematocrit) and lowered temperature, and this increased viscosity (η) increases resistance (R). Thus:

$$R \propto \eta$$

As the length of the resistance pathway (i) increases, the greater the resistance. Thus:

$$R \propto l$$

The most important factor that influences resistance to blood flow is the vessel radius (r). The relationship between the radius and resistance is rendered as:

$$l/r^4 \propto R$$

Thus, the vessel radius has a very important effect on resistance, such that when the radius is halved, resistance increases by a factor of 16. Vessel radius is by far the most important variable that normally alters resistance. The relationship between viscosity, length, and vessel radius has been quantified in Poiseuille's law, which states that:

$$R = 8\eta l/\pi r^4$$

and:

$$Q = \Delta P \pi r^4/8\eta l$$

Although this relationship was originally developed to describe the flow of a homogeneous fluid through a rigid tube, it approximates the flow of the circulation under most circumstances.

The **circulation** is not a single circuit but is composed of a number of circuits arranged in series as well as in parallel. The total peripheral resistance is a summation, and is *not* an arithmetic sum, of all the resistances in the circulation, however. The type of circuit determines its contribution to total peripheral resistance.

A circuit composed of three resistance elements (R_1, R_2, and R_3) arranged in series is illustrated in Figure 21-13A. P_1 is greater than P_2 which is greater than P_3, and flow (Q) through the circuit proceeds from left to right. The sum of the individual driving pressures (P_1, P_2, P_3) across each circuit is equal to the driving pressure across the entire system (P_T), that is:

$$\Delta P_T = \Delta P_1 + \Delta P_2 + \Delta P_3$$

Ohm's law as it pertains to circulation holds that $\Delta P = QR$, so by substitution the following is true:

$$Q_T R_T = Q_1 R_1 + Q_2 R_2 + Q_3 R_3$$

This is a closed circuit, so flow through each resistance element is equal and also equal to the total flow, such that:

$$R_T = R_1 + R_2 + R_3$$

Thus, for a circuit that has resistances arranged in series, the total resistance is equal to the sum of the component resistances.

For a circuit composed of resistances arranged in parallel (see Fig. 21-13B), the total resistance is determined differently. The total flow through the circuit is equal to the sum of flows through each resistance unit, and the driving pressure is the same across each resistance.

$$Q_T = Q_1 + Q_2 + Q_3$$

$$\Delta P_T = \Delta P_1 + \Delta P_2 + \Delta P_3$$

By substitution:

$$\Delta P_T / R_T = \Delta P_1 / R_1 + \Delta P_2 / R_2 + \Delta P_3 / R_3$$

and:

$$1/R_T = 1/R_1 + 1/R_2 + 1/R_3$$

As evident from the arithmetic, for resistances arranged in parallel, the reciprocal of the total resistance is equal to the sum of the reciprocals of the component resistances.

It is obvious that resistances arranged in parallel are more efficient than resistances arranged in series because the pump does not have to work as hard to generate a large driving pressure. A parallel system has the advantage not only of efficiency but also of continuity. This means that one of the resistances can be eliminated from the circuit altogether without significantly affecting

Fig. 21-13. (A) Three circuits in a series. Each circuit has a resistance. The driving pressure for circuit 1 is $P_1–P_2$; for circuit 2, $P_2–P_3$; and for circuit 3, $P_3–P_4$. (B) Three circuits in parallel. Each circuit has a resistance. The driving pressure for each circuit is the same as the driving pressure for the entire system and is equal to $P_1–P_2$. (See text for in-depth description.)

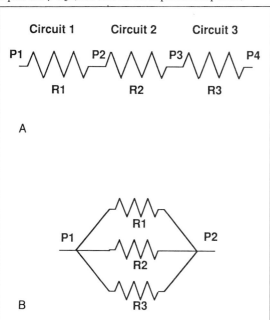

blood flow to other portions of the circulation; pressure is the regulated variable. Even though less efficient, series resistance units are sometimes necessary, however. Two prominent examples are the portal venous connections between the gastrointestinal tract and liver and between the hypothalamus and anterior pituitary. Toxins that enter the body by means of the gastrointestinal tract go into the portal vein and are effectively removed by the liver before they are exposed to the rest of the body. In the hypothalamopituitary circulation, blood first perfuses the hypothalamus, which secretes hormonal substances that regulate hormone secretion by the anterior pituitary (Chapter 47).

The total peripheral vascular resistance is the sum of all resistances in the systemic circulation that the left ventricle must overcome to eject blood. For the systemic circulation, then, if the driving pressure is 90 mm Hg (aortic − right atrial pressures) and cardiac output is 5 liters per minute, the total peripheral vascular resistance is 18 mm Hg · min/L. Total vascular resistance across the pulmonary circuit is much less; driving pressure (16 − 2 mm Hg [pulmonary arterial − pulmonary venous]) divided by cardiac output (5 liters per minute, or 2.4 mm Hg · min/L). This difference in systemic and pulmonary resistances accounts for why the left and right ventricles differ in wall thickness and mass.

PERIPHERAL BLOOD FLOW

Distribution

The proportion of the cardiac output that various areas of the peripheral circulation receive is shown in Figure 21-1 and the blood flow each organ receives is given in Table 21-1.

The distribution of peripheral blood flow is balanced between the needs of the body and the needs of the tissues. Each cell must receive an adequate supply of oxygen and nutrients, and under normal circumstances this is not difficult to achieve. However, there are situations in which blood flow to most organs ceases or is quite reduced, while that to the brain and heart is preserved. Obviously, this cannot persist for too long, but represents an important survival mechanism in emergency situations. In this section, the factors that regulate blood flow to the brain, muscle, and skin are discussed.

Brain

Preserving blood flow to the brain is the most important function of the circulation, because cerebral tissue cannot survive on anaerobic metabolism and begins to die if blood flow is arrested for more than a few minutes. Several mechanisms maintain **cerebral blood flow,** one of which is **arterial blood gases**. There is a positive curvilinear relationship between cerebral blood flow and arterial carbon dioxide tension (Fig. 21-14A). There is a large-gain control in this relationship, in that small increases in arterial blood carbon dioxide tension produce large increases in cerebral blood flow. It is generally believed that this response is mediated in part through changes in the extracellular pH. The increase in cerebral blood flow when carbon dioxide tension or hydrogen ion concentrations increase prevents wide oscillations in pH and carbon dioxide concentrations in the cerebrospinal fluid (CSF). This is an important feature, because pH changes alter the membrane potentials and function of nerve cells. In addition, reduced oxygen availability causes cerebral vasodilation. Figure 21-14B illustrates the relationship between arterial oxygen content, cerebral blood flow, and cerebral oxygen delivery. As arterial oxygen content decreases, cerebral blood flow increases exponentially. This is an important relationship, because cerebral oxygen delivery (the product of cerebral blood flow and arterial oxygen content) stays constant. This response appears to be locally regulated because it is observed in animals in which innervation of blood vessels has been interrupted surgically or pharmacologically or hormonal responses have been inhibited. This is an important local vascular response that maintains cerebral oxygen delivery.

Cerebral blood flow is also controlled by autoregulation of blood flow in the face of widely vary-

Table 21-1. Approximate Blood Flow to Organs

Organ	Mass(kg)	Basal Blood Flow		Maximum Blood Flow	
		L/min	ml/min/100 gm	L/min	ml/min/100 gm
Brain	1.5	0.75	50	2.1	140
White	0.9	0.23	25	0.9	100
Gray	0.6	0.60	100	2.0	333
Heart	0.3	0.24	80	1.2	400
Muscle	30	1.0	3–5	18	60
Skin	2	0.2	3–5	3–4	150–200
Gastrointestinal tract	2.5	1.0	40	5.0	200
Liver*	1.4	0.5	40	3.0	200
Kidney	0.3	1.2	400	1.4	450
Bone	27	0.8	3	4	15

*These values denote hepatic arterial contribution only.

ing perfusion pressures of from 60 to 150 mm Hg (see Chapter 20). When arterial blood pressure decreases below 60 mm Hg and cerebral blood flow falls, the brain tissue begins to become ischemic. This response, particularly in the cardiovascular regulatory portions of the brain, can elicit powerful stimulation of the peripheral sympathetic nervous system, which produces generalized vasoconstriction and an increase in arterial blood pressure. This response is effective down to arterial blood pressures of 15 to 20 mm Hg, and is so powerful that blood flow to other areas can drop to zero in an effort to preserve cerebral blood flow. This reflex response is called the **central nervous system ischemic response**. Coronary blood flow is also maintained (see Chapter 20) because the sympathetic activation simultaneously increases myocardial work by increasing heart rate and contractility. As the myocardial blood flow is regulated predominantly by metabolic factors, intense sympathetic nervous system activation tends to increase myocardial blood flow.

A special instance of the central nervous system ischemic response is the **Cushing's reflex**. The brain floats in CSF and is encased in a nonexpandable skull. When the CSF pressure increases, as it can during infection such as meningitis, which produces swelling of cerebral

tissue, or as the result of trauma, the cerebral vessels become compressed. Peripheral sympathetic nerves are then activated, and arterial blood pressure increases until cerebral perfusion pressure is restored and cerebral blood flow is returned to normal. This is another extremely powerful reflex that can produce profound hypoperfusion of nearly all other parts of the circulation.

Cerebral blood vessels are innervated, but these neural mechanisms modify cerebral blood flow only weakly. Any effect nerve stimulation has on cerebral blood flow can be effectively overruled by other factors that regulate cerebral blood flow.

Skeletal Muscle

The blood flow to resting skeletal muscle is relatively low; normally only 3 to 4 ml/min/100 gm of muscle and only 10 percent of the capillary beds are perfused. This is sufficient to meet the basal metabolic needs of resting muscle. Blood vessels in skeletal muscle are innervated; they constrict in response to alpha-adrenergic stimulation and dilate in response to beta-adrenergic or cholinergic stimulation. When skeletal muscle is not working and blood flow is needed elsewhere, neural mechanisms constrict muscle vessels to

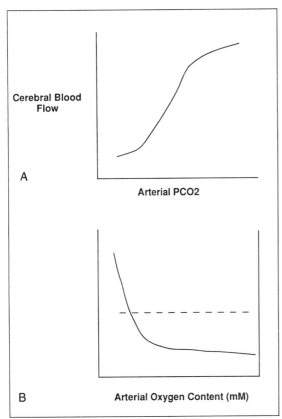

Fig. 21-14. Relationship between (A) cerebral blood flow and carbon dioxide tension (PCO_2) and (B) cerebral blood flow (*solid line*), cerebral oxygen delivery (*dashed line*), and arterial oxygen content.

divert blood to the needed areas. When skeletal muscle is active, however, neural influences on blood flow are overridden by powerful local metabolic and vascular control mechanisms. Strenuous exercise can increase blood flow to muscle by as much as 25-fold, to a maximum of about 80 ml/min/100 gm in which *all* capillary beds are perfused. The primary regulators of skeletal muscle blood flow during exercise are **metabolic factors**. When muscle contracts, it necessarily consumes oxygen and nutrients (primarily glucose and lactate) and produces carbon dioxide and other metabolic waste products. These re-

sultant changes—a decrease in oxygen tension and increases in the concentrations of carbon dioxide, lactic acid, hydrogen ions, potassium ions—directly increase muscle blood flow. The predominant effector has not been clearly identified, and it is probable that all changes make some contribution to the change in blood flow. As the increase in blood flow reverses these changes (that is, the increased perfusion washes out these substances), tissue concentrations return to normal. With aerobic exercise, blood flow is maintained at a steady level, albeit one higher than normal, in pace with the increase in metabolic rate. On average, blood flow is matched to the metabolic rate, although this may vary from moment to moment because the muscle can depend on anaerobic metabolism for only short periods. Under anaerobic conditions, muscle blood flow remains elevated until the concentrations of all effector substances return to normal. Thus, for a given period of exercise, the increase in metabolic rate is matched by an increase in blood flow, even if the exercise is maximal for a time and blood flow remains elevated into the period of recovery from exercise.

Skin

Blood flows to the skin to nourish the epidermis and to help regulate body temperature. A brief description of the **anatomy of cutaneous circulation** is important to an understanding of the physiology of cutaneous blood flow (Fig. 21-15). Blood enters the skin in the dermis through arterioles. The arterioles branch into metarterioles and these give rise to capillary loops that radiate out toward the epidermal layer. The capillary loops then drain into venules that drain into a venous plexus located in the dermal layer. In certain areas of the body (soles of feet, palms, ears, lips, and nose), blood that enters a cutaneous artery can either flow through the capillary loops via the arterioles or bypass the capillaries by flowing through short communications between the venous plexus and artery, the arteriovenous anastomosis. These vessels are richly inner-

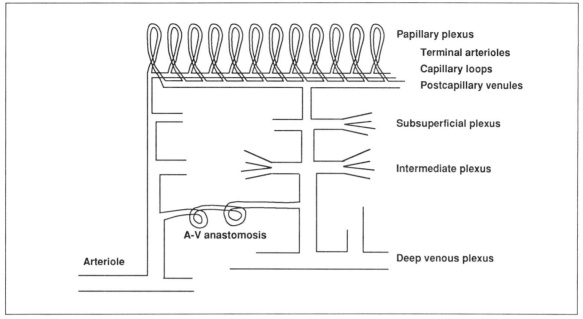

Fig. 21-15. The microcirculation in plantar skin. (See text for detailed explanation.) (*A-V* = arteriovenous.)

vated with sympathetic nerve fibers. Sympathetic nerve stimulation constricts arterioles and arteriovenous anastomoses; intense stimulation can reduce cutaneous blood flow to nearly zero.

Blood flows through the capillaries much more slowly than through the arteriovenous anastomosis. This is an important feature of cutaneous blood flow and its role in temperature regulation. When the body is cold, the sympathetic nervous system is activated and arterioles and arteriovenous anastomoses constrict. Blood flow then bypasses the cutaneous circulation and returns quickly to the body's core. The heat in the blood is thus retained and body temperature is maintained. When the body is warm, the sympathetic activation of the skin is inhibited and the arterioles dilate. A major portion of blood flows through the capillaries and heat is dissipated in the venous plexus in the dermis and epidermal skin layers, and is ultimately lost to the atmosphere.

REGIONAL CIRCULATION IN THE FETUS AND THE TRANSITION AT BIRTH

The Fetal Circulation

Course of Blood Flow

The placenta is the site of nutrient, oxygen, and waste exchange between the mother and fetus, such that the placenta functions as the gastrointestinal tract, lungs, and kidneys for the fetus. **Placental exchange** poses a unique problem of blood flow distribution for the fetus, in that blood rich in oxygen and nutrients must be distributed to developing organs and waste products must be eliminated adequately.

Blood rich in oxygen and nutrients returns to the fetal body from the placenta in the umbilical veins (Fig. 21-16). Blood flows from the umbilical veins to the portal sinus to enter, in approximately equal portions, either the hepatic microvasculature or the **ductus venosus**, a conduit

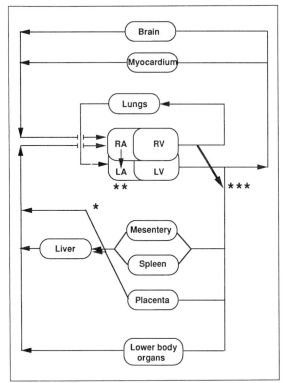

Fig. 21-16. Course of circulation in the fetus (***** = ductus venosus; ****** = foramen ovale; ******* = ductus arteriosus; *LA* = left atrium; *LV* = left ventricle; *RA* = right atrium; *RV* = right ventricle.)

vessel that connects the umbilical vein and inferior vena cava. Well-saturated blood derived from the umbilical vein joins poorly saturated blood derived from the venous drainage of lower-body organs in the inferior vena cava. These bloodstreams do not mix but streamline alongside each other, with the well-saturated stream medial to the poorly saturated stream. About half of the inferior vena caval return flows through the right atrium to the right ventricle. The remaining half, composed primarily of well-saturated blood, flows through the communication between the right and left atria, the **foramen ovale**, to the left ventricle. Thus, shunting of umbilical venous blood across the ductus venosus and foramen ovale delivers high concentrations of oxygen and nutrients primarily to upper-body

organs, which include the brain and heart of the fetus.

Venous drainage from upper-body organs returns to the heart in the superior vena cava. This blood, along with half of the return from the inferior vena cava, is ejected by the right ventricle. Because pulmonary vascular resistance is quite high in the fetus, most of the right ventricular output bypasses the pulmonary circulation and flows through the **ductus arteriosus**, a large communication between the main pulmonary trunk and descending aorta. A major portion of blood flow in the descending aorta is directed toward the placenta. In this way, venous drainage from upper- and lower-body organs is shunted toward the placenta, where wastes are eliminated and oxygen and nutrients are acquired.

There are important differences between the fetal and postnatal circulations. The **fetal circulation** is a parallel circuit and the **postnatal circulation** is a series circuit. In the fetus, left and right ventricular outputs are not equal; the right ventricular output exceeds the left ventricular output by a factor of two, and they perfuse different portions of the peripheral circulation. By virtue of the large vascular shunts in the fetal circulation, pressures in the left and right ventricles are roughly equivalent, as are pressures in the aorta and pulmonary artery. Right atrial pressure is greater than left atrial pressure, so blood flows through the foramen ovale from the right to left atrium.

Distribution of Blood Flow to Fetal Organs

The term used to refer to cardiac output in the fetus is **combined ventricular output**. It is the sum of the right and left ventricular outputs, and is used rather than *cardiac output* (equal to either the right or left ventricular output) because of the parallel arrangement of the fetal circulation. The proportions of combined ventricular output distributed to peripheral organs is different in the fetus than in the adult, because organs that are relatively inactive during fetal life (the kidneys, lungs, and gastrointestinal tract) receive a low proportion of total blood flow (Table 21-2).

Table 21-2. Percentage of Total Output of Blood from the Heart to Peripheral Organs

Organ	Fetus	Adult
Placenta	40	0
Lungs	5	100*
Brain	10	14
Heart	4	4
Skin, muscle, bone	33	26
Kidneys	3	25
Gastrointestinal tract	3	21
Other	2	10

*Remember that this is a series circuit.

The Transition at Birth

Immediately after birth, several events occur. Pulmonary ventilation and gas exchange is initiated and the placenta is removed from the circulation. The environmental temperature decreases as the newborn leaves the warm, humid amniotic cavity for a drier, cooler environment. These events produce marked cardiovascular changes. With the initiation of pulmonary ventilation, pulmonary vascular resistance decreases dramatically. Factors responsible for this include the establishment of an air–liquid interface that reduces surface tension and an increase in the oxygen content of the blood, which is a vasodilator in the pulmonary circulation. These factors appear to be mediated by local activation of arachidonic acid metabolism and increased prostacyclin production. When pulmonary vascular resistance decreases, pulmonary blood flow increases and pulmonary arterial pressure decreases. As a consequence, blood flow through the ductus arteriosus slows and pulmonary venous return to the heart accelerates. Left atrial pressure increases and the foramen ovale closes functionally soon after birth. These changes reverse the blood pressure gradient across the foramen ovale, so that left atrial pressure exceeds right atrial pressure. Elimination of the placental circulation increases the systemic vascular resistance. The decrease in pulmonary vascular resistance and increase in systemic vascular resistance causes a separation of blood pressures in the aorta and pulmonary artery. With these changes, the parallel fetal circulation is transformed into a series circulation in the newborn. The foramen ovale and ductus venosus close within a few days of birth, and the ductus arteriosus closes within seven to ten days.

At birth, oxygen consumption increases by a factor of two to three, because of the necessity of maintaining body temperature in a colder environment. In addition, heart rate and cardiac output increase. Normal thyroid function before birth and sympathetic stimulation at birth is necessary for these changes to occur.

NATIONAL BOARD TYPE QUESTIONS

Select the one most correct answer for each of the following

1. During exercise, blood flow to many tissues is reduced. Blood flow to the brain is unchanged, and blood flow to the heart and skeletal muscles is greatly increased. This is evidence for:
   A. A very selective distribution of sympathetic vasoconstrictor fibers
   B. The activity of the sympathetic vasodilator system
   C. Selective vasoconstriction in large distributing arteries
   D. The actions of the muscle pump to increase flow through muscle and brain
   E. Generalized vasoconstriction and local metabolic regulation of blood flow

2. An agent that increases vascular compliance will:
   A. Decrease the volume of blood in the arteries
   B. Decrease the volume of blood in the veins
   C. Increase arterial blood pressure
   D. Increase right atrial pressure
   E. Decrease venous return

3. Which is the highest value?
   A. Mean arterial pressure
   B. Systolic arterial pressure
   C. Diastolic arterial pressure

D. Pulse pressure

E. Mean circulatory pressure

Use the following choices to answer questions 4, 5, and 6.

A. Aorta

B. Arteries

C. Arterioles

D. Capillaries

E. Venules and veins

4. Which area of the circulation has the greatest cross-sectional area?

5. Which area of the circulation normally contains the most blood volume?

6. Which area of the circulation offers the greatest resistance to blood flow?

ANNOTATED ANSWERS

1. E. Exercise is a good example of how blood flow distribution to peripheral organs is a balance between the needs of the body as a whole and the needs of individual organs. During exercise, blood flow to working muscle (the heart and skeletal muscle) is increased by local metabolic regulatory factors. If the exercise is strenuous, blood flow to the skin also increases in direct proportion to the increase in heat production. Blood flow to other organs, with the exception of the brain, decreases due to sympathetic vasoconstrictor mechanisms.

2. E. An increase in vascular compliance means that the blood vessels will accomodate a larger volume of blood and that blood pressure will either not change or will decrease.

3. B. Arterial blood pressure normally oscillates between 120 mm Hg (systolic pressure) and 80 mm Hg (diastolic pressure). The mean arterial pressure in this case is a weighted average and is approximately 93 mm Hg. Pulse pressure is the difference between 80 and 120 mm Hg, or 40 mm Hg. The mean circulatory pressure is the equilibrium pressure of the circulation when blood flow is zero. Under normal circumstances, the mean circulatory pressure is 7 mm Hg.

4. D.

5. E.

6. C.

The aorta and large arteries are the conducting or distributing portion of the circulation. They are compliant, but not too compliant, and are responsible for converting intermittent blood flow ejected from the left ventricle to continuous blood flow to the peripheral organs. The arterioles are the major site of resistance in the circulation, and the greatest pressure decrease occurs there. The capillaries have the largest total cross-sectional area because of their enormous number. The venules and veins constitute the collection portion of the circulation. The highly compliant venules and veins serve important blood reservoirs.

BIBLIOGRAPHY

Rudolph, A. M., Iwamoto, H. S., and Teitel, D. F. Circulatory changes at birth. *J. Perinat. Med.* 16:9–21, 1988.

Shepherd, J. T., and Abboud, F. M. eds. *Handbook of Physiology, Section 2: The Cardiovascular System, Volume III: Peripheral Circulation and Organ Blood Flow,* Part 1. New York: Oxford University Press, 1984.

22 Microcirculation and Lymphatic Circulation

PATRICK D. HARRIS AND GARY L. ANDERSON

Objectives

After reading this chapter, you should be able to:

Describe vasomotion

List the factors important in determining diffusion

Explain the meaning of *reflection coefficient* and *permeability*

Explain the Starling hypothesis in terms of water movement from the capillaries

List the normal values for arteriole, capillary, and venule hydrostatic and colloid osmotic pressures

Explain how capillary pressure is regulated by arteriole and venule pressures and resistances

Describe lymphatic drainage

Explain the problems associated with tissue edema

Various organs and tissues have a microcirculation made up of small blood vessels (arterioles, capillaries, and venules) that can be seen clearly only under a microscope. Blood travels into a tissue through the muscular arterioles and then into the capillaries, which have a single layer of endothelial cells through which oxygen and nutrients pass to the nearby tissues. Arterioles, which range from 10 to 150 μm in diameter, regulate the distribution of blood flow to various groups of capillaries. Some of the smallest arterioles (called *metarterioles*) can serve as thoroughfare channels that bypass the capillary beds by shunting blood directly into the venules.

Because different sizes of arterioles can dilate or constrict somewhat independently of each other, flow patterns throughout the microcirculation can vary in speed and direction and from continuous to intermittent. There are cross connections between arterioles that permit flow to change directions, but, most often, flow through the arterioles is rapid, continuous, and in one direction. In contrast, capillary flow can be highly variable, with long periods (up to 30 minutes) when flow ceases in small groups of adjacent capillaries.

VASOMOTION

The smaller arterioles (less than 40 μm in diameter) exhibit **vasomotion,** which is characterized by periods of one to ten seconds when flow fluctuates because of rhythmic changes in the arteriole diameter. Arteriole diameter is determined by a balance between the contractile activity of the smooth muscle in the arteriole wall and the intraluminal pressure, which holds open (distends) the arteriole. Vasomotion can occur because of the **passive collapse** or **stretch** of the arteriole wall with a respective decrease or increase in intravascular pressure. More often, however, vasomotion is an **active response** brought about by contraction (increased tension) or relaxation (decreased tension) of the smooth muscle in the arteriole wall. At times, vasomotion can be very pronounced, with the arteriole relaxing completely and then closing completely during contraction. In resting skeletal muscle, vasomotion is common in small arterioles and occurs at a rate of 25 to 40 contractions per minute.

Blood flow from the capillaries (which are 0.5 to 1 mm long) collects in venules. The smaller venules (10 to 40 µm in diameter) have an endothelial cell layer surrounded by minimal adventitia and occasional contractile cells, called **pericytes**. These venules are involved in **transvascular exchange,** which is the transport of fluid and large molecules (macromolecules) across the vascular wall. During **inflammation,** large quantities of plasma proteins (in particular albumin) leave the bloodstream and enter the interstitial fluid through large gaps (up to 1 µm long and 0.4 µm wide) between endothelial cells in the venule wall. The larger venules and veins are principally collecting and storage vessels for the return of blood to the heart. The large venules have smooth muscle, and some (primarily in the skin) exhibit a regular rhythmic vasomotion of 12 to 20 contractions per minute.

MICROVASCULAR FUNCTION

Arterioles, capillaries, and venules are specialized components (called *effectors*) of the control systems that regulate several microvascular functions. These microvascular functions consist of two major processes: (1) **delivery** (called *nutrient flow*) of blood flow containing oxygen and other nutrients *to the capillaries;* and (2) **exchange** of fluid and solutes *across the capillaries* or venule walls.

THE EXCHANGE FUNCTION

The exchange of water and solutes between capillaries and the tissue cells is coordinated by several factors, one important one being the number of capillaries with flow (that is, the amount of capillary wall area available for exchange). More metabolically active tissues have a greater number of capillaries per gram of tissue, or **capillary density**. For example, heart, muscle, brain, and glandular tissues have many more capillaries per gram of tissue than do bone and cartilage. A large capillary wall area allows increased exchange between the capillary blood and the interstitial space. For example, approximately 300 ml of water diffuses every minute across the capillary walls in each 100 gm of skeletal muscle because skeletal muscle has a high capillary density.

A second important factor in transvascular exchange is the **lipid solubility** of the material to be exchanged. Molecules like water, sodium, chloride, and glucose readily diffuse through holes (pores) in the capillary wall, but these molecules do not diffuse as easily through endothelial cell membranes because they are not very soluble in lipid, which is a major component of the cell membrane. Lipid-soluble substances, especially dissolved gases such as oxygen and carbon dioxide, also diffuse through capillary pores, but there is a much higher exchange of these substances through the endothelial cell membranes because they are very soluble in the lipid of the cell membrane. Thus, oxygen and carbon dioxide diffuse very rapidly between the blood and the tissue interstitium, while other materials, such as water, move more slowly.

A third important factor in the transvascular exchange of material is the **free diffusion coefficient** for that material being exchanged between the blood and the interstitium. This coefficient is a measure of how fast material placed at the center of a container will disperse or move throughout water in the container. Material that consists of small molecules and molecules with very little net electric charge have very high free diffusion coefficients and disperse rapidly.

A fourth important factor in the transvascular exchange of material is the **relative concentrations of the material** in the blood and the tissue interstitium. A large difference in concentration between these two sites produces a higher rate of transvascular exchange.

All of the above factors combine to determine transvascular exchange by a process called **diffusion**. Figure 22-1 illustrates diffusion for a material (Q) that is moving from inside the vessel to the outside (the interstitial space). The rate of material movement (quantity per second) is represented by the expression dQ/dt. The amount of capillary wall area is defined as the product of circumference and length ($2\pi rl$) of the vessel. The difference between the concentration for Q inside the blood vessel (C_i) and for Q outside the vessel (C_o) represents the **driving force** (ΔC) for the movement of Q across the vessel wall. The ease with which Q can cross the vessel wall is called the **permeability** (P) of the vessel wall to Q. Permeability (expressed as centimeters per second) is determined by the lipid solubility and the free diffusion coefficient for Q. A high lipid solubility and a high free diffusion coefficient yield a high permeability. Overall, the rate of Q movement by diffusion is defined as:

$$\frac{dQ}{dt} = (2\pi rl)\,(P)\,(\Delta C)$$

Blood vessel permeability and thus the rate of diffusion is not the same for all blood vessels or for all tissues. For example, liver capillaries have a very high permeability even to large-molecular-weight solutes such as albumin (molecular weight, 69,000). Skeletal muscle, on the other hand, has capillaries that are not permeable to albumin. The brain has capillaries that restrict the diffusion of almost all solutes, and the transport of many substances across the endothelial cells of brain capillaries is accomplished only by other processes, called *pinocytosis* and *vesicular transport*. This special barrier property of the brain microcirculation is referred to as the **blood—brain barrier**.

Generally, the venous end of a capillary is more permeable to small lipid-insoluble solutes, primarily because the venous end appears to have more pores than the arteriole end. However, even the most liberal estimates indicate that pores occupy less than 1 percent of the total capillary surface area. Thus, lipid-insoluble materials diffuse slowly, even at the venous end of the capillary, and lipid-soluble substances, which can diffuse through the cell membrane as well as through pores in the capillary wall, move readily across the vessel wall at both the arterial and venous ends.

There are several very important features about the control of microcirculation that must be included in a consideration of the factors that determine the rate of transvascular diffusion for any material. Lipid-insoluble materials, such as glucose, albumin, and small charged solutes (for example, sodium and chloride), are generally stored outside of the bloodstream in many tissues. In contrast, lipid-soluble materials, such as oxygen and carbon dioxide, are essentially not stored in tissues. Thus, lipid-soluble materials can be depleted quickly when blood flow to tissues is severely reduced.

The arrival of lipid-soluble materials in the tissue interstitium is limited primarily by the amount of these materials that enter the capillary

Fig. 22-1. The important components in determining diffusion of substances across the vessel wall. (See text for further explanation.)

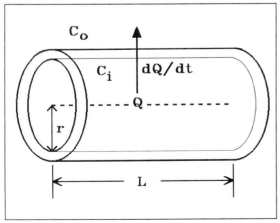

through blood flow, and not by diffusion of these materials across the blood vessel wall. This means that lipid-soluble materials are **flow-limited** in their entry and exit from the tissue interstitium, and the interstitial level of these materials is primarily regulated by the blood flow in the microvessels of these tissues.

The arrival of lipid-insoluble materials such as albumin and glucose in the tissue interstitium is primarily **limited by the diffusion** of these materials across the blood vessel wall, and not by the quantity that enters the capillary through blood flow. The interstitial level of these materials is primarily regulated through tissue control of the microvessel wall permeability to these materials. There is one major exception. Water is a lipid-insoluble material, but the interstitial water level is not regulated by tissue control of vessel permeability to water. The mechanism of water exchange is described in the next section.

TRANSVASCULAR EXCHANGE OF WATER

The net transfer of water from the capillaries to the interstitial space is only minimally influenced by diffusion, because the driving force for water is almost zero. Instead, the transvascular exchange of water between the blood and the interstitium occurs primarily through the **bulk flow of water**. The pumping action of the heart creates hydrostatic pressure in the blood vessels, and the difference between the hydrostatic pressure in the capillaries and that in the tissue interstitium forces water through pores in the capillary wall. The amount of bulk flow is determined by the magnitude of the hydrostatic pressure difference across the capillary wall and by the pore resistance to bulk flow.

In addition to the hydrostatic pressure, the presence of molecules other than water produce **osmotic** or **oncotic pressures** that influence the bulk flow of water. Plasma contains dissolved protein (mainly albumin) which, because of its large size (molecular weight, 69,000), cannot readily diffuse through the small capillary pores. The presence of protein in blood plasma produces a

capillary oncotic pressure, also called **plasma colloid osmotic pressure,** which keeps water from leaving the capillaries. In contrast, proteins in the tissue interstitium exert a **tissue colloid osmotic pressure** that draws water out of the capillaries.

The net water movement across the capillary wall is determined by the sum of these four separate forces; the hydrostatic forces within and without the capillary and the colloid osmotic forces within and without the capillary. The **net outward force** is a hydrostatic force. The **net inward force** is a colloid osmotic force. The combined effect of these forces on transvascular water movement is described by the following equation, which was developed by Starling in 1896:

$$\dot{Q}(H_2O) = CFC \, [(CHP - THP) - \sigma \, (COP - TOP)]$$

Where $\dot{Q}(H_2O)$ is the transvascular water flow, in milliliters per minute per 100 gm of tissue; CFC is the capillary filtration coefficient, in milliliters per minute per millimeter mercury per 100 gm of tissue; CHP is the capillary hydrostatic pressure, in millimeters mercury; THP is the tissue (interstitial) hydrostatic pressure, in millimeters mercury; σ is the reflection coefficient (no unit of measure) for the movement of proteins (colloids) across the capillary wall; COP is the capillary (colloid) osmotic pressure, in millimeters mercury; and TOP is the tissue (colloid) osmotic pressure, in millimeters mercury.

The **capillary filtration coefficient** is represented by the product of surface area and permeability, because it is determined by the length and diameter of the capillaries (surface area) and the number and size of the pores through which water can exit the blood vessel (water permeability). This coefficient is sometimes expressed as a **hydraulic conductivity factor**.

The **protein reflection coefficient** is the inverse of the protein permeability of the vessel wall. A molecule that "reflects" from the capillary wall and does not cross the capillary wall into the interstitium has zero permeability and a maximum reflection coefficient of 1. A molecule that does not reflect from the capillary wall but easily passes through the capillary wall into the interstitium has a high permeability and a reflection coefficient of zero. The normal reflection coefficient for plasma proteins at the capillary wall is almost 1.0, indicating low permeability and a high reflection coefficient, as these plasma proteins are almost totally reflected at the capillary wall.

The **plasma colloid osmotic pressure** is about 28 mm Hg. Approximately 19 mm Hg of this is provided by the plasma proteins. Because the plasma proteins are negatively charged, they exert an electric force on sodium and other positively charged ions. Thus, these positive ions are somewhat retained in the vascular space and add 9 mm Hg to the plasma protein effect on osmotic pressure. Some protein does leak into the interstitial space, primarily through the venules. With time, this protein is returned to the circulation by the lymphatic system. This protein leakage means that the interstitial space contains a small amount of soluble protein that exerts approximately 4.5 mm Hg of **tissue colloid osmotic pressure**.

Capillary hydrostatic pressure varies from tissue to tissue; it is particularly high in kidney glomerular capillaries (45 mm Hg) and low in intestinal (10 mm Hg) and lung (8 mm Hg) capillaries. In skeletal muscle, capillary pressure is nearly 40 mm Hg at the arterial end and 12 mm Hg at the venous end. The interstitial fluid space has a hydrostatic pressure that may be slightly positive or even slightly negative, depending on factors such as tissue structure, muscle activity, and interstitial fluid volume. Over time, the tissue hydrostatic pressure is normally small and assumed to be approximately zero.

The combined hydrostatic and oncotic pressures determine a net force for water movement across the capillary wall. If the net force is positive, water moves out of the capillary blood, called **filtration**. If the net force is negative, water moves into the capillary blood, called **reabsorption**. Filtration actually occurs near the arterial end of the capillary and reabsorption near the venule end, because the balance of hydrostatic and osmotic forces is different at various locations along the capillary.

As shown in Figure 22-2, the **net hydrostatic pressure** (the difference between the capillary and tissue hydrostatic pressure, shown as a *solid line*) decreases in a linear fashion from 40 to 12 mm Hg along the length of capillary. In contrast, the **net colloid osmotic pressure** (the difference between the capillary and tissue osmotic pressures, shown as a *dashed line*) changes in a curvilinear fashion along the length of the capillary. At the arterial end of the capillary, where the net hydrostatic pressure exceeds the net oncotic pressure, net filtration occurs and water leaves the capillary. This increases the concentration of protein within the capillary, as the capillary wall is not very permeable to protein. The higher protein concentration near the midpoint along the capillary length produces a higher plasma colloid osmotic pressure at that point. At the venule end of the capillary, where the net hydrostatic pressure is less than the net oncotic pressure, there is net reabsorption of water and water enters the capillary to restore the colloid osmotic pressure to normal.

In skeletal muscle, there is slightly more filtration than reabsorption throughout the whole capillary, with the excess filtration becoming lymph drainage from that tissue area. However, glomerular capillaries in the kidney have filtration along most of their length and the capillaries in the intestine and lung have reabsorption along most of their length. Thus, capillaries with high hydrostatic pressures (such as those in the

Fig. 22-2. The distribution of pressure and the sites of net filtration and net reabsorption of fluids within capillaries.

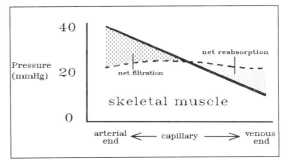

kidney) are primarily filtering capillaries, and capillaries with low hydrostatic pressures (such as those in the intestine and lungs) are primarily reabsorptive capillaries. Despite these differences, the primary physiologic mechanism underlying the transcapillary exchange of water in all capillaries is the **capillary hydrostatic pressure**.

DETERMINANTS OF CAPILLARY PRESSURE

Capillary hydrostatic pressure is controlled by the arterioles, venules, and the heart, according to the following formula:

$$CHP = \frac{R_V}{R_A} P_A + P_V$$

Where CHP is the pressure at the midpoint in the capillary microvasculature; R_V is the resistance of the venule microvasculature; R_A is the resistance of the arteriole microvasculature; P_A is the pressure at the beginning of the arteriole microvasculature (similar to mean arterial pressure); P_V is the pressure at the end of the venule microvasculature (similar to central venous pressure); and F is the flow through the microvasculature. This is diagrammed in Figure 22-3.

The capillary pressure will increase whenever arterial pressure increases (as in hypertension), venous pressure increases (as in congestive right-heart failure), venule resistance to flow increases

Fig. 22-3. The important factors in determining capillary pressure. (See text for further explanation.)

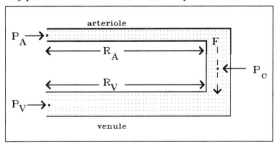

(as in obstructive venous clots), or arteriole resistance to flow decreases (as in arteriole vasodilation).

The most important point is that capillary pressure is much more sensitive to changes in venous pressure than to changes in arterial pressure. The ratio of venule to arteriole resistance (R_V/R_A) is normally about 0.1. Thus, arterial pressure (P_A) must increase 10 mm Hg to cause a 1-mm Hg increase in capillary hydrostatic pressure. In contrast, a 1-mm Hg increase in venous pressure will cause about a 1-mm Hg increase in capillary hydrostatic pressure. This means that capillary hydrostatic pressure is somewhat protected from changes in arterial pressure, but is highly influenced by even small changes in central venous pressure.

LYMPHATIC DRAINAGE

Capillary filtration predominates slightly over capillary reabsorption and this produces **tissue lymph flow**. About 90 percent of the net filtration is reabsorbed directly into the blood of capillaries or postcapillary venules. The lymphatics drain the remaining 10 percent of the filtered fluid. Of course, changes in capillary pressure vary this balance between filtration and reabsorption and thus alter lymph flow. When capillary pressure decreases significantly, as in hemorrhagic shock, reabsorption can occur along the entire capillary to stop lymph flow in any tissue. Some tissues, such as skin, have a specialized vascular smooth muscle (called *precapillary sphincters*) at the entrance to the capillary bed. When the precapillary sphincters in these tissues close completely, capillary pressure decreases to the venous pressure level, which triggers reabsorption along the entire capillary.

The **lymphatics** form a closed, separate fluid-flow system composed of endothelium-lined vessels that closely parallel the larger arterioles and venules. Lymph capillaries originate as blind-ended sacs close to blood capillaries and are anchored by fine filaments to the surrounding tissue. The lymphatic sacs have one-way valves that open when tissue pressure exceeds lymph pressure. Lymph vessels contract to propel the lymph fluid into the larger collecting lymph vessels which eventually connect to the thoracic duct, where the lymph fluid empties into the venous circulation. The total volume of lymph fluid returned to the circulation is approximately three to four liters per day. This lymph circulation system is especially important for transporting fluid containing chylomicrons absorbed from the intestine and plasma proteins that leak from the microcirculation.

EDEMA

Edema is the accumulation of excess fluid in the interstitial or extravascular space. Edema is clinically detectable only when considerable excess fluid accumulates in the soft tissue spaces. One cause of edema is **reduced concentrations of plasma protein,** which reduce colloid osmotic pressure; this increases net filtration at the arteriole end of the capillary and reduces net reabsorption of filtered fluid at the venule end. This oncotic pressure problem can result from liver disease, as the liver normally synthesizes many of the plasma proteins. Another cause of edema is reduced outflow of lymph brought about by blockage of lymph vessels or by increased pressure within the vena cava, which prevents lymph fluid from leaving the thoracic lymph duct.

Histamine release during inflammation can substantially dilate arterioles in order to increase capillary hydrostatic pressure and net filtration. Histamine release also decreases the reflection coefficient for plasma proteins so that protein leak from the vascular space into the interstitium is increased. This leakage happens in the small postcapillary venules (not in capillaries) where histamine causes large gaps (up to 1.0 μm in length and 0.4 μm in width) to form between venule endothelial cells. Plasma proteins and other large molecules can easily escape through these large gaps into the interstitium, and this reduces the protein reflection coefficient and increases filtration. Pronounced edema eventuates when net filtration increases such that the amount of fluid entering the interstitium exceeds the lymphatic drainage capacity.

Bacterial toxins and **burns** also cause increases in capillary permeability, which result in edema. Edema is particularly serious in the lungs because it widens the distance between the lung capillaries and the alveolar air sacks, and this increases the diffusion distance for the movement of oxygen from the lung air into the red blood cells. In this situation of **pulmonary edema,** there is not enough time for oxygen to travel from the alveolus to the capillary as the individual red blood cells pass through the lung microcirculation. Thus, the blood leaving the lungs is only partially oxygenated.

SUMMARY

The microcirculation consists of small arterioles, capillaries, and venules that serve as the exchange sites for water, nutrients, and the waste products from cell metabolism. Arterioles help control the amount of blood flow to the capillaries by dilating or contracting in response to signals from the central nervous system or from certain endocrine organs. Arterioles also exhibit vasomotion (rhythmic contractions) that helps regulate capillary blood flow. The exchange of solutes between blood and interstitial fluid occurs when materials move across the vessel wall by diffusion, which partly depends on the permeability of the capillary and venule walls. Small lipid-soluble substances (like oxygen) diffuse rapidly, while large lipid-insoluble substances (like albumin) diffuse very slowly. The balance between hydrostatic and colloid osmotic pressures across the vessel wall regulate the exchange of water. High-pressure capillaries (such as glomerular capillaries in the kidney) filter fluid out of the capillary and low-pressure capillaries (such as intestinal villus capillaries) reabsorb fluid back into the bloodstream. Edema arises when the net filtration of fluid into the interstitium exceeds the capacity of the lymph system to return fluid to the circulation. Agents such as histamine, which increase venule wall permeability to plasma proteins (such as albumin), promote net fluid filtration and often result in edema.

NATIONAL BOARD TYPE QUESTIONS

Select the one most correct answer for each of the following

1. Edema will most likely result from each of the following conditions *except*:
   A. High venous pressure
   B. Lymphatic obstruction
   C. Low plasma protein concentration
   D. Increased mean arterial pressure
   E. Increased capillary hydrostatic pressure
2. Capillary filtration will be increased by each of the following *except*:
   A. Increased plasma protein concentration
   B. Increased venous hydrostatic pressure
   C. Increased arterial hydrostatic pressure
   D. Increased periods of precapillary sphincter dilation
   E. Increased arteriole diameters
3. If the mean capillary hydrostatic pressure is 35 mm Hg, the plasma protein osmotic pressure is 24 mm Hg, the tissue hydrostatic pressure is 2 mm Hg, the tissue oncotic pressure is 3 mm Hg, and the plasma protein reflection coefficient is 1, there will be a net force for the production of interstitial fluid of:
   A. 16 mm Hg
   B. 12 mm Hg
   C. 10 mm Hg
   D. 6 mm Hg
   E. None of the above
4. Capillaries and small venules are the site of solute exchange with interstitial spaces because:
   A. Their radius is small
   B. The hydrostatic pressure difference

across the wall of these vessels exceeds that of other systemic vessels
   C. The velocity of flow is lowest in these vessels
   D. Flow to capillaries is regulated by arteriole diameters
   E. None of the above
5. Which of the following is the major route by which oxygen diffuses from blood to the interstitium?
   A. Transcellular (across endothelial cell walls)
   B. Paracellular (between endothelial cell junctions)
   C. Pores within endothelial cells
   D. Carried by vesicles
   E. None of the above

ANNOTATED ANSWERS

1. D. Increases in mean arterial pressure have little involvement in increasing capillary pressure. Only about one-tenth of the arterial pressure increase is transmitted into the capillary. Thus, moderate increases in central venous pressure can readily cause systemic edema, but moderate increases in arterial pressure do not. Increases in capillary pressure are caused mainly by increases in venous pressure.
2. A. Increasing the protein concentration in blood increases the plasma colloid osmotic pressure, and this exerts a force that draws fluid from the interstitial space into the capillary blood to cause reabsorption of the tissue fluid.
3. B. The hydrostatic filtration pressure is 33 mm Hg (35 − 2 mm Hg) and the osmotic reabsorptive pressure is 21 mm Hg (24 − 3 mm Hg). This yields a *net* filtration pressure of 12 mm Hg (33 − 21 mm Hg).
4. E. Solute exchange occurs in the capillaries and venules because of the large surface area and thin walls of these vessels. These vessels

23 Regulation of the Cardiovascular System

ALVIN S. BLAUSTEIN AND RICHARD A. WALSH

Objectives

After reading this chapter, you should be able to:

Outline the general organization of the systems that control circulation and explain the importance of these systems

Explain the basic properties of control systems, principally feedback and gain

Identify the major sensor, integrating, and effector components involved in reflex control

Explain the regulatory mechanisms involved in postural changes and Valsalva's maneuver

Explain and give examples of the autocrine, paracrine, and endocrine control of arterial pressure and volume

Describe the mechanisms responsible for the distribution of cardac output during exercise

The major function of the circulatory system is to ensure that all cells receive the substrates required for nutrition, respiration, growth, repair, and excitation, and that potentially toxic metabolic wastes are removed. The circulation also integrates organ function by transporting regulatory substances, and assists in defending the organism against environmental challenges by serving as a reservoir and vehicle for the immune and coagulation systems. To meet these requirements, the circulation must be able to respond to the local needs of individual tissues, to the more generalized demands of exercise and reproduction, and to physical factors such as injury, temperature changes, or gravity. Integrated control of the circulation sustains constant tissue perfusion of vital organs such as the heart and

brain in response to acute, potentially adverse shifts in blood volume. Prolonged changes in blood volume and cardiac or vascular growth and development are the result of chronic adaptive mechanisms.

Effective control of cardiac output and arterial pressure requires coordinated interaction between the vasculature and heart. This is accomplished by three major regulatory systems—the nervous system, a system of peptides that affects vascular tone and is important in volume regulation, and local factors in the heart, endothelium, and organ circulations (Table 23-1). These systems adjust the volume and resistance in various circulatory compartments, and modify the force and rate of cardiac contraction and relaxation.

Table 23-1. Elements of Circulatory Control

Sensors and Integrators	Effectors
Detectors	Vascular smooth muscle
Baroreceptors	Arteries
Chemoreceptors	Arterioles
Cardiac receptors	Venules
Osmoreceptors	Veins
Ergoreceptors	
Thermoreceptors	Microcirculation
	Endothelial junctions
Endothelium	
Relaxing factor(s)	Volume regulators
Endothelins	Kidney
Prostacyclin	Hypothalamus
Prostaglandins	Neurohypophysis
	Adrenal cortex
Vasoactive peptides	
Angiotensins	Specialized tissues
Atrial natriuretic	Adrenal medulla
peptide	
Vasopressin	Heart
Vasoactive intestinal	Siroatrial or atrioven-
peptide	tricular nodes
Opioids	His-Purkinje system
Substance P	Atrial muscle
	Ventricular muscle
Catecholamines	
Norepinephrine	
Epinephrine	
Dopamine	
Specialized substances	
Histamine	
Serotonin	
Kinins	
Autonomic nervous	
system	
Cortical centers	
Vasomotor areas	
Hypothalamus	
Neurohypophysis	
Sympathetic ganglia	

BASIC FEATURES OF CIRCULATORY CONTROL

Blood flow is distributed to many organs with individual requirements that may vary by as much as 25 times over the basal demand. Flow is regulated by local factors that match need to perfusion. These local perfusions form the major determinant of cardiac output through their combined effect on venous return. This local control usually involves physical factors, metabolic products, and peptides that regulate function through either **autocrine** or **paracrine mechanisms**. Autocrine control occurs when a substance elaborated by a cell regulates the function of that cell; paracrine control results when such substances regulate the function of neighboring cells. To maintain **arterial pressure** in the face of diverse organ requirements that redistribute cardiac output and blood volume, the autonomic nervous and endocrine systems communicate with target organs systemically rather than locally. **Neural components** consist of widely distributed receptors and neuroeffector junctions in the heart and blood vessels, and integrating areas in the brain. **Peptide** and **steroid hormones** regulate vascular tone and intravascular volume. Neuroendocrine interactions are responsible for much of the cardiovascular endocrine regulation, which involves effects distant from their site of hormone elaboration.

The primary goal of circulatory control is to maintain an arterial pressure sufficient to provide the energy for perfusion without damaging vital organs or vessels. Regulation of pressure or volume involves a **sensor,** the control system or integrator, and **effector organs** that form a *loop* to maintain the variable within physiologic limits (Fig. 23-1A). Changes in arterial pressure or volume represent **inputs,** or perturbations, to the control system and initiate **outputs,** or corrections, which restore pressure or volume to normal. When the control system suppresses the perturbation, it exhibits **negative feedback,** and this is characteristic of most control systems in the body. Thus, when systemic pressure or volume falls, adjustments increase resistance and cardiac function, and mobilize or retain fluid. The relationship between the input and output characterizes the ability of the control system to maintain homeostasis, and is called the **gain**. The gain is defined as the ratio of the **correction** (the degree to which adjustments compensate) to the **error** (the deviation from homeostasis remaining after adjustment). For example, if arte-

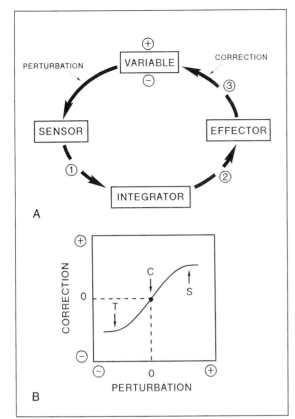

Fig. 23-1. Characteristics of a negative feedback system. (A) The basic elements of a control system (loop) regulating a variable are the *sensor*, which detects a perturbation, the *integrator*, which interprets the sensor's signal and "instructs" an *effector*, which in turn corrects the variable. Site *1* represents a potential opening in the loop commonly used to study integrating systems by controlling the signal to the integrator. A second system may also contribute a signal to the same integrator, modifying the response. Site *2* is used to isolate the effector and study the range and intensity of its responses to putative or known mediators. Here responses may be elicited in other systems as well. At Site *3*, the specific nature and characteristics of the effector's output can be observed, leading, for example, to the discovery of changes in effector sensitivity. The effector's output may also perturb another system, as when sympathetic renal vasoconstriction reduces renal blood flow and initiates alterations in volume regulation. (B) The behavior of a control system can be expressed as the relationship between the perturbation (i.e., the degree to which the controlled variable is disturbed) and the correction. Most often this relation-

rial pressure falls acutely from a mean of 80 mm Hg to a mean of 50 mm Hg, and subsequent compensation returns the pressure to 70 mm Hg, the gain would be the ratio of 20 (the correction of 50 to 70) to 10 (the difference between 70 and 80).

For most control systems, the relationship between the magnitudes of the output, or response, and the input is neither constant nor linear. Instead, at one extreme, the system may be relatively insensitive to the input, which must achieve a **threshold** before an output is elicited; at the other extreme, there is a **saturation zone,** whereby the output is maximum regardless of the perturbation. Between these, the system exhibits a steep relationship between input and output, such that any displacement from equilibrium quickly activates a brisk corrective response (see Fig. 23-1B). For the regulation of both arterial pressure and volume, either can be affected by a number of inputs that open the simple loop between the sensor and the effector. In addition, the pressure and volume loops interact such that changes in one of the variables may initiate a correction in the other system (Fig. 23-1A; sites *2* and *3*). These systems function as open-loop, negative-feedback systems. In experiments, a loop is deliberately opened, usually at the sensor (Fig. 23-1A, site *1*), so that the investigator can control the perturbations detected by the receptor and determine the response of the system.

In almost all cases, **positive feedback** in biologic systems is maladaptive and results when control systems are inadequate or ill-suited to overcome perturbations. In this event, the

ship is sigmoid. The *threshold* (*T*) is the smallest disturbance that will initiate any correction; the *equilibrium point* (*C*) is the usual operating point of the system, and need not be in the middle as it is on this diagram. However, it generally lies along the steepest portion of the response curve. The *saturation plateau* (*S*) represents the maximum output of the control system, regardless of the magnitude of the perturbation. In a system with negative feedback, if the perturbation shifts the equilibrium point upward (+), a negative correction (−) will move the system back toward C.

original disturbance is not corrected but accentuated. For example, in chronic congestive heart failure there is increased intravascular volume and frequently ventricular dysfunction. The sympathetic division of the autonomic nervous system is activated and stimulates the heart to contract more frequently and forcefully. Resistance vessels also constrict, reducing renal blood flow and initiating mechanisms responsible for salt and water retention. The result is further circulatory overload and eventually edema formation. However, interactions between positive and negative feedback loops appear to be important in normal short-term cardiovascular adaptation.

OVERVIEW OF CIRCULATORY FUNCTION

For the most part, the circulation constitutes a **closed system** of blood vessels whose pressure and volume are tightly regulated. The most important elements of this control system are summarized in Figure 23-2. The fluid in the vascular system is in **dynamic equilibrium** with interstitial fluid bathing the cells, which leaves and enters the circulation through the capillaries and lymphatics. The heart connects the venous and arterial sides of the circulation and imparts the mechanical energy required for all fluid transfers. The amount of blood leaving the heart following each contraction is called the **stroke volume** and the amount flowing through the arterial tree per unit of time is the **cardiac output**.

The cardiac output is delivered to the peripheral tissues by the **large arteries** such as the aorta. These conductance vessels have relatively little smooth muscle in their walls and are not significantly affected by vascular control. They do, however, contain mechanoreceptors (the carotid sinus and aortic arch baroreceptors) that initiate circulatory reflexes important in controlling systemic arterial pressure. The elastic tissue of the aorta and its branches convert pulsatile cardiac flow into a continuous, steady-state flow optimal for perfusion of the smaller arteries and arterioles. These smaller vessels are surrounded

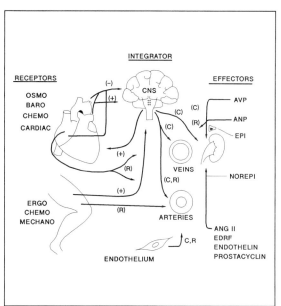

Fig. 23-2. Elements of circulatory control for systemic arterial pressure and volume. On the left are the *receptors,* which sense a variety of variables to which there are circulatory responses. Generally, their outputs are proportional to the perturbation, though for each variable there are some receptor types that are activated only at physiologic extremes. Much of the *integration* and *coordination* takes place in the central nervous system (*CNS*), where neural and hormonal elements are in direct contact. The response is determined by a balance of *effectors,* listed on the right, which ultimately depend on chemical mediators acting at target organs. These target organs include circulatory elements such as the heart and blood vessels, as well as volume regulators such as the kidney and neurohypophysis. (*AVP* = arginine vasopressin; *ANP* = atrial natriuretic peptide; *EPI* = epinephrine; *NE* = norepinephrine; *Ang II* = angiotensin-II; *EDRF* = endothelium-derived relaxing factor; (−) = inhibitory; (+) = excitatory; *C* = constrict; *R* = relax)

by layers of smooth muscle cells in direct contact with endothelium on the luminal side, and are richly innervated on the adventitial side. Both the endothelium and neural connections provide an important regulatory input that determines the tension in the smooth muscle encircling the vascular lumen. This, in turn, changes the cross-sectional area of these vessels. The **effective** cross-sectional area in the muscular arteries, arterioles, and venules is the most important determinant of steady-state peripheral resistance. By the time blood reaches the microcirculation, the loss of energy across the arterial resistance has caused the mean pressure to decline from between 80 and 100 mm Hg to approximately 20 mm Hg near the arteriolar end of the capillaries, and to between 5 and 10 mm Hg in the small veins distal to the venules. The pressure drop that occurs from artery to vein is the most important source of energy for venous return. It is likely that some **tissue metabolites** like adenosine or locally synthesized **vasoactive substances** like prostaglandins or kinins exert their effects near the arteriolar and venulocapillary junctions. In this way, tissue factors can "feed back" to regulate local flow.

In contrast to arteries, **veins** are highly distensible and, together with the venules and venous sinuses, contain 60 to 65 percent of the blood volume, one-third of which is in the splanchnic veins. By regulating the functional cross-sectional area of the venous compartment, the body can control the amount of blood available for translocation from the venous to the arterial side of the circulation (the cardiac output). An increase in the venomotor tone decreases venous capacitance and redistributes blood volume, thus increasing cardiac output; a decrease in venomotor tone has the exact opposite effect. The large veins pass between major body compartments and are affected by local external pressures, such as normal positive pressures exerted by exercising skeletal muscles and the normally negative pressures in the thoracic cavity. Because venous pressures are relatively low and capacitance (volume–pressure relationship) is large, such external forces may variably facilitate or inhibit venous return, with only small changes in venous pressure adding dynamic modifiers to the more static behavior of the vascular system.

Ultimately, the amount of venous return is largely influenced by the **pressure difference** between the venules and right atrium. Because dynamic factors are superimposed on more static behavior, the heart must be able to adjust to momentary changes in venous return arising during altered respiration or posture, or the more sustained increases produced by exercise. Intrinsic cardiac mechanisms, such as the **length–tension relationship** (Chapter 19), allow the heart to vary the **stroke volume** with each beat; **neurohormonal influences**, including catecholamines, mediate changes in the cardiac rate and force of contraction that can be sustained somewhat longer.

NEURAL INFLUENCES ON CIRCULATORY CONTROL

Overview of the Autonomic Nervous System

The most pervasive and well-understood integrating regulatory system controlling cardiac output and arterial pressure is the **autonomic nervous system**. This system directly influences vasomotor tone and cardiac function (heart rate and contractility) through its two major divisions, sympathetic and parasympathetic, which act on specific adrenergic and cholinergic receptors, respectively. It also affects systemic volume and peripheral resistance by modulating the release of certain **peptide hormones**. **Neural control** involves the assimilation of inputs from the cerebral cortex (motor, visual, labyrinthine, olfactory) and specialized sensors (mechanoreceptors, chemoreceptors, osmoreceptors, and thermoreceptors) and their integration into several regions (hypothalamus, pons, and medullary), to conduct efferent nerve impulses to the periphery over the sympathetic and parasympathetic pathways. The dynamic balance between these two systems determines the **net response**.

The central **sympathetic outflow** converges on preganglionic neurons located primarily in the lateral horn or intermediolateral cell column of the spinal cord. The axons of these neurons are small and myelinated, exiting the spinal cord through ventral roots. These axons often travel and ramify within the chain of paravertebral ganglia before synapsing with the postganglionic neurons in one or more sympathetic ganglia.

There are three large ganglia at the cervical level: superior, middle, and inferior. The inferior ganglion often merges with the first thoracic ganglion to form the stellate ganglion, which innervates the heart. Otherwise, the paravertebral ganglia are segmental from the thoracic through lumbar segments. Most often, the postganglionic fibers exit to join visceral or spinal nerves via unmyelinated rami. Occasionally, preganglionic axons continue peripherally and synapse either in prevertebral ganglia (celiac or mesenteric) or, in a unique case, with the chromaffin cells of the adrenal medulla. In general, preganglionic synapses use the neurotransmitter **acetylcholine** (ACh) while postganglionic synapses use **norepinephrine** (NE).

Parasympathetic outflow commences in the motor nuclei of cranial nerves III, VII, IX, and X, or from sacral neurons in the spinal gray matter. The cranial preganglionic neurons are myelinated and synapse with postganglionic fibers on the target organ; those from the sacral segments emerge by means of ventral roots traveling in the pelvic nerves. Neurotransmission at all synapses in the parasympathetic division depends on **acetylcholine**.

Central integrating components of the autonomic nervous system are organized longitudinally, so that axons from both the central and peripheral sites converge on the sympathetic and parasympathetic neurons throughout the hypothalamus, pons, and medulla. All these regions contain synapses important to cardiovascular autonomic regulation. It is not possible to organize these into discrete anatomic–functional centers because of the complexity of their interconnections; however, there are some general descriptions that apply to their organization (Fig. 23-3).

The **hypothalamus** is located on the floor of the third cerebral ventricle and connects the endocrine functions of the pituitary, the vasomotor areas in the pons and medulla, and the higher cortical centers. This organization facilitates patterned cardiovascular responses to behavioral states, such as aggression and sex, and provides one of the places where the cardiovascular and volume control systems communicate. The hypothalamus can modify thirst and modulate serum osmolarity by controlling the secretion of **arginine vasopressin** (a hormone that modifies the renal control of urine osmolarity) from the posterior pituitary. Cardiovascular responses to thermal stress are also likely to originate in the hypothalamus.

The overall organization of the **vasomotor areas** is complex, but there appear to be three functionally overlapping anatomic zones that interact extensively (see Fig. 23-3). These include a **vasoconstrictor** area in the upper anterolateral medulla; the **vasodilator** area in the lower anterolateral medulla; and the **sensory** area which integrates the vasoconstrictor and vasodilator areas and is located bilaterally in the nucleus tractus solitarii of the posterolateral medulla and lower pons. Regions that seem more devoted to modifying heart rate than vasomotor tone are located in the thalamus, posterior and posterolateral regions of the hypothalamus, and dorsal region of the medulla (the traditional cardioaccelerator region). Accelerator regions are located primarily on the right sides of both the hypothalamus and medulla.

The heart and blood vessels are under continuous stimulation by sympathetic and parasympathetic impulses. Changes reflect an altered balance between these two divisions, which generally behave in a roughly reciprocal fashion. This reciprocal behavior represents interaction among the higher centers. The effects of the two divisions are further modified at the effector organ by the distribution and density of the innervation.

Effects of Sympathetic Neural Control

Input resulting in the stimulation or withdrawal of the sympathetic nervous system represents the most potent control of the **peripheral circulation**. Broadly categorized, fibers travel either in specific sympathetic nerves (those innervating the viscera and heart) or join the paravertebral sympathetic chain, synapsing in various secondary ganglia which give rise to spinal nerves (in-

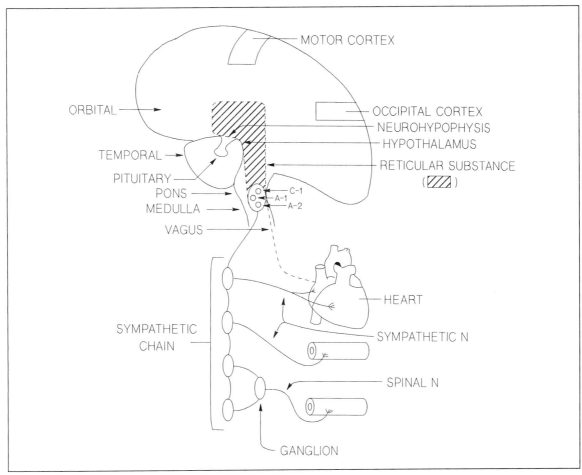

Fig. 23-3. Relationship between vasomotor centers and major target organs. Input from the cortex and peripheral or central sensors converges on the *hypothalamus* and *vasomotor areas* in the pons and medulla, where preganglionic autonomic neurons are concentrated. This figure focuses on inputs arising from other cortical centers, particularly behavioral (*temporal lobe*), motor (*motor cortex*), and visual (*occipital cortex*). There are some connections from these centers to the hypothalamus, which also has osmoreceptors and thermoreceptors. The hypothalamus has outputs to the *pituitary*, as well as to the pons and medulla. Control over the circulation by the vasomotor centers is largely the result of balancing the sympathetic (*sympathetic chain and spinal sympathetic nerves*) and parasympathetic (*vagus*) neural outflow to the heart, resistance vessels, and kidney, where renal renin release is modulated. (*C-1* = vasoconstrictor area; *A-1* = vasodilator area in the lower anterolateral medulla; *A-2* = sensory area which integrates C-1 and A-1.)

nervating peripheral vessels) (see Fig. 23-3). The **vascular nerves** terminate on small arteries, arterioles, venules, and veins, allowing neural modulation of resistance and vascular volume. **Cardiac nerves**, many of which descend from the

stellate ganglia, innervate both the atria and ventricles.

Most often, reflex sympathetic stimulation causes vasoconstriction by releasing NE from sympathetic nerve endings, and is an example of

paracrine function. If the sympathetic nerve to a limb is stimulated, local vascular resistance increases, blood flow shows a corresponding decrease, capillary pressure falls (absorbing local interstitial fluid), and blood volume is displaced from the limb. NE exerts its effect by occupying specific (alpha) receptors on vascular smooth muscle and acts through a second messenger, cyclic adenosine $3':5'$ monophosphate (cAMP). In a metabolically active organ, local influences may override the autonomic ones. The inhibition of sympathetic outflow allows vessels to dilate and respond to local humoral and myogenic stimuli.

Although most reflex sympathetic stimulation produces vascular constriction, a subset of fibers release ACh rather than NE. These fibers arise in the motor cortex of the cerebrum, then pass through the hypothalamus and medulla before joining other sympathetic nerves. For the most part, these neurons innervate the vasculature of skeletal muscle and provoke an **anticipatory increase** in local blood flow prior to exercise. Another secondary outcome of the sympathetic stimulation of these fibers is the release of epinephrine from the adrenal medullae. Unlike NE, epinephrine stimulates both alpha and beta receptors. In low quantities, epinephrine causes vasodilation and cardiac stimulation; at higher concentrations, vasoconstriction predominates.

Direct cardiac effects of reflex sympathetic stimulation include increased heart rate, more forceful and rapid cardiac ejection, and faster relaxation and filling. Its effect on the pacemakers and conduction tissue initiate the **increased heart rate**. Beta-adrenergic stimulation of the sinoatrial node shifts the pacemaker activity to the pacemaker cells, with faster intrinsic rates of depolarization. It causes hyperpolarization of the membrane and faster spontaneous depolarization, probably secondary to increased activity of the sarcolemmal Na^+-K^+ pump and a larger inward K^+ current. The long-lasting **inward Ca^{2+} current** is also greater. The result is an action potential with increased amplitude, faster upstroke, and shorter duration. **Greater contractility** is the consequence of higher cytosolic calcium concentration (enhanced release from

the sarcoplasmic reticulum) and enhanced calcium responsiveness of the contractile proteins. Finally, **faster relaxation** is due to enhanced phosphorylation of both troponin I and phosphalamban. Troponin I abruptly terminates contraction by allowing the thin filaments to release bound calcium, and phosphalamban permits the more rapid transport of calcium to the sarcoplasmic reticulum.

Reflex sympathetic stimulation is important in the increase in cardiac output necessary during exercise or other forms of stress. These cardiac stimulating effects of the sympathetic nervous system increase the metabolic requirements of heart muscle, and this forms the basis for using graded exercise stress to evaluate cardiac reserve in disease states. As recently observed in experimental heart failure models, the failure of cardiac contractility to increase may occur because the beta receptor is uncoupled from its guanine nucleotide–binding regulatory proteins (G proteins) causing a blunted increase in intracellular adenylate cyclase activity following sympathetic stimulation. Sympathetically mediated actions are largely responsible for maintaining systemic arterial pressure and vital organ perfusion during hypovolemic states and cardiac dysfunction.

Effects of Parasympathetic Neural Control

The parasympathetic nervous system has a **cranial division,** supplying the blood vessels of the head and viscera, and a **sacral division** with fibers innervating the vessels of the genitalia, bladder, and large intestine. Because these fibers supply only a small percentage of the resistance vessels, ordinarily the parasympathetic component of the autonomic nervous system has little involvement in the regulation of arterial pressure. It does, however, play an important role in modulating the cardiac rate. Fibers traveling in the **vagus nerve** innervate the sinoatrial and atrioventricular nodes as well as the atrial musculature. Changes in heart rate arise from the shift to P cells, with slower intrinsic rates of depolarization and changes in membrane depolarization secondary to ACh stimulation. The ACh-activated K^+ channels are opened, the outward K^+ current increases, and the pacemaker current is inhibited. ACh is coupled to the operation of channels by cholinergic receptors and G-regulatory proteins. When the vagus nerve is stimulated, heart rate and the force of atrial contraction both decline. Because of effects on atrioventricular conduction, the coordination of atria and ventricles is often disrupted. This combination of effects may lower cardiac output by 40 to 50 percent. The effects of vagal stimulation are evident following external massage of the carotid sinus, which stimulates the glossopharyngeal afferent limb of the baroreceptor reflex and modifies efferent parasympathetic outflow. This slows cardiac rate and atrioventricular conduction.

Other Neurovascular Transmitters

Although most neurovascular communications are transmitted by NE or ACh, there are other neurotransmitters involved in cardiovascular reflexes. **Substance P** is a neurotransmitter peptide that is widely distributed in the brain and peripheral nervous system. Its cardiovascular regulatory potential is suggested by its relatively high concentration in the vasomotor areas (particularly the nucleus tractus solitarii and dorsal motor nucleus of the vagus), where it may interact with the opiate peptide system. In addition, it is present in the nerves innervating virtually every vascular bed, where its release triggers vasodilation through a specific receptor. **Opioids** such as the enkephalins and endorphins are also widely distributed in the brain and spinal cord. While infusion of these neurotransmitters produces transient vasodilation, they are thought to cooperate with other neurotransmitters operating in the same synaptic cleft to modulate the responses. They appear to be most involved in the behavioral responses to pain and exercise. **Vasoactive intestinal polypeptide** is found in the brain (in descending order of concentration: cerebral, cerebellar, basilar vertebral, and spinal arteries), gut, salivary glands, uterus, and skeletal muscle. It is a potent vasodilator and also

increases heart rate above that obtained with sympathetic stimulation only.

Baroreceptors and Control of Arterial Pressure

A complex set of afferent inputs to the central cardiovascular areas arise from both within and without the brain. Among these are the peripheral sensors such as the baroreceptors, chemoreceptors, and cardiac mechanoreceptors. The **baroreceptor system** consists principally of the carotid sinus and aortic arch mechanoreceptors, central vasomotor integrating areas, and autonomic efferents, and exerts short-term control over arterial pressure. In vivo, the baroreceptor system operates as an open loop, with negative feedback and its components buffering potentially large changes in arterial pressure (for instance, those produced by changes in posture). Under resting conditions, the system is static; however, it can be modified by periodic or transient perturbations, such as respiration or exercise, and therefore it also has dynamic characteristics. The neural outflow from the vasomotor centers modulates the smooth muscle tone of resistance vessels, the force of myocardial contraction, and the heart rate, which buffer changes in systemic arterial pressure or blood volume. The sympathetic efferent flow is not uniformly distributed to all resistance beds but exhibits a characteristic pattern. Such patterns add a layer of complexity to the integrative activities at the different levels of the nervous system.

The **carotid sinus** baroreceptor is located in a segmental enlargement at the bifurcation of the common carotid artery. The receptors are in the adventitia of the sinus wall and are innervated by a branch of the glossopharyngeal nerve, which carries afferent traffic to the nucleus tractus solitarii in the medullary area of the brainstem. **Strain energy density** (the force required to produce an incremental stretch) in the wall of the sinus is linearly related to pressure over a wide range of values from 50 to 250 mm Hg. Over much of this range, the physical distortion produced by an increase in pressure also has a direct linear relationship to afferent nerve activity. The rate of afferent nerve discharge is largely influenced by the mean arterial pressure, and to a lesser extent by pulse pressure (Fig. 23-4). Therefore, for any given mean pressure, a narrower pulse pressure decreases afferent activity. The relationship between sinus pressure and nerve discharge exhibits **hysteresis;** that is, it differs

Fig. 23-4. Baroreceptor nerve responses to alterations in mean and phasic pressures. (A) The discharge rate of the baroreceptor nerve increases as mean arterial pressure increases. These discharges generally inhibit central vasomotor centers. At lower pressures, the effects of phasic (pulsatile) variations in arterial pressure are superimposed on the effects of mean pressure, but, at higher systemic pressures, this variation is no longer evident, indicating receptor saturation. (B) The difference between baroreceptor nerve responses to pulsatile and constant pressure stimuli. When the input is phasic, the discharge rate in the nerve remains higher than that when the input is constant, indicating that receptors adapt under these conditions. Thus, phasic stimuli are more effective in maintaining baroreceptor function than are continuous ones (Modified from: Berne, R. M., and Levy, M. N., eds. *Physiology*, 2nd ed. St. Louis: C. V. Mosby, 1988, Pp. 519–520.)

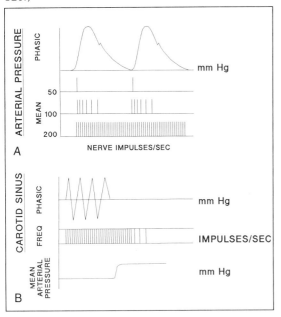

slightly depending on whether measurements are made during an increase or decrease in pressure. Because the rate of nerve discharge is closely related to the stretch associated with dimensional changes in the sinus wall, physiologic (catecholamines) and pathologic (hypertension or atherosclerosis) factors that modify the distensibility of the carotid sinus also change the relationship between intraluminal pressure and stretch.

The **aortic arch** baroreflex system is similar to that of the carotid sinus. Nerve endings are concentrated at the junction between the adventitia and media, and serve as stretch receptors with their afferent impulses traveling through both myelinated and unmyelinated fibers in the vagus nerve. The threshold of pressure stimulation for aortic receptors is about 90 mm Hg, compared with 60 mm Hg for the carotid receptors (Fig. 23-5-). Therefore, the carotid sinus is still important in modulating blood pressure and heart rate at lower pressures, a feature especially important for maintaining cerebral perfusion in an upright posture. This difference in characteristics of the two systems allows the aortic reflex to buffer the carotid reflex. If the afferent nerves from the aortic arch receptors are transected (that is, the loop is opened at a sensor) and pressure in the vascularly isolated carotid sinus is lowered from 100 to 50 mm Hg, systemic arterial pressure rises from 85 to 150 mm Hg. This demonstrates the unopposed action of the carotid sinus reflex at lower pressures. With the aortic nerves intact, arterial pressure rises only to 120 mm Hg, illustrating the buffering action of the aortic on the carotid sinus reflex.

As a demonstration of this, if blood pressure increases, the carotid and aortic receptors are activated by the deformation caused by changes in local wall stress. This produces an increase in afferent impulses traveling through the vagus and glossopharyngeal nerves, respectively. These impulses reach synapses in the central vasomotor centers of the pons and medulla, inhibiting sympathetic efferent nerve activity to heart and resistance vessels and veins (particularly splanchnic). Parasympathetic outflow to the heart then increases. The net effects are cardiac

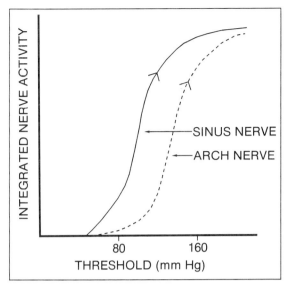

Fig. 23-5. Difference in carotid sinus and aortic arch baroreceptor responses. As pressure sensed by the receptors is increased from 50 to 200 mm Hg, the afferent nerve discharges detected from the carotid sinus and aortic arch exhibit differences. The carotid baroreceptor has an almost immediate threshold response, which is steepest up to pressures of 160 mm Hg, where there is saturation. In contrast, the threshold of the aortic baroreceptor is between 80 and 90 mm Hg, and saturation occurs at 200 mm Hg. This allows the two receptor systems to interact, particularly at lower pressures.

slowing and a fall in blood pressure, due both to arterial and venous dilation and to the decreased force of cardiac contraction, which correct the increase in blood pressure.

Chemoreceptors

Arterial chemoreceptors are located in the carotid arteries and aortic arch in the same regions as baroreceptors. They are formed by a varying number of specialized type I and type II cells. The **type I cells** form synapses with nerve fibers and a group of four to six cells is encircled by a type II cell. This array is close to an extensive capillary network. The type I cells are excitable and act as receptors that release neurotransmitters (for ex-

ample, NE or substance P) which activate the afferent nerves. Type II cells are inexcitable, and though essential for sensor function, their function is not well understood. Hypoxia or decreased pH slow mitochondrial electron transport and this changes the ratio of ADP to ATP. This leads to mitochondrial calcium release, a rise in the cytosolic calcium level, and a decrease in the frequency of action potential formation in type I cells.

The **carotid bodies** are innervated by a branch of the glossopharyngeal nerve, and the **aortic bodies** are supplied by a branch of the vagus. The fibers from each may be either myelinated or unmyelinated. Nerve discharges are stimulated by decreases in pH or the oxygen tension (P_{O_2}), or by increases in the carbon dioxide tension (P_{CO_2}) and temperature. P_{O_2} and P_{CO_2} interact during afferent nerve activity. For any given arterial P_{O_2}, the number of discharges increases at higher P_{CO_2}; conversely, for any given P_{CO_2}, the number of impulses increases with lower P_{O_2}. No entirely satisfactory overview of the role of chemoreceptors in cardiovascular regulation has been formulated, unlike the prominent role that has been identified for them in ventilatory regulation.

Systemic hypoxia produces profound cardiovascular adjustments, mediated by peripheral chemoreceptors, the direct effects of hypoxemia on the central nervous system, and the effects of increased ventilation on lung inflation and p_{CO_2}. The carotid and aortic body chemoreflexes are responsible for the reflex systemic arterial hypertension, which is mediated by sympathetic outflow from the vasomotor areas. Increases in heart rate, contractility, and cardiac output are most likely due to the combined effects of central nervous system hypoxia and increased ventilation, as chemoreceptor denervation does not abolish the cardiac responses, while lung denervation largely prevents or reverses them.

Cardiac Mechanoreceptors

Mechanoreceptors in the heart possess both vagal and sympathetic afferents. Those with afferents traveling in the vagus have sensors in the atria and ventricles; those with sympathetic afferents have receptors on the pulmonary veins and coronary vessels.

Vagal Afferents

Atrial A and B receptors are distinguished by their location at the **venoatrial junctions,** as well as by their function. Type A receptors seem to react primarily to heart rate, but they adapt (gradually decrease their rate of discharge after an initial increase) to long-term changes in atrial volume. Type B receptors respond to short-term changes in atrial volume, increasing discharges during atrial distention. The afferent nerves from both receptors are largely myelinated. **C fibers** are nonmyelinated vagal fibers and arise from receptors scattered throughout the atria. These discharge with a low frequency and respond with increased discharge to increases in atrial pressure. The A and B receptors are thought to mediate the increase in heart rate associated with atrial distention, known as the **Bainbridge reflex**. In contrast, activation of the **atrial C fibers** is generally vasodepressor in nature, with cardiac slowing and peripheral vasodilation. There is great variability in the net response to atrial distention among mammalian species.

Ventricular mechanoreceptors also have afferents with myelinated and unmyelinated fibers. In most mammals, the receptors are found in both ventricles. Discharge in the myelinated fibers decreases periodically with inspiration and increases with a rise in ventricular pressure. Ventricular C fibers are located primarily in the **epicardium** and discharge more rapidly in response to increases in both systolic and diastolic pressure. They exhibit a sharp threshold, discharging only at high systolic pressures, but progressively increase as diastolic pressures rise from 5 to 20 mm Hg. Ventricular distention can produce powerful depressor reflexes during both bradycardia and hypotension, called the **Bezold-Jarisch reflex**. C fibers appear to be more important than myelinated ones in conducting the relevant afferent nerve traffic. The central connections for this reflex are in the nucleus tractus solitarii, which has both sympathetic and parasympathetic synapses. Activation of cardiac C fibers also induces marked relaxation of the stomach by means of vagal noncholinergic fibers and is part of a more generalized activation of the vomiting reflex.

Sympathetic Afferents

The sympathetic afferents are less well understood than the vagal afferents. **Atrial fibers** appear to increase activity with rises in atrial pressure and volume, and can respond to phasic changes in atrial volume. **Ventricular fibers** may be either myelinated or unmyelinated and show modest increases in discharge rate when end-diastolic pressure (unmyelinated) or ventricular systole (myelinated) is elevated. Receptors on the coronary vessels discharge more frequently as blood flow or intracoronary pressures fall. The increased discharge rate during interrupted coronary flow may be important during myocardial ischemia. Recently, a class of receptors has been identified that are activated by low pressures in the atria and ventricle.

The importance of **cardiac mechanoreceptors** in the short-term regulation of arterial pressure is unknown, although they are much less involved than the baroreceptors. There is evidence that these atrial and ventricular receptors may affect the release of vasopressin and the renal release of renin by modifying selected efferent sympathetic outflow.

Circulatory Control During Postural Changes

The circulatory changes associated with standing up from a recumbent position illustrate how rapid control by neural mechanisms is integrated. Upon standing, **gravity** initially displaces blood from the thorax and splanchnic beds to the lower extremities and buttocks within a few heartbeats. The **labyrinthine organs** are also stimulated due to the change in position. The magnitude of the displaced blood helps to determine the ultimate response. There is a reduction in cardiac volumes, with atrial pressures falling from 2 or 3 mm Hg to -8 to -10 mm Hg, and a resulting decline in cardiac filling. **Stroke volume** and **cardiac output** fall rapidly. The reduction in central volume stimulates atrial and ventricular receptors, increasing traffic through the sympathetic afferents and decreasing traffic through the vagal afferents. In addition, the **stretch** on the carotid sinus and aortic arch baroreceptors decreases, reducing the frequency of afferent discharge in the glossopharyngeal and vagus nerves. The result is both disinhibition and direct stimulation of vasoconstrictor and cardiac accelerator regions of the vasomotor areas.

Sympathetic outflow from the nucleus tractus solitarii and vasoconstrictor regions increases to offset the potential decrease in systemic arterial pressure produced by upright posture. The first detectable changes are venous and arterial vasoconstriction that lead to decreased venous caliber and capacitance, particularly in the splanchnic veins which are nearly maximally constricted in the upright position, and increased peripheral resistance; arterial pressure also increases. Shortly thereafter, the heart rate and rate of contractility are accelerated, but despite this, stroke volume and cardiac output are reduced predominantly by the effects on preload (decreased cardiac volumes) that result from venous pooling.

An inability to sustain blood pressure when standing (orthostatic hypotension) may be caused by neuropathies that impair receptor or effector function (failure to vasoconstrict or to increase heart rate), by drugs that block the autonomic nervous system, or by reduced intravascular volume.

Circulatory Changes During Valsalva's Maneuver

Valsalva's maneuver, which involves straining to expire against a closed glottis, is often performed during defecation and parturition. It is a useful way to evaluate the integrated autonomic control of the circulation and demonstrates how neural and intrinsic cardiac control systems compensate for a physical stress (Fig. 23-6). At the onset of the maneuver, intrathoracic pressure increases abruptly, compressing the great vessels, raising the aortic pressure, and displacing blood volume out of the chest (**Phase I**). In **Phase II,** the increased intrathoracic pressure narrows veins at the thoracic inlet and obstructs systemic venous return. Intracardiac volumes decrease by 25 to 30 percent, reducing cardiac preload and hence stroke volume; arterial pressure falls and pulse pressure narrows. Low-pressure cardiac receptors are stimulated and this leads to sympathetic afferent stimulation of vasoconstrictor regions. Baroreceptor impulses originating from the carotid sinus and aortic arch decrease because of lowered mean aortic pressure and pulse pressure. The resulting sympathetic outflow increases heart rate and contractility, and constricts peripheral arteries (increasing resistance) and veins (increasing venous pressure outside the chest). As a result, cardiac volume, stroke volume, cardiac output, and arterial pressure stabilize and may even increase. At the moment the strain is released (**Phase III**), intrathoracic pressure falls, relieving constraining effects on the thoracic vena cava, heart, and aorta. Immediately, there is a transient fall in systemic pressure. The sympathetically mediated vascular responses initiated and sustained during straining then produce a large venous pressure gradient between the extrathoracic capacitance vessels and the underfilled low-pressure right atrium, driving an explosive venous return. Within a few seconds, intracardiac volumes return to premaneuver levels. This recruitment of the cardiac length–tension mechanisms coupled with sympathetic cardiac support increases stroke volume, and translocates the increased venous return into the constricted arterial circulation. The sudden transfer of this additional volume raises arterial and pulse pressures above premaneuver levels (**Phase IV**). This increases wall stress in the heart and baroreceptor regions. Afferent nerve traffic from these sensors increases, inhibiting central outflow and leading to recovery.

When there is **autonomic dysfunction,** the

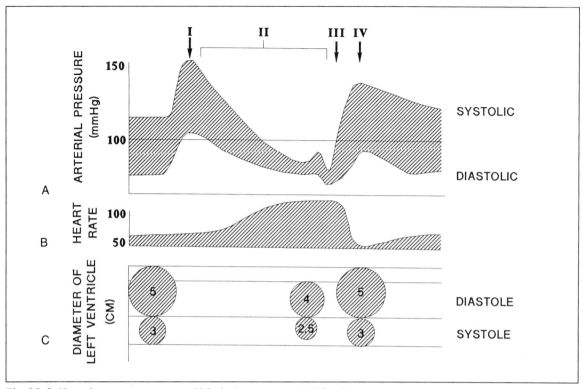

Fig. 23-6. Hemodynamic responses to Valsalva's manuever. (A) The changes in systemic arterial pressure. (B) The heart rate responses. (C) Representative left ventricular minor axis dimensions during diastole and systole, as measured by reflected ultrasound. During the strain phase (*II*), arterial pressures and pulse pressures fall, heart rate increases, and cardiac volumes decline. After release of the strain (*IV*), there is a brief overshoot in arterial pressure which reflects the persistent effects of circulatory adjustments made during phase II. (*Phase I* = onset of maneuver; *Phase III* = release of strain.)

tachycardia of Phase II and overshoot in arterial pressure of Phase IV may be absent. In the event of ventricular dysfunction or obstruction to left ventricular inflow, Phases II, III, and IV are missing, and there is a classic square-wave response, whereby the Phase I response is sustained until release.

LOCAL INFLUENCES AND CIRCULATORY CONTROL

Vascular Smooth Muscle

Vascular smooth muscle can constrict tonically and change its tension in response to both **cen-trally dispatched signals** and **local factors**. These characteristics permit flexible and precise circulatory control. Unlike the organized contractile units in striated muscle, **actin** and **myosin** are distributed throughout the smooth muscle cell, aligned more or less parallel to the long axis of the cell. A characteristic action potential is *not* required for excitation–contraction coupling, because graded changes in membrane potential initiate varying degrees of constriction. The control of **phosphorylation** and **intracellular calcium** concentration is required for actin–myosin reactions and are critical elements in the precise regulation of vessel caliber. The control of phosphorylation is closely linked to myoplasmic calcium concentration, which is in turn influenced by four factors: (1) passive calcium influx related to transmembrane gradients; (2) influx through voltage-dependent channels in the plasma membrane; (3) influx through receptor-operated channels; and (4) intracellular release from the sarcoplasmic reticulum (Chapter 15). G-regulatory proteins couple receptors to other membrane proteins, which induce the diverse cellular responses. The actions of G-regulatory proteins are discussed fully in Chapters 6 and 15.

Vasodilator Mechanisms

Atrial natriuretic peptide (ANP) is a recently discovered endocrine hormone synthesized in the heart, and found predominantly in granules in the atrial appendage. It appears that this hormone is synthesized as a precursor, then cleaved by a serum protease to a molecule with an active region containing a minimum of 23 amino acids. **Release of ANP** is promoted by atrial stretch, beta-adrenergic stimulation, and increased heart rate. The granules fragment and the small packets are transported along microtubules to the cell surface, where the hormone is released.

ANP is a direct-acting vasodilator that antagonizes the vasoconstrictor **angiotensin-II,** but cannot overcome high doses of NE. ANP stimulates membrane-bound cyclic guanosine $3':5'$-monophosphate (cGMP) as a second messenger to induce its vasodilating effects. At least part of its ability to produce natriuresis and diuresis derives from its **vasodilating capabilities**. In vivo, ANP promotes arterial and venous dilation. When infused systemically, ANP reliably reduces arterial pressure almost immediately in a dose-dependent manner. Because the drug also dilates veins, cardiac output falls and peripheral resistance is therefore unchanged. Renin secretion is suppressed and plasma levels decline, perhaps due to the accelerated sodium delivery to the macula densa. ANP also blocks the effects of angiotensin-II on aldosterone release, as well as

lowers angiotensin levels. About one hour after the start of the infusion, arterial pressure returns to pre-infusion levels, reflecting the influence of other compensatory mechanisms. These effects are most important as part of the short-term regulation of **volume** and **pressure**. Levels of ANP are elevated during heart failure; however, its effects are offset by potent vasoconstrictor mechanisms and stimuli for sodium retention.

The **kinins** are a group of polypeptide vasodilators. As with other vasoactive peptides, they are synthesized and circulate in the form of larger, inactive molecules in the blood and tissue fluid. Bioconversion to active molecules occurs locally, where concentrations may be a good deal higher than systemic levels. The best understood of these is **bradykinin,** which is formed from **kallikrein**. When activated, kallikrein forms an intermediate, kallidin, that is converted to bradykinin. Bradykinin has only local effects on the circulation, because it is quickly inactivated by **carboxypeptidases** or converting enzyme. Bradykinin is an important mediator of the inflammatory response influencing endothelial permeability and plays a regulatory role in the control of local circulations in the skin, kidney, and salivary glands. **Histamine, serotonin,** and **prostaglandins** have highly specific functions but no known systemic role in circulatory control.

The **endothelium** is also an important modulator of smooth muscle activity. Its ability to communicate with smooth muscle is especially important in understanding how circulatory substances which have specific receptors or those released from the vessel wall regulate smooth muscle tone (Table 23-2). Its role in mediating smooth muscle relaxation was first recognized when the usual vasodilator response to ACh observed in larger arteries was abolished by denuding the endothelium (Fig. 23-7). The search for an **endothelium-derived relaxing factor** (EDRF) that diffused into smooth muscle resulted in the identification of **nitric oxide,** generated from L-arginine by nitric oxide–synthase. This represented the first, but not necessarily only, EDRF. Nitric oxide has a half-life of only a few

Table 23-2. Endothelium-Derived Relaxing Factor–mediated Vasodilators

Substance	Receptor
Acetylcholine	Muscarinic
Norepinephrine/ epinephrine	Alpha$_2$
Histamine	H$_2$
Vasopressin	V$_1$
Bradykinin	
Platelet aggregation	
ADP	Purinergic
Serotonin	S$_1$
Shear stress	

ADP = adenosine diphosphate.

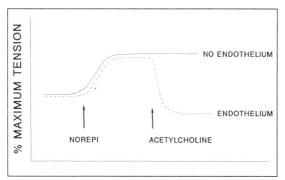

Fig. 23-7. Endothelium-dependent relaxation of vascular smooth muscle. This is a typical example of the effects of endothelial-derived relaxing factor. Two strips or rings of vascular smooth muscle are bathed in a nutrient buffer, one with the endothelium intact and the other with the endothelium gently abraded or digested. The vessel is then preconstricted, in this example with norepinephrine (NOREPI), and then exposed to the test agent. If tension falls in the vessel with intact endothelium but not in the vessel without endothelium, this indicates endothelial-induced relaxation. This is an example of paracrine function which is important in the local control of the circulation. Acetylcholine is the standard against which other substances are compared.

seconds and is quickly inactivated by the superoxide anion. It exerts paracrine control on vasodilation by stimulating the accumulation of cGMP, a breakdown product of guanosine triphosphate.

In the microcirculation, EDRF may mediate the vasodilation associated with **inflammation** and **trauma,** which are local hyperemic responses. EDRF is also released in normal vessels in response to **platelet aggregation** and **shear stress** exerted by flowing blood. When diseases such as atherosclerosis disrupt the endothelium, they may interfere with the synthesis of EDRF, or the ability to vasodilate in response to a number of stimuli. Recently, the excessive vasodilation associated with **bacteremic shock** has been linked to uncontrolled production of nitric oxide. **Prostacyclin** is a vasodilator derived from the family of prostaglandins synthesized in endothelial cells. It also inhibits platelet aggregation and works synergistically with EDRF. Their combined effects may wash away platelet plugs, formed during physiologic vascular trauma. Because it is not stored and has a short half-life, prostacyclin's effects are only local and promote smooth muscle relaxation. In certain disease states, such as diabetes mellitus and atherosclerosis, endothelial cells have a decreased capacity to synthesize prostacyclin and respond poorly to EDRF. Nicotine inhibits its synthesis.

Vasoconstrictors

The endothelium is also the source of substances that initiate smooth muscle contraction, including **thromboxane A$_2$, prostaglandin H$_2$,** and the **superoxide anion,** all of which are short acting. A group of active peptides, **endothelins,** has recently been isolated. They are formed by enzymatic cleavage from a larger, inactive precursor and constrict both arterial and venous smooth muscle in all vascular beds, acting through a unique receptor in the media. Endothelins open a voltage-sensitive Ca^{2+} channel that activates inositol-tris-phosphate and diacylglycerol. This further increases intracellular Ca^{2+} concentrations. Pretreatment with Ca^{2+}-channel blockers prevents the vasoconstrictor actions of endothelin, demonstrating its dependence on the extracellular Ca^{2+} level. The constriction produced by endothelin follows a brief vasodilation, and is slow in onset and sustained. It is reversed by stimulation of cAMP or by glyceryl trinitrate, which increases cGMP levels much like EDRF does.

Endothelin causes arterial constriction, and thus arterial resistance is augmented. Its venoconstrictor actions decrease capacitance and increase cardiac preload. It is also a potent antinatriuretic. However, its effects on arterial resistance generally predominate, and cardiac output falls during endothelin infusions. In the heart, endothelin has an **inotropic effect,** at least on atrial muscle, and stimulates the secretion of ANP. The role of endothelin in cardiovascular control is currently unknown, as are details about the regulation of its synthesis and release. It is known to be released in response to raised thrombin levels, suggesting that it may be important in the evolution of responses to hemorrhage. Its slow secretion and onset of action, as well as its sustained action, are not consistent with the properties needed for rapid circulatory regulation. Recently, it has been noted to possess the characteristics that mitogen exhibits in smooth muscle cells, fibroblasts, and mesangial cells, suggesting a role in systemic arterial hypertension.

Endothelial cells, as well as cardiac and renal cells, contain intrinsic **renin—angiotensin systems. Angiotensin-II** is a powerful vasoconstrictor peptide. It is formed by the action of the enzyme renin on its precursor, angiotensinogen. The product, angiotensin-I, is not vasoactive but is cleaved to angiotensin-II by a converting enzyme, for which potent pharmacologic inhibitors have been synthesized. Angiotensin-II is also a **salt-retaining hormone** when released into the circulation, stimulating the synthesis and release of aldosterone. It also has cardiac **inotropic properties**. In these capacities, it functions as an endocrine hormone. Its release is modulated by the sympathetic division of the autonomic nervous system, representing another interaction of the pressure and volume control systems. Angiotensin appears to be involved chiefly in the intermediate and long-term regulation of **systemic pressure** and **volume**. Local angiotensin systems may play a role in the regulation of

organ perfusion and serve as a **growth factor** for myocytes and fibroblasts. These regulatory functions are examples of **autocrine regulation**. Regulation of growth may be critical to cardiovascular adaptation in hypertension.

Arginine vasopressin (AVP) and its precursor **neurophysin** are synthesized in the magnicellular neurons of the supraoptic, paraventricular, and suprachiasmatic nuclei of the hypothalamus. Neurosecretory granules are transported axonally to the neurohypophysis, where they are released from nerve endings into the systemic circulation. AVP is also a neurotransmitter found in central regions involved in circulatory control, including the nucleus tractus solitarii and dorsal motor nucleus of the vagus. AVP also projects to the cerebral cortex and spinal cord. Thus, AVP exhibits features of both **paracrine** and **endocrine control**.

AVP has **two major circulatory roles**: (1) to maintain arterial pressure in the presence of reduced blood volume; and (2) to regulate osmolality in the face of an increased plasma solute concentration. Its actions on arterial pressure reflect its interaction with vasoconstrictor V_1 receptors, and its central action augments the baroreceptor reflex by modifying sympathetic outflow and accentuating the bradycardia. In contrast to its constricting effects in the periphery, AVP induces vasodilation in cerebral vessels, and this depends on EDRF. AVP's role in maintaining volume is mediated through arterial and atrial receptors, which inhibit systemic AVP secretion when activated.

EXERCISE: INTEGRATION OF CENTRAL AND LOCAL REGULATION

Exercise is the most common and powerful stimulus to which circulation must adjust. Circulatory alterations begin even before activity, when cortical centers influence vasomotor areas, leading to parasympathetic withdrawal. This increases heart rate and myocardial contractility. At the same time, sympathetic cholinergic vasodilation increases blood flow to the muscle, and alpha vasoconstriction reduces blood flow to the skin, kidney, and splanchnic regions. As a result, peripheral resistance and arterial pressure initially increase.

Once **exercise** begins, there is an important and progressive redistribution of cardiac output, summarized in Figure 23-8. Local changes in skeletal muscle become the most important de-

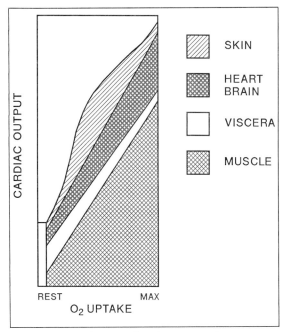

Fig. 23-8. Distribution of cardiac output during exercise. Exercise activates both local and central circulatory control mechanisms. Local vasodilation, which is mediated by increased oxygen flux as well as adenosine and potassium levels, distributes cardiac output to exercising muscles in proportion to their metabolic needs. At the same time, the heart requires an incremental percentage of overall cardiac output to increase the coronary blood flow so that it can supply the energy needed to sustain cardiac output. Cardiac performance is enhanced both by the fall in systemic arterial resistance and by the effects of the sympathetic nervous system. To dissipate metabolic heat, skin vessels dilate at all but the highest levels of exercise, when that blood volume may be required to further augment cardiac output. (Modified from: Berne, R. M., and Levy, M. N., eds. *Physiology*, 2nd ed. St. Louis: C. V. Mosby, 1988, P. 562.)

terminants of venous return. Central mechanisms maintain arterial pressure by increasing peripheral resistance, mobilizing venous reservoirs, improving cardiac performance, and distributing cardiac output to the heart and working skeletal muscles. The increase in skeletal muscle blood flow is brought about by several proposed **vasodilator influences**: (1) increased oxygen flux into muscle cells from tissue fluid, thus decreasing the amount of oxygen available for smooth muscle contraction; (2) effects of adenosine introduced through specific vascular receptors; (3) effects of potassium released from the working cells; and (4) local changes in pH due to lactic acid and CO_2 accumulation. Secondary factors such as the duration of each muscle contraction (causing external vascular compression) and capillary recruitment (opening of capillaries serving previously inactive muscle units) also affect local blood flow.

The **mass sympathetic discharge** that accompanies exercise increases the heart rate, the strength of its contraction, and speed of relaxation. These effects are mediated by both sympathetic nerves and epinephrine release from the adrenal medulla. Most peripheral arterioles are constricted in nonmuscular organs; the venules and capacitance veins also constrict. These changes have **several outcomes**. **First,** volume is redistributed from the peripheral into the central circulation, as veins constrict and capillary pressure in nonworking tissues falls, transporting fluid into the vasculature. Even though some fluid moves into the working tissues, the net result is that more volume is available for sustaining venous return and cardiac output. **Second,** venoconstriction increases the pressure gradient for return of blood to the heart and, together with muscular and respiratory pumps, augments venous flow. **Third,** systolic arterial pressure rises. This represents the combined influence of localized arteriolar constriction (which increases resistance in nonworking tissues but decreases overall peripheral resistance) and the rapid translocation of blood from the venous to the arterial circulation. This rapid transfer of blood is possible because of sympathetic cardiac stimulation.

The increase in mean arterial pressure is due largely to heightened systolic pressure and sustains the driving force for perfusion of constricted beds, while permitting the powerful autoregulatory beds of the heart and kidney to remain well perfused. In addition, the rise in pressure augments blood flow through the vasodilated circulation of the working tissue.

Under **extreme conditions,** the muscles cease to receive sufficient oxygen to maintain aerobic performance. The reason for this may be an inability to sustain adequate oxygenation of the blood (high altitude or lung diseases), or failure to deliver the required cardiac output caused by volume depletion, cardiac ischemia, failure, or arrhythmia. Clinical exercise testing is designed to determine whether such limiting factors are present. When exhaustion is reached, arterial pressure may fall and endogenous vasoconstrictors, such as epinephrine and AVP, are secreted to overcome local vasodilatation. Volume loss, increased body core temperature, failed local energy reserves, and emotional fatigue lead to collapse. A complete discussion of exercise physiology is presented in Chapter 60.

ADAPTATION

When a physiologic or pathologic stress is applied repeatedly, there are prolonged changes in the circulation. Ideally, these are associated with a sustained ability to maintain the organism in the fully functional state and are considered adaptive. An example of physiologic adaptation is observed in inhabitants at **high altitudes** (chronic hypoxia) who have increased blood and red cell volumes, increased capillarity of tissue circulations, and growth of the pulmonary arterial and right ventricular muscle in response to hypoxic vasoconstriction. Another example of physiologic adaptation is **physical training**. This produces decreased peripheral arterial resistance, increased cardiac volumes and muscle mass, and improved oxygen use in peripheral tissues.

Examples of changes in circulation brought about by patholic conditions are systemic hypertension and chronic heart failure. In **systemic hypertension,** vascular smooth muscle mass increases and there is cardiac growth (hypertrophy). In **chronic heart failure** due to muscular or valvular disease, there are long-term changes in the circulatory volume, cardiac mass, and baroreceptor sensitivity. In these two pathologic conditions, the chronic changes may not adequately compensate for the prolonged perturbations caused by the disease. Eventually there is clinically apparent deterioration after a long, relatively asymptomatic period of compensation. A detailed discussion of circulatory adaptation and growth is not appropriate to this text, but it is important to appreciate the existence of systems dedicated to controlling adaptations.

SUMMARY

The circulation maintains local environmental conditions essential for cellular function. Control of the circulation is important for the integration of the local blood flow requirements largely related to metabolic demands. Control of the systemic arterial pressure provides the energy for perfusion of all tissues. Three important systems interact to exert and coordinate control. The first is the **autonomic nervous system,** consisting of peripheral receptors, central integrating vasomotor centers, and neuroeffector junctions on the heart and blood vessels. The second comprises the **local regulators** in the tissues. These include intrinsic cardiac mechanisms such as the length–tension relationship, which permit the heart to adjust to momentary changes in venous return, and paracrine regulation of local circulations. The third system consists of the **endocrine regulators** of systemic vascular volume. These systems overlap (for example, in the hypothalamus) and interact extensively to permit acute and long-term regulation of arterial pressure and blood volume. At the same time, they preserve the autonomy of organ and tissue blood flow, which is the cornerstone of circulatory flexibility and adaptation. The control systems involved in circulatory regulation exhibit **negative feedback,** responding with adjustments that oppose the perturbations from equilibrium values.

These systems are activated by ordinary activities, such as postural changes, exercise, and instinctive behaviors. They are also triggered by disease states or extraordinary stresses, such as trauma or loss of intravascular volume. They are capable of sustaining changes and bridging the gap between adjustment and adaptation.

NATIONAL BOARD TYPE QUESTIONS

Select the most appropriate answer for each of the following

1. Control systems are vital homeostatic mechanisms. Which statement best describes the behavior of those systems that maintain systemic arterial pressure?
    A. Interacting, closed-loop systems, exhibiting a direct positive relationship between perturbations and corrections
    B. Isolated, closed-loop systems, exhibiting an inverse but direct relationship between perturbation and correction
    C. Interacting, open-loop systems, exhibiting a sigmoid inverse relationship between perturbation and correction
    D. Isolated, open-loop systems, exhibiting a direct positive relationship between perturbation and correction
    E. A single, closed-loop system, exhibiting a sigmoid inverse relationship between perturbation and correction
2. Which of the following scenarios would not initiate a positive-feedback cycle?
    A. Following standing from a supine position, baroreceptor activation slows the heart rate and decreases peripheral resistance
    B. Following trauma causing blood loss, volume and pressure sensors increase heart rate and peripheral resistance in skeletal muscle beds; in addition, renal vascular resistance falls, ANP is released from the heart, and AVP secretion is inhibited
    C. When a person is escaping from a predator, local mechanisms increase blood flow to the rapidly moving limbs and heart and

mobilize tissue fluid; at the same time, central mechanisms reduce blood flow to the kidneys and splanchnic beds

Match the effector substance with the major mechanism by which it exerts control over target tissues. There is only one correct answer for each neurohormone.

3. Aldosterone
4. NE
5. Angiotensin-II

6. Arginine vasopressin

7. EDRF

A. Endocrine only
B. Paracrine only
C. Endocrine and autocrine
D. Endocrine, paracrine, and autocrine

E. Paracrine and autocrine

8. Local regulatory mechanisms are most important in the circulatory control in which scenario?
    A. Change in posture
    B. Systemic hypoxia
    C. Moderate exercise
    D. Valsalva's maneuver
    E. Weightless environment
9. Which of the following interventions is least likely to increase cardiac output: (i) stimulation of sympathetic nerves from the stellate ganglia; (ii) selective stimulation of the central vasoconstrictor centers; (iii) rapid intravenous infusion of compatible blood; (iv) pressure applied to the vascularly (not neurally) isolated carotid sinus; or (v) a shot of adrenaline.
    A. i only
    B. ii and iv
    C. v only
    D. i, iii, and v
    E. All increase output

ANNOTATED ANSWERS

1. C. The control of systemic arterial pressure requires the interaction of systems regulat-

ing peripheral arterial resistance and capacitance as well as blood volume. Inputs from one of these systems (for example, volume loss) results in adjustments in both systems, restoring first arterial pressure and then blood volume. In general, over the physiologic range there is an inverse sigmoid relationship between the perturbation and the correction, so that the output of the control system opposes the perturbation. This restores the system toward its equilibrium or operating point, and represents negative feedback.

2. C. In both A and B the perturbation produces a "correction" that is in the same direction, thus accentuating the perturbation. In A, standing displaces blood from the central to peripheral veins. Vasodilation and cardiac slowing would cause this blood to pool, further lowering blood pressure and stimulating more vasodilation and cardiac slowing. In B, the increase in heart rate and peripheral resistance would sustain arterial pressure, but renal vasodilation, and the activation of stimuli that promote urine flow (and therefore volume loss), would further reduce blood volume and possibly overwhelm the other adjustments. Therefore, the volume control system is exhibiting positive feedback. C is the usual response during exercise.

3. A. Aldosterone is an endocrine hormone produced in the adrenal cortex in response to angiotensin-II. This hormone is released into the bloodstream and acts primarily at the renal tubules, influencing Na^+-K^+ exchange.

4. E. NE is a catecholamine secreted from secretory granules into synaptic clefts. Its effects are exerted by means of receptors on target tissues innervated by the sympathetic nerves (paracrine). It does have some autocrine function, in that it is involved in feedback to presynaptic receptors on the nerve terminal.

5. D. Angiotensin-II acts as an endocrine hormone in the regulation of blood volume and systemic vascular resistance, as a paracrine substance in the regulation of local organ blood flow, and as an autocrine factor in the regulation of cell growth.

6. C. AVP is elaborated by the neurohypophysis and performs as an endocrine hormone with regard to its action on the V_2 receptors of the renal collecting ducts. It also exerts paracrine control as a neurotransmitter in the central nervous system and promotes cerebral vasodilation by activating V_1 receptors.

7. B. Nitric oxide is produced in endothelial cells and acts on neighboring smooth muscle cells; it offers an example of paracrine regulation.

8. C. Muscular activity produces vasoactive metabolites, increased local oxygen flux and potassium release, and changes in H^+ flux. These factors produce local vasodilation and increased blood flow to the working muscle and through the coronary arteries to supply the myocardium. There are also systemic circulatory adjustments, but these local factors redistribute a disproportionate amount of the increased cardiac output to the working tissues, where vascular resistance is lowest.

9. B. The stellate ganglia are part of the cervical sympathetic chain and have efferent connections to the heart. Stimulation increases heart rate and contractility through release of NE at the neuroeffector junctions on the heart. Rapid volume infusion increases venous return and activates the heart's length–tension mechanism. This increases stroke volume and, at least over the short term, cardiac output. Adrenaline (epinephrine) is a catecholamine that stimulates cardiac contractility and heart rate. It also induces vasodilation in selected vascular beds, thereby lowering peripheral resistance. The combination of these effects increases cardiac output. In contrast, vasoconstriction increases arterial vascular resistance, raising blood pressure and therefore ventricular afterload. It may also activate the baroreceptor reflexes, thus withdrawing sympathetic cardiac stimulation. Both

of these mechanisms decrease cardiac output. Carotid sinus pressure activates the baroreceptor reflexes, slowing heart rate, decreasing atrial contractility, and reducing cardiac output transiently.

BIBLIOGRAPHY

Laragh, J. H., and Atlas, S. Atrial natriuretic hormone: a regulator of blood pressure and volume homeostasis. *Kidney Int.* 34(Supp 25); S64–71, 1988.

Shepherd, J. T. Reflex control of arterial pressure. *Cardiovasc. Res.* 16:357–383, 1982.

Shepherd, J. T., and Abboud, F. M., eds. *Handbook of Physiology, Volume III: The Peripheral Circulation, Part 2: Cardiovascular Reflexes and Circulatory Integration.* New York: Oxford University Press, 1984.

Vane, J. R., Anggard, E. E., and Gotting, R. M. Regulatory Functions of vascular endothelium. *N. Engl. J. Med.* 323:27–36, 1990.

Vanhoutte, P. M. Endothelium-derived relaxing and contracting factors. *Adv. Nephrol.* 19:3–16, 1990.

Yanagisawa, M., and Masaki, T. Endothelin: a novel endothelium-derived peptide. *Pharmacol.* 38:1877–1883, 1989.

VI Hemostasis and Blood Coagulation

24 Platelet Function and Blood Coagulation

ROBERT F. HIGHSMITH

Objectives

After reading this chapter, you should be able to:

Define hemostasis and describe how the hemostatic mechanism operates in relation to thrombosis and hermorrhage

Explain why the hemostatic mechanism is a "potential" one and describe the adequate stimuli necessary to elicit the response

Illustrate how the hemostatic mechanism may be stimulated by abnormal processes, leading to thrombosis

Explain why, when the hemostatic mechanism is unable to respond to a stimulus such as vascular injury fatal hemorrhaging may result

List the basic components of the normal hemostatic response to vessel injury

Describe the process by which platelets interact with an injured vessel wall to form a temporary hemostatic plug

Delineate those factors that mandate the transformation of the temporary hemostatic plug to a definitive mass of fibrin

Describe the production of coagulation factors and the role of vitamin K in their synthesis

Describe the major enzymatic pathways of blood coagulation and the mechanisms that initiate them

Contrast the kinetics of thrombin evolution via the extrinsic pathway versus the intrinsic pathway

Explain the importance of both enzymatic pathways for achieving normal coagulation

List the diverse effects of thrombin and explain its pivotal role in the overall hemostatic response to vessel injury

---

All organs and tissues of the body depend on blood for oxygen delivery, exchange of gases and nutrients, biologic waste removal, and hormonal communication. This system in turn demands the existence of a very competent mechanism to ensure that the circulating blood stays in the vascular compartment (no bleeding) and remains fluid at all times (no clotting).

Injury to vessels transporting blood under considerable pressure can lead to life-threatening bleeding, or hemorrhage. Fortunately, an elaborate host-defense mechanism is normally operative that, in most cases, depending on the site and severity of the trauma, can completely halt the extravasation of blood from the vascular compartment. This mechanism, termed **hemostasis,** represents the concert of events responsible for the rapid repair of any break in the vascular endothelium without compromising the fluidity of the blood. These events greatly assist in maintaining the internal environment, and thus represent a classic example of a homeostatic control system.

The hemostatic mechanism is a **potential system**, designed to reseal the endothelial surface in response to injury. However, this potential for rapid localized hemostasis within a fluid medium is not without risk, for essentially the same processes that stop bleeding, if abnormal, can produce **hemorrhage** or intravascular clotting—**thrombosis**. Because it is a potential system, the hemostatic mechanism must be triggered. If the mechanism fails to respond to the stimulus or if one or more of its components does not function, hemorrhage may develop. Conversely, if the hemostatic mechanism triggers spontaneously or is not regulated properly, thrombosis may arise. Fortunately, there are a number of checks and balances that collectively permit the timely arrest of bleeding after injury without compromising blood fluidity.

OVERVIEW AND COMPONENTS OF HEMOSTASIS

A competent hemostatic mechanism involves the proper triggering and interplay of the following: (1) blood vessel constriction; (2) platelet aggregation and fusion; (3) activation of blood coagulation factors; and (4) regulation and limitation of coagulation.

When a vessel is injured, blood vessels constrict rapidly and platelets aggregate to form a **temporary hemostatic plug**. These events involve smooth muscle shortening in the vessel wall and alterations in platelet biochemistry, and constitute the **primary hemostatic response**. The third and fourth steps, representing the **secondary phase of hemostasis**, are also activated by vessel injury, but are temporally sequenced to permit localized stoppage of bleeding. Thus, activation of coagulation first produces deposition of a blood clot, or insoluble **fibrin**, that anchors

the hemostatic plug and allows for definitive hemostasis to proceed. In addition, numerous other regulatory factors and systems are present that limit the coagulation process to the site of vessel injury.

The relative importance of each of these components in the overall hemostatic response is determined by where in the vasculature the injury occurs, the caliber or size of the vessel, and the pressure differential between the inside and outside of the vessel. For example, vasoconstriction dominates in halting bleeding at the arteriolar and venular level because of the relatively high proportion of smooth muscle in these vessels. On the other hand, platelet aggregation and fusion are particularly important when there is injury at the capillary level, because of lessened mechanical opposition of lower blood flow and pressure as well as the lack of smooth muscle in these vessels. Bleeding from damaged larger arteries can be profuse, due to the elevated outward-driving pressure. The formation of fibrin is mandatory for effective hemostasis in these vessels. Obviously, none of these components can halt bleeding from a laceration or rupture of the major conduit arteries. In this event, counterbalancing pressures must be applied and immediate surgical repair of the damaged vessel undertaken. The extent of bleeding is also affected by the pressure in the vascular area surrounding the injury. For example, bleeding into joints and skeletal muscle is constrained by the elevated tissue pressures in these areas, whereas hemorrhage in the gastrointestinal tract is often difficult to control.

The rest of this chapter deals with the major events of hemostasis—namely, the roles of blood vessel constriction, platelet activation, and blood coagulation in the normal hemostatic response to vessel injury.

VASCULAR AND PLATELET RESPONSES TO INJURY

Blood Vessel Constriction

When **arterioles** (small resistance vessels) are severed, the immediate control of hemorrhage is accomplished very efficiently by **vasoconstriction**. The importance of vasoconstriction in effective hemostasis is exemplified by the profuse bleeding that may result from trauma to **hemangiomas** (vascular tumors) or areas of **vasculitis** (inflamed vessels), conditions in which the contractile responses in blood vessels are limited. Although capillaries do not have smooth muscle fibers, some control of bleeding is probably gained through contraction of precapillary sphincters. The type of injury, whether transecting or crushing, determines the efficiency of vasoconstriction in halting blood loss. Although larger arteries and veins do contract when injured, the reflexive constriction is usually not enough to minimize bleeding in areas where there are elevated pressures and rapid blood flow.

The mechanisms underlying vasoconstriction are poorly understood, because of the considerable technical difficulty in studying injured vessels with intact neural inputs. However, the response appears to have two components: a **rapid reflexive** contraction, lasting 10 to 30 seconds which depends on sympathetic neural inputs, and a **slower** (60 minutes or so) **myogenic** component that may be mediated by the release of local vasoactive substances. Possible chemical mediators are **serotonin** (5-hydroxytryptamine) and **thromboxane A₂,** which are released from aggregating platelets, or the recently discovered potent vasoconstrictor **endothelin,** which is produced by damaged endothelial cells.

Role of Platelets in Hemostasis

Normal Platelet Production and Structure

Platelets are anuclear, discoid cells manufactured from the cytoplasm of megakaryocytes in the bone marrow. Substantial reserves of mature platelets are not available in the marrow. Instead, platelet production is regulated to meet circulating demands through humoral stimulation by thrombopoietin, in a fashion analogous to red blood cell regulation by erythropoietin. Platelets normally circulate at a concentration of about 200,000 to 400,000 platelets per microliter of blood and have a half-life of about 9 to 10 days.

Under normal conditions, about two-thirds of the total platelets are in the systemic circulation, with the remainder sequestered in the spleen. **Splenomegaly** (enlargement of the spleen) can cause increased splenic pooling of platelets and a decreased availability of circulating platelets, known as **thrombocytopenia,** thus increasing the tendency for bleeding. **Platelet turnover,** which is calculated from the platelet count divided by the platelet survival time, is a direct estimate of the rate of platelet removal from the circulation and is an important variable in understanding arterial thrombosis (see Chapter 5). In normal subjects, the platelet turnover is about 25,000 to 35,000 platelets per microliter of blood per day. Thus, if circulating platelet levels are suddenly depleted, caused by increased platelet consumption or destruction, it may take approximately 5 to 7 days for the platelet count to be restored to normal.

Platelets are the smallest of the circulating formed elements, with an average diameter of 3 to 4 μm, thickness of about 1 μm, and cell volume of 7 to 10 fl. The plasma membrane has numerous invaginations and is continuous with an **open canalicular system** that serves as a pathway for both the uptake of extracellular calcium and release of intracellular material (Fig. 24-1). There is also a **dense tubular system** consisting of smooth endoplasmic reticulum that does not communicate with the extracellular space but interdigitates with the open canalicular system. Together, these two tubular systems appear to make up the **Ca²⁺ regulatory site** for platelets. The platelet is bounded by a trilaminar plasma membrane, which is particularly rich in phospholipids, cholesterol, and glycolipids. The **membrane phospholipids** are particularly important as cofactors in the coagulation process and as a source of arachidonic acid in the production of thromboxane A₂. Just beneath the plasma membrane is a circumferential band of microtubules that serve as a cytoskeleton.

Platelets contain a variety of proteins common to contractile cells, including actin, myosin, tropomyosin, calmodulin, myosin light-chain kinase, filamin, and troponin. Energy for platelet

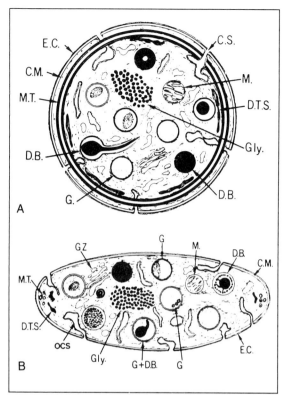

Fig. 24-1. (A) Equatorial section and (B) cross section of platelet ultrastructure as revealed by electron microscopy. *EC* = exterior coat; *CM* = trilaminar unit membrane; *CS*, in A, *OCS* in B = open canalicular system; *MT* = microtubules; *Gly* = glycogen; *M* = mitochondria; *G* = alpha granules; *DB* = dense bodies; *DTS* = dense tubular system; and *GZ* = Golgi apparatus. (From: White, J. G., and Gerrard, J. M. Ultrastructural features of abnormal platelets: a review. *Am. J. Pathol.* 83:591, 1976. With permission from the American Association of Pathologists.)

contraction is provided by ATP through both anaerobic glycolysis and mitochondrial oxidative metabolism. Platelets contain several organelles. The largest is the electron **dense body** or **granule,** which contains ADP, calcium, and serotonin. Other organelles include the alpha granules, small mitochondria, and glycogen stores. The **alpha granules** contain platelet-specific proteins, such as beta-thromboglobulin,

platelet-derived growth factor, and platelet factor 4, as well as proteins also found in plasma (fibrinogen, albumin, and fibronectin). When stimulated, platelet secretion involves activation of the contractile machinery and release of storage granule contents into the open canalicular system and ultimately into the extracellular fluid space.

Platelets and the Response to Vessel Injury

Platelets are essential for normal hemostasis and have several important functions. These are: (1) maintenance of **vessel integrity** by sealing minor disruptions of the vascular endothelial surface; (2) formation of **hemostatic plugs** following vessel injury and disruption of the endothelium; (3) provision of **procoagulant activity** and phospholipid, which are required for efficient formation of fibrin, the end product of the coagulation pathway; and (4) provision of **platelet-derived growth factor,** which may be an important mitogen in the wound healing process and in the development of atherosclerosis.

Platelet Adhesion

Normally platelets circulate freely and do not adhere to the endothelial surface. Although the reason for this is not clear, it may be due to the basal secretion of **prostacyclin,** a potent inhibitor of platelet aggregation, by endothelial cells. However, when the endothelial surface is disrupted, platelets accumulate rapidly at the site of vessel injury in a process termed **platelet adhesion** (Fig. 24-2). This initiates platelet involvement in hemostasis and takes only seconds. This reaction is characterized by the formation of pseudopodia in the membranes of those platelets in contact with connective tissue elements in the subendothelial space. Platelets have a particularly high affinity for **collagen** and, under experimental conditions, will selectively migrate to this subendothelial protein.

The process of platelet adhesion requires a plasma cofactor protein, named **von Willebrand factor** (vWF) after the Finnish physician who first

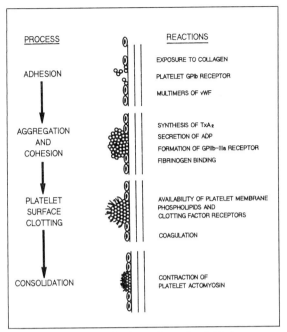

PROCESS

ADHESION

AGGREGATION
AND
COHESION

PLATELET
SURFACE
CLOTTING

CONSOLIDATION

REACTIONS

EXPOSURE TO COLLAGEN

PLATELET GPIb RECEPTOR

MULTIMERS OF vWF

SYNTHESIS OF TxA₂
SECRETION OF ADP
FORMATION OF GPIIb-IIIa RECEPTOR
FIBRINOGEN BINDING

AVAILABILITY OF PLATELET MEMBRANE
PHOSPHOLIPIDS AND
CLOTTING FACTOR RECEPTORS

COAGULATION

CONTRACTION OF
PLATELET ACTOMYOSIN

Fig. 24-2. Stages in the formation of a stable hemostatic plug in response to vessel injury. TxA_2 = thromboxane A_2; vWF = von Willebrand factor; ADP = adenosine diphosphate; $GPIb$, $GPIIb$-$IIIa$ = glycoproteins. (Modified from: Rapaport, S. I. Hemostasis. In: West, J.B., ed. *Physiological Basis of Medical Practice.* Baltimore: Williams & Wilkins, 1985, P. 412.)

described the inherited bleeding disorder associated with its deficiency. (It is frequently referred to as **factor VIII-vWF,** because factor VIII, one of the coagulation proteins, circulates as a complex with vWF, and, at one time, the two factors were thought to be the property of a single molecule.) vWF is synthesized by endothelial cells and is absorbed by circulating platelets and by exposed subendothelial tissue. A platelet membrane glycoprotein, **GP1b,** serves as the surface receptor for polymers of vWF. Hereditary absence of this glycoprotein receptor, known as the **Bernard-Soulier syndrome,** or absence of vWF, called **von Willebrand's disease,** results in defective platelet adhesion and a severe bleeding tendency. These characteristics indicate that both vWF and platelet membrane glycoproteins

are key elements in the interaction of the platelet membrane with the subendothelium.

Platelet Activation and Aggregation

Blood vessel disruption and the exposure of subendothelial tissue also results in the activation and aggregation of platelets. These overlapping events involve: (1) changes in platelet shape; (2) the release of ADP and serotonin from the dense granules; (3) liberation and oxidation of arachidonic acid from the platelet membrane; (4) formation of thromboxane A_2; and (5) the simultaneous formation of small amounts of thrombin by means of coagulation, with subsequent thrombin-induced changes in platelet function.

Collagen and the initial **thrombin** formed after vessel injury are the primary triggers of platelet activation and aggregation. After binding to specific glycoprotein receptors on the platelet membrane, these substances activate membrane phospholipases and liberate arachidonic acid (Fig. 24-3). The arachidonate in the platelet (but not in the endothelial cell) is rapidly converted to **thromboxane A₂.** Besides its potent vasoconstrictor ability, thromboxane A_2 mobilizes Ca^{2+} from various intracellular storage sites, particularly the dense tubular system. The resulting elevation in cytosolic Ca^{2+} levels is most likely the final mediator of platelet aggregation and release. Ca^{2+} then complexes with calmodulin and this leads to phosphorylation of platelet myosin, contraction of actomyosin, and release of intracellular granule contents (for example, ADP and serotonin). The elevated Ca^{2+} level also activates membrane phospholipases that further amplify the process.

As noted in Fig. 24-3, **aspirin** is a potent inhibitor of **cyclooxygenase,** the enzyme that converts arachidonate to the endoperoxides. A brief exposure to aspirin acetylates and inactivates cyclooxygenase. Furthermore, because the platelet is incapable of de novo synthesis of proteins and enzymes, inactivation of cyclooxygenase by aspirin is irreversible for the life span of the platelet.

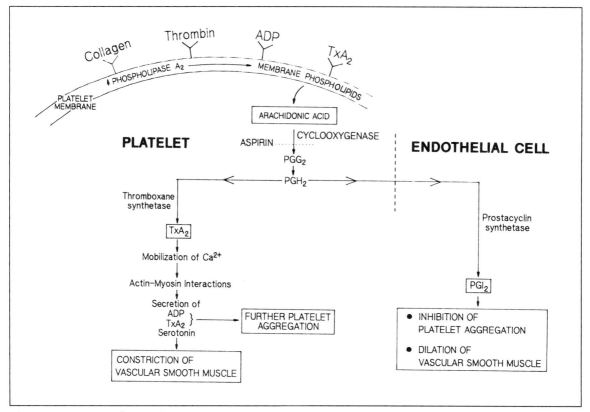

Fig. 24-3. Oxidation of arachidonic acid in the platelet in response to platelet aggregating agents. (*ADP* = adenosine diphosphate; *TxA₂* = thromboxane A₂; *PGG₂* and *PGH₂* = cyclic endoperoxides; *PGI₂* = prostacyclin.) Metabolism of arachidonate via the lipoxygenase pathway, leading to the formation of leukotrienes is not shown; its potential role in the platelet aggregation response is not clearly known. To the right of the dashed line is depicted the metabolism of endoperoxides by the endothelial cell.

Although aspirin does prolong the bleeding-time test of platelet function, it does not usually cause a bleeding tendency. Thus, unknown alternative mechanisms must operate in the platelet that induce aggregation and secretion and compensate for an aspirin-induced defect in the synthesis of thromboxane A₂. However, ingestion of aspirin by someone who has another abnormality that affects platelet number or function can precipitate a profound bleeding tendency.

The ADP released by platelets in contact with collagen or thrombin induces further platelet aggregation, and thereby increases the size of the platelet plug. ADP initiates aggregation by bind-ing to specific receptors on the platelet. Receptor occupation in turn mobilizes fibrinogen-binding sites, consisting of a complex between the platelet-membrane glycoproteins GPIIb and GPIIIa. The subsequent bridging between platelets is mediated by the Ca²⁺-dependent binding of fibrinogen molecules at the platelet membrane. Thus, normal platelet aggregation requires fibrinogen and Ca²⁺, and does not occur when GPIIb and GPIIIa are abnormal, such as witnessed in cases of **Glanzmann's thrombasthenia**. Finally, serotonin is also released by collagen-stimulated and thrombin-stimulated platelets, and causes vasoconstriction in regions containing smooth muscle fibers. The net effect of these reactions is a contracted blood vessel and a propagating platelet mass that may be sufficient (without fibrin formation via coagulation) to halt bleeding. However, depending on the severity and site of the injury, anchoring of the temporary hemostatic plug with fibrin may be necessary. In addition, the released ADP is slowly metabolized and the adherent platelets tend to disaggregate.

BLOOD COAGULATION

Overview

Curiosity and interest in the phenomenon by which liquid blood is transformed into a gel-like mass when removed from the body dates back to the days of Plato and Aristotle. It is only within the last few decades, however, that our understanding of the mechanisms and regulation of **blood coagulation** has reached the point where it can be applied in the treatment of coagulation disorders. The overall purpose of blood coagulation is to seal damaged vessels and to anchor platelet plug formation by forming a definitive insoluble fibrous meshwork, **fibrin,** from a soluble precursor in plasma, **fibrinogen**.

In simplest terms, the clotting of blood involves the enzymatic activation of several plasma proteins possessing highly specific functions, known as **coagulation factors**. This process consists of a series of cascading proteolytic reactions in which an inert or minimally active precursor is converted to an active proteolytic enzyme, a **serine protease,** which then cleaves and activates the next inert precursor. Despite considerable sequence identity in the coagulation factors, they also possess highly specific and unique substrate binding sites.

The cascading nature of the activation of coagulation allows for numerous points of control and for **amplification,** whereby the activation of a few molecules of initial protease translates into the formation of many molecules of product. In many of the reactions, Ca^{2+} is required and phospholipid (derived primarily from platelet membranes) and other substances serve as required cofactors. These cofactors are very important, for they (1) accelerate the reaction rates so that ample product is formed rapidly; (2) provide a "surface" upon which the reactions can occur, protected from inhibition and blood flow dilution; (3) confer specificity to the reactions by aligning the proteases with the correct substrates; and (4) are supplied primarily at the damaged tissue site, and thereby localize the coagulation process.

Nomenclature

Both Roman numerals and common names are used to identify the factors involved in blood coagulation (Table 24-1). Several years ago, an international research committee assigned Roman numerals to the various factors. Although most of the number designations are now commonly accepted, some of the synonyms are still used. For example, fibrinogen, prothrombin, tissue thromboplastin, and Ca^{2+} are rarely referred to as factors I, II, III, and IV, respectively. The numerals were assigned to reflect the order of the factors' discovery, and not the reaction sequence.

Most of the coagulation factors exist in plasma as inactive precursors, termed **procoagulants**. After activation by a process called **limited proteolysis,** these activated factors are identified by appending an "a" to the factor number (for example, inactive factor X is converted to factor Xa during the clotting process). Fibrinogen is an exception, for it is converted to an insoluble and nonenzymatic fibrin molecule that is the major structural protein of the blood clot.

Production of Coagulation Factors and the Role of Vitamin K

All of the factors circulate in plasma in concentrations ranging from about 1 to 150 µg/ml, except fibrinogen, which is about a thousand-fold greater at 2.5 mg/ml (See Table 24-1). Thus, under normal circumstances, there is virtually an inexhaustible supply of fibrinogen as the final substrate for coagulation (and thrombosis). The liver manufactures all the factors except factor

Table 24-1. Glossary of Coagulation Proteins

Factor	Synonym	Plasma Concentration[a]		Half-life (days)
		(μg/ml)	(nM)	
I	Fibrinogen	2,500	7,000	5
II	Prothrombin	100–150	2,000	3
III	Tissue thromboplastin, tissue factor	0	0	—
V	Proaccelerin, labile factor	7	1	1.5
VII	Proconvertin, stable factor	0.5	10	0.2
VIII	Antihemophilic factor	5–10	—[b]	0.5
IX	Christmas factor	4	70	1
X	Stuart factor	8	140	2
XI	Plasma thromboplastin antecedent	4	25	2.5
XII	Hageman factor	29	360	2
XIII	Fibrin-stabilizing factor	8	25	7
Prekallikrein	Fletcher factor	45	510	?
HMWK	Fitzgerald factor	70	583	?

[a] Values derived from ranges reported in literature.
[b] Value depends on which estimate of molecular weight is used.
HMWK = high-molecular-weight kininogen.

VIII, the antihemophilic factor, which is most likely produced in endothelial cells or megakaryocytes, or both places. (Recall that factor VIII circulates as a complex with vWf.) Therefore, with the exception of factor VIII, the plasma concentrations of all clotting proteins decline dramatically in severe liver disease, thus increasing the risk of a hemorrhagic condition.

Vitamin K is a fat-soluble vitamin necessary for the synthesis of certain Ca^{2+}-binding domains in factors II (prothrombin), VII, IX, and X (Fig. 24-4, as well as protein C (discussed in Chapter 25). Vitamin K is needed for the hepatic post-translational formation of gamma-carboxyglutamic acid residues (GLA) that are formed by the addition of a second carboxyl group at the gamma-carbon of selected glutamic acid residues. The additional carboxylation takes place at ten to twelve specific glutamyl residues in the amino-terminal portion of the molecules. Because of their tertiary conformation, pairs of GLA residues form a unique negative-charge density and a Ca^{2+} binding site that permits the vitamin K–dependent factors to bind to phospholipid via Ca^{2+} bridges.

Vitamin K deficiencies can lead to severe bleeding tendencies and ineffective coagulation. When this happens, normal amounts of vitamin K–dependent proteins are synthesized, but they lack the unique GLA residues and are therefore ineffective during the coagulation process. (As will be discussed, this same mechanism is the underlying principle in the treatment of venous thrombosis using vitamin K antagonists, such as warfarin, which render certain key coagulation factors ineffective and thereby block clot formation.) Vitamin K deficiencies can be brought about by inadequate dietary intake or by fat malabsorption in the intestine. The vitamin is also produced endogenously by gastrointestinal bacteria and then absorbed. Thus, unless there is malabsorption, vitamin K deficiency can only occur because of either inadequate oral intake or the use of broad-spectrum antibiotics that block endogenous production.

Most of the clotting factors have very short in-

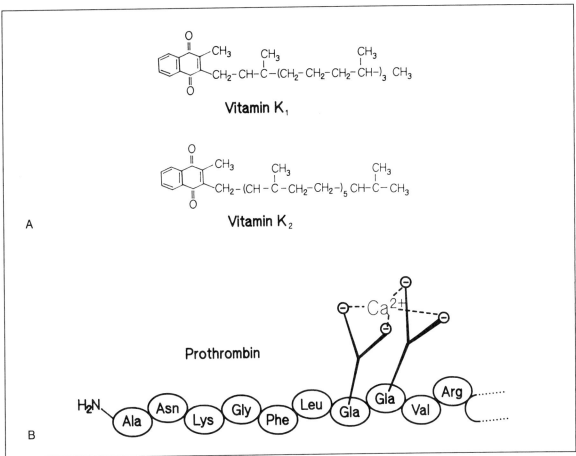

Fig. 24-4. Structure and action of vitamin K. (A) Two forms of vitamin K, one from dietary sources (plant, Type K_1) and the other from bacteria (type K_2). Vitamin K is a napthoquinone derivative with a hydrophobic side-chain that differs in the two forms. (B) An example of the vitamin K-dependent, posttranslational insertion of gamma-carboxy-glutamic acid residues (*GLA*) into the amino-terminal sequence (*Ala, Asn, Lys,* etc.) of prothrombin. Several similar insertions are made into each of the vitamin K-dependent clotting factors. The resulting GLA residues form a negatively charged region that serves as a Ca^{2+}-binding domain through which the factors bind to phospholipid surfaces. (Modified from: Ogston, D. *The Physiology of Hemostasis.* Cambridge: Harvard University Press, 1983; and from: Thompson, A.R., and Harker, L. A. *Manual of Hemostasis and Thrombosis,* 3rd ed. Philadelphia: F. A. Davis, 1983, P. 119.)

travascular half-lives, relative to other plasma proteins; these times range from just a few hours (factors VII, VIII, and IX) up to seven days (factor XIII) (see Table 24-1). Therefore, a patient with a bleeding disorder attributed to a deficiency in a factor that is rapidly metabolized often requires repeated infusions of factor concentrates in order to maintain stable and normal levels. Experimental kinetic studies have indicated that activation and consumption of coagulation factors do not account for their rapid catabolism. The liver is particularly effective in clearing the circulation of activated procoagulants, thereby reducing the tendency for dissemination of the clotting process.

Enzymatic Pathways of Coagulation

Historical Perspective

The first clotting protein, discovered by Malpighi in the 17th century was **fibrin**—the insoluble ropelike meshwork of a blood clot that is formed rapidly when blood is exposed to foreign surfaces. It was soon recognized that fibrin cannot exist as such in the normal circulation, and a soluble precursor of the molecule, now known as **fibrinogen,** was correctly predicted. Further information about the mechanism by which fibrinogen is converted to fibrin eluded detection until the late 1800s, when some key observations were made by Buchannan and Schmidt. These pioneering scientists noted that a substance was formed during the clotting process (was present in serum but not in plasma) which could hasten the conversion of fibrinogen to fibrin and thus the coagulation of freshly drawn blood. This process was later confirmed to be activated by a proteolytic enzyme now known as **thrombin**. This discovery led to the obvious postulation that thrombin must exist in circulating blood as an inactive precursor, termed **prothrombin**.

Only within the last few decades have we understood the mechanisms responsible for converting prothrombin to thrombin. During this period, numerous coagulation factors were discovered to be essential components of normal hemostasis. Nearly all the factors were detected by studies on patients with bleeding disorders caused by an inherited deficiency of the factor. In fact, many of the hemorrhagic diseases and the deficient factors involved were named after the family or persons afflicted (for example, Hageman factor and Hageman trait; Christmas factor and Christmas disease). Although some of these common names are still used, the factors are now generally referred to by their Roman numeral designations.

Mechanisms of Thrombin Evolution

The reactions leading to the formation of thrombin from prothrombin can be simplified and divided (for teaching purposes only) into three major enzymatic pathways—the **intrinsic, extrinsic,** and **common** pathways (Fig. 24-5). All of the components required for the intrinsic pathway to operate exist in circulating blood, hence the name "intrinsic." However, the extrinsic pathway is initiated by a factor not normally present in the bloodstream but supplied by damaged tissue, hence the "extrinsic." Activation of either of these two reaction chains precipitates activation of factor X, which in turn converts prothrombin to thrombin, followed by obligatory conversion of fibrinogen to fibrin. That portion of the overall reaction sequence shared by both the intrinsic and extrinsic pathways, beginning at the factor X step, is the common pathway. Although the intrinsic and extrinsic pathways can be preferentially activated in the test tube for diagnostic purposes, in vivo they both operate in response to vessel injury and both must function for normal hemostasis to take place. In other words, a factor deficiency in one pathway cannot be fully compensated for by activation of the other intact pathway.

The Intrinsic Pathway   The intrinsic pathway, leading to the activation of factor X, is initiated by blood coming in contact with a negatively charged foreign surface and involves six clotting proteins: factor XII, plasma prekallikrein (PK), high-molecular-weight kininogen (HMWK), factor XI, factor IX, and factor VIII. The initial stage of this pathway involves the first four factors, and is sometimes referred to as the **contact phase** of coagulation because the early reactions are triggered by contact with a foreign surface (Fig. 24-6). This phase has some unique characteristics compared with other reactions in the coagulation process. For example, the reactions during the contact phase do not depend on Ca^{2+} and the factors involved are avidly absorbed onto the foreign surfaces where the reactions occur. The factors involved in contact activation

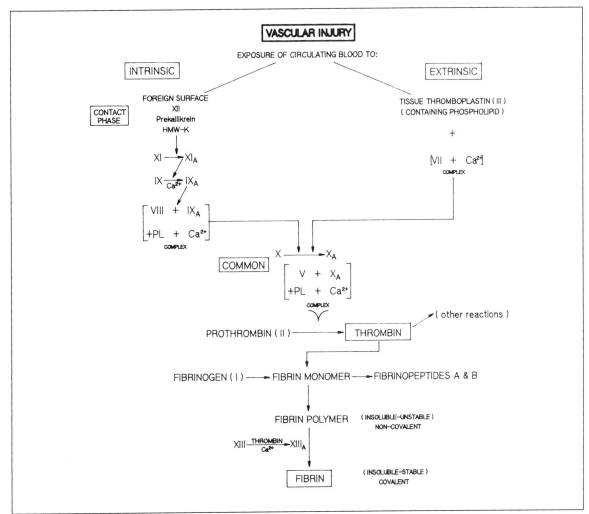

Fig. 24-5. The intrinsic, extrinsic, and common enzymatic pathways of blood coagulation, (See text for detailed description.) (*HMW-K* = high-molecular-weight kininogen; *PL* = phospholipid.)

of blood coagulation may also participate in other host defense mechanisms, including fibrinolysis, kinin generation in the inflammatory response, and complement activation. Finally, although the contact phase can be readily demonstrated in vitro and tested, the pathophysiologic significance of these initial reactions remains questionable because individuals with defects in these factors are largely asymptomatic. In fact, of all the contact factors, only severely depressed levels of factor XI are associated with abnormal bleeding.

Experimental studies have revealed that, when factor XII, PK, HMWK, and factor XI are properly assembled on a negatively charged surface, factor XI is rapidly activated to XIa (see Fig. 24-6). The enzyme responsible for catalyzing the formation of XIa is activated-factor XII (XIIa). The precise sequence of events in the initial activation of factor XII is not fully understood. However,

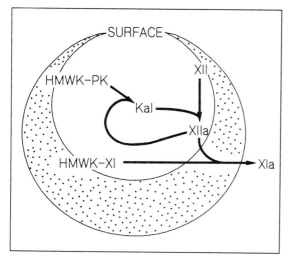

Fig. 24-6. The contact phase of blood coagulation, showing the reactions leading to the activation of factor XI when blood is exposed to a negatively charged foreign surface. (See text for the other explanation.) (*HMWK* = high-molecular-weight kininogen; *PK* = prekallikrein; *Kal* = kallikrein.)

reciprocal-activation reactions are the most likely source of the response. For example, only a few molecules of XIIa, formed perhaps by a conformational realignment of active sites in response to the negatively charged surface, convert PK to kallikrein, which in turn activates additional factor XII. HMWK has no known enzymatic function, but serves as a necessary cofactor in several of these reactions. Both PK and factor XI exist in plasma as a complex with HMWK. Thus, HMWK probably serves as a carrier molecule, bringing PK and factor XI close to the surface-bound factor XII. In this way, HMWK enhances the ability of factor XIIa to activate not only factor XI, but also PK to kallikrein, which in turn generates additional factor XIIa.

In the presence of Ca^{2+}, factor XIa next activates factor IX by limited proteolysis at two sites. Once activated, factor IXa converts inactive factor X to its active form (Xa) in the presence of factor VIII, phospholipids, and Ca^{2+}. This reaction takes place efficiently only when the factors first form a multimolecular complex on the phospho-

lipid surface, with factors IXa and X bound at their GLA residues to the phospholipid via Ca^{2+} bridges. In vivo, the phospholipid is provided by the platelet membrane and factor VIII serves as a cofactor in this process. (Both the availability of platelet phospholipid and the binding of factor VIII to the lipid matrix are greatly enhanced by thrombin.) The formation of the multimolecular assembly on the phospholipid surface accelerates the reaction by targeting and immobilizing the substrates and enzymes as well as minimizing the effect of blood-flow dilution.

The Extrinsic Pathway An alternative mechanism for the activation of factor X by means of the extrinsic pathway is simultaneously initiated when blood is exposed to underlying tissue (see Fig. 24-5). The combination of a ubiquitous tissue lipoprotein, termed **tissue factor** or **tissue thromboplastin** (factor III), with factor VII, a plasma protein, triggers rapid activation of factor X to Xa. Factor VII, a vitamin K–dependent clotting factor present in very low levels in plasma, binds to the phospholipid portion of factor III via its GLA residues and Ca^{2+}. The resulting complex then develops an active site in the factor VII portion to form factor VIIa, which then converts factor X to Xa by a mechanism identical to that described for the factor IXa–factor VIII–Ca^{2+}–phospholipid complex.

Feedback between Intrinsic and Extrinsic Pathways Recent experimental evidence has indicated that **feedback mechanisms** and interactions may occur at several points between the intrinsic and extrinsic pathways. For example, the activation of factor XII, besides triggering the intrinsic pathway, may activate factor VII in the extrinsic pathway. Also, the factor VIIa–factor III–Ca^{2+} complex may activate factor IX in the intrinsic pathway. However, these findings were obtained using purified factors and substrates; so the physiologic significance of these interactions remains to be elucidated.

The Common Pathway In a fashion similar to the activation of factor X by factor IXa, the activation of prothrombin by factor Xa requires factor V, phospholipid, and Ca^{2+} as cofactors. (As with factor VIII in the activation of factor X, both the availability of phospholipid from platelet membranes and the binding of factor V to the lipid matrix are greatly enhanced by thrombin.) The vitamin K–dependent proteins, prothrombin and factor Xa, bind to the platelet phospholipid via GLA residues and Ca^{2+} bridges. Factor V, a very large molecule (250 to 300 kDa), also binds to the platelet phospholipid and to prothrombin. Formation of this multimolecular complex on the phospholipid surface requires factor V as a nonenzymatic cofactor that efficiently aligns the substrate (prothrombin) and the enzyme (factor Xa) for optimal interaction. Factor Xa induces two sequential proteolytic cleavages in prothrombin. First, an amino-terminal fragment and a carboxy-terminal fragment, called **prethrombin,** are released. Cleavage at a second site on prethrombin yields the final active protease of coagulation, **thrombin**.

Formation and Stabilization of Fibrin

Fibrinogen is a very large glycoprotein (340 kDa) that circulates in plasma at high concentrations of between 2 and 4 mg/ml. It is a dimeric molecule with two identical halves, each consisting of three polypeptide chains, the $(A\alpha)_2$, $(B\beta)_2$, and $(\gamma)_2$ chains (Fig. 24-7 A). The chains are connected by disulfide bonds, which also bridge the two halves through their Aα and γ chains near the amino terminus. The intact molecule has a trinodular structure, as revealed by electron microscopy. The distal nodules are composed of the carboxy segments of the β and γ chains, with the α chains extending outward from this region. The central nodule is composed primarily of the amino-terminal regions of each set of three chains.

Fibrinogen is converted to fibrin in three distinct states. In the **first stage,** thrombin, in a highly specific fashion, cleaves fibrinogen at the amino-terminal end of each α and β chain to yield four small fragments, termed **fibrinopeptides** (two fibrinopeptides A and two fibrinopeptides B) (see Fig. 24-7 B and C). The A peptide (16 amino

acids) is released before the B peptide (14 amino acids). The action of thrombin thus exposes new amino acid residues in the central nodule region that serve as binding sites for subsequent polymerization of the molecule. Once the fibrinopeptides have been cleaved from fibrinogen, the resulting molecule is called a **fibrin monomer**.

In the **second stage,** molecules of fibrin monomer polymerize in a nonenzymatic fashion by end-to-end and side-to-side binding to form a **fibrin polymer** (Fig. 24-8). The binding of monomers to form fibrin polymers is brought about by weak, noncovalent interactions. If there are enough fibrin polymers, the interactions cause the formation of fibrin strands that are insoluble in plasma. This type of fibrin polymer, in the absence of factor XIIIa, is readily soluble in denaturing agents such as urea and monochloroacetic acid, and is also very susceptible to hydrolysis by the fibrinolytic enzyme **plasmin** (see discussion of fibrinolysis in Chapter 25).

The **final stage** of coagulation is stabilization of the fibrin molecule by factor XIII, which is activated to its enzymatic form, **factor XIIIa,** by limited proteolysis with thrombin in the presence of Ca^{2+}. Factor XIIIa is a **transglutaminase** that catalyzes the formation of covalent bonds between adjacent side-chains of the fibrin polymer (Fig. 24-9). Specifically, amide linkages form rapidly between a gamma-carbonyl group of glutamine on one gamma chain with an epsilon-amino group of lysine in an adjacent gamma chain, with the chains oriented in an antiparallel fashion. Although slower, similar linkages are formed between adjacent alpha chains. The net effect is an increase in the mechanical ridigity of fibrin and the production of a tightly cross-linked clot that is insoluble in denaturing agents. The stabilized fibrin molecule is highly resistant to the action of plasmin, the fibrinolytic enzyme. Recent evidence indicates that this property probably stems from the fact that the primary inhibitor of plasmin, **alpha₂-antiplasmin,** is also cross-linked to fibrin by factor XIIIa.

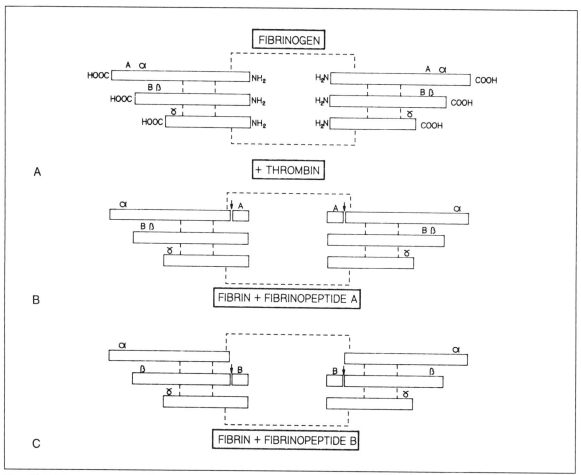

Fig. 24-7. The conversion of fibrinogen to fibrin monomer by thrombin. (A) Intact fibrinogen. (B and C) In the presence of thrombin, two pairs of fibrinopeptides *A* and *B* are sequentially removed from fibrinogen to yield monomeric fibrin. (*Broken lines* = disulfide bonds.) (Modified from: Ogston, D. *The Physiology of Hemostasis.* Cambridge: Harvard University Press, 1983, P. 90.)

Coagulation Kinetics and the Key Roles of Thrombin

Thrombin is central to the overall process of coagulation. In addition to converting fibrinogen to fibrin and activating factor XIII, it also potentiates factors V and VIII in their binding to phospholipid-Ca^{2+} matrices. Also important is thrombin's ability to indirectly furnish phospholipid from the platelet by virtue of its potent effect on the platelet membrane (see Fig. 24-3). The

phospholipid so derived provides a surface where coagulation reactions can proceed rapidly and efficiently, most notably those involving factors V and VIII.

Thrombin evolution via the extrinsic pathway is initially much faster than that via the intrinsic pathway, due to the presence of fewer rate-limiting reaction sequences. The efficient operation of the intrinsic pathway requires phospholipid at the factor IXa–VII-Ca^{2+} stage. However, once

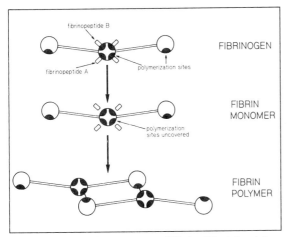

Fig. 24-8. The polymerization of fibrin. The trinodular structure of fibrinogen and fibrin is represented as spheres. The central nodule, containing the thrombin cleavage sites, is composed primarily of the NH_2-terminal ends of each set of the three polypeptide chains, while the distal nodules comprise the carboxy-terminal portions of the chains. After release of the fibrinopeptides by thrombin, polymerization sites are exposed in the central nodule region. Spontaneous polymerization then occurs between this portion of the molecule, with other sites located in the distal nodules of adjacent fibrin monomers. (Modified from: Ogston, D. *The Physiology of Hemostasis.* Cambridge: Harvard University Press, 1983, P. 91.)

phospholipid is available (because of thrombin acting on platelet membranes), the intrinsic pathway becomes fully operative. Furthermore, **amplification** occurs while the intrinsic pathway is being activated, in that the formation of a few molecules of activated product lead to the activation of many more molecules in the next step. Thus, in response to vessel injury, a small amount of thrombin is rapidly manufactured, primarily (but not exclusively) from activation of the extrinsic pathway. Through its action on platelets and factors V and VIII, thrombin, in a positive-feedback fashion, then allows unimpeded operation of the intrinsic pathway, explosive generation of more thrombin, and subsequent conversion of fibrinogen to fibrin. A theoretical view of these events is depicted in Figure 24-10. To reiterate, both pathways must be operative in order to achieve the critical thrombin concentration necessary to convert enough fibrinogen to fibrin, and hence to ensure effective hemostasis and cessation of bleeding.

Figure 24-10 also illustrates the need to restore the thrombin evolution curve to baseline values, by limiting the formation of thrombin once hemostasis is effective. This is particularly important, because fibrinogen, the substrate for thrombin, normally exists in very high concentrations, and therefore itself cannot limit the coagulation process. Hence, two very important questions that will be addressed in Chapter 25 are: (1) how is the coagulation of blood restricted to the site of vessel injury? and (2) how is this potentially explosive system regulated so that coagulation does not disseminate or take place in the absence of vessel injury (thrombosis)?

SUMMARY

When a vessel is injured, a series of host-defense reactions is initiated at the interface of the blood and vessel wall to arrest bleeding and achieve effective hemostasis. Platelets play a key role in this process by forming a temporary hemostatic plug, which, in addition to smooth muscle contraction, may suffice to halt bleeding from small vessels. The biochemical mechanisms underlying platelet activation and the aggregation response to collagen and thrombin have been largely elucidated. Platelet membranes also contribute phospholipid, which is an important cofactor in the coagulation response. Coagulation is initiated via two pathways when blood is exposed to foreign surfaces and to tissue thromboplastin. A series of cascading reactions, featuring limited proteolysis, amplification, and feedback, produces the insoluble fibrin matrix. Thrombin, platelet phospholipid, and Ca^{2+} play multiple key roles in coagulation. The explosive generation of thrombin, with subsequent formation of fibrin, is a high-risk system that must be carefully regulated to ensure that coagulation is restricted to the site of vessel injury.

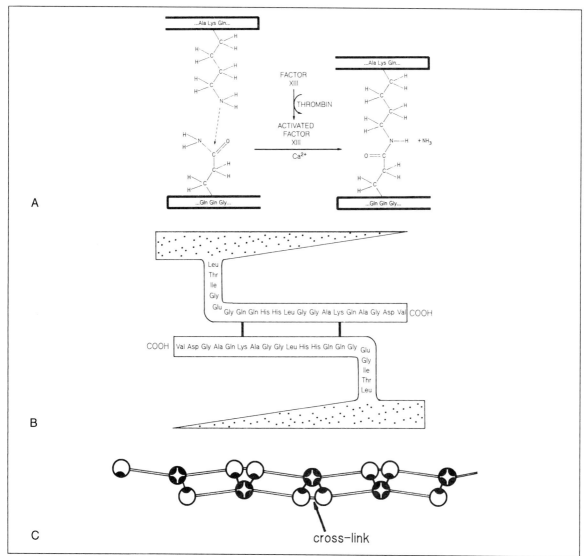

Fig. 24-9. Stabilization of fibrin by activated factor XIII. (A) Factor XIII is activated by thrombin and, in the presence of Ca^{2+}, catalyzes peptide-bond formation between glutamine and lysine residues of adjacent molecules of fibrin. (B) The antiparallel arrangement of adjacent gamma chains and the location of the covalent linkages catalyzed by factor XIIIa. (C) Spatial arrangement of trinodular fibrin molecules after cross-linking with factor XIIIa. (A and B modified from: Doolittle, R. F. Fibrinogen and fibrin. *Sci. Am.* 245:126-135, 1981; C modified from: Ogston, D. *The Physiology of Hemostasis.* Cambridge: Harvard University Press, 1983, P. 91.)

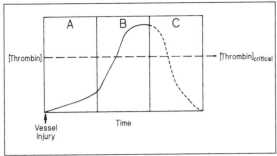

Fig. 24-10. Hypothetical thrombin concentration as a function of time after vessel injury. (A) Thrombin evolution primarily activated by the extrinsic pathway. (B) Thrombin evolution primarily activated by the intrinsic pathway. Both pathways must be operative in order to achieve the critical thrombin concentration necessary to generate fibrin in amounts sufficient for effective hemostasis. (C) Regulation of thrombin evolution and plasma levels by regulatory mechanisms.

NATIONAL BOARD TYPE QUESTIONS

Select the one most correct answer for each of the following

1. Platelets:
   A. Are multinucleated cells with a half-life of about 10 days
   B. Actively bind and internalize serotonin, thereby inactivating this vasoactive substance
   C. Undergo a dramatic change in shape and function when stimulated by collagen or thrombin, a process inhibited by cyclooxygenase antagonists such as aspirin
   D. Contain a single nucleus, numerous mitochondria, and intracellular stores of glycogen
   E. At a count of 250,000 per microliter of blood would produce a profound bleeding tendency
2. When platelets contact subendothelial collagen,
   A. The collagen fibrils are rapidly degraded by platelet-derived thrombin
   B. The platelet-release reaction is initiated,
   whereby the platelets swell, the cell membrane disintegrates, and the intracellular contents are released
   C. They are activated by tissue thromboplastin (factor III) and the clotting sequence begins
   D. ADP is released from the collagen fibrils and causes further platelet aggregation
   E. Antithrombin is released from the platelet, inactivates thrombin, and thereby restricts the coagulation process to the site of vascular injury
3. Which of the following is a normal function of thrombin?
   A. Functions as a transglutaminase that directly and independently cross-links adjacent fibrin polymers
   B. Directly causes the polymerization of soluble fibrin monomers to form fibrin polymers that are insoluble in urea
   C. Provides phospholipid through its effect on platelet membranes
   D. Can activate the clot-dissolving (fibrinolytic) system, in combination with factor XIII and Ca^{2+}
   E. Potentiates the biologic activity of factor VII
4. Which of the following statements concerning vitamin K is *correct?*
   A. Vitamin K is a fat-soluble vitamin necessary for normal platelet function
   B. The synthesis of Ca^{2+}-binding GLA residues in factors II (prothrombin), VII, IX, and X requires vitamin K
   C. Thrombin's action on platelet membranes requires vitamin K
   D. The only source of vitamin K is from dietary intake
   E. Although the synthesis of platelets does not require vitamin K, their ability to aggregate in response to collagen strongly depends on vitamin K
5. Which of the following statements concerning coagulation factors is *correct?*
   A. All of the factors are synthesized in the liver

B. The concentration of the factors in plasma ranges from about 1 to 150 μg/ml, with the exception of fibrinogen which is about a thousand-fold greater at 2.5 mg/ml

C. In severe liver disease, the plasma levels of all factors are equally suppressed

D. Activation and consumption are the primary pathways by which the factors are normally catabolized

E. The intravascular half-lives for each of the factors do not exceed one day

ANNOTATED ANSWERS

1. C. Cyclooxygenase inhibitors block the platelet aggregation response.
2. B. These are the normal reactions that occur when platelets come in contact with collagen.
3. C. Thrombin is a potent stimulus initiating platelet aggregation, release, and disintegration of the plasma membrane, thereby providing phospholipid.
4. B. This is the only correct statement concerning vitamin K.
5. B. This is the only correct statement concerning coagulation factors.

BIBLIOGRAPHY

Bowie, E. J. W., and Sharp, A. A., eds. *Hemostasis and Thrombosis.* London: Butterworth, 1985.

Doolittle, R. F. Fibrinogen and fibrin. *Sci. Am.* 245:126-135,1981.

Ogston, D. *The Physiology of Hemostasis,* Cambridge: Harvard University Press, 1983.

Rapaport, S. I. Hemostasis. In: West, J. B., ed. *Physiological Basis of Medical Practice,* llth ed. Baltimore: Williams & Wilkins, 1985, Pp. 409-432.

Ratnoff, O. D., and Forbes, C. D., eds. *Disorders of Hemostasis.* Orlando, FL: Grune & Stratton, 1984.

Thompson, A. R., and Harker, L. A. *Manual of Hemostasis and Thrombosis,* 3rd ed. Philadelphia: F. A. Davis, Co. 1983.

25 Regulation of Coagulation and Pathophysiology of Hemostasis

ROBERT F. HIGHSMITH

Objectives

After reading this chapter, you should be able to:

List the major regulatory factors that localize and restrict the coagulation process

Explain the role of blood flow dilution and consumption of coagulation factors in the regulation of coagulation

Compare and contrast the counterbalancing effects of arachidonic acid metabolism in the platelet versus that in the endothelial cell

Describe the role of protease inhibitors in the regulation of coagulation and predict the consequences of their absence

Describe the components of the fibrinolytic enzyme system and indicate how it is physiologically and therapeutically activated

Explain the mechanism by which fibrin and fibrinogen degradation products are formed and define their effects on coagulation

Describe the thrombomodulin–protein C anticoagulant system and discuss its importance in the regulation of coagulation

Summarize and integrate the reactions of the entire hemostatic mechanism, including how it is initiated and regulated

Define *hemorrhage* and *thrombosis* in terms of what components of the hemostatic mechanism are affected

Describe the basic principles underlying the most common tests of platelet and coagulation factor function and identify the specific causes of abnormal bleeding based on the results of these tests

Compare arterial and venous thrombosis, including the calculation of platelet and fibrinogen turnover

Explain the mechanisms of action of heparin, warfarin, and plasminogen activators and their use in the treatment of thrombosis

This chapter is divided into two major sections. The first presents the mechanisms regulating blood coagulation and a summary of the integrated reactions of hemostasis. The second part is devoted to the pathophysiology of hemostasis, including a discussion of the most common tests of hemostatic function and examples of specific hemorrhagic and thrombotic disorders.

REGULATION OF COAGULATION

Under normal circumstances, blood remains fluid at all times; platelet fusion and fibrin deposition occur only at sites of vessel injury and only in amounts sufficient to halt bleeding. Because of the virtually unlimited supply of platelets and fibrinogen and the dire consequences of dissemination of the platelet–fibrin mass, an abundance of mechanisms has evolved to regulate hemostasis. Even more astounding is how effective hemostasis can take place despite such a wealth of inhibitory and regulatory elements.

Stimulus Localization and Blood Flow

The endothelium lines the entire vasculature in a continuous fashion; it is also nonthrombogenic and normally does not activate platelets or the coagulation process. This unique property of endothelial cells is in large part due to their basal secretion of **prostacyclin** (PGI$_2$), a vasodilator and potent inhibitor of platelet aggregation, as well as the production of substances that inhibit or oppose coagulation (activators of fibrinolysis, antithrombin, and protein C). Thus, hemostasis is a potential process that is only triggered by local perturbations of the endothelium, such as those caused by vessel injury. In addition, the cofactors necessary for effective hemostasis are provided locally, do not circulate, and serve as sinks for binding certain activated coagulation factors.

One of the most important activities that localize and restrict hemostasis is the rapid movement of fresh, "unactivated" blood through the vessels, causing substantial dilution of the procoagulants activated at the site of injury. In addition, the shear forces and turbulence created by flowing blood mechanically oppose clot formation. The importance of blood flow in opposing coagulation is clearly illustrated by the increased incidence of deep venous thrombosis in bedridden patients, when the beneficial effects of blood flow are minimal. The injection of activated coagulation factors into experimental animals also does not produce thrombosis unless a vessel segment is occluded shortly after the infusion.

Consumption and Catabolism of Coagulation Factors

During the clotting process, several components in plasma are consumed. For example, the cell-free fluid that can be obtained from a clot after coagulation (**serum**) is depleted of fibrinogen, prothrombin, factors V, VIII, and XIII, as well as platelets. This consumption constitutes a local means of limiting coagulation, as the levels of these components must be replenished before clotting can proceed. In the event of minor vessel injury, the circulating levels of these components are not affected and they are replenished by the delivery of fresh plasma to the injury site. However, in conditions involving massive systemic coagulation, such as **disseminated intravascular coagulation,** circulating levels of these factors may be depleted and this fosters a bleeding tendency. In this instance, replenishment requires de novo synthesis of new protein and platelets, or, in severe cases, transfusion of whole blood, plasma, or factor concentrates.

Because of their high molecular weight, coagulation factors are not filtered by the kidney. It is the liver that has a major role in both the synthesis and degradation of most coagulation factors. Thus liver disease is frequently accompanied by marked alterations in the coagulation system. In terms of the regulation of coagulation, the liver can remove activated coagulants from the bloodstream, but not their unactivated procoagulant forms. For example, experimental infusion of activated coagulant proteins into the systemic circulation rapidly induces formation of thrombi following venous occlusion, but not if these proteins are infused into the portal vein. Isolated liver perfusion experiments have also indicated that factor Xa is selectively removed during passage through the liver but not factor X. Finally, reticuloendothelial cells such as macrophages efficiently bind and remove soluble fibrin polymers but do not bind the precursor, fibrinogen.

Platelet–Endothelial Cell Interactions

As mentioned in Chapter 24, very few platelets normally adhere to the luminal surface of the vessel wall. The nonthrombogenic nature of this surface is in large part attributable to the basal secretion from the endothelium of the products of arachidonic acid metabolism. In both the platelet and endothelial cell, activation of **phospholipase A₂** liberates arachidonate from membrane phospholipids, which is then rapidly oxidized by cyclooxygenase to form endoperoxides (see Fig. 24-3). In the platelet, **thromboxane synthetase** converts the endoperoxide, PGH₂, to the potent mediator of platelet aggregation and vasoconstriction, **thromboxane A₂**. However, in the endothelial cell, oxidation of arachidonic acid generates PGI₂ because of the **PGI₂ synthetase** present in this cell. In direct contrast to thromboxane A₂, PGI₂ is a potent inhibitor of platelet aggregation and a vasodilator. Thus, by virtue of its basal secretion of PGI₂, the endothelial cell counterbalances the release of thromboxane A₂ by stimulated platelets. Furthermore, it has been shown in vitro that endoperoxides released from aggregating platelets may be taken up by endothelial cells, metabolized, and released as PGI₂. Therefore, as activated platelets come into contact with an intact endothelial cell capable of producing PGI₂, the transferal of endoperoxide substrates from the platelet to the endothelial cell may represent another means of limiting the growth of platelet hemostatic plugs.

Inhibition of Activated Coagulation Factors

Human plasma contains an abundance of naturally occurring **protease inhibitors** that effectively block the activity of proteolytic enzymes, particularly those possessing serine residues at the active site (Table 25-1). These inhibitors collectively ensure that any "spillover" of serine proteases (for example, activated coagulant proteases) at the site of vessel injury is rapidly neutralized in the systemic circulation. In vitro, the inhibitors display considerable overlap in their specificity toward different proteases. However, the diseases associated with inherited deficiencies of each inhibitor are very specific, suggesting distinct in vivo roles for each inhibitor.

Antithrombin (sometimes referred to as antithrombin III or heparin cofactor) is particularly important in the overall regulation of coagula-

Table 25-1. Plasma Protease Inhibitors

Inhibitor	Plasma Concentration (mg/dl)[a]	Deficiency State
Alpha₁-antitrypsin	200–400	Pulmonary emphysema
Alpha₂-macroglobulin	150–350	?
Antithrombin (antithrombin III or heparin cofactor)	18–30	Thrombosis
C1 inactivator (C1 esterase inhibitor)	15–35	HANE
Alpha₂-antiplasmin (alpha₂-plasmin inhibitor)	5–7	Hemorrhage
Protein C	0.4	Thrombosis

[a] Values derived from Ratnoff and Forbes, 1984, P. 41. HANE = hereditary angioneurotic edema.

tion. This inhibitor forms a stable covalent complex with thrombin (as well as with factors XIIa, XIa, Xa, IXa, and kallikrein), so as to completely and irreversibly block the serine residues of these enzymes at the active site. If the anticoagulant **heparin** is present (and perhaps heparin-like mucopolysaccharides in the vessel wall), the rate of inhibition of thrombin by antithrombin is greatly accelerated. Thus, antithrombin is a necessary cofactor in the anticoagulant activity of heparin. (The molecular basis of these interactions is discussed later in this chapter.) The importance of antithrombin in the regulation of coagulation is exemplified by the severe thrombotic episodes that affect patients with hereditary deficiencies of this inhibitor.

With the exception of alpha₂-macroglobulin, hereditary defects in each of the other antiproteases are also closely associated with specific clinical syndromes. Although the manifestations of these diseases vary widely, a common feature is unopposed proteolysis. Like antithrombin deficiency, the other syndromes illustrate the importance of inhibitors as regulators, not only for blood coagulation but for other proteolytic systems as well.

Activation of Fibrinolysis

Fibrinolysis is the naturally occurring process that dissolves fibrin clots. Like coagulation, the fibrinolytic enzyme system is a potential one that must be activated (Fig. 25-1). When an endogenous or exogenous activator substance is present, the single-chain inactive zymogen, **plasminogen,** is converted by limited proteolysis to a two-chain active serine protease, **plasmin**.

If plasmin is formed in plasma devoid of clots, it is rapidly and irreversibly inhibited by alpha$_2$-antiplasmin in a fashion similar to the inhibition of thrombin by antithrombin (a 1 to 1 covalent complex of inhibitor and enzyme). Under unusual circumstances, such as alpha$_2$-antiplasmin deficiency or repeated activation of plasminogen by the administration of exogenous activators, free plasmin can be generated in plasma, with consequential widespread proteolysis. Although rare, this can lead to the breakdown of several proteins necessary for normal coagulation, such as factors V and VIII as well as fibrinogen, thereby precipitating a hemorrhagic condition. However, because of the high affinity of both plasminogen activator and plasminogen for fibrin, the physi-

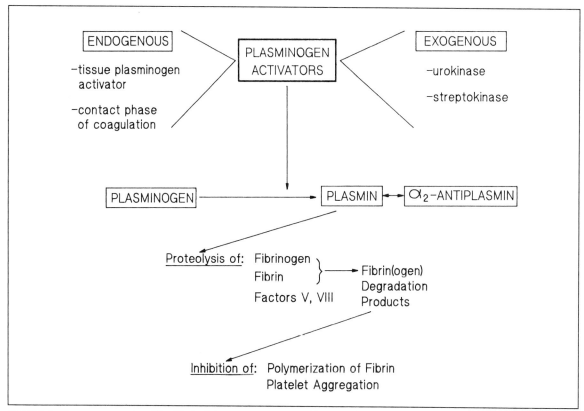

Fig. 25-1. The components of the fibrinolytic enzyme system. (See text for elaboration.)

ologic activation of plasmin most likely takes place in situ (within the fibrin clot), in an environment protected from inhibition.

When activated on the fibrin surface, plasmin hydrolyzes peptide bonds in fibrin to yield progressively smaller fragments, called **fibrin degradation products**. If generated in plasma in amounts exceeding inhibitor concentrations, plasmin also cleaves fibrinogen in a similar fashion to produce **fibrinogen degradation products**. Fibrin and fibrinogen are cleaved by plasmin in a sequential and asymmetric fashion, yielding four products: fragments X, Y, D, and finally E (Fig. 25-2). Fibrin that has been cross-linked by factor XIIIa is highly resistant to lysis by plasmin, but the degradation products that are formed contain combinations of the fragments, such as two D fragments and one E fragment.

The degradation products of fibrin and fibrinogen inhibit hemostasis in several ways. For example, the aggregation of platelets is greatly impeded by fragments X, Y, and D. These cleavage products, but not the smaller fragment E, effectively compete with the D domain of intact fibrinogen for binding to a platelet membrane receptor. The fragments are thus highly inhibitory, as binding of intact fibrinogen to a specific receptor site (glycoprotein IIb–IIIa) on the platelet membrane is a prerequisite for the normal platelet aggregation response. In addition, the larger fragments, X and Y, interfere with the polymerization of fibrin by disrupting the interaction of intact monomers.

Plasminogen activators (PAs) play an obvious key role in initiating fibrinolysis, both from a physiologic and therapeutic standpoint. The most commonly recognized PAs include the factor XIIa–kallikrein–high-molecular-weight kininogen system, urokinase, streptokinase, and tissue-type PA (see Fig. 25-1). It has long been known that coagulation is somehow temporally linked to the inflammatory response and to the initiation of clot lysis. Although the common trigger for these host-defense reactions may exist at the factor XIIa step of intrinsic coagulation

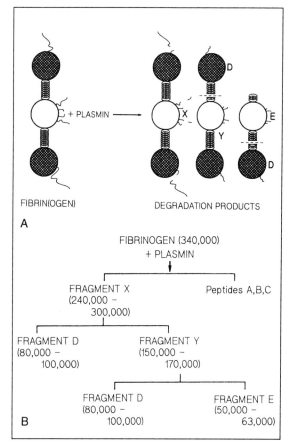

Fig. 25-2. Stages in the degradation of fibrin and fibrinogen by plasmin. (A) The trinodular structure of fibrin and fibrinogen and the plasmin cleavage sites. (B) Approximate molecular weights (in daltons) of the major fragments of fibrin and fibrinogen following degradation by plasma. (See text description.) (A modified from: Rapaport, S. I. Hemostasis. In: West, J. B., ed. *Physiological Basis of Medical Practice,* 11th ed. Baltimore: Williams & Wilkins, 1985, P. 428; B modified from: Ogston, D. *The Physiology of Hemostasis.* Cambridge: Harvard University Press, 1983, P. 148.)

(based on findings from in vitro experiments using highly purified proteins), the physiologic and pathologic significance of this pathway is not known. **Urokinase,** a urinary protein produced by the renal tubular epithelium, is a potent PA that has been used successfully as a thrombolytic

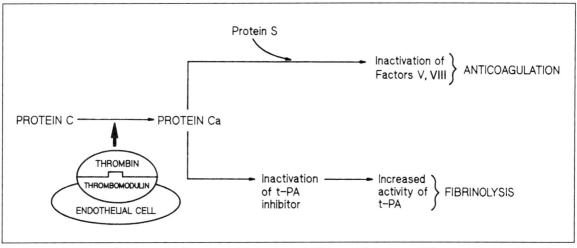

Fig. 25-3. Thrombin–thrombomodulin activation of protein C. (See text description.) (*Ca* = activated protein C; *t-PA* = tissue-type plasminogen activator.)

coagulation reactions occur. Thrombin not only has positive-feedback effects that cause the explosive activation of the coagulation cascade, but also initiates negative-feedback mechanisms that control its own generation (for example, the protein C–thrombomodulin anticoagulant system). Finally, there are numerous regulatory mechanisms that either operate constantly or are triggered by coagulation, and these ensure that hemostasis is restricted to the site of vessel injury.

PATHOPHYSIOLOGY OF HEMOSTASIS

The primary focus of the preceding sections was on the normal physiology of hemostasis. Because numerous texts on the clinical disorders of hemostasis are readily available (see Bibliography for additional reading), relatively little detail will be presented here. Instead, this section will be devoted to the **conceptual** aspect of the underlying causes and diagnoses of the most common hemorrhagic and thrombotic disorders, within the framework of the material presented thus far. The commonly used screening tests of hemosta-

sis are presented to test the reader's understanding of the underlying mechanisms, but the clinical management and treatment of hemostatic disorders are not covered.

The patient with a bleeding or thrombotic disorder can confront the clinician with a diagnostic challenge of varying complexity. A thorough patient history is critical and nearly always revealing. In most cases, the history and physical examination, coupled with a few basic screening tests, will readily suggest the most common problem. Frequently, however, emergency interventions must be initiated to properly manage life-threatening hemorrhage or thrombosis. In either event, a thorough knowledge of the basic hemostatic mechanisms and of the screening tests is required.

Hemorrhagic Disorders

Nearly all of the hemostatic mechanisms discussed so far must be triggered. If the stimulus occurs, such as a break in the endothelium, and any of the components of the mechanism do not respond normally (for example, blood vessel abnormality, platelet or coagulation factor levels

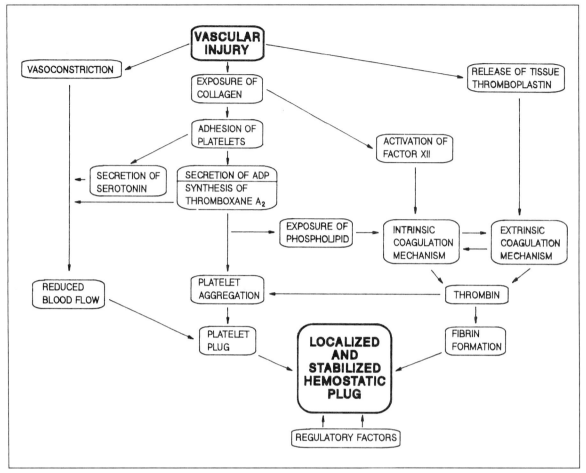

Fig. 25-4. Summary of the integrated hemostatic response to vessel injury. (See text description.)

below normal or abnormal in quality because of a genetic defect or some other abnormality), a hemorrhagic condition could develop. In fact, if the defect is severe, spontaneous hemorrhage may arise when there is no apparent trauma. Following are brief descriptions of several commonly used tests that assess abnormalities in platelet number and function, as well as screen for coagulation factor deficiencies.

Platelet Screening Tests
Platelet Count   Measuring the concentration of platelets in blood is a customary part of the hemostasis screening profile, particularly because defective platelet plug formation is commonly caused by a lowered platelet count, known as **thrombocytopenia**. The concentration of platelets in anticoagulated blood can be estimated crudely by visual inspection of Wright-stained peripheral smears of whole blood. The normal ratio of platelets to red blood cells is approximately 1 to 20. Platelet concentration is

more accurately determined by chamber counting under phase-contrast microscopy or by electronic particle counting. The normal platelet count is 250,000 ± 60,000 per microliter of blood.

Bleeding Time   Because the platelet count cannot reveal abnormal platelet function, it is often necessary to screen for both qualitative and quantitative platelet defects. This is accomplished by the template **bleeding time test,** which screens the overall competence of platelets in forming hemostatic plugs in response to a standardized incision. To perform the test, a blood pressure cuff is applied to maintain venous pressure at 40 mm Hg and a standardized incision is made on the volar surface of the forearm using a template. Bleeding from the wound stops when sufficient platelets have aggregated and functioned properly to hold against the standardized back pressure. The wound is blotted with filter paper at 30-second intervals, and once blood no longer stains the paper the time elapsed determines the bleeding time.

As with all tests of hemostatic function, normal control times must be carefully determined for each laboratory. Under usual circumstances, and with platelet counts above 100,000 normal platelets per microliter, the normal bleeding time is 4.5 ± 1.5 minutes. Prolongation is borderline when bleeding time is approximately 8 to 10 minutes, and significant platelet dysfunction or thrombocytopenia is implied when values exceed 10 minutes. To reiterate, the bleeding time can be prolonged due to either a qualitative or quantitative platelet defect, and the screening test must be done in conjunction with a platelet count to differentiate between these possibilities. Blood coagulation factor defects are also not assessed in the bleeding time test because cessation of bleeding in this test depends primarily on platelet number and function, as the vessel injury is at the capillary level.

Coagulation Screening Tests   The **partial thromboplastin time** (PTT) measures the overall competency of the intrinsic pathway of coag-

ulation (see Fig. 24-5.) Partial thromboplastins (such as cephalin or Inosithin) are added to citrated plasma and the time it takes for coagulation to occur is measured after the addition of phospholipid and Ca^{2+}. A variant of this test, the activated PTT, eliminates the varying degree of contact activation that affects the PTT test. In the activated PTT, the contact factors are maximally activated by first mixing the plasma with particulates, such as kaolin, Celite, or powdered glass. The activated PTT and PTT preferentially screen the intrinsic pathway because factor III (tissue thromboplastin) is not available to initiate the extrinsic pathway, while the reagents used (a particulate foreign surface, phospholipid, and Ca^{2+}) permit clotting to proceed in the test tube via the intrinsic pathway. (One should remember, however, that both pathways are needed for normal hemostasis in vivo. The role of platelets in providing phospholipid is bypassed in both this test and the prothrombin time test because synthetic phospholipid is added as a reagent.) The test therefore bypasses the early stage of the extrinsic pathway.

Normal values for the activated PTT are about 35 to 45 seconds, depending on the commercial source of the reagents and the person performing the tests. The activated PTT will be prolonged if a circulating anticoagulant is present, or if any of the factors involved in the intrinsic pathway are either significantly below normal levels or are defective in quality.

The **prothrombin time** (PT) assesses the overall competency of the extrinsic pathway of coagulation (see Fig. 24-5). To perform the test, synthetic tissue thromboplastin (factor III), containing phospholipid is added to citrated plasma, and the coagulation time is measured after Ca^{2+} is added. The PT screens the extrinsic pathway because the degree of contact activation of the intrinsic pathway is minimal, while the factor III added as a reagent rapidly interacts with factor VII in the plasma sample and clotting proceeds via the extrinsic pathway. This test therefore bypasses the early stages of the intrinsic pathway.

Normal values for the PT are approximately 10 to 15 seconds, again depending on the person performing the test and the commercial source of the thromboplastin. The PT will be prolonged if a circulating anticoagulant is present or if any of the factors in the extrinsic pathway are significantly below normal levels or are defective in quality.

The **thrombin time** (TT) evaluates the integrity of the final stage of coagulation when soluble fibrinogen is converted to the insoluble fibrin polymer. To perform the test, a standard dilute solution of thrombin is added to citrated plasma and the coagulation time is then measured. The test screens only the last stage of coagulation, because exogenous thrombin is used as a reagent and therefore the patient's ability to generate thrombin by means of either the intrinsic or extrinsic pathway is bypassed.

Normal values for the TT are approximately 18 to 20 seconds. The TT will be prolonged if plasma fibrinogen levels are significantly low, or if heparin, fibrin and fibrinogen degradation products, or a defective fibrinogen molecule is present. The test may be used to monitor systemic fibrinolytic and anticoagulant therapy and is more sensitive to alterations in fibrinogen content or quality than are the PTT and PT. Thus, patients with a mild fibrinogen abnormality may have a normal or slightly prolonged PT and PTT, but a considerably prolonged TT.

The end point for all of the three coagulation tests (PTT, PT, and TT) is the appearance in plasma of the insoluble fibrin polymer. However, these tests cannot distinguish whether the fibrin formed is a weakly associated or covalently bonded polymer. For this reason, the tests are insensitive to factor XIII deficiencies. To **screen for factor XIII** defects, the solubility of the patient's clot must be evaluated. If the clot is a covalently linked polymer (that is, normal factor XIII), it is insoluble in 5 M urea or 1 percent monochloroacetic acid. When factor XIII is lacking, however, the defective clot is readily soluble in these reagents. Thus, clot solubility serves as a means for identifying factor XIII defects.

Tables 25-2 and 25-3 classify hemostatic disorders based on their screening test profiles. Considerable diagnostic discrimination between

Table 25-2. Evaluation of Hemostatic Deficiencies by the Prothrombin Time and Partial Thromboplastin Time

Deficiency	Partial Thrombo- plastin Time	Prothrombin Time
Prekallikrein	P	N
HMWK	P	N
Factor XII	P	N
Factor XI	P	N
Factor IX	P	N
Factor VIII	P	N
Factor VII	N	P
Factor X	P	P
Factor V	P	P
Prothrombin	P	P
Fibrinogen	P	P
Factor XIII	N	N

HMWK = high-molecular-weight kininogen; P = prolonged; N = normal.

Table 25-3. Classification of Hemostatic Deficiencies Based on Coagulation Screening Tests

Deficiency	Partial Thrombo- plastin Time	Prothrom- bin Time	Thrombin Time
Intrinsic pathway	P	N	N
Extrinsic pathway	N	P	N
Common pathway	P	P	N
Fibrinogen defect	N[a]	N[a]	P

[a] Prolonged if fibrinogen defect is severe.
P = prolonged; N = normal.

hemostatic disorders is afforded by just the three coagulation screening tests. However, a **definitive diagnosis** is often possible only through the use of specific immunochemical factor assays. Alternatively, the ability of the patient's plasma to correct the prolonged test times of artificially depleted plasmas or plasmas from patients with specific factor deficiencies can be assessed, and this usually permits identification of the abnormality.

Examples of Specific Hemorrhagic Disorders

Abnormalities of Platelet Plug Formation

Hemorrhagic disorders can be broadly classified according to which phase of the hemostatic mechanism is affected. For example, abnormalities in the formation of temporary hemostatic plugs can be due to vascular defects, producing an inability of the vessel to constrict properly (inflammation, hemangioma, and so on), or, more likely, due to quantitative or qualitative platelet defects. These potential causes of bleeding are only minimally assessed by the bleeding time and platelet count. Although hereditary thrombocytopenia is extremely rare, **acquired thrombocytopenia** is the most common cause of impaired hemostatic function. A decrease in platelet number can be due to decreased production (bone marrow disease), enhanced platelet destruction, seen in autoimmune reactions such as **idiopathic thrombocytopenic purpura,** or increased pooling of platelets caused by **splenomegaly**.

Qualitative platelet defects can be either hereditary or acquired (for example, drug-induced or secondary to other diseases). Examples of hereditary platelet abnormalities include a lack of membrane receptors necessary for adherence to collagen, as seen in **Bernard-Soulier disease,** as well as a deficiency of those receptors that normally mediate platelet aggregation, a condition known as **Glanzmann's thrombasthenia**. **Von Willebrand's disease** is a common hereditary hemostatic disorder that is unusual in that it is inherited as an autosomal dominant trait (nearly all others are autosomal recessive or sex-linked recessive disorders), and is manifested as a combined qualitative platelet defect and coagulation factor defect. Platelets from patients with von Willebrand's disease do not adhere normally to injured vessel walls. This is caused by the decreased concentration of a plasma protein (von Willebrand factor), believed to form a part of the factor VIII complex, that is necessary for normal platelet adhesion. Thus, affected patients display a prolonged bleeding time, decreased platelet adhesiveness, and a lowered factor VIII level. The depression in the factor VIII level in von Willebrand's disease represents a decrease in both immunological levels of factor VIII (as measured with an antibody) and functional levels of factor VIII (as measured with a clotting assay). It is important to distinguish these two activities of factor VIII, for in hemophilia, another hereditary disorder involving factor VIII (as will be discussed), immunologic levels of factor VIII are normal but functional levels are depressed.

Abnormalities of Coagulation

Coagulation defects that produce a bleeding tendency can result from decreased production or increased destruction of the coagulant proteins, the presence or absence of inhibitory substances, or the production of an abnormal (dysfunctional) coagulant molecule. The PT, PTT, TT, and clot solubility tests are used routinely to screen the competency of the coagulation factors (see the section "Coagulation Screening Tests" and Tables 25-2 and 25-3).

The most common congenital bleeding disorder is caused by factor VIII deficiency, and is referred to as hemophilia A or classic hemophilia. This hereditary defect is a sex-linked recessive disorder affecting only males; it accounts for about three-fourths of all congenital defects and occurs in about 1 in 10,000 persons. Factor IX deficiency, known as hemophilia B or Christmas disease, is about a fourth as frequent as hemophilia A. The coagulant defect in classic hemophilia is a functional deficiency of factor VIII, as measured with clotting assays. The problem is not caused by a deficient rate of production, but by the production of an abnormal factor VIII molecule. Thus, factor VIII antigen levels are normal but the functional levels are low.

The severity of bleeding in hemophiliacs is roughly proportional to the severity of the defect. Patients with less than 1 percent of normal factor VIII activity are rare and usually suffer episodes of spontaneous bleeding. Most patients have factor levels in the range of 15 to 40 percent and do not have spontaneous bleeding. Instead, these patients have hemorrhagic complications in response to trauma (surgery, tooth extractions, exercise-related injuries, and the like), and effective hemostasis requires the deposition of fibrin. A major complication of hemophilia is joint bleeding, or **hemarthrosis,** which, if severe and repetitive, can lead to crippling synovitis and arthritis.

The diagnosis of severe hemophilia is easily done by noting the history and inheritance of the disease, coupled with a prolonged activated PTT, normal PT, and normal bleeding time. Less severe or mild cases can be difficult to distinguish from von Willebrand's disease. The characteristics of these two disorders that allow for their differential diagnoses are presented in Table 25-4. The primary characteristics that are distinct are the inheritance patterns, bleeding time, and factor VIII antigen levels. In addition, gastrointestinal bleeding is a frequent complication of von Willebrand's disease and hemarthrosis is a common feature of hemophilia.

Table 25-4. Characteristics of von Willebrand's Disease and Classic Hemophilia

Characteristic	von Willebrand's Disease	Hemophilia
Common bleeding site	Gastrointestinal	Joints
Inheritance	Autosomal dominant	X-linked recessive
Prothrombin time	N	N
Partial thromboplastin time	P	P
Platelet count	N	N
Bleeding time	P	N
Factor VIII (antigen)	Low	N
Factor VIII (clotting)	Low	Low

P = prolonged; N = normal.

THROMBOTIC DISORDERS

Thrombosis develops when the hemostatic mechanism is abnormally triggered or when it is not restricted or regulated properly. In comparison to hemorrhagic disease, thrombosis of either the venous or arterial vasculature is very common and, in fact, is a major cause of morbidity and mortality in Western civilizations. In many cases, death from thrombosis is particularly tragic, because it may occur without any apparent clinical warning, and frequently at a relatively young age. Thrombosis represents a pathologic occlusion or narrowing of a blood vessel caused by inappropriate intravascular coagulation. These events may lead to decreased delivery of oxygen and nutrients to the affected area, or areas, tissue ischemia, and, finally, cellular death—**infarction**. Its consequential severity is determined by where it is formed or where it travels to if it is dislodged—**embolization**. For example, small deposits of fibrin can often be tolerated in areas such as skeletal muscle which possess good collateral circulation; however, thrombi dislodged in the venous system may embolize to the lung, where they become **pulmonary emboli,** and clots forming in the coronary circulation may produce **myocardial infarction.** Both conditions can be life-threatening.

Arterial Thrombosis

Thrombi developing on the arterial side of the circulation are composed primarily of platelets. Relatively little fibrin deposition and trapping of erythrocytes are apparent in the early stages of arterial thrombosis, owing to the high pressures and rapid flow characteristic of this circulation. Although the precise causes remain uncertain, the formation of arterial thrombi appears to be precipitated by an abnormal vessel wall with damaged or disrupted endothelium. Thus, areas of the arterial tree that frequently contain lesions of the vessel wall are particularly susceptible to thrombus formation. For example, atherosclerotic plaque formation, stenotic valves, and chemical or mechanical trauma to the endothelium may initiate thrombosis.

Atherosclerosis of the coronary arteries, or **ischemic heart disease,** is the leading cause of mortality in the United States. Furthermore, it is now evident that thrombosis (often accompanied by vasospasm) is the most common cause of myocardial infarction in atherosclerotic coronary arteries. The initial event in arterial thrombosis appears to be the formation of a platelet mass that adheres to either the vessel wall lesion or the damaged endothelium. The platelet mass then enlarges and intrudes into the lumen of the vessel. Although occlusion can occur at the site of the lesion, more commonly embolization of the platelet mass causes occlusion downstream at the arteriolar or capillary level. After occlusion, the static blood flow then fosters the deposition of fibrin. Depending on the particular circulation involved, these events can lead to tissue and cellular ischemia, with myocardial or cerebral infarctions the most common.

Recent experimental evidence indicates that the platelet is not only the primary component of the arterial thrombus, but may also be involved in the development of the atherosclerotic lesion. During platelet aggregation at the site of damaged endothelium, a substance called **platelet-derived growth factor** (PDGF) is released from the alpha granules, and it is a potent mitogen for the growth of vascular smooth muscle cells in vitro. It is theorized that, in vivo, the mitogenic stimulus provided by PDGF promotes growth and proliferation of the underlying smooth muscle cells cells, causing this cell type to invade the intimal region and produce plaque formation. In experimental animal models of atherosclerosis, reducing the platelet count markedly retards the development of lesions. A similar phenomenon has been observed for animals with qualitative platelet defects.

Arterial thrombi can be diagnosed and localized using a variety of means, including noninvasive blood flow measurements and autoradiographic scanning. A cardinal feature is increased platelet consumption and platelet turnover (see the section "Fibrinogen and Platelet Turnover"). Unfor-

tunately, there is no effective treatment for arterial thrombosis. Anticoagulant therapies, which are quite effective in treating venous thrombosis, are largely unsuccessful in preventing arterial thrombosis, as this primarily involves platelets and not coagulation per se. Use of platelet suppressive agents, such as aspirin, has met with limited success. Obviously, much more research is needed to design highly specific therapeutic agents and measures that prevent the development of atherosclerotic lesions.

Venous Thrombosis

Thrombi that form on the venous side of the circulation primarily contain fibrin, along with trapped erythrocytes and some platelets. These thrombi can develop when there are no apparent lesions of the vessel wall and appear to be provoked by stasis or venous pooling of blood. Because even the normal venous circulation is characterized by low pressures and flow, any events that favor venous pooling of blood and stasis will further predispose the subject to spontaneous activation of the coagulation process. For this reason, venous thrombosis can pose a major threat for bedridden patients, older individuals with sedentary life-styles, or those whose occupations require limited movement. Additional predisposing conditions include surgery, pregnancy, congestive heart failure, varicose veins, and obesity. The fact that static blood flow is such a common predisposing feature of venous thrombosis suggests that some degree of coagulation may be ongoing and that adequate blood flow accompanying physical activity may be a key to prevention. Blood flow dilution is also an important factor for localizing and restricting the coagulation process. Another serious threat posed by deep venous thrombosis is the potential for embolization; in this event, the dislodged fibrin circulates and is deposited in the lung (pulmonary embolus).

Widely available methods for **detecting venous thrombosis** include contrast and isotopic venography, Doppler flow measurements, and radiolabeled fibrinogen uptake. The **contrast**

venogram is perhaps the most standard laboratory procedure. Perfusion lung scans or pulmonary angiograms are required to establish the diagnosis of pulmonary embolism. A major feature of venous thrombosis is increased fibrinogen consumption and turnover (see the next section, "Fibrinogen and Platelet Turnover"). In contrast to the treatment of arterial thrombosis, venous thrombosis is very amenable to anticoagulant therapy.

Fibrinogen and Platelet Turnover

The turnover of fibrinogen and platelets in arterial and venous thrombosis is a dynamic and meaningful parameter in understanding the major differences between these two conditions.

The concentration of any substance ([S]) in whole blood is yielded by the product of its turnover times the mean survival time:

$$[S]_{\text{whole blood}} = \text{turnover}_s \times \text{survival time}_s$$
$$\text{(Eq. 25-1)}$$

Analysis of this simple equation illustrates why the measurement of static blood concentrations of substances, such as fibrinogen or platelets, is of limited value. For example, during thrombosis, the survival time for platelets or fibrinogen is suddenly lowered due to consumption of these substances. In response to this consumption, feedback regulation becomes operative and more fibrinogen or platelets are produced (that is, turnover is increased). If the increase in turnover rate counterbalances the decrease in survival, the whole blood concentration will be unchanged. Thus, static blood concentrations often do not reflect major changes in the metabolism of substances. A much more useful and dynamic measure is to estimate the actual turnover of the substances of interest. The turnover of fibrinogen or platelets can be calculated from the mean survival times of radiolabeled fibrinogen or platelets following infusion into patients. Thus, by rearranging Equation 25-1, the turnover of any substance (T_s) is given by the blood concentration of S ([S]) divided by the mean survival time:

$$T_s = [S]/\text{survival time}_s \qquad \text{(Eq. 25-2)}$$

Under normal conditions, platelet turnover is approximately 2.5 times 10^4 platelets per microliter per day (2.5 x 10^5 platelets per microliter, divided by 10 days), and fibrinogen turnover is about 0.5 mg per milliliter per day (2.5 mg per milliliter, divided by 5 days). The range of normal turnover values and those occurring during a theoretical thrombotic episode in the arterial and venous circulations are shown in Figure 25-5. In ongoing **arterial thrombosis,** there is an increased platelet consumption and therefore decreased survival time as platelet thrombi are formed in the arterial circulation. To maintain normal circulating platelet levels, the platelet turnover is increased. Because fibrinogen con-

Fig. 25-5. Platelet and fibrinogen turnover under normal conditions and following arterial or venous thrombosis. (See text explanation.)

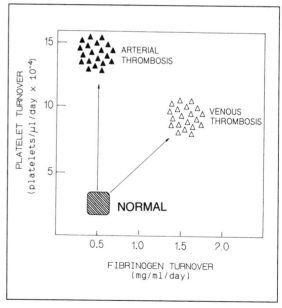

sumption in arterial thrombosis is largely unaffected (unless there is severe and widespread occlusion), the fibrinogen turnover remains unchanged.

In ongoing **venous thrombosis,** there is an increase primarily in fibrinogen consumption, and therefore a decrease in the fibrinogen survival time as fibrin thrombi are formed in the venous circulation. To maintain normal circulating fibrinogen levels, the fibrinogen turnover accelerates. In severe cases in which many platelets are trapped in the fibrin thrombi, the platelet turnover may increase as well. It should be noted that Figure 25-5 is hypothetical and is presented only to help in understanding the basic concepts. In fact, such measurements are performed only in experimental settings; the actual clinical presentations of arterial and venous thrombosis are quite distinct, for which other more definitive diagnostic procedures are available.

Anticoagulants in the Treatment of Venous Thrombosis

Heparin and **warfarin** (Coumarin) are two widely used anticoagulants that differ markedly in their mechanisms of action. Heparin is a highly sulfonated, negatively charged mucopolysaccharide isolated from animal lungs and intestinal mucosa. It also exists in human mast cells, and structurally similar substances are found in the extracellular matrix of the endothelium. Heparin alone has little anticoagulant activity. However, in the presence of the plasma cofactor, antithrombin, heparin is a potent inhibitor of the coagulation process. As depicted in Figure 25-6, heparin binds to antithrombin and induces a conformational change in the inhibitor molecule, so that the reactive site of antithrombin is more favorably positioned for interaction with the active-site serine residue of thrombin. Thus, the interaction of heparin and antithrombin produces nearly instantaneous inhibition of thrombin. As indicated in the discussion of antithrombin, nearly all other activated coagulant enzymes are similarly inhibited by anti-

thrombin, with heparin enhancing their rates of interaction as well.

Heparin therapy results in substantial prolongation of the clotting times of the three major screening tests (PTT, PT, and TT), with the TT and the PTT being the most sensitive in monitoring this anticoagulant. Intravenous heparin is indicated in **acute venous thrombosis** to prevent **recurrent clotting** or to halt the extension of existing thrombi while endogenous fibrinolysis is allowed to proceed. This anticoagulant is also used during various surgical procedures to prevent clotting. Low-dose heparin, given subcutaneously, has been used with some success to prevent venous thromboembolism. Although heparin is widely used with success, several factors make it a less than ideal anticoagulant. For example, it is not effective when given orally and needs to be injected. Repeated infusions may be required, as it is rapidly metabolized with a half-life of about 90 minutes. There is also considerable variability in the response to heparin in different patients and within the same patient, and secondary bleeding complications may arise.

Warfarin (coumarin) is an orally effective anticoagulant that is commonly used for the long-term prevention of **venous thrombosis**. It was originally discovered as the active ingredient (bishydroxycoumarin) in spoiled sweet clover that induced a hemorrhagic condition after ingestion by cattle. This finding rapidly led to the use of coumarin-like derivatives as rodenticides, and, surprisingly, in the prevention of thromboembolism.

The coumarin nucleus is chemically similar to that of vitamin K; both compounds most likely compete for a common receptor site in the liver. Thus, warfarin antagonizes the action of vitamin K in the hepatic synthesis of the vitamin K–dependent clotting factors (prothrombin, factors VII, IX, and X). As discussed in Chapter 24, vitamin K allows for the posttranslational insertion into these proteins of gamma-carboxyglutamic acid residues (GLA) that are essential for their clotting function—particularly for their ability to

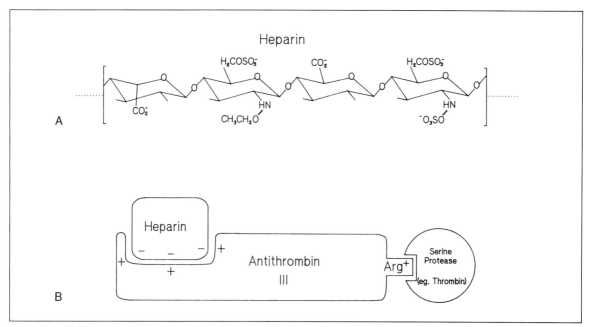

Fig. 25-6. (A) A tetrasaccharide portion of the heparin molecule. (B) The mechanism of action of heparin as an anti-coagulant. Heparin binds to antithrombin, inducing a conformational change in the inhibitor, so that the reactive-site arginine residues are more favorably positioned for covalent binding to the active-site serine residues of proteases such as thrombin. (Modified from: Thompson, A. R., and Harker, L. A. *Manual of Hemostasis and Thrombosis*, 3rd ed. Philadelphia: F. A. Davis, 1983, P. 152.)

bind Ca^{2+} and phospholipid. The effect of warfarin then is to induce an abnormal structure of certain key clotting factors (for example, they contain no GLA residues, do not bind Ca^{2+} and phospholipid avidly, and therefore do not function properly during coagulation). In the presence of warfarin, the newly synthesized factors are immunologically and metabolically almost identical to their normal counterparts but markedly abnormal in function. Therefore, following warfarin therapy, the levels of the altered vitamin K–dependent factors are simply a reflection of their normal half-life, which ranges from about 5 to 7 hours for factor VII to about 60 hours for prothrombin (Fig. 25-7). Because warfarin's anticoagulant effect is due to a metabolic action, about 3 to 5 days of therapy are required before effective anticoagulation can be achieved and the functional levels of the vitamin K–dependent factors are adequately suppressed.

Conceptually, warfarin therapy induces a bleeding tendency in patients who are prone to clotting. Therefore, patients undergoing warfarin treatment must be carefully monitored to ensure proper adjustment of the dose, and thus to minimize the risk of hemorrhage and maximize the inhibition of clotting. This is normally accomplished by beginning with low doses of warfarin and daily monitoring of the clotting time of the patient's plasma, until the degree of prolongation is acceptable. The PTT is normally used for this because it is very sensitive to changes in factor VII, the vitamin K–dependent protein whose functional levels decline most rapidly in response to warfarin. In the event of "overshoot," warfarin's effects can be reversed by administering vitamin K; however, reestablishing normal levels of functional vitamin K–dependent factors requires several days and may be accompanied by a rebound tendency for thrombosis to occur.

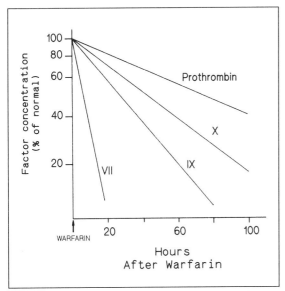

Fig. 25-7. Functional levels of prothrombin and factors VII, IX, and X following warfarin administration.

Fibrinolytic Therapy

As discussed earlier in this chapter, activation of the plasminogen–plasmin system by infusing activator compounds, such as streptokinase and urokinase, has met with some success in the short-term treatment of patients with ongoing thrombosis. However, both compounds have major drawbacks and place the patient at a significant risk of bleeding. Of particular promise is the use of recombinant PA in fibrinolytic therapy. This agent appears to mimic the endogenous PA of endothelial cell origin. It binds specifically to the components of the clot matrix, activates plasminogen in situ, and thereby elicits clot dissolution in an environment largely protected from inhibition by alpha$_2$-antiplasmin. Early clinical trials of t-PA indicate that it is an effective thrombolytic agent in certain conditions and carries a low risk of causing secondary bleeding.

In the past, thrombolytic therapy was only used for treating certain types of thrombosis. For example, anticoagulants have been successfully used in the treatment of patients with venous thrombosis, and, because of the risk of bleeding with urokinase or streptokinase, these activators have not been widely used in such patients. However, the plasminogen activators have been used very effectively in acute situations when the immediate benefit may outweigh the risk (for example, massive pulmonary embolism). In addition, these agents, as well as t-PA, appear to be quite effective in reopening clotted coronary arteries, if given by coronary catheterization within a few hours of the onset of the thrombotic event. By quickly reestablishing vessel patency, these agents can preserve the ischemic myocardium.

SUMMARY

An abundance of regulatory factors exist or become operative in order to localize and restrict hemostasis to the site of vessel injury. For example, the endothelium is normally nonthrombogenic and produces substances with antiplatelet and vasodilatory actions. In addition, certain key coagulation factors are consumed during the clotting process and their activated forms are rapidly diluted by blood flow, cleared by the liver, or effectively inactivated by the plasma protease inhibitors such as antithrombin. Activation of fibrinolysis and the protein C–thrombomodulin proteolytic systems represent two additional major anticoagulant mechanisms.

Although relatively rare, hemorrhagic conditions can develop spontaneously or when the components of the hemostatic mechanism do not respond to the stimulus of vessel injury. As assessed by simple screening tests, hemorrhagic disorders can result from blood vessel dysfunction or from either a qualitative or quantitative defect in the platelets or coagulation proteins. Thrombosis is a common disorder in which the hemostatic mechanism is abnormally triggered or improperly regulated. Arterial and venous thrombosis differ considerably in their etiology and therapy, as well as in the extent of platelet involvement versus activation of coagulation.

NATIONAL BOARD TYPE QUESTIONS

Select the one most correct answer for each of the following

1. Which of the following statements is *true* concerning von Willebrand's disease and classic hemophilia?

   A. Both are autosomal dominant diseases that affect only males

   B. The bleeding time of von Willebrand's disease is prolonged and the PT of hemophilia is normal

   C. Both have a decreased platelet count

   D. Von Willebrand's disease represents a combined deficiency of an abnormal platelet count and a factor VII deficit, while hemophilia exhibits only a prolonged bleeding time

   E. Both are bleeding disorders involving decreased factor VIII antigen (immunologic) levels but normal factor VIII functional levels

2. Which of the following statements is *false?*

   A. An antithrombin-deficient patient would tend to have thrombotic complications more readily than a normal person

   B. Heparin is an effective anticoagulant in blood drawn into a test tube as well as when injected into the bloodstream

   C. The bleeding time of a normal person after ingesting aspirin is more prolonged than the bleeding time of a hemophiliac

   D. The anticoagulant action of warfarin (coumarin) requires several days to become effective, both in the bloodstream and in blood drawn into a test tube

   E. Thrombotic disorders are more common than are primary inherited bleeding disorders

3. Compared with venous thrombosis, which one of the following best describes arterial thrombosis?

   A. Is best treated with vitamin K antagonists

   B. Occurs primarily in areas with static blood flow

   C. Exhibits a decreased mean platelet survival time

   D. Is frequently accompanied by an increase in fibrinogen turnover

   E. Is frequently treated with orally effective anticoagulants such as coumarin

4. A patient with a bleeding disorder had a normal platelet count, normal TT, and normal PT. Which of the following defects is *not possible* (that is, can be ruled out as a possible diagnosis)?

   A. Qualitative platelet defect

   B. Classic hemophilia

   C. Factor XIII defect

   D. Vitamin K deficiency

   E. Factor IX defect

5. A person has the following laboratory values: platelet count, 250,000 platelets/μl; mean platelet survival time, 10 days; fibrinogen turnover, 0.4 mg/ml/day; and plasma fibrinogen level, 2.0 mg/ml. Which of the following is *true?*

   A. The bleeding time is significantly prolonged

   B. The person most likely has ongoing venous thrombosis

   C. The mean fibrinogen survival time is 50 days

   D. The platelet turnover is 2.5 x 10⁴ platelets/μl/day

   E. The person most likely has ongoing arterial thrombosis

6. Which of the following is characteristic of the most common hereditary coagulation defect?

   A. The PT, TT, and bleeding time are normal

   B. An autosomal dominant mode of inheritance

   C. A prolonged bleeding time

   D. A combined deficiency of factor VIII and a platelet defect

   E. A normal PTT

7. Which of the following tend to restrict intravascular coagulation?

   A. Hepatic clearance of activated coagulation factors but not their inactive forms

   B. Thromboxane production by endothelial cells

C. Inhibition of the fibrinolytic enzyme system

D. A reflex splenomegaly immediately after intravascular coagulation

E. Consumption of vitamin K during the coagulation process

ANNOTATED ANSWERS

1. B. Patients with von Willebrand's disease have a qualitative defect plus an abnormal factor VIII level; hence, they have a prolonged bleeding time. Hemophilia represents a pure factor VIII defect that is not screened for in the PT; hence the PT is normal.

2. D. Warfarin is a vitamin K antagonist whose anticoagulant action is directed at the hepatic synthesis of certain coagulation proteins. Its effect in vivo therefore requires several days. It is ineffective in vitro.

3. C. As platelet thrombi are formed during arterial thrombosis, platelet consumption is increased and this increases the platelet survival time.

4. D. Because a bleeding time, PTT, and clot solubility test were not performed, the defects in A, B, C, and E are still possible and would yield the test profile given. However, vitamin K deficiency would yield a prolonged PT and therefore could be ruled out as a possible diagnosis.

5. D. The values are approximately normal. The platelet turnover is 250,000 platelets/μl/10 days, or 2.5×10^4 platelets/μl/day.

6. A. The most common hereditary coagulation defect is hemophilia A, which represents the synthesis of an abnormal factor VIII molecule. The PT, TT, and bleeding time are all normal, only the PTT is prolonged.

7. A. The clearance of activated coagulants by the liver is an important regulatory factor that tends to localize and restrict the coagulation process. The other events do not accomplish this.

BIBLIOGRAPHY

Bowie, E. J. W., and Sharp, A. A., eds. *Hemostasis and Thrombosis.* London: Butterworth, 1985.

Doolittle, R. F. Fibrinogen and fibrin. *Sci. Am.* 245:126-135,1981.

Ogston, D. *The Physiology of Hemostasis,* Cambridge: Harvard University Press, 1983.

Rapaport, S. I. Hemostasis. In: West, J. B., ed. *Physiological Basis of Medical Practice,* llth ed. Baltimore: Williams & Wilkins, 1985, Pp. 409-432.

Ratnoff, O. D., and Forbes, C. D., eds. *Disorders of Hemostasis.* Orlando, FL: Grune & Stratton, 1984.

Thompson, A. R., and Harker, L. A. *Manual of Hemostasis and Thrombosis,* 3rd ed. Philadelphia: F. A. Davis, Co. 1983.

VII Respiratory Physiology

26 Introduction and Anatomy

DOUGLAS K. ANDERSON

Objectives

After reading this chapter, you should be able to:

Describe the basic anatomy and functions of the pulmonary system

Define the lung volumes and capacities

Describe the primary functions of the muscles of respiration

The respiratory system is responsible for the exchange of oxygen and carbon dioxide by the body. The gas-exchanging organs are the **lungs**. In addition to gas exchange, the lungs have a variety of other functions, including contributing to the body's acid–base balance, speech, metabolism of certain vasoactive compounds, and defense against infection. In this section, however, discussion will be limited to the gas-exchange function of the pulmonary system.

Metabolizing cells require a continuous supply of oxygen and continuously produce carbon dioxide. To provide adequate oxygen to the cells and remove sufficient amounts of carbon dioxide from them, the lungs must be ventilated and perfused at all times. Gas exchange is limited if the lungs (or a region of the lungs) are ventilated but not receiving adequate blood flow or vice versa. The purpose of this section is to describe the steps required to move oxygen from the air to the mitochondria and carbon dioxide from the cells to the external environment. Discussion will focus on lung mechanics, ventilation, gas exchange and transport, pulmonary blood flow, the regulation or control of ventilation, and the matching of ventilation to blood flow. However, before considering the physiology of respiration, it is necessary to understand the anatomy of the lungs and airways, and the remarkable way the structure of the lungs accomplishes their primary function of gas exchange.

ANATOMY OF THE LUNGS

Airways and Alveoli

Air passes into the **tracheobronchial tree** (airways) through the nose and mouth. It is warmed to body temperature, humidified, and, if entering through the nose, filtered. The airways bring the inspired air to the gas-exchange region of the lungs; these are the respiratory bronchioles, alveolar ducts, and alveoli. From the trachea to the alveolar sacs, there are approximately 23 generations of sequential branching (Fig. 26-1). The first 16 or so generations constitute the **conducting zone**. The blood flow to this portion of the airways is primarily to provide nutrients to the smooth muscle of the airways that constitute the conducting zone. Because no gas is exchanged in these airways, the conducting zone constitutes the **anatomic dead space.** Each generation of branching increases the collective

415

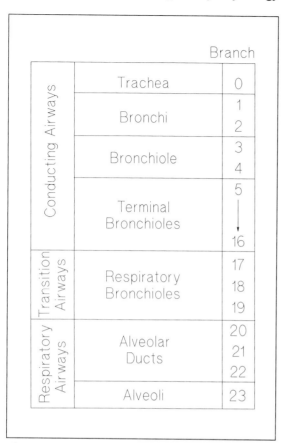

Fig. 26-1. Diagram of the airways. The first 16 levels of branching constitute the conducting zone. Branches 17 to 23 are the transition and respiratory zone where gas exchange occurs. (Modified from: West, J. B. *Respiratory Physiology* 4th ed. Baltimore: Williams & Wilkins, 1990, P. 6.)

respiratory zone, comprising respiratory bronchioles (with a smooth muscle coat and occasional alveoli), alveolar ducts (with smooth muscle sphincters only), and alveolar sacs. It is in the respiratory zone, principally the alveolar sacs, that gas exchange occurs. There are approximately 300 million alveoli in the lung, each with an average diameter of 0.3 mm.

These multiple generations of branching translate into an enormous area for gas exchange (50 to 100 m²; average, 70 m²). This large contact area between alveolar gas and pulmonary capillary blood, coupled with a very thin alveolar–capillary barrier (approximately 0.5 μm), highly soluble respiratory gases (oxygen and carbon dioxide), and driving pressures for both oxygen and carbon dioxide between the alveoli and pulmonary capillary blood endow the respiratory system with the ideal characteristics for exchanging gas by passive diffusion.

The movement of air from the nose to the terminal bronchioles is accomplished by **bulk flow**. At the level of the terminal bronchioles where the total cross-sectional area is great, the forward velocity of bulk air flow is negligible and **gas diffusion** becomes the predominant process. The distance from the terminal bronchiole to alveolar sacs is small (5 mm), so the time required to equilibrate alveolar gas with inspired air is also small (less than 1 second). However, low air velocity in the terminal portions of the conducting zone causes suspended particles such as dust and asbestos to be deposited there.

Gas-Exchange Interface

The pulmonary capillaries are extremely short and heavily interconnected, effectively presenting a sheet of blood to the alveolar wall for gas exchange (Fig 26-2). There are approximately 250 to 300 billion pulmonary capillaries in the lung, or about a thousand capillaries per alveolus. Each capillary also makes contact with several alveoli (being sandwiched between them), thereby taking maximum advantage of the large available surface area (Fig. 26-3). Alveolar gas must cross **six layers** to enter erythrocytes in pul-

cross-sectional area of the airways, while reducing the radius of each individual airway and the velocity of air flow within that airway. Smooth muscle innervated by autonomic fibers can vary the airway diameter of the conducting zone, but this smooth muscle also responds to certain chemicals and drugs.

Around the sixteenth or seventeenth branching generation (between the terminal and respiratory bronchioles), there is a **transitional zone**. Here, the walls thin out but retain their smooth muscle coat. Generations 17 to 23 constitute the agent. However, because it is excreted by the kidney and does not circulate in blood, it most likely has no role in intravascular fibrinolysis. **Streptokinase,** an exogenous PA, is a product of hemolytic streptococci and, like urokinase, has been used as an activator of plasminogen in the treatment of thrombosis. Although both urokinase and streptokinase therapy have met with some success, these fibrinolytic agents have major drawbacks, such as the cost of production, immunogenicity, and a tendency to foster residual bleeding and strokes because of the high dosages frequently employed.

The most important physiologic PA is **tissue-type PA** or (t-PA). Because it is produced by vascular endothelial cells, this protein is in an ideal position to interact with circulating plasminogen and to trigger fibrinolysis following clot formation at the vessel wall. t-PA is unique among the PAs in that it possesses a very high affinity for fibrin and fibrinogen as well as plasminogen; this enables it to elicit plasmin formation within the fibrin matrix—an environment that protects the fibrinolytic enzyme from rapid inhibition by the plasma protease inhibitors. Thus, the t-PA molecule can evoke highly specific and localized fibrinolysis. Recombinant t-PA is now available for treating thrombotic complications, and most clinical trials indicate that it is an effective thrombolytic agent. How the secretion and activity of this important PA are regulated remains uncertain, but recent research has revealed that endothelial cells also express at least two highly specific inhibitors of t-PA. Although this finding adds yet another level of complexity to the overall regulation of the fibrinolytic system, the t-PA inhibitor molecules are probably very important in coordinating fibrinolysis with coagulation.

The Thrombomodulin–Protein C Anticoagulant System

Protein C is another vitamin K–dependent plasma protein which, when activated, has a profound anticoagulant action. When formed during the coagulation process, thrombin is not only bound and inhibited by antithrombin to prevent systemic coagulation, but also binds locally to **thrombomodulin,** an endothelial cell membrane protein. The resulting thrombin–thrombomodulin complex activates circulating protein C to Ca (Fig. 25-3). Protein Ca has two activities that together promote anticoagulation and stimulate fibrinolysis. First, protein Ca inactivates factor VIII and, in the presence of **protein S,** another vitamin K–dependent protein, also hydrolyzes factor V. Inactivation of these two factors effectively blocks further coagulation. In addition, because protein Ca is free to circulate and has a relatively long half-life in plasma, it can also limit systemic clotting. Second, protein Ca inactivates the inhibitor of t-PA produced by the endothelium, thereby allowing t-PA to initiate fibrinolysis. The importance of the protein C system as an endogenous anticoagulant is illustrated by the severe recurrent venous thrombosis in patients suffering from genetic deficiencies of this protein.

THE INTEGRATED REACTIONS OF HEMOSTASIS

An overall summary of the hemostatic response to vessel injury is given in Figure 25-4. When a vessel is injured, **smooth muscle contraction** along with **platelet aggregation** and fusion initiate temporary hemostatic **plug formation** that, in some cases, may be sufficient to halt bleeding. In most situations, however, **fibrin** must be deposited at the site to accomplish effective hemostasis. Coagulation is initiated by the exposure of blood to tissue thromboplastin (extrinsic) and the activation of factor XII by negatively charged foreign surfaces (intrinsic). The subsequent formation of thrombin converts soluble **fibrinogen** to insoluble fibrin polymer. **Thrombin** plays a pivotal role in numerous aspects of coagulation and its regulation. For instance, thrombin stabilizes the fibrin molecule and further enhances hemostasis by stimulating factors V and VIII, as well as platelets, by providing the membrane phospholipid on which

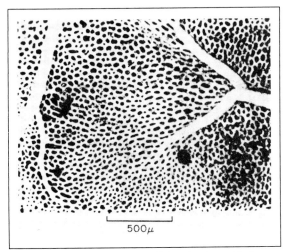

Fig. 26-2. Photomicrograph of the alveolar wall of a frog showing the short, dense capillary segments that create an essentially continuous sheet of blood. (Reproduced with permission from: Maloney, S. E., and Castle, B. L. *Respir. Physiol.* 7:150, 1969.)

monary capillaries. These are: (1) a fluid layer containing surfactant (see Chapter 27) on the inner surface of the alveolus; (2) the alveolar epithelium; (3) an interstitial space filled with fluid; (4) the capillary endothelium; (5) plasma in the capillaries; and (6) the erythrocyte membrane. The first four layers (the so-called alveolar–capillary barrier) collectively average 0.5 μm in thickness.

The fluid-filled interstitial space is in contact with **lymphatics,** thereby furnishing a conduit for draining interstitial fluid from the lung. This drainage helps to keep the alveoli dry. Although a general widening of the interstitial space, either by fibrosis (due to such agents as asbestos and coal dust) or interstitial edema, can, theoretically, reduce diffusional gas exchange, thickening of the interstitial space is generally not considered clinically important in reducing gas diffusion across the alveolar–capillary barrier.

Lung Volumes and Capacities

An understanding of the various volumes and capacities (the sum of two or more volumes) that

are used to describe lung function is integral to a discussion of pumonary physiology (Fig 26-4A). Measurable physiologic lung volumes and capacities can provide an index of pulmonary function, but are arbitrarily defined and differ among individuals according to age, body type, conditioning, and gender. Consequently, these volumes are not necessarily of diagnostic value, but can aid in differentiating between the two major types of lung disorders—obstructive and restrictive—and in quantifying the extent of the abnormality. All lung volumes, capacities, and ventilations are expressed as BTPS (body temperature, ambient pressure, saturated with water vapor). All gas volumes in blood (for example, oxygen consumption and carbon dioxide production) are expressed as STPD (standard temperature and pressure, dry gas). Ventilation and gas volumes in blood are covered in Chapter 28.

The **total lung capacity** (*TLC*) is the total volume of air in the lungs when they are maximally inflated. The **residual volume** (*RV*) represents the volume of air left in the lungs after a maximal expiration, that is, the lungs do not empty to a completely airless state on exhalation. The RV changes with age and gender. In young adults, the RV accounts for approximately 20 percent of the TLC, increasing in older individuals to about 30 percent. These age-related changes stem from alterations in the elastic recoil of the lung and mobility of the chest wall.

The **vital capacity** (*VC*) is the maximum volume of air that can be exhaled after a maximal inspiration. The VC constitutes 80 percent of the TLC and has three components. The first is the **tidal volume** (*VT*), which is that volume of air inspired and expired with each normal breath. VT can increase during exercise to approach the VC. During normal breathing, VT is approximately 500 ml. The second component of VC is the **inspiratory reserve volume** (*IRV*). This is the quantity of air that can be inhaled into the lungs from a normal end-tidal inspiratory position. The third component is the **expiratory reserve volume** (*ERV*), which is the amount of air that can be exhaled from the lungs from a normal end-tidal expiratory position. Both the IRV and

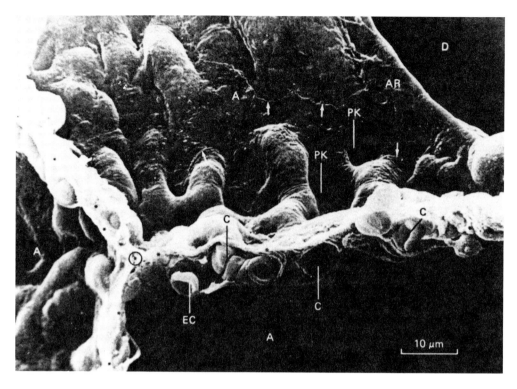

Fig. 26-3. Scanning electron photomicrograph of pulmonary capillaries (*C*) in cross section. Erythrocytes (*EC*) can be seen inside the capillaries. (*A* = alveolus; *PK* = pore of Kohn; *AR* = alveolar opening into an alveolar duct; *D* = alveolar duct.) (Reproduced with permission from: Levitzky, M. G. *Pulmonary Physiology* 3rd ed. New York: McGraw-Hill, 1991. P. 6.)

ERV decrease as the V_T increases and approaches the VC; in other words, the V_T increases at the expense of the IRV and ERV.

The **functional residual capacity** (*FRC*) is that volume of air in the lungs at the end of a normal expiration, that is, the FRC is the resting volume of the lung and is the sum of RV and ERV (see Fig. 26-4A). At the FRC, the inward elastic recoil of the lung is in equilibrium with the outward elastic recoil of the chest wall.

The FRC, like some of the other lung volumes and capacities, can differ with age, gender, and body type. However, these lung volumes and capacities can also be considerably affected by mechanical problems that alter the elastic recoil properties of the lung (restrictive lung diseases) or by conditions that block the airways (obstructive lung diseases). Conditions that limit the mobility of the chest wall can also affect the lung volumes and capacities. Asthma is an example of **obstructive disease**. This disorder is caused by a narrowing of the airways due to spasm of the bronchial smooth muscle, bronchial wall edema, or increased mucus production. Air is then trapped in the lungs distal to the obstruction, resulting in increased FRC, RV, and TLC but decreased VC (see Fig. 26-4B). Other obstructive lung diseases, such as emphysema and chronic bronchitis, also increase the resistance to air flow and can produce similar changes in the FRC, RV, TLC, and VC. Conversely, **restrictive lung diseases,** such as pulmonary fibrosis, increase the

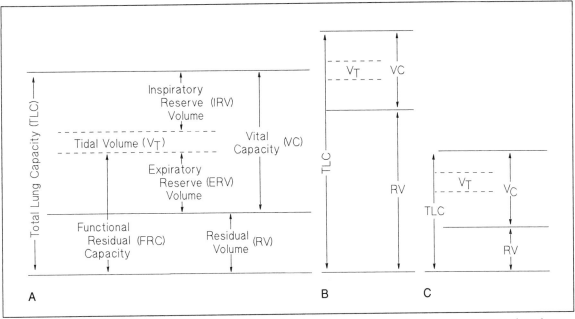

Fig. 26-4. (A) Lung volumes and capacities. (B) Effects of obstructive lung disease (e.g., emphysema) on these lung volumes. (C) Effects of restrictive lung disease (e.g., pulmonary interstitial fibrosis) on these same lung volumes and capacities.

"stiffness" of the lungs. Because these restrictive disorders augment lung elasticity (that is, decrease lung compliance and increase lung recoil; see Chapter 27), the lungs cannot expand normally and VC, TLC, FRC, and RV all decrease (see Fig. 26-4C).

Respiratory Muscles

The respiratory muscles change the dimensions of the thoracic cavity, causing it to function as a pump. Changing the size of the thoracic cavity alters the dimensions of the lungs. Increasing the volume of the thoracic cavity increases the volume of the lungs, causing air to be inhaled. Reducing the volume of the thoracic cavity decreases the lung volume, causing air to be expelled.

Different muscles accomplish inspiration and expiration. The most important **inspiratory muscle** is the **diaphragm,** which is innervated by the phrenic nerve. Contraction of the diaphragm enlarges the vertical dimension of the thoracic cavity by pushing the abdominal contents down and increases the transverse dimension by pushing the margins of the ribs laterally. Contraction of the **external intercostals** stiffens the chest wall and prevents these muscles from being pulled inward during inspiration, thereby reducing the dimensions of the thoracic cavity. During exercise and hyperventilation, the **accessory muscles of inspiration** are also recruited. These include the **scalene** and **sternomastoids,** both of which enlarge the thoracic cavity by elevating the sternum.

At rest or even during moderate exercise, expiration is a **passive** event, mediated by the return of the diaphragm to its relaxed position and by the elastic recoil of the lungs. With increased levels of ventilation, expiration becomes active and involves the muscles associated with expiration. The **abdominal musculature** (the rectus abdominis, the internal and external oblique, and the transverse abdominal) are the primary expiratory muscles. The **internal intercostals** are also active during forced expiration.

SUMMARY

All metabolizing tissues require oxygen and produce carbon dioxide, and the respiratory system is responsible for their transport. There are a number of steps involved in the movement of oxygen from the air to the tissues and carbon dioxide from the tissues to the external environment. These include the movement of air into and out of the lungs, the matching of the pulmonary blood flow to ventilation, the neural control of ventilation, the movement of oxygen and carbon dioxide between the lungs and the pulmonary circulation, and the transport of respiratory gases between the lungs and tissues.

NATIONAL BOARD TYPE QUESTIONS

Select the one most correct answer for each of the following

1. The VC is the:
   A. Volume of air remaining in the lungs after a maximum expiration
   B. Total volume of air in the lungs when they are maximally inflated
   C. Volume of air inspired and expired with each normal breath
   D. Amount of air that can be inspired from the end-tidal expiratory position
   E. Maximum volume of air that can be exhaled after a maximum inspiration
2. Which of the following patterns of lung volumes and capacities are most likely in a patient with a restrictive lung disease?
   A. Increased FRC, RV, and TLC; decreased VC
   B. Decreased FRC, RV, TLC, and VC
   C. Decreased TLC and RV; increased FRC and VC

D. Increased VC and RV; decreased FRC and TLC
   E. Increased FRC, RV, TLC, and VC
3. Near the terminal bronchioles:
   A. Gas exchange occurs
   B. The movement of air is accomplished by bulk flow
   C. There is no smooth muscle
   D. The movement of air is by diffusion
   E. There is no cartilaginous support
4. The primary muscle of inspiration is the:
   A. Diaphragm
   B. Sternomastoids
   C. Rectus abdominis
   D. External oblique
   E. Internal intercostals
5. Which of the following statements is true?
   A. Gas exchange occurs in the conducting zone of the airways
   B. The alveolar–capillary barrier presents a formidable obstacle to gas exchange
   C. The collective cross-sectional area of the airways decreases with each generation of airway branching
   D. The conducting zone of the airways constitutes the anatomic dead space
   E. The VC is the sum of RV and V_T

ANNOTATED ANSWERS

1. E. By definition, the maximum volume of air that can be exhaled after a maximum expiration is the VC
2. B. Because the lungs become "stiffer" and more difficult to inflate in restrictive lung diseases, the VC, TLC, FRC, and RV are all decreased.
3. B. In the airways down to the terminal bronchioles, movement of air is produced by bulk flow.
4. A. The principal muscle of inspiration is the diaphragm.
5. D. The conducting zone of the airways is ventilated but not perfused. Consequently, there

is no gas exchange in this region of the airways and the conducting zone is the anatomic dead space.

BIBLIOGRAPHY

Cherniack, N. S., Altose, M. D., and Kelsen, S. G. The Respiratory System. In: Berne, R. M., and Levy, M. N., eds. *Physiology,* Section VI. St. Louis: C. V. Mosby, 1983.

Comroe, J. H. *Physiology of Respiration,* 2nd ed. Chicago: Year Book Medical Publishers, 1974.

Levitzky, M. G. *Pulmonary Physiology,* 3rd ed. New York: McGraw-Hill, 1991.

Mines, A. H. *Respiratory Physiology,* 2nd ed. New York: Raven Press, 1986.

Taylor, A. E., Rehder, K., Hyatt, R. E., and Parker, J. C., *Clinical Respiratory Physiology.* Philadelphia: W. B. Saunders, 1989.

West, J. B. *Respiratory Physiology,* 4th ed. Baltimore: Williams & Wilkins, 1989.

27 Mechanics of Respiration

DOUGLAS K. ANDERSON

Objectives

After reading this chapter, you should be able to:

Define and discuss the relationships among the pressures responsible for the movement of air into and out of the lungs

Describe the pressure–volume changes in one respiratory cycle

Define the "resistances" to ventilation, that is, those forces that oppose lung expansion and air flow

Define and discuss the factors that affect lung compliance

Define and understand the determinants of air flow resistance

The mechanics of breathing encompass those factors that influence lung elasticity and air flow. They require an understanding of the pressures exerted to overcome the elastic recoil of the lungs and chest wall and the resistance to air flow that together effect volume changes in the lungs. Elastic properties of the lung and chest wall are termed **static** because they are defined during periods of zero air flow. The properties of airway resistance are studied during periods of air flow, and are therefore called **dynamic**.

PRESSURES

For air to move into and out of the lungs, a pressure gradient between the alveoli and the atmosphere must be created. The **atmospheric,** or barometric, **pressure** exists at the nose and mouth and by convention is defined as **0 cm H_2O**. **Alveolar pressure** is the pressure in the alveoli. When there is no air flow, the alveolar pressure equals the atmospheric pressure, or 0 cm H_2O.

When the alveolar pressure falls below the atmospheric pressure, a **pressure gradient,** extending from the atmosphere to the alveoli, is created and air moves into the lungs. Conversely, when the alveolar pressure exceeds the atmospheric pressure, the pressure gradient travels from the alveoli to the atmosphere and air flows out of the lungs. The pressure that establishes these gradients is the **intrapleural pressure,** which is inside the thoracic cavity but surrounding the lungs in the thin, serous fluid-filled space between the parietal and visceral pleura. The cohesive forces of the fluid molecules cause the two pleural surfaces to adhere to each other. At rest, the lungs are partially inflated and, because of their elastic nature, tend to collapse. The chest wall is drawn inward and, because it also has elastic properties, tends to recoil outward. Because pressure is equal to the force per unit area, these opposing forces act on the pleural surfaces to produce a subatmospheric intrapleural pressure which averages between -3 to

−5 cm H_2O at rest. The intrapleural pressure becomes more subatmospheric as the forces trying to pull the two pleural layers apart increase. Because the alveolar pressure always exceeds the pressure in the intrapleural space, the lungs tend to expand into this region of lower pressure. It is this subatmospheric intrapleural pressure that keeps the lungs inflated. Stated differently, the **transpulmonary pressure** (here defined as the transmural pressure across the alveolar wall, which is equal to the difference between the alveolar and intrapleural pressures) is the distending pressure of the alveoli. As long as the intrapleural pressure is less than the alveolar pressure, the alveoli will be partially inflated. When the transpulmonary pressure is increased by making the intrapleural pressure more subatmospheric (or by raising alveolar pressure), the alveoli expand further and their volume increases.

PRESSURE–VOLUME RELATIONSHIPS DURING ONE INSPIRATION–EXPIRATION CYCLE

Figure 27-1 illustrates the volume, pressure, and air flow changes that take place during a normal quiet breath. The tidal volume (V_T), air flow, and intrapleural pressure are all measured directly. Alveolar pressure can only be calculated. To measure the intrapleural pressure, a pressure-sensitive balloon transducer is passed into the esophagus. Because the thin-walled esophagus traverses the intrapleural space, pressure changes in this space are transmitted to the esophageal balloon.

At the end of expiration and just before the start of inspiration, air flow is zero because there is no driving pressure (that is, the atmospheric pressure equals the alveolar pressure, which by definition is 0 cm H_2O). Intrapleural pressure is subatmospheric, averaging 5 cm H_2O (actually −5 cm H_2O) below atmospheric pressure.

Inspiration is initiated when the inspiratory muscles (primarily the diaphragm) contract and

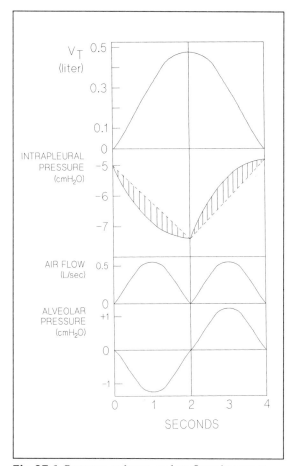

Fig. 27-1. Pressure, volume, and air flow changes in one inspiration–expiration (breathing) cycle. (V_T = tidal volume.) (Modified from: Comroe, J. H. *Physiology of Respiration* 2nd ed. Chicago: Year Book Medical Publishers, 1974. P. 102.)

thus enlarge the thoracic cavity. With the increasing thoracic volume, the intrapleural pressure becomes more subatmospheric and the lungs expand. With this increased lung volume, alveolar pressure decreases and becomes subatmospheric (−1 to −2 cm H_2O), causing air to flow into the lungs down this pressure gradient. Air continues to fill the lungs until the alveolar pressure again equals the atmospheric pressure. At this point, air flow ceases, inspiration ends, and the volume of air that has entered the lungs is the V_T.

When inspiration terminates, the diaphragm relaxes and the lungs recoil. As they recoil, the intrapleural pressure becomes less subatmospheric and lung volume is reduced, such that it returns toward the **functional residual capacity** (FRC) or resting volume. This decreased lung volume compresses the alveolar gas, causing alveolar pressure to exceed the atmospheric pressure, which causes air to flow down this gradient from the lungs to the atmosphere. When the alveolar pressure again equals the atmospheric pressure (when the V_T has been removed from the lungs), air flow ceases, expiration ends, and the cycle is repeated.

"RESISTANCES" TO VENTILATION

Certain factors oppose lung expansion and air flow. These consist of the **elastic recoil** properties of the lung (which includes surface tension forces), the **resistance to air flow** produced when air moves through the airways, and **tissue resistance**. Of the three, tissue resistance (which is caused primarily by the sliding of the lung tissues over each other as the lungs inflate and deflate) is minor and will not be considered in this discussion.

Figure 27-1 describes two pathways by which **intrapleural pressure** is changed during a normal breathing cycle. The *dashed line* shows the course intrapleural pressure would take if there were no air flow. This is the pressure change necessary for overcoming the elastic recoil forces of the lung only. The *solid line* describes the changes in intrapleural pressure during air flow. This line diverges from the *dashed line* because it is the sum of the pressure changes needed to overcome both the lung elastic recoil and the resistance to air flow. Thus, the difference between the two lines is the pressure change required to overcome the air flow resistance. The remainder of this chapter concentrates on the factors that produce the elastic properties of the lungs and that contribute to or change the resistance to air flow. These concepts are important because most lung disorders involve changes in lung elasticity or air flow.

COMPLIANCE

Compliance is defined as a unit change in the lung volume per the unit change in pressure, expressed as liters per centimeter of water, and is an index of the "expandability" of the lungs. This calculation is termed **static compliance** if the pressure and volume measurements are made when there is no air flow. An increase in the volume change per unit change in pressure indicates **increased lung compliance**. This results from reduced lung elasticity or elastic recoil, and means the lungs are easier to inflate. A decrease in the volume change per change in pressure indicates **reduced lung compliance**. This means the lungs are "stiffer," in that they display increased elasticity or elastic recoil and are harder to inflate. Thus, compliance is the inverse of elasticity or elastic recoil.

PRESSURE—VOLUME (COMPLIANCE) CURVE OF EXCISED LUNGS

The pressure—volume relationship of the lung can be demonstrated in lungs in vitro (Fig. 27-2). The lungs can be inflated in two ways: by placing the lungs in an airtight jar and reducing the pressure in the jar around the lungs; or by applying positive pressure through the trachea directly into the lungs. In either situation, the pressure inside the lungs exceeds the outside or surrounding pressure and the lungs inflate. By measuring the change in volume that occurs with each change in pressure, compliance curves like those in Figure 27-2 can be generated. There are several important features of these curves. Even though the lungs have been removed from the body, they are not airless—a small volume of **minimal air** remains. Starting at this point of maximum deflation, increasing the pressure inside the lungs (or decreasing the pressure around the lungs) causes the lung volume to increase. This is not a linear increase but is S—shaped. At a very

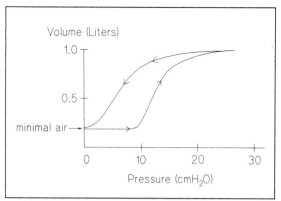

Fig. 27-2. Pressure–volume (compliance) curve of air-inflated excised lungs. Direction of *arrowheads* indicates inflation and deflation curves. (Modified from: West, J. B. *Respiratory Physiology* 4th ed. Baltimore: Williams & Wilkins, 1990. P. 90.)

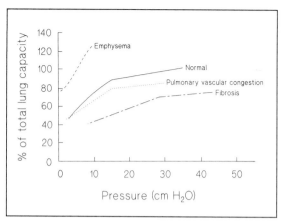

Fig. 27-3. Effects of disease on lung compliance.

low lung volume, compliance is low, such that it takes a relatively large change in pressure to achieve a change in volume. This occurs because, at this low volume, most of the alveoli in these excised lungs have collapsed and the cohesive forces at the liquid–liquid interface in these alveoli must be overcome. Thus, the initial pressure is expended in opening these closed alveoli, without a substantial change in volume.

Once open, the lungs distend easily. Between 5 to 20 cm H_2O, there is a relatively large change in the volume per unit change in pressure, indicating that the lungs are quite compliant over this pressure range. Above 20 cm H_2O, however, compliance again declines, because at high lung volumes the maximum distensibility of the lung is being reached (the lung is near maximally stretched) and further increases in pressure will produce only small volume increases.

The deflation curve for excised lungs differs from the inflation curve. This **hysteresis loop** results primarily from changes in the surface tension at different lung volumes, which will be discussed later in this chapter.

Anything that changes the lung's elastic properties alters compliance. Figure 27-3 shows how various pathologic states affect compliance. (The curves in this figure correspond to the deflation curve in Figure 27-2.) If the lungs become more compliant (an increased change in volume per unit change in pressure), the compliance curve shifts to the left of normal. Conversely, if the lungs become stiffer or less compliant, the compliance curve shifts to the right, because the change in lung volume per unit change in pressure is reduced. **Emphysema** is a condition in which lung tissue is lost. Consequently, the elasticity or elastic recoil of the lung is decreased and the lungs become more distensible (compliance increases). In **fibrosis,** connective tissue infiltrates the interstitial space. This makes the lung tissue stiffer, causing increased lung elasticity and decreased compliance. Thus, these lungs are difficult to distend. Compliance is also reduced if the pulmonary vasculature becomes engorged with blood. This can result from increased pulmonary venous pressure due to left heart failure. In fact, any condition that decreases the distensibility of the lung, such as alveolar edema or atelectasis (collapsed alveoli), decreases compliance.

REGIONAL DIFFERENCES IN VENTILATION

Ventilation (the volume of air inspired and expired per unit time; see Chapter 28 for a more complete discussion) is not uniform thoughout the lung. Rather, there is a continuum from the top to the bottom of the lung, with the lowest level of ventilation at the apex and the highest at the base. Thus, more air per unit of time is brought into and removed from the lower lung than from the upper lung. This difference in ventilation is illustrated in Figure 27-4.

Intrapleural pressure is not uniform around the lung because the lungs are suspended in the thoracic cavity, causing their full weight to be supported at the apex. Consequently, the forces tending to separate the parietal and visceral pleura are greater at the lung apex than at its base. As a result, intrapleural pressure increases from the top to the bottom of the lung. In other words, pressure in the intrapleural space is more subatmospheric at the apex of the lung than at its base. Because of this **intrapleural pressure gradient** from the apex to the base of the lung, there is a corresponding transpulmonary pressure gradient. The higher transpulmonary pressure at the apex of the lung causes the alveoli in this region to be more expanded than those at

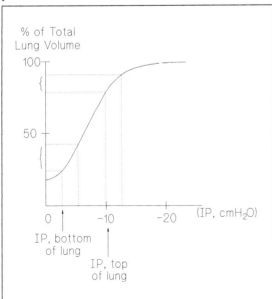

Fig. 27-4. The source of differences in ventilation from the apex to the base of the lung. (*IP* = intrapleural pressure.)

the base. Thus, for a given change in the intrapleural pressure, the alveoli on or near the upper flat portion of the compliance curve (those that are near the top of the lung) will distend less than the alveoli toward the base of the lung, which are on the steep part of the compliance curve. Consequently, over time more gas is moved into and out of the alveoli at the base of the lung than at the apex.

SPECIFIC COMPLIANCE

From previous discussions, it should be obvious that compliance varies with lung volume. Compliance is high at lower lung volumes, whereas it is low at higher lung volumes. Compliance also varies with lung size; it is higher in large individuals or animals with a larger total lung capacity (TLC) than in smaller individuals or species with a smaller TLC. Surgical removal of a lung or portion of a lung also reduces compliance. Thus, to determine if changes in compliance are due to alterations in the elastic properties of the lung and not in the size or volume of the lung, compliance must be normalized to some lung volume that generally approximates the FRC. A person with a normal compliance of 0.2 liter per cm H_2O and FRC of 2.5 liters would have a specific compliance of 0.08 cm H_2O^{-1}. **Specific compliance** is relatively consistent across species and normal individuals with widely different lung sizes, but changes in disease states that affect the elastic properties of the lungs.

PRESSURE–VOLUME (COMPLIANCE) CURVE OF THE LUNGS AND CHEST WALL

As mentioned earlier in this chapter, both the lungs and chest wall have elastic properties. Normally, the elastic recoil forces of the lungs and chest wall oppose each other, in that the lungs are inflated and recoil inward toward their equilibrum or resting position while the chest wall is pulled inward and recoils out toward its equilibrium position. Indeed, when the intrapleural space is opened to the atmosphere (pneumothorax) and the intrapleural pressure equals the atmospheric pressure, the lungs collapse and the chest wall springs outward.

The interaction between the lungs and chest wall over a wide range of volumes and pressures is demonstrated in Figure 27-5. A **relaxation pressure curve** for the entire respiratory system (lungs and chest wall) can be generated if the airway pressure is measured at different lung volumes while the respiratory muscles are relaxed. If the intrapleural pressure is also measured, the individual recoil pressures of the lungs and chest wall can be calculated. Thus, the relaxation pressure is the algebraic sum of the recoil pressures of the lung and chest wall.

If no external force is applied to the system (when the respiratory muscles are relaxed), the relaxation pressure is zero at the point when the inward recoil of the lungs is exactly counterbalanced by the outward recoil of the chest wall. This is the **resting respiratory position** and the volume of air remaining in the lungs is the FRC. With inspiration, the lungs inflate and move away from their equilibrium position, whereas the chest wall moves toward its resting point. Consequently, during this portion of inspiration, the lung elastic forces are the only ones that the respiratory muscles need to overcome. When the chest wall reaches its resting position (at about 70 percent of vital capacity), the relaxation pressure is entirely due to the lung elastic recoil pressure. Above this point, the chest wall is beyond its equilibrium position and the inspiratory muscles must then overcome the recoil of both the lungs and the chest wall. Thus, **maximum inspiration** (and vital capacity) is determined by the strength of the inspiratory muscles in relation to the recoil forces of the lung and chest wall.

With **maximum expiration,** the lungs approach their resting position and the relaxation pressure is created almost entirely by the elastic recoil pressure of the chest wall. At this point, there are minimal lung recoil forces and the expiratory muscles have compressed the chest wall as far as possible from its equilibrium point. This is when the lungs contain minimal air.

If the elastic recoil of either the lungs or chest wall changes without a corresponding change in the other, the FRC is altered and the relaxation pressure curve is shifted to either the left or right of normal. For example, with the loss of elastic tissue in **emphysema,** there is diminished elastic recoil force in the lungs without an accompanying loss of recoil in the chest wall. Thus, the lungs are more compliant and distended (FRC is increased) and the relaxation pressure curve is shifted to the left. Conversely, in **pulmonary fibrosis,** the elastic recoil force of the lungs, but not of the chest wall, is increased. This causes the lung compliance or distensibility to be decreased. Consequently, the FRC is decreased and the relaxation pressure curve is shifted to the right.

Fig. 27-5. Pressure–volume curves of the lung and chest wall. The relaxation pressure is the algebraic sum of the recoil pressures of the lungs and chest wall. (Modified from: Slonim, N. B., and Hamilton, L. H. *Respiratory Physiology* 2nd ed. St. Louis: Mosby, 1971. P. 52.)

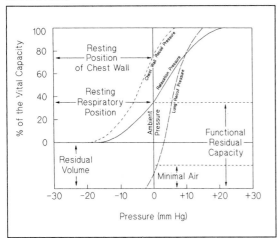

SURFACE TENSION

Figure 27-6 depicts the surface tension created when a liquid (such as water) is exposed to air. The *open circles* represent water molecules and the *solid squares* are gas molecules. In liquid, the molecular attracting forces acting on each

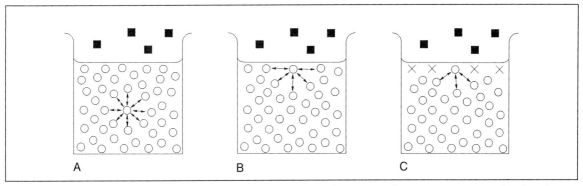

Fig. 27-6. Schematic depiction of the concept of surface tension. The *open circles* represent water molecules, the *closed squares* represent air molecules, and the *x's* represent surface active or surfactant molecules. (A) The molecular forces acting on each water molecule are equal in all directions. (B) The air–water interface. (C) Action of surfactant molecules. (Modified from: Conroe, J. H. *Physiology of Respiration* 2nd ed. Chicago: Year Book Medical Publishers, 1974. P. 107.)

individual water molecule are equal in all directions (see Fig. 27-6A). At the air–water interface, however, the water molecules beneath the surface exert stronger attracting forces on the surface water molecules than do the gas molecules (see Fig. 27-6B). Thus, there are stronger forces pulling the surface water molecules down than there are pulling them up. As a result of this imbalance of attractive forces, the surface shrinks to its smallest possible area. The resulting force between the surface molecules is the **surface tension**.

There is a thin film of fluid that lines the alveoli. Because this alveolar fluid has an air interface, it creates a surface tension. This surface tension of the alveolar fluid–air interface is depicted in Figure 27-7. Both the elastic recoil forces and the surface tension forces tend to collapse the alveolus. Thus, alveolar size is determined by the alveolar pressure (which expands the alveolus) that balances the sum of the elastic recoil and surface tension forces.

As indicated previously, surface tension plays a major role in the overall **compliance** of the lungs. The contribution of surface tension to lung compliance can be seen when the pressure–volume (compliance) curves in air (see Fig. 27-2) are contrasted with those from saline-inflated lungs (Fig. 27-8). When filled with air, the lungs'

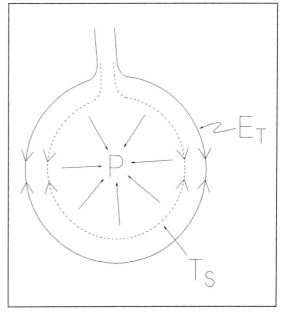

Fig. 27-7. The forces acting at the alveolar liquid–air interface. (E_T = tissue tension; T_s = surface tension; and P = intra-alveolar pressure. *Arrowheads* indicate the direction of the forces.) (Modified from: Levitzky, M. G. *Pulmonary Physiology* 3rd ed. New York: McGraw-Hill, 1991. P. 25.)

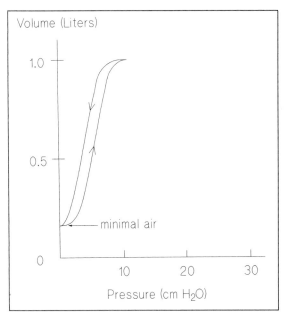

Fig. 27-8. Pressure–volume (compliance) curve for saline-inflated excised lungs. (Compare with Figure 29-2.) (Modified from: West, J. B. *Respiratory Physiology* 4th ed. Baltimore: Williams & Wilkins, 1990. P. 93.)

inflation curve differs from their deflation curve (a hysteresis loop) (see Fig. 27-2). When inflated with saline, hysteresis is abolished and the lungs have much greater compliance than when they are inflated with air (see Fig. 27-8). When the lungs are filled with saline, the air–liquid interface is eliminated and this abolishes the surface tension forces. This leaves elastic recoil as the only force opposing lung distention, and demonstrates that surface tension constitutes a substantial portion of the total recoil force of the lung.

PULMONARY SURFACTANT

Surfactants are surface–active molecules that reduce surface tension forces in the lung. These molecules accumulate on the surface of the alveolar fluid and interdigitate between the water molecules (see Fig. 27-6C). This separates the surface water molecules and reduces the attractive forces between them, thus decreasing surface tension.

Characteristics of Pulmonary Surfactant

Surfactant is a complex phospholipid that is a combination of dipalmitoyl phosphatidylcholine (DPPC), as well as other lipids and proteins. It is formed in type II alveolar cells and is rapidly synthesized and turned over. DPPC has two long-chain saturated fatty acids (palmitate) esterified to the first and second carbons of a glycerol molecule. A choline base is located on the third glycerol carbon. The fatty acids are nonpolar and hydrophobic, whereas the charged choline base is polar and hydrophylic. Thus, DPPC orients perpendicularly to the air–water interface, such that the polar end is dissolved in the water and the nonpolar fatty acids project toward the alveolar lumen (Fig. 27-9).

Physiologic Importance of Pulmonary Surfactant

Surfactant lowers the surface tension in the alveoli, thereby increasing lung compliance. This reduces the effort needed to expand the lungs with each breath.

Surfactant also promotes the stability of alveoli and helps keep them from collapsing. The lung is a collection of 300 million basically spherical alveoli of different sizes, all communicating with each other. There is little variation in alveolar pressure throughout the lung, meaning that the pressure is essentially the same in all open and

Fig. 27-9. The air–water interface in the presence of molecules of the surfactant dipalmitoyl phosphatidyl-choline (*DPPC*).

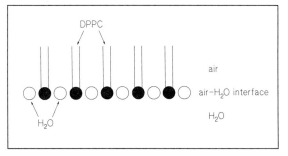

functioning alveoli. According to the law of Laplace, the distending pressure in a sphere is equal to 2T/r, where T is the wall tension and r is the radius of the sphere. Thus, if wall tension remains constant, the distending pressure increases if the radius decreases, and vice versa. When this law is applied to alveoli, the wall tension is the tension in the wall of an alveolus (the sum of elastic and surface tension; see Fig. 27-7) and the radius is the alveolar radius. Because the distending pressure is essentially the same in the millions of alveoli of different radii, the wall tension must change commensurate with the change in radius; otherwise, small alveoli will empty into larger alveoli and then collapse. Surfactant is responsible for changing the alveolar wall tension. When the radius is small (during expiration), surfactant molecules are packed tightly together, widely separating the water molecules and reducing the wall tension. When the radius is large, the surfactant molecules are scattered, allowing the water molecules greater access to each other and thereby increasing the alveolar wall tension.

Surfactant helps keep the alveoli dry. Surface tension also tends to draw fluid into the alveoli, causing alveolar edema. Because surfactant reduces the surface tension forces, the affinity for fluid to move into the alveoli is diminished.

Respiratory Distress Syndrome of Newborns

Respiratory distress syndrome of newborns is a serious condition that mainly affects premature infants, and is caused by inadequate synthesis of surfactant. In lungs of such infants, surface tension forces are high, compliance is low, and there are large regions of collapsed alveoli (atelectasis) and edema. These infants must expend a tremendous amount of energy just to inflate their lungs. Treatment of the disease is limited. Positive-pressure ventilation is required, and encouraging results have been obtained with the aerosol administration of bovine surfactant.

RESISTANCE TO AIR FLOW

As discussed previously, there are two primary factors that must be overcome in order to move air into and out of the lungs; these are the elastic recoil forces of the lungs and chest wall, and airway resistance. Of the two, airway resistance is the most important. The factors governing air flow through the airways are the same as those regulating the flow of any fluid through tubes.

Physical Factors Determining the Resistance to Air Flow

There are two primary types of air flow in the airways—laminar and turbulent. In **laminar flow,** the stream lines travel parallel to the sides of the tubes. Laminar flow occurs in all airways whenever air flow is low, but generally it is limited to the smaller airways because air flow in these regions is usually low. In **turbulent flow,** the flow in the stream lines is agitated. Turbulent flow can develop at the branch points of airways or when there are irregularities in the airways caused by mucus, tumors, or foreign bodies, even at low flow rates. The flow in the larger airways is also turbulent when air flow velocity is elevated.

Laminar flow through straight, smooth tubes is governed by **Poiseuille's law**:

$$V = Pr^4/8nl \qquad \text{Eq. 27-1}$$

Where V is the air flow, P is the pressure difference (the driving pressure), r is the radius of the tube, n is the viscosity of the fluid, and l is the length of the tube.

Resistance to air flow (R) is equal to the pressure difference divided by flow, or:

$$R = P/V \qquad \text{Eq. 27-2}$$

Combining and rearranging Equations 27–1 and 27–2 yields:

$$R = 8nl/r^4 \qquad \text{Eq. 27-3}$$

Equation 27–3 shows that the airway radius is the most important determinant of the resistance to air flow. For example, halving the radius increased resistance sixteen-fold, but doubling the length of the airway only doubles airway resistance.

Turbulent flow alters the relationship between the driving pressure, resistance, and air flow. When air flow becomes turbulent, the driving pressure has to increase (by the square of the flow) in order to maintain air flow constant. Thus:

$$P = V^2R \qquad \text{Eq. 27-4}$$

For much of the tracheobronchial tree, air flow is **transitional,** in that there is a combination of turbulent and laminar flows. Thus, for transitional flow:

$$P = VR_1 + V^2R_2 \qquad \text{Eq. 27-5}$$

For normal individuals, the second (turbulent) component is relatively small during quiet breathing and Equation 29–5 reduces to:

$$P = VR_1 \qquad \text{Eq. 27-6}$$

Principal Sites of Airway Resistance

Total airway resistance is the sum of the resistances from the nose and mouth and from the tracheobronchial tree. The nose and mouth account for a substantial portion of total airway resistance.

For the tracheobronchial tree, the total cross-sectional area of the airways increases from the larger to the smaller airways, even though the individual radii are decreasing (Fig. 27-10). The chief site of airway resistance in the tracheobronchial tree is at the medium-sized segmental bronchi, where the radius of the individual bronchi is decreased but the total cross-sectional area is not yet substantially increased. The least resistance to air flow is in the very small and numerous terminal bronchioles, with their tiny individual radii but enormous cross-sectional area.

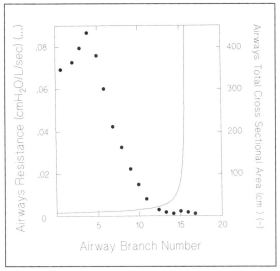

Fig. 27-10. Airway resistance (*solid circles*) and total cross-sectional area (*solid line*) plotted as a function of airways branch number. Airway resistance decreases and the total cross-sectional area of the airways increases going from the trachea to the smaller airways. (Modified from: West, J. B. *Respiratory Physiology* 4th ed. Baltimore: Williams & Wilkins, 1990. P. 105.)

Determinants of Airway Resistance

Lung Volume

When lung volumes increase, airway resistance decreases (Fig. 27-11). Lung volume modifies airway resistance in two ways. First, lung tissue is tethered to the airways. With increasing lung volume the **airways are "pulled open,"** thereby decreasing resistance. Second, increasing the lung volume elevates the **transmural pressure** across the airways and this increases their radii and decreases resistance. Thus, because lung volume is greater at the apex than at the base of the lung (see Fig. 27-4), airway caliber is larger at the top than at the bottom of the lung, causing airway resistance to increase progressively from the top to the base of the lung.

Bronchial Smooth Muscle

Airways from the trachea down to the alveolar ducts are supplied with smooth muscle, which is

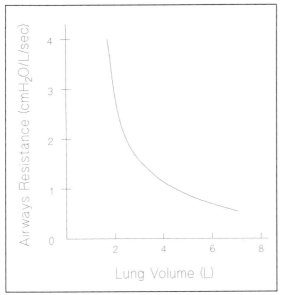

Fig. 27-11. Airway resistance plotted as a function of lung volume. Resistance decreases as lung volume increases. (Modified from: West, J. B. *Respiratory Physiology* 4th ed. Baltimore: Williams & Wilkins, 1990. P. 106.)

controlled by the autonomic nervous system. **Adrenergic sympathetic** activity or sympathomimetic compounds relax bronchial smooth muscle, whereas **cholinergic parasympathetic** activity or parasympathomimetic agents constrict bronchial smooth muscle. This is the opposite of what takes place in vascular smooth muscle, because beta$_2$ receptors (adrenergic receptors that mediate smooth muscle relaxation) predominate in bronchial smooth muscle.

Irritants such as smoke and dust and certain chemicals such as histamine can cause **reflex constriction** of the airways. These substances may stimulate the **rapidly adapting receptors** in the airways. In addition, emboli that lodge in the pulmonary circulation will cause airways in the poorly perfused areas to constrict. This appears to be a direct effect of decreased alveolar carbon dioxide tension on the smooth muscle of airways and may be one mechanism involved in balancing ventilation and perfusion (see Chapter 31).

SUMMARY

To move air into and out of the lungs, pressures must be generated that overcome the elastic recoil of the lungs and chest wall as well as the resistance to air flow. Atmospheric, alveolar, and intrapleural pressures are the important pressures in creating air flow. The distensibility of the lungs is defined in terms of changes in volume per unit change in pressure, and is called *compliance.* Lung compliance also includes surface tension forces. Surface tension is reduced and varied in the lung through the influence of surfactant, a surface active molecule which is a complex phospholipid. Resistance to air flow is determined by the pressure difference divided by the air flow. The radii of the individual airways decrease going from the larger to the smaller airways, but the total cross-sectional area of the tracheobronchial tree increases. Consequently, the least resistance to air flow is generated in the terminal bronchioles and the greatest air flow resistance in the tracheobronchial tree arises in the medium-sized segmental bronchi. The caliber and, hence, resistance of the airways can be determined passively by changes in lung volume and actively by changes in the autonomic activity to the smooth muscle of the airways.

NATIONAL BOARD TYPE QUESTIONS

Select the one most correct answer for each of the following

1. Intrapleural pressure is:
   A. The pressure inside the lungs
   B. The difference between alveolar and atmospheric pressures
   C. Generally greater than atmospheric pressure
   D. Generally less than atmospheric pressure
   E. Generally slightly greater than alveolar pressure
2. During normal tidal breathing (0.6 liter), a person shows an intrapleural pressure

change of 3 cm H_2O. If his FRC is 2.5 liters, his specific compliance (in cm H_2O^{-1}) is:

A. 0.83
B. 0.2
C. 0.033
D. 0.33
E. 0.08

3. Pulmonary surfactant is responsible for:

A. Changing radius of the alveoli by changing alveolar pressure

B. Changing alveolar pressure by keeping the surface tension constant

C. Helping to equalize the alveolar pressure throughout the lung by changing the surface tension in alveoli of different sizes

D. Changing alveolar pressure by changing the radius of the alveoli

E. Changing the surface tension by changing alveolar pressure

4. Resistance to air flow is:

A. Highest at the top of the lung
B. Lowest in the small terminal bronchioles
C. Lowest at the nose and mouth
D. Increased by sympathetic stimulation
E. Decreased at low lung volumes

5. Lung compliance is increased:

A. When surfactant levels decrease
B. In patients with emphysema
C. By pulmonary congestion
D. In patients with pulmonary interstitial fibrosis
E. When airway resistance is increased

ANNOTATED ANSWERS

1. D. The pressure between the lungs and chest wall (the intrapleural pressure) is subatmospheric

2. E. 0.6 liter/3 cm H_2O/2.5 liters = 0.08 cm H_2O.

3. C. Surfactant varies the surface tension, which helps equalize pressure in alveoli of different sizes.

4. B. The lowest resistance to air flow is at the terminal bronchioles with their enormous cross-sectional area.

5. B. Patients with emphysema lose lung elastic tissue. Consequently, they have increased lung volume with a given change in intrapleural pressure, which, by definition, is an increase in compliance.

BIBLIOGRAPHY

Cherniack, N. S., Altose, M. D., and Kelsen, S. G. The Respiratory System. In: Berne, R. M., and Levy, M. N., eds. *Physiology*, Section VI. St. Louis: C. V. Mosby, 1983.

Comroe, J. H. *Physiology of Respiration*, 2nd ed. Chicago: Year Book Medical Publishers, 1974.

Levitzky, M. G. *Pulmonary Physiology*, 3rd ed. New York: McGraw-Hill, 1991.

Mines, A. H. *Respiratory Physiology*, 2nd ed. New York: Raven Press, 1986.

Taylor, A. E., Rehder, K., Hyatt, R. E., and Parker, J. C. *Clinical Respiratory Physiology*. Philadelphia: W. B. Saunders, 1989.

West, J. B. *Respiratory Physiology*, 4th ed. Baltimore: Williams & Wilkins, 1989.

28 Ventilation, Gas Exchange, and Transport

DOUGLAS K. ANDERSON

Objectives

After reading this chapter, you should be able to:

Understand the basic properties of gases

Define minute ventilation, alveolar ventilation, and dead space

Describe the relationship among alveolar ventilation, alveolar oxygen and carbon dioxide partial pressures, oxygen consumption, and carbon dioxide production

Explain the source of the simplified alveolar gas equation

Describe the factors that govern the diffusion of gases through the alveolar–capillary membrane

Discuss the equilibration times for oxygen and carbon dioxide across the alveolar–capillary membrane

Discuss the factors responsible for the transport of oxygen by the blood

Define oxygen content and capacity and oxyhemoglobin saturation

Define the reasons for the shape of the oxyhemoglobin dissociation curve and discuss its physiologic significance

Delineate the mechanisms that control carbon dioxide transport by the blood and describe the carbon dioxide dissociation curves for whole blood

The principal function of the lung and chest wall system is the conveyance of air to the alveolar–capillary interface to supply oxygen to and remove carbon dioxide from the pulmonary capillary blood. Ventilation responds to changes in the metabolic demand of the body and the environmental gas composition and pressure. O_2 and CO_2 movement across the alveolar–capillary barrier is accomplished by diffusion and is governed by **Fick's law**. In the blood, O_2 is transported to the tissues attached primarily to hemoglobin, whereas CO_2 is returned to the lungs principally in the form of bicarbonate. The entire process and mechanisms responsible for moving O_2 from the air to the tissues and CO_2 from the tissues to the atmosphere is the topic of this chapter. Preliminary to this, the behavior of individual gases, mixtures of gases, and gases in liquid will be summarized.

PROPERTIES OF GASES

The factors that determine the pressure of a gas are described by the **general gas law:**

435

$$P = nRT/V \qquad \text{Eq. 28-1}$$

Where P represents pressure; V, volume; n, the number of gas molecules; R, the gas constant; and T, temperature. This equation shows that pressure is directly related to temperature and inversely related to volume. This means that at a constant volume, if temperature is raised, pressure increases. Further, if the volume containing the gas is increased at a constant temperature, the pressure falls.

Dalton's law states that the partial pressure of a single gas (P_g) in a mixture of gases is equal to the product of the total pressure (P_T) and the mole fraction (f_g) of the single gas (that is, the fraction of the total number of gas molecules that are molecules of the single gas).

$$P_g = P_T \times f_g \qquad \text{Eq. 28-2}$$

For example, the pressure of dry air at sea level (P_{ATM}) is 760 mm Hg and O_2 (fO_2) constitutes 20.95 percent of this. Thus, the partial pressure of O_2 (PO_2) at sea level is:

$$PO_2 = P_{ATM} \times fO_2$$
$$PO_2 = (760)(0.2095)$$
$$PO_2 = 159 \text{ mm Hg}$$

In a mixture of gases, the total pressure of the mixture (P_T) is equal to the sum of the partial pressures of the individual gases composing the mixture:

$$P_T = P_1 + P_2 + P_3 + \ldots \qquad \text{Eq. 28-3}$$

Each gas acts as if the others did not exist.

Air is a mixture of gases, including nitrogen, O_2, CO_2, and certain inert (I) gases. Thus for air:

$$P_{ATM} = P_{N_2} + P_{O_2} + P_{CO_2} + P_I \qquad \text{Eq. 28-4}$$

The above applies to *dry* gas mixtures. Inspired atmospheric air is warmed and humidified as it passes through the nose and mouth. The water vapor pressure (P_{H_2O}) of a saturated gas varies with temperature. At body temperature, P_{H_2O} is

47 mm Hg. Because water vapor is a gas and exerts a partial pressure, the partial pressure of inspired gases in the airways is:

$$P_{ATM} = P_{N_2} + P_{O_2} + P_{CO_2} + P_{H_2O} + P_I \qquad \text{Eq. 28-5}$$

Thus, the PO_2 of inspired air (PIO_2) in the airway is:

$$PIO_2 = (P_{ATM} - P_{H_2O}) \times fO_2$$
$$PIO_2 = (760 - 47)(0.2095)$$
$$PIO_2 = (713)(0.2095)$$
$$PIO_2 = 149 \text{ mm Hg}$$

Thus, the PO_2 has been reduced by approximately 6 percent, which equals the amount of water vapor added to the inspired air at the nose and mouth. The partial pressures of the other gases are also lessened by the partial pressure of water vapor, so that the sum of the partial pressures of all the gases that make up the inspired air does not exceed P_{ATM}.

As one ascends to higher altitudes, P_{ATM} declines but the proportions of the gases that make up air do not change. For example, if P_{ATM} is 349 mm Hg at 20,000 feet (6,000 meters) above sea level, then:

$$PO_2 = (349 - 47)(0.2095)$$
$$PO_2 = 63 \text{ mm Hg}$$

When a gas (such as O_2 or CO_2) is exposed to a liquid, it will dissolve in that liquid. The amount or concentration of gas (C_g) dissolved is the product of the partial pressure of the gas (P_g) in the gas phase and the solubility of the gas (S) in the liquid, such that:

$$C_g = P_g \times S \qquad \text{Eq. 28-6}$$

At equilibrium, the partial pressure of a gas in a liquid equals its partial pressure in a gas mixture that is exposed to the liquid. The amount of gas dissolved in a liquid depends not only on its partial pressure but also on its solubility. Thus, if a gas is very soluble, its concentration in a liquid may be high even though its partial pressure is low. For two gases with the same partial pressure, the one with the greatest solubility will be the most concentrated in the liquid. CO_2 is about 24-fold more soluble in water than O_2. So, at any given partial pressure, about twenty-four times more CO_2 than O_2 will be dissolved.

Some **important pressures** (given in millimeters mercury) are: alveolar O_2 (PAO_2), 100 to 105; pulmonary vein O_2 (systemic arterial; PaO_2), 95 to 100; alveolar CO_2 ($PACO_2$), 40; pulmonary vein CO_2 (systemic arterial; $PaCO_2$) 40; inspired Po_2, 152; pulmonary artery O_2 (systemic vein), 40; inspired PCO_2, 0.3; pulmonary artery CO_2 (systemic vein), 45; expired Po_2, 120; tissue Po_2, 40 or less; expired PCO_2, 32; and tissue PCO_2, 47 or more.

In the respiratory system, pressure is the driving force and gases move down the pressure gradients. For O_2, the gradient is directed from the atmosphere to the tissues; for CO_2, the gradient is directed from the tissues to the atmosphere.

In the normal lung, the equilibrium of O_2 and CO_2 between each alveolus and its investing capillaries is very rapid and complete, in that the partial pressures in both the alveoli and capillaries are essentially the same. However, for the lung as a whole and for systemic arterial gas pressures, equilibration is not complete (PAO_2 exceeds PaO_2 by about 5 to 8 mm Hg). This is caused by physiologic ventilation–perfusion mismatching in the lung and arteriovenous shunts (for example, nutrient circulation to the upper airways). (Ventilation–perfusion inequalities and shunting are discussed in Chapter 31.)

PaO_2 and $PaCO_2$ levels represent a balance between what comes into the lung in inspired air and what is removed (or added) from pulmonary capillary blood. The factors influencing the **levels of inspired gases** include the atmospheric pressure, the percentage of O_2 and CO_2 in the inspired air, and the level of alveolar ventilation ($\dot{V}_A$). (*Note that the dot over the letter indicates some volume per unit time—a flow.*) Determinants of what is added or removed by the blood include the Po_2 and PCO_2 in blood (which are functions of the metabolic rate) and the rate or volume of the pulmonary blood flow ($\dot{Q}$). Normally these factors do not change singly but act in concert; for example, if metabolic rate is altered, so is $\dot{V}_A$ and $\dot{Q}$. This maintains PaO_2 and $PaCO_2$ as close as possible to optimal values (100 mm Hg and 40 mm Hg, respectively) over a broad range of atmospheric pressures, PIO_2 levels, and degrees of physical activity.

VENTILATION

As noted in Chapter 26, all lung volumes, capacities, and ventilation are expressed as BTPS (body temperature, ambient pressure, saturated with water vapor); gas volumes in blood are expressed as STPD (standard temperature and pressure, dry gas). **Ventilation** is the volume of gas that is inspired or expired per unit time. **Minute ventilation** ($\dot{V}_E$) is the volume of gas inspired and expired per minute (in liters per minute, BTPS) and is the product of the tidal volume (V_T) and the frequency of breathing (n):

$$\dot{V}_E = V_T \times n \qquad \text{Eq. 28-7}$$

The V_T is all the air that enters the lungs with each breath. It includes the volume of air that enters the alveoli (V_A) and the air that remains in the conducting zone of the airways (where no gas is exchanged with pulmonary capillary blood), called the **anatomic dead space** (V_D). In healthy individuals V_D is estimated to be equal to the body weight in pounds. Thus:

$$V_T = V_D + V_A \qquad \text{Eq. 28-8}$$

The **physiologic dead space** is defined as any part of the lung that is ventilated but not perfused. This wasted ventilation (as it is more appropriately termed) increases in certain pathologic conditions, such as pulmonary emboli. In normal individuals, the ratio of V_D to V_T is approximately 0.3. Like other lung volumes, respiratory dead space is expressed as BTPS.

Alveolar ventilation ($\dot{V}_A$) is the volume of fresh gas that reaches the gas exchange zone (alveoli) each minute (liters per minute, BTPS). The $\dot{V}_A$ is of key importance because it represents the amount of fresh inspired air that is available for gas exchange. $\dot{V}_A$ equals the frequency of breathing times the difference between V_T and V_D. Thus, the magnitude of $\dot{V}_A$ depends on the V_T, the V_D, and the frequency of breathing:

$$\dot{V}_A = (V_T - V_D) \times n \qquad \text{Eq. 28-9}$$

ALVEOLAR VENTILATION, CARBON DIOXIDE, AND OXYGEN

The purpose of $\dot{V}_A$ is to supply the O_2 removed from the blood by the tissues and to remove from the alveoli the CO_2 added to blood by the tissues. Thus, $\dot{V}_A$ matches O_2 intake with O_2 consumption ($\dot{V}_{O_2}$), and CO_2 elimination with CO_2 production ($\dot{V}_{CO_2}$) and, in the process, maintains alveolar and arterial blood gas levels essentially constant. Consequently, there is a relationship between the P_{AO_2}, $\dot{V}_{O_2}$, and $\dot{V}_A$ as well as between P_{ACO_2}, $\dot{V}_{CO_2}$, and $\dot{V}_A$. In all of the following relationships, $\dot{V}_{O_2}$ and $\dot{V}_{CO_2}$ are expressed as STPD and $\dot{V}_A$ as BTPS.

Carbon Dioxide

The relationship between $\dot{V}_A$ and P_{ACO_2} is given by:

$$P_{ACO_2} = P_{ICO_2} + \dot{V}_{CO_2}/\dot{V}_A \times K \qquad \text{Eq. 28-10}$$

Where K is a proportionality constant equal to 0.863 if $\dot{V}_{CO_2}$ is at STPD and $\dot{V}_A$ is expressed as BTPS. Because the partial pressure of inspired CO_2 (P_{ICO_2}) is essentially zero (approximately 0.3 mm Hg at sea level), this term can be eliminated, reducing the equation to:

$$P_{ACO_2} = \dot{V}_{CO_2}/\dot{V}_A \times K \qquad \text{Eq. 28-11}$$

At a constant $\dot{V}_{CO_2}$ (200 ml/min), P_{ACO_2}, and thus arterial PCO_2 (P_{aCO_2}), varies inversely with $\dot{V}_A$ (Fig. 28-1). Consequently, if $\dot{V}_{CO_2}$ is unchanged, doubling $\dot{V}_A$ will halve P_{ACO_2} and P_{aCO_2}, and halving $\dot{V}_A$ will double P_{ACO_2} and P_{aCO_2}.

Generally, $\dot{V}_A$ and $\dot{V}_{CO_2}$ do not change independently. Rather, an increase in metabolic rate accelerates $\dot{V}_{CO_2}$. The excess CO_2 produced will stimulate the central and peripheral chemore-

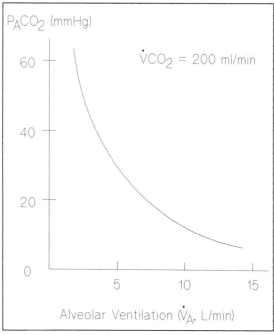

Fig. 28-1. Alveolar CO_2 tension (P_{ACO_2}) plotted as a function of alveolar ventilation ($\dot{V}_A$) at a constant production of ($\dot{V}_{CO_2}$). As $\dot{V}_A$ increases, P_{ACO_2} declines. (Modified from: Cherniack, N. S., Altose, M. D., and Kelsen, S. G. The respiratory system. In Berne, R. H., and Levy, M. N., eds., *Physiology*. St. Louis: Mosby, 1983. P. 685.)

ceptors (see Chapter 30 for details on respiration regulation). This causes $\dot{V}_A$ to increase appropriately, resulting in a relatively constant P_{ACO_2} and P_{aCO_2} over a broad range of activity (metabolic) levels.

Oxygen

The relationship for $\dot{V}_A$ and P_{AO_2} is expressed as follows:

$$P_{AO_2} = P_{IO_2} - \dot{V}_{O_2}/\dot{V}_A \times K \qquad \text{Eq. 28-12}$$

Where K is again 0.863. The P_{IO_2} is approximately 150 mm Hg and, unlike P_{ICO_2}, cannot be ignored. At constant levels of $\dot{V}_{O_2}$ (approximately 250 ml/min), both P_{AO_2} and arterial PO_2 (P_{aO_2}) increase with increasing $\dot{V}A$ and approach the PIO_2 (Fig. 28-2). Again, the $\dot{V}A$ does not normally change in the face of a constant $\dot{V}O_2$ but is regulated to meet metabolic demands, such that the PAO_2 and PaO_2 remain relatively unaltered over a wide metabolic range.

SIMPLIFIED ALVEOLAR GAS EQUATION

Equations 28-11 and 28-12 can be combined to form the simplified alveolar gas equation:

$$PAO_2 = PIO_2 - PACO_2/R + F \qquad \text{Eq. 28-13}$$

Where R is the respiratory exchange ratio (an index of the metabolic rate), equal to $\dot{V}CO_2/\dot{V}O_2$, and F is a correction factor of approximately 1 to

Fig. 28-2. Alveolar CO_2 tension (PAO_2) plotted as a function of ($\dot{V}A$) at a constant O_2 consumption ($\dot{V}O_2$). As $\dot{V}A$ increases, PAO_2 rises approaching the level of inspired O_2. (Modified from: Cherniack, N. S., Altose, M. D., and Kelsen, S. G. The respiratory system. In Berne, R. H., and Levy, M. N., eds., *Physiology*. St. Louis: Mosby, 1983. P. 686.)

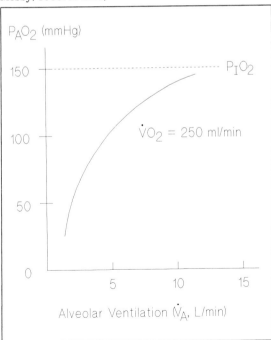

3 mm Hg. In its complete form, the **alveolar gas equation** is expressed as:

$$PAO_2 = PIO_2 - PACO_2 (FIO_2 + 1 - FIO_2/R) \qquad \text{Eq. 28-14}$$

Where FIO_2 (the fraction of inspired oxygen) is 0.2095. Equation 28-14 indicates that PAO_2 depends on PIO_2 and the respiratory exchange ratio. Also, at a given PIO_2 and a constant respiratory exchange ratio, PAO_2 and $PACO_2$ are inversely related. For example, if $\dot{V}A$ increases, PAO_2 will rise and $PACO_2$ will fall. **Alveolar hypoventilation** occurs when $\dot{V}A$ is less than what the metabolic rate requires. Thus, PAO_2 and PaO_2 will decrease and $PACO_2$ and $PaCO_2$ will increase. Conversely, **alveolar hyperventilation** arises when $\dot{V}A$ exceeds the metabolic demands. In this event, PAO_2 and PaO_2 increase and $PACO_2$ and $PaCO_2$ decrease.

GAS EXCHANGE (DIFFUSION)

Generally, PaO_2 and $PaCO_2$ can be used interchangeably with PAO_2 and $PACO_2$, respectively, in the alveolar gas equation because of the rapid equilibration of O_2 and CO_2 across the alveolar–capillary barrier. These gases move between the alveoli and pulmonary capillary blood through the process of **diffusion.** This transfer of gases is governed by **Fick's law,** which states that the volume of gas ($\dot{V}_g$) transferred across the alveolar–capillary membrane per unit of time is *directly related* to the:

1. Driving pressure across the alveolar–capillary membrane, that is, the difference in the partial pressure of the gas between the alveoli and capillary blood ($PA - PC$)
2. Area of the membrane (A)
3. Solubility (S) of the gas

and is *inversely related* to the:

4. Length of the diffusion pathway (the thickness of the membrane [T])
5. Square root of the molecular weight (MW) of the gas.

Usually solubility and molecular weight are incorporated into a diffusion constant (D):

$$D \propto S/\sqrt{MW} \qquad \text{Eq. 28-15}$$

The expression for Fick's law is then:

$$V_g = A \times D \times (P_A - P_C)/T \qquad \text{Eq. 28-16}$$

The solubility of CO_2 in water and its molecular weight are about twenty-four and 1.4 times greater than those for O_2, respectively. This combination causes the rate of CO_2 diffusion through both the tissue fluid and into the liquid surface to be about twenty-fold higher than that of O_2.

As discussed previously, the lung is well designed to transfer gases between the alveoli and pulmonary capillary blood. There is a large surface area for gas exchange (about 70 square meters) that is primarily due to the subdivision of the lung into approximately 300 million alveoli. In addition, the diffusion pathway is short (averaging 0.5 μm) and there are transmembrane pressure gradients in the correct directions for movement of both O_2 and CO_2.

Transfer of gases across the alveolar–capillary barrier can be either diffusion or perfusion limited. A gas that does not equilibrate across the alveolar–capillary barrier, such that its partial pressure gradient is maintained when the blood is in transit through a pulmonary capillary, is said to be **diffusion limited**. Stated differently, the diffusion-limited transfer of a gas depends on the physical properties of the alveolar–capillary barrier (such as the thickness of the barrier and the area for gas exchange) and not on the blood flow rate in the pulmonary capillary. The movement of carbon monoxide across the alveolar–capillary barrier is diffusion limited.

Conversely, a gas that equilibrates rapidly across the alveolar–capillary barrier is **perfusion limited**. In this case, the partial pressure for the gas in the pulmonary capillary blood becomes virtually equal to that in the alveoli very soon after blood has entered the pulmonary capillary. Once its partial pressure gradient across the alveolar–capillary barrier is eliminated, no additional dif-

fusion of this gas can occur during the remaining time blood is in transit through the pulmonary capillary. Thus, the amount of this gas transferred depends on the blood flow and not on the diffusion properties of the alveolar–capillary barrier. Nitrous oxide is a gas that is perfusion limited.

Equilibration of Oxygen

Under normal conditions, O_2 transfer is **perfusion limited**. This is demonstrated in Figure 28-3, which plots P_{O_2} in blood as a function of the blood transit time in a single pulmonary capillary. It takes blood an average of 0.75 second to

Fig. 28-3. Capillary partial pressure of O_2 (P_{O_2}) plotted as a function of the mean transit time for a red blood cell to pass through a pulmonary capillary. (A) The normal O_2 equilibration curve between the alveolus and the pulmonary capillary. O_2 equilibration is essentially complete by the time the red blood cell is one-third of its way through the capillary. (B) The O_2 equilibration curve if the alveolar–capillary barrier becomes moderately thickened. (C) The O_2 equilibration curve between the alveolus and pulmonary capillary if this thickening becomes severe. (Modified from: West, J. B. *Respiratory Physiology* 4th ed. Baltimore: Williams & Wilkins, 1990. P. 25.)

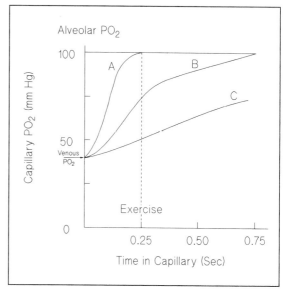

traverse one pulmonary capillary. *Curve A* in Figure 28-3 shows that, under resting conditions, there is essentially complete equilibration between alveolar and capillary P_{O_2} once blood has passed only one-third of the way through the capillary. Thus, the rate of blood flow through the capillary can be substantially increased, but essentially complete equilibration between capillary and alveolar P_{O_2} still takes place. For example, during heavy exercise, the increased cardiac output can reduce the capillary transit time by as much as threefold, to 0.25 second. This two-thirds reduction in the transit time does not appreciably alter the alveolar–capillary equilibration of O_2, and the end-capillary P_{O_2} will remain near resting levels. Thus, in normal individuals, O_2 transfer is perfusion limited even during exercise. In some well-conditioned athletes, the capillary transit time during exercise becomes so short (less than 0.25 second) that O_2 equilibration is incomplete. In this special situation, O_2 transfer becomes diffusion limited.

Theoretically, if the alveolar–capillary barrier thickens because of disease (for example, interstitial pulmonary fibrosis or edema), O_2 transfer could be impeded and blood may need to traverse almost the entire length of the capillary to achieve equilibrium (see Fig. 28-3, *curve B*). At rest, the end-capillary P_{O_2} in such a person would be near normal, but the reduced capillary transit time that occurs with exercise would impose a significant diffusion limitation on O_2 transfer, and end-capillary P_{O_2} would fall. If the thickening is extreme, then O_2 equilibration may not occur even at rest (see Fig. 28-3, *curve C*).

This perfect matching between one alveolus and capillary for O_2 does not occur throughout the whole lung because of the ventilation and perfusion inequalities and shunting that typically exist even in the healthy lung. (This subject is covered in Chapter 31.)

Equilibration of Carbon Dioxide

The equilibration of CO_2 across the alveolar–capillary barrier occurs at about the same rate as oxygen, in that equilibration is complete when blood is slightly over one-third of the way through the capillary (Fig. 28-4, *curve A*). Thus, the transfer of CO_2 like that of O_2 is perfusion limited over a wide range of activity levels. In addition, if thickening of the alveolar–capillary barrier is sufficiently severe, CO_2 transfer could become diffusion limited and end-capillary P_{CO_2} levels would rise (see Fig. 28-4, *curve B*). O_2 and CO_2 have essentially the same equilibration times even though CO_2 is twenty-four times more soluble in water and diffuses twenty times faster through water than does O_2. This is due primarily to two reasons. First, the pressure gradient for CO_2 across the alveolar–capillary barrier is substantially less than that for O_2 (5 mm Hg versus 60 mm Hg, respectively). Second, CO_2 is transported in blood primarily as bicarbonate and carbamino compounds. It takes a finite period to reverse

Fig. 28-4. Equilibration of CO_2 across the alveolar–capillary barrier. (A) CO_2 equilibration is essentially complete by the time the red blood cell is a little over one-third of its way through the capillary. (B) With thickening of the alveolar–capillary barrier, CO_2 equilibration may be incomplete. (P_{CO_2}, P_{O_2} = partial pressures of CO_2 and O_2, respectively.) (Modified from: West, J. B. *Respiratory Physiology* 4th ed. Baltimore: Williams & Wilkins, 1990. P. 29.)

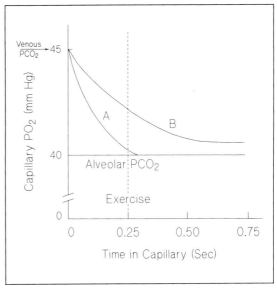

these reactions in the lung to form dissolved CO_2, so that it can diffuse from blood into the alveoli. Together, these two factors are primarily responsible for the slightly longer P_{CO_2} equilibration time.

GAS TRANSPORT BY THE BLOOD

Oxygen

O_2 is transported in the blood in two forms: **physically dissolved** and attached to **hemoglobin**.

At 37°C, each milliliter of plasma takes up 0.00003 ml of O_2 for each 1-mm Hg increase in P_{O_2}. Thus, there is 0.3 ml of dissolved O_2 in every 100 ml of blood at 37°C with a P_{O_2} of 100 mm Hg. If this were the only way blood could transport O_2, the cardiac output would have to be 83 liters per minute and every O_2 molecule would have to be extracted from the blood to meet an O_2 consumption in *resting* humans of 250 ml/min. Obviously, O_2 carriage by the blood solely in this form could not meet even the most basic metabolic requirements of the body and an additional form of O_2 transport is needed. Hemoglobin provides this transport.

Hemoglobin is a remarkable compound made up of four polypeptide chains, each with a **heme** moiety attached. The heme molecule consists of a porphyrin ring with one atom of ferrous iron in the center. Each of these four iron molecules can reversibly combine with one molecule of O_2. The iron molecules stay in the ferrous state when they bind with O_2, so this reaction is more correctly called **oxygenation** than oxidation. Thus, the presence of hemoglobin in the blood permits much greater quantities of O_2 to be transported than in the dissolved state alone. O_2 attached to hemoglobin (oxyhemoglobin) accounts for about 98 percent of O_2 transported by blood. Hemoglobin permits blood to absorb sixty-five times as much O_2 as plasma at a P_{O_2} of 100 mm Hg.

Definition of Oxygen Content and Capacity and Oxyhemoglobin Saturation

The **O_2 content** of blood is the total amount of O_2 carried in blood, that is, the O_2 content is the sum of the O_2 combined with hemoglobin and the O_2 dissolved in plasma. Thus, the O_2 content varies with the total amount of hemoglobin in blood and with the P_{aO_2}.

The **O_2 capacity** of blood is defined as the maximum amount of O_2 that can be carried by hemoglobin. One gram of hemoglobin that is fully loaded with O_2 combines with 1.34 ml of O_2. Whole blood in healthy adults contains approximately 15 gm of hemoglobin (Hb) per 100 ml of blood. Thus:

$$\begin{aligned}
O_2 \text{ capacity} &= 1.34 \text{ ml } O_2 \times \text{gm Hb/vol. of blood} \\
&= 1.34 \text{ ml } O_2/\text{gm Hb} \times 15 \text{ gm Hb}/100 \\
&\qquad \text{ml of blood} \\
&= 20.1 \text{ ml } O_2/100 \text{ ml of blood}
\end{aligned}$$

$$\text{Eq. 28-17}$$

This calculation reveals that the O_2 capacity varies with the hemoglobin content of blood.

The **percentage saturation of hemoglobin with O_2** is the amount of O_2 actually combined with hemoglobin divided by the O_2 capacity times 100. This quantity ($\%HbO_2$) can be plotted as a function of the P_{aO_2} yielding the oxyhemoglobin dissociation curve.

The Oxyhemoglobin Dissociation Curve

Plotting the dissolved O_2 content against P_{aO_2} yields a straight line, but the relationship between P_{aO_2} and $\%HbO_2$ saturation is not linear but is S-shaped (Fig. 28-5) The slope is steep when P_{aO_2} is in the lower ranges and is essentially flat at higher P_{aO_2} values. The S-shaped oxyhemoglobin dissociation curve indicates that the amount of O_2 bound to hemoglobin is relatively constant over a fairly wide range of higher P_{aO_2} values (flat part of the curve), but, at lower P_{aO_2} levels, there are gradually larger changes in the amount of O_2 bound to hemoglobin for a given change in P_{aO_2} (steep part of curve). This occurs because hemoglobin is made up of four subunits that load or unload their attached O_2 molecules with different affinities. As P_{aO_2} decreases from 100 mm Hg, there is little change in the $\%HbO_2$ saturation until P_{aO_2} reaches about 60 to 70 mm

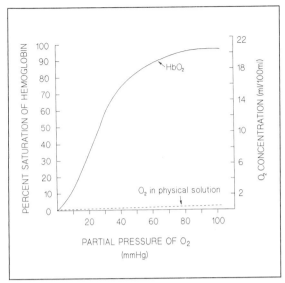

Fig. 28-5. Oxyhemoglobin (HbO$_2$) dissociation curve (*solid line*) compared with the amount of O$_2$ in physical solution (*dashed line*). Partial pressure of O$_2$ in blood, pH, and temperature is held constant at 40 mm Hg, 7.4, and 37°C, respectively. (Modified from: West, J. B. *Respiratory Physiology* 4th ed. Baltimore: Williams & Wilkins, 1990. P. 70.)

Hg. Thus, it takes a substantial decline in PaO$_2$ to offload the first O$_2$ molecule, causing the upper portion of the curve to be relatively flat. The unloading of the remaining three O$_2$ molecules is progressively easier, and this accounts for the steep part of the curve.

Physiologic Significance of the Shape of the Oxyhemoglobin Dissociation Curve

The plateau portion of the oxyhemoglobin dissociation curve provides a "reserve" for the availability of O$_2$. PaO$_2$ (and consequently PaO$_2$) can lower substantially with little change in the %HbO$_2$ saturation. This means that PaO$_2$ can vary widely (from 100 to 60 mm Hg, or about 40 mm Hg) and yet essentially the same amount of O$_2$ will attach to hemoglobin.

The steep part of the oxyhemoglobin dissociation curve enables tissues to extract relatively large amounts of O$_2$ from blood with relatively small changes in the PaO$_2$. Consequently, on the steep part of the curve, almost 60 percent of the O$_2$ attached to hemoglobin can be offloaded with a 40-mm Hg change in PaO$_2$ (that is, from 60 to 20 mm Hg). This allows for large quantities of O$_2$ to be offloaded from hemoglobin when the PaO$_2$ is low, thereby making it available to metabolically active tissues where it is needed.

Factors Affecting the Oxyhemoglobin Dissociation Curve

The oxyhemoglobin dissociation curve shown in Figure 28-5 only illustrates PaO$_2$ as affecting the onloading and offloading of O$_2$ from hemoglobin. Actually, when arterial blood reaches the tissue where the PO$_2$ is less than 40 mm Hg, the curve deviates to the right. This rightward shift is due to the metabolic activity at the tissue level with a resulting change in temperature, PCO$_2$, and pH (Fig. 28-6). An increase in temperature and PCO$_2$, and decrease in pH, shift the oxyhemoglobin dissociation curve to the right. Conversely, a decrease in temperature and PCO$_2$ and an increase in pH shift the curve to the left. This effect of PCO$_2$ and pH on the oxyhemoglobin dissociation curve is called the **Bohr effect**. A rightward shift of the curve means that, at a given PaO$_2$, there is less O$_2$ bound to hemoglobin (more offloading of O$_2$); a leftward shift means that, at a given PaO$_2$, there is more O$_2$ bound to hemoglobin. As blood passes through the capillaries of metabolizing tissue, PO$_2$ and pH declines and PCO$_2$ and temperature increases, caused by the metabolically active tissue. The shift in the oxyhemoglobin dissociation curve resulting from the changes in pH, PCO$_2$, and temperature facilitates the unloading of O$_2$ from hemoglobin. Thus, additional O$_2$ (exceeding that provided by the decline in PO$_2$ from arterial to venous blood) is made available to the tissues and the curve deviates to the right (that is, an extra 3 to 5 percent of O$_2$ is removed from hemogloblin).

CO$_2$ and H$^+$ combine with sites on hemoglobin, changing its configuration and facilitating the offloading of oxygen. Both CO$_2$ and H$^+$ bind more avidly to partially unsaturated hemoglobin. Thus, the initial unloading of O$_2$ from hemoglobin

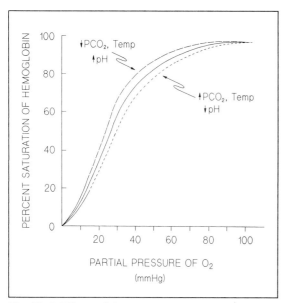

Fig. 28-6. Effects of changing partial pressure of CO_2 (Pco_2), pH, and temperature on the dissociation of O_2 from hemoglobin. Increases in pH and decreases in Pco_2 and temperature shift the oxyhemoglobin dissociation curve to the left. Conversely, a decrease in pH and increases in temperature and Pco_2 shift the curve to the right.

is caused by a decline in PaO_2. However, once partially unsaturated hemoglobin appears, H^+ and CO_2 begin to bind, thereby enhancing the further off-loading of O_2 along with the decline in PaO_2. Thus, in actively metabolizing tissue (such as in skeletal muscle during exercise) where there are additional increases in Pco_2 and temperature along with a lower PaO_2 and pH, the amount of O_2 offloaded from hemoglobin in this tissue is greater than at rest. The obvious purpose is to provide added O_2 to the tissues when the metabolic rate is increased.

2, 3-Diphosphoglycerate (2,3-DPG) is a metabolite of anaerobic glycolysis and is highly concentrated in red blood cells (15 mol per gram of hemoglobin). Like CO_2 and H^+, 2,3-DPG facilitates the unloading of O_2 from hemoglobin. It is generally believed that increases in 2,3-DPG levels can induce a rightward shift of the oxyhemoglobin dissociation curve. However, measurable in-

creases in 2,3-DPG seem to occur only under certain conditions such as at hypoxia (produced by high altitudes) or exercise, and this rise develops only after some delay. For example, 2,3-DPG does not become elevated until after about one hour of vigorous exercise, and there may be no increase at all in some well-trained athletes.

What then is the role (if any) of 2,3-DPG under normal resting conditions? Without a substantial change in concentration, can 2,3-DPG contribute to the onloading and offloading of O_2 from hemoglobin? As previously noted, 2,3-DPG is highly concentrated in red blood cells at all times. Similar to the actions of the H^+ and CO_2, the binding of 2,3-DPG to hemoglobin facilitates the release of O_2 by changing the configuration of the hemoglobin molecule. When hemoglobin is essentially completely oxygenated, most of the 2,3-DPG–binding sites are covered. However, when O_2 is released from hemoglobin in the tissues (initiated by decreased PaO_2) and partially reduced hemoglobin appears, binding sites for 2,3-DPG are uncovered. The 2,3-DPG molecules attach to these sites, causing a change in the molecular configuration of hemoglobin and thus enhancing the unloading of O_2. In the lung, hemoglobin is reoxygenated, 2,3-DPG–binding sites are covered, and the 2,3-DPG molecules (along with CO_2 and H^+ from other sites on the hemoglobin molecule) are offloaded from hemoglobin. Thus, the loading and unloading of O_2 from hemoglobin is a dynamic process that depends on the PaO_2, the levels of CO_2 and 2,3-DPG, pH, and temperature.

Effects of Anemia and Carbon Monoxide on Oxygen Transport

The curves in Figure 28-7 are not oxyhemoglobin dissociation curves (%HbO$_2$ saturation), but represent plots of the *amount* of O_2 bound to hemoglobin (oxyhemoglobin content) as a function of PaO_2. *Curve A* depicts data from a normal subject, whereas *curve C* illustrates data from an **anemic** person (a hemoglobin concentration of 6 gm/100 ml of blood; normal, 15 gm/100 ml). Thus, the oxyhemoglobin content is reduced in this person, which means the total amount of O_2 carried by

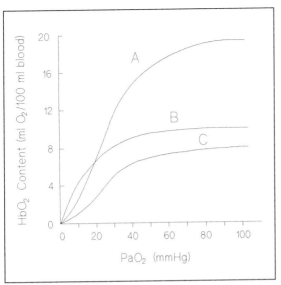

Fig. 28-7. Effects of anemia (C) and carbon monoxide poisoning (B) on the oxyhemoglobin (*HbO₂*) content (the volume of O₂ bound to hemoglobin). (A) The normal curve. Note that both anemia and carbon monoxide poisoning reduce the HbO₂ content but for different reasons (see text). (Modified from: Cherniack, N. S., Altose, M. D., and Kelsen, S. G. The respiratory system. In Berne, R. H., and Levy, M. N., eds. *Physiology*. St. Louis: Mosby, 1983. P. 691.)

hemoglobin is decreased because the concentration of hemoglobin is reduced. However, at a PaO₂ of 100 mm Hg, the hemoglobin in an anemic individual is still about 97 percent saturated. Thus, even though the hemoglobin levels are depressed, the remaining hemoglobin is saturable. In anemic blood, as PaO₂ levels are reduced, substantially more O₂ is unloaded from hemoglobin than normal (rightward shift of *curve C* from *curve A*). This represents a compensation for the lowered O₂ content in the blood of anemic individuals. This prolonged reduction in O₂ levels is likely to produce lactic acidosis (decreased pH) and to increase 2,3-DPG levels (anaerobic shift in metabolism) in the blood, both of which facilitate the offloading of O₂.

Curve B describes the oxyhemoglobin in a person exposed to **carbon monoxide**. The affinity of carbon monoxide for the O₂-binding sites on

hemoglobin is more than two hundred times greater than that of O₂. Because carbon monoxide competes with O₂ for binding sites on hemoglobin, exposure to carbon monoxide reduces the oxyhemoglobin content. In this event, the %HbO₂ saturation is reduced, such that the hemoglobin levels are normal but carbon monoxide is bound to about 60 percent of the O₂-binding sites. Carbon monoxide binding causes a leftward shift of the oxyhemoglobin content curve. Because most of the O₂-binding sites are occupied by carbon monoxide molecules that do not respond to declining PaO₂ levels, the O₂ molecules that remain on hemoglobin are more avidly bound and unload slower than normal, resulting in the leftward shift of *curve B* from *curve A*.

Carbon Dioxide

Like O₂, the movement of CO₂ between compartments proceeds down its pressure gradient; from tissue PCO₂ to blood plasma to red blood cells to lungs, in descending order. There are three forms of CO₂ transport: (1) physically dissolved; (2) as bicarbonate; and (3) in combination with blood proteins as carbamino compounds. These three mechanisms of CO₂ transport at the tissue are depicted in Chapter 38, (see Fig. 38-2). The reverse process occurs in the lungs.

Because CO₂ is about twenty-four times more soluble than O₂, **dissolved CO₂** is a more important form of CO₂ transport than is dissolved O₂. Approximately 5 to 10 percent of the CO₂ is transported in the dissolved form. At 37°C each milliliter of blood takes up 0.0007 ml of CO₂ for each 1-mm Hg increase in PCO₂, which is about twenty times greater than the rate for O₂.

Some of the dissolved CO₂ reacts with water, forming carbonic acid, which immediately breaks down to **bicarbonate** and H⁺:

$$CO_2 + H_2O \rightleftharpoons H_2CO_3 \rightleftharpoons H^+ + HCO_3$$

This reaction is very slow in plasma. However, in the red blood cell it is 13,000 times faster because of the presence of the enzyme carbonic anhydrase, which catalyzes the hydration of CO₂ to carbonic acid. The H^+ that is formed is buffered by the partially oxygenated hemoglobin. The bicarbonate that is produced diffuses out of the red blood cells into plasma, and this prevents its accumulation in the red blood cell, which would slow or halt the hydration of CO_2. Thus, even though most of the bicarbonate is formed in red blood cells, much of it diffuses out and is transported in plasma. Between 60 to 65 percent of the CO_2 in the blood is in the form of bicarbonate.

With the leakage of bicarbonate from erythrocytes into plasma, chloride ions from plasma move into the red blood cell to maintain electric neutrality. This exchange of anions is known as the **Hamburger** or **chloride shift**. Buffering the H^+ reduces the polyvalent anionic charge on hemoglobin. This loss of intracellular negative charge is replaced by the monovalent anions, bicarbonate and chloride. Consequently, the total number of ions in the cell increases, elevating the internal osmotic pressure. Water moves into red blood cells to preserve the **osmotic equilibrium**. This causes a slight swelling of erythrocytes in the venous blood and a consequental increase in the venous hematocrit relative to that of arterial blood.

Carbamino compounds are formed from the reaction of CO_2 with terminal amine groups, *primarily* on hemoglobin (although CO_2 will also form carbamino compounds with plasma proteins, but to a much lesser extent). This is depicted by the following structural formula:

$$Hb\text{-}NH_2 + CO_2 \rightleftharpoons Hb - N \overset{\displaystyle H^+}{\underset{\displaystyle COO^-}{}} + H^+$$

Carbamino compounds account for about 30 percent of the CO_2 transported in blood.

Carbon Dioxide Elimination by the Lungs

The pressure gradients for O_2 and CO_2 across the alveolar–capillary barrier initially stimulate the simultaneous movement of these two gases between the alveoli and blood. As CO_2 enters the alveoli from plasma, the **law of mass action** demands that the reaction for the hydration of CO_2 be shifted to the left. This process plus the binding of O_2 with hemoglobin causes both CO_2 and H^+ to dissociate from hemoglobin; the released H^+ then combines with bicarbonate to reform CO_2. Basically, the movement of CO_2 and O_2 at the lungs is the reverse of their movement at the tissues (see Fig. 38-2).

Carbon Dioxide Dissociation Curve

The CO_2 dissociation curve relates changes in the **CO_2 content** of whole blood to changes in the blood P_{CO_2}, not the changes in binding of CO_2 to some carrier molecule such as hemoglobin (that is, it is not a percentage saturation curve like the oxyhemoglobin dissociation curve) (Fig. 28-8). Several points distinguish this curve. First, the CO_2 dissociation curve is not S-shaped like the oxyhemoglobin dissociation curve, but is essentially linear in the physiologic $PaCO_2$ range (from 40 to 50 mm Hg). Second, reducing the O_2 saturation of hemoglobin causes a leftward shift of the CO_2 dissociation curve. Consequently, as O_2 is unloaded from hemoglobin, the CO_2 content of blood is **increased,** primarily because unsaturated or deoxyhemoglobin is a weaker acid than oxyhemoglobin. Thus, as hemoglobin becomes progressively more unsaturated, it more readily binds the H^+ formed by the dissociation of carbonic acid, thereby allowing more CO_2 to be transported as bicarbonate. In addition, as hemoglobin desaturates, CO_2 forms carbamino compounds with the deoxyhemoglobin, which also adds to the CO_2 content. Thus, just as O_2 dissociation from hemoglobin does not physiologically obey one curve, neither does CO_2 dissociation from whole blood. There is an additional change in CO_2 content resulting from the difference in the oxyhemoglobin concentration between the lungs and the microvasculature of the tissues. Finally, at any partial pressure, the CO_2 content of whole blood is more than twice that for O_2, primarily because of the greater solubility of CO_2 in blood.

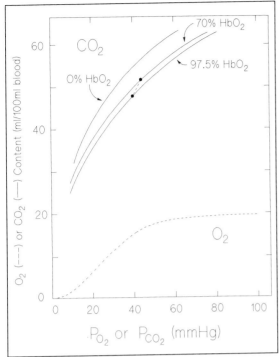

Fig. 28-8. CO_2 (*solid lines*) and O_2 (*dashed line*) plotted as a function of partial pressures of CO_2 (Pco_2) or O_2 (Po_2), respectively. CO_2 content is increased with desaturation of hemoglobin and decreased with increasing saturation of hemoglobin. In the physiologic range, the CO_2 content curve is essentially linear, as compared with the oxyhemoglobin dissociation curve. Note that the CO_2 content of whole blood is approximately twenty times higher than the O_2 content. (Modified from: Mines, A. H. *Respiratory Physiology* 2nd ed. New York: Raven Press, 1986. P. 77.)

SUMMARY

Alveolar ventilation supplies the O_2 needed for metabolism and eliminates the CO_2 produced by the tissues. Consequently, the level of alveolar ventilation is responsive to the metabolic demand. The transfer of O_2 and CO_2 across the alveolar–capillary barrier occurs by means of diffusion down their respective partial pressure gradients. O_2 is primarily transported in the blood from the lungs to the tissues, bound to hemoglobin. The $\%HbO_2$ saturation is predominantly determined by the PaO_2 but is also significantly influenced by CO_2 and 2,3-DPG levels, pH, and temperature.

CO_2 is conveyed from the tissues to the lungs in three forms: physically dissolved, attached to hemoglobin as carbamino compounds, and as bicarbonate. Of the three, bicarbonate is the most important, accounting for up to 65 percent of the total amount of CO_2 transported. The CO_2 content of the blood is a function of the Pco_2 and the $\%HbO_2$ saturation.

NATIONAL BOARD TYPE QUESTIONS

Select the one most correct answer for each of the following

1. A healthy female weighing 120 lbs (54kg) with V_T of 400 ml per breath at a rate of 12 breaths per minute has a $\dot{V}A$ (liters per minute) of:
   A. 4.80
   B. 3.00
   C. 6.24
   D. 3.36
   E. 4.33

2. A normal individual is flying in an open-cockpit airplane at 6,000 ft (1,800 meters). At this altitude, his PiO_2 is 120 mm Hg and his $\dot{V}A$ is increased. Assuming his respiratory exchange ratio stays at 0.8, what would his $PaCO_2$ be (in mm Hg) if his PaO_2 is restored to 100 mm Hg by the increased $\dot{V}A$ (ignore the correction factor)?
   A. 32
   B. 16
   C. 25
   D. 42
   E. 56

3. Most CO_2 in the blood is transported in the form of:
   A. Carbamino compounds combined with plasma proteins
   B. Carbamino compounds combined with hemoglobin

C. Physically dissolved in plasma and the cytosol of red blood cells

D. Carbonic acid in plasma

E. Bicarbonate in plasma

4. An otherwise normal person who is anemic (6 gm of hemoglobin per 100 ml blood) would be expected to have:

A. A lower O_2 content than normal

B. A lower %HbO_2 saturation than normal

C. A lower PaO_2 than normal

D. A higher O_2 capacity than normal

E. A leftward shift of the oxyhemoglobin dissociation curve

5. All of the following occur during passage of blood through tissue capillaries *except:*

A. O_2 diffuses from the red blood cell into the tissues

B. CO_2 is released from hemoglobin

C. Chloride moves from the plasma into the red blood cell

D. Bicarbonate diffuses from the red blood cell into the plasma

E. CO_2 diffuses from the tissues into capillary blood

BIBLIOGRAPHY

Cherniack, N. S., Altose, M. D., and Kelsen, S. G. The Respiratory System. In: Berne, R. M., and Levy, M. N., eds. *Physiology,* Section VI. St. Louis: C. V. Mosby, 1983.

Comroe, J. H. *Physiology of Respiration,* 2nd ed. Chicago: Year Book Medical Publishers, 1974.

Levitzky, M. G. *Pulmonary Physiology,* 3rd ed. New York: McGraw-Hill, 1991.

Mines, A. H. *Respiratory Physiology,* 2nd ed. New York: Raven Press, 1986.

Taylor, A. E., Rehder, K., Hyatt, R. E., and Parker, J. C. *Clinical Respiratory Physiology,* Philadelphia: W. B. Saunders, 1989.

West, J. B. *Respiratory Physiology,* 4th ed. Baltimore: Williams & Wilkins, 1989.

ANNOTATED ANSWERS

1. D. (0.40 liter/breath − 0.12 liter/breath) × 12 breaths/min = 3.36 liter/min.

2. B. Rearrange the simplified alveolar gas equation and solve for $PaCO_2$, such that (120 mm Hg − 100 mm Hg) × 0.8 = 16 mm Hg.

3. E. 65 percent of the CO_2 in blood is transported in the plasma as bicarbonate.

4. A. The O_2 content in the blood of anemic individuals is reduced because the concentration of hemoglobin in the blood is reduced.

5. B. At the tissue level, CO_2 diffuses into blood and forms carbamino compounds with hemoglobin; it is not released from hemoglobin.

29 Pulmonary Blood Flow

DOUGLAS K. ANDERSON

Objectives

After reading this chapter, you should be able to:

Define and compare the pressures within the pulmonary vascular system to those in the systemic circulation

Describe the factors affecting pulmonary vascular resistance

Explain the effects of gravity on the distribution of blood flow in the lung

---

The pulmonary circulation is basically a low-pressure, low-resistance, highly compliant system. Its function is to accept the **entire cardiac output** for gas exchange, and not to regulate the supply of blood to individual organs to meet metabolic demands, as the systemic circulation does. The walls of the pulmonary vessels are thinner and contain less smooth muscle than the walls of the systemic vessels. In addition, the pulmonary capillary segments are very short, forming a dense meshwork that essentially yields a continuous sheet of blood flow in the alveolar wall.

PRESSURES IN THE PULMONARY VASCULAR SYSTEM

Pressure in the pulmonary artery is about 25 mm Hg systolic and 8 mm Hg diastolic, with a mean of about 14 mm Hg [calculated as (systolic pressure − diastolic pressure)/3 + diastolic pressure]. Pressure in the left atrium is about 5 mm Hg, resulting in a pressure drop across the pulmonary circulation of approximately 9 mm Hg. In the systemic circulation, pressure in the aorta averages 93 mm Hg, with systolic and diastolic

pressures of 120 and 80 mm Hg, respectively. If the pressure in the right atrium averages 2 mm Hg, then the pressure gradient across the systemic circulation is about 91 mm Hg, or ten times greater than that in the pulmonary circulation.

PULMONARY VASCULAR RESISTANCE

Pulmonary vascular resistance is 1.8 mm Hg/liter/min and **systemic vascular resistance** is 18 mm Hg/liter/min if the cardiac output is 5 liters/min. Thus, the pulmonary circulation is a highly distensible system, with a resistance that is only about ten percent of the systemic vascular resistance. Although there is some active regulation of the pulmonary circulation, changes in the pulmonary vascular resistance are primarily passive, and consist of responses to changes in lung volume and pulmonary arterial and venous pressures.

PASSIVE CHANGES IN THE PULMONARY VASCULAR RESISTANCE

Effects of Lung Volume

The pulmonary circulation is composed principally of two types of vessels: extra-alveolar and intra-alveolar. The larger arteries and veins constitute the **extra-alveolar** vessels. They lie outside the alveoli, are tethered to the elastic tissue of the lung, and are exposed to the intrapleural pressure. The **intra-alveolar** vessels (the pulmonary capillaries) lie between the alveoli. These two groups of vessels respond differently to changes in lung volume, with both contributing to the overall pulmonary vascular resistance.

At lower lung (alveolar) volumes, the intra-alveolar vessels are near maximally open, and therefore their resistance to blood flow is minimal (Fig. 29-1, *curve B*). However, with increasing lung volume, these intra-alveolar vessels are compressed by the distended alveoli and this progressively increases their resistance to blood flow (see Fig. 29-1, *curve B*). Conversely, at low lung volumes, the caliber of the extra-alveolar vessels is small because the transmural pressure gradient across the walls of these vessels is reduced due to the lesser subatmospheric pressure in the intrapleural space. Consequently, vascular resistance in the extra-alveolar vessels is high at low lung volumes (see Fig. 29-1, *curve C*). With increasing lung volume, the intrapleural pressure becomes more subatmospheric, elevating the transmural pressure gradient across the extra-alveolar vessels. This increased transmural pressure, coupled with the added radial traction on the extra-alveolar vessels imposed by the surrounding lung tissue as it expands, causes these vessels to distend and thereby decreases their vascular resistance (see Fig. 29-1, *curve C*).

Because the greatest cross-sectional area exists in the millions of intra-alveolar vessels, increasing vascular resistance in these vessels offsets decreased resistance in the extra-alveolar vessels. Thus, total pulmonary vascular resistance is heightened at higher lung volumes when vascular

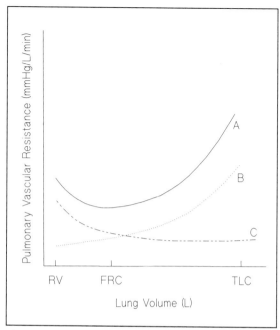

Fig. 29-1. Pulmonary vascular resistance plotted as a function of lung volume. (A) The total pulmonary vascular resistance, which is the sum of the resistance contributed by the intra-alveolar vessels (B) and extra-alveolar vessels (C). The functional residual capacity (*FRC*) is that lung volume with the least total vascular resistance. (*RV* = residual volume; *TLC* = total lung capacity.) (Modified from: Taylor, A.E., et al. *Clinical Respiratory Physiology*. Philadephia: Saunders, 1989. P. 75.)

resistance in the intra-alveolar vessels is high (see Fig. 29-1, *curve A*). Total pulmonary vascular resistance is lowest at the functional residual capacity when there is sufficient lung inflation to open the extra-alveolar vessels with minimal closing of the intra-alveolar vessels (see Fig. 29-1, *curve A*).

Recruitment and Distention of Capillaries

As previously indicated, the pulmonary circulation is remarkably compliant. Pulmonary vascular resistance declines as the pressure in the pulmonary circulation rises. This is depicted in Figure 29-2, which is a plot of the pulmonary

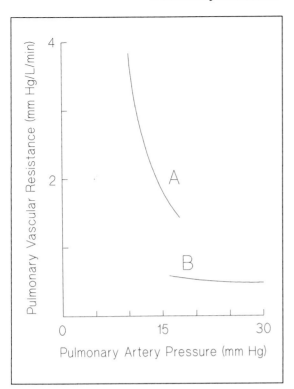

Fig. 29-2. Pulmonary vascular resistance plotted as a function of pressure in the pulmonary artery at normal (A) and elevated (B) left atrial pressures. (Modified from: Taylor, A. E., et al. *Clinical Respiratory Physiology.* Philadephia: Saunders, 1989. P. 74.)

distended due to the venous back pressure. As a consequence, little further decrease in resistance can occur (see Fig. 29–2, *curve B*).

ACTIVE REGULATION OF PULMONARY VASCULAR RESISTANCE

Although changes in pulmonary resistance are achieved mainly by passive factors, resistance can be actively modified by **neural, chemical,** and **humoral** influences. The pulmonary blood vessels are innervated by both sympathetic and parasympathetic fibers but more sparsely than are the systemic vessels. The density of innervation is greatest in the larger vessels, decreasing with vessel size. **Sympathetic** stimulation constricts the pulmonary blood vessels (or at least increases the tone of the larger blood vessels), whereas **parasympathetic** stimulation causes vasodilation.

There are a variety of **vasoactive compounds** that affect pulmonary vascular resistance. The **vasoconstricting** agents include: arachidonic acid, leukotrienes, thromboxane A_2, prostaglandin F_2, angiotensin-II, serotonin, epinephrine, and norepinephrine. The **vasodilating** compounds are: acetylcholine, bradykinin, and prostacyclin.

Alveolar Hypoxia

Local alveolar hypoxia (a decrease in the alveolar oxygen tension in a restricted region of the lung) produces vasoconstriction of the vessels to that portion of the lung. This is an important reaction that shifts blood away from poorly ventilated alveoli to better-ventilated ones, thereby matching perfusion with ventilation (see Chapter 31). It is the reduced oxygen tension in the alveoli (and not in the blood) that is thought to cause vasoconstriction of the precapillary small muscular arteries leading to the hypoxic region. This vasoconstriction may be due to the direct effects of oxygen on the vascular smooth muscle or may be mediated by some vasoactive agent or agents.

vascular resistance as a function of pressure in the pulmonary artery at normal (*curve A*) and elevated (*curve B*) left atrial pressures. At normal pressures, approximately half of the pulmonary capillaries are closed. With increasing pulmonary arterial pressures, these previously closed capillaries open, in what is known as **recruitment,** and as vascular pressure continues to rise, these patent microvessels become **distended** (see Fig. 29–2, *curve A*). The net effect is an increase in the total cross-sectional area of the pulmonary capillaries, resulting in decreased pulmonary vascular resistance. If the left atrial pressure is elevated, increasing pulmonary arterial pressure has little effect on the pulmonary vascular resistance because all capillaries are open and fully

Effects of Gravity on Pulmonary Blood Flow

As indicated in Figure 29-3, pulmonary blood flow decreases from the bottom to the top of the lung in upright individuals. This occurs because gravity creates a gradient of vascular pressures from the top to the bottom of the lung, such that the pressure is lower at the apex than at the base of the lung. Thus, the intravascular pressure in the lung is unlike the alveolar pressure which is essentially constant throughout the lung.

Fig. 29-3. The effect of gravity on the distribution of blood flow in the upright lung. Blood flow is highest at the base of the lung and decreases toward the apex. (See text for detailed description.) (Pa = arterial pressure; Pv = venous pressure; PA = intra-alveolar pressure.) (Modified from West, J. B. *Respiratory Physiology* 4th ed. Baltimore: Williams & Wilkins, 1990. P. 40.)

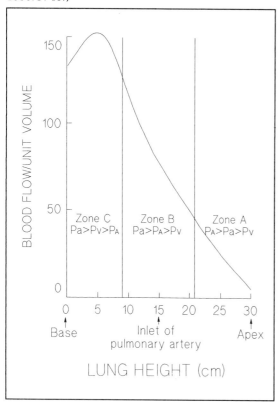

The distance from the apex to the base of the lung is approximately 30 cm. If, for example, the pulmonary artery enters the lung at its midpoint, then the top of the lung is 15 cm above the inlet of the pulmonary artery. If the pulmonary artery has a pressure of 20 cm H_2O, the arterial pressure at the top of the lung is 5 cm H_2O (the arterial pressure at the lung apex equals the pressure at the inlet of the pulmonary artery [20 cm H_2O] less the height of a column of blood extending from the inlet to the top of the lung [15 cm]).

As illustrated in Figure 29-3, there is a *potential* region (*zone A*) at the top of the lung where arterial, capillary, and venous pressures are all less than the alveolar pressure. In this situation, the blood vessels would be completely collapsed and no blood would flow through this region of the lung. If this region continued to be ventilated, it would become **alveolar dead space,** such that it would be ventilated but not perfused. Such a zone does not usually exist; vascular pressure is normally higher than alveolar pressure, so that the capillary flow commonly occurs at the top of the lung in standing individuals. However, such a zone can exist if the pulmonary arterial pressure is reduced (for example, by hemorrhage) or if the alveolar pressure is increased (for example, by positive-pressure ventilation).

In the erect position, pulmonary arterial pressure increases toward the base of the lung, approaching the pressure in the pulmonary artery near its inlet (see Fig. 29-3, *zone B*). Below the inlet, arterial pressure in the lung exceeds the pressure in the pulmonary artery (see Fig. 29-3, *zone C*). Thus, in this example, the arterial pressure would be 20 cm H_2O at the inlet of the pulmonary artery and 35 cm H_2O at the bottom of the lung, 15 cm below the inlet.

In *zone B*, the arterial pressure is greater than the alveolar pressure, but the alveolar pressure may still slightly exceed the venous pressure. In this case, flow is determined by the difference between the arterial and alveolar pressures and not the difference between the arterial and venous pressures as in most other organs. Thus, venules may be partially compressed in the upper portions of *zone B*, thereby restricting blood flow.

However, because the arterial pressure is increasing in *zone B*, the pressure difference (difference between arterial and alveolar pressure) for flow becomes greater and blood flow increases in the lower regions of this zone. In addition, with the increasing arterial pressure in *zone B*, capillary recruitment can take place.

In *zone C*, both the arterial and venous pressures exceed the alveolar pressures, and flow is determined, as in most organs, by the difference between the arterial and venous pressures. In *zone C*, it is likely that most capillaries are open and the increases in blood flow are due chiefly to distention of the microvessels.

The gradient in pulmonary blood flow from the top to the bottom of the lung is also caused by the higher lung (alveolar) volumes at the top of the lung. These higher volumes tend to compress the capillaries and increase resistance in the apical region, as compared to the base of the lung where alveolar volume is less (see Fig. 29–1, *curve B*). However, as previously noted, at very low lung volumes, resistance in the extra-alveolar vessels increases because the transmural pressure is reduced across the wall of these vessels and the radial traction on these vessels is decreased (see Fig. 29–1, *curve C*). This increased resistance in the extra-alveolar vessels is responsible for the decline in blood flow at the base of the lung (the lowest portion of *zone C*).

When an individual is supine, the pressure differences between the apex and base of the lung are abolished because the lung is not as wide as it is tall. Consequently, pulmonary blood flow is more homogeneous throughout the supine lung than throughout the upright lung.

SUMMARY

The pulmonary circulation represents a low-pressure, low-resistance, highly distensible system. It serves to accept the entire cardiac output for gas exchange. Although there is some active regulation of the pulmonary circulation, pulmonary vascular resistance is determined primarily by passive events, including lung volume and the pulmonary arterial and venous pressures. In addition, gravitational effects on pulmonary arterial pressure cause blood flow to be nonuniform throughout the upright lung, such that blood flow is greatest at the base and progressively decreases toward the apex. These passive changes in blood flow are accomplished primarily by the recruitment and distention of capillaries.

NATIONAL BOARD TYPE QUESTIONS

Select the one most correct answer for each of the following

1. The pulmonary circulation:
   A. Is a low-pressure, high-resistance system
   B. Is a high-pressure, low-resistance system
   C. Is a high-pressure, high-resistance system
   D. Is a low-pressure, low-resistance system
   E. Has no vascular smooth muscle, as have systemic vessels
2. Pulmonary vascular resistance is:
   A. Decreased by increasing pulmonary arterial or venous pressure
   B. Highest at low lung volumes
   C. Highest at the functional residual capacity of the lung
   D. Lowest at high lung volumes
   E. Increased by stimulation of parasympathetic nerves.
3. Pulmonary blood flow is:
   A. Lowest at the base of the lung because of gravitational effects, with a resultant low vascular pressure
   B. Highest at the inlet of the pulmonary artery
   C. Dependent on pulmonary arterial pressure
   D. Highest at the apex of the lung
   E. Primarily regulated by autonomic nervous system control of the pulmonary vascular smooth muscle
4. Which of the following produces vasodilation of the pulmonary vasculature?
   A. Epinephrine
   B. Alveolar hypoxia
   C. Arachidonic acid

D. Angiotensin-II

E. Acetylcholine

5. Vascular resistance in the pulmonary circulation is highest in the:

A. Extra-alveolar vessels at low lung volumes

B. Extra-alveolar vessels at the functional residual capacity

C. Extra-alveolar vessels at high lung volumes

D. Intra-alveolar vessels at high lung volumes

E. Intra-alveolar vessels at low lung volumes

ANNOTATED ANSWERS

1. D. Pressure and vascular resistance in the pulmonary circulation is one-tenth of the pressure and resistance in the systemic circulation.

2. A. Increasing pulmonary arterial or venous pressure recruits and distends pulmonary capillaries, thus decreasing pulmonary vascular resistance.

3. C. At normal left atrial pressures, increasing the pressure in the pulmonary artery recruits closed capillaries and, with increasing pressure, distends already open capillaries, resulting in decreased pulmonary vascular resistance.

4. E. As in the vascular smooth muscle of systemic smooth muscle, acetylcholine vasodilates the pulmonary vasculature. (Recall that acetylcholine vasoconstricts the smooth muscle of the airways.)

5. D. The millions of intra-alveolar vessels located between alveoli provide the greatest cross-sectional area in the pulmonary circulation. At high lung volumes, these vessels are compressed. Thus, at high lung volumes, the highest pulmonary vascular resistance is in the intra-alveolar vessels.

BIBLIOGRAPHY

Cherniack, N. S., Altose, M. D., and Kelsen, S. G. The Respiratory System. In: Berne, R. M., and Levy, M. N. eds. *Physiology,* Section VI. St. Louis: C. V. Mosby, 1983.

Comroe, J. H. *Physiology of Respiration,* 2nd ed. Chicago: Year Book Medical Publishers, 1974.

Levitzky, M. G. *Pulmonary Physiology,* 3rd ed. New York: McGraw-Hill, 1991.

Mines, A. H. *Respiratory Physiology,* 2nd ed. New York: Raven Press, 1986.

Taylor, A. E., Rehder, K., Hyatt, R. E., and Parker, J. C. *Clinical Respiratory Physiology.* Philadelphia: W. B. Saunders, 1989.

West, J. B. *Respiratory Physiology,* 4th ed. Baltimore: Williams & Wilkins, 1989.

30 Control of Ventilation during Wakefulness

SHAHROKH JAVAHERI AND DOUGLAS K. ANDERSON

Objectives

After reading this chapter, you should be able to:

Give an overview of the control of breathing and distinguish between homeostatic and behavioral functions of the respiratory system

Distinguish various elements of the respiratory system involved in the control of breathing

Describe the role of the brain in the control of breathing

Describe the homeostatic mechanisms that maintain the oxygen and carbon dioxide partial pressures and H^+ concentration

Explain how hypoxemia and hypercapnia differ in their control of breathing

Name and describe the role of the intrathoracic receptors in the control of breathing

Understand the importance of upper airway respiratory muscles in regulating upper airway patency

Describe how the control of breathing differs during wakefulness and sleep

---

The major function of the respiratory system is to maintain normal blood partial pressures of the two vital respiratory gases, O_2 and CO_2, along with the H^+ concentration $[H^+]$. This important regulatory function is referred to as the **homeostatic** (chemostatic or metabolic) **function** of the respiratory system, and is achieved by adjusting ventilation to the metabolic need (O_2 consumption $[\dot{V}O_2]$ and CO_2 production $[\dot{V}CO_2]$) of the organism.

The breathing apparatus, however, is also used for nonhomeostatic (behavioral) functions, such as **phonation.** With the ever-occurring variations in metabolic needs (chemostatic function) and behavioral functions, particularly during wakefulness, the act of breathing becomes complex and must be governed precisely by a hierarchy of control systems.

The metabolic and behavioral functions of the respiratory system are regulated by the central nervous system (CNS), where **respiratory rhythmogenesis** arises. The **central pattern generators** produce the **automatic respiratory rhythm** that underlies the periodic cycles of inspiration and expiration. About seventy years ago, it was demonstrated that respiratory rhythm can persist after removal of the entire brain above the brainstem, but automatic breathing ceases after transection of the brainstem at the medullary-spinal level (first cervical level). However, even today, precise histologic localization of the respiratory centers in the pontomedullary area remains ill-defined.

COMPONENTS OF THE RESPIRATORY SYSTEM THAT CONTROL BREATHING

Control of respiration involves three components: sensors, controllers, and effectors (Fig. 30-1). In this complicated system, the **controllers** are "the respiratory centers" and they harbor multiple important functions. These are:

1. Respiratory rhythmogenesis—central pattern generation
2. Neural translation of the central rhythm to the motor output that drives the respiratory muscles
3. Adjustment of the rhythm and motor output to meet the metabolic needs (homeostatic functions)

Fig. 30-1. The components of the respiratory system that control breathing. In a broad sense, these include sensors, controllers, and effectors. The respiratory centers receive afferent information from multiple sites (sensors). Axons from the medullary respiratory neurons decussate below the obex of the medulla oblongata and descend in the reticulospinal (*RS*) tracts in the ventrolateral spinal cord to synapse with spinal motor neurons of the phrenic and intercostal nerves that innervate the thoracic inspiratory and expiratory pump muscles. The voluntary breathing system, which originates in the cerebral cortex, descends via the corticospinal (*CS*) tract to the spinal motor neurons innervating the respiratory pump muscles and via the corticobulbar tract innervating the upper airway respiratory muscles, by way of the cranial nerves (*CN*). In this diagram, ventilation is determined by the balance between upper airway resistance (determined by state of activation of muscles of upper airways) and the pressure generated by the pump muscles. Arterial blood partial pressures of O_2 and CO_2 (PaO_2 and $PaCO_2$) and the H^+ concentration ($[H^+]$) are the controlled variables and influence the activity of the chemoreceptors.

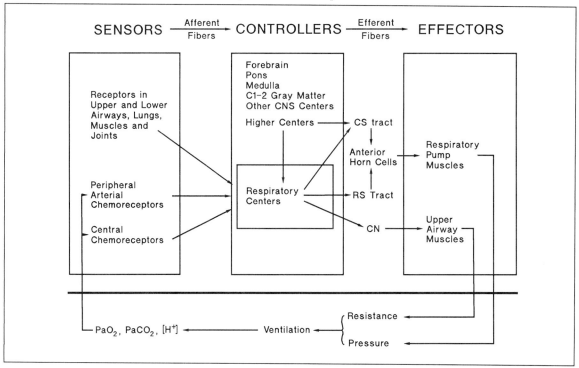

4. Adjustment of the rhythm and motor output to meet behavioral and voluntary functions (nonhomeostatic functions)
5. Most efficient use of the respiratory system to achieve various functions with minimal energy expenditure

The respiratory centers are the sites of respiratory rhythmogenesis, the basic function underlying the act of automatic breathing. The centers receive multiple inputs from a variety of **sensors** located at various sites (Table 30-1). These sensors perceive changes in a wide variety of variables, such as the partial pressure of CO_2 (Pco_2) and O_2 (Po_2), $[H^+]$, and lung inflation, and transmit this information (in the form of increased or decreased activity) to the controllers. After processing this information, the controllers alter their level of activity, which is conveyed to multiple **effectors** (see Fig. 30-1). The most basic effectors are the respiratory muscles, including the thoracic inspiratory pump muscles and the muscles of the upper airways. As noted in Chapter 26, the main inspiratory muscle is the **diaphragm.** The major respiratory muscles of the upper airways are the so-called **upper airway dilators,** which regulate the cross-sectional area and resistance of the upper airway. These mus-

Table 30-1. Sensors (Receptors) with Afferent Input to the Respiratory Centers

From within the brain
  Central chemoreceptors
  Hypothalamic (temperature) receptors
  Forebrain centers (voluntary functions)
From outside the brain
  Peripheral arterial chemoreceptors (primarily carotid
    bodies)
  Upper airway receptors
    Nasal
    Pharyngeal
    Laryngeal
  Pulmonary receptors
    Stretch receptors
    Irritant receptors
    C fibers
  Respiratory muscle receptors
  Costovertebral joint receptors

cles contract in phase with inspiration, increasing the tone in the compliant upper airways and thereby preventing the walls of the upper airway from being drawn inward by the subatmospheric intraluminal pressure generated by the diaphragm.

Although the respiratory centers receive information from various receptors over a variety of neural pathways (see Fig. 30-1), the output information is conveyed to the diaphragm via the phrenic nerves, to the intercostal muscles via the intercostal nerves, and to the muscles of the upper airways via the cranial nerves (see Fig. 30-1). Thus, air flows through the upper airway into the lungs where gas exchange occurs.

This chapter will review the metabolic and behavioral functions of the respiratory system and consider some of the more important reflex arcs involved in the control of breathing. Various sensors, respiratory centers, and muscles of the upper airway will be emphasized. The respiratory thoracic pump muscles were covered in Chapter 26. Control of breathing during sleep is beyond the scope of this book and is only briefly considered.

RESPIRATORY CENTERS AND NEUROGENESIS OF BREATHING

Over seventy years ago, results of brain transection experiments indicated that the **central automatic respiratory centers** were located in the brainstem and consisted of a **pneumotaxic center** in the rostral pons, an **apneustic center** in the caudal pons, and the **medullary centers** (Fig. 30-2). Transections made above the **upper pons** (*level 1*) caused no change in the respiration of an anesthetized animal, and this localized the site of respiratory centers to the pons and medulla. Sectioning the **vagi** with or without this transection caused slowing and deepening of respiration. This pattern of breathing is due to removal of inhibitory afferent input from pulmonary stretch receptors (to be discussed). A transection made at the **midpons** (*level 2*) caused some slowing and deepening of respiration with the vagi intact. When the vagi were

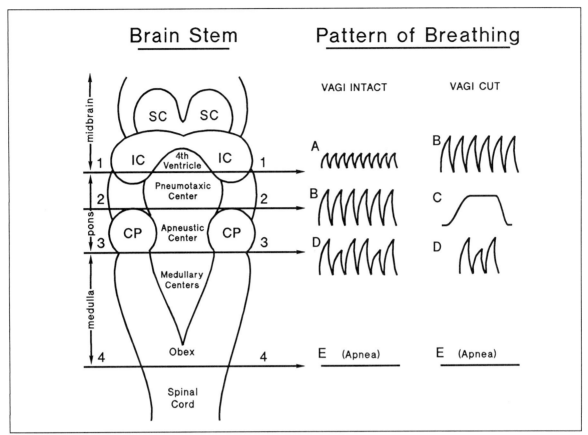

Brain Stem

Pattern of Breathing

Fig. 30-2. Lumsden's experiments with brainstem transections at levels 1 to 4 (see text). The diagram to the left is a dorsal view of the pontomedullary area. The breathing patterns (labeled *A* to *E*) represent those that appear following complete transections at each level. These patterns are: normal breathing (*A*), slow deep breathing (*B*), apneustic breathing (*C*), gasping (*D*), and apnea (*E*). (*SC* and *IC* = the superior and inferior colliculi respectively; *CP* = the cerebellar peduncles.) (Modified from: Comroe, J. H., ed. *Physiology of Respiration*. Chicago: Year Book Medical Publishers, 1975.)

sectioned, this pontine transection produced a series of prolonged inspirations, punctuated by very brief expirations, termed *apneustic breathing*. This respiratory pattern supposedly arises because a center (called the *apneustic center*) in the lower pons assumes control of the breathing pattern. Because apneustic breathing was unmasked by vagal and upper pontine sections, it was concluded that normal activity in the apneustic center was periodically inhibited by impulses traveling in the vagus and by impulses from the pneumotaxic center in the upper pons.

Removal of the **pontine apneustic center** by transection between the pons and medulla (*level 3*) produced a gasping, spasmodic respiratory pattern, characterized by brief and usually maximal inspiratory efforts that terminated abruptly (see Fig. 30-2). Sectioning between the **medulla** and **spinal cord** (*level 4*) causes all spontaneous respirations to cease (apnea) in the resting expiratory position (at the functional residual capacity). This indicated that the **medullary centers** were critically important for spontaneous respiration.

STRUCTURE AND FUNCTION CHARACTERISTICS OF RESPIRATORY CENTERS

Although these early studies localized the respiratory centers to the pontomedullary areas, detailed anatomic localization had to await the advent of electrophysiologic techniques. Microelectrodes placed in different parts of the brainstem have been used to explore the regions involved in **rhythmogenesis.** These studies revealed the presence of **respiration-related neurons** that link oscillation to respiratory output, such as phrenic nerve motor activity. Although the basic neural mechanism of rhythmogenesis is still unclear, these studies have identified multiple neural aggregates located bilaterally in the various parts of the brainstem (Fig. 30-3). Thus, the **pneumotaxic center** corresponds to two ros-

tral pontine nuclei, the **parabrachialis medialis** and **Kolliker-Fuse nuclear complex.** The **apneustic center** is considered to be a diffuse entity in the pontine reticular formation, and the **medullary centers** consist of at least two separate groups of nuclei, the **dorsal** and **ventral respiratory groups** (see Fig. 30-3).

Dorsal Respiratory Group

The dorsal respiratory group is located in the dorsomedial part of the medulla, corresponding to the ventrolateral nucleus of the tractus solitarii (see Fig. 30-3). This respiratory group contains mainly **inspiratory-related neurons.**

The axonal projections of the neurons of the **nucleus tractus solitarii** terminate in the cervical and thoracic anterior spinal motor neurons of the phrenic and intercostal nerves, and exhibit discharge patterns similar to those of the phrenic and intercostal nerves.

Fig. 30-3. Medulla and part of the spinal cord, showing the two main aggregates of respiration-related neurons, the dorsal (*DRG*) and ventral (*VRG*) respiratory groups. These nuclei are spread symmetrically and bilaterally. The VRG with its various subdivisions, including the most caudal part (C-1–C-2 of the cervical segments) are depicted. (*nTS* = nucleus tractus solitarii; *nA* = nucleus ambiguus; *nPA* = nucleus para-ambigualis; *nRA* = nucleus retroambigualis.) (Modified from: Euler, C. V. Brain stem mechanisms for generation and control of breathing pattern. In: Cherniak, N. S., Widdicombe, J. G., eds. *Handbook of Physiology, The Respiratory System,* Volume II. New York: Oxford University Press, 1986.)

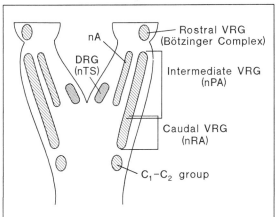

Ventral Respiratory Group

The ventral respiratory group is more diffuse than the dorsal one, and consists of multiple neural aggregates extending longitudinally from the rostral to caudal medulla and possibly up to the second cervical segment of the spinal cord. In a broad sense, the ventral respiratory group consists of the **Bötzinger complex,** in the vicinity of the nucleus retrofacialis, the **nucleus ambiguus** and **para-ambigualis,** and the **nucleus retroambigualis** (see Fig. 30-3). The ventral respiratory group contains neurons that are active during both inspiration and expiration. The expiration-related neurons project to expiratory intercostal motor neurons.

There may also be inspiratory mechanisms at work in the gray matter of the first and second cervical segments of the spinal cord. If this is confirmed, these C-1–C-2 inspiratory neurons (which project to the phrenic motor neurons) would represent the most caudal extension of the ventral respiratory group (see Fig. 30-3).

Pontine Respiratory Group

The respiration-related neurons in the pontine respiratory group include the **parabrachialis medialis** and **Kolliker-Fuse nuclei** in the dorsolateral rostral pons, which constitute the so-called **pneumotaxic center.** These pontine nuclei transmit information to the **inspiratory-off switch** (IO-S), which has not been fully defined histologically, but determines the inspiratory time. This mechanism also receives information from the pulmonary stretch receptors via the vagus nerves. Both vagal stimulation (lung inflation) and input from the pontine respiratory group activate the IO-S neurons, causing termination of inspiration. In contrast, the combination of vagotomy and lesioning of the pontine respiratory group (transection at level 2; see Fig. 30-2) abolishes the IO-S function, resulting in uninterrupted inspiratory activity that lasts for long periods—apneustic breathing. The IO-S mechanism is also stimulated by increased body (or hypothalamic) temperature, and this initiates tachypnea. This mechanism (panting) is used to dissipate heat in some animals when they become hyperthermic.

SIMPLIFIED MODEL OF RESPIRATORY RHYTHMOGENESIS

Although the respiratory centers of rhythmogenesis have been localized to the brainstem region, it has not been possible to precisely locate the basic rhythm generator, the **central pattern generator,** and the mechanisms underlying respiratory rhythmogenesis remain ill-defined. A simplified working model is presented (Fig. 30-4).

Two important features of this model are the **magnitude** of the central inspiratory activity, which determines the intensity of the desire to inspire and is reflected in the inspiratory flow rate, and the **duration** of the central inspiratory activity, which determines the inspiratory time. Because flow equals the pressure divided by resistance, in our model, **inspiratory air flow** is the

function of contractions of the diaphragm (generating pressure) and of the muscles of the upper airway (determining upper airway resistance). The **inspiratory time** is the function of the IO-S mechanism, which, when turned on, stops central inspiratory activity, thereby terminating inspiration. **Expiration** begins passively through the elastic recoil of the respiratory system. As long as the IO-S is active (the expiratory time), inspiration is not resumed. At very high levels of ventilation, expiration can also become active, in that the expiratory muscles contract rhythmically to augment expiration. This indicates the presence of central expiratory activity (not shown in the model).

In this model, the **central inspiratory activity integrator** receives chemoreceptor input as well as other inputs. The output from this integrator feeds both the IO-S mechanism and all the motor neurons of the phrenic, intercostal, and cranial nerves (these innervate the upper airway muscles). Activity in the IO-S mechanism determines inspiratory time. It is the interaction between inspiratory flow (designated here as mean inspiratory flow [MIF]) and inspiratory time (TI) that determines the tidal volume (VT), as $V_T = MIF \times T_I$. Ventilation, therefore, is directly related to mean inspiratory flow, which reflects central inspiratory activity. Stimuli such as increased P_{CO_2} augment the central inspiratory activity by means of raised chemoreceptor activity (see Fig. 30-4), thereby increasing mean inspiratory flow and ventilation.

RECEPTORS IN THE UPPER AIRWAYS AND CHEST WALL

Aside from the receptors in the tracheobronchial tree and the chemoreceptors, which will be discussed, many other receptors transmit afferent input to the pontomedullary centers. Examples of these receptors include those in the **upper respiratory tract** (including the nose, pharynx, and larynx), the **respiratory muscles** (muscle spindles and tendon organs), and the **costovertebral joints.** The respiratory effects of these reflex arcs are complex and will not be discussed

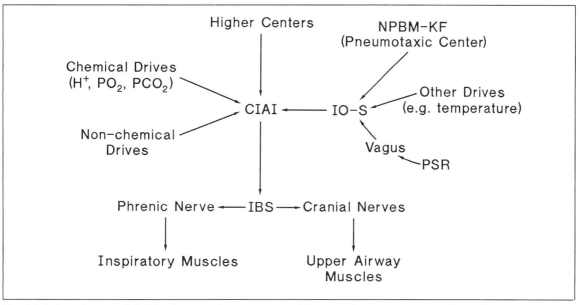

Fig. 30-4. A working model of the functional organization of the central pattern generator. Central inspiratory activity integrator (*CIAI*) neurons are active during inspiration, sending afferent information to inspiratory bulbospinal (*IBS*) neurons. Efferent neural signals reach the inspiratory muscles via different pathways, as shown. Afferent input from the pulmonary stretch receptors (*PSR*) feeds into the inspiratory-off switch (*I-OS*) center, which also receives afferent information from the parabrachialis medialis and Kolliker-Fuse nuclear complex (*NPBM-KF*). The I-OS, in turn, inhibits the CIAI neurons, terminating inspiration. Various drive inputs may stimulate or inhibit the CIAI neurons or the I-OS center. CIAI neurons send appropriate information to IBS neurons, which may stimulate or inhibit the motor neurons of the cranial and spinal nerves. (P_{O_2} and P_{CO_2} = partial pressures of O_2 and CO_2, respectively.)

in detail. Some examples are the mammalian diving reflex, induced when water is applied to the nose or face, and the cough reflex or apnea, provoked by stimulation of laryngeal receptors. **Apnea** is of major interest because of the sudden infant death syndrome and sleep apnea syndrome.

RECEPTORS IN THE TRACHEOBRONCHIAL TREE AND THE LUNG PARENCHYMA

There are at least three different groups of afferent end organs in the wall of the lower airways and lung parenchyma that may influence the breathing pattern. These are the **pulmonary stretch receptors,** the **rapidly adapting irritant receptors,** and **C fibers.**

Pulmonary Stretch Receptors

The pulmonary stretch receptors are located in the smooth muscle of the trachea and larger bronchi. The afferent nerves originating from these receptors pass primarily in the vagus nerves into the brain (see description of the IO-S mechanism).

Airway distention is the major stimulus of the pulmonary stretch receptors. Many of these receptors discharge spontaneously when functional residual capacity is reached, their activity markedly increasing with airway distention. This increase in activity ebbs slowly with time, hence they are called **slowly adapting receptors.**

In the **Hering-Breuer reflex,** lung inflation inhibits inspiration and prolongs expiration. The pulmonary stretch receptors are stimulated by lung inflation and this increases their afferent activity in the vagus nerves. This activates the IO-S mechanism, which terminates inspiration, thereby limiting further lung inflation. For this reason, sectioning the vagus nerves in anesthetized, spontaneously breathing animals produces a prolonged inspiratory time. This reflex is active in animals and infants, but is not of major importance in adult humans except during deep breathing.

Rapidly Adapting Irritant Receptors

Nerve endings in the epithelium of large airways probably serve as the rapidly adapting irritant receptors. These receptors are stimulated by **intraluminal irritants** such as cigarette smoke and by **inflammatory mediators** such as histamine and prostaglandins. However, they quickly adapt to stimulation and their activity returns to basal levels despite continued stimulation, hence their name. Information from the rapidly adapting irritant receptors travels through the vagal fibers to the brain, where the efferent limb of this reflex (the motor fibers in the vagus) activates contraction of the airway smooth muscle (bronchospasm). This reflex is therefore important in asthma. Rapidly adapting irritant receptors also mediate the cough reflex. **Coughing** is an attempt to expel the intraluminal irritants that stimulate the irritant receptors.

C Fibers and Juxtacapillary Receptors

The C fibers and juxtacapillary (J) receptors are present both in bronchi (bronchial C fibers) and in lung parenchyma (pulmonary C fibers or J receptors). The J receptors are located in the interstitium of the alveolar wall juxtaposed to capillaries. J receptors are presumably stimulated by "distention" of the interstitial space, as may occur in pulmonary interstitial edema and

fibrosis. The **reflex arc** is otherwise similar to that described for the pulmonary stretch receptors. Stimulation of the J receptors is believed to account for the **rapid breathing** (tachypnea) seen in a variety of cardiopulmonary disorders such as pneumonia, pulmonary edema, embolism, and fibrosis.

PERIPHERAL ARTERIAL CHEMORECEPTORS

The peripheral arterial chemoreceptors include the carotid and aortic bodies. The **carotid bodies** are located in the tissue between the internal and external carotid arteries (Fig. 30-5) and the **aortic bodies** are located at the arch of the aorta. The carotid bodies should not be mistaken for the carotid sinuses, which are baroreceptors that detect changes in blood pressure and are located in the wall of the internal carotid arteries. The carotid bodies are highly vascularized with a rich blood flow. Their main nerve supply is provided by the **carotid sinus nerve,** which is a branch of the glossopharyngeal nerve.

In most species, including humans, the carotid bodies are the predominant arterial chemoreceptors mediating the hyperventilatory response to decreased arterial O_2 tension (PaO_2), known as **hypoxemia.** In anesthetized animals, hypoxemia causes ventilatory depression if the carotid bodies are absent. This hypoxic ventilatory depression is probably due to the depressant action of hypoxemia on the brain, mediated by the release of multiple neurotransmitters such as **adenosine. Theophylline,** an adenosine antagonist, partially reverses the hypoxic ventilatory depression.

Ultrastructurally, there are two main cell types in the carotid bodies, the type I and II cells. The **type I** cells are presumably the chief cells involved in **chemotransduction.** The carotid bodies respond to changes in blood PO_2, $[H^+]$, and K^+ concentration ($[K^+]$). Because the systemic circulation furnishes their blood supply and because of their proximity to the lung and heart, the carotid bodies detect changes in PaO_2, PCO_2,

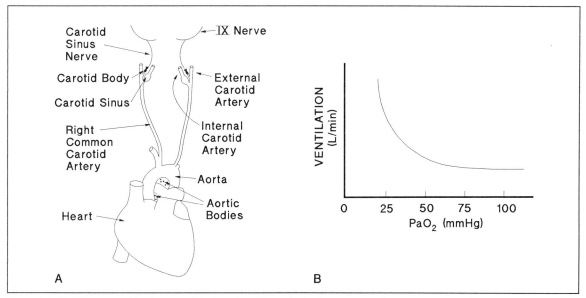

Fig. 30-5. (A) The carotid bodies and (B) the related hyperbolic hypoxic ventilatory response. The arterial O_2 tension (PaO_2) must decrease considerably before notable changes in ventilation occur; however, any further decrease in PaO_2 results in progressively increasing ventilation.

and [H+] relatively quickly, resulting in the immediate ventilatory response.

This negative-feedback loop is illustrated in Figure 30-5. Hypoxemia, increased plasma [H+] (acidemia), and heightened [K+] (hyperkalemia) stimulate the carotid bodies, thus accelerating impulses to the brainstem respiratory centers from these chemoreceptors. Added activity from the respiratory centers to the thoracic and upper airway muscles causes increased ventilation that delivers more O_2 to and removes excess CO_2 from the lungs. The arterial CO_2 tension ($PaCO_2$) therefore decreases (which in turn reduces the carbonic acid level and [H+]), and PaO_2 increases, partially relieving the initial stimulus—the hypoxemia or acidemia. **Hyperkalemia** also excites these receptors, which may play a role in mediating the hyperpneic response of exercise. **Hypercapnia** may also stimulate the peripheral chemoreceptors independent of associated changes in [H+].

The importance of the **carotid bodies** in Po_2 homeostasis is demonstrated in people staying at high altitudes, where, despite considerable drops in the atmospheric Po_2, compensatory hyperventilation maintains adequate Po_2 for survival. The hypoxemia associated with pulmonary disease also stimulates the peripheral chemoreceptors in an attempt to increase alveolar ventilation and PaO_2. However, PaO_2 must fall considerably (below approximately 55 mm Hg) before the receptors are sufficiently stimulated (see Fig. 30-5). The hyperbolic nature of the changes in the rate of discharge of afferent fibers from the carotid body illustrate this, as do the hyperbolic changes in ventilation when Po_2 is reduced (see Fig. 30-5). Another important point is that PaO_2 represents the stimulus to the carotid bodies and not the O_2 content of the arterial blood. This is why anemia and carbon monoxide inhalation are not associated with compensatory hyperventilation.

CENTRAL CHEMORECEPTORS

The central chemoreceptors are situated somewhat superficially on the ventrolateral aspect of the medulla. At least **three areas,** designated *M,* *S,* and *L* for the scientists that identified these areas, have been defined (Fig. 30-6). It is believed that **areas M** and **L** are chemosensitive and that their output originates from **area S,** which is not chemosensitive. Recent evidence suggests that there may also be some chemoreceptors deeper in the medulla.

The main stimulus to the central chemoreceptors is the [**H**+] of the brain extracellular fluid that bathes these receptors. These receptors are separated from blood by the blood–brain barrier, which is formed by the tight junctions between the endothelial cells of the cerebral capillaries. Because the blood–brain barrier resists ionic diffusion, it takes several minutes for changes in plasma [H+] to be reflected in the brain extracellular fluid. In metabolic acidosis or alkalosis, the magnitude of steady-state changes in brain extracellular fluid [H+] is much smaller than the changes in plasma [H+]. This difference exists because the blood–brain barrier is also equipped with **transport carriers** that can move various ions into and out of the extracellular fluid, thereby regulating its ionic composition. **Central chemoreceptors** are stimulated by increased [H+] in extracellular fluid during metabolic aci-

dosis; conversely, their activity diminishes when extracellular fluid [H+] decreases during metabolic alkalosis. This information is signaled to the respiratory centers (see Fig. 30-1), and initiates compensatory changes in ventilation and $PaCO_2$, which in turn mitigate the changes in [H+] induced initially by the metabolic acid–base perturbation (negative-feedback loop of the central chemoreceptors).

Brain extracellular fluid [H+] rises rapidly with **acute increases in PaCO**$_2$ because CO_2 is a lipid-soluble molecule that diffuses quickly across the blood–brain barrier. In the extracellular fluid, CO_2 is hydrated to form carbonic acid, which dissociates to H+ and HCO_3^- (the bicarbonate radical). It is this H+ that stimulates the central chemoreceptors. However, recent findings have suggested that PCO_2 may exert an independent stimulatory effect on the central chemoreceptors separate from the H+.

The sensitivity or gain of the central chemoreceptors to changes in [H+] appears to exceed that of the peripheral chemoreceptors, such that most of the steady-state ventilatory response to changes in [H+] (metabolic or respiratory) is mediated by the central chemoreceptors. Studies in some animal species have revealed that the hyperventilation caused by severe metabolic acidosis is almost equivalent before and after denervation of the peripheral chemoreceptors.

Fig. 30-6. The operation of the negative-feedback system that mediates the compensatory hyperventilation initiated by hypoxemia. (*PCR* = peripheral arterial chemoreceptors; $\dot{V}A$ = alveolar ventilation; $PaCO_2$ and PaO_2 = arterial CO_2 and O_2 tension, respectively. *Arrows* indicate increases or decreases in the level of activity.)

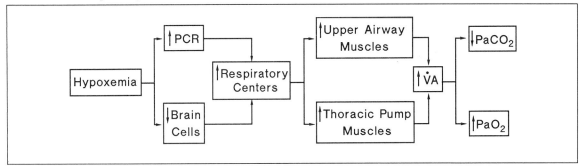

SUMMARY OF THE REFLEX ARCS OF THE PERIPHERAL AND CENTRAL CHEMORECEPTORS

Both the peripheral and central chemoreceptors are part of a **negative-feedback system** that functions primarily to correct disturbances in Po_2, $[H^+]$, and Pco_2. These **reflex arcs** are important elements of the metabolic (homeostatic) function of the respiratory system. The peripheral and central chemoreceptors are stimulated or depressed by altered $[H^+]$ in their chemical environment, initiated by metabolic and respiratory acid–base disturbances. Increases in Pco_2 may be independent of $[H^+]$ in activating peripheral and central chemoreceptors. Decreased PaO_2 stimulates the peripheral chemoreceptors and has a central depressant effect. Because of their strategic location, peripheral chemoreceptors respond quickly to changes in the **arterial blood chemistry** (Po_2, $[H^+]$, and $PaCO_2$). The central chemoreceptors are believed to be more sensitive (higher gain) than the peripheral chemoreceptors to changes in $[H^+]$ and $PaCO_2$, and to mediate most of the ventilatory response to changes in $[H^+]$ stemming from either respiratory or metabolic sources.

CONTROL OF VENTILATION DURING EXERCISE

A fascinating phenomenon is the mechanism, or mechanisms, that accentuate ventilation during exercise. PaO_2 and $PaCO_2$ remain relatively constant despite increasing levels of physical activity that heighten $\dot{V}o_2$ and $\dot{V}co_2$. These increases are reflected by changes in the mixed venous (pulmonary arterial) blood Po_2 (which falls) and Pco_2 (which increases) during exercise. Alveolar ventilation also increases in proportion to $\dot{V}co_2$, such that their ratio, which equals $PaCO_2$ (see Chapter 28), remains constant. **Multiple hypotheses** have been suggested to account for the hyperpnea of exercise and the tight match of $\dot{V}co_2$ to alveolar ventilation. These include **neurogenic drives** originating from central (cortical or hypothalamic) and peripheral (proprioceptors and muscle

spindles stimulated by motion and tension) sources, **humoral factors,** involvement of the **central and peripheral chemoreceptors,** and **cardiodynamic factors** (increased blood flow to the lung containing excess CO_2). However, none has been unanimously accepted.

During heavy exercise, alveolar ventilation exceeds the immediate need of $\dot{V}co_2$ and this reduces $PaCO_2$. The point where the rise in alveolar ventilation departs from $\dot{V}co_2$ is called **anaerobic threshold** because of the coincidental development of plasma metabolic lactacidosis. It is believed that the rise in plasma $[H^+]$ stimulates the carotid bodies, and this increases alveolar ventilation beyond that required by the metabolic activity. Other studies have suggested that the rise in ventilation may, in part, be related to hyperkalemia.

VOLUNTARY AND BEHAVIORAL CONTROL OF BREATHING

During deep anesthesia and slow-wave sleep, breathing primarily accommodates metabolic needs (homeostasis). During wakefulness, however, the breathing apparatus subserves both homeostatic and nonhomeostatic (voluntary) functions (see Fig. 30-1). During **wakefulness,** in keeping with priorities imposed by the prevailing circumstances, the pattern of breathing is adjusted to meet the functional requirements. The **voluntarily controlled functions** and behaviors span one's lifetime and include sucking, swallowing, defecation, crying, singing, trumpeting, laughing, phonation and speech, cough, control of posture, breath holding, and hyperventilation. We control inspiratory and expiratory times and air flow rates to accommodate these voluntary functions.

During these voluntary and behavioral acts, PaO_2 and $PaCO_2$ and $[H^+]$ are disturbed; these represent the controlled variables of the metabolic respiratory control system. Although these disturbances are tolerated to some extent, the metabolic control system soon intercedes. Breath-holding is a specific example: if the subject first hyperventilates (lowers P_{CO_2}) or breathes pure O_2, it can be prolonged.

As depicted in Figure 30-1, the **descending neural signals** that drive the cranial nerves (innervating muscles of upper airways) and the **spinal respiratory motor neurons** (innervating the thoracic pump muscles) consist of those arising from voluntary (for example, cortical) and involuntary (pontomedullary) sources. The corticospinal (pyramidal) tract may transmit the voluntary information directly from the cortex to the respiratory motor neurons. Alternatively, information from that cortex may pass through pontomedullary respiratory centers before descending to the motor neurons via the reticulospinal tract (extrapyramidal tract).

PaCO₂ HOMEOSTASIS

The normal resting $PaCO_2$ is determined where the metabolic hyperbola (Chapter 28; Fig. 28-1) intercepts the curve of the CO_2 response (Fig. 30-7). Because of the characteristic linear relationship of P_{CO_2} to ventilation (linear gain), small changes in P_{CO_2} produce immediate compensatory changes in ventilation (see Fig. 30-7). This is in contrast to the PaO_2, which can decrease considerably before evoking significant compensatory ventilatory changes (see Fig. 30-5). In addition, because the metabolic rate is tightly linked to alveolar ventilation, as $\dot{V}_{CO_2}$ increases (exercise, ingestion of food high in carbohydrates), alveolar ventilation increases proportionately and $PaCO_2$ remains constant. For these reasons, $PaCO_2$ homeostasis appears to be under tight control.

RESPIRATORY MUSCLES OF UPPER AIRWAYS

The nose, mouth, pharynx, and larynx that constitute the upper airways serve as the conduit for air flow into the trachea. The upper airways contribute to the **anatomic dead space** because they do not take part in gas exchange. However, **flow-resistive changes** in the upper airways can greatly influence ventilation, and hence gas ex-

change. An example of this is upper airway occlusion, which may occur during sleep and is the underlying cause of **obstructive sleep apnea.** This occlusion occurs because the walls of upper airways are not rigid but contain soft tissues such as muscles and vessels. If the muscle tone is poor and the vascular bed is congested, this can profoundly reduce the cross-sectional area and increase resistance. During inspiration when the inspiratory muscles generate negative suction, a compliant upper airway is prone to collapse.

Of particular importance are those groups of **upper airway muscles** that phasically contract to stiffen the upper airway during each inspiration, preventing its collapse. Activation of these dilator muscles has been shown to precede diaphragmatic activation, which allows the upper airways to withstand the negative intraluminal pressure. These muscles are also stimulated by **hypoxemia** and **hypercapnia,** in a manner similar to that described for the respiratory pump muscles (see Fig. 30-6).

An important dilator muscle is the **genioglossus** which pulls the tongue ventrally when it contracts. Relaxation of this muscle during sleep, along with gravitational force exerted on the tongue in the supine position, displaces the tongue backward, causing pharyngeal obstruction and sleep apnea.

SUMMARY

The respiratory centers are located in the pontomedullary area of the brain and produce the respiratory rhythmogenesis underlying the act of breathing. The neural signals travel to muscles of the upper airways (via cranial nerves) and to the thoracic pump muscles (via phrenic and intercostal nerves).

The generation of rhythmic neural signals by the respiratory centers depends on afferent inputs from multiple receptors (sensors) located both inside and outside the brain. The carotid bodies constitute the primary peripheral arterial chemoreceptors that mediate the hyperventilatory response to hypoxemia. The increased ven-

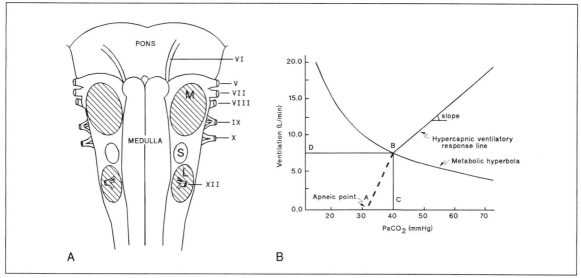

Fig. 30-7. (A) The central chemoreceptors (L = Loeschcke; S = Schlafcke; M = Mitchell) on the ventral aspects of the medulla. (B) The related hypercapnic ventilatory response. Areas M and L are presumably CO_2 sensitive. Note the linear ventilatory response to changes in the arterial CO_2 tension ($PaCO_2$). The slope of the line (reflecting gain in the system) defines the ventilatory response to CO_2. The average slope ranges between 1 and 3 liters/min/1-mm Hg rise in the partial pressure of CO_2. Also shown is the metabolic hyperbola. C is the resting $PaCO_2$ (40 mm Hg) and D is the resting ventilation. A is the so-called apneic point, or that $PaCO_2$ below which ventilation ceases.

tilation caused by hypercapnia is mediated mostly by the central chemoreceptors in the medulla oblongata. These receptors are stimulated by increased $[H^+]$ in the extracellular fluid bathing them. This accounts for most of the compensatory increase in ventilation (and fall in arterial PCO_2) during metabolic acidosis, and the compensatory fall in ventilation (and rise in PCO_2) during metabolic alkalosis. Normally ventilation is maintained proportional to the metabolic rate, such that $PaCO_2$ and PaO_2 remain constant. This is how O_2 and CO_2 levels remain constant during exercise despite several-fold increases in the metabolic rate ($\dot{V}CO_2$). The primary function of the respiratory system is to keep PO_2, PCO_2, and $[H^+]$ as close to normal as possible. This is referred to as *metabolic* or *homeostatic function*. Although breathing is chiefly automatic (homeostatic), nonhomeostatic factors (behavioral control) also influence ventilation. The respiratory apparatus is used to achieve important behavioral acts such as speech, swallowing, and control of posture. Normally, there is a complex interaction between the behavioral and homeostatic control of breathing.

NATIONAL BOARD TYPE QUESTIONS

Select the one most correct answer for each of the following

1. Which of the following receptors mediate the hyperventilatory response to hypoxemia?
   A. Central chemoreceptors
   B. Pulmonary stretch receptors
   C. Carotid bodies
   D. Irritant receptors

Diabetic ketoacidosis develops in an otherwise healthy man. He is seen in an emergency room where an arterial blood sample is obtained to assess the severity of the metabolic acidosis.

Choose the correct answer for questions 2 through 5.

2. As a result of metabolic acidosis, $PaCO_2$ will:
   A. Not change
   B. Decrease
   C. Increase

3. This change in $PaCO_2$ means that alveolar ventilation has _____ in relation to $\dot{V}CO_2$.
   A. Increased
   B. Not changed
   C. Decreased

4. Which of the following receptors are involved in mediating the above ventilatory response?
   A. The arterial and central chemoreceptors
   B. Pulmonary stretch receptors
   C. Hypothalamic receptors
   D. Irritant receptors

5. Compared with PaO_2, $PaCO_2$ is normally tightly controlled because:
   A. Hypercapnic ventilatory response is linear, with an average slope of 1 to 3 liters/min per each 1-mm Hg rise in PCO_2
   B. The hypoxic ventilatory response is linear
   C. The hypercapnic ventilatory response is hyperbolic
   D. All of the above

ANNOTATED ANSWERS

1. C. The central chemoreceptors and other brain cells may actually be depressed by hypoxemia. Irritant receptors are stimulated by chemical substances in the intraluminal airway and not by hypoxemia. Pulmonary stretch receptors are mechanoreceptors.

2. B. Diabetic ketoacidosis produces a compensatory fall in $PaCO_2$ because of the rise in alveolar ventilation.

3. A. The rise in ventilation exceeds the increased $\dot{V}CO_2$ and this is why $PaCO_2$ decreases.

4. A. Ventilation rises because the increase in $[H^+]$ stimulates the arterial and central chemoreceptors.

5. A. The hypercapnic ventilatory response is linear, with an average slope of 1 to 3 liters/min per each 1-mm Hg rise in PCO_2. Therefore, as soon as PCO_2 increases, ventilation also increases, and this mitigates the rise in PCO_2. In contrast, PO_2 must decline considerably before ventilation is notably increased, because the hypoxic ventilatory response is hyperbolic.

BIBLIOGRAPHY

Berger, A. J., Mitchell, R. A., and Severinghaus, J. W. Regulation of respiration. *N. Engl. J. Med.* 297: 91–97; 138–143; 194–201, 1977.

Dempsey, J. A., and Forster, H. V. Mediation of ventilatory adaptations. *Physiol. Rev.* 62: 262–346, 1982.

Euler, C. V. Brain stem mechanisms for generation and control of breathing pattern. In: Cherniak, N. S., Widdicombe, J. G., eds. *Handbook of Physiology, The Respiratory System*, Volume II. New York: Oxford University Press, 1986.

Javaheri, S., and Kazemi, H. Metabolic alkalosis and hypoventilation in man. *Am. Rev. Resp. Dis.* 136: 1011–1016, 1987.

31 Matching of Ventilation with Blood Flow

DOUGLAS K. ANDERSON

Objectives

After reading this chapter, you should be able to:

Define the normal alveolar ventilation to perfusion ratio and its extremes

Describe the physiologic and pathologic factors that cause mismatching of ventilation with blood flow

Describe the effects on the blood gases of alveolar ventilation to perfusion

Low arterial oxygen tension (PaO_2) is frequently seen in patients with cardiopulmonary disease. The most frequent cause of this hypoxemia is uneven matching of ventilation and blood flow (alveolar ventilation/perfusion ratio [$\dot{V}A/\dot{Q}$] inequality). To optimize the efficiency of gas exchange and achieve normal blood gas levels, the alveoli must be *both* adequately ventilated *and* perfused. Thus, not only must the alveoli freely inspire O_2 and expire CO_2 (ventilation) but sufficient blood must perfuse these alveoli to transport O_2 to the tissues and return CO_2 from the tissues to the alveoli for removal.

THE ALVEOLAR VENTILATION/ PERFUSION RATIO

Figure 31-1 illustrates the matching of ventilation and perfusion, the two extremes of $\dot{V}A/\dot{Q}$ mismatching and its effect on alveolar and blood gases. *A* represents the normal situation in which $\dot{V}A$ and $\dot{Q}$ are matched ($\dot{V}A/\dot{Q}$ = 0.8 to 1.2). *B* is the extreme circumstance of **venous to arterial**

shunting of blood where $\dot{Q}$ is normal but there is no $\dot{V}A$ ($\dot{V}A/\dot{Q}$ = 0). In this situation, the alveolar and blood partial pressures of O_2 (PO_2) and CO_2 (PCO_2) approach those of venous blood. *C* is the extreme condition of **alveolar dead space** in which $\dot{V}A$ is normal but there is no $\dot{Q}$ ($\dot{V}A/\dot{Q}$ = ∞). Under these circumstances, alveolar PCO_2 and PO_2 approach the levels in inspired air. Thus, *B* and *C* represent the extremes of a continuum of possible $\dot{V}A/\dot{Q}$ ratios in the lung.

FACTORS RESPONSIBLE FOR MISMATCHING OF $\dot{V}A/\dot{Q}$

Normal or Physiologic Factors

Figure 31-2 summarizes the normal changes that occur in blood flow, ventilation, and $\dot{V}A/\dot{Q}$ ratios proceeding from the top to the bottom of the lung. As previously discussed, in the normal, healthy, upright lung, ventilation and blood flow decrease from the base to the apex. However, as indicated by the slope of the lines in Figure 31-2, the rate

469

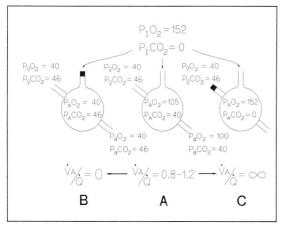

Fig. 31-1. The spectrum of possible alveolar ventilation to perfusion ($\dot{V}A/\dot{Q}$) ratios in the lung. (A) The normal situation, with normal alveolar and blood gas values. (B) The lowest possible $\dot{V}A/\dot{Q}$ ratio—blood flow with no ventilation or a shunt. Alveolar and blood gas values in this situation approach those seen in venous blood. (C) The highest possible $\dot{V}A/\dot{Q}$ ratio, in which there is wasted ventilation or alveolar dead space with alveolar gas values that approach those in inspired air. ($P\bar{v}O_2$ = mean venous O_2 tension; PIO_2 = inspiratory O_2 tension; PAO_2 = alveolar O_2 tension; PaO_2 = arteriolar O_2 tension.)

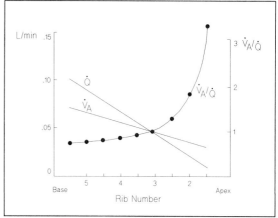

Fig. 31-2. Plot of ventilation, blood flow, and alveolar ventilation to perfusion ($\dot{V}A/\dot{Q}$) ratios as a function of the height of the lung. Although both ventilation and perfusion decrease progressively from the bottom to the top of the lung, perfusion exceeds ventilation in the lower regions of the lung and vice versa toward the top of the lung. Consequently, $\dot{V}A/\dot{Q}$ ratios increase from the bottom to the top of the lung, in that low $\dot{V}A/\dot{Q}$ ratios are normally found toward the bottom of the lung and high $\dot{V}A/\dot{Q}$ ratios are seen in the upper portion. (Modified from West, J. B. *Respiratory Physiology* 4th ed. Baltimore: Williams & Wilkins, 1990. P. 61.)

of the decrease in blood flow exceeds that for ventilation. Consequently, there is a gradient of low to high $\dot{V}A/\dot{Q}$ ratios from the bottom to the top of the lung. In the lower portions of the lung, blood flow exceeds ventilation, causing low $\dot{V}A/\dot{Q}$ ratios. In the upper regions of the lung, ventilation surpasses blood flow, resulting in high $\dot{V}A/\dot{Q}$ ratios. When blood flow and ventilation are essentially equal, $\dot{V}A/\dot{Q}$ ratios are normal. Thus, in the normal upright lung, $\dot{V}A/\dot{Q}$ ratios can range from under 0.5 to over 3.

Regions of the lung with low $\dot{V}A/\dot{Q}$ ratios have low PO_2 and high PCO_2, relative to normal. Areas of the lung with high $\dot{V}A/\dot{Q}$ ratios have relatively high PO_2 levels and low PCO_2 levels. This consequental mismatching of $\dot{V}A/\dot{Q}$ from the bottom to the top of the lung causes a range of values for PO_2 (90 to 130 mm Hg) and PCO_2 (35 to 42 mm Hg) in the effluent blood from the various regions of the lung. Even though O_2 may essentially be com-

pletely equilibrated across the alveolar–capillary barrier for the individual units, the range of $\dot{V}A/\dot{Q}$ ratios (and hence, range of alveolar PO_2 [PAO_2] and arterial PO_2 [PaO_2] values) common in the healthy lung contributes to the difference normally seen between the PO_2 in mixed alveolar gas and the PO_2 in systemic arterial blood—$P(A–a) O_2$ (see Chapter 28 for a description of some of the important pressures).

Another factor that contributes to the overall disparity between the PAO_2 and the PaO_2 is the presence of anatomic venous-to-arterial shunts. Normally a small percentage (2 to 5 percent) of the venous return is not exposed to the gas exchange surfaces of the lung but instead passes directly into the systemic arterial circulation. This includes nutrient blood flow coming from the upper airways and collected by the bronchial veins and coronary venous blood that drains directly into the left ventricle via the thebesian veins. This "dumping" of unoxygenated blood into the systemic circulation further lowers the PaO_2, causing additional deviation of the PaO_2 from the PAO_2.

Pathologic Factors

There are a number of conditions that can restrict ventilation and blood flow, producing abnormally large mismatching of ventilation with perfusion. This **$\dot{V}A/\dot{Q}$ mismatching** leads to low $\dot{V}A/\dot{Q}$ ratios and to subsequent **hypoxemia.**

Hypoventilation can eventuate from either uneven resistance to air flow within the lung or uneven lung compliance. Examples of **uneven resistance** to air flow include bronchoconstriction (asthma), collapse of airways (emphysema), narrowing of airways (bronchitis), and compression of airways (tumors, edema). Examples of **uneven lung compliance** consist of an increase in elastic recoil (fibrosis), loss of elastic recoil (emphysema), and insufficient surfactant levels (infant respiratory distress syndrome).

Pulmonary blood flow can be restricted by constriction or compression of blood vessels (tumors, edema), obliteration of vessels (emphysema, fibrosis), or blockage (embolization or thrombosis) of some part of the pulmonary circulation.

When pulmonary arterial blood flows to the systemic circulation without any effective contact with alveolar gas, this creates a **venous-to-arterial shunt.** In the same way that normal venous-to-arterial shunts lower PaO_2, pathologic shunts can cause significant $\dot{V}A/\dot{Q}$ mismatching (low $\dot{V}A/\dot{Q}$ units) and hypoxemia. Interatrial septal defects are examples of pathologic anatomic shunts. Physiologic or functional shunts can arise with any disease process causing atelectasis or a consolidation of alveolar spaces that continue to be perfused. In the respiratory distress syndromes of both adults and children, alveoli either collapse from high surface-tension forces or become filled with edema fluid, hemorrhage, or cellular debris. A physiologic shunt exists for as long as there is perfusion to these areas, and hence hypoxemia develops.

EFFECTS OF $\dot{V}A/\dot{Q}$ INEQUALITY ON BLOOD GASES

Figure 31-3 depicts two lung units, one with a low and the other with a high $\dot{V}A/\dot{Q}$ ratio. The levels of PO_2 and PCO_2 in inspired air and in the pulmonary arterial blood supplying these units are normal. If equal volumes of blood are entering and leaving both units, then the low $\dot{V}A/\dot{Q}$ unit is underventilated and the high $\dot{V}A/\dot{Q}$ unit is overventilated.

Oxygen

Capillary blood leaving the low $\dot{V}A/\dot{Q}$ unit will be hypoxemic, in that the PaO_2 percentage of oxyhemoglobin ($\%HbO_2$), and O_2 content will all be reduced. If the PaO_2 from this unit is, for example, 60 mm Hg, the O_2 content would be approximately 18.0 ml of O_2 per 100 ml of blood with a $\%HbO_2$ of about 89 percent. Ventilation in the other unit is increased relative to its blood flow, such that the $\dot{V}A/\dot{Q}$ ratio is elevated and blood leaving this unit has an abnormally high PaO_2, $\%HbO_2$, and O_2 content of 125 mm Hg, 99 percent, and 20.2 ml of O_2 per 100 ml of blood, respectively. When the blood from these two units merges, the approximate PaO_2, $\%HbO_2$, and O_2 content of the mixed effluent blood would be 75 mm Hg, 94 percent, and 19.1 ml of O_2 per 100 ml of blood, respectively. Thus, for O_2, the high $\dot{V}A/\dot{Q}$ units only partially compensate for the low $\dot{V}A/\dot{Q}$ units, in that the O_2 values in the mixed effluent blood are lower than the normal values of PaO_2, 100 mm Hg; $\%HbO_2$, 97.4 percent; and O_2 content, 19.8 ml of O_2 per 100 ml of blood.

The reason the units with high $\dot{V}A/\dot{Q}$ ratios do not compensate for the low units is that the units with high $\dot{V}A/\dot{Q}$ ratios add relatively little O_2 to the blood, compared with the decrement in O_2 caused by the units with low $\dot{V}A/\dot{Q}$ ratios, because of the nonlinear shape of the oxyhemoglobin dissociation curve (see Fig. 28-5). Because this curve is almost horizontal when PO_2 exceeds 80 mm Hg, raising the PO_2 to 125 mm Hg would elevate the saturation of hemoglobin from approximately 97.4 percent (at a normal PO_2 of about 95 to 100

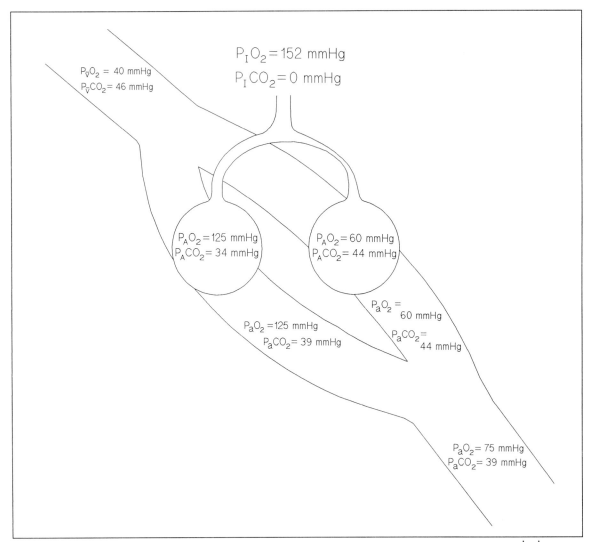

Fig. 31-3. Depiction of two lung units, with equal blood flow to and from both units. One unit has a low $\dot{V}_A/\dot{Q}$ ratio; the other has a high $\dot{V}_A/\dot{Q}$ ratio. Levels of venous blood and inspired gases are normal. ($P\bar{v}O_2$ = mean venous O_2 tension; PiO_2 = inspiratory O_2 tension; PAO_2 = alveolar O_2 tension; PaO_2 = arteriolar O_2 tension.) (Modified from: Cherniack, N. S., Altose, M. D., and Kelsen, S. G. The respiratory system. In Berne, R. H., and Levy, M. N., eds., *Physiology*. St. Louis: Mosby, 1983. P. 700.)

mm Hg) to around 99 percent. This would increase the volume of O_2 carried by hemoglobin by about 0.32 ml of O_2 per 100 ml of blood. In the low $\dot{V}_A/\dot{Q}$ unit (with a PO_2 of 60 mm Hg), the volume of O_2 carried by hemoglobin is about 17.8 ml of O_2 per 100 ml of blood. This would represent an O_2 decrement of approximately 1.68 ml of O_2 per 100 ml of blood, rendered as: 19.48 ml of O_2 per 100 ml of blood (the volume of O_2 combined with hemoglobin at a PaO_2 of 100 mm Hg) minus 17.8 ml of O_2 per 100 ml of blood. Thus the high $\dot{V}_A/\dot{Q}$ unit, which adds about 0.32 ml of O_2 per 100 ml of blood, cannot offset the loss of 1.68 ml of O_2 per 100 ml of blood from the low $\dot{V}_A/\dot{Q}$ unit. Consequently, the PaO_2, $\%HbO_2$, and O_2 content of the mixed effluent blood all fall below normal levels.

Carbon Dioxide

In subjects with $\dot{V}_A/\dot{Q}$ mismatching, there is often a normal or even subnormal $PaCO_2$. This can arise because the shape of the CO_2 dissociation curve for whole blood, unlike the oxyhemoglobin dissociation curve (compare Figures 28-5 and 28-9), is essentially linear over a wide range of PCO_2 values. Consequently, the PCO_2 and CO_2 in mixed effluent blood are essentially the average of their respective contents from high and low $\dot{V}_A/\dot{Q}$ units. For the two lung units depicted in Figure 31-3, blood leaving the low $\dot{V}_A/\dot{Q}$ unit has elevated PCO_2 and CO_2 contents (approximately 44 mm Hg and 52.1 ml of CO_2 per 100 ml of blood, respectively) but blood from the high $\dot{V}_A/\dot{Q}$ unit has low PCO_2 and CO_2 contents (approximately 34 mm Hg and 45.3 ml of CO_2 per 100 ml of blood, respectively). Because the CO_2 contents and PCO_2 values from these two units can be averaged (the CO_2 content/PCO_2 relationship is linear; see Fig. 28-8), the mixed effluent blood has a PCO_2 of 39 mm Hg and a CO_2 content of 47.8 ml of CO_2 per 100 ml of blood; both values are essentially normal. Thus, under these circumstances, the decreased PCO_2 and CO_2 content in the high $\dot{V}_A/\dot{Q}$ unit compensates for the increase in PCO_2 and CO_2 content in the blood from the low $\dot{V}_A/\dot{Q}$ unit.

In a disease state, the number of low $\dot{V}_A/\dot{Q}$ units will be increased, causing PaO_2 to fall and $PaCO_2$ to rise. Stimulation of the central and peripheral chemoreceptors by either the elevated $PaCO_2$ or the hypoxic stimulation of the peripheral chemoreceptors augments alveolar ventilation (creating high $\dot{V}_A/\dot{Q}$ ratios) in those regions of the lung that are normal or not yet significantly diseased. In this case, systemic blood is likely to be hypoxemic with a near-normal $PaCO_2$. However, as the disease worsens, the area of the lung with low $\dot{V}_A/\dot{Q}$ ratios will enlarge. Because this occurs at the expense of the normal to high (healthy) $\dot{V}_A/\dot{Q}$ regions, the lung becomes progressively less able to

provide adequate O_2 levels and to expel CO_2, thus intensifying the hypoxemia and CO_2 retention.

COMPENSATORY MECHANISMS MATCHING VENTILATION AND PERFUSION

As mentioned in Chapter 29, alveolar hypoxia, that is, low **alveolar Po_2** in poorly ventilated alveoli, causes **arteriolar constriction,** which increases the resistance to blood flow to these poorly ventilated alveoli. This event in turn redistributes blood flow to well-ventilated alveoli. In addition, alveolar hypocapnia, that is, low **alveolar Pco_2** in overventilating alveoli or alveoli receiving a decreased blood flow, **constricts the small airways** leading to these alveoli. This increased airway resistance reduces the ventilation to these alveoli, resulting in a redistribution of gas to other alveoli with adequate blood flow, thereby creating a better match between ventilation and perfusion.

SUMMARY

To achieve maximal gas exchange and normal blood gas levels, alveoli must be both ventilated and perfused. Ventilation with little or no perfusion is called *alveolar dead space* and results in PaO_2 and $PaCO_2$ levels that approach their respective levels in inspired air. Perfusion without ventilation is termed *venous-to-arterial shunt* and causes alveolar gas pressures to approximate those in venous blood. Shunts and alveolar dead space represent the extreme limits of low to high $\dot{V}_A/\dot{Q}$ ratios, respectively. Because the progressive decline in blood flow from the bottom to the top of the lung exceeds that for ventilation, there is a normal mismatching of ventilation with perfusion, and hence a range of $\dot{V}_A/\dot{Q}$ ratios from the base to the apex of the lung. This $\dot{V}_A/\dot{Q}$ mismatching plus normal shunting causes aortic Po_2 levels that are slightly lower than alveolar values. For O_2, units with high $\dot{V}_A/\dot{Q}$ ratios cannot compensate for units with low $\dot{V}_A/\dot{Q}$ ratios because the shape of the oxyhemoglobin dissociation curve is not linear. Thus, high $\dot{V}_A/\dot{Q}$ units cannot replace the

O_2 lost by the low $\dot{V}_A/\dot{Q}$ units. Conversely, CO_2 retained by units with low $\dot{V}_A/\dot{Q}$ ratios can be eliminated by the units with high $\dot{V}_A/\dot{Q}$ ratios, because the CO_2 dissociation curve for whole blood is essentially linear in the physiologic range. Thus, as long as sufficient healthy lung tissue remains, high $\dot{V}_A/\dot{Q}$ units can compensate for low $\dot{V}_A/\dot{Q}$ units and thereby maintain normal or near-normal blood levels of CO_2. However, as increasing areas of the lung become diseased, the ratio of high $\dot{V}_A/\dot{Q}$ units to low $\dot{V}_A/\dot{Q}$ units decreases, and CO_2 is retained.

NATIONAL BOARD TYPE QUESTIONS

Select the one most correct answer for each of the following

1. In the normal lung:
   A. Low $\dot{V}_A/\dot{Q}$ ratios are found primarily at the apex
   B. Low $\dot{V}_A/\dot{Q}$ ratios are found primarily at the base
   C. Both $\dot{V}_A$ and $\dot{Q}$ are higher at the apex than at the base
   D. Normal $\dot{V}_A/\dot{Q}$ mismatching occurs because $\dot{V}_A$ is higher at the base than at the apex, whereas $\dot{Q}$ is higher at the apex than at the base
   E. At the apex, $\dot{Q}$ normally exceeds $\dot{V}_A$

2. With respect to a normal PaO_2 of 100 mm Hg and $PaCO_2$ of 40 mm Hg, which of the following changes would be exhibited by blood coming from a low $\dot{V}_A/\dot{Q}$ unit?
   A. Low PaO_2 and high $PaCO_2$
   B. High PaO_2 and low $PaCO_2$
   C. Low PaO_2 and low $PaCO_2$
   D. High PaO_2 and high $PaCO_2$
   E. No change

3. Which of the following statements is true, assuming equal areas of the lung are affected?
   A. Units with high $\dot{V}_A/\dot{Q}$ ratios cannot compensate for the loss of CO_2 caused by units with low $\dot{V}_A/\dot{Q}$ ratios

   B. High $\dot{V}_A/\dot{Q}$ ratios are exhibited by the lower half of the lung and low $\dot{V}_A/\dot{Q}$ ratios by the upper half
   C. Increased PCO_2 in the alveoli with high $\dot{V}_A/\dot{Q}$ ratios can increase ventilation in low $\dot{V}_A/\dot{Q}$ units via stimulation of the central chemoreceptors
   D. Units with high $\dot{V}_A/\dot{Q}$ ratios cannot add enough O_2 to blood to offset the O_2 loss caused by units with low $\dot{V}_A/\dot{Q}$ ratios
   E. In $\dot{V}_A/\dot{Q}$ mismatching, there is always a low PaO_2 and an elevated $PaCO_2$

4. In a person breathing room air, which of the following combinations of partial pressures (in mm Hg) would most likely be caused by an area of the lung with a $\dot{V}_A/\dot{Q}$ ratio of 0.4?
   A. $PO_2 = 100$; $PCO_2 = 40$
   B. $PO_2 = 115$; $PCO_2 = 65$
   C. $PO_2 = 115$; $PCO_2 = 25$
   D. $PO_2 = 65$; $PCO_2 = 65$
   E. $PO_2 = 65$; $PCO_2 = 25$

5. Which of the following statements is true?
   A. An area of the lungs with a high $\dot{V}_A/\dot{Q}$ ratio will have a high PaO_2 and $PaCO_2$
   B. Alveolar volume, airway resistance, and pulmonary blood flow are all higher in the lower half of the lung compared with the upper half
   C. In the normal lung, alveolar ventilation progressively decreases from the base to the top
   D. A pathologic shunt is any area of the lung ventilated but not perfused
   E. An area of the lung that is perfused but not ventilated is considered alveolar dead space

ANNOTATED ANSWERS

1. B. At the base of the lung, pulmonary blood flow exceeds alveolar ventilation, resulting in a reduced $\dot{V}_A/\dot{Q}$ ratio.

2. A. In low $\dot{V}_A/\dot{Q}$ units, ventilation is reduced

relative to blood flow. This hypoventilation lowers the P_{O_2} and elevates P_{CO_2} in the blood, leaving the low $\dot{V}_A/\dot{Q}$ units.

3. D. Because the oxyhemoglobin dissociation curve is S-shaped, the addition of O_2 by the high $\dot{V}_A/\dot{Q}$ ratios (on the flat part of the curve) cannot offset the decrement in O_2 caused by the low $\dot{V}_A/\dot{Q}$ ratios (on the sloped portion of the curve).

4. D. A $\dot{V}_A/\dot{Q}$ ratio of 0.4 is below the normal range of 0.8 to 1.2. Thus, one would expect a lowered P_{O_2} and an elevated P_{CO_2} from this region of the lung.

5. C. Like pulmonary blood flow, alveolar ventilation is highest at the base of the lung and lowest at the apex. This gradient of ventilation and perfusion proceeding from the top to the bottom of the lung is responsible for the normal mismatching of ventilation with perfusion seen in the normal lung.

BIBLIOGRAPHY

Cherniak, N. S., Altose, M. D., Kelsen, S. G. The Respiratory System. In: Berne, R. M., and Levy, M. N., eds. *Physiology*, Section VI. St. Louis: C. V. Mosby, 1983.

Comroe, J. H. *Physiology of Respiration*, 2nd ed. Chicago: Year Book Medical, 1974.

Levitzky, M. G. *Pulmonary Physiology*, 3rd ed. New York: McGraw-Hill, 1991.

Mines, A. H. *Respiratory Physiology*, 2nd ed. New York: Raven Press, 1986.

Taylor, A. E., Rehder, K., Hyatt, R. E., and Parker, J. C. *Clinical Respiratory Physiology*. Philadelphia: W. B. Saunders, 1989.

West, J. B. *Respiratory Physiology*, 4th ed. Baltimore: Williams & Wilkins, 1989.

VIII Renal and Acid-Base Physiology

32 Elements of Renal Function

ROBERT O. BANKS

Objectives

After reading this chapter, you should be able to:

List the functions of the kidneys

List the major anatomic structures in the kidney and describe their functional organization

List and define the three major mechanisms at work in the formation of urine

Cite the formula for renal clearance

Explain why the clearance of a substance that is freely filtered and not reabsorbed or secreted equals the glomerular filtration rate

Describe the application of the Fick principle to the kidney and explain why the clearance of a substance that is totally cleared by the kidney equals the renal plasma flow

Define the filtration fraction and give normal values

Distinguish between total renal plasma flow and effective renal plasma flow

Explain renal extraction

Define filtered load and fractional excretion, and calculate the filtered Na^+ load per day and the fractional Na^+ excretion

One of Claude Bernard's many contributions to physiology was formulating the concept of constancy of the "milieu interieur," an idea later encompassed by Walter Cannon in his term **homeostasis,** defined as maintenance of the steady-state. In this context, the kidney represents the primary organ of homeostasis in its regulation of both the volume and composition of many constituents of the extracellular fluid space. Although this section on renal physiology primarily focuses on the homeostatic role of the kidney, the excretion of metabolic waste products and the regulation of **red blood cell synthesis** through the **production of erythropoietin** represent two other important renal functions.

ANATOMY OF THE KIDNEY

The anatomy of the kidney is summarized in Figures 32-1 through 32-3. There are two clearly defined zones in the kidney—an outer region, referred to as the **cortex,** and an inner region, called the **medulla.** In the human kidney, the medulla is divided into a number of conical areas called the **renal pyramids.** Within each pyramid, the medulla can be further partitioned into an **outer medulla,** containing an **outer** and **inner stripe,** and an **inner medulla.** The apex of the inner medulla is often referred to as the **papilla** and protrudes into the pelvic space.

As is illustrated in Figures 32-1 and 32-2, the kidney is highly vascular. Even though the

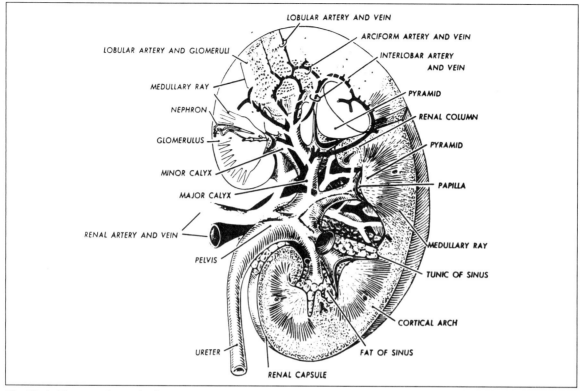

Fig. 32-1. The vascular and tubular networks in the human kidney. Lobular arteries and veins are also referred to as *intralobular vessels; arciform* and *arcuate* are also synonymous. (Reproduced with the permission of Oxford University Press. From: Smith, H. W. *Principles of Renal Physiology.* New York: Oxford University Press, 1956.)

combined kidney weight constitutes a relatively small percentage (0.5 percent) of the total body weight, approximately 20 percent of the entire cardiac output perfuses the kidney under resting conditions. Because the normal cardiac output is about 5 liters/min, the combined renal blood flow (RBF) to the right plus left kidney is close to 1 liter/min.

The major sites of **resistance to blood flow** in the kidney are located in the afferent and efferent arterioles. Of the total renal vascular resistance (defined by the ratio of the mean arterial pressure minus the mean renal venous pressure divided by the RBF), most (greater than 90 percent) is due to the combined resistance of the afferent and efferent arteriole. This **portal system** allows for very efficient regulation of the pressure in the

glomerular capillaries, which are interposed between the two arteriolar segments of the renal vasculature. As will be further discussed in Chapter 33, the pressure in glomerular capillaries is one of the major determinants of the rate of glomerular filtration (GFR).

The functional unit of the kidney is the **nephron,** and there are approximately 1.2 million of these tubular structures in each kidney. Within the closed end of each tubule is the filtering unit of the nephron, the **renal corpuscle.** The renal corpuscle is composed of the glomerular capillaries, generally referred to as the **glomerulus,** and the surrounding tubular element, known as **Bowman's capsule.**

Glomeruli are located only in the cortex. There are, however, three general types of nephrons:

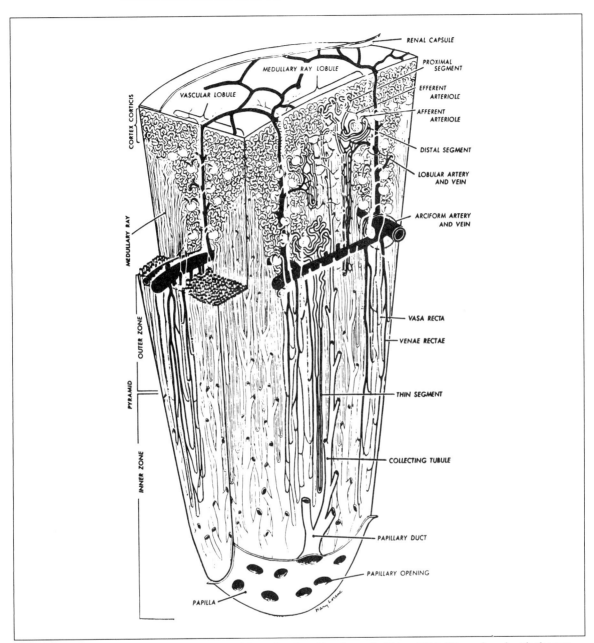

Fig. 32-2. A higher-magnification depiction of the subject matter shown in Figure 32-1. (Reproduced with the permission of Oxford University Press. From: Smith, H. W. *Principles of Renal Physiology.* New York: Oxford University Press, 1956.)

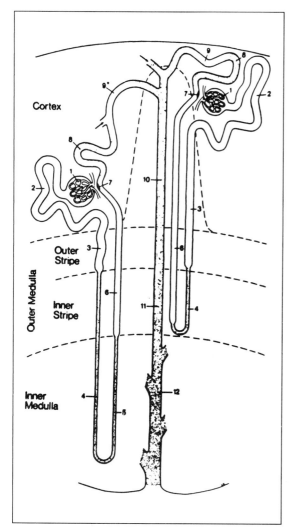

Fig. 32-3. An illustration, not drawn to scale, of the long-looped and short-looped nephrons. (*1* = The renal corpuscle; *2* = the proximal convoluted tubule; *3* = the proximal straight tubule; *4* = the descending limb of Henle's loop; *5* = the thin ascending limb; *6* = the thick ascending limb; *7* = the macula densa; *8* = the distal convoluted tubule; *9* = the connecting tubule; *9** = the connecting tubule of the juxtamedullary nephron that forms an arcade; *10* = the cortical collecting duct; *11* = the outer medullary collecting duct; *12* = the inner medullary collecting duct.) A medullary ray in the cortex is delineated by a *dashed line.* (Reproduced with permission of: Kritz, W., and Bankir, L. A standard nomenclature for structures of the kidney. *Am. J. Physiol.* 254: F1–F8, 1988.)

those with glomeruli originating in the superficial regions of the cortex, referred to as **superficial** or **outer cortical nephrons;** those with glomeruli originating in the **midcortical regions;** and those with glomeruli in the **juxtamedullary region** (see Fig. 32-3). In the human kidney, approximately 15 percent of the total nephron population is of the juxtamedullary variety.

There are four major tubular segments of the nephron, the **proximal tubule,** the intermediate tubule or **loop of Henle,** the **distal tubule,** and the **collecting duct** (see Fig. 32-3). Each of these can be further divided. The proximal tubule has a convoluted portion, or the **proximal convoluted tubule,** and a straight segment, also known as the **pars recta.** Three cell types form the epithelial lining of these proximal structures. The **S_1 segment,** composed of S_1-type cells, makes up the beginning and middle portions of the convoluted proximal tubule. S_1 cells are characterized by numerous microvilli and mitochondria. The **S_2 segment** forms the remaining portion of the convoluted tubule and the initial portion of the straight segment. S_2 cells have fewer microvilli and mitochondria than do S_1 cells. Finally, the remaining portion of the pars recta is called the **S_3 segment;** S_3 cells have few microvilli and mitochondria.

The **loop of Henle** is divided into the descending and ascending limbs. These loops descend to varying depths into the medulla of the kidney. In general, those that originate from glomeruli in superficial regions of the kidney descend only to the outer medulla, the so-called **short-looped nephrons;** those of the juxtamedullary variety descend into the inner medulla, often to the tip of the papilla, the so-called **long-looped nephrons.** The lumen of both the descending and ascending limbs of Henle's loop are lined with relatively thin epithelial cells, and are therefore referred to as the **thin descending** and **thin ascending limbs.** These epithelial cells contain very few mitochondria and microvilli.

The initial segment of the distal tubule is the thick ascending limb of the outer medulla, referred to as the **medullary thick ascending limb.** In the cortex, the thick ascending limb is referred to as the **cortical thick ascending limb,** each of which comes into contact with the afferent arteriole of its parent glomerulus, forming a complex known as the **juxtaglomerular apparatus.** This unit is composed of differentiated epithelial cells of the tubule (macula densa cells) and afferent arteriole (juxtaglomerular cells) plus interposed lacis cells.

The terminal segment of the distal tubule (beyond the macula densa) is the **distal convoluted tubule.** The distal convoluted tubules of superficial nephrons return to the surface of the kidney and represent, in addition to the proximal convoluted tubules of these nephrons, one of the tubular regions accessible to **micropuncture.**

The **collecting duct** is made up of several segments, and, although there are minimal *morphologic* differences between them, there are important *physiologic* differences. The first segment is the **connecting tubule** and consists of two cell types: **connecting tubule** cells and **intercalated cells.** The transition to the cortical collecting duct is arbitrarily defined by the first appearance of collecting duct cells. The cortical collecting duct is located primarily in the medullary rays of the kidney. At the transition between the cortex and medulla, the tubule is referred to as the **outer medullary collecting duct.** Similarly, the collecting duct in the inner medulla is referred to as the **inner medullary collecting duct.** There are two primary cells in the collecting duct—**collecting duct** or **principal cells** and the **intercalated cells.**

There are important anatomic and, therefore, physiologic, relationships between the vascular and tubular elements of the kidney. Thus, efferent arterioles of the superficial cortex, and to a lesser extent those in the midcortical region, form **peritubular capillary networks** that are closely associated with the proximal and distal convoluted tubules of the parent glomerulus. By contrast, efferent vessels of juxtamedullary glomeruli form loop-type structures, called **vasa recta,** which descend to varying depths in the medulla. Consequently, the postglomerular blood of juxtamedullary nephrons does not perfuse proximal or distal convoluted tubules, and therefore can-

not be affected by transport-related events in these regions of the nephron. Instead, these proximal and distal tubular segments are perfused with blood from the efferent arterioles of glomeruli that originate in the midcortex.

GENERAL ASPECTS OF URINE FORMATION

The physiologic basis of renal function can be summarized by three primary events: **filtration, reabsorption,** and **secretion.**

Filtration

Filtration takes place only in the glomerulus and depends on the balance between **hydrostatic** and **colloid osmotic pressures** in the glomerular capillary and in Bowman's space. In other words, filtration is regulated by the **Starling forces** governing fluid movement across the capillary wall (see Chapters 1 and 20 for a review of Starling forces). The rate of filtration in the glomerulus is usually expressed as milliliters per minute or as liters per day.

Reabsorption

Reabsorption represents the **movement of water and solute** from the tubular lumen to the peritubular capillary network. Reabsorption of water is passive, but that of solute may be active or passive depending on the particular solute and nephron segment involved.

Secretion

Secretion represents the **net addition** of solute to the tubular lumen and almost always is an **active transport process.** Secretion is often mistakenly equated with excretion, which merely refers to the solute contained in the final urine.

RENAL CLEARANCE

The clearance principle, as applied to the kidney, is exceedingly useful and important for determining several variables of renal function in both the experimental and clinical setting. At first glance, the concept is deceptively simple, yet it is often difficult to fully appreciate the physiologic implications of a clearance value. Therefore, the concept of clearance will be approached by analyzing its definition, its mathematical formula, and the clearance of several solutes, the renal handling of which are well known.

Clearance is defined as the volume of plasma required to supply a given amount of a substance excreted in the urine per unit of time; it can also be defined as the volume of plasma cleared of a given substance per unit of time. In general, for clearance to be measured, the subject must be in a steady-state condition. The **formula for clearance** is:

$$C_x = \frac{(U_x)(V)}{P_x}$$

Where C_x is the clearance of substance x, U_x is the urine concentration of substance x, V is the urine flow rate, and P_x is the arterial plasma concentration of substance x. To satisfy the condition of a steady-state, the urine and plasma concentrations of substance x and the urinary flow rate must not change during the collection period.

The **units of clearance** are defined by the following equation:

$$C = \frac{(U)(V)}{P}$$

$$C = \frac{(\text{mass/volume})(\text{volume/unit time})}{\text{mass/volume}}$$

C = volume/unit time (usually expressed as milliliters per minute)

Examples of Clearance

Inulin Clearance

Inulin is a polymer of fructose and is one of the solutes used to measure the volume of the **extracellular fluid** (see Chapter 1). It is an inert substance that freely crosses most capillaries yet does not traverse the cell membrane. Consequently,

inulin is freely filtered in the kidney, not reabsorbed or secreted. Given these characteristics, the inulin clearance therefore equals the GFR, for the reasons illustrated in Figure 32-4. As this figure shows, the amount of inulin entering the 2.4 million nephrons per unit of time, in a steady-state, must equal the amount of inulin leaving these nephrons. Because inulin is freely filtered, the inulin concentration in the glomerular capillary plasma equals its concentration in the fluid in Bowman's space, both of which are equal to the inulin concentration in arterial plasma. Therefore, the amount of inulin entering the nephron per unit of time is given by the product of the inulin concentration in arterial plasma, times the volume flow of fluid entering the nephron—the GFR. In other words:

Fig. 32-4. This figure represents the sum of all the nephrons. Since inulin is neither reabsorbed nor secreted, the amount of inulin filtered must equal the amount excreted. These facts, coupled with the fact that inulin is also freely filtered, provide the basis for equating the clearance of inulin with the glomerular filtration rate (*GFR*). (P_{inulin} = inulin concentration in arterial plasma.)

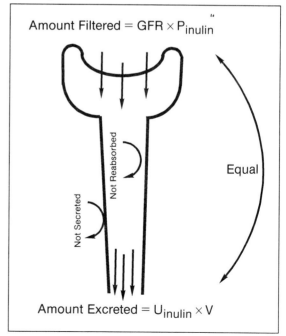

Amount Filtered = GFR × P_{inulin}

Not Reabsorbed

Not Secreted

Equal

Amount Excreted = U_{inulin} × V

$(GFR)(P_{inulin}) = (U_{inulin})(V)$
or $GFR = (U_{inulin})(V)/P_{inulin}$

Thus, the clearance of inulin, or of any freely filtered, non-reabsorbed and nonsecreted solute equals the GFR.

A healthy 20-year-old person who weighs 70 kg has a GFR of about 120 ml/min or about 70 ml per minute per square meter of body surface area. Males and females with similar weights have similar GFR values. The GFR decreases with aging, such that an average 70-kg, 60-year-old person has a GFR of about 100 ml/min. By contrast, at birth the GFR, corrected for body surface area, is relatively low and averages only about 20 ml min^{-1} m^{-2}. The GFR reaches the adult value of 70 ml min^{-1} m^{-2} by 1 to 3 years of age.

Creatinine Clearance

In the human kidney, besides being freely filtered, a small amount of creatinine (Cr) is secreted into the lumen of the proximal tubule. Consequently, the amount of creatinine excreted slightly exceeds the amount filtered. The difference is relatively small and averages only about 10 percent in the normal kidney (at low GFR values, the secreted component becomes a larger proportion of the amount excreted). Because creatinine is an endogenous substance and its clearance only exceeds the GFR by approximately 10 percent, creatinine clearance is an important clinical assessment. Moreover, because the production rate of creatinine is relatively constant (about 1.2 mg/min in a 70-kg person) and is not affected by renal disease, creatinine clearance and the plasma creatinine concentration provide important information regarding the **status of glomerular function.** The usefulness of the plasma creatinine concentration as a predictor of glomerular function is illustrated in the following analysis:

Creatinine production rate = a constant rate (k) of approximately 1.2 mg/min
Creatinine production rate = creatinine excretion rate
Therefore, $k = (U_{Cr})(V) = (GFR)(P_{Cr})$

Given a normal GFR of 120 ml/min, the average plasma creatinine concentration is therefore about 0.01 mg/ml, or 1 mg/dl.

To illustrate the importance of an increasing plasma concentration of creatinine, assume that a patient has an initial plasma creatinine concentration of 1 mg/dl, but several months later this increases to 10 mg/dl. If the patient does not have muscular dystrophy or some other disease that can alter the production rate of creatinine, then the patient's GFR would have decreased to one-tenth of its initial value. In other words, an estimate of the patient's GFR can be obtained by simply determining the plasma concentration of creatinine (it is important to note that, following a change in the GFR, 4 to 5 days must elapse before a new steady-state plasma creatinine value can be attained). If the plasma creatinine concentration is elevated, and because the plasma creatinine concentration is only an estimate of the GFR due to the assumptions inherent in creatinine analysis (to be discussed), collection of a timed urine sample (usually for 24 hours) or evaluation of the clearance of an exogenously administered glomerular marker can then be used to verify the status of glomerular function.

As mentioned, there are some qualifications and assumptions regarding the predictive nature of the plasma creatinine concentration as it relates to the GFR. Thus, a number of factors affect the plasma creatinine concentration besides the GFR and its endogenous production rate. These factors include the degree of tubular secretion, as well as the dietary intake of creatinine and its extrarenal excretion. Other solutes in plasma also interfere with the chemical assay normally used for creatinine. For these reasons, it is important to measure the renal clearance of creatinine or an exogenous glomerular marker when renal disease is anticipated.

Urea Clearance

Urea is an end product of hepatic protein metabolism. It is routinely measured in blood as the **blood urea nitrogen** (BUN) and its clearance is normally 40 to 50 percent of the GFR, indicating that there is net passive reabsorption of urea. Urea clearance varies with the urine flow rate, reaching approximately 60 percent of the GFR at high urine flow rates. In addition, the production rate of urea is not constant, but varies with the amount of protein in the diet and when diseases affect liver function. Therefore, compared to plasma creatinine values, the BUN is less reliable as an indicator of glomerular dysfunction. Nonetheless, the BUN, which is normally about 10 to 20 mg/dl (about 6 mM/liter), does increase markedly in patients with renal disease.

Clearance of Organic Acids and Bases

The **Fick principle,** a concept discussed in conjunction with the determination of cardiac output (see Chapter 17), also provides the basis for one of the methods used to estimate **renal plasma flow** (RPF). If a solute, y, is excreted by the kidney but not metabolized and does not enter the lymphatics or red blood cells, then the amount of y entering and leaving the kidney in a steady-state will be the same:

$$(RPF_a)(P_{ay}) = (U_y)(V) + (RPF_v)(P_{vy})$$

Where P_{ay} and P_{vy} are the arterial and renal venous plasma concentrations of y, respectively, and a and v are the arterial and venous values, respectively.

Therefore, if RPF_a equals RPF_v, then:

$$RPF = (U_y)(V)/(P_{ay} - P_{vy})$$

This equation does not yield a clearance but illustrates that RPF can be calculated if the amount of a substance excreted and the difference in the arterial and venous concentrations across the kidney are known. Moreover, a number of weak organic acids and bases are avidly secreted by the proximal tubule, resulting in a very low renal venous concentration of these solutes. One of the most widely studied weak organic acids is **p-aminohippuric acid** (PAH). If the concentration of PAH in renal venous plasma is zero, then the above equation reduces to a clearance:

$$RPF = (U_{PAH})(V)/P_{aPAH}$$

This equation is the clearance of PAH (C_{PAH}). Thus, if a given solute is avidly secreted by the kidney enough that the renal venous concentration of the material is zero, the clearance of that solute will equal the RPF.

If the renal venous concentration of PAH is zero, then the **renal extraction** (E), where E = $P_a - P_v / P_a$, for PAH would be 1.0 or 100 percent. The renal extraction of all the organic acids and bases is, in fact, less than 100 percent, with that for PAH averaging about 90 percent. Therefore, the clearance of PAH is less than the actual RPF by approximately 10 percent. Some investigators have referred to the clearance of PAH as an indicator of the **effective renal plasma flow.**

The average PAH clearance and extraction values for a healthy 20-year-old person weighing 70 kg are 540 ml/min and 0.9, respectively. Therefore, the average RPF (given by the clearance value of PAH divided by the extraction value) is 600 ml/min. As the normal hematocrit (hct) is approximately 40 percent, the RBF is 1,000 ml/min (if one neglects the non-red blood cell elements in blood, then RBF = RPF/1 − hct). RPF decreases with age, reaching about 420 ml/min at age 60.

Given that the GFR is normally about 120 ml/min and the RPF is 600 ml/min, only about 20 percent of the total renal plasma flow is actually filtered. This relationship between RPF and GFR is referred to as the **filtration fraction** (FF), and is equal to the ratio of the GFR to the RPF:

$$FF = GFR/RPF$$

The filtration fraction is not a fixed value, and, as will be discussed further in Chapter 33, the factors that affect GFR and RPF ultimately affect the filtration fraction.

Examples of weak organic acids and bases secreted by the proximal tubule are listed in Table 32-1.

Glucose and Albumin Clearance

The clearances of glucose and albumin are defined, as previously explained, by the amount of

Table 32-1. Some Representative Weak Organic Acids and Bases

Organic Acids	Organic Bases
p-Aminohippurate	Thiamine
Salicylic acid	Serotonin
Phenol red (phenolsul-	Histamine
fonphthalein)	Guanidine
Probenecid	Procaine
Nitrofurantoin	N'-methylnicotinamide
Penicillin G	
Chlorothiazide	
Furosemide	
Diodrast[a]	
Uric acid	

[a]Injectate form of iodopyracet.

glucose and albumin excreted divided by the plasma glucose and albumin concentrations, respectively. Under normal conditions, virtually no glucose or albumin is excreted, and therefore the clearance of these solutes is zero. This illustrates one of the **significant limitations** of the clearance approach, namely, that the clearance data alone does not permit firm conclusions to be obtained about the renal handling of a given solute. Thus, the clearances of both glucose and albumin are zero, yet, as will be discussed, glucose is freely filtered but avidly reabsorbed by a Na^+-coupled, secondary active reabsorptive process in the early portions of the proximal tubule. By contrast, the **permeability and selectivity** of the renal corpuscle primarily account for the low clearance of albumin. Other methodologies, such as **micropuncture** and **microperfusion** of isolated tubular segments are used to elucidate tubular events.

Sodium Clearance

The clearance of Na^+ is equal to $(U_{Na})(V)/P_{Na}$. Under normal conditions, the amount of Na^+ ingested is equal to the amount lost from the body (see Chapter 36 for a complete discussion of Na^+ balance). Although the amount of Na^+ ingested varies considerably among individuals, the average person consuming a Western diet ingests about

150 mEq/day. Because approximately 95 percent of this Na^+ (or about 140 mEq/day) is excreted in the urine (5 percent is lost in feces and sweat) and as the plasma Na^+ concentration is about 140 mEq/liter, the clearance of Na^+ is roughly 1 liter/day (140 mEq per day/140 mEq per liter), or only about 0.7 ml/min.

Based on the above analysis for Na^+, it is apparent that only a small fraction of the filtered Na^+ load, defined as the product of the GFR and the plasma Na^+ concentration, is actually excreted, that is, most of the filtered Na^+ is reabsorbed. Thus, the fractional excretion (FE) of any freely filtered substance, x, is given by:

$$FE_x = \text{amount of x excreted/amount of x filtered}$$
or
$$FE_x = (U_x)(V)/(GFR)(P_x) = C_x/GFR = C_x/C_{inulin}$$

The fractional excretion of Na^+ would be less than 1 percent (140 mEq/day excreted compared with approximately 25,000 mEq/day filtered).

SUMMARY

A synopsis of the renal processing of the substances discussed in this chapter is presented in Table 32-2. The differences in the renal handling of these representative solutes translate into a spectrum of renal clearances, differing from one solute to another and varying with the status of renal function. The clearance of a substance such as inulin that is freely filtered, not reabsorbed or secreted, is equal to the GFR. In the clinical setting, the clearance of creatinine is especially useful for estimating glomerular function. If a particular solute has a clearance that is greater than inulin's and if the solute is freely filtered and not synthesized by the kidney, this indicates that the solute undergoes **net secretion** by the kidney (there could be a combination of active or passive reabsorption in addition to the active secretion). Similarly, if another solute has a clearance less than inulin and is also freely filtered, **net reabsorption** must predominate (again a combination of active or passive reabsorption as well as active secretion could occur).

Table 32-2. Partition of Delivered Substances in the Kidney[a]

Solute	In Renal Artery	After Filtration		After Reabsorption		After Secretion	
		Blood	Filtrate	Blood	Urine	Blood	Urine
RBC, albumin	100	100	0	100	0	100	0
Glucose	100	80	20	100	0	100	0
Inulin and creatinine	100	80	20	80	20	80	20
PAH	100	80	20	80	20	10	90
H_2O and Na^+	100	80	20	99	1	99	1

[a]Expressed as percentage of total amount entering the kidney.
RBC = red blood cells; PAH = *p*-aminohippuric acid.

Clearance must represent a virtual volume. For example, in the illustration for Na^+, 0.7-ml "packets" of plasma, devoid of Na^+, do not circulate. Rather, it is as if 0.7 ml of plasma were completely cleared of Na^+ each minute. In actuality, the excreted Na^+ derives from the entire extracellular fluid, or more precisely, from the total quantity of exchangeable Na^+ in the body. Inulin clearance can be equated with an important variable of renal function—the GFR. Similarly, the clearances of PAH and other weak organic acids and bases approximate another important variable of renal function—the RPF.

NATIONAL BOARD TYPE QUESTIONS

Select the one most correct answer for each of the following

1. If the GFR is 120 ml/min and the urine flow rate is 0.5 ml/min, the urine to plasma inulin ratio would be:
   A. 10/1
   B. 24/1
   C. 100/1
   D. 240/1
   E. 1,000/1
2. In a steady-state, the clearance of inulin is equal to the GFR because the:
   A. Volume of plasma cleared equals 100 ml/min
   B. Amount of inulin infused equals the amount of inulin excreted

C. Amount of inulin excreted minus the amount of inulin filtered is zero
D. Amount of inulin excreted equals the GFR
E. Concentration of inulin in urine equals the concentration in plasma

3. Which of the following statements regarding inulin is *false* (assuming that the amount of inulin infused is always equal to the amount of inulin excreted)?
   A. The clearance increases as the GFR increases
   B. The amount of inulin excreted increases as the GFR increases
   C. The clearance of inulin increases as the plasma inulin concentration increases
   D. The amount of inulin filtered increases as the GFR increases
   E. At a constant GFR, the amount of inulin excreted increases as the plasma concentration of inulin increases
4. Which of the following statements regarding the renal handling of creatinine in the human is true? In a steady-state:
   A. The amount of creatinine produced is greater than the amount of creatinine excreted
   B. The amount of creatinine excreted is less than the amount of creatinine filtered
   C. The amount of creatinine produced equals the amount of creatinine excreted even when the GFR is decreased
   D. The plasma concentration of creatinine remains constant even when the GFR is decreased

E. The clearance of creatinine equals the RPF

5. Based on the following data, which statement is correct?

Oxygen consumption for the whole body = 250 ml/min

Mixed arterial oxygen concentration = 20 ml/dl of whole blood

Mixed venous oxygen concentration = 15 ml/dl of whole blood

RBF/cardiac output = 0.3

GFR/RPF = 0.2

Arterial hematocrit = 0.5

Arterial plasma Na^+ concentration = 140 mEq/ml

Urine Na^+ concentration = 280 mEq/ml

Urine flow rate = 0.75 ml/min

A. The RBF is 1,200 ml/min

B. The GFR is 120 ml/min

C. The fractional excretion of Na^+ is 2.0 percent

D. The fraction of the filtered Na^+ reabsorbed is 94 percent

E. The urine/plasma inulin concentration ratio is 200/1

ANNOTATED ANSWERS

1. D. Calculated using the clearance formula: C = (U)(V)/P.

2. C. When the amount of the substance filtered equals the amount of the substance excreted, the clearance of that substance equals the GFR.

3. C. The clearance of inulin is not affected by the plasma inulin concentration; because inulin clearance equals the GFR, only changes in the GFR alter inulin clearance.

4. C. As the GFR decreases, the plasma concentration of creatinine increases, but the amount of creatinine excreted remains constant and equal to the amount of creatinine produced.

5. E. This calculation is based on the Fick principle, the fraction of the cardiac output perfusing the kidneys and the clearance of inulin.

BIBLIOGRAPHY

Beeuwkes, R., III. The vascular organization of the kidney. *Annu. Rev. Physiol.* 42: 531–542, 1980.

Kritz, W., and Bankir, L. A standard nomenclature for structures of the kidney. *Am. J. Physiol.* 254: F1–F8, 1988.

33 Renal Hemodynamics and Glomerular Filtration

ROBERT O. BANKS

Objectives

After reading this chapter, you should be able to:

List and describe the forces involved in the formation of the glomerular ultrafiltrate

Describe the composition of the glomerular ultrafiltrate

List and describe the physiologic mechanisms involved in the regulation of the glomerular filtration rate and renal blood flow

Explain how autoregulation controls renal blood flow and the glomerular filtration rate and cite possible mechanisms

State the normal adult values for the glomerular filtration rate and renal blood flow in humans

List the major metabolic substrates and describe the unique nature of renal metabolism

Describe the renal handing of low-molecular-weight and high-molecular-weight proteins

As noted in Chapter 32, the kidneys represent only about 0.5 percent of the body weight, yet receive approximately 20 percent of the cardiac output. This relatively large blood flow per gram of tissue (about 4 ml min^{-1} gm^{-1}) is one of the highest blood flow values for any of the organs in the body and undoubtedly reflects the need to support the high filtering capacity of the kidneys rather than to provide for any unique nutrient requirements. The high flow is the result of a relatively low renal vascular resistance. **Renal vascular resistance** is a physiologic variable that is affected by changes in renal sympathetic nerve activity (the kidney is richly innervated); by the influence of a number of hormones, autacoids, and paracrine substances; and by changes in the renal arterial blood pressure (autoregulation of renal blood flow [RBF]).

The profile of the vascular blood pressure values in the renal circulation is illustrated in Figure 33-1. As noted in Chapter 32 and shown in this figure, the major resistance sites—the vascular regions with the largest decreases in pressure—are located in the afferent and efferent arterioles.

Compared with other capillary beds, the pressure in the glomerular capillary is high. This high pressure accounts for the **net ultrafiltration** pressure in the glomerulus, as will be discussed. In addition, the high vascular resistance of the efferent arteriole produces a relatively low pressure in the peritubular capillaries; this low capillary pressure in the postglomerular vessels fosters a force that favors the reabsorption of solutes and water from the renal interstitial fluid space back into the vascular space.

Approximately 90 to 95 percent of the total

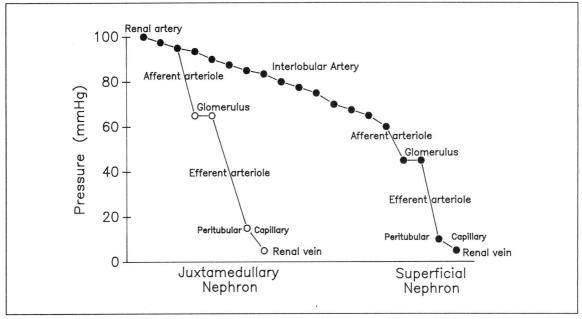

Fig. 33-1. Blood pressure values in arteries, arterioles, capillaries, and veins of the kidney. The largest decreases in pressure, reflecting the highest intrarenal vascular resistance, occur in the afferent and efferent arterioles.

postglomerular RBF perfuses the renal cortex. The remaining 5 to 10 percent perfuses the renal medulla via the vasa recta, but only a relatively small fraction of this (less than 1 percent) perfuses the inner medulla. The low inner medullary blood flow helps conserve the **medullary osmolar gradient,** a phenomenon that will be described in Chapter 35.

GLOMERULAR DYNAMICS

Structure

As is illustrated in Figure 33-2, the glomerular architecture is ideally suited to support its primary function—the formation of an ultrafiltrate of plasma. Each **glomerulus** is composed of 20 to 40 capillary loops with an estimated total surface area of as much as 15,000 cm²/100 gm. A high glomerular **ultrafiltration coefficient** (K_f, which is the product of the **hydraulic permeability** and the **filtration surface area**), in con-

junction with the favorable net filtration pressure that exists between the glomerular capillary and Bowman's space, results in the ultrafiltration of relatively large volumes of plasma (180 liters/day).

Within the glomerulus, **podocytes** with projected foot processes create *filtration slits* that are about 250 Å wide. These slits may contribute to the **perm-selectivity** of the glomerular unit. However, more recent evidence has demonstrated that **fixed negative charges** (probably afforded by the basement membrane) are important in determining the glomerular perm-selectivity. Filtration normally approaches zero for substances with the size and charge of albumin (molecular weight, 60,000; diameter, 36 Å). Indeed, the loss of negative charges in the glomerulus may account for the proteinurea often associated with glomerulopathies. Figure 33-3 shows how both the charge and size affect the filtration and subsequent excretion of different sizes of neutral and charged dextran molecules. For any given molecular radius, the fractional clearance of cationic

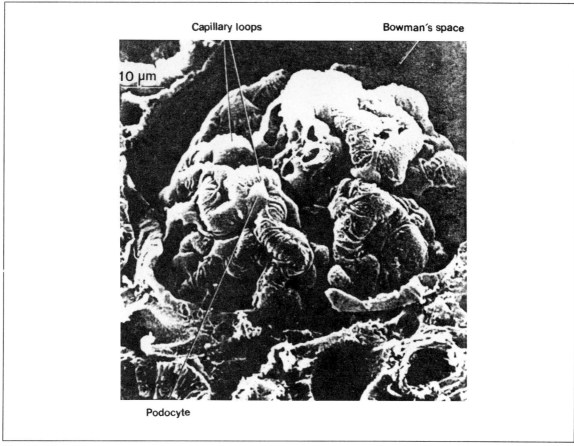

Fig. 33-2. A scanning electron photomicrograph of a glomerulus. (From Spinnelli, F., Wirz, H., and Brucher, C. Fine structure of the kidney revealed by scanning electron-microscopy. Basel: Ciba-Geigy, 1972.)

dextran molecules exceeds that of neutral molecules, but the fractional clearance of anionic molecules is less. By contrast, in rats with experimentally induced nephrotoxic serum nephritis, the fractional clearance of anionic dextran molecules markedly increases and resembles that of the neutral dextrans; the excretion of neutral dextran molecules is similar in normal and experimental rats.

Physical Forces

As noted in Chapter 32, **glomerular filtration** is determined by the balance of **Starling forces** that exist between the capillary lumen and Bowman's space. Accordingly:

$$GFR = K_f[(HP_{cap} + COP_{BS}) - (HP_{BS} + COP_{cap})]$$

Eq. 33-1

Where K_f is the ultrafiltration coefficient, equal to the capillary surface area times the hydraulic (water) permeability coefficient, HP_{cap} is the mean (average) hydrostatic pressure along the length of the capillary, COP_{cap} is the mean (average) colloid osmotic pressure along the length of the glomerular capillary, and HP_{BS} and COP_{BS} are the hydrostatic and colloid osmotic pressures in

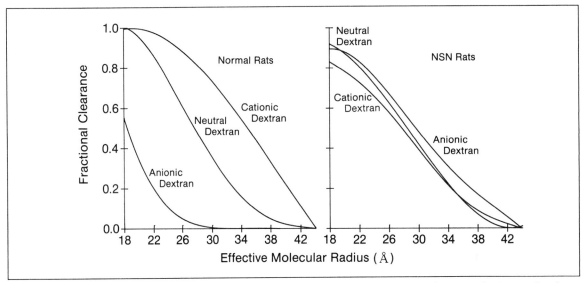

Fig. 33-3. The fractional clearance ($C_{dextran}/C_{inulin}$) of different-sized neutral, cationic, and anionic dextran molecules in normal rats and in rats with experimentally induced nephrotoxic serum nephritis (*NSN*). (Modified from: Bohrer, M. P., et al. Permselectivity of the glomerular capillary wall. *J. Clin. Invest.* 61:72–78, 1978).

Bowman's space, respectively (the term *hydro-static pressure* as used in these equations actually refers to the total pressure, which is the sum of the hydraulic plus the hydrostatic pressure.)

Because the **colloid osmotic pressure** in Bowman's space is normally zero, the above equation reduces to:

$$GFR = K_f [(HP_{cap} - HP_{BS}) - COP_{cap}] \qquad \text{Eq. 33-2}$$
$$\text{or } GFR = K_f (\Delta P - COP_{cap}) \qquad \text{Eq. 33-3}$$

Where $\Delta P = HP_{cap} - HP_{BS}$.

The **hydrostatic** and **colloid osmotic pressure** in afferent and efferent arterioles have been measured directly in some species of animals, such as the Munich-Wistar rat and the squirrel monkey, that have glomeruli on the surface of the kidney. Because most animals have glomeruli that originate below the surface of the kidney, afferent and efferent vessels cannot be micropunctured. A typical profile for the hydrostatic pressure difference ($HP_{cap} - HP_{BS}$) and the colloid

osmotic pressure along the length of the glomerular capillary in the Munich-Wistar rat is shown in Figure 33-4A. Since both the hydrostatic and colloid osmotic pressure change along the length of the capillary, the ultrafiltration pressure is constantly decreasing as it approaches the efferent arteriolar end of the capillary. As is illustrated in Figure 33-4A, the **net ultrafiltration pressure** is delineated by the area between the curves for the net hydrostatic pressure and the colloid osmotic curves along the length of the glomerular capillary. Thus Equation 33-3 reduces to:

$$GFR = K_f (PUF) \qquad \text{Eq. 33-4}$$

Glomeruli characterized by data similar to those illustrated in Figure 33-4A are said to be in **filtration pressure equilibrium,** whereby the net ultrafiltration pressure is zero at the end of the capillary. Glomeruli that possess this equilibrium have a net ultrafiltration pressure that is highly dependent on the rate of plasma flow through the glomerulus, in that increases in RPF augment the net ultrafiltration pressure.

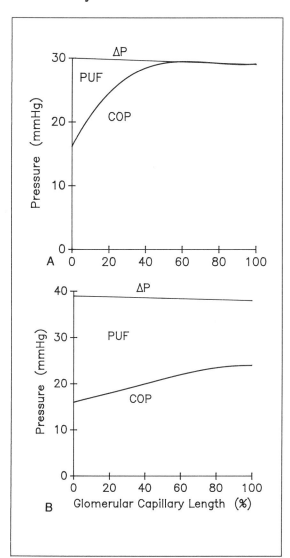

Fig. 33-4. Typical pressure profiles for ΔP and COP along the length of a glomerular capillary of rat (A) and dog (B). (ΔP is the difference in hydrostatic pressure between the capillary and Bowman's space and COP is the colloid osmotic pressure in plasma.) These data depict examples of filtration-pressure equilibrium (A) and disequilibrium (B). Average hydrostatic pressure values in Bowman's space are 11 mm Hg in the rat and 21 mm Hg for the dog. It should be noted that not all investigators have observed filtration-pressure equilibrium in the rat.

By contrast, in glomeruli exhibiting **filtration-pressure disequilibrium,** whereby there is a net ultrafiltration pressure remaining at the end of the glomerular capillary, this pressure is less sensitive to increases in RPF. Indirect evidence has suggested that the glomeruli in the dog (and probably human) kidney are characterized by filtration-pressure disequilibrium. The estimated profiles for glomerular capillary pressures in the dog are illustrated in Figure 33-4B.

Values for K_f in the renal vascular bed are one to two orders of magnitude greater than those in other vascular beds, such as in skeletal muscle. A number of factors may affect the K_f, and therefore potentially influence the GFR. As was the case for RPF's effect on the GFR, the status of filtration-pressure equilibrium versus disequilibrium also determines whether changes in K_f alter the GFR. Specifically, under conditions of *disequilibrium,* increases in K_f augment the GFR but changes in K_f during filtration-pressure *equilibrium* do not markedly affect the GFR.

REGULATION OF RENAL BLOOD FLOW AND THE GLOMERULAR FILTRATION RATE

Changes in either the afferent or efferent arteriolar resistance, or both, affect renal hemodynamics and, if there are resultant changes in glomerular capillary pressure, also affect the GFR. Because the glomerulus is a **portal system,** the potential for varying glomerular capillary pressure is great. As discussed previously, depending on whether filtration-pressure dysequilibrium or equilibrium dominates, changes in K_f or RPF will also affect the filtration rate. Finally, as illustrated in Equation 33-1, the plasma colloid osmotic pressure is one of the determinants of the GFR and any changes will alter the filtration rate.

Hormonal Control

Many endogenous and exogenous substances affect RBF and the GFR, and some of these are listed in Table 33-1.

Administration of vasoconstrictors or activation

Table 33-1. Some Vasoactive Substances and
Conditions that Affect Renal Blood Flow

Vasoconstrictors	Vasodilators
Norepinephrine (alpha receptors)	Acetylcholine and other muscarinic agonists
Epinephrine	Bradykinin
Angiotensin-II	Prostaglandins E_2 and I_2
Endothelin	Histamine (H_1 and H_2 receptors)
Vasopressin (V_1 receptors)	Isoproterenol
Serotonin	Dopamine (D_2 receptors)
	Papaverine
	Hyperosmotic solutions (e.g., mannitol)
	Ca^{2+}-channel antagonists
	Adenosine (also induces a transient constriction)
	Sodium nitroprusside
or conditions such as:	*or conditions such as:*
Hemorrhage (mediated by hormones and increased renal nerve activity)	Expansion of the extracellular fluid (mediated by hormones and decreases in renal nerve activity)
Elevated ureteral pressure (chronic)	Elevated ureteral pressure (acute) (may be mediated by histamine and prostaglandins)
Strenuous exercise (mediated by hormones and increases in renal nerve activity)	

of the sympathetic nervous system reduces the
GFR. By contrast, most vasodilators cause little
or no change in the GFR. This has been attributed to secondary factors that operate during infusions of the vasodilators, such as decreased
efferent arteriolar resistance with a resultant increase in tubular pressure, and to agonist-related
decreases in K_f.

As discussed in Section V on "Cardiovascular Physiology," a number of **vasodilators** exert
their actions by means of an endothelium-dependent series of events. Although it is more
difficult to assess the potential role of the
endothelium at the organ level, it is likely that
the endothelium dependence of certain vasodilators such as acetylcholine, histamine, and prostaglandins, as well as independence of such
agents as nitric oxide and nitroprusside, also applies to vasodilator-induced changes in RBF.

Some of the **vasoactive agents** listed in Table
33-1 also exert direct tubular actions. Thus, for
example, angiotensin-II, which is one of the most
potent endogenous biologic vasoconstrictors,
also stimulates proximal tubular Na^+ reabsorption. The **renin-angiotensin-aldosterone** system is important in the control of both vascular
reactivity and volume (see Chapter 36).

Another important factor that increases RBF
and GFR is **protein intake**. Glomerular hyperfiltration and elevated RBF are deleterious, causing glomerular sclerosis, and are of particular
concern when there is reduced renal mass such
as exists in various renal diseases. Dietary protein restriction can, under some conditions, offset the progressive loss of glomerular function,
and some have advocated this to be an important
therapeutic measure in treating renal disease.

Sympathetic Nervous System

Adrenergic postganglionic fibers innervate
both vascular and tubular elements in the kidney, including the afferent and efferent arteriole,
the proximal tubule, and the loop of Henle. The
functional importance of the renal nerves has
been the focus of intense investigation for many
years. Nonetheless, the role of renal nerves in normal homeostatic functions of the kidney has not
been fully elucidated. On the other hand, a number of conditions are associated with activation
or inhibition of renal nerve activity, and consequently markedly affect renal function. One of
these conditions is **hemorrhage**. The loss of vascular volume leads to a decrease in the mean ar-

terial blood pressure and subsequent activation of the carotid sinus and aortic arch reflexes. Heightened renal sympathetic nerve activity activates alpha-adrenergic receptors in afferent and efferent vessels, thereby increasing renal vascular resistance. Both RBF and the GFR decrease in this pathologic condition.

AUTOREGULATION OF RENAL BLOOD FLOW AND GLOMERULAR FILTRATION RATE

One of the important physiologic variables that does not normally influence renal hemodynamics or the glomerular filtration rate is **renal arterial blood pressure** (RAP). It should be noted that the experimental manipulation used in studies that evaluate the effects of perfusion pressure on renal hemodynamics only affects RAP. This maneuver is generally accomplished by placing an adjustable clamp on the renal artery of an experimental animal. Under these conditions, the mean arterial blood pressure (aortic blood pressure) is constant. As discussed previously, changes in arterial blood pressure activate the sympathetic nervous system, and consequently affect RBF and the GFR. Changing only the RAP avoids changes in the systemic arterial blood pressure and secondary changes in the renal hemodynamics.

Using the above methodology, changing RAP from approximately 180 to 80 mm Hg causes relatively small changes in RBF and GFR. Typical autoregulatory profiles for RBF and GFR are illustrated in Figure 33-5. As the RAP is reduced, because flow is constant, there must be a decrease in renal arterial resistance, as illustrated by Ohm's law. Current data indicate that the **afferent arteriole** is the primary site of pressure-induced changes in renal vascular resistance, but under some conditions the efferent arteriole may also be involved. Thus, as RAP is lowered, there is a relatively large decrease in the afferent arteriolar resistance. Under some conditions, particularly those associated with **high plasma renin concentrations,** such as low dietary Na+ intake, reductions in RAP may also be associated with small, angiotensin-II–mediated increases in efferent arteriolar resistance. As a consequence of afferent, and possibly efferent, arteriolar resistance changes within the autoregulatory range, glomerular capillary pressure remains relatively constant.

Autoregulation can be demonstrated in the isolated, perfused kidney, proving that renal nerves are not required for this phenomenon to occur. There are two major theories that have been advanced to account for autoregulation, and either one or a combination of both could be involved. The first is the **myogenic theory,** which proposes that changes in RAP directly affect smooth muscle activity. For example, a decrease in RAP leads to a decrease in stretch of the afferent arteriole causing the smooth muscle to relax and thereby decreasing vascular resistance. The second theory is the **distal tubular feedback theory,** or the so-called tubuloglomerular feedback proposal. It is well established that, if the composition or delivery rate of a tubular solute entering the distal nephron is changed, this is detected by some mechanism which in turn changes the resistance of preglomerular elements (the afferent arteriole). If the flow entering the distal nephron is reduced, this causes a decrease in afferent arteriolar resistance, an increase in RBF, and an increase in the net ultrafiltration pressure. The sensor mechanism in the distal nephron appears to be the **macula densa** but the nature of the triggering signal (changes in concentration of Na+ or Cl− or total solute) in the tubular lumen remains unknown. Evidence does not favor a major role for neural factors, adrenergic receptors, or the prostaglandin system. It has been reported that histamine H_1 receptor antagonists block RBF and GFR autoregulation, suggesting that histamine may contribute to this phenomenon. In addition, the increase in efferent arteriolar resistance that may occur when RAP is reduced in Na+-depleted animals appears to be related to activation of the renin-angiotensin system.

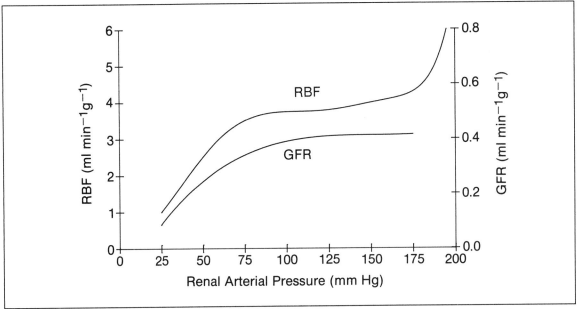

Fig. 33-5. Typical renal blood flow (*RBF*) and glomerular filtration rate (*GFR*) autoregulatory curves in the dog.

COMPOSITION OF THE GLOMERULAR FILTRATE

The ultrafiltrate of plasma that is formed in the glomerulus is about 94 percent water and 6 percent solute and is virtually protein-free. **Diffusable nonelectrolytes** (for example, glucose and urea) have the same concentration in filtrate as in plasma. The **Gibbs-Donnan effects** of charged, impermeant proteins produce small differences in the concentrations of filtered ions between plasma and ultrafiltrate. For example, the concentration ratio of monovalent anions in the filtrate to plasma is about 1.05; it is about 0.95 for monovalent cations.

With regard to **proteins,** the concentration of albumin in plasma is about 40 gm/liter. The concentration of albumin in the early proximal tubule of the rat is about 15 *mg*/liter or less. More than 99 percent of these small amounts of filtered albumin is reabsorbed by the proximal tubule. Low-molecular-weight proteins (10,000 to 50,000 daltons) are filtered to varying degrees and are also reabsorbed in the proximal tubule.

In **disease states** that primarily interfere with glomerular function, large quantities of high-molecular-weight proteins are filtered and excreted (protein excretion may exceed 20 gm/day). However, because low-molecular-weight proteins are filtered to some extent but reabsorbed in the proximal tubule, tubular disorders are more likely to be associated with impaired protein reabsorption and a consequential excretion of larger quantities of low-molecular-weight proteins (a so-called tubular proteinurea).

Some high-molecular-weight proteins are excreted by healthy individuals under certain conditions. These include so-called **exercise proteinurea** and **postural proteinurea,** whereby more protein is excreted by subjects while upright than when recumbent.

METABOLISM

The metabolic cost of the work performed by the kidney is high. **Oxygen consumption** by the kidney per gram of tissue weight is higher than that for any other organ, primarily because of the large amount of Na$^+$ that must be reabsorbed (approximately 1 mole of O_2 per 28 Eq of Na$^+$ reabsorbed). Considering the very high blood flow per gram of kidney, the arteriovenous O_2 content difference is very low (approximately 1.7 vol/dl) and is the lowest for any organ.

There is a linear relationship between renal O_2 consumption and the GFR. The reason for this close correlation is that an increase in the GFR is associated with an increase in the filtered Na$^+$ load and a resultant increase in Na$^+$ reabsorption. Because Na$^+$ reabsorption is an energy-dependent process, O_2 consumption by the kidney also increases.

Another unique aspect of O_2 dynamics in the kidney relates to the status of **medullary oxygenation**. Because the **vasa recta** are organized in a countercurrent fashion (see Chapter 35 for details), O_2 diffuses down a partial pressure gradient from the descending into the ascending limb; a CO_2 gradient in the opposite direction reverses the net movement of that gas from the ascending into the descending limb of the vasa recta. These events result in a relatively low medullary partial pressure for O_2 coupled with a relatively high partial pressure for CO_2.

The primary substrates utilized by the kidney are **free fatty acids** (particularly palmitic acid), **citrate**, and **lactate**. The kidney consumes very little glucose, and is in fact a gluconeogenic organ.

SUMMARY

The kidney is a highly vascular organ, receiving approximately 20 percent of the cardiac output. Thus, under normal conditions, blood flow to the kidneys is about 1 liter/min. Because the hematocrit is about 40 percent, the renal plasma flow is roughly 600 ml/min. Of this, approximately 20 percent, or 120 ml/min, enters Bowman's space in the form of an ultrafiltrate of plasma, which is referred to as the GFR. This ultrafiltrate is formed through the difference in Starling forces between the lumen of the glomerular capillary and Bowman's space. A number of important physiologic variables affect the RBF and GFR. RBF is primarily determined by the vascular resistance in the afferent and efferent arterioles. Because the glomerular capillary is positioned between these two arteriolar units, the glomerular capillary pressure, and hence the GFR, is also affected by the relative balance of afferent and efferent arteriolar resistance. In addition, the GFR is influenced by changes in the colloid osmotic pressure and by changes in the K_f of the glomerular capillary unit. A number of hormones, autacoids, and neural factors influence these physiologic variables, and thereby alter the RBF and GFR. On the other hand, changes in RAP in the range of approximately 80 to 180 mm Hg, do not markedly alter RBF or GFR, a phenomenon known as autoregulation. The mechanism responsible for autoregulation is unknown but appears to involve both a myogenic component and one that depends on a distal tubule feedback mechanism. The ultrafiltrate normally contains all the small diffusible nonelectrolytes and electrolytes in approximately the same concentrations as in arterial plasma. Because most of the ultrafiltrate is eventually reabsorbed along the nephron by active transport events or by processes that depend on these energy-consuming systems, the metabolic cost to the kidney is relatively large.

NATIONAL BOARD TYPE QUESTIONS

Select the one most correct answer for each of the following

1. Which of the following is true with regard to renal metabolism?
   A. Glucose is the major metabolic substrate
   B. The arteriovenous O_2 content difference is large across the kidney and, in this regard, is similar to the coronary circulation
   C. Increases in the GFR generally lead to increased O_2 consumption
   D. O_2 consumption per gram of kidney is very low compared with that of the liver
   E. Anaerobic glycolysis predominates in the cortex

2. A glomerular capillary unit is characterized, in each of the following conditions, by filtration pressure equilibrium. Which of the following would increase the filtration rate?
   A. A decrease in K_f
   B. An increase in arterial colloid osmotic pressure
   C. A decrease in afferent arteriolar resistance
   D. An increase in tubular pressure
   E. A 10-mm Hg decrease in the mean arterial blood pressure

3. Renal vascular resistance:
   A. Is equal to $1/R_{afferent\ art.} + 1/R_{efferent\ art.}$
   B. Is relatively high compared with other organs
   C. Decreases during hypovolemia
   D. Is lower in hypertensive individuals than in normotensive individuals
   E. Decreases when renal perfusion pressure decreases from 120 to 80 mm Hg

4. The fluid in Bowman's space:
   A. Contains some protein, but primarily that with a molecular weight greater than 20,000 daltons
   B. Would show a fall in pressure if the tubular flow rate in the early distal nephron was increased
   C. Has a glucose concentration higher than that in efferent arteriolar plasma leaving the glomerulus
   D. Has a pressure that is inversely related to the mean arterial pressure
   E. Has a Na^+ concentration lower than that of plasma

5. The ratio of the clearance of anionic dextran molecules to the clearance of inulin ($C_{dextran}/C_{inulin}$) will:
   A. Decrease as the plasma inulin concentration increases
   B. Increase as the molecular radius of the anionic dextran molecule is increased
   C. Be higher compared to neutral dextran molecules of similar size
   D. Attain a value of zero at a lower molecular radius than would neutral dextran molecules of similar size

E. Decrease during conditions such as nephrotoxic serum nephritis

ANNOTATED ANSWERS

1. C. An increase in GFR is associated with an increase in the filtered load of Na^+, an increase in Na^+ reabsorption, and a corresponding increase in O_2 consumption.

2. C. A decrease in the afferent arteriolar resistance would increase the glomerular capillary pressure and also the RPF, both of which augment the ultrafiltration pressure (increases in RPF cause an increase in PUF when the capillary unit is in filtration-pressure equilibrium).

3. E. RBF displays autoregulation between mean renal perfusion pressures of approximately 80 and 180 mm Hg. Even if efferent arteriolar resistance increases, because RBF remains constant when renal perfusion pressure is reduced from 120 to 80 mm Hg, there must be a net decrease in the renal vascular resistance.

4. B. Increases in the distal tubular flow rate cause an increase in afferent arteriolar resistance and therefore reduce the GFR and tubular pressure in Bowman's space.

5. D. Compared with neutral dextran molecules, the excretion (and clearance) of similar-size anionic dextran molecules is less (see Fig. 33-3).

BIBLIOGRAPHY

Arendshorst, W. J., and Gottschalk, C. W. Glomerular ultrafiltration dynamics: euvolumia and plasma volume-expanded rats. *Am. J. Physiol.* 239:F171–186, 1980.

Brenner, B. M. Nephron adaptation to renal injury or ablation. *Am. J. Physiol.* 249:F324–F337, 1985.

Brenner, B. M., and Humes, H. D. Mechanisms of glomerular ultrafiltration. *N. Engl. J. Med.* 297:148–154, 1977.

Knox, F. G., et al. Regulation of glomerular filtration and proximal tubular reabsorption. *Circ. Res.* 36–37:I107–I118, 1975.

Navar, L. G. Renal autoregulation: perspectives from

whole and single nephron studies. *Am. J. Physiol.* 234:F357–F370, 1978.

Oken, D. E. Does the ultrafiltration coefficient play a role in regulating glomerular filtration in the rat? *Am. J. Physiol.* 256:F505–F515, 1989.

Pollack, D. M., and Banks, R. O. Perspectives on renal blood flow autoregulation. *Proc. Soc. Exp. Biol. Med.* 198: 800–805, 1991.

Silva, P. Renal fuel utilization, energy requirements, and function. *Kidney Int.* 32(suppl 22):S-9–S-14, 1987.

Wright, F. S., and Briggs, J. P. Feedback control of glomerular blood flow, pressure and filtration rate. *Physiol. Rev.* 59:958–1006, 1979.

34 Proximal Tubule Function

ROBERT O. BANKS

Objectives

After reading this chapter, you should be able to:

Distinguish between active and passive transport systems in the kidney

Describe both reabsorptive and secretory transport maximums with examples of each

Calculate the reabsorptive and secretory transport maximums, if given data

Explain renal threshold

Define the significance of a transport maximum value relative to the normal plasma concentration of a given substance

Characterize the renal handling of amino acids

Identify the major events that occur in the proximal tubule, including the nature and quantity of fluid reabsorption (H_2O, HCO_3^-, Cl^-, Na^+, K^+, glucose, amino acids, Ca^{2+}, HPO_4, and so on

Explain the osmolarity of tubular fluid in the proximal tubule

State the fraction of water and Na^+ reabsorbed in the proximal tubule

Describe the phenomenon of glomerular–tubular balance and the possible mechanisms involved

Interpret the physiologic significance of increases and decreases in tubular fluid to plasma ratios

---

The clearance of various substances such as creatinine (for estimating the glomerular filtration rate [GFR] and *p*-aminohippuric acid (PAH) (for estimating the renal plasma flow [RPF]) provides useful information concerning the function of the whole kidney. However, most clearance measurements offer little information about either the specific tubular mechanisms or the nephron segments involved in the renal handling of individual solutes. Thus, other methods must be used to elucidate the physiologic events that occur in specific regions of the nephron. Two major techniques are **micropuncture** and the **in vitro perfusion** of isolated tubular segments.

MICROPUNCTURE

The technique of micropuncture was developed in the 1920s. In it, nephrons are impaled with micropipettes and samples of tubular fluid are then withdrawn and analyzed (only nanoliter quantities are obtained, which therefore necessitate microanalytic procedures). There are several in vivo micropuncture approaches and these are:

1. Sampling of freely flowing tubular fluid, whereby a sample of tubular fluid is collected without interrupting tubular flow.
2. Occluding the nephron at the site of puncture

by injecting an oil droplet; the tubular fluid near the oil droplet can then be collected at a rate equal to the tubular flow rate.

3. Placing a large oil droplet in a single nephron; the oil droplet is then "split" by injecting a volume of fluid of choice into the droplet; after the small volume of fluid has equilibrated, it is then recollected and analyzed.

4. Placing an oil droplet in a nephron to occlude flow, followed by insertion of a perfusion pipette into the nephron just distal to oil and a collection pipette farther downstream; the tubular segment is then perfused with a fluid of choice.

One of the limitations of micropuncture is that the approach can, primarily, only be used on proximal and distal convoluted tubules of superficial nephrons, though some segments of medullary structures can be punctured in species that have very long papillae.

In Vitro Perfusion of Isolated Tubular Segments

To perform the in vitro perfusion of isolated segments, the segments of the nephron are microdissected, attached to perfusion and collection pipettes, and then suspended in an "extracellular" bathing fluid. Fluids of choice can be placed either in the medium or in the perfusate. The advantage of this approach is that many sections of the nephron, from both superficial and deep nephrons, can be studied. Virtually all major divisions of the nephron have, in fact, been evaluated using this technique.

PROXIMAL TUBULAR FUNCTION

Sodium Reabsorption

Approximately 60 to 70 percent of the filtered Na^+ is reabsorbed in the proximal tubule. These and other values for the fractional solute reabsorption in a given segment of the nephron are calculated from the **filtration rate of the single nephron** (the single-nephron glomerular filtration rate [SNGFR]), the tubular fluid to plasma concentra-

tion ratio (TF/P ratio) of the solute in question, the TF/P ratio of inulin at that same point, and the tubular fluid flow rate (V). The following analysis illustrates how this is derived.

If f is the fraction of a filtered solute, x, remaining at any point, y, along the nephron, because the amount of x at any point y equals $(TF_x)(V_{TF})$ and the amount of x filtered equals $(SNGFR)(P_x)$, then:

$$f = \frac{(TF_x)(V_{TF})}{(SNGFR)(P_x)} \qquad \text{Eq. 34-1}$$

or:

$$\frac{TF_x}{P_x} = f\left[\frac{SNGFR}{V_{TF}}\right] \qquad \text{Eq. 34-2}$$

From Equation 34-5, it will be seen that:

$$\frac{SNGRF}{V_{TF}} = \frac{TF_{inulin}}{P_{inulin}}$$

Therefore,

$$\frac{TF_x/P_x}{TF_{inulin}/P_{inulin}} = f \qquad \text{Eq. 34-3}$$

In other words, the TF/P ratio of a freely filtered solute divided by the TF/P inulin ratio equals the fraction of the initial filtered quantity remaining at a particular point along the nephron. The TF/P Na^+ ratio at all points along the proximal tubule is 1.0. By contrast, the TF/P ratio for inulin at the end of the proximal tubule is 3:1. Consequently, at the end of the proximal tubule:

$$\frac{TF_{Na}/P_{Na}}{TF_{inulin}/P_{inulin}} = \frac{1}{3}$$

Or one-third of the filtered Na^+ remains at this point.

The nature of Na^+ and other solute reabsorption in the proximal tubule is illustrated in Figure 34-1. **Na^+ reabsorption** accounts for a major

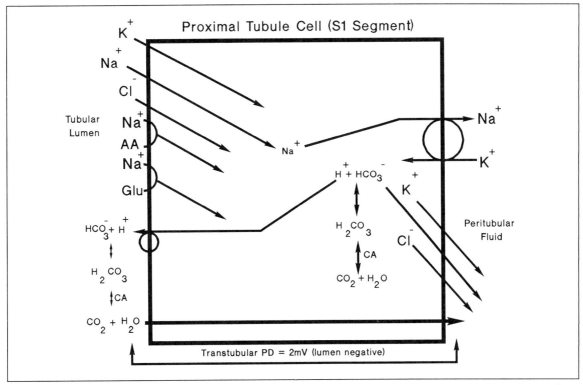

Fig. 34-1. Some of the major transport-related events in the S_1 segment of the proximal tubule. (*AA* = amino acids; *Glu* = glucose; *PD* = potential difference; *CA* = carbonic anhydrase.)

portion of the fluid reabsorbed in the proximal tubule (see the section "Volume Reabsorption"). As shown in Figure 34-1, the driving force for Na^+ reabsorption from the lumen of the tubule into the proximal tubular cell is generated by a favorable electrochemical gradient for Na^+. A low intracellular fluid concentration of Na^+ is maintained by the action of a **Na^+—K^+ ATPase pump** located on the basolateral surface of the proximal tubular cell. The entry of Na^+ into proximal tubular cells is accompanied by any one of a number of **secondary transport processes,** including Cl^-, amino acids, and glucose reabsorption as well as H^+ secretion.

Osmolarity of Proximal Tubule Fluid

An important aspect of proximal tubule function is that reabsorption occurs **isosmotically;** in other words, the tubular fluid to plasma osmolar ratio is 1.0 at all points along the proximal tubule. The fact that the TF/P osmolar ratio is unity in the proximal tubule shows that the water permeability of this segment of the nephron is very high. As will be discussed, current evidence indicates that the osmolar concentration of tubular fluid in the proximal tubule may be a few (3 to 5) milliosmoles per liter less than plasma's, indicating that there are "trivial" degrees of luminal hypotonicity in this segment.

Organic Solute Reabsorption

Most (more than 90 percent) of the filtered glucose, amino acids, and HCO_3^- is reabsorbed by **Na$^+$-dependent symport processes** in the early portions of the proximal tubule (see Figure 34-1). The reabsorption of **glucose** and **amino acids** takes place within the first 20 to 30 percent of the proximal tubule. Because these nonelectrolytes are reabsorbed with Na$^+$, the process is **electrogenic** and this generates a small, lumen-negative, transtubular potential averaging about 2 mV in this region of the nephron. The **reabsorption of HCO$_3^-$** is secondary to and depends on a Na$^+$–H$^+$ antiport process (see Figure 34-1). The reabsorption of HCO_3 occurs in the early portions of the proximal tubule and is a preferential process, compared with the **reabsorption of Cl$^-$**. Consequently, the tubular fluid Cl$^-$ concentration increases in the S$_l$ segment of the proximal tubule. The TF/P Cl$^-$ ratio for the last 75 percent of the proximal tubule is about 1.2:1. In addition, because the Cl$^-$ permeability of the latter portions of the proximal tubule is relatively high, a small lumen-positive, transtubular potential averaging 2 mV develops in the last 75 percent of the proximal tubule. Details of HCO_3^- reabsorption and of the factors that influence the reabsorption of this important anion are covered in Chapter 39.

Potassium and Calcium Reabsorption

Roughly 60 to 70 percent of the filtered K$^+$ and Ca^{2+} is reabsorbed in the proximal tubule by means of combined active and passive diffusion, including solvent drag-related events mediated by paracellular pathways.

Volume Reabsorption

Approximately 60 to 70 percent of the filtered volume load is reabsorbed by the time it reaches the end of the proximal tubule. In other words, if the GFR is 120 ml/min, 70 percent or 84 ml/min is returned to the circulation via the peritubular capillaries surrounding the proximal tubule. Consequently, the volume flow entering the loop of Henle is approximately 36 ml/min (120 − 84 ml/min).

Estimates of the **fractional volume reabsorption** in the proximal tubule are determined from micropuncture samples of tubular fluid taken from segments near the end of the proximal tubule. Specifically, comparing inulin concentrations in tubular fluid samples with those in plasma allows for calculation of fractional volume reabsorption. Because inulin is freely filtered, not reabsorbed or secreted, in the steady-state, the amount of inulin entering a single nephron must equal the amount of inulin leaving that single nephron and must equal the amount of inulin passing at any point along that nephron, or:

$$(SNGFR)(P_{inulin}) = (TF_{inulin})(V_{TF}) \qquad \text{Eq. 34-4}$$

Therefore,

$$\frac{P_{inulin}}{TF_{inulin}} = \frac{V_{TF}}{SNGFR} \qquad \text{Eq. 34-5}$$

The P/TF inulin ratio therefore represents the fraction of the initial filtered volume remaining in the nephron at that point.

The last accessible micropuncture site is about 60 to 70 percent of the total proximal tubule length. Tubular fluid samples obtained from this region have inulin concentrations that are approximately twice those of plasma, such that the TF/P inulin ratio is 2. Consequently, the projected value for TF/P inulin ratios at the end of the proximal tubule is 3:1. This means that one-third of the initial filtered volume remains and that two-thirds (hence 60 to 70 percent) was reabsorbed.

Glomerulotubular Balance

Fractional volume reabsorption in the proximal tubule is relatively constant and is independent of fluctuations in the GFR. This phenomenon is referred to as **glomerulotubular** (G-T) **balance** and serves to minimize large changes in the delivery rate of fluid to the loop of Henle and remaining portions of the distal nephron that would result from changes in the GFR. For example, under normal conditions, characterized by a GFR of 120 ml/min, 70 percent or 84 ml/min is reabsorbed in the proximal tubule, leaving 36 ml/min to enter the loop. If for some reason there was a 20 percent increase in the GFR and the proximal tubule reabsorbed a constant volume rather than a constant fraction, the following would occur: 144 − 84 ml/min = 60 ml/min entering the loop. Thus, a 20 percent increase in the GFR could lead to a 67 percent increase in the delivery rate of fluid to the loop (from 36 to 60 ml/min). However, in the presence of a constant fractional reabsorption of fluid (invoking the G-T balance), the following would obtain: (144 ml/min)(0.7) = 101 ml/min reabsorbed, or only 43 ml/min entering the loop. Consequently, G-T balance markedly attenuates changes in the loop delivery rate that arise during increases in the GFR.

The **mechanism of the G-T balance** is unknown but may stem from two factors: (1) increases in **Starling reabsorptive forces** in the peritubular capillaries (this hypothesis depends on there being a greater increase in the GFR than in the RPF, such that the colloid osmotic pressure of postglomerular blood would increase and there would be a consequential increase in the reabsorptive force at the level of the tubule) or (2) increases in the **filtered load of organic solute** (glucose and amino acids) that would in turn be reabsorbed in the early portions of the proximal tubule and therefore promote the reabsorption of water.

POSSIBLE MECHANISMS OF FLUID REABSORPTION IN THE PROXIMAL TUBULE

Standing Osmolar Gradient

For many years, fluid reabsorption in the proximal tubule was thought to eventuate from a "standing" osmolar gradient in the lateral intercellular space of proximal tubular cells, and the highest sodium chloride concentration was thought to exist near the tight junction. This osmolar gradient in the lateral intercellular fluid space, coupled with the high water permeability of the proximal tubule, was thought to generate the osmotic force for water reabsorption. However, some recent investigators have questioned whether this model adequately accounts for the water reabsorption by the proximal tubule. Based on experimental data and theoretical considerations, the mechanisms of **luminal hypotonicity** and **anion asymmetry** were formulated to explain water reabsorption in the proximal tubule.

Luminal Hypotonicity

The theory of luminal hypotonicity is based on the fact that the water permeability of the proximal tubule is relatively high. Thus, when the osmolality differences between the tubular lumen and renal interstitial fluid space are comparatively small, theoretically, large volumes of water could be reabsorbed. The theory derives from the concept that the **active transport of Na⁺** across the basolateral membrane (via the Na⁺−K⁺ ATPase pump) and the resultant **diffusion gradient** for the entry of Na⁺ (cotransported with other solutes such as HCO_3^-, Cl^-, and organic solute) from the tubular lumen into the cell generates tubular fluid that is slightly hypotonic compared to plasma; the osmolality difference may only be 3 to 5 mOsm/kg of H_2O. Coupled with the high water permeability of the proximal tubule, this **"trivial" osmolar difference** may be sufficient to account for the relatively large volume of water reabsorbed in this segment of the nephron. A difference of 3 to 5 mOsm/kg of H_2O between the tubular fluid and plasma is within the statistical variation of osmolality measurements and is in accord with the previous statement that the osmolar TF/P ratio is always 1.0 along the proximal tubule.

Anion asymmetry

As previously noted, the preferential reabsorption of HCO_3^- in the early portions of the proximal tubule augments luminal Cl^- concentrations, whereby there is an axial decrease in the HCO_3^- concentration and a corresponding increase in the Cl^- concentration. In addition, the later portions of the proximal tubule become appreciably more permeable to Cl^- than to HCO_3^-. Therefore, HCO_3^- in the peritubular fluid becomes an effective **osmotic reabsorptive force** in this region. A favorable concentration gradient for passive Cl^- reabsorption in this segment also causes Na^+ and water reabsorption throughout the later segments of the proximal tubule.

Finally, **Starling forces** in the peritubular capillary favor the net reabsorption of fluid from the interstitial fluid space; blood in the peritubular capillary is postglomerular, and therefore has a low hydrostatic pressure and high colloid osmotic pressure.

REABSORPTIVE AND SECRETORY TRANSPORT MAXIMUMS

Many substances are reabsorbed or secreted by active transport systems. **Active transport events** are typically characterized by high-energy dependency and saturation kinetics, including both competitive and noncompetitive inhibition.

Reabsorptive Transport Maximums

As already noted, glucose is rapidly removed from the early portions of the proximal tubule (within the first 20 to 30 percent of the proximal tubule the glucose concentration falls to virtually zero). The Na^+-dependent reabsorptive process for glucose exhibits both competitive inhibition (it displays a reabsorptive transport maximum $[T_m]$) and noncompetitive inhibition with agents such as the **glucosides,** an example of which is phlorhizin. The **renal handling of glucose** can be partitioned as follows: (1) amount of glucose filtered per unit of time; (2) amount of glucose reabsorbed per unit of time; and (3) amount of glucose excreted per unit of time. Moreover, the amount of glucose excreted $(U_{glu})(V)$ equals the difference between the amount of glucose filtered and the amount of glucose reabsorbed, or:

$$(U_{glu})(V) = (GFR)(P_{Glu}) - \text{ amount of glucose reabsorbed}$$

or:

$$\text{Amount of glucose reabsorbed} = (GFR)(P_{Glu}) - (U_{Glu})(V)$$

If the **plasma glucose concentration** (P_{Glu}) varies over a wide range, a **glucose "titration" curve** can be plotted, similar to that shown in Figure 34-2. Several facts are illustrated in this figure. **First**, as the plasma glucose concentration increases, the filtered load of glucose (the product of the plasma glucose concentration and the GFR) increases in a linear fashion; this assumes that the GFR is constant, which, in practice, is difficult to achieve because increases in the plasma glucose concentration are usually associated with changes in the GFR. **Second,** at low plasma glucose concentrations, all the filtered glucose is reabsorbed, such that none is excreted. **Third,** the amount of glucose reabsorbed reaches a maximum value, the so-called **T_m for glucose** (T_{mG}). In the human, the T_{mG} is 300 to 400 mg/min.

The **renal threshold,** defined as the plasma glucose concentration at which glycosuria (glucose in the urine) is observed, is about 180 to 200 mg/dl. The renal threshold is considerably lower than that calculated because of the phenomenon of **splay**. Thus, given an average T_{mG} of 360 mg/min and an average GFR of 120 ml/min, the predicted renal threshold would be calculated as follows:

$$(GFR)(P_{Glu}) = T_{mG}$$
$$(120 \text{ ml/min})(P_{Glu}) = 360 \text{ mg/min}$$
$$P_{Glu} = 3 \text{ mg/ml, or } 300 \text{ mg/dl}$$

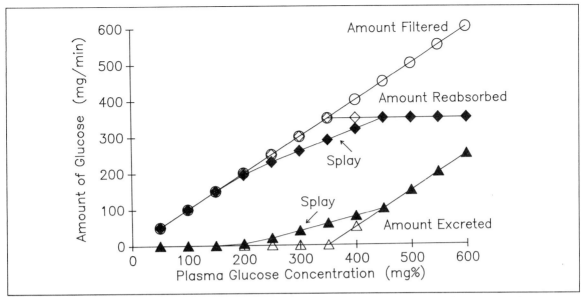

Fig. 34-2. Amounts of glucose filtered, reabsorbed, and excreted are shown as a function of the plasma glucose concentration. The transport maximum for glucose (T_{mG}) is given by the plateau value for the amount of glucose reabsorbed. Splay refers to the fact that glucose is excreted below the plasma value that would be predicted based on the glomerular filtration rate and the T_{mG}.

There could be a number of factors that influence splay, including a heterogeneity of T_{mG} values in different nephrons. Thus, some nephrons have lower T_{mG} values than others, and consequently the reabsorptive capacity is exceeded at relatively low plasma glucose concentrations.

The effect on the glucose clearance when the plasma glucose concentration is changed is illustrated in Figure 34-3. At low plasma glucose concentrations the glucose clearance is zero. However, once the renal threshold is exceeded, glucose is excreted and its clearance value can be calculated. As the plasma glucose concentration continues to increase, the clearance increases and approaches the GFR. In other words, at high plasma glucose concentrations, the amount of glucose excreted approaches the amount filtered.

Because the T_{mG} is high relative to the amount filtered, the kidney normally reabsorbs all of the filtered glucose. Thus, the kidney does not participate in regulating the plasma glucose concentration.

Glycosuria can be of **extrarenal origin,** as exemplified by diabetes mellitus. In this disorder, which is due either to a lack of sufficient pancreatic insulin secretion or to resistance to the hormone, the plasma glucose concentration increases, the amount of glucose filtered exceeds the T_{mG}, and the additional glucose is excreted. Glycosuria can also be of **renal origin,** so-called renal glycosuria. For example, glycosuria is associated with some forms of congenital diseases involving defects in the glucose reabsorptive mechanisms. **Fanconi's syndrome** is one such disorder. Besides glycosuria, features of Fanconi's syndrome generally include aminoaciduria, phosphaturia, and bicarbonaturia. Glycosuria is also seen in patients with heavy metal poisoning (lead induces a Fanconi-like syndrome) and when glycosuria is experimentally induced with drugs such as phlorhizin.

Other substances that have reabsorptive T_m's are amino acids, organic anions, phosphate, and sulfate.

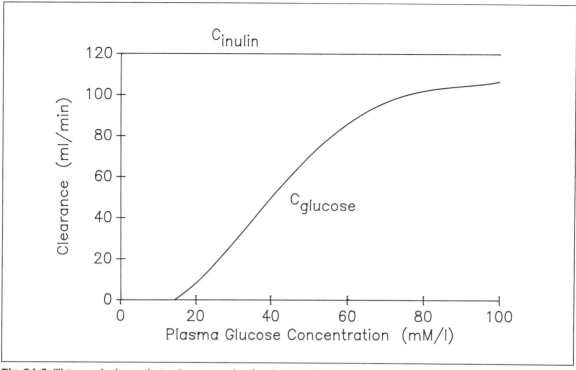

Fig. 34-3. This graph shows that a clearance value for glucose ($C_{glucose}$) can be obtained above the renal threshold and that the $C_{glucose}$ approaches the glomerular filtration rate at plasma glucose concentrations well above threshold. (C_{inulin} = clearance for inulin.)

Amino Acids

Amino acids are freely filtered but completely reabsorbed in the early portions of proximal tubules through the mediation of Na^+-dependent processes. Several independent or partially overlapping transport mechanisms exist for the following groups of amino acids: **acidic** amino acids—glutamic and aspartic acids; **iminoglycine** amino acids—glycine, proline, and hydroxyproline; **basic** amino acids—cysteine, lysine, arginine, and ornithine; and **neutral** amino acids—alanine, valine, leucine, isoleucine, methionine, serine, threonine, histidine, tyrosine, phenylalanine, and tryptophan.

The T_m for amino acids is high relative to the filtered load. Consequently, the kidney normally reabsorbs all the filtered amino acids and there-fore does not regulate their plasma concentrations.

Organic Anions

Among the **organic anions, T_m** values exist for citrate, acetoacetate, uric acid (undergoes both reabsorption and secretion), and proteins.

Phosphate

Phosphate exists in blood as a **lipid fraction** (8 mg/dl), an **ester-phosphate fraction** (1 mg/dl), and an **ionized inorganic phosphate fraction** (3 mg/dl or 1 mM/liter). In **plasma**, which contains only an **ionized fraction**, about 10 to 20 percent of the total phosphate content is bound to plasma proteins; it is the ionized phosphate in the plasma compartment that is stabilized by renal function.

The major site of **phosphate reabsorption** is in the proximal tubule. Its reabsorption is closely coupled to Na^+ reabsorption. Phosphate has a low T_m (0.1 mM min^{-1} 1.73 m^{-2}), such that relatively small changes in the plasma phosphate concentration produce large changes in the amount excreted. In addition, the T_m of phosphate is not fixed. One of the primary factors that affect phosphate reabsorption is **parathyroid hormone** (PTH). PTH inhibits phosphate reabsorption by accelerating the adenylate cyclase activity in the proximal convoluted tubule, the cortical ascending limb, the distal convoluted tubule, and the branching collecting tubule.

PTH also inhibits Ca^{2+} reabsorption in the proximal tubule, but in the distal nephron, it increases Ca^{2+} reabsorption with an attendant decrease in Ca^{2+} excretion. Finally, PTH also activates **renal 25-(OH) D_3-1$\propto$-hydroxylase,** the enzyme that converts 25-(OH)D_3 to 1,25-(OH)$_2D_3$, the biologically active form of vitamin D. The overall regulation and systemic effects of PTH are discussed in Chapter 52.

Sulfate

The normal plasma concentration of inorganic sulfate is 1 to 1.5 mM/liter. The reabsorptive T_m for sulfate is about the same as that for phosphate.

Secretory Transport Maximum

Many weak organic acids and bases (for a partial list see Table 32-1) are characterized by secretory T_ms. In the human, the T_m values for PAH, Diodrast, and phenol red average 80, 57, and 36 mg min^{-1} 1.73 m^{-2}, respectively. The rapid uptake of various organic acids by the kidney serves as the basis for their use in renal and urologic evaluations.

The **renal handling of organic acids** can be divided into three phases: (1) amount of organic acid filtered per unit of time; (2) amount of organic acid secreted per unit of time; and (3)

amount of organic acid excreted per unit of time. Using PAH as an example: the amount of PAH filtered plus the amount secreted equals the amount excreted.

If increasing amounts of PAH are infused into healthy subjects, a **PAH "titration" curve** can be obtained, similar to that in Figure 34-4. As this figure shows, given a constant GFR, the amount of PAH filtered increases linearly as the plasma PAH concentration increases. The total amount of PAH excreted exceeds the amount filtered by the amount secreted, but the amount secreted attains a maximum value, the T_m of PAH.

Increasing the plasma concentration of PAH (or of other weak organic acids and bases) reduces the clearance of PAH (and of the other weak acids or bases), and this is illustrated in Figure 34-5. As the plasma concentration of PAH increases, the clearance of PAH decreases and approaches the GFR. In other words, because the amount of PAH secreted at high plasma PAH concentrations is constant (it reaches the T_m value) whereas the amount of PAH filtered (and excreted) continues to increase as the plasma PAH concentration increases, the secreted component becomes an increasingly smaller fraction of the total PAH excreted. Thus, at high plasma PAH concentrations, the amount of PAH excreted approaches the amount filtered, and consequently the clearance of PAH approaches the GFR.

SUMMARY

The largest fraction (about 60 to 70 percent) of the filtered volume and of the filtered Na^+ is reabsorbed in the proximal tubule. In addition, most of the filtered organic solute (glucose, amino acids, and bicarbonate) is reabsorbed by Na^+-dependent, cotransport processes, primarily in the beginning regions of the proximal tubule. A portion of the Na^+ that is reabsorbed in the proximal tubule is exchanged for intracellular H^+, most which are reabsorbed as water during the hydrogenation of the filtered bicarbonate. The primary factor that initiates Na^+ reabsorption in the proximal tubule, and consequently the reabsorption of water and other solutes, is the passive entry of

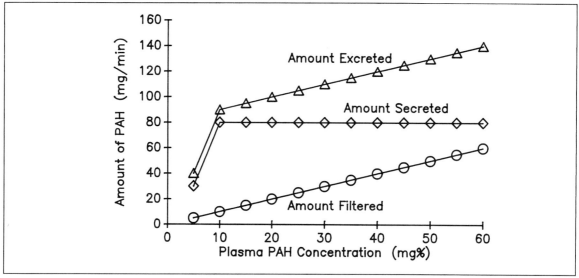

Fig. 34-4. Amounts of *p*-aminohippurate (*PAH*) filtered, secreted, and excreted are shown as a function of the plasma PAH concentration. The transport maximum for PAH represents the plateau value for the amount of the weak organic acid secreted.

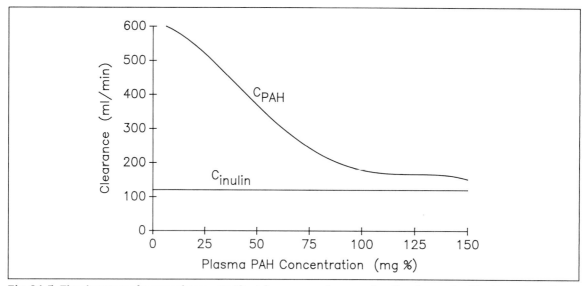

Fig. 34-5. The clearance of *p*-aminohippurate (C_{PAH}) decreases and approaches the glomerular filtration rate as the plasma PAH concentration increases. (C_{inulin} = clearance for inulin.)

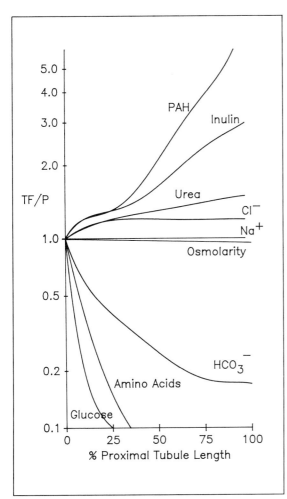

Fig. 34-6. Tubular fluid (*TF*) to plasma (*P*) concentration ratios for some of the major solutes are plotted as a function of distance along the proximal tubule. (*PAH* = *p*-aminohippurate.)

tively secreted into the proximal tubule and possess secretory T_m values.

A summary of the major events that take place in the proximal tubule is well illustrated when the TF/P ratios of different solutes are plotted as a function of the fractional length of the proximal tubule (Fig. 34-6). The tubular fluid and *not* the arterial concentration is the variable that changes as a function of distance along the tubule. An understanding of the data illustrated in Figure 34-6 should further clarify the major concepts underlying proximal tubule function.

NATIONAL BOARD TYPE QUESTIONS

Select the one most correct answer for each of the following

1. With regard to the renal handling of glucose, which of the following statements is true?
   A. The renal extraction of glucose at plasma glucose concentrations well above threshold approaches that of inulin
   B. The renal threshold for glucose could be calculated if one knew the T_{mG} and the GFR
   C. At high plasma glucose concentrations, the clearance of glucose approaches the RPF
   D. Glucose is usually excreted after a high carbohydrate meal
   E. The concentration of glucose in the efferent arteriole is higher than that in the afferent arteriole

2. Using the following data, which statement about proximal tubular function is true?

	Arterial Plasma Concentration	End Proximal Fluid Concentration
Na⁺(mEq/L)	140	140
Cl⁻(mEq/L)	100	120
Inulin (mg/ml)	10	40

   A. One-fourth of the filtered Na⁺ was reabsorbed

Na⁺ into the tubular cell. The Na⁺ gradient from the tubular lumen into the cell is generated by the Na⁺-K⁺ ATPase pump located on the basolateral surface of the tubular cell. Many of the filtered solute molecules, such as glucose, amino acids, sulfate, and phosphate, are reabsorbed by transport processes that have a T_m, and consequently display saturation kinetics. Similarly, a number of weak organic acids and bases are ac-

B. More than 75 percent of the filtered Cl⁻
was reabsorbed

C. 70 percent of the filtered water was reabsorbed

D. 70 percent of the filtered Cl⁻ was reabsorbed

E. An agent that inhibits fluid reabsorption
in the proximal fluid is likely present

3. In the early portions of the proximal tubule:
A. The TF/P ratio for urea is less than 1.0
B. The TF/P Na⁺ ratio is 1.2 to 1.3
C. The TF/P Cl⁻ ratio increases
D. The TF/P creatinine ratio is less than 1.0
E. The TF/P glucose ratio ranges from 1.2 to
1.3

4. A sample of tubular fluid at the end of a superficial proximal tubule would normally:
A. Have creatinine and chloride concentrations higher than those in plasma
B. Contain no glucose but have large
amounts of amino acids
C. Have a pH less than 6.0 and little HCO₃⁻
D. Contain urea at a concentration less than
that in plasma
E. Contain endogenous weak organic bases
such as choline, at concentrations slightly
less than those in plasma

5. The TF/P inulin ratio at a micropuncture site
near the end of a proximal tubule is 2.5:1. If
the single-nephron GFR is 50 nl/min, at the
point of micropuncture:
A. 30 nl/min would have been reabsorbed
B. The TF/P Na⁺ ratio would be 2.5:1
C. 50 percent of the initial filtered volume
would remain
D. The TF/P urea ratio would be 2.5:1
E. The TF/P creatinine ratio would be 2.0:1

ANNOTATED ANSWERS

1. A. Because the clearance of glucose approaches the GFR at high plasma glucose
concentrations, the renal extraction of glucose approaches the renal inulin extraction.

2. D. This calculation is based on the TF/P Cl⁻
ratio divided by the TF/P inulin ratio.

3. C. The TF/P Cl⁻ ratio increases from 1.0 to
1.2 in the early segments of the proximal
tubule (due to preferential organic solute
reabsorption with Na⁺) and remains elevated
in the S_2 and S_3 portions of the tubule.

4. A. Besides the increase in tubular fluid Cl⁻
concentrations (see Question 3), as fluid is
reabsorbed, the tubular fluid creatinine concentration increases.

5. A. This calculation is based on Equation 34-
4; the tubular flow rate is 20 nl/min, which
means that 30 nl/min must have been reabsorbed.

BIBLIOGRAPHY

Andreoli, T. E., and Shafer, J. A. Effective luminal hypotonicity: the driving force isotonic proximal tubular fluid absorption. *Am. J. Physiol.* 236:F89–96,
1979.

Diamond, J. M., and Bossert, W. H. Standing-gradient
osmotic flow. A mechanism for coupling water and
solute transport in epithelia. *J. Gen. Physiol.*
50:2061–2083, 1967.

Knox, F. G., and Haramati, A. Renal regulation of phosphate excretion. In: Seldin, D. S., and Giebisch, G.,
eds. *The Kidney.* New York: Raven Press, 1985.

Larson, T. S., et al. Renal handling of organic compounds. In: Massry, S. G., and Glassock, R. J. eds.
Textbook of Nephrology. Baltimore: Williams & Wilkins, 1989.

35 Urinary Concentration and Dilution: Loop of Henle Function

ROBERT O. BANKS

Objectives

After reading this chapter, you should be able to:

Describe the transport and permeability characteristics of the descending and ascending loop of Henle and the distal nephron as they pertain to the generation of the medullary osmolar gradient

Explain the role of the vasa recta in maintaining the osmolar gradient

Describe the roles of antidiuretic hormone and urea in the generation of the medullary osmolar gradient

---

The ingestion of solute and water varies widely from day to day, yet the osmolality of the body fluids remains relatively constant at approximately 280 mOsm/kg H_2O. Maintaining constant osmolality in the body fluid depends on the ability of the kidney to excrete urine with a wide range of osmolar concentrations. The human kidney can produce urine with osmolalities ranging from about 50 mOsm/kg H_2O (specific gravity, 1.002) to about 1,200 mOsm/kg H_2O (specific gravity, 1.032). Osmolality is often measured by **freezing point depression** (1 mOsm/kg H_2O decreases the freezing point by 1.86°C) or by **lowering of the vapor pressure**. **Specific gravity** is not a very precise evaluation of the total solute concentration because it varies with the composition of the solution (for example, the presence of proteins in a solution markedly affects the specific gravity determination but has little effect on the osmolality of the solution).

Events that take place in the renal medulla, by and large, account for the ability to excrete, on one hand, a urine that can be about one-sixth the osmolality of body fluids to one that is as much as four times as concentrated.

For many years it has been recognized that an **osmolar gradient** exists between the corticomedullary junction and the tip of the papilla. Tissue samples from the cortex have an osmolality of 280 mOsm/kg H_2O whereas the osmolality at the tip of the papilla is 1,200 mOsm/kg H_2O. This is illustrated in Figure 35-1, which plots the osmolar constituents in kidneys from dogs following 24 hours of water deprivation. As can be seen, most of the hypertonic fluid in the outer medulla is composed of NaCl whereas urea contributes significantly to the osmolar gradient in the inner medulla. Tissue samples obtained from the tip of the papilla display an osmolality of about 1,200 mOsm/kg H_2O, consisting of approximately 600

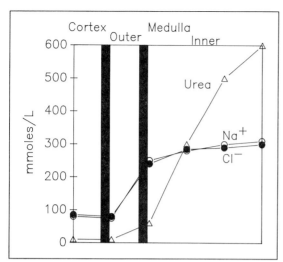

Fig. 35-1. The concentrations of Na^+, Cl^-, and urea are shown in the cortex, outer medulla, and inner medulla of dehydrated dogs. The osmolar gradient in the outer medulla is primarily generated by NaCl, whereas both NaCl and urea contribute to the gradient in the inner medulla. (Adapted from: Ullrick, K. J., and Jarausch, K. H. Untersuchungen zum Problem der Harkonzentrierung und Harnverdünnung *Pflugers Arch. ges. Physiol.* 262:537–550, 1956.)

mOsm/kg H_2O (300 mM) of NaCl and 600 mOsm/kg H_2O (600 mM) of urea.

THE COUNTERCURRENT MULTIPLICATION THEORY

The formulation of a countercurrent multiplication theory to explain the origin of the osmolar gradient in the renal medulla emerged during the 1940s and 1950s and evolved largely from the theoretical and experimental studies of a group of Swiss investigators, including Kuhn, Ryffel, Wirz, and Hargitay. The phenomena associated with fluid systems in which there is counter flow, with transfer of solute (or energy) between the two limbs, served as a foundation for the theory. An example of a countercurrent multiplication system is shown in Figure 35-2, which illustrates the mechanism by which **countercurrent flow** can generate a **thermal gradient** within a system.

Among the requirements for the generation of a gradient are: **adequate time** for temperature equilibration between the ascending and descending limbs, a **heat source,** and **adequate insulation** of the system. Application of these countercurrent principles to the kidney to explain the origin of the osmolar gradient is illustrated in Figure 35-3. **Active transport of solute** (sodium), in the absence of simultaneous water flow, from the thin and thick ascending limb of Henle's loop was hypothesized to be the initiating step for the osmolar gradient. **NaCl** was then thought to diffuse into the descending limb of Henle's loop and recycle into the medulla. The **vasa recta,** which also exhibit counter flow, would function as passive exchangers; **solute** reabsorbed in the inner medulla would move down its concentration gradient from the ascending into the descending vasa recta and minimize dissipation of the gradient. **Water** and **salt** would enter the vasa recta because of the favorable balance of Starling forces (high colloid osmotic pressure and low hydrostatic pressure) between the renal interstitium and the capillary lumen.

Since the initial formulation of the countercurrent multiplication theory to explain the medullary osmolar gradient, a number of subsequent experimental observations have been consistent with the model. First, the **water permeability** of the ascending limb of Henle's loop (both the thin and the thick segment) has been found to be virtually zero under all conditions, including those characterized by large changes in the plasma concentration of **antidiuretic hormone** (ADH), the major hormone affecting the water permeability of several portions of the collecting duct (see below and Chapter 37). Second, fluid at any level in the ascending limb is **hypotonic** relative to the renal interstitium at that level. Indeed, samples of tubular fluid from the early distal tubule are markedly hypotonic to plasma, with osmolalities in the range of 100 to 150 mOsm/kg H_2O. Third, **active solute reabsorption** in the thick ascending limb is well established. In contrast to other regions of the nephron, the transport process in the thick ascending limb is a **Na^+-Cl^-,**

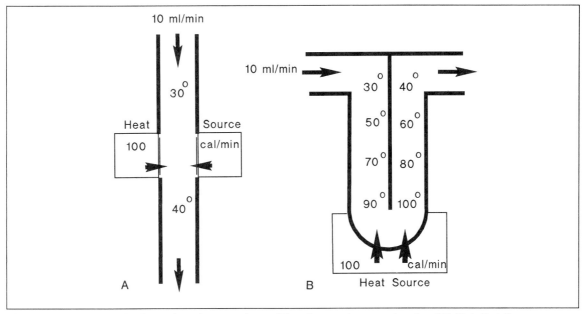

Fig. 35-2. Generation of a thermal gradient in a linear (A) and a countercurrent system (B). Note that the energy added to both the linear system and the countercurrent system is the same and that, in the steady-state, the temperature outflow from both systems is 40°C (during establishment of the thermal gradient in B, there would be a period when the temperature outflow would be less than 40°C).

K$^+$-Cl$^-$ cotransport event (see Fig. 35-4). This anion–cation transport accounts for the selective action of the so-called **loop diuretics** in this region of the nephron (see Chapter 37).

Although these experimental observations agree with the countercurrent multiplication theory, other findings cannot be accounted for by the model. First, **the tubular architecture** of the medulla is very complex, with a great deal of dissociation between the ascending and descending limbs of the loop of Henle. Second, the thin ascending limb of Henle's loop does not appear to **actively transport Na$^+$ or Cl$^-$**. Morphologic evaluations of this region have revealed thin epithelial cells with few mitochondria. Most physiologic studies have also failed to demonstrate active transport in this region, although some investigators have cited evidence of a modest amount of active transport. A **net solute extraction** along the entire ascending limb (thin and thick portion) is critical for the generation of an osmolar gradient within the entire medulla. Third, the

countercurrent multiplication model does not provide a **role for urea** in the generation of the gradient. It has been known for many years that protein deprivation compromises the ability to elaborate a concentrated urine, and this deficit can be rectified by administering urea to the experimental animal.

COUNTERCURRENT MULTIPLICATION WITHOUT ACTIVE TRANSPORT IN THE THIN ASCENDING LIMB

More recently, a countercurrent multiplication model with **passive solute reabsorption** in the thin ascending limb of Henle's loop has been proposed to explain the origin of the osmolar gradient. This theory is based on the transport and permeability properties of the descending limb of Henle's loop, the thin ascending limb, the thick ascending limb, the distal convoluted tubule, and

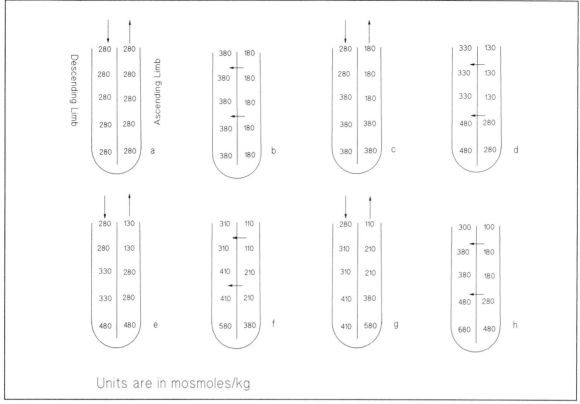

Fig. 35-3. Countercurrent multiplication in the loop of Henle. The model assumes that an osmolar difference of 200 mOsm/liter can be generated between the descending and ascending limbs when flow is stopped. (Adapted from: Pitts, R. F. *Physiology of the Kidney and Body Fluids,* 3rd ed. Chicago: Year Book Medical Publishers, 1974.)

the cortical, outer medullary, and inner medullary collecting ducts. To appreciate how this theory might explain the origin of the medullary osmolar gradient, the physiologic properties of each of these nephron segments will be summarized.

Major Permeability and Transport Events

Descending Limb of Henle's Loop

Most studies utilizing perfusion of isolated descending limb of Henle's loop segments from rabbits, rats, and hamsters have shown that this portion of nephron has (1) a **high H₂O permeability;** (2) a very **low solute permeability,** even

to urea; and (3) **no active transport**. Because of the low solute permeability of this nephron segment, equilibration between tubular fluid and the renal interstitium takes place by means of water extraction. Therefore, the concentration of NaCl, the major solute entering the loop (leaving the proximal tubule) increases to about 600 mm/liter at the turn (at the base of the loop). Of note, the fluid outside the nephron in the renal interstitium contains 600 mm urea and 300 mm NaCl.

There is some controversy over the solute permeability of the thin descending limb (point 2 above). Some investigators have presented evidence that there is significant entry of solute into the descending limb, which is not consistent

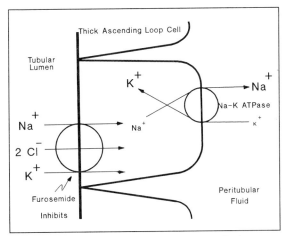

Fig. 35-4. A model illustrating the nature of solute transport in the thick ascending limb of Henle's loop.

with and, in fact, contradicts the countercurrent multiplication model which proposes passive reabsorption in the thin ascending limb.

Thin Ascending Limb

Experiments on isolated thin ascending limb segments have shown that this section of the nephron is impermeable to water, moderately permeable to urea, highly permeable to NaCl with evidence of a carrier-mediated, facilitated diffusion of Cl⁻, and incapable of active solute transport. As indicated in the previous section, there is evidence that a modest degree of **active transport** occurs in the thin ascending limb. Because of these permeability properties of this segment of the nephron and the fact that there is a **favorable NaCl gradient** between the tubular lumen (600 mm NaCl) and the renal interstitium (300 mm NaCl), NaCL passively diffuses into the renal interstitium. Because water cannot follow (this segment of the nephron is always impermeable to water), fluid inside the tubule becomes hypotonic relative to the renal interstitium.

Thick Ascending Limb

The two primary features of the thick ascending limb are that it is **impermeable to water** and it

exhibits a large amount of **solute transport**. As already noted, the cotransport of Na⁺–K⁺–2Cl⁻ predominates in this region of the nephron. The transtubular potential is lumen-positive relative to the plasma. This Na⁺–K⁺–2Cl⁻ cotransport can be inhibited by ethacrynic acid and furosemide, two of the so-called **loop diuretics**. Because the reabsorption of solute in this region of the nephron initiates the medullary osmolar gradient, if these agents are administered in doses sufficient to block all solute reabsorption in the thick ascending limb, the entire osmolar gradient in the renal medulla is abolished.

The physiologic variables of the remaining nephron segments are also important for an understanding of the countercurrent multiplication model with passive solute reabsorption in the thin ascending limb of Henle's loop, and are described in the following sections.

Distal Convoluted Tubule

The distal convoluted tubule is characterized by low water and urea permeability in both the presence and absence of ADH.

Cortical Collecting Duct

The cortical collecting duct has a low urea permeability in both the presence and absence of ADH. Conversely, the water permeability in this segment varies and depends on the presence or absence of ADH. Thus, there is a high water permeability when ADH is present but this region is impermeable to water when the hormone is lacking.

Outer Medullary Collecting Duct

The water and urea permeability properties of the outer medullary collecting duct are the same as those of the cortical collecting duct.

Inner Medullary Collecting Duct

The water permeability of the inner medullary collecting duct is also regulated by ADH. In contrast to other segments, however, this segment displays variable urea permeability; it is high in the presence of ADH and low in its absence.

MODEL FOR GENERATION OF THE OSMOLAR GRADIENT WITHOUT ACTIVE TRANSPORT IN THE THIN ASCENDING LIMB

Incorporating the facts given in the previous sections, the following analysis reflects some of the sequential phases postulated to occur during the generation of the osmolar gradient, and this is also illustrated in Figure 35-5. For purposes of this discussion, it is assumed that there is initially no osmolar gradient in the medulla and that there is a high plasma concentration of ADH.

The first stage in the process is the **cotrans-port of Na$^+$—K$^+$—2Cl$^-$** in the thick ascending limb, a phase that has been referred to as the **single effect mechanism**. Because the thick ascending limb of Henle's loop is not permeable to water, this transport process delivers a hypotonic fluid to the distal convoluted tubule. During this initial phase, the tubular fluid entering the distal tubule would be hypotonic relative to plasma, but not as low as that observed in the steady-state condition.

Because of the high titer of ADH in plasma, as the hypotonic tubular fluid progresses through the **cortical collecting duct**, a region of the nephron surrounded by an isosmotic (280 mOsm/kg H$_2$O) fluid, **tubular fluid equilibrates with the interstitium** through the extraction of water. However, since this segment of the nephron is always impermeable to urea, the concen-

Fig. 35-5. A model of countercurrent multiplication without active transport in the thin ascending limb of Henle's loop. (See text for details.)

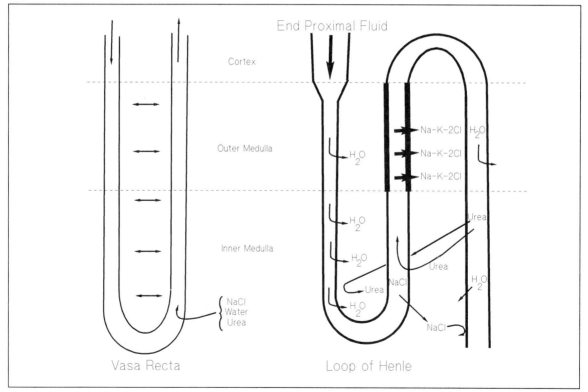

tration of urea in tubular fluid leaving this region is elevated. As the tubular fluid flows into the **outer medullary collecting ducts,** it enters a region of the nephron where it encounters **elevated osmolarity** surrounding the tubule; this is due to the solute extrusion from the thick ascending limb. Since the collecting duct in the outer medulla is also permeable to water when ADH is present, but not to urea, the tubular flow rate is further reduced and the concentration of urea further increased.

Tubular fluid then enters **the inner medulla,** a region where the collecting ducts exhibit an ADH-dependent permeability to both water and urea; consequently, urea diffuses into the medullary interstitium.

In the **descending limb of Henle's loop,** the elevated osmolarity in the renal interstitium of the outer medulla also prompts the passive extraction of water from this segment. As tubular fluid in the descending limb enters the inner medulla it encounters a region where there is an elevated urea concentration in the interstitium, and thus the reabsorption of water continues. Because the fluid at the **end of the proximal tubule** is primarily composed of NaCl, the reabsorption of water in the descending limb promotes delivery of a hypertonic NaCl solution to the turn of the loop. As this tubular fluid enters the **thin ascending limb,** which is permeable to salt but impermeable to water, NaCl passively diffuses out of the nephron and down a **NaCl concentration gradient**. Thus, there is **net solute loss** in the thin segment in the absence of active transport. A portion of the **urea** that is cycled into the medullary interstitium diffuses into the ascending limb and therefore is recycled through the distal nephron. This entire process continues until a **steady-state** is reached, wherein the total amount of solute and water entering the medulla equals the total amount leaving the medulla. However, as was the case with the thermal model depicted in Figure 35-2, a quantity of solute is trapped in the medulla with a resultant osmolar gradient directed from the outer medulla to the tip of the papilla.

Water enters the **medullary interstitium** by means of two sources—the descending limb of Henle's loop and the collecting duct. The effective **osmotic pressure** across the inner medullary collecting duct is a function of the difference in the NaCl concentration between the renal interstitium and fluid in the tubule; urea does not promote water reabsorption because, in the presence of ADH, the collecting duct is permeable to urea. Water entering the medullary interstitium is reabsorbed into the **vasa recta** because the Starling forces in these vascular loops favor net fluid reabsorption; blood in the vasa recta has traversed the glomerulus and the efferent arteriole, and consequently has an elevated colloid osmotic pressure and reduced hydrostatic pressure. **Solute** (primarily NaCl and urea) that enters the capillaries of the vasa recta with water subsequently diffuses down its concentration gradient from the ascending into the descending vascular limb, thereby minimizing solute loss from the medulla; the vasa recta thereby function as passive countercurrent exchangers.

While the **energy input** for generating the osmolar gradient is furnished by the $Na^+-K^+-2Cl^-$ transport in the outer medulla, the system is critically dependent on the subsequent events associated with the cycling of urea. The model explains why, for example, the inner medullary gradient decreases in the absence of ADH or during protein starvation. Both conditions would foster decreased cycling of urea in the system.

Finally, as already noted in this chapter, there is evidence that the passive movement of solute out of the thin ascending limb of Henle's loop cannot account, quantitatively, for the osmolar gradient in the inner medulla, even under the most optimal conditions with no urea entering the descending limb of Henle's loop. Thus, other events may also participate in the generation of the medullary osmolar gradient.

OTHER MAJOR REABSORPTIVE EVENTS IN THE LOOP OF HENLE

Approximately 20 percent of the **filtered Ca²⁺** is reabsorbed in Henle's loop through a combination of active (primarily in the thick ascending limb) and passive transport processes. Parathyroid hormones accentuate the active component of Ca^{2+} reabsorption in the thick ascending loop, whereas the loop diuretics (such as furosemide) decrease Ca^{2+} reabsorption in this region. About 20 percent of the **filtered K^+** is also reabsorbed in the loop through a combination of passive and active processes. Finally, virtually all the **ammonia** secreted in the proximal tubule is reabsorbed through a combination of active and passive transport processes in the loop (see also Chapter 39 for details).

SUMMARY

The ability to excrete urine with an osmolality ranging from 50 to 1,200 mOsm/kg H_2O depends on a number of factors, central to which is the presence of an osmolar gradient in the medulla. The osmolar gradient primarily reflects the reabsorption of $Na^+-K^+-2Cl^-$ in the thick ascending limb of Henle's loop and the subsequent transport-related and permeability-related events in the combined distal tubule and collecting duct system. The net effect of this process is that the concentration of urea in the collecting duct increases with a simultaneous cycling of urea into the inner medulla. An elevated urea concentration in the inner medulla fosters water reabsorption from the thin descending limb of Henle's loop and an increased NaCl concentration in the tubular lumen. As tubular fluid enters the thin ascending limb, the elevated NaCl concentration inside the nephron relative to the renal interstitium provides a favorable concentration gradient for NaCl reabsorption; this occurs in the absence of concomitant water flow because the entire ascending limb of Henle's loop is impermeable to water under all conditions. The vasa recta function as passive countercurrent exchangers.

NATIONAL BOARD TYPE QUESTIONS

Select the one most correct answer for each of the following

1. During dehydration:
   A. The volume flow out of the vasa recta would normally exceed the inward flow
   B. The inner medullary collecting ducts are impermeable to water
   C. The water permeability of the ascending limb of Henle's loop is relatively high due to the presence of ADH
   D. Recycling of urea in the medulla would be less than that during water diuresis
   E. Urine would be hypotonic relative to plasma

2. During water diuresis:
   A. The plasma ADH concentration is high
   B. The water permeability of the cortical diluting segment (the cortical thick ascending limb) is high
   C. The osmolar gradient in the inner medulla decreases
   D. The inulin concentration at the turn of a long loop of Henle is less than that leaving the loop
   E. The urea concentration in the inner medulla increases

3. Which of the following statements about the function of long loops of Henle is true?
   A. Fluid entering the loop (leaving the proximal tubule) is hypertonic
   B. Fluid at the turn is hypertonic compared with the interstitium at that level
   C. Water permeability of the descending limb is low
   D. Fluid leaving the loop (entering the early distal tubule) is hypotonic to plasma
   E. ADH modulates the water permeability of the ascending limb

4. In a long loop of Henle:
   A. Water is extracted from the ascending limb but solute is lost from the descending limb
   B. Water permeability is unaffected by ADH
   C. Only small amounts of filtered K^+ are reabsorbed (less than 5 percent)
   D. The creatinine concentration in the tubular fluid at the turn is about half that in fluid leaving the loop

E. NaCl reabsorption predominates in the descending limb

5. Under conditions of hydropenia (water deprivation), in the lumen of a descending limb of Henle's loop of a juxtamedullary nephron, the concentration of which of the following *decreases* toward the turn of the loop?

A. Na^+
B. K^+
C. Water
D. Urea
E. Ammonia

ANNOTATED ANSWERS

1. A. During dehydration, water is reabsorbed from the thin descending limb as well as from the collecting duct; this water is absorbed into the vasa recta, and is thereby returned to the general circulation.

2. C. The osmolar gradient in the medulla depends on urea recycling, a process that is suppressed by low ADH concentrations, as is the case during water diuresis.

3. D. Samples of tubular fluid in the early segments of the distal tubule are, under normal conditions, markedly hypotonic (100 to 150 mOsm/liter) to plasma.

4. B. ADH has no effect on the water permeability of the loop of Henle.

5. C. The descending limb is permeable to water but not to solute. Therefore, as the tubule descends into the hypertonic renal medulla, water is extracted from the tubular lumen.

BIBLIOGRAPHY

Andreoli, T. E., et al. Questions and replies: renal mechanisms for concentrating and diluting processes. *Am. J. Physiol.* 235:F1–F11, 1978.

Jamison, R. L. The renal concentrating mechanism. *Kidney Int.* 32(suppl 21):S-43–S-50, 1987.

Kokko, J., and Rector, F. C., Jr. Countercurrent multiplication system without active transport in inner medulla. *Kidney Int.* 2:214–223, 1972.

Moore, L. C., and Marsh, D. L. How descending limb of Henle's loop permeability affects hypertonic urine formation. *Am. J. Physiol.* 239:F57–F71, 1980.

Stephenson, J. L. Concentration of urine in a central core model of the renal counterflow system. *Kidney Int.* 2:85–94, 1972.

36 The Distal Nephron: Homeostatic Mechanisms

ROBERT O. BANKS

Objectives

After reading this chapter, you should be able to:

Describe the factors that are known to influence Na+ excretion, including the glomerular filtration rate, aldosterone, and third factors such as plasma oncotic pressure, renal arterial pressure, atrial natriuretic factor, and other hormones

Cite the regulatory mechanisms for aldosterone release, the site of aldosterone action, how much of the filtered Na+ load is controlled by aldosterone, and its mechanism of action

Describe the renin-angiotensin system and the factors that control renin release

Explain how Na+ is reabsorbed in the distal tubule

Describe the renal handling of K+, including the factors that affect its distribution between the extracellular and intracellular fluid spaces

One of the major functions of the kidney is to **regulate extracellular fluid (ECF) volume** and **osmolality**. As NaCl is the major osmotic constituent of the ECF, Na+ balance is a primary determinant of ECF volume. In other words, ECF volume regulation is essentially synonymous with Na+ regulation as illustrated by the existence of a **hygienic safety range** of Na+ intake of approximately 50 to 200 mEq/day. A very low salt intake is associated with **hypotension**, particularly orthostatic-induced hypotension, and an **inability to control for volume losses** that arise from sweating or injuries. By contrast, high salt intakes (usually exceeding 250 mEq/day) can precipitate Na+ and volume retention, resulting in **hypertension**. There are also individuals who have a genetic predisposition to salt retention; such people have a reduced ability to excrete salt, and develop hypertension at Na+ intake values that are within the normal range.

SODIUM HOMEOSTASIS

The total amount of exchangeable Na+ in the body is about 40 mEq/kg of body weight. As just noted, the amount of Na+ ingested each day varies widely from individual to individual. When there is very low salt intake (less than 50 mEq/day), the kidneys excrete urine that is virtually free of Na+. However, because the average person consumes about 150 mEq/day, to remain in Na+ balance, 150 mEq/day must be lost from the body. Unless large quantities of perspiration are formed, most of this ingested salt is excreted by the kidneys (about 5 percent is lost in the feces). Consequently, as discussed in Chapter 32, a representative value for Na^+ clearance would be:

$$C_{Na} = \frac{(U_{Na})(V)}{P_{Na}} = \frac{140 \text{ mEq/day}}{140 \text{ mEq/liter}} = 1 \text{ liter/day}$$

Where C refers to clearance, U to the urine concentration, V to the urine flow rate, and P to the arterial plasma concentration, all in relation to Na^+.

Given a daily glomerular filtration rate (GFR) of 180 liters/day, the filtered Na^+ load is yielded by: $(GFR)(P_{Na}) = (180 \text{ liters/day})(140 \text{ mEq/liter}) = 25,200 \text{ mEq/day}$.

Thus, as more than 25,000 mEq of Na^+ are filtered each day, but only 140 mEq/day are excreted, the **excreted component** represents less than 1 percent of the filtered Na^+ load and more than 99 percent is reabsorbed. In other words:

$$FE_{Na} = \frac{\text{amt. } Na^+ \text{ excreted } = (U_{Na})(V)}{\text{amt. } Na^+ \text{ filtered } = (GFR)(P_{Na})} \qquad \text{Eq. 36-1}$$

$$= \frac{140 \text{ mEq/day}}{25,200 \text{ mEq/day}} = 0.01$$

Where FE refers to the fractional excretion, in this case, of Na^+.

Another equation that illustrates the basis for the fractional excretion of any freely filtered solute, x, is:

$$FE_x = C_x/GFR \qquad \text{Eq. 36-2}$$

Approximately 70 percent of the **filtered Na^+ load** is reabsorbed in the proximal tubule, 20 percent in the loop of Henle, 5 percent in the distal convoluted tubule and cortical collecting duct, and 5 percent in the medullary collecting duct. **Aldosterone,** the primary mineralocorticoid secreted by the adrenal cortex, is one of the major hormones that regulate Na^+ excretion. Aldosterone primarily affects the fraction of the filtered Na^+ load reabsorbed in the cortical collecting duct region of the nephron. Thus, in the absence of aldosterone, roughly 5 percent of the filtered Na^+ load is excreted; if sufficient amounts of the hormone are present, more than 99 percent of the filtered Na^+ is reabsorbed.

The quantity of Na^+ regulated by aldosterone is substantial. Given a filtered Na^+ load of 25,000 mEq/day, 1,260 mEq/day (5 percent of the filtered Na^+ load) would be lost if the mineralocorticoid was lacking. Thus, in the absence of aldosterone, an intake of 1,260 mEq of Na^+ each day (representing more than 70 gm of NaCl) would be required to achieve Na^+ balance. A daily intake of that quantity of Na^+ is not possible, and, as the loss of large amounts of salt and water produces profound hypotension, untreated **hypoaldosteronism** is fatal. By contrast, patients with **hyperaldosteronism** retain salt and water, are usually hypertensive and hypokalemic, and suffer from metabolic alkalosis.

Na^+ Reabsorption in the Aldosterone-Sensitive Region of the Nephron

The paradigm for **solute reabsorption** and **secretion** in the aldosterone-sensitive region of the nephron, the connecting segment, and the cortical collecting duct is illustrated in Figure 36-1. A significant fraction of the Na^+ reabsorbed in this region is exchanged for intracellular K^+ and H^+. Recent data indicate that the secretion of K^+ and H^+ may actually take place in different cell types in this region of the nephron. Thus, in principal cells, Na^+ passively enters the intracellular fluid (ICF) compartment, creating a lumen-negative potential. This potential promotes both **Cl^- reabsorption** via a paracellular pathway and **K^+ secretion** across the luminal membrane. By contrast, the **H^+ secretion** occurs in intercalated cells by means of a H^+-ATPase pump situated on the luminal membrane, a process that is not coupled to Na^+ reabsorption (Na^+ is not reabsorbed by the intercalated cells). Both cell types are stimulated by aldosterone. H^+ secretion by intercalated cells is also stimulated by **acidemia**. Thus, not only does aldosterone stimulate Na^+ reabsorption but the hormone enhances the secretion and, therefore excretion, of K^+ and H^+ as well.

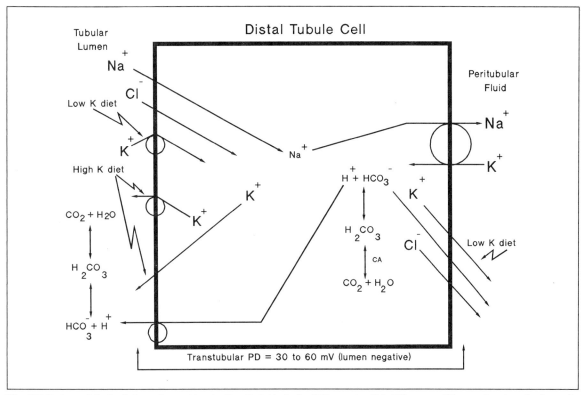

Fig. 36-1. A model of solute reabsorption in the distal tubule. (*PD* = potential difference; *CA* = carbonic anhydrase.)

Consequently, not only do **fluid volume imbalances** arise with both hyperaldosteronism and hypoaldosteronism, but there are also changes in acid-base and K^+ balance. **Hypoaldosteronism** is usually associated with a metabolic acidosis and hyperkalemia, whereas **hyperaldosteronism** is often characterized by metabolic alkalosis and hypokalemia. Finally, as shown in Figure 36-1, when K^+ is depleted, a net reabsorption of K^+ occurs in this region of the nephron; this reabsorptive process may be driven by a K^+-ATPase pump located on the luminal surface of intercalated cells.

Mechanism of Action of Aldosterone

When aldosterone is administered to an aldosterone-deficient animal, there is a delay of 30 to 90 minutes before changes in electrolyte excretion appear. The reason for this **delay** is that aldosterone, like other adrenal steroids, modifies the functional activity in its target tissue by changing the rate of a series of metabolic reactions. Aldosterone diffuses down its concentration gradient into cells of the cortical collecting tubule, where it is bound to a cytoplasmic or, more likely, a nuclear receptor protein. The steroid–receptor complex increases RNA transcription and ultimately results in the synthesis of new proteins. Currently it is unclear how these newly synthesized proteins alter cellular function. They may stimulate the peritubular membrane Na^+–K^+ ATPase pump by furnishing additional energy or may directly facilitate Na^+ entry into the cell by increasing the Na^+ permeability of the luminal membrane; recent studies have indicated that the latter mechanism predominates. Along with the increased Na^+ permeability of the luminal membrane, a process that can be blocked with the diuretic amiloride, there is an increase in the ICF Na^+ concentration and a secondary increase in Na^+–K^+ ATPase activity in the basolateral membrane.

If aldosterone is administered for several days to a healthy subject, the amount of Na^+ excreted initially decreases but then gradually returns to control values within 3 to 5 days, a phenomenon referred to as **aldosterone escape**. Interestingly, K^+ excretion does not escape, and therefore K^+ depletion develops if the mineralocorticoid infusion is continued. Evidence has been reported that, during aldosterone escape, the amount of Na^+ reabsorbed in the pars recta of the proximal tubule or in the loop of Henle is reduced, which offsets the enhanced reabsorption of Na^+ in the aldosterone-sensitive region of the nephron. The reduced reabsorption of Na^+ in the pars recta or loop of Henle may be related to changes in a number of other **non-aldosterone factors** that regulate Na^+ reabsorption. These factors would be activated by the aldosterone-induced expansion of the ECF space (see discussion of third factors at the end of this chapter).

Regulation of Aldosterone Secretion

The secretion of aldosterone from the **zona glomerulosa** in the **adrenal cortex** is regulated by several physiologic variables. The two primary factors are the plasma concentrations of **angiotensin-II** and of K^+; increases in either of these substances stimulate the secretion of aldosterone from the adrenal cortex. **Adrenocorticotropic hormone** exerts a permissive effect on the zona glomerulosa, in that relatively low levels of it are required for angiotensin-II and K^+ to impose their regulatory actions. Other hormones also affect aldosterone release. For example, **atrial natriuretic factor** (ANF), which will be discussed, inhibits and **endothelin** stimulates aldosterone secretion. Finally, there is evidence that **plasma osmolality** may also directly affect aldosterone secretion; increases in plasma osmolality are associated with decreases in aldosterone secretion.

The Renin-Angiotensin System

Angiotensin-II is the end product of a biologic cascade system initiated by the release of **renin,** a proteolytic enzyme produced by the **juxtaglomerular apparatus** in the kidney. Several factors, all related to arterial blood pressure, participate in the regulation of renin release from the kidney: (1) direct effects of **mean arterial blood pressure,** probably mediated by changes in wall tension in the afferent arteriole (a decrease in mean arterial blood pressure causes a decrease in wall tension of afferent arteriole and an increase in renin release); (2) altered **renal sympathetic nerve activity** (the juxtaglomerular apparatus is richly innervated, and increased sympathetic nerve activity, acting via beta-adrenergic receptors, causes an increase in renin release); and (3) changes related to the **GFR** (a decrease in the GFR prompts a decrease in the Na^+ delivery rate to the macula densa, which in turn is associated with increased renin release).

The renin substrate is a circulating alpha$_2$ globulin, **angiotensinogen** (produced in the liver), and the biochemical product is a decapeptide, **angiotensin-I**. A converting enzyme that exists in relatively large amounts in the lungs but is also found in the kidneys removes two C-terminal amino acids from angiotensin-I, forming the biologically active compound, angiotensin-II. Angiotensin-II is degraded to a biologically inactive peptide, angiotensin-III, by angiotensinase.

Angiotensin-II is one of the primary agonists for **aldosterone release,** though angiotensin-II has a number of other important biologic actions. It serves as a **potent vasoconstrictor,** which is central to its role in the long-term regulation of arterial blood pressure. Angiotensin-II also directly promotes **Na^+ reabsorption** in the proximal tubule. This action complements the effects of aldosterone on Na^+ reabsorption in the cortical collecting duct. In addition, angiotensin-II stimulates the *thirst center*.

Sodium Excretion and the GFR

In addition to the renin-angiotensin-aldosterone system, a second major factor that can influence the excretion rate of Na^+ is the GFR. Despite glomerulotubular balance, an elevated GFR invariably leads to increased Na^+ excretion. However, whether the GFR functions as a physiologic variable in the regulation of Na^+ excretion is unclear. As has been noted, a number of hormones affect the GFR, and therefore potentially regulate Na^+ excretion by means of this mechanism. Certainly, however, adjustments are made in the quantity of Na^+ reabsorbed by the nephron. Because less than 1 percent of the filtered Na^+ load is normally excreted, relatively small changes in the amount of Na^+ reabsorbed translate into relatively large changes in the amount of Na^+ lost or retained.

Third Factors

Numerous studies have suggested that there are many **non-aldosterone**, **non-GFR factors** involved in the regulation of Na^+ excretion. These regulatory events are often referred to as **third factors**. Descriptions of some of these factors, which may be interrelated and involve similar mechanisms, follow.

Atrial Natriuretic Factor

Extracts of the atrial, but not ventricular, myocardium contain a potent natriuretic factor. A family of biologically active peptides has been identified that derive from a common 152–amino acid precursor with at least 1 percent of the total messenger RNA activity specific for this factor. The circulating peptide appears to be a 28–amino acid peptide containing a disulfide ring. The factor is stored in so-called **atrial-specific granules** of atrial myocytes. Bolus injections of ANF elicit a very rapid and transient increase in the **urine flow rate** and **Na^+ excretion**. ANF also suppresses **cardiac output**, with a subsequent period of hypotension. Although ANF relaxes smooth muscle *in vitro*, it does not cause renal or mesenteric vasodilation (a transient, small increase in RBF has been reported by some investigators following pharmacologic doses of the peptide).

The mechanism by which ANF induces **natri-uresis** is still unknown. It has been shown that increases in Na^+ excretion produced by the peptide can be dissociated from changes in the GFR and RBF (high doses of ANF increase the GFR). Therefore, a direct tubular action of ANF has been proposed. It is of note that ANF does not inhibit $Na^+–K^+$ ATPase and, therefore, is not the elusive ouabain-like substance thought to be involved in the regulation of Na^+ reabsorption (to be discussed).

There is evidence that **atrial-stretch activation** of alpha$_1$ receptors and other events stimulate ANF release. Along these lines, ANF may participate in the phenomenon of **immersion diuresis and natriuresis**. Submersion in water is known to augment interstitial fluid pressure and plasma volume, with a resultant increase in atrial stretch and ANF release.

ANF receptors have been found in a number of tissues. In the kidney, these receptors reside in the **glomerulus** and **medullary collecting tubule**. There are also receptors in the **adrenal gland** (ANF inhibits aldosterone release) and the **anterior pituitary gland**.

ANF is thought to participate in the natriuretic response to acute changes in plasma or ECF volume. Long-term increases in Na^+ intake do not appear to alter ANF secretion. Other factors, such as a ouabain-like factor, could be involved in the renal response to elevations in dietary Na^+ intake.

Ouabain-like Factor

A ouabain-like hormone has been proposed to account for the inhibition of Na^+ reabsorption in the distal nephron (through decreases in $Na^+–K^+$ ATPase) observed during **chronic volume expansion**. The postulated hormone is often referred to as the *natriuretic* hormone. Extracts of plasma or urine from individuals or animals with expanded ECF volumes contain a natriuretic substance. Indeed, anephric patients have high concentrations of this substance. The factor is often distinguished by its ability to reduce Na^+ transport across isolated epithelial membranes (ANF does not affect Na^+ transport in isolated epithelia). Both a low-molecular-weight (1,000 daltons) and higher-molecular-weight (10,000 to 30,000 daltons) substance have been isolated. When the larger-molecular-weight substance is injected into test animals, natriuresis occurs within about one hour and lasts for more than two hours. By contrast, the low-molecular-weight substance evokes a rapid but short-lived natriuresis. Several reports indicate that this natriuretic factor originates in the brain.

Renal Nerves

The kidney is richly innervated with **sympathetic nerves,** and many physiologic and pathologic conditions have an impact on renal nerve activity. It has been known for many years that cutting the renal nerves produces a "denervation diuresis and natriuresis." Along these lines, there is evidence that **norepinephrine** stimulates Na$^+$ reabsorption in the proximal tubule and loop of Henle. Conversely, changes in the plasma concentration of hormonal and humoral agents can compensate for the renal nerves, as transplanted, denervated kidneys can function normally.

Pressure Natriuresis and Diuresis

Increases in **arterial pressure,** mediated by some unknown mechanism, provoke increased Na$^+$ excretion. There is evidence that pressure natriuresis is related to changes in proximal tubule function.

One group of investigators has suggested that **abnormal pressure natriuresis** is one of the major factors that contributes to the maintenance of hypertension. In other words, if renal function is normal, hypertension cannot be sustained, such that elevated pressure causes natriuresis and diuresis, with a resultant decrease in vascular volume and, consequently, blood pressure.

Physical Forces

Changes in **Starling forces** may be involved in the natriuresis observed in some settings. For example, diminished colloid osmotic pressure may decrease fluid reabsorption along the nephron. Such a mechanism may contribute to the reduced fluid reabsorption observed in the proximal tubule during saline expansion (fractional reabsorption can decrease from a control value of 70 percent to as low as 50 percent following expansion).

Changes in the distribution of blood flow or the GFR in the kidney have also been suggested to influence Na$^+$ excretion. One aspect of intrarenal hemodynamics currently being studied is the possibility that augmented medullary and papillary blood flow reduces the osmolar gradient and hence increases Na$^+$ excretion, but conclusive evidence that these events are important is lacking.

Other Hormones

There are a number of known hormones or humoral substances with natriuretic properties. These include **antidiuretic hormone** (ADH), **oxytocin, histamine,** and some **prostaglandins**. Whether the natriuretic effect of these substances is physiologic or pharmacologic remains to be determined.

Summary Scheme for Sodium Balance

The kidney varies the amount of Na$^+$ excreted to ensure a constant circulating blood volume. Thus, depending on the dietary Na$^+$ intake, Na$^+$ excretion can range from very low to relatively large amounts. The ingestion of salt heightens body fluid osmolality, and, because the primary volume of distribution of NaCl is in the ECF, the increase in osmolality shifts water from the ICF into the ECF. The increase in body fluid osmolality also stimulates the release of ADH from the posterior pituitary gland (see Chapter 37 for details), which in turn activates the thirst center in the hypothalamus. If water is ingested, it is retained until normal osmolality is achieved. This sets up an isotonically expanded ECF, a decrease in colloid osmotic pressure, and an increase in arterial or venous blood pressure, or both. Third factor events are therefore activated, including the pressure natriuretic phenomenon, the release of ANF, and perhaps the ouabain-like natriuretic hormone. In addition, the renin-angiotensin system is inhibited, with a resultant decrease in the aldosterone output from the adrenal glands. Na^+ excretion rate increases, water returns to the cells, and the ECF volume is reduced isosmotically by a proper plasma concentration of natriuretic factors and ADH.

The adjustments associated with changes in Na^+ intake are rather slow compared with those that occur during changes in water consumption. Several hours (more on a day-to-day scale) must elapse before Na^+-induced changes in ECF volume are manifested, but increases in water intake with accompanying decreases in body fluid osmolality result in changes in renal function that take only minutes to occur (see Chapter 37).

POTASSIUM HOMEOSTASIS

The **total body content** of K^+ is about 3,500 to 4,000 mEq (50 to 55 mEq/kg). Because the **normal ECF concentration** of K^+ is to 4 to 5 mEq/liter, only about 60 mEq of the total body K^+ content are in the ECF. High ICF concentrations of K^+ are necessary for many biochemical reactions to take place. In addition, the concentration ratio of K^+-ICF to K^+-ECF is critical for maintaining **resting membrane potentials,** and therefore the excitability of nerve and muscle cells.

Distribution of Potassium between the ECF and ICF

A number of both physiologic and pathologic factors affect the distribution of K^+ between the ICF and ECF and are therefore important in an analysis of K^+ balance. These factors include the acid-base status of the individual, the plasma concentration of several hormones, the activity of the Na^+-K^+ ATPase pump, and the rate of cell breakdown.

Acid-Base Status

During **acidosis**, H^+ shifts into cells in exchange for ICF K^+ (and for some ICF Na^+), and consequently the plasma K^+ concentration increases. The opposite occurs during **alkalosis**. For every 0.1 pH unit change, there is an approximate 0.6-mEq/liter change in the plasma K^+ concentration.

Hormones

Several hormones, including **insulin, catecholamines,** and **aldosterone,** promote the cellular uptake of K^+. Along these lines, it is of interest to note that insulin, in conjunction with glucose plus sodium bicarbonate, is often used for treating acute hyperkalemia. The liver is particularly important in this regulation, as about 70 percent of the insulin-induced K^+ uptake takes place in this organ.

Disease States

An **increase in cell breakdown,** such as occurs in severe trauma, provokes K^+ release into the ECF, and potentially, depletes K^+ stores. **Chronic diseases** such as heart failure also lead to decreases in the total body K^+ content. Whether the plasma K^+ concentration is altered depends on subsequent renal adjustments.

Renal Excretion of Potassium

The **dietary intake** of K^+ in a healthy adult is 40 to 120 mEq/day. Under normal conditions, most of the ingested K^+ is excreted in the urine (small amounts of K^+ are also lost in feces and sweat). Furthermore, a significant portion of the K^+ excreted derives from K^+ secretion in the combined distal tubule and cortical collecting duct system (about 90 percent of the filtered K^+ is reabsorbed in the proximal tubule plus the loop of Henle). As a result of these K^+ reabsorptive events, the K^+ concentration entering the distal tubule is relatively low (less than 1 mEq/liter)

Potassium Regulatory Factors

There are three primary factors that determine the rate of K$^+$ secretion: the **concentration gradient for K**$^+$ between the cortical collecting duct cell and the tubular lumen, the **transepithelial potential difference** between the tubular cell and the lumen of the nephron, and the **K$^+$ permeability** of the luminal membrane. Each of these events is summarized below.

The Cell-to-Lumen Potassium Gradient

An exaggerated difference between the K$^+$ concentration in the cell and the tubular lumen leads to an increase in the K$^+$ secretion rate and vice versa. The following factors affect the K$^+$ concentration in the cell and the tubular fluid.

At least three factors affect the **cell K$^+$ concentration** in the combined distal tubule and cortical collecting duct; these include the plasma K$^+$ concentration, the arterial pH, and the plasma concentration of aldosterone.

An increase in K$^+$ intake will raise the **plasma K$^+$ concentration** and stimulate Na$^+$-K$^+$ ATPase–mediated active transport. This occurs in all cells in the body, including those in the cortical collecting duct, and thereby increases the gradient for K$^+$ secretion into the tubular lumen.

Since a decrease in **arterial pH** shifts K$^+$ out of cells, K$^+$ concentrations in the distal tubular cells drop during acidosis. Consequently, the K$^+$ gradient between distal tubular cells and the tubular lumen decreases and K$^+$ secretion diminishes. Although hyperkalemia can develop during the acute phase of metabolic acidosis, K$^+$ wasting and a resultant hypokalemia normally arise in chronic acidosis. The opposite takes place during alkalosis.

Aldosterone stimulates the peritubular membrane Na$^+$-K$^+$ ATPase pump (probably by means of the increased Na$^+$ entry across the luminal membrane) in cortical collecting cells, thereby enhancing the distal tubular cell concentration of K$^+$.

The **tubular fluid concentration** of K$^+$ generally varies inversely with the rate of flow entering the cortical collecting duct. Thus, an accelerated distal tubular flow rate increases the K$^+$ gradient between the cell and the lumen (or the transtubular potential difference, as will be discussed) and results in increased K$^+$ secretion. Many diuretics (see Chapter 37) enhance K$^+$ excretion, mediated in part by these mechanisms. Similarly, isotonic expansion of the ECF space reduces fluid reabsorption in the proximal tubule, thereby increasing the distal flow rate and K$^+$ secretion. Volume depletion has the opposite effect.

The Transepithelial Potential Difference

The transtubular potential in the distal nephron is very high (about 30 to 50 mV lumen-negative relative to plasma) and is related to the transport of Na$^+$ from the lumen into the plasma. Increases in the transtubular potential difference favor K$^+$ secretion. Agents that affect the permeability of the luminal membrane to Na$^+$ alter the transepithelial potential and secondarily influence K$^+$ secretion. For example, the diuretics **amiloride** and **triamterene** (see Chapter 37) reduce the Na$^+$ permeability of the luminal membrane, which in turn reduces the transepithelial potential difference and decrease the K$^+$ secretion rate.

The Potassium Permeability of the Luminal Membrane

Changes in the permeability of the luminal membrane alter the K$^+$ secretion rate. **Aldosterone** appears to enhance the luminal membrane permeability to both Na$^+$ and K$^+$. Thus, increases in plasma aldosterone concentrations are associated with augmented K$^+$ secretion (and excretion).

CALCIUM REABSORPTION IN THE DISTAL NEPHRON

About 10 to 15 percent of the **filtered Ca^{2+}** is reabsorbed in the distal convoluted tubule and another 5 percent in the collecting duct. These are both active transport events that are stimulated by parathyroid hormone (PTH) secretion

(PTH stimulates Ca^{2+} reabsorption in these regions through the operation of a cAMP-dependent processes). Thus, under normal conditions, about 1 percent of the filtered Ca^{2+} load is excreted.

SUMMARY

It is essential that Na^+ and K^+ balance is maintained. The importance of balancing the intake and output of Na^+ is primarily dictated by the fact that changes in Na^+ balance markedly affect ECF volume. K^+ balance is critical because of the impact of ECF K^+ concentrations on the resting membrane potential. The amount of Na^+ excreted is largely determined by the amount reabsorbed (or, more accurately, not reabsorbed) from the initial filtered Na^+ load. Under normal conditions, relatively small fractions (less than 1 percent) of the filtered Na^+ load are excreted. Changes in the Na^+ reabsorption within the aldosterone-sensitive region of the nephron, located primarily in the cortical collecting duct, determine to a large extent the amount of Na^+ excreted. In addition, as most of the filtered K^+ is reabsorbed before it reaches the aldosterone-sensitive region of the nephron, the amount of K^+ secreted in this region determines the amount of K^+ excreted. Aldosterone release from the adrenal gland is regulated primarily by the plasma concentrations of K^+ and angiotensin-II. Angiotensin-II production is controlled by renin output from the kidney, which in turn is elevated by decreases in the renal perfusion pressure, increases in the renal sympathetic nerve tone, or decreases in distal tubular flow rate or any combination of these. Other factors, such as ANF and renal arterial pressure, also participate in the regulation of Na^+ balance.

NATIONAL BOARD TYPE QUESTIONS

Select the one most correct answer for each of the following

1. K^+ secretion in the distal nephron:
   A. Increases with an aldosterone antagonist
   B. Increases on a low K^+ diet
   C. Increases during respiratory acidosis
   D. Increases during treatment with diuretics which act proximal to the distal nephron
   E. Decreases after several days of aldosterone treatment (aldosterone escape)

2. An inhibitor of aldosterone would:
   A. Enhance K^+ and Na^+ reabsorption
   B. Enhance K^+ and Na^+ excretion
   C. Enhance H^+ and Na^+ excretion
   D. Enhance Na^+ reabsorption but reduce K^+ secretion
   E. Enhance Na^+ excretion but lower H^+ secretion

3. Licorice contains an aldosterone-like steroid, glycyrrhizic acid, so individuals ingesting large amounts of licorice would have:
   A. A reduced inulin space
   B. Hyperkalemia (high plasma K^+ concentration)
   C. Enhanced Na^+ excretion
   D. Metabolic alkalosis
   E. Hypotension

4. Renin release from a kidney:
   A. Increases as arterial pressure increases
   B. Decreases as renal sympathetic nerve activity increases
   C. Is enhanced during a high salt diet
   D. Increases if the renal artery is constricted (stenosis)
   E. Decreases during treatment with a converting-enzyme inhibitor

5. Aldosterone:
   A. Secretion is inhibited by elevated plasma K^+ concentration
   B. Binds to membrane receptors of cells in the distal nephron
   C. Stimulates cAMP production in cells of the distal nephron
   D. Increases K^+ excretion
   E. Secretion decreases as the plasma concentration of adrenocorticotropic hormone increases

ANNOTATED ANSWERS

1. D. Diuretic agents that inhibit Na^+ reabsorption proximal to the aldosterone-sensitive region of the nephron accentuate K^+ excretion. The mechanism is elicited by a diuretic-induced decrease in the tubular fluid K^+ concentration or an increase in the transtubular potential difference in the aldosterone region, with a resultant increase in K^+ secretion rate.

2. E. Inhibitors of aldosterone reduce Na^+ reabsorption as well as K^+ and H^+ secretion.

3. D. Aldosterone and aldosterone-like drugs stimulate H^+ secretion, and therefore can foster metabolic alkalosis.

4. D. Renin release from the kidney increases as renal arterial pressure decreases, renal sympathetic nerve tone increases, and distal tubular flow rate decreases.

5. D. Aldosterone stimulates K^+ secretion.

BIBLIOGRAPHY

Anger, M. S., et al. Water and sodium metabolism. In: Massry, S. G., and Glassock, R. J., eds. *Textbook of Nephrology*. Baltimore: Williams & Wilkins, 1989.

Folkow, B. Salt and hypertension. *News in Physiological Sciences* 5:220–224, 1990.

Gottschalk, C. W., Moss, N. G., and Colindres, R. E. Neural control of renal function in health and disease. In: Seldin, D. S., and Giebisch, G., eds. *The Kidney*. New York: Raven Press, 1985.

Guyton, A. C. Renal function curve—a key to understanding the pathogenesis of hypertension. *Hypertension* 10:1–6, 1987.

Sanson, S. C., et al. Potassium homeostasis. In: Massry, S. G. and Glassock, R. J., eds. *Textbook of Nephrology*. Baltimore: Williams & Wilkins, 1989.

37 Water Balance and Diuretics

ROBERT O. BANKS

Objectives

After reading this chapter, you should be able to:

State the osmolality of tubular fluid in each major segment of the nephron during dehydration and water diuresis

Cite the fraction of water and Na^+ reabsorbed in each major segment of the nephron

Explain the cellular mechanism of action of antidiuretic hormone

Describe the production and storage site of antidiuretic hormone and the factors that control its release

State the formula for free water clearance

Define osmolar and free-water clearance

Give the range of urine osmolality

Describe the significance of positive and negative free-water clearance values

Describe the mechanisms of action of various diuretics

Body fluid osmolality is closely regulated by balancing the intake and loss of water. The intake of water is the sum of (1) that contained in the food we eat; (2) that derived from the oxidation of ingested food, or so-called metabolic water; and (3) that which we drink. The amount taken in with drinking can vary substantially from individual to individual, but certain minimum requisite amounts are imposed by (1) insensible water loss; (2) water lost in the feces; and (3) an osmotic ceiling on the urine osmolality. The **thirst center** assures that this minimum daily intake of water is met.

WATER DEPRIVATION

The **maximum water deprivation** that is compatible with life varies among species: the max- imum weight loss from desiccation in humans is about 20 percent; this rises to 40 percent in dogs and cats and up to 60 percent in earthworms. A 10 percent weight loss through desiccation in humans usually compromises mental and physical functioning. Problems of dehydration are particularly critical in **infants,** as roughly 70 percent of an infant's weight is water (compared with 60 percent in adults). Thus, an infant weighing 5 kg contains approximately 3.5 liters of water. Given a fluid intake of about 800 ml/day, this represents a turnover of roughly 25 percent of the infant's total body water. A 70-kg adult, on the other hand, ingests about 1 to 2 liters of fluid per day, constituting less than 5 percent of the total body water. Additional water loss in an infant (diarrhea, sweating, or vomiting) more rapidly becomes a life-threatening situation than in an adult.

MINIMUM DAILY WATER INTAKE

The following example is an estimate of the **minimum daily intake of water** through drinking necessary for a 70-kg individual. The calculation will vary, depending on such factors as diet, temperature, exercise, and humidity. This example assumes the individual ingests 1.5 kg of food.

Water intake

From food	900 ml
From oxidation of food	400 ml
Total water ingested without drinking	1,300 ml

Water output

Insensible water loss (skin and lungs, or about 0.5 ml/kg-hr)	840 ml
In feces	100 ml
Minimum urine volume	750 ml
Total water lost	1,690 ml

The value for the minimum daily urine volume is based on the total osmoles that must be excreted and the maximum urine osmolality. In a **normal diet,** about 150 mEq of sodium chloride and 50 mEq of K^+ plus anion are ingested daily for a total of 400 mOsm of ions that must be excreted. An additional 500 mOsm of solute, primarily urea, must also be excreted. Assuming a maximum urine osmolality (U_{os}) of 1,200 mOsm/kg H_2O, the minimum 24-hour urine volume (V) would be:

Total solute excreted = $(U_{os})(V)$ = 900 mOsm/day
$(U_{os})(V)$ = (1,200 mOsm/kg H_2O)(V) = 900 mOsm/day
V = 750 ml/day

For this particular individual, 1,690 less 1,300 ml, or 390 ml, represents the obligatory water intake that must be ingested by drinking. Most individuals exceed this minimum daily intake, and, consequently, the osmolality of urine collected during 24 hours would be less than 1,200 mOsm/kg H_2O in these people. The exact value of course depends on the quantity of water consumed, but generally ranges between 500 and 800 mOsm/kg H_2O.

ANTIDIURETIC HORMONE

The amount of water excreted by the kidney is controlled by **antidiuretic hormone** (ADH), which is also known as *vasopressin.* ADH regulates the water permeability of the **cortical** and **medullary collecting duct system.** If ADH is lacking, this region of the nephron is virtually impermeable to water, and, thus, water inside the nephron at the turn of Henle's loop would be excreted (the ascending limb of Henle's loop is always impermeable to water).

Mechanism of Action of ADH

The antidiuretic action of ADH is mediated by **V_2 receptors** in the basolateral membrane of the collecting duct. Activation of these receptors stimulates adenylate cyclase activity, and consequently increased conversion of ATP to cyclic AMP (cAMP). The mechanism by which cAMP increases water permeability is unclear. cAMP is known to stimulate kinases, and, cAMP-induced changes in microtubules and microfilaments have also been reported. cAMP is converted to an inactive form (5′ AMP) by **phosphodiesterse.** These events are summarized in Figure 37-1.

The above series of events can be modulated by different agents at several points. For example, **calcium** can inhibit both the binding of ADH to the V_2 receptor and adenylate cyclase activity. **Prostaglandin E_2** and **bradykinin** also suppress adenylate cyclase activity and the ADH-induced **hydro-osmotic response** (the effect of bradykinin may be mediated by an increase in prostaglandin E_2 synthesis). **Aldosterone** potentiates ADH, probably through inhibition of phosphodiesterase activity. Finally, **parathyroid hormone** also potentiates ADH.

ADH increases **vascular resistance** by activating **V_1 receptors** on smooth muscle cells. The

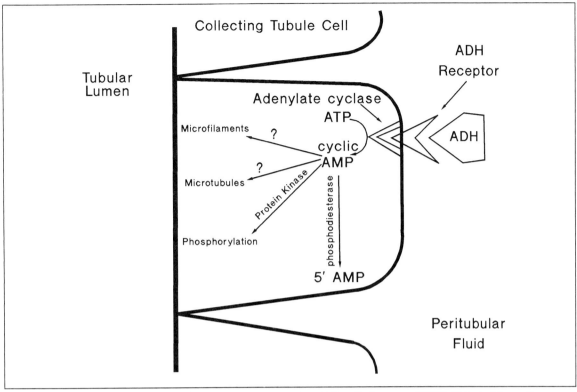

Fig. 37-1. The major events in the hydro-osmotic response to antidiuretic hormone (*ADH*). (Adapted from: Dousa, T.P., and Valtin, H. Cellular actions of vasopressin in the mammalian kidney. *Kidney Int.* 10:46–63 1976.)

effects of this are coupled to activation of the phosphoinositide pathway.

Synthesis of ADH

ADH is a **peptide hormone** synthesized in the **supraoptic** and **paraventricular nuclei** of the hypothalamus and transported down the axons of these cells in association with a larger (10,000 daltons) protein, **neurophysin. Oxytocin** is also produced in this area and undergoes a similar series of events.

ADH is stored in large granules of the posterior lobe of the pituitary gland. With appropriate stimulation and depolarization of these fibers, these granules release their contents by **reverse endocytosis** (exocytosis). Both ADH and neuro-

physin appear to be discharged into the circulation.

Secretion of ADH

Two major stimuli evoke ADH release: **hypertonicity** of the body fluids and **volume depletion.** A relatively small decrease (about 1 percent) in the body fluid osmolality below the set point, generally about 280 mOsm/kg H_2O, is associated with a decline in the plasma ADH concentration to virtually undetectable levels (less than 1 ng/ml). By contrast, as the plasma osmolality exceeds 280 mOsm/kg H_2O, the plasma ADH concentration increases, reaching values of about 10 ng/ml at 300 mOsm/kg H_2O.

Volume (pressure) **receptors** that influence

ADH release are located in several regions of the circulation. **Low-pressure receptors** in the **left atrium** are among the most important, but input may also derive from baroreceptors in the aortic and carotid regions. Elevated atrial or arterial pressure does not markedly limit ADH secretion, but a decrease in these pressures is a potent stimulus for its release. Many **other factors** can also affect the release of ADH; angiotensin-II, nicotine, and pain stimulate ADH release, while ethanol and glucocorticoids inhibit its release.

ADH has a relatively short half-life of 5 to 10 minutes once it enters the circulation, because it is enzymatically degraded in plasma and rapidly cleared by the kidneys (ADH may be secreted by the proximal tubule).

QUANTITATION OF WATER EXCRETION: FREE WATER CLEARANCE

One way to quantitate the amount of **solute-free water** that is either excreted or retained by the kidney is to evaluate the **free-water clearance.** (The term *solute-free water* refers to the quantity of water that must be added or extracted from a urine sample to render it isosmotic to plasma). Free-water clearance (CH_2O) is defined as follows:

$$C_{H_2O} = V - C_{os} \qquad \text{Eq. 37-1}$$

Where

$$C_{os} = \frac{(U_{os})(V)}{P_{os}} \qquad \text{Eq. 37-2}$$

All the above expressions are given in milliliters per minute. U_{os} refers to osmolality, V to the urine flow rate, C_{os} to the osmolar clearance, and P_{os} to the plasma osmolality.

Free-water clearance is, in some respects, a misnomer because it is not an actual clearance, but, as this definition illustrates, represents the difference between the urine flow rate and the osmolar clearance. **Osmolar clearance,** a conventional clearance, stands for the volume of plasma completely cleared of total solute per unit of time.

The following examples illustrate the physiologic implications of free-water clearance values under three different conditions and are based on the evaluation of the 70-kg subject cited previously in this chapter.

Condition 1: Excretion of Isotonic Urine

In this condition, the individual is producing urine with an osmolality of 280 mOsm/kg H_2O and has a plasma osmolality of 280 mOsm/kg H_2O. Regardless of the urine flow rate, the free-water clearance in this condition would be zero. Thus, when the urine osmolality equals the plasma osmolality:

$$C_{H_2O} = V - \frac{(U_{os})(V)}{P_{os}} = 0$$

In other words, this person would be neither excreting nor retaining excess water. Rather a **zero free-water clearance** represents an isosmotic reduction in the body fluid volume.

Condition 2: Excretion of Hypotonic Urine

When urine osmolality is less than plasma osmolality, the free-water clearance becomes positive. This condition, known as **water diuresis,** arises when ADH release from the pituitary is inhibited; solute-free water, relative to plasma, is thus excreted and body fluid osmolality increases.

If the 70-kg individual were to rapidly ingest (within 5–10 min) one liter of water, and had an initial plasma osmolality of 280 mOsm/kg H_2O, the plasma osmolality would fall to 273 mOsm/kg H_2O after water absorption (this calculation is based on a total body water content of 60 percent, or 42 liters, prior to water ingestion). ADH concentrations in plasma would rapidly decline, resulting in the formation of a large volume of hypotonic urine. At peak diuresis, the subject's urine flow rate might be about 12 ml/min with a urine osmolality of 50 mOsm/kg H_2O. Thus, the free-water clearance at peak diuresis would be:

$$C_{H_2O} = V - \frac{(U_{os})(V)}{P_{os}}$$

$$= 12 \text{ ml/min} -$$
$$\frac{(50 \text{ mOsm/kg})(12 \text{ ml/min})}{273 \text{ mOsm/kg}}$$

$$= 12 \text{ ml/min} - 2.2 \text{ ml/min}$$

$$= + 9.8 \text{ ml/min}$$

In essence, the kidney is excreting the solute contained in 2.2 ml of plasma in 12 ml of urine. Thus, in this example, an additional 9.8 ml of solute-free water is excreted each minute.

Solute-free water is generated primarily in the **ascending loop of Henle,** but to some degree, also in the **distal** and **collecting tubules,** where solute is extracted from tubular fluid without simultaneous water reabsorption. The maximum positive free-water clearance is about 15 ml/min.

Condition 3: Excretion of Hypertonic Urine

This calculation will utilize the values given at the beginning of this chapter. Thus, urine osmolality is 1,200 mOsmoles/kg H_2O with a urine flow rate of 750 ml/day or 0.52 ml/min. The free-water clearance would therefore be calculated as follows:

$$X_{H_2O} = 0.52 \text{ ml/min} -$$
$$\frac{(1,200 \text{ mOsm/kg})(0.52 \text{ ml/min})}{300 \text{ mOsm/kg}}$$

$$= 0.52 \text{ ml/min} - 2.08 \text{ ml/min}$$

$$= - 1.56 \text{ ml/min}$$

In this example, the free-water clearance is negative and this connotes a state of **dehydration.** In other words, by excreting the solute contained in 2.08 ml of plasma in only 0.52 ml of urine, each minute 1.56 ml of solute-free water is conserved. The total amount of solute excreted is approximately the same as that for the subject in condition 2; the individual must excrete 900 μOsm/day, or approximately 600 μOsm/min, since the only difference between the two conditions is the amount of water consumed.

A free-water clearance of $- 2$ ml/min is approximately the maximum negative free-water clearance that the human kidney can produce. Because it can be awkward to deal with negative values, when free water clearance becomes less than zero, the term **T^{c}H$_2$O** is often used and stands for the difference between the osmolar clearance and the urine flow rate. The superscript c indicates that negative free water is formed in the collecting duct.

PATHOLOGIC CONDITIONS ASSOCIATED WITH ADH SECRETION

Excess ADH Secretion

The **syndrome of inappropriate secretion of ADH** is characterized by chronic hypo-osmolality and hyponatremia because of the excess water retention and increased Na^+ excretion involved. Thus, elevated ADH concentrations result in a reduced urine output, which in turn provokes increased blood volume and blood pressure with a resultant pressure-induced natriuresis.

The inappropriate release of ADH may be of **pituitary origin** (related to, for example, CNS infections, certain drug treatments, or other endocrine imbalances) or of **non-pituitary origin** (a frequent complication of lung carcinomas). The major symptoms of hypo-osmolality are neurologic and stem from the development of **cerebral edema.**

ADH Deficiency

ADH deficiency is a feature of two conditions: **central diabetes insipidus,** in which patients cannot secrete ADH, and **nephrogenic diabetes insipidus,** in which the plasma concentrations of ADH are elevated, but the kidney cannot conserve water because of tubular defects. A variety of factors can precipitate the nephrogenic form. The disorder can be inherited (a rare condition), drug related, or a complication of renal disease.

Hyperosmolality and hypernatremia do not

develop in diabetes insipidus if water intake is sufficient. The symptoms of hypernatremia are primarily neurologic and range from muscle weakness to convulsions and coma.

DIURETICS

Pharmacologic agents are often used to promote the excretion of excess salt and water. Although a complete description of diuretics is beyond the scope of this textbook, Table 37-1 lists some of the major diuretics and gives the site and mechanism of action of these agents.

Carbonic anhydrase is an important enzymatic component in the reabsorption of sodium bicarbonate (see Chapter 39). Agents such as **acetazolamide** (Diamox) that inhibit carbonic anhydrase activity elicit a modest diuretic response that involves the excretion of an alkaline urine (sodium bicarbonate). These agents are primarily used to treat edematous patients with metabolic alkalosis.

Osmotic agents, such as mannitol, are often used to reduce intracranial pressure and help prevent acute renal failure. Their primary mechanism of action relates to the fact that they are filtered but not reabsorbed. Thus, when infused in sufficient quantities, the osmotic agent reduces the obligatory water reabsorption along the nephron.

The so-called **loop diuretics,** such as furosemide and ethacrynic acid, inhibit the Na^+-K^+-2 Cl^- transport process in the ascending limb of Henle's loop. These agents are widely used for the treatment of edema and hypertension. As more than 20 percent of the filtered Na^+ is reabsorbed in the loop of Henle, these agents can promote the excretion of relatively large amounts of Na^+. Loop diuretics also enhance the excretion rate of K^+. This results from inhibition of the Na^+-K^+- $2Cl^-$ transport process and the secondary increase in tubular flow rate through the cortical collecting duct with a subsequent increase in K^+ secretion (see Chapter 36 for the factors that affect K^+ secretion in the cortical collecting duct). Finally, because relatively large amounts of Ca^{2+} are reabsorbed in the loop of Henle, treatment

with loop diuretics also markedly augments Ca^{2+} excretion.

The major site of action of the **thiazides** is in the early portions of the distal tubule. Chlorothiazide is one such agent. Although the biochemical mechanism of action of the thiazides is unknown, the primary diuretic effect appears to result from the inhibition of a Na^+-Cl^- cotransporter located on the luminal membrane of the distal convoluted tubule. The resultant increase in the tubular flow rate entering the collecting duct also elevates K^+ secretion. In contrast to the loop diuretics, the thiazides promote Ca^{2+} reabsorption, and therefore reduce Ca^{2+} excretion. The thiazides are used routinely for treating edema and hypertension.

Finally, the **K^+-sparing diuretics** represent a class of several agents that cause a modest increase in Na^+ and Cl^- excretion without elevating K^+ excretion. Examples of these diuretics are spironolactone, amiloride, and triamterene. Spironolactone is a competitive aldosterone-receptor antagonist, whereas the other two agents lessen the permeability of the luminal membrane to Na^+ in the cortical collecting duct. These agents are generally administered in conjunction with a loop diuretic or a thiazide to prevent the hypokalemia often associated with the loop diuretic or the thiazide.

SUMMARY

Body fluid osmolality (normally about 280 mOsm/liter) is regulated within a relatively narrow range. Water balance is accomplished primarily by means of changes in the amount of water excreted or retained by the kidney and by appropriate increases in the amount of water ingested through stimulation of the thirst center. Decreased body fluid osmolality is rapidly offset by decreased ADH secretion from the pituitary. As ADH concentrations fall, the water permeability of the collecting duct declines, the urine flow rate then increases, and urine osmolality drops. By contrast, water deprivation leads to an increase in body fluid osmolality because of the continued insensible water loss. Consequently, the plasma

Table 37-1. A Summary of Major Classes of Diuretics

Drug	Major Site of Action	Mechanism	Comments
Acetazolamide (Diamox)	Proximal tubule	Carbonic anhydrase inhibitor (interferes with HCO_3^- reabsorption)	Na^+, HCO_3^-, K^+*, and H_2O mainly excreted
Osmotic diuretics (mannitol; glucose in uncontrolled diabetes mellitus)	Primarily loop of Henle, also proximal tubule	Is a non-reabsorbable solute. If infused in sufficient amounts: (1) retains H_2O in the nephron and (2) increases blood flow to the medulla and decreases the osmolar gradient	Excretion of H_2O in excess of NaCl; K^+* also excreted; diuresis is not affected by ADH
Loop diuretics (ethacrynic acid, furosemide)	Thick ascending limb of Henle's loop	Inhibits $Na^+-K^+-2Cl^-$ reabsorption (can abolish the osmolar gradient; cannot concentrate or dilute urine. Thus lower C_{H_2O} and T^cH_2O)	Large amounts of Na^+, Cl^-, and K^+* can be excreted (as much as 40% of the filtered Na^+ and H_2O)
Thiazides	Cortical diluting segment (also known as the cortical thick ascending limb) and distal convoluted tubule	May inhibit a Na^+-Cl^- cotransporter on the luminal membrane; lower C_{H_2O} but no effect on T^cH_2O	5–8% of the filtered Na^+ excreted. K^+*, Cl^-, and H_2O also excreted
Spironolactone	Aldosterone-sensitive region	Aldosterone antagonist	Slow onset of action (12 + hr). Na^+, Cl^-, and H_2O are excreted. K^+ is retained (a K^+-sparing diuretic)
Triamterene and amiloride	Aldosterone-sensitive region	Not aldosterone antagonists: agents block Na^+ entry into CCD cells	Na^+, Cl^-, and H_2O excreted. K^+-sparing diuretics

*Any agent that interferes with Na^+ reabsorption proximal to the site of K^+ secretion will enhance K^+ excretion. In the presence of these drugs, the distal tubular flow rate increases, the lumen concentration of K^+ decreases, or the transtubular potential becomes more negative, thereby favoring the K^+ gradient between the distal tubular cells and the lumen. K^+ secretion (and, therefore, K^+ excretion) increases.
ADH = antidiuretic hormone; C_{H_2O} = free-water clearance; T^cH_2O = (see text, p. 539); CCD = cortical collecting duct.

concentration of ADH increases, water excretion decreases (urine osmolality increases), and the thirst center is stimulated. Water excretion by the kidney can be quantitated by applying the concept of free-water clearance, which represents the difference between the urine flow rate and the osmolar clearance. Positive values indicate that the kidney is producing a urine that is diluted (compared to plasma), and is thereby concentrating the body fluids. Negative values are obtained when the kidney elaborates a concentrated urine, and is therefore diluting the body fluid osmolality. Finally, a free-water clearance of zero signifies that the kidney is neither concentrating nor diluting body fluid osmolality, and an isosmotic urine is being produced.

NATIONAL BOARD TYPE QUESTIONS

Select the one most correct answer for each of the following

1. Antidiuretic hormone:
   A. Affects water permeability of the thick but not the thin ascending loop
   B. In excess can lead to hypernatremia
   C. Inhibits adenylate cyclase activity in the collecting duct
   D. Would be relatively ineffective in reversing a mannitol (osmotic) diuresis
   E. Action on the kidney can be blocked by a phosphodiesterase inhibitor
2. ADH-dependent changes in the water permeability of the kidney:
   A. Can be potentiated by phosphodiesterase inhibitors such as theophylline
   B. Are mediated by messenger RNA
   C. Are primarily localized to the ascending loop of Henle
   D. Are a cGMP-related event
   E. Are always associated with ADH-dependent changes in the urea permeability of the cortical collecting duct
3. On a normal diet, about 900 mOsm/day are excreted. Assuming the maximum urinary osmolar concentration is 1,200 mOsm/liter, the minimum daily urine volume (in milliliters per day) would be:
   A. 75
   B. 120
   C. 240
   D. 500
   E. 750
4. The following data were obtained from individuals A and B:

	A	B
Urine osmolality (mOsm/kg H_2O)	100	500
Urine flow rate (ml/min)	4.0	0.8
Plasma osmolality (mOsm/kg H_2O)	280	280

Which of the following statements is true?
   A. The osmolar clearance is the same in both subjects but the free-water clearances differ
   B. The plasma concentration of ADH in person B is probably less than that in person A
   C. Person A could be dehydrated and receiving furosemide
   D. The cAMP concentrations in the collecting ducts of person A are greater than those in person B
   E. The urine creatinine concentration in person B is less than that in person A
5. Free-water clearance:
   A. Is always positive when the urine to plasma osmolar ratio is greater than 1
   B. Is always negative when the urine to plasma osmolar ratio is less than 1
   C. Is zero when the urine to plasma osmolar ration is 2 and the urine flow rate is 2 ml/min
   D. Approaches zero during a maximum loop diuretic–induced diuresis
   E. Becomes more negative as ADH concentrations decrease

ANNOTATED ANSWERS

1. D. Increasing the filtered load of a non-reabsorbable solute causes an osmotic-induced retention of water in the tubule. Consequently, osmolar gradients across the nephron are reduced, with a resultant attenuation in the action of ADH.
2. A. The water permeability–related actions of ADH are mediated by increases in cAMP activity; since phosphodiesterase degrades cAMP, an inhibitor of phosphodiesterase potentiates ADH activity.
3. E. Calculated from the equation for free-water clearance [$C_{H_2O} = V - (U_{os})(V)/P_{os}$].
4. A. The osmolar clearance, given by $(U_{os})(V)/P_{os}$, is the same in both individuals.

5. D. Loop diuretics inhibit the Na^+–K^+–$2Cl^-$ transport in the thick ascending limb of the loop of Henle. Thus, the ability to either concentrate or dilute urine is attenuated and the final urine approaches isotonicity.

BIBLIOGRAPHY

Abramow, M., Beauwens, R., and Cogan, E. Cellular events in vasopressin action. *Kidney Int.* 32(suppl 21):S-56-S-66, 1987.

Robertson, G. L. Physiology of ADH secretion. *Kidney Int.* 32(suppl 21):S-20-S-26, 1987.

Suki, W. N. Renal actions and uses of diuretics. In: Massry, S. G., and Glassock, R. J., eds. *Textbook of Nephrology.* Baltimore: Williams & Wilkins, 1989.

38 Acid-Base Balance: Biochemistry, Physiology, and Pathophysiology

JOHN H. GALLA AND SHAHROKH JAVAHERI

Objectives

After reading this chapter, you should be able to:

Describe the overall purpose and burden of acid-base homeostasis

Identify H⁺ concentration as a dependent variable and the major independent variables that determine it in biologic solutions: strong ions, weak acids, and partial presence of CO_2

Explain the difference between the H⁺ concentration and pH

Describe the nature of physiochemical buffers and how they participate in acid-base balance

Understand why CO_2 is an important physiologic buffer and how it behaves in vivo

List the four simple disorders of acid-base balance and understand their pathophysiologic concepts

Describe the concepts of respiratory or metabolic compensation for primary perturbations and of mixed disturbances, and recognize the presence of compensation

---

In this chapter, acid-base balance is introduced with an overview of its homeostasis followed by discussions of its biochemistry and physiologic and pathophysiologic states. Emphasis is on modern quantitative principles espoused by Stewart (1983).

GENERAL BACKGROUND

Purpose of Acid-Base Homeostasis

The *milieu interieur* of the body is regulated to maintain the **H⁺ concentration** ([H⁺])* of its var-

*Square brackets around a substance denote concentration.

ious compartments within a very narrow range, for the most fundamental reasons. From a biochemical standpoint, the configuration or tertiary structure of many proteins is dependent on the ambient [H⁺]. In particular, the activities of many enzymes are exquisitely sensitive to the [H⁺]. Thus, if [H⁺] is perturbed sufficiently from normal, the structure of circulating substances and those on cell membranes and cells is likely to be altered in such a way that would prevent them from entering into reactions, interacting with receptors, engaging in O_2 exchange, and so on. From this perspective, it is obvious that, without the exquisite regulation of [H⁺],

metabolic processes would rapidly become deranged
or cease.

Burden of Homeostasis

The regulation of $[H^+]$ of the various body fluid compartments is a dynamic process in which $[H^+]$ is continually perturbed by normal metabolism and promptly corrected to normal. Each day a person consumes carbohydrates, fats, and proteins that are used to produce energy, maintain cellular structure, and restore excreted substances. Most of these substrates are eventually catabolized in mitochondria during normal aerobic metabolism:

$$C_6H_{12}O_6 + 6O_2 \rightarrow 6CO_2 + 6H_2O + \text{heat}$$
$$\text{Eq. 38-1}$$

In all, about 15,000 to 20,000 mmols of CO_2 are generated each day and must be excreted by the lungs to maintain the constancy of the interior milieu, **homeostasis.** The protein consumed, about 1 gm per kilogram of body weight per day, yields about 70 mEq of inorganic acid. For example, organic sulfur, a fixed or nonvolatile acid, is oxidized to inorganic sulfur:

$$\text{cysteine} + O_2 \rightarrow \text{urea} + H_2O + CO_2 + SO_4^{2-} + 2H^+$$
$$\text{Eq. 38-2}$$

Fixed acid is excreted by the kidneys.

ACID-BASE BIOCHEMISTRY

Nature of $[H^+]$

The $[H^+]$ in biologic fluids is most commonly expressed by its negative logarithm to the base 10, or **pH.** Because of this logarithmic notation, the changes in $[H^+]$ expressed as pH can be deceptive, as an increase in $[H^+]$ actually represents a decrease in pH and a small change in pH may in fact indicate a large change in $[H^+]$. Because of the long-standing and widespread use of pH to denote $[H^+]$, both expressions are used interchangeably in this chapter.

$[H^+]$ is a **variable** that *depends* on a number of independent variables in any given solution. Thus, $[H^+]$ can be described by a set of equations that incorporates these variables and is based on **physicochemical principles,** including the laws of electroneutrality and conservation of mass and the equilibrium reactions. To understand how the $[H^+]$ of a solution is determined, the simplest solution will be considered first and then ones of increasing complexity.

In the case of **pure water,** $[H^+]$ is described by two relationships. First, by its **dissociation constant** (K_w), which in turn varies directly only according to the independent variable, temperature:

$$[H^+] \times [OH^-] = K'_w \qquad \text{Eq. 38-3}$$

Where $K'_w = K_w \times [H_2O] = 10^{-14}$. $[H_2O]$ behaves like a constant in this expression; because its concentration is so large (55 M) compared with the other factors, it is incorporated into K'_w. This relationship, which must always be satisfied, shows that $[H^+]$ and $[OH^-]$ behave in a reciprocal manner.

Second, as in all solutions, **electroneutrality** must be preserved. Thus, for pure water:

$$[H^+] = [OH^-] \qquad \text{Eq. 38-4}$$

Equations 38-3 and 38-4 must be satisfied and, when they are solved simultaneously, express the $[H^+]$ for water:

$$[H^+] = \sqrt{K'_w} = 1 \times 10^{-7} \text{ M} \qquad \text{Eq. 38-5}$$

As seen in this expression, the pH of water at 25°C is 7.00. As the temperature increases, K'_w increases and, consequently, $[H^+]$; the pH decreases.

Effect of Strong Ions on $[H^+]$

Biologic solutions are not pure water. Most contain a wide array of substances, including electrolytes. Those electrolytes that for all practical purposes **dissociate completely,** such as Na^+, K^+, Ca^{2+}, Mg^{2+}, Cl^-, and SO_4^{2-}, are classified as **strong ions.** Some organic anions of sufficiently strong acids, such as lactate, also behave like strong ions. Through a series of algebraic expressions that include the law of electroneutrality and equilibrium reactions (similar to those described for H_2O), the effect of strong ions is incorporated into the expression for $[H^+]$. Consider, for example, only Na^+ and Cl^- in a solution. First, electroneutrality must be satisfied:

$$([Na^+] + [H^+]) - ([Cl^-] + [OH^-]) = 0 \quad \text{Eq. 38-6}$$

Solved simultaneously with Equation 38-3, which must always be satisfied, yields:

$$[H^+]^2 + ([Na^+] - [Cl^-]) \times [H^+] - K'_w = 0$$
$$\text{Eq. 38-7}$$

Rearranged in the quadratic form, an expression is yielded that incorporates these strong ions:

$$[H^+] = \sqrt{K'_w + ([Na^+] - [Cl^-])^2/4} - $$
$$([Na^+] - [Cl^-])/2 \qquad \text{Eq. 38-8}$$

In this equation, $[H^+]$ is a dependent variable and a function of the difference between the charges of the strong electrolytes and K'_w, both of which are *independent* variables. For all substances that behave like strong ions, this difference can be expressed as the **strong ion difference** (SID), which is the sum of all the strong cations minus the sum of all the strong anions in the solution, and is the single most important determinant of $[H^+]$ or $[OH^-]$ in this solution. The essential feature of SID is the **difference in charge** and not the ion species. For example, if Ca^{2+} and SO_4^{2-} are added to the NaCl solution, SID is expressed as:

$$([Na^+] + [Ca^{2+}]) - ([Cl^-] + [SO_4^{2-}]) = SID$$
$$\text{Eq. 38-9}$$

In solutions with a positive SID, such as various body fluids, $[H^+]$ is small and $[OH^-]$ predominates. For example, Equation 38-6 can be rearranged to solve for $[OH^-]$:

$$([Na^+] - [Cl^-]) + [H^+] = [OH^-] \qquad \text{Eq. 38-10}$$

Because Equation 38-3 must always be satisfied, if $[Na^+]$ is 140 mM and $[Cl^-]$ is 100 mM, the SID is $(+)40$ mM. Thus, $[OH^-]$ must be approximately 40 mM but $[H^+]$ is only 1×10^{-12} mM.

Regardless of whether SID is positive or negative, electroneutrality must be preserved; in the example shown in Equation 38-10, the difference between $[H^+]$ and $[OH^-]$ is equal to SID. When SID is substituted, Equation 38-8 is restated as:

$$[H^+] = \sqrt{K'_w + ([SID]/2)^2} - [SID]/2 \quad \text{Eq. 38-11}$$

Thus, as SID in physiologic solutions is several orders of magnitude greater than K'_w, SID dominates the resultant $[H^+]$.

The effect of strong ions on the $[H^+]$ of a solution is conventionally demonstrated by titration with a strong acid or base, such as hydrochloric acid or sodium hydroxide (Fig. 38-1). In this example, the initial point of addition, indicated by the zero point on the X axis, is arbitrarily set at a $[H^+]$ of 10^{-4} mM (pH 7.00), at the midpoint of the Y axis; the initial point of addition can be set anywhere. The addition of a **strong acid** introduces strong anions and that of a **strong base** introduces strong cations, as shown along the ordinate of Figure 38-1.

Although H^+ or OH^- is also added to the solution during titration, *equal* changes in strong ions evoke *different* changes in $[H^+]$ and $[OH^-]$, depending on the initial or ambient $[H^+]$ of the solution. For example, if 2 mEq of HCl is added at pHs of approximately 12, 10.5, or 1.5, the effect on the final $[H^+]$ is dramatically different, as can be seen by plotting these values on the titration curve (see Fig. 38-1). In a strongly alkaline solution (pH 12), the addition of 2 mEq of HCl has a trivial effect on $[H^+]$ but $[H^+]$ increases linearly in a strongly acidic solution (pH 1.5). The reason for this, of course, is that, in alkaline solutions, H^+ reacts with OH^- to form H_2O while the added Cl^- remains an anion. In an acidic solution, H^+ does not react with OH^- because virtually none is present and each milliequivalent of H^+ added remains a proton; Cl^- remains an anion. The

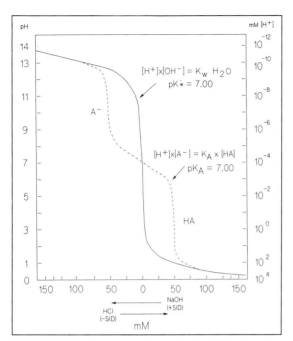

Fig. 38-1. Titration curves for water *(solid line)* and a solution buffered with a weak acid *(HA; dashed line).* K_w = dissociation constant; pK_w = negative logarithm of K_w; K_a, pK_a = weak acid dissociation constant; *SID* = strong ion difference.)

intermediate effect is observed at pH 10.5 because of the limited availability of OH^- around the neutral point for pure water (pH 7.00).

The impact of titration on changes in pH (see Fig. 38-1, the left scale of the Y axis) can be more difficult to visualize. In contrast to the $[H^+]$, the effect of the 2 mEq of HCl on pH is most dramatic at the neutral point but is numerically similar at both high and low initial pHs. This is, of course, simply due to the fact that pH is a logarithmic function that obscures the dramatic difference in the effect of HCl.

Effect of Weak Ions on [H+]

Not all solutes dissociate completely as strong ions do. Biologic solutions, in particular, may contain substantial concentrations of **incompletely dissociated,** or "weak," acids or bases,

that add another level of complexity to the solution. Weak acids, for example, contribute at least three quantities to a solution—HA, A^-, and H^+ for monovalent compounds—as described by the relationship:

$$[H^+] \times [A^-] = K_A \times [HA] \qquad \text{Eq. 38-12}$$

or

$$pH = pK_A + \log [A^-]/[HA] \qquad \text{Eq. 38-13}$$

where K_A is the weak-acid dissociation constant, also expressed as pK_A; its negative logarithm (HA) is any weak acid and A^- is its conjugate base. **Proteins** and **phosphates** are the major weak acids in the body.

According to the **law of the conservation of mass:**

$$[A_{TOT}] = [HA] + [A^-] \qquad \text{Eq. 38-14}$$

Thus, though the total amount of weak acid $([A_{TOT}])$ is constant (Equation 38-14), the proportion that exists as A^- and HA may change (Equation 38-12). When solving for $[H^+]$, the relationships described by Equations 38-12 and 38-14 must be satisfied along with those described by Equation 38-3 and the electroneutrality equation, which now includes $[A^-]$. At this juncture in our discussion, $[H^+]$ is a function of four independent variables: K'_w, SID, K_A, and A_{TOT}.

Buffers

Weak acids can act as chemical buffers under the appropriate conditions. A **buffer** is a substance that can give or accept protons in a manner that tends to minimize changes in $[H^+]$ when substances, such as strong ions, which might otherwise produce large changes in $[H^+]$, are introduced into solutions. A buffer exists in solution as a weak *acid* (proton donor) (HA) and a **conjugate** as a weak *base* (proton acceptor) (A^-) pair in the relationship:

$$[HA] \leftrightarrow [H^+] + [A^-] \qquad \text{Eq. 38-15}$$

For polyvalent buffers, multiple such relationships exist and in a complex solution;

$$[H_2A] \leftrightarrow [H^+] + [HA^-] \leftrightarrow 2[H^+] + [A^{2-}];$$
$$H_3A \ldots \text{etc.} \qquad \text{Eq. 38-16}$$

As the influence of SID on the $[H^+]$ of the solution depends on the ambient $[H^+]$, even more so does the **buffering power.** The effect of adding strong ions to a solution containing a monovalent buffer, HA, with a pK_A arbitrarily set at 7.00 is shown by the *dashed curve* in Figure 38-1. At the alkaline end of the titration curve, where only HA is present, as well as at the acidic end, where only A^- exists, the solution behaves like water, as if no buffer were present. However, within about ± 1 pH until of the pK_A (7.00, in this case), the addition of nearly 50 mM of strong ion has little effect on $[H^+]$, indicating that the solution is buffered. Figure 38-1 illustrates two important principles regarding buffers:

1. The **pK'** is the pH at which the [HA] equals $[A^-]$, that is, $[A^-]/[HA] = 1.0$.
2. At the pK', the change in $[H^+]$ elicited by the addition of strong ions is least, or, restated, the *most* effective buffering lies in the range of the linear portion (the midpoint) of the titration curve for that buffer.

Isohydric Principle

The **buffering power** of any complex solution (the sum of the effect of all buffers present) can also be determined by titration. When several buffers exist in a solution, all of the buffer pairs are in equilibrium at the same $[H^+]$, which in turn depends on the type and concentration of the buffer pairs:

$$\begin{aligned} pH &= pK'_1 + \log [HCO_3^-]/[H_2CO_3] \\ &= pK'_2 + \log [HPO_4^{2-}]/[H_2PO_4^-] \\ &= pK'_3 \qquad \text{Eq. 38-17} \end{aligned}$$

This important principle states that, by analyzing one buffer pair, the pH of the system can be determined and hence the status of all other buffer pairs in the solution. This is known as the **isohydric principle.** However, even though $[H^+]$ may be used to determine the status of all buffer pairs, this does *not* mean that $[H^+]$ is an independent variable.

Effect of CO_2 on $[H^+]$

Finally, and of particular importance in biologic solutions, the addition of CO_2, a volatile weak acid, to a solution introduces another independent variable and several molecular species:

$$[\text{dissolved } CO_2] = P_{CO_2} \times S_{CO_2} \qquad \text{Eq. 38-18}$$

$$[\text{dissolved } CO_2] + H_2O \leftrightarrow H_2CO_3 \leftrightarrow H^+ + HCO_3^- \qquad \text{Eq. 38-19}$$

$$[\text{dissolved } CO_2] + OH^- \leftrightarrow HCO_3^- \leftrightarrow H^+ + CO_2^{2-} \qquad \text{Eq. 38-20}$$

The partial pressure of CO_2 (P_{CO_2}) and its solubility coefficient (S_{CO_2}) determine its concentration according to **Henry's law;** S_{CO_2}, in turn, depends on temperature and the other constituents of the solution. Combining these relationships and their dissociation constants algebraically yields two expressions that describe the relationships of all these **molecular species:**

$$[H^+] \times [HCO_3^-] = K_2 \times P_{CO_2} \qquad \text{Eq. 38-21}$$
$$[H^+] \times [CO_3^{2-}] = K_3 \times [HCO_3^-] \qquad \text{Eq. 38-22}$$

The behavior of CO_2 in the physiologic environment will be discussed later, but this expression shows that HCO_3^- is also a dependent variable.

Summary of Biochemical Considerations

Three independent variables (SID, A_{TOT}, and CO_2) and six variables that depend on them and are pertinent to biologic solutions have now been considered. Plasma SID represents the difference between the sum of all the strong cations and anions; A_{TOT} includes the plasma proteins and phosphates; and P_{CO_2} is set by alveolar ventilation. When the six expressions for these variables (Equations 38-3, 38-11, 38-12, 38-14, 38-21, and 38-22, as well as electroneutrality, which now includes $[HCO_3^-]$ and $[CO_3^{2-}]$) are solved simultaneously with the known plasma values for $[Na^+]$, $[K^+]$, $[Cl^-]$, as well as total protein content, and the arterial P_{CO_2} ($PaCO_2$) with dissociation constants at 37°C, a fourth order polynominal yields a value for $[H^+]$ of 40 nEq/liter or a pH of 7.40. This affirms the validity of this approach.

PHYSIOLOGIC pH

Nature and Maintenance of Physiologic pH

These principles of acid-base biochemistry must now be considered in the context of a physiologic system that places certain constraints on the physicochemical system. The normal $[H^+]$ of the mammalian arterial blood at 37°C is 40 nM (10^{-9}M), or a pH of 7.40 with a range of about 7.36 to 7.44. The range of extracellular $[H^+]$ that is compatible with life is roughly 100 to 16 nM (pH 7.00 to 7.80), although people may survive more extreme transient deviations. Intracellular pH is about 7.00, whereas the neutral pH (when $[H^+] = [OH^-]$) at 37°C is 6.70; neutral pH is lower with fever because K'_w increases with increasing temperature. Under normal dietary conditions as well as within a broad range of acid or base loads, both **physicochemical** and **physiologic buffering systems** tightly maintain pH in the normal range.

Buffering of CO_2

As the CO_2 produced by normal body catabolism enters the capillary blood, it is buffered primarily and initially by hemoglobin within erythrocytes (RBCs) (Fig. 38-2). Although this CO_2 is initially uncharged, it freely enters the RBC, where within milliseconds carbonic anhydrase catalyzes its combination with H_2O to form H^+ and H_2CO_3. Within microseconds, H_2CO_3 dissociates to form H^+ and HCO_3^-, the HCO_3^- exchanging for plasma Cl^- through anion antiporters (band 3 protein)

in the RBC membrane (the chloride shift). In this manner, about 80 percent of the CO_2 generated is transported as HCO_3^- in the plasma en route to the lungs for excretion.

CO_2 is also rapidly and nonenzymatically bound to the alpha-amino groups of deoxyhemoglobin (Hgb) to form carbamino groups and produce within the RBC:

$$Hgb \cdot R\text{-}NH_2 + CO_2 \leftrightarrow Hgb \cdot R\text{-}NHCOO^- + H^+$$
$$\text{Eq. 38-23}$$

Hemoglobin has ionized groups with strengths that change in keeping with the state of hemoglobin oxygenation. Thus, when oxyhemoglobin loses its O_2, it can accept H^+, which combines with the NH moiety on the imidazole ring:

$$O_2 \cdot Fe \cdot Hgb\text{-}NH + H^+ \leftrightarrow O_2 + Fe \cdot Hgb\text{-}NH_2^+$$
$$\text{Eq. 38-24}$$

This allows more CO_2 to react with H_2O and be carried in the plasma as HCO_3^-. Thus, at a constant P_{CO_2}, the CO_2 content of deoxygenated hemoglobin is greater than that of oxygenated blood. This is known as the **Haldane effect.**

Physiologic Buffering

The fixed acids, CO_2/HCO_3^- buffer pair, hemoglobin within RBCs, plasma proteins, and the phosphate buffer systems—all participate in the **initial physicochemical buffering** in the blood. The CO_2/HCO_3^- buffer system has a major role, and its relationship to $[H^+]$ is described by the **Henderson-Hasselbalch equation,** which is a logarithmic rearrangement of Equation 38-21:

$$pH = pK' + \log [HCO_3^-]/[H_2CO_3] \quad \text{Eq. 38-25}$$

in which H_2CO_3 is in equilibrium with dissolved CO_2 and is equal to the $PaCO_2$ multiplied by 0.03, which is the solubility coefficient at 37°C in plasma. Thus, at normal acid-base equilibrium:

$$
\begin{aligned}
pH &= 6.1 + \log 24 \text{ mM}/(40 \times 0.03) \text{ mM} \\
&= 6.1 + \log 24 \text{ mM}/1.2 \text{ mM} \\
&= 6.1 + \log 20 \\
&= 7.40 \quad\quad\quad\quad\quad\quad\quad \text{Eq. 38-26}
\end{aligned}
$$

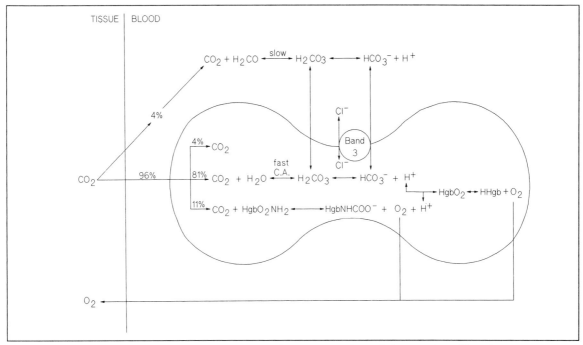

Fig. 38-2. CO_2 distribution in tissue capillary blood and transport and metabolism in the red blood cells. (CA = carbonic anhydrase; *Hgb* = hemoglobin; *Band 3* = anion exchanger. Percentages indicate the approximate distributions.)

Where 6.1 is the pK' of the CO_2/HCO_3^- buffer pair, 24 mM is the normal plasma $[HCO_3^-]$, and 40 mm Hg is the normal $PaCO_2$.

Based on the characteristics of ideal buffers, described earlier in this chapter, the CO_2/HCO_3^- pair should be a poor buffer system because of its pK of 6.1. Theoretically, the $H_2PO_4^-/HPO_4^{2-}$ buffer pair should be the better system, in that the linear portion of its titration curve lies within the 7.00 to 7.80 pH range. This paradox regarding the **primacy of the CO_2/HCO_3^- system** is due to two of its salient features: (1) the prompt excretion of CO_2 by the lungs and its maintenance at 40 mm Hg by the adjustment of ventilation to CO_2 production; and (2) its abundance in the body.

To illustrate the operation of the physicochemical buffer system, the response to a **fixed acid load** is considered. If a strong acid is added to a closed, unbuffered system, the pH plummets, as seen for water (see Fig. 38-1). If a simplified plasma containing $NaHCO_3$ is incorporated into the system, there is prompt physicochemical buffering of the added acid. For example:

$$5H^+ + 5Cl^- + 10Na^+ + 10HCO_3^- \rightarrow 10Na^+ + 5Cl^- + 5HCO_3^- + 5H_2CO_3 \qquad \text{Eq. 38-27}$$

Although the H_2CO_3 rapidly dissociates to CO_2 and H_2O, the degree of buffering would be limited if the CO_2 were trapped in a closed system.

However, because the physiologic buffering system is open, the CO_2 released in the reaction can be excreted by the lungs. Thus, PCO_2 decreases and this reduces the $[H_2CO_3]$ and $[H^+]$ (Equations 38-19 and 38-20). Therefore, the $[HCO_3^-]/[H_2CO_3]$ ratio become closer to its normal value at pH 7.40 (see Equation 38-26). More importantly, the **increased arterial $[H^+]$** rapidly signals the respiratory centers to increase ventilation, which further decreases $PaCO_2$ and brings the $[HCO_3^-]/[H_2CO_3]$ ratio and all other buffer pairs even closer to their normal values.

Within minutes, further buffering takes place in soft tissue and bone, where proteins and organic phosphates are the principal participants. Finally, over the course of a few days, the kidney excretes the excess acid, thereby increasing plasma SID and reclaiming HCO_3^- used during initial buffering. At this point, the body $[H^+]$ returns to normal.

Under normal **steady-state conditions**—a diet consisting of 70 gm of protein per day which generates about 70 mEq of acid—approproximately 30 mM H^+ is excreted as titratable acids (PO_4, SO_4, and so on) and the remainder excreted as ammonium by the NH_4^+/NH_3 urinary buffer system. In the renal response to reclaiming the HCO_3^- expended in the titration of acid loads, the **NH_4^+/NH^3 buffer** pair in the urine is quantitatively more important (see Chapter 39 for details).

DISORDERS OF ACID-BASE BALANCE

When the control mechanisms that maintain acid-base balance either become deranged or are overwhelmed by other pathologic processes, clinical disorders of acid-base balance occur. These can precipitate **acidosis,** an abnormal condition that tends to retain acid or lose base, or **alkalosis,** in which base is retained or acid is lost. These disorders are further classified as either **respiratory,** in which the clearance of CO_2 through the lungs is the primary disturbance, or **metabolic,** in which the extrapulmonary organs are the sites of the primary disturbance.

The primary disorder itself should be differentiated from the actual $[H^+]$ of the blood. **Acidemia** and **alkalemia** refer to a blood $[H^+]$ above or below the normal limits, respectively; blood pH, of course, would be the inverse of this. The distinction between, for example, acidosis and acidemia is important because more than one primary disorder can co-exist in a patient; thus,

coexisting acidosis and alkalosis may produce a normal arterial $[H^+]$. These are called **mixed disturbances.** To diagnose these disorders clinically, the following more simplified approach than that given in the foregoing theoretical analysis is applied.

Classification

The four basic acid-base disorders can be defined in terms of the perturbations of the variables of the **Henderson-Hasselbalch equation,** and are summarized in Table 38-1, with the primary disturbance indicated in bold type. Although the plasma $[HCO_3^-]$ is a dependent variable, it reflects changes in SID, which would otherwise be difficult and cumbersome to calculate.

For the respiratory disorders, in which CO_2 is primarily disturbed, the blood $[HCO_3^-]$ and $[H^+]$ are altered by changes in $PaCO_2$ and a set of compensatory responses. In the metabolic disorders, the plasma SID is altered primarily and this is reflected in changes in the plasma $[HCO_3]$. A change in the $PaCO_2$ compensates for this. All of these variables are commonly determined by clinical laboratories.

Compensation

A primary disturbance in the respiratory system activates a metabolic response that serves to minimize or buffer the magnitude of the disturbed $[H^+]$ from its normal concentration and vice versa. The net response by the buffer systems that are not primarily disturbed is called the **compensatory response.**

The physicochemical buffering responses occur rapidly and include the participation of buffers in the interstitial and intracellular compartments as well as the intravascular compartment. In the respiratory compensation of primary metabolic disorders, changes in plasma $[H^+]$ and brain extracellular fluid $[H^+]$ affect the neuronal output from peripheral chemoreceptors (the carotid bodies) and central chemoreceptors (mostly located near the ventral surfaces of the medulla oblongata). These receptors then cause the med-

Table 38-1. Characteristics of Primary Acid-Base Disorders*

Type	pH	$[H^+]$	$PaCO_2$	$[HCO_3^-]$	SID
Respiratory acidosis	Decreased	Increased	**Increased**	Increased	Increased
Respiratory alkalosis	Increased	Decreased	**Decreased**	Decreased	Decreased
Metabolic acidosis	Decreased	Increased	Decreased	Decreased	**Decreased**
Metabolic alkalosis	Increased	Decreased	Increased	Increased	**Increased**

$PaCO_2$ = arterial CO_2 tension; SID = strong ion difference.
*Bold type indicates the primary disturbance (see text).

ullary respiratory centers to stimulate respiratory muscles via the phrenic and intercostal nerves, resulting in increased alveolar ventilation (see details in Chapter 30). This **respiratory response** is complete within minutes.

In contrast to these two rapid responses, the **maximum compensatory metabolic response** by the kidneys for primary respiratory disorders takes from two to five days to effect changes in SID and HCO_3^-. This response buffers the change in $[H^+]$ induced by the rise in $PaCO_2$. The renal mechanisms involved in this response are covered in Chapter 39.

Because these compensatory responses are physiologic, they are **normal** and **predictable,** in that they can be described in **mathematical terms.** The magnitudes of the responses have been derived empirically from clinical observations and are not otherwise discoverable. While different studies show variability in the compensation formulas, probably because of the populations or species studied and the analytical methods used, reasonably widely accepted formulas are offered (Table 38-2), expressed in terms of the variables of the Henderson equation.

Because both the physicochemical and respiratory buffering responses are rapid, the duration of the metabolic disorder is not considered and only one factor is provided for each disorder (Table 38-2). In contrast, because the renal compensatory response in the primary respiratory disorders is slow, the duration of the disorder must be considered in order to determine the appropriateness of compensation, making factors for both acute and chronic disorders necessary. To illustrate the use of these factors, in a simple

Table 38-2. Factors for Predicting Compensation in Simple Acid-Base Disorders

Disorder	Compensation Coefficient*
Metabolic acidosis	$(-)$ 11 to 15 mm Hg $PaCO_2/$ $(-)$ 10 mM $[HCO_3^-]$
Metabolic alkalosis	$(+)$ 6 to 7 mm Hg $PaCO_2/$ $(+)$ 10 mM $[HCO_3^-]$
Respiratory acidosis	
Acute	$(+)$ 1 mM $[HCO_3^-]/$ $(+)$ 10 mm Hg $PaCO_2$
Chronic	$(+)$ 3.5 mM $[HCO_3^-]/$ $(+)$ 10 mm Hg $PaCO_2$
Respiratory alkalosis	
Acute	$(-)$ 2.5 mM $[HCO_3^-]/$ $(-)$ 10 mm Hg $PaCO_2$
Chronic	$(-)$ 5.0 mM $[HCO_3^-]/$ $(-)$ 10 mm Hg $PaCO_2$

*The use of the coefficient to calculate change assumes the following normal values: serum $[HCO_3^-]$ = 24 mM and $PaCO_2$ = 40 mm Hg.

metabolic acidosis with a serum $[HCO_3^-]$ of 14 mM, the change in serum $[HCO_3^-]$ from normal is $(-)$10 mM $(24 - 14 = 10)$ and the predicted change in $PaCO_2$ is $(-)$13 mm Hg, or an absolute value of 27 mm Hg $(40 - 13 = 27)$.

Figure 38-3 is a composite acid-base map that depicts these relationships. In it, $PaCO_2$ is plotted against serum $[HCO_3^-]$ and also shown are isopleths for $[H^+]$ that were calculated from the Henderson-Hasselbalch equation. This is a variant of the $pH–HCO_3$ diagram popularized by Davenport, in which $PaCO_2$ is plotted as isopleths, but is preferred because $[H^+]$ is clearly a dependent variable in this format. The *heavy dashed lines*

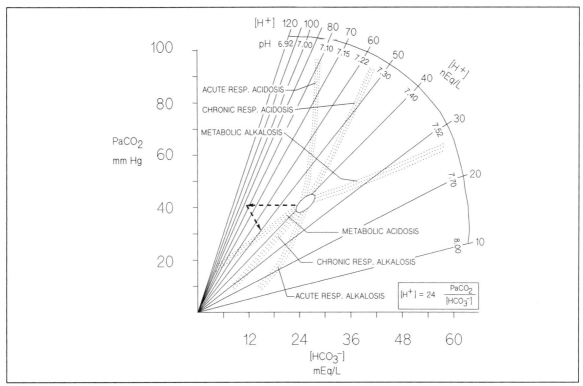

Fig. 38-3. Relationship of arterial CO_2 tension ($PaCO_2$) with HCO_3^- concentration ($[HCO_3^-]$); isopleths for the H^+ concentration ($[H^+]$) are also shown. The darkened lines indicate the ranges of values expected for the primary acid-base disorders, which are appropriately compensated for. The changes in the three variables expected for a metabolic acidosis with and without respiratory compensation are shown.

describe a simple effect of respiratory compensation in metabolic acidosis. If the serum $[HCO_3^-]$ decreases from about 26 to 11 mEq/liter without respiratory compensation, the arterial $[H^+]$ would be 90 nEq liter (pH, 7.05); with respiratory compensation, the $PaCO_2$ falls to 23 mm Hg and the resultant $[H^+]$ is only 65 nEq/liter (pH, 7.26). The *shaded areas* denote the confidence limits for the four basic types of simple disorders; the effect of the duration of both acute and chronic respiratory disorders on compensation is also shown. As a corollary to these compensation formulas, if a particular primary disorder does not conform to the predicted compensation, this implies the existence of at least one other primary disorder, a mixed disorder.

Clinical Diagnosis of Acid-Base Disorders

For clinical purposes, the **arterial blood-gas analysis** provides both $PaCO_2$ and pH or $[H^+]$, or both, from which the plasma $[HCO_3^-]$ may be calculated. In addition, the serum total CO_2 content, which represents the sum of the concentrations of HCO_3^- and the different forms of CO_2 in the blood, the latter normally about 1.2 mM, is usually directly determined. The application of a few simple steps yields the correct acid-base diagnosis.

First, the prudent diagnostician can validate the accuracy of laboratory data by comparing the calculated plasma $[HCO_3^-]$ with the directly determined serum total CO_2 concentration. If these values agree within ± 3 mM, the data can be considered accurate; if not, the data are internally inconsistent and one of the values is wrong.

If the plasma [HCO_3^-] has not been calculated from the blood-gas analysis, it can be done easily by applying the **Henderson-Hasselbalch equation.** To use the arithmetic form of this equation (the Henderson equation), pH units must be converted to [H^+] (Table 38-3). $PaCO_2$ and [H^+] are then substituted in the following equation, which is a rearrangement of Equation 38-21:

$$[H^+] \text{ (nM)} = 24 \times PaCO_2 \text{ (mm Hg)}/[HCO_3^-] \text{ (mM)}$$
$$\text{Eq. 38-28}$$

The clinical laboratory data must always conform to this equation. This equation is helpful clinically because the two variables, PCO_2 and [H^+], can be measured reliably and reproducibly in easily available and comparatively inexpensive apparatus. (Unfortunately, some have interpreted the equation to mean that [H^+] and [HCO_3^-] are independent variables; indeed, some textbooks on acid-base indicate that changing [H^+] primarily affects [HCO_3^-] and vice versa. From the foregoing discussion it should be clear that only one independent variable is being measured—PCO_2. The other independent variables that determine [H^+] and [HCO_3^-] are not a part of the equation.)

The values for three components of the Henderson-Hasselbalch equation are then examined to determine the dominant type of acid-base disorder (see Table 38-2). For example, predominant metabolic acidosis would be implied by decreases in plasma [HCO_3], $PaCO_2$, and pH. Next, to distinguish between a simple (single) or mixed (multiple) disorder, a simple disorder is first assumed and this is tested by applying the appropriate compensation factor. If, in this example of a metabolic acidosis, the predicted compensatory respiratory response is calculated to have a $PaCO_2$ of 20 mm Hg and 20 mm Hg is measured, this confirms a single disorder and the existence of only a metabolic acidosis. If, on the other hand, 30 mm Hg is measured, this higher than predicted $PaCO_2$ indicates a mixed disorder that includes both a primary metabolic acidosis and a primary respiratory acidosis. Two or more disorders can co-exist in a patient, but a consideration of the diagnostic challenge of such cases is beyond the objectives of this text.

Metabolic Acidosis

Metabolic acidoses arise either because of a change in the proportions of the measured anions (Cl^- and HCO_3^-) or accumulation of other strong acids not usually measured in the blood; both represent decreases in SID. When there is a **change in the proportions of measured anions,** chloride is retained and HCO_3^- is excreted abnormally, as seen in renal tubular acidosis and diarrhea. **Strong acids can accumulate** because of ingestion, excess production beyond the excretory capacity, or impaired excretion per se, such as caused by ketoacidosis (acetoacetate and beta-hydroxybutyrate), lactic acidosis, uremia, and toxins.

In **lactic acidosis,** for example, the SID decreases because lactate, with a pK of 3, behaves

Table 38-3. Approximation of [H^+] to Corresponding pH Units

To approximate the [H^+] corresponding to a given pH unit, begin with the datum: pH 7.00 equals 100 nM [H^+]. Multiply this value by 0.8 to obtain an estimate of the [H^+], corresponding to the next higher 0.1 pH unit. Repeat this operation to obtain the [H^+] corresponding to successive 0.1-increments in pH units. Interpolate for values between succcessive 0.1-pH units.

pH and [H^+]

pH	7.00	7.10	7.20	7.30	7.40	7.50	7.60	7.70	7.80
	$=$	$\approx$	$\approx$	$\approx$	$\approx$	$\approx$	$\approx$	$\approx$	$\approx$
[H^+]	100	80	64	52	40	32	25	20	16

like a strong anion and $[HCO_3^-]$ decreases as it reacts with and buffers the increase in $[H^+]$. In **renal tubular acidosis,** the SID decreases with an increase in plasma $[Cl^-]$ and a direct decrease in $[HCO_3^-]$. Anions (A^-), which buffer the rise in $[H^+]$, decrease to form HA (Equation 38-15). Respiratory compensation occurs with increased ventilation to excrete CO_2 and lower $[PaCO_2]$, which, as an independent variable, offsets the increase in $[H^+]$.

Metabolic Alkalosis

Metabolic alkaloses are characterized by an **increase in SID** and the **tendency to accumulate base in the extracellular fluid.** In one major group of disorders, the **chloride-depletion alkaloses,** SID increases because of the loss of chloride from the body, which leads to retention of base, primarily as HCO_3^-. Loss of gastric fluid and chloruretic diuretic use are examples of conditions that cause this form of alkalosis. In the other major group, the **potassium-depletion alkaloses,** the increase in SID arises from a shift of acid into the intracellular compartment; these disorders usually occur when there is mineralocorticoid excess such as aldosterone-producing adenomas or severe hypertension. In both types of alkalosis, plasma proteins and other buffers release H^+ according to the formula given in Equation 38-15. With the fall in $[H^+]$, a decrease in ventilation produces respiratory compensation that raises the $PaCO_2$, which in turn ameliorates the initial decrease in $[H^+]$.

Respiratory Acidosis

The central event in respiratory acidoses is **diminished clearance of CO_2** or hypoventilation, which produces hypercapnea, or an elevated $PaCO_2$ (see Chapter 28). Among the main causes of CO_2 retention, the most common are **disorders of the lung parenchyma** that interfere with CO_2 excretion because of ventilation and perfusion mismatch (emphysema, asthma, and adult respiratory distress syndrome). **Neuromuscular diseases,** such as poliomyelitis and Guillain-

Barré syndrome, which impair respiratory pump muscles, and **drugs,** such as morphine and anesthetics, which suppress the neuronal circuitry of breathing, cause global hypoventilation and elevate $PaCO_2$.

An increase in PCO_2, as an independent variable, increases $[H^+]$ (Equation 38-19). However, a number of biochemical reactions are initiated that minimize the rise in $[H^+]$. Weak acids, such as hemoglobin and proteins, titrate $[H^+]$ according to Equation 38-15. SID increases (becomes more positive) as Cl^- moves into RBCs in exchange for HCO_3^- (chloride shift) (see Fig. 38-2) and as Na^+ and K^+ move from the intracellular into the extracellular fluid. In **acute respiratory acidosis,** both the rise in SID and the reactive weak acids (fall in A^-) elevate plasma $[HCO_3^-]$ and reduce $[H^+]$.

In **chronic respiratory acidosis,** in which high PCO_2 levels persist for a few days, SID becomes more negative with a further rise in plasma $[HCO_3^-]$ and a further drop in $[H^+]$. The change in SID is due to an increase in renal Cl^- excretion, which leads to a fall in the plasma $[Cl^-]$ and the generation of new bicarbonate. The excretion of H^+ occurs through the synthesis of NH_3, which forms NH_4^+ (see Chapter 39).

Respiratory Alkalosis

The central event in respiratory alkalosis is the **excess clearance of CO_2** (an increase in alveolar ventilation disproportionate to CO_2 production) which results in reduced $PaCO_2$ (hyperventilation). The most common causes of respiratory alkalosis are **pregnancy, high altitude, lung diseases,** and **liver diseases.** Other causes include endotoxinemia, hysteria, and any condition associated with severe hypoxemia.

As an independent variable, the decrease in $PaCO_2$ lowers $[H^+]$, as the reaction (Equation 38-19) is shifted to the left by the accelerated CO_2 excretion through the lungs. Compensatory mechanisms mitigate such decreases in $[H^+]$. SID becomes less positive as Cl^- and lactate move into the extracellular fluid and Na^+ and K^+ move out. Lactate concentrations increase because of the heightened activity of certain glycolytic enzymes secondary to intracellular alkalosis. Organic anions and plasma proteins (A^-) increase in accordance with Equation 38-15. In the face of a **low** **PaCO$_2$** that persists for three to five days, renal adaptive mechanisms are maximized that further raise the plasma [Cl$^-$]. This then elicits a fall in [HCO$_3$$^-$] and a rise in [H$^+$] (see Chapter 39).

SUMMARY

The [H$^+$] of the body is tightly regulated by and is a function of three independent variables: the SID, the A$_{TOT}$ and nature of the weak acids (buffers), and Pco$_2$ set by alveolar ventilation. The simultaneous solution of the expressions for these independent variables and their associated relationships permit calculation of the in vivo [H$^+$]. For physiologic buffering, nonvolatile (fixed) acids are affected by rapidly responsive physicochemical buffers, mainly proteins and phosphates, in the extracellular and intracellular spaces, and by changes in the alveolar ventilation that adjust Pco$_2$. CO$_2$, the major product of catabolism and a volatile acid, is buffered in the blood primarily by hemoglobin and promptly excreted by the lungs. A defect or overload in any of these mechanisms disrupts acid-base balance, and such disorders are classified as acidosis or alkalosis and further defined as respiratory or metabolic, depending on the nature of the defect. These four acid-base disturbances can be understood in terms of the major independent variables.

NATIONAL BOARD TYPE QUESTIONS

Select the one most correct answer for each of the following

1. A comatose patient is seen in the emergency room with an arterial blood pH of 7.20, a PaCO$_2$ of 72 mm Hg, and a plasma [HCO$_3$$^-$] of 27 mM. This implies which acid-base disorder(s)?
   A. Metabolic acidosis
   B. Acute respiratory acidosis
   C. Chronic respiratory acidosis
   D. Both A and B
   E. Data are internally inconsistent

2. Which of the following blood constituents does not independently influence arterial pH?
   A. Chloride
   B. Protein
   C. Bicarbonate
   D. Phosphate
   E. Hemoglobin

3. A patient who has been vomiting after having consumed junk food for several days has an arterial blood pH of 7.50, a PaCO$_2$ of 48 mm Hg, and a plasma [HCO$_3$$^-$] of 37 mM. Which acid-base disorder(s) is (are) likely?
   A. Metabolic alkalosis
   B. Chronic respiratory alkalosis
   C. Acute respiratory acidosis
   D. Both A and C
   E. Data are internally inconsistent

4. A 35-year-old woman with a known history of psychiatric problems is comatose when brought to the emergency room. There are no signs of trauma. Physical examination reveals: blood pressure, 85/50; pulse, 94 bpm and regular; and respirations, 8 to 10 per minute and shallow. The arterial blood pH is 7.05, PaCO$_2$ is 60 mm Hg, and serum total CO$_2$ 15 mM. Which acid-base disorder(s) is (are): likely?
   A. Metabolic acidosis
   B. Respiratory acidosis
   C. Both A and B
   D. Metabolic alkalosis
   E. Data are internally inconsistent

5. A 48-year-old male smoker has had cough with sputum production for 20 years and dyspnea on exertion for 10 years. Five years ago, cor pulmonale was diagnosed and treated. Repeated bouts of respiratory infection and exacerbation of respiratory insufficiency characterized his course. His arterial pH was 7.37; PaCO$_2$, 50 mm Hg, and serum [HCO$_3$], 28 mM. The likely acid-base disorder(s) present is (are):

A. Metabolic acidosis

B. Acute respiratory acidosis

C. Chronic respiratory acidosis

D. Chronic respiratory acidosis and metabolic alkalosis

E. Data are internally inconsistent

ANNOTATED ANSWERS

1. B. From Table 38-3, pH 7.20 approximates [H^+] of 64 nM. From Equation 38-28, 24 × 72/64 = 27 mM, which makes the data consistent. From Table 38-1, the increases in [H^+], $PaCO_2$, and plasma [HCO_3^-] conform to the pattern for respiratory acidosis. From Table 38-3, the small 3-mM increment in plasma [HCO_3^-] for the 32–mm Hg increment in $PaCO_2$ conforms to the predicted compensation for an acute respiratory acidosis. This is expected because full metabolic compensation for respiratory disorders requires 2 to 5 days to take effect. Thus, there is no additional disorder.

2. C. Bicarbonate is *dependent* on PCO_2, the SID, and A_{TOT}. Cl^-, as a strong ion, varies independently according to intake and renal excretion and does not participate in reactions; it is a major determinant of SID. Protein and phosphate concentrations are determined by intake and cellular metabolism; as weak acids, their total concentrations vary independently of changes in the acid-base balance. The concentration of hemoglobin is regulated by multiple elements controlling the erythron and operating independently of acid-base balance.

3. A. From Table 38-3, pH 7.50 approximates [H^+] of 32 nM. From Equation 38-28, 24 × 48/32 = 36 mM, which means the data are consistent. From Table 38-1, the decrease in [H^+] and the increases in $PaCO_2$ and plasma [HCO_3^-] conform to the pattern for metabolic

alkalosis. From Table 38-2, the 8-mm Hg increment in $PaCO_2$ for the 13-mm increment in plasma [HCO_3^-] conforms to the compensation for a metabolic alkalosis. Thus, no additional disorder is present.

4. C. From Table 38-3, pH 7.05 approximates [H^+] of 90 nM. From Equation 38-28, 24 × 60/90 = 16 mM; the data are therefore consistent. From Table 38-1, the increases in [H^+] and $PaCO_2$ conform to the pattern for respiratory acidosis, but the increase in [H^+] and the decrease in plasma [HCO_3^-] conform to the pattern for metabolic acidosis. From Table 38-3, the 9-mM decrement in plasma [HCO_3^-] would predict a 12-mm Hg decrement in $PaCO_2$ of from 40 to 28 mm Hg. Instead, a 20-mm Hg increment in $PaCO_2$ has occurred; the $PaCO_2$ is higher than that predicted for a simple metabolic acidosis. Thus, a mixed disorder is present, involving both primary metabolic acidosis and primary respiratory acidosis.

5. C. From Table 38-3, pH 7.37 approximates [H^+] of 43 nM. From Equation 38-28, 24 × 50/43 = 28 mM; the data are therefore consistent. From Table 38-1, the increase in [H^+] and the increases in $PaCO_2$ and plasma [HCO_3^-] conform to the pattern for respiratory acidosis. From Table 38-2, the 4-mM increment in plasma [HCO_3^-] for the 10-mm Hg increment in $PaCO_2$ conforms to the compensation predicted for a chronic respiratory acidosis. Thus, no additional disorder exists.

BIBLIOGRAPHY

Davenport, H. W. *The ABC of Acid-Base Chemistry*, 6th ed. rev. Chicago: University of Chicago Press, 1974.

Seldin, D. W., and Giebisch, G. *The Regulation of Acid-Base Balance.* New York: Raven Press, 1989.

Stewart, P. A. Modern quantitative acid-base chemistry. *Can. J. Physiol. Pharmacol.* 61:1444-1461, 1983.

39 Renal Acid-Base Regulation and Micturition

ROBERT O. BANKS

Objectives

After reading this chapter you should be able to:

Describe the role of the kidneys in regulating extracellular pH

Explain the process involved in urinary acidification: the events in both the proximal and distal tubule, formation of titratable acidity, ammonia production, importance of carbonic anhydrase, and production of new HCO_3^-

Define nonionic diffusion and explain how it influences the excretion of ammonium and other weak bases and acids

List the factors that influence HCO_3^- excretion

Compare the renal participation in acid-base balance during acute and chronic alkalosis and acidosis

Describe the process of micturition

A large portion of the diet in many Western countries consists of protein, and persons on such a high-protein diet metabolize some 50 to 100 mEq of so-called nonvolatile or fixed inorganic and organic acids per day. In contrast to the H^+ that derives from the hydration of CO_2 and is eliminated through the lungs, the nonvolatile acids must first be buffered by bicarbonate and hemoglobin and then excreted by the kidney. Nonvolatile acid production can increase substantially under both normal and pathologic conditions. Thus, lactic acid production is significantly enhanced during heavy exercise and hypoxia. In other conditions, such as uncontrolled diabetes mellitus, the production of nonvolatile acids can increase by as much as tenfold. Diets rich in vegetables and fruits generate nonvolatile products that are alkalis, but these products must also be excreted by the kidney.

The kidney participates in the **overall regulation of acid-base balance** in two ways: (1) it reclaims virtually all of the filtered HCO_3^- and (2) it excretes acid. **Acid excretion** is accomplished through these routes. The first is excretion of free H^+; because the minimum pH of the urine is about 4.5, this avenue accounts for less than 0.5 percent of the total H^+ excreted per day. The second route is excretion of titratable acid and the third is ammonia production and the resultant ammonium excretion. Adjustments in the third route of acid excretion primarily explain the increased H^+ excretion seen during a metabolic acidosis. The enhanced production of ammonia by the kidney during acidemia takes four to five days to reach a maximum (as will be discussed).

BICARBONATE REABSORPTION

When the plasma HCO_3^- concentration is less than 26 to 28 mEq/liter, virtually all the **filtered HCO_3^- is reabsorbed.** The bulk of this reabsorption occurs in the **early proximal tubule** (within the first 20 to 30 percent of the total proximal tubule) by means of H^+ secretion (see Fig. 34-1). Most of the H^+ secreted in the proximal tubule derives from **Na^+—H^+ exchange.** However, approximately 20 percent of the HCO_3^- reabsorption in the proximal tubule can continue following inhibition of Na^+ transport, suggesting that H^+ secretion in the proximal tubule can also be mediated by a **H^+-ATPase, electrogenic process.**

Secreted H^+ combines with filtered HCO_3^-, thereby generating H_2CO_3. Because the brush border of the proximal tubule contains **carbonic anhydrase,** the H_2CO_3 instantaneously dissociates into CO_2 and H_2O. These products diffuse passively across the luminal membrane and, in the presence of carbonic anhydrase in the cell, are converted back into H_2CO_3 and then to H^+ plus HCO_3^-. The HCO_3^- then moves passively across the peritubular membrane (evidence suggests that this step may be a carrier-mediated, $H^+ = 3HCO_3^-$ cotransport process). The net effect of this series of events is the reabsorption of $NaHCO_3$.

The small amount of filtered HCO_3^- that leaves the proximal tubule (about 10 to 20 percent of the filtered load) is reabsorbed in the **distal tubule** by a similar mechanism (see Fig. 38-1). The major difference, however, is that there is little if any carbonic anhydrase in the brush border of the distal tubule (it is only found in the cells of the distal tubule). On the other hand, the luminal pH of the distal tubule can fall as low as 4.5 (versus 7.0 in the proximal tubule) because a H^+-ATPase pump exists on the luminal membrane. This relatively high H^+ concentration shifts the H^+ plus HCO_3^- reaction to the right (to H_2CO_3 and to CO_2 plus H_2O), and therefore promotes reabsorption of essentially all HCO_3^- in the distal tubule. The reabsorption of HCO_3^- across the basolateral membrane in the distal tubule appears to be mediated by a Cl^--HCO_3^- exchanger. Thus, on a normal protein diet with 50 to 100 mEq of fixed acid formed each day, virtually all the filtered HCO_3^- is reabsorbed (180 liters/day $\times$ 24 mEq/liter = 4,320 mEq/day filtered; 5 mEq/day is normally excreted and 4,315 mEq/day is reabsorbed). It is important to note that 4,315 mEq of H^+ was secreted in order to reabsorb the 4,315 mEq of HCO_3^-. This H^+, however, is not excreted. Furthermore, the HCO_3^- consumed during the initial plasma buffering of the fixed acid has not been replaced; instead HCO_3^- is replaced through the formation of titratable acid and excretion of NH_4^+, as will be discussed.

HCO_3^- reabsorption is affected by a number of factors (Table 39-1). Two of these factors, the plasma HCO_3^- concentration and the extracellular fluid volume, are illustrated in Fig. 39-1. As the plasma HCO_3^- concentration is increased by infusing solutions of $NaHCO_3$, there is a resulting increase in the extracellular fluid volume as well as the plasma HCO_3^- concentration. Under those conditions, as is illustrated by the lines marked "expanded" in Fig. 39-1, the amount of HCO_3^- reabsorbed attains a plateau value at high plasma HCO_3^- concentrations. That is, HCO_3^- reabsorption appears to be characterized by a transport maximum. However, when the plasma HCO_3^- is increased with minimal expansion of the extracellular fluid space, as illustrated by the lines marked "nonexpanded" in Fig. 39-1, virtually all the filtered HCO_3^- is reabsorbed, even at high plasma HCO_3^- concentrations. The difference between HCO_3^- reabsorption in the nonexpanded and the expanded states may be related to the fact that reabsorptive events in the proximal tubule decrease during expansion of the extracellular fluid space and increase during volume contraction; these phenomena may be related to changes in the balance of Starling reabsorptive forces between the postglomerular plasma and the renal interstitial fluid space.

Table 39-1. Factors Affecting HCO_3^- Reabsorption

Factors that Increase HCO_3^- Reabsorption	Factors that Decrease HCO_3^- Reabsorption	Mechanism
MAJOR FACTORS		
Volume contraction	Volume expansion	Probably via changes in proximal Na^+ reabsorption
Increased filtered HCO_3^- load	Decreased filtered HCO_3^- load	Also may be via proximal Na^+ reabsorption
Hypercapnia	Hypocapnia	Possibly via changes in cell pH
Hypokalemia	Hyperkalemia	Possibly via changes in cell pH
	Carbonic anhydrase inhibition	Carbonic anhydrase is essential for HCO_3^- reabsorption in proximal tubule
MINOR FACTORS		
Hypercalcemia	Hypocalcemia	A combination of direct tubular effects, hemodynamic effects, and external buffering of acid govern the minor factors
PTH deficiency	PTH excess	
Vitamin D	Vitamin D deficiency	
Thyroid hormone	Phosphate depletion	
Glucose	Hypothyroidism	
	Basic amino acids (lysine, arginine)	
	Maleic acid	

PTH = parathyroid hormone.

EXCRETION OF TITRATABLE ACID

The ability of a weak acid to participate in the formation of titratable acid depends on the quantity of the buffer present and on its weak-acid dissociation constant (pK_a) (maximum buffering occurs within ± 1 pH unit of the pK_a). **HPO_4** (pK_a, 6.8) is the major urinary titratable acid buffer. **Uric acid** (pK_a, 5.75) and **creatinine** (pK_a, 4.97) are present but in smaller amounts. (In uncontrolled diabetes mellitus the large quantities of keto acids produced, such as beta-hydroxybutyrate [pK_a, 4.8], contribute to the urinary buffering.) The formation of titratable acid is illustrated in Figure 39-2. It is important to note that the production of titratable acid generates "new HCO_3^-," which is the HCO_3^- that replaces the buffer base consumed in the initial buffering process of the fixed acid.

AMMONIA PRODUCTION AND ACID EXCRETION

The sum of ammonium and titratable acidity minus the amount of excreted HCO_3^- equals the **net acid excretion.** As already noted, the net acid excretion for a normal protein diet is about 50 to 100 mEq/day. Approximately 30 mEq of this H^+ burden is normally excreted as NH_4^+. This NH_4^+ excretion is also advantageous, because it also elicits the generation of "new" HCO_3^-.

As is illustrated in Figure 39-3, **ammonia** is synthesized in tubular cells in several regions of the nephron. Recent evidence, however, indicates that most is generated in the proximal tubule. Ammonia is derived from the metabolism of amino acids, particularly (but not solely) from **glutamine.** The **biochemical events** that characterize ammonia production via glutamine metabolism are illustrated in the following reaction:

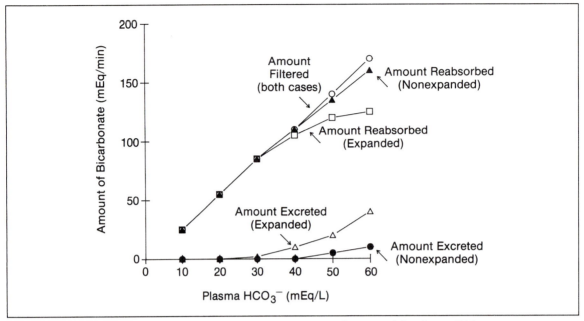

Fig. 39-1. Amounts of HCO_3^- filtered, reabsorbed, and excreted as a function of the plasma HCO_3^- concentration. Data from two groups of rats are illustrated, one group that was volume expanded during the process of increasing the plasma HCO_3^- concentration and a second group that was not. As can be seen, the nonexpanded group reabsorbed virtually all the filtered HCO_3^-, even at high plasma HCO_3^- concentrations. (Adapted from: Purkerson, M. L. et al. On the influence of extracellular fluid volume expansion on bicarbonate reabsorption in the rat. *J. Clin. Invest.* 48:1754–1760, 1969.)

$$\begin{array}{c} NH_3 \\ glutamine \rightarrow \uparrow \rightarrow glutamic\ acid \\ glutaminase \end{array}$$

$$\begin{array}{c} NH_3 \\ glutamic\ acid \rightarrow \uparrow \rightarrow alpha\text{-}ketoglutarate \\ glutamate\text{-}dehydrogenase \end{array}$$

Ammonia can readily diffuse into both the tubular lumen and plasma. There is evidence, however, that the transfer of ammonia and ammonium from the cell into the lumen may involve both an **active component** and a component dependent on a **Na$^+$-NH$_4^+$ antiport.** In addition, nonionic diffusion contributes significantly to the trapping of ammonia in the lumen of the tubule, as will be discussed. There is also evidence that a substantial portion of the ammonia secreted in the proximal tubule is reabsorbed (probably as NH_4^+) in the ascending limb

of the **loop of Henle;** most of the NH_4^+ excreted appears to derive from subsequent secretion into the collecting duct.

NONIONIC DIFFUSION

The NH_3 molecule is **lipid soluble** and can readily diffuse across the peritubular and luminal membranes but the charged component, NH_4^+, cannot. This phenomenon is termed **nonionic diffusion.** Because the pH of tubular fluid is lower than that of plasma, there is a preferential movement of NH_3 into the tubular lumen (the reaction, $NH_3 + H^+ \rightarrow NH_4^+$, provides a "sink" for the continued movement of NH_3 into the lumen). As noted in Figure 39-3, the secretion of NH_3 and subsequent excretion of NH_4^+ not only produces the net loss of H^+ but, as indicated, also leads to the generation of "new" HCO_3^-.

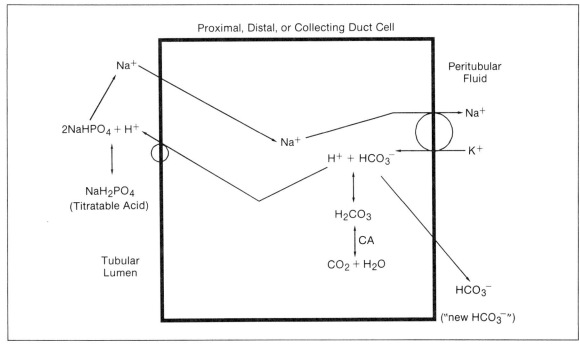

Proximal, Distal, or Collecting Duct Cell

Peritubular Fluid

Na^+

$2NaHPO_4 + H^+$

NaH_2PO_4
(Titratable Acid)

Na^+

$H^+ + HCO_3^-$

H_2CO_3

CA

$CO_2 + H_2O$

Na^+

K^+

HCO_3^-

("new HCO_3^-")

Tubular
Lumen

Fig. 39-2. The formation of titratable acid by tubular cells. During this process, dibasic weak acids (such as HPO_4^{2-}) are converted into the monobasic form. Consequently, Na^+ and newly synthesized HCO_3^- are reabsorbed; the "new" HCO_3^- replenishes that consumed during the initial buffering of the fixed acid. (*CA* = carbonic anhydrase.)

Nonionic diffusion is an important phenomenon that affects the excretion of all **weak acids and bases.** These, of course, can be endogenous substances, such as NH_3, or pharmacologic compounds. Thus, the rate of excretion of many drugs depends on (and can be regulated by adjusting) the **urinary pH.** For example, the excretion rate of aspirin (pK_a, 3.0) is relatively low when the urinary pH is between 4 and 5. By contrast, during treatment with acetazolamide (which causes a large increase in HCO_3^- excretion and a resultant urinary pH in the alkaline range), the clearance of aspirin approaches the glomerular filtration rate. In other words, weak acids are trapped in the compartment with the lower concentration of H^+, whereas weak bases are concentrated in the compartment with the higher concentration of H^+.

The major urinary adjustment during chronic **metabolic acidosis** is an increase in ammonia production (Fig. 39-4). As is illustrated in this figure, the production rate of **ammonia** during acidosis can increase markedly; as much as a tenfold increase in ammonia production has been reported. The increase in NH_3 production during acidosis takes four to five days to reach a maximum. This delay may be due to slow changes in the factors regulating entry of glutamine into the mitochondria or a delayed increase in the enzymatic activity involved in glutamine metabolism. These facts are further illustrated in Figure 39-5, which shows the net acid excretion in human subjects following ingestion of NH_4Cl. NH_4Cl is converted to ammonia and HCl in the liver, and this precipitates metabolic acidosis. As can be seen, virtually all the additional acid load is excreted through increased ammonia production, a process that requires several days before a steady-state is reached.

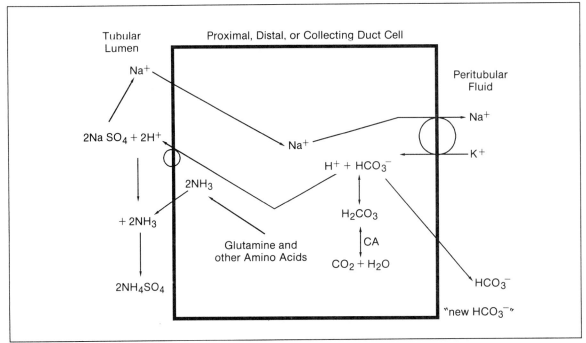

Fig. 39-3. The production of ammonia and its subsequent excretion. This process also replenishes the HCO_3^- consumed during the initial buffering of the fixed acid. (*CA* = carbonic anhydrase.)

Fig. 39-4. Urinary excretion of ammonia plus ammonium in control and chronically acidotic dogs. The urinary pH in both groups was initially about 5.0 but was rendered progressively alkaline by infusing $NaHCO_3$. (Data from: Pitts, R. F. Renal excretion of acid. *Fed. Proc.* 7:418–426, 1948.)

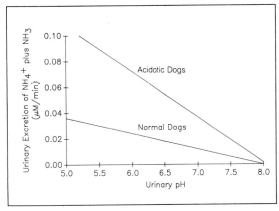

MICTURITION

Because of the positive pressure exerted in **Bowman's space,** tubular fluid moves along the entire nephron. Once in the ureter, urine is moved into the bladder under the combined influence of **gravitational forces** and **contractions of the ureteral musculature**. Urine is prevented from refluxing into the ureter by a valvular flap at the junction of the ureter with the bladder.

The **capacity of the bladder** can exceed 400 ml. As the bladder is initially distended, the first 30 to 50 ml of urine causes an increase in bladder pressure of about 10 to 20 cm H_2O. Continued filling then produces little change in pressure until the bladder capacity is reached. When the bladder volume is approximately 200 ml, afferent impulses in the pelvic nerves create the sensation of distention and initiate the desire to urinate. Voluntary control can be maintained during continued filling of the bladder, but, as filling exceeds

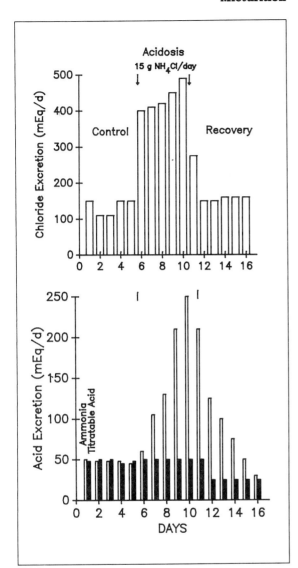

Fig. 39-5. Renal excretion of Cl⁻ and acid (ammonia and titratable acid plotted separately) in human subjects before, during, and after an acid load (15 gm of NH₄Cl per day; NH₄Cl is converted to HCl and ammonia by the liver). Cl⁻ excretion increases rapidly but ammonia excretion (the primary urinary buffer that increases in response to the acid load) requires several days to attain a steady-state. (Data from: Pitts, R. F. Renal excretion of acid. *Fed. Proc.* 7:418–426, 1948.)

400 ml, pressure can increase to 100 cm H₂O and involuntary micturition will result.

Initiation of micturition is normally a voluntary act and may be related to the willful relaxation of the external sphincter through inhibition of the pudendal nerves. The subsequent micturition reflex is an **automatic spinal reflex.** Activation of parasympathetic fibers causes stimulation of the detrusor muscle and an increase in intravesical pressure of up to 25 to 50 cm H₂O.

The bladder is also innervated by **sympathetic** (hypogastric) **fibers,** but their function is not entirely clear. Activation of beta-adrenergic fibers to the fundus causes relaxation whereas activation of alpha-adrenergic fibers in the neck of the bladder and in the urethra causes contraction of those structures.

SUMMARY

Most people who consume a typical Western diet produce 50 to 100 mEq/day of so-called nonvolatile or fixed acid, which must be excreted by the kidney. In the process of excreting acid, the kidney reabsorbs virtually all the filtered HCO₃⁻. Additional HCO₃⁻, so-called new HCO₃⁻, is also generated by the kidney as a by-product of acid excretion. Net acid excretion is accomplished both through the formation and excretion of titratable acid and through the secretion of ammonia and subsequent excretion of ammonium (very small amounts of H⁺, less than 0.5 percent of the daily H⁺ burden, are excreted as free cations). Under normal conditions, approximately half the acid is excreted as titratable acid and half as ammonium. Eighty to ninety percent of the filtered bicarbonate is reabsorbed in the proximal tubule, and the remaining fraction is reabsorbed in the distal tubule. As H⁺ is secreted (either in exchange for Na⁺ or via a proton pump), the subsequent combination of H⁺ with filtered HCO₃⁻ generates CO₂ and H₂O. (In the proximal tubule, carbonic anhydrase in the brush border facilitates this process; in the distal tubule, a high luminal H⁺ concentration shifts the reaction to CO₂ plus H₂O). The production of titratable acid (the process of hydrogenation of the conjugate base) is primarily accomplished by a phosphate buffer. Ammonia is produced in tubular cells, principally from glutamine. Ammonia can diffuse into both the lumen of the tubule and the vascular space; however, because of the phenomenon of nonionic diffusion and the presence of a H^+ gradient between the tubular lumen and plasma, larger amounts of the gas are trapped in the lumen as ammonium. During acidosis, the compensatory increase in acid excretion is primarily brought about by increases in ammonia production, but several days must elapse before enhanced synthesis of ammonia achieves a new steady-state.

Micturition is normally a voluntary process that appears to be initiated by the deliberate relaxation of the external sphincter. The desire to urinate is evoked as the bladder fills to about 200 ml. Once initiated, the micturition reflex is an automatic spinal reflex that involves activation of parasympathetic nerves and contraction of the detrusor muscle.

NATIONAL BOARD TYPE QUESTIONS

Select the one most correct answer for each of the following

1. HCO_3^- reabsorption in the proximal tubule:
   A. Occurs primarily in the latter third of this segment
   B. Increases during treatment with acetazolamide
   C. Increases during volume contraction
   D. Increases during hyperventilation
   E. Is cotransported with potassium
2. Net acid excretion:
   A. Increases as the amount of protein in the diet decreases
   B. Can be calculated if the amount of HCO_3^- reabsorbed is known
   C. Can be calculated from the glomerular filtration rate, the plasma pH, and the urinary pH
   D. Can be calculated from the amount of

ammonium, titratable acid, and HCO_3^- excreted
   E. Is primarily regulated by the amount of titratable acid formed
3. During metabolic acidosis:
   A. HCO_3^- excretion increases
   B. The major increase in H^+ excretion is brought about by an increase in the excretion of titratable acid
   C. A carbonic anhydrase inhibitor would correct the low plasma pH
   D. The pH in the proximal tubule lumen would still be around 7.0, or unchanged from the normal value
   E. The excretion of K^+ increases
4. An individual has a fixed acid load of 100 mEq/day. Assuming that this person is in acid-base balance, has normal plasma electrolyte concentrations, and a normal glomerular filtration rate, which of the following statements is true?
   A. The individual is secreting approximately 100 mEq/day of H^+
   B. Most (more than 90 percent) of the acid would be excreted as ammonium
   C. Very little ammonia would be produced by the kidney
   D. Most (more than 95 percent) of the filtered HCO_3^- would be reabsorbed
   E. About the same amount of acid would be excreted by the lungs
5. Expansion of the extracellular fluid space with isotonic saline is associated with:
   A. Decreased HCO_3^- reabsorption
   B. Increased fractional Na^+ reabsorption in the proximal tubule
   C. Increased aldosterone secretion
   D. Decreased atrial natriuretic peptide secretion
   E. Decreased fractional Na^+ excretion

ANNOTATED ANSWERS

1. C. The volume of the extracellular fluid is one of the major factors affecting HCO_3^- reab-

sorption; there is an inverse relationship between the two.

2. D. Net acid excretion is given by the sum of titratable acid plus ammonium minus HCO_3^- excretion; it is much less than the total amount of H^+ secreted by the nephron, as most of the H^+ is involved in the reabsorption of HCO_3^-.

3. D. In contrast to the distal tubule, which can generate a large pH gradient between the lumen and plasma, the lowest pH in the lumen of the proximal tubule is about 7.0.

4. D. The fixed acids that are produced from protein metabolism are rapidly buffered by HCO_3^- and other buffers. All the filtered HCO_3^- must be reabsorbed from the filtrate to conserve that which has not been consumed in this process.

5. A. A number of factors (see Table 39-1) affect HCO_3^- reabsorption, including the expansion and contraction of the extracellular fluid space.

BIBLIOGRAPHY

Good, D. W., and Knepper, M. A. Ammonia transport in the mammalian kidney. *Am. J. Physiol.* 248:F459-F471, 1985.

Rector, F. C., Jr. Sodium, bicarbonate, and chloride absorption by the proximal tubule. *Am. J. Physiol.* 244: F461-F471, 1983.

IX Gastrointestinal Physiology

40 Gastrointestinal Nervous System

EDWARD S. REDGATE

Objectives

After reading this chapter, you should be able to:

Identify the principal pathways involved in the sympathetic innervation of the gastrointestinal tract, the responses of target tissue to sympathetic discharges, the neurotransmitters responsible for sympathetic activity, and the distribution of tonic and reflexive sympathetic control

Identify the pathways of somatic and parasympathetic innervation of the esophagus and how they control swallowing and esophageal peristalsis

Explain how vagal excitory and inhibitory fibers control the circular muscle tone of the lower esophageal sphincter and the receptive relaxation response of the proximal stomach, and their role in the small and large intestines

Describe how the intrinsic innervation of secretory epithelium and external musculature is organized and how this innervation controls the activities of these tissues

---

The remarkable ability of the gastrointestinal tract to transfer nutrients, salts, and water across the mucosal epithelium is achieved through the coordinated action of several effectors:

1. The **motor apparatus** physically reduces the size of the food particles
2. The manifold **secretions** of the mucosal epithelium and glands chemically reduce the food particles to more elemental compounds
3. The **motor apparatus** mixes and propels the resulting **chyme** through the tube
4. The **vasoconstrictor muscles** direct appropriate changes in blood flow to the above structures

The coordination of this symphony of events is acomplished by the **autonomic nervous system,** in conjunction with **endocrine and paracrine secretions.** This chapter focuses on the role of the autonomic nervous system and Chapter 41 on the role of the gastrointestinal hormones. The autonomic nervous has two separate outputs to the gut, extrinsic and intrinsic, and these converge on the final common pathway neurons. The **extrinsic innervation** originates at various levels of the cerebrospinal axis and consists of the parasympathetic and sympathetic fibers. The **intrinsic innervation** consists of ganglia located in an intramural plexus called the **enteric nervous system.** This accounts for the innervation of the segments of the gastrointestinal tract lying between the midesophageal level and the internal anal sphincter but not for the innervation of the upper esophageal musculature and the external anal sphincter. The upper esophagus and the external anal sphincter are composed of striated muscle. The **upper esophagus** is not innervated by the autonomic nervous system but by special visceral efferent fibers in the vagal nerves; the **external anal sphincter** is innervated by sacral somatic efferent fibers in the pudendal nerves.

The **degree of dependence** of various parts of the gastrointestinal system on these sources of innervation varies considerably. For example, at the oral and anal ends of the gastrointestinal tract where the upper esophagus and external anal sphincters are composed of striated muscle, denervation causes a loss of function of the striated muscle. The muscle in the rest of the gastrointestinal tract is smooth muscle innervated by the enteric nervous system, which can continue to function after surgical separation from the central nervous system.

There are two well-established enteric motor neuron pathways to gastrointestinal smooth muscle: **cholinergic** and **nonadrenergic noncholinergic** (NANC). This combination of intrinsic nerves and smooth muscle is able to function independently of the innervation by extrinsic nerves, but the degree of independence is not uniform. In certain structures, such as the lower esophageal sphincter, the stomach musculature, and the ileocecal and internal anal sphincters, extrinsic innervation exerts significant control over the motility of the sphincter. This is not the case in the small intestine.

In view of these variations in the origin and degree of neural control, the innervation of the gastrointestinal tract will be described starting with the sympathetic and parasympathetic innervation and then proceeding to the intrinsic or enteric innervation.

EXTRINSIC INNERVATION

Sympathetic Control

The sympathetic outflow to the gastrointestinal tract originates in the **intermediolateral cell column** of the *thoracolumbar spinal cord.* As shown in Figure 40-1, the preganglionic neurons project cholinergic fibers into the splanchnic nerves, which converge on noradrenergic neurons in the prevertebral ganglia. In view of species variation and the lack of precise knowledge, it is not possible to be entirely specific in describing the pathways of sympathetic fibers innervating the gastrointestinal tract, but it is clear that most of the sympathetic outflow to it originates in ganglia, as shown in Table 40-1.

Noradrenergic fibers from these prevertebral ganglia activate vasoconstrictors and inhibit secretomotor neurons, relax nonsphincter circular muscle, and contract sphincter circular muscle. The prevertebral ganglion cells are the site of convergence of **two inputs:** one from the extrinsic preganglionic fibers, as already described, and the other from sensory afferents originating in the muscle coats and mucosa. The influence of the sympathetic system is mainly manifested in a tonic activity in **splanchnic blood vessels,** and in the inhibition of **secretions by the mucosal epithelium.** The sympathetic activity on the gastrointestinal blood vessels constricts arterioles and the larger veins, thereby decreasing splanchnic vessel diameter and increasing resistance to blood flow. The vasoconstrictor tone is mediated by the release of **norepinephrine** in combination with the **neuropeptide Y** (NPY). Norepinephrine acts at adrenergic receptors and NPY potentiates the constrictor action of norepinephrine. The tonic activity of the sympathetic innervation mediated by the adrenergic fibers containing somatostatin acts on the secretomotor innervation of the mucosal epithelial cells to decrease water and electrolyte secretion. Loss of this sympathetic inhibitory control causes marked secretion by the intestinal mucosal epithelium, which may result in diarrhea.

In regard to the sympathetic innervation of the **external muscle layers,** there are only a few sympathetic nerve fibers supplying the smooth muscle of the nonsphincter regions. In the **sphincter regions** (lower esophageal, pyloric, ileocecal, and internal anal), noradrenergic innervation is normally more dense than in adjacent nonsphincter regions. The sphincter muscle is supplied with alpha-excitatory and beta-inhibitory adrenergic receptors. It is the **alpha-adrenergic receptors** of the muscle that are physiologically activated

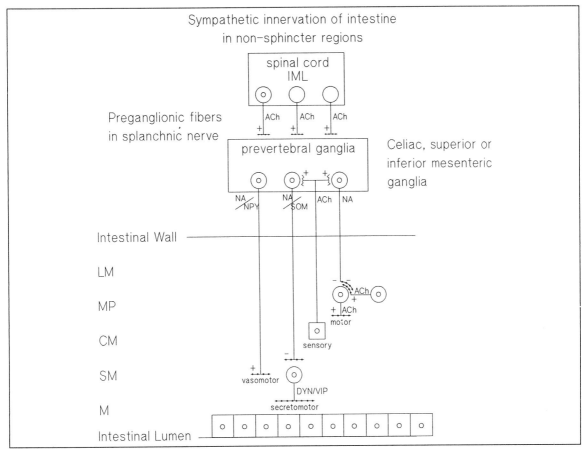

Fig. 40-1. Sympathetic innervation of the intestine in nonsphincter regions. The preganglionic sympathetic neuron output from the thoracolumbar spinal cord innervates ganglion cells in prevertebral ganglia. These postganglionic sympathetic neurons innervate vascular smooth muscle, secretomotor neurons in the submucosal plexus, and motor neurons in the myenteric plexus. The sympathetic output is excitatory to the vascular smooth muscle but inhibitory to the secretomotor and motor neurons. (*LM* = longitudinal muscle; *MP* = myenteric plexus; *CM* = circular muscle; *ACh* = acetylcholine; *NA* = norepinephrine; *SOM* = somatostatin; *SM* = submucosal plexus; *M* = mucosa; *IML* = intermediolateral cell column; *NPY* = neuropeptide Y; *DYN* = dynorphin; *VIP* = vasoactive intestinal peptide.)

by the release of norepinephrine from adrenergic nerves, causing contraction of the sphincter muscle. There is evidence suggesting that the tonic sympathetic discharge to sphincter muscle originates in the spontaneously active preganglionic neurons of the lumbar spinal cord or in ganglion cells of the prevertebral ganglia. However, if the entire sympathetic chain is removed in laboratory animals, no apparent changes

in the digestive process or lasting changes in sphincter activity are observed. From these experimental observations, it appears that the intact parasympathetic innervation and intrinsic reflexes are sufficient and that the sympathetic supply is not essential to the digestive process.

In the **nonsphincteric regions,** the major influence of adrenergic neurons on the motility of

Table 40-1. Innervation of the Gastrointestinal System by Postganglionic Sympathetic Fibers

Ganglion of Origin	Destination
Superior cervical	Esophagus
Middle	Esophagus
Stellate	Esophagus
Thoracic	Esophagus
Celiac	Lower esophageal sphincter, stomach, pylorus, proximal duodenum, sphincter of Oddi
Superior mesenteric	Caudal duodenum, jejunum, ilium, ileocecal sphincter, cecum
Inferior mesenteric	Transverse colon, descending colon, sigmoid colon, rectum, internal anal sphincter

the external muscle coats is not exerted directly on the muscle but indirectly by inhibiting the activity of cholinergic ganglion cells in the myenteric plexus which excite muscle contractions. These adrenergic nerve terminals form a dense plexus among the ganglion cells of the myenteric plexus. Norepinephrine presynaptically suppresses both acetylcholine and serotonin release in the intrinsic interneuronal circuitry. While these adrenergic nerves do not discharge spontaneously, they may be induced to discharge reflexly, especially in a protective manner, such as when the muscle wall is threatened by excessive distention or in response to peritoneal irritation. This may lead to a state of **ileus** in which there is no motor activity in the intestinal musculature.

Parasympathetic Control

The role of the parasympathetic innervation of the gastrointestinal tract is different from that of the sympathetic innervation. While excitation of the sympathetic innervation constricts arteries and large veins, inhibits secretion, constricts sphincters, and inhibits nonsphincteric muscle activity, excitation of the parasympathetic inner-

vation not only tends to **oppose all these actions,** but also sets in motion coordinated sequences of motility, secretion, and blood flow changes that promote **digestion.** The characteristics of these coordinating actions are unique for each segment of the gastrointestinal tract, so each segment must be considered separately.

Extrinsic Innervation of the Esophagus

The motor fibers innervating the upper striated muscle as well as the lower smooth muscle portion of the esophagus travel in the **vagus nerve,** CN X (Fig. 40-2; Table 40-2). The upper striated muscle portion of the esophagus is supplied by **special visceral efferent** fibers in the vagus nerve originating in the **nucleus ambiguus** of the medulla; the lower smooth muscle part is supplied by **general visceral efferent** fibers originating in the

Fig. 40-2. Major neural pathways involved in the control of swallowing. The inputs from pharyngeal and esophageal sensory receptors traveling in cranial nerves V, IX, and X project to centers in the forebrain and medulla oblongata that control swallowing. The output from the medullary swallowing center over cranial nerves V, VII, X, and XII controls striated muscles of the pharynx and upper esophagus (branchiomotor); the output from the dorsal motor nucleus of cranial nerve X (vagus) controls the lower esophagus. (*CN V* = trigeminal nerve; *CN VII* = facial nerve; *CN XII* = hypoglossal nerve; *CN X* = vagus; *NST* = nucleus tractus solitarii; *NA* = nucleus ambiguus; *RF* = reticular formation; *DMN X* = dorsal motor nucleus X.)

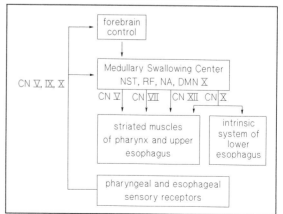

Table 40-2. Esophageal Peristalsis

Components	Primary Peristalsis	Secondary Peristalsis
Mechanoreceptor location	Palate, pharynx, epiglottis	Esophageal wall
Stimulus	Swallowing	Esophageal distention
Afferent nerves		
Upper esophagus	Trigerminal, glossopharyngeal, and vagus nerves	Vagus nerve
Lower esophagus	Vagus nerve	Vagus nerve, myenteric plexus
Type of motor discharge	Motor program	Local reflex discharge
Output pathway		
Upper esophagus	SVE in vagus nerve	SVE in vagus nerve
Lower esophagus	GVE in vagus nerve	GVE in vagus nerve, enteric nerves
Type of muscle		
Upper esophagus	Striated	Striated
Lower esophagus	Smooth	Smooth
Effect of vagotomy		
Upper esophagus	Loss of primary peristalsis	Loss of secondary peristalsis
Lower esophagus	Loss of primary peristalsis	No loss

SVE = special visceral efferent; GVE = general visceral afferent.

dorsal motor nucleus of the vagus. The discharge of the vagal motor neurons is controlled by **afferent fibers** of pharyngeal and esophageal sensory receptors that mainly travel in the superior laryngeal branch of the vagus and synapse in the **nucleus of the solitary tract.** There are also **descending polysynaptic pathways** to these nuclei from neurons in the forebrain that may also play a role in initiating swallowing. A **central mechanism** appears to control **swallowing,** and this consists of the nucleus of the solitary tract, the vagal nuclei, and adjacent areas of the medullary reticular formation. When triggered, this central mechanism programs a sequence of motor neuron discharges controlling **relaxation** of the upper and lower esophageal sphincters as well as **peristalsis** of the striated and smooth muscle parts of the esophageal body. When the motor discharge program of the swallowing center is engaged, an organized sequence of events occurs in which the tonic activity of the cricopharyneal muscle is inhibited and the phar-

yngoesophageal sphincter relaxes, permitting food to pass into the esophageal body. When a **food bolus** is formed and propelled from the pharynx through the upper esophageal sphincter into the esophagus, sensory receptors are stimulated, which initiate waves of peristaltic contractions in the esophagus (Fig. 40-3).

Esophageal peristalsis is evoked in two ways (see Table 40-2). The **swallowing** itself initiates one type of esophageal peristalsis called **primary peristalsis.** The **local distention of the esophageal** muscle wall elicits a second type of peristalsis called **secondary peristalsis.** Secondary peristalsis is excited by sensory input produced by distention at each level of the esophagus, and it empties the esophagus if primary peristalsis fails to. Swallowing takes place when the swallowing center is activated by descending signals from the forebrain or reflexly when sensory receptors in specific areas of the palate, pharynx, and epiglottis are stimulated. These areas are innervated by the **maxillary branch** of the trigeminal

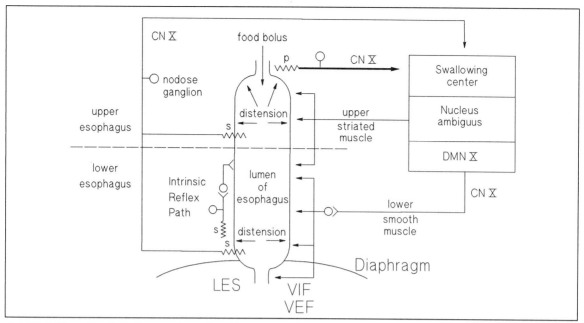

Fig. 40-3. Neural control of upper striated and lower smooth muscle regions of the esophagus. Passage of a food bolus from the pharynx into the esophagus stimulates sensory receptors, initiating primary peristalsis. Local distention of the esophagus initiates secondary peristalsis. (*VEF* = vagal excitatory fiber; *VIF* = vagal inhibitory fiber; *LES* = lower esophageal sphincter; *zigzag lines* = mechanoreceptor symbol; *p* = primary peristalsis; *s* = secondary peristalsis; *CN X* = vagus; *DMN X* = dorsal motor nucleus.)

nerve, the glossopharyngeal nerve, and the superior laryngeal nerve (see Fig. 40-2). When the swallowing center is activated, **interneuronal circuits** are triggered to release the programmed sequence of coordinated motor events of primary peristalsis in the upper striated muscle esophagus as well as in the lower smooth muscle esophagus (see Fig. 40-3). Evidence indicates that the **centrally programmed sequence of motor events** of primary peristalsis may be played out without further afferent support. As a result, primary peristalsis may may continue without interruption even if the bolus is diverted through a slit in the esophagus, because it depends on an intact swallowing center and efferent vagal nerves and not on esophageal distension. In contrast, secondary peristalsis consists of a succession of locally initiated reflexes originating in esophageal receptors. It relies on afferent and efferent neural pathwaysin the vagal nerves of the upper striated

muscle esophagus and on the long vagal reflex pathways plus the local enteric system reflex pathway in the lower smooth muscle esophagus. Therefore, if a bolus of food descends partway through the esophagus but then deviates through an esophageal slit, peristalsis of the secondary type ceases below the slit. Peristalsis of the primary type continues to complete its movement to the lower end of the esophagus, but is diminished. As a food bolus descends in an intact esophagus, secondary peristalsis reinforces the force and velocity of the primary peristalsis, so that the distention created by the bolus produces a larger response than just that from a "dry swallow."

Cervical vagotomy causes loss of function of the striated muscle esophagus because it denervates both the striated muscle innervated by motor neurons of the nucleus ambiguus and the sensory receptors. In contrast, **motor innerva-**

tion of the smooth muscle esophagus is supplied by preganglionic fibers originating in the dorsal motor nucleus of the vagus, which terminate in the enteric system of the lower esophagus. While vagotomy severs the preganglionic innervation of the lower esophagus, it does not interrupt the local reflex pathways in the enteric system of the lower esophagus. Secondary peristalsis may be evoked through this local reflex pathway. This information helps us to understand how peristaltic motor events may persist in clinical cases when a neuromuscular blocking agent for skeletal muscle is administered. After **skeletal neuromuscular blockade** (nicotinic cholinergic receptor antagonist), the oropharynx and upper striated muscle esophagus are paralyzed but primary and secondary peristalsis persist in the lower smooth muscle esophagus (muscarinic cholinergic receptors).

Parasympathetic Innervation of the Lower Esophageal Sphincter

Unlike the parasympathetic innervation of other visceral structures, such as that of the heart and bladder, the parasympathetic innervation of the lower esophageal sphincter consists of **two types of preganglionic fibers,** as shown in Figure 40-4. One set of preganglionic fibers in the vagus nerves is believed to synapse with NANC neurons of the enteric system, which inhibit a permanent myogenic tone of the lower esophageal sphincter smooth muscle and are called **vagal inhibitory fibers** (VIFs). The other set of preganglionic fibers is believed to synapse with cholinergic enteric neurons, which increase tone in the smooth muscle of the sphincter and are called **vagal excitory fibers** (VEFs). As a food bolus travels through the esophagus, the distention of the lower esophageal muscle wall stimulates the neural circuitry of the swallowing center and activates reflex responses of the VEFs and VIFs. This causes the lower esophageal sphincter to relax as a result of the increased discharge of VIFs and decreased discharge of VEFs. Once the bolus passes into the stomach, the sphincter closes because the VEF and VIF discharges reverse as VIF discharge de-

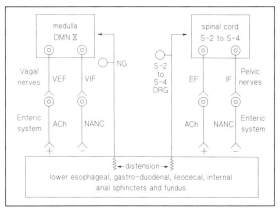

Fig. 40-4. Parasympathetic excitatory and inhibitory fiber innervation of the gastric fundus and gastrointestinal sphincters. The myogenic tone of the lower esophageal sphincter, gastric fundus, gastroduodenal, ileocecal, and internal anal sphincters is controlled by parasympathetic excitatory and inhibitory fibers: these outputs are found in vagal nerves *(VEF, VIF)* above the transverse colon, and for pelvic nerves *(EF, IF)* below the transverse colon. (*DMN X* = dorsal motor nucleus X; *ACh* = acetylcholine; *NANC* = nonadrenergic, noncholinergic; *DRG* = dorsal root ganglion; *S-2 to S-4* = sacral spinal segments 2 to 4; *NG* = nodose ganglion.)

creases and VEF discharge increases. Because both VEFs and VIFs exist in the vagal nerves, sectioning or stimulation of these nerves produces variable results. However, vagotomy may decrease or even abolish the ability to relax the lower esophageal sphincter due to the loss of the VIFs. Failure of the lower esophageal sphincter smooth muscle to relax completely is seen in the disease **esophageal achalasia.** Among the many consequences of lower esophageal sphincter achalasia are difficulty in swallowing, excessive distention of the esophagus due to accumulation of food, and regurgitation and aspiration of food particles.

Extrinsic Control of the Stomach

Distention is the most important local stimulus controlling gastric motility. There are two major patterns of **gastric motility:** the postprandial mode and the interdigestive mode. The

postprandial mode is initiated by the entry of food into the stomach. When a volume of food passes from the esophagus into the stomach, the intraluminal gastric pressure tends to rise but this is quickly limited by a reflex relaxation response in the proximal stomach. This response is an extrinsic reflex that relies on vagal afferent and efferent fibers that serve to limit the **intraluminal gastric pressure.** The relaxation response restricts the **gastroduodenal pressure gradient** which tends to expel gastric contents through the pylorus into the duodenum and also reduces the possibility of gastroesophageal reflux.

In the **receptive relaxation response** of the proximal stomach, the VEF and VIF type of preganglionic vagal fibers of the stomach alter their discharge rate reciprocally in a manner similar to that exhibited by the lower esophageal sphincter (see Fig. 40-4). The gastric distention also activates **extrinsic reflexes** mediated by vagal afferent and efferent fibers as well as **intrinsic reflexes,** which increase the mixing and grinding movements of the gastric corpus and antrum. As the **proximal stomach** maintains an appropriate intragastric pressure, the **antrum** is loaded and gastric chyme is pumped by the antrum into the **duodenum.** Duodenal distention activates extrinsic VEF and VIF reflex pathways similar to those that limit the gastroduodenal pressure gradient. In each of the already mentioned pathways, the VEFs synapse with excitatory cholinergic neurons while the VIFs synapse with inhibitory NANC fibers. The reflexes are blocked by vagotomy or by treatment with a ganglionic blocker of the nicotinic type. The **neuromuscular junction** of postganglionic fibers (VEF pathway) is blocked by treatment with a cholinergic blocker of the muscarinic type. When vagotomy blocks the receptive relaxation response, the reservoir function of the stomach is impaired. Intragastric pressure increases when a volume of food enters the stomach and the emptying of liquids into the duodenum is accelerated. In contrast, the motility of the corpus of the stomach and especially the antrum is reduced, so that the emptying of solids is delayed.

Besides the extrinsic reflexes responding to distention that are activated by receptors in the muscular coats, there are extrinsic reflexes responding to **chemical conditions** such as pH, or hypertonicity or hypotonicity, which involve receptors in the mucosa. Besides controlling gastric motility, extrinsic reflexes also influence the **secretion of HCl** and **pepsinogen** by the gastric secretory epithelium. Proximal gastric vagotomy decreases the secretion of HCl and pepsinogen but preserves antral motility, and has been used to manage duodenal ulcers caused by acid secretion mediated by vagal innervation. There are **chemoreceptors** in the duodenum that respond to glucose, pH, fats, fatty acids, and hypertonicity. There are also mechanoreceptors in the muscular layers that react to stretch. Stimulation of these receptors activates vagal reflexes that inhibit gastric emptying. These enterogastric reflexes also appear to employ VEF-type and VIF-type preganglionic vagal fibers and are extinguished by vagotomy.

In the **interdigestive period** that follows stomach emptying, a distinct pattern of motility emerges in the stomach and small intestine called the **migrating motor** (or myoelectric) **complex.** It is characterized by a motor pattern in which any solids remaining in the gastrointestinal tract are swept toward the colon. This motor activity appears to be regulated by a combination of neural and humoral factors, as **ganglionic** or **neuromuscular blockade** suppresses it and the intestinal neuropeptide **motilin** initiates it.

Extrinsic Innervation of the Small and Large Intestine

The extrinsic parasympathetic innervation of the small and large intestine consists of fibers that travel in the vagus and pelvic nerves (see Fig. 40-4). The **vagal nerves** innervate the small intestine and the ascending and transverse colon; the **pelvic nerves** innervate the descending colon and the rectum. The vagal motor neurons originate in the **dorsal motor nucleus of the vagus,** and the afferent fibers traveling in the vagal nerves have cell bodies in the **ganglion nodosum.** The pelvis nerve motor neurons originate in the **intermediolateral** and **ventral horn**

areas of sacral spinal cord segments S-2 to S-4, and the cell bodies of the pelvic afferent axons are located in the S-2 to S-4 dorsal root ganglia. The activity of the parasympathetic innervation of the small intestine is believed to resemble that of the esophagus and stomach, in that there is an excitatory cholinergic fiber pathway and an inhibitory NANC fiber pathway, but the density of preganglionic vagal fibers decreases distally. There are more vagal fibers going to the esophagus and stomach than to the duodenum, and the vagal innervation of the distal small intestine and colon has not been well described. Electrical stimulation or sectioning of the mixture of excitatory and inhibitory fiber pathways in the vagal nerves may elicit inconsistent results. However, evidence suggests that a tonic VEF exists that stimulates intestinal motility.

Extrinsic Innervation of the Colon and Internal Anal Sphincter

The same general pattern of extrinsic nervous control as that described for the esophagus, stomach, and small intestine is seen in the colon and internal anal sphincter (see Fig. 40-4). There are both excitatory and inhibitory fiber pathways in the vagal and pelvic nerve regions of the large intestine, where transmission depends on acetylcholine or NANC signals, respectively. Stimulation of the vagal innervation of the colon elicits contraction of both circular and longitudinal muscles, and stimulation of the pelvic nerve results in relaxation of the internal anal sphincter.

ENTERIC NERVOUS SYSTEM

Intrinsic Innervation by the Enteric Nervous System

The enteric nervous sytem consists of the **ganglionated plexuses,** which lie in the wall of the gastrointestinal tract and extend from the esophagus to the internal anal sphincter. The two major plexuses of the enteric system are the myenteric and submucosal. As shown in Figure 40-5, the **myenteric plexus** is located between the longitudinal and circular layers of the external musculature and the **submucosal plexus** is situated on the submucosal surface of the circular muscle layer. The ganglia of these plexuses are extensively interconnected. These plexuses receive sensory input and contain networks of interneurons that connect with motor and secretomotor neurons. There are **sensory inputs** of mucosal origin that respond to chemical and osmotic stimulation as well as inputs activated by radial stretch of the external musculature. There are also **excitatory** and **inhibitory inputs** from the parasympathetic division of the autonomic nervous system. These inputs synapse with elaborate nerve networks of interneurons that coordinate different patterns of motility and secretomotor activity. These connections are not completely understood but encouraging progress is being made. It is known that the enteric system contains a **large number of neurons,** between 10^7 and 10^8, which is comparable to the neuron population in the spinal cord. In view of this, the enteric system is considered a major part of the nervous system, and one which contains very complex circuits. This view has been reinforced by the discovery of a diversified population of neurons containing many different neurotransmitter and neuromodulator substances, most of which are neuropeptides also found in the central nervous system.

There are distinct target organs for the two major plexuses. The **major outputs** of the myenteric plexus are directed to the **circular** nd **longitudinal muscles.** In certain parts of the gastrointestinal tract, it is possible to distinguish between a plexus innervating the circular muscle and one for the longitudinal muscle. In contrast, the major output of the submucosal plexus is to the **secretory epithelium** and **enteroendocrine cells** (hormone-secreting cells of the mucosa).

Innervation of the Secretory Epithelium

The secretory activity of the intestinal epithelium responds to both neural and hormonal signals. The chemical and osmotic properties of food and distention of the gut activate neural pathways in the submucosal plexus that control the secretion

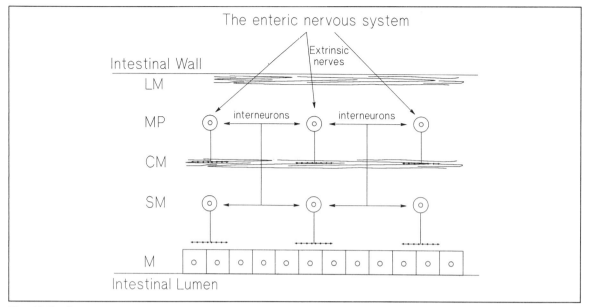

Fig. 40-5. The enteric nervous system contains two main plexuses, the myenteric and the submucosal. The myenteric plexus mainly innervates the external muscle layer and the submucosal plexus, the secretory epithelium. The neurons of each of these plexuses are interconnected and the plexuses are interconnected. The plexuses also receive both sympathetic and parasympathetic extrinsic innervation. (*LM* = longitudinal muscle; *MP* = myenteric plexus; *CM* = circular muscle; *SM* = submucosal plexus; *M* = mucosa.)

of the intestinal epithelium. The sensory information is encoded in action potentials that evoke neurotransmitter release in the submucosal plexus. As shown in Figure 40-6, **acetylcholine** is one of these neurotransmitters, and it acts at nicotinic cholinergic receptors to excite interneurons and secretomotor neurons. The secretomotor neurons may also release acetylcholine that in turn excites muscarinic receptors on epithelial cells, or these secretomotor neurons may release NANC neurotransmitters, such as vasocative intestinal peptide (VIP), dynorphin, substance P, cholecystokinin, and gastrin releasing peptide, either separately or in combination. The secretomotor neuron activity may trigger the secretion of substances such as pepsinogen, mucus, gastrin, and electrolytes. The activity of the intrinsic secretory pathways is modulated by input from both the extrinsic parasympathetic and sympathetic nerves. Stimulation of the sympathetic nerves inhibits neuronal activity in the

submucosal ganglia, decreases secretion by the secretory epithelium, and enhances absorption. Removal of the sympathetic input by paralysis or denervation of the sympathetic innervation causes enhanced secretion, reduced absorption, and possibly diarrhea.

Intrinsic Innervation of the External Musculature

Almost all of the axons supplying the **circular muscle** originate in the myenteric plexus and most of the axons supplying the **longitudinal muscle** appear to arise from the myenteric or a closely related plexus. A concept that has emerged from recently acquired information is that the coordination of the longitudinal and circular muscle of the intestine cannot be explained solely by reflex-type activity. Instead, it is proposed that: (1) the interneuronal networks in the myenteric plexus contain preprogrammed circuits that

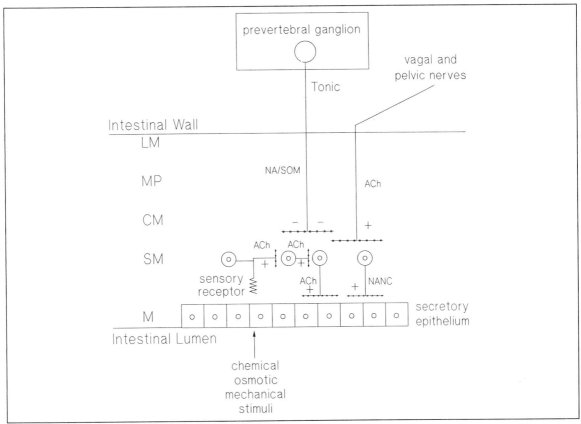

Fig. 40-6. The secretory epithelium is innervated by cholinergic and nonadrenergic, noncholinergic *(NANC)* secreto-motor neurons. The activity of these secretomotor neurons is coordinated by sensory receptor input to the enteric system, and by extrinsic sympathetic and parasympathetic innervation. There is a tonic inhibitory sympathetic input and an excitatory parasympathetic input. (*LM* = longitudinal muscle; *MP* = myenteric plexus; *CM* = circular muscle; *SM* = submucosal plexus; *M* = mucosa.)

selectively activate each of the various motility patterns; and (2) with the exception of sphincter motility, motility patterns such as segmentation, peristalsis, the migrating motor complex, and ileus (see Chapter 41) appear to be generated by circuits in the enteric system and do not depend on long reflex pathways involving the central nervous system.

The preprogramed **patterns of motility** result from the combined function of both the enteric neurons and the external musculature. The circular layer of the external musculature is a **syncytium** that is continuously excited by the rhythmic discharges of pacemaker smooth muscle cells. Its contractile activity generates many of the mechanical forces that mix and propel the luminal contents. The specific intrinsic neural circuits responsible for the motility patterns are not known, but recently acquired information has permitted insight into the neural and muscular mechanisms that generate some of the various patterns of motility (Table 40-3). A major function of the intrinsic innervation is inhibition of the spread of pacemaker-initiated excitation throughout the circular muscle syncytium. The intrinsic nerves produce this inhibition, and this

Table 40-3. Examples of Responses of Innervated Circular Muscle of Intestine to Selected Agents

Agent	Proposed Action	Circular Muscle Response
Tetrodotoxin	Blocks Na+-dependent action potentials and abolishes spikes	Spasm of nonsphincter circular muscle
VIP	Inhibits circular muscle contractile activity	Sphincter muscle relaxes
VIP antiserum	Inactivates VIP released by inhibitory enteric neurons	Spasm of nonsphincter circular muscle and achalasia of sphincter muscle
Opiate-like substances	Decrease spread of signals producing sequential contraction of circular muscle in succeeding segments	Continuous segmentation
Serotonin	Increases spread of signals producing sequential contraction of circular muscle in succeeding segments	Peristalsis, hyperpropulsion

VIP = vasoactive intestinal peptide.

effect can be shown when an isolated strip of small intestine is exposed to **tetrodotoxin** (TTX), which blocks Na+-dependent action potentials. TTX abolishes NA+ spikes and prevents the release of an inhibitory transmitter (probably VIP or nitric oxide) from inhibitory NANC intrinsic motor neurons. In the absence of the inhibitory transmitter, the strip of external musculature, which prior to TTX treatment was relaxed, develops unpatterned contractile spasms.

The **inhibitory intrinsic motor neurons** exist at all levels of the gut, and, in nonsphincter regions, they continuously suppress the myogenic contractile activity of circular muscle. In a sphincter region, such as the lower esophageal sphincter, the inhibitory motor neurons do not continuously suppress the myogenic contractile activity of the circular muscle of the sphincter, but are brought into action reflexly during passage of the luminal contents. The evidence sup-

porting **VIP** serving as the neurotransmitter for the inhibitory neuron is that: (1) VIP-containing fibers innervate the lower esophageal sphincter; (2) VIP is released during lower esophageal sphincter muscle inhibition (relaxation); (3) VIP application relaxes circular muscle; and (4) injection of **VIP antiserum** blocks neurally induced relaxation of the circular muscle. Inhibitory and excitatory synaptic discharges of intrinsic interneurons maintain control over the inhibitory motor neurons. Inhibitory interneuronal discharges suppress the release of VIP or nitric oxide and this permits increased myogenic contractions of the circular muscle. The process of inhibiting an inhibitory neuron is called **disinhibition.** It is believed that the disinhibition of such motor neurons plays a major role in generating the stereotyped motility patterns of **circular muscle.** The **longitudinal muscle** differs in this respect. Instead of receiving inhibitory fibers, it is innervated by a cholinergic excitatory pathway.

Since preprogrammed patterns of motility can be evoked by the application of certain putative neurotransmitters to isolated gut preparations, it is believed that the **interneuronal circuits** involved are "hard-wired" programs, comparable to those known to control stereotyped patterns of motor behavior in invertebrates that are evoked by command neurons. For example, in the intestine, **opiate-like substances** evoke segmentation and suppress peristalsis, while serotonin does the opposite. This evidence has been assembled in a model in which opiate-like substances and serotonin activate circuits controlled by command neurons. These **command neurons** cause stereotyped movements by activating certain neuronal networks formed by the enteric nervous system. Other examples of **stereotyped motor behavior** controlled by command neurons are ileus and circular muscle spasm. In the case of **ileus,** the command neurons elicit a continuous discharge of the intrinsic inhibiting motor neurons, causing circular muscle relaxation. Conversely, a **circular muscle spasm** occurs if the intrinsic inhibitory motor neurons are absent or inactive. Inhibitory motor neuron inactivity precipitates the spread of pacemaker-driven contractile activity throughout the circular muscle syncytium.

This model explains the derangements in motility seen in Hirschsprung's disease and achalasia. In **Hirschsprung's disease,** spasms of contractile activity of circular muscle occur when enteric ganglion cells are lacking in the myenteric plexus. It is believed that the circular muscle spasms are due to the loss of intrinsic inhibitory motor neurons. There is also a loss of myenteric ganglion cells in **achalasia** plus an inability to adequately relax the sphincter.

SUMMARY

Although our knowledge of gastrointestinal innervation is still far from complete, the properties of certain components of the innervation are beginning to emerge. The major source of innervation is the autonomic nervous system, while the contribution of the somatic nervous system is limited to important controls over the striated muscles involved in swallowing and defecation. With regard to vasomotor and secretomotor control and innervation of the external muscle layers, the effectors from the midesophageal area to the internal anal sphincter receive innervation from the intrinsic and extrinsic parts of the autonomic nervous system. The intrinsic part is called the *enteric nervous system* and is composed of a diverse array of neurons, most containing neuropeptides. These neurons form an intramural network responsible for producing different coordinated patterns of motor and secretomotor activity.

The extrinsic part of the autonomic nervous system consists of the parasympathetic and sympathetic divisions. The parasympathetic aspect innervates the smooth muscle of the external muscle layers and contains antagonistic outputs consisting of cholinergic excitatory and NANC inhibitory neurons. The circular muscle of the gut is a spontaneously excited electrical syncytium that is activated by pacemaker smooth muscle cells whose contractile activity is held in check by the tonic discharge of inhibitory NANC neurons. Release from this inhibitory tone of NANC neurons, or disinhibition, elicits increased contractile activity. In contrast, the cholinergic innervation of circular muscle is tonically excitatory. These two opposing pathways to circular muscle, one tonically inhibitory and the other tonically excitatory, are extrinsically innervated by preganglionic vagal fibers to that part of the gut orad to the transverse colon. These outputs are the VIFs and VEFs.

Neural control of gut motility is thought to function in the following manner. The extrinsic VIF and VEF pathways control the relaxation and contraction of the lower esophageal sphincter, the receptive relaxation response of the proximal stomach, and the function of the gastroduodenal and ileocecal sphincters. The signals delivered by the parasympathetic VIF and VEF discharges appear to represent a variety of command signals that selectively activate different preprogrammed patterns of gut motility. Other examples of command signals are those that evoke segmentation, peristalsis, the migrating motor complex, and ileus. In addition to the NANC fibers controlling smooth muscle, there are a variety of NANC neurons that function as secretomotor neurons. These NANC secretomotor neurons receive excitatory cholinergic and inhibitory adrenergic input.

The overall activity of the sympathetic division of the autonomic nervous system innervating the gut suppresses digestion. Adrenergic neuron discharges inhibit the excitatory cholinergic input to circular muscle in nonsphincter regions but excite sphincter muscles to contract, thereby lessening propulsion of intraluminal contents. The inhibition of propulsion, contraction of sphincters, and suppression of secretions transforms the activity of the gut from a propulsive to absorptive mode.

NATIONAL BOARD TYPE QUESTIONS

Select the one most correct answer for each of the following

1. The sympathetic innervation of the gut:
   A. Acts presynaptically to release acetylcholine in the myenteric ganglia
   B. Activates alpha receptors, causing relaxation of sphincter muscles
   C. Tonically relaxes splanchnic vascular smooth muscle
   D. Tonically inhibits secretion by mucosal epithelium
   E. Co-releases NPY which inhibits norepinephrine activity
2. Esophageal body reflex activity:
   A. Is not controlled by long reflex pathways
   B. Is coordinated by a swallowing center in the glossopharyngeal nucleus
   C. Persists in the upper esophagus after exposure to curare
   D. Persists in the lower esophagus after atropine administration
   E. Is stimulated by local distention

3. Relaxation of the lower esophageal sphincter:
   A. Is stimulated by gastric receptors
   B. Involves increased VIF and decreased VEF discharge
   C. Is evoked by acetylcholine release from cholinergic fibers
   D. Is independent of myogenic tone
   E. Is controlled by local reflex pathways
4. The receptive relaxation response of the stomach:
   A. Is blocked by sympathetic denervation
   B. Propels gastric chyme into the duodenum
   C. Is stimulated by low gastric pH
   D. Inhibits contractions of the antrum
   E. Prevents reflux into the esophagus
5. The intrinsic innervation of the secretory epithelium:
   A. Contains tonically active inhibitory NANC neurons
   B. Originates in neuron cell bodies located in the myenteric plexus
   C. Responds to the discharge of postganglionic sympathetic neurons with increased epithelial cell secretions
   D. Responds to osmotic stimulation of mucosal receptors with increased epithelial cell secretion
   E. Responds to vagal cholinergic input with decreased epithelial cell secretion

---

ANNOTATED ANSWERS

1. D. The sympathetic innervation of the gut acts presynaptically to inhibit acetylcholine release in the myenteric ganglia, activates alpha receptors, causing contraction of sphincter muscles, tonically constricts blood vessels, and inhibits secretion. NPY potentiates norepinephrine.
2. E. Esophageal body reflex activity is controlled by long reflex pathways (primary peristalsis) that are coordinated by a swallowing "center" located in nucleus tractus solitarii,

vagal nuclei, and reticular formation. Curare blocks the nicotinic receptors of the upper esophageal striated muscle and atropine blocks the muscarine receptors of the lower esophageal smooth muscle. Local distention stimulates the short reflex pathways (secondary peristalis).

3. B. Relaxation of the lower esophageal sphincter involves increased VIF and decreased VEF discharge. It is not controlled by acetylcholine, relies on the presence of myogenic tone, and is controlled by long reflex pathways stimulated by esophageal receptors.

4. E. The receptive relaxation response is not controlled by sympathetic innervation, opposes gastric emptying, is not stimulated by low pH, does not inhibit antral motility, and decreases gastroesophageal reflux.

5. D. The intrinsic innervation of the secretory epithelium does not contain tonically active inhibitory NANC neurons, originates in the neuron cell bodies of the submucosal plexus, is inhibited by sympathetic input and excited by parasympathetic input, and contains local circuits that respond to osmotic stimuli by activating epithelial cell secretion.

BIBLIOGRAPHY

Furness, J. B., and Costa, M. *The Enteric Nervous System*. New York: Churchill Livingstone, 1987.

Johnson, L. R., ed. In: *Physiology of the Gastrointestinal Tract*, 2nd ed. New York: Raven Press, 1987.

Schultz, S. G., Wood, J. D., and Rauner, B. B., eds. In: *Handbook of Physiology*, Section 6: The Gastrointestinal System. Bethesda, MD: The American Physiological Society, 1989.

Wood, J. D. Neurophysiological Theory of Intestinal Motility. *Jpn. J. Smooth Muscle Res.* 23: 143–186, 1987.

41 Gastrointestinal Hormones

EDWARD S. REDGATE

Objectives

After reading this chapter, you should be able to:

Define endocrine, paracrine, autocrine, and neurocrine

Identify what criteria must be met to establish the endocrine nature of a chemical messenger

Explain the roles of gastrin, cholecystokinin, secretin, and gastric inhibitory polypeptide, and describe the factors controlling their release

---

Communication between cells in the gastrointestinal tract is subserved not only by the neural reflex pathways described in Chapter 40 but also by a host of chemical messengers. Virtually every function of the gut appears to be influenced by one or more of these chemicals, called **hormones,** which are produced by specialized cells found throughout the system. There are four functional categories of hormones: endocrine, paracrine, autocrine, and neurocrine. Hormones are classified according to their **cells of origin** and the **mode of delivery** to target cells. **Endocrine hormones** are secreted into the circulation and the target cells are equipped with receptors that recognize the hormones. **Paracrine** and **autocrine hormones** are secreted into the interstitial fluid surrounding the secretory cells and their action is restricted either to nearby cells or to their cells of origin, respectively. **Neurotransmission** may be regarded as a special form of paracrine hormone action, in which hormone release (a neurocrine) is restricted to the synaptic region. While neurocrine hormones comprise a diverse array of molecular types (amines, amino acids, peptides), the gut endocrine and paracrine hormones known

at this time are exclusively **peptides.** Because neurocrines were described in Chapter 40, endocrine and paracrine hormones will be the focus of this chapter.

The **study of gastrointestinal peptides** entered a new era with the advent of certain immunologic, genetic, and chemical techniques: immunocytochemistry, radioimmunoassay, methods for studying gene expression and peptide synthesis, and techniques for isolating and culturing functionally intact peptide-secreting cells. Application of these techniques has revealed that significant structural similarities exist among certain gastrointestinal peptides, and these have been grouped into families (Table 41-1), such as the **gastrin** and **secretin** families. As shown in Table 41-1, there is an abundant variety of peptides in the gastrointestinal tract, suggesting the presence of extensive hormonal controls, but at this time the hormonal status of only a few is known. A peptide must fulfill **certain criteria** before it can be considered a true hormone. The following must be demonstrated in order to establish that a gastrointestinal peptide is indeed an endocrine hormone: (1) it is released into the

Table 41-1. Gastrointestinal Peptide Families

Family	Major Members	Principal Biologic Functions
Gastrin	Gastrin	+ gastric acid, + trophic to mucosa
	Cholecystokinin	+ pancreozymin, + gallbladder
Secretin	Secretin	+ pancreas and biliary bicarbonate
	Glucagon	− intestinal motility
	Vasoactive intestinal peptide	+ pancreatic and intestinal secretion
	Gastric-inhibitory polypeptide	− gastric acid
	PHI-27	+ pancreatic secretion
Pancreatic polypeptide	Pancreatic polypeptide	− pancreatic secretion
	Peptide YY	− pancreatic secretion
	Neuropeptide Y	+ vasoconstriction
Opioids	Enkephalin	− intestinal transit
	Beta-endorphin	− intestinal transit
Tachykinin-bombesin	Substance P	+ smooth muscle
	Substance K	+ smooth muscle
	Gastrin-releasing peptide	+ gastrin release
Orphan peptides	Somatostatin	− gastric acid
	Neurotensin	− gastric acid

+ = stimulates; − = inhibits.

circulation in response to a physiologic stimulus, such as feeding; (2) it produces a response that can be mimicked when the peptide is infused at concentration levels that are physiologic; and (3) the response to the physiologic stimulus can be antagonized by infusion of a specific antiserum or a selective antagonist. Of the several dozen substances currently being investigated as possible gastrointestinal endocrine hormones (see Table 41-1), only four peptides are known to comply with these criteria: **gastrin, cholecystokinin** (CCK), **secretin,** and **gastrin-inhibitory polypeptide** (GIP).

Gastrin, CCK, secretin, and GIP are found in the **enteroendocrine cells** scattered throughout the mucosa in certain regions of the gut. **Gastrin** is localized in the mucosa of the **gastric antrum,** while **CCK, secretin,** and **GIP** dwell in the mucosa of the **proximal small intestine.** As a rule, the **endocrine cells** releasing the peptide hormones are oriented in the mucosa such that one surface of the cell is exposed to the contents of

the gastric or intestinal lumen where chemicals in the chyme may stimulate or inhibit the cells, and the opposite surface is exposed to interstitial fluid in contact with the gastrointestinal blood vessels and the circulation. The stimulus and response relationships of these four hormones are summarized in Table 41-2.

GASTRIN

Gastrin is a peptide produced by **G cells** of the gastric antrum and duodenum. It is the most potent known stimulant of **gastric acid secretion** by parietal cells. When stimulating gastric acidity, gastrin relaxes proximal gastric stomach muscle, thus retarding gastric acid emptying into the duodenum. There are two major forms of gastrin in plasma, **G-17** and **G-34** (the numbers denote the fact that they are 17 and 34 amino acid polypeptides, respectively). Most of the plasma G-17 originates in the pyloric antral mucosa, while half of the plasma G-34 originates in

Table 41-2. Stimulus–Response Relationship of Gut Peptides

Variable	Gastrin	CCK	Secretin	GIP
Stimuli	Feeding, gastric distention, digested protein	Digested proteins and fats	pH 4.5, digested fat	H⁺, osmolarity, fats
Cell of origin	G cell	I cell	S cell	GIP cell
Responses	+ acid, − emptying, + mucosal growth	− emptying	− acid, − emptying	− acid, − emptying
Stomach				
Pancreas		+ enzymes, + bicarbonate*	+ bicarbonate, + enzymes*	+ insulin
Liver		+ bicarbonate*	+ bicarbonate	
Gallbladder		+ contraction		

CCK = cholecystokinin; GIP = gastrin-inhibitory polypeptide or glucose-dependent insulinotropic peptide; + = stimulates; − = inhibits; * = CCK + secretin synergism.

the antrum and half arises from the duodenum. In the pyloric antral region, **G cells** are scattered throughout the pyloric antral glands and are oriented in the typical gut endocrine cell fashion, whereby the **apical border** extends to the gastric lumen and the **basal granule—containing portion** contacts interstitial fluid. When stimulated by feeding, the plasma concentration of gastrin increases within a few minutes, peaks 20 to 30 minutes later, and consists of roughly equal portions of G-17 and G-34.

The **physiologic stimulus** for gastrin release is **feeding.** The mechanisms controlling release are complex and consist of neural reflex pathways with cholinergic and nonadrenergic, noncholinergic (NANC) neurotransmission that acts directly on somatostatin cells and G cells, respectively. Direct effects are exerted by **somatostatin** (a paracrine) as well as by the **gastric acidity** and the **amino acid** and **peptide content** of the gastric chyme, as shown in Figure 41-1.

Neural activity in the vagal pathways to G cells commences with the smell and taste of food, and is continued by local and vagovagal reflexes elicited by gastric distention when food enters the stomach. In addition, the products of protein digestion act directly on G cells. Undigested protein does not have any particular effect on G cells, but amino acids, especially tryptophan, and poly-

peptides are potent stimuli while an elevated H⁺ concentration is inhibitory below pH 4.5. A model showing the major factors that govern gastrin release is illustrated in Figure 41-1. The major **neural input** to the G cells is vagal, and the **major chemicals** in the chyme that modulate gastrin release are the **products of protein digestion** and the gastric acidity. **One vagal pathway** to the G cells consists of cholinergic preganglionic fibers to NANC postganglionic fibers in the intrinsic system. This NANC stimulatory pathway releases gastrin-releasing polypeptide (GRP), a member of the bombesin family. The gastrin acts on parietal cells, causing HCl secretion. A **second vagal pathway** to the G cells is indirect. Postganglionic cholinergic fibers activated by the vagal input terminate on somatostatin cells, which are in direct contact with G cells. Release of acetylcholine inhibits somatostatin release. **Somatostatin** is a paracrine hormone that inhibits G-cell secretion. Thus, electrical field stimulation of either of these vagal inputs to the G-cell control mechanism stimulates gastrin release: one by releasing an excitatory NANC transmitter (GRP) and the other by inhibiting an inhibitory paracrine hormone (somatostatin). Additional controls arise from the contact of the apical surface of the G cells with the contents of the gastric lumen. The products of protein digestion directly stimulate the G cells

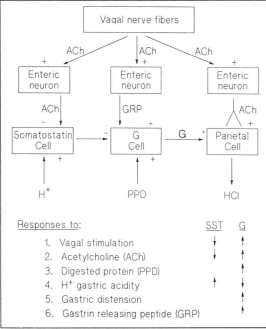

Responses to: SST   G

1. Vagal stimulation
2. Acetylcholine (ACh)
3. Digested protein (PPD)
4. H⁺ gastric acidity
5. Gastric distension
6. Gastrin releasing peptide (GRP)

Fig. 41-1. Endocrine, neurocrine, and paracrine factors controlling HCl secretion by parietal cells. Parasympathetic innervation of epithelial cells involved in controlling HCl secretion. Parietal cell secretion of HCl is stimulated by a cholinergic vagal pathway. The vagal pathway also excites gastrin (G) release (an endocrine) and inhibits somatostatin (SST) release (a paracrine-inhibiting gastrin release). H^+ feedback increases somatostatin secretion and the presence of amino acids increases gastrin secretion. *(Ach* = acetylcholine; *PPD* = products of protein digestion; *GRP* = gastrin-releasing peptide.)

to release gastrin, while the H^+ from gastric HCl stimulate somatostatin release, which in turn inhibits gastrin release.

In summary, the evidence indicates that gastrin release is not only the result of excitatory stimulation by NANC neurocrine hormone and the products of protein digestion, but also the combined result of the inhibitory action of the paracrine hormone, somatostatin, and gastric luminal H^+ concentration. Finally, gastrin performs an important role in gut function, besides controlling gastric acid secretion. It imposes essential **trophic actions** on the oxyntic (acid-se-

creting) gastric mucosa, the mucosa of the small and large intestines, and possibly also the exocrine pancreas.

CHOLECYSTOKININ

Cholecystokinin (CCK) is a multifunctional endocrine hormone possessing **cholecystokinetic** (stimulating gallbladder contraction), **pancreozymic** (stimulating pancreatic enzyme secretion), and **pancreatic growth–promoting** (trophic) actions. It is the single major factor controlling gallbladder contraction and pancreatic enzyme secretion. The **CCK molecule** is very heterogeneous, consisting of a large group of peptides that are products of preproCCK. Based on current knowledge, CCK-8, CCK-22, and CCK-33 appear to be the dominant forms of CCK in the plasma. Structurally and functionally CCK is related to gastrin but has a **higher affinity** for the receptors that stimulate gallbladder contraction and pancreatic enzyme secretion. CCK is produced by **I cells,** which are distributed in the duodenum and upper jejunum. In their **cholecystokinetic** mode, CCK peptides potently contract gallbladder muscles and relax the sphincter of Oddi, so that bile is ejected into the duodenum. CCK also inhibits gastric emptying by relaxing the proximal stomach and stimulating the pyloric sphincter; in this role, CCK acts as an **enterogastrone,** meaning a hormone of intestinal origin that inhibits gastric activity. In their **pancreozymic** function, CCK peptides directly stimulate pancreatic acinar cells to release pancreatic enzymes. The most important stimulus for CCK release is when the products of fat and protein digestion, such as amino and fatty acids, are brought in contact with the mucosa of the upper small intestine.

When the **pancreatic acinar cells** are under the sole influence of CCK, the amount of enzyme secreted is relatively small. However, when another hormone, secretin, is present, enzyme secretion is heightened. This is an example of **potentiation,** in that secretin alone does not stimulate pancreatic enzyme secretion but augments the action of CCK. This potentiation of the action of CCK by secretin is physiologically important for achieving optimal levels of pancreatic enzymes for the digestion of dietary nutrients. The converse is also true, in that CCK potentiates the action of secretin. **Secretin** is a hormone that originates from the duodenal mucosa, and regulates hepatic and pancreatic secretion of bicarbonate ion. In the pancreas, CCK enhances the secretin-induced secretion of bicarbonate by pancreatic duct cells but by itself cannot do this.

SECRETIN

Secretin is the endocrine hormone whose extraction from intestinal mucosa and intravenous infusion by Bayliss and Starling in 1902 sparked the concept of **endocrine regulation** and coining of the term *hormone,* from the Greek word meaning "I arose to activity." Secretin exhibits **sequence homology** with several other gastrointestinal peptides, including vasoactive intestinal peptide (VIP), GIP, PHI-27, and pancreatic glucagon. Secretin is secreted by the S cells, which are distributed mainly in the upper small intestine. The major stimulus for secretin release is the **presence of H**$^+$ in the upper small intestine. When gastric chyme is discharged into the initial 2 to 3 cm of the duodenum, the pH of the chyme is often below the threshold (estimated to be 4.5) for stimulating secretin release. A significant amount of secretin release is also stimulated by the products of **fat digestion,** such as fatty acids. However, the **presence of HCl** is the major stimulus. **Secretin** primarily controls the secretion of water and bicarbonate by pancreatic duct cells and potentiates the action of CCK in stimulating the secretion of enzymes by pancreatic acinar cells. Secretin also stimulates secretion of the aqueous bicarbonate-rich component of hepatic bile.

Acid in the duodenum releases not only secretin but also CCK. These two hormones potentiate each others actions on the exocrine pancreas. By itself, neither secretin nor CCK in the amounts released by a meal is potent enough to substantially stimulate exocrine pancreatic secretion of water and bicarbonate. Secretin may also interact

with CCK to delay gastric emptying, that is, it performs the function of an enterogastrone.

GASTRIC INHIBITORY POLYPEPTIDE

GIP participates in two physiologic regulations: it inhibits **gastric acid secretion** and stimulates **insulin release.** GIP's inhibition of gastric acid secretion accounts for why the introduction of certain food substances into the small intestine suppresses gastric acid secretion, making GIP another enterogastrone. These inhibitory responses arise after duodenal distention or after gastric acid, fats, or hyperosmolar solutions have entered the duodenum. It appears that GIP is one of several enterogastrones secreted by the upper small intestine which limit the amount of gastric acid in the intestine. The interactions of these gastrointestinal peptides are complex and may involve mediation by locally released somatostatin.

GIP is one of the hormones secreted by the gastrointestinal tract that participates in the control of **insulin secretion** by the endocrine pancreas. Although plasma glucose levels constitute the most important determinant of insulin release, blood levels of certain amino acids, gastrointestinal hormones, such as GIP, and the autonomic nervous system are also involved. GIP is an **anticipatory type** of controller of insulin secretion, in that GIP is released as gastric chyme enters the duodenum. Because of this, insulin secretion rises earlier and to a greater extent than it would if the plasma glucose level were the only controller.

SUMMARY

The abundance of peptides in the gastrointestinal mucosa suggests that they are extensively involved in the endocrine and paracrine regulation of gastrointestinal function, but only four of these peptides are well-established hormones: gastrin, CCK, secretin, and GIP. Gastrin exists in typical endocrine cells (G cells) scattered in the glands of the antral mucosa and, when released into the circulation, stimulates gastric acid secretion. Feeding activates two types of vagal pathways—cholinergic and NANC—which stimulate G cells to release gastrin. Food entry into the stomach further modulates gastric acid secretion by evoking local and vagovagal pathway responses to the distention and to the presence of the products of protein digestion (excitatory) and the H^+ concentration (inhibitory). Gastrin also has a trophic effect on the oxyntic gastric mucosa and the mucosa of the small and large intestines. CCK, secretin, and GIP reside in the endocrine cells of the mucosa of the upper small intestine. CCK controls gallbladder contraction, sphincter of Oddi relaxation, pancreatic enzyme secretion, secretion of the aqueous component of bile, and gastric emptying. CCK release is most potently stimulated by the presence of protein and fat digestion products in the upper small intestine. Secretin is mainly released when gastric or fatty acids make contact with the mucosa of the upper intestine. This constitutes an endocrine hormone—controlling secretion of an aqueous bicarbonate fluid by duct cells of the pancreas and liver. Secretin and CCK interact to potentiate each other. GIP is an enterogastrone secreted by the mucosal endocrine cells of the upper small intestine and inhibits gastric activity when gastric acid, fats, or hyperosmotic solutions are discharged into the small intestine. A major action of GIP is stimulation of insulin secretion.

NATIONAL BOARD TYPE QUESTIONS

Select the one most correct answer for each of the following

1. Match the diagrams in Figure 41-2 with the appropriate secretory-to-target cell mode of action.
   ____ Paracrine
   ____ Neurocrine
   ____ Endocrine
   ____ Autocrine
2. Which of the following is a valid criterion for an endocrine hormone?

A. It is released into the interstitial fluid in response to a physiologic stimulus and acts directly on an adjacent target cell
B. It produces a response that can be mimicked by several other substances at lower concentrations
C. Response to the substance is diminished by general anesthesia
D. The response can be antagonized by a specific antiserum
3. Gastrin is a hormone that:
A. Originates in the antral and duodenal mucosa
B. Is secreted by mucosal cells whose apical borders contact the submucosa
C. Is not released until food distends the stomach
D. Is released by the action of secretin on the antral mucosa
E. Stimulates gastric emptying
4. CCK is a hormone that:
A. Stimulates gastric emptying
B. Has a trophic effect on the antral mucosa
C. Is pancreozymic
D. Contracts the sphincter of Oddi
E. Is released when carbohydrates enter the duodenum
5. Secretin is a hormone that:
A. Is released from the duodenal mucosa by the action of products of protein digestion
B. Interacts with CCK to inhibit biliary bicarbonate secretion
C. Acts as an enterogastrone
D. Causes bile salt secretion
E. Has growth-promoting effects on the intestinal mucosa

ANNOTATED ANSWERS

1. A, neurocrine; B, endocrine; C, autocrine; D, paracrine.
2. D. An endocrine hormone should be released in response to a physiologic stimulus, carried by the circulation to a distant target cell, and antagonized by a specific antiserum.

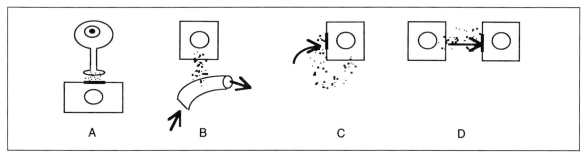

Fig. 41-2.

3. A. Gastrin originates in the antral and duodenal mucosa, where it is secreted by G cells whose apical borders are in contact with the contents of the gastric lumen. Gastrin is released by the smell, taste, chewing, and swallowing of food, as well as by gastric distention. It also inhibits proximal stomach contraction and thus gastric emptying.

4. C. CCK is an enterogastrone and is pancreozymic; it relaxes the sphincter of Oddi and is released when the products of fat and protein digestion are present in the duodenum.

5. C. Secretin is released from S cells in the duodenal mucosa in response to gastric and fatty acids. It acts as an enterogastrone, potentiates the action of CCK, stimulates biliary bicarbonate and not bile salt secretion, and has a growth-promoting effect on the pancreas.

BIBLIOGRAPHY

Del Valle, J., and Yamada, T. The gut as an endocrine organ. *Ann. Rev. Med.* 41:447, 1990.

Johnson, L. R., ed. *Physiology of the Gastrointestinal Tract,* 2nd ed. New York: Raven Press, 1987.

Schultz, S. G., Wood, J. D., and Rauner, B. B. In: *Handbook of Physiology,* Section 6: The Gastrointestinal system. New York: Oxford University Press, 1989.

42 Gastrointestinal Motility

FRANK C. BARONE AND JOSEPH D. FONDACARO

Objectives

After reading this chapter, you should be able to:

Identify the anatomic features of the gastrointestinal tract that provide for motility, specifically the longitudinal and circular smooth muscle layers

Describe the major types of motility patterns seen in the gastrointestinal tract and the primary functions of these patterns

Discuss the motility patterns seen in specific areas of the gastrointestinal tract and the significance of the migrating motor complexes

Describe the various phases of swallowing and the anatomic components (and their role) important to each phase

Explain how the electrical and mechanical events are related to gastric motility

Describe the pattern of motility unique to the small and large intestines and the mechanisms of defecation

Identify various diseases of gastrointestinal motility

In the normal physiologic setting, the human gastrointestinal tract propels nutrients, liquids, other ingested materials, and secretions in an aboral direction from the mouth to anus. This movement, termed **motility,** is involuntary, with the exception of the oral phase of swallowing. Motility is accomplished by discrete and coordinated mechanical movements of each organ in the tract. The term *motility* is applied also to the mechanical contractions of the gastrointestinal tract that are nonpropulsive and that function to mix ingested food with the secretions of mucosal cells that line various digestive organs. Both the propulsive and nonpropulsive motility patterns are a result of coordinated contractions of discrete sets of longitudinal and circular muscles (the intestinal or visceral smooth muscle). The

gallbladder can also contract in response to appropriate stimulation, so that bile is efficiently secreted into the duodenum. The **mouth, pharynx,** and **esophagus** provide a passageway for ingested materials to enter the stomach, where food is stored and mixed with the various components of gastric juice (including salivary enzymes). **Gastric motility** provides for "metered" emptying of this now partially digested material into the upper small intestine, which is a pattern unique to this organ and regulated by both intrinsic and extrinsic nervous and humoral factors. **Intestinal motility** allows for nonpropulsive (mixing) and propulsive contractions so that efficient digestion and absorption can occur. **Colonic motility,** which is normally slower than that exhibited by the small bowel, is also both nonpropulsive and propulsive to ensure the maximum absorption of water and salt and the efficient elimination of waste. This chapter will review the anatomic features of the gastrointestinal musculature, the various specific motor functions of each major organ in the gastrointestinal tract, and some of the clinical manifestations of abnormal motility.

ANATOMIC CONSIDERATIONS

The basic anatomy of the gastrointestinal wall, including the location of the longitudinal and circular musculature, is illustrated in Figure 42-1. (This chapter will only cover the neural elements of the gastrointestinal tract. Chapter 40 contains a more detailed description of the gastrointestinal nerves and their role in the regulation of the functions of this system.) The outermost layer of

muscle is composed of **longitudinal fibers** and beneath this is the **circular muscle layer.** Between these muscle layers is the **myenteric nerve plexus,** which is a component of the enteric nervous system and provides for intrinsic modulation of gastrointestinal function. The **submucosal layer** is located beneath the circular smooth muscle and is composed of stromal tissue that contains blood vessels, the submucosal nerve plexus (the other branch of the enteric nervous system), lymphatic channels, and various cell types, including those that comprise the gastrointestinal immune system. Finally, the **mucosal layer** of epithelial cells lines the lumen of the major organs of the tract.

Different motility patterns are exhibited throughout the gastrointestinal tract (Table 42-1). Tonic or sustained contractions are characteristic of the **sphincteric muscle areas** of the gut. The **upper and lower esophageal sphincters,** the **pyloric** and **ileocecal valves,** and the **internal anal sphincter** are all sphincter mus-

Fig. 42-1. The histologic arrangement of the layers of tissues of the gastrointestinal wall.

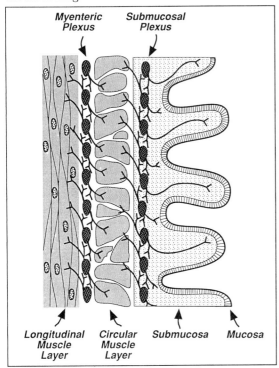

Myenteric Plexus Submucosal Plexus

Longitudinal Muscle Layer   Circular Muscle Layer   Submucosa   Mucosa

Table 42-1. Types of Gastrointestinal Motility Patterns and the Major Organs or Structures Involved

Type of Contractions	Organs and/or Structure
Tonic	Upper and lower esophageal sphincters, pyloric valve, sphincter of Oddi (associated with the common bile duct), ileocecal valve, internal anal sphincter
Propulsive peristalsis	Esophagus, lower two-thirds of stomach, small intestine, rectum
Migrating motor complex	Fasting or empty small intestine
Mass movement	Ascending, transverse, and descending colon
Nonpropulsive segmentation	Small intestine
Haustration	Ascending, transverse, and descending colon

cles that maintain a tonic state of contraction. Peristalsis or peristaltic contractions propel luminal contents aborally and are seen in the **esophagus, distal half** of the stomach, the small intestine, and the rectum. **Peristalsis** is characterized by waves of alternating constriction and relaxation such that a band of constriction propels the food bolus from behind and the area in front relaxes, thus facilitating forward passage of the bolus. After the bolus has passed, sphincteric tone is re-established in the area of constriction.

In the **colon,** there is a much different pattern of propulsive activity called **mass movement. Peristalsis** is manifested again in the **rectum.** There are also nonpropulsive motor patterns, termed **segmentation,** seen in the small bowel and haustrations of the colon, which serve to mix digested food (small intestine) and feces (colon) with other luminal contents and maximize exposure of material to absorptive surfaces. These mixing, nonpropulsive patterns are absent throughout the gastrointestinal tract during fasting and when the bowel is empty.

Fasting actually reveals another discrete motility pattern of the gut that migrates from the stomach to terminal ileum at a speed of about 5 cm per minute and occurs about every 90 minutes. Termed the **migrating motor complex,** it consists of little or no activity that changes suddenly to maximum propagated contractile activity, and functions to cleanse the lumen of indigestible material.

MOTILITY OF THE MOUTH AND PHARYNX

Chewing begins the process of gastrointestinal motility, but is not necessary for the digestion of food. It is normally voluntary, but does have an involuntary component, in that food placed into the mouth of an unconscious person may initiate a chewing reflex. Chewing simply fragments ingested food, mixes it with salivary enzymes, and lubricates it to facilitate swallowing.

Swallowing or **deglutition** begins the obvious motility of the gastrointestinal tract, and, though initiated voluntarily, becomes completely involuntary within one second. Its prime function is to get food or liquid into the stomach without forcing it into the nasopharynx or trachea. Because so many different mechanisms participate in the process, swallowing is separated into three phases: the oral phase, the pharyngeal phase, and the esophageal phase. These are illustrated in Figure 42-2.

In the **oral phase,** the tongue selects a portion of food and pushes it back against the soft palate, which is slightly elevated (see Fig. 42-2A). As the tongue pushes the food farther back, the soft palate closes the nasopharynx and food enters the upper pharynx. In the **pharyngeal phase,** the presence of food causes the upper pharyngeal constrictor muscle to contract, ensuring closure of the nasopharynx and initiating sequential contractions of the middle and lower pharyngeal constrictor muscles (see Fig. 42-2B). This wave of contraction propels the food toward the esophagus. As food approaches the esophagus, the epiglottis is lowered and the glottis is elevated slightly so that the larynx is covered, preventing food from entering the trachea. At the same time, the upper esophageal sphincter momentarily relaxes, facilitating the entry of food into the esophagus (see Fig. 42-2C).

During the **esophageal phase,** peristalsis is initiated as the wave of pharyngeal contractions passes through the upper esophageal sphincter. This closes the sphincter behind the bolus of food and normal peristaltic contractions transport the food down the esophageal tube (see Fig. 42-2D). As this propulsive wave reaches the lower esophageal sphincter, the area of relaxation preceding the bolus momentarily relaxes the sphincter and food easily enters the upper stomach. As the propulsive wave passes, sphincteric tone is re-established in the area of constriction behind the food bolus and the lower esophageal sphincter closes, preventing reflux of food from the stomach into the esophagus. These three phases ordinarily take about 10 seconds, depending on the consistency of the food consumed. Liquids pass much more quickly.

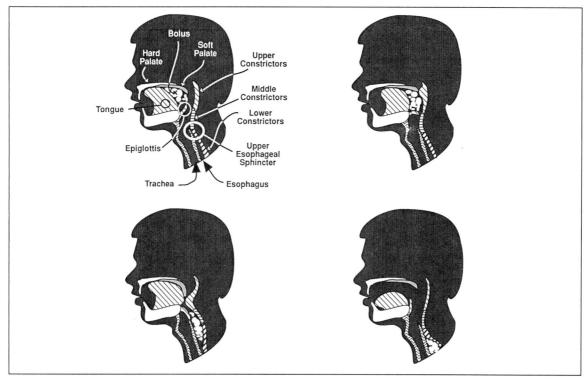

Fig. 42-2. The mechanisms of swallowing, illustrating the oral, pharyngeal, and esophageal components. (Modified from: Weisbrodt, N. W. Esophageal Motility. In: Johnson, L. R., ed. *Gastrointestinal Physiology,* 3rd ed. St. Louis: Mosby, 1985.)

ESOPHAGEAL MOTILITY

The esophagus is about 22 cm long. The **extrinsic innervation** of the esophagus involves the vagus nerve synapsing with the myenteric plexus neurons that control peristalsis. These myenteric plexus neurons innervate other myenteric neurons above in the outer longitudinal smooth muscle layer and below in the inner circular smooth muscle layer. The **upper third** of the esophagus is composed of striated muscle that receives direct input from the vagal efferent fibers to produce coordinated peristaltic waves. The **middle third** of the esophagus contains mixed smooth and striated muscle with a myenteric plexus that receives vagal input. The intrinsic nervous system receives input from sensory receptors and is also stimulated by stretch (or distention). Con-

sequently, if the thoracic vagus nerve is cut, normal peristalsis still occurs in the lower two-thirds of the esophagus.

In the **lower esophagus,** which contains only smooth muscle, neuronal impulses pass from the vagus nerves to myenteric neurons. From there, impulses pass to other myenteric cells and to the inner and outer muscle layers; these regulate peristalsis. **Primary peristalsis** (swallow reflex) in the esophagus is triggered by swallowing in the pharynx and continues even in the absence of a bolus in the esophageal lumen (dry swallow). **Secondary peristalsis** is stimulated by distention of the esophagus produced by an intraluminal bolus, an event similar to that seen in other parts of the gut (though for shorter distances). Secondary peristalsis is mediated entirely by the **enteric nervous system,** and is achieved through the coordinated contraction and relaxation of the muscle layers. The peristaltic reflex is initiated by intraluminal distention which in turn stimulates stretch receptors in the esophageal wall and initiates a progressive neurally mediated sequence. This reflex is also seen in the small intestine. Reflex relaxation (or lack of contractions in an organ such as the esophagus, which has no resting intraluminal tension or tone) precedes the bolus. The simultaneous contraction of circular muscle and relaxation of longitudinal muscle behind the bolus then further propels it forward. As the first portion of circular muscle relaxes, the next portion contracts and the process is repeated. It has the appearance of a contractile wave that moves slowly (3 cm per second) down the esophagus.

Figure 42-3 summarizes the roles of the vagus nerve, myenteric nerves, and intrinsic muscle excitability in the regulation of esophageal motility. **Vagal nerve stimulation** relaxes the upper and lower esophageal sphincters. The lower esophageal sphincter also relaxes immediately when a swallow is initiated in the pharynx or when the lower esophagus is distended; the enteric nervous system mediates these events. Peristalsis in the esophagus can be evoked by a swallow or by an intraluminal bolus, and occurs along the upper esophagus through a direct vagal–hindbrain–vagal reflex arc. In the lower esophagus, peristalsis can be independent of the vagus and is mediated by the enteric nervous system. Intrinsic smooth muscle excitability and tone shut the esophageal sphincters and keep them in a tonic state.

Specific diseases can arise because of abnormal esophageal motility. **Gastroesophageal reflux disease,** and the resulting esophagitis, is caused by incomplete closure of the lower esophageal sphincter. (decreased resting pressure) or when there are increased transient relaxations of the sphincter not associated with swallowing. Gastric contents then reflux into the esophagus and cause mucosal irritation, inflammation, and eventual erosion and bleeding of the esophagus. The initial symptoms are heartburn, indigestion, and a sour taste in the mouth. It is aggravated by the ingestion of coffee, citrus juice, and ethanol, and by cigarette smoking. **Achalasia** represents a failure of the lower esophageal sphincter to relax completely during swallowing. Food therefore does not pass easily from the esophagus into the stomach; vomiting and weight loss are the major symptoms. **Diffuse esophageal spasm** consists of repetitive, high-amplitude contractions that can be painful and produce aperistalsis.

GASTRIC MOTILITY

The major functions of the stomach are to **store food** temporarily, **continue digestion** by chemically reducing food particle size, and **regulate emptying** into the duodenum. The stomach is divided into three distinct regions: the fundus, the body, and the antrum, which terminates at the pylorus or gastroduodenal junction (Fig. 42-4). The upper portion of the **fundus** lacks phasic motility, but exhibits minimal tonic pressure and a high compliance for storage. Food can accumulate in the fundus with a large volume change, but no pressure change. This phenomenon is called **receptive relaxation** and is mediated by the vagus nerves via inhibitory neurons of the myenteric plexus. The **body** of the stomach displays a motility that mixes and grinds the food with gastric juice and propels the food and liquid toward the antrum and pyloric area for regulated emptying.

Located at approximately the midpoint of the greater curvature of the body of the stomach is an area of rapid spontaneous depolarization known as the **gastric pacemaker.** It establishes the maximum rate of gastric contractile activity for this and the more distal areas of the stomach. The gastric pacemaker generates a well-defined electrical event that can be recorded as intrinsic electrical activity from both the pacemaker area and more distal regions of the stomach (Fig. 42-5). The gastric smooth muscle elicits two types of electrical activity: slow waves and spike potentials. **Slow waves** are slow depolarizations occurring at a frequency of three to six per minute and are often called the **basic electrical rhythm.**

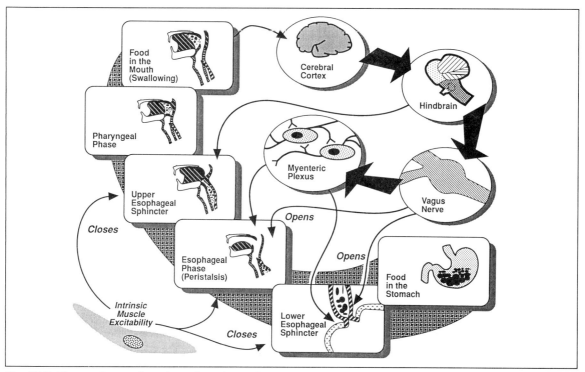

Fig. 42-3. The regulation of esophageal motility and esophageal sphincteric activity. (See text for detailed description.)

Fig. 42-4. The areas of the stomach and their associated structures.

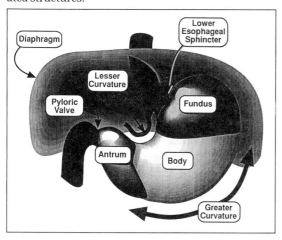

Fig. 42-5. The basic electrical rhythm of the stomach, with slow wave and spike potential activity illustrating the relationship of spike potentials to muscle contractions.

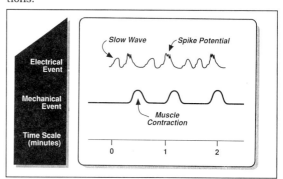

Spike potentials are periodic fast waves of depolarization that most often follow a slow wave and, when they do, always initiate contractions of the stomach musculature. Spike potentials are produced by the action of excitatory or disinhibitory enteric nervous regulation. Because muscle contractions always follow spike potentials associated with a slow wave, the slow wave or basic electrical rhythm controls the maximum rate of contraction of the stomach musculature, which is about three to six contractions per minute.

Spike potentials and the subsequent contractions of the gastric musculature are elicited directly by vagal nerve influence as well as indirectly through the myenteric plexus; distention of the stomach as it fills with food and liquid also evokes them. The hormone **gastrin,** released by antral G cells in response to vagal influence and gastric distention, also accentuates the occurrence of spike potentials and contractions. These contractions are primarily peristaltic, with some nonpropulsive contractions manifested during peak digestion. These propulsive peristaltic waves continually move partially digested food toward the pylorus and are the prime mechanism for gastric emptying.

Gastric emptying after feeding takes longer for solids than for liquids—about three hours versus one hour, respectively—because of the time it takes for the distal stomach to reduce solids to a fluid consistency composed of particles less than 2 mm in diameter. Solid particles of food are broken up by gastric peristaltic contractions that move the particles against a closed pylorus. Only a small amount of partially digested material can move into the duodenum before the pylorus closes. A major portion of this wave rebounds back over the distal stomach and mixes the food with gastric secretions, thus promoting further digestion. The **pylorus** limits the size of particles entering the duodenum and appears to prevent the reflux of duodenal contents into the stomach. **Undigestible particles** remain in the stomach until fasting **migrating motor complexes** move them through the bowel.

Following eating, the primary determinant of emptying is **volume;** the larger the meal volume, the greater the stretch on the gastric wall and the greater the force and frequency of gastric contractions. **Isosmotic gastric contents** empty faster than do hypo-osmotic or hyperosmotic contents because of the feedback inhibition exerted by duodenal osmoreceptors. In addition, gastric contents with a **low pH** (higher H+ concentration) empty slowly because of a combined neural and hormonal mechanism. Sensory receptors are activated that transmit an inhibitory reflex to the stomach and **secretin** is released from cells in the duodenum, which reduces the tone of the proximal stomach and decreases antral peristalsis. Intraluminal duodenal nutrients also appear to reduce gastric emptying through the release of hormones. **Cholecystokinin** (CCK) release is elicited by fat and proteins and **gastric inhibitory peptide** (GIP) release is evoked by glucose. The two hormones appear to mediate gastric emptying by reducing gastric contractions and intragastric pressure. Both are secreted by the duodenal mucosa. An intact **vagal innervation** appears to be important, but not essential, for regulating all aspects of gastric emptying, as loss of vagal integrity modifies the gastric and duodenal neuroneuronal and neurohumoral mechanisms involved. Figure 42-6 illustrates the

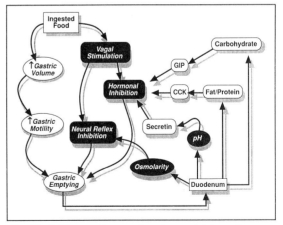

Fig. 42-6. The regulation of gastric motility and gastric emptying. (*GIP* = gastric inhibitory peptide; *CCK* = cholecystokinin.)

mechanism of gastric emptying and the feedback regulatory process involved.

Clinical problems can result from **altered gastric emptying. Hypertrophic pyloric stenosis** in infants is caused by thickening of the pyloric sphincter with accompanying spasticity. It prevents adequate emptying of the stomach. These infants appear hungry, eat and vomit excessively, and lose weight. A **chronic ulcer** in the pyloric channel in an adult may cause scarring, which can narrow the pyloric channel enough to obstruct it. Its symptoms include vomiting, dehydration, and acid-base and electrolyte imbalances. **Motility disorders** of the stomach typically foster delayed gastric emptying of solid food (termed *gastroparesis*) that is manifested by symptoms of anorexia, persistent fullness after meals, nausea, vomiting, and pain. These can be associated with abnormal smooth muscle electrical activity, smooth muscle contractile activities (dystrophies), and altered enteric (diabetic gastroparesis) or extrinsic (vagotomy) nervous innervation. Delayed gastric emptying is occasionally the primary cause of **gastroesophageal reflux disease.** Vagotomy eliminates fundic-receptive relaxation and accommodation, accelerates the emptying rate of liquids, and results in early satiety.

Vomiting (emesis), which under some conditions may not be abnormal, is usually infrequent but can indicate a serious disorder. Protracted or frequent vomiting can lead to metabolic alkalosis and electrolyte imbalances. Before vomiting, one senses the inevitable with the feeling of nausea. This reflects a diffuse discharge from both sympathetic and parasympathetic nerves. A patient about to vomit exhibits **sympathetic responses** that include dilatation of pupils, pallor, sweating, cold skin, increased heart rate, and increased respiration. **Parasympathetic responses** consist of increased salivation; more pronounced motor activity in the esophagus, stomach, and duodenum; and relaxation of the upper and lower esophageal sphincters. Duodenal contents can be forced into the stomach under these conditions. During vomiting, a person retches, takes a deep breath, and the glottis closes. The pylorus closes

with no gastroesophageal motility, the abdominal muscles contract, and the stomach is squeezed between the diaphragm and the abdominal muscles, thereby rapidly emptying the stomach. The gastric contents are propelled up the esophagus and out the mouth. Emesis is stimulated in the brain through sensory stimulation or injury to the viscera or head. It may also be provoked by unusual labyrinth stimulation and by chemical stimulation of receptors on cells in the chemoreceptive trigger zone, located on the floor of the fourth ventricle in or near the area postrema, a circumventricular organ situated outside the blood-brain barrier.

MOTILITY OF THE SMALL INTESTINE

The small intestine is commonly divided into three proximal-to-distal segments: the **duodenum, jejunum,** and **ileum.** It has a total length of approximately 6 to 8 meters in the average adult human and functions to absorb water, electrolytes, nutrients, and other nondietary components (such as drugs) necessary to maintain health. To do this, the intestinal contents must be moved in a manner that not only brings them **in contact** with the intestinal mucosa but also **propels** them along this tubular organ. Although the **slow-wave** and **spike-potential** mechanisms of contractility as well as fasting and fed motor patterns resemble those of the stomach, slow waves are controlled by several pacemaker areas. Their rate is highest in the duodenum (about 12 per minute), and this decreases in a graded manner down to the terminal ileum (about 8 per minute).

Propulsive motility in the small intestine is accomplished primarily by peristalsis. The mechanisms involved have been studied more in the small bowel than in any other part of the gut, and the salient features of this reflex are given in the section "Esophageal Motility." Peristaltic contractions do not travel the entire length of the small intestine, except for the propagated contractile waves that arise at certain times during the migrating motor complex. Vagal innervation, which is important for gastric peristaltic amplitude, is not necessary for intestinal peristalsis. Instead, the **enteric nervous system** (mainly the myenteric plexus), **intrinsic smooth muscle** excitability, and certain gastrointestinal **hormones** modulate the peristaltic activity of the small bowel. The purpose of the fasting pattern or migrating motor complex in the small bowel is similar to that for the stomach; it cleanses the bowel of nonabsorbable and indigestible material. During the **fed motility pattern** in the small intestine, segmentation and occasional peristaltic reflex contractions serve to mix chyme with digestive enzymes. Furthermore, the nonpropulsive, nonpropagating contractions allow contact with the mucosal surface to promote absorption of nutrients. An illustration of **segmentation** is provided in Figure 42-7. Segmentation is also regulated by the enteric nervous system, intrinsic smooth muscle excitability, and gastrointestinal hormones. These controls are absent during the fasting state.

The **biliary tract,** consisting of the hepatic duct, gallbladder, common bile duct, sphincter of Oddi, exhibits characteristic motility patterns. This system stores and delivers bile to the duodenum during feeding to provide for the efficient digestion and absorption of lipids. The flow of bile to the duodenum is regulated primarily by CCK, which is released from duodenal mucosal cells by the presence of dietary fats and amino acids. **CCK** contracts the gallbladder and relaxes the sphincter of Oddi, thus promoting the flow of bile to the upper small bowel.

Disorders of small intestinal motility can lead to constipation or diarrhea, abdominal distention and pain, or nausea and vomiting. Much basic research needs to be done on the small intestine to better understand the causes and consequences of motility dysfunction. However, it is believed that these symptoms are due to altered slow-wave and spike-potential changes in extrinsic neuronal or hormonal regulation, or because of a smooth muscle disorder. For example, constipation often arises during **pregnancy** and may be associated with decreased slow-wave frequency produced by elevated progesterone levels. **Hyperthyroidism** elevates slow-wave frequency

Fig. 42-7. The segmental motor pattern of the small intestine. (A) In this sequence, represented by the *solid line,* short segments of bowel are alternately contracted and relaxed (*black arrows*). (B) As segmentation occurs, represented by the *shaded area,* areas of contraction relax and areas of relaxation contract (*white arrows*). This moves the intestinal contents back and forth, mixing them with digestive enzymes and exposing them to the absorptive surface.

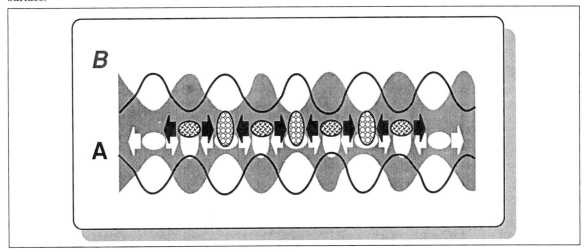

and produces diarrhea. Constipation is observed often in **hypothyroidism,** which decreases slow-wave frequency. **Intestinal pseudoobstruction** is characterized by dilated, nonfunctioning intestinal segments and has been attributed to both smooth muscle and enteric neuronal deficits. **Scleroderma** produces intestinal dilation because of a failure of enteric nervous system regulation (absent response to luminal distention). **Bacterial overgrowth** of the bowel has been associated with an abnormal or deficient fasting migrating motor complex of the small intestine. Also, in the later stages of **scleroderma** and in some forms of **bowel obstruction,** there is a loss of smooth muscle that further contributes to bowel dilatation.

In terms of the biliary tract, **gallbladder dysmotility** (poor CCK-induced emptying) is observed in the presence of cholelithiasis (gallstones). Also, sphincter of Oddi dysfunction, termed **biliary dyskinesia,** can produce pain due to a paradoxical CCK-induced increase in sphincter pressure.

COLONIC MOTILITY

The ileocecal sphincter (a zone about 4 cm long) is an area of elevated pressure that separates the colon from the small intestine. It possesses those characteristics that qualify it to be a sphincter. That is, it has increased tone that promotes aboral flow but it can contract to prevent retrograde flow. The colon, or large intestine, is roughly 1.5 meters long and is divided anatomically into the ascending, transverse, and descending segments. The colon's function is to conserve water and electrolytes by means of absorption and to form, store, and eliminate waste. Although the basic histological features of the colonic wall resemble those of the other portions of the gastrointestinal tract (see Fig. 42-1), colonic longitudinal muscle is concentrated in three bundles of **taeniae coli.** These bundles cause the areas of longitudinal muscle in the large bowel wall to protrude during circular muscle contractions, forming **sacculations** or **haustra.** The purpose of haustrations or haustral contractions

is similar to that of segmentation in the small bowel; they allow mixing of the contents and expose them to the surface mucosa for absorption. Likewise, they are nonpropulsive and nonpropagating. Taeniae coli are not found in the sigmoid colon, that area of the intestine between the descending colon and the rectum. At the terminal rectum, the circular smooth muscle coat thickens to form the **internal anal sphincter,** which is surrounded by striated muscle originating from the pelvic floor and forming the **external anal sphincter. Sympathetic innervation** of the colon is furnished via lumbar colonic nerves. **Cranial parasympathetic** (vagal) **innervation** is provided to the right colon and **sacral parasympathetic innervation** supplies the entire colon via pelvic nerves. Unlike the relatively short **transit time** through the small intestine (several hours), colonic transit is measured in days. Much mixing occurs in the right (ascending) colon, where slow waves of contraction move in the oral direction to delay transit and promote mixing and absorption of water and electrolytes.

The nature of these **slow waves of contraction** in the human colon is controversial. They cannot always be recorded and, when they are, appear to have a frequency of between three and ten per minute. As in the upper gastrointestinal tract, slow waves may be linked to spike potentials that initiate contractions. However, three distinct types of spike activity and related contractile events take place in the colon. **Short spike bursts** arise most frequently; these are associated with a single slow wave and appear to mediate haustral contractions. **Long spike bursts** occur with several slow waves, move in either direction, and appear to accelerate colonic motor patterns, consisting of both haustral activity and propagated, peristaltic-like contractions called *mass movement.* **Migrating spike bursts** are rapid oscillations of long duration. They are not related to slow waves, which usually begin in the transverse and descending colon, and appear to mediate mass movements. **Mass movement** constitutes the major propulsive motor activity in the colon. Shortly after eating, when mass movement is usually initiated, a relatively large bolus of stored material, kept in the lower ascending colon, is carried some distance up this colonic area. While haustral contractions keep the bolus firm, a segment may be broken off and, during a subsequent mass movement, be moved rapidly around the hepatic flexure into the transverse colon. Subsequent propulsive activity progressively transports this segment (or a fraction thereof) through the transverse and descending colon into the **sigmoid colon.** There, normal peristalsis is re-established, which moves the bolus into the **rectum,** where it is stored until defecation. The normal **colonic transit time** in an average adult human is from one to three days.

In the **ascending colon,** slow waves tend to move in a retrograde direction, which normally retards flow and facilitates the mixing and absorption of water and electrolytes. The **final stages** of fluid absorption and feces formation and storage take place in the descending colon. In the transverse and descending colon, slow waves of contraction move in the expected aboral direction. The lower sigmoid colon, rectum, and anal sphincters collectively provide for defecation and fecal continence.

Defecation is a complex act that involves the colon, rectum, anal sphincters, and striated muscles of the pelvic floor, abdominal wall, and diaphragm. The **rectosigmoid area** stores fecal matter until mechanoreceptors in the wall of the rectum are stimulated. This initiates reflex neuronal impulses and the internal anal sphincter muscle (an involuntary muscle) relaxes, signaling the urge to defecate. Voluntary contraction of the striated external anal sphincter muscle inhibits the reflex, causing the rectal wall to relax (receptive relaxation) and the internal anal sphincter to contract or close. At the appropriate time and place, the external sphincter is voluntarily relaxed (aided by a sitting or squatting posture), with additional pressure furnished through voluntary contraction of the striated muscles. Propulsive contraction of the rectum and sigmoid colon, coupled with relaxation of both sphincters, forces fecal material out of the anal canal.

Disorders of colonic motility can lead to changes in the consistency and frequency of de-

fecation (including diarrhea and constipation), pain or distention when colonic wall tension is increased, and incontinence. As in the small intestine, more basic research is needed, to provide further insight into the causes of the specific motility dysfunctions responsible for these colonic disorders. However, clinical colonic motor disorders are known to be associated with defects in the intrinsic and extrinsic innervation, slow waves, spike potentials, and smooth muscle functioning. For example, damage to the **cervical or thoracic spinal cord** abolishes colonic compliance with reflex defecation that is ordinarily elicited by distention, because sensation and voluntary control are now missing. **Sacral nerve damage** interrupts important reflex pathways and produces an atonic colon; this causes a dramatic loss of colonic sensation, with only intrinsic autonomic reflex defecation response remaining, and also impairs colonic emptying. In **Hirschsprung's disease** (aganglionosis or congenital megacolon), there is complete loss of the enteric nervous system, specifically the myenteric plexus, in a segment of the lower bowel. Because a major role of the enteric nervous system is to inhibit gut myogenic activity, the aganglionic segment is permanently contracted (or, if only the internal anal sphincter is involved, does not relax). As in the small intestine, **scleroderma** initially produces a neuronal abnormality (loss of postprandial motor response), but, as the disease progresses, the muscle eventually cannot respond. **Irritable bowel syndrome** refers to a condition of altered bowel function, constipation or diarrhea, and abdominal pain but without any detectable organic disorder. Affected patients are more sensitive to distention of the rectosigmoid area and exhibit a lower threshold for pain to this stimulation. Although not well understood, irritable bowel syndrome is linked to a heightened colonic responsiveness to many types of stimulation, including stress, and such patients can also exhibit abnormal motility patterns of the small bowel in response to food. **Diverticular disease** is associated with herniation of the mucosa or submucosa (diverticulum) through the muscle wall of the sigmoid colon, and appears to be caused by increased intraluminal pressure. Microperforation of the mucosa in a diverticulum may lead to inflammation, called **diverticulitis. Fecal incontinence** can be caused by sphincteric muscle damage inflicted by trauma or inflammatory diseases, such as Crohn's disease or ulcerative colitis. **Internal anal sphincter dysfunction** can be the aftermath of sclerosis, neuromuscular disease such as diabetic visceral neuropathy, and inflammatory bowel disease. **External anal sphincter dysfunction** often results from neuropathies or dermatomyositis.

SUMMARY

The gastrointestinal tract possesses a unique arrangement of smooth muscle layers that are responsible for the distinct patterns of motility observed during and after consumption of a meal. These patterns are both propulsive and nonpropulsive (or mixing) in nature. Propulsive motility patterns move food along and are seen throughout the tract, from the mouth to anus. Peristalsis is the major propulsive activity in the esophagus, stomach, small intestine, and rectum, whereas mass movement is the propulsive mechanism in the colon. Segmentation in the small bowel and haustrations in the colon represent the major nonpropulsive forces responsible for mixing the intestinal contents with secretions from both the mucosa and the accessory organs of the tract. They also optimize contact of the luminal contents with the absorptive surfaces. Extrinsic and intrinsic nerves, as well as locally released hormones, regulate these motility patterns according to the volume and chemical composition of the intestinal contents. Diseases of motility are quite varied, and are determined by which of the many variables regulating motility is compromised.

NATIONAL BOARD TYPE QUESTIONS

Select the one most correct answer for each of the following

1. Of the following motility patterns seen in the gastrointestinal tract, which is considered a totally voluntary process?
   A. Rhythmic segmentation
   B. Esophageal peristalsis
   C. Chewing
   D. Mass movement
   E. Swallowing

2. Which of the following events in the gastrointestinal tract results in the aboral propulsion of contents?
   A. Slow waves
   B. Peristalsis
   C. Haustrations
   D. Segmentation
   E. Contraction of sphincters

3. The migrating motor complex:
   A. Is recorded in the small and large intestines
   B. Occurs randomly throughout the gastrointestinal tract
   C. Is induced by sphincteric relaxation
   D. Is only observed in the fasting state
   E. Is the primary mechanism for moving food through the gastrointestinal tract

4. The major function of the colon is:
   A. The absorption of water and electrolytes, thus reducing the volume of fluid excreted
   B. To absorb dietary nutrients
   C. To reabsorb bile salts
   D. Chylomicron formation
   E. To absorb vitamin B_{12}

5. The basic electrical rhythm of the stomach
   A. Is a mechanical force
   B. Determines the slowest rate of peristaltic contractions
   C. Always causes contractions
   D. Determines the maximum rate of peristaltic contractions
   E. Has nothing to do with contractions

ANNOTATED ANSWERS

1. C. Rhythmic segmentation, esophageal peristalsis, and mass movements of the colon are

clearly involuntary. Swallowing is initiated by a voluntary movement. However, the pharyngeal and esophageal phases of swallowing render it a not entirely voluntary motor process. Chewing is the only totally voluntary motor process.

2. B. Slow waves are an electrical phenomenon of smooth muscle membrane depolarization. Haustrations and segmentation are motor patterns that mix food (or chyme) with other intestinal contents. Sphincteric contractions can actually impede the aboral movement of intestinal contents. Peristalsis is the only true propulsive motor pattern.

3. D. The migrating motor complex is only recorded in the small intestine and represents an orderly (not random) occurrence. The complex is independent of sphincteric activity and is not considered to represent the primary motor pattern for propelling food following a meal. It is only observed in the fasting state.

4. A. The major function of the colon is the absorption of water and electrolytes. Only in certain pathologic situations are nutrients, bile salts, or vitamin B_{12} present in the colon. The colonic mucosa is considered unequipped to effectively absorb these substrates. Likewise, chylomicron formation does not take place in the colonic mucosal cells.

5. D. Peristaltic contractions of the stomach always result when a spike potential is generated. In turn, a spike potential is always associated with a slow wave of the basic electrical rhythm. Because of this interdependency, the basic electrical rhythm determines the maximum rate of peristaltic contractions in the stomach. In this regard, it does influence contractions. The basic electrical rhythm is an electrical, not mechanical, event but it does not always cause contractions as do spike potentials.

BIBLIOGRAPHY

Camilleri, M., and Phillips, S. F. Disorders of small intestinal motility. *Gastroenterol. Clin. North Am.* 18:405-424, 1989.

Davenport, H. W. *Physiology of the Digestive Tract; An Introductory Text,* 3rd ed. Chicago: Year Book Medical Publishers, 1971.

Hendrix, T. R., Castell, D. O., and Wood, J. D. American Gastroenterological Association undergraduate teaching project, Unit 10B, alimentary tract motility: stomach, small intestine, colon, and biliary tract. Timonium, MD: Milner-Fenuicle, Inc, 1987.

Huizinga, J. D. Electrophysiology of human colon motility in health and disease. *Clin. Gastroenterol.* 15:879-901, 1986.

Johnson, L. R., ed. *Gastrointestinal Physiology,* 3rd ed. St. Louis: Mosby, 1985.

Johnson, L. R., ed. *Physiology of the Gastrointestinal Tract,* 2nd ed. New York: Raven Press, 1987.

Minami, H., and McCallum, R. W. The physiology and pathophysiology of gastric emptying in humans. *Gastroenterology* 86:1592-1610, 1984.

Paterson, W. G., and Goyal, R. K: Oesophageal motility and its disorders. *Cur. Op. Gastroenterol.* 1:549-559, 1985.

Read, N. W., and Timms, J. M: Defecation and pathophysiology of constipation. *Clin. Gastroenterol.* 15:937-965, 1986.

Sernka, T. J., and Jacobsen, E. D: *Gastrointestinal Physiology—The Essentials,* 2nd ed. Baltimore: Williams and Williams, 1983.

43 Gastric Secretion

JOHN CUPPOLETTI AND DANUTA H. MALINOWSKA

Objectives

After reading this chapter, you should be able to:

Describe the structure of the stomach, gastric glands, and secretory cells

Explain the secretory function of the stomach

List the receptors that control gastric HCl secretion

Cite the second-messenger systems that control gastric HCl secretion

List the membrane transport proteins involved in gastric HCl secretion

Explain the mechanisms of gastric HCl secretion

Understand the rationale behind the use of antisecretory agents for treating ulcers and gastric secretory disease

Describe the sites and mechanisms of action of gastric antisecretory agents

Gastric secretions include pepsin, hydrochloric acid, and mucus. In the stomach, HCl exists at a concentration of 0.15 M and is lethal to most organisms that would otherwise thrive in the digestive tract. HCl, in conjunction with pepsin, is normally required for the initial stages of protein digestion. However, normal levels of HCl may interfere with ulcer healing, and treatment of ulcers relies on the use of antisecretory agents. Mucus has been suggested to play a role in protecting the stomach from gastric secretions.

STRUCTURE OF THE GASTRIC MUCOSA

The structure of the stomach is shown in Figure 43-1. It is divided into three main regions: the **fundus, corpus** (body), and **pyloric antrum.** The surface of the stomach, or **mucosa,** is highly invaginated and is composed of gastric glands (see

Fig. 43-1A). These invaginated regions are termed **gastric pits.** A diagram of a **gastric gland** is shown in Figure 43-1B. Near the top or neck of the gland are the surface mucous cells that spill out to form the surface of the mucosa. The mucous neck cells are located around the middle of the gastric gland. The **parietal** (oxyntic cells) and **peptic cells** (chief cells) exist in approximately equal amounts and compose the bulk of the base of the gland. The parietal cells secrete HCl and the peptic cells secrete pepsin.

GASTRIC JUICE

Gastric juice is slightly hyperosmotic (325 mOsm/liter versus 300 mOsm/liter for plasma). It is slightly enriched in K^+ (approximately 10 mM versus 5 mM for plasma) and is low in Na^+ (2 to 4 mM versus 140 mM for plasma). The concentration of H^+ is 160 to 170 mM and Cl^- has a

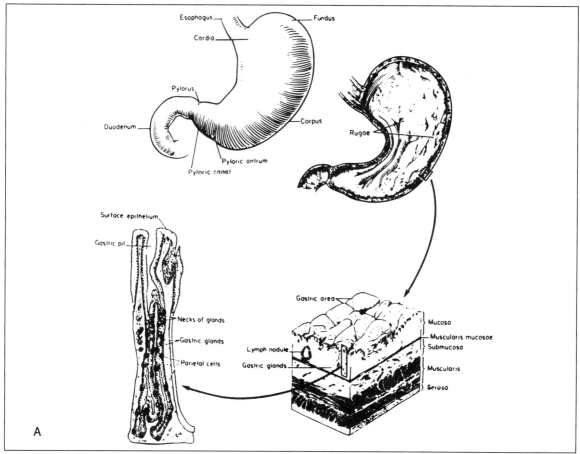

Fig. 43-1. (A) Upper: diagram of the stomach showing main regions. Lower: cross-section of the entire stomach wall from the corpus region, and a gastric pit with gastric glands. (B) Diagram of a gastric gland from the corpus region of a mammalian stomach, illustrating the different cell types present. (From: Ito, S. Functional gastric morphology. In: Johnson, L.R., ed. *Physiology of the Gastrointestinal Tract.* New York: Raven Press, 1987, Pp. 817-851.)

concentration of about 180 mM. HCl constitutes the bulk of gastric secretion (about 170 mM). Gastric juice has an approximate pH of 1; the pH of serum is about 7. This dramatic difference in pH represents a million-fold gradient of H^+ across the gastric mucosa.

FINE STRUCTURE OF THE PARIETAL CELL

The parietal cell is a large pyramidal-shaped cell that is rich in mitochondria. This large popula-

tion of mitochondria is consistent with the cell's involvement in the active transport of enormous quantities of HCl, a process that requires a large expenditure of energy. Parietal cells also contain large quantities of **cytoskeletal proteins,** including actin, tubulin, and cytokeratins. Another characteristic of parietal cells is that, at rest, the cytoplasm is filled with smooth-surfaced, elongated membranes, called **tubulovesicles** (Fig. 43-2A). As shown in Figure 43-2B, when acid secretion is stimulated, these tubulovesicles rearrange, perhaps with the help of the cytoskeleton,

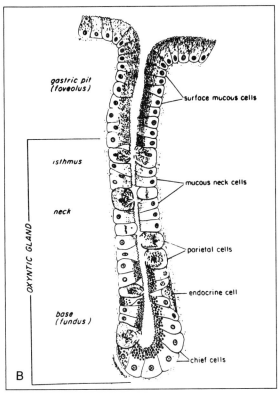

Fig. 43-1. (continued)

to form large intracellular canals, known as **intracellular canaliculi** or **apical membrane,** that are lined with microvilli. These canals communicate with the lumen of the gland and are the intracellular site of gastric acid secretion. Acid formed in these extracytoplasmic spaces flows into the lumen of the gland and eventually into the stomach. The canalicular or apical membranes contain the proteins that use metabolic energy in the form of ATP for the production of HCl.

REGULATION OF GASTRIC ACID SECRETION: RECEPTORS

Histamine, cholinergic agents (acetylcholine, or its stable analogue, carbachol), and gastrin all **stimulate acid secretion.** Somatostatin, epi-

dermal growth factor, beta-adrenergic agonists, and enteroglucagon **inhibit acid secretion,** but this action appears to be an indirect one, as specific receptors for these agents have not been identified on the parietal cells. Two types of histamine receptors exist. There are H_1 **receptors,** which are blocked by diphenhydramine and other H_1 receptor antagonists, but they do not participate in the regulation of HCl secretion. H_2 **receptors,** which are blocked by cimetidine and ranitidine, but not H_1 blockers, are also present. **Histamine** is a potent stimulant of gastric acid secretion, and H_2 receptor antagonists (for the chemical structure of cimetidine, see Fig. 43-4A), which prevent histamine from binding to the H_2 receptors of the parietal cell basolateral membrane, are widely used for reducing acid secretion in treating gastric ulcers. Histamine receptors do not exist on peptic cells, and therefore histamine does not stimulate pepsin secretion. **Vagal stimulation** leads to stimulation of acid secretion through the release of acetylcholine. This secretory response can be inhibited by atropine. Thus the parietal cells contain a **muscarinic receptor.**

The peptide hormone **gastrin** is a potent stimulant of gastric acid secretion. Gastrin receptors are present on the parietal cells of some species, including the dog and rat, but are apparently absent in other species, including rabbit and human. Nevertheless, gastrin is a potent stimulant of acid secretion in humans, and its effect appears to be mediated by the release of histamine from another cell type in the gastric glands. Histamine then interacts with the parietal cell H_2 receptor to stimulate acid secretion. Thus, parietal cells contain receptors for both histamine (H_2) and acetylcholine (muscarinic), and some species possess receptors for gastrin (Fig. 43-3). H_2 blockers are effective *antiulcer agents* because they reduce HCl levels, thereby promoting healing.

INTRACELLULAR SECOND MESSENGERS

Stimulants of acid secretion act by binding to receptors on the **basolateral membrane** of the

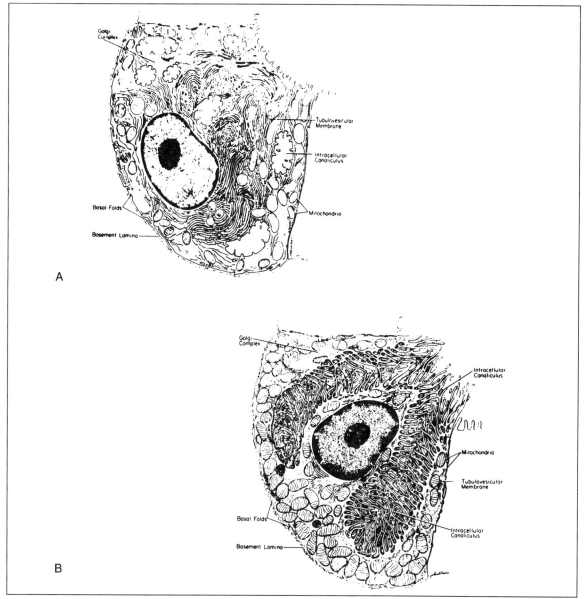

Fig. 43-2. The morphologic changes that occur in the parietal cell upon stimulation of acid secretion. (A) Resting, nonsecreting parietal cell. The cytoplasm contains numerous mitochondria and is filled with tubulovesicles. (B) Stimulated, acid-secreting parietal cell. The tubulovesicles have disappeared and an extensive, expanded intracellular canaliculus lined with microvilli is apparent. (From: Ito, S. Functional gastric morphology. In: Johnson, L.R., ed. *Physiology of the Gastrointestinal Tract.* New York: Raven Press, 1987, Pp. 817-851.)

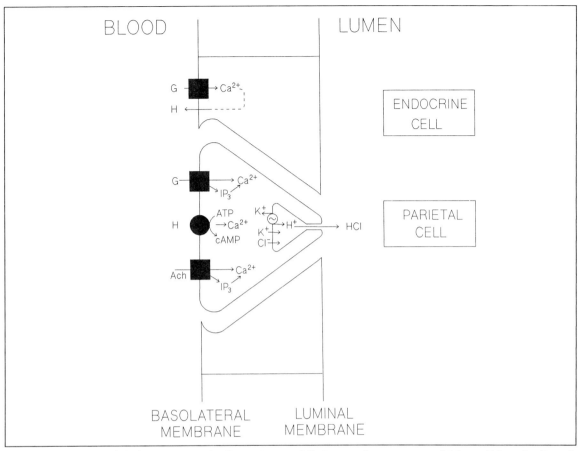

Fig. 43-3. Summary of the known parietal cell receptors, and their second messengers, which result in activation of acid secretion. (*H* = histamine receptor; *G* = gastrin receptor; *ACh* = acetylcholine receptor; *IP*$_3$ = inositol-tris-phosphate; *cAMP* = cyclic AMP).

parietal cell. Second messengers are thereby released, which act on target systems in the cell. The second messengers that are known to be active in the parietal cell following hormone release are **cyclic AMP** (cAMP), **Ca²⁺,** and **inositol-tris-phosphate.**

Histamine binding to the H₂ receptors of the parietal cells activates **adenylate kinase.** This enzyme generates cAMP from ATP. Increased intracellular **cAMP levels** only occur with histamine stimulation. cAMP binds to cAMP-dependent protein kinase; this kinase is inactive when cAMP is absent. When cAMP binds to the

regulatory subunit of the cAMP-dependent protein kinase, the active catalytic subunit is released. This catalytic subunit then phosphorylates a variety of target proteins in the parietal cell. Some of these targets are **cytoskeletal proteins,** which may be involved in the characteristic morphologic rearrangement that accompanies acid secretion. Other potential targets of the kinase are the **ion channels** of the parietal cell's apical and basolateral membrane and enzymes that participate in the production of metabolic energy.

Histamine, muscarinic agents, and gastrin act

to raise intracellular **free Ca**$^{2+}$ levels in the parietal cell. This increase arises from the release of Ca^{2+} from intracellular stores and from increased Ca^{2+} influx across the basolateral membrane. There is a Ca^+-dependent and phospholipid-dependent protein kinase (protein kinase C) in the parietal cell that may play a role in regulating the morphologic or secretory changes that occur when Ca^{2+} is released, but the cellular systems that are affected by this protein remain to be elucidated.

Inositol-tris-phosphate is released by gastrin and muscarinic agonists. It and diacylglycerol are the products of hydrolysis mediated by a membrane-associated phospholipase. A major target for inositol tris-phosphate is a Ca^{2+}-release channel present in the membranes of intracellular vesicles that store Ca^{2+}. Diacylglycerol is an activator of the Ca^{2+}-dependent and phospholipid-dependent protein kinase (protein kinase C).

Studies on the parietal cell have revealed that, just as there are overlaps among the second messengers released by the stimulants of acid secretion, so there are also links at this level. As more is learned about the targets of the second messengers and the control of the various cellular systems which must be activated in order to increase metabolic output, maintain ionic homeostasis, rearrange the cell morphology, and finally increase acid secretion, overlap at the level of the second-messenger targets is also expected. Figure 43-3 summarizes second messengers that operate in the parietal cell.

DETERMINANTS OF HCl PRODUCTION

Stimulation of acid secretion increases oxygen use. The mitochondria use this extra oxygen for the production of ATP and the oxygen is not directly involved in HCl production, as occurs in a variety of oxidation-reduction reactions. Rather, studies on detergent-permeabilized parietal cells have shown that HCl production by the stimulated parietal cell requires only K^+, Cl^-, and ATP. Thus, HCl is formed in an enzyme reaction catalyzed by a K^+-stimulated and H^+-transporting

ATPase—the gastric (H, K)-ATPase, also referred to as the **gastric proton pump.** The gastric (H, K)-ATPase protein is a close relative of (Na, K)-ATPase (see Chapter 3). The (H, K)-ATPase is not inhibited by cardiac glycosides, but is inhibited by a newly discovered antisecretory agent, omeprazole (Fig. 43-4B), which is active in the acid environment of the apical membrane and which inhibits acid secretion through covalent modification of the (H, K)-ATPase. This class of antisecretory agents is especially effective in the management of Zollinger-Ellison syndrome,which is characterized by an overgrowth of parietal cells and excess acid production. Omeprazole provides an effective treatment for

Fig. 43-4. Structures of two clinically used inhibitors of gastric acid secretion. (A) Cimetidine (Tagamet) is a histamine H_2-receptor antagonist. It contains an imidazole ring and is chemically related to histamine. (B) Omeprazole (Prilosec) is a substituted benzimidazole sulfoxide. In an acid environment it becomes active and inhibits the gastric (H, K)-ATPase. (C) For comparison, this panel shows the structure of the agonist, histamine.

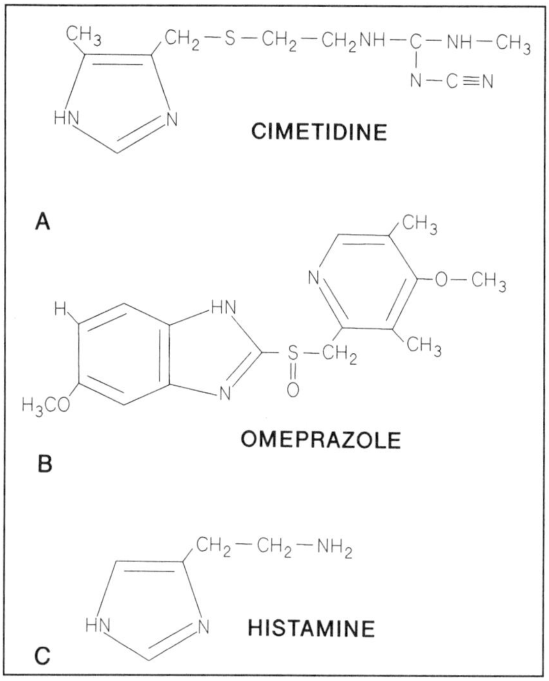

Zollinger-Ellison syndrome because, unlike H_2 blockers, it acts directly on the (H, K)-ATPase.

As shown in Figure 43-5, the process by which the apical (secretory) membrane produces HCl requires the following:

1. ATP hydrolysis and the production of ADP and inorganic phosphate by the (H, K)-ATPase; this requires that K^+ be present at a site on the extracytoplasmic surface of the (H, K)-ATPase.
2. A mechanism for moving K^+ from the cytosol of the parietal cell into the stomach to satisfy the K^+ requirement of the (H, K)-ATPase.
3. Cl^- must accompany the K^+ that leaves the cell in order to balance the ionic charge and provide equivalents of Cl^- for HCl secretion.

The proteins that catalyze the movement of K^+ and Cl^- are likely to be ion channels, and these channels are probably regulated in gastric HCl secretion. The transport of K^+ across the fully elaborated secretory membrane, along with Cl^-, furnishes the K^+ required for the (H, K)-ATPase and the Cl^- needed for HCl production. Inhibition of either the K^+ or Cl^- pathway or the (H, K)-ATPase diminishes acid secretion. Thus, when the (H, K)-ATPase is in the tubulovesicles in the resting parietal cell and the K^+ and Cl^- channels

Fig. 43-5. The components of the parietal cell apical (or secretory) membrane that elicit HCl production. (1) (H, K)-ATPase; (2) K^+ channel, and (3) Cl^- channel.

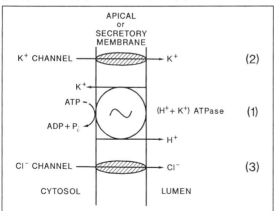

are inactive, acid is not secreted. When acid secretion is stimulated by any potent agonist, (H, K)-ATPase appears in the elaborated canalicular membrane, the K^+ and Cl^- channels become active, and HCl is produced.

ION TRANSPORT AT THE BASOLATERAL MEMBRANE

The H^+ that appears in the stomach derives from the splitting of water: $H_2O \rightarrow H^+ + OH^-$. Thus, an hydroxyl ion is produced in the cytosol for each proton transported into the lumen of the stomach. If OH^- were allowed to accumulate in the parietal cell, the pH of the cell would rapidly become basic. Fortunately, the cell is well equipped with carbonic anhydrase, which catalyzes the production of HCO_3^-, and with Cl^-–HCO_3^- exchangers in the basolateral membrane, which rapidly move HCO_3^- out of the cell and into the blood. The Cl^-–HCO_3^- exchanger is inactive at rest and is stimulated by H^+ secretion. At rest, the Na^+–H^+ exchanger in the basolateral membrane appears to be the primary factor in regulating intracellular pH.

The (Na, K)-ATPase of the parietal cell basolateral membrane serves to maintain low concentrations of Na^+ and high concentrations of K^+ in the parietal cell; Ca^{2+}-ATPases in the basolateral and intracellular membrane maintain a low intracellular Ca^{2+} concentration.

PEPSIN SECRETION

The peptic (chief) cell, situated in the base of the gastric gland, is responsible for the production of pepsin, the major hydrolytic enzyme in the stomach. Pepsinogen, which is the inactive precursor of pepsin, is kept in the storage granules of the peptic cell. Pepsin secretion is stimulated by cholinergic, muscarinic agents and by beta-adrenergic agents. Exocytosis releases pepsinogen into the stomach, where autocatalytic cleavage converts it into pepsin in the presence of HCl. Pepsin is only active at the acidic pH of the stomach, and therefore does not play a role in further digestion in the intestine. Other enzymes that are released by the pancreas act in the intestine.

SUMMARY

The gastric glands are made up primarily of surface cells, mucous cells, peptic (chief) cells, and parietal cells (oxyntic cells). The peptic cells are responsible for pepsin secretion and the parietal cells for HCl secretion. The basolateral membrane of parietal cells contains receptors for histamine and acetylcholine. In some species, there are also gastrin receptors. In humans, gastrin appears to be activated through the release of histamine from another type of cell, near the parietal cells. The basolateral membrane also contains membrane transport proteins, including (Na, K)-ATPase and the Cl^-–HCO_3^- exchanger. The Cl^-–HCO_3^- exchanger is important for removing the equivalents of OH^- that are produced with H^+. Upon stimulation of acid secretion, there is a morphologic rearrangement of the parietal cell, wherein intracellular membranes (tubulovesicles) are transformed into intracellular canals (intracellular canaliculi), which are the sites of acid secretion and which directly communicate with the gland lumen. The secretory membrane contains the (H, K)-ATPase. This protein is a K^+-dependent ATP hydrolase that acts in concert with K^+ and Cl^- pathways to produce concentrations of HCl as high as 0.16 M in the liter quantities of gastric juice that are produced several times each day. Gastric ulcers can be controlled by the inhibition of HCl production. Potential sites of regulation of acid secretion include the basolateral membrane receptors, the gastric proton pump, and the ion channels that control the gastric proton pump. Regulation of these ion channels may underlie the physiologic control of gastric acid secretion.

NATIONAL BOARD TYPE QUESTIONS

Select the one most correct answer for each of the following

1. Gastric acid secretion can be stimulated by:
   A. Histamine only
   B. Phospholipids
   C. Histamine, gastrin, and acetylcholine
   D. Cimetidine and atropine
   E. Angiotensin
2. The enzyme responsible for gastric acid secretion is:
   A. The HCl cotransporter
   B. K^+ phosphatase
   C. The Na^+–H^+ exchanger
   D. H^+-K^+ ATPase
   E. A redox system
3. Gastric acid secretion:
   A. Occurs across the parietal cell apical membrane
   B. Occurs across the parietal cell basolateral membrane
   C. Occurs across the peptic cell secretory membrane
   D. Results from the concerted action of an apical membrane Cl^-–HCO_3^- exchanger and a Na^+–proton exchanger
   E. Originates in the surface mucous cells
4. The effects of histamine on the gastric parietal cell are:
   A. Increased acid secretion and intracellular Na^+ concentration
   B. Increased cellular respiration, and decreased intracellular cAMP and increased cGMP levels
   C. Increased acid secretion, and increased intracellular cAMP and decreased intracellular Ca^{2+} levels
   D. Increased pepsin secretion and intracellular Ca^{2+} concentrations
   E. Increased acid secretion, cellular respiration, intracellular cAMP levels, and intracellular Ca^{2+} concentrations
5. Control of gastric HCl production:
   A. Can be achieved by H_1 antagonists such as diphenhydramine
   B. Can be achieved in the Zollinger-Ellison syndrome with the use of H_2 antagonists
   C. Is not thought to be important to the healing of ulcers

D. Cannot be achieved by commonly used antiulcer agents

E. Can be achieved by therapeutic agents that act at membrane receptors or at the secretory membrane

ANNOTATED ANSWERS

1. C. Histamine, gastrin, and acetylcholine all stimulate acid secretion.
2. D. (H, K)-ATPase is the enzyme responsible for acid secretion.
3. A. Acid secretion occurs across the parietal cell apical membrane.
4. E. Histamine stimulates acid secretion, which results in increased cellular respiration, and increased intracellular cAMP and intracellular Ca^{2+} levels.
5. E. H_2 blockers bind to the histamine (H_2) receptor at the basolateral membrane; ome-

prazole acts by inhibiting the gastric (H, K)-ATPase at the secretory membrane. Because control of gastric HCl is required for normal healing of the stomach, H_2 blockers are used as antiulcer agents. In the Zollinger-Ellison syndrome, the more potent (H, K)-ATPase inhibitors are necessary to reduce HCl levels.

BIBLIOGRAPHY

Forte, J. G., and Wolosin J. M. HCl secretion by the gastric oxyntic cell. In: Johnson, L.R., ed. *Physiology of the Gastrointestinal Tract.* New York: Raven Press, 1987, Pp. 853-863.

Ito, S. Functional gastric morphology. In: Johnson, L. R., ed. *Physiology of the Gastrointestinal Tract.* New York: Raven Press, 1987, Pp. 817-851.

Soll, A. H., and Berglindh, T. Physiology of isolated gastric glands and parietal cells: receptors and effectors regulating function. In: Johnson, L. R., ed. *Physiology of the Gastrointestinal Tract.* New York: Raven Press, 1987, Pp. 883-909.

44 Pancreatic Exocrine Secretion

JOSEPH D. FONDACARO

Objectives

After reading this chapter, you should be able to:

Identify the important anatomic features of the pancreas

Describe the relationship between the neural, vascular, and cellular components of the pancreas

Discuss the role of the exocrine secretions in normal digestive tract physiology

Explain the action of glycolytic, proteolytic, and lipolytic enzymes on the dietary components of a normal meal

Describe the integral role of pancreatic HCO_3^- secretion in the digestive process, acid buffering, and duodenal mucosal protection

Discuss the neural and hormonal regulation of enzyme and HCO_3^- secretions and the function of second messengers in acinar and duct cells

---

The pancreas is an unusual organ in that it is a major component of two physiologic systems: **the gastrointestinal tract** and the **endocrine system.** The exocrine secretions of the pancreas are essential for the normal digestion and absorption of dietary nutrients. Furthermore, a component of pancreatic secretion provides an important protective mechanism for the duodenal mucosa. Thus, given the vital importance of this organ to the normal digestive process, the roles of the pancreas have significant clinical and physiologic relevance. Acute and chronic pancreatitis, pancreatic carcinomas, and cystic fibrosis are major diseases that compromise pancreatic exocrine function, while diabetes mellitus is considered the major metabolic disease caused by loss of pancreatic endocrine secretion. In this chapter, we will explore the major anatomic considerations that are important and perhaps unique to the pancreas, the vital endocrine and exocrine secretions of this organ and their function, and the intricate role of nerves and humoral agents in the regulation of pancreatic exocrine secretion.

ANATOMIC CONSIDERATIONS

The pancreas is a **retroperitoneal organ** that can be described as an "accessory" organ of the gastrointestinal tract. It is strategically located between the antral portion of the greater curvature of the stomach and the duodenal bulb, and extends horizontally in a tapered fashion almost to the spleen. From this location, pancreatic exocrine secretions are delivered to the **proximal duodenum,** such that maximal efficiency can be obtained from their presence in the gut lumen.

The **arterial blood supply** to the pancreas is derived from branches of the celiac and superior mesenteric arteries, the major contributor being the pancreaticoduodenal artery. Arteries and arterioles course through the stroma of the pancreas and terminate in capillary beds close to the exocrine-secreting cells, called **acini** (Fig. 44-1). Likewise, arterioles and capillaries surround the **islets of Langerhans** that contain the endocrine-secreting cells of the pancreas. Venous blood from the organ drains into the **portal circulation.**

The **efferent motor nerves** to the pancreas are supplied by both sympathetic and parasympathetic branches of the autonomic nervous system. **Postganglionic sympathetic fibers** arise from the celiac and superior mesenteric plexuses and travel to the pancreas along the arterial blood vessels. These adrenergic nerves release **norep-**

inephrine and are believed to influence pancreatic function mainly through the modulation of arterial blood flow. **Preganglionic parasympathetic fibers** from the vagus nerve terminate in the pancreas. **Postganglionic cholinergic fibers** release **acetylcholine** and are believed to be important in regulating pancreatic function. These fibers terminate on pancreatic acinar and islet cells to regulate their secretion (see Fig. 44-1). Furthermore, **cholinergic muscarinic receptors** have been identified on the acinar cell membrane, supporting the belief that direct parasympathetic influence plays a vital role in pancreatic exocrine secretion.

Recently, important **neurocrine influences** on pancreatic function have been identified, and

Fig. 44-1. Functional anatomy of the pancreas. The pancreas contains both the endocrine- and exocrine-secreting functional units. (*ACh* = acetylcholine.)

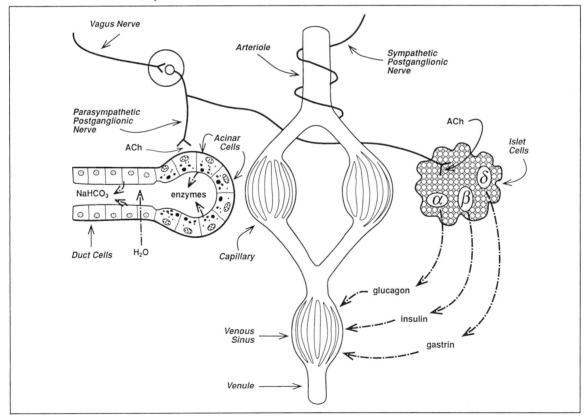

these have taken the form of nonadrenergic, noncholinergic nerves—the **peptidergic nerves.** These nerves have been shown to directly influence pancreatic acinar and duct cell secretion. They also operate through the modulation of sympathetic and parasympathetic fibers. Cell surface receptors have also been identified on the end organs. The **peptide neurotransmitters** that are considered important in regulating pancreatic function are vasoactive intestinal peptide (**VIP**), somatostatin, and enkephalin. **VIP** stimulates the pancreatic secretion of mainly fluid and electrolytes; **somatostatin** inhibits VIP's effect on pancreatic secretion. **Enkephalins** suppress pancreatic fluid and enzyme secretion, most likely by modulating other neural pathways. **Afferent sensory fibers** run from the pancreas to the central nervous system and are generally nociceptive in nature.

Major Cell Types

There are three major groups of cells that make up the pancreas: acinar cells, duct cells, and the cells of the islets of Langerhans (see Fig. 44-1). The **acinar cells** are the enzyme-secreting cells of the pancreas and are arranged in numerous blind-end glandlike structures called **acini.** Secretions from these cells collect in the acinar duct and travel through a network of converging ducts to the **main pancreatic duct.** From there, they are emptied into the duodenum. Acinar cells are characterized by their pyramidal shape and by their dense cytoplasmic granules containing the digestive enzymes to be secreted, many of which are in an inactive precursor form (see Fig. 44-1). Occasionally, a second type of acinar cell, called a **centroacinar cell,** is encountered. Although these cells are somewhat similar in shape to the more peripheral acinar cell, they are located near the center of the acinus and secrete water and electrolytes into the acinar space.

Duct cells are fairly homogeneous, and, as their name implies, they line the inner surface of the branching pancreatic ducts (see Fig. 44-1). Somewhat more columnar in shape, these cells are the major contributors of water and electrolytes, the most important of which is HCO_3^-, found in pancreatic juice.

Both the acinar and duct cells are richly supplied with capillaries that supply the raw materials for synthesis and secretion. Abundant **parasympathetic motor nerve pathways** deliver the appropriate stimulation for acinar cell secretion. Islet cells elaborate endocrine secretions and will not be discussed here.

EXOCRINE SECRETION

The product of pancreatic exocrine secretion, often referred to as **pancreatic juice,** contains both organic and inorganic components. The **organic components** are those major enzymes necessary for the digestion of dietary nutrients and are synthesized and secreted by acinar cells. HCO_3^- is the chief **inorganic component** and is secreted into pancreatic juice by duct cells. **Water** is also a major component of pancreatic juice and is supplied primarily by the duct cells through an osmotic response to HCO_3^- secretion. The normal adult human pancreas is capable of elaborating approximately 1.5 liters of pancreatic juice per day, which is a remarkable feat for an organ that constitutes barely 0.1 percent of the total body weight.

Organic Components

The **digestive enzymes** in pancreatic juice are proteins synthesized in acinar cells (Fig. 44-2). This synthesis takes place in the **ribosomes** lining the **rough endoplasmic reticulum.** As enzymatic proteins are produced, they are transferred to the lacunae of the rough endoplasmotic reticulum and eventually concentrate and bud off as vacuoles. These vacuoles migrate to the **Golgi complex,** where they are enveloped in membranes to form **zymogen granules** containing the inactive precursor (or proenzyme) form of the active enzyme. The **proenzymes** are stored in these granules in the apical portion of the acinar cells until appropriate stimulation of the cells triggers secretion by a mechanism known as exocytosis. **Exocytosis** is thought to consist of the

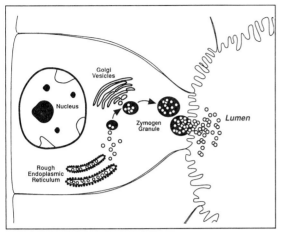

Fig. 44-2. Synthesis and release of digestive enzymes by an acinar cell of the pancreas. (Modified from: Johnson, L. R., ed. *Physiology of the Gastrointestinal Tract,* 2nd ed. New York: Raven Press, 1985.)

following events: (1) migration of the zymogen granule toward the inner surface of the apical membrane; (2) fusion of the granule membrane with that of the cell, a Ca^{2+}-requiring step; and (3) elimination of the bilayers and release of enzyme into the acinar space. This entire secretory process is the result of a series of biochemical events known as **stimulus–secretion coupling.** Utilizing intracellular second-messenger systems, this cascade of events links the stimulation of the acinar cell at the basolateral membrane by specific secretagogues to the release of digestive enzymes at the apical membrane.

Distinct receptors have been identified on the basolateral membrane of the acinar cell as being involved in the secretory process. Separate receptors bind the hormones cholecystokinin and gastrin and the neurotransmitter acetylcholine. Upon binding, receptor–ligand interaction activates phosphatidylinositol hydrolysis, thus elevating the intracellular Ca^{2+} concentration. Ca^{2+}, utilizing either a Ca^{2+}–calmodulin complex or the Ca^{2+}, phosphatidyl serine–dependent protein kinase C, triggers exocytosis. Another receptor on the basolateral membrane of the acinar cell binds the hormone **secretin.** When secretin

binds to its receptor, the signal for secretion is communicated by means of the adenylate cyclase–cyclic AMP pathway to the apical cell border, resulting in secretion. Secretin potentiates cholecystokinin's action on the enzyme secretion of acinar cells. For all receptors, signal transduction via the second-messenger systems links events at the basolateral membrane to the apical membrane, an intriguing aspect of biochemical regulation.

The **major digestive enzymes** in pancreatic juice are listed in Table 44-1, which shows that there are enzymes for the three classes of dietary nutrients. The **proteolytic enzymes** are secreted in an inactive proenzyme form, which prevents the autodigestion of pancreatic tissue. Once inside the lumen of the duodenum, **enterokinase,** an enzyme secreted by the duodenal mucosa, converts trypsinogen to its active form **trypsin** (Fig. 44-3). This conversion sets in motion the conversion of the other three proenzyme forms to their active moieties, as well as the autocatalysis of trypsin on its own precursor (see Fig. 44-3). Finally, because secretion of these enzymes by acinar cells is accompanied by the secretion of water and electrolytes (presumably by centroacinar cells), the substance leaving the acinar space is essentially an ultrafiltrate of plasma.

Pancreatic **alpha amylase** hydrolyzes 1,4-glycosidic bonds within dietary carbohydrate mol-

Table 44-1. Major Digestive Enzymes Found in Human Pancreatic Juice

Glycolytic
  Pancreatic alpha amylase
Lipolytic
  Pancreatic lipase
  Colipase (2)*
Proteolytic
  (Pro)carboxypeptidase (4)
  Trypsin(ogen) (3)
  Chymotrypin(ogen)

*The numbers in parentheses represent the number of different isoforms of the enzyme found in pancreatic juice.

All of the major proteolytic enzymes are secreted in precursor forms, as indicated by the appropriate prefix or suffix in parentheses.

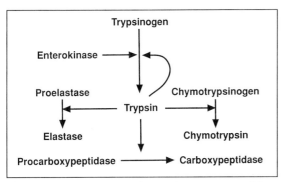

Fig. 44-3. The activation of proteolytic enzymes of the pancreas.

Fig. 44-4. The digestive action of pancreatic alpha amylase on dietary polysaccharide.

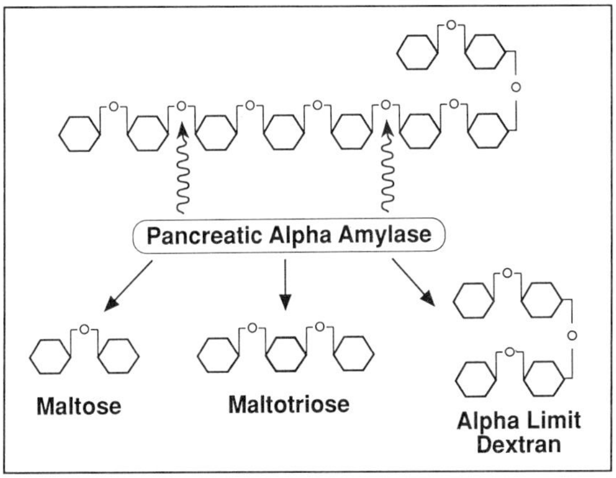

ecules (Fig. 44-4). This enzyme does not affect 1,6 linkages or terminal 1,4 bonds. Thus, the end-products of pancreatic amylase activity are **maltose** (a disaccharide consisting of two glucose molecules), **maltotriose** (a trisaccharide made up of three glucose molecules), and **alpha-limit dextrans** (smaller branch-chain polysaccharides). This is the primary step required for the digestion of **dietary carbohydrates.** However, the resulting molecules are still too large for intestinal absorption, and additional breakdown is needed. This is accomplished by specific enzymes located in the **brush-border membranes** of the small intestinal mucosal cells, which re-

duce these molecules to glucose and other simple sugars.

Pancreatic **lipase** and **co-lipase** reduce dietary triglycerides to simpler molecules to enable their absorption. The **triglyceride molecule** is made up of three fatty acid moieties coupled by ester linkages to the glycerol backbone (Fig. 44-5). The lipases preferentially cleave the 1 and 1' ester linkages to eventually yield a 2-monoglyceride and two free fatty acid moieties. Occasionally the reaction includes the ester bond at the "2" position, yielding **free glycerol,** which is readily absorbed. All of this lipolytic activity requires the emulsifying action of bile salts in order to solubilize triglycerides in the water environment of the intestinal lumen, so that the hydrophilic enzymes can digest the individual molecules. Once liberated, fatty acids and 2-monoglycerides partition readily into **bile salt micelles,** which are transported through the aqueous phase of the bowel lumen to the absorbing enterocytes.

Once activated, the **proteolytic enzymes** secreted by the pancreas cleave peptide linkages within the complex structure of dietary protein. **Trypsin** and **chymotrypsin** are endopeptidases that target internal peptide linkages, yielding dipeptides and tripeptides and other small-peptide chains. Many of these can be absorbed by the intestinal cells. **Carboxypeptidase** is an exopeptidase that attacks the ends of a peptide chain, liberating free amino acids. In addition,

Fig. 44-5. The digestive action of pancreatic lipase and co-lipase on dietary triglyceride.

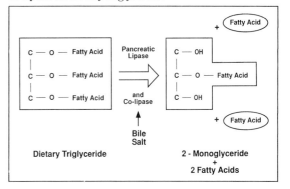

exopeptidases secreted by the intestinal mucosa also act on small peptides to yield free amino acids. These simple amino acid molecules are readily absorbed by the intestine (see Chapter 46).

Inorganic Components

As **fluid** leaves the acini and proceeds down the pancreatic ducts, its ionic composition changes. The degree to which it is modified depends on the secretory rate, a phenomenon that will be discussed later in this chapter. The major inorganic components of pancreatic juice are **water** and **electrolytes,** which are furnished mainly by **duct cells.** These cells are somewhat smaller than acinar cells and do not contain zymogen granules. Instead, they possess a specific transport mechanism in their apical and basolateral membranes that facilitates the secretion of certain electrolytes, of which HCO_3^- is the primary constituent.

As **pancreatic juice** exits from the acinar region and travels down the ducts, the initial concentrations of the major **electrolytes,** Na^+, K^+, Cl^-, and HCO_3^-, reflect that of the extracellular fluid. Upon stimulation of the duct cells, as ordinarily occurs during and immediately after a meal, there are specific changes in the electrolyte composition of the pancreatic juice emptied into the duodenum. The Na^+ and K^+ levels remain relatively constant, but the concentration of HCO_3^- increases significantly and that of Cl^- falls. Thus, the pH of the pancreatic juice entering the duodenum is **alkaline** (approximately 7.5 to 8.0). This marked increase in the pH enables pancreatic juice to buffer the extremely acid gastric juice. This **buffering** of gastric juice by pancreatic secretion is considered a major protection against erosion of the duodenal mucosa and the formation of duodenal ulcers.

Another important function of pancreatic HCO_3^- secretion concerns the action of pancreatic **digestive enzymes.** These enzymes have an optimal pH well above that of gastric juice. Thus, the buffering of gastric juice optimizes the activity of these enzymes in the small intestine. It has been shown that the increase in pancreatic juice HCO_3^- content is proportional to the rate of stimulation of the pancreas (Fig. 44-6). Thus, as food enters the digestive tract, the degree of intensity of neural and hormonal stimulation of the pancreas is increased, resulting in an elevated HCO_3^- concentration in pancreatic juice. Furthermore, the Cl^- concentrations in the pancreatic juice vary inversely with stimulation of secretion, and thus the HCO_3^- concentration (see Fig. 44-6). This suggests a coupling of the transport systems responsible for the duct-cell secretion of these electrolytes.

Duct-cell HCO_3^- is provided by several different sources (Fig. 44-7). **HCO_3^- dissolved in plasma** is freely diffusible down its concentration gradient through the basolateral membrane of duct cells into the cytoplasm. Carbon dioxide can diffuse from plasma into the cell as well as be produced by cellular metabolism. This carbon dioxide combines with water in the **hydration reaction** catalyzed by carbonic anhydrase to form the unstable carbonic acid, which immediately dissociates into H^+ and HCO_3^-. These **two sources** provide the HCO_3^- that is secreted into the duct lumen by two transport mechanisms. HCO_3^- can **freely diffuse** down its concentration gradient across the apical membrane of the cell into the duct lumen. HCO_3^- can also be exchanged one-to-one for luminal Cl^- by an **apical membrane anion exchanger** (see Fig. 44-7). This is thought to represent a specialized apical membrane Cl^-–HCO_3^- exchange protein, and its existence accounts for the reciprocal relationship between the luminal Cl^- and HCO_3^- concentrations as the secretory rate increases. Finally, the H^+ produced by the hydration reaction is transported out of the duct cell by a **basolateral cation exchanger,** the Na^+–K^+ exchange mechanism which is energized by the basolateral (Na, K)-ATPase–mediated Na^+ pump. The Cl^- that accumulates in the cell from the apical Cl^-–HCO_3^- exchange diffuses through the basolateral membrane into the blood. These specialized transport systems of the duct cell membranes ensure that HCO_3^- is efficiently secreted into pancreatic juice, and that the venous blood leaving the pancreas has a lower pH than the arterial blood entering the pancreas. Water is freely diffusible via para-

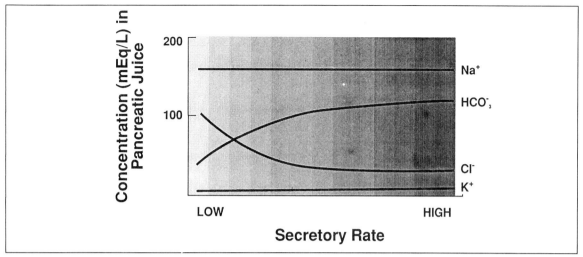

Fig. 44-6. Concentration of major ions in pancreatic juice as compared with the rate of stimulation of secretion. (Modified from Johnson, L. R., ed. *Physiology of the Gastrointestinal Tract,* 2nd ed. New York: Raven Press, 1987.)

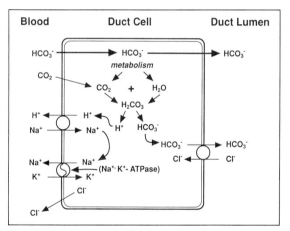

Fig. 44-7. A composite of the transport process involved in pancreatic duct cell HCO_3^- secretion.

cellular pathways in order to maintain osmotic balance, while the ionic composition of luminally discharged pancreatic juice may vary greatly from that of the ultrafiltrate of plasma first produced by acinar cell secretion.

CONTROL OF PANCREATIC EXOCRINE SECRETION

The secretion of pancreatic juice is under combined **hormonal** and **neural control. Parasympathetic postganglionic fibers** of the vagus nerve and the gastrointestinal hormones **cholecystokinin** and **secretin** are the major mediators of regulation. It has been useful in our understanding of these processes to divide the **regulation** of pancreatic exocrine secretion into three phases: **cephalic, gastric,** and **intestinal.** Although these phases may overlap considerably in time and may actually interact, there are distinguishing features that coincide with the location of food in certain regions of the gastrointestinal tract (Table 44-2).

Cephalic Phase

The cephalic phase of pancreatic secretion is triggered by the central nervous system recognition and integration of the sight, smell, and taste of food. This phase has been demonstrated in humans when subjects either just see and smell appetizing foods or chew food without swallowing it. These stimuli lead to increased pancreatic

Table 44-2. Control of Pancreatic Exocrine Secretion

Phase	Stimulus	Response
Cephalic	Sight, smell, and taste of food	Vagal efferent nerves release ACh which stimulates parietal cells to secrete acid and acinar cells to secrete digestive enzymes
Gastric	Presence of food in the stomach	Gastropancreatic reflex and gastrin stimulate acinar cells; balloon distention causes secretion of enzymes
Intestinal	Acid and partially digested food in duodenum	Vagovagal reflex stimulates enzyme secretion; release of CCK and secretin from duodenal mucosa stimulates enzyme and HCO_3^- secretion, respectively

ACh = acetylcholine; CCK = cholecystokinin.

enzyme and HCO_3^- secretion. These findings have also been demonstrated in a variety of animal studies. It has been shown experimentally that the degree of enzyme secretion in the cephalic phase is approximately 50 percent of the maximum response induced by the administration of cholecystokinin and secretin. Furthermore, when duodenal acidification accompanies "sham feeding," the secretory response is markedly enhanced.

The cephalic phase of pancreatic secretion is mediated primarily by the **vagus nerve,** with efferent cholinergic fibers directly and indirectly innervating the pancreatic acinar cells. The vagus may also indirectly influence duct-cell secretion by modulating the peptidergic nerves that innervate these cells. Several lines of evidence support the role of the vagus in this phase. **First,** vagal fibers are the **primary source** of central nervous system cholinergic nerves to the pancreas. **Second,** in animal studies, electrical stimulation of vagal efferent fibers elicits a **pancreatic secretion** that is similar in composition to that produced during sham feeding. **Third,** the exogenous administration of **cholinergic agonists** activates a similar secretory response. **Fourth,** the pancreatic secretory response to sham feeding can be inhibited or abolished by the administration of **anticholinergic drugs**. Sham feeding has also been shown to cause in-

sulin release. Because sham feeding stimulates antral gastrin release and gastric secretion as well, gastric acid emptying into the duodenum and circulating gastrin levels are suspected to enhance the pancreatic secretory response.

Gastric Phase

It is apparent from the previous discussion that the cephalic and gastric phases overlap considerably. However, physiologic events occur in the stomach that can modulate pancreatic secretion.

Besides vagally mediated gastrin release and gastric acid secretion, **balloon distention** of the stomach markedly enhances pancreatic secretion. Secretion is inhibited when the anticholinergic agent **atropine** is administered, but only truncal **vagotomy** produces long-term blockade of the response. The results of these studies suggest that the gastric phase of pancreatic secretion is mediated by a **cholinergic nerve reflex,** most likely involving vagal afferent and efferent fibers (a vagovagal reflex). However, other evidence also suggests the involvement of **gastrin** in this phase. **Distention** of both the proximal and distal portions of the stomach stimulates pancreatic secretion. **Vagotomy** abolishes the response to proximal distention but not that to distal distention, implying that gastrin release elicited by distal distention evokes the pancreatic response. Furthermore, the intravenous administration of an extract of antral mucosa in experimental animals provokes pancreatic secretion.

Although serum gastrin levels have not been examined during these distention studies, it is assumed that the gastric phase is mediated by the combination of a vagally mediated gastropancreatic reflex and antral gastrin release.

Intestinal Phase

The intestinal phase of pancreatic secretion appears to be somewhat more complex than the cephalic and gastric phases, because of increasing evidence that certain partially digested dietary substrates, as well as hormones and neural reflexes, participate in this phase. However, studies in the intestine are easier to conduct because a variety of test substances can be perfused or instilled into the small intestine without activating central nervous system or gastric responses. Gastric acid, monoglycerides, amino acids, peptides, fatty acids, and other substrates individually and in combination stimulate pancreatic exocrine secretion in appropriate amounts and composition to promote the efficient digestion of dietary nutrients and the buffering of gastric effluent.

Acidification of the proximal duodenum is a major stimulant of pancreatic exocrine secretion. This was observed initially as early as 1902 in pioneer experiments conducted by Bayliss and Starling. They observed that when acid was brought into contact with the duodenal mucosa of experimental animals, it elicited a "watery" secretion from the pancreatic duct. This occurred whether the pancreas was left intact or denervated. They then observed that a crude homogenate of duodenal mucosa, when injected intravenously, evoked the same response. They postulated that a substance released by the duodenal mucosa in response to gastric acid circulates in the blood to stimulate pancreatic secretion. They named this substance secretin, thus giving rise to the science of endocrinology. It was later established that the major components of acid-induced pancreatic secretion are

Na^+, HCO_3^-, and water, and that very little protein is secreted by the pancreas in response to duodenal acidification. It has also been established that the specialized cells of the duodenal mucosa, the **amine precursor uptake and decarboxylation cells,** synthesize and release secretin. Although secretin is the major hormonal mediator of pancreatic HCO_3^- secretion, in the normal response to a meal, it is known that cholecystokinin can augment the action of secretin on pancreatic duct cells.

The **duct-cell secretory response** to secretin is mediated by cAMP, which serves to increase the production of HCO_3^- by duct-cell carbonic anhydrase and to stimulate Cl^-–HCO_3^- exchange. The major function of **pancreatic HCO_3^-** is to buffer gastric acid and create an optimal pH environment for digestive enzyme activity. Though some HCO_3^- is contributed by bile and duodenal mucosal cell secretion, these sources provide only about 10 to 15 percent of the total amount. Thus, pancreatic HCO_3^- is the primary source of buffer for neutralizing the gastric acid entering the duodenum.

Within the stomach, the action of gastric acid, pepsin, lingual lipase, and salivary amylase begins the **initial breakdown** of solid food into smaller particles. Once these partially digested products are emptied into the duodenum, they become **important stimulants** for the intestinal phase of pancreatic exocrine secretion. Peptides, amino acids, fatty acids, and monoglycerides have all been shown to elicit pancreatic enzyme secretion. Cholecystokinin, which is released by the specialized amine precursor uptake and decarboxylation cells of the duodenal mucosa, is the major mediator of this response. Some evidence indicates a **neural reflex** involving duodenal and pancreatic cholinergic nerves. It remains unclear what the specific mechanism is that causes these partially digested substances to stimulate the intestinal cells to release hormones or initiate the neural reflex. The cholecystokinin-induced and cholinergic-initiated responses of the acinar cells are brought about by a rise in the intracellular Ca^{2+} level, which acts as the second-messenger system. It has been demonstrated that secretin can augment the action of cholecystokinin and acetylcholine at the acinar cell through a cAMP-mediated response.

SUMMARY

The pancreas possesses a unique anatomic arrangement of nerves, blood vessels, and exocrine secretory units called *acini.* This arrangement allows for the efficient delivery of nutrients, which serve as the raw materials required for enzyme synthesis and extrinsic regulatory influences (neurotransmitters and hormones) necessary for the optimal functioning of this organ. Acinar cells synthesize and secrete digestive enzymes with glycolytic, proteolytic, and lipolytic properties. The proteolytic enzymes are secreted in precursor forms and are activated by the pH in the intestinal lumen. Pancreatic duct cells secrete sodium bicarbonate, which osmotically stimulates the movement of water into pancreatic juice and buffers the gastric contents entering the duodenum. Enzyme and HCO_3^- secretion is governed mainly by cholecystokinin and secretin, respectively; these are hormones that are secreted by the duodenal mucosa in response to the presence of food. The parasympathetic nervous system also regulates pancreatic secretion via the vagus nerve.

NATIONAL BOARD TYPE QUESTIONS

Select the one most correct answer for each of the following

1. Diversion of the pancreatic duct away from the small intestine could cause:
   A. Steatorrhea (fat malabsorption)
   B. Essential amino acid deficiency
   C. Duodenal ulceration
   D. Altered carbohydrate metabolism
   E. All of the above
2. A physiologic stimulation of pancreatic exocrine secretion occurs in response to:
   A. Gallbladder contraction
   B. Duodenal acidification
   C. Release of pepsinogen
   D. Decreased cAMP levels in the acinar cells
   E. Colonic distention
3. Pancreatic HCO_3^- and fluid secretion would be *most* stimulated by:
   A. Perfusion of the duodenal lumen with a neutral solution containing free amino acids
   B. Distention of the gastric antrum with a neutral solution
   C. Smelling a steak meal
   D. Perfusion of the duodenal lumen with an acidic solution containing partially hydrolyzed protein
   E. Intravenous infusion of insulin
4. The enzyme carbonic anhydrase is important in the secretion of:
   A. Cholecystokinin
   B. Secretin
   C. Procarboxypeptidase
   D. Pancreatic HCO_3^-
   E. Co-lipase
5. As the rate of pancreatic secretion increases, one would expect the Cl^- concentration in the secretion to:
   A. Increase
   B. Decrease
   C. Remain unchanged
   D. Parallel the Na^+ concentration
   E. Parallel the K^+ concentration

ANNOTATED ANSWERS

1. E. Steatorrhea is a condition in which there is excess lipid in the stool. In pancreatic duct diversion, this excess results from the inability of the intestine to absorb undigested dietary lipid because pancreatic lipase and co-lipase are absent from the small bowel lumen. Likewise, with diversion of the pancreatic duct away from the bowel, protein digestion is inhibited, and thus essential amino acids are unavailable for absorption and assimilation. Finally, by diverting pancreatic juice, the duodenal mucosa is exposed to unbuffered gastric acid as the

stomach empties. This scenario would quickly cause erosion and ulceration of the duodenal mucosa. Since the digestion of dietary carbohydrates would also be markedly compromised, glucose and other simple sugars would not be available for absorption and metabolism by the body.

2. B. Acidification of the duodenum stimulates the release of secretin from the duodenal mucosa, which in turn stimulates pancreatic HCO_3^- secretion. Gallbladder contractions occur in response to cholecystokinin and do not influence pancreatic secretion. Pepsinogen, secreted by the stomach, is a proteolytic enzyme precursor and does not stimulate pancreatic secretion. Decreasing cAMP levels in acinar cells would inhibit pancreatic secretion, and colonic distention plays no role in this process.

3. A. An acidic duodenum is the major stimulus of pancreatic HCO_3^- and water secretion. Likewise, partially hydrolyzed protein can influence pancreatic secretion via cholecystokinin release by the duodenum. Perfusion of the duodenum with a neutral solution inhibits pancreatic HCO_3^- and fluid release.

Distention of the stomach or smelling a steak would initiate enzyme secretion but not fluid or HCO_3^- secretion. Also, the rate of pancreatic secretion is independent of plasma insulin levels.

4. D. HCO_3^- liberated by the action of carbonic anhydrase in pancreatic duct cells is the source of HCO_3^- for secretion. Cholecystokinin, secretin, procarboxypeptidase, and co-lipase are all peptides, and do not depend on carbonic anhydrase activity.

5. B. Because HCO_3^- secretion by pancreatic duct cells is energized by a Cl^-–HCO_3^- exchanger, the luminal Cl^- concentration would decrease as the rate of pancreatic secretion increases. Na^+ and K^+ concentrations remain relatively constant.

BIBLIOGRAPHY

Johnson, L. R., ed. *Gastrointestinal Physiology*, 3rd ed. St. Louis: C. V. Mosby, 1985.

Johnson, L. R., ed. *Physiology of the Gastrointestinal Tract*, 2nd ed. New York: Raven Press, 1987.

Sernka, T. J., and Jacobson, E. D., eds. *Gastrointestinal Physiology—The Essentials*, 2nd ed. Baltimore: Williams and Wilkins, 1983.

45 Liver and Biliary System

JAMES E. HEUBI

Objectives

After reading this chapter, you should be able to:

Describe the structural and functional relationships of hepatic secretion

Identify the major elements of bile acid and bilirubin handling by the liver

List the principal factors regulating biliary secretion

List the major components of bile

Identify the components of the enterohepatic circulation of bile acids, and how bile acid pool sizes are regulated

Discuss how the biliary tract responds to a food stimulus

Explain how gallstones are formed

The functional integrity of the liver is essential for the maintenance of life. The liver is responsible for key components of **intermediary metabolism,** and its product, bile, facilitates fat absorption and cholesterol excretion. **Bile** is a complex mixture of exocrine secretions and excretory products of the liver. It consists of varying amounts of water, cholesterol, lecithin, bile salts, bile pigments, protein, and inorganic salts of plasma. Bile is formed by the liver cells, secreted into the bile canaliculi, and transported to the gallbladder, where it is concentrated and stored until appropriate stimulation initiates its delivery to the duodenum. The excretory products of bile, such as **bile pigments,** are eventually excreted in the feces; other components such as **bile acids** are important in the overall digestive and absorptive functions of the intestine. Although gallbladder contraction and the expulsion of large amounts of bile into the duodenum are triggered by the presence of food in the intestine,

bile formation by the liver is relatively independent of digestive events. The anatomic and functional aspects of the liver, the formation and components of bile, bilirubin formation and excretion, and the regulation of biliary secretion as it relates to digestion will be the focus of this chapter.

HEPATIC STRUCTURE

The important microscopic features of the hepatic architecture are shown in Figure 45-1. The basic unit of hepatic structure is the **lobule.** Within it, the liver cells, **hepatocytes,** are grouped around a central vein and radiate out in a spokelike fashion. The **sinusoids** extend from the central vein to the portal tracts, or **triads,** which include a portal vein, hepatic artery, and small bile ducts. **Bile canaliculi** lie between adjacent hepatocytes and drain bile into the bile ducts at the periphery of the lobule. This strategic

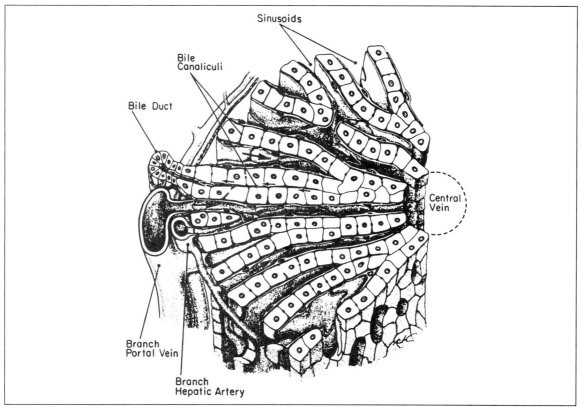

Fig. 45-1. A portion of the hepatic lobule. The cords of hepatocytes radiate out from the central vein. Branches of the portal vein, hepatic artery, and bile ducts located in the periphery make up the portal triad. The *large arrows* show the flow of blood from the portal to central veins and the *small arrows* indicate canalicular bile flow. (Reproduced with permission from: Bloom, W., and Fawcett, D. W. *A Textbook of Histology,* 10th ed. Philadelphia: W. B. Saunders, 1975.)

placement of hepatocytes is well adapted to their functions. The **venous sinusoids** are not only lined with the usual vascular endothelium but also with specialized cells of the reticuloendothelial system, called **Kupffer's cells. Blood flows** from the gastrointestinal tract through the **portal vein** to the liver. From the portal tracts, it traverses the short sinusoids between cords of liver cells, one to two cells thick, and the Kupffer's cells. Because of the large fenestrations between the endothelial cells lining the sinusoids, each hepatocyte is in intimate contact with sinusoidal blood. This arrangement promotes the efficient extraction of many compounds from the plasma.

Blood is then returned to the general circulation via the **central veins,** which empty into the **hepatic vein.**

The bile **canaliculi** are approximately 1 μm in diameter and are surrounded by hepatocytes. They are closed at the end nearest the central vein and have no membrane or wall (Fig. 45-2). Therefore, the plasma membrane of the hepatocyte that borders the canalicular lumen serves as the **canalicular membrane. Secretions** from the hepatocytes are collected in the canalicular lumen and eventually transported to an intricate system of bile ductules and ducts that are lined by epithelial cells. The **ductal system** from the right

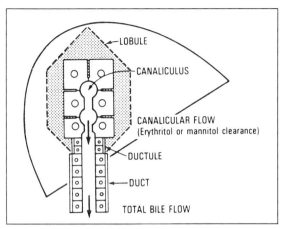

Fig. 45-2. The biliary tract. (Reproduced with permission from: Schiff, L., and Schiff, E. R. *Diseases of the Liver,* 6th ed. Philadelphia: Lippincott, 1987.)

and left lobes empties bile by means of the right and left hepatic ducts, which meet to form the **common hepatic duct.** The **cystic duct** drains the gallbladder and the **common hepatic duct** joins with the cystic duct to form the **common bile duct,** which enters the duodenum at the duodenal papilla where the sphincter of Oddi is located (see Fig. 45-7).

SPECIAL ASPECTS OF THE STRUCTURE AND FUNCTION RELATIONSHIPS

The bile canaliculus has a high surface to volume ratio. The **surface area** for the human liver, excluding microvilli, is about 10.5 square meters. The **intercellular junctions** are important for secretion. These include the **tight junction, intermediate junction, desmosome,** and **gap junctions.** The unidirectional blood flow from the portal to hepatic veins creates a **translobular concentration gradient** for each of a number of solutes. Likewise, **biliary secretion** into the canaliculus behaves in a similar fashion (Fig. 45-3). Canalicular bile flows in the opposite direction of sinusoidal plasma flow. Consequently, the concentration of solutes in the canaliculi near the

portal venous end is higher than that at the hepatic venous end. This causes bile to contain high concentrations of organic solutes.

HEPATIC FUNCTION

The liver is an essential organ for a number of reasons. It performs functions necessary for the functioning of the **cardiovascular** and **immune systems,** and it secretes important substances into the gastrointestinal tract. It is pivotal in the regulation of **metabolism;** these roles include protein synthesis, the storage of multiple vitamins and iron, and the degradation of hormones, drugs, and toxins.

The liver regulates carbohydrates, lipid, and protein metabolism. **Glycogen** synthesized during a glucose excess can be metabolized and released, called **glycogenolysis,** during prolonged fasts to maintain a relatively constant blood glucose concentration. The liver is also capable of synthesizing **glucose** through the conversion of other substances, including amino acids, referred to as **gluconeogenesis. Lipids** absorbed from the intestine are packaged as **chylomicrons** and carried to the systemic circulation via the lymphatic system. The **triglycerides** of chylomicrons are hydrolyzed by lipoprotein lipase located on endothelial surfaces. Cholesterol-rich chylomicron remnants are removed by the liver and degraded. The **hepatocytes** synthesize **very-low-density lipoproteins,** which are a major source of cholesterol and triglycerides for use by body cells. Organelles within the hepatocytes are responsible for beta-oxidation of fatty acids, which provides energy, and their metabolic endproducts, **ketone bodies** (acetoacetate, beta-hydroxybutyrate, and acetone). The liver is centrally involved in **protein breakdown** and **synthesis.** When proteins are catabolized, their constituent amino acids are deaminated to form ammonia. Within the liver, **ammonia** is converted to urea by metabolism in the urea cycle. Ninety-five percent of all **major plasma proteins** is synthesized by the liver, including albumin, apoproteins (the protein component of lipoproteins), fibrinogen, and those proteins involved in blood clotting.

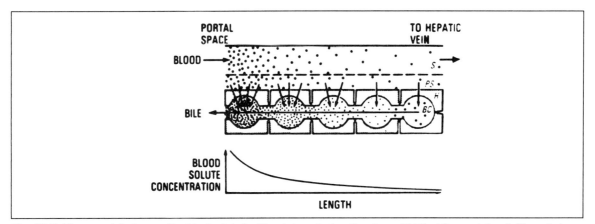

Fig. 45-3. The translobular gradient is depicted with solute entering the sinusoid (S) at the portal end. The solute concentration (*dots*) declines as it passes through the pericanalicular space (*PS*), into hepatocytes (*H*), and into the bile canaliculus (*BC*). Bile flows in the opposite direction to blood flow, so that solute is most concentrated in the canaliculi nearest the portal space. (Reproduced with permission from: Schiff, L., and Schiff, E. R. *Diseases of the Liver*, 6th ed. Philadelphia: Lippincott, 1987.)

The liver is a major storage site for **iron.** In addition, fat-soluble vitamins, including vitamins A, D, and K, as well as vitamin B_{12}, are stored in the liver.

Drug, hormone, and toxin degradation is an important hepatic function. The liver participates in the conversion of many compounds to their inactive forms. Conjugation with glucuronic acid, glycine, or glutathione is a common pathway in the degradation or detoxification that occurs in the liver.

Approximately 1,400 ml of **blood** passes through the liver each minute. Of this, nearly 1,000 ml comes from the **portal circulation,** which drains the organs of the gastrointestinal tract and the spleen. The remaining 400 ml consists of arterial blood derived from the **hepatic artery.** This total blood flow represents about 25 to 28 percent of the cardiac output. **Hepatic arterial blood** supplies nutrients to the structural components of the liver, including the biliary tract, and eventually joins the hepatic sinusoids to mix with portal blood. Blood from the **hepatic sinusoids** eventually drains into the hepatic vein, which empties into the inferior vena cava cephalad to the liver. The liver performs two major vascular functions: **storage** and **filtration.** The liver is

somewhat expandable and compressible, and therefore can store as much as 500 ml of blood. Two factors influence the amount of blood stored in the liver: **right atrial pressure** and **portal venous pressure.** As these pressures increase, the liver expands and more blood is stored. Conversely, as blood volume shrinks, as occurs during hemorrhage, the liver contracts and returns more blood to the systemic circulation.

The one liter per minute of blood delivered to the liver by the portal system filters through the venous sinusoids. These reticuloendothelial cells are capable of phagocytizing nearly 99 percent of the bacteria and other foreign substances in the blood. Thus, it serves as an important line of defense between the gastrointestinal tract and the systemic circulation, which helps to prevent systemic infection and antigen sensitization.

BILE COMPOSITION

Inorganic electrolytes exist in bile at concentrations similiar to those of plasma. Biliary calcium concentrations may exceed those in the plasma because of binding to bile acids and micelle formation. For the same reason, biliary bicarbonate concentrations are likely to be higher than those in plasma. The **major organic components** of bile are conjugated bile acids, phospholipids, cholesterol, and bile pigments. There are low levels of proteins and hormones.

The **bile acid concentration** in bile ranges from 2 to 45 mM. Typically bile acids are present in the form of glycine or taurine conjugates, with a small fraction existing as free bile acids. Bile acids are **amphipathic molecules** that form **micelles** (macromolecular aggregates) above a critical micellar concentration, which in human bile is approximately 2 to 4 mM. Bile acids in hepatic and gallbladder bile greatly exceed this critical concentration; therefore, bile acids in bile and in the duodenal and jejunal lumen exist as micelles, with small amounts present as **monomers.**

The **phospholipid** concentration in human bile ranges from 0.3 to 11 mM; that for **cholesterol** is 1.6 to 8.3 mM. Although the levels of organic compounds in bile may vary according to the location in the biliary tract and intestinal lumen (because of reabsorption of inorganic solutes, especially by the gallbladder), the ratios of these compounds remain relatively fixed. Bile acids are especially important in relation to the other two organic compounds, for two reasons: (1) hepatocyte secretion of cholesterol and phospholipids relies on bile acid secretion; and (2) the solubility of cholesterol and phospholipids largely depends on bile acids. Because of the formation of mixed micelles containing cholesterol, phospholipid, and bile acids, solubilized concentrations of both cholesterol and phospholipid can far exceed their normal aqueous solubilities. In normal humans, bile may be supersaturated with cholesterol during a portion of the day and cholesterol crystals may precipitate out of bile. In conjunction with compounds that form a nidus (including proteins and calcium), cholesterol crystals may aggregate to foster the development of cholesterol **gallstones.**

The typical concentrations of **bile pigments** range from 0.3 to 3.2 mM. In human bile, about 80 percent of the total pigments occur as bilirubin diglucuronide, while bilirubin monoglucuronide accounts for 20 percent. Small quantities of unconjugated bilirubin and other conjugates may also be present, but are of minimal importance. **Conjugation** is essential for the elimination of bilirubin from the body. This process, catalyzed by UDPglucuronyl transferase, involves esterification of bilirubin to a glycosidic compound, glucuronic acid. The conjugation of bilirubin, with the resultant formation of glucuronides, makes it more hydrophilic, and promotes its excretion in urine and bile while impeding intestinal and gallbladder absorption.

Protein concentrations in bile are extremely low, usually 300 to 3,000 mg/liter in human bile. Serum albumin is the most abundant protein; other proteins exist in an inverse relationship to their molecular weight.

SECRETION

Osmotic filtration is considered the major bile flow–generating mechanism. Two routes have been proposed for the passage of material from the sinusoids to the canaliculi: the **transcellular** and **paracellular pathways.** There is clear evidence that the transcellular pathway is the principal route, with mounting evidence of a minor role for the paracellular pathway.

Bile acid–dependent flow, linked directly to bile acid transport through the hepatocyte and into the canicular lumen, provides the major impetus for bile flow. The other component, **bile acid–independent flow,** accounts for a smaller fraction of bile flow.

Recent research has characterized a **sodium-coupled transport system** of bile acids from the sinusoidal blood supply into the hepatocyte (Fig. 45-4). **Taurocholate uptake** depends on an inwardly directed sodium gradient that may be diminished in the presence of other bile acids, amino acids, ouabain, and furosemide. The sodium gradient is continuously maintained by (Na, K)-ATPase. Little is known about **intracellular transport.** Bile acids appear to bind to cytosolic proteins (Y[1] and glutathione-S transferase). **Vesicular transport,** which may be affected by microfilament or microtubule inhibitors, is possibly the principal intracellular transport

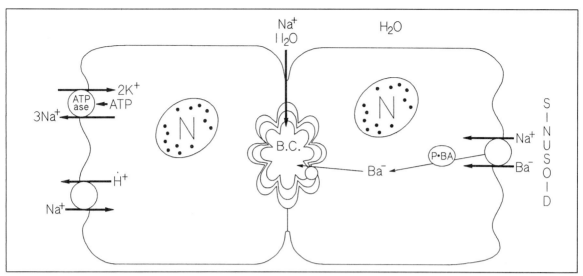

Fig. 45-4. Bile acid (BA^-) movement across the hepatocyte into the bile canaliculus (BC). Bile acids are transported across the sinusoid membrane by a Na^+-dependent transport process. Within the cytosol, bile acids are bound to protein (P) and carried to the canaliculus, where a Na^+-independent transport process facilitates movement into the canaliculus. (Adapted from: Sellinger, M., and Boyer, J. L. *Progress in Liver Disease,* Volume IX. Philadelphia: W. B. Saunders, 1990.)

pathway. During transcellular passage, unconjugated bile acids are conjugated to taurine and glycine. Canalicular transport carriers facilitate the passage of bile acids into the canalicular lumen. Unlike transport at the sinusoidal membrane, the **canalicular passage** of bile acids is a sodium-independent process, driven by a negative transmembrane potential. The **osmotic activity** of transported bile acids promotes the movement of water and other solutes into the canalicular lumen. Movement across the paracellular pathway is poorly understood, but also appears to be powered by osmotic filtration.

A **linear relationship** has been demonstrated to exist between bile flow and bile acid excretion. At low rates of bile acid excretion, bile flow is considered **bile acid–independent flow.** The most reasonable explanation for this flow is that it arises from the movement of inorganic electrolytes. About 30 to 60 percent of the basal bile flow can be attributed to either bile acid–dependent or bile acid–independent flow.

Bilirubin uptake and **secretion** by the he-

patocyte merits particular attention. Bilirubin in the serum derives from the breakdown of hemoglobin plus other nonhemoglobin substances. Normally unconjugated bilirubin is rapidly removed from the circulation by the liver and serum bilirubin concentrations are less than 1 mg/dl. Bilirubin is complexed with albumin in the **systemic circulation.** It then dissociates from the albumin and is transported into the cell by a carrier-mediated system shared with other organic anions but not with bile acids. In the **hepatocyte,** bilirubin binds to cytosolic proteins, ligandin, and Z protein. Within the cell, bilirubin is conjugated and thereafter secreted across the canaliculus into bile by a carrier-mediated transport system like the one on the sinusoidal membrane. Bilirubin secretion is enhanced by bile acids, even though it is excreted by a different pathway. Bile salts may form a "**micellar sink**" in the bile canaliculus, which augments passive diffusion of the pigment from the cell into the bile.

Bilirubin **glucuronides** are deconjugated by bacterial or intestinal glucuronidases. Thereafter, bacteria further metabolize them into urobilinogens and urobilins. Unconjugated bilirubin and urobilinogen undergo an enterohepatic circulation. In the adult this is of minimal significance; however, in the newborn, absorption of unconjugated bilirubin may cause "physiologic jaundice." Reabsorbed conjugated bilirubin and urobilinogen may be excreted through the kidney. This excretion depends on the size of the fraction of both substances that is unbound to albumin.

REGULATION OF BILIARY SECRETION

Choleresis (the hepatic secretion of bile) is stimulated by increasing the flux of bile acids across the hepatocyte. Non-micelle-forming bile acids (such as dehydrocholate) or those with a high critical micellar concentration have a greater choleretic effect than do the more physiologic bile acids. This finding suggests that bile acid–dependent flow is influenced by the number of osmotically active solute particles made available by passive diffusion and as solvent drugs. Pathophysiologic conditions tht might accentuate bile acid flux through the hepatocyte and enhance bile flow consist of increases in the bile acid pool size, including alterations in intestinal transit and gallbladder contractility. Additional compounds, including organic anions (Bromasulphalein, rose bengal, endocyanine green), stimulate choleresis by a mechanism similar to that which operates for the bile acids. In contrast to the effects described for bile acids, certain bile acids, including taurolithocholic acid and sulfated forms of lithocholic acid, may inhibit bile flow, presumably through either toxic injury to the hepatocytes, with attendant reduced bile flow, or through alteration of the bile acid–independent bile flow.

Bile acid–independent bile flow is enhanced by drugs such as phenobarbital and pharmacologic doses of corticosteriods. Inhibitors of sodium transport, including ouabain, ethacrynic acid, and amiloride, all diminish bile acid–independent bile flow. Estrogens suppress bile flow, predominantly by reducing bile acid–independent flow; however, direct effects of estrogens on plasma membranes may contribute to these flow reductions. Reduced circulating levels of thyroid hormone and the antipsychotic drug, chlorpromazine, both lead to inhibition of bile acid–independent flow.

BILE DUCT SECRETION AND ABSORPTION

Evidence of ductal secretion, which accounts for 30 percent of basal bile flow in humans, has derived from studies examining the choleretic effect of secretin. **Secretin infusion** stimulates increases in flow and changes in bile composition (raised HCO_3^- level and pH, and lowered concentrations of bile acids). The secretory activity of bile ductules and ducts may explain the choleresis that arises in certain diseases which involve an increased number of bile ducts or dilatation of the biliary tree. Bilie ductules and ducts are capable of reabsorption. The relative importance of secretion and absorption probably varies during the day and is not well understood.

BILIARY SECRETION IN HUMANS

Much of the information regarding **solute transport** has been furnished by studies in nonhuman mammalian models. In humans, secretion of **canalicular bile** averages $11\mu l$ per μmol of bile acid and approximately 15 μmol of **bile acid** is secreted each minute, yielding a mean bile acid–dependent flow of 0.15 to 0.16 ml per minute. The estimated canalicular bile acid–independent flow is 0.16 to 0.17 ml per minute and ductular secretion is 0.11 ml per minute. This gives an estimated bile output of 600 ml per day.

ENTEROHEPATIC CIRCULATION OF BILE ACIDS

Bile acids serve at least three major purposes: (1) they **enhance bile flow** during secretion across the biliary canaliculi; (2) they form aggregates, called

mixed micelles, in the upper small intestine, which solubilize the water-insoluble products of lipolysis, thereby facilitating their absorption; and (3) they are major regulators of **sterol metabolism.** In health, the **enterohepatic circulation** is localized to the liver and biliary tract, the intestinal tract, and portal and peripheral circulation. In disease states, such as **cholestasis,** the pool shifts from the intestinal tract and biliary tract into the liver and peripheral circulation. In the liver, cholesterol is converted to the highly polar bile acids, **cholic acid** (3α, 7α, 12α-trihydroxy-5β-cholanic acid) and **chenodeoxycholic acid** (3α, 7α-dihydroxy-5β-cholan-24-oic acid), which are termed **primary bile acids** (Fig. 45-5). The initial step of bile acid synthesis involves the **7α-hydroxylation** of the sterol nucleus of cholesterol, which is effected by the rate-limiting

Fig. 45-5. Primary bile acids synthesized in the liver from cholesterol and regulated by the rate-limiting enzyme, cholesterol 7-α-hydroxylase. Also shown are the secondary bile acids produced by bacterial 7-α-dehydroxylation after deconjugation.

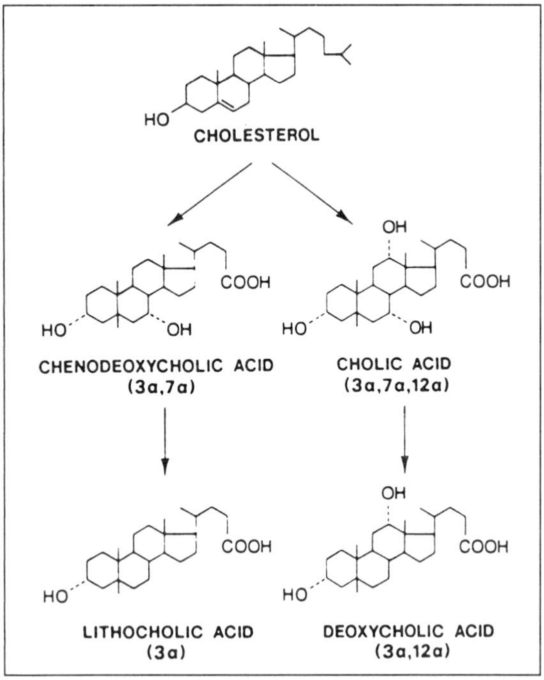

enzyme of bile acid synthesis, **cholesterol 7α-hydroxylase.** This enzyme is regulated by a mechanism of feedback inhibition based on the quantities of bile acids returning to the liver via the portal vein. After hepatic synthesis, bile acids are conjugated with glycine or taurine as *N*-acyl conjugates. All but a small fraction of bile acids are conjugated before excretion by the hepatocyte.

During **fasting,** most of the bile acid pool is in the gallbladder; however, some bile acids are always entering the intestinal lumen and produce basal intraluminal concentrations and small but measurable concentrations in the serum because of continuous intestinal absorption. When a **meal** is consumed, the gallbladder contracts at least once and often twice, whereupon most of the bile salt pool is emptied into the small intestinal lumen (Fig. 45-6). It is crucial to remember that bile acids are not secreted into the intestine in isolation but rather as **mixed micelles** containing bile acids, cholesterol, and phospholipids. Within the **upper small bowel,** bile salts form mixed micelles in combination with the lipolytic products, fatty acids, and monoglycerides. The lipolytic products are absorbed and bile acids are then reabsorbed in two ways. An **active transport system** localized to the distal ileum is the principal pathway for reabsorption. **Passive transport by nonionic diffusion** may occur anywhere along the small and large intestine; however, such absorption is markedly curtailed because the dissociation constants of many bile acids are low and they tend to remain ionized at the luminal pH. Therefore, only unconjugated bile acids or glycine conjugates of dihydroxy bile acids are likely to be un-ionized and converted into a state suitable for passive absorption. In addition, luminal nutrients tend to inhibit the passive uptake of bile acids.

After bile acids are transported out of the enterocyte, they are carried to the liver, where transport systems very efficiently remove them from the portal blood. A small fraction of the bile acids in portal blood spill over into the systemic circulation, leading to a predictable postprandial rise in serum bile acids during each of the six to

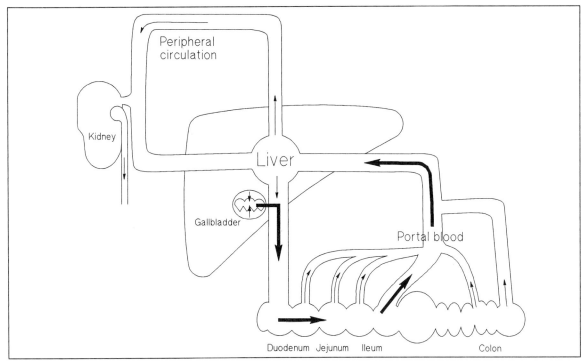

Fig. 45-6. The enterohepatic circulation of bile acids. When the gallbladder contracts, bile acids are expelled into the duodenal lumen. Passive absorption is shown by *small arrows* directed along the entire length of the small and large intestine. The *large arrow* going from the ileum into the portal blood depicts the active transport of bile acids by a carrier-mediated transport system. Bile acids return to the liver via the portal vein. The liver efficiently removes all but a small portion of bile acids that spill over into the systemic circulation. This produces a small, but measurable, postprandial rise in the serum level of bile acids. A small quantity of bile acids is filtered and excreted by the kidneys.

ten enterohepatic cycles of bile acids that occur during a day.

Most of the conjugates of cholic and cheno-deoxycholic acid excreted in bile are reabsorbed without alteration by intraluminal bacteria. Approximately one-fourth of the primary bile acids are deconjugated by bacteria in the terminal ileum and colon. Most of these free bile acids are reabsorbed and reconjugated in the liver with either glycine or taurine. Additional modifications to the steroid nucleus after deconjugation may also be accomplished within the gut lumen. The most common **biotransformation** made by bacteria is 7α-dehydroxylation, with the formation of **secondary bile acids.** The **primary bile acids,** cholic and chenodeoxycholic acid, are

transformed into the secondary bile acids, de-oxycholic acid (3α, 12α-dihydroxy-5β-cholanic) and lithocholic acid (3α-hydroxy-5β-cholanic), respectively (see Fig. 45-5). Approximately one-quarter to one-third of the pool of primary bile acids is converted to secondary bile acids each day. The **deoxycholic acid** produced is absorbed relatively efficiently (about one-third to one-half) and conjugated with either glycine or taurine. The **glycine** and **taurine** conjugates are then recycled and reabsorbed with efficiencies similar to the conjugates of chenodeoxycholic acid. Only about one-fifth of the newly formed lithocholic acid is absorbed, because it is very insoluble and adsorbs to colonic bacteria. If reabsorbed, lith-ocholic acid is also conjugated with glycine or taurine; however, most conjugates are sulfated in the "3" position to form sulfated **lithocholate conjugates. Sulfated conjugates** are poorly absorbed and pass into the colon, whereas **nonsulfated** lithocholate conjugates are efficiently reabsorbed.

With efficient recycling and synthesis, the biliary bile acids in humans generally comprise cholic (36 percent), chenodeoxycholic (36 percent), deoxycholic (24 percent), and lithocholic (1 percent) acids, in a ratio of 3 to 4:1 of glycine-to-taurine conjugates. Depending on diet and the methodology used, normal adults excrete 22 to 650 mg of bile acids per day. Fecal bile acids in healthy humans reflect the effects of major bacterial transformation. Most are unconjugated and lack the 7α-hydroxyl group. Hepatic bile acid synthesis is regulated by feedback inhibition and by the hepatic synthesis of chenodeoxycholic acid and cholic acid balances that compensate for daily stool losses, thereby creating a steady-state and keeping the primary bile acid pool sizes constant.

REGULATION OF THE ENTEROHEPATIC CIRCULATION OF BILE ACIDS

The regulation of bile acid pool size is carefully controlled by the liver through a **feedback inhibition system.** The rate-limiting enzyme of bile acid synthesis, **cholesterol 7α-hydroxylase,** is controlled by the return of bile acids to the liver through the portal vein. The bile acid pool size may be depleted, leading to reduced intestinal concentrations of bile acids and impaired fat solubilization, when the enterohepatic circulation is interrupted by ileal disease (Crohn's disease), ileal resection, or small intestinal bypass. When excess bile acids are lost in the feces, the liver enhances synthesis up to tenfold in an attempt to maintain a steady bile acid pool size. If the loss exceeds the hepatic ability to synthesize bile acids, the pool will diminish. Minor modifications of pool size may result when the small intestinal transit time is decreased or the gallbladder is surgically removed. In contrast, pool sizes may increase when gallbladder emptying is impaired, as observed in celiac disease or pregnancy, or when the small intestinal transit time is increased.

BILIARY TRACT PHYSIOLOGY

The **smooth muscle** of the common bile duct contracts rhythmically and may enhance bile flow within the biliary tree. The **sphincter of Oddi,** located at the junction of the common bile duct and duodenum, also undergoes rhythmic contractions. During **fasting,** bile flows predominantly into the gallbladder because of a pressure gradient created by the contracted sphincter of Oddi (Fig. 45-7A). The **gallbladder** is a saccular organ with a capacity of 15 to 60 ml in adults. The bile collected during the **interdigestive phase** is concentrated predominantly by the reabsorption of water. Bile acids, bilirubin, cholesterol, and phospholipids are concentrated threefold and tenfold in the gallbladder. When a meal is ingested, the gallbladder contracts and the sphincter of Oddi relaxes. Initially during the **cephalic** and **gastric phases** of digestion, this is mediated by cholinergic stimulation mediated by the vagus nerve (see Chapter 44). The greatest gallbladder emptying occurs during the **intestinal phase**. Products of digestion, especially fatty acids and amino acids, stimulate the release of **cholecystokinin** (CCK) by the I cells in the upper small intestinal mucosa. Circulating CCK causes strong contraction of the gallbladder smooth muscle and relaxation of the sphincter of Oddi (see Fig. 45-7B). CCK appears to be the major hormone mediating gallbladder contraction; **gastrin,** in doses far exceeding the physiologic range, may also stimulate contraction, but secretin has little effect. Intraduodenal acid and alcohol stimulate sphincter of Oddi contraction as does morphine.

GALLSTONES

The prevalence of gallstones or **cholelithiasis** increases with age. Gallstones rarely develop in infants and children. There are two major types of

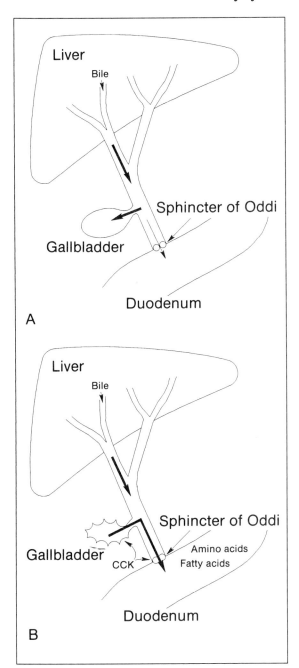

A

B

gallstones: **calcium bilirubinate** (pigment) and **cholesterol.** About 85 percent of stones in adults in the United States are composed predominantly of cholesterol. Most gallstones that arise in childhood are pigment stones resulting from hemolytic disorders. A sharp increase in the frequency of cholelithiasis is found in women after puberty and a gradual increase is noted in males throughout adulthood.

Cholesterol gallstone formation is the consequence of several interrelated events, including: (1) hepatic secretion of bile that is supersaturated with cholesterol; (2) nucleation of cholesterol monohydrate crystals; and (3) impaired gallbladder emptying. The bile of virtually all subjects in whom cholesterol stones develop is supersaturated with cholesterol, but many adults without stones have bile supersaturated with cholesterol during at least a portion of the day. Unlike normal individuals, those with gallstones exhibit rapid nucleation of cholesterol crystals. A delicate balance appears to exist between the promoters and inhibitors of crystal nucleation, and this determines whether gallstones form. Gallbladder stasis causes retention of crystals, with subsequent enlargement of the calculi.

Conditions that affect any of the steps described will foster stone formation. **Cholesterol supersaturation** is found commonly in obese patients, in patients who have undergone ileal resection, and in patients taking clofibrate, an agent used to treat hyperlipidemia, or estrogens. Specific conditions that alter the nucleating factors (glycoproteins and mucin) include **total parenteral nutrition** and **cystic fibrosis. Gallbladder stasis** eventuates when patients are treated with total parenteral nutrition, and is also seen during pregnancy and in celiac disease. **Pigment stones**

Fig. 45-7. The fasting (A) and postprandial (B) biliary tract. During fasting, a small amount of bile is excreted into the duodenum but most is stored in the gallbladder because of tonic contraction of the sphincter of Oddi. With a meal stimulus, cholecystokinin (*CCK*) is secreted by the small intestinal mucosa, which causes the sphincter of Oddi to relax and the smooth muscle of the gallbladder to contract.

are commonly associated with hemolytic disorders, biliary infection, and alcoholic cirrhosis.

Until recently, the only therapy for symptomatic gallstones was **cholecystectomy.** Recently introduced alternatives to surgery include **sonication** (lithotripsy), **chemical solubilization** with organic solvents infused through biliary catheters, and **oral bile acid** therapy.

SUMMARY

Hepatic structure and function are closely related. Bile flow is governed principally by the transport of bile acids across the sinusoidal and canalicular membranes. Likewise, bilirubin uptake and excretion are mediated by transport proteins on both the sinusoidal and canalicular membranes. The enterohepatic circulation of bile acids furnishes an efficient means to ensure adequate biliary concentrations of bile acids for solubilizing hydrophobic compounds such as cholesterol and phospholipid (thereby minimizing the risk of gallstone formation). It also ensures adequate intraluminal bile acid concentrations that promote efficient solubilization of the lipolytic products of intraluminal digestion. Meal-stimulated, hormonally mediated (CCK) contraction of the gallbladder and relaxation of the sphincter of Oddi allows delivery of the concentrated bile and this facilitates the formation of intraluminal micelles containing bile acids, fatty acids, phospholipids, monoglycerides, and fat-soluble vitamins.

NATIONAL BOARD TYPE QUESTIONS

Select the one most correct answer for each of the following

1. Bile secretion depends on the efficient movement of bile acids through the hepatocyte. Which of the following is *not* characteristic of this process?
   A. Bile acids traverse the hepatocyte cytosol unbound to protein
   B. A sodium-dependent transport system facilitates movement of bile acids across the sinusoidal membrane
   C. The canalicular bile acid carrier, which is not sodium dependent, facilitates movement across that membrane
   D. Amino acids, ouabain, and furosemide inhibit transport across the sinusoidal membrane by dissipating the sodium gradient
   E. Phenobarbital increases bile acid–independent bile flow

2. Bile acids are a major component of bile. Which of the following is correct regarding bile acids?
   A. Bile acids are usually unconjugated in bile
   B. Above a certain concentration, bile acids form macromolecular aggregates called *micelles*
   C. Bile acids play a minor role in solubilizing phospholipids and cholesterol in bile
   D. The ratios between bile acids, phospholipids, and cholesterol in bile vary markedly over broad concentration ranges
   E. Gallstones form when the concentration of cholesterol in bile goes below the limits of its solubilization by bile acids

3. When the enterohepatic circulation of bile acids is interrupted by ileal disease or resection, what is the usual consequence?
   A. Very low fecal bile acid loss
   B. Reduction of the bile acid concentrations in the upper small bowel, with consequental fat malabsorption
   C. Increased serum cholesterol concentration
   D. Reduced hepatic bile acid synthesis
   E. The cholesterol solubility in bile is unchanged

4. Bilirubin metabolism in humans is characterized by the following:
   A. Bilirubin is not protein-bound in the systemic circulation
   B. Bilirubin passively diffuses across the sinusoidal membrane into the hepatocytes
   C. Bilirubin binds to the cytosolic proteins, ligandin and Z protein, within the hepatocyte

D. Bilirubin diffuses out of the hepatocyte into the canaliculi

E. A, B, and D are true

5. The enterohepatic circulation of bile acids promotes efficient recycling of bile acids with adequate small bowel concentrations maintained to ensure solubilization of lipolytic products. The following is true about bile acids:

A. They are amphophilic compounds that form micelles above a critical concentration, called the *critical micellar concentration*

B. The primary bile acids are cholic and deoxycholic acid

C. Secondary bile acids are formed by bacterial dehydroxylation of the primary bile acids only after they are deconjugated

D. The glycine-to-taurine conjugation ratio of bile acids in bile is typically 1:2

E. A and C are correct

tration is reached, and are essential for solubilizing phospholipids and cholesterol.

3. B. If the enterohepatic circulation of bile acids is interrupted, reduced bile acid pools cause low upper intestinal concentrations and consequental fat malabsorption.

4. C. Bilirubin is bound to albumin in the systemic circulation. It dissociates from albumin and is transported by a carrier-mediated mechanism across the sinusoidal membrane. Within the cell, bilirubin is bound to cytosolic proteins, ligandin and Y^1. Exit from the cell is facilitated by a carrier-mediated transporter at the canalicular membrane.

5. E. The primary bile acids, cholic and chenodeoxycholic acid, are deconjugated and dehydroxylated to form the secondary bile acids, deoxycholic and lithocholic acids. Typically, the ratio of glycine-to-taurine conjugation of bile acids is 3 to 4:1.

ANNOTATED ANSWERS

1. A. Bile acids are generally bound to proteins within the cytosol. Although controversy exists regarding their relative importance, Y^1 and glutathione-S transferase are the principal proteins involved in intracellular bile acid movement.

2. B. Bile acids form macromolecular aggregates called *micelles* when a critical concen-

BIBLIOGRAPHY

Erlinger, S. *Secretion of Bile in Diseases of the Liver,* 6th ed. Schiff, L., and Schiff, E. R., eds. Philadelphia: Lippincott, 1987, Pp. 77–101.

Hofmann, A. F. The Enterohepatic Circulation of Bile Acids in Man. In: Stollerman, G. H., ed. *Advances in Internal Medicine.* Chicago: Year Book Medical Publishers, 1976, Pp. 501–534.

Sellinger, M., and Boyer, J. L. Physiology of Bile Secretion and Cholestasis. In: Popper, H., and Schaffner, F., eds. *Progress in Liver Disease,* Volume IX. Philadelphia: W. B. Saunders, 1990, Pp. 237–260.

46 Intestinal Absorption

JOSEPH D. FONDACARO

Objectives

After reading this chapter, you should be able to:

Characterize the volume load introduced into the human intestine every 24 hours

Describe how the intestinal mucosa is designed anatomically and functionally to handle this fluid challenge

Describe the mechanisms of intestinal water and electrolyte (especially Na^+ and Cl^-) transport and absorption, and the intracellular pathways that regulate these membrane processes

Discuss the principles of active, energy-dependent, Na^+-coupled transport and how the inwardly directed Na^+ electrochemical gradient energizes this transport process

Describe the regulatory roles of the autonomic and enteric nervous systems and the influence of various hormones on intestinal absorption

Explain the mechanisms of carbohydrate, protein, and lipid digestion and the absorption of the end-products of these digestive processes

Describe the role of bile salts in the digestion and absorption of dietary lipids

---

The mammalian intestine performs several important functions, all of which are integral components of the total digestive process. The intestine, along with the pancreas, secretes digestive enzymes and mixes these enzymes with dietary components, thus providing small, simple molecules for efficient absorption. The intestine, along with the pancreas and the biliary tract, secretes HCO_3^- that buffers the strong acid of gastric juice and thereby protects its delicate epithelium from erosion, damage, and ulceration. In addition, the musculature of the intestine provides for the efficient, coordinated, and timely movement of nonabsorbed substrates in an aboral direction for eventual elimination. One of the most striking aspects of intestinal function is the absorption of water, electrolytes, and dietary nutrients. In the normal-functioning mammalian intestine, nearly all of the dietary nutrients and approximately 95 to 98 percent of the water and electrolytes that enter the upper small bowel are absorbed. This chapter will review the major components of intestinal electrolyte and nutrient absorption, those processes whose net result is the movement of electrolytes, water, and metabolic substrates into the blood for distribution and use throughout the body.

ANATOMIC CONSIDERATIONS

Before exploring how the bowel is equipped functionally to handle the process of absorption, it is important to understand how the bowel is anatomically designed to serve this function. Figure 46-1 is a diagram showing how the intestinal structure amplifies the total surface area to promote efficient absorption. If one envisions the

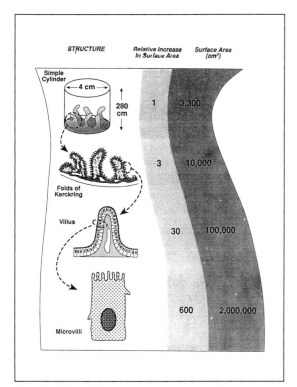

Fig. 46-1. Amplification of intestinal surface area.

small bowel as a **cylinder** with a length of approximately 280 cm and diameter of about 4 cm, the total internal surface area yielded by these dimensions is approximately 3,300 cm². Using this as a reference point, the **epithelial folds of Kerckring** alone amplify the surface area by a factor of three, to yield an area of about 10,000 cm². Next, the numerous villi of the intestinal surface further augment the surface area by a factor of ten, to yield an area approximately thirty times that of the reference point, or about 100,000 cm². Finally, the **microscopic surface of the luminal border** of the villous cells contains a microvillar membrane that contributes an additional twentyfold increase in surface area, such that it is now six hundred times that of the reference standard, or approximately 2,000,000 cm². In graphic terms, the total surface area rendered by the bowel anatomy for the purpose of absorption is about the size of a singles tennis court.

FLUID AND ELECTROLYTE ABSORPTION

The mammalian intestine is the primary site of fluid and solute absorption within the gastrointestinal tract. The small intestine passively and efficiently absorbs large volumes of fluid by actively transporting smaller quantities of osmotically active solutes. To appreciate the enormous challenge confronting the transport mechanisms of the intestine, it is important to consider the total fluid volume handled by the intestines on a daily basis.

Table 46-1 gives a breakdown of the volume of fluid entering the small intestine from all the potential contributors and how this volume is handled. The healthy adult ordinarily consumes approximately 2.0 liters of fluid a day in the diet. In addition, during digestion of dietary substrates, certain organs within the gastrointestinal tract secrete fluid that contains their particular contribution to the digestive process. In a 24-hour period, the salivary glands, stomach, pancreas, biliary system, and the intestine itself secrete about 7.0 liters of fluid. This yields a total

Table 46-1. Daily Total Fluid Volume (in liters) Handled by the Intestines

Ingested fluid		2.0	
Secreted fluid			
Saliva	1.0		
Gastric juice	2.0		
Pancreatic juice	0.5		
Bile	1.5		
Intestinal secretion	2.0		
Total secretions		7.0	
Total fluid input			*9.0*
Absorption			
Small intestine	7.5		
Colon	1.4		
Total fluid absorbed			*8.9*
Total fluid excreted			*0.1*

fluid volume of roughly 9.0 liters entering the duodenum daily. Ordinarily, the small intestine absorbs about 7.5 liters, or 83 to 85 percent, of this fluid. Combined with the roughly 1.4 liters absorbed by the large bowel, the total fluid absorption represents approximately 98 percent of the total fluid intake, leaving approximately 100 ml of fluid excreted daily from the gastrointestinal tract.

Cellular Transport Process

The transport mechanisms of the intestinal epithelium are an integral aspect of fluid and electrolyte transport that takes place in the intestine. The wide variety of absorptive and secretory processes exhibited by the intestine are mediated by a single layer of epithelial cells, the **intestinal mucosa,** which lines the entire luminal surface of the intestinal tract. This mucosal layer possesses fingerlike projections or elevations, called **villi,** and pitted areas, called **crypts** (Fig. 46-2).

The mucosal cells in these two regions are anatomically different. The **villous cell** has a characteristic microvillous or brush-border luminal membrane, while the **crypt cell** has no outstanding or distinguishing microanatomic features. Furthermore, these cells are functionally, as well as structurally, different (see Fig. 46-2). The villous cells are almost exlusively absorptive and are equipped with the necessary mechanisms to accomplish this task. The crypt cells are primarily **secretory** and possess those mechanisms consistent with this role.

The scientific literature on electrolyte transport and ion-coupled transport processes is enormous. Therefore, it is not possible to give an exhaustive review here, and only the transport of Na^+, K^+, Cl^-, H^+, and HCO_3^- is emphasized in the following discussion.

As just mentioned, the villous cell is the primary site of absorption. There are three basic pathways for **Na^+ absorption** across the intestinal mucosa (Fig. 46-3). It can enter the villous

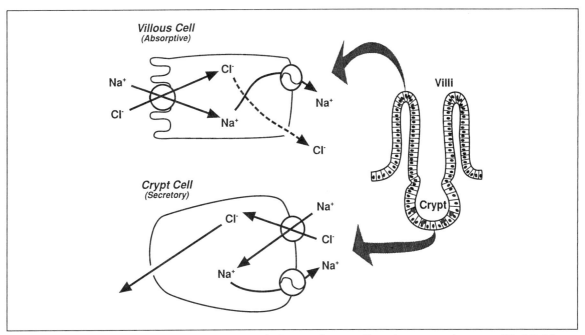

Fig. 46-2. Anatomic and functional differentiation of villous and crypt cells in the intestinal mucosa. (Modified from: Field, M., Fordtran, J. S., Schultz, S. G., eds. *Secretory Diarrhea*. Bethesda, MD: American Physiological Society, 1980.)

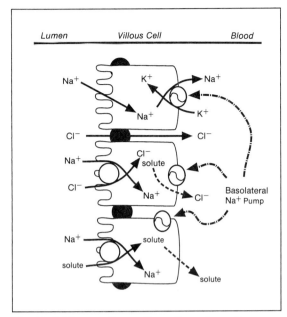

Fig. 46-3. Three pathways for sodium absorption across the small intestinal mucosa. (See text for detailed description.)

cell at the luminal membrane through a Na$^+$-conducting pathway, the so-called **Na$^+$ channel.** This pathway is sensitive to the drug amiloride, and Na$^+$ entry is accomplished by an inwardly directed diffusional gradient for the cation. The second mechanism for Na$^+$ absorption is a **coupled NaCl cotransport** process. This luminal membrane carrier must bind Na$^+$ and Cl$^-$ before it delivers either ion to the cell interior. The driving force for this mechanism is again the inwardly directed Na$^+$ gradient. Finally, Na$^+$ can be absorbed by **Na$^+$-solute–coupled cotransport.** The **solutes** known to be cotransported with Na$^+$ are glucose and amino acids (to be discussed later in this chapter), certain bile acids, and some water-soluble vitamins. Once inside the cell, the solute is freely diffusible down a concentration gradient through the basolateral membrane of the cell for eventual discharge into the portal bloodstream. The feature common to all of these Na$^+$ absorptive processes is the driving force for Na$^+$ entry—the **inwardly directed Na$^+$ gra-**

dient, which is maintained by the basolateral **Na$^+$ pump.** This Na$^+$ pump is energized by the hydrolysis of ATP which is catalyzed by (Na, K)-ATPase, and exchanges 3 Na$^+$ out for 2 K$^+$ in. Thus, following a meal, the intracellular Na$^+$ concentration is maintained very low relative to the extracellular environment, in particular the intestinal lumen. The movement of electrolytes and the transported solutes causes the concomitant movement of water into the bloodstream in an osmotically obligatory fashion.

Of the three modes of entry, it has been determined that **coupled NaCl cotransport** is the primary contributor to Na$^+$ absorption. Because of this, investigators have focused on further elucidating the mechanism. Figure 46-4 illustrates several important features of the NaCl cotransport process. First, it is believed that the cotransport of Na$^+$ and Cl$^-$ occurs as a result of two separate but functionally parallel exchange pro-

Fig. 46-4. The Na$^+$–H$^+$ and Cl$^-$–HCO$_3^-$ exchange mechanisms of coupled NaCl absorption.

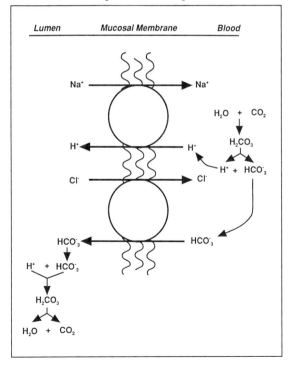

cesses residing on the luminal membrane: **Na⁺—proton exchange** and **Cl⁻—HCO₃⁻ exchange.** Additionally, in certain areas of the intestine, these exchange processes are thought to be involved in regulating and maintaining appropriate intracellular pH, since an ion exchanger can function in both directions. The absorption of Cl⁻ into the bloodstream can proceed by two routes: (1) **passive diffusion** from the cell across the basolateral membrane; and (2) a **paracellular shunt pathway.** Finally, using specific membrane vesicle preparations, it has been demonstrated that the luminal and basolateral membranes may possess a Na⁺—proton exchanger, perhaps serving also as a means of maintaining a constant intracellular pH.

The process of **electrolyte secretion** is a function of the crypt cell. For many years, scientists studying intestinal transport focused exclusively on absorption, and almost totally neglected the equally important function of intestinal secretion. It has been within the last two decades that intestinal electrolyte secretion has emerged as a subject of intense investigation.

Figure 46-5 shows a typical crypt cell and illustrates several important transport processes characteristic of this cell type. As noted previously, these cells are secretory, with Cl⁻ secretion being the primary event. This is acomplished by the luminal Cl⁻ conductive pathway, or **Cl⁻ channel.** Cl⁻ secretion is believed to be maintained at a **basal rate** under normal physiologic conditions. However, in certain **disease states** re-

sulting in **secretory diarrhea,** this Cl⁻ secretion may be dramatically enhanced in the small intestine and is considered the primary focus of the pathophysiology. The **outwardly directed Cl⁻ gradient** is believed to be maintained by a **Na⁺—K⁺—2Cl⁻ cotransport** mechanism located on the basolateral membrane. The functioning of this cotransport process and the subsequent "loading" of the cell with Cl⁻ is energized by the inwardly directed Na⁺ gradient, maintained by the basolateral Na⁺ pump. **Alterations** in the Na⁺ pump, the Na⁺—K⁺—2Cl⁻ cotransporter, or the basolateral K⁺ channel can seriously affect Cl⁻ secretion. Thus, these three basolateral transport processes can be regarded as "functionally" coupled to Cl⁻ secretion.

Intracellular Regulation of Transport

Like most physiologic events, these transport processes are regulated by both intracellular and extracellular mechanisms. As is true in other systems, the **intracellular regulation** of intestinal electrolyte transport is made up of a complex series of biochemical events. It is well accepted that modulation of the **cyclic nucleotide pathway** involving cAMP and of the intracellular Ca²⁺ concentration alters luminal membrane transport in the villous and crypt cells. Both of these pathways are referred to as **second-messenger systems** and are involved in **signal transduction.**

Figure 46-6 illustrates the role of **adenylate cyclase and cAMP** in regulating ion transport. Various substances that may alter transport (for example, hormones) bind to specific cell surface receptors. This stimulus triggers several biochemical events within the cell. A stimulatory guanine nucleotide—binding protein activates the enzyme adenylate cyclase. The exact role of this and other **G proteins** in intracellular regulatory pathways is currently being elucidated. Adenylate cyclase catalyzes the conversion of ATP to cAMP. An elevated cAMP level, in turn, activates a cAMP-dependent protein kinase. The activity of this protein kinase, presumably through a phosphorylation reaction, inhibits coupled NaCl absorption in the villous cell and stimulates

Fig. 46-5. Transport processes in crypt cells.

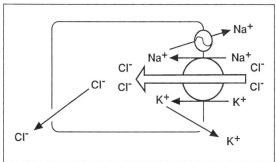

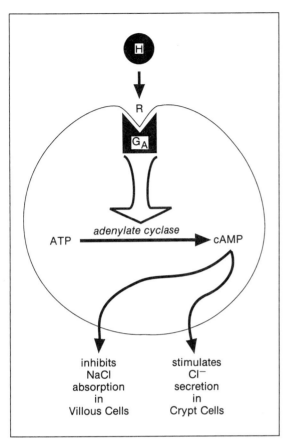

Fig. 46-6. The role of adenylate cyclase and cyclic AMP in the regulation of intestinal electrolyte transport. (H = hormone; B = receptor; G_A = guanine nucleotide regulatory protein.)

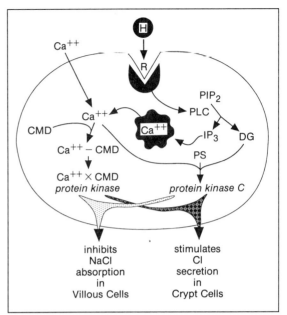

Fig. 46-7. The role of the calcium regulatory pathways in the control of intestinal electrolyte transport. (CMD = calmodulin; H = hormone; R = receptor; PIP_2 = phosphatidylinositol-bis-phosphate; IP_3 = inositol-tris-phosphate; PS = phosphatidylserine; DG = diacylglycerol.)

Cl⁻ secretion in the crypt cell. Agents that lower cAMP levels have been reported to produce an opposite effect on these transport processes. A similar pathway can be defined for the involvement of the cyclic nucleotide **cGMP** in **villous cell absorption.**

The **signal transduction system** involving **intracellular Ca²⁺** is somewhat more complex. It is believed that two distinct Ca²⁺ regulatory pathways reside in these transporting enterocytes (Fig. 46-7). While the **Ca²⁺—calmodulin** (CMD) pathway has been known for some time, only re-

cently have the **Ca²⁺-phospholipid—dependent protein kinase C** (PKC) pathway and the involvement of the inositol phospholipids in the intracellular regulation of transport been appreciated. Both of these pathways are triggered by a rise in the intracellular Ca²⁺ concentration, which can be achieved in two ways. The **first** is through the interaction of hormones with receptors, which are linked functionally to Ca²⁺ channels: Ca²⁺ influx is enhanced and the level of cytosolic free Ca²⁺ is therefore increased. In the **second** route, a certain hormone-receptor binding may stimulate phospholipase C, thereby accelerating the conversion of phosphatidylinositol to phosphatidylinositol-bis-phosphate (PIP₂), which is broken down to inositol-tris-phosphate (IP₃) and diacylglycerol. IP₃ stimulates the release of Ca²⁺ from intracellular storage sites, thus elevating the cytosolic free Ca^{2+} level. Ca^{2+} may interact with CMD and, through the stimulation of a Ca^{2+}-CMD–dependent protein kinase, alter transport processes, as shown in Figure 46-7. In the presence of phosphatidylserine, diacylglycerol and cytosolic Ca^{2+} stimulate PKC, which presumably has similar effects on the transport of electrolytes. More recent findings using specific inhibitors of Ca^{2+}-CDM and of PKC suggest that the primary role of Ca^{2+}-CDM is regulating NaCl absorption, while PKC primarily alters Na^+–H^+ exchange and Cl^- secretion.

Finally, important interactions between the membrane **phospholipids** and **arachidonic acid metabolism** have been proposed and observed. In particular, inhibitors of arachidonic acid metabolism to prostaglandins have been shown to suppress PKC-mediated secretion. These studies suggest that **prostaglandins,** which are known stimulators of intestinal secretion, are perhaps intimately involved in the combined phosphatidylinositol plus PKC regulatory pathway.

While these intracellular second-messenger regulator systems are not unique to intestinal mucosal cells, a basic appreciation for their role in intestinal transport may elucidate certain disease conditions, such as secretory diarrhea and other transport disorders of the gastrointestinal tract.

ELECTROLYTE AND FLUID ABSORPTION IN THE COLON

The mammalian large intestine plays a vital role in the fluid and electrolyte composition of stool. The colon **absorbs** Na^+, Cl^-, and water and **secretes** small quantities of K^+ and HCO_3^-. The **transport mechanisms** for electrolytes in the colon are similar to those of the small intestine but the colon lacks nutrient transport processes. This poses a potentially serious clinical problem when the small intestine is affected by **nutrient malabsorption.** Unabsorbed dietary substrates entering the colon can provide an osmotic challenge to the movement of fluid across the colonic mucosa, setting in motion a potential diarrheal condition. Fortunately, the colonic epithelium absorbs Na^+ more efficiently than does the small intestinal mucosa. Furthermore, the colon responds more readily to **aldosterone,** by increasing Na^+ absorption, than does the small bowel.

Although the healthy human colon absorbs approximately 1,400 ml of fluid per day, it has the capacity to absorb more than three times that volume, or about 4,400 ml. This absorptive "reserve" has been referred to as **colonic salvage** and is of utmost importance in the regulation of stool water excretion and management of potential diarrheal conditions (Table 46-2). In malabsorptive diseases of the small intestine, the amount of fluid delivered to the colon is markedly increased. In many cases, colonic salvage can compensate for this by absorbing more fluid and preventing diarrhea of small bowel origin (see Table 46-2, column 2). In some cases, such as **cholera,** the fluid volume delivered to the colon (excess fluid from the small intestine) exceeds the absorptive reserve, leading to fluid loss (see Table 46-2, column 3). Similarly, in diseases of the colon, such as certain **inflammatory conditions,** colonic fluid absorption may be compromised. While the volume of fluid entering the colon may be essentially normal, the absorptive capacity of the colon is markedly reduced, resulting in diarrhea of colonic origin (see Table 46-2, column 4).

Fluid absorption in the colon is determined by the **absorption of electrolytes,** primarily Na^+ and Cl^-. In the colon, **Na^+ absorption** involves electrogenic Na^+ transport (via Na^+ channels) and neutral NaCl cotransport. These two mechanisms of Na^+ absorption are driven by a **lumen-to-cell Na^+ concentration gradient,** which is maintained by the basolateral Na^+–K^+ pump and resembles that of the small bowel. The resultant accumulation of K^+ in the absorbing cell drives **K^+ secretion** across a luminal membrane K^+ channel. **Cl^- absorption** is accomplished by a passive Na^+-independent diffusional process along an inwardly directed Cl^- gradient and by a Cl^-–HCO_3^- exchanger, resulting in **HCO_3^- secretion.** In the normal colon, electrolyte

Table 46-2. Colonic Salvage and Failure of this Mechanism in Diarrheas of the Small Bowel and Colon*

	Health	Small Bowel Disease		Colonic Disease
Fluid volume to the colon	1,500	4,500	6,000	1,500
Colonic fluid absorption	1,400	4,400	4,400	700
Stool water	100	100	1,600	800
Diarrhea	No	No	Yes	Yes

*All quantities are given in milliliters.

absorption exceeds secretion, resulting in **net water absorption.**

NEURAL AND HORMONAL REGULATION OF TRANSPORT

The various processes of intestinal electrolyte transport are precisely coordinated with the other intestinal functions of motility and secretion to establish an efficient and integrated sequence to the total digestive process. Extrinsic and intrinsic nervous innervation from the autonomic and enteric nervous systems, respectively, along with various hormonal influences, ensure this coordination once food is present in the digestive tract.

Figure 46-8 depicts a full-thickness section through the small intestine, consisting of the outer longitudinal muscle layer, the circular muscle, the submucosal space, and the mucosal cell layer. The **submucosal nerve plexus** in the submucosal space is a component of the enteric nervous system of the gastrointestinal tract. As shown, **intramural reflex pathways** can regulate Cl⁻ secretion. Sensory receptors in the mucosal region detect various characteristics of luminal contents (texture and fluidity). This afferent information is integrated and transferred in the submucosal plexus through the release of acetylcholine, which evokes a rapid excitatory response in the motor neurons innervating the crypt region. This **cholinergic stimulation** of the crypt cells accelerates Cl⁻ secretion, most

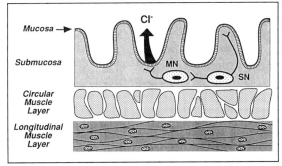

Fig. 46-8. An intramural reflex pathway that regulates chloride secretion in the crypts. (*SN* = sensory neurons; *MN* = motor neurons.) (Modified from: Cooke, H. J. Neural and Humoral Regulation of Small Intestinal Elecrolyte Transport. In: Johnson, L. R., ed. *Physiology of the Gastrointestinal Tract,* 2nd ed. New York: Raven Press, 1987.)

likely mediated by a Ca^{2+}, phospholipid, and PKC pathway.

Extrinsic neural influence may also play a role in this regulatory scheme (Fig. 46-9). These pathways are somewhat more complex, in that they include the myenteric plexus of the enteric nervous system and some external nerves of the autonomic nervous system. The previous discussion described how submucosal nerves can influence Cl⁻ secretion. However, it is known that other influences may be involved in the secretory response. Submucosal nerves can be modulated by nerves of the myenteric plexus, which release either **serotonin** (which is stimulatory) or **enkephalins** (which are inhibitory). Furthermore,

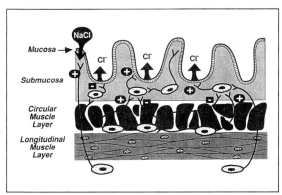

Fig. 46-9. The role of the intrinsic and extrinsic nervous systems in the regulation of intestinal electrolyte transport. (Modified from: Cooke, H. J. Neural and Humoral Regulation of Small Intestinal Electrolyte Transport. In Johnson, L. R., ed. *Physiology of the Gastrointestinal Tract,* 2nd ed. New York: Raven Press, 1987.)

extrinsic sympathetic fibers can influence the activity of enteric, submucosal nerves or provide direct input to the epithelial cells. In this case, **norepinephrine,** the neurotransmitter released by these nerves, has a dual action. Norepinephrine inhibits submucosal nerve activity and reduces Cl⁻ secretion, while directly stimulating villous cells to increase NaCl absorption. This action has been mimicked by alpha$_2$ adrenergic receptor agonists, which have been proposed as a class of compounds for antidiarrheal therapy.

The hormone **somatostatin** may also be involved as a neurotransmitter in the sympathetic neural influence on transport, but this interaction has not been fully elucidated. Because of the well-known action of somatostatin in reducing fluid and electrolyte losses in diarrhea, a very effective somatostatin analog has been developed and has shown a limited but successful effect in treating the fluid and electrolyte loss secondary to secretory diarrhea.

It is now well established that several naturally occurring **gastrointestinal hormones** and **paracrine substances** can stimulate electrolyte (NaCl) and fluid absorption (Table 46-3). For the most part, however, the cellular mechanisms that

Table 46-3. Endocrines, Paracrines, and Neurocrines that Influence Electrolyte Transport

Stimulators of Absorption

Aldosterone	Glucocorticoids
Dopamine	Neuropeptide Y
Enkephalins	Norepinephrine
Epinephrine	Somatostatin

Stimulators of Secretion

Acetylcholine	Neurotensin
Bombesin	Prostaglandins
Bradykinin	Serotonin
Histamine	Substance P
Leukotrienes	Vasopressin
Motilin	Vasoactive intestinal polypeptide

mediate the absorptive response are unknown. In contrast, there are many hormones and paracrine substances within the gastrointestinal tract and elsewhere that promote fluid and electrolyte secretion (see Table 46-3). The mechanism of action of many of these secretagogues is known, in that, generally speaking, they alter cyclic nucleotide or Ca^{2+} metabolism within the enterocytes, leading to secretion.

CARBOHYDRATE AND PROTEIN DIGESTION AND ABSORPTION

The digestion and absorption of dietary carbohydrates and proteins takes place in the **small intestine.** These are extremely efficient processes, in that essentially all of the carbohydrates and proteins consumed are absorbed. **Carbohydrates** are the most important component of the diet from a caloric point of view and constitute about 45 to 50 percent of the typical Western diet. Most caloric intake for the majority of people is obtained from carbohydrates, and therefore they represent our "cheapest" source of energy. In contrast, **proteins** are the "expensive" food in the diet. Meats, fish, eggs, and dairy products are high in protein. A limited amount of protein is obtained from vegetables, and, in many countries, vegetables are the main source of protein.

While vegetables are a much cheaper source of protein than are meats, the residents of these countries are at risk of developing diseases of protein deficiency, such as **kwashiorkor.** In the United States, meat is readily available, and on the average, protein constitutes about 25 percent of the total caloric intake. Both carbohydrates and proteins must be digested to liberate the simple molecules of hexoses and amino acids, respectively, and thus make absorption possible.

Carbohydrates

Dietary carbohydrates are either digestible or indigestible (Table 46-4). The common sources of **digestible carbohydrates** are starches (complex sugars), table sugar, fruits, and milk. The **indigestible carbohydrates** are found in legumes in the form of oligosaccharides (for example, raffinose) and in some vegetables, fruits, and grains as the polysaccharides cellulose, hemicellulose, pectin, and gums. These are often referred to as **dietary fiber** and their value to human health has been emphasized of late.

Starches are complex sugar polymers consisting of long chains of glucose molecules linked by a 1,4-glycosidic bond and branching side chains of glucose connected to the main chain by a 1,6

Table 46-4. Important Dietary Carbohydrates

Food Source	Carbohydrate
Digestible Carbohydrates	
Starch	Glucose
Milk and milk products	Glucose, galactose
Sugarcane	Glucose, fructose
Vegetables	Fructose
Fruits	Fructose
Honey	Fructose
Corn syrup	Fructose
Indigestible Carbohydrates	
Legumes	Raffinose
Vegetables, fruits	Cellulose, hemicellulose, pectin, gums

bond. The **digestion of starch** to simple hexose molecules involves a two-step process: a luminal phase and a brush-border phase (Fig. 46-10). The **luminal phase** actually begins in the mouth with the action of saliva on ingested food. **Saliva** is a mixture of secretions from three pairs of glands, the parotid, submandibular, and sublingual salivary glands. Secretion from the **parotid glands** is watery; it serves to lubricate food and wash away and solubilize food particles from the taste buds, causing food to be tasted as it is consumed. This secretion also **buffers** potentially injurious agents, such as hot or acidic liquids, and lubricates the structures of the mouth for the production of clear and distinct speech. **Sublingual** and **submandibular glands** secrete a more viscous fluid, which also lubricates food, and it contains the enzyme *salivary amylase.* This enzyme initiates the digestion of carbohydrates, which is continued in the esophagus and for a short time in the stomach. It is estimated that as much as 75 percent of the carbohydrates consumed is digested down to disaccharides before it reaches the small intestine. Most of the luminal phase occurs in the upper small intestine as pancreatic alpha amylase is secreted into the lumen. The **complex starch polymer** is ultimately reduced to maltose, maltotriose, and alpha-limit dextrans by these enzymes (see also Chapter 44).

These three products are dispersed in the **bulk-water phase** and the **unstirred water layer**

Fig. 46-10. The luminal and brush border phases of intestinal starch digestion.

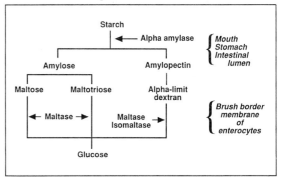

within the intestine and are eventually exposed to the brush-border membrane of the absorbing enterocyte. These cells secrete disaccharidases and trisaccharidases that reduce these substrates to their simplest form—**glucose** (see Fig. 45-10). Once in this form, carbohydrates can be absorbed into the enterocyte by mechanisms that will be discussed below, and are eventually discharged into the bloodstream.

Sucrose (cane sugar) and **lactose** (milk sugar) are disaccharides that require only a brush-border phase of digestion. Sucrose is reduced to glucose and fructose and lactose to glucose and galactose by the action of the disaccharidases **sucrase** and **lactase,** respectively, which are secreted by the brush-border membrane of the enterocytes. These carbohydrates are now in a form that facilitates their absorption.

Carbohydrate-induced diarrhea most often results from lactose intolerance because of a genetic deficiency of the brush-border enzyme, lactase. This syndrome, called **hypolactasia,** is the most frequent genetic deficiency in humans and affects approximately one-half of the world's population. This is an example of an **osmotic diarrhea,** and thus the primary factor governing the amount of fluid lost is the amount of lactose ingested. These osmotic diarrheas are generally of colonic origin.

Hexose

Glucose, the major hexose derived from carbohydrate digestion, is absorbed by two processes: passive diffusion and active, carrier-mediated transport. The **passive diffusion** absorption of glucose depends on concentration and can occur through the cells as well as by a paracellular route. As long as the glucose concentration in the lumen remains higher than that in the blood, glucose will diffuse down its concentration gradient and be absorbed. There is some evidence to suggest that the passive diffusion of glucose through the intestinal mucosal cells is carrier-mediated, which would constitute a specialized diffusional system called **facilitated diffusion.** Regardless of the diffusion mechanism, approx-

imately 80 percent of luminal glucose is absorbed by diffusion. The remaining 20 percent is absorbed by an **energy-dependent, carrier-mediated, Na$^+$—glucose cotransport process,** which is maintained for as long as glucose exists in the intestinal lumen. However, following a meal, this active transport of glucose is "masked" by the overwhelming amount of passive diffusion taking place in response to the high luminal glucose concentration. As the luminal glucose concentration falls to about 5 mM and below, active glucose transport becomes dominant and eventually sequesters all remaining glucose in the blood.

The mechanism of active glucose transport is the now classically defined **Na$^+$-coupled, energy-dependent, uphill transport** found in many epithelial cell systems, both in humans and other animal species (Fig. 46-11). In this scheme, Na$^+$ and glucose bind to an apical membrane transport protein and are deposited in the cytosol because of a conformational change in the transport protein, and because the affinity of the protein for the substrates changes from high to low. Once inside the enterocyte, glucose accumulates to a level that exceeds its concentration in the interstitial space and blood. Glucose then diffuses down this concentration gradient, out of the cell, and into the blood. Following a meal, this cell-to-blood gradient is maintained for some time, because the active transport of glucose from the lumen sequesters most, if not all, of the luminal glucose and stimulates villous blood flow, thus preventing glucose from collecting in the tissue. Cytosolic Na$^+$ is pumped from the interior of the cell by the basolateral (Na, K)-ATPase—mediated Na$^+$ pump. This low intracellular Na$^+$ concentration creates the inwardly directed Na$^+$ gradient and energizes the transport of glucose (and other substrates, as will be discussed). In regard to glucose, this system is often referred to as **secondary active transport,** since the active (energy-consuming) step involves Na$^+$ extrusion, and energy from the ATPase reaction is not directly used in moving the glucose.

There are several lines of experimental evidence pointing to the active transport of glucose by the

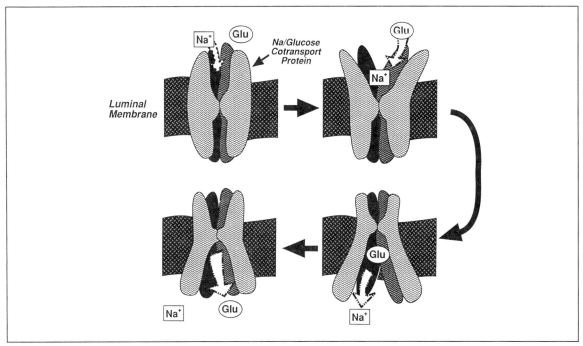

Fig. 46-11. Active Na$^+$-coupled glucose (*Glu*) transport by the intestinal villous cell carrier protein. (Modified from: Stevens, B. R., et al. Intestinal brush border membrane Na$^+$/glucose cotransporter function *in situ* as a homotetramer. *Proc. Natl. Acad. Sci.* 87:1456–1460, 1990.)

intestinal mucosa. If the glucose concentration is high in the blood and low in the intestinal lumen, glucose does not accumulate in the lumen, but continues to be absorbed from the lumen. Furthermore, when the glucose concentration in the blood is equal to that in the lumen, there is still net movement of glucose from the lumen to blood. If Na$^+$ is removed from the lumen, active glucose transport is suppressed. In addition, **metabolic poisons** that inhibit (Na, K)-ATPase block glucose transport. Finally, active glucose transport exhibits the **saturation kinetics** typical of a carrier-mediated transport process, with a rate of absorption equal to one-half the maximum (the T$_m$ or K$_m$) occurring at about 5 to 10 mM.

Galactose is also actively transported by the glucose carrier system, and has been shown experimentally to be a competitive inhibitor of glucose transport. However, the carrier system has a somewhat lower affinity for galactose than for glucose. **Fructose** is not actively transported by intestinal cells but is absorbed by a carrier-mediated, facilitated diffusion system. The distinguishing feature of a facilitated diffusional system is that, because it is carrier mediated, equilibrium is achieved sooner than it would be in a simple non-carrier–mediated, passive-diffusion system. Carrier-mediated fructose absorption does not require the input of energy.

Proteins

Like carbohydrates, dietary proteins must be reduced to their simplest forms before they can be absorbed by the intestinal mucosa. This digestive process begins in the stomach, through the action of the gastric enzyme **pepsin,** and is continued in the upper small bowel by the proteolytic enzymes of the **pancreas.** Cytosolic dipeptidases from the brush borders of the villous cells and from enterocytes mediate the final breakdown of small peptides into individual **amino acids. Hydrolytic products** of protein digestion (amino acids, dipeptides, and tripeptides) are each capable of being absorbed intact across the luminal membrane of the intestine into the enterocyte and eventually into the bloodstream.

In the intestinal lumen, amino acids are actively transported into the blood by active **Na$^+$-dependent, carrier-mediated** systems similar to those for glucose. These transport systems possess all the characteristics of the glucose system except that they are specific to amino acid transport. Furthermore, there appear to be separate carrier systems specific for certain classes of amino acids; that is, the neutral, dibasic, acidic, proline, and phenylalanine-methionine groups. All the characteristics of an active carrier-mediated process exist in these groups, such as competitive inhibition, Na$^+$ dependency, saturation kinetics, and metabolic energy. Some amino acids, such as phenylalanine and the basic amino acids, are absorbed primarily through facilitated diffusion from the lumen to blood.

LIPID DIGESTION AND ABSORPTION

Approximately 95 to 98 percent of the total dietary lipids consumed is in the form of **triglyceride.** The remaining 2 to 5 percent is made up of phospholipids, cholesterol, and cholesterol esters. Upon oxidation in the tissues, lipids yield more than **twice as much energy** per gram than do carbohydrates or proteins and furnish up to 40 percent of our total daily caloric intake. Lipids also have a **greater satiety** value than do carbohydrates or proteins, as they tend to remain in the stomach longer and are digested more slowly. Lipids are important sources of **carbon** for the biosynthesis of cholesterol and other steroids. Many plant triglycerides provide essential **fatty acids.**

Dietary triglycerides must also be broken down into **simpler molecules,** monoglycerides, fatty acids, and glycerol, to facilitate efficient absorp-tion. A very small fraction of dietary triglyceride is digested in the mouth and stomach by **lingual lipase,** which is secreted by the salivary glands. However, most triglycerides are digested in the upper small intestine. To accomplish this, **two obstacles** must be overcome. First, the lumen of the upper small bowel consists of an aqueous medium and triglycerides are not soluble in water. Second, digestive enzymes are proteins and are dissolved in the aqueous environment of the intestinal lumen. Thus, triglycerides must be solubilized in the aqueous phase before digestion can occur. This requires a coordinated and efficient series of mechanical and chemical interactive steps that are, for convenience sake, divided into **intraluminal** and **intracellular events.**

Intraluminal Events

As just noted, triglyceride must be solubilized in the aqueous environment of the lumen in order for digestion to occur. This is summarized in Figure 46-12. This is accomplished by the action of **bile salts,** which are introduced into the lumen by biliary secretion. Bile salts are **amphipathic** molecules, in that they have both hydrophilic and lipophilic properties. Thus, they are soluble in both the aqueous and lipid mediums and can solubilize or emulsify lipids in water. Bile salts are also involved in lipid absorption, which will be discussed.

Once triglycerides are emulsified in the aqueous phase, pancreatic lipase catalyzes their further breakdown. **Co-lipase** serves two important functions: it **complexes** the lipase to the bile salt and the triglyceride to ensure digestion and it also lowers the **optimum pH** of the enzyme to near the prevailing pH (6 to 7) of the intestinal lumen. The **pancreatic lipase** cleaves the ester linkage of the fatty acid and glycerol at the "1" and "1'" position, liberating two free fatty acids and 2-monoglyceride. Occasionally the ester bond at the "2" position is cleaved, leaving a free glycerol molecule. These resulting lipid substrates are slightly more soluble in the aqueous medium of

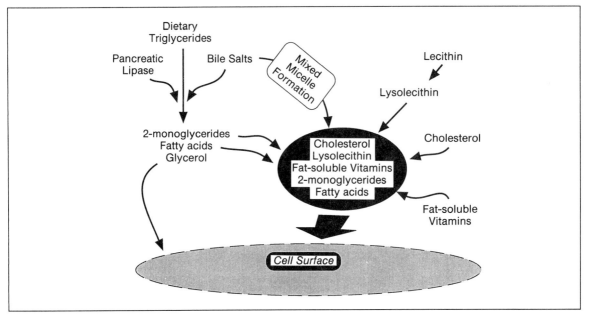

Fig. 46-12. Intraluminal events in intestinal lipid digestion and absorption.

the lumen than are triglycerides, but still require the action of bile salts for complete solubility.

At this point, the **bile salt concentration** in the lumen becomes important. Because of their physiochemical nature, bile salts form **micelles** above a certain concentration, called the **critical micellar concentration** (CMC). Following a meal, the luminal bile salt concentration is well above the CMC and the bile salt monomers aggregate to form micelles, which are spherical structures with a diameter of approximately 50 Å. In this form, they contain no other substrates and are referred to as *simple micelles.* The orientation of the bile salt monomers in the micellar structure is such that their lipophilic end faces inward, creating a **hydrophobic core.** The end-products of triglyceride digestion, namely the monoglycerides and free fatty acids, are sequestered with these micelles, such that they partition easily into the lipid core. These micelles that contain lipid products in their core are called *mixed micelles.* This not only ensures their solubilization in the water medium of the lumen but pro-

vides a vehicle for their delivery to the absorbing enterocyte. **Triglycerides** do not partition readily into these micelles and stay in the lumen to be emulsified and digested. **Glycerol,** which is also liberated by triglyceride digestion, does not partition into micelles but is freely dispersed through the water environment, eventually to be absorbed into the enterocytes by simple diffusion.

Other **dietary lipids** partition into bile salt micelles. Cholesterol, cholesterol esters, the lipid-soluble vitamins (A, D, E, and K), and the end-products of phospholipid digestion, the lyso-phospholipids, are all sequestered into micelles. The bile-salt mixed micelles maintain the solubility of these lipids in the predominantly water environment of the intestinal lumen.

Before discussing lipid absorption and to clarify the integral role bile salt micelles play in this process, certain features of the **aqueous environment** within the lumen of the small intestine will be described. The peristaltic contractions of the intestinal smooth muscle gently move the luminal contents in an aboral direction. In the aqueous medium, this movement assumes the characteristics of **laminar flow.** Thus, the flow rate is greatest in the center and decreases toward the luminal wall. This centrally moving fluid is referred to as the **bulk-water phase.** Immediately adjacent to the luminal wall of mucosal cells, the flow rate is essentially zero and is referred to as the **unstirred water layer.** The thickness of this unstirred water layer varies inversely with the frequency of segmental and propulsive contractions of the visceral smooth muscle of the intestine. Nevertheless, it does represent an appreciable **diffusional barrier** and slows the transit of absorbable substrates to the apical cell surface. This would pose a very significant hindrance to lipid absorption, if it were not for the fact that these absorbable lipids are solubilized within the micelles. Therefore, these lipid-laden micelles gain access to the absorbing surfaces in the same way other substrates do. Through the mixing and churning action created by segmental contractions, mixed micelles contact and adhere to the absorbing cell surface, such that lipids freely diffuse out of the lipophilic micellar core and into the lipid matrix of the apical membrane of the enterocyte. Neither bile salt micelles nor any monomeric bile salt molecules diffuse into the cell but are recycled back into the bulk-water phase. The precise mechanisms of micellar adherence, dispersion, and recycling are unknown.

Intracellular Events

The fatty acids sequestered in the apical membrane of enterocytes must be transferred to the cytosol and ultimately into the bloodstream, and this process is depicted in Figure 46-13. The first intracellular event requires the assistance of a cofactor that promotes the solubilization of these lipids in the aqueous cytosol. A protein with a high affinity for these fatty acids exists in the cytosol of the enterocytes. This protein, with a molecular weight of 12,000, is called **fatty acid–**

binding protein (FABP). It binds fatty acids and transports them into the cytosol, where they are re-esterified into triglycerides. Because some fatty acids are known to be cytotoxic, FABP not only transports fatty acids but also protects the cell from any adverse effects.

Once fatty acids enter the cell, they are first activated to **acetyl coenzyme A** (CoA), a reaction that requires Mg^{2+}, ATP, and the enzyme acetyl CoA synthetase. This reaction is rendered as follows:

$$\text{fatty acid } + \text{ CoA} \xrightarrow[Mg^{2+}]{ATP} \text{ acetyl CoA}$$

In this form, they enter one of the two **biosynthetic pathways,** which brings about their re-esterification and the resynthesis of triglycerides. These pathways are the **2-monoglyceride** and the **alpha-glycerol phosphate** pathways. It has been suggested that the 2-monoglyceride pathway is the predominant route of triglyceride resynthesis in humans. Phospholipids likewise are exclusively resynthesized through the alpha-glycerol phosphate pathway.

Before exiting the cell, these reformed triglycerides, along with cholesterol, cholesterol esters, and various lipoproteins, are packaged into another lipid-carrying particle called a **chylomicron,** which is derived from beta-lipoprotein synthesis in the enterocyte. The phospholipids to be absorbed, along with about 70 percent of the absorbable free cholesterol, form a monolayer on the surface membrane of the chylomicrons. Chylomicrons vary in size from 700 to 6,000 Å, depending on how much lipid they contain. The exact mechanism that accomplishes this packaging of lipids in these spheres is unknown. The final intracellular event in **lipid absorption** is the diffusion of chylomicrons from the cytosol, through the basolateral membrane, and into the lymphatic channel of the villus, known as the central **lacteal.** From here they are delivered to the systemic circulation through the thoracic duct.

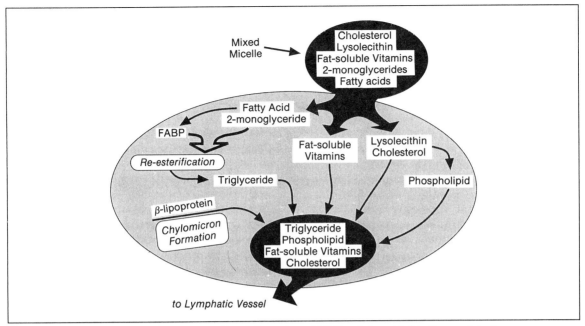

Fig. 46-13. Intracellular events in intestinal lipid absorption. (*FABP* = fatty acid–binding protein.)

ENTEROHEPATIC CIRCULATION AND ABSORPTION OF BILE SALTS

Bile acids are secreted by hepatocytes and collect in the bile canaliculi of the liver parenchyma. In the **interdigestive phase,** most bile salts are stored in the gallbladder and a very small amount leaks into the intestine. During and immediately after a **meal,** the gallbladder empties bile salts into the common bile duct and eventually into the intestinal lumen. In the **small intestine,** bile salts perform their important actions on lipid digestion and absorption, which are described in the preceding section. Bile salts are eventually absorbed by the small intestine and returned to the liver by means of the **portal circulation.** Upon reaching the **liver,** they are sequestered by **hepatocytes** and recycled through the system. Approximately 95 to 98 percent of the secreted bile salts are reabsorbed and recirculated in this pathway, called the **enterohepatic circulation** (see Fig. 46-13). The remaining 2 to 5 percent of

unabsorbed bile salts enters the colon and is ultimately excreted in the feces. This loss of bile salts is easily replenished by the de novo hepatic synthesis of bile salts. Thus, the enterohepatic circulation and hepatic synthesis allow the body to maintain a constant bile salt pool of 3 to 5 gm. This pool cycles five to ten times a day; thus, it is not unusual for a single bile salt molecule to recycle three times during one meal.

There are **several important factors** that affect the enterohepatic circulation and the maintenance of a bile salt pool. They include: (1) the relative impermeability of the upper small bowel to bile salts; (2) the active transport and efficient absorption of bile salts by the terminal ileum; (3) a patent portal blood flow; (4) active sequestration of bile salts from the portal blood by hepatocytes; (5) secretion of bile salts by the liver and storage in the gallbladder; and (6) the delivery of bile salts to the intestinal lumen via the biliary system. It is easy to see how a disturbance in any of these components, such as may result from liver disease or inflammatory bowel disease, will not only seriously affect the bile salt pool but will markedly compromise lipid digestion and absorption.

The **intestinal absorption of bile salts** constitutes an interesting aspect of the enterohepatic circulation and the functionality of the small intestine. The **upper two-thirds of the small intestine** is relatively impermeable to bile salts and very little absorption takes place there. Thus, although there is still some lipid in this region, more than enough solubilization will be occurring because of the adequate quantities of bile salts, ensuring efficient digestion and the absorption of dietary lipids.

As bile salts move into the **lower one-third of the small bowel,** the mucosa becomes more permeable to these substrates and the monomeric bile salts are absorbed through diffusion, which continues throughout the remaining small intestine. As the luminal concentration falls below the CMC, micelles disperse until all the bile salts exist as monomers. In the **terminal ileum** of humans and most other mammalian species, bile salts are rapidly and efficiently absorbed by an **active Na⁺-dependent cotransport system** that is similar to the glucose and amino acid active transport systems operating in the upper small bowel. Thus, by the time a fraction of intestinal contents reaches the ileocecal valve, nearly 98 percent of the bile salt content has been absorbed into the portal blood.

VITAMINS

Vitamins from various animal and plant sources are dietary nutrients that are essential for normal human metabolism. These organic compounds are not manufactured by the body, and thus efficient intestinal absorptive mechanisms must ensure their adequate daily intake. Vitamins are classified in many ways, but, in terms of absorption, they are classified according to whether they are lipid-soluble or water-soluble.

The **lipid-soluble vitamins A, D, E, and K** are absorbed with other lipid-soluble nutrients and use the **bile-salt mixed micelle** as a vehicle for their solubilization and transport to the cell surface. The mechanisms involved are the same as those described for lipid absorption. **Water-soluble vitamins** are absorbed by processes similar to those that operate for sugars and amino acids. While few details regarding water-soluble vitamin absorption are available, a few specific mechanisms of absorption of some of these essential substrates are well known.

Vitamin B (thiamine), **vitamin C,** and **folic acid** are absorbed primarily by passive diffusion. At lower concentrations, these vitamins can be actively transported by an active Na⁺-dependent process. This process is thought to be carrier-mediated, requiring metabolic energy derived from the hydrolysis of ATP.

Vitamin B$_{12}$ absorption represents a special case. To achieve efficient absorption this vitamin requires a glycoprotein called **intrinsic factor,** which is secreted by the parietal cells of the stomach. Once intrinsic factor complexes with vitamin B$_{12}$, the complex binds to specific receptors on ileal enterocytes and is internalized. Receptor binding of the intrinsic factor–B$_{12}$ complex requires Ca^{2+} or Mg^{2+} and an alkaline pH. The internalization process is not energy dependent and presumably occurs through **pinocytosis.**

A marked decrease in vitamin B$_{12}$ intake or any disease state that diminishes the secretion of intrinsic factor will prevent or inhibit intrinsic factor–B$_{12}$ complexing or binding, or will compromise ileal absorption, leading to vitamin B$_{12}$ deficiency. Because this vitamin is important for red blood cell formation and metabolism, if left untreated, vitamin B$_{12}$ deficiency will lead to **pernicious anemia.**

SUMMARY

The human intestine is equipped anatomically and functionally to handle the nearly 9.0 liters of fluid delivered to it daily. The countless numbers of folds and villi endow the human intestine with an absorption surface area of nearly 2,000,000 cm^2. The epithelial lining absorbs electrolytes through the influence of passive diffusion, ion exchangers, and energy-dependent pumps, creating osmotic gradients for efficient water absorption. Intricate intracellular mechanisms mediated by Ca^{2+} and cAMP precisely regulate these processes. Many, if not most, of these mechanisms are influenced extrinsically by the autonomic nervous network and intrinsically by the enteric nervous system, as well as a variety of locally released hormones. Carbohydrates, proteins, and lipids are digested by specific enzymes that are furnished by a variety of sources, including the salivary glands, stomach, pancreas, and the intestinal mucosal cells. These many digestive enzymes ensure that dietary nutrients are the right size for efficient absorption. Specific transport mechanisms also exist for the efficient absorption of these digestive end-products. Bile salts play a dual role in lipid digestion and absorption, first by emulsifing triglycerides to facilitate their breakdown by lipases, and second by solubilizing free fatty acids, cholesterol, and fat-soluble vitamins for efficient absorption. All of the processes described herein are integral components of the total digestive process.

NATIONAL BOARD TYPE QUESTIONS

Select the one most correct answer for each of the following

1. Essential to the absorption of long-chain fatty acids from the lumen of the intestine into the villous cell is:
   A. Mixed micelle formation
   B. Na^+-dependent cotransporter
   C. Carboxypeptidase activity
   D. Isomaltase
   E. Increased beta-lipoprotein activity
2. With regard to the intestinal Na^+ transport, which one of the processes described below is correct?
   A. Elevated cAMP concentration in the enterocyte promotes NaCl absorption in the intestine
   B. NaCl absorption in the gut is secondary to water flux
   C. Luminal membrane Cl^- conductance balances or neutralizes Na^+ absorption

D. (Na, K)-ATPase activity is essential for maintaining the inward Na^+ electrochemical gradient
E. Increased intracellular Ca^{2+} concentration increases Na^+ absorption.
3. The end-products of carbohydrate and protein digestion have the following characteristic in common:
   A. They require chylomicrons in order to exit the mucosal cell
   B. They enter the bloodstream after being transported through the lymphatic vessels of the intestine
   C. They require Na^+ for active transport
   D. They require mixed micelles for efficient absorption
   E. They only pass through paracellular routes of absorption
4. Which of the following would be expected in a person with complete lactase deficiency, following ingestion of a milk meal?
   A. Increased electrolyte and water absorption in the small intestine
   B. Decreased sucrose digestion
   C. Increased glucose absorption in the colon
   D. The small intestine contents flowing into the colon will contain lactose
   E. A decrease in maltose absorption
5. The absorption of water by the intestinal epithelial cells:
   A. Is an active energy-dependent process
   B. Is a passive diffusional process secondary to active solute transport
   C. Is coupled to Cl^- secretion
   D. Is driven by peristaltic "squeezing" of the water into capillaries
   E. Requires micelles to create temporary "pores" in the epithelial cell membranes

---

ANNOTATED ANSWERS

1. A. Bile-salt mixed micelles are essential for the absorption of most lipids and lipid-soluble vitamins from the intestinal lumen into villous cells. A Na^+-dependent cotransporter

has not been described for lipid absorption. Carboxypeptidase and isomaltase are proteolytic and glycolytic enzymes, respectively. Beta-lipoproteins are utilized during chylomicron formation for the triglyceride "exiting" from the villous cell into the lymphatic channel.

2. D. Elevated cAMP levels in the mature enterocyte inhibit NaCl absorption. While NaCl can follow water across most membranes, water flux follows NaCl absorption in the intestine. Cl$^-$ absorption does not parallel Na$^+$ absorption, which primarily explains the electrical potential difference maintained across the intestinal mucosal layer. Increasing intracellular Ca^{2+} concentrations would decrease Na$^+$ absorption by intestinal cells. It is essential to all the processes of Na$^+$ absorption that the basolateral (Na, K)-ATPase –mediated Na$^+$ pump maintains the lumen-to-cell electrochemical Na$^+$ gradient.

3. C. Simple sugars and amino acids require a Na$^+$-dependent cotransport for active energy-utilizing absorption. Chylomicrons and mixed micelles are required for the absorption of dietary lipids and are not used by carbohydrate and protein end-products. Only absorbed lipids are carried by lymphatic channels to the bloodstream after absorption from the intestine. While paracellular routes may be recruited for absorption by these nutrients, the transcellular movement from lumen to blood is well established as an integral part of the total absorptive process.

4. D. A person lacking lactase will be unable to digest the milk sugar, lactose. Thus, the undigested lactose from a milk meal would likely enter the colon undigested in such a person. Water absorption also would be decreased. The digestion of sucrose and maltose and the absorption of glucose and electrolytes are all essentially unaffected.

5. B. The movement of water across any membrane is a passive diffusional process and is a result of the osmotic forces generated by solute transport acting on it. Water absorption by active transport has never been described. Water absorption is not coupled to Cl$^-$ secretion; in fact, water would diffuse into the lumen of the intestine during Cl$^-$ secretion. Peristalsis does not influence water absorption by squeezing or forcing water through the intestinal wall. There is no evidence to support a role for micelles in intestinal water absorption.

BIBLIOGRAPHY

Field, M., Fordtran, J. S., and Schultz, S. G., eds. *Secretory Diarrhea*. Bethesda, MD: American Physiological Society, 1980.

Fondacaro, J. D. Intestinal electrolyte transport and diarrheal disease. *Am J. Physiol.* 250(13): G1–G8, 1986.

Fondacaro, J. D. Intestinal absorption of bile acids. In: A. Kuksis, ed. *Fat Absorption*. Boca Raton, FL: CRC Scientific Press, 1986

Johnson, L. R., ed. *Physiology of the Gastrointestinal Tract*, 2nd ed. New York: Raven Press, 1987.

Stevens, B. R., et al. Intestinal brush border membrane Na$^+$/glucose cotransporter function *in situ* as a homotetramer. *Proc. Natl. Acad. Sci.* 87:1456–1460, 1990.

X Endocrine Physiology

47 The Hypothalamus and Neuroendocrinology

LAWRENCE A. FROHMAN

Objectives

After reading this chapter, you should be able to:

List the different neuroregulatory systems in which the hypothalamus participates and distinguish their differences

Describe the origin of the blood supply to the pituitary and its functional importance

List the hypophysiotropic hormones of the hypothalamus and the pituitary hormones that they regulate

Identify the mechanisms used by hypothalamic hormones to alter pituitary hormone secretion

Describe the individual hypothalamic–pituitary–target gland axes, the hormones involved, and the functions they serve

Characterize the role of cytokines in neuroendocrine activation

---

Considerable regulatory control of hormonal secretion and metabolic processes is exerted by the central nervous system (CNS). The integration of this control occurs in the **ventromedial hypothalamus,** known as the **hypophysiotropic area,** and consists of three major components.

The first is a **neuroendocrine system** involving clusters of peptide-secreting and monoamine-secreting neurons in the anterior and medial (periventricular) ventral hypothalamus, whose products are transported along nerve fibers to terminals in the outer layer of the median eminence. The releasing and inhibiting hormones are secreted into capillaries of the hypothalamic–hypophyseal portal vascular system and transported to the pituitary where they regulate the secretion of the anterior pituitary hormones.

The second component of this integration process is a **neurohypophysial pathway** from selected nuclear regions in the anterior hypothalamus that traverses the floor of the ventral hypothalamus and pituitary stalk and terminates in specialized neuronal elements, called *pituicytes*, located in the posterior pituitary. This system is responsible for **osmoregulation** mediated by the secretion of **vasopressin** (antidiuretic hormone) and for **parturition and nursing,** which are mediated by the secretion of **oxytocin.** A more detailed discussion of this system is found in Chapter 49.

The third aspect is a **neurometabolic pathway** that traverses the base of the brain and autonomic nervous system pathways of the spinal cord and terminates in the liver, gastrointestinal tract, pancreas, adrenal medullae, and adipose tissue. This pathway consists of bidirectional fibers and is involved in **substrate regulation.** Its primary effects are on **substrate regulation** (for example, glucose, fatty acids, and amino acids) and on **metabolic homeostasis,** consisting of the control of food intake (satiety), temperature (thermoregulation), and body fat stores (nutrient regulation).

ANATOMIC

Hypothalamus

The neuronal **perikarya** (cell bodies) of the neuroendocrine system are distributed throughout the mediobasal hypothalamus. Although the hypophysiotropic hormone–secreting neurons receive input from many brain regions in response to external environmental changes, they maintain their secretory activity even in the absence of extrahypothalamic input, indicating that their most important homeostatic signals are derived from the circulation. Neuronal perikarya that secrete thyrotropin-releasing hormone (TRH), corticotropin-releasing hormone (CRH), and somatotropin release–inhibiting factor (somatostatin; SRIF) are located in the **anterior hypothalamus,** in the region of the **paraventricular nucleus.** Those secreting growth hormone–releasing hormone (GRH) are located almost exclusively in the **arcuate nucleus,** which also contains the perikarya of gonadotropin-releasing hormone (GnRH) neurons. Dopamine-secreting neurons arise from the **tuberoinfundibular tract,** and those releasing a recently identified prolactin-inhibiting factor exhibit an identical distribution to those secreting GnRH.

The releasing and inhibiting hormones are **axonally transported** and stored in nerve terminals in the outer layer of the **median eminence,** where their concentrations are many times greater than those in other hypothalamic regions. When appropriately stimulated, the nerve terminals release their contents into the **portal capillaries,** which transport the hormones to the **pituitary** (Fig. 47-1). Because portal blood flow is not compartmentalized, the various cell types

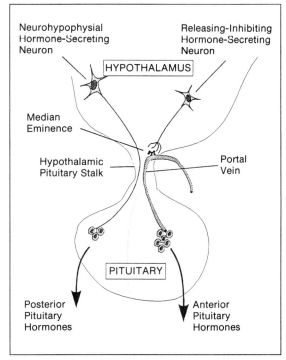

Fig. 47-1. The hypothalamic–pituitary neuroendocrine and neurohypophysial systems.

in the anterior pituitary have access to all of the releasing and inhibiting hormones. Thus, **specificity of action** is achieved by the presence of specific hormone receptors on individual cell types rather than by anatomic segregation.

Neurons composing the neurohypophysial system are anatomically more distinct, with neuronal perikarya located in the paraventricular and supraoptic nuclei of the **anterior hypothalamus.** The **cell bodies** are large (magnocellular) in contrast to the releasing hormone–secreting cells (parvicellular). The posterior pituitary hormones are synthesized in cell bodies as part of a precursor molecule, which also contains their carrier proteins, known as **neurophysins.** During transport from the hypothalamus to the posterior pituitary, the precursor molecules are converted into mature hormones, and, in the pituicytes of the posterior pituitary, they are packaged into **secretory granules,** which are released in response to appropriate stimulation of the receptors on the cell bodies in the hypothalamus. The posterior pituitary is consequently a truly integral part of the brain.

The **neurometabolic functions** of the hypothalamus are closely tied to the sympathetic and parasympathetic components of the **autonomic nervous system.** The hypothalamus is divided into a **medial sympathetic** and a **lateral parasympathetic zone,** though the neuronal perikarya involved with a specific function cannot always be localized to one precise nuclear region. Those neurons participating in the inhibitory **control of food intake** (satiety) tend to be located more medially, while those concerned with the **stimulation of appetite** are located more laterally. This distribution helps explain why destructive lesions of the hypothalamus, which frequently occur in the midline, are more likely to cause obesity than anorexia or starvation, because they abolish satiety control. In addition, some fiber tracts from these hypothalamic regulatory centers cross the midline, which protects the organism from interruption of normal regulatory control unless there is bilateral hypothalamic destruction.

Portal Vascular System

The blood supply of the anterior pituitary is distinct from that of the posterior pituitary and the CNS, in that it is not directly connected with the systemic arterial system. All **arterial blood** flows first through the hypothalamic arteries, which give rise to an extensive capillary plexus in the outer layer of the median eminence, juxtaposed to the nerve terminals of the hypophysiotropic hormone–secreting neurons. The **blood–brain barrier** in this region of the brain is incomplete, permitting access of charged particles and proteins in circulating plasma to the interstitial areas of the ventral hypothalamus and nerve terminals in the median eminence. This allows the nerve terminals and their neuronal perikarya to respond to changes in circulating hormones and metabolic signals, in addition to CNS-derived signals, for hormone release. The portal capillaries

combine to form a series of portal veins that descend through the portal stalk and reach the anterior pituitary, where they again give rise to a series of capillaries that bathe the pituitary cells. **Venous drainage** from this system enters the posterior pituitary and eventually reaches the systemic veins in the region of the petrosal sinus. Some blood flow travels in a reverse direction, from the posterior pituitary to the anterior pituitary, though the physiologic significance of this circulation with respect to anterior pituitary hormone secretion is unknown. Similarly, some of the anterior pituitary venous drainage seems to ascend the pituitary stalk, though it does not appear to reach the level of the median eminence.

HYPOTHALAMIC HORMONES
Structure and Synthesis

The hypothalamic hormones that regulate **anterior pituitary hormone** secretion and their overall effects on the individual pituitary hormones are shown in Figure 47-2. Although each pituitary hormone was originally thought to be regulated by one hypothalamic hormone, it is

Fig. 47-2. Hypothalamic hormone and pituitary hormone relationships. Hormones boxed in interrupted lines have not been fully characterized. (*CRF* = corticotropin-releasing factor; *VP* = vasopressin; *GnRH* = gonadotropin-releasing hormone; *GRH* = growth hormone–releasing hormone; *SRIF* = somatostatin; *TRH* = thyrotropin-releasing hormone; *PRF* = prolactin-releasing factor; *PIF* = prolactin-inhibiting factor; *DA*-dopamine, *Prl* = prolactin; *ACTH* = adrenocorticotropic hormone; *LH* = luteinizing hormone; *FSH* = follicle-stimulating hormone; *GH* = growth hormone; *TSH* = thyroid-stimulating hormone.)

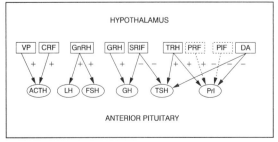

now known that several types of patterns exist and control of most of the pituitary hormones involves multiple hypothalamic hormones (some of which remain to be identified). In addition, there are stimulatory and inhibitory hypothalamic hormones for several pituitary hormones, and single hypothalamic hormones have effects on multiple pituitary hormones.

The predominant effect of the hypothalamic hormones on the pituitary, with a single exception, is **stimulatory.** Therefore, disruption of the integrity of the hypothalamic–pituitary interconnection decreases the secretion of pituitary hormones. The control of **prolactin** secretion is the exception; elimination of the hypothalamic input increases hormone release.

All of the hypothalamic hormones, again with a single exception, are **single-chain peptides** with sequence lengths ranging from 3 to 44 amino acids (Fig. 47-3). While the shorter peptides (TRH and GnRH) slow almost no species variation, each of the larger peptides exhibits either variable chain length (SRIF and GRH) or a small to moderate variation among species (GRH and CRH). Two candidates for prolactin-releasing and prolactin-inhibiting factors are actually contained within the structure of a larger precursor that also gives rise to other neuropeptides. **Vasoactive intestinal peptide** (VIP) stimulates prolactin release and has the same precursor as peptide histidine–isoleucine. A putative prolactin-inhibiting peptide is located on the C-terminal extension of GnRH (GnRH-associated peptide). In addition, a partially characterized peptide in the neurointermediate lobe of the pituitary also exhibits potent prolactin-releasing effects. The single nonpeptide hypophysiotropic hormone is **dopamine.** This catecholamine, besides its important role as a neurotransmitter, is the most important physiologic inhibitor of prolactin secretion.

Each of the peptide hormones is derived from a **precursor molecule** that is synthesized in a manner similar to that for other proteins. There is a single copy of the hypothalamic hormone on each precursor (except for TRH, where multiple copies are present), and processing of the precursor is believed to occur in the neuronal perikarya and also during transport along the axon fibers. **Differential processing** of the precursor is believed to account for the generation of those hormones that exist in multiple forms. While the processing enzymes remain to be identified, they are not unique to the specific hypothalamic neurons, as GRH, CRH, TRH, and SRIF have each been observed in non-neural tissue under normal conditions (placenta, pancreatic islets, C cells of thyroid) and in certain pathologic states (ectopic hormone secretion).

Regulation of Secretion: Role of Biogenic Amines and Neuropeptides

The hypothalamic hormone–secreting neurons have been called **transducer cells** because they possess both neural and endocrine characteristics. Although they respond to classic neurotransmitter-mediated signals, they release peptide hormones into a regional, or in some cases, systemic vascular system. The neurotransmitter signals used for communication throughout the CNS—monoamines and neuropeptides—also regulate hypothalamic hormone secretion. Their effects may occur at axodendritic interactions and at axoaxonic connections on the releasing hormone–containing nerve terminals. Multiple neurotransmitters and neuropeptides may participate in the release of a single hypothalamic hormone, and intermediary neurons may also be involved.

The **monoamines** involved in the neuroregulation of hypothalamic hormones include catecholamines (dopamine, norepinephrine, epinephrine), indolamines (serotonin, melatonin), acetylcholine, gamma-aminobutyric acid (GABA), histamine, and glutamine. A detailed discussion of their synthesis and action is beyond the scope of this chapter. However, it is important to remember that agents used to modify their function, including synthesis blockers, receptor agonists and antagonists, reuptake inhibitors, and degradative enzyme inhibitors, are all ca-

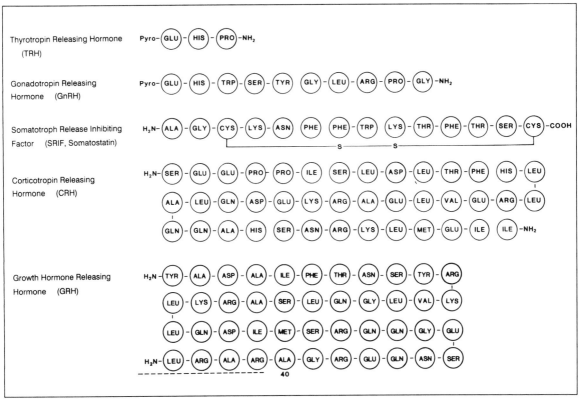

Fig 47-3. Structures of hypothalamic-releasing and -inhibiting hormones. The sequences of thyrotropin-releasing hormone, gonadotropin-releasing hormone, and somatostatin (somatotropin release–inhibiting factor) are identical in all mammalian species studied. The sequence of human corticotropin-releasing hormone (CRH) has been deduced from that of its complementary DNA and differs by 7 amino acids from ovine CRH. Two forms of growth hormone–releasing hormone have been identified in humans, which differ only by the additional carboxyl-terminal tetrapeptide, indicated by the interrupted line. Sequences in most large mammals show small variations, while those in rodents differ by more than 30 percent. (From: Frohman, L. A., and Krieger, D. T. Neuroendocrine Physiology and Disease. In: Felig, P. F., et al., eds. *Endocrinology and Metabolism*, New York: McGraw-Hill, 1987, P. 200.)

pable of modifying hypothalamic hormone secretion and pituitary function.

Numerous **neuropeptides** have been identified in the hypothalamus and are capable of influencing hypothalamic hormone secretion. They are listed in Table 47-1. Many of these peptides are also widely distributed in extraneural tissue, particularly in the gastrointestinal tract where they exert independent actions. They have existed throughout evolution, however, a lot of them as far back as unicellular organisms; this

underscores their importance in intercellular communication. Overall, they appear to be crucial in a number of integrative systems relating to homeostatic mechanisms, such as nutrition, growth, and reproduction, and their actions in the CNS complement those exhibited in extraneural tissues. The hypothalamic-releasing hormones themselves are included in this category, because there is now abundant evidence that they also serve as neurotransmitters and neuromodulators within the CNS.

Table 47-1. Neuropeptides with Effects on Hypothalamic Releasing and Inhibiting Hormones

I. Gastroenteropancreatic peptides
  Cholecystokinin
  Gastrin
  Secretin
  Motilin
  Substance P
  Neurotensin
  Gastrin-releasing peptide
  Insulin
  Glucagon
  Pancreatic polypeptide
 	
II. Hypothalamic hormones
  Somatostatin
  Thyrotropin-releasing hormone
  Growth hormone-releasing hormone
  Corticotropin-releasing hormone
  Vasopressin
	III. Endorphin/enkephalin peptides
	  Methionine/leucine enkephalin
	  Dynorphin
	  Beta-endorphin
	  Alpha melanocyte–stimulating hormone
	
	IV. Others
	  Neuropeptide PYY
	  Neuropeptide PHI/PHM
	  Calcitonin
	  Calcitonin gene–related peptide
	  Angiotensin
	  Bradykinin
	  Galanin

Pattern of Hypothalamic Hormone Secretion

Hypothalamic hormone levels in portal vessels have been measured in several animal species, and the results have been extrapolated to humans, based on a similarity of responses to the hormones among the various species. There is good evidence for the **pulsatile secretion** of GnRH, GRH, SRIF, and CRH (Fig. 47-4). The pulse frequency varies between species but, at least for GnRH, is closely entrained with the pulsatile pattern of luteinizing hormone (LH) secretion by the pituitary. CRH secretion also appears to be entrained to that of adrenocorticotropic hormone (ACTH), though the interrelationship of GRH and SRIF pulses to growth hormone (GH) pulses is a bit more complex. Alterations in the pulsatile pattern of hormone secretion have been observed in patients with certain types of neuroendocrine disorders, such as **anorexia nervosa.**

Transport and Metabolism of Hypothalamic Hormones

The extremely short distance between the median eminence and pituitary eliminates necessity for carrier or binding proteins for the releasing hormones. There is no need for the hormones to exist in the peripheral circulation. Reports of their presence there are frequently artifactual (produced by nonspecific factors in the radioimmunoassay), are attributable to cross-reactivity with partially metabolized hormones, or are the result of their secretion from extraneural sites. A large number of peptidases exist in plasma that rapidly destroy the releasing hormones. They have become the focus of much interest, in conjunction with the development of long-acting synthetic hormone analogues. Because of these considerations, the biologic effects of hypothalamic hormones on the pituitary can be regarded as a single-pass phenomenon.

Mechanism of Action of Hypothalamic Hormones

The hypophysiotropic hormones bind to **high-affinity receptors** on anterior pituitary cells and affect both hormone secretion and cellular function through a variety of mechanisms. Pituitary hormone secretion is stimulated through the participation of one or more second-messenger systems; these include adenylate cyclase–cyclic AMP

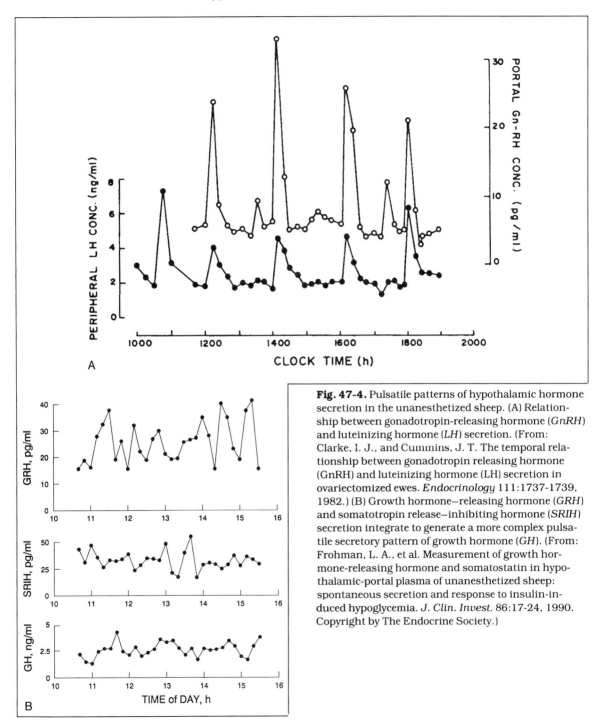

Fig. 47-4. Pulsatile patterns of hypothalamic hormone secretion in the unanesthetized sheep. (A) Relationship between gonadotropin-releasing hormone (*GnRH*) and luteinizing hormone (*LH*) secretion. (From: Clarke, I. J., and Cummins, J. T. The temporal relationship between gonadotropin releasing hormone (GnRH) and luteinizing hormone (LH) secretion in ovariectomized ewes. *Endocrinology* 111:1737-1739, 1982.) (B) Growth hormone–releasing hormone (*GRH*) and somatotropin release–inhibiting hormone (*SRIH*) secretion integrate to generate a more complex pulsatile secretory pattern of growth hormone (*GH*). (From: Frohman, L. A., et al. Measurement of growth hormone-releasing hormone and somatostatin in hypothalamic-portal plasma of unanesthetized sheep: spontaneous secretion and response to insulin-induced hypoglycemia. *J. Clin. Invest.* 86:17-24, 1990. Copyright by The Endocrine Society.)

(cAMP), Ca^{2+}—calmodulin, and phosphatidyli-nositol–protein kinase C. Individual **releasing hormones** appear to utilize different mechanisms to release the same pituitary hormone. In addition, the releasing hormones are important stimuli for enhancing the **gene expression** of the hormones, whose release they stimulate, and the effects (where studied) appear to be independent. In the **somatotroph,** this effect is mediated by cAMP. Releasing hormones also stimulate overall **RNA and protein synthesis,** enhance expression of selected **proto-oncogenes,** and stimulate [^{3}H]**thymidine incorporation** into DNA, leading to cellular hyperplasia and even tumor formation.

The mechanism of inhibiting hormone action is less clearly understood but, in part, relates to an inhibition of adenylate cyclase activity, enhanced phosphodiesterase activity, and impaired transmembrane Ca^{2+} transport. In addition, there appear to be other effects in the late stages of hormone secretion, possibly involving exocytosis.

The actions of all hypothalamic hormones are influenced by **target gland hormones,** including glucocorticoids, sex hormones, thyroxine, inhibin, and a number of growth factors. These hormones modify the number of hypothalamic hormone receptors and also exert effects at post-receptor sites.

INDIVIDUAL HYPOTHALAMIC–PITUITARY HORMONE SYSTEMS

Each of the hypophysiotropic–pituitary hormone systems consists of closed-loop feedback systems mediated by blood-borne signals, including those from target organs. Superimposed on these target organs are open-loop signals that are mostly of CNS origin and mediated by neurotransmitters; these signals reflect alterations in environment (temperature, light), stress (physical and psychic), and intrinsic rhythmicity (ranging from short-term or ultradian to seasonal periodicity). Consequently, both internal and external environmental factors exert major forces on the activity of these systems.

Hypothalamic-Pituitary-Adrenal Axis

CRH is a 41–amino acid peptide, whose release is affected by nearly all neurotransmitters. The effect of **stress,** the major external stimulus for CRH secretion, is mediated through **nicotinic cholinergic receptors. Serotonin** also stimulates CRH release, but cholinergic interneurons are involved because atropine can block this response. **Norepinephrine** and **GABA** inhibit cholinergic stimulation of CRH secretion. **Melatonin** and **enkephalins** also inhibit CRH release, and melatonin is likely responsible for the circadian pattern of ACTH secretion that is entrained to the light-dark cycle.

CRH stimulates the release of **ACTH,** which in turn stimulates glucocorticoid and mineralocorticoid secretion by the adrenal cortex. ACTH secretion is inhibited by glucocorticoids through both rapid (minutes) and delayed (hours) feedback mechanisms. Both types of feedback operate at the level of the pituitary and hypothalamus, by inhibiting the response to and secretion of CRH, respectively. Longer-term effects of glucocorticoids include the inhibition of ACTH and CRH gene expression. In addition, ACTH inhibits CRH release by a short-loop feedback effect.

There is considerable interaction between the feedback influence of glucocorticoids and neurotransmitters. For example, **phenytoin,** a membrane stabilizer, decreases CNS sensitivity to glucocorticoid feedback, thereby diminishing the ACTH response to **metyrapone** (an adrenal enzyme inhibitor that reduces circulating glucocorticoid levels), but also enhances pulsatile ACTH secretion, though does not affect the ACTH response to vasopressin or stress. **Vasopressin** stimulates ACTH release, both directly and through a CNS-mediated mechanism, and also potentiates the effects of CRH.

Hypothalamic-Pituitary-Gonadal Axis

GnRH is a decapeptide (10 amino acids) that stimulates the release of both LH and follicle-stimulating hormone (FSH). **Dopamine** and

serotonin are the primary neurotransmitters that affect GnRH secretion; the former is stimulatory and the latter is inhibitory. A central inhibitory role of CRH has also been proposed. The regulation of the reproductive hormone axis is complex, varying with age and sex. Prior to **midpuberty,** FSH secretion is greater and is more responsive to GnRH than is LH secretion. After this, the pattern is reversed, concomitant with the development of a greater CNS sensitivity to the inhibitory feedback effects of gonadal steroids. The sleep-related pulsatile secretion of LH and synchronization of LH and FSH pulses begin in the **late prepubertal period** (ages 7 to 9 years) and are responsible for the nocturnal rises in testosterone (boys) and estradiol (girls) that initiate the clinical manifestations of puberty. There is a simultaneous decrease in the sensitivity of the CNS to steroid feedback effects which culminates in the **cyclic preovulatory gonadotropin surge** that results in onset of **cyclic ovulation.** In men, the nocturnal gonadotropin surges are replaced by a somewhat random pulsatile pattern throughout the day; a more regular 90-minute pattern is observed in mature women.

During **menopause,** when ovarian follicles disappear and the secretion of the major ovarian hormones decreases, FSH and LH secretion is enhanced, though their inherent pulsatility is preserved. A similar increase in gonadotropins is observed in **men in the eighth and ninth decades,** along with decreasing testicular function. **Stress** causes a transitory increase in LH in men, followed by a prolonged decrease and accompanied by reduced testosterone levels. In women, stress is associated with hypothalamic **anovulation.**

Gonadal steroids regulate the tonic secretion of LH and FSH by a negative-feedback mechanism that operates at both the hypothalamic and pituitary levels, with testosterone more potent than estrogen. **Inhibin,** a peptide produced by the germinal epithelium, selectively inhibits FSH secretion, probably by impairing the effects of GnRH. The effects of estrogen on cyclic gonadotropin secretion, however, are opposite those on tonic secretion. The rising **estradiol** levels in the preovulatory period are actually responsible for the surge of LH that triggers **ovulation.** These stimulatory feedback effects also occur at both the hypothalamic and pituitary levels, though the pituitary appears to be a more important site of action.

Hypothalamic-Pituitary-Thyroid Axis

TRH is a tripeptide that is the primary regulator of thyroid-stimulating hormone (TSH), though the inhibitory effects of SRIF and dopamine are also of physiologic significance under certain conditions. TRH secretion is stimulated by norepinephrine and dopamine, and inhibited by serotonin. It is not known whether the release of TRH is pulsatile. TSH secretion enhances thyroid hormone secretion, and the feedback effects, primarily mediated by triiodothyronine, predominantly affect the pituitary through an inhibition of TRH action. Longer-term effects also take place in the hypothalamus through inhibition of TRH gene expression. A reduction in circulating thyroid hormone levels translates into an acute rise in TSH secretion, for which TRH is not required. However, prolonged and sustained TSH secretion does require TRH. TRH stimulates TSH biosynthesis and release, and, in the absence of TRH, TSH glycosylation is altered, resulting in a biologically less potent molecule.

Rapid changes in environmental conditions necessitating increased metabolic activity (for example, cold exposure) provoke a noradrenergic-mediated CNS stimulus to increased TRH secretion. This effect is readily demonstrable in infants, but not in adults, who possess other more important mechanisms of thermogenesis, such as shivering and the mobilization of free fatty acids through activation of the autonomic system.

Somatostatin inhibits TSH secretion but has a relatively minor role in its physiologic regulation under normal conditions, as it is much less effective in inhibiting TSH than GH secretion. Nevertheless, when SRIF secretion is increased (for instance, because of increased GH levels), TSH production can be suppressed. Dopaminergic inhibition of TSH secretion is most readily demonstrable when TSH secretion is elevated. The overall importance of dopaminergic control, however, remains to be determined.

Hypothalamic-Pituitary-Somatotroph-Liver Axis

The secretion of **GH** is regulated by a releasing (GRH) and an inhibiting (somatostatin; SRIF) hormone. GRH exists in both 40– and 44–amino acid forms, which have identical biologic activity. The **neurotransmitter regulation** of GRH is extensive, with stimulatory effects from acetylcholine, norepinephrine, and epinephrine (through the alpha-adrenergic receptor), as well as from dopamine, serotonin, and enkephalin. Beta-noradrenergic receptors are inhibitory, and the effect of GABA may be either inhibitory or stimulatory. SRIF secretion is stimulated by dopamine and inhibited by acetylcholine. Within the hypothalamus, GRH and SRIF have a reciprocal influence on each other's secretion. GH secretion is characterized by a pulsatile pattern, frequently beneath the limits of detectable measurement, superimposed on which are occasional surges of secretion that are related to the postabsorptive state and in association with deep sleep (electroencephalograph stages III to IV). GH secretion changes dramatically with age. Extremely high levels are seen neonatally, which decrease within a few weeks of life. Pulses that are indistinguishable from those of adults, except for increased frequency, are seen during puberty. After the fourth decade, GH secretion diminishes, and, after the sixth decade, significant pulsatile secretion is uncommon during either waking or sleeping periods.

Nutrient status profoundly affects GH secretion. Increases in amino acid levels, decreases in free fatty acid levels, and hypoglycemia all stimulate GH secretion; hyperglycemia inhibits GH release. GH secretion is stimulated by exercise and stress. **Thyroxine, glucocorticoids,** and

sex steroids all affect GH secretion, though in a complex manner. In general, both **androgens** and **estrogens** stimulate GH responses to many stimuli while **glucocorticoids** inhibit the same responses. However, the timing of hormone administration in relation to testing is of crucial importance. GH secretion is sexually dimorphic, in that women generally have higher baseline levels and lower pulse peaks than do men. Thyroid hormone deficiency tends to decrease GH secretion, though the effect in humans is much less pronounced than in rodents, and is related to species differences in the effects of thyroid hormone on the GH gene.

A **closed-loop feedback system** also exists for GH secretion, and is mediated by both GH itself and the major GH-dependent growth factor, insulin-like growth factor I (somatomedin C; IGF-I). IGF-I and GH both stimulate SRIF release and inhibit GRH release. In addition, IGF-I inhibits basal and GRH-stimulated GH secretion by the pituitary. Each of these effects occurs at hormone levels that are achieved under physiologic conditions. Although most circulating IGF-I originates from the liver, the feedback (and also growth-promoting) effects may also occur through a paracrine action, since IGF-I is produced by many tissues.

Hypothalamic-Lactotroph-Breast Axis

The secretion of **prolactin** is under an inhibitory CNS tone which is chiefly supplied by dopamine. There have been claims of other peptide prolactin–inhibiting factors, though none have received widespread acceptance. Both TRH and VIP have physiologic prolactin-releasing activity, though their relative importance in the overall control of prolactin is controversial. Serotonin stimulates prolactin release through effects on a prolactin-releasing factor, most likely VIP. Melatonin and histamine have stimulatory effects within the CNS, as do enkephalins and GABA. The effects of these two latter substances are related to their inhibitory role on the tuberoinfundibular dopaminergic system.

The secretion of prolactin is enhanced by **tactile stimulation of the breast** via sensory-receptors on the nipple and areola, fibers from which travel through the intercostal nerves to the spinal cord. Prolactin levels increase during pregnancy as a result of estrogen stimulation of the **lactotroph.** The rapid decline in estrogen and progesterone levels after parturition establishes an environment for lactation from the estrogen-primed breast. The suckling stimulation of prolactin secretion is mediated by VIP, though it is not responsible for the **milk letdown reflex,** which depends on **oxytocin.**

Prolactin is also a **stress-responsive hormone,** and increased levels are seen after surgical or emotional stress, exercise, and insulin-induced hypoglycemia. Thyroid hormone exerts an inhibitory effect on prolactin secretion, which is increased in thyroid hormone deficiency.

NEUROIMMUNOENDOCRINOLOGY

The immune system is one of the major intracellular communication systems. Considerable interaction between the immune and neuroendocrine systems has recently been recognized, as well as the regulatory effects they exert on one another. The regulatory signals of the immune system, the **lymphokines** and **monokines,** affect the hypothalamic pituitary system, and **neuroendocrine peptides** such as ACTH and beta-endorphin have been shown to be produced by immunologically competent cells (lymphocytes). Although an understanding of how these two systems interact is still in its infancy, certain important concepts are recognized.

It is now clear that certain monokines or lymphokines are capable of stimulating the hypothalamic–pituitary–adrenal axis. **Interleukin-1** (IL-1), a product of activated macrophages, stimulates CRH secretion from the hypothalamus, both after systemic administration and in vitro. Because IL-1 is released during infection and is believed to serve as an endogenous pyrogen, it is a likely candidate as the mediator (or one of the mediators) of the associated **hypothalamic—pi-**tuitary—adrenal stress response. How it interacts with the CNS is unclear; it may stimulate CRH directly or through interneurons. The mechanisms of a feedback system exist, as activation of this axis increases glucocorticoid production by the adrenals, which in turn suppresses many immunologic functions. The same compound has been shown to stimulate somatostatin release and decrease TSH release by an action imposed at the hypothalamic or pituitary level.

Cytokines other than IL-1 have also been reported to affect neuroendocrine function. **IL-6,** another cytokine produced by T cells, monocytes, and endothelial cells, can stimulate the release of GH, prolactin, and LH directly from the pituitary. In addition, a **thymic factor** (thymosin fraction 5) stimulates GnRH release from the hypothalamus and, after systemic administration, ACTH release. Although the role of these agents in neuroendocrine function under normal physiologic conditions remains to be clarified, the interrelationship of the neuroendocrine and immune systems is clearly important and may serve as a vehicle for regulating pathophysiologic responses to stimuli that provoke immunologic mechanisms.

SUMMARY

The CNS regulates the endocrine system through specific neuroanatomic pathways that involve the hypothalamus and a series of specific releasing and inhibiting hormones that control the secretion of the anterior pituitary. Each hormone is involved in a complex integrated feedback system that maintains a level of basal hormone secretion, and the feedback systems are influenced by signals derived from perturbations of the external environment, inherent biologic rhythms, and circulating nutrients. The most important of endocrine functions—growth, reproduction, and the response to stress—all require the integrative functions of this system. In addition, the hypothalamus has a major involvement in non-pituitary—mediated homeostatic mechanisms, including neurometabolic regulation, thermoregulation, autonomic nervous system function, and immunoregulation.

NATIONAL BOARD TYPE QUESTIONS

Select the one most correct answer for each of the following

1. The tonic influence of the CNS is stimulatory for all of the following pituitary hormones except:
   A. Corticotropin
   B. FSH
   C. GH
   D. Prolactin
   E. TSH

Feedback inhibition of the endocrine axis that includes the hormone in column A is mediated by which hormone in column B?

A	B
2. GnRH	A. Cortisol
3. TRH	B. IGF-I
4. CRH	C. Estradiol
5. GRH	D. Triiodothyronine

6. The secretion of all of the following are increased during stress except for:
   A. GH
   B. TSH
   C. IL-1
   D. ACTH
   E. Prolactin

7. The hypothalamic-pituitary portal system is important in the regulation of all of the following hormones except for:
   A. GH
   B. ACTH
   C. Vasopressin
   D. Prolactin
   E. TSH

8. IL-1 has been implicated as a mediator of the stress response for which of the following hypothalamic hormones?
   A. TRH
   B. CRH
   C. SRIF
   D. GnRH
   E. GRH

ANNOTATED ANSWERS

1. D. Loss of hypothalamic control of prolactin secretion of the anterior pituitary results in increased secretion. Thus, transection of the pituitary stalk produces hyperprolactinemia together with diminished secretion of all other pituitary hormones.

2. C. Estradiol is a potent inhibitor of GnRH secretion, and this action forms the basis for the ovulation-inhibiting effects of oral contraceptives.

3. D. Triiodothyronine has a strong inhibitory effect on TRH secretion as well as on the expression of the gene for the TRH precursor.

4. A. Cortisol, representing the glucocorticoid class of hormones, is a potent inhibitor of CRH secretion. Therapy with cortisone blocks CRH secretion and action on the pituitary.

5. B. GRH secretion is inhibited by GH and also by IGF-I, which may have direct effects and effects mediated through somatostatin secretion.

6. B. TSH secretion is not stimulated during stress. It is frequently inhibited, probably because of increased glucocorticoid secretion.

7. C. Vasopressin is synthesized in the paraventricular nucleus and transported to the posterior pituitary via neural fibers in the pituitary stalk. The other hormones are regulated by hypothalamic hormones that are secreted into portal vessels.

8. B. IL-1 has been shown to stimulate the release of CRH and is believed to mediate the effects of several types of bacterial endotoxins.

BIBLIOGRAPHY

Bateman, A., Singh A., Kral, T., and Solomon, S. The immune-hypothalamic-pituitary-adrenal axis. *Endocrinol. Rev.* 10:92–100, 1989.

Frohman, L. A., and Krieger, D. T. Neuroendocrine physiology and disease. In: Felig, P. A., Baxter, J., Broadus, A. E., and Frohman, L. A, eds. *Endocri-nology and Metabolism.* New York: McGraw-Hill, 1987, Pp. 185-246.

Krieger, D. T., and Hughes J. C., eds. *Neuroendocrinology.* Sunderland, MA: Sinauer Associates, 1980.

Reichlin, S. Neuroendocrinology. In: Wilson, G. D. and Foster, D. W., eds. *Textbook of Endocrinology,* Philadelphia: W. B. Saunders, 1992, Pp. 135–220.

48 The Anterior Pituitary

LAWRENCE A. FROHMAN

Objectives

After reading this chapter, you should be able to:

Describe the anatomy and blood supply of the pituitary gland

List the cell types of the anterior pituitary and the hormones they secrete

Cite the families of pituitary hormones and describe their characteristics

List the individual pituitary hormones, their general chemical characteristics, and their target glands

Discuss the major actions of each of the hormones of the anterior pituitary

Describe the types of tests used for evaluating the secretion of the individual pituitary hormones

ANATOMY

The pituitary gland is located in the **sella turcica,** a saddle-shaped cavity at the base of the skull that forms an integral part of the sphenoid bone. It is bounded anteriorly by the midline **tuberculum sellae** and the **anterior clinoid processes,** which project posteriorly from the sphenoid wings. Posteriorly, the sellar boundaries consist of the **dorsum sellae,** the lateral portions of which form the **posterior clinoid processes.** The lateral boundaries of the pituitary consist of the nonosseous medial wall of the cavernous sinus, through which travel the internal carotid artery together with the third, fourth, and sixth cranial nerves. The roof of the pituitary consists of a thickened reflection of the dura mater, the **diaphragma sellae,** which is attached to the clinoid processes. The **pituitary stalk** and its portal blood vessels pass through a foramen in this membrane. The outer layer of the dura mater extends into the pituitary sella to form

its **periosteum,** thereby making the pituitary extradural and not in contact with the cerebrospinal fluid. The sella shape varies from ovoid to sphenoid and it weighs about 0.5 to 0.7 gms. About two-thirds of the total weight comprises the anterior lobe, and the rest, the posterior lobe. The pituitary can increase up to twice its normal size during **pregnancy,** primarily because of hyperplasia of one cell type, the **lactotroph.**

The **blood supply** of the pituitary originates from the **internal carotid artery,** through interconnecting branches of the circle of Willis, and the three hypophysial arteries. The branches supplying the stalk and the posterior pituitary are derived directly from these arteries, whereas the anterior pituitary has no direct arterial supply. Rather, its entire vascular supply originates in a portal system that first bathes the outer layer of the median eminence, then coalesces into long and short portal veins that travel primarily on the anterior surface of the pituitary stalk, and finally

681

terminates in a dense plexus of sinusoidal capillaries within the substance of the anterior pituitary. **Venous drainage** from the anterior pituitary enters the posterior pituitary and from there to the cavernous sinus or the petrosal sinus. Some retrograde flow up the portal veins has been demonstrated, though it is currently believed that such flow does not reach the median eminence and thus does *not* provide a mechanism for the direct feedback of pituitary hormones on the hypothalamus, as had been proposed.

The **nerve supply** of the anterior pituitary is sparse and almost entirely consists of postganglionic fibers of the sympathetic nervous system that accompany and terminate on arteriolar vessels. Their exact function is unknown, though they may regulate pituitary blood flow, and, in this manner, regulate the extent of exposure of the anterior pituitary to hypothalamic hormones. They may also play a pathophysiologic role in the development of pituitary ischemia and necrosis that occurs during parturition, when there is severe blood loss.

DEVELOPMENTAL ASPECTS

The anterior pituitary (adenohypophysis, *pars anterior*) is derived from an ectodermal evagination of the oropharynx, **Rathke's pouch,** that fuses with an outpouching of the third ventricle, early in fetal development. The latter eventually develops into the **neurohypophysis** (*pars nervosa*), or posterior lobe. The posterior portion of Rathke's pouch is less well developed, particularly in humans, and is called the **pars intermedia,** or **intermediate lobe.** Recent evidence has indicated unique cell types and hormone precursor processing in this pituitary region. Remnants of Rathke's pouch may persist within the pituitary as a cleft or small colloid-filled cysts, or along the tract from which it arose. Cells in this region may assume the characteristics of mature anterior pituitary cells, secrete hormones, and even undergo tumor formation with hormone hypersecretion. **Pituitary hormones** exist as early as four to five weeks of gestation and can respond

prenatally to some hypothalamic-releasing hormones. However, the full feedback regulatory system is not believed to be fully established until postnatal life.

Cells of the **anterior pituitary** develop the capacity to secrete **adrenocorticotropic hormone** (ACTH), **growth hormone** (GH), **prolactin, thyroid-stimulating hormone** (TSH), **luteinizing hormone** (LH), and **follicle-stimulating hormone** (FSH). Those in the **posterior pituitary** secrete **vasopressin** and **oxytocin,** while cells of the **intermediate lobe** secrete **ACTH, melanocyte-stimulating hormone** (MSH), **beta-lipotropin** (β-LPH), and **endorphins.** Within the anterior pituitary, cells are arranged in a sinusoidal or rosette formation, and all cell types are intermingled. More importantly, however, is the association seen at the **microscopic level** between somatotrophs and thyrotrophs and between gonadotrophs and lactotrophs. There is increasing evidence for intercellular communication between the various cell types in the anterior pituitary, which may have an important physiologic function.

PITUITARY CELL TYPES

The cells of the anterior pituitary can be divided into several subtypes, based on the class of hormones they secrete. They include the **somatomammotropic group,** responsible for the secretion of GH and prolactin; the **glycoprotein-secreting cells,** which secrete TSH, LH, and FSH; the **corticotroph cells,** which secrete ACTH; and the other hormones derived from the **proopiomelanocortin** (POMC) **molecule,** and the **folliculostellate cells,** recently identified to secrete an endothelial-cell growth factor. There are additional cells types, but their function is not known. The various cell types were originally identified by their **staining characteristics,** based on pH-dependent histochemical stains (acidophils, basophils, and chromophobes). **Morphologic differences** in granule size, when viewed under the electron microscope, also exist among the various cell types. However, precise identification of the cell types is currently determined through the use of specific immunohistochemical stains that indicate the presence of specific hormones.

Somatomammotrophic Family

The somatomammotrophic family includes three cell types: somatotrophs, lactotrophs, and somatomammotrophs. The **somatotrophs** were originally identified as acidophilic cells and their function as GH-secreting cells was originally based on their predominance in tumors associated with acromegaly and gigantism. They are the most common cell type and are located primarily in the lateral wings of the anterior pituitary. The **lactotrophs** are also acidophilic, though they stain less intensely than do somatotrophs. They are slightly less numerous than are somatotrophs, secrete prolactin, and are situated more peripherally. The lactotrophs increase greatly in number during pregnancy, under the influence of estrogen, and may constitute up to 75 percent of the mass of the anterior pituitary. Lactotrophic tumors are associated with increased prolactin secretion. The **somatomammotroph,** which constitutes about 1 to 2 percent of the anterior pituitary cells, can secrete both GH and prolactin, and is believed to serve as a precursor cell for both somatotrophs and lactotrophs. It is a more primitive cell type, stores less hormone, and retains the capacity to undergo replication in tissue culture, at least in rodents. Tumors derived from this cell type exhibit hypersecretion of both GH and prolactin.

Glycoprotein Hormone–Secreting Family

The glycoprotein hormone–secreting family consists of two cell types: thyrotrophs and gonadotrophs. The **thyrotrophs** are intensely basophilic, are polyhedral, and tend to be concentrated in the anterior portion of the pituitary, near the midline. They secrete TSH and normally represent about 5 to 6 percent of the anterior lobe cells, though marked increases in size and number occur after thyroidectomy. A single gonadotroph cell is responsible for both LH and FSH secretion. Although gonadotrophs constitute only 3 to 4 percent of the anterior lobe cells, they increase in number after castration. During pregnancy, they decrease as a consequence of the high levels of circulating steroids secreted by the corpus luteum in response to chorionic gonadotropin production and by the placenta during the second and third trimesters.

Corticotroph Family

There are two types of cells in the corticotroph family, which are basophilic or chromophobic. One type is located in the **medial region** of the anterior lobe and the other in the **junctional region** between the anterior and posterior lobes. Anterior lobe corticotrophs are sparsely granulated, in contrast to those in the junctional area that have large electron-dense granules. Both types produce peptides that are derived from the common POMC precursor. However, the specific peptides found in each of the cell types vary greatly, based on differences in the processing enzymes present. **ACTH,** the most physiologically important member of this hormone family, is secreted primarily by the cells in the anterior lobe. In states of **glucocorticoid insufficiency,** anterior lobe corticotrophs proliferate, while those in the junctional area do not.

Folliculostellate Cells

A small number of cells exhibit a stellate shape and send processes into the perivascular spaces. A peptide hormone, **endothelial cell growth factor,** with mitogenic activity specific for endothelial cells, has recently been isolated from these cells, and they must therefore be added to the list of anterior pituitary hormone–secreting cells.

ANTERIOR PITUITARY HORMONES

There are three separate families of anterior pituitary hormones: corticotropin-lipotropin, glycoprotein, and somatomammotropin. A comparison of their chemical characteristics is given in Table 48-1. Additional discussion of the neuroendocrine regulation of each of the hormones is provided in Chapter 47.

Corticotropin-Related Peptides

ACTH and its related family of peptides are derived from a **single precursor molecule,** POMC, which has a molecular weight of about 29,000 and is glycosylated. A stepwise processing of this molecule occurs, with cleavage into three major fragments (Fig. 48-1). The amino-terminal end has an uncertain biologic function, the midportion fragment contains ACTH, and the carboxyl-terminal fragment contains β-LPH. ACTH is not further processed in the anterior lobe corticotrophs but is further cleaved to α-MSH and corticotropin-like intermediate lobe peptide in the junctional zone cells (and also in the central nervous system). β-LPH is also further processed in these same cells to other endorphin-related peptides. Although the structures of β-MSH and met-enkephalin are contained within that of β-LPH, each of these compounds has a different mechanism of synthesis, with that for met-enkephalin originating from a separate precursor.

Adrenocorticotropin

ACTH is a 39–amino acid, single-chain peptide with all of its biologic activity contained in the first 24 residues. Its primary site of action is the **adrenal cortex,** where it stimulates the secretion of glucocorticoids, mineralocorticoids, and androgenic steroids. ACTH binds to specific receptors on the adrenocortical cell plasma membranes and exerts its action by cyclic AMP–mediated effects that enhance cholesterol side-chain cleavage, resulting in its conversion to **pregnenolone.** In addition, ACTH has true trophic effects on the adrenal cortex, stimulating DNA and protein synthesis and leading to **adrenocortical hyperplasia.** ACTH also possesses **extraadrenal effects,** including stimulation of lipolysis in adipose cells,

insulin release from pancreatic beta-cells, release of GH from somatotrophs, and enhancement of amino acid and glucose transport into muscle. However, with the exception of patients who have ACTH-secreting tumors, it is unlikely that circulating ACTH levels are ever achieved which are sufficient to produce these effects. ACTH has **pigmenting action** which, though less potent than that of MSH, is of likely physiologic importance in humans, particularly in states of ACTH hypersecretion.

ACTH is measured primarily by **radioimmunoassay** (RIA), and levels in plasma vary from less than 1 pg/ml (less than 0.2 pM) to 80 pg/ml (18 pM). During stress, ACTH levels may reach 200 to 300 pg/ml (45 to 67 pM). ACTH is rapidly cleared from the circulation, with a **plasma half-life** of 3 to 9 minutes. The **daily secretion rate** is approximately 25 µg/day. ACTH secretion is **episodic,** but a diurnal rhythm is also observed, with highest levels in the early morning and lowest levels in the late evening (Fig. 48-2). Pulses of ACTH secretion immediately precede those of cortisol, though the pulsatility of cortisol is more difficult to detect because of its more prolonged half-life in the circulation.

Beta-Lipotropin and Beta-Endorphin

β-LPH and beta-endorphin are released from the pituitary in equimolar ratios to those of ACTH, in response to nearly all types of stimuli, as would be expected based on their presence in the same precursor molecule. **Plasma levels** of these two peptides, however, do not parallel those of ACTH because of their more prolonged half-life in the circulation. β-LPH is cleared primarily by the kidneys, and levels rise disproportionate to those of ACTH during renal failure. The preservation of normal ACTH levels under these circumstances indicates that β-LPH exerts little feedback effect on the pituitary. Although biologic effects of β-LPH (pigmentation) and beta-endorphin (analgesia) are well established, the levels of these peptides in the peripheral circulation are too low for these effects to occur in man.

Table 48-1. Anterior Pituitary Hormones

Class	Members	Molecular Weight	Amino Acids	Carbohydrate	Other Features
Corticotropin-lipotropin	ACTH	4,500	39		All members derived from a single precursor
	α-MSH	1,800	13		N terminal has 13 amino acids of ACTH; in humans, found only in fetal life and in tumors
	β-LPH	11,200	91		
	β-Endorphin	4,000	31		C-terminal (amino acids 61 to 91) portion of β-LPH
Glycoprotein	LH	29,000	Alpha subunit, 89; beta subunit, 115	1% sialic acid	All have two subunits; the alpha subunit is identical or nearly identical and the beta subunit confers biologic specificity
	FSH	29,000	Alpha subunit, 89; beta subunit; 115	5% sialic acid	
	TSH	29,000	Alpha subunit, 89; beta subunit, 112	1% sialic acid	
	Chorionic gonadotropin[a]	46,000	Alpha subunit, 92; beta subunit, 139	12% sialic acid	
Somatomammotropin	Growth hormone	21,800	191		All single-chain proteins with 2 or 3 disulfide bridges
	Prolactin	22,500	198		
	Placental lactogen	21,800	191		

ACTH = adrenocorticotropic hormone; MSH = melanocyte-stimulating hormone; LPH = lipotropin; LH = luteinizing hormone; FSH = follicle-stimulating hormone; TSH = thyroid-stimulating hormone.

[a]Of placental origin and included for comparison purposes.

[b]Carbohydrate-containing forms of prolactin have recently been identified.

Adapted from: Frohman, L. A. Diseases of the anterior pituitary. In: Felig, P., et al., eds. *Endocrinology and Metabolism*, 2nd ed. New York: McGraw-Hill, 1987.

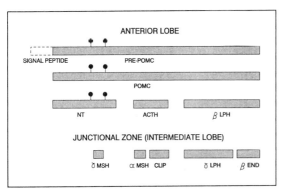

Fig 48-1. The posttranslational processing of proopiomelanocortin *(POMC)*, the precursor of adrenocorticotropic hormone *(ACTH)*. After removal of the signal peptide from the POMC precursor *(pre-POMC)*, the glycosylated (indicated by *closed circles*) POMC is further processed in the anterior lobe to an amino-terminal peptide *(NT)*, ACTH, and beta-lipotropin *(β-LPH)*. In the junctional zone, each of these peptides is further cleaved; gamma-melanocyte-stimulating hormone (γ-MSH) is derived from NT; α-MSH and corticotropin-like intermediate lobe peptide *(CLIP)* are derived from ACTH; and γ-LPH and beta-endorphin *(β-END)* are derived from β-LPH.

Glycoprotein Hormones

The pituitary glycoprotein hormones consist of TSH, LH, and FSH. In addition, there is a placental glycoprotein hormone, **chorionic gonadotropin,** that is structurally similar to LH and exhibits many of its biologic characteristics. The glycoprotein hormones are composed of **two subunits,** alpha and beta, each consisting of a protein core with branched carbohydrate side chains that constitute from 15 to 30 percent of the hormones' molecular mass. The carbohydrates are essential for the hormones' biologic activity and also markedly influence their stability in plasma. Within a particular species, the alpha subunits of the pituitary glycoprotein hormones are identical but the beta subunits vary, providing the biologic specificity of action. Even the beta subunits exhibit some homology, however. Up to 50 percent of the amino acids may be identical or the product of a single DNA nucleotide change, and it is therefore not surprising that an overlap

in biologic function can be demonstrated when there are extremely high levels in the circulation, and that the hormones are not species specific. The subunits by themselves do not possess intrinsic biologic activity. The beta subunit of chorionic gonadotropin, in particular, is identical to that of LH, with the exception of a 30-amino acid extension at the carboxyl terminal. Besides secreting the intact glycoprotein hormones, each of the cell types also secretes the free alpha subunit. In general, the free subunit has an altered carbohydrate content and is unable to combine with the beta subunit, in contrast to the alpha subunit that is prepared by chemically dissociating the intact hormone. The control of alpha-subunit glycosylation, which occurs postranslationally, depends on the presence of the appropriate hypothalamic-releasing hormone (for example, in states of TRH deficiency, TSH is synthesized with an altered carbohydrate content). The individual subunits are under separate genetic control, and the rate-limiting step in hormone biosynthesis appears to depend on the beta subunit.

Thyroid-Stimulating Hormone

The actions of DTSH on the thyroid cells parallel those of ACTH on the adrenal cortex. TSH binding leads to the **activation of adenylate cyclase,** which mediates a number of processes, including iodide transport and binding and thyroglobulin synthesis and proteolysis. TSH also **stimulates RNA** and **protein synthesis,** thus increasing thyroid size and vascularity. TSH is measured by RIA, and ultrasensitive assays are now readily available that can distinguish between low, normal, and high levels. **Normal TSH concentrations** in plasma range from 0.5 to 6 μU/ml. In prolonged and severe thyroid hormone deficiency, TSH levels may exceed 100 μU/ml; in conditions of thyroid hormone excess, levels are undetectable (less than 0.1 to 0.15 μU/ml). The half-life of TSH in circulation is approximately 75 minutes and the daily secretion rate is 100 to 200 mU. Secretion rates of up to ten times the normal rate may occur in hypothyroidism.

Besides its neuroendocrine control, TSH secre-

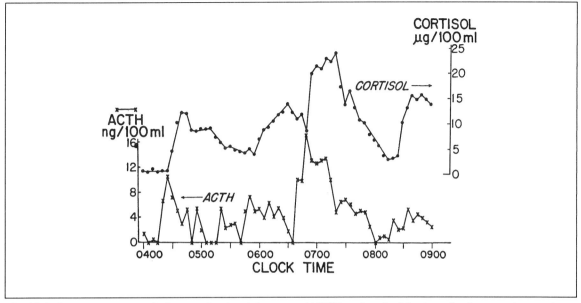

Fig. 48-2. Secretory pattern of adrenocorticotropic hormone *(ACTH)* and cortisol during a five-hour, early morning period in a normal man. The more pronounced pulsatility of ACTH is, to a large extent, a consequence of its more rapid disappearance from plasma than is cortisol. (From: Gallagher, T. F., et al., ACTH and cortisol secretory pattern in man. *J. Clin. Endocrinol. Metab.* 36:1058, 1973. Copyright by The Endocrine Society.)

tion is also affected by thyroid hormone, glucocorticoids, estrogen, and GH. **Glucocorticoids** suppress both basal and stimulated TSH secretion and **estrogens** enhance the TSH response to TRH. This is reflected in the higher TSH responses in women as compared to men, and during the late follicular phase of the menstrual cycle in women, when estradiol levels are increased. It is not known whether this phenomenon contributes to the greater incidence of goiter in women.

Luteinizing Hormone and Follicle-Stimulating Hormone

LH and FSH are the two pituitary hormones that regulate **gonadal function. FSH** stimulates ovarian follicular growth, testicular growth, and spermatogenesis. **LH** promotes ovulation and follicular luteinization, stimulates testicular Leydig cell function, and enhances the production of sex steroids from both the ovary and testis.

FSH promotes ovarian growth and maturation of the primordial follicle, while LH stimulates the production estrogens by the thecal cells and progesterone by the corpus luteum, by enhancing the conversion of cholesterol to pregnenolone. In the testis, FSH acts on Sertoli cells, where, together with testosterone, the production of an androgen-binding protein is stimulated. The target cell of LH in the testis is the Leydig cell, where testosterone production is stimulated. The androgen-binding protein transports testosterone in high concentrations into the tubular cells to stimulate spermatogenesis. Details of the actions of LH and FSH are provided in Chapters 54 and 55.

Gonadotropin concentrations are primarily determined by RIA, in which there is some degree of cross-reactivity between the hormone and its subunits, though this is not of practical importance. More significant, however, is the cross-reactivity between LH and chorionic gonadotropin. Although most chorionic gonadotropin RIAs are specific for the unique portion of the molecule, the reverse is not generally true.

Plasma levels of LH and FSH vary with the **menstrual cycle.** Plasma FSH levels rise slightly and then decline during the early follicular phase, at which time LH levels are stable or slightly increasing. Increasing estrogen stimulation at **midcycle** causes an abrupt increase in the LH level, which together with an increase in FSH levels, triggers **ovulation.** Both hormone levels decline during the **luteal phase.** In men, FSH and LH levels are similar to those in women during the follicular phase. No identifiable cyclic function exists in either the hypothalamus or testis, and the primary feedback effect on LH secretion is mediated through testosterone, acting at both the hypothalamic and pituitary levels. The negative feedback on FSH secretion in both sexes occurs primarily through gonadally produced **inhibin,** acting at the pituitary. The hormone levels increase in response to age-associated decreases in gonadal function in both sexes. This occurs at menopause in women; in men, it is generally seen after age 70. The half-life of LH in plasma is approximately 50 minutes, whereas that of FSH is three to four hours and that of chorionic gonadotropin even longer. The difference is attributed to the varying content of sialic acid in the three hormones, since desialation markedly shortens the half-life of the hormone in plasma.

The secretion of LH and FSH is also influenced by **prolactin,** which exerts inhibitory effects directly, or is mediated through dopamine or other neurotransmitters in the central nervous system. The LH response to gonadotropin-releasing hormone (GnRH) is also inhibited by glucocorticoids.

Somatomammotropic Hormones

The three members of this family are **GH, prolactin,** and a placental hormone, **chorionic somatomammotropin,** also known as **placental lactogen** (PL). Placental lactogen is structurally very similar to GH (83 percent homology), but also displays considerable homology with prolactin and is believed to be the phylogenic ancestor of the two pituitary hormones. In contrast, GH and prolactin only exhibit 16 percent homology. The GH and PL family actually consists of five separate genes. Of the three GH genes, only one is expressed in the pituitary; a variant gene is expressed in the placenta and exhibits biologic activity. Only one of the two PL genes is expressed. Each hormone is a single-chain peptide with a molecular weight of between 21,000 and 22,000, and one, prolactin, exists in a slightly higher molecular weight, glycosylated form, with a reportedly reduced biologic activity. Despite these differences, each hormone has intrinsic growth-promoting and lactogenic properties. GH and prolactin both exist as larger molecules in the circulation. Some of the heterogeneity in size is attributable to noncovalent homodimerization through disulfide interchange, and some is the consequence of association with binding proteins. There are a number of GH variants, including proteolytically modified forms, electrophoretic variants, and a smaller 20-kDa molecule that is lacking amino acids 32 to 46, the result of differential messenger RNA splicing. This smaller GH variant binds to the GH receptor and exhibits effects that are comparable to those of the regular 22-kDa GH.

Growth Hormone

GH has a major role in **promoting linear growth** and in the regulation of **metabolism.** Administration of GH to patients deficient in the hormone produces a positive nitrogen balance, decreased urea production and body fat stores, increased muscle mass, and enhanced carbohydrate utilization. GH produces **biphasic effects** on circulating levels of glucose, amino acids, and fatty acids, with an initial decrease followed by a return to normal levels, or even an increase. At the **cellular level,** GH stimulates the uptake and incorporation of amino acids into protein, enhances RNA synthesis, accelerates glucose uptake, and antagonizes the lipolytic effect of catecholamines. These acute effects disappear within three to four hours of continuous exposure and are replaced by a series of **insulin antagonistic effects,** including enhanced triglyceride lipolysis, increased sensitivity to catecholamine-mediated lipolysis, and impairment of glucose uptake and utilization as a consequence of inhibition of pyruvate decarboxylation. These effects form the basis for the **diabetogenic effects** of GH. In the pancreatic beta cell, GH exhibits multiphasic effects on insulin secretion. An acute direct stimulatory effect is followed by a secondary inhibitory effect and then a prolonged stimulation of insulin secretion, secondary to the impairment of carbohydrate metabolism. It is this insulin antagonistic effect that is of greatest importance in the development of diabetes in states of GH hypersecretion.

GH also stimulates the production of several tissue growth factors, particularly **insulin-like growth factor I** (IGF-I), also known as **somatomedin C.** IGF-I is a peptide with a molecular weight of 7,500 that structurally resembles proinsulin and binds to the insulin receptor. Separate IGF-I receptors exit in many tissues, including cartilage cells, where the peptide stimulates sulfate and amino acid uptake and their incorporation into proteoglycan. IGF-I can thus be considered a mediator of many GH effects and acts as an **autocrine** or **paracrine growth factor.** It is also released into the circulation, primarily from the liver and kidney, and may act as a **true hormone.** Many of the actions originally attributed to GH are in fact caused by IGF-I. Furthermore, IGF-I and GH act **synergistically** in cell differentiation and cell growth. GH serves to commit a precursor cell (such as a fibroblast or prechondrocyte) to a specific pathway of differentiation and IGF-I enhances its growth and replication. In addition, GH enhances the activity of another growth factor, **epidermal growth factor,** by stimulating production of its receptor. Overall, the capacity of tissues such as the liver, heart, and kidney to undergo hypertrophy depends on GH.

GH in the circulation is measured by RIA. Because GH is secreted in an episodic manner, with values ranging from undetectable levels (less than 1 ng/ml or less than 45 pM) to levels of from 30 to 50 ng/ml (1.4 to 2.3 nM), mean values can only be determined by integrating a series of repeated measurements. In adults, the mean levels are generally under 2 ng/ml (less than 100 pM). About 70 percent of **GH secretion** occurs during the night in association with deep sleep (Fig. 48-3). Values in women during the reproductive years are slightly greater, both basally and in response to stimuli. The **integrated secretory rate** during the adolescent growth spurt is greater and after the sixth decade is considerably lower. The largest secretory bursts of GH take place at night, in association with deep sleep (electroencephalograph stages III to IV). The episodic pattern of GH secretion is also important in modulating its metabolic actions, as described, since the nearly complete absence of GH effects during trough periods is important in maintaining the anabolic as compared to the insulin antagonistic actions of the hormone on peripheral tissues. In female rodents, the less pulsatile pattern of GH secretion is also responsible for the sexually dimorphic pattern of certain hepatic enzymes and at least one epidermal growth factor (EGF) receptor. The extent of this phenomenon in humans is unknown. GH is cleared from plasma primarily by the liver but also by the kidney. The **plasma half-life** of GH is 20 minutes and the 24-hour secretion rate in adults ranges from 300 to 400 $\mu g/m^2$.

Prolactin

Prolactin chiefly stimulates the **synthesis of milk constituents** (for example, casein, lactalbumin, lipids, and carbohydrates). There are prolactin receptors on the alveolar surface of mammary cells and also in the liver and kidney. Breast development during puberty is not determined by prolactin secretion, and a pathologic increase in prolactin secretion does not, by itself, increase breast size. During **pregnancy,** however, prolactin, together with estrogen, progesterone, and PL, promote breast enlargement and initiate milk production. The abrupt decrease in placental estrogen and progesterone immediately following parturition initiates lactation. Continued prolactin secretion is necessary for maintaining **lactation** in the postpartum period, and the return of prolactin levels to normal is delayed

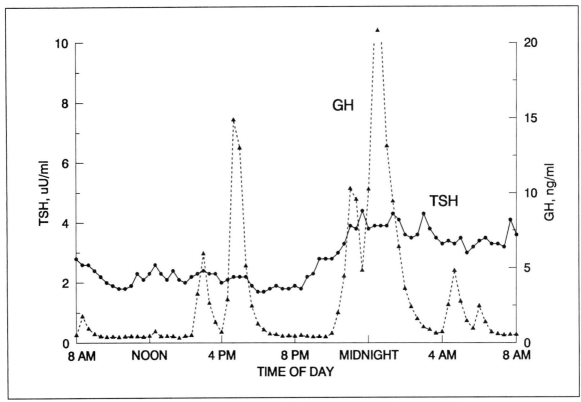

Fig. 48-3. Pulsatile pattern of growth hormone *(GH)* and thyroid-stimulating hormone *(TSH)* secretion in a normal man. The largest pulses of GH are seen in association with electroencephalograph stages III to IV sleep (deep sleep). TSH levels, while increased during the night, exhibit a different pattern.

in women who nurse for prolonged periods. Although prolactin displays other effects in lower species, including behavioral (maternal activity) and water metabolism, these activities have not been confirmed in humans.

Prolactin is measured by RIA and **normal levels** are lower than 15 ng/ml (under 0.7 nM) in men and less than 20 ng/ml (under 0.9 nM) in women. Prolactin levels do not change significantly during the menstrual cycle, though levels in pregnancy increase to 200 ng/ml (9 nM) at term. The **half-life** of prolactin in plasma is approximately 50 minutes, with removal primarily accomplished by the liver and kidney.

CLINICAL EVALUATION OF ANTERIOR PITUITARY HORMONES

The evaluation of anterior pituitary hormones involves the assessment of spontaneous secretion, the pituitary secretory reserve using one or more stimuli, and their ability to respond to suppressive agents. Not all tests are appropriate for each hormone and the clinical setting in which hormone dysfunction is observed or suspected determines which type of testing is indicated. The general endocrine physiologic principle followed is that **stimulation tests** are performed when hormone levels are in the basal state and **suppression tests** are performed when hormone levels are elevated or borderline elevated. In general, measurements of hormones in plasma have replaced urine measurements, thereby simplifying and shortening the sample collection procedures. However, urine collections still have considerable merit in evaluating overall secretory activity (during a 24-hour period).

ACTH measurements are generally performed in conjunction with those of cortisol. In cases of suspected **primary adrenal insufficiency,** a single plasma ACTH level is often diagnostic, as it will be elevated if the disease originates in the adrenal gland (primary) and normal or low in hypothalamic pituitary disease (secondary). Differentiation of the latter two conditions can frequently be accomplished by testing with corticotropin-releasing hormone (CRH) (which stimulates the pituitary directly) or with **insulin hypoglycemia** (which requires the participation of the combined hypothalamic–pituitary unit), or both methods can be used. A positive response to CRH, coupled with no response to insulin, points to the central nervous system as the source of the problem. A single **injection of ACTH** along with the measurement of plasma cortisol levels is also a simple screening procedure for the adequacy of endogenous ACTH secretion, as the unstimulated adrenal cortex undergoes atrophic changes and cannot respond normally. The axis can also be tested with **metyrapone,** an inhibitor of adrenal 11-hydroxylase. The drug inhibits cortisol biosynthesis, thereby reducing the feedback effects on CRH and ACTH and leading to a rise in the ACTH level. However, this type of stimulus differs from that of insulin hypoglycemia, which evaluates stress-induced ACTH release, a pathway largely independent of feedback regulation.

The best assessment of **suspected increased ACTH production** is by measuring the 24-hour urinary cortisol excretion. As a second step, the suppressibility of ACTH and cortisol by dexamethasone (a synthetic glucocorticoid) is used to determine the integrity of feedback inhibition.

TSH levels are the most useful indicator of TSH secretory capacity when evaluating a patient with **hypothyroidism.** If the disease originates in the **thyroid** (primary), the TSH levels will be elevated. If it is of **pituitary** (secondary) or **hypothalamic** (tertiary) **origin,** the level will be low or (inappropriately) normal. Differentiation between hypothalamic and pituitary disease can generally be accomplished by TRH stimulation, which customarily produces an enhanced or prolonged increase in TSH levels in hypothalamic disease and a reduced or absent response in pituitary disease. On occasion, however, the TSH response may not be exaggerated in hypothalamic hypothyroidism. A suspected **autonomous origin** to the elevated TSH levels (caused by a TSH-secreting tumor) is assessed by suppression with thyroxine or triiodothyronine, using the concept of normal feedback inhibition. If suppression occurs only at abnormally high thyroid hormone levels, this indicates thyroid hormone resistance, which can arise from a genetic or drug-induced condition.

LH and **FSH secretions** are assessed by measuring basal levels in conjunction with gonadal steroid levels and in response to GnRH challenge. If there is **generalized gonadal failure,** testosterone or estradiol levels are reduced and LH and FSH levels are elevated. **Selective gonadal failure** (such as Sertoli cell dysfunction with impaired spermatogenesis) will result in a disproportionate increase in FSH levels with only minimal changes in the LH levels. If LH and FSH levels are not elevated in the presence of decreased levels of gonadal steroids, the disturbance is either pituitary or hypothalamic in origin. GnRH stimulation may be useful in distinguishing between the two, but, in contrast to TRH, a single injection of GnRH may not stimulate the LH and FSH release from a pituitary gland that has not been exposed to endogenous GnRH for a long time. Repeated stimulation at frequent intervals (every three hours for five to seven days) is often necessary to prime the gonadotrophs to respond.

GH secretion is tested if there is a suspected deficiency in children with growth failure or in patients with destructive lesions (usually tumors) of the pituitary gland. It is also assessed in patients with suspected GH-secreting tumors causing **gigantism** or **acromegaly,** a disorder characterized by the enlargement of hands and feet, coarsening of facial features, soft tissue overgrowth, and metabolic disturbances. These abnormalities are caused by GH overproduction. Adequacy of GH secretion is determined by one of several methods: repeated measurement of plasma hormone levels at frequent (every 20 minutes) intervals to demonstrate pulsatility and determine overall 24-hour integrated secretion values, frequent nighttime sampling to demonstrate sleep-associated GH secretion, or stimulation with one or more agents such as growth hormone–releasing hormone (acting directly at the pituitary), insulin hypoglycemia, arginine infusion, or the alpha-adrenergic agonist clonidine, all of which act within the central nervous system.

GH overproduction is frequently revealed simply by measuring the basal hormone level in plasma, if sufficiently high. The inability of GH to be suppressed by hyperglycemia (after ingestion of glucose) indicates autonomous secretion of the hormone, typically seen in GH-secreting tumors.

Prolactin secretion is assessed by the measurement of hormone levels under basal conditions. Reduced hormone levels are rarely of clinical diagnostic importance, because they appear only in association with panhypopituitarism. **Hyperprolactinemia,** in contrast, is an important problem and is documented by repeated measurement of the hormone on separate occasions (because prolactin is a stress-responsive hormone and may be transiently elevated in anxious patients). Although agents are available that can both stimulate (TRH, dopamine antagonists) and suppress (dopamine agonists) prolactin levels, they are rarely needed for clinical purposes.

SUMMARY

The anterior pituitary, once called the "master gland," is the primary target through which the central nervous system controls endocrine function. It is composed of several different cell types, each of which produces one or more hormones that stimulate target endocrine glands or other

sites, such as the liver, breast, skeletal muscle, or bone. The secretion of each of the pituitary hormones is regulated by hypothalamic-releasing or hypothalamic-inhibiting hormones, or both, and by feedback signals derived from the target glands or organs affected by the pituitary hormone. Measurement of pituitary hormones in circulation is used to assess pituitary function and involves both dynamic and static measurements.

NATIONAL BOARD TYPE QUESTIONS

Select the one most correct answer for each of the following

Match the hormone in column A with the most appropriate attribute in column B.

A	B
1. GH	A. Derived from a precursor containing other hormones
2. TSH	B. Stimulates casein synthesis
3. ACTH	C. Some effects mediated through IGF-I
4. Prolactin	D. Secretion suppressed by inhibin
5. FSH	E. Secretion decreased by thyroxine

6. Included among the effects of growth hormone are all, except which one of the following?
   A. Stimulation of insulin secretion
   B. Enhanced amino acid uptake
   C. Enhanced glucose uptake
   D. Enhanced triglyceride lipolysis
   E. Increased conversion of amino acids to glucose

7. Which of the following anterior pituitary cells secretes more than one hormone?
   A. Somatotrophs
   B. Corticotrophs
   C. Lactotrophs
   D. Gonadotrophs
   E. Thyrotrophs

8. The glycoprotein alpha subunit is secreted by all of the following cells, except the:
   A. Thyrotroph

B. Trophoblast
C. Lactotroph
D. Gonadotroph

9. All of the following statements are correct except:
A. The GH and PL gene family consists of five genes
B. The GH variant gene is not expressed
C. The PL gene is expressed only in the placenta
D. The 22-kDa GH and the 20-kDa variant GH are derived from the same gene
E. Prolactin is secreted by both somatomammotrophs and lactotrophs

ANNOTATED ANSWERS

1. C. Some of the growth-promoting effects of GH stem from a direct action of the hormone; others are mediated through the production of one of several growth factors, IGF-I being best understood.
2. E. TSH secretion is affected by the level of circulating thyroid hormones and is decreased when levels are elevated. Measurement of TSH levels is useful for assessing endogenous thyroid function.
3. A. ACTH is derived from a precursor, proopiomelanocortin, which serves a similar role for both β-LPH and beta-endorphin. Beta-endorphin is located on the carboxyl terminal of the ACTH precursor.
4. B. The major effect of prolactin is on the breast, where it stimulates the production of milk proteins, the most prominent of which is casein.
5. D. The primary inhibitor of FSH secretion is inhibin, a polypeptide that is secreted by the ovary and testis and blocks the effects of GnRH.
6. E. GH has many metabolic effects on carbohydrate metabolism that regulate the uptake and use of substrates such as glucose. It does not, however, participate in gluconeogenesis.
7. D. The gonadotrophs are the only type of anterior pituitary cell that secretes more than one hormone (LH and FSH). Although corticotrophs secrete β-LPH and beta-endorphin as well as ACTH, there is as yet no convincing evidence that these compounds serve as hormones when secreted from the pituitary.
8. C. Each of the cell types listed, with the exception of the lactotroph, secretes a glycoprotein hormone consisting of a common alpha subunit and a hormone-specific beta subunit. The lactotroph secretes prolactin, which also exists in a glycosylated form, but does not contain the alpha subunit.
9. B. The GH variant gene is expressed, but only in the placenta. It exhibits growth-promoting effects similar to those of GH.

BIBLIOGRAPHY

Daniels, G. H., and Martin, J. B., Neuroendocrine regulation and diseases of the anterior pituitary and hypothalamus. In: Braunwald, E. et al, eds., *Harrison's Principles of Internal Medicine.* New York: McGraw-Hill, 1987, Pp. 1694-1717.

Frohman, L. A. Diseases of the anterior pituitary. In Felig, P. A., et al, eds. *Endocrinology and Metabolism,* New York: McGraw-Hill, 1987, Pp. 247-337.

Thorner, M. O., et al. The anterior pituitary. In: Wilson, J. D., and Foster, D. W., eds. *Williams Textbook of Endocrinology.* Philadelphia: W. B. Saunders, 1992, Pp. 221-310.

49 The Posterior Pituitary

LAWRENCE A. FROHMAN

Objectives

After reading this chapter, you should be able to:

Discuss the components of the hypothalamic–neurohypophysial system and the hormones they secrete

Describe the mechanisms of biosynthesis, transport, storage, and secretion of the neurohypophysial hormones

List the sites of vasopressin activity and describe its mechanism of action

Describe the major mechanisms involved in the regulation of vasopressin secretion and their interaction

List the sites and describe the mechanism of action of oxytocin

The posterior pituitary is a component of two major and independent neuroendocrine systems that produce the hormones vasopressin and oxytocin, each of which has multiple functions. The most important action of vasopressin relates to water conservation; oxytocin is important in uterine contraction at parturition and also in milk secretion. The anatomy, biochemistry, and secretion of these systems and their hormones will be considered together, and the mechanism of action and regulation of secretion will be described separately.

ANATOMY

The posterior pituitary or **neurohypophysis** is an integral part of the central nervous system. It extends ventrally from the **median eminence** of the hypothalamus and is attached to the caudal border of the **anterior pituitary.** The intrasellar portion of the neurohypophysis is known as the **pars nervosa,** and the portion above the dia-

phragma sellae is the **infundibulum. The blood supply** of the posterior pituitary is derived directly from the arterial circulation by branches of the intracavernous portion of the internal carotid artery and the posterior communicating arteries. Venous drainage is to the jugular vein via the cavernous and petrosal sinuses.

The overall physiologic units of the posterior pituitary consist of neuronal cell bodies, the **perikarya,** which are located in two nuclear groups in the anterior hypothalamus: the **paraventricular nucleus** (PVN) and the **supraoptic nucleus** (SON). The neuronal perikarya are relatively large and have been termed *magnocellular neurons.* From the perikarya in the SON, axonal fibers travel through the ventral hypothalamus and pituitary stalk, then terminate in bulbous extensions called **pituicytes** in the pars nervosa. These neurosecretory cells are responsible for most, if not all, of the vasopressin and oxytocin that is secreted into plasma. A second system orginates in the PVN and sends fibers to the median eminence, where the released vasopressin participates in the control of adrenocorticotropic hormone secretion, to the medulla and spinal cord, where it is involved in regulating autonomic nervous system function, and to the amygdala, other hypothalamic nuclei, and the wall of the third ventricle, the presumed sites of origin for vasopressin and oxytocin in the cerebrospinal fluid.

BIOCHEMISTRY, SYNTHESIS, AND SECRETION

There are multiple neuropeptides in the posterior pituitary that probably have important physiologic functions, though the most well characterized and studied are **vasopressin** and **oxytocin**. Each is a **nonapeptide** consisting of a six-member disulfide-containing ring and a carboxylamidated three-residue tail (Fig. 49-1). The structures of the two peptides differ by only two amino acids. Both are derived from a single precursor present in nonmammalian vertebrates, **arginine vasotocin,** that exhibits similar biologic properties. Extensive structure–function studies of vasopressin have indicated that the arginine in the "8" position is critical to the pressor activity but not to the antidiuretic activity of vasopressin. Substitution of D-arginine in this position, along with removal of the terminal amino group of cysteine (desamino-8-D-arginine vasopressin [DDAVP] produces a highly potent and long-acting antidiuretic peptide possessing virtually no

pressor activity. This agent is currently the drug of choice in treating disorders of vasopressin deficiency (diabetes insipidus).

Vasopressin and oxytocin are synthesized in the SON and PVN perikarya as a large **prohormone** (molecular weight, 21,000 to 23,000) containing an amino-terminal neuropeptide, followed by a **neurophysin** (a binding protein specific for each neuropeptide) with a molecular weight of 10,000, and a **carboxyl-terminal peptide.** The precursor is proteolytically processed, and **neurosecretory granules** are formed that contain the neuropeptide bound to its neurophysin. Although vasopressin and oxytocin are made in separate cell bodies, the biosynthetic processes are similar. The granules are then transported down the axon by a mechanism that is not yet defined, and are then stored in pituicytes in the posterior pituitary or in nerve terminals in the central nervous system.

Hormone secretion is accomplished through the classic stimulus–secretion coupling mechanism described for many neurohormones and neurotransmitters. The stimulus, which is applied to the perikarya in the SON or PVN, propagates an electrical impulse down the axon, leading to depolarization of the cell membrane; increased calcium permeability, leading to rapid entry of calcium; and exocytosis of the secretory granule. Thus, the neurophysin is secreted in equimolar amounts to the neuropeptide. The relatively low binding affinity and pH optimum, however, promote nearly complete dissociation of the complex in plasma.

In addition to vasopressin and oxytocin, **endothelin,** an endothelial cell–derived peptide with potent vasoconstricting properties, has recently been identified in the posterior pituitary and shown to be secreted under conditions that favor vasopressin secretion. Its role in the response to osmotic regulation remains undefined.

Fig. 49-1. Chemical structures of vasopressin and oxytocin, the major hormones of the posterior pituitary.

VASOPRESSIN

NH$_2$ - Cys – Tyr – Phe – Gln – Asn – Cys – Pro – Arg – Gly – NH$_2$
|_____ s – s _____|

OXYTOCIN

NH$_2$ - Cys – Tyr – Ile – Gln – Asn – Cys – Pro – Leu – Gly – NH$_2$
|_____ s – s _____|

VASOPRESSIN

Mechanism of Action

The primary effect of vasopressin is to enhance the formation of **concentrated** (hypertonic) **urine.** It also increases vascular smooth muscle tone, enhances intestinal motility, and stimulates the production of clotting factors, including factor VIII and von Willebrand's factor. Two separate **vasopressin receptors** have been identified, one (V_1) that mediates the pressor effect and a second one (V_2) that mediates antidiuresis and the effects of clotting factors.

In the vasopressin-responsive **renal epithelial cells,** receptor binding is coupled to a guanine-nucleotide regulatory protein (G_s) that stimulates cyclic AMP formation and protein phosphorylation. This heightens water permeability in the collecting duct and the medullary thick ascending loop of Henle. If vasopressin is lacking, the collecting tubules are almost completely impermeable to water. In its presence, there is a concentration-dependent increase in the water-specific channels or pores that accentuates the transport of solute-free water through the luminal membranes. The renal handling of salt and water metabolism is discussed in Chapter 37. A feedback process that modulates the effects of vasopressin is mediated by **prostaglandin E,** the synthesis of which is stimulated by the hormone. Prostaglandin E, in turn, inhibits the effects of vasopressin on adenylate cyclase.

The **pressor effects** of vasopressin in humans are much greater than those necessary for maximal antidiuresis, and there is currently no convincing evidence that the hormone has any physiologic role in regulating blood pressure in humans. Similarly, any physiologically important effects on intestinal motility and clotting factors have yet to be proved. Nevertheless, the use of vasopressin, or more recently DDAVP, in the treatment of milder forms of **hemophilia** has been valuable in reducing the need for plasma transfusions.

Plasma Levels and Metabolism

Plasma levels of vasopressin in normally hydrated, nonstressed individuals are less than 2 pg/ml. Levels are highest in the early morning and lowest in late afternoon. Most of the peptide is inactivated in the liver, though plasma enzymatic degradation also occurs, particularly during pregnancy, when plasma vasopressinase activity increases. In contrast, plasma neurophysin levels are considerably greater than those of vasopressin, because of a much slower metabolic clearance of the protein.

Regulation of Secretion

Osmoregulation

Under physiologic conditions, the most important regulator of vasopressin secretion is the **osmotic pressure of plasma**. This control is mediated by a group of highly specific neurons called **osmoregulators** that are located in the anterior hypothalamus, near but clearly distinct from the SON. There are undoubtedly bimodal inhibitory and stimulatory inputs from the osmoregulatory neurons, but the net result resembles that of a "**set point**" or threshold, **mechanism**. Plasma osmolality normally fluctuates within a very narrow range of between 275 to 290 mOsm/kg. Within this range, plasma vasopressin levels are near or beneath the lower limit of detection (less than 2 pg/ml). When the osmolality level is exceeded (the critical value varies among individuals), there is a steep and proportionate rise in vasopressin levels, such that a 1 percent change in osmolality triggers vasopressin release by an amount (1 pg/ml) sufficient to alter urine osmolality (Fig. 49-2). The set point, though constant in an individual for long periods, may vary by up to 5 mOsm/kg in conjunction with other factors, such as age, hemodynamic alterations, and pregnancy. The rates of change in vasopressin secretion in response to alterations in osmolality are also constant in individuals and appear to be determined genetically. However, they may be modified if blood volume, blood glucose, or calcium levels are

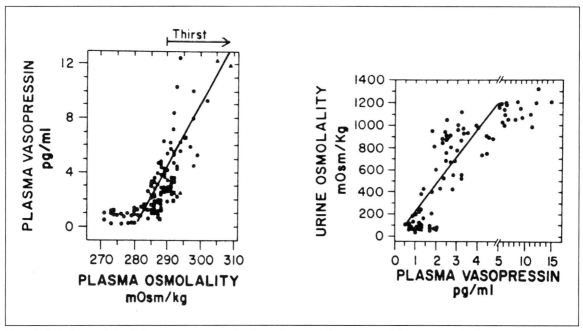

Fig. 49-2. Relationship of plasma vasopressin levels to plasma and urine osmolality in normal adults. A very narrow range of change in the plasma osmolality results in increased vasopressin secretion and near-maximal antidiuresis, even before thirst perception. (From: Robertson, G. L. Posterior pituitary. In: Felig, P. F., et al, eds. *Endocrinology and Metabolism.* New York: McGraw-Hill, 1987. Pp. 341.)

disturbed, or by certain drugs, such as lithium. The sensitivity of the osmoregulatory mechanism also varies with the **solute** involved. While **Na**$^+$ is the most potent solute and, together with its anions, constitutes 95 percent or greater of the plasma osmotic pressure, **sugars** such as sucrose and mannitol also have similar stimulatory effects. One exception is **urea**, which is almost entirely ineffective in stimulating vasopressin secretion, for reasons that are still unclear. Osmoreceptors respond to changes in the glucose level, but only when entry of this solute into cells is impaired, such as in uncontrolled diabetes.

Hemodynamic Regulation

Changes in **blood volume** and **blood pressure** profoundly affect vasopressin secretion, inde-

pendent of osmoregulation. These changes are recognized by **baroreceptors** (pressure-sensitive receptors) in both the low-pressure (cardiac atria) and high-pressure (carotid bifurcation and aortic arch) regions of the circulatory system. They travel via parasympathetic fibers in the vagus and glossopharyngeal nerves, which synapse in the brainstem with fibers that project to the PVN and SON. Their impulses appear to be chiefly inhibitory under nonstressed basal conditions. When blood pressure decreases, the plasma vasopressin level rises exponentially in proportion to the degree of hypotension. A reduction of more than 10 percent is generally necessary before plasma vasopressin levels are increased; a 20 to 30 percent decrease in blood pressure elevates levels to many times that needed for maximal antidiuresis. A blood volume decrease of more than 7 percent is required in order to increase plasma vasopressin levels and levels are markedly elevated when there is a 20 percent decrease. Increases in blood volume or blood pressure produce corresponding decreases in the vasopressin levels.

Because the variation in body water content is very small under normal physiologic conditions, the osmoregulatory mechanisms would appear to supply the primary control of vasopressin secretion. However, fluctuations in blood pressure during exercise or stress are enough to provoke baroreceptor regulatory input that may well be comparable to that of the osmoreceptor in regulating vasopressin secretion (Fig. 49-3). Yet, these two mechanisms are distinct, as shown by individuals with disturbed baroreceptor activation of vasopressin secretion who respond normally to stimulation of the osmoreceptor.

Fig. 49-3. Stimulus–response relationship between plasma vasopressin levels and percentage change in osmolality, blood volume, and blood pressure in normal adults. (From: Robertson, G. L. Posterior pituitary. In: Felig, P. F., et al, eds. *Endocrinology and Metabolism.* New York: McGraw-Hill 1987, P. 344.)

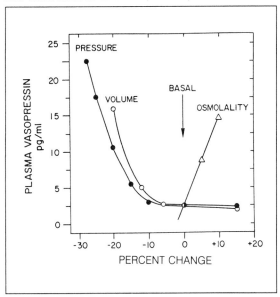

Neuroregulation

The most potent neurogenic stimulus for vasopressin secretion is **nausea.** This pathway originates from the **area postrema** in the medulla and is mediated by **dopaminergic fibers.** Antidopaminergic agents that are effective in alleviating nausea also inhibit vasopressin secretion. Many physiologic and pathologic processes associated with increased vasopressin secretion and antidiuresis, such as vasovagal reactions, acute motion sickness, and acute hypoxia, are probably mediated by their associated emetic stimuli.

Stress and **increased temperature** also enhance vasopressin secretion, though it is not clear whether these effects are primary or are mediated by associated changes in blood pressure or the effective blood volume. Clinically, high fevers may be associated with vasopressin secretion that is osmotically inappropriate. **Severe hypoxemia** may also stimulate vasopressin secretion, even in the absence of nausea. **Central glucopenia** is another potent stimulus for vasopressin secretion. Although the stimulus is not specific for vasopressin, it may be physiologically important, for example, in diabetics who are receiving excessive amounts of insulin. **Pain** also constitutes a stimulus for vasopressin release, but the response is not specific.

Angiotensin

Angiotensin-II is a potent stimulus for vasopressin secretion. It has a **central site of action** and also stimulates the **thirst mechanism.** The nature of this stimulation is not known, though its role is believed to be physiologic because angiotensin antagonists inhibit vasopressin release, implying the presence of an endogenous stimulatory tone.

OXYTOCIN

Mechanism of Action

The biologic activity of oxytocin in **females** is confined to the **uterus** and **breasts.** There are specific myometrial receptors that bind the hormone, leading to membrane depolarization and myometrial contraction. These effects are enhanced by estrogen and are important during parturition. Although the initiation of labor is largely independent of oxytocin, the hormone appears to participate in the final expulsion of the fetus and placenta when the cervix is fully dilated.

Oxytocin also produces contraction of the myoepithelial cells of the **breast alveoli,** causing expulsion of milk from the secretory channels into the large sinuses connected to the nipple. Although oxytocin does not participate in milk production, which is largely regulated by prolactin, it is critical to the secretory process, and was originally called the "milk letdown" factor.

In **males,** oxytocin participates in the process of **sperm ejection** into semen upon stimulation of the reproductive organs, though the importance of this action is uncertain. The hormone also exhibits slight antidiuretic activity (about 0.5 to 1 percent), but it is doubtful that this represents a physiologic role.

Regulation of Secretion

The regulation of oxytocin secretion is less well understood than that of vasopressin. The major stimulus appears to be **suckling,** which is mediated through nerve fibers in the nipple. **Distention of the female genital tract,** particularly during parturition, also serves as possible stimulus for oxytocin release. Estrogen administration heightens the secretion of the oxytocin-associated neurophysin, and was originally believed to elevate oxytocin release as well. However, the specificity of assays used to demonstrate these effects has recently been questioned, and the possibility of another, yet unidentified, posterior pituitary hormone has been raised. **Pain** is also a stimulus for the secretion of oxytocin, though this response is not specific.

NATIONAL BOARD TYPE QUESTIONS

Select the one most correct answer for each of the following

1. All of the following are important in the control of vasopressin secretion under physiologic conditions, except:
   A. Plasma osmolality
   B. Blood volume
   C. Blood pressure
   D. Blood urea
   E. Body temperature
2. The actions of vasopressin include all of the following, except:
   A. Increases water permeability of the collecting duct
   B. Increases vascular smooth muscle tone
   C. Increases intestinal motility
   D. Increases factor VIII production
   E. Increases sodium transport in the ascending loop of Henle
3. Oxytocin has all of the following actions, except:
   A. Myometrial contraction
   B. Initiation of labor
   C. Mammary myoepithelial contraction
   D. Fetal and placental expulsion
   E. Sperm ejection into semen

ANNOTATED ANSWERS

1 D. Even though urea penetrates the blood–brain barrier very slowly, it is a poor stimulator of vasopressin. This observation strongly implies that the osmoreceptor is located outside the blood-brain barrier.

2 E. It is precisely because vasopressin increases water reabsorption in the loop of Henle in the absence of solute that excretion of concentrated urine occurs.

3 B. Although oxytocin appears to have a role in the expulsion of the fetus and placenta

from the uterus, it is not involved in the initiation of labor, which appears to be affected by increased activity in the fetal adrenal.

BIBLIOGRAPHY

Baylis P. H. Vasopressin and its neurophysin. In: DeGroot, L. J., et al, eds. *Endocrinology.* Philadelphia: Saunders, 1989, Pp. 213–229.

Reeves, W. B., and Andreoli, T. E. The posterior pituitary and water metabolism. In: Wilson, J. D., and Foster, D. W., eds. *Williams Textbook of Endocrinology.* Philadelphia: W. B. Saunders, 1992, Pp. 311–356.

Robertson, G. L. Posterior pituitary. In: Felig, P. A., eds. *Endocrinology and Metabolism.* New York: McGraw-Hill 1987, Pp. 338–385.

50 The Thyroid

TIMOTHY C. WILLIAMS

Objectives

After reading this chapter, you should be able to:

Outline thyroid development and describe its fully formed structure

Describe the basic steps in thyroid hormone synthesis, storage, secretion, and metabolism

Explain how thyroid hormones interact with cells

Summarize some of the important biologic effects of thyroid hormones

Describe thyroid hormone regulation

The thyroid synthesizes, stores, and secretes two iodine-containing hormones, **3,5,3′,5′-tetraiodothyronine** (T_4), also known as thyroxine, and **3,5,3′-triiodothyronine** (T_3), that are important regulators of development and metabolism (Fig. 50-1). These hormones are necessary for normal central nervous system development during fetal growth and the first several years of life. They are also critical for normal bone growth and sexual maturation during childhood and adolescence. Throughout life, they are important in maintaining metabolic homeostasis in virtually every cell, tissue, and organ throughout the body. Thus, adequate levels of thyroid hormones are essential for the ongoing normal function of almost every system and process in the body.

The thyroid also produces **calcitonin,** a hormone involved with calcium homeostasis (discussed in Chapter 51).

EMBRYOLOGY AND ANATOMY

The initial event in thyroid development is a **ventral evagination of endoderm** in the floor of the oropharynx in the region where the tongue forms. This subsequently becomes bifid and migrates caudad. After this tissue reaches the lower neck region, the remaining structure connecting it and the mouth, called the **thyroglossal duct,** is obliterated. Tissue is also incorporated from the lower lateral branchial pouch regions (ultimobranchial bodies) into the developing lobes of the thyroid that gives rise to **parafollicular C cells** which produce calcitonin.

The fully developed thyroid consists of **two lobes,** each situated just lateral to the trachea, below the larynx and above the suprasternal notch. The two lobes are connected across the midline by the **thyroid isthmus,** which is located just below the cricoid cartilage anterior to the trachea. This configuration is achieved by the end of the third gestational month, when thyroid hormone production first occurs. The normal adult thyroid weighs about 15 to 20 gm and each lobe measures about 4 x 2 cm.

Microscopically, the thyroid consists of **follicles.** These are spherical structures composed

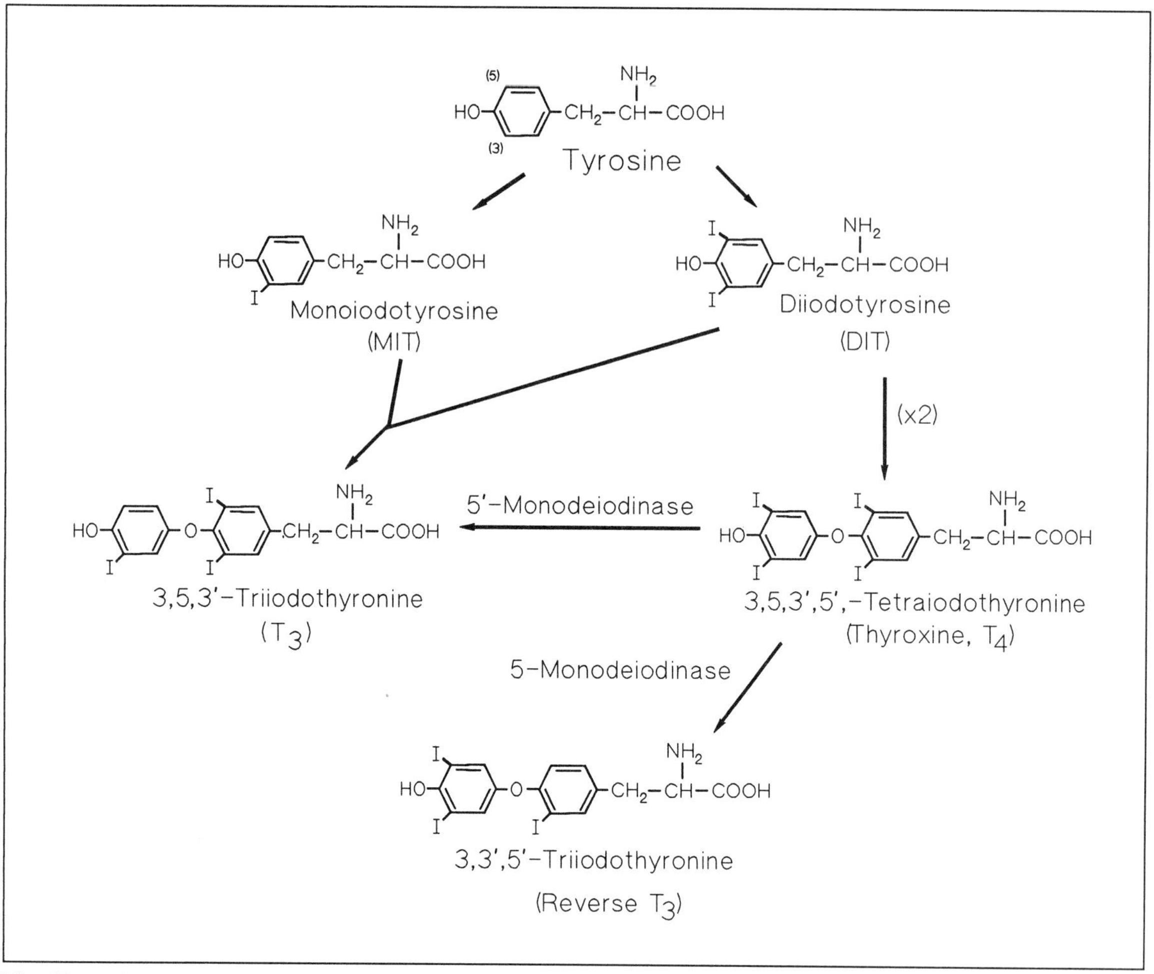

Fig. 50-1. Thyroid hormone synthesis and structures of thyroid hormones and their precursors.

of a central core of **colloid** (stored thyroid hormone) and an outer rim made up of a single layer of **follicular epithelial cells** that produce thyroid hormones. Interspersed among the follicles are connective tissue, blood vessels, nerves, and parafollicular C cells.

THYROID HORMONES

Biosynthesis

Thyroid hormone synthesis depends on an adequate supply of **dietary iodine;** this iodine is reduced to **iodide** (I^-) in the gastrointestinal tract and is rapidly absorbed. Although the recommended daily intake of iodine is at least 150 μg, at least 50 to 75 μg is required to prevent goiter formation due to iodine deficiency.

Follicular cells acquire and concentrate I^- from plasma along their basal membranes in an active-transport process called **I⁻ uptake** or **trapping** (Fig. 50-2). This energy-dependent and saturable process is linked with (Na, K)-ATPase. The I⁻ is rapidly oxidized and incorporated into **tyrosyl** residues located in thyroglobulin, a glycoprotein produced by follicular cells, in a process termed **organification** (see Fig. 50-2). **Thyroglobulin** is located in vesicles produced by the **Golgi apparatus.** It appears that I⁻ oxidation and organification occur primarily in vesicles that are near or bound to apical cell membranes. Thyroglobulin has a molecular weight of approximately 660,000 and contains over one hundred tyrosyl residues; however, only about 20 to 25 percent of these residues are available for iodination. Multiple tyrosyl residues in each thyroglobulin molecule are iodinated, either at one or two positions, to form a number of both **monoiodotyrosine** (MIT) and **diiodotyrosine** (DIT) residues (see Figs. 50-1 and 50-2). One MIT and one DIT or two DIT residues combine to form T_3 and T_4, respectively, in a process known as **coupling.** This process depends on the conformation of the thyroglobulin molecule and is catalyzed by **thyroid peroxidase,** a heme-containing and vesicle membrane–bound glycoprotein. This enzyme also catalyzes both I⁻ oxidation and organification and requires hydrogen peroxide as an oxidizing agent. Hydrogen peroxide is probably generated by NADPH–cytochrome c reductase.

Storage

Iodinated thyroglobulin containing both T_3 and T_4 as well as MIT and DIT is extruded from inner or apical membranes of follicular cells by the exocytosis of vesicle contents into the central colloid region where it is stored (see Fig. 50-2). About ten to fifteen times more T_4 than T_3 is stored by this process, and significant reservoirs of thyroid

Fig. 50-2. Thyroid hormone synthesis and secretion by a thyroid follicular cell. (*MIT* = monoiodotyrosine; *DIT* = diiodotyrosine; *TG* = thyroglobulin; T_4 = thyroxine: T_3 = 3,5,3′ triiodothyronine.)

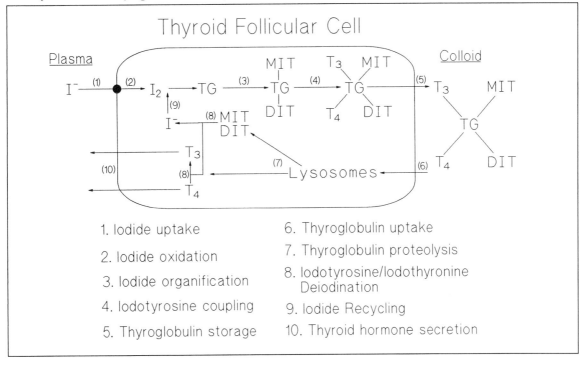

1. Iodide uptake
2. Iodide oxidation
3. Iodide organification
4. Iodotyrosine coupling
5. Thyroglobulin storage
6. Thyroglobulin uptake
7. Thyroglobulin proteolysis
8. Iodotyrosine/Iodothyronine Deiodination
9. Iodide Recycling
10. Thyroid hormone secretion

hormones are maintained under usual circumstances that guard against thyroid hormone deficiency and provide a readily available and large supply of hormone without the need for new hormone synthesis.

Secretion

Thyroid hormone secretion begins with the **endocytosis** of colloid into small vesicles at the apical borders of follicular cells (see Fig. 50-2). These colloid vesicles containing iodinated thyroglobulin combine in the follicular cytoplasm with lysosomes containing protease activity. As these **phagolysosomes** are transported toward basal regions of the follicular cells, the amount of colloid decreases because of its proteolysis to T_4, T_3, MIT, and DIT. Thyroid hormones are eventually secreted at **basal cell membranes,** gaining access to the plasma while an intracellular deiodinase removes I^- from MIT and DIT for recycling within cells (see Fig. 50-2). In addition, a monodeiodinase (MDI) also converts some T_4 to T_3 within the follicular cells (see Figs. 50-1 and 50-2).

Transport

More than 99 percent of the thyroid hormones circulate in blood bound to **plasma proteins,** thus, only a small portion is unbound or free. The major plasma proteins that bind thyroid hormones are synthesized in the liver and include **thyroxine-binding globulin** (TBG), **albumin,** and **thyroxine-binding prealbumin** (TBPA). TBG binds about 60 percent of the circulating thyroid hormones, whereas albumin and TBPA each bind about 20 percent.

The total plasma concentration of T_4 is about sixty times greater than that of T_3; however, the percentage of free T_3 is fifteen times greater than that of T_4 because of differing affinities between the hormones for binding to these plasma proteins. This results in a fourfold greater free-plasma concentration of T_4 than of T_3.

Only the **free fraction** of these hormones is readily available for cellular uptake and use. In addition, it is the free concentration of thyroid hormones rather than the bound or total concentration that is regulated by the **hypothalamus** and **pituitary.**

Cellular Interaction

The interaction of thyroid hormones with cells is complex and incompletely understood. It appears that these hormones, in particular T_3, exert their major effects through binding to specific **intranuclear receptors.** These nuclear receptors are **oncogene products** that resemble those for steroid hormones and vitamins A and D. This interaction alters DNA transcription into messenger RNA, and the eventual translation into various proteins that have myriad effects on cellular function and structure. The affinity of these nuclear receptors is approximately tenfold greater for T_3 than for T_4, which explains to some extent why T_3 is three to four times more potent that T_4 in its metabolic effects. The cellular effects of T_4 may largely stem from its intracellular conversion to T_3. In addition, T_3 binds to sites on the cell-surface membranes and mitochondria, thus indicating that it may have multiple sites of action.

Actions and Effects

Thyroid hormones have effects on almost every cell in the body. Therefore, almost every bodily function or process is to some extent influenced by these hormones. **Some important cellular constituents** that are stimulated by thyroid hormones include beta-adrenergic receptors, (Na, K)-ATPase, myocardial myosin ATPase, and a plethora of enzymes that regulate the production and degradation of metabolic fuels.

Thyroid hormones **stimulate** the rate of **hepatic glucose production, glucose utilization** in tissues, **lipolysis, hepatic fatty acid oxidation** and **ketone production, hepatic cholesterol** and **triglyceride synthesis, cholesterol** and **triglyceride metabolism,** and the **synthesis of various proteins** that are important structural and regulatory components of cells. These myriad metabolic effects produce an overall increase in oxygen consumption and combined calorigenesis and thermogenesis. Thyroid hormones also augment the function of the central and sympathetic nervous systems and the heart, which is at least in part mediated by their regulation of catecholamine receptors. More specifically, thyroid hormones have been demonstrated to increase the number of **beta-adrenergic receptors.** In addition, thyroid hormones modulate the secretion of other hormones, particularly **growth hormone** and **sex steroids.** Another important process regulated by thyroid hormones is the development of the **central nervous system,** which may be mediated by the stimulation of nerve growth factor production.

Metabolism

Both T_4 and T_3 are metabolized in extrathyroidal sites by **deiodination.** 5-MDI removes an I^- from the outer ring of T_4 and converts it into the more potent T_3 (see Fig. 50-1). There are two types of 5-MDI: **type 1,** which is found primarily in the liver and kidney and is responsible for producing most of the circulating T_3, and **type 2,** which is found primarily in the pituitary and brain. The regulation of these two types of enzymes is different and, in some respects, divergent. Hypothyroidism, starvation, diabetes mellitus, renal failure, and a variety of drugs decrease type 1 activity, whereas hyperthyroidism and overnutrition increase type 1 activity. Hypothyroidism increases and hyperthyroidism decreases brain and pituitary type 2 activity, which may serve to protect the central nervous system from the metabolic consequences of thyroid hormone deficiency or excess. However, **pituitary thyrotrophs** are unique because hypothyroidism produces increased TSH secretion because of decreased intracellular T_3 availability. Chronic diseases, nutritional status, and drugs have little if any effect on type 2 activity. In contrast, 5-MDI removes an inner-ring I^- from T_4, which converts it to the metabolically inactive reverse T_3 (see Fig. 50-1).

About 40 percent of the T_4 produced daily is metabolized by each **MDI pathway** and the remainder is excreted in bile. T_3 and reverse T_3 are subsequently further deiodinated to metabolically inactive substances. These metabolic transformations can occur in a variety of tissues, though most take place in the liver and kidney.

Regulation

Secretion of thyroid hormones is primarily regulated by **TSH,** also called *thyrotropin.* This glycoprotein hormone, with a molecular weight of 28,000, is synthesized and secreted by **thyrotroph cells** of the **anterior pituitary**. TSH stimulates almost all the processes involved in thyroid hormone synthesis and secretion. In man, this hormone is the only known physiologic stimulator of thyroid hormone secretion.

TSH secretion, in turn, depends on an interaction between the **free plasma thyroid hormone concentrations** (a classic feedback-inhibition loop) and **hypothalamic factors** (Fig. 50-3). Both T_4, either directly or by its conversion to T_3 within pituitary thyrotrophs, and T_3 inhibit TSH secretion, which is mediated by an interaction with thyrotroph nuclear receptors. There is evidence that TSH synthesis and the number of cell-surface **thyrotropin-releasing hormone** (TRH) receptors are both diminished by these events. The predominant **hypothalamic effect** on TSH secretion is stimulatory and is mediated by TRH, a tripeptide. In addition, both **somatostatin** (also known as *somatotropin release–inhibiting hormone*) and **dopamine,** which are produced in the hypothalamus, as well as **glucocorticoids** have been demonstrated to suppress TSH secretion. However, a physiologic role for these substances is questionable. The various hypothalamic factors that influence TSH secretion are synthesized and secreted by neurons in the hypothalamus whose axons terminate in or around the median eminence. The factors gain access to the pituitary by means of the **hypothalamic–pituitary portal vessel system** and interact with specific cell-surface receptors on thyrotrophs. A feedback effect of thyroid hormones

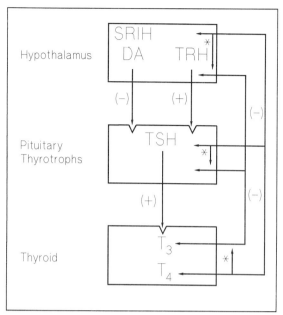

Fig. 50-3. Regulation of thyroid hormone secretion. (* = 5'-monodeiodinase; + = stimulation; − = inhibition; *SRIH* = somatotropin release–inhibiting hormone; *DA* = dopamine; *TRH* = thyrotropin-releasing hormone; *TSH* = thyroid-stimulating hormone; T_3 = 3,5,3' triiodothyronine; T_4 = thyroxine.)

on TRH synthesis has also been demonstrated recently.

Thus, pituitary thyrotrophs and their interaction with the thyroid are analogous to a thermostat and its relationship to a furnace, an appropriate and convenient analogy since thyroid hormones are thermogenic. **Thyrotrophs** (the thermostat) are intrinsically set to produce an amount of TSH (the circuit-linking the thermostat and furnace) that will maintain **euthyroidism.** If the thyroid (furnace) fails to produce a sufficient amount of circulating free thyroid hormones (heat), the decrement is perceived by the thyrotrophs. This prompts increased TSH secretion that stimulates the thyroid to secrete more thyroid hormones until euthyroidism is re-established. Conversely, if there is a surfeit of circulating free thyroid hormones, the increment is perceived by thyrotrophs. This in turn triggers

decreased TSH secretion and thus decreased stimulation of the thyroid. Thyrotrophs are extremely sensitive to alterations in the free circulating concentrations of thyroid hormones, with relatively small increments or decrements translating into marked reciprocal alterations of TSH secretion. There is also evidence that I⁻ itself may to some extent regulate thyroid hormone production and secretion. In general, relative I⁻ deficiency leads to increased hormone production, and vice versa.

THYROID DISEASES AND TESTS

Diseases of the thyroid can be divided into anatomic and functional disorders, but these categories are not mutually exclusive, and in fact often occur together. **Anatomic abnormalities** consist of developmental disorders such as agenesis or dysgenesis, solitary nodules that may be cysts, infections, inflammations, benign or malignant neoplasms, and diffuse enlargement or goiter. **Functional disorders** include insufficient thyroid hormone secretion (hypothyroidism or myxedema) and excess thyroid hormone secretion (hyperthyroidism or thyrotoxicosis).

Anatomic thyroid disease usually causes few problems unless there is thyroid dysfunction or there is marked enlargement that causes cosmetic deformity or impingement on other important structures, such as the trachea, recurrent laryngeal nerves, or esophagus. In contrast, functional thyroid disease often results in marked problems, although it is rarely fatal except in extreme situations. Common symptoms of **hypothyroidism** in adults include cold intolerance, constipation, weight gain, dry skin, loss and coarsening of hair, fatigue, slow or impaired mentation, and excess menstruation. Hypothyroidism causes additional problems in children. When it arises during fetal or neonatal life, a condition called **cretinism,** it results in poor intellectual development and growth. Hypothyroidism that develops during later childhood is associated with decreased linear growth and abnormal sexual maturation. Symptoms of **hyperthyroidism** include heat intolerance, diarrhea, weight loss, diaphoresis, muscle weakness, nervousness, and absent or scant menses.

The initial evaluation of any thyroid disease includes a careful history and physical examination. Further evaluation of **structural abnormalities** may include a **radioiodine scan** of the thyroid. This procedure takes advantage of the fact that the thyroid is the only tissue in the body that significantly takes up and concentrates I^-. To perform this test, a small dose of iodine 123 is administered by mouth. Twenty-four hours later, a scintillation counter is placed over the neck region to detect the amount and distribution of the radioiodine uptake. Another useful test for assessing anatomic abnormalities, particularly solitary nodules, is **fine-needle aspiration biopsy** of the area to process tissue for cytopathologic examination. Evaluation of **functional thyroid disease** usually involves the measurement of plasma concentrations of T_3, T_4, and TSH by **radioimmunoassay.** In general, hyperthyroidism exhibits elevated T_3 and T_4 levels and decreased or undetectable TSH levels. Hypothyroidism is usually associated with the opposite. However, there are extremely rare pituitary thyrotroph tumors that secrete excess amounts of TSH in an unregulated fashion, and this causes hyperthyroidism in conjunction with increased TSH levels. In addition, a disease process in the hypothalamus or pituitary that impairs TRH or TSH secretion, respectively, may lead to hypothyroidism with decreased or inappropriately low TSH levels.

SUMMARY

The thyroid produces two hormones, T_3 and T_4, that are essential for normal neurologic development, linear bone growth, and proper sexual maturation. These hormones also affect almost all cells in the body by regulating their production of various regulatory and structural proteins. Thus, thyroid hormones are necessary for integrating normal development and proper bodily function throughout life.

Thyroid hormones are synthesized by thyroid follicular cells in a series of steps that include I^- uptake, oxidation, and organification. This last step involves the incorporation of I^- with tyrosyl residues located in thyroglobulin, a large glycoprotein. This forms iodotyrosines that are subsequently coupled to create T_3 and T_4, which are stored in colloid located in the central core of the follicles.

Thyroid hormone secretion involves the reuptake of stored hormones into the apex of follicular cells and their transport to the base, where they are secreted and discharged into the plasma. They circulate in plasma, bound predominantly to several proteins, with only a small fraction that is free or available for cellular interaction. Thyroid hormone secretion is regulated by TSH produced in pituitary thyrotrophs whose secretion, in turn, depends on the amount of circulating free thyroid hormones. This involves a classic feedback-inhibition loop and the stimulatory influence of hypothalamic TRH.

The interaction of thyroid hormones with cells is predominantly mediated by their binding to specific nuclear receptors. This results in altered DNA transcription and messenger RNA translation into a variety of regulatory and structural proteins that mediate the effects of these important hormones.

NATIONAL BOARD TYPE QUESTIONS

Select the one most correct answer for each of the following

1. Which of the following is vital for thyroid hormone synthesis?
   A. Thyroxine-binding globulin
   B. I^-
   C. TRH
   D. 5-MDI$^-$
   E. Tryptophan
2. Which of the following is the most important physiologic regulator of thyroid hormone secretion?
   A. Dopamine
   B. Somatostatin
   C. Tyrosine

D. Thyroglobulin

E. Thyrotropin

3. Which of the following is not synthesized in thyroid follicular cells?

A. Calcitonin

B. Thyroglobulin

C. T_4

D. T_3

E. Thyroid peroxidase

4. Which of the following is true with respect to the interaction of thyroid hormone with cells?

A. Thyroid hormones bind primarily to specific cell-surface receptors on target cells

B. Thyroid hormone interacts with the thyroid hormone receptors to activate adenylate cyclase, producing cyclic AMP as a secondary messenger

C. Modulation of the calcium channels and intracellular calcium concentrations is vital

D. Thyroid hormones alter DNA transcription by binding to specific nuclear receptors

E. Stimulatory and inhibitory subunits of cell-surface receptors are intricately involved with the effects of thyroid hormones on cells

Select the response most closely associated with each numbered item. (The heading may be used once, more than once, or not at all.)

A. Thyroid peroxidase

B. Colloid

C. Thyroglobulin

D. TSH

E. 5-MDI

5. A glycoprotein containing tyrosyl residues that is necessary for thyroid hormone synthesis

6. An important storage form of thyroid hormones

7. Catalyzes I^- oxidation and organification as well as iodotyrosine coupling

8. A glycoprotein important in the regulation of thyroid hormone secretion

---

ANNOTATED ANSWERS

1. B. I^- is an essential component of thyroid hormones. In fact, thyroid hormones are the only known biologically active compounds that contain iodide.

2. E. Thyrotropin or TSH is the only known physiologic regulator of thyroid hormone secretion.

3. A. All the substances listed are synthesized in thyroid follicular cells except calcitonin, which is produced by parafollicular C cells.

4. D. Thyroid hormones, like steroid hormones, interact with cells by binding to specific nuclear receptors and this results in altered DNA transcription.

5. C. Thyroglobulin is a glycoprotein containing tyrosyl residues and is necessary for thyroid hormone synthesis.

6. B. Colloid, located in the central core of the follicles, is the storage form of thyroid hormones.

7. A. Thyroid peroxidase is an enzyme that catalyzes I^- oxidation and organification and also iodotyrosine coupling.

8. D. TSH, or thyrotropin, is a glycoprotein that is the only known physiologic regulator of thyroid hormone secretion.

BIBLIOGRAPHY

DeGroot L. J., et al, eds. *The Thyroid and Its Diseases*, 5th ed. New York: J. Wiley & Sons, 1984.

Ingbar, S. H., and Braverman L. E., eds. *The Thyroid: A Fundamental and Clinical Text*, 5th ed. Philadelphia: J. B. Lippincott, 1986.

51 Calcium and Phosphate Homeostasis

TIMOTHY C. WILLIAMS

Objectives

After reading this chapter, you should be able to:

Explain the importance of calcium and phosphate in structure and function

Describe calcium and phosphate distribution

Outline parathyroid and parafollicular C cell development and anatomy

Characterize the basic steps in parathyroid hormone, calcitonin, and vitamin D synthesis, storage, secretion, and metabolism

Describe how parathyroid hormone, calcitonin, and vitamin D interact with cells

Outline some important biologic effects of parathyroid hormone, calcitonin, and vitamin D

Describe how parathyroid hormone, calcitonin, and vitamin D are involved in integrated calcium and phosphate homeostasis

The physiologic importance of calcium and phosphate can be divided into two categories, **structural** and **functional.** Both substances are crucial components for normal **bone mineralization** that is necessary for the structural integrity of the skeletal system. **Phosphate,** in the form of phosphoproteins and phospholipids, is also a vital constituent of cellular structure. **Extracellular calcium** has a critical role in neuromuscular function and coagulation, while **intracellular calcium** acts as both a regulator of a variety of cellular processes, such as enzymatic reactions and secretion, and a secondary messenger for external regulators such as certain peptide hormones. **Intracellular phosphate** is an essential component of a large number of biologic substances, including enzyme cofactors, nucleic acids, regulatory phosphoproteins, glycolytic intermediates, and molecules that store chemical energy such as ATP. Thus, calcium and phosphate are not only important for imparting structural skeletal strength but also for maintaining the proper structure and function of every cell in the body.

CALCIUM AND PHOSPHATE DISTRIBUTION

The normal adult contains approximately 1,000 gm of calcium and 600 gm of phosphate. Most calcium (99 percent) and phosphate (85 percent) exists in bone as **hydroxyapatite,** a crystalline substance that also contains hydroxyl ions. Therefore, only about 1 percent of calcium and 15 percent of phosphate are found in

extraskeletal sites, including extracellular fluid and cellular cytosol.

In **plasma,** about 50 percent of **calcium** is bound to proteins, primarily **albumin,** and a small portion is complexed with other ions. Thus, less than half of the circulating calcium is freely ionized, but it is this ionized fraction that is physiologically important and regulated by the interactions of **parathyroid hormone** (PTH), **calcitonin** (CT), and **vitamin D.** In contrast, only about 10 percent of the **plasma phosphate** is protein-bound and most circulates in a **free ionic form** or is **complexed with cations.** In addition, plasma phosphate concentrations are not tightly regulated within a narrow range, like calcium, but may vary by up to 50 percent within a day.

Within **cells,** approximately 99 percent of **calcium** is complexed with phosphate in the mitochondria, and a small portion is also bound to plasma membranes and the endoplasmic reticulum. Therefore, only about 1 percent is free in cytosol. Thus, the cytosolic concentration of calcium is a hundredfold less than that in the mitochondria, but it is 10,000-fold less than that in plasma. These calcium concentration gradients are maintained by **energy-requiring transport systems** that pump calcium from cytosol into the internal organelles or the extracellular space. In contrast, **phosphate** is widely distributed in cells as a constituent of a variety of structural and functional molecules, such as phosphoproteins, phospholipids, nucleic acids, and intermediates of carbohydrate metabolism. Nevertheless, the phosphate concentration in the mitochondria is about fivefold greater than that in cytosol and the cytosolic phosphate concentration is about half that of plasma. Little is known about how these relatively small phosphate gradients are maintained.

Because of the critical role of calcium and phosphate in structure and function, complex interrelated systems have evolved to maintain the homeostasis of these substances. These systems include the **parathyroids** that secrete PTH, the **parafollicular C cells** that secrete CT, and the **liver** and **kidneys** that produce vitamin D.

PARATHYROIDS
Embryology and Anatomy

The two pairs of parathyroids develop from **dorsal endoderm** that is associated with the bilateral third and fourth branchial pouches. Tissue from the third pouch, which also gives rise to the **thymus,** migrates caudad during development to become the **inferior pair** of parathyroids, with each gland near the lower poles of the thyroid lobes. Tissue from the **fourth branchial pouches,** which gives rise to the **superior pair** of parathyroids, develops in close association with the fifth branchial pouches (ultimobranchial bodies) or lateral thyroid anlagen. This tissue remains relatively stationary during development, such that each of the superior glands is situated dorsal and lateral to the thyroid near the level of the isthmus.

In the **adult,** the parathyroids are oval and measure about 6 mm in length and 4 mm in width, with each gland weighing approximately 40 mg. The **superior glands** are usually located laterally on the dorsal aspect of the thyroid near the level of the isthmus. The **inferior glands** are near the lower poles of the thyroid lobes in a more ventral position. Although there are usually four parathyroids, as few as two or as many as eight have been reported. In addition, the anatomic location of the inferior glands is variable, and they may be located more caudad and even in the anterior mediastinum.

Microscopically, the parathyroids are encapsulated by **fibrous connective tissue.** The glandular parenchyma is partitioned into lobules and these lobules are formed into cords by a highly vascular **fibroconnective tissue stroma** that also contains nerves, lymphatics, and fat cells. The most abundant epithelial cell of the parathyroid parenchyma is the **chief cell** which synthesizes PTH. These cells are polyhedral, contain clear cytoplasm, and are arranged in an interconnecting trabecular pattern. The other less abundant cell type is the **oxyphil,** which is larger and contains eosinophilic granules and a small hyperchromatic nucleus. These cells are scattered singly or in small aggregates throughout the gland. The function of these cells is unknown.

Parathyroid Hormone

Biosynthesis, Storage, Secretion, Circulation, and Metabolism

PTH is a **single-chain polypeptide** consisting of 84 amino acids with no disulfide links; its molecular weight is about 9,300. In humans there is a single PTH gene located on the short arm of **chromosome 11.** The gene codes for messenger RNA (mRNA) that is translocated from the nucleus to the cytoplasmic rough endoplasmic reticulum, where it is translated into a larger **prepro-PTH** polypeptide that is rapidly cleaved to produce pro-PTH. This **pro-PTH** is transferred to the **Golgi apparatus,** where it is cleaved to PTH that is then packaged into membrane-bound secretory vesicles for storage in the cytoplasm. These secretory vesicles are subsequently transferred to and fuse with the plasma membrane, extruding PTH from the cells by **exocytosis.**

Another substance, called **parathyroid secretory protein,** is also produced by the chief cells. This is a large glycoprotein that exists as a dimer of identical subunits, with a molecular weight of about 70,000. It is synthesized, stored, and secreted along with PTH, but has no known physiologic function.

The **circulation** of intact PTH in plasma is relatively brief (minutes) because of its rapid metabolism by the kidney and liver and loss in urine. The initial site of cleaveage is in the region of the amino acids 33 to 37, resulting in a longer carboxyl-terminal fragment and a shorter amino-terminal fragment. Because the biologically active portion of the peptide that interacts with PTH receptors resides within the first 27 amino acids of the amino terminal, these amino-terminal fragments may retain biologic activity but the carboxyl-terminal fragments are inert. Amino-terminal fragments, like PTH, also have a short plasma half-life due to renal excretion and degradation. In contrast, carboxyl-terminal fragments are slowly eliminated from the plasma by renal excretion and degradation, resulting in a long plasma half-life (hours) for these biologically inactive substances.

Cellular Interaction

PTH interacts with target cells in bone and kidney by binding to specific cell-surface receptors in the same way as other polypeptide hormones. This interaction leads to activation of membrane-bound adenylate cyclase that generates cyclic AMP (cAMP) from ATP. The cAMP subsequently interacts with regulatory subunits of protein kinase, liberating the catalytic subunit of the enzyme. The enzyme then catalyzes the phosphorylation of proteins that mediate various hormonal effects, but the mechanisms involved are poorly understood. In addition, other secondary messengers such as calcium may participate. It is unclear whether PTH receptors are identical in various PTH-responsive cells.

Effects

PTH has direct effects on bone and kidney and indirect effects on intestine. In **bone,** the major action is **mobilization of calcium** and **phosphate** from bone to extracellular fluid, accomplished in a biphasic fashion. Initially calcium and phosphate are released from areas of bone that are in **rapid equilibrium** with the extracellular fluid. This happens within minutes and the effect rapidly dissipates once PTH is removed. It is likely mediated through the stimulation of osteolysis by preformed osteoclasts (cells that resorb mineralized bone). About 12 hours later, there is **a delayed effect** that is characterized by marked resorption of bone due to the recruitment of new highly active osteoclasts. Within several days, there is also increased osteoblast activity (cells that synthesize bone matrix) that results in **remodeling of bone,** in conjunction with the increased osteoclast activity. These delayed effects may persist for several days after hormone stimulation is stopped.

Osteoclasts do not have PTH receptors; they only exist on osteoblasts. The effects of PTH on osteoclasts are therefore likely mediated by stimulation of osteoblastic production of a substance that is secreted locally within bone, that subsequently stimulates osteoclasts.

PTH has three major actions on the **kidney:** (1) it **increases calcium reabsorption** in distal nephrons; (2) it **decreases phosphate reabsorption** in the proximal nephrons; and (3) it stimulates 1-alpha hydroxylase, a mitochondrial mixed-function oxidase, in the proximal tubules that converts **25-hydroxyvitamin D** (25-OHD) to the more potent **1,25-dihydroxyvitamin D** (1,25-OH$_2$D). PTH also decreases proximal tubular reabsorption of sodium, bicarbonate, and water and increases magnesium reabsorption in the ascending limb.

PTH indirectly stimulates the intestinal absorption of calcium and phosphate. This is mediated by PTH's enhancement of the renal production of 1,25-OH$_2$D, the most potent vitamin D metabolite.

Regulation

PTH secretion is primarily regulated by **serum ionized calcium (Ca^{2+}) concentrations:** increases inhibit and decreases stimulate PTH release. The mechanisms that govern this are poorly understood, however. Preliminary data suggest that a **secondary messenger** within the chief cells, such as cAMP, may perform this function. Relatively small increments or decrements in the serum Ca^{2+} level translate into marked reciprocal changes in the PTH secretion, and this offers protection against perturbations of serum Ca^{2+} levels. PTH secretion is also modulated by several other factors. **Hypomagnesemia** inhibits and beta-adrenergic agonists stimulate PTH release. However, the importance of magnesium and catecholamines in the physiologic regulation of PTH secretion is questionable.

PARAFOLLICULAR C CELLS

Embryology and Anatomy

Parafollicular C cells that produce CT are embryologically derived from **neural crest cells** that migrate into the ventral portion of the fifth branchial pouches (ultimobranchial bodies). This tissue is incorporated into the lateral lobes of the thyroid during development, and this accounts for their intrathyroidal location. These cells are a minor constituent of the total thyroid mass and are usually seen as single cells or small clusters of cells scattered among follicles in the medial part of the lateral lobes.

Calcitonin

Biosynthesis, Storage, Secretion, Circulation, and Metabolism

CT is a 32–amino acid, single-chain polypeptide with a single disulfide link at the amino terminal and a proline-amide at the carboxyl terminal; its molecular weight is about 3,400. The gene for CT resides on the short arm of **chromosome 11** and codes for three peptides: **CT, katacalcin,** and **CT gene–related peptide** (CGRP). There are actually two CT genes in this region, alpha and beta, that are near the PTH gene.

The gene codes for two precursors: **CT–katacalcin** and **CGRP.** The **mRNA** for these alternative precursors is produced by a single transcript that is differentially processed in various tissues. CGRP is produced in the thyroid along with CT–katacalcin, but only CGRP is produced in appreciable quantities in **extrathyroidal sites** such as the central nervous system, where it probably functions as a neurotransmitter. The mRNA for **katacalcin** flanks the mRNA for CT, and these are cotranslated in the thyroid, leading to the synthesis, storage, and secretion of both peptides in equimolar ratios. However, the physiologic significance of katacalcin is unknown.

Biosynthesis of CT and **katacalcin** proceeds much like that described for PTH. The processed mRNA is translocated to the cytoplasmic rough endoplasmic reticulum, where it is translated into a larger **prepro-CT–katacalcin polypeptide.** This large precursor is probably reduced to a smaller **pro-CT–katacalcin polypeptide** before transfer to the Golgi apparatus. This substance is then processed into CT and katacalcin, and these are subsequently packaged into **membrane-bound secretory vesicles.** These substances are co-secreted by the binding of secretory vesicles to the plasma membrane, with the subsequent exocytosis of contents into the extracellular space. The processed mRNA for **CGRP** undergoes the same series of events, and CGRP may also be co-secreted with these two peptides or packaged into separate secretory vesicles.

CT in plasma may exist in several forms; as a **monomer, dimer,** or a **smaller metabolite.** Little is known about either the biologic significance of these various forms or their metabolism, though the kidney appears to be the principal site of clearance.

Cellular Interaction

CT interacts with cells in bone and kidney by binding to specific cell-surface receptors in the same way as do other peptide hormones. Like PTH receptors, it appears that **cAMP** is an important secondary messenger, but little else is known about this interaction. Certain areas of **nephrons** contain **CT receptors,** primarily where PTH receptors and actions are lacking. In addition, **osteoclasts** but not osteoblasts have been demonstrated to possess CT receptors.

Effects

The principal physiologic effect of CT is to **lower serum calcium** and **phosphate concentrations.** This appears to be predominantly mediated through the inhibition of osteoclastic activity, with a resulting reduction in both bone resorption and the release of calcium and phosphate into the extracellular fluid. In the **kidney,** CT has also been shown to accelerate calcium, phosphate, and sodium excretion and to stimulate 1-alpha-hydroxylase activity in the proximal nephrons.

Regulation

The primary regulator of CT secretion is the **Ca²⁺ concentration** in serum. **Hypocalcemia** results in suppressed secretion, whereas **hypercalcemia** leads to increased secretion. A number of other factors have also been demonstrated to influence secretion of this hormone; these include estrogen, 1,25-OH₂D, gastrin, and cholecystokinin.

VITAMIN D

Biosynthesis, Storage, Secretion, Circulation, and Metabolism

Vitamin D is a fat-soluble substance that is both a **vitamin** and a **classic steroid hormone.** It can be synthesized in humans or can be derived from animal and plant sources through the diet. **Cholecalciferol**, also called *vitamin D₃*, is produced in animal and human skin, and **ergocalciferol,** also called *vitamin D₂*, is produced in plants. Both are produced nonenzymatically by ultraviolet irradiation. The precursor in animals and humans is 7-**dehydrocholesterol** and the precursor in plants is a similar substance, **ergosterol** (Fig. 51-1). These two forms of vitamin D are actually **prohormones,** with little if any biologic activity, and their **active metabolites** are equipotent. They will both be referred to here as *vitamin D.*

Once vitamin D is synthesized in skin or absorbed from the intestine, it is transported by the blood to the liver, where it is hydroxylated by a microsomal mixed-function oxidase at the C-25 position to produce 25-OHD, a weakly active form of the vitamin (see Fig. 51-1). This substance is transported to the kidney via the blood, where it is hydroxylated at the C-1 position in the proximal nephrons to produce the most potent form of the vitamin, **1,25-OH₂D** (see Fig. 51-1). It may also be hydroxylated in the proximal nephrons by a mitochondrial enzyme at the C-24 position to form 24,25-OH₂D, a weakly active compound of unknown biologic importance. Both 1-alpha-hydroxylase and 24-hydroxylase activity have been demonstrated in other tissues, but their physiologic significance is questionable.

Vitamin D and its metabolites are stored in **adipose tissue** and **muscle,** but this storage is not closely regulated. Rather, it depends simply on the supply of vitamin available from the exogenous and endogenous sources.

All forms of vitamin D circulate in plasma

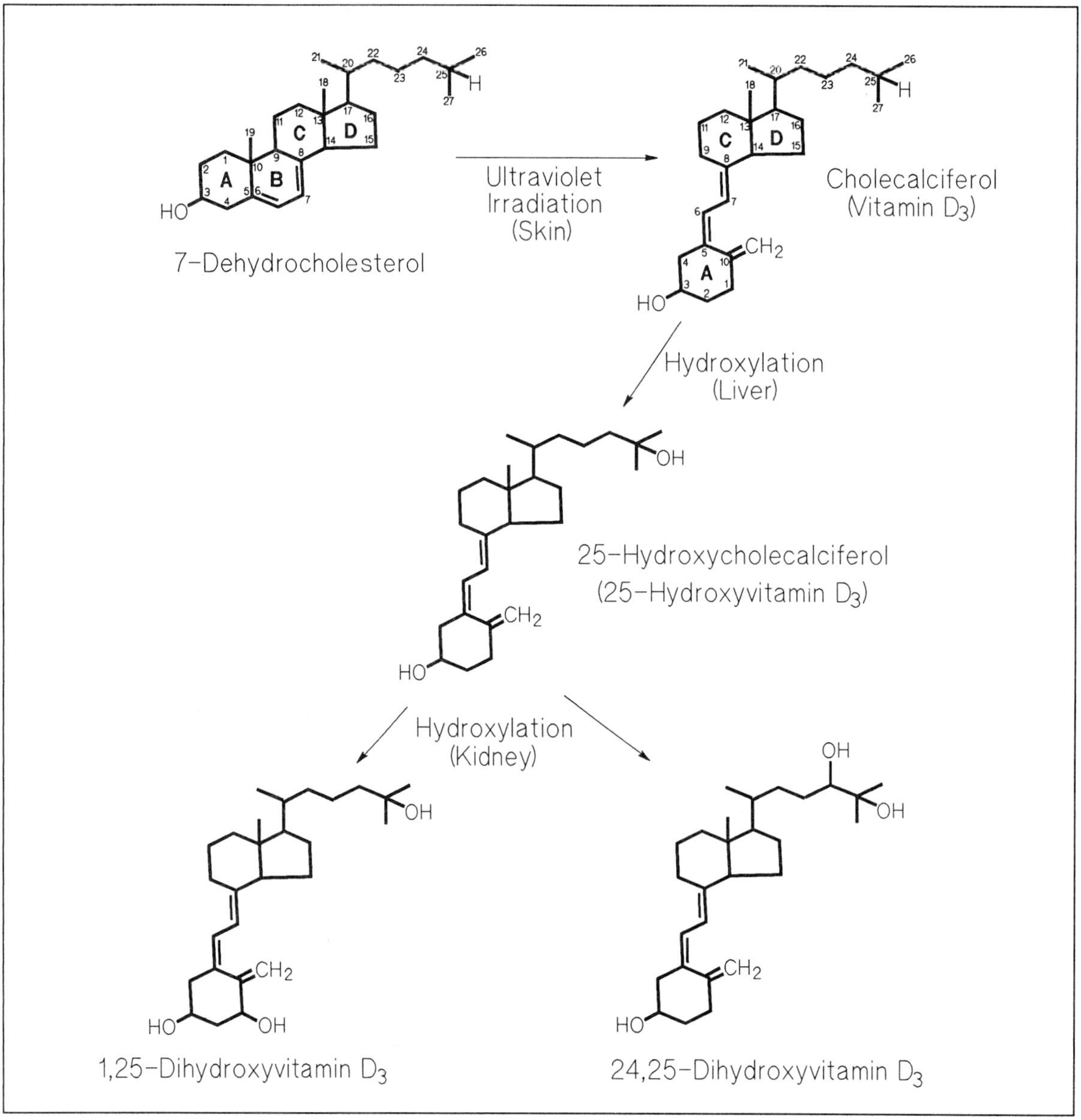

Fig. 51-1. Biosynthesis and structures of cholecalciferol (vitamin D_3) and its metabolites. Biosynthesis and structures of ergocalciferol (vitamin D_2) and its metabolites (not shown) are similar.

bound to a specific protein, called **vitamin D–binding protein.** This is an alpha-globulin with a molecular weight of about 56,000 that contains a single binding site capable of binding all vitamin D metabolites, but with differing affinities. It has a greater affinity for 25-OHD than for vitamin D or 1,25-OH$_2$D. At least in part this explains why 25-OHD is the major circulating form of vitamin D in plasma. Although 25-OHD circulates in plasma at a concentration 1,000-fold greater than that of 1,25-OH$_2$D, 1,25-OH$_2$D is several-fold more potent than 25-OHD and is considered responsible for most vitamin D action.

Hydroxylated vitamin D metabolites are primarily cleared by side-chain oxidation in the liver, with excretion of these products in bile. There is a significant enterohepatic recirculation of these metabolites, such that as much as 20 percent is reabsorbed.

Cellular Interaction

Vitamin D interacts with target cells in the intestine, bone, and kidney, in the same fashion as do other steroid and thyroid hormones. It enters cells and binds to specific nuclear receptors, and this leads to alterations in the transcription of mRNA from DNA. The mRNA is then translocated to the cytoplasmic rough endoplasmic reticulum, where it is translated into proteins that mediate the hormone's effects. One such protein, **calcium-binding protein,** appears to have an important role in these actions, at least with respect to calcium absorption in the intestine.

Effects

The primary physiologic effects of vitamin D are on the intestine, bone, and kidney. In the **intestine,** which is the source of new calcium and phosphate for maintaining normal structure and function after birth, 1,25-OH$_2$D enhances the absorption of both calcium and phosphate. This occurs predominantly in the small bowel. A small amount of **calcium** can be passively absorbed but most is absorbed by an active saturable process that depends on stimulation by vitamin D. In contrast, **phosphate absorption** is also somewhat passive but dependent on a vitamin D–stimulated active transport process that has a high capacity and is essentially unsaturable under normal conditions. The amount of phosphate absorbed is therefore determined by the amount available through diet. This less restricted control of phosphate absorption probably explains why serum phosphate concentrations vary widely, depending on oral intake. Both calcium and phosphate absorption require uptake into intestinal cells from the luminal brush border, transport and processing through cells, and finally secretion into the extracellular space by an active transport process.

The effects of vitamin D on **bone** are both direct and indirect. The principal effect is to provide adequate amounts of calcium and phosphate for **mineralization,** which is mediated indirectly by its enhancement of gastrointestinal absorption of these substances. **1,25-OH$_2$D** also augments bone resorption alone and in response to PTH. This action is probably mediated by stimulation of osteoclastic activity, but the mechanism is poorly understood. One theory is that 1,25-OH$_2$D induces monocytes to differentiate into osteoclasts rather than stimulating existing osteoclasts, since vitamin D receptors have been found in monocytes that are progenitors of osteoclasts but not in osteoblasts or mature osteoclasts.

In the **kidney**, 1,25-OH$_2$D decreases 1-alpha-hydroxylase and increases 24-hydroxylase activity in the proximal nephrons. Vitamin D has also been reported to influence calcium and phosphate reabsorption, but this remains a controversial finding. In addition, vitamin D receptors and actions have been demonstrated in a variety of other tissues, but their physiologic significance is unknown.

Regulation

The regulation of vitamin D production is complex. The **amount of vitamin D** in the diet and exposure to **ultraviolet irradiation** determine how much prohormone is available for 25-hydroxylation in the liver. Excess vitamin D,

25-OHD, and 1,25-OH$_2$D are all reported to diminish the activity of this hepatic enzyme. Subsequent 1-alpha hydroxylation in the kidney is also inhibited, whereas 24-hydroxylase is stimulated by 1,25-OH$_2$D. PTH and hypophosphatemia both stimulate renal 1-alpha-hydroxylase whereas hyperphosphatemia inhibits it.

INTEGRATED CALCIUM AND PHOSPHATE HOMEOSTASIS

The primary regulator of serum Ca^{2+} is **PTH.** The **parathyroids** are exquisitely sensitive to alterations in the level of serum Ca^{2+}. A rise or fall precipitates a **marked reciprocal alteration** in PTH secretion that initiates a series of events that attempt to restore the Ca^{2+} level to normal (Fig. 51-2). When the Ca^{2+} level decreases, increased PTH secretion stimulates calcium and inhibits phosphate reabsorption and enhances the production of 1,25-OH$_2$D in the nephrons. 1,25-OH$_2$D, in turn, stimulates the gastrointestinal absorption of both calcium and phosphate. PTH, either alone or in combination with 1,25-OH$_2$D, increases the bone resorption of calcium and phosphate. These complex interrelated events increase the Ca^{2+} concentration. Increases of serum Ca^{2+} precipitate opposite effects that attempt to lower the Ca^{2+} level. Whether or not CT is significantly involved in these homeostatic mechanisms in humans is controversial because excesses or deficiencies of this hormone in humans are not associated with Ca^{2+} disturbances. Thus, it seems likely that CT may have a modulatory rather than primary regulatory role.

The **kidneys** are the principal regulators of the **serum phosphate** level but phosphate concentrations are not tightly regulated like the Ca^{2+} levels. As previously mentioned, intestinal phosphate absorption is a high-capacity process that is loosely regulated, and therefore the amount of phosphate absorption depends on the amount in the diet. The kidneys maintain a **basal intrinsic phosphate reabsorptive capacity,** and thus serum phosphate concentrations vary widely depending on oral intake, under usual circumstances.

Extreme decrements in the serum phosphate level stimulate renal 1-alpha-hydroxylase activity and the increased production of 1,25-OH$_2$D that subsequently leads to increased intestinal absorption as well as bone resorption of both phosphate and calcium (Fig. 51-3). The resultant increment in the serum Ca^{2+} concentration suppresses the PTH secretion that in turn enhances renal phosphate reabsorption. These events work in concert to raise the serum phosphate level. States of marked hyperphosphatemia trigger opposite events that attempt to restore the serum phosphate to normal. There is also evidence that the kidneys are able to autoregulate their intrinsic phosphate reabsorptive capacity, such that it increases with decrements and decreases with increments of serum phosphate.

DISORDERS OF CALCIUM AND PHOSPHATE

Disorders of calcium and phosphate can be divided into those involving **mineralization of the skeleton** and **functional abnormalities** associated with perturbations of the serum Ca^{2+} and phosphate concentrations. Both types of disorders are commonly encountered in clinical medicine.

Decreased bone mass, referred to as **osteoporosis** or **osteopenia,** is probably the most common disorder of calcium and phosphate. It is encountered most often in postmenopausal women, and can be prevented to some extent by **estrogen replacement.** Although it is usually **idiopathic,** it may be caused by bone neoplasms, hyperparathyroidism, and a number of other diseases. It may be asymptomatic or associated with extreme pain and deformity due to vertebral compression fractures. Hip and wrist fractures due to trauma, even trivial injury, also often occur. A less common problem of bone mineralization is called **osteomalacia** when it occurs in adults and **rickets** in growing children. In these disorders, the bone matrix is produced normally but it is improperly mineralized. They are usually caused by **vitamin D deficiency**.

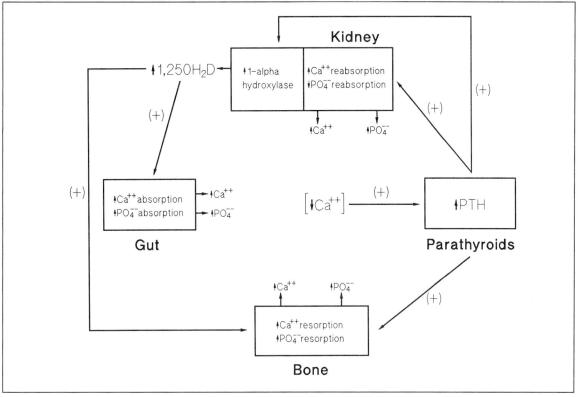

Fig. 51-2. Integrated Ca^{2+} homeostasis. The responses to mild decreases of serum Ca^{2+} concentrations are shown; opposite responses to mild increases occur. (+ = stimulation; *PTH* = parathyroid hormone.)

Disorders of serum calcium, and to a lesser extent phosphate, can cause profound dysfunction in a variety of processes because of the critical role these substances have in maintaining **cellular homeostasis.** For example, both hypercalcemia and hypocalcemia cause severe **neurologic dysfunction** and hypophosphatemia produces **extreme muscle weakness.** Such disorders may be caused by excess or deficient intake of calcium, phosphate, or vitamin D. They may also arise because of abnormally increased or decreased PTH secretion and renal dysfunction. In addition, both malignancy and granulomatous diseases often cause hypercalcemia. The etiology of hypercalcemia in the malignancy is usually the autonomous production of a peptide by the tumor that is similar to PTH, called *PTH-related peptide,* which can interact with PTH receptors. In **granulomatous diseases,** it is the unregulated elaboration of 1-alpha-hydroxylase in granulomas that results in increased 1,25-OH₂D production.

Evaluation of mineralization defects may include x-ray studies of the bone, radioisotopic bone scans, measurement of bone density by photon absorptiometry, and measurement of serum calcium, phosphate, PTH, and vitamin D concentrations, depending on the clinical situation. The **evaluation of serum calcium and phosphate disorders** obviously includes measurement of these substances, and also may include various bone assessments mentioned previously as well as determination of the plasma PTH and vitamin D levels, depending on the clinical setting.

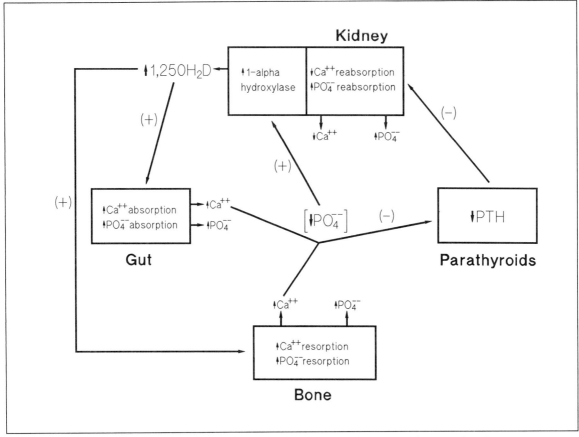

Fig. 51-3. Integrated phosphate homeostasis. The responses to marked decreases of serum phosphate concentrations are shown; opposite responses occur to marked increases. (+ = stimulation; − = inhibition; *PTH* = parathyroid hormone.)

SUMMARY

The maintenance of normal calcium and phosphate homeostasis is essential for the proper mineralization of the skeleton and for the proper structure and function of all cells. Thus, complex interrelated systems have evolved that work in concert to regulate and ensure that these substances are maintained in appropriate concentrations and proportions throughout the body. The major factors that contribute to this regulation and that provide a delicate check-and-balance system are PTH, CT, and vitamin D.

NATIONAL BOARD TYPE QUESTIONS

Select the one most correct answer for each of the following

1. Which substance is an important regulator of the serum Ca^{2+} level in humans?
   A. CT
   B. Katacalcin
   C. CGRP
   D. Parathyroid secretory peptide
   E. PTH
2. Which organ or tissue is primarily respon-

sible for producing 1,25-OH$_2$D
A. Skin
B. Liver
C. Bone
D. Kidney
E. Intestine

Select the response most closely associated with each numbered item. (The headings may be used once, more than once, or not at all.)
A. PTH
B. 25-OHD
C. CT
D. 1,25-OH$_2$D
E. 7-Dehydrocholesterol

3. The most potent form of vitamin D
4. Produced by chief cells
5. Produced by parafollicular C cells

Select the response most closely associated with each numbered item.
A. Vitamin D
B. PTH
C. Both
D. Neither

6. Peptide(s) that regulate(s) the serum Ca^{2+} level
7. Hormone(s) that primarily interact(s) with nuclear receptors
8. Hormone(s) essential for normal calcium and phosphate homeostasis

ANNOTATED ANSWERS

1. E. PTH is the most important of the substances listed in regulating the serum Ca^{2+} level.
2. D. 1,25-OH$_2$D is produced in the kidneys by 1-alpha hydroxylation of 25-OHD.
3. D. 1,25-OH$_2$D is the most potent form of vitamin D.
4. A. PTH is produced in the chief cells of parathyroids.
5. C. CT is produced in parafollicular C cells of the thyroid.
6. B. PTH is a peptide hormone that regulates the Ca^{2+} levels.
7. A. Vitamin D is a steroid hormone that interacts primarily with nuclear receptors.
8. C. PTH and vitamin D are hormones that are both essential for normal calcium and phosphate homeostasis.

BIBLIOGRAPHY

Aurbach, G. D., Marx, S. J., and Spiegel, A. M. Parathyroid hormone, calcitonin, and the calciferols. In: Wilson, J. D., and Foster, D. W., eds. *Textbook of Endocrinology*, 7th ed. Philadelphia: W. B. Saunders, 1985, Pp. 1137–1217.

Stewart, A. F., and Broadus, A. E. Mineral metabolism. In: Felig, P., et al, eds. *Endocrinology and Metabolism*, 2nd ed. New York: McGraw-Hill, 1987, Pp. 1317–1453.

52 Physiology of the Adrenal Gland

LAWRENCE M. DOLAN

Objectives

After reading this chapter, you should be able to:

Describe the significance of the anatomic relationship between the adrenal cortex and the medulla

Define the major branch points in steroid hormone synthesis and their significance

List the major hormones produced by each section of the adrenal gland

Define the major factors that regulate the production and release of each adrenal hormone

State the action of each adrenal hormone

Describe the degradation of each adrenal hormone

Compare the known differences in steroid hormone receptors

Apply the general principles of adrenal function tests to the diseases described

---

The **adrenal glands** are paired structures located superior to the upper pole of each kidney and lateral to the lower thoracic and upper lumbar vertebrae (Fig. 52-1). Each gland is composed of an **outer cortex** that completely encircles the **inner medulla**. The cortex and medulla are distinct structural units.

Three groups of **arteries** perfuse each gland. The **superior group** arises from the inferior phrenic artery, the **middle group** directly from the aorta, and the **inferior group** from the renal artery. These arteries enter the **adrenal capsule** and form a **thin plexus**. Blood vessels from the plexus form a **sinusoidal circulation**, which nurtures the cortex. After traversing the cortex, the blood vessels reunite to form the **plexus reticularus** at the cortical-medullary junction. A few arterioles pass directly through the cortex to supply the plexus reticularus. Thus, the blood flow to the medulla includes not only arterial

sources but also blood that has traversed the cortex. Blood from the plexus reticularus then bathes the cells of the medulla. The complex blood supply to the adrenal glands protects the glands from infarction and provides the appropriate milieu for the production of epinephrine.

EMBRYOLOGY

Two separate germ lines contribute to the formation of the adrenal gland. The **cortex** arises from mesoderm found medial to the genital ridge, and is present by the fourth week of gestation. The **medulla** is of ectodermal origin and evolves from neural crest cells that invade the "cortical" mesoderm during the fourth week of gestation. Thus, the intimate anatomic association between the cortex and medulla is established early in development.

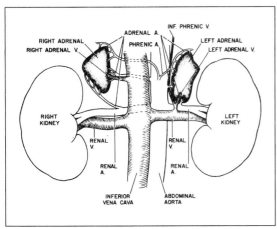

Fig. 52-1. Diagram of the adrenal glands, their location, and their blood supply. (*A* = artery; *V* = vein.) (Adapted from: Felig, P. *Endocrinology and Metabolism*. New York: McGraw-Hill, 1981, P. 387, with permission.)

HISTOLOGY

The **fetal adrenal cortex** consists of three concentric layers (Fig. 52-2). The outer layer, the **true cortex**, is beneath the connective tissue capsule. The middle layer, the **fetal cortex**, constitutes 80 percent of the adrenal gland at birth. The fetal cortex involutes during the first three weeks of life, leaving only a remnant of connective tissue surrounding the inner layer, the **medulla**.

After the first month of life, the true cortex dominates and comprises approximately 90 percent of the adrenal gland's weight. The **true cortex** is composed of three concentric layers. Directly beneath the capsule is the **zona glomerulosa**, which constitutes 15 percent of the cortex. The **zona fasciculata**, immediately below the zona glomerulosa, makes up 75 percent of the cortex. The **zona reticularis**, the inner layer of the adult cortex, surrounds the medulla. The medulla occupies approximately 10 percent of the adrenal gland. The cortex produces three types of **steroid hormones**: mineralocorticoids, glucocorticoids, and androgens. The medulla manufactures **catecholamines**.

STEROID HORMONES
Synthesis

Cholesterol is the common precursor of cortical steroid hormones. A unique cascade of biochemical alterations within each cortical zone produces steroid hormones with different biologic properties (Fig. 52-3). The **zona glomerulosa** synthesizes aldosterone. **Aldosterone**, the primary mineralocorticoid, contributes to sodium retention and K^+ and H^+ excretion by the kidney, gastrointestinal tract, and salivary and sweat glands. The **zona fasciculata** synthesizes cortisol, a glucocorticoid. **Cortisol** plays a major role in metabolic homeostasis of cells. It also helps to maintain the circulating blood sugar level by diminishing peripheral glucose use and increasing hepatic glucose production from amino acids and free fatty-acid precursors. It also participates in the response to physical and emotional stress and facilitates macrophage, B-cell, and T-cell functions. The zona reticularis manufactures androgens and physiologically insignificant amounts of estrogens. **Dehydroepiandrosterone** (DHEA), **dehydroepiandrosterone-sulfate** (DHEA-S), and **androstenedione** are the major androgens. The androgens contribute to **virilization** (pubic and facial hair development) and **somatic growth**.

The adrenal cortex does not store a large concentration of any of the steroid hormones. Therefore, an increased demand for steroid hormones requires a readily available supply of cholesterol. The adrenal gland has three sources of cholesterol: (1) increased uptake from circulating low-density and high-density lipoproteins, which is the primary source; (2) endogenous synthesis by the gland; and (3) hydrolysis of stored cholesterol esters.

The rate-limiting step in the production of cortical steroid hormones is the conversion of cholesterol to pregnenolone (see Fig. 52-3). Subsequent alterations in the activity of the enzymatic pathways dictate the type and concentration of steroid hormone produced.

Figure 52-3 details the production of aldosterone in the zona glomerulosa. The unique features

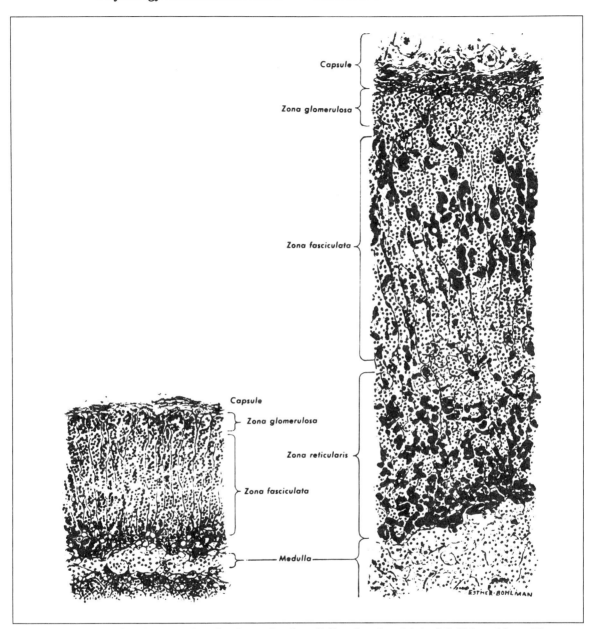

Fig. 52-2. Histology of the human adrenal at 6 months of age (left) and in the adult (right). (Adapted from: Bloom and Fawcett, *A Textbook of Histology*. Philadelphia: W. B. Saunders Company, 1962, P. 361, with permission.)

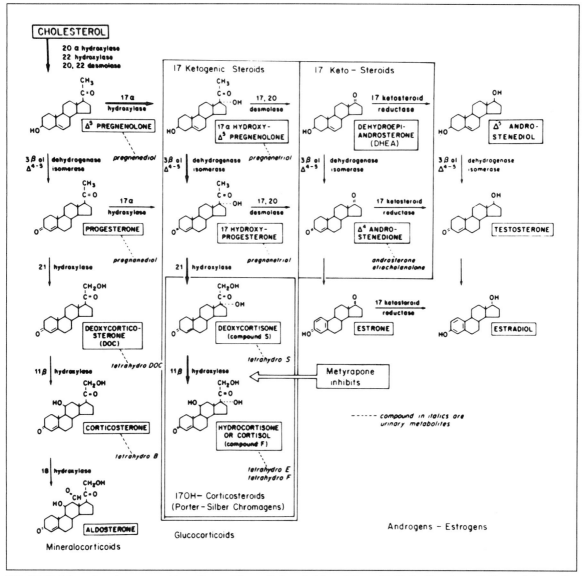

Fig. 52-3. Adrenal cortex enzymatic pathway. (From: Dolan, L. M. and Carey R. M. In: Vaughan, E. D., and Carey, R. M., eds. *Adrenal Disorders.* New York: Thieme Medical Publishers, 1989, P. 85, with permission.)

of the mineralocorticoid enzymatic pathway are the presence of 18-hydroxylase and the absence of 17α-hydroxylase. This lack of 17α-hydroxylase prevents the zona glomerulosa from synthesizing glucocorticoids, androgens, or estrogens. 18-Hydroxylase catalyzes the final step in aldosterone

production but is not found in other cortical layers. This special enzymatic milieu makes the zona glomerulosa the sole source of aldosterone.

The presence of 17α-hydroxylase in the zona fasciculata and reticularis permits both pregnenolone and progesterone to enter the glucocorticoid and androgen pathways. The subsequent enzymatic conversion of these substances results in the production of cortisol, DHEA, DHEA-S, and androstenedione, as well as the estrogens. Figure 52-3 documents the details of the biochemical pathways involved.

Secretion and Metabolism

Mineralocorticoids

The renin–angiotensin system and changes in the serum concentrations of potassium, sodium, and adrenocorticotropin hormone (ACTH) regulate aldosterone secretion. The **renin–angiotensin system** is the major factor governing aldosterone synthesis and secretion. **Renin**, a proteolytic enzyme, is synthesized, stored, and secreted by the juxtaglomerular apparatus of the kidney (Fig. 52-4). It is released in response to reduced systemic pressure, increased sympathetic nervous system activity, and decreased

Fig. 52-4. The juxtaglomerular apparatus. (Adapted from: Davis, J. O. What signals the kidney to release renin? *Circ. Res.* 28:301, 1971, with permission of the American Heart Association.)

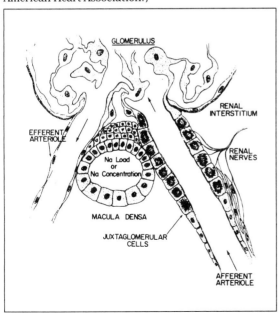

serum sodium or potassium concentrations. Conversely, a rise in systemic blood pressure, decreased sympathetic nervous system activity, or increased angiotensin-II concentrations suppress renin release.

Renin cleaves **angiotensinogen**, an alpha$_2$ globulin produced by the liver, to form the decapeptide, **angiotensin**-I (Fig. 52-5). Angiotensin-I is converted to **angiotensin-II** by a converting enzyme that removes two amino acids from the carboxyl terminal of angiotensin-I. The converting enzyme is found in numerous organs, including the lung, kidney, and liver, as well as throughout the systemic vascular bed. Amino-terminal cleavage of angiotensin-II by aminopeptidases produces **angiotensin-III**.

Angiotensin-II has two major functions. It is the **most potent vasoconstrictor** found in mammals and stimulates the **production and secretion of aldosterone** from the cells of the zona glomerulosa. Angiotensin-II has two known influences on the **mineralocorticoid enzymatic pathway**. It increases the conversion of cholesterol to **pregnenolone** and the conversion of corticosterone to **aldosterone** (see Fig. 52-3). Angiotensin-III is only 25 percent as effective as its parent compound, both as a vasoconstrictor and as a stimulator of aldosterone secretion. The role of angiotensin-III in steroidogenesis is unclear.

Angiotensin-II binds to a surface-cell receptor and this initiates its biologic action. The exact intracellular messenger for angiotensin-II has not been identified, but there is mounting evidence that a rise in intracellular calcium concentration is the second messenger. Intracellular calcium concentration can be altered by directly affecting the calcium flux across membranes or by varying the activity of phospholipase C. **Phospholipase C** (a membrane enzyme) catalyzes the breakdown of phosphoinositol-4, 5-bis-phosphate (a membrane phospholipid) to form both inositol-trisphosphate (IP$_3$) and diacylglycerol (DAG). IP$_3$ promotes calcium flux from the endoplasmic reticulum to the intracellular space. DAG enhances calcium binding to protein kinase C, and this enhances protein kinase C activity. Activation of

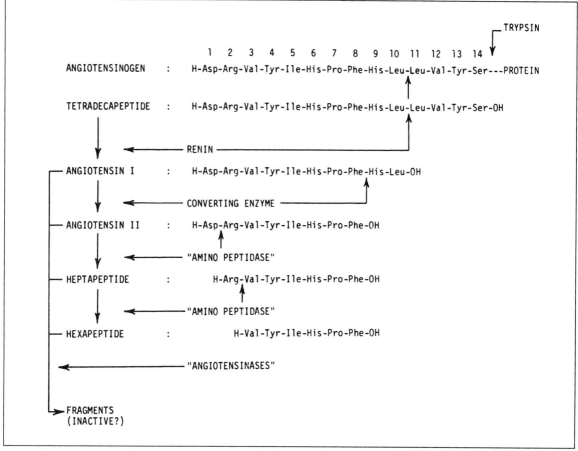

Fig. 52-5. The renin–angiotensin system. (Adapted from: Genest, J. *Hypertension*. New York: McGraw-Hill, 1977, P. 141, with permission.)

protein kinase C is associated with the phosphorylation and dephosphorylation of other intracellular proteins that are believed to mediate the biologic effect of angiotensin-II. The intracellular messenger for angiotensin-III has not yet been determined.

Angiotensin-II and angiotensin-III are both inactivated by cleavage to smaller peptides by angiotensinases, endopeptidases, and carboxypeptidases. These peptidases are found in serum and the vascular beds of the kidney, liver, and lung.

Other factors that directly regulate aldosterone secretion include serum potassium, sodium, and ACTH concentrations. Small alterations in the serum potassium level (0.1 mEq/liter) and large changes in the serum sodium concentration (greater than 10 mEq/liter) can affect aldosterone secretion. In both instances, there is an inverse relationship between the electrolyte concentration and the rate of aldosterone secretion.

The final known mediator of aldosterone secretion is **ACTH**. In normal individuals, an acute injection of ACTH results in a prompt increase in the serum aldosterone concentration. Continuous infusion produces a rise in the serum aldosterone level, which is maintained for 24 hours. Subsequently, the concentration falls to preadministration levels despite the continued administration of ACTH.

Aldosterone is poorly bound to serum proteins and has a serum half-life of 20 to 30 minutes. Aldosterone is metabolized by the liver, primarily to **tetrahydroaldosterone**. **Tetrahydroaldosterone**, which comprises 30 to 40 percent of the aldosterone excreted, and an **18-glucuronide** by-product are the major metabolites.

Glucocorticoid

ACTH, a 39-amino acid polypeptide secreted by the pituitary gland, exerts a major influence on adrenal cortical function (Fig. 52-6). It stimulates cortisol release, increases the uptake of cholesterol from serum lipoproteins, preserves and enhances the ability of the enzymatic cascade to produce cortisol, and maintains the adrenal cortex through a trophic action.

The serum **cortisol** concentration increases within three minutes of the administration of ACTH. Because only a small amount of cortisol is stored in the adrenal gland, the rise in its serum concentration is primarily achieved by increased hormone production. ACTH appears, however, to be a **regulator** rather than an initiator of cortisol synthesis and secretion. This concept is supported by the following evidence. Low serum cortisol concentrations are found in subjects who have undergone total hypophysectomy, and the serum cortisol concentration does not increase after an acute administration of ACTH in such subjects. In this situation, priming with an infusion of ACTH for 48 to 72 hours is required before the serum cortisol concentration increases.

ACTH also has a trophic action on the **adrenal gland**. When administered to both intact and hypophysectomized animals, it causes adrenal cortical hypertrophy. Cortical atrophy develops after total hypophysectomy. If one of the adrenal

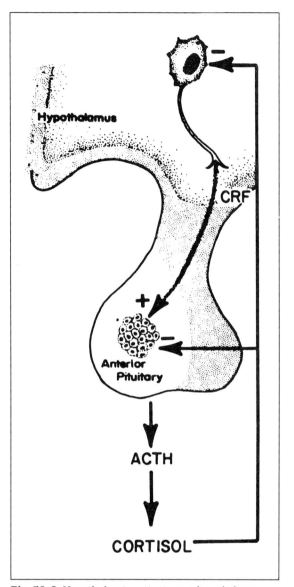

Fig. 52-6. Hypothalamic–pituitary–adrenal glucocorticoid axis. (*CRF* = corticotropin-releasing factor; *ACTH* = adrenocorticotropin hormone.) (Adapted from: Felig, P., et al, eds., *Endocrinology and Metabolism*. New York: McGraw-Hill, 1981, P. 216, with permission.)

glands is removed, the remaining adrenal gland exhibits compensatory hypertrophy. In hypophysectomized animals, compensatory hypertrophy is not observed unless exogenous ACTH is administered.

ACTH secretion is regulated by an inherent circadian rhythm, stress, and the circulating cortisol concentration. The morning serum cortisol concentration varies between 8 and 25 μg/dl, with an approximate 50 percent decrease by late afternoon. The **circadian rhythm** is established shortly after infants begin sleeping through the night. The serum cortisol concentration also rises in response to **physical or emotional stress**. Both the diurnal rhythm and the stress response of ACTH are mediated by the release of a hypothalamic peptide, **corticotropin-releasing hormone** (CRH). The **cerebral cortex** and **brainstem** regulate the release of CRH from the hypothalamus. Increased serum cortisol concentration inhibits the release of both CRH and ACTH (see Fig. 52-6).

ACTH binds to a surface-cell receptor to initiate ACTH's biologic effect. The exact intracellular mechanism whereby ACTH stimulates cortisol synthesis and release is unknown. The administration of ACTH, however, activates adenylate cyclase with attendant increased production of cyclic AMP (cAMP). cAMP subsequently activates protein kinase A, resulting in the phosphorylation of specific proteins. The rise in cAMP levels precedes heightened steroidogenesis. cAMP derivatives can also stimulate steroid production. These associations suggest that the generation of cAMP is intimately involved in glucocorticoid production.

At physiologic concentrations, 80 percent of cortisol is bound to corticosteroid-binding globulin (CBG), 15 percent to albumin, and the remaining 5 percent to other serum glycoproteins. Physiologic and pharmacologic changes alter the CBG level. **Pregnancy** and **hyperthyroidism** increase its concentration, while **liver disease** and **nephrotic syndrome** lower the concentration. These conditions cause a change in the total serum cortisol concentration but do not significantly influence the **free cortisol concentra-**

tion, which is the metabolically active component.

The half-life of serum cortisol is 60 to 120 minutes, and the primary site for cortisol metabolism is the **liver.** To do this, the double bond in the cholesterol ring (see Fig. 52-3) is reduced, and this forms the inactive compound, dihydrocortisol. Sixty to seventy percent of the metabolites of cortisol are conjugated to glucuronide, while a small fraction is bound to sulfates. The addition of glucuronide and sulfate increases the solubility of the metabolites and enhances renal excretion.

Androgens

DHEA, **DHEA-S**, and **androstenedione** are the principal androgens secreted by the adrenal cortex, but the production and release of these compounds are not well understood. Acute administration of ACTH elicits a rise in DHEA. Serum DHEA concentration mimics the **circadian variation** seen in serum cortisol concentration. Prolonged ACTH administration raises both the DHEA and DHEA-S concentrations.

ACTH is not the only regulator of androgen secretion. The circulating DHEA-S concentration increases at adrenarche, which occurs one year before the onset of puberty, and is unassociated with significant changes in ACTH or cortisol secretion. This dissociation of adrenarche both from puberty or a change in cortisol secretion makes it unlikely that either the gonadotropins or ACTH mediate adrenarche. Investigators have suggested two regulators of adrenarche: (1) a **maturational shift** in the pathway of adrenal steroid biosynthesis (increased activity of 17, 20-desmolase activity [see Fig. 52-3]) or (2) increased secretion of a yet to be isolated hormone, referred to as **adrenal adrenarche–stimulating factor**. Investigators have suggested both a pituitary and an extrapituitary source of this factor.

Circulating androgens are weakly bound to circulating proteins. The androgens have several potential metabolic fates. **Degradation** primarily takes place in the liver but can also occur in the kidney. DHEA and androstenedione are converted principally to sulfate or glucuronide conjugates and then excreted in the urine. DHEA can, however, be directly excreted in the urine.

Receptors

In contrast to the peptide hormone receptors, which are found on the surface of the cell membrane, steroid hormone receptors are found in either the **cytoplasm** or **nucleus** (Fig. 52-7). The lipophilic properties of the steroid hormones allow their passage through the cell membrane and access to the receptor. After the hormone binds to the receptor, the receptor undergoes a conformational change that reveals a DNA-binding site. If the receptor–hormone complex is in the cytosol, it translocates to the nucleus. In the nucleus, the receptor–hormone complex binds to chromatin by means of the DNA-binding site. This process initiates or inhibits the transcription of specific genes producing messenger RNA (mRNA). mRNA is then translated into specific proteins. Changes in the concentrations of induced proteins mediate the biologic effect of the steroid hormone. When transcription is completed, the receptor–hormone complex dissociates from the chromatin and each other. The receptor returns to its original environment (nucleus or cytoplasm) and configuration (hidden DNA-binding site).

Mineralocorticoid Receptor

The gene for the human mineralocorticoid receptor is present on **chromosome 4**. This receptor

Fig. 52-7. Activation of the glucocorticoid receptor (*GR*). (*CBG* = corticosteroid-binding globulin.) (Adapted from: DeGroot, L. *Endocrinology*, 2nd ed. Philadelphia: W. B. Saunders Co, 1989, P. 1558, with permission.)

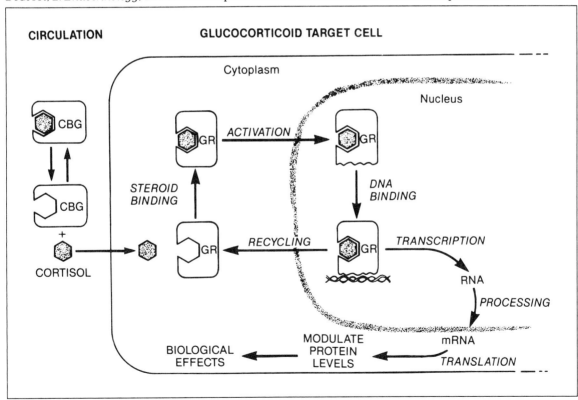

has been cloned and sequenced, revealing a 984–amino acid protein with a molecular weight of 107 kDa. The mineralocorticoid receptor is found primarily in the cytoplasm. The specific intracellular messengers that mediate the biologic action of the mineralocorticoids have not been defined.

Recent work with the mineralocorticoid receptor has revealed some intriguing data. In vitro, the receptor binds aldosterone and certain glucocorticoids at physiologic concentrations with a high affinity. These data are consistent with the significant homology that exists between the mineralocorticoid and glucocorticoid receptors at their steroid hormone–binding sites. However, in vivo experiments using tissues responsive to mineralocorticoids demonstrate binding of mineralocorticoid but not glucocorticoid to the mineralocorticoid receptor. These data suggest that the tissue response to mineralocorticoid is dictated by factors other than the mineralocorticoid receptor alone. Two possible mechanisms have been suggested. First, CBG does not bind aldosterone, but may selectively inhibit the binding of glucocorticoid to the mineralocorticoid receptor in mineralocorticoid-responsive tissues. Second, 11β-hydroxysteroid dehydrogenase, an enzyme that converts cortisol to its 11-keto analogue, may be present in the mineralocorticoid-responsive tissues. The 11-keto analogue of cortisol does not bind to the mineralocorticoid receptor. Thus, degradation of glucocorticoid may take place locally in tissues that are mineralocorticoid responsive. Further work is required to clarify these issues.

Glucocorticoid Receptor

The human glucocorticoid receptor gene is found on **chromosome 5**. The receptor, a 777–amino acid protein, has a molecular weight of 94 kDa and is found in both the cytoplasm and nucleus. The significant homology between the glucocorticoid and mineralocorticoid receptor and its possible physiologic significance have been discussed. The specific intracellular proteins generated after mRNA translation that mediate the biologic actions of glucocorticoids have not been identified.

Androgen Receptor

The gene for the androgen receptor is found on the X chromosome, and the human form of the receptor has a molecular weight of 98 kDa. Human androgen receptors are found in both the nucleus and cytoplasm. The specific intracellular proteins that mediate the biologic actions of androgen have not been identified.

CATECHOLAMINES

Synthesis

Catecholamines are naturally occurring compounds consisting of a catechol (3,4-dihydrophenyl) nucleus linked to an amine (Fig. 52-8). The catecholamines are synthesized by the sequential conversion of dietary tyrosine to epinephrine. **Tyrosine hydroxylase** is the major regulator of catecholamine production (see Fig. 52-8). **Dopamine**, **norepinephrine**, and **epinephrine** are the catecholamines found in humans. Each is stored in high concentrations in separate vesicles.

During **gestation** and the **newborn period**, the primary end-product of the adrenal medulla is norepinephrine. With **advancing age**, there is a dramatic decline in the norepinephrine-to-epinephrine ratio, associated with a marked rise in **phenylethanolamine-N-methyl transferase** (PNMT) activity, the enzyme that catalyzes the conversion of norepinephrine to epinephrine (see Fig. 52-8). PNMT activity requires high local concentrations of glucocorticoid. Thus, the production of epinephrine by the medulla is linked to the anatomic and vascular relationship between the cortex and medulla, which ensures that the medullary cells are bathed with a high concentration of cortisol.

Secretion and Metabolism

The major stimulus for catecholamine release from the adrenal medulla is **preganglionic sym-**

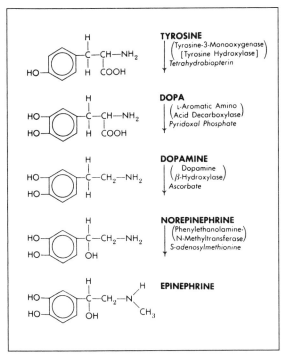

Fig. 52-8. Enzymatic pathway and cofactors for catecholamine synthesis. (Adapted from: *Goodman and Gilman's The Pharmacological Basis of Therapeutics*, 8th ed. Copyright 1990, Pergamon Press, Inc., P. 102, with permission of MacMillan Publishing Co.)

pathetic nervous system activation. Stress, change in posture or temperature, asphyxia, hypotension, low blood sugar levels, and sodium depletion are all factors that activate the sympathetic nervous system. Direct stimulation of the medullary cells by histamine, acetylcholine, and angiotensin-II also produces catecholamine release, but the physiologic significance of these factors is unknown.

Serum catecholamines are rapidly removed from the circulation and have a serum half-life of less than 20 seconds. This brief half-life is the combined result of rapid uptake of catecholamines by tissues and inactivation in the vascular system and liver. The major process of catecholamine inactivation is believed to be uptake by tissues. The process has been divided into two categories: uptake$_1$ and uptake$_2$. **Uptake$_1$**

consists of the removal of serum catecholamines, primarily norepinephrine, by postganglionic sympathetic neurons, with subsequent sequestration in neuronal storage vesicles. **Uptake$_2$** principally involves the accumulation of epinephrine by extraneuronal tissues. Epinephrine is then degraded by catecholamine-*o*-methyltransferase and monoamine oxidase to form inactive metabolites. The primary metabolites are vanillylmandelic acid, metanephrine, normetanephrine, and their glucuronide and sulfate derivatives.

Receptors

The catecholamine receptors consist of several similar, but distinct, types of surface-cell receptors defined by their physiologic action and pharmacologic specificity. The subtypes include alpha$_1$, alpha$_2$, beta$_1$, and beta$_2$ adrenergic receptors. In general, norepinephrine and epinephrine have an equipotent effect on the alpha$_1$, alpha$_2$, and beta$_1$ receptors. Epinephrine has a more potent effect on the beta$_2$ receptor.

The **alpha$_1$ receptors** are primarily found on vascular smooth muscle and their activation results in **vasoconstriction**. The alpha$_1$ receptor activates phospholipase C, a membrane enzyme, and this leads to the generation of intracellular second messengers, DAG and IP$_3$ (see the section on mineralocorticoid secretion and metabolism for details). Both products enhance the intracellular calcium concentration.

The **alpha$_2$ receptors** are found principally on presynaptic nerve terminals. Activation inhibits the production of cAMP and attenuates norepinephrine release by the sympathetic nervous system. Postganglionic alpha$_2$ receptors are also found in gastric smooth muscle cells and the beta cells of pancreatic islets. Stimulation causes **decreased gastric motility** and **attenuated insulin secretion**.

The **beta$_1$ receptors** are found primarily on cardiac pacemakers and muscle cells. Stimulation increases cAMP production with increased **chronotropic** and **inotropic function**.

The **beta$_2$ receptors** are found principally on

respiratory smooth muscle cells, uterine smooth muscle, salivary glands, and the liver. Stimulation increases cAMP production, with resultant respiratory and uterine smooth muscle dilatation, increased secretion by the salivary glands, and enhanced glycogenolysis and gluconeogenesis.

MAJOR CLINICAL DISORDERS

The major clinical disorders of the adrenal gland can be defined as pathologic states reflecting either an excessive or deficient production of the adrenal hormones. Following are examples of diseases involving each hormone.

Mineralocorticoid

Primary **hyperaldosteronism**, a rare cause of mild to moderate hypertension, is the result of the autonomous, unilateral, and excessive secretion of aldosterone by a tumor. **Hypokalemia** is a cardinal sign. Total body potassium depletion results in excessive H+ secretion by the cortical collecting duct, resulting in metabolic alkalosis. Diagnostic tests reveal autonomous mineralocorticoid secretion. Surgical removal of the tumor is required for effective treatment.

Deficient aldosterone production can result from a number of enzymatic deficiencies in steroidogenesis (see Fig. 52-3) or from destruction of the zona glomerulosa or the adrenal cortex (Addison's disease). Isolated **mineralocorticoid deficiency** causes intravascular volume depletion, hyponatremia, hyperkalemia, and shock. In **Addison's disease** (mineralocorticoid and glucocorticoid deficiency), the patient also exhibits signs of insufficient cortisol production (to be discussed). The lack of aldosterone response to ACTH (see the section "Adrenal Function Tests") is diagnostic. Effective treatment requires exogenous mineralocorticoid replacement.

Glucocorticoid

Cushing's disease is characterized by the inability of physiologic concentrations of serum glu-

cocorticoid to suppress ACTH secretion, resulting in excessive cortisol secretion. The signs and symptoms include growth failure, cutaneous hyperpigmentation (ACTH excess), moon facies, purple striae, mild hypertension, trunkal obesity, and thin skin. Diagnosis requires the suppression of glucocorticoid secretion by a pharmacologic, but not physiologic, dose of glucocorticoid. Treatment usually requires the surgical removal of the pituitary lesion.

Glucocorticoid deficiency can result from an enzymatic deficiency in steroidogenesis (congenital adrenal hyperplasia) or destruction of the adrenal cortex (Addison's disease). Decreased glucocorticoid production provokes excessive ACTH secretion. Although five types of congenital adrenal hyperplasia have been described, the most common is **21-hydroxylase deficiency** (Fig. 52-3 and 52-9). The position of the deficient enzyme in the steroidogenic cascade (see Fig. 52-1) dictates the steroid hormones produced and the accompanying clinical signs and symptoms. The **clinical picture** can include hyperpigmentation, shock due to cardiac dysfunction, mineralocorticoid deficiency, and excessive virilization of a genetic female or undervirilization of a genetic male. **Addison's disease** is characterized by glucocorticoid deficiency (cardiac dysfunction, anorexia, fatigue, weakness) and mineralocorticoid deficiency (as previously described). Treatment of congenital adrenal hyperplasia and Addison's disease requires the administration of glucocorticoid and mineralocorticoid.

Androgen

Excessive androgen production is characterized by premature or exaggerated virilization. This can result from certain forms of congenital adrenal hyperplasia (as described) or an adrenal tumor. This latter condition requires surgical removal of the tumor.

An isolated deficiency of adrenal androgen production has no known clinical consequence.

Catecholamines

A prime example of excessive catecholamine production is a **pheochromocytoma**. The typical symptoms of this rare cause of intermittent hypertension are headaches, palpitations, excessive perspiration, and paroxysm (pallor, anxiety, nausea, and weakness). Surgical removal is the standard form of therapy.

Epinephrine plays a secondary role in the acute response to **hypoglycemia**. Some individuals with diabetes mellitus do not produce enough epinephrine, and are thus at risk for asymptomatic, prolonged hypoglycemia.

ADRENAL FUNCTION TESTS

Diagnostic tests that assess adrenal integrity document either basal or dynamic function. **Baseline tests** employ single or serial measurements of the circulating concentration of a hormone or a timed urine collection. The former test assesses a single moment or series of moments in time. The latter gives an integrated view of baseline function over a longer period. Either method can only be used as a screening process, but baseline tests are not very sensitive for diagnosing adrenal insufficiency.

Dynamic adrenal function tests attempt to either stimulate or suppress the basal level of the adrenal hormone of interest. Stimulation tests are required when decreased production is suspected. Suppression tests are in order if the converse is entertained.

The interpretation of both baseline and dynamic tests must take into consideration the clinical status of the patient. Dietary sodium intake and the patient's posture significantly affect mineralocorticoid metabolism. Stress can prevent glucocorticoid suppression. **Androgen tests** are influenced much less by the clinical status of the patient. **Catecholamines values**, however, are significantly influenced by exercise, stress, posture, and a number of medications. Before performing any adrenal function tests, the patient's clinical status should match that of the reference population to avoid the risk of misinterpretation of the results.

Mineralocorticoid

Basal serum aldosterone concentrations yield little useful diagnostic information. **Hypokalemic metabolic alkalosis** can be an early clue to excessive aldosterone production. A 24-hour urine collection for determination of the metabolites of aldosterone also offers little insight into the status of a patient unless the values are markedly elevated.

Manipulation of aldosterone secretion is, however, an effective diagnostic tool. The most useful stimulation test exploits the response to ACTH administration. If the adrenal gland is present and the enzymatic cascade intact, there is a prompt rise in aldosterone secretion following the acute administration of ACTH. Some investigators have suggested that the adrenal gland must also be exposed to low levels of ACTH in order to respond to exogenous ACTH. Clinical experience, however, suggests that the enzymatic system of the zona glomerulosa is maintained by the renin–angiotensin system, even in subjects who suffer from ACTH deficiency.

If excessive aldosterone production is suspected, suppression tests are in order. A number of methods, including oral and intravenous sodium loading, mineralocorticoid and glucocorticoid administration, and changes in posture, can be used to identify and define the pathophysiologic process.

Glucocorticoid

Basal glucocorticoid function tests give some insight into the status of the CRH-ACTH–adrenal axis. A low 8:00 A.M. serum cortisol value raises the question of glucocorticoid deficiency. Lack of a circadian rhythm or increased urinary concentrations of free cortisol, 17-hydroxycorticosteroids, 17-ketogenic steroids, or metabolites of a precursor of cortisol (see Fig. 53-3) suggest either excessive ACTH production or autonomous activity in the zona fasciculata. Only dynamic testing can identify the pathologic process.

Three **stimulation tests** are commonly employed: stress, metyrapone, and ACTH. **Insulin-induced hypoglycemia** is the most widely used stress test. An appropriate rise of serum cortisol in response to hypoglycemia documents an intact CRH–ACTH–adrenal axis. A subnormal response fails to identify which part of the axis is deficient. Because of the risks associated with hypoglycemia, **metyrapone** is preferred by some to evaluate the CRH–ACTH–adrenal axis. Metyrapone inhibits the final step in the production of cortisol, rendering the patient glucocorticoid deficient (see Fig. 52-3). A normal response is an increase in ACTH and the immediate precursor of a cortisol, deoxycortisone. As with insulin-induced hypoglycemia, a subnormal response does not identify the specific part of the axis that is defective.

A normal response to **exogenous ACTH** documents a functionally intact adrenal gland that has been primed with endogenous ACTH, but the test does not directly document CRH and ACTH sufficiency. Investigators have compared the response of individuals to ACTH challenge with those undergoing insulin-induced hypoglycemia. An intact response to ACTH has been highly correlated but not absolutely associated with a normal response to hypoglycemia. Thus, despite the high correlation between the two tests, one cannot be absolutely sure that the CRH–ACTH portion of the axis is intact based solely on a normal response to ACTH.

Suppression tests exploit the response to different doses of **dexamethasone** (an extremely potent glucocorticoid) to identify the source of either excessive ACTH (pituitary, ectopic) or cortisol production. In normal individuals, endogenous glucocorticoid production is suppressed with low-dose dexamethasone (see Fig. 52-6). In subjects with excessive ACTH production of pituitary origin, cortisol production is inhibited only with high-dose dexamethasone. Individuals who show no suppression with either low or high doses of dexamethasone have either an ectopic (nonpituitary) source of excess ACTH production

or autonomous production of cortisol by the adrenal gland. The concentration of basal circulating ACTH and radiographic studies can help to differentiate ectopic ACTH production from an adrenal tumor.

Androgen

Androgen function tests are principally employed to identify individuals with excessive androgen production caused by a physiologic (enzymatic defect in steroidogenesis, increased production of ACTH by the pituitary) or autonomous (tumor) process. Elevated basal serum levels of DHEA-S, androstenedione, or 17-ketosteroids can be used to screen individuals with suspected excessive production (see Fig. 52-3). The dexamethasone suppression test differentiates between physiologic and autonomous processes.

Medulla

Baseline medullary function tests can be used to identify individuals with excessive catecholamine production. Both single-serum and 24-hour urine concentrations are employed.

Dynamic tests include the administration of glucagon, phentolamine, or clonidine. **Glucagon** stimulates the catecholamine release. A patient with excessive catecholamine production will exhibit a rapid and dramatic increase in blood pressure. The **phentolamine** test blocks the $alpha_1$-adrenergic effect of catecholamines. Subjects synthesizing an excessive quantity of catecholamine will display a fall in systemic pressure. **Clonidine** inhibits centrally mediated adrenergic function. As a result, a subject with autonomous catecholamine production will not show decreased sympathetic nervous system activity or serum catecholamine concentration.

SUMMARY

The basic anatomic and physiologic relationships within the adrenal cortex and between the adrenal cortex and medulla have been described. This chapter also highlights the important physiologic branch points in steroidogenesis. The characteristics of adrenal hormone receptors have been described and compared. Finally, selected diseases have been briefly described and tests of adrenal function discussed to help the student place the basic anatomy and physiology of the adrenal gland into a clinical context.

NATIONAL BOARD TYPE QUESTIONS

Select the one most correct answer for each of the following

1. Acute ACTH administration stimulates the release of which hormone?
   A. Norepinephrine
   B. Aldosterone
   C. Thyroid hormone
   D. Epinephrine
2. The unique enzymatic cascade for promoting aldosterone synthesis by the zona glomerulosa includes:
   A. The presence of cortisol reductase
   B. The presence of 17,20-desmolase
   C. The absence of 17α-hydroxylase
   D. The presence of tyrosine hydroxylase
3. The removal of epinephrine principally includes:
   A. Catecholamine uptake$_1$
   B. Storage in preganglionic sympathetic neurons
   C. Degradation by aminopeptidases
   D. Catecholamine uptake$_2$
4. The production of epinephrine depends on:
   A. An intact zona glomerulosa
   B. DHEA
   C. High local cortisol concentration
   D. Angiotensin-II production
5. The adrenal glands:
   A. Have a single arterial blood supply
   B. Are derived from two germ layers
   C. Store large concentrations of steroid hormones
   D. Synthesize steroid hormones from a common precursor, alanine
6. Steroid hormone receptors are:
   A. Present on surface-cell membranes
   B. Present in the nucleus and cytoplasm
   C. Stimulate cAMP production
   D. Bind catecholamines

ANNOTATED ANSWERS

1. B. All adrenal cortex steroid hormones are released by acute ACTH stimulation, whereas epinephrine, norepinephrine, and thyroid hormone are not.
2. C. The absence of 17 -hydroxylase limits steroidogenesis in the zona glomerulosa to aldosterone production.
3. D. Catabolism of epinephrine is preceded by uptake$_2$ into non-neuronal tissues.
4. C. Cortisol induces PNMT, an enzyme required for epinephrine synthesis.
5. B. The adrenal glands are derived from two germ layers (mesoderm, ectoderm) and synthesize steroid hormones from a common precursor, cholesterol.
6. B. Steroid hormones diffuse freely through the cell membrane and bind to receptors in the cytoplasm and nucleus.

BIBLIOGRAPHY

Bondy, P. K. Disorders of the adrenal cortex. In: Wilson, J. D. and Foster, D. W., eds. *William's Textbook of Endocrinology*, 7th ed. Philadelphia: W. B. Saunders, 1985, Pp. 816–890.

DeQuattro, V. Catecholamines and adrenal disorders. In: DeGroot, L. J., ed. *Endocrinology*, 2nd ed. Philadelphia: W. B. Saunders, 1989, Pp. 1717–1800.

Dolan, L. M. and Carey, R. M. In: Vaughan, E. D., and Carey, R. M., eds. *Adrenal Disorders*. New York: Thieme Medical Publishers, 1989, Pp. 81–145.

53 The Endocrine Pancreas and Glucose Regulation

MICHAEL A. BERK

Objectives

After reading this chapter, you should be able to:

List the major pancreatic hormones and describe their major function, or functions

Cite the factors that influence pancreatic hormone secretion

Explain the functional role of the portal circulation in relationship to glucose homeostasis

Describe insulin synthesis and secretion, and its action on target tissues

Characterize the roles of the autonomic nervous system, liver, and counterregulatory hormones in the maintenance of euglycemia and the prevention of hypoglycemia

---

The hormones of the endocrine pancreas play a major role in regulating the plasma and tissue levels of carbohydrates, lipids, and amino acids in both the fed and fasted states. They regulate substrate delivery for the synthesis of glycogen, lipids, and proteins in the **fed state**, as well as influence the catabolism of these entities to provide substrates for the more immediate energy requirements of the organism in the **fasted state**. Although the exocrine function of the pancreas is a major component of digestion, this chapter focuses on the endocrine component.

ANATOMY AND EMBRYOLOGY

The **adult pancreas** is the fusion product of two duodenal buds, referred to as the *dorsal* and *ventral pancreas*. In most humans, the major excretory duct of the pancreas enters the duodenum via the ampulla of Vater, after emptying into the common bile duct. It is supplied by both the **systemic and portal circulation**. The exocrine and endocrine tissues are histologically distinct, in that the latter are found in the islets of Langerhans.

The **pancreatic parenchyma** derives from a network of tubules that form from the endoderm. Early in the fetal period, the acini (exocrine) begin to develop. The islets of Langerhans develop from groups of cells that separate from these tubules and lie between the acini. These structures comprise less than 20 percent of the pancreatic volume. Insulin secretion usually begins at about twenty weeks' gestation.

ENDOCRINE (ISLET) CELLS

The endocrine cell types and their chief hormonal secretions are listed in Table 53-1. The alpha cells

Table 53-1. Endocrine Pancreas Cell Types

Cell Type	Hormone Secreted	Percentage of Total Islet Cells
Alpha	Glucagon	15
Beta	Insulin	65
Delta	Somatostatin	20
	Gastrin	

Adapted from: Felig, P., et al, eds. In: *Endocrinology and Metabolism*, 2nd ed. New York: McGraw-Hill, 1987.

secrete **glucagon** as their primary product. **Insulin** is secreted by the beta cells, and the delta cells secrete both **somatostatin** and **gastrin**. All of these hormones are polypeptides and share synthetic and secretory characteristics. Insulin will be used as a model to describe synthesis and secretion in more detail. Gastrin will not be discussed, as its primary function is to regulate gastric retion rather than promote homeostasis.

PORTAL CIRCULATION

The hormones released by the islet cells are discharged primarily into the portal circulation. All of these hormones have physiologic effects on the liver and are in part extracted by it. For this reason, the concentrations of these hormones are higher in the portal circulation than in the peripheral (systemic) circulation, and estimates of their secretion rates and kinetics have therefore only recently been possible. In addition, studies in humans, which may require the peripheral (systemic) administration of hormones, must often achieve high systemic concentrations of hormone to ensure that a physiologic portal concentration is reached.

GLUCOREGULATION

The absorptive state that exists after a balanced meal elevates the plasma concentration of carbohydrates, lipids, and amino acids. Homeostatic mechanisms, primarily hormonal, maintain the plasma concentrations of these fuels in a physiologic range (for example, 60 to 100 mg/dl for glucose in humans), so that current energy needs can be met, with storage of fuel for use during the postabsorptive and fasting states.

NERVOUS SYSTEM

The autonomic nervous system participates in glucoregulation. **Beta-adrenergic stimulation** causes insulin secretion and **alpha-adrenergic stimulation** activates glucagon secretion. There is also considerable evidence for a "glucose sensor" in the central nervous system that triggers autonomic responses to plasma glucose levels, especially during hypoglycemia, and this will be discussed later. The "sensor" is located in the hypothalamus, though its exact mechanism of action and the neurotransmitters that mediate the response are unknown.

INSULIN SYNTHESIS AND SECRETION

Insulin is a polypeptide hormone consisting of 51 amino acids in two chains (A and B) joined by two disulfide bridges with a molecular weight of approximately 5,800. Most of the amino acids have been preserved throughout mammalian evolution. For example, human insulin differs from pork insulin by only one amino acid and from beef insulin by three. Insulin is synthesized in the beta cells as a single-chain prohormone, **proinsulin**, with a molecular weight of about 9,000. Proinsulin is actually a conversion product of a larger molecule, **preproinsulin**, which is processed to proinsulin immediately after its messenger RNA–directed synthesis. Once in the **Golgi apparatus**, proinsulin is cleaved into equimolar amounts of insulin and **C peptide**, the inactive cleavage product. This process continues in the cell's secretory granules, so that the secretion of insulin and C peptide into the portal circulation is also equimolar.

A rising plasma glucose concentration elicits a rapid rise in the plasma insulin level that lasts for minutes, followed by a more gradual and persistent rise that lasts for two to three hours. The rapid or first phase reflects the release of previously synthesized insulin from storage granules, and the more prolonged response or second phase is thought to represent the secretion of newly synthesized insulin. Thus, after ingestion of a glucose load, there is a rapid rise in the plasma glucose level, followed by a gradual decline to fasting concentrations within two to three hours. As the plasma glucose concentration falls, a change from the fed to the fasted state occurs. At this point, insulin secretion dissipates and other mechanisms that maintain euglycemia and prevent hypoglycemia are activated.

INSULIN ACTION

Paracrine

Insulin secretion acts in a paracrine fashion to suppress glucagon secretion.

HEPATIC

One of the primary actions of insulin is to suppress glucose output by the liver. Glycogen synthesis is increased and its degradation is inhibited by insulin. In addition, its action on muscle and adipose tissue suppresses the release of amino acids for synthesis into glucose, via the Cori cycle, and the release of free fatty acids for conversion to ketone bodies. Thus, glucose output by the liver is both directly and indirectly inhibited by the action of insulin, while glycogen storage is promoted.

ADIPOSE AND MUSCLE

Insulin promotes the storage of triglycerides in adipose tissue, a process known as **lipogenesis**. The absence of insulin results in **lipolysis**, which provokes an increase in the free fatty-acid concentrations and production of ketone bodies by the liver. These actions occur through the insulin-mediated stimulation of **lipoprotein lipase** and inhibition of **hormone-sensitive lipase**.

In **muscle**, insulin promotes protein synthesis; its absence results in protein breakdown, with increased amino acid delivery to the liver to provide substrates for the Cori cycle.

Insulin also augments the intracellular potassium concentration, usually at pharmacologic doses. Lack of insulin causes an increased potassium concentration in the extracellular space but decreased total body potassium, as will be discussed.

INSULIN RECEPTORS

Insulin acts on specific protein receptors located on the cell surface. Binding of insulin to its specific receptor activates postreceptor events, leading to the activation of enzymatic processes that promote glucose entry into cells, protein synthesis, lipogenesis, and glycogenesis, depending on the tissue where insulin is acting. The insulin action at the receptor level seems to be divided into two different molecular events. Insulin activates a **tyrosine kinase** that phosphorylates its own receptor, known as **autophosphorylation**.

In **obesity** as well as **non-insulin-dependent diabetes mellitus** (to be discussed), there are fewer insulin receptors. This condition is thought to be secondary to the down-regulation of the insulin receptor produced by increased concentrations of circulating insulin. In more severe cases of non-insulin-dependent diabetes mellitus, the postreceptor events are also disturbed.

DIABETES MELLITUS

Diabetes mellitus is defined as a syndrome of defective carbohydrate, protein, and lipid metabolism, resulting in **hyperglycemia** and **chronic vascular complications**. Only the former will be dealt with in this chapter.

The majority of people with diabetes mellitus can be classified into two groups, those who are **insulin-dependent** and those who are **non-insulin-dependent**.

Insulin-Dependent Diabetes Mellitus

Insulin-dependent diabetes is an autoimmune disease characterized by total destruction of the pancreatic islets resulting in a complete absence of insulin. This deficiency fosters a variety of metabolic disturbances, which, if untreated, can lead to coma and death. Lack of insulin, as is seen in uncontrolled or untreated insulin-dependent diabetes mellitus, leads to extracellular hyperglycemia and intracellular glucopenia (Fig. 53-1). The **extracellular hyperglycemia** causes increasing serum glucose concentrations and osmotic diuresis. The osmotic diuresis leads to renal sodium and free water excretion, causing volume depletion, dehydration, and total depletion of the body's sodium content.

Intracellular glucopenia activates a variety of compensatory mechanisms designed to restore normal intracellular glucose concentrations. The

Fig. 53-1. Metabolic consequences of insulin deficiency that eventually lead to the development of diabetic ketoacidosis.

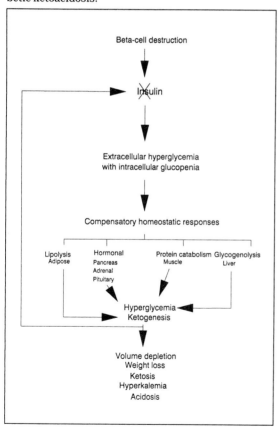

specific cellular trigger for these events is not known. These compensatory mechanisms involve all major nutrient groups. **Glycogenolysis** is increased in the liver and muscle. **Proteolysis** is accelerated in muscle to provide substrate for the Cori cycle. **Lipolysis** is heightened in adipose tissue, thus increasing the concentration of serum free fatty acids, which are converted to the ketone bodies, acetoacetate and beta-hydroxybutyrate. These ketone bodies serve as alternate fuel sources. They are excreted by the kidney but, because they are negatively charged, must be accompanied by positively charged ions. This results in dissipation of the bicarbonate, ammonia, and protein buffer systems, which leads to metabolic acidosis. Metabolic acidosis then triggers compensatory respiratory alkalosis as well as the intracellular buffering of H^+ by potassium. Thus, potassium leaves the cell in exchange for H^+, resulting in elevated extracellular potassium concentrations. The excess potassium is excreted by the kidney, leading to total depletion of the body's potassium store.

In addition to the above processes, insulin deficiency activates hormonal mechanisms, all of which result in hyperglycemia (see the section "Glucose Counterregulation"). Thus, one can discern an overall picture of hyperglycemia, hyperkalemia with total body potassium depletion, metabolic acidosis with respiratory alkalosis, volume depletion, and total body sodium depletion.

Non-Insulin-Dependent Diabetes Mellitus

Non-insulin-dependent diabetes is the result of both **defective insulin secretion** by the islets of Longerhans and **resistance to insulin action** in insulin target tissues. It is not clear whether the pancreatic defect precedes the peripheral defect, or vice versa. Figure 53-2 summarizes the events leading to hyperglycemia in non-insulin-dependent diabetes. Peripheral insulin resistance arises usually because of a decreased number of insulin receptors. Because this phenomenon also occurs in obese individuals who do not have diabetes, and as most patients with non-insulin-

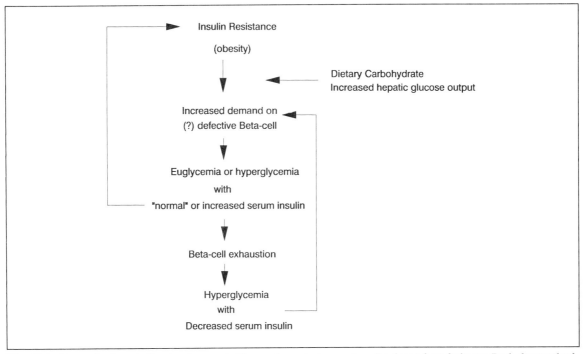

Fig. 53-2. Events leading to the development of hyperglycemia in non-insulin-dependent diabetes. Both diminished insulin secretion and peripheral insulin resistance contribute to this pathophysiologic process.

dependent diabetes are overweight, obesity appears to be a key factor in the development of the disease. The hyperglycemia that occurs as the result of insulin resistance elicits increased secretion of insulin, resulting in raised concentrations of plasma insulin. However, this increase may then down-regulate insulin receptors, further exacerbating hyperglycemia. It appears that the physiologic response to insulin resistance (increased insulin secretion) in some individuals is defective, so that hyperglycemia persists. The attempt by the defective beta cell to secrete more insulin leads to beta-cell exhaustion and worsening hyperglycemia. Increased glucose output by the liver is also seen. These events culminate in hyperglycemia with osmotic diuresis and volume depletion. Weight loss occurs when the hyperglycemia persists. However, because insulin is present, the ketonemia and acidosis seen with insulin-dependent diabetes are rarely observed.

GLUCOSE COUNTERREGULATION

The central nervous system neither synthesizes nor stores glucose, its primary fuel. The body, therefore, has developed redundant compensatory mechanisms that maintain euglycemia in the face of physiologic events, such as fasting and exercise, that can result in hypoglycemia. These actions involve the dissipation of insulin action, the autonomic nervous system, and hormonal mechanisms, referred to as *counterregulatory hormones*.

Dissipation of Insulin

The dissipation of insulin is the most obvious way of preventing hypoglycemia in humans. Insulin has a blood half-life of approximately 15 to 30 minutes, though its biologic effect in target tissues is somewhat longer. Thus, when the plasma glucose concentration falls, insulin secretion diminishes. Experimental studies in humans have demonstrated that, even when serum insulin levels remain elevated, other mechanisms that raise the plasma glucose concentration become active to prevent a further decrement and also to increase plasma glucose to euglycemic concentrations. These mechanisms are primarily mediated by hormones.

Counterregulatory Hormones

In vivo experimental studies in animals and humans have demonstrated that hormonal mechanisms begin acting within minutes to prevent hypoglycemia in the face of insulin-induced hypoglycemia, exercise, and late after glucose ingestion. The hormones involved in this process are **glucagon** (pancreas), the **catecholamines**—epinephrine (adrenal medulla) and norepinephrine (central and peripheral nervous systems)—**growth hormone** (anterior pituitary), and **cortisol** (adrenal cortex). The **counterregulatory hormones** can be divided into those that **act within minutes** (glucagon and the catecholamines) and those that **act hours later** (cortisol and growth hormone). Another pancreatic hormone, **somatostatin**, should also be mentioned. It inhibits growth hormone, insulin, and glucagon secretion and has proved to be a useful pharmacologic tool in in vivo and in vitro studies of the actions of the those hormones. Its significance in glucose counterregulation in the intact organism under physiologic conditions is not clear.

Glucagon

Glucagon is synthesized primarily in the **alpha cells** of the pancreatic islets. Like insulin, it is synthesized as a larger prohormone and processed by the cell to the active form. Although large forms of glucagon, molecular weight of approximately 9,000, exist in the circulation, it is the **species with a molecular weight of 3,500** that is active in glucose counterregulation. Glucagon is secreted into the portal circulation and

exerts its primary effect on the liver, where it enhances glycogenolysis, gluconeogenesis, and protein breakdown, thereby providing an immediate supply of glucose. Glucagon may also exert a paracrine effect and limit insulin secretion by the alpha cell. As glucagon (and other counterregulatory hormones) increases the plasma glucose level, however, insulin secretion occurs, suggesting that the glucose stimulus predominates over any direct effect hormone secretion might have on insulin secretion.

Catecholamines

Both of the plasma catecholamines, epinephrine and norepinephrine, increase the plasma glucose concentration. **Epinephrine** is secreted by the **adrenal medulla** into the venous circulation, so that it acts as a true hormone. **Norepinephrine** serves primarily as a neurotransmitter. Thus its concentration in plasma under physiologic conditions reflects its "spillover" from the nerve terminals into the circulation, and does not accurately indicate its concentrations in the tissues. Studies have shown that epinephrine is the primary counterregulatory catecholamine with respect to the plasma glucose concentration. Norepinephrine may exert a counterregulatory effect through the adrenergic stimulation of hepatic glucose production. Epinephrine, acting through beta-adrenergic receptors and cyclic AMP, raises the plasma glucose level by inhibiting insulin-mediated glucose use by the peripheral tissues and, to a lesser extent, increasing hepatic glucose production. Its effects occur within minutes. It also augments lipolysis, the heart rate, and systolic blood pressure, but diminishes diastolic blood pressure. The increase in heart rate together with other adrenergic effects, including diaphoresis and tremor, provide "warning" symptoms of hypoglycemia in humans.

Growth Hormone

Growth hormone is secreted by the **anterior pituitary** in response to **hypoglycemia**. It is secreted within minutes of the hypoglycemic stimulus, but exerts its effects hours later. It acts to increase the plasma glucose concentration by inhibiting insulin-mediated glucose use by peripheral tissues. It also increases lipolysis.

Cortisol

Cortisol is secreted by the **adrenal cortex**. Its onset of action to raise plasma glucose levels occurs hours after its secretion. Cortisol increases gluconeogenesis in liver by means of muscle protein catabolism that furnishes the gluconeogenic substrate. It also interferes with insulin-mediated glucose utilization by the peripheral tissues.

From the above discussion it can be inferred that the counterregulatory responses to hypoglycemia are redundant. Studies in both animals and humans have shown that **glucagon** is the primary counterregulatory hormone that mediates the acute response to hypoglycemia. **Epinephrine** appears to become critical when glucagon secretion is absent, as arises in some disease states such as diabetes mellitus. This hierarchy also holds late after glucose ingestion and during exercise. It is less clear if the later-acting hormones, growth hormone and cortisol, are important in the counterregulatory response to hypoglycemia.

SUMMARY

The plasma glucose concentration is maintained within very narrow limits in normal humans. This regulation involves the interaction of the liver, central nervous system, peripheral tissues, and hormonal regulators from the endocrine pancreas, pituitary, and adrenal cortex. Insulin is the major hormone that lowers the plasma glucose level into the normal range, while enhancing glycogenesis, lipogenesis, and protein synthesis. These processes also serve to lower the hepatic glucose output and increase insulin-mediated glucose uptake by the peripheral tissues.

Redundant mechanisms exist to increase the plasma glucose level in the face of hypoglycemia. These include dissipation of insulin action, counterregulatory hormone secretion, and hepatic au-

toregulation. The counterregulatory hormones augment glycogenolysis, lipolysis, and gluconeogenesis. These processes result in increased hepatic glucose output and decreased insulin-mediated glucose uptake.

NATIONAL BOARD TYPE QUESTIONS

Select the one most correct answer for each of the following

1. If beta-adrenergic blockade is pharmacologically induced in a volunteer who is then given a large dose of insulin, what would be the expected plasma glucose response over the next two hours?
   A. Plasma glucose concentration would decrease
   B. Plasma glucose concentration would decrease then increase
   C. Plasma glucose concentration would increase
   D. Plasma glucose concentration would increase then decrease

2. The same patient as in Question 1 is studied using the same protocol, except that he is also given somatostatin along with beta-adrenergic blockade prior to insulin administration. What would be the expected plasma glucose response over the next two hours?
   A. Plasma glucose concentration would decrease
   B. Plasma glucose concentration would decrease than increase
   C. Plasma glucose concentration would increase
   D. Plasma glucose concentration would increase than decrease

3. Which peripheral plasma hormone measurement most accurately reflects its secretion?
   A. Insulin
   B. Norepinephrine
   C. Glucagon
   D. Cortisol

4. A patient with insulin-dependent diabetes is being treated for diabetic ketoacidosis. After

several hours of treatment with intravenous insulin and fluids, her blood glucose level is 150 mg/dl and her potassium content is 3.7 mEq/l. However, her serum ketones are still positive and her serum bicarbonate remains at 12. Continued therapy would involve:

A. Continued intravenous insulin

B. Glucose added to the intravenous solution

C. Continued intravenous potassium

D. Subcutaneous insulin

5. A patient presents with recurrent syncope and has a documented blood glucose level of 20 mg/dl. To distinguish an insulin-secreting tumor from exogenous (surreptitious) insulin administration, one would measure the:

A. Serum potassium concentration

B. Serum insulin concentration

C. Serum glucagon concentration

D. Serum C-peptide concentration

6. A patient with a glucagon-secreting tumor would show the following characteristics:

A. Normal to elevated serum insulin concentration

B. Elevated serum glucose concentration

C. Elevated plasma glucagon concentration

D. Elevated liver glycogen content

ANNOTATED ANSWERS

1. B. The plasma glucose level would decrease in response to insulin, then increase as a result of, primarily, glucagon secretion.

2. A. The plasma glucose concentration would decrease and not return to normal during the time specified because of blockade of both epinephrine and glucagon.

3. D. Insulin and glucagon are secreted into the portal circulation and extracted by the liver, and therefore peripheral plasma measurements do not accurately reflect secretion. Norepinephrine measurements in plasma do not accurately reflect levels at the nerve terminals.

4. C. Insulin needs to be continued to inhibit lipolysis and thereby ketogenesis. Glucose is necessary to prevent hypoglycemia and provide a substrate for insulin. Insulin shifts potassium to the intracellular space, so its administration should be continued.

5. B. Insulin and C peptide are secreted in equimolar amounts. Because exogenous insulin contains virtually no C peptide, its concentrations would be low, while insulin concentrations would be high in a patient taking insulin. An insulin-secreting tumor would elevate the concentrations of both.

6. C. The glucagon concentration would be elevated secondary to tumor production. This would augment the plasma glucose level, which would promote insulin secretion. Glycogen stores would likely be depleted.

BIBLIOGRAPHY

Felig, P., et al, eds.: *Endocrinology and Metabolism*, 2nd ed. New York: McGraw-Hill, 1987.

Cryer, P. E. Glucose counterregulation in man. *Diabetes* 30:261, 1983.

Rizza, R. A., Cryer, P. E., and Gerich, J. E. Role of glucagon, epinephrine and growth hormone in human glucose counterregulation: effects of somatostatin and adrenergic blockade on plasma glucose and glucose flux rates following insulin-induced hypoglycemia. *J Clin Invest* 64:62, 1979.

XI Reproductive Physiology

54 Sexual Differentiation, Development, and Maturation

ANDREW R. LaBARBERA

Objectives

After reading this chapter, you should be able to:

Identify the factors that determine gonadal sex

Identify the factors that determine phenotypic sex, that is, the sex of the internal and external genitalia

Describe the extent of prenatal gametogenesis in females and males

Explain how the reproductive tracts of males and females differentiate

Explain the role of steroid hormones in the differentiation of the reproductive tract in males and females

Explain the role of steroid hormones in the sexual differentiation of the brain

Describe the origin of breasts in males and females, including prenatal and postnatal breast development

Describe the changes in the secretion of hypothalamic, pituitary, and ovarian hormones during puberty

---

Functionally, the reproductive system includes the gonads, the internal genitalia of the reproductive tract, the external genitalia, and the hypothalamic–pituitary unit. Differentiation of the reproductive systems of both females and males commences early in fetal life. Development of the primary sex organs, the **gonads**, including the **germ cells**, begins before the secondary or accessory sex organs start to develop, but development of primary and secondary sex organs is not completed until puberty. The control mechanisms in the hypothalamic–pituitary unit, though functional during childhood, remain quiescent until puberty.

EARLY DEVELOPMENT OF THE GONADS

The **embryonal gonad** consists of three elements, each with a different origin. **First**, the **germ cells** are derived from the primitive ectodermal cells of the inner cell mass. They do not arise from the same cells as the somatic cells of the gonads. Primordial germ cells can be identified as early as the fifth day of gestation in the blastocyst. In the fourth week of gestation, large, spherical primordial germ cells in the endodermal yolk-sac epithelium begin to migrate to the coelomic epithelium of the genital or gonadal ridges as the dorsal area of the yolk sac is incorporated into the embryo. The **supporting cells** of this coelomic epithelium constitute the **second element** of the embryonic gonad. These epithelial cells, which later give rise to the Sertoli cells of the testis and the granulosa cells of the ovary, penetrate the underlying mesenchymal cells to form the irregularly shaped primitive sex cords. **Stromal** or **interstitial cells** that are derived from the original mesenchymal cells of the gonadal ridge constitute the **third element** of the embryonic gonad. The formation of the early embryonic gonad is complete by the fifth to sixth week of gestation. At the end of the sixth week, the gonad of either sex is still both indifferent and bipotential. In other words, the gonad is neither male nor female and can develop in either direction as determined by the chromosomal composition.

GENETIC DETERMINATION OF GONADAL SEX

The chromosomal sex determines gonadal sex, which in turn determines the phenotypic sex (Fig. 54-1). The presence of a Y chromosome determines whether testes will develop. Regardless

Fig. 54-1. Genetic determination of the sex of the offspring.

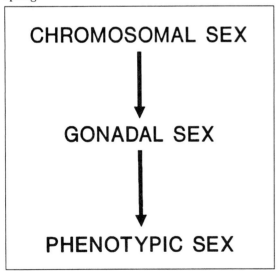

CHROMOSOMAL SEX

GONADAL SEX

PHENOTYPIC SEX

of the number of X chromosomes that exist, if there is a Y chromosome, testes will develop at least partially; if a Y chromosome is not present, ovaries will develop. All of the testis-determining genes must be present in order for testes to develop completely. These genes normally are located on the short arm of the Y chromosome near the centromere.

Complete sexual differentiation and development requires either two normal X chromosomes for **genetic females** or one X chromosome and one normal Y chromosome for **genetic males**. Normal sexual differentiation also requires certain autosomal genes in both genetic males and females. The testis-determining genes prevent activation of the autosomal genes for ovarian development during gestation. If there are two X chromosomes in addition to a Y chromosome, then testes will develop but spermatogenesis will be impaired. In the absence of a Y chromosome, the phenotype will be female regardless of the number of X chromosomes, but two X chromosomes are necessary for the complete differentiation and function of ovaries. Sexual anomalies can occur when the X and Y chromosomes fail to segregate completely during meiosis. Some of the most common disorders of sexual differentiation are summarized in Table 54-1. Disorders such as **Turner's syndrome** and **Klinefelter's syndrome** result from the abnormal segregation of sex chromosomes during meiosis. Individuals with disorders such as male or female pseudohermaphroditism possess normal complements of sex chromosomes but have defects of sex hormone synthesis or action.

DIFFERENTIATION AND DEVELOPMENT OF THE GONADS

In the **seventh week of gestation**, gonadal development in males and females diverges (Fig. 54-2). In **males** with a normal XY chromosomal complement, the primitive sex cords do not degenerate. Rather, the primitive sex cords continue to develop under the influence of the testis-determining genes. These cords penetrate deep

Table 54-1. Disorders of Sexual Differentiation

Gonadal dysgenesis (Turner's syndrome)
    Karyotype: 45,XO (loss of second sex chromosome)
    Phenotype: female
    Manifestation: short stature, primary amenorrhea, sexual infantilism, and elevated gonadotropin concentrations

Klinefelter's syndrome
    Karyotype: XXY (extra sex chromosome)
    Phenotype: male
    Manifestation: small firm testes, azoospermia, gynecomastia, elevated gonadotropin levels, and mental and social impairment

Male pseudohermaphroditism
    Karyotype: XY (normal)
    Phenotype: male/female
    Manifestation: testes present, and deficiencies of androgen biosynthesis, androgen action (testicular feminization), or mullerian-inhibiting hormone formation cause failure of virilization

Female pseudohermaphroditism
    Karyotype: XX (normal)
    Phenotype: female/male
    Manifestation: ovaries present, normal reproductive tract, and excess androgen production (congenital adrenal hyperplasia) results in virilization

into the medulla to form the medullary cords. Near the **hilus** of the testis, the cords disperse into strands of cells that later develop into the **rete testis**. In contrast to the ovary, the testis cords, or spermatogenic cords, become separated from the surface epithelium by a dense layer of connective tissue, the **tunica albuginea**. No cortical cords develop in the testes. By the **fourth month of gestation**, the testis cords are continuous with the rete testis and consist of primitive germ cells and **sustentacular cells of Sertoli** (Sertoli cells), which originate from the surface of the gonad. At puberty, each cord develops a lumen to become a **seminiferous tubule**. The **mesenchymal cells** between the cords give rise to the **interstitial cells of Leydig** (Leydig cells), so that by the fourth to sixth months these androgen-producing cells which influence differentiation of the male reproductive tract are particularly abundant.

Sertoli cells begin to secrete the **histocompatibility Y antigen** in the seventh week of gestation. The role of this antigen is controversial. It has been thought to be involved in male sexual differentiation. However, in some species, the equivalent histocompatibility antigen is associated with the female sex chromosome and is produced by the female and not the male.

In **normal females** with two X chromosomes, the ovaries begin to develop around the **twelfth week of gestation**. The primitive sex cords are arranged in clusters containing primordial germ cells and are located in the medullary portion of the gonad. The medullary cords eventually degenerate to be replaced by stromal elements, which later form the medulla of the ovary. The surface epithelium proliferates, giving rise to the cortical cords, which penetrate into the underlying mesenchyme. By the **fourth month**, the cortical cords consist of clusters of epithelial cells surrounding primordial germ cells. The epithelial cells of the cortical cords later develop into the cells of the ovarian follicle.

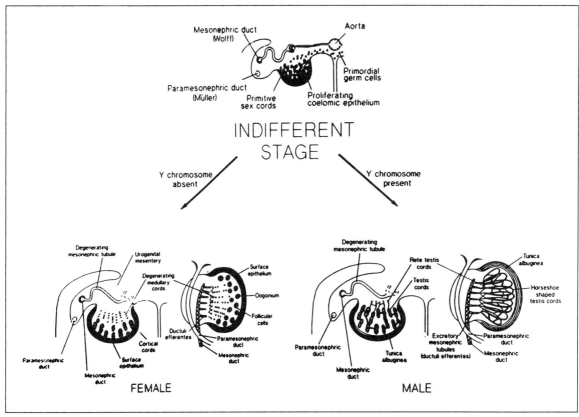

Fig. 54-2. Differentiation of the gonads. (Top) Indifferent gonads at six weeks of gestation. (Bottom left) Around the twelfth week, the indifferent gonads differentiate into ovaries if no Y chromosome is present. (Bottom right) Around the seventh week, the gonads begin to differentiate into testes if a Y chromosome is present.

PRENATAL GAMETOGENESIS

Human somatic cells normally contain 46 chromosomes. **Oocytes** (female gametes) and **spermatozoa** (male gametes) must each contain only half the normal complement of chromosomes (N or **haploid** number), so that after fertilization the **zygote** will contain the normal **diploid**, or 2N, number of chromosomes. The number of chromosomes is reduced from the 2N number in the primordial germ cells to the N number in the mature gametes by **meiosis**, a process of chromosomal reduction division unique to germ cells. During meiosis, genes derived from the maternal chromosomes can be exchanged with genes de-

rived from paternal chromosomes prior to segregation of chromosomes into daughter cells. Thus, the mature gamete contains genes from both the mother and father.

Oogenesis

The production of female germ cells begins during embryonic life and is completed during the adult reproductive period (Fig. 54-3). In the female, **oogenesis**, the process by which a primordial germ cell develops into an ovum, begins soon after the primordial germ cell arrives in the gonad during **embryogenesis**. The primordial germ cells undergo numerous **mitotic divisions**,

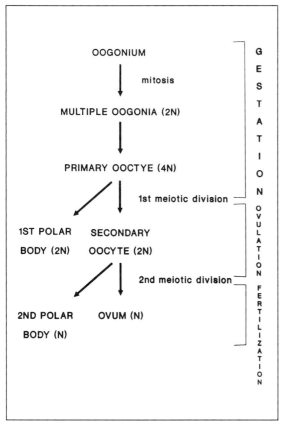

Fig. 54-3. Oogenesis. Production of female germ cells begins during embryonic life and is completed during the adult reproductive period. Oogonia proliferate by mitosis, differentiate into larger primary oocytes, replicate their DNA, and enter the first meiotic division. Meiosis begins during prenatal development in all oocytes, is arrested in the dictyotene stage of the first meiotic prophase, and does not resume until shortly before ovulation in individual oocytes. One primary oocyte gives rise to one ovum and two nonfunctional polar bodies.

so that by the end of the third month of gestation the fetal ovary contains clusters of **oogonia** derived from a single germ cell and surrounded by a layer of epithelial cells. During the fourth month, oogonia begin to differentiate into the larger **primary oocytes**, replicate their DNA, and enter the **prophase** of the first meiotic division. At this time, the primary oocytes are **tetraploid**,

that is, they have the 4N quantity of DNA. Oogonia continue to divide so that by the **fifth month of gestation** the ovary contains the maximum number of primary oocytes, approximately seven million. Meiosis in all of the female germ cells begins during prenatal development, is arrested at the dictyotene stage of prophase of the first meiotic division, and does not resume until shortly before ovulation in individual oocytes.

Degeneration of many primary oocytes and oogonia by atresia commences during the fifth month of gestation, while development continues. Most of the oogonia degenerate by the seventh month. Individual surviving primary oocytes, however, are surrounded by the flat epithelial cells, thus forming the **primordial follicles**. At birth, the ovary contains between 700,000 and two million primordial follicles. During childhood most of the oocytes become atretic, so that at puberty only approximately 40,000 oocytes remain in the ovary. Throughout this entire period, the oocytes remain arrested in the first meiotic division.

Spermatogenesis

Only the initial stages of germ cell production occur during embryonic life in the male (Fig. 54-4). Unlike the female, in whom meiosis begins during fetal life, meiosis in male germ cells does not begin until **puberty** and then continues throughout life. At birth, the primitive sex cords of the testis contain large **spermatogonia** surrounded by supporting cells, which become the Sertoli cells. The maturation divisions of male germ cells differ from those of the female germ cells in that a **primary spermatocyte** gives rise to four **spermatids**, whereas a primary oocyte results in a single mature oocyte and two nonfunctional polar bodies. **Spermatogenesis**, the process of differentiation of the primordial germ cells in the male, does not begin until puberty. Differentiation of male germ cells in the testis occurs in the **Sertoli cells**, which are epithelial cells lining the lumen of the **seminiferous tubules**.

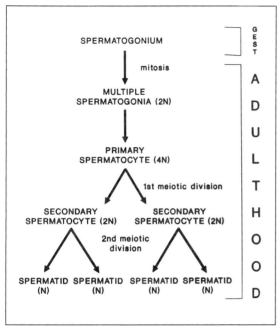

Fig. 54-4. Spermatogenesis. Only the initial stages of germ cell production in the male occur during embryonic life, so that at birth the primitive sex cords consist of large spermatogonia surrounded by supporting cells that become the Sertoli cells. After puberty, spermatogonia continue to proliferate by mitosis and undergo meiosis. One primary spermatocyte gives rise to four spermatids, each of which will mature into a spermatozoon.

DIFFERENTIATION AND DEVELOPMENT OF THE REPRODUCTIVE TRACT

Until the eighth week of gestation, the urogenital tracts and external genitalia of both males and females are indifferent and indistinguishable (Fig. 54-5). Initially, each sex has a **dual paired-duct system** consisting of a **mesonephric duct** and a **paramesonephric duct** within each **mesonephric kidney**. The mesonephric, or **wolffian**, duct initially connects networks of capillaries in the mesonephros and extends caudally to the urogenital sinus. The paramesonephric, or **mullerian**, duct develops at approximately six weeks as an evagination in the coelomic epithelium on

the anterolateral surface of the urogenital ridge. This evagination becomes tubular and is associated with the wolffian duct at the caudal end. The wolffian ducts are required for the formation of the mullerian ducts. The wolffian ducts terminate in the **urogenital sinus**, the lower portion of which eventually contributes to formation of the **external genitalia**.

The indifferent external genitalia are discernible as a **genital eminence**, which is a rounded mass extending from the umbilicus to the tail. The genital eminence consists of a genital tubercle flanked by genital swellings. The urogenital sinus opens between the genital swellings and is surrounded by genital or cloacal folds, which develop into the urethral folds.

Male

Shortly after formation of the testis cords in the sixth to seventh weeks of gestation, the mullerian ducts begin to regress in response to **mullerian-inhibiting hormone,** a large dimeric glycoprotein of approximately 140,000 molecular weight, which is produced by the Sertoli cells of the fetal testes. As the number of Leydig cells of the fetal testes increases, increasing amounts of the androgenic hormone **testosterone** are produced in response to **human chorionic gonadotropin**, which is synthesized and secreted by the **placenta**. **Androgens** are not necessary for regression of the mullerian duct; rather, they independently stimulate development of the wolffian duct system, with **virilization** of the urogenital sinus and external genitalia. Androgen production peaks at 12 to 13 weeks of gestation and then slowly declines. The capacity of the fetal testis to produce androgen corresponds to the induction of the **3β-hydroxysteroid dehydrogenase** enzyme system. The crucial role of testicular secretions in male sexual differentiation is summarized in Figure 54-6.

The **wolffian ducts** begin to develop into the male reproductive tract soon after the mullerian ducts begin to regress (see Fig. 54-5). Under the influence of androgenic hormones, the **epididymis** forms from the area of the wolffian duct

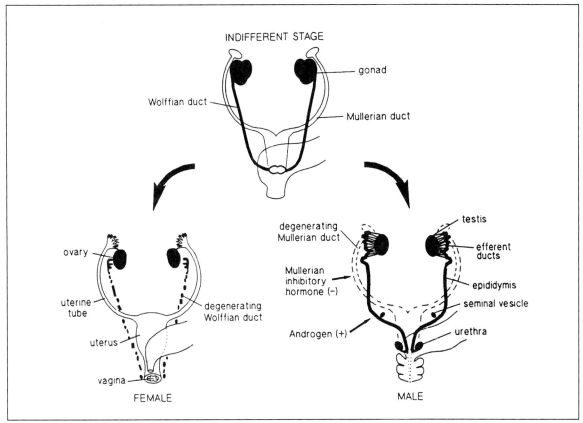

Fig. 54-5. Differentiation of the reproductive tract. Until the seventh week of gestation, each sex has a bipotential, dual paired-duct system. In the male, the wolffian (mesonephric) ducts develop into the male reproductive tract and the mullerian (paramesonephric) ducts regress. Conversely, in the female, the mullerian ducts develop into the female reproductive tract and the wolffian ducts regress.

closest to the testis; the muscular **vas deferens** forms from the middle portion of the duct; an outbudding of the lower portion of the duct develops into the **seminal vesicle**; and the portion of the wolffian duct closest to the urogenital sinus forms the **ejaculatory duct**. The **prostate gland** and the membranous portions of the **urethra** arise from the urogenital sinus.

Development of the **external genitalia**, which is androgen-dependent, occurs primarily during the first trimester of pregnancy (Fig. 54-7). The genital tubercle enlarges and elongates to form the **male phallus**. By the time of birth, the male genital tubercle is much larger than that of the

female. The genital swellings migrate posteriorly and fuse to form the **scrotum**. The urethral folds close around the urogenital or urethral groove to form the **urethra of the penis**. Much later in fetal development, the testes descend.

Cells of some **androgen-responsive tissues** in the male fetus, such as the rete testis, epididymis, vas deferens, and seminal vesicles, respond to **testosterone**. Testosterone binds to receptors in target cells and modulates the expression of genes necessary for growth and differentiation of these tissues. Other androgen-responsive tissues, such as the urethra, prostate, scrotum, and penis, contain an enzyme, **5α-reductase,** which

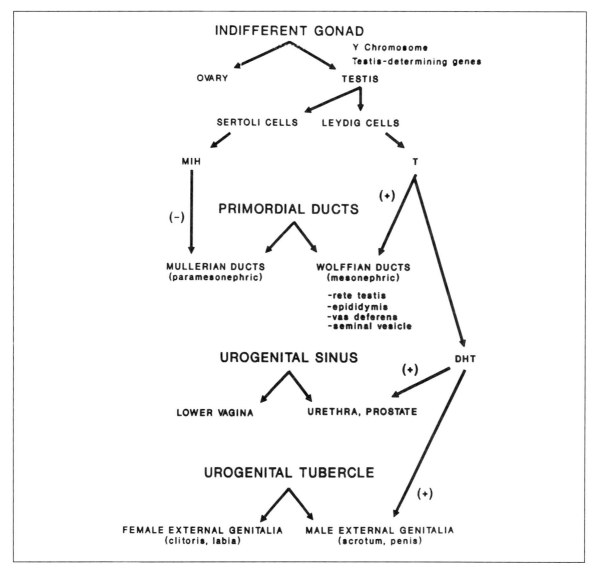

Fig. 54-6. Control of sexual differentiation. In the sixth to seventh weeks of gestation, gonadal differentiation diverges. Under the influence of testis-determining genes on the Y chromosome in males, the gonads differentiate into testes. Mullerian-inhibiting hormone (*MIH*) secreted by the Sertoli cells of the testes causes regression of the mullerian ducts. Testosterone secreted by the Leydig cells of the testes causes development of the male reproductive tract. Some male sex accessory tissues convert testosterone to dihydrotestosterone (*DHT*) via the 5α-reductase enzyme, in which case DHT is the active androgen. In the absence of a Y chromosome, the gonads differentiate into ovaries. In the absence of androgens, the wolffian ducts degenerate. In the absence of MIH, the mullerian ducts develop into the female reproductive tract.

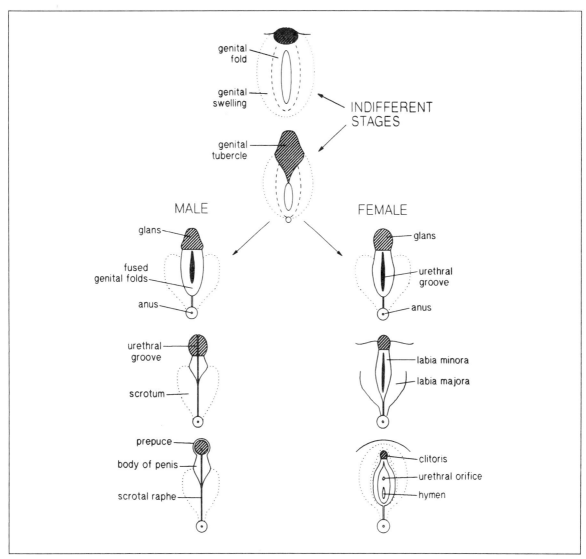

Fig. 54-7. Differentiation of the external genitalia. In the presence of androgen (male), the urogenital tubercle develops into the male external genitalia (scrotum, penis). In the absence of androgen (female), the urogenital tubercle develops into the female external genitalia (labia, clitoris).

converts testosterone to **dihydrotestosterone** (DHT). These tissues contain receptors for DHT. A deficiency of 5α-reductase and, consequently, inability to convert testosterone to DHT, results in incomplete differentiation of the male internal and external genitalia.

Female

In the absence of fetal testicular androgen secretion in females, the paired mesonephric or wolffian ducts regress automatically near the end of the second month of gestation (see Fig. 54-5). The paramesonephric, or mullerian, ducts differentiate into the main genital tract of the female. The cranial portion of the mullerian duct develops into the **fallopian tube**, whereas the middle and caudal portions of the duct develop into the **uterus** and **uterine canal**, respectively. The uterine canal is the precursor of the **uterine corpus** and the **cervix**. The thick **uterovaginal plate**, which develops at the region of contact between the wolffian and mullerian ducts and the urogenital sinus, enlarges due to proliferation of **endodermal cells**. The cells in the center of this plate eventually break down to form the lumen of the vagina, which remains separated from the urogenital sinus by the **hymen**, a thin tissue plate.

In the female, the genital tubercle elongates slightly and develops into the **clitoris**, which is much smaller than the penis in the male (see Fig. 54-7). The genital swellings enlarge to form the **labia majora**, whereas the urethral folds do not fuse but give rise to the **labia minora**. The urogenital groove remains open to form the **vestibule**.

It is not known whether **estrogens** are involved in the differentiation of the **external genitalia** in the female. The **fetal ovary** is capable of producing estrogen early in gestation. However, estrogens in the fetal circulation are derived from the mother, the placenta, and the fetal adrenal cortex and ovary. Thus, removal of the fetal ovaries by castration would not eliminate the principal source of estrogens. The cells of the **primitive sex accessory structures** do possess androgen receptors, so that excessive androgen production does cause profound virilization of the external genitalia of the female. Exposure to **progestins**, which possess androgenic activity, results in varying degrees of virilization, depending on the androgen activity of the steroid.

SEXUAL DIFFERENTIATION OF THE BRAIN

Gonadal steroid hormones are important in the regulation of **sexual behavior**. Besides directing differentiation of the reproductive tract, steroids appear to play a role in the differentiation of the neural systems that regulate sexual function and behavior in the adult. **Receptors for estrogens and androgens** have been identified in the anterior pituitary, hypothalamus, midbrain, amygdala, and cerebral cortex of adult females and males, respectively. The neural consequences of estrogen or androgen binding have not yet been determined.

The hypothalamic–pituitary unit of the adult female produces hormones, called **gonadotropins**, that stimulate the gonads in a cyclic pattern, with a periodicity of approximately 28 days. This cyclicity is determined by the changing sensitivities of the hypothalamic–pituitary unit to estrogens and progestins (see Chapter 55). Gonadotropin secretion in the male differs from that in the female in that it is not cyclic.

Evidence obtained in laboratory animals indicates that **androgens** secreted by the testes during the periods of fetal and neonatal development are responsible for the acyclic pattern of gonadotropin secretion characteristic of males. Administration of testosterone to neonatal female rats produces acyclic gonadotropin secretion and sterility, or **masculinization**, when they become adults. Conversely, castration of neonatal male rats eliminates androgen production and the adults exhibit cyclic female gonadotropin secretion. In some species, testosterone is first converted to estradiol before exerting its effect, and, in these species, the administration of estradiol to neonatal males also produces masculinization of the hypothalamic–pituitary axis.

BREAST DEVELOPMENT

The development of breasts is identical in males and females before puberty. Mammary glands begin as bilateral **mammary lines**, which are bandlike thickenings of the ectodermal epidermis. In a seven-week-old embryo, these lines extend from the forelimbs to the hindlimbs on the ventral surface. The mammary lines disappear shortly after formation, except for a single small portion on each side in the thoracic region. These remaining **mammary buds** penetrate the underlying mesenchyme and remain virtually unchanged until the fifth month. Each mammary bud then gives rise to 15 to 25 separate sprouts, which form small outbuddings. The **epithelial sprouts** proliferate and canalize throughout the remainder of gestation to form the **lactiferous ducts**, while the small outbuddings form the **small ducts** and **alveoli** of the mammary glands. The lactiferous ducts open into a small epithelial pit, which develops into the nipple connecting the lactiferous ducts to the outside.

Just before birth, the breast undergoes a brief period of growth, but then the mammary glands regress because of low levels of estrogens and progestins. The breast remains quiescent until puberty. Further growth is proportional to the growth of the remainder of the body.

With the increased concentrations of estradiol and progesterone associated with puberty in females, the breasts grow and develop. The **areolae** enlarge and become more pigmented. The amount of connective tissue and, more significantly, the amount of adipose tissue increase, and the glands become more vascularized. The ducts enlarge to form rudimentary lobules, and alveolar structures enlarge. Fully developed alveoli, however, only appear during pregnancy.

The **growth and function of the mammary glands** in the **nonpregnant state** depend primarily on estrogens and progestins, but growth hormone, thyroid hormones, and adrenal corticosteroids have permissive effects. Progesterone by itself has little effect on the mammary gland. Ductal breast tissue alternately proliferates and regresses during each menstrual cycle. Proliferative changes are maximal late in the luteal phase, after exposure to peak concentrations of progesterone. Breast development and function during and after pregnancy are discussed in Chapter 57.

NEUROENDOCRINE CONTROL AND PUBERTY

The reproductive systems of males and females appear to remain dormant during childhood. The **hypothalamic–pituitary–gonadal axis** is at least partly functional during this period. Removal of the gonads, which produce small amounts of steroid hormones, produces a rise in plasma gonadotropin levels. Moreover, administration of very small amounts of steroid hormones suppresses the already low levels of gonadotropins. Thus, the small amounts of steroid hormones produced by the gonads during childhood are able to suppress the secretion of gonadotropins by the hypothalamic–pituitary unit. This mechanism is referred to as **negative feedback**.

Males, at approximately 10 years of age, and females, at approximately 11 years, begin to undergo **puberty**, which is the transition from the juvenile state to adulthood. Puberty generally lasts two to four years. In females, puberty culminates in the onset of **menstruation** between the ages of eleven and sixteen. During puberty, the sexually immature child is transformed into a sexually mature adolescent. The hypothalamic–pituitary–gonadal axis gradually begins to functon in an adult manner, so that the pituitary secretes increased amounts of the **gonadotropins**, **follicle-stimulating hormone** (FSH), and **luteinizing hormone** (LH). Increased gonadotropin secretion augments secretion of **sex hormones** by the gonads. As a result, **secondary sex characteristics** appear and mature; the adolescent growth spurt occurs; fertility is acheived; and profound psychologic changes are observed. These changes constitute **sexual maturation**. The signal for puberty is unknown, but the **timing** of this transitional state may be related to general bodily growth.

Sexual maturation is characterized by an increase in gonadal hormone secretion, referred to as **gonadarche**. The first recognizable hormonal change associated with puberty is elevated adrenal androgen production, referred to as **adrenarche**. The signal for adrenarche has not been identified. The source of increased gonadal activity seems to arise in the **brain** and **hypothalamus**, since both the anterior pituitary gland and the gonads are capable of adult function during childhood if appropriately stimulated. The transformation from an immature to a mature state is due to maturation of the neural mechanisms in the brain that modulate secretion of gonadotropin-releasing hormone (GnRH) by the hypothalamus; this hormone stimulates the anterior pituitary gland to secrete gonadotropic hormones. During puberty, secretion of GnRH by the hypothalamus increases. The control of GnRH secretion is thought to reside in a GnRH oscillator or pulse generator, which is the central neural regulator of GnRH secretion. THis **GnRH pulse generator** has not been identified, nor is it known whether it is separate from or the same as the GnRH-secreting hypothalamic neurons.

Changes in Gonadotropins during Prenatal and Postnatal Development

Puberty can be considered the **final phase of development** of the hypothalamic–pituitary–gonadal axis that begins during fetal life. The concentrations of the gonadotropins, FSH and LH, in the fetal circulation of both males and females increase markedly around the middle of pregnancy (Fig. 54-8). Peak concentrations are achieved between 100 and 150 days of gestation, and then decline as the hypothalamic–pituitary unit becomes very sensitive to the negative-feedback inhibition by gonadal steroids. The **magnitude of the rise** in fetal concentrations of FSH and LH is more dramatic in females (see Fig. 54-9) than in males (see Fig. 54-8) because the hypothalamic–pituitary unit is much more sensitive to the negative feedback effects of circulating androgens than to those of circulating estrogens.

FSH and LH in fetal blood are secreted by the fetal pituitary gland, apparently in response to GnRH released by the fetal hypothalamus. The fetal pituitary does respond to exogenous synthetic GnRH. Moreover, the hypothalamic–hypophysial circulation and the hypothalamic nuclei exist in midgestation. The **bell-shaped time course** of gonadotropin secretion by the fetus (see Fig. 54-8) is thought to be due to the maturation of hypothalamic GnRH and pituitary FSH and LH secretory capacities (increased gonadotropins) before the hypothalamic–pituitary

unit matures and is able to respond to the negative feedback effects of fetal and placental androgens and estrogens (decreased gonadotropins). Immediately after birth, neonatal gonadotropin concentrations rise temporarily because of the sudden absence of inhibitory placental sex steroids (see Fig. 54-9). FSH and LH are elevated in the plasma of male infants to levels that equal or exceed those seen during prepubertal development; elevated gonadotropin secretion is associated with elevated testosterone levels. Plasma LH levels in female infants resemble those in male infants, but plasma FSH levels remain elevated for several years, often in the range of those seen in castrated women. This sex-related difference is thought to indicate that the hypothalamic GnRH pulse generator in female infants operates at a slower frequency than that in the adult. Gonadotropin levels decline and the hypothalamic–pituitary unit, though seemingly mature in both males and females, remains quiescent until the beginning of puberty.

Male

The **testis** grows very little during childhood. At the onset of **puberty**, it begins to increase from a volume of approximately 2 ml to the adult volume of 12 to 25 ml. The **Leydig cells**, which synthesize and secrete testosterone, increase dramatically in number, with a consequential increase in the circulating testosterone levels. Daytime concentrations of testosterone in plasma increase from 0.2 ng/ml before puberty to approximately 6 ng/ml after puberty. The **seminiferous tubules** increase in diameter and in tortuosity, due to the differentiation of Sertoli cells and proliferation of spermatogonia. **Pubertal changes** in males include an increase in muscle mass, broadening of the shoulders, and thickening of the vocal cords, which results in the voice breaking.

Female

During childhood, the **ovary** increases linearly in size. The **follicles** continue to grow, but all be-

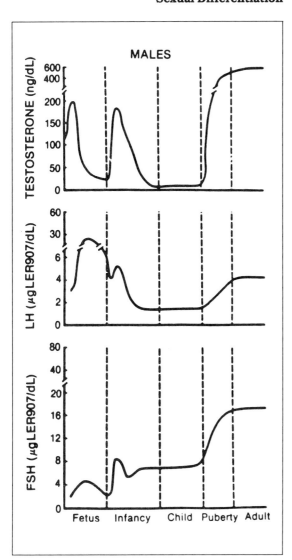

MALES

Fig. 54-8. Plasma levels of testosterone, luteinizing hormone (*LH*), and follicle-stimulating hormone (*FSH*) during prenatal and postnatal development in the male. The bell-shaped time course of fetal gonadotropin secretion probably reflects maturation of hypothalamic gonadotropin-releasing hormone (*GnRH*) and pituitary FSH and LH secretory capacities (increased gonadotropins) prior to maturation of the ability of the hypothalamic–pituitary unit to respond to the negative-feedback effects of fetal and placental androgens and estrogens (decreased gonadotropins). Immediately after birth, neonatal gonadotropin concentrations rise temporarily due to the sudden absence of negative feedback by inhibitory sex steroids from the placenta. Gonadotropin levels decline and the hypothalamic–pituitary unit remains quiescent until the beginning of puberty. At puberty, plasma gonadotropin levels begin to rise to adult levels, apparently due to gradually diminishing sensitivity of the hypothalamic–pituitary unit to the negative-feedback effects of androgens. (From: Winter, J. S. D., et al. *J. Clin. Endocrinol. Metab.* 42:679, 1976, copyright by The Endocrine Society.)

come atretic by the time the **antrum** develops. The growing follicles are capable of **steroidogenesis**, as estradiol concentrations in ovarian venous blood are higher than those in the peripheral circulation. Between the ages of 8 and 10 years in girls, morning **estradiol** concentrations in serum begin to reach those seen in adult women during the follicular phase of the menstrual cycle; the secretion of androgens, such as dehydroepiandrosterone, by the adrenals and

ovaries also increases. In this early pubertal period, labial hair appears, due to increased androgen levels, and the breasts enlarge, due to increased estrogen levels. In mid to late puberty, the pituitary acquires the ability to secrete a burst of gonadotropins in response to GnRH after exposure to elevated levels of estradiol. This ability of estrogen to amplify the pituitary responsiveness to GnRH, which occurs just before ovulation, is termed **positive feedback**. Development of this control mechanism is gradual. As positive feedback develops, the female begins to experience cyclic reproductive function, the **menstrual cycle**. The first menstrual period, **menarche**, usually takes place between the ages of twelve and thirteen, but the **first ovulation** generally does not occur until at least six months later. **Regular ovulatory menstrual cycles** commence up to several years later. During the transition period, the female experiences **anovulatory cycles** in which no ovum is released. **Pubertal changes** in the female include rapid growth of the pelvis and increased fat formation over the shoulders, pelvis, buttocks, and thighs.

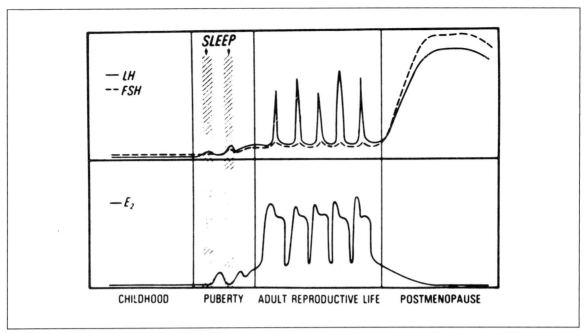

Fig. 54-9. Plasma levels of estradiol (E_2), luteinizing hormone (*LH*), and follicle-stimulating hormone (*FSH*) during the postnatal life of the female. Pituitary gonadotropin and ovarian steroid levels remain low during childhood. At puberty, transient elevations of gonadotropins are first observed during sleep. As the sensitivity of the hypothalamic–pituitary unit to negative feedback inhibition by ovarian estrogens diminishes and the sensitivity to the positive-feedback effect of estrogen increases, gonadotropins and estrogens begin to be secreted in a cyclic manner. Cyclic function continues until menopause, when ovarian function declines. Decreased estrogen levels in the postmenopausal period lead to a rise in the plasma gonadotropin concentrations because of release of the hypothalamic–pituitary unit from the negative-feedback inhibition by estrogen. (From: Rebar, R. W. Normal physiology of the reproductive system. In: *Endocrine and Metabolism Continuing Education and Quality Control Program.* American Association for Clinical Chemistry, Inc.: November 1982.)

SUMMARY

The reproductive system includes the gonads, the internal genitalia of the reproductive tract, the external genitalia, and the hypothalamic–pituitary unit. The development and differentiation of the male and female reproductive systems begins *in utero*. Phenotypic sex, or the sex of the internal and external genitalia, is determined by the sex of the gonad, which in turn is determined by the chromosomal sex or the complement of X and Y chromosomes. The reproductive tracts differentiate from a bipotential system consisting of two paired duct systems, the mullerian ducts and the wolffian ducts. Breast development prior to puberty is similar for males and females, but diverges at puberty when estrogens and progestins stimulate further development in females. Puberty, the period during which the sexually immature child is transformed into a sexually mature adolescent, results from the maturation of central neural positive and negative feedback mechanisms that regulate the hypothalamic secretion of GnRH, which, in turn, regulates secretion of the anterior pituitary hormones, FSH and LH.

NATIONAL BOARD TYPE QUESTIONS

Select the one most correct answer for each of the following

1. The sex of the gonads is determined by:
   A. Plasma levels of testosterone and estradiol
   B. The complement of X and Y chromosomes
   C. The presence of mullerian-inhibiting hormone
   D. Plasma levels of the gonadotropins, FSH and LH, in the fetal circulation
   E. None of the above
2. Development of male external genitalia requires:
   A. That there be a single Y chromosome
   B. Prior regression of the wolffian ducts
   C. That there be no estrogen in the fetal circulation
   D. Androgens
   E. None of the above
3. Breast development:
   A. Is identical in males and females before puberty
   B. Requires estrogens and progestins for growth and function to occur in the adult female
   C. Begins with the formation of bandlike thickenings of the ectodermal epidermis
   D. During female puberty involves enlargement of the areolae, lactiferous ducts, and alveoli, and an increase in the amounts of adipose and connective tissues
   E. All of the above
4. The mullerian-inhibiting hormone is:
   A. Required for differentiation of the wolffian (mesonephric) ducts
   B. Synthesized and secreted by Sertoli (sustentacular) cells
   C. Synthesized and secreted by interstitial cells of Leydig
   D. Secreted in response to human chorionic gonadotropin
5. LH and FSH levels in the fetal circulation increase around the middle of pregnancy because:
   A. The hypothalamus of the mother secretes increased amounts of GnRH
   B. The fetal pituitary is sensitive to the positive-feedback effects of steroid hormones
   C. The fetal hypothalamic–pituitary unit is not yet sensitive to the negative-feedback effects of steroid hormones
   D. The placenta has acquired the capacity to synthesize LH and FSH
   E. All of the above

ANNOTATED ANSWERS

1. B. Gonadal sex is determined by chromosomal sex.
2. D. The development of male external genitalia depends on the presence of androgens.
3. E.
4. B. Mullerian-inhibiting hormone is synthesized and secreted by Sertoli cells of the testis.
5. C. LH and FSH levels in the fetal circulation increase around the middle of pregnancy because the fetal hypothalamic–pituitary unit is not yet sensitive to the negative-feedback effects of steroid hormones.

BIBLIOGRAPHY

Knobil, E., and Neill, J. D., eds. *The Phsiology of Reproduction*. New York: Raven Press, 1988.
Sadler, T. W. *Langman's Medical Embryology*. Baltimore: Williams & Wilkins, 1985.

55 The Female Reproductive System

ANDREW R. LaBARBERA

Objectives

After reading this chapter, you should be able to:

List the general functions of the female reproductive system in the nonpregnant state

State the components of the female reproductive system and their specific functions

Describe the development of ovarian follicles, mature ova, and corpora lutea

Explain the synthesis, secretion, and effects of ovarian steroid hormones

Explain how positive and negative steroid feedback regulates gonadotropin secretion

Describe the physiologic basis of the menstrual cycle and the coordinated cyclic changes in the functions of the hypothalamus, pituitary, ovaries, and reproductive tract

Explain how sexual function facilitates the transport of spermatozoa through the reproductive tract

Identify the causes and effects of menopause, the beginning of reproductive senescence

---

The reproductive system of the female functions to produce offspring. It produces germ cells for sexual reproduction; provides an environment for the transport of the male germ cells, the spermatozoa, to the fallopian tubes for fertilization; provides an appropriate environment for the development of the embryo; and produces milk for the nourishment of the young offspring. Although the reproductive organs are established during the embryonic and fetal period, they do not reach full maturity until puberty. The active reproductive period begins with **menarche,** the first menstruation, at puberty and lasts until **menopause**, which is the cessation of reproductive function when the supply of ovarian follicles is exhausted.

Approximately once a month throughout the reproductive period of the female, the ovary cyclically produces a follicle containing a **mature gamete**, or **ovum**. The cycle, which is referred to as the **menstrual cycle**, is controlled by gonadotropic hormones secreted by the pituitary gland in response to the hypothalamic hormone **gonadotropin-releasing hormone** (GnRH). Secretions of the hypothalamic–pituitary unit are modulated by both positive- and negative-feedback signals from the ovaries. The components of the reproductive system undergo simultaneous cyclic changes, and each cycle begins with the shedding of the vascularized luminal epithelium of the uterus, called **menses**. The functions of the hypothalamus, pituitary, ovary, reproductive tract, and mammary glands in the nonpregnant and pregnant states are integrated and controlled by neural and hormonal signals. Female reproductive function during the pregnant and postpartum states is discussed in Chapter 57.

OVARY

The ovary serves **four functions** critical to reproduction: (1) the cyclic production of gametes; (2) secretion of hormones that prepare the reproductive tract to receive and nurture the conceptus and influence the development of secondary sexual characteristics; (3) secretion of hormones that participate in conditioning the mammary glands for lactation; and (4) feedback regulation of hypothalamic–pituitary secretion.

Follicle

The **basic functional unit** of the ovary is the follicle, which consists of an immature female gamete, the **oocyte**, surrounded by one or more layers of specialized follicular cells. These cells secrete autocrine, paracrine, and endocrine factors that modulate the functions of the oocyte and other cells within the follicle, affect the structure and function of the female accessory sex organs, and modulate the actions of the hypothalamic–pituitary unit. The structure of each follicle is related to its stage of development (see Fig. 55-1). Follicular growth and differentiation are mediated by the two anterior pituitary gonadotropins, **follicle-stimulating hormone** (FSH) and **luteinizing hormone** (LH), although FSH is capable of stimulating complete development of the follicles by itself. The ovary contains both nongrowing and growing follicles; the nongrowing follicles greatly outnumber the growing ones throughout most of the female's lifespan.

The onset of puberty and the consequential increase in the secretion of GnRH by the hypothalamus and of FSH and LH by the pituitary marks the beginning of the ongoing development of follicles beyond the **primary stage**. Follicles do not begin to progress to the **preovulatory stage** until several months after menarche, however. In each menstrual cycle, several follicles are recruited to develop into **secondary follicles**. Normally, only a single follicle in each cycle becomes **dominant**, completes the process of differentiation and maturation, and is then **ovulated**. The remainder undergo degeneration, or **atresia**.

Evidence from other species indicates that **folliculogenesis**, beginning with the recruitment of primordial follicles from the pool of developing follicles to formation of the dominant follicle, is a continuous process that appears to span three to four menstrual cycles. During the first cycle, follicles develop to the **secondary preantral stage;** during the second to fourth cycles, they become **tertiary antral follicles;** and during the fourth cycle they become mature **preovulatory follicles.** Follicular development requires physiologically effective levels of gonadotropins, as development beyond the primordial follicle stage is rarely seen in hypogonadotropic hypogonadal women.

Primordial Follicle

The nongrowing follicles are the primordial follicles formed during gestation. Each primordial follicle consists of an oocyte with a diameter of approximately 25 μm, surrounded by a single layer of flat, squamous epithelial cells which are the precursors of the **granulosa cells.** The primordial follicle is surrounded by a **basement membrane.** The **oocyte** is maintained in the arrested **dictyotene stage** of the first meiotic prophase by the putative **oocyte maturation inhibitor,** presumably secreted by the epithelial cells. Beginning in fetal life and continuing throughout childhood, primordial follicles start to enlarge and develop by mitosis, but then degenerate through atresia. Consequently, the number of primordial follicles in the ovary is reduced from one to two million at birth to between 40,000 and 400,000 at puberty. The recruitment and atresia of follicles continue throughout the reproductive lifespan, so that the pool of available follicles is continually decreasing.

Primary Follicle

Early in **folliculogenesis,** the oocyte and surrounding follicle grow in tandem. The **oocyte** begins to increase in size to approximately 100 μm,

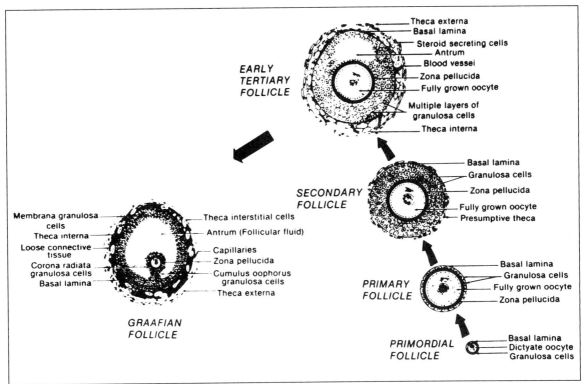

Fig. 55-1. Developmental stages of ovarian follicles and the corpus luteum. During gestation, multiple primordial follicles are formed and consist of an oocyte surrounded by a single layer of granulosa cells and a basement membrane. Transformation of a primordial follicle into a primary follicle is characterized by enlargement of the oocyte and a change in the shape of the granulosa cells from flat to cuboidal. Primary follicles enlarge by mitosis of the granulosa cells to form secondary follicles. The oocyte achieves its maximal size of 120 μm. Under the influence of follicle-stimulating hormone (FSH) and estradiol, the follicle continues to enlarge by mitosis of the granulosa cells into a tertiary follicle, and the *antrum*, a fluid-filled cavity, forms. FSH and estradiol continue to stimulate the follicle to enlarge and become responsive to luteinizing hormone (LH). The LH surge causes the follicle to rupture and expel the cumulus–oocyte complex. The residual granulosa cells hypertrophy and differentiate (*luteinization*) to form the corpus luteum. (From: Erickson, G. F., et al. The ovarian androgen-producing cells: a review of structure/function relationships. *Endocr. Rev.* 6:371, 1985, copyright by The Endocrine Society.)

and the surrounding epithelial cells enlarge and acquire the cuboidal shape of **granulosa cells.** The newly formed granulosa cells of this primary follicle acquire receptors for FSH and proliferate more rapidly. A glycoproteinaceous material begins to form the **zona pellucida,** which eventually surrounds the entire oocyte.

Secondary Follicle

Granulosa cells continue to proliferate and the oocyte reaches its maximal size to become a secondary follicle. Once two to three layers of granulosa cells form, **mesenchymal cells** migrate to the **basal lamina** and align themselves in parallel around the entire follicle to form the beginnings of the **theca interna** and **theca externa. Capillary networks** form around the theca cells of the secondary follicles. At the end of this stage, the follicle is approximately 400 μm in diameter and consists of a fully grown 120 μm–diameter oocyte surrounded by a thick zona pellucida, up to eight layers of stratified, columnar granulosa cells, a basal lamina, and the presumptive theca. The oocyte has the capacity to complete the first step of meiotic maturation, consisting of germinal vesicle breakdown and progression to **metaphase I.**

Tertiary Follicle

As the follicle enlarges, fluid accumulates among some of the granulosa cells, forming the **antrum,** which is a cavity filled with follicular fluid. At this stage, the follicle is a **tertiary antral follicle.** The number of granulosa cells increases a further 100-fold to 1,000-fold. **Gap junctions** form between the granulosa cells, permitting the intercellular transfer of molecules of up to 1,200 molecular weight. Some of the fibroblastlike cells of the **theca interna** transform into the large, epithelial **theca–interstitial cells.** Theca cells have **LH receptors,** enabling them to respond to LH. LH induces synthesis of the cholesterol side-chain cleavage, 3β-hydroxysteroid dehydrogenase–Δ^{5-4} isomerase and 17α -hydroxylase enzyme complexes in theca cells, resulting in increased production of androstenedione and testosterone, the obligatory androgen precursors for **estrogen biosynthesis.** FSH acts on granulosa cells to induce synthesis of the **aromatase enzyme complex,** which converts androgens to estrogens, resulting in increased plasma concentrations of estrogen. The effects of FSH on granulosa cells are amplified by **insulin-like growth factor I** (IGF-1), formerly referred to as *somatomedin C*; it is a peptide that is produced in granulosa cells as well as other bodily tissues. Estrogen acts within the follicle to enhance the responsiveness of granulosa cells to FSH. It also progressively sensitizes the pituitary gland to GnRH, accounting for the positive-feedback effect of estrogen on the hypothalamic–pituitary unit.

Graafian Follicle

As the granulosa cells proliferate further under the influence of FSH and estradiol and the volume of follicular fluid increases, the follicle enlarges to become a preovulatory **graafian follicle.** The granulosa cells differentiate into several ultrastructurally **specialized subpopulations.** Those immediately surrounding and in contact with the oocyte through the zona pellucida form the **corona radiata.** The coronal cells are in turn surrounded by and in contact with the cells of the **cumulus oophorus.** These cells send projections into the oocyte, leading to putative important interactions between the oocyte and the surrounding cells. The granulosa cells lining the follicle aromatize androgens synthesized in the theca to estrogens in increasing amounts. Consequently, plasma concentrations of estradiol continue to increase exponentially. FSH induces a 20-fold to 100-fold increase in LH receptors as well as an increase in LH-responsive adenylyl cyclase activity in the granulosa cells, thus preparing them to respond to the preovulatory surge of LH. The theca–interstitial cells in the theca interna continue to proliferate, until there are approximately eight layers of theca–interstitial cells in a mature graafian follicle. As the follicle matures, these cells produce increasing amounts of androgen under the influence of LH. The cells of the theca externa are interspersed with blood vessels and smooth muscle cells innervated by sympathetic nerves. Nerve fibers and blood vessels do not penetrate the basement membrane to innervate the granulosa cell layer. The oocyte of a mature graafian follicle has the capacity to proceed to metaphase II of meiosis and to complete meiotic maturation after fertilization. Normal oocyte maturation requires adequate levels of estradiol in the follicle. An imbalance in the ratio of androgen to estrogen is associated with the **polycystic ovarian syndrome.** In this disorder, follicles become very large (cysts) but do not differentiate normally and do not produce normal ova.

Follicular Fluid

The composition of the fluid in the follicular antrum changes as the follicle develops and the granulosa and theca cells mature. Initially antral follicles contain a solution of **proteoglycans,** predominantly chondroitin sulfate, which is rapidly diluted by fluid derived from plasma. The **osmolality and electrolyte composition** of follicular fluid are nearly identical to those of plasma. The protein content ranges from 50 to 100 percent of that of serum. FSH and LH levels reflect those of serum. Steroid hormone and precursor levels are determined by the rates of synthesis and diffusion from the theca and granulosa cells as well as concentrations of steroid-binding proteins, which heighten steroid levels relative to plasma. **Numerous enzymes** have been detected in follicular fluid, including peptidases, phosphatases, nucleotidase, hyaluronidase, plasmin, and collagenase. Plasmin and collagenase are thought to be involved in weakening the follicle prior to rupture, or ovulation.

Corpus Luteum

The preovulatory surge in the LH level, which is initiated by GnRH after the pituitary gonadotropins have been sensitized to GnRH by high levels of estradiol, causes ovulation of the **cumulus–oocyte complex.** Prior to ovulation, the granulosa and theca cells lining the follicle begin a process of cytodifferentiation termed **luteinization.** The granulosa cells hypertrophy, increasing from a diameter of 8 to 10 μm prior to ovulation to a diameter of 30 to 35 μm after ovulation. The luteinized granulosa cells, or **luteal cells**, fill much of the follicular cavity to form the **corpus luteum** (see Fig. 55-1). Ultrastructurally, luteal cells resemble steroid-secreting cells, in that they possess smooth endoplasmic reticulum, abundant tubular mitochondrial cristae, and numerous cytoplasmic lipid droplets containing cholesterol esters. Theca–interstitial cells also luteinize to form **theca–lutein cells,** acquiring an ultrastructure similar to that of the granulosa cells; they are much smaller, however, reaching a diameter of only about 15 μm. After ovulation and

the formation of a **fibrin clot** in the follicular cavity, the **basal lamina** breaks down and the cavity is invaded by blood vessels from the theca interna. An **unidentified angiogenic factor** appears to participate in the formation of the blood vessels in the corpus luteum. **Luteinization** is maximal 5 to 6 days after ovulation, when the corpus luteum becomes maximally functional.

The **primary function** of the corpus luteum appears to be to secrete **progesterone,** which is necessary for preparing the **uterine endometrium** to accept the blastocyst for implantation and maintaining the fetal–placental unit during early pregnancy. If fertilization of the ovum and implantation of the blastocyst do not occur, the corpus luteum degenerates through the process of **luteolysis,** which is evident histologically eight days after ovulation. Progesterone production begins to decline around the tenth day after ovulation.

Low levels of LH are necessary for luteal function. Administration of LH during the luteal phase increases plasma progesterone levels and can prolong the lifespan of the corpus luteum for at least several days. Conversely, administration of antiserum to LH during the luteal phase can cause premature menstruation. The effects of LH on luteal cells are mediated by LH receptors coupled to adenylyl cyclase. Although both LH receptors and LH-responsive adenylyl cyclase activity decline after the LH surge through the processes of down-regulation and desensitization, respectively, they reappear during the early luteal phase and become maximal in the midluteal phase.

Regulation of the corpus luteum in women is poorly understood. The human corpus luteum produces **androstenedione** and **estradiol** as well as **progesterone** and **17α-hydroxyprogesterone.** LH can increase both progestin and estrogen production, but FSH, whose receptors are maximal in the early luteal phase, increases only estrogen production. Plasma estrogen levels and luteal aromatase activity both increase prior to luteolysis. Furthermore, exogenous estrogen inhibits luteal progesterone production. Although **prostaglandin** $F_{2\alpha}$ ($PGF_{2\alpha}$) can cause luteal regression in primates, removal of the uterus, a principal source of $PGF_{2\alpha}$, has no effect on the lifespan of the corpus luteum in humans. Luteal cells contain receptors for **oxytocin,** a peptide of the posterior pituitary, but the role of oxytocin in regulating luteal function has not been established.

Secretory Products

The **ovary** produces several substances that regulate ovarian function itself, control the hypothalamic-pituitary secretion of gonadotropins, and condition the accessory reproductive organs for pregnancy. These include steroid hormones, inhibin, growth factors, prostaglandins, proteoglycans, and proteolytic enzymes.

Secretion of steroid hormones is a principal function of the follicle, interstitial cells, and corpus luteum of the ovary. The **follicle** produces progestins, androgens, and estrogens, whereas the **corpus luteum** produces mainly progestins and some estrogen. The **steroid hormones** are synthesized from **cholesterol** (Fig. 55-2), which is derived from **three sources:** (1) circulating low-density (LDL) and high-density lipoproteins; (2) pre-formed cholesterol stored in the ovarian cell; and (3) *de novo* synthesis from acetate within the ovarian cell. In the avascular antral and preovulatory follicles, cholesterol is probably synthesized *de novo*, as the basement membrane of the follicle excludes cholesterol-carrying LDLs from the granulosa cell layer. After **vascularization** associated with **luteinization,** luteal cells use circulating LDLs as the primary source of cholesterol. LH increases the number of LDL receptors in luteal-cell membranes, leading to increased uptake of cholesterol from the LDLs. Cholesterol is stored in lipid droplets in ovarian cells as esters of **long-chain fatty acids.** Intracellular concentrations of free cholesterol are maintained by three enzymes that are regulated by hormones: (1) **cholesterol ester synthetase** (acyl coenzyme A : cholesterol-acyl transferase), which esterifies free cholesterol; (2) **cholesterol esterase** (sterol ester hydrolase), which catalyzes the release of cholesterol; and (3) **HMG CoA reductase** (3-hydroxy-3-methylglutaryl coenzyme A-reductase), which is the rate-limiting enzyme in cholesterol biosynthesis. LH increases luteal cell cholesterol concentrations by increasing both HMG CoA reductase and cholesterol esterase activities.

Progestin

Progestins, or C_{21} steroids, are important both as hormones and as precursors for the **biosynthesis of androgens and estrogens. Progesterone** is the most important progestogenic hormone, but **pregnenolone** is most significant as a precursor for other steroid hormones. Progesterone production (1 to 3 mg per day) and plasma progesterone levels (0.05 μg/dl) are lowest during the follicular phase of the menstrual cycle until just before ovulation. They rise in response to preovulatory LH stimulation of granulosa cells in the graafian follicle and then decline. After ovulation and luteinization, progesterone production by the luteal cells increases to 10 to 40 mg per day, with plasma progesterone levels reaching 0.5 to 2.5 μg/dl in the midluteal phase.

The **biosynthesis of progesterone** from cholesterol begins in the **mitochondria,** where pregnenolone is synthesized (see Fig. 55-3). The **rate-limiting reaction** is cleavage of the cholesterol side chain by means of the C_{27}–side-chain-cleavage cytochrome P_{450} enzyme, which is induced in granulosa cells by FSH. Pregnenolone is converted to progesterone through the action of the microsomal enzyme 3β-hydroxysteroid dehydrogenase–Δ^{5-4} isomerase, which is induced by FSH and LH. Progesterone is cleared from the circulation rapidly, with a metabolic clearance rate of 2,100 liters per day, and excreted in the urine. Approximately 10 percent of the total progesterone in the circulation is excreted as the conjugated metabolite **pregnanediol glucuronide.**

Progesterone acts on the **reproductive tract,** the **mammary glands** of the breast, and on the **hypothalamic–pituitary unit** and exhibits the following actions:

1. Increases secretion of glycogen-rich mucus by the uterine endometrium

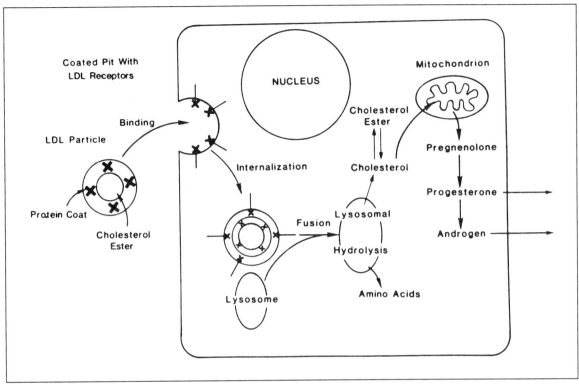

Fig. 55-2. Source of ovarian cellular cholesterol. Ovarian cells utilize cholesterol, the precursor of steroid hormones, derived from three sources: (1) circulating low-density (*LDL*) and high-density (*HDL*) lipoproteins; (2) pre-formed cholesterol stored within the cell; or (3) *de novo* synthesis from acetate. Circulating lipoprotein is the preferred source. (From: Ross, G. T., and Schreiber, J. R., Jr. The ovary. In: Yen, S. S. C., and Jaffe, R. B., eds. *Reproductive Endocrinology. Physiology, Pathophysiology and Clinical Management*, 2nd ed. Philadelphia: W. B. Saunders, 1986. P. 125.)

2. Decreases contractility of the uterine myometrium
3. Increases consistency of the cervical mucus
4. Decreases GnRH pulse frequency
5. Stimulates development of the lobules and alveoli of the mammary glands

During the **luteal phase** of the menstrual cycle, progesterone promotes active secretion by the mucosa of the fallopian tubes and by the glands of the uterine endometrium in preparation for possible implantation of a blastocyst. It also decreases the frequency of contractions of the uterine myometrium to minimize expulsion of the ovum. The prolonged elevated progesterone levels during **pregnancy** cause the smooth muscle of the uterine myometrium to relax and the uterus to expand. Progesterone also changes the **secretory activity of the cervix,** causing the cervical mucus to decrease in volume and become thicker in consistency. In the breast during pregnancy and postpartum lactation, progesterone synergizes with estrogen and the lactogenic hormones, placental lactogen and prolactin, to stimulate development of the **lobuloalveolar system.** Progesterone plays an important role in regulating **FSH and LH secretion** by the pituitary gland. It decreases secretion of the gonadotropins, primarily by decreasing the frequency of **hypothalamic GnRH pulses.**

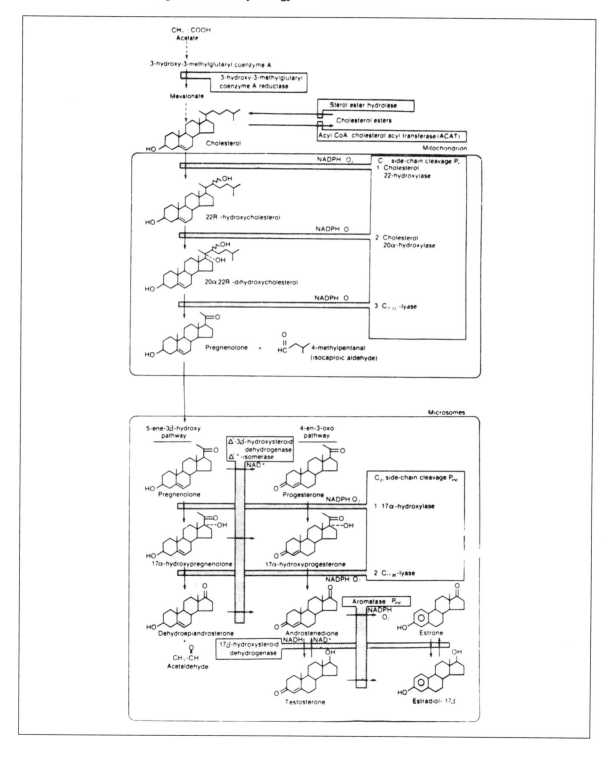

The effects of progesterone are mediated by specific high-affinity intracellular receptors located at the nuclear membrane of the cell. **Receptor binding** makes the receptor capable of binding to nuclear acceptor sites associated with DNA and nuclear proteins. Binding of progesterone-receptor complexes to DNA heightens the synthesis of the messenger RNAs (mRNAs), which direct the numerous biosynthetic processes.

Androgen

Ovarian androgens, or C_{19} steroids, are important in females as **substrates for estrogen biosynthesis** in granulosa cells. Adrenal androgens are important for the development of female axillary and pubic hair that occurs at **adrenarche,** which is maturation of the adrenal function at puberty. **Androstenedione** and **testosterone,** which are secreted by theca and interstitial cells, are the most important ovarian androgens. Androstenedione production in women averages 3 mg per day. Only approximately half of the circulating androstenedione is produced by the **ovary;** most of the remainder is produced by the **adrenal glands.** Concentrations of androstenedione in **follicular fluid** are 100 to 500 times greater than those in plasma, and reach a maximum of approximately 74 μg/dl during the midfollicular phase, after which they decline. The androstenedione concentration in plasma is between 40 and 240 ng/dl, with higher concentrations during the luteal phase than during the follicular phase.

Pregnenolone is converted to androgens by a series of microsomal enzymes (see Fig. 55-3). The **rate-limiting step** is C_{21} side-chain cleavage, which involves the 17α-hydroxylase : $C_{17,20}$-lyase enzyme complex. LH induces this enzyme system in the theca–interstitial cells of the follicle. In the human ovary, the **preferred pathway** for androgen biosynthesis is from pregnenolone to dehydroepiandrosterone, which is converted to androstenedione. In **premenopausal women**, androstenedione production (3 mg per day) predominates over testosterone production (300 μg per day); in postmenopausal women, more testosterone than androstenedione is produced. Androstenedione and testosterone are metabolized to 17-ketosteroids and excreted in the urine as **conjugated steroid glucuronosides.** Androstenedione functions primarily as the substrate for estrogen biosynthesis in the granulosa cells. However, the mechanism by which it migrates through the basement membrane to the granulosa cells has not been established. Androgens also appear to modulate **steroidogenesis** in the human ovary. Androstenedione and testosterone can both inhibit progesterone biosynthesis in the granulosa cells. Furthermore, atretic follicles have higher levels of androgens in the follicular fluid than do normally developing follicles.

Estrogen

Estradiol-17β and **estrone** quantitatively are the most important estrogens, often referred to as C_{18} steroids. Estrogens are synthesized in follicular granulosa cells from androgens derived from the theca–interstitial cells (Fig. 55-4). Conversion of androstenedione to estrone and of testosterone to estradiol-17β is catalyzed by the **rate-limiting cytochrome P$_{450}$-dependent**

Fig. 55-3. Biosynthesis of ovarian steroid hormones. Synthesis of progesterone from cholesterol begins in the mitochondria, where pregnenolone is synthesized via the C_{27}–side-chain-cleavage cytochrome P$_{450}$ system. Pregnenolone is converted to progesterone via the microsomal enzyme 3β-hydroxysteroid dehydrogenase/Δ^{5-4} isomerase. Pregnenolone is converted to androgens by microsomal enzymes. The rate-limiting step is C_{21}–side-chain cleavage, which involves the 17α -hydroxylase/$C_{17,20}$-lyase system. In the human ovary, the preferred pathway is from pregnenolone to dehydroepiandrosterone, which is converted to androstenedione. Androstenedione is converted to estrone via the aromatase cytochrome P$_{450}$ system, which converts the A ring of the steroid nucleus to an aromatic ring. Estrone is converted to 17β-estradiol by 17β-hydroxysteroid dehydrogenase. (From: Gore-Langton, R. E., and Armstrong, D. T. Follicular steroidogenesis and its control. In: Knobil, E., and Neill, J. D., eds. *The Physiology of Reproduction*, vol. 1. New York: Raven Press, 1988. P. 334.)

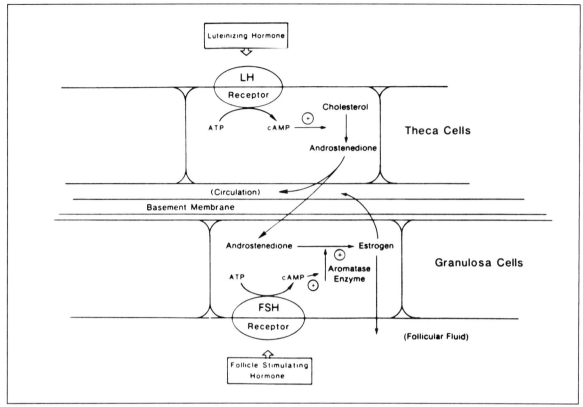

Fig. 55-4. Compartmentalization of estrogen biosynthesis in the follicle. Luteinizing hormone (*LH*) stimulates theca cells to synthesize androstenedione from cholesterol. Androstenedione diffuses both into the circulation and across the basement membrane into granulosa cells. Androstenedione is converted to estrone and then to estradiol by the aromatase enzyme system, which is induced in granulosa cells by follicle-stimulating hormone (FSH). (From: Ross, G. T., and Schreiber, J. R., Jr. The ovary. In: Yen, S. S. C., and Jaffe, R. B., eds. *Reproductive Endocrinology. Physiology, Pathophysiology and Clinical Management,* 2nd ed. Philadelphia: W. B. Saunders, 1986. P. 124.)

aromatase enzyme complex, which is induced in granulosa cells by FSH. Estrone is converted reversibly to estradiol-17β by the **17β-hydroxysteroid dehydrogenase** enzyme present in granulosa cells. During the **early follicular phase** of the menstrual cycle, estrone and estradiol are secreted in nearly equal amounts, 60 to 170 μg per day. As the **dominant follicle** enlarges in the **latter half of the follicular phase,** estradiol secretion increases to 400 to 800 μg per day, resulting in plasma estradiol concentrations of 25 to 40 ng/dl. The **corpus luteum** secretes about 250 μg of estradiol per day, which is approxi-

mately four times the secretion of estrone. Only 2 to 3 percent of **circulating estradiol** is free, in that it is not bound to plasma-binding proteins. Approximately 38 percent is bound to **testosterone estradiol—binding globulin (TeBG),** a beta-globulin produced in the liver with a higher affinity for testosterone than for estradiol. The remaining 60 percent of circulating estradiol is bound to **albumin.**

Numerous **metabolites of estrogens** are excreted in both the urine and bile. The **liver** plays a major role in the metabolism and excretion of estrogens. Most of the metabolites of estrogens are sulfated or conjugated, or both, to glucosi-duronic acid in the liver. The two major metabolites of both estradiol and estrone are **estrone glucuronide** and **free unconjugated estriol,** but **catechol estrogens** are also excreted in the urine. Physiologically, estradiol is more than ten times as potent as estrone and more than eighty times as potent as estriol. The **metabolic clearance rates** of estradiol and estrone are 1,350 and 2,210 liters per day, respectively.

Estrogen has numerous and diverse effects on the organs and tissues of the reproductive system (Table 55-1). These effects are observed during the **follicular phase** of each menstrual cycle and are reversed during the **luteal phase.** Estrogen stimulates **cellular proliferation** and **growth of tissues** of the reproductive tract. **At puberty,** estrogen causes an increase in size of the fallopian tubes, uterus, vagina, and external geni-

Table 55-1. Actions of Estradiol

1. Stimulate enlargement of the uterus and vagina to adult size
2. Stimulate growth of the ovarian follicles
3. Stimulate growth and cellular proliferation of the uterine endometrium (increase protein synthesis, proliferation of glands, and blood supply)
4. Increase motility of the uterine myometrium and fallopian tubes
5. Increase sensitivity of the myometrium to oxytocin
6. Increase the quantity and decrease the viscosity of cervical mucus
7. Promote thickening (cornification) of the vaginal epithelial mucosa
8. Increase GnRH receptors in gonadotropes
9. Increase the pituitary content of follicle-stimulating hormone, luteinizing hormone, and prolactin
10. Decrease the gonadotropin-releasing hormone pulse amplitude
11. Stimulate growth of the mammary ducts
12. Increase general protein anabolism (to a lesser degree than androgens)
13. Stimulate growth and closure of the epiphyses
14. Increase synthesis of plasma proteins, such as testosterone-estradiol–binding globulin, in the liver
15. Cause moderate retention of sodium, chloride, and water in the kidneys

talia; conversely, estrogen deprivation results in atrophy of these organs. In the **fallopian tubes,** estrogen increases proliferation of cells, including the ciliated epithelial cells of the mucosal lining that promote movement of the ovum toward the uterus. The **endometrium** of the uterus is profoundly affected by estrogens, which cause thickening due to proliferation of the stromal cells, endometrial glands, and blood vessels. Estrogens also increase the number of **progesterone receptors** in endometrial cells; endometrial sensitivity to progesterone requires prior exposure to estrogen. The **lining of the cervix** produces copious amounts of a thin, watery mucus after exposure to estrogen, and the cells lining the vagina proliferate, transform from cuboidal to stratified, and become more secretory.

Within the **ovary,** estrogen enhances its own production and amplifies the effects of FSH on granulosa cell function and follicular development. Estrogen synthesized in the granulosa cells can inhibit synthesis of androgen in the theca cells, thereby coordinating production of substrate and product. In the **breasts,** estrogen causes deposition of fat and growth of stromal tissues and ducts. Estrogen is responsible for breast enlargement but not for milk production.

Estrogens cause deposition of fat in subcutaneous tissues other than those of the breast. This effect, which is most pronounced in the buttocks and thighs, results in a lower specific gravity of the female body than the male body. Estrogens have dual effects on the **skeleton.** They augment osteoblastic activity, leading to rapid growth at puberty, but they also cause closure of the epiphysial plates, which prevents further growth of the long bones. Estrogen receptors have been identified in osteoblastic cells.

Ovarian function is regulated indirectly by estrogens by means of their positive-feedback and negative-feedback effects on the hypothalamic–pituitary unit. Estrogen suppresses the pituitary secretion of FSH and LH by the hypothalamic–pituitary unit, and thus the gonadotropins are maintained at a relatively low level by this negative-feedback effect. **Decreased plasma estradiol levels,** such as occur after menopause or removal of the ovaries, result in greatly elevated plasma gonadotropin levels. The mechanism of this negative-feedback effect of estrogen on the hypothalamic–pituitary unit has not been established definitively. However, it is thought that estrogen decreases **the amplitude of GnRH pulses,** since estrogen decreases the amplitude of pituitary LH pulses. Paradoxically, rising estrogen levels also cause the **surge of LH and FSH** in the late follicular phase prior to ovulation. As the dominant follicle grows and secretes increasing amounts of estrogen, plasma levels of estrogen rise. Elevated estrogen concentrations progressively sensitize the pituitary gland to the relatively constant pulsatile hypothalamic GnRH secretion, culminating in the preovulatory LH surge. This positive-feedback effect of estrogen is due both to increased pituitary receptors for GnRH and to increased contents of FSH and LH in the gonadotropes. The paradoxical positive-feedback and negative-feedback effects of estrogens are related to the plasma levels and the duration of exposure to these steroids.

Estradiol, like progesterone, binds to specific high-affinity receptor proteins that are only loosely associated with the nucleus of the target cell. Binding of estradiol to its receptor transforms the **estradiol-receptor complex** to a form that binds tightly to **nucleoprotein acceptor sites** associated with DNA. Estradiol thus regulates gene transcription leading to the synthesis of mRNAs, which direct the formation of specific proteins responsible for estrogenic effects.

Inhibin

Inhibin is a glycoprotein with an apparent molecular weight of 32,000 and consists of two dissimilar subunits that are covalently linked by disulfide bonds. The larger alpha subunit has an apparent molecular weight of 20,000. The smaller beta subunit, which has an apparent molecular weight of 13,000, exists in two forms, β_A and β_B. Both $\alpha\beta_A$, known as **inhibin A,** and $\alpha\beta_B$, known as **inhibin B,** are synthesized in ovarian granulosa cells. Both **FSH** and **testosterone** stimulate granulosa cells to synthesize and secrete

inhibin. In **males,** inhibin is produced by Sertoli cells in the testis. The initial half-life of inhibin in the circulation is approximately 15 minutes.

Inhibin preferentially suppresses the **synthesis and secretion of FSH** by gonadotropes in the anterior pituitary. Circulating FSH levels are inversely related to circulating inhibin levels. **Inhibin production,** which is low at the beginning of the menstrual cycle, increases late in the follicular phase, reaches a peak prior to the midcycle surge of LH and FSH, decreases slightly, and then increases further in the midluteal phase to levels approximately twice the levels at midcycle. As the corpus luteum declines late in the luteal phase, inhibin levels decrease and FSH levels increase at the same time as the onset of the next menstrual cycle.

NEUROENDOCRINE REGULATION OF THE OVARIES

Regulation of ovarian function occurs through the secretion of two **gonadotropic hormones** synthesized in the anterior pituitary gland in response to hypothalamic stimulation (Fig. 55-5). The structurally similar glycoproteins, FSH and LH, are the most important reproductive peptide hormones produced by the anterior pituitary in the nonpregnant state. Synthesis and secretion of FSH and LH are regulated by **hypothalamic** and **ovarian hormones. Prolactin,** which is also produced in the anterior pituitary, and **oxytocin,** which is secreted by the posterior pituitary, are important during and after pregnancy and are discussed in Chapter 57.

Follicle-Stimulating Hormone

FSH is a heterodimeric protein with a molecular weight of 32,600. It consists of two dissimilar subunits strongly associated by noncovalent interactions. The **alpha subunit** is identical to the alpha subunit of LH, thyrotropin (TSH), and human chorionic gonadotropin (hCG), and confers species specificity on the hormone. The beta **subunit** confers hormonal specificity on FSH. Each subunit has two N-linked, branched oligo-

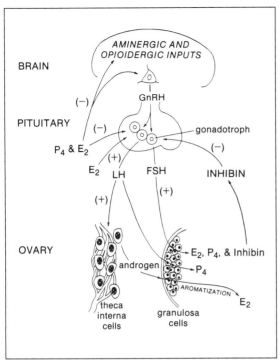

Fig. 55-5. Major feedback mechanisms regulating the hypothalamic–pituitary–ovarian axis. The hypothalamus secretes gonadotropin-releasing hormone (*GnRH*) in a pulsatile manner. GnRH diffuses into the hypothalamic–hypophysial portal vessels, which transport it to the anterior pituitary. Gonadotropes in the pituitary secrete follicle-stimulating hormone (*FSH*) and luteinizing hormone (*LH*) in response to GnRH. LH stimulates the theca cells to produce androgen. FSH stimulates the granulosa cells to differentiate and acquire the capacity to aromatize androgens to estrogens and to produce inhibin. As the follicles enlarge, they produce more estrogen and inhibin. Estrogen progressively sensitizes the pituitary to GnRH (positive feedback); inhibin decreases FSH secretion by gonadotropes (negative feedback). Estrogen and progesterone both modulate hypothalamic secretion of GnRH (negative feedback).

saccharide moieties that are involved in the receptor-mediated activation of adenylyl cyclase. These carbohydrate side chains of FSH have terminal sialic acid (*N*-acetylneuraminic acid) residues that prolong the half-life of the gonadotropin in the circulation.

Specialized cells, **gonadotropes,** in the anterior pituitary gland synthesize, store, and secrete both FSH and LH. The **mRNAs** for the alpha and beta subunits of FSH are encoded by separate genes on **chromosomes 6 and 11,** respectively. The peptides are synthesized in the endoplasmic reticulum and glycosylated in the Golgi apparatus. **Synthesis of FSH** is regulated primarily by GnRH and inhibin. **GnRH** has a permissive effect on the synthesis of the beta subunit of FSH. Synthesis of the beta subunit is rate-limiting for the production of FSH, whereas synthesis of the alpha subunit occurs at a relatively constant rate. GnRH also stimulates secretion of mature, fully glycosylated FSH from secretory granules. **Inhibin,** whose secretion varies during the menstrual cycle, opposes the effects of GnRH; it inhibits both the synthesis and release of FSH.

Circulating levels of FSH reflect a steady-state determined by rates of secretion and disappearance. FSH is secreted by the pituitary in a pulsatile manner in response to pulses of GnRH, though discrete pulses of FSH in the circulation are difficult to discern because of the relatively long half-life of FSH in the circulation. The initial **half-life of FSH** is between three and four hours. Both the **liver** and **kidney** are important in the clearance and excretion of FSH. FSH in the pituitary is turned over once per day, and this accounts for the amount secreted and metabolized. Both FSH synthesis and secretion are inhibited by a direct action of inhibin on pituitary gonadotropes.

The effects of FSH on follicular granulosa cells are mediated by membrane-bound receptors that are coupled functionally to the membrane-bound adenylyl cyclase enzyme system. Activation of the cyclase in response to stimulation by FSH involves the interaction of the FSH-receptor complex with a stimulatory **guanine nucleotide regulatory protein** (G$_s$), which must bind both magnesium and guanosine triphosphate in order to be activated. **Cyclic AMP** (cAMP), which is generated from ATP by the action of cyclase, activates one or more cAMP-dependent protein kinases, which catalyze the phosphorylation of proteins. Theoretically, FSH could exert its effects by regulating either enzyme activities or substrate levels.

FSH stimulates numerous processes associated with **differentiation** in the granulosa cells, leading to growth and development of the ovarian follicles. These include increased lactic acid production; induction of aromatase, the enzyme complex that catalyzes conversion of androgens to estrogens; induction of the cholesterol side-chain cleavage enzyme, which is rate-limiting in the synthesis of pregnenolone, a precursor of progesterone; induction of LH receptors; induction of LH-sensitive adenylyl cyclase; increased activity of plasminogen activator; enhanced synthesis of proteoglycans; and increased synthesis of inhibin.

Luteinizing Hormone

Structurally, LH is very similar to FSH. It is a heterodimeric protein with a molecular weight of 29,400 and consisting of two dissimilar, noncovalently associated subunits. The **alpha subunit,** which is identical to that of FSH, confers species specificity on the hormone. The **beta subunit** confers hormonal specificity on LH. LHβ has a single N-linked, branched oligosaccharide moiety, and less than half the sialic acid content of FSH, so that the half-life of LH in the circulation is much shorter than that of FSH.

LH is synthesized, stored, and secreted by the same gonadotropes of the anterior pituitary gland that produce FSH, though some gonadotropes may produce only LH or only FSH. The gene for the **mRNA** of the beta subunit of LH is located on **chromosome 19.** Synthesis, glycosylation, cisternal packaging, and storage of LH resemble those for FSH, except that the terminal sialic acid–galactose disaccharide of FSHβ is replaced by sulfated *N*-acetyl-D-galactosamine in LHβ. **Synthesis of LH** is regulated primarily by GnRH and estrogens. GnRH increases the number of translatable mRNAs for the beta subunit, so that synthesis of the beta subunit, which is **rate-limiting** for production of the hormone, is enhanced. **GnRH** also stimulates secretion of mature, fully glycosylated LH from secretory granules. **Estrogens** increase the pituitary content of LH and potentiate gonadotrope responsiveness to GnRH. The effect of the rising plasma estrogen concentrations on the pituitary response to GnRH in the late follicular phase of the menstrual cycle accounts for the dramatic increase in LH secretion prior to ovulation, referred to as the **midcycle LH surge**. LH is secreted in a **pulsatile manner** in response to pulses of GnRH (Fig. 55-6). **Progesterone** indirectly inhibits secretion of LH by decreasing the GnRH pulse frequency. This effect is evident during the **luteal phase** of the menstrual cycle (Fig. 55-6).

The **liver** and the **kidney** are involved in the **clearance and excretion of LH.** Because LH contains less sialic acid than FSH, LH is cleared from the circulation more rapidly than FSH. The initial **half-life** of LH is approximately 20 minutes. The rapid clearance of LH from the circulation contributes to the pronounced pulsatile nature of plasma concentrations of the hormone.

The effects of LH on follicular granulosa cells, theca cells, and interstitial cells are mediated by **membrane-bound receptors** that are coupled functionally to cAMP-dependent protein kinase by the intracellular mediator cAMP, which is produced through activation of the membrane-bound adenylyl cyclase enzyme system. As with other cyclase systems, the LH-responsive production of cAMP is modulated by magnesium and guanosine triphosphate by means of a guanine nucleotide regulatory (G_s) protein.

LH is a major regulator of **steroid biosynthesis in the ovary.** It stimulates androgen production in the theca and interstitial cells throughout follicular development, as well as estradiol and progesterone production in mature, differentiated granulosa cells of preovulatory follicles. Unlike FSH, which regulates steroidogenesis by controlling the concentrations of rate-limiting biosynthetic enzymes, LH regulates the **availability of steroidogenic substrates** by enhancing the mobilization, transport, and metabolism of cholesterol. LH increases the activity of the **cholesterol ester hydrolase,** an enzyme which de-esterifies cholesterol and provides free cholesterol for steroidogenesis.

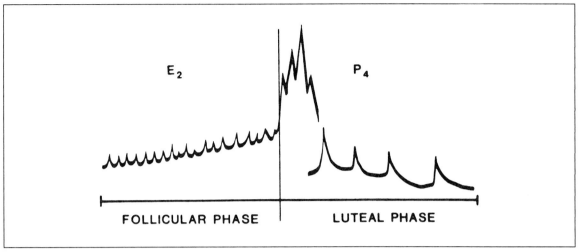

Fig. 55-6. Changing pattern of pulsatile luteinizing hormone (LH) secretion during the menstrual cycle. During the estrogen-dominated follicular phase, LH is secreted approximately once every 90 minutes. Progesterone secreted during the luteal phase decreases the frequency of LH pulses. (From Yen, S. S. C. The human menstrual cycle. In: Yen, S. S. C., Jaffe, R. B., eds. *Reproductive Endocrinology. Physiology, Pathophysiology and Clinical Management,* 2nd ed. Philadelphia: W. B. Saunders, 1986. P. 220.)

Gonadotropin-Releasing Hormone

Secretion of FSH and LH by the gonadotropes of the anterior pituitary gland is controlled by neurons with cell bodies in the **mediobasal hypothalamus** (MBH), particularly the arcuate nucleus–median eminence region. These neurons secrete GnRH into the extracellular space. The hormone diffuses into the hypothalamic–hypophysial portal vessels, which carry it to the anterior pituitary gland. Either surgical separation of the MBH from the remainder of the central nervous system or ablation of the arcuate nucleus causes an abrupt decline in circulating levels of both FSH and LH.

GnRH, which is sometimes referred to as *LH-releasing hormone,* is a decapeptide (10 amino acids) with a molecular weight of 1,182 that stimulates both the synthesis and release of FSH and LH by gonadotropes of the anterior pituitary. It is synthesized in neurons that form **loose networks,** rather than distinct nuclear clusters, in the MBH, which contains the arcuate nucleus and the periventricular structures adjacent to it.

Release of GnRH from nerve endings in the median eminence and diffusion into the portal circulation is pulsatile; secretion of pituitary gonadotropins is correspondingly pulsatile. The initiation of each LH pulse is synchronous with a marked increase in **multiunit activity,** the pulse generator, in the MBH. The relationship between the vollies of electrical activity and the initiation of LH pulses is absolute. The GnRH pulse generator, which has not been identified, can be suppressed by **alpha-adrenergic** and **antidopaminergic blocking agents,** as well as by **morphine** and the endogenous opiate **beta-endorphin.** Inhibition of LH secretion by beta-endorphin can be reversed by administration of the opiate antagonist **naloxone. Opioid peptides** appear to mediate stress-induced inhibition of reproductive function. GnRH pulse frequency, as indicated by the frequency of LH pulses in peripheral plasma, is more rapid in the estrogen-dominated **follicular phase** of the menstrual cycle than in the progesterone-dominated **luteal phase**. It is thought that **estrogen** exerts its negative-feedback effect on gonadotropin secretion by decreasing the **amplitude of GnRH pulses. Progesterone** is thought to exert its negative-feedback effect by decreasing the **frequency** of **GnRH pulses**. In the absence of the ovaries, artificial reduction of the GnRH pulse frequency below the physiologic rate leads to increased plasma FSH and decreased plasma LH levels, probably because the gonadotropin pulses are larger when the GnRH stimulation occurs more slowly.

MENSTRUAL CYCLE

During the reproductive lifespan of the adult female, between menarche and menopause, the reproductive system cyclically undergoes a series of structural and functional changes that result in gamete production and preparation of the reproductive tract for implantation of a fertilized zygote, should fertilization occur. This recurring menstrual cycle usually lasts 28 to 32 days. By convention, timing of the cycle begins either with menses (Fig. 55-7) or with the ovulatory surge of LH. During the menstrual cycle, an ovarian follicle develops to maturity, ovulates, and then luteinizes to form a corpus luteum. The cells of the follicle and the corpus luteum secrete different amounts of the steroid hormones estradiol and progesterone, which produce characteristic changes in the reproductive tract. The absence of cyclic reproductive function coupled with failure to menstruate is termed **amenorrhea.**

Follicular Phase

The follicular phase, which begins with the onset of menstrual bleeding, is when a **single ovarian follicle** becomes dominant and matures and the cells of the uterine endometrium rapidly proliferate. It is also referred to as the *preovulatory* or *proliferative* phase. At the beginning of this phase, **estradiol** and **inhibin** production is low. The GnRH pulse generator produces a pulse of GnRH approximately once every 90 minutes. Pituitary gonadotropes respond to each pulse of GnRH by releasing a pulse of either LH or FSH,

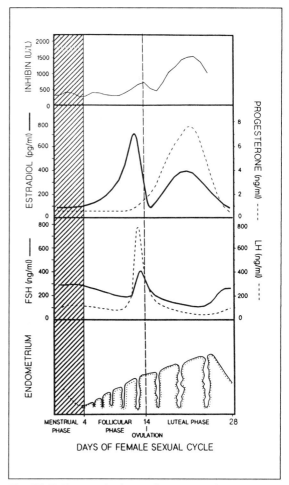

Fig. 55-7. Temporal relationship among changes in pituitary, ovarian, and endometrial function throughout the menstrual cycle. The beginning of menses is considered day 0. (*FSH* = follicle-stimulating hormone; *LH* = luteinizing hormone.)

or both. The **plasma FSH level** begins to rise late in the luteal phase of the previous menstrual cycle and continues into the early follicular phase, due to a decline in inhibin secretion. **Plasma LH levels** are low during the **early follicular phase.** LH and FSH are carried by the circulation to the ovary where: (1) LH stimulates the **theca—interstitial cells** to differentiate and produce increased androgen; and (2) FSH stimulates the

granulosa cells to proliferate and then differentiate.

As multiple follicles enlarge and the granulosa cells differentiate under the influence of FSH and estrogen, increasing amounts of **thecal androgen** are aromatized to estrogen, causing plasma levels of estradiol to rise. The estrogen has a **negative-feedback effect** on the hypothalamus, thereby modulating the amplitude rather than the frequency of LH and FSH pulses. **FSH secretion** is also negatively modulated by inhibin. However, **inhibin levels** do not rise until the **late follicular phase.** Thus, the midfollicular phase decline in FSH secretion may primarily reflect the negative-feedback effect of estrogen. Alternatively, the pituitary may become progressively sensitized to inhibin. Estrogen acts within a follicle to sensitize granulosa cells to FSH, so that a follicle producing large quantities of estrogen can continue to develop despite declining plasma FSH levels. In this way, the most mature follicle appears to become the **dominant follicle.** As the follicle grows and matures, plasma estrogen concentrations increase exponentially with time. In the late follicular phase, the plasma levels of androgens and the progestin 17α-hydroxyprogesterone increase. The **uterine endometrium** increases in size and vascularity, and the pituitary gland gradually becomes sensitized to estrogen. After plasma estradiol concentrations have exceeded 150 to 200 pg/ml for approximately 36 hours, the negative-feedback effect of estrogen is reversed and the pituitary secretes a surge of LH and FSH. This preovulatory surge of gonadotropins appears to be due to the direct effects of estrogen on pituitary gonadotropes, rather than to a change in the GnRH pulse frequency or amplitude.

Ovulation

The preovulatory surge of LH in plasma, which lasts approximately 48 hours from onset to the return of pre-surge levels, induces several profound, cAMP-mediated changes in the **graafian follicle.** Maturation of the **oocyte** resumes, so that by the time the ovulated oocyte reaches the fallopian tube, meiosis will have reached the stage of extruding the **first polar body.** As the granulosa and theca cells begin to luteinize in response to LH-induced increases in the intracellular cAMP content, they secrete increased amounts of estradiol and progesterone. As a result, plasma concentrations of estradiol rise sharply and plasma progesterone levels continue to increase. The follicular cells also secrete **plasminogen activator,** which leads to an increase in the activity of the proteolytic enzyme **collagenase** in the follicular fluid. Collagenase digests collagen fibers in the follicular wall, resulting in **increased distensibility.** In response to LH, secretion of the prostaglandins, PGE_2 and $PGF_{2\alpha}$, by granulosa and theca cells is augmented. PGE_2 enhances plasminogen activator production. **Histamine** (from mast cells), **norepinephrine** (from noradrenergic neurons in the follicular wall), and $PGF_{2\alpha}$ appear to facilitate rupture of the follicle by increasing vascular permeability, hyperemia, and, consequently, intrafollicular pressure. The combined effect of decreased tensile strength of the follicular wall and an intrafollicular pressure of 15 to 20 mm Hg triggers **follicular rupture,** with expulsion of the **cumulus—oocyte complex. Ovulation** occurs 1 to 2 hours before the plasma progesterone levels peak, or 34 to 35 hours after the onset of the LH surge.

Luteal Phase

After ovulation, a **fibrin clot** forms in the cavity of the ruptured follicle. The granulosa and theca cells lining the wall of the follicle continue to luteinize, and, as the corpus luteum forms, plasma progesterone and estradiol concentrations continue to rise. The **lifespan of the corpus luteum** determines the length of the luteal phase of the menstrual cycle. The luteal phase is more constant than the follicular phase, and usually lasts approximately 14 days. **Progesterone** decreases the frequency of GnRH pulses, so that the frequency of LH pulses decreases, but the concomitant increase in LH pulse amplitude causes little change in the mean plasma concentrations of LH.

Continued secretion of progesterone, which inhibits follicular development, depends on LH.

The excitability of the uterine myometrium decreases and the secretory activity of the endometrium increases, both in response to progesterone. This altered intrauterine environment facilitates **implantation of the blastocyst.**

Menses

If the cycle is nonfertile and pregnancy does not ensue, the corpus luteum ceases to function, degenerates, and involutes in a process termed **luteolysis.** This process usually begins 14 to 15 days after ovulation. **Progesterone levels** decline to levels seen during the follicular phase of the cycle. Three days following the initiation of luteolysis, the endometrial lining of the uterus is shed in menstruation. Approximately one day before menstruation, plasma FSH levels rise and **folliculogenesis** resumes. **Menses,** which consists of endometrial vasospasm and ischemic necrosis culminating in desquamation and bleeding, usually lasts one to four days. Approximately 35 ml each of blood and serous fluid are usually lost.

REPRODUCTIVE TRACT

The reproductive tract (Fig. 55-8), which is the duct system consisting of the fallopian tubes, uterus, cervix, and vagina, facilitates: (1) reception and transport of the spermatozoa; (2) reception and transport of an ovum; (3) fertilization of the ovum by a single spermatozoan; (4) embryogenesis and fetal development; and (5) delivery of an infant. The components of the reproductive tract cyclically undergo a series of dramatic structural and functional changes. During the proliferative **follicular phase** of the cycle, the reproductive tract is dominated by **estrogen;** during the secretory **luteal phase,** by **progestin.**

Fallopian Tubes

The paired fallopian tubes, or **oviducts,** consist of (1) an **infundibulum** with the open ostium surrounded by fimbriae, one of which is attached to the ovary; (2) an outer, dilated **ampulla,** which curves over the ovary; and (3) a constricted **isthmus,** which includes an intramural portion that extends into the fundus of the uterus. Each section of the fallopian tube is composed of **three layers:** (1) an outer **serosal surface,** which is continuous with the peritoneum; (2) a middle **muscular layer;** and (3) an inner **mucosal layer,** which comprises epithelium and connective tissue. The **epithelium,** which consists of ciliated columnar cells and secretory, nonciliated columnar cells, is continuous with the uterine mucosa. The **isthmus** has a thick muscular wall and a thin mucosa, whereas the convoluted **ampulla** has a relatively thin muscular wall but a thick mucosa consisting of numerous papillary folds that markedly increase the surface area of the tubal lumen.

The fallopian tube serves as a **conduit for the ovulated ovum** and the **spermatozoa,** and as a source of **nutrients** for the gametes and early preembryo. It is the normal site of **fertilization.** The **ciliated cells** at the open fimbriated end (ostium) of the oviduct direct the ovum into the infundibulum and down through the ampulla toward the isthmus. **Estrogens** promote growth, proliferation, and ciliogenesis in the epithelium of the fallopian tubes during the follicular phase of the menstrual cycle. In addition, estrogens elevate the levels of progesterone receptors. Estrogen and progestin together condition the cells of the muscular layers to contract in a coordinated manner to promote transport of the ovum to the uterus.

The **composition of oviductal fluid** is regulated by the secretory epithelial cells, which secrete fluid against a pressure gradient. **Sodium** is the principal cation and **chloride** the principal anion, and the concentrations of both are higher than those in plasma. In addition, oviductal fluid is enriched in **potassium**, so that the ratio of potassium to sodium is higher than that in plasma. **Electrolyte concentrations** fluctuate during the menstrual cycle, with the sodium and chloride concentrations decreasing when the estrogen content is lowest. The **protein concentration** is highest in the immediate

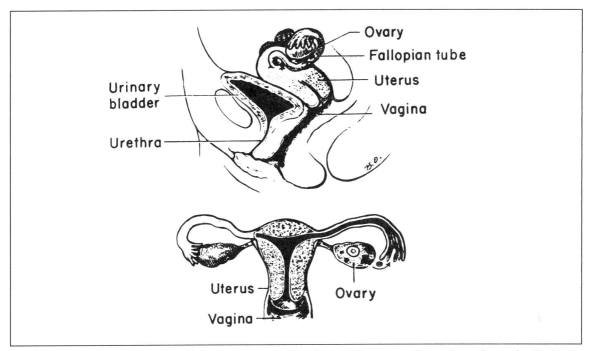

Fig. 55-8. Female reproductive tract, showing the anatomic relationships among the ovaries and reproductive tract. (From Guyton, A. C. *Textbook of Medical Physiology*, 6th ed. Philadelphia: W. B. Saunders, 1981. P. 1005.)

postovulatory period. For **metabolism,** human oviductal epithelial cells utilize the glycolytic pathway and produce **lactic acid,** which is present at concentrations of 1.0 to 8.4 gm/dl, compared with concentrations of only 1.6 to 3.6 mg/dl for **glucose. Bicarbonate ion,** which stimulates oxygen uptake by the spermatozoa, has concentrations of 110 to 220 mg/dl. Both lactic acid and bicarbonate are important for the cleavage of zygotes in the fallopian tubes.

Uterus

The uterus is a muscular organ consisting of a **body,** or **corpus,** the top of which is the **fundus,** or **dome,** which connects with the two fallopian tubes; and the **neck,** or **isthmus,** which leads into the cervix. Like the oviducts, the uterus is composed of three **tissue layers:** (1) the **external serosa,** which is continuous with the folds over

the urinary bladder and rectum; (2) the middle **muscular layer,** the **myometrium,** which consists of three vascularized layers of smooth muscle and connective tissue; and (3) the inner **mucosal layer,** the **endometrium,** which lines the uterine lumen. The functions of the endometrium and myometrium during implantation, pregnancy, and parturition are discussed in Chapter 57.

Endometrium

The endometrium plays a major role in pre-pregnancy reproductive function. Its secretions affect the spermatozoa en route through the uterine cavity to the oviducts, and provide nutrients to the zygote during the days prior to implantation. The endometrium must be appropriately conditioned and highly vascularized in order for implantation to occur. The endometrial layer of the uterus contains **secretory glands, stromal** tissue, **blood vessels,** and an **epithelium** that covers the luminal surface.

The endometrium undergoes a characteristic sequence of changes in response to ovarian steroid hormones during each menstrual cycle. **Optimal growth** and **maturation** of the endometrium require sequential exposure to these ovarian hormones in definite ratios. During the **proliferative phase,** when plasma estrogen levels are increasing, the thickness of the endometrium increases to 3 to 5 mm because of the proliferation of glandular and superficial epithelium, stroma, and blood vessels. The stroma becomes edematous and the glands become very tortuous. During the **secretory phase** after ovulation, when both progesterone and estrogen are secreted by the corpus luteum, the thickness of the endometrium increases to 4 to 6 mm. **Glycogen-containing vacuoles** appear at the base of the glandular epithelial cells, and **lipid** and **glycogen** accumulate in the stromal cells. The glandular cells secrete fluid into the glandular lumina. The stroma becomes loose and more edematous, facilitating implantation of the blastocyst. In the latter part of the secretory phase, lymphocytes invade the stroma. If implantation does not occur, levels of ovarian hormones drop precipitously, polymorphonuclear leukocytes migrate into the stroma, and the secretory phase is followed by desquamation of the endometrium and menses, the genital bleeding due to sloughing of the endometrial lining of the uterus. Only a thin layer of endometrial stroma remains after menstruation. New epithelial cells appear on the luminal surface of the endometrium within 3 to 7 days after the beginning of menstruation and the proliferative phase ensues. If implantation does occur, the endometrium gives rise to the **decidua** (Chapter 57).

Myometrium

Contractions of myometrial smooth muscle fibers are crucial in the expulsion of the fetus at parturition. It is important that the uterus not contract prematurely during either the preimplantation or postimplantation phases of pregnancy. **Progesterone** directly influences myometrial contractility. It increases the membrane potential and the liminal stimulus required to elicit contractions of the myometrial smooth muscle. Thus, during the secretory phase of the menstrual cycle when implantation would occur, the myometrium is quiescent. If implantation takes place and pregnancy ensues, the high plasma levels of progesterone maintain the myometrium in a quiescent state, so that the developing embryo and fetus are not expelled. After the 38th week of pregnancy, progesterone levels fall and the electrical activity of the myometrium increases.

Cervix

The cervix consists of an endocervical canal that communicates with the uterus via the **internal os** and with the vagina via the **external os.** It is composed mainly of connective tissue with scattered smooth muscle fibers. The **vaginal portion** of the cervix is covered with layers of squamous cells that are continuous with the vaginal mucosa. The **epithelium** of the endocervical canal consists of tall, secretory columnar cells that respond to estrogens by increasing in height and by accumulating columns of cervical mucus at their apical end. After ovulation, a portion of the endocervical cell is sloughed off.

The mucus secreted by the endocervical cell is rich in **glycogen, glycoproteins,** and **glycosaminoglycans.** Both before and after ovulation, the cervical mucus is highly viscous and sparse. Around the time of ovulation, however, the mucus becomes thin and increases sevenfold to facilitate the rapid transport of sperm to the fallopian tubes.

Vagina

The vagina is the outermost section of the female reproductive tract. It is composed of **three layers** of tissue: (1) the **luminal mucosa,** which is continuous with the cervical mucosa and the cutaneous epithelium of the labia; (2) the bilayered **muscular layer;** and (3) the outer **adventitia,** which consists of connective tissue continuous with the connective tissue of the bladder, rectum, and other pelvic structures. The cells of the vaginal mucosa proliferate from mitoses in the cells of the basal layer. Estrogens, androgens, and progestins enhance this thickening of the vaginal epithelium, although **estradiol** is the most potent steroid. The cells transform from cuboidal to stratified, making the lining of the vagina much more resistant to infection and trauma than that of a non-estrogen-primed epithelium. Progestins also bring about desquamation of the superficial layers of epithelial cells.

SEXUAL FUNCTION

Female sexual function promotes the reception and transport of spermatozoa in the female reproductive tract. The physiologic processes of the female sexual act are influenced by both **psychic factors,** which are not well understood but are responsible for the "sex drive," and by **reflexes** that involve the local stimulation of components of the reproductive organs. Neuronal signals triggered by local stimulation, such as massage or irritation, of the perineal region, the genitalia, and the urinary tract are transmitted via the pudendal nerve and sacral plexus to the brain. The **clitoris** is especially sensitive to physical stimulation.

The female **sexual response** involves physiologic responses that can be divided into **four phases**. These sexual responses, which begin at puberty and are amplified by estrogens, can occur throughout a woman's life even in an estrogen-deficient state. The **first phase, excitement or arousal,** occurs in response to psychogenic or neurogenic stimulation and includes (1) **erection** due to gradually increasing vasocongestion and muscular tension in the genitalia; and (2) **lubrication** of the vagina. During this phase, signals from the parasympathetic nerves cause arteriolar dilation and venous constriction in the erectile tissues that surround the introitus and project into the clitoris. As a result, blood rapidly accumulates in the erectile tissue, enabling the introitus to tighten around the male penis. At the same time, parasympathetic signals to the bilat-

eral **Bartholin's glands** beneath the labium minus lead to increased secretion of mucus into the vagina. This mucus, along with other vaginal secretions, provides lubrication during sexual intercourse, so that the sexual act produces a satisfactory massaging sensation in the woman, rather than irritation. This **massaging sensation** is necessary for the female to progress to the **second phase** of the sexual response, the **plateau phase.** Continued satisfactory stimulation causes excitement to increase in intensity to a plateau, where a high state of sexual sensitivity and receptivity is maintained. The length of this phase can vary greatly among individuals. The plateau phase culminates in **orgasm,** the **third phase** of the sexual response.

Orgasm occurs when the sexual organs receive appropriate psychic signals from the brain, at the same time that they are receiving local stimulation of maximum intensity. Because psychologic and physical intensity are maximal, orgasm is referred to as the *climax*. During orgasm, the **perineal muscles,** which receive neural signals similar to those that cause ejaculation in the male, contract rhythmically. At the same time, intense sexual signals transmitted to the brain lead to increased **muscular tension** throughout the body. Orgasm culminates in rapid release from the developed vasocongestion during intense muscular tension, and it may be accompanied by secretion of copious amounts of vaginal fluid. The **fourth phase** of sexual response, **resolution,** is often characterized by relaxation and sleep. Many women, however, return to the first phase of arousal, remain responsive to sexual stimulation, and may experience plateau and orgasm repeatedly.

MENOPAUSE

Menopause, defined as the last menstrual period, signals the **cessation of reproductive function** in women and marks the end of a gradual diminution of the number of follicles in the ovary. It usually occurs around the age of 51. About seven years before menopause, the menstrual cycle length begins to be more variable as the follicular phases shorten and ovulations do not occur. Long, short, normal, and anovulatory cycles can be interspersed during the **perimenopausal period.**

Nearly complete **loss of oocytes** is a major factor in the onset of menopause. Exhaustion of the supply of primordial follicles leads to cessation of ovarian follicular production of estrogen and progestin, as well as of inhibin, and a consequential decrease in the plasma levels of these hormones. The **hypothalamic–pituitary unit** also becomes less sensitive to the negative-feedback effects of estrogen. Consequently, plasma **FSH levels** rise several fold to levels seen in castrated women because of elimination of negative feedback. The excessive secretion of **LH** associated with menopause is thought to be due to the altered microheterogeneity of the hormone, causing secretion of forms of LH with relatively low biologic activity. The **neuroendocrine mechanisms** that control FSH and LH secretion remain intact, since surges of FSH and LH, similar to normal preovulatory surges, can be induced by the administration of estrogen and progestin. The **pituitary response to GnRH** is not impaired by menopause, although artificially induced hypersecretion of GnRH can lead to depletion of pituitary stores of the hormone.

Certain **physiologic changes** eventuate from the decreased secretion of ovarian steroid hormones. These include hot flashes or flushes, atrophy of the genitourinary tract, atrophy of the epithelial glands and ducts of the breasts, and bone loss. Tissues that depend on estrogen and progestin regress after menopause. Bone loss, or **osteoporosis,** results from both direct and indirect effects of estrogen on bone, but is not solely a function of plasma estrogen concentrations. Similarly aged males, who normally have much lower plasma estrogen levels than females, also experience a certain degree of bone loss. Estrogen appears to have permissive effects on calcium absorption and to enhance the proliferation and differentiation of osteoblasts, the cells responsible for bone formation.

Hot flashes, which are irregularly occurring thermogenic episodes that cause peripheral vasomotor dilation and sweating, are associated with **transient elevations** in the plasma levels of GnRH, LH, and catecholamines. Although the cause of hot flashes is unknown, administration of estrogen prevents them. In some women, hot flashes subside within five years of menopause; in others, they continue unabated for many years.

Genital atrophy encompasses the vulva, vagina, urethra, uterus, and fallopian tubes. With estrogen depletion, the size of these organs, their vascularity, and their secretions diminish. Administration of estrogen will not reverse atrophy of the external genitalia, though it can reverse atrophy of the internal genitalia.

SUMMARY

The reproductive system of the female consists of paired ovaries, the reproductive tract or internal genitalia, external genitalia, and the breasts. The ovaries produce germ cells containing the haploid number of chromosomes. Ovarian cells of the follicle, corpus luteum, and interstitium produce hormones that condition the reproductive tract to receive the male germ cells and to provide a beneficial environment for fertilization, implantation of the blastocyst, and pregnancy. Reproductive function begins at puberty when the hypothalamic–pituitary unit matures functionally. During the reproductive period, a mature ovum is produced by the ovary each month. The cyclic production of a female gamete and the concomitant changes in the reproductive tract are referred to as the *menstrual cycle*. The cyclic reproductive processes are controlled by interacting neural and hormonal signals originating from the hypothalamic–pituitary–ovarian axis. As the pool of primordial follicles in the ovary is depleted, reproductive function declines and ceases at *menopause*.

NATIONAL BOARD TYPE QUESTIONS

Select the one most correct answer for each of the following

1. Progesterone:
   A. Decreases myometrial contractility
   B. Decreases the liminal stimulus necessary for contraction of myometrial smooth muscle
   C. Binds to plasma membrane–bound receptors that then transport the hormone to the nucleus of the cell
   D. Prepares endometrial cells to respond to estrogen by increasing estrogen receptors
   E. None of the above

2. Which of the following is *not* a precursor in the synthesis of 17β-estradiol in the ovary?
   A. Androstenedione
   B. Testosterone
   C. Cholesterol
   D. Pregnenolone
   E. Estriol

3. FSH and LH:
   A. Are both required for the development of ovarian follicles
   B. Secretion by the anterior pituitary gland is modulated by inhibin
   C. Are glycosylated heterodimeric proteins that bind to membrane-bound receptors
   D. Concentrations in the plasma peak during the early follicular phase of the menstrual cycle
   E. None of the above

4. Excitement or arousal, the first phase of the female sexual response,
   A. Includes rapid accumulation of blood in the erectile tissue of the introitus and clitoris
   B. Includes increased secretion of mucus into the vagina in response to parasympathetic signals to Bartholin's glands
   C. Occurs in response to psychogenic or neurogenic stimuli
   D. Is amplified by estrogen
   E. All of the above

5. Menopause, the onset of reproductive senescence, occurs because:
   A. The uterus is no longer capable of sustaining a fetus
   B. Pituitary gonadotropes synthesize excessive quantities of FSH and LH
   C. Excessive plasma inhibin levels suppress FSH production
   D. The pool of ovarian follicles is depleted
   E. All of the above

ANNOTATED ANSWERS

1. A. Progesterone decreases myometrial contractility by increasing the membrane potential and the liminal stimulus needed to elicit contractions of myometrial smooth muscle.
2. E. Estriol is a metabolite of estradiol.
3. C.
4. E.
5. D.

BIBLIOGRAPHY

Becker, K., et al, eds. *Principles and Practice of Endocrinology and Metabolism.* Philadelphia: J. B. Lippincott, 1990.

Knobil, E., and Neill, J. D., eds. *The Physiology of Reproduction.* New York: Raven Press, 1988.

Odell, W. D., and Moyer, D. L. *Physiology of Reproduction.* Saint Louis: C. V. Mosby, 1971.

Yen, S. S. C., and Jaffe, R. B., eds. *Reproductive Endocrinology. Physiology, Pathophysiology and Clinical Management,* 2nd ed. Philadelphia: W. B. Saunders, 1986.

56 The Male Reproductive System

ANDREW R. LaBARBERA

Objectives

After reading this chapter, you should be able to:

List the components of the male reproductive system and their functions

Describe the synthesis, secretion, and effects of testicular steroid hormones

Explain how the hypothalamus and anterior pituitary gland regulate male reproductive function

Explain how negative feedback by gonadal steroids and inhibin modulates gonadotropin secretion

Describe spermatogenesis and the maturation of spermatozoa

Explain how the reproductive tract functions to deliver mature spermatozoa to the female reproductive system

---

The male reproductive system, which consists of the paired testes, the reproductive tract or internal genitalia, and the external genitalia, functions to produce germ cells for sexual reproduction and to deliver them to the female reproductive tract. The organs of the male reproductive system, which are established during **gestation,** reach full maturity during puberty. Production of **spermatozoa,** the male germ cells, differs from that of germ cells in the female in two major respects: (1) development beyond the primordial germ cell stage does not occur until puberty in the male; and (2) beginning at puberty, the development of spermatozoa is continuous. **Spermatogenesis,** the process by which primordial germ cells form into spermatozoa, and the **production of androgens,** the male sex hormones, both occur in the testes and are controlled by the hypothalamic–pituitary unit. Gonadotropin secretion and at least some spermatogenesis

usually continue until death in the male, though, between the ages 40 to 50, testosterone secretion and male sexual function begin to decline.

TESTIS

The paired testes serve at least three functions critical to successful propagation of the species: (1) production of **gametes;** (2) secretion of **hormones** that cause differentiation of the brain and reproductive tract during fetal and neonatal development, maintain the structure and function of the reproductive tract in the sexually mature adult, participate in the regulation of metabolism and skeletal growth, and promote development of secondary sexual characteristics; and (3) **feedback regulation** of hypothalamic–pituitary hormone secretion. The testes consist primarily of the **seminiferous tubules** that are separated by interstitial tissue (Fig. 56-1). Each testis is

789

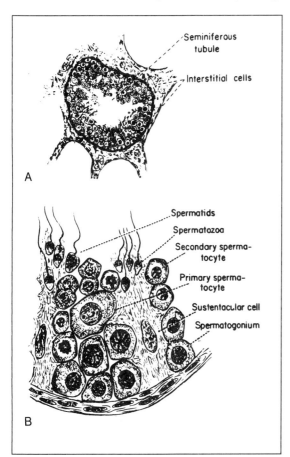

Fig. 56-1. Cross-section of a seminiferous tubule. (A) Relationship between the tubule and the surrounding stroma containing interstitial cells of Leydig. (B) Relationship between the germ cells and the Sertoli (sustentacular) cells. Spermatogonia are located near the outer basement membrane of the tubule. As spermatogenesis progresses, the germ cells move steadily closer to the lumen and become engulfed by the Sertoli cells. (From Guyton, A. C. *Textbook of Medical Physiology,* 6th ed. Philadelphia: W. B. Saunders, 1981. P. 993.)

surrounded by a thick fibrous capsule, the **tunica albuginea,** and is located outside the body in the **scrotal sac.** If the testes do not descend into the scrotum properly (cryptorchidism), usually during the sixth to eighth month of gestation, spermatogenesis and androgen production are

not normal. While the cause of **cryptorchidism** is not known, the disorder is frequently associated with abnormal fetal testicular androgen production.

Interstitial Cells of Leydig

The spaces between the seminiferous tubules are filled with interstitial tissue composed of connective tissue stroma, blood and lymph vessels, nerves of the testicular parenchyma, mast cells, macrophages, and the interstitial cells of Leydig, or Leydig cells. The androgen-secreting Leydig cells, which develop from the mesenchymal cells of the stroma rather than from epithelial cells, are large cells with a diameter of up to 20 μm. Ultrastructurally, these cells, which are scattered through the stroma surrounding the seminiferous tubules, are typical steroid-secreting cells with abundant smooth endoplasmic reticulum, mitochondria, and Golgi stacks and lipid droplets containing cholesterol esters. The principal function of the Leydig cells appears to be to produce **androgens,** though some estrogen is also secreted.

Leydig cells differentiate and begin to secrete androgens during the **seventh week of fetal life** in response to **chorionic gonadotropin** (hCG), which is produced by the placenta. During childhood, when gonadotropin secretion is low, the Leydig cells regress to an undifferentiated state. When plasma LH levels rise again at **puberty,** the Leydig cells differentiate again and acquire the characteristics of steroid-producing cells. They are maintained by LH during adulthood.

Androgen
Androgenic steroid hormones are the **male sex hormones.** They perform numerous functions throughout the body (Table 56-1). These include: (1) differentiation of the **male reproductive tract** and **brain** during fetal life; (2) stimulation of **testes descent** during the last two months of gestation; (3) stimulation of maturation and maintenance of the **reproductive tract,** includ-

Table 56-1. Actions of Androgens

1. Stimulate differentiation of the male reproductive tract
2. Stimulate sexual differentiation of the male brain
3. Stimulate maturation of the external genitalia
4. Increase size of larynx and thickness of vocal cords
5. Increase hair growth (facial, axillary, pubic) and hair recession (balding)
6. Increase libido and sexual potency
7. Increase aggressive behavior
8. Maintain spermatogenesis in conjunction with FSH
9. Maintain male reproductive tract and accessory sex glands
10. Decrease luteinizing hormone, and to a lesser extent follicle-stimulating hormone, secretion by the hypothalamic–pituitary unit
11. Stimulate bone growth and closure of epiphyses
12. Increase protein anabolism (increase protein synthesis, decrease protein catabolism), leading to increased linear body growth, nitrogen retention, and muscular development

ing the internal and external genitalia, at puberty and during adulthood; (4) maintenance of **spermatogenesis** in the adult testes; (5) **negative-feedback regulation of LH secretion** by the pituitary; and (6) facilitation of **sexual drive** and aggressive behavior. Androgens also stimulate development of the male **secondary sexual characteristics,** including: (1) **growth of facial, chest, axillary, and pubic hair,** as well as hair recession and balding; (2) hypertrophy of the **laryngeal mucosa** causing enlargement of the larynx and deepening of the voice; (3) development of **increased musculature** due to increased protein deposition and nitrogen retention; and (4) enhancement of **linear growth** through stimulation of bone growth. In addition, androgens are very potent **anabolic hormones** that accelerate metabolism.

Testosterone is the major androgen synthesized by the Leydig cells and is secreted at a rate of approximately 6 to 7 mg per day. Ninety-five percent of plasma testosterone in men (600 to 700 ng/dl) is secreted by the testes. Other potent androgens produced by the testes include **5α-**dihydrotestosterone (DHT; 0.3 mg per day) and **androstenedione** (2.4 mg per day), which serves mainly as a precursor for estrogen biosynthesis. DHT and androstenedione are present in plasma at concentrations of 50 to 60 ng/dl and 150 ng/dl, respectively.

Testosterone is **synthesized from cholesterol,** which is preferentially derived from circulating **low-density lipoproteins** (LDL), but which can be synthesized de novo from **acetate** in the Leydig cells (Fig. 56-2). Cholesterol derived from LDL is stored in lipid droplets in the cytoplasm. It is transported to the mitochondria, where it is converted to **pregnenolone** through the action of the C_{27}–side-chain-cleavage P_{450} enzyme. Pregnenolone is then transported to the microsomal compartment (smooth endoplasmic reticulum). In man, pregnenolone is converted to testosterone via the Δ^5 pathway, involving the intermediates 17α-hydroxypregnenolone and dehydroepiandrosterone. The conversion of pregnenolone, which is a C_{21} steroid, to dehydroepiandrosterone, which is a C_{19} steroid, occurs through the action of the C_{21}–side-chain-cleavage P_{450} enzyme, which possesses both 17α-hydroxylase and $C_{17,20}$-lyase activities. LH enhances androgen biosynthesis primarily by stimulating transport of cholesterol from the outer to the inner mitochondrial membrane. LH may also increase the amounts or activities of the steroidogenic enzymes.

Testosterone, which has a low solubility in an aqueous medium, is carried in the blood bound to plasma proteins. The principal androgen transport protein is testosterone-estradiol–binding globulin (TeBG), a 94,000-molecular-weight glycoprotein with one steroid-binding site per molecule. Thirty percent of the testosterone in plasma is bound to TeBG; 67 percent circulates bound to albumin and other plasma proteins; and only 3 percent is free. Testosterone bound to TeBG does not appear to be available for metabolism or for uptake by target tissues, as the rate of testosterone metabolism correlates directly with the amount of free plus albumin-bound testosterone. The biologically relevant testosterone

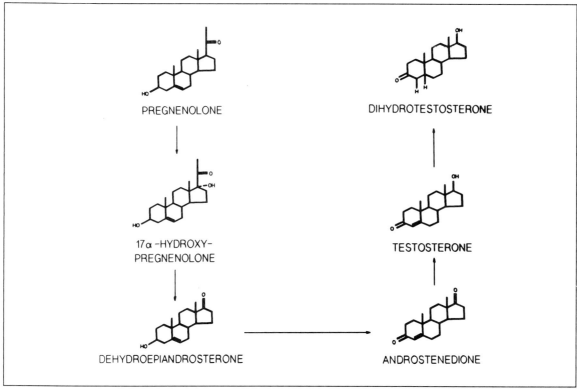

Fig. 56-2. Biosynthesis of testicular androgens in Leydig cells. Testosterone is synthesized from cholesterol, which is transported to the mitochondria where it is converted to pregnenolone through the action of the C_{27}–side-chain-cleavage P_{450} enzyme. Pregnenolone is transported to the microsomal compartment (smooth endoplasmic reticulum). Pregnenolone is converted to testosterone via the Δ^5 pathway, involving the intermediates 17α-hydroxypregnenolone and dehydroepiandrosterone. The conversion of pregnenolone, which is a C_{21} steroid, to dehydroepiandrosterone, which is a C_{19} steroid, occurs through the action of the C_{21}–side-chain-cleavage P_{450} enzyme, which has both 17α-hydroxylase and $C_{17,20}$-lyase activities.

is that which is free; many species do not have TeBG. The plasma concentration of TeBG is enhanced by both estrogen and thyroid hormone, whereas testosterone itself decreases plasma TeBG.

Testosterone has a **plasma half-life** of approximately 15 minutes. It can be metabolized either to **active compounds** in specific target tissues or to **inactive compounds,** which are excreted. Examples of the first instance include: (1) aromatization to estradiol in the brain, breast, and other tissues; (2) conversion by 5α-reductase in the prostate, urethra, scrotum, and penis, and

in skin to DHT, a more potent androgen than testosterone; and (3) conversion by 5β-reductase to compounds that possess little androgenic activity, but that stimulate production of red blood cells in bone marrow. In the second situation, androgens are metabolized to inactive sulfates and glucuronide conjugates in the liver, and then excreted in either the bile or urine. The mean metabolic clearance rate for testosterone is approximately 1,000 liters per day.

Testosterone diffuses from the plasma into target cells, where it either binds immediately to an intracellular receptor or is first converted to DHT or estradiol. Testosterone and DHT bind to specific high-affinity intracellular **androgen receptors** that appear to be loosely associated with the nuclear membrane. Estradiol also binds to specific high-affinity receptors similarly situated within the cell. The **androgen-receptor complex** interacts with specific nuclear acceptor sites on **chromatin.** This binding to chromatin triggers RNA synthesis, which in turn leads to increased synthesis of the proteins necessary for growth and differentiated function.

Seminiferous Tubules

Each mature testis contains 20 to 25 meters of highly convoluted, anastomosing **seminiferous tubules** per gram of tissue. The ends of these tubules open into the **rete testis,** a network of intercommunicating channels which link the seminiferous tubules with the efferent ducts. The **cytoarchitecture** of the tubular epithelium is characterized by a **basement membrane** lined with sustentacular (Sertoli) cells and spermatogonia. The **Sertoli cells** are surrounded by **spermatogonia,** which are progressively more developed going from the basement membrane to the tubular lumen. The luminal surface of the Sertoli cell surrounds **spermatids** that are maturing into spermatozoa.

Sertoli Cells

The Sertoli cells, which are tall, columnar cells that extend centripetally from the basement membrane to the lumen, constitute the nongerminal component of the epithelium of the seminiferous tubule. These cells are linked by tight junctional complexes to form the **blood–testis barrier,** which limits the movement of fluid and nutrients between the interstitial space of the testis and the lumen of the seminiferous tubule. **Lipid-soluble substances,** such as steroids, are able to traverse the barrier and enter the lumen of the seminiferous tubule at a rate that is related to their lipid solubility. Presumably lipid-soluble substances diffuse passively through the Sertoli cells. **Glucose** is transferred across the barrier

by facilitated diffusion. The significance of the blood–testis barrier is unknown, but it may serve to prevent the body's immune system from recognizing the haploid germ cells. If the barrier is interrupted, as happens in vasectomy, the male can become immune to his own spermatozoa. Findings from animal studies indicate that the barrier does not form until puberty. The shape, volume, and ultrastructure of Sertoli cells change to accommodate the changing size and shape of the surrounding germ cells at various stages of development.

Sertoli cells have several functions: (1) provision of **nutrients** to the germ cells; (2) synthesis of **multiple proteins** that are secreted into the luminal fluid and are also present in the serum, including ceruloplasmin, acidic glycoprotein, and transferrin; (3) synthesis of **estrogen;** (4) production of **androgen-binding protein** (ABP); (5) **phagocytosis** of damaged germ cells; and (6) synthesis and secretion of **inhibin.**

Inhibin

Sertoli cells synthesize and secret **inhibin,** a heterodimeric protein that selectively inhibits the synthesis and secretion of **follicle-stimulating hormone** (FSH) by pituitary gonadotropes and apparently is identical to the inhibin synthesized by granulosa cells. The structure and synthesis of inhibin by ovarian granulosa cells are discussed in Chapter 55. The concentrations of FSH and inhibin in the circulation are in a **dynamic equilibrium.** FSH stimulates Sertoli cells to produce inhibin, which, in turn, negatively modulates further FSH production. If one or both testes are removed or if the seminiferous tubules are damaged, the plasma FSH concentration rises and cannot be suppressed by administration of testosterone.

Androgen-Binding Protein

ABP is a dimeric protein that binds androgens with high affinity. The two protomers have molecular weights of 48,000 and 46,000. ABP is found both in the luminal fluid and blood, and is structurally very similar to the serum transport protein, TeBG, also referred to as *sex hormone–binding globulin*. Because ABP is concentrated in the luminal fluid and in the epididymis, its putative physiologic role is to carry testosterone within the Sertoli cell and from the testis to the epididymis in order to maintain high testosterone concentrations within these androgen-dependent tissues.

Estrogen

The principal estrogens in the plasma of men are **estradiol** and **estrone**. Only 10 to 20 percent of circulating estrogens are synthesized by the Sertoli cells in the testes, however. The remainder originates through the extragonadal conversion of testosterone and androstenedione by means of **cytochrome P$_{450}$-dependent aromatase,** which is present in numerous tissues. The **ratio of estrogens to androgens** appears to be more important than the absolute levels of the steroids, because decreased plasma testosterone levels with normal plasma estrogen levels result in **feminization**.

Spermatogenesis

Development of the male germ cells occurs continuously and repeatedly in the epithelium lining the seminiferous tubules of the mature testis. Areas of active spermatogenesis are interspersed with resting epithelia, so that areas of progressively more developed germ cells succeed each other along the length of the tubule. As the germ cells develop, they move from the outer wall to the luminal surface, so that the least developed cells line the walls of the tubule and the most developed cells line the tubular lumen.

Spermatogonia are the least differentiated cells of the germ cell population. They are very large with prominent nuclei and few cytoplasmic organelles and are located around the walls of the seminiferous tubules. These cells, which are derived from primordial germ cells, **divide by mitosis,** both to form the pool of cells that will undergo meiosis and also to replenish themselves. **Three progressive stages of sperma-** **togonia** can be distinguished: (1) **type A, dark spermatogonia,** which have darkly staining, vacuolated nuclei and are in close contact with the basement membrane of the wall of the tubule; (2) **type A, pale spermatogonia,** which have less darkly staining nuclei and have less contact with the basement membrane; and (3) **type B spermatogonia,** which contain peripheral nuclear chromatin aggregations and have the least contact with the basement membrane. The type B spermatogonia give rise to the **primary spermatocytes.**

Prior to prophase of the first meiotic division, primary spermatocytes, which resemble type B spermatogonia, double the DNA content of their 46 chromosomes, so that they have 92 daughter chromatids with the 4N DNA content. DNA replication in primary spermatocytes represents the final synthesis of DNA in spermatogenesis.

Primary spermatocytes undergo the first meiotic division to yield two secondary spermatocytes, each with the N, or haploid, number of chromosomes; each chromosome has two chromatids, so that the secondary spermatocyte has the 2N DNA content. **Secondary spermatocytes,** which are spherical cells with a diameter of 10 to 12 μm, have a short lifespan. They rapidly undergo the second meiotic division, during which the chromatids of each chromosome separate to the two daughter cells by processes similar to those of mitosis. The final result of these divisions is four spermatids, each with 23 unpaired chromosomes, the N or haploid number. Each spermatid has 22 autosomes and either an X or a Y chromosome.

Spermiogenesis

The spherical spermatids, which are considerably smaller than the secondary spermatocytes, mature into spermatozoa through the process of spermiogenesis. Neither mitosis nor meiosis is involved in the transformation of the spermatid into an elongated, highly organized cell. Spermatids are located near the luminal surface of the seminiferous tubule and during the early phase of spermiogenesis they become engulfed by Sertoli cells.

During spermiogenesis, both the **cytoplasm and nucleus of the spermatid continue to differentiate** and the **shape changes from spherical to elongated.** The Golgi apparatus aligns itself on one side of the cell nucleus and begins to form small vesicles that appear as **dense bodies** (pro-acrosomic granules). These granules coalesce to form the **head cap** and the **acrosome,** which contains glycosaminoglycans and the enzymes acid phosphatase and nucleoside phosphatase. In the nucleus, **basic proteins** such as histones and protamine combine with DNA to produce a semicrystalline chromatin structure. The **nucleoplasm** fills with tightly packed dense granules. The **midpiece** and **flagellum** form from the centrioles on the sperm head opposite the site where the acrosome forms. Most of the cytoplasm migrates to the area around the midpiece, forms a residual body, and pinches off from the cell.

Mature Spermatozoon

The mature spermatozoon consists of a **head,** which is ovoid and flattened anteriorly and comprises the nucleus and acrosome, and a **tail,** which is composed of the neck, midpiece, principal piece, and end piece. The **condensed nuclear elements** from the spermatid are located in the **head.** The **acrosomal portion** of the head contains **Golgi elements and enzymes,** such as hyaluronidase and proteinases, that are responsible for penetration of the ovum. The **mitochondria** from the spermatid form a mitochondrial sheath around the axial filament complex in the midpiece. The **ultrastructure of the axial filament complex** of the tail, which extends from neck to end, resembles that of cilia with the same number and arrangement of longitudinal tubules. The flagellum has some additional coarse actin fibers. The **centrioles,** which are involved in spindle formation in the fertilized ovum, are located in the neck of the tail.

Fully developed, but nonmotile, mature spermatozoa are released from the Sertoli cells into the tubular lumen. A fluid drive generated by the Sertoli cells washes the spermatozoa out of the tubules and into the rete testis and from there into the epididymis.

The **spermatogenic cycle,** the frequency with which spermatogenesis begins in a given section of the seminiferous tubule, takes 16 days in humans. **Spermatogenesis,** from the time a spermatogonium begins to replicate its DNA to the time mature spermatozoa are released from the Sertoli cells into the tubular lumen, requires 74 days, or approximately 4.5 spermatogenic cycles.

Maintenance of spermatogenesis requires testosterone, the principal androgenic steroid hormone produced by the interstitial cells of Leydig. Androgen production, in turn, is regulated by the anterior pituitary hormone luteinizing hormone (LH), which formerly was referred to as *interstitial cell–stimulating hormone.* FSH, which is required for the development of the seminiferous epithelium and initiation of the mitotic phase of spermatogenesis, is not necessary for subsequent stages of spermatogenesis. Thus, spermatogenesis in the adult male can be maintained with testosterone alone.

NEUROENDOCRINE REGULATION OF THE TESTES

Testicular function of the adult male is controlled by the **hypothalamic–pituitary unit** through the action of the gonadotropins, FSH and LH, on the testes (Fig. 56-3). In the male, FSH and LH act on cell types whose secretions have separate negative-feedback effects on the secretion of the corresponding gonadotropins. **Gonadotropin-releasing hormone** (GnRH) (see Chapter 55), a hypothalamic hormone, stimulates pituitary gonadotropes to secret FSH and LH. Because GnRH is secreted in a pulsatile manner, FSH and LH are also secreted in this fashion. **FSH** stimulates Sertoli cells that produce inhibin, which in turn specifically inhibits FSH secretion; **LH** stimulates Leydig cells that produce testosterone, which in turn specifically inhibits LH secretion. **Inhibin** has a negative-feedback effect on the **pituitary gonadotropes,** whereas **testosterone** appears to exert its negative-feedback effect

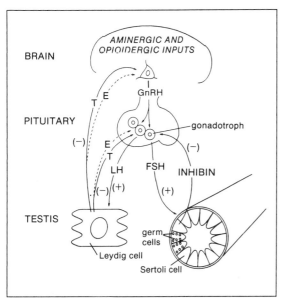

Fig. 56-3. Hypothalamic–pituitary–testicular axis. The anterior pituitary synthesizes and secretes follicle-stimulating hormone (*FSH*) and luteinizing hormone (*LH*) in response to hypothalamic gonadotropin-releasing hormone (*GnRH*), which is secreted in a pulsatile manner. FSH stimulates the Sertoli cells, which synthesize and secret inhibin. LH stimulates the Leydig cells to produce testosterone. Inhibin and, to a lesser extent, testosterone have a negative-feedback effect on FSH secretion, whereas testosterone has a negative-feedback effect on LH secretion. Inhibin blocks FSH synthesis and secretion in the pituitary gonadotrope, whereas testosterone slows the frequency of GnRH secretion and may have a direct effect on the pituitary, leading to a decrease in the frequency of LH pulses. Both FSH and testosterone are necessary for spermatogenesis in the seminiferous tubules.

primarily on the **hypothalamus,** and to a lesser extent on the pituitary. When the plasma testosterone level drops, the frequency of pulsatile GnRH secretion increases. Thus, the plasma FSH and LH concentrations are maintained in dynamic equilibrium with the plasma inhibin and testosterone levels, respectively. In contrast with testosterone, **estradiol** exerts a negative feedback on LH secretion by decreasing the amplitude of the pituitary discharge of LH.

Follicle-Stimulating Hormone

The structure, synthesis, and secretion of FSH are discussed in Chapter 55. FSH is secreted by the **basophilic cells** of the anterior pituitary gland in response to stimulation by the hypothalamic hormone GnRH. FSH secretion in males neither fluctuates much during the day, nor does it show the cyclic variation seen in females. **Testosterone,** the principal male sex steroid, only partially suppresses FSH secretion. Rather, the primary negative modulator of FSH production is inhibin, a product of the Sertoli cell which is stimulated by FSH.

The role of FSH in adult testicular function is not understood completely. It is clear, however, that FSH and testosterone must be present for **development of the seminiferous epithelium.** Once the tubules are established, FSH is not required. If the epithelium regresses, FSH must be administered together with testosterone in order to restore it. FSH enhances production of a number of Sertoli cell proteins in culture, including **transferrin** (an iron-binding protein), **plasminogen activator, fibroblast-growth factor,** and **insulin-like growth factor I** (somatomedin C). The long-term administration of pharmacologic doses of testosterone or androgenic steroids, such as those used by body builders, will suppress FSH secretion to the extent that spermatogenesis ceases.

The effects of FSH on Sertoli cells are mediated by membrane-bound receptors. The **FSH receptor,** a high-molecular-weight, multimeric complex containing sialylated glycoprotein, is coupled to adenylyl cyclase, the membrane-bound enzyme that catalyzes the conversion of ATP to cyclic AMP. Cyclic AMP regulates protein kinases that in turn regulate the expression of genes for enzymes and substrates involved in Sertoli cell function.

Luteinizing Hormone

The structure, synthesis, and secretion of LH are discussed in Chapter 55. LH is secreted in a pulsatile manner by **basophilic cells** in the anterior pituitary gland in response to pulsatile stimulation by GnRH. Unlike FSH, the synthesis and secretion of LH are under the **negative-feedback control of testosterone,** which is produced by the Leydig cells in response to stimulation by LH. **Testosterone** inhibits pituitary LH secretion by reducing the frequency of LH pulses, presumably by decreasing the frequency of hypothalamic GnRH pulses. **Estradiol** also decreases LH secretion but it acts by reducing pituitary sensitivity to GnRH. Inhibin has no effect on LH synthesis or secretion.

The physiologic role of LH is to **maintain testosterone production by Leydig cells.** The effects of LH can be achieved by administration of testosterone, which alone is necessary for maintaining spermatogenesis in the differentiated seminiferous epithelium. As in the ovary (Chapter 55), the effects of LH on Leydig cells are mediated by specific high-affinity, membrane-bound receptors that are coupled to the membrane-bound enzyme adenylyl cyclase. LH stimulates **steroidogenesis** in the Leydig cells by enhancing the activity of **cholesterol ester hydrolase,** or cholesterol esterase, which results in increased stores of cholesterol for androgen biosynthesis in the Leydig cell. It also enhances transport of cholesterol into the mitochondria.

Gonadotropin-Releasing Hormone

Both FSH and LH are synthesized and secreted in response to GnRH, which is secreted in a pulsatile manner similar to that in females (Chapter 55). In males, the rate of GnRH secretion increases and hypothalamic GnRH content decreases after **castration,** which interrupts the negative-feedback effects of steroids. Treatment with either testosterone, DHT, or estradiol reverses the effects of castration, indicating that GnRH secretion is modulated by both testosterone and estradiol. As in females, GnRH secretion is also modulated by input from higher centers in the brain. Thus, **neurogenic and psycho-** genic influences such as stress can decrease the secretions of the hypothalamic–pituitary unit and, consequently, testicular function.

Prolactin

The anterior pituitary of men secretes prolactin, but the significance of its secretion is not clear. Prolactin does cause a reduction in the conversion of testosterone to DHT, so it is possible that the hypogonadism frequently observed in hyperprolactinemic males results from decreased 5α-reductase activity.

REPRODUCTIVE TRACT

The male reproductive tract is a duct system with associated secretory glands (Fig. 56-4). It transports spermatozoa from the testes to the urethral opening in the penis. Each testis has an individual duct system consisting of **efferent** or **collecting ducts,** the **epididymis,** the **vas deferens,** and the **ejaculatory duct.** The **ejaculatory ducts** from the two testes drain into the **urethra.** The **secretory glands** include the **seminal vesicles,** the **prostate gland,** the **bulbourethral** or **Cowper's glands,** and the **glands of Littré.** The efferent ducts, epididymis, vas deferens, and seminal vesicles develop from the **wolffian ducts,** and thus share a common mesonephric origin with the kidneys. In the **reproductive tract,** spermatozoa mature and the composition of the seminal fluid is altered to enhance survival of the spermatozoa after ejaculation.

Ducts

Efferent Ducts

The **rete testis** drains into 15 to 20 efferent or collecting ducts that penetrate the **tunica albuginea** at the upper portion of the testis. Each efferent duct passes upward and becomes highly coiled to form a compact, cone-shaped structure. These multiple coiled tubules are surrounded by

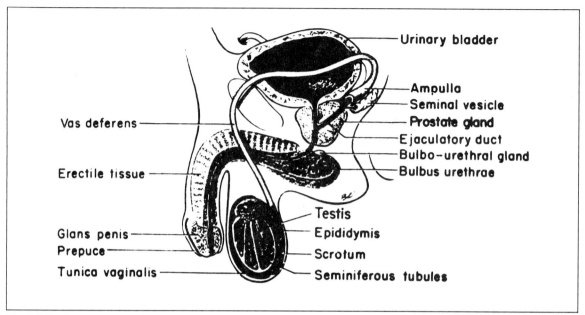

Fig. 56-4. The testis, reproductive tract, and external genitalia of the male. In each testis, spermatozoa released from the Sertoli cells into the lumen of the seminiferous tubule flow into the rete testes, which drain into the collecting ducts. The collecting ducts become highly coiled to form the head (caput) of the epididymis. The tortuous body (corpus) of the epididymis extends down along the posterolateral border of the testis into the tail (cauda) of the epididymis, where the duct is less convoluted. In the epididymis, spermatozoa acquire the ability to fertilize ova and the potential for motility. The epididymis empties into the vas deferens, which enlarges to form the ampulla before entering the prostate gland. The portion of the vas deferens just beyond the point where the seminal vesicle empties into it is the ejaculatory duct, which penetrates the prostate and traverses the gland to join the urethra. The urethra opens at the end of the penis. At the back of the urinary bladder, outcroppings of the vas deferens from the testes form the paired seminal vesicles. The single prostate surrounds the urethra as it emerges from the urinary bladder and the two ejaculatory ducts. (From: Guyton, A. C. *Textbook of Medical Physiology,* 6th ed. Philadelphia: W. B. Saunders, 1981, P. 992.)

connective tissue and form the **head,** or **caput,** of the epididymis.

The **efferent ducts,** which develop under the influence of testosterone, but not DHT, consist of a thin layer of smooth muscle, a basement membrane, and an epithelial layer lining the lumen. The **epithelium** has alternating groups of ciliated high-columnar cells and nonciliated low-columnar cells. The latter cells are thought to be secretory and contribute to the absorption of the large volume of fluid entering from the rete testis. Particulate matter is removed from the fluid in the ducts by **endocytosis.**

Epididymis

Each epididymis consists of head (caput), body (corpus), and tail (cauda). The **body,** which consists of the tortuous ductus epididymidis, extends down along the posterolateral border of the testis. In the **tail,** the ductus epididymidis becomes less convoluted and leads into the vas deferens. The **lumen of the epididymis** is lined with a pseudostratified columnar epithelium containing both small basal cells and tall columnar cells. The tall cells form the continuous even surface, however.

The **transit time from caput to cauda** appears to be 5 to 6 days. During this time, spermatozoa acquire both the ability to fertilize ova and the potential for motility. Spermatozoa do not actually become motile until ejaculation and exposure to oxygen or lactic acid. They are capable of surviving longer in the epididymis than in any other segment of the reproductive tract, and the **cauda** is the major site of storage of spermatozoa in the duct system.

The **tall cells,** which have nonmotile stereocilia, secrete ions, nutrients, proteins, glycoproteins, enzymes, and other substances necessary for both the survival and maturation of spermatozoa. As in the efferent ducts, the cells of the epididymal epithelium also absorb large quantities of fluid. Epididymal fluid is **enriched in potassium,** compared with its concentration in whole semen. It also contains high concentrations of **glyceryl phosphorylcholine,** which must be metabolized to glyceryl phosphate by glycerophosphorocholine esterase in order to be used as an energy substrate by spermatozoa. The luminal fluid gradually becomes acidified as it moves from the rete testis to the cauda epididymidis.

The **epididymis** is an androgen-dependent tissue. The epithelial cells lining the adult male epididymis contain 5α-reductase, which metabolizes testosterone to DHT, and they contain abundant androgen receptors that bind DHT preferentially. Removal of the testes causes atrophy of the epididymis.

Vas Deferens

The vas deferens, or ductus deferens, is a continuation of the epididymis. This duct, whose luminal epithelium has important absorptive and secretory properties, begins where the epididymis straightens and reverses direction to ascend along the posterior border of the testis toward the inguinal canal. Inside the abdominal cavity, the vas deferens is associated with blood vessels and nerves to form the **spermatic cord,** which remains outside the peritoneum, however. The vas deferens enlarges to form the **ampulla** just before it enters the **prostate gland.** The am-

pulla serves as a secondary storage site for spermatozoa.

Ejaculatory Duct

The portion of the vas deferens just beyond the point where the seminal vesicle empties into it is termed the *ejaculatory duct*. This duct, which is a common duct of both the testis and the seminal vesicle, penetrates the upper surface of the prostate gland and traverses the gland to join the urethra.

Urethra and Penis

The urethra continues through the prostate on its way from the bladder to the penis. The urethra opens at the end of the penis. Ordinarily, the penis is flaccid, but erotic influences elicit involuntary autonomic reflexes that cause the penis to increase in size and become erect. This phenomenon, **erection,** enables the penis to perform the male sexual act.

Sex Accessory Glands

Seminal Vesicles

At the back of the urinary bladder, an outcropping of the vas deferens forms the seminal vesicle, which appears to be a blind, lobulated, elongated sac. In the adult male, each seminal vesicle is a coiled tube that is approximately 15 cm long and has little storage capacity. The seminal vesicle is composed of **three layers:** (1) an outer, fibrous connective tissue coat containing elastic fibers; (2) a middle muscular layer; and (3) an inner mucous membrane that is highly folded, greatly increasing the surface area of the epithelium. The epithelial lining consists mainly of tall columnar cells.

The **secretions of the seminal vesicles** are alkaline and rich in fructose, citric acid (the major anion), inositol, sorbitol, 19-hydroxylated prostaglandins, and reducing substances. Most of the fructose in semen is secreted by the seminal vesicles to be used as an energy source for spermatozoa motility after ejaculation. The fluid is also rich in potassium (20 mM) but poor in sodium, and contains ascorbic acid, inorganic phosphorus, and acid-soluble phosphorus.

Prostate Gland

The prostate gland is a multilobular organ, approximately 4 cm in diameter, that surrounds the urethra as it emerges from the urinary bladder and the two ejaculatory ducts. It is firm and surrounded by a thin capsule containing both smooth muscle fibers and connective tissue. The prostate gland is composed of numerous individual glands that are made up of alveoli lined primarily with tall columnar epithelial cells. These glands drain via a system of branching ducts and tubules into the prostatic urethra, and are embedded in the stroma of the organ, which consists of smooth muscle and fibrous connective tissue.

The prostate gland is sensitive to and dependent on the androgen DHT, which is the principal regulator of growth, differentiation, and function of the prostate gland. If the testes are removed, the secretory cells of the prostate gland shrink. Prostatic cells contain 5α-reductase, which converts testosterone to DHT.

The secretion of the prostate gland in the adult male is a slightly acidic (pH, 6.5), colorless, thin liquid. It contains acid phosphatase, magnesium, calcium, citric acid, spermine, and fibrinolysin, as well as several strong proteolytic enzymes. Fructose is not secreted by the prostate gland. Sodium, rather than potassium, is the main cation and large quantities of zinc are secreted.

Semen

Semen is composed of spermatozoa and seminal plasma. The **seminal plasma** consists of fluid secretions from the vas deferens, epididymis, seminal vesicles, prostate gland, and mucous glands. A concentration of spermatozoa in the semen that is too high or low can cause decreased motility. The volume of the normal human ejaculate is 2 to 6 ml; spermatozoa are normally present at concentrations of 40 to 100 million per

milliliter. Seminal plasma is enriched in potassium, zinc, citric acid, fructose, phosphorylcholine, spermine, amino acids, prostaglandins, and the enzymes acid phosphatase, diamine oxidase, lactic dehydrogenase, β-glucuronidase, α-amylase, and seminal proteinase. The pH normally ranges between 7.5 and 7.8, because the prostatic secretions partially neutralize the more basic secretions of the other segments of the reproductive tract.

The average ejaculate volume is 3 to 3.5 ml, and the major contributions to its volume are: seminal vesicles, 1.5 to 2 ml; prostate gland, 0.5 ml; bulbourethral glands and glands of Littré, 0.1 to 0.2 ml. During emisson and ejaculation, the secretions of these glands are released sequentially, with the mucous secretions of the bulbourethral glands first and the secretions of the seminal vesicles last.

Semen has a **milky appearance** imparted by the prostatic secretions; it is **viscous** because of the secretions from the mucous glands and seminal vesicles. Semen also contains **prostatic clotting enzymes** that activate fibrinogen secreted by the seminal vesicles to produce a weak coagulum in the ejaculate. The **coagulum,** which keeps the ejaculated sperm immobile, dissolves 15 to 20 minutes after ejaculation from lysis by fibrinolysin. After ejaculation, spermatozoa only survive for 24 to 72 hours, although they can live for weeks in the male reproductive tract.

MALE SEXUAL FUNCTION

Fertility in the male refers to the at least occasional ability to produce enough semen at ejaculation containing a sufficient number of normal healthy spermatozoa to impregnate a fertile woman. **Potency** refers to the ability to engage in intercourse, which depends on penile erection. The **male sexual response** consists of three components: erection, emission, and ejaculation. Sexual function is controlled by the **central nervous system,** which processes both **tactile stimuli** from the organs of the reproductive tract as well as **psychic stimuli.** Thoughts of the sexual act can cause erection, emission, and ejaculation.

Erection

Erection is defined as penile rigidity or **tumescence.** In men, the penis contains erectile tissue located in three corporal bodies, the **two dorsal corpora cavernosa** and the **single ventral corpus spongiosum,** through which passes the urethra. Erection involves filling of the cavernous spaces of the three corpora with blood in response to psychic and tactile stimuli. **Parasympathetic impulses** transmitted to the penis from the sacral plexus simultaneously elicit dilation of the arteries and constriction of the veins in the penis. Consequently, arterial blood fills the erectile tissue under high pressure, causing the penis to become hard and elongated.

The **glans penis,** which has a network of sensory end-organs that transmit sexual sensations to the central nervous system via the pudendal nerve and the sacral plexus, is very sensitive and important for initating and maintaining erection before and during the male sexual act. Impulses from the anal epithelium, scrotum, and other perineal structures, as well as from internal organs of the genitourinary tract such as the prostate, bladder, urethra, and seminal vesicles, aid in amplifying sexual sensation.

The parasympathetic impulses also cause secretion of a small amount of mucus by the glands of Littré and the bulbourethral glands. This mucus helps lubricate the penis during intercourse, although most lubrication is provided by secretions of the female reproductive tract. The pain sensation that can result from inadequate lubrication can lead to loss of erection.

Emission

Emission is defined as the deposition of seminal fluid components from the vas deferens, seminal vesicles, and prostate gland into the posterior urethra. When sexual stimulation peaks, sympathetic impulses originating in the spinal cord and passing through the hypogastric plexus initiate contractions of the epididymis, vas deferens, and ampulla to expel the sperm into the urethra. This is followed rapidly by contractions of the muscular layer of the prostate gland and then by contractions of the seminal vesicles. The combined contractions expel the prostatic and seminal fluids and force the sperm forward. The sperm and the prostatic and seminal secretions mix with the mucus secreted by the bulbourethral glands to form the semen.

Ejaculation

Ejaculation is the passage of semen through the urethra and its expulsion from the urethral meatus. Filling of the urethra with semen triggers sensory signals that travel through the pudendal nerves to the spinal cord, which, in turn, transmits nerve impulses to the skeletal muscles surrounding the erectile tissue of the penis. These rhythmic nerve impulses stimulate rhythmic, wavelike contractions of the muscle that increase the pressure and expel the semen from the urethra.

SUMMARY

The reproductive system of the male consists of paired testes, the reproductive tract or internal genitalia, and external genitalia. Male germ cells, or spermatozoa, containing the haploid number of chromosomes are produced in the seminiferous tubules. Interstitial cells produce the male sex hormones, or androgens, that maintain the structure and function of the internal genitalia. Reproductive function begins at puberty, when the hypothalamic–pituitary unit matures functionally. The production of spermatozoa differs from the production of ova in the female in two major respects: (1) development beyond the primordial germ cell stage does not occur until puberty; and (2) beginning at puberty, development of spermatozoa is a continuous process. Testicular function is controlled by the gonadotropic hormones FSH and LH. FSH stimulates Sertoli cells, which nurture the developing germ cells; LH stimulates the interstitial cells of Leydig, which synthesize and secrete testosterone, the principal androgen. Male reproductive function is controlled by interacting neural and hormonal signals of the hypothalamic–pituitary–testicular axis.

NATIONAL BOARD TYPE QUESTIONS

Select the one most correct answer for each of the following

1. Testosterone, the principal male sex steroid hormone, is:
   A. Synthesized from estradiol-17β
   B. Produced primarily in the Sertoli cells of the testes in response to FSH
   C. Not required for the initiation and maintenance of spermatogenesis
   D. Converted to DHT by 5α-reductase in the cells of the prostate gland and some other reproductive tissues
   E. Present in the plasma as a free hormone not bound to plasma proteins

2. Inhibin is a male hormone that:
   A. Is a lipophilic steroid
   B. Is synthesized in the interstitial cells between the seminiferous tubules
   C. Suppresses production of FSH but not of LH
   D. Reduces the frequency of pulsatile secretion of GnRH by the hypothalamus
   E. Is produced in response to testosterone

3. Spermatogenesis, the process by which spermatogonia develop into spermatozoa:
   A. Is a cyclic process that begins approximately every 28 days throughout the testes
   B. Requires FSH and estrogen
   C. Occurs in the interstitium surrounding the seminiferous tubules
   D. Yields male gametes with the diploid number of chromosomes
   E. Involves two meiotic divisions that produce four spermatids from a single primary spermatocyte

4. Sexual potency in the male refers to the ability to:
   A. Engage in intercourse, which depends on erection of the penis
   B. Produce mature spermatozoa
   C. Produce 2 to 6 ml of semen per ejaculate
   D. Synthesize and secrete testosterone in the testes
   E. All of the above

5. In the male, FSH:
   A. Binds to receptors on primary spermatocytes, causing them to enter meiosis
   B. Concentrations in plasma are independent of hypothalamic GnRH secretion
   C. Stimulates Sertoli cells to synthesize inhibin and androgen-binding protein
   D. Concentrations in plasma decrease after removal of the testes (castration)
   E. All of the above

ANNOTATED ANSWERS

1. D.
2. C. Suppresses synthesis and secretion only of FSH.
3. E. Involves two meiotic divisions which result in four spermatids, each with the N number of chromosomes and the N complement of DNA, from a single primary spermatocyte.
4. A.
5. C.

BIBLIOGRAPHY

Ham, A. W. *Histology*, 7th ed. Philadelphia: J. B. Lippincott, 1974.

Knobil, E., and Neill, J. D., eds. *The Physiology of Reproduction*. New York: Raven Press, 1988.

Odell, W. D., and Moyer, D. L. *Physiology of Reproduction*. Saint Louis: C. V. Mosby, 1971.

Yen, S. S. C., and Jaffe, R. B., eds. *Reproductive Endocrinology. Physiology, Pathophysiology and Clinical Management*, 2nd ed. Philadelphia: W. B. Saunders, 1986.

57 Fertilization, Pregnancy, and Lactation

ANDREW R. LaBARBERA

Objectives

After reading this chapter, you should be able to:

Explain how spermatozoa travel through the female reproductive tract to the fallopian tubes, the usual site of fertilization

List the cellular events that take place during fertilization of the ovum and the early development of the zygote and pre-embryo

Explain how the blastocyst implants in the endometrial lining of the uterus

Characterize the development and function of the placenta in nurturing the fetus

Describe the regulation of maternal and fetal metabolism during pregnancy

Explain how parturition is initiated and proceeds to expel the fetus

Describe the process of lactation and the postpartum changes in maternal function

---

Male gametes, which are delivered to the female reproductive tract during coitus, migrate to the fallopian tubes, where the ovum is fertilized. The zygote that results from fusion of a spermatozoon and an ovum migrates to the uterus and implants in the uterine wall. As the conceptus develops, a physical barrier known as the **placenta** forms between the mother and fetus to regulate the exchange of nutrients, gases, and wastes. The fetal-placental unit secretes numerous hormones which regulate metabolism in both the mother and the fetus during the period of development which is known as **gestation** or **pregnancy**. After approximately 39 weeks of gestation, the fetus and the placenta are expelled from the uterine cavity by the process known as **parturition**. The newborn is nurtured by breast milk obtained from the mother through suckling. The offspring reaches maturity at puberty.

GAMETE TRANSPORT IN THE FEMALE REPRODUCTIVE TRACT

Fertilization, which is the fusion of a male gamete (spermatozoon) and a female gamete (ovum), takes place in one of the oviducts, or fallopian tubes, of the female reproductive tract. Therefore, spermatozoa must be delivered into the female tract and travel to the site of fertilization. The ovulated egg likewise must travel from the ovary to the site of fertilization.

803

Ovum

As discussed in Chapters 54 and 55, the adult ovary contains primary oocytes that are arrested in the **dictyotene stage of prophase** of the first meiotic division. In this stage, a primary oocyte has replicated its DNA, so that it contains the **diploid** (2N) number of chromosomes—44 autosomes and 2 sex chromosomes—but the **tetraploid** (4N) quantity of DNA, because each chromosome has two daughter chromatids. The **oocyte** enlarges during childhood, but the nucleus remains unchanged. Meiosis resumes in the primary oocyte of the mature graafian follicle during the preovulatory surge in plasma luteinizing hormone (LH), approximately 36 hours before ovulation (Fig. 57-1). The first meiotic division, which is completed several hours prior to ovulation, results in **two daughter cells.** The larger cell is the **secondary oocyte** and the smaller cell is the **first polar body,** which lies between the **zona pellucida** and the **vitelline membrane** of the secondary oocyte. Each cell has the haploid (N) number of chromosomes and the diploid quantity (2N) of DNA. Further nuclear maturation in the secondary oocyte is halted unless it is fertilized. Simultaneous with the resumption of oocyte maturation, the cells of the **cumulus oophorus** secrete copious amounts of **proteoglycan** that is rich in hyaluronic acid, and the cumulus expands.

After ovulation, the ovum and its surrounding mass of cumulus oophorus cells are transported along the surface of the ovary and through the **ostium,** the opening at the fimbriated end of one of the fallopian tubes. This transport, which requires several minutes, is facilitated by the beating motion of the cilia lining the fimbria. Contractions of the oviductal musculature direct the ovum into the ampulla, where it remains for approximately three days. The ovum is kept in the ampulla primarily by constriction of the ampullary–isthmic sphincter.

Spermatozoa

During **coitus**, which usually occurs within 10 minutes of **intromission** in the human, semen is ejaculated into the vagina close to the external os of the cervix. The alkaline semen buffers the acidic (pH, less than 5.0) vaginal fluid to provide a temporarily favorable environment for the spermatozoa, which must undergo a sequence of incompletely defined spontaneous biochemical changes, known as **capacitation,** in order to become **hypermotile** and able to fertilize an ovum. Spermatozoa can be capacitated *in vitro* by incubation at 37°C.

Within one minute of ejaculation, the seminal plasma coagulates. The **coagulum,** which keeps the spermatozoa in the vagina until they become motile, is subsequently broken down during the next 20 to 30 minutes by proteolytic enzymes in the ejaculate. Usually the spermatozoa are maximally motile and the coagulum is completely liquefied within one hour of ejaculation.

Motile sperm migrate through the cervical mucus at a rate of 2 to 3 mm per minute. Sperm are propelled through the female reproductive tract by their own flagellar action, as well as by the combined effects of vaginal and uterine contractions and the negative vaginal pressure that occurs after orgasm. The contractions may be amplified by the copious amounts of 19-hydroxylated prostaglandins present in semen. The first sperm reach the fallopian tubes within five minutes of ejaculation, but the number in the tubes does not peak until four to six hours after ejaculation. The **uterotubal junction** is a major barrier to the ascent of the spermatozoa, which must reach the ampulla of the fallopian tube where fertilization occurs. Fewer than 200 of the 20 to 500 million motile sperm are in the fallopian tubes at any one time. Spermatozoa traverse the tubes and continue out the infundibular end into the peritoneal cavity. Motile sperm are usually present in the female reproductive tract for 48 to 60 hours but have been found for up to 85 hours after intercourse.

FERTILIZATION

Mature ova remain fertile for only up to 15 to 18 hours after ovulation, and, if fertilization does not occur, the ovum degenerates. Often an ovum

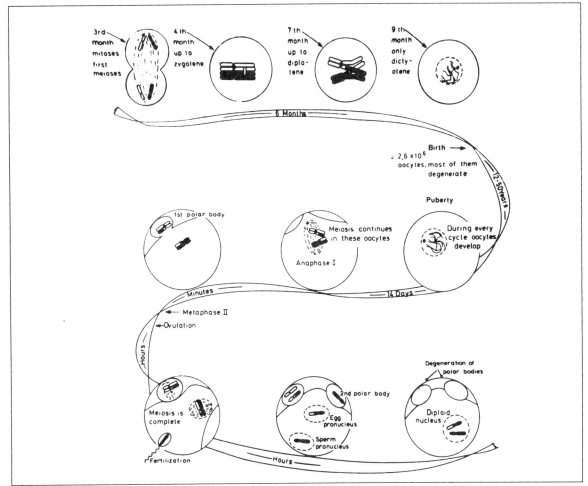

Fig. 57-1. Meiosis in the oocyte. Meiosis begins during early fetal life, but is arrested at the dictyotene stage of the first meiotic prophase by the ninth month. Oocytes degenerate throughout gestation and childhood. Beginning at puberty, the ovulatory surge of luteinizing hormone stimulates oocytes to resume meiosis, extrude the first polar body, and progress to metaphase II. Fertilization causes meiosis to resume, nuclear maturation to be completed, and the second polar body to be extruded. Membranes form separately around the chromosomes from the mother (female pronucleus) and the father (male pronucleus). The two pronuclei fuse after several hours, and the first cleavage division begins. (From Bresch, C., and Haussman, R. *Klassiche und moleculare Genetik*, 3rd ed. New York: Springer-Verlag, 1972.)

is fertilized by a spermatozoon that preceded it into the ampulla of the fallopian tube. Capacitated spermatozoa lose their potential to fertilize ova within approximately 24 hours after intercourse.

The anterior portion of the sperm nucleus is covered by a membrane-bound, caplike structure termed the **acrosome,** and contains a variety of hydrolytic enzymes including hyaluronidase and acrosin. When a spermatozoon encounters the cumulus—oocyte mass in the fallopian tube, the structure of the outer plasma membrane of the acrosome breaks down and the outer acrosomal membrane becomes fenestrated in a process termed the **acrosome reaction** (Fig. 57-2). This process, which appears to be induced by the cumulus oophorus or the zona pellucida, or both, but which can occur in their absence, requires a high extracellular calcium concentration that favors a massive calcium influx. The **hydrolytic enzymes** released during the acrosome reaction facilitate passage of the spermatozoon through the glycoprotein-rich matrix surrounding the cells of the cumulus oophorus and through the zona pellucida of the ovum, which is rich in sulfated glycoprotein.

After penetrating the zona pellucida, the spermatozoon lies in the **perivitelline space** adjacent to the plasma membrane of the ovum (See Fig. 57-2). The plasma membrane of the equatorial portion of the sperm head fuses with the **oolemma** of the ovum, and microvilli of the ovum membrane surround the sperm head. The putative sperm receptor on the oolemma has not been identified. Fusion of the two gamete membranes activates the ovum, which causes a series of biochemical events characterized by **exocytosis** of cortical granules and **resumption of meiosis.** Release of the contents of the cortical granules seems to prevent penetration of the ovum by more than one spermatozoon, known as **polyspermy.**

Fertilization causes the **egg nucleus,** which was arrested at metaphase of the second meiotic division, to complete maturation (see Fig. 57-1). The resulting **female pronucleus** and the second polar body, which is extruded into the perivitelline space, both have the haploid number of chromosomes and the haploid quantity of DNA. Meanwhile, the **sperm nucleus,** which is embedded in the cytoplasm of the ovum, swells and decondenses to form the **male pronucleus.** DNA synthesis proceeds synchronously in both pronuclei of the zygote to duplicate the chromosomes. Fertilization is complete when the membranes of the pronuclei, which are in close contact with each other, break down, the chromosomes mingle, and mitosis and cleavage of the **one-cell zygote** to a **two-cell pre-embryo** occur, approximately 24 hours after fertilization (see Fig. 57-2).

Fig. 57-2. Fertilization. (A) A mature ovum surrounded by cells forming the corona radiata is ovulated. (B) The cumulus oophorus and corona radiata expand. (C) Sperm bind to the membrane of the oocyte, the sperm and ovum membranes fuse, and the sperm head is engulfed by the ovum. (D) The sperm nucleus decondenses to form the male pronucleus. (E) DNA synthesis proceeds in both male and female pronuclei of the zygote in order to duplicate the chromosomes. The pronuclear membranes break down, the chromosomes mingle, and mitosis and cleavage of the one-cell zygote to a two-cell pre-embryo occur. (From Guyton, A. C. *Textbook of Medical Physiology*, 6th ed. Philadelphia: W. B. Saunders, 1981, P. 1022.)

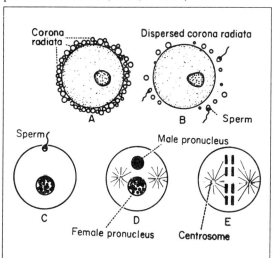

PRE-IMPLANTATION DEVELOPMENT

The **zygote** is the same size as the unfertilized ovum, approximately 100 to 125 μm in diameter, and is still invested by the zona pellucida. Development through the early cleavage stages progresses as the pre-embryo descends through the fallopian tube, assisted by ciliary motion and muscular contraction. The cryptoid surface of the oviductal lining and the spastic contractions of the estrogen-dominated isthmus portion of the tube impede movement of the pre-embryo into the uterus, so that the pre-embryo is retained in the fallopian tube for approximately three days. A rising plasma progesterone level causes the tone of the smooth muscle in the oviductal wall to decrease, so that the isthmus relaxes and the pre-embryo can pass through the uterotubal junction into the uterus.

While in the fallopian tube, the pre-embryo undergoes **cleavage division.** The cells, or **blastomeres,** of the two-cell, four-cell, and eight-cell pre-embryos all are **totipotential.** In other words, at these stages of development, each cell is capable of developing into a complete human being. At the **eight-cell stage,** the cells undergo **compaction,** in which the individual blastomeres become less prominent and begin to display polarity. Cells around the periphery develop microvilli on their outer surfaces. At the 16-cell **morula stage,** the cells on the inside of the preembryo begin to develop into the **inner cell mass,** which eventually develops into the **fetus;** the cells on the outside begin to develop into the **trophectoderm,** which gives rise to the **extra-embryonic tissues,** including the placenta and membranes. Six days after conception, when the morula consists of approximately 64 cells, a cavity, or **blastocoele,** begins to appear opposite the inner cell mass. As cells surrounding the blastocoele degenerate, the cavity enlarges and becomes filled with fluid. At this stage, the inner cell mass, termed the **blastocyst,** is clearly distinguishable from the trophectoderm, which is composed of a single layer of cells. The trophectodermal cells form giant cells that have tight junctions capable of excluding large molecules and which completely encircle the blastocyst.

IMPLANTATION OF THE BLASTOCYST

Approximately seven days after fertilization, the blastocyst, which consists of approximately 200 cells, loses the zona pellucida in a process termed **hatching** and implants in the wall of the uterus. Implantation, also termed **nidation,** depends on prior conditioning of the endometrium by progesterone, which causes the stromal cells to swell and accumulate glycogen, protein, and lipids intracellularly. Administration of antiprogestins prevents implantation. The blastocyst is capable of secreting **human chorionic gonadotropin** (hCG), which stimulates cells of the corpus luteum to produce **progesterone.**

The blastocyst attaches at its embryonic pole to the wall of the uterine fundus, and the microvilli of the trophectodermal cells interdigitate with the microvilli on the luminal surface of the endometrial cells. The trophectodermal, or **trophoblast,** cells then invade through the basement membrane underlying the endometrial epithelium, aided by proteolytic enzymes secreted by the blastocyst, and establish an **implantation site** in the endometrial stroma. The stromal cells then enlarge, or **decidualize,** in that they become **transcriptionally active** and surround the blastocyst. The decidual response of the uterine endometrium also involves heightened **vascular permeability** around the nidation site, which is thought to be mediated by the local production of **histamine** and **prostaglandins.** After implantation, the embryo develops rapidly to form distinct placental and fetal structures.

DEVELOPMENT AND FUNCTION OF THE PLACENTA

The trophoblast differentiates into **two layers** within eleven days of fertilization. The inner layer, the **cytotrophoblast,** consists of numerous individual cells. The outer layer, the **syncytiotrophoblast,** is much thicker and resembles a continuous mass of cytoplasm containing many nuclei. This **syncytium** contains small spaces or **lacunae,** which enlarge and become continuous with each other. Early in pregnancy, the lacunae form **sinuses** that contain blood from the maternal uterine veins and venous sinuses eroded by the trophoblast. As the lacunae enlarge, the strands of trophoblast left between them form fingerlike projections, termed **villi.** Each villus has a core of cytotrophoblast covered with an irregular layer of syncytiotrophoblast cells. The villi become vascularized by **fetal blood vessels,** so that the fetal capillary endothelium is close to the trophoblast cells and fetal blood flows by the sixteenth day after fertilization. Later in pregnancy, the trophoblast erodes the maternal spiral arteries, so that maternal blood flows into the intervillous spaces surrounding the villi. The placenta forms a **barrier** that permits the exchange of nutrients, gases, and metabolic wastes without maternal and fetal blood ever mixing.

The fully developed **fetal–placental–uterine** unit is illustrated in Figure 57-3. **At term,** the placenta is a disk-shaped, ovoid structure that is approximately 18 times 20 cm in diameter and 2 cm thick. It consists of: (1) the **inner amnion,** which is made up of a single layer of ectodermal epithelium that completely encloses the embryo; (2) the **middle chorion,** which surrounds the amnionic sac and includes the villi and trophoblast; and (3) the **decidua** of the maternal endometrium. The **umbilical cord,** which arises from the center of the disk, carries fetal blood between the fetus and the chorionic villi.

TRANSFER OF NUTRIENTS AND GASES

The placenta functions principally to provide a barrier for **nutrients** to diffuse from the maternal blood supply to the developing fetus and for **excretory products** to diffuse from the fetal to the maternal blood supply. The placenta is also a **protective barrier** against microorganisms. The two principal elements in the **transplacental transfer** of substances are: (1) the **maternal-facing trophoblast microvillus membrane;** and (2) the **fetal-facing basal trophoblast membrane.** These membranes regulate the transfer of glucose, lactate, certain amino acids, oxygen, carbon dioxide, electrolytes, and several minerals between the mother and fetus.

The **net transfer** of a substance from the maternal to the fetal circulation is the **net transplacental flux** (J_{net}), which follows the **Fick principle** and can be defined as:

$$J_{net} = J_{mf} - J_{fm}$$

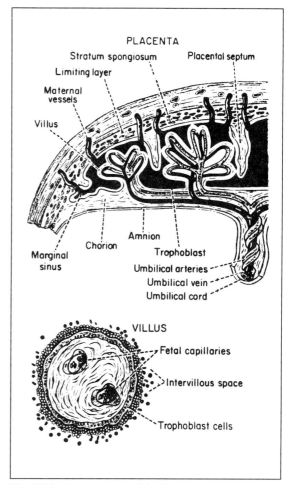

Fig. 57-3. Cross-section of the fetal–placental–uterine unit. (Top) Relationship between the fetal and maternal structures that allows gases and nutrients to be exchanged without the blood supplies mixing. (Bottom) Cross-section of a villus that is bathed by maternal blood and through which course fetal capillaries. (From: Guyton A. C. *Textbook of Medical Physiology,* 6th ed. Philadelphia: W. B. Saunders, 1981, P. 1024.)

Where J_{mf} is the flux from the maternal to the fetal compartment and J_{fm} is the flux from the fetal to the maternal compartment. The **unidirectional fluxes** in each direction for a particular substance are complex processes because they are affected by: (1) the **rate of delivery** of the substance to the placenta by means of the circulation; (2) the **difference in the concentrations** of the substance between the maternal and fetal circulations; (3) the **interaction of the substance** with the different components of the placental barrier; and (4) the **relationship between the maternal and fetal blood flows.**

Small Organic Molecules

D-Glucose is transferred from the mother to fetus by **facilitated diffusion.** There is a small but significant difference between the maternal (4.4 mmol/liter) and fetal (3.6 mmol/liter) plasma glucose concentrations. The **fetal arterial plasma glucose concentration** is a function of the maternal arterial or venous plasma glucose concentration up to approximately 20 mmol/liter, at which concentration the **transfer mechanism** becomes saturated. **Glucose transfer kinetics** are complicated by active glucose metabolism in the placenta itself. Moreover, some of the glucose that is metabolized in the placenta is supplied by the fetus.

Transfer of **L-lactate** across the placenta also occurs by **facilitated diffusion.** Presumably lactate is transferred as **lactic acid,** since there is evidence of the cotransport of H^+. An artificially induced H^+ gradient can precipitate the net transfer of lactic acid against its concentration gradient. **Carriers that bind lactate** have been demonstrated on both the maternal and fetal sides of the placenta. **Pyruvic acid** competes with lactic acid for the carriers. The placenta produces lactate, which is primarily released into the maternal circulation at midgestation. Near term, placental lactate is released into both the maternal and fetal circulations, so that there is no net transplacental transfer of lactate.

Many **amino acids** are transferred across the placenta by **energy-dependent active transport.** It appears that they are transported actively across the maternal-facing plasma membrane of the trophoblast cell. They then diffuse across the fetal-facing plasma membrane into the fetal circulation. **Neutral amino acids** are transported by three systems. The **A (alanine) system** transports alanine, glycine, proline, serine, threonine,

and glutamine. The **L (leucine) system** transports leucine, isoleucine, valine, and phenylalanine, as well as alanine, serine, threonine, and glutamine. The **ASC (alanine–serine–cysteine) system** preferentially transports alanine, serine, threonine, and glutamine. Transport mechanisms for **basic and acidic amino acids** have not been demonstrated. Measurements of **human fetal umbilical venous–arterial plasma concentration** differences indicate that there is **net uptake** by the fetus of at least alanine, glycine, leucine, isoleucine, phenylalanine, and histidine. In contrast to other amino acids, **glutamate** does not cross the placenta at physiologic concentrations, but is removed from the fetal circulation even though its concentrations in umbilical venous plasma exceed those in maternal plasma. The placenta releases **glutamine** into the umbilical circulation. **Glycine** and **leucine** appear to be produced in the placenta, since they are secreted into the umbilical circulation, but removal from the maternal circulation is undetectable.

The human placenta is permeable to **free fatty acids** and **ketones.** Maternal concentrations of most free fatty acids are greater than fetal concentrations, and they appear to cross the placenta by **simple diffusion. Arachidonic acid,** however, is found in higher concentrations in fetal plasma than in maternal plasma. **Free fatty acids,** which are transferred from mother to fetus, are rapidly esterified to **triglycerides** in the fetal liver and do not cross the placenta. **Cholesterol,** which is carried in lipoprotein particles in the circulation, is transferred from mother to fetus through the uptake of **low-density lipoprotein particles** by trophoblast cells. **High-density lipoprotein** is not taken up by these cells. The total concentration of lipoprotein-bound cholesterol in maternal venous plasma (230 mg/dl) is approximately four times that in mixed umbilical plasma (60 mg/dl) at term.

Ions

The transfer of **ions** and **minerals** between mother and fetus varies greatly, both among the solutes and among species. In humans, it appears that the placenta transfers most sodium from the maternal circulation (135 mmol/kg water) to the fetal circulation (137 mmol/kg water) by **passive diffusion.** A small amount may be transported actively by cotransport with other solutes. The concentration of potassium is higher in the fetal circulation (5.7 mmol/kg H_2O) than in the maternal circulation (3.8 mmol/kg H_2O), but the transfer mechanism is not understood well. Chloride appears to be able to diffuse across the placenta with concentrations in maternal plasma (107 mmol/kg H_2O) resembling those in fetal plasma (109 mmol/kg H_2O). Both calcium and **phosphorus** concentrations in fetal plasma exceed those in maternal plasma, and evidence suggests that both are transferred across the placenta by **active transport.**

Vitamins

Most **water-soluble vitamins** are thought to be transferred across the placenta by **specialized transport mechanisms. Lipid-soluble vitamins,** in contrast, appear to **diffuse freely** across the placenta. In humans, transport of water-soluble **vitamin C** is energy dependent and requires sodium and (Na, K)-ATPase. It is not clear whether fetal plasma concentrations of vitamin C are significantly higher than maternal plasma concentrations. **Vitamin D_3** and its metabolites are present in higher concentrations in the mother than in the fetus. Administration of large quantities of 1,25-dihydroxycholecalciferol (1,25-OHD or **vitamin D**) elevates fetal concentrations, but there does not appear to be net transfer of 1,25-OHD, since the placenta itself is capable of converting 25-hydroxycholecalciferol to 1,25-OHD. **Vitamin B_{12}** is present in higher concentrations in the fetus than in the mother and is transferred across the placenta by **active transport.** Other water-soluble vitamins found in higher concentrations in fetal plasma than in maternal plasma, and which appear to be transferred by **active transport,** include **vitamin B_6, biotin, folate, nicotinate, pantothenate, riboflavin,** and thiamine.

Maternal plasma concentrations of lipid-soluble **vitamin A** (retinol) are more than 2.5 times those in fetal plasma at term. Retinol appears to diffuse passively through the placenta because of its **lipophilicity,** and then binds to **retinol-binding protein.** Retinol-binding protein also crosses the placenta, but to a lesser extent than retinol. As more than 90 percent of retinol in human fetal serum is complexed with either retinol-binding protein or prealbumin, the fetal–maternal transfer rate of retinol is only one-fifth the maternal–fetal transfer rate across the placenta.

Proteins

The placenta is relatively impermeable to proteins, with the exception of certain immunoglobulins and retinol-binding protein. Proteins such as **albumin** and **small peptide hormones** are not able to cross the placenta, suggesting that permeability of the placenta to hydrophilic molecules is limited by the small pores. Protein hormones such as **hCG** and **chorinoic somatomammotropin** (hCS) are produced by trophoblast cells and secreted into the maternal circulation. The fact that these hormones are excluded from the fetal circulation indicates the existence of a barrier between the placenta and fetus. Fetally produced **α-fetoprotein** is normally present in maternal plasma at low concentrations, indicating that it can cross the placenta at least to a small extent. It appears that the **placental capillary wall** is the effective barrier to transplacental transfer of large or medium-sized molecules, whereas the **fetal surface of the trophoblast** is limiting for small proteins.

Immunoglobulin G (IgG) is transferred from mother to fetus in quantitatively important amounts. **Fc receptors** for IgG exist on the microvillous membranes of trophoblast cells. IgG, but not other classes of immunoglobulins, is taken up by trophoblast cells through endocytosis.

Maternal antibodies can be both beneficial and deleterious to the fetus.

Respiratory Gases

Exchange of O_2 and CO_2 between the mother and fetus across the placenta occurs by **simple diffusion,** as it does in the lungs. O_2 dissolved in maternal blood in the placental sinuses diffuses through the cells of the villi and placental capillaries and into the fetal blood because the **partial pressure of O_2** (Po_2) of **maternal blood** in the placental sinuses is approximately 50 mm Hg, whereas the Po_2 of blood in the **umbilical vein** leaving the placenta is only approximately 30 mm Hg. The fetus is able to obtain sufficient O_2 at such a low Po_2 because: (1) fetal blood has approximately 50 percent more O_2-carrying hemoglobin than maternal blood; (2) at a given Po_2, fetal hemoglobin can carry 20 to 30 percent more O_2 than adult hemoglobin, so that at pH 7.4 fetal hemoglobin is 80 percent saturated at a Po_2 of 34 mm Hg (Fig. 57-4); and (3) hemoglobin can carry more O_2 at a low partial pressure of CO_2 (Pco_2) than at a high Pco_2 known as the **Bohr effect.** Transfer of O_2 and CO_2 depends in part on the **blood flow rates** in the uterine and umbilical circulations and on the **geometric relationships** between these two circulations. CO_2 exchange is also affected by the **equilibrium between carbonic acid and bicarbonate.** The **Fick principle** can be used to determine O_2 uptake by the uterus. **Uterine blood flow** (F; in milliliters per minute) and the O_2 contents (at standard temperature and pressure [STP]) of blood simultaneously drawn from a maternal artery (A;ml_{STP}/ml blood) and from the uterine vein (V;ml $_{STP}$/ml blood) are measured and O_2 uptake by the uterus is calculated by the formula:

$$O_2 \text{ uptake} = (A - V) \times F$$

Maternal uptake by the uterus of the pregnant sheep is approximately 48 ml_{STP}/min. Similarly, O_2 uptake by the fetus can be determined by measuring umbilical blood flow (f, in milliliters per minute) and the O_2 contents of blood simultaneously drawn from the umbilical artery (a; ml_{STP}/ml blood) and the umbilical vein (v; ml_{STP}/ml blood). O_2 uptake by the fetus is yielded by the formula:

$$O_2 \text{ uptake} = (v - a) \times f$$

In the pregnant sheep, O_2 uptake by the fetus is 26.5 ml_{STP}/ min. It is not possible to perform all of the above measurements during pregnancy in humans. However, maternal blood flow rates to the human placenta (124 ml/min/kg uteroplacental unit) are much lower than those to the sheep placenta (288 ml/min/kg utero–placental–fetal unit). **Fetal blood flow** to the placenta in both

Fig. 57-4. Oxyhemoglobin dissociation curves for maternal and fetal blood at pH 7.4. Fetal hemoglobin is able to carry 20 to 30 percent more O_2 than maternal hemoglobin. At pH 7.4, fetal hemoglobin is 80 percent saturated at a partial pressure of O_2 (Po_2) of 34 mm Hg. (From: Meschia, G. In: Creasey, R. K., and Resnik, R., eds. *Maternal-Fetal Medicine: Principles and Practice.* Philadelphia: W. B. Saunders, 1989, P. 309.)

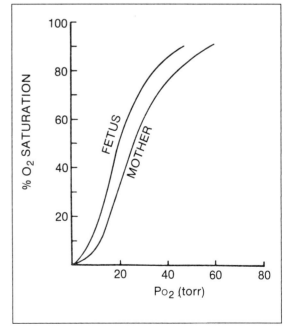

humans (120 ml/min/kg fetus at 35 weeks of gestation) and sheep (256 ml/min/kg fetus) is similar to **maternal blood flow.**

CO_2 formed from O_2 metabolism must be transported by the maternal circulation to the lungs, where it is eliminated. The placental barrier membrane is highly permeable to CO_2. For CO_2 to diffuse from the fetus to the mother, the P_{CO_2} of the fetal circulation must be higher than that of the maternal circulation. The P_{CO_2} of umbilical arterial blood is approximately 48 mm Hg compared with 40 mm Hg in maternal arterial blood. Thus, CO_2 is able to diffuse across the placenta along a **pressure gradient.** Alterations in maternal respiration, and consequently in maternal arterial P_{CO_2}, are rapidly reflected in the fetal circulation.

ENDOCRINE CONTROL OF MATERNAL AND FETAL METABOLISM

Because the fetus must be nourished by the mother, the placenta secretes factors that regulate the maternal and fetal environments. **Placental factors** include hCG, hCS, estrogens, progestins, growth factors, and a number of peptides that are either identical or similar to peptides of the hypothalamic–pituitary unit. At least hCG, estrogens, and progestins are required for maintenance of pregnancy. In addition, secretion of **anterior pituitary hormones** increases to such an extent that the **pituitary gland** enlarges by approximately 50 percent during the first trimester of pregnancy and by more than 100 percent by term. **Pituitary hormones** include prolactin, adrenocorticotropin, and thyrotropin. Changes in maternal plasma concentrations of hormones are illustrated in Figure 57-5. The importance of these hormones in maintaining pregnancy is discussed in the following sections.

Chorionic Gonadotropin

hCG is a glycosylated heterodimeric protein with a molecular weight of 38,600 produced by the syncytiotrophoblast. The **alpha subunit** synthe-

sized in the placenta is nearly identical to the alpha subunits of follicle-stimulating hormone (FSH), luteinizing hormone (LH), and thyroid-stimulating hormone (TSH) synthesized in the pituitary. The beta subunit of hCG is considerably larger than the beta subunits for the other hormones; it consists of 145 amino acids and has a molecular weight of 24,000. hCGβ also differs from the beta subunits of FSH, LH, and TSH in the type of **carbohydrate residues** and its **lack of sulfation.** Deglycosylation of hCG increases **receptor-binding potency** by approximately twofold, with concomitant inhibition of hCG's ability to activate adenylyl cyclase.

As with FSH and LH, a single copy of the gene for the alpha subunit is located on **chromosome 6.** In striking contrast to LH, there are seven genes or **pseudogenes** for hCGβ on **chromosome 19.** Only one of the genes is transcribed, however. Translation of the messenger RNAs (mRNAs) for the alpha and beta subunits appears to be similar to that for the pituitary gonadotropins, because the placental cells contain free alpha subunits and intact hCG, but little if any free hCGβ. Synthesis, glycosylation, and packaging are accomplished by mechanisms similar to those for the pituitary. The synthesis and secretion of the hCG syncytiotrophoblast appear to be enhanced by gonadotropin-releasing hormone (GnRH) synthesized in the cytotrophoblast.

Recent evidence suggests that one of the fragments of hCGβ, termed the *beta core fragment,* which is detected in both plasma and urine, may be secreted by the placenta and by hCG-producing neoplasms. The clinical utility of this peptide in monitoring treatment of hormone-producing neoplasms is under investigation.

hCG is detectable in the circulation within eight days of conception. Maternal peak plasma concentrations of 1,000 to 2,000 IU/dl are achieved around the fourteenth week of pregnancy, but fetal plasma concentrations are much lower, approximately 3 IU/dl. hCG-stimulated **luteal progesterone secretion** peaks around the fourth week after conception, much earlier than the peak of hCG, suggesting that the corpus luteum becomes at least partially refractory to hCG

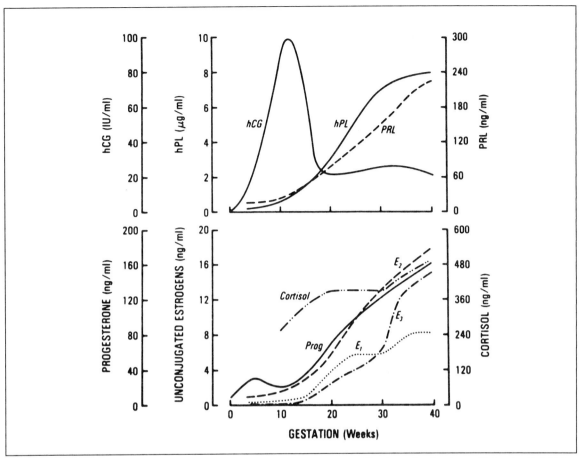

Fig. 57-5. Maternal serum concentrations of hCS (or hPL), progesterone (*Prog*), cortisol, estrone (E_1), estradiol (E_2), and estriol (E_3) throughout pregnancy. (From: Rebar, R.W. Maternal serum hormone concentrations during pregnancy. In: Creasy, R. K., and Resnik, R., eds. *Maternal-Fetal Medicine: Principles and Practice.* Philadelphia: W. B. Saunders, 1989, P. 156.)

stimulation. During early pregnancy, the placenta is gradually able to secrete sufficient quantities of progesterone so that the corpus luteum is no longer needed. Consequently, hCG levels decline and the corpus luteum regresses, since the placenta is fully developed as a steroidogenic organ by the twentieth week of pregnancy.

hCG is secreted during the first trimester of pregnancy to stimulate the corpus luteum to produce progesterone, 17α-hydroxyprogesterone, and relaxin. If fertilization and implantation occur, hCG prevents the corpus luteum from de-

clining during the menstrual cycle, so that progestins can be synthesized and secreted by the corpus luteum until the placenta can assume this function. **Exogenous hCG** can prolong the life of the corpus luteum if fertilization and implantation do not occur. An exponential increase in hCG during the first few weeks of pregnancy is an indicator that the placenta is developing normally.

hCG stimulates testosterone biosynthesis in the **fetal testes,** if present. Maximal levels of testosterone in the blood of male fetuses occur during the eleventh to seventeenth weeks after conception, coinciding with peak hCG concentrations. Moreover, hCG stimulates both testosterone and DNA synthesis in fetal testes during this period, thus providing the androgens that stimulate male sexual differentiation which takes place around this time.

Regulatory roles for hCG in other tissues have been suggested but not definitively proved. These include **regulation of steroidogenesis** in both the fetal adrenal cortex and the placenta. By virtue of its structural similarity to LH, hCG can interact with LH receptors and exert LH-like effects when administered artificially. hCG only has physiologic significance during pregnancy, however.

Metabolic clearance of hCG from the maternal circulation, like that of LH and FSH, depends on both the liver and kidney. The hormone has a relatively long biphasic rate of disappearance, on the order of 24 to 36 hours, with a first component of 6 to 8 hours. The long half-life has been attributed to its high content of sialic acid. In contrast to intact hormone, the half-life of the free alpha subunit is approximately 13 minutes, whereas the half-life of the free beta subunit is approximately 41 minutes. Completely desialylated hCG has a plasma half-life of just a few minutes and low biologic activity *in vivo* because of rapid hepatic uptake.

hCG binds to **specific membrane-bound receptors** in cells of the corpus luteum. Because **luteal cells** result from luteinization of granulosa cells, the hCG receptor is identical to the LH receptor of preovulatory granulosa cells. hCG receptors gradually appear on the surface of luteal cells after disappearance (down-regulation) of the LH receptors, which is induced by the midcycle surge of LH. **Signal transduction** for hCG is nearly identical to that for LH with only minor quantitative differences in the characteristics of hCG-receptor–G_s–adenylyl cyclase interactions.

Chorionic Somatomammotropin

hCS, also referred to as *placental lactogen,* is a **single-chain polypeptide** consisting of 191 amino acids with a molecular weight of 23,279.

It is synthesized and secreted by the syncytiotrophoblast. Structurally, it closely resembles both growth hormone (somatotropin) and prolactin. The genes for hCS are located on **chromosome 17.** The hormone is present in syncytiotrophoblasts by the second week after conception and is detectable in maternal plasma as early as the fourth week (0.7 to 0.9 µg/dl. Unlike hCG, the concentration of hCS increases throughout pregnancy in direct proportion to the weight of the placenta. Expression of hCS mRNA in the syncytiotrophoblast does not change during pregnancy, suggesting that the progressive increase in hCS secretion is due to the proliferation of syncytiotrophoblast cells. At term, **maternal plasma concentrations** normally range from 0.5 to 1.5 mg/dl. hCS concentrations in the **fetal circulation** are very low, averaging only approximately 0.003 mg/dl. Following delivery of the placenta, hCS disappears from the maternal circulation within 10 to 12 minutes.

hCS is one of the principal regulators of **maternal homeostasis** during pregnancy. It causes alterations in **maternal intermediary metabolism** that increase the availability of glucose to the fetus. hCS **enhances insulin secretion** by the pancreas in response to a glucose load; however, it also **increases insulin resistance,** such that it diminishes the effects of insulin on tissues such as liver. The net result is impaired glucose tolerance and elevated plasma glucose levels. hCS also **enhances lipolysis.** Thus, it increases concentrations of nonesterified fatty acids, ketones, and glycerol in the blood. Finally, hCS may **inhibit secretion of growth hormone** by the pituitary, which is depressed during pregnancy. Although the hormone has lactogenic properties and may have mammotropic effects, it does not appear to be involved in milk production in humans.

No secretagogues that regulate hCS secretion by human placental cells have been identified. However, arachidonic acid and phospholipase A_2, which catalyzes cleavage of arachidonic acid from membrane phospholipids, stimulate: (1) intracellular calcium mobilization; (2) hydrolysis of phosphoinositides to diacylglycerols and inositol phosphates; and (3) release of hCS. **Diacylglycerols** increase protein kinase C activity and the synthesis and secretion of hCS, and **inositol triphosphate** mobilizes calcium. Thus, **arachidonic acid** might mediate the synthesis and secretion of hCS by an unidentified secretagogue, by causing hydrolysis of phosphoinositides to diacylglycerols and inositol phosphates, which in turn increase protein kinase C activity and calcium mobilization, leading to increased synthesis and secretion of hCS.

Steroid Hormones

Throughout gestation, from implantation to parturition, progestins, androgens, and estrogens are important to both the mother and fetus. Steroid hormones are synthesized, metabolized, and excreted by mechanisms that involve the mother, placenta, and fetus (Fig. 57-6).

Progesterone is synthesized by the maternal–placental unit. Maternal plasma concentrations of progesterone and urinary excretion of its metabolite, **pregnanediol,** do not change following fetal death **in utero,** indicating that progesterone biosynthesis is independent of fetal steroidogenesis. Maternal plasma progesterone concentrations range from $4\mu g/dl$ during the **first trimester** to 16 $\mu g/dl$ during the **third trimester.** At term, approximately 90 percent of the 250 mg produced each day by the placenta is secreted into the maternal circulation, and 10 percent is delivered to the fetal circulation. During the first six weeks of gestation, maternal plasma progesterone is produced by the **corpus luteum.** During the eighth to ninth weeks, progesterone production shifts to the placenta, and this is known as the **luteal–placental shift.**. As in the corpus luteum, progesterone is synthesized from cholesterol obtained by the uptake of low-density lipoprotein by trophoblast cells. **Cholesterol** is converted to pregnenolone through the action of cytochrome P_{450} enzyme–dependent side-chain cleavage and hydroxylation by mitochondrial enzymes. **Pregnenolone** is converted to progesterone by the microsomal enzymes 3β-hydroxysteroid dehydrogenase and $\Delta^{5,4}$-isomerase.

Progesterone synthesis is stimulated by beta-adrenergic agonists, whose effects are mediated by cyclic AMP and are inhibited by GnRH. Since the placenta does not possess 17-hydroxylase, maternal plasma concentrations of 17α-hydroxyprogesterone are very low in midpregnancy. In early pregnancy maternal 17α-hydroxyprogesterone is produced in the **corpus luteum.** After the thirty-second week of pregnancy, maternal plasma concentrations of 17α-hydroxyprogesterone increase due to production by the **fetal adrenal gland.**

Progesterone has several actions crucial to the initiation and maintenance of pregnancy: (1) it prepares the uterine endometrium for **implantation;** (2) it promotes **decidualization** of the endometrial cells to provide nourishment for the early embryo; (3) it acts **synergistically with relaxin** to reduce uterine contractility and sensitivity to oxytocin during pregnancy; and (4) it promotes development of the **mammary glands** prior to lactation.

Androgens are synthesized by the maternal and fetal adrenal glands but not by the placenta, which lacks the 17-hydroxylase and 17,20-desmolase necessary to convert C_{21} steroids (progestins) to C_{19} steroids (androgens). The placenta takes up both **dehydroepiandrosterone** (DHEA) and **dehydroepiandrosterone-sulfate** (DS) from the maternal and fetal circulations. Sulfate is cleaved from the DS to yield DHEA, which, in turn, is converted to **androstenedione** and **testosterone.** Maternal plasma levels of DS and DHEA remain virtually unchanged throughout pregnancy, whereas levels of androstenedione and testosterone increase by twofold to threefold. Androstenedione and testosterone serve as substrates for **aromatase,** which converts the androgens to estrogens.

Estrone and **estradiol** are synthesized in syncytiotrophoblast cells from androstenedione and testosterone through the action of the cytochrome P_{450}–dependent aromatase. Estradiol concentrations in maternal plasma increase throughout pregnancy, reaching levels of 2 to 3 $\mu g/dl$. **Estriol** is not formed from estradiol or estrone because the placenta lacks 16-hydroxylase. How-

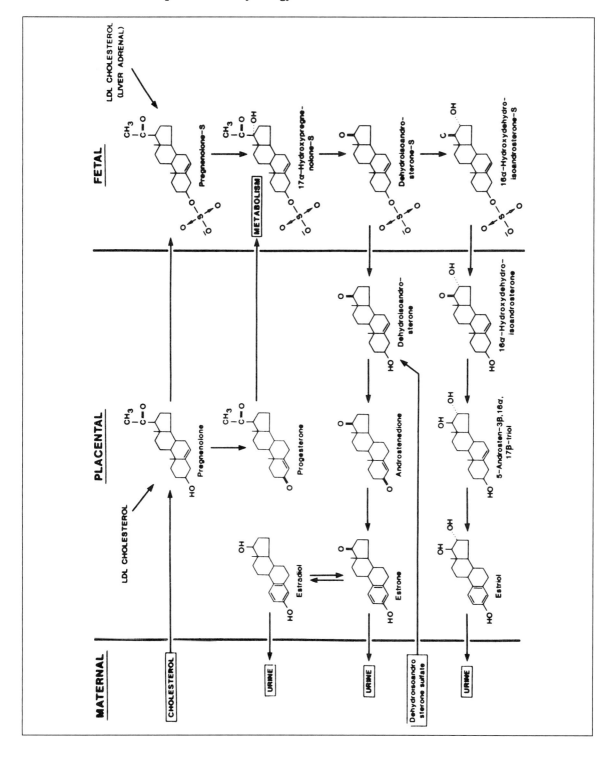

ever, copious amounts of estriol are formed from 16α-hydroxydehydroepiandrosterone, which originates from the fetal adrenal gland. Concentrations of estriol in maternal plasma are very low during the first trimester, but then are similar to estrone levels for the duration of pregnancy. **Estrogens** are very important in the mother during pregnancy. They promote **relaxation of the pelvic ligaments** and **elasticity of the pubic symphysis,** so that the fetus can pass through the birth canal at parturition. Estrogens also are necessary for **enlargement of the breasts** and for **growth of the ductule system** and **lobule—alveolar tissue** of the breasts. Finally, estrogens cause enlargement of both the **uterus** and of the **external genitalia** of the mother.

Other Placental Factors

Several peptides that are identical to hypothalamic peptides exist in the placenta. **GnRH** is synthesized in the cytotrophoblast, which is close to the synchtiotrophoblast, the site of synthesis of hCG and hCS. GnRH is able to stimulate the placental production of hCG. GnRH synthesis in the placenta may be inhibited by **inhibin,** which is also present in the cytotrophoblast.

Both **adrenocorticotropin** (ACTH) and **corticotropin-releasing hormone** (CRH), as well as the ACTH-related **proopiomelanocortin—derived peptides** such as beta-lipotropin, alpha-melanocyte—stimulating hormone, and beta-endorphin, are synthesized in the placenta. The presence of placental ACTH in the maternal circulation contributes to the mother's relative resistance to negative-feedback suppression of ACTH by glucocorticoids. Levels of ACTH in maternal plasma are similar in the first and third trimesters, but levels of CRH in both maternal

and fetal plasma rise progressively during pregnancy.

The human placenta produces both a **chorionic thyrotropin** (hCT) and **thyrotropin-releasing hormone** (TRH), but their physiological importance has not been established. Activities that resemble somatostatin-releasing hormone and growth hormone releasing—hormone have been identified in the placenta.

Prolactin

Prolactin is a **single-chain polypeptide** produced by specialized cells (lactotropes) of the anterior pituitary gland. In humans, the hormone is also synthesized in the decidua but not in the placenta. Structurally, prolactin is very similar to growth hormone and hCS. The glycosylated form has a molecular weight of 25,000, whereas the nonglycosylated form has a molecular weight of 23,000. The gene that encodes the mRNA for prolactin is located on **chromosome 6. Preprolactin,** the precursor of PRL, is synthesized on ribosomes in the endoplasmic reticulum, processed to the mature form, and then stored in secretory granules.

Prolactin is secreted in males and in both pregnant and nonpregnant females. During **pregnancy,** maternal plasma prolactin concentrations gradually begin to rise in the first trimester and, by term, reach levels ten times those in nonpregnant women, or approximately 12 µg/dl. This increase coincides with increased plasma estrogen concentrations. **After parturition,** basal plasma concentrations return to nonpregnant levels.

Synthesis and secretion of prolactin by pituitary lactotropes are regulated by hypothalamic factors and by estrogen. The hypothalamus secretes **dopamine,** a neurotransmitter, which

Fig. 57-6. Synthesis of progestins, estrogens, and androgens in the maternal—placental—fetal unit. Progesterone is synthesized by the maternal ovary until the eighth or ninth week, at which time the placenta is able to secrete sufficient progesterone to maintain pregnancy. Estrogens are synthesized in the placenta from dehydroepiandrosterone-sulfate (dehydroisoandrosterone-sulfate) synthesized in the fetal adrenal gland. (From: Solomon, S. The placenta as an endocrine organ: Steroids. (LDL = low-density lipoprotein.) In: Knobil, E., and Neill, J. D., eds. *The Physiology of Reproduction.* New York: Raven Press, 1988, P. 2086.)

inhibits pituitary, but not decidual, prolactin secretion. There is substantial physiologic evidence for a **prolactin-releasing factor,** but it has not yet been identified. **TRH** stimulates prolactin secretion. Thus, individuals with **primary hypothyroidism** (elevated TRH and TSH levels) can have hyperprolactinemia. **Estrogens** enhance prolactin production by increasing the amount of preprolactin mRNA. Estrogen also increases the size (hypertrophy) and number (hyperplasia) of lactotropes in the pituitary. In both nonpregnant and pregnant states, maternal prolactin secretion is pulsatile, with the pulse magnitude greater during sleep. **Chronic excessive secretion of prolactin** (hyperprolactinemia) in nonpregnant women can lead to abnormal production of milk (galactorrhea) and absence of menses (amenorrhea).

The **fetal pituitary** also synthesizes, stores, and secretes prolactin, beginning early in gestation. Secretion is highest, however, late in gestation when levels in fetal plasma exceed those in maternal plasma. Although the physiologic role of prolactin in the fetus is unknown, evidence suggests that it **promotes fetal lung maturation** by increasing levels of pulmonary **lecithin,** a component of surfactant. Prolactin exists in very high concentrations in amniotic fluid, which has concentrations fivefold to tenfold greater than those in maternal serum. Decidual prolactin synthesis can account for its presence in amniotic fluid. Prolactin is not synthesized in the trophoblast or amnion. Because experimental intraamniotic injection of prolactin in monkeys has been observed to cause a 50 percent decrease in amniotic fluid volume, prolactin may be an **osmoregulator** in the amniotic fluid compartment of the fetal–placental unit. Prolactin promotes growth of the **mammary glands,** initiates **postpartum milk delivery,** and **maintains milk secretion** during lactation.

MATERNAL PHYSIOLOGY

During pregnancy, maternal physiology is reorganized to accommodate both the metabolic demands of the fetus, and to prepare the mother for delivery of the fetus and postpartum nourishment of the neonate. Reorganization of maternal function occurs first in response to **maternal reproductive hormones** and then in response to hormones of the **fetal–placental unit.** Major changes take place in water and electrolyte balance, cardiovascular function, respiration, and metabolism.

Maternal blood volume expands by approximately 40 percent during pregnancy. The plasma volume begins to increase early in pregnancy and then more rapidly during the second trimester; it increases only slightly during the third trimester. An increased plasma supply is required to furnish the utero–fetal–placental unit with enough blood and to protect the mother against the blood loss associated with delivery. A human mother loses an average of **500 ml of blood** during a vaginal delivery. The expansion of maternal blood volume appears to be brought about through the retention of both water and minerals by the kidneys. During pregnancy, an extra **500 to 900 mEq of sodium, 300 mEq of potassium,** and **30 mEq of calcium** are retained. Because a proportionately greater amount of water, 4 to 6 liters, is retained, the **sodium concentration** in the maternal circulation decreases. Although the mechanism of pregnancy-induced electrolyte and water retention has not been precisely defined, estrogens and progesterone both augment plasma concentrations of **aldosterone,** the adrenal cortical hormone that promotes sodium retention and an increase in total body water. **Red blood cell volume** in the maternal circulation begins to increase during the second trimester, with the greatest increase occurring in the third trimester.

Cardiac output increases during pregnancy in order to maintain arterial blood pressure and achieve adequate uterine blood flow in the face of **decreased peripheral vascular resistance.** Maternal resting cardiac output reaches a peak by the twentieth week of pregnancy at which time it has increased by an average of 40 percent, from 5 to 7 liters per minute. Increased cardiac output is due to both **increased heart rate** and **increased stroke volume.** The increase in cardiac output is necessary to compensate for the proportionately greater decline in peripheral vascular resistance, which reaches its lowest level in midpregnancy.

Blood flow to the uterus increases as pregnancy advances and has been estimated to approach 1,200 ml per minute, or 17 percent of the maternal cardiac output, in humans near term. Because uterine blood flow increases more slowly than the size of the fetal–placental unit, the **rate of O_2** extraction from uterine blood increases during pregnancy.

Maternal O_2 consumption increases during pregnancy to meet the energy requirements for maternal and fetal metabolism and synthesis of new tissue. The **maternal metabolic rate** is higher at term than in the nonpregnant state. Basal O_2 consumption increases by approximately 51 ml O_2 per minute, or 20 percent of the total maternal O_2 uptake at term. O_2 uptake is achieved through an increase in minute ventilation, which is augmented in turn through increased tidal volume without increased respiratory rate. The plasma bicarbonate concentration is also reduced to offset the respiratory alkalosis that would result from this hyperventilation.

The average **net energy cost** of pregnancy has been estimated to be 77,000 kcal. It comprises: (1) the caloric value of maternal and fetal tissues formed during pregnancy; (2) the energy required for the biosynthetic processes themselves; (3) the energy required to maintain newly formed tissues; and (4) the energy required for the transport and exchange of nutrients and gases among the mother, placenta, and fetus. Maternal metabolism fulfills these energy requirements, not only for the pregnancy itself, but also to ensure that the fetus has adequate energy stores to survive the immediate postnatal period and that the mother has adequate energy stores to nourish the fetus in the event of diminished food intake.

PARTURITION

A mature fetus is expelled from the uterus, normally at term, by a process referred to as **parturition.** The mechanism of parturition in humans is not completely understood, and this is due, in part, to the major differences among mammalian species in terms of the types and quantities of hormones secreted near term. It is clear that the **initiation of parturition,** or **labor,** is not a function of the maturity of the fetus itself, because labor frequently occurs before (*preterm*) the fetus is fully developed. Rather, it appears to result from the maturation of a complex system of communication among various endocrine organs.

Regulatory Factors

Parturition is the culmination of processes that take place throughout pregnancy to prepare the reproductive tract for expulsion of the fetus. Near term, the **uterus becomes progressively more excitable. Rhythmic contractions** of the uterus increase in frequency and intensity to a level where the fetus is forcefully expelled from the uterus. Estradiol, progesterone, relaxin, prostaglandins, oxytocin, and catecholamines all play roles in the ongoing or short-term regulation of cellular processes prior to parturition. In contrast to other species, adrenal glucocorticoids do not appear to play a major role in human parturition. The **primary signal** that initiates parturition has not been established.

Estradiol and Progesterone

Estradiol and progesterone have opposite permissive effects on uterine contractility. The **plasma concentrations** of both hormones increase during the first seven months of pregnancy, but, during the last two months, progesterone concentrations stabilize and only the estrogen concentrations continue to increase. As a woman approaches term, levels of **estrone sulfate,** which can be converted to estradiol and estrone in the chorion and decidua, increase in the amniotic fluid. Estrogens themselves inhibit the synthesis of progesterone from pregnenolone. Thus, the estrogen–progesterone ratio increases during the two months preceding parturition.

Estrogens also promote parturition in part by

inhibiting production of **PGI₂,** an inhibitory prostaglandin, and enhancing production of stimulatory prostaglandins such as **PGE₂** and **PGF₂ₐ** in the decidua. Estrogens increase the concentration of oxytocin receptors in the myometrium, resulting in increased myometrial sensitivity to oxytocin in late pregnancy. They also promote the formation of gap junctions between myometrial cells, fostering increased electrical coupling among cells.

Relaxin

During pregnancy, a 23,000-molecular-weight polypeptide is produced in the corpus luteum, placenta, and decidua of the uterus. This polypeptide, **preprorelaxin,** is processed by proteolytic enzymes to yield the active relaxin molecule, which has a molecular weight of 6,300. Structurally, it resembles insulin, in that it consists of two dissimilar chains possessing some homology with insulin and insulin-like growth factors. The two relaxin chains are held together by **disulfide bonds** located in positions similar to those in the insulin molecule. The mRNA for the preprorelaxin molecule is encoded by two genes located on the short arm of **chromosome 9,** though only one of the genes is known to be actively transcribed during pregnancy.

The **corpus luteum** in the ovary is the major source of relaxin in the peripheral circulation during pregnancy. Uterine decidual tissue and placental tissue produce lesser amounts of the hormone, which may act locally within the uterus. Peripheral blood levels of relaxin are elevated throughout most of gestation, but they are highest near the end of the first trimester, reaching concentrations of greater than 0.1 μg/dl and then declining to and remaining at less than 0.05 μg/dl during the twenty-fifth to thirty-ninth weeks of gestation. Corpora lutea produce relaxin in response to **hCG**. Circulating levels of relaxin are directly related to the amount of circulating hCG and to the number of corpora lutea in the ovary. Relaxin is not detectable in plasma in the nonpregnant state.

Relaxin has several effects on the reproductive tract, although its importance in humans has not been established as well as its role in other mammalian species. Relaxin initially appears to **inhibit myometrial contractility** throughout pregnancy in order to maintain the uterus in a quiescent state, so that the fetus is not expelled prematurely. It later causes **relaxation of the pelvic ligaments,** producing separation of the pubic symphysis. In women, pubic separation is detectable by the end of the first month of pregnancy, is near maximal by the fifth to seventh month, and is then relatively constant during the last two to three months. Relaxin is also one of the factors that enhances **softening of the cervix** in preparation for passage of the fetus. This softening, or "ripening", of the cervix involves an increase in water content, decrease in collagen content, and increase in the total glycosaminoglycan content, which includes a decrease in sulfated glycosaminoglycans. The effects of relaxin on the pubic symphysis and the cervix require **prior sensitization by estrogen**.

Oxytocin

Oxytocin is a **nonapeptide** (nine amino acids) that stimulates secretion of PGE₂ and PGF₂ₐ from the decidua. It is synthesized by neurons whose cell bodies are located in the supraoptic and paraventricular nuclei of the **hypothalamus.** The hormone is transported through the axons bound to the protein **neurophysin I** and delivered to nerve endings in the neurohypophysis, or posterior pituitary gland, where it is secreted. Because it is undetectable in plasma until after the beginning of parturition, its probable function is to enhance rather than initiate parturition. Uterine responsiveness to oxytocin is enhanced by estrogen, which could account for the twelvefold increase in myometrial oxytocin receptors between the thirteenth and seventeenth weeks of pregnancy and term.

Prostaglandins

Prostaglandins, also termed **eicosanoids,** are thought to constitute the most **important signal for parturition** because they: (1) increase in concentration in amniotic fluid at parturition; (2) administration of prostaglandin synthase inhibitors such as aspirin suppresses uterine activity and prolongs pregnancy; and (3) the myometrium is very sensitive to exogenous prostaglandins. PGE_2 and $PGF_{2\alpha}$ can augment myometrial contractility throughout pregnancy. They appear to modulate calcium fluxes in myometrial smooth muscle fibers. $PGF_{2\alpha}$ inhibits calcium uptake by the sarcoplasmic reticulum, so that cytoplasmic calcium concentrations increase. Elevated intracellular calcium levels activate **myosin light-chain kinase** (MLCK), a critical enzyme in myometrial contraction. MLCK catalyzes phosphorylation of the myosin light chain. The high-energy phosphate bond of phosphorylated myosin is hydrolyzed, yielding energy for the cross-linking of myosin and actin, which shortens the myometrial smooth muscle cell.

PGE_2 is synthesized in both the amnion and decidua; little, if any, is synthesized in the chorion. $PGF_{2\alpha}$ can be formed in the decidua from PGE_2 supplied by the amnion, arachidonic acid released from the amnion, and the *de novo* synthesis of arachidonic acid in the decidua. Arachidonic acid is liberated from its intracellular esterified storage form by the action of phospholipase A_2, a lysosomal enzyme whose regulation by a variety of stimulators is only beginning to be understood.

Mechanism

Delivery of the fetus, or labor, is effected by the sustained, rhythmic, and coordinated contractions of uterine smooth muscle, and can be divided into three phases. The **first phase** extends from the first uterine contractions to complete dilatation of the cervix. The **second phase** extends from complete cervical dilatation to birth of the infant. The **third phase** extends from delivery of the infant to delivery of the placenta. The three phases last a total, on average, of 14 hours for a woman's first delivery.

Unlike other muscular contractions, the uterine contractions of labor are painful. They occur approximately ten minutes apart during the first stage of parturition; the interval decreases to as little as two minutes during the second stage. Each contraction typically lasts sixty seconds. During labor, the cervix shortens and dilates, so that the diameter of the external cervical os, the opening to the uterus, enlarges from a few millimeters to approximately 10 cm. As labor progresses, the fetal membranes rupture and amniotic fluid is released. Delivery of the infant follows as the upper portion of the uterus thickens and continues to contract strongly.

LACTATION

Postpartum growth, development, and survival of the newborn require nourishment which is provided for a variable time in the mother's milk. Milk is produced in the mammary glands of the appropriately conditioned breasts. It is delivered to the infant in response to the mechanical and neural stimuli that constitute the **suckling reflex.**

Development of the Breasts and Mammary Glands

Throughout gestation, the mammary glands grow and develop in response to the coordinated actions of pituitary, ovarian, thyroid, and adrenal hormones (Fig. 57-7). During the **first and second trimesters,** estrogens, growth hormone, and glucocorticoids synergize to stimulate growth of the **mammary ductal system.** Stromal tissue increases, but there is some loss of interstitial adipose tissue. The terminal portions of the mammary glands enlarge, the breasts become more vascularized, and the nipples enlarge. During the **second and third trimesters,** the lobuloalveolar epithelium of the mammary gland differentiates and acquires the capacity to synthesize milk in a process termed **lactogenesis. Lubuloalveolar development** requires estrogen, progesterone, and prolactin. Lactogenesis during pregnancy is minimal, however, and lactation is absent because estrogens and progesterone inhibit the actual formation of milk.

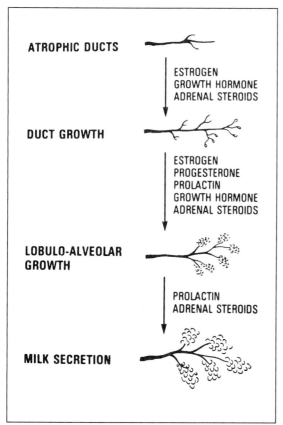

ATROPHIC DUCTS

↓ ESTROGEN
GROWTH HORMONE
ADRENAL STEROIDS

DUCT GROWTH

↓ ESTROGEN
PROGESTERONE
PROLACTIN
GROWTH HORMONE
ADRENAL STEROIDS

LOBULO-ALVEOLAR
GROWTH

↓ PROLACTIN
ADRENAL STEROIDS

MILK SECRETION

Fig. 57-7. Hormonal regulation of mammary gland development. During the first and second trimesters of pregnancy, estrogens, adrenal steroids, and growth hormone synergize to stimulate growth of the mammary ducts. During the second and third trimesters, the lobuloalveolar epithelium differentiates and acquires the capacity to secrete milk. Estrogen, progesterone, and prolactin are required at this stage, as well as glucocorticoids and growth hormone. Milk production does not begin until estrogen and progesterone levels decline after parturition. (From Rebar, R. W. Maternal serum hormone concentrations during pregnancy. In: Creasy, R. K., and Resnik, R., eds. *Maternal-Fetal Medicine: Principles and Practice.* Philadelphia: W. B. Saunders, 1989, P. 155.)

Synthesis and Secretion of Milk

In the final month of pregnancy, the parenchymal cells of the mammary glandular alveoli hypertrophy because of the intracellular accumulation of

colostrum which is a hyaline, eosinophilic, proteinaceous, low-fat secretion. It is composed of: 87 percent water, 1.3 percent fat, 3.2 percent lactose, 7.9 percent protein (principally alpha lactalbumin, lactoferrin, and immunoglobulin A), and 0.6 percent mineral-containing ash. Colostrum has a higher protein content but lower carbohydrate level than mature human milk. Within two to three days after parturition and the decline of estrogen and progesterone levels in maternal plasma, the fat content of the mammary secretion increases abruptly and it becomes **milk.**

Human milk is an emulsion of fat in water that is isotonic with plasma. It has the following composition: 87 percent water, 4.5 percent fat, 6.8 percent lactose, 0.9 percent casein, 0.4 percent lactalbumin and other proteins, and 0.2 percent mineral-containing ash. It has an energy content of 60 to 75 kcal/dl, and milk production can reach 1.5 liters per day. Milk is not secreted before parturition because estrogen and progesterone secreted by the placenta block the lactogenic effects of prolactin and hCS. At birth, the mammary glands are freed from the inhibitory effects of the estrogen and progesterone. The lactogenic properties of prolactin precipitate synthesis of milk rather than colostrum. Other hormones, such as glucocorticoids, growth hormone, and parathyroid hormone, appear to be important in regulating the composition of milk.

Prolactin

Prolactin stimulates several important biochemical processes in milk synthesis, and thus is the most important hormone for **lactogenesis.** Prolactin stimulates cells of the mammary glandular epithelium to synthesize the milk proteins **casein** and **alpha-lactalbumin.** Administration of the dopamine agonist **2-bromo-α-ergocryptine** in the immediate postpartum period causes prolactin levels to drop and breast engorgement and lactation to cease. The first step in the interaction of prolactin with target cells is binding to cell-surface receptors, which elicits increased expression of the genes for milk proteins and RNA synthesis. However, the intermediate steps in prolactin action are not known. Similarly, it is also not known how estrogens and progesterone block the stimulatory effects of prolactin on lactogenesis.

Basal prolactin levels in maternal plasma gradually return to nonpregnant levels within several weeks of birth. Stimulation of the nipples by the infant's suckling, however, causes a transient tenfold increase in prolactin secretion, which lasts approximately one hour each time suckling occurs. These repeated prolactin surges provide the stimulus for continued lactogenesis. If the prolactin surges are blocked or absent, they disappear within a few days. Thus, if the mother does not nurse, prolactin surges gradually disappear and lactogenesis ceases.

Suckling and Milk Ejection

Milk is continuously synthesized and secreted into the lumina of the alveoli (Fig. 57-8). Milk does not flow readily from the alveoli to the duct system, so that milk does not leak out of the nipple. The 12 to 20 lactiferous ducts, or **galactophores,** leading into the nipple are dilated to form sinuses. Most milk, however, is stored in the alveoli. Milk is ejected, or **let down,** from the alveoli only in response to suckling by the infant. The mammary glands contain contractile tissue consisting of **myoepithelial cells** that surround the alveoli and small ducts and **smooth muscle** that encircles the larger milk ducts and blood vessels. The **myoepithelial cells** generate the major force for milk ejection.

The **suckling reflex,** which controls milk ejection, involves **neurogenic and hormonal signals,** including the hormone **oxytocin** (Fig. 57-9). **Mechanical stimulation** of the nipple generates impulses that are transmitted to the spinal cord through the dorsal roots of the spinal nerves. The **neural pathways** terminate in the paraventricular and supraoptic nuclei of the hypothalamus. **Oxytocin** synthesized by neurosecretory cells in these nuclei and secreted from the posterior lobe of the pituitary gland stimulates the myoepithelial cells surrounding each alveolus to contract, thus expelling milk. Milk ejection

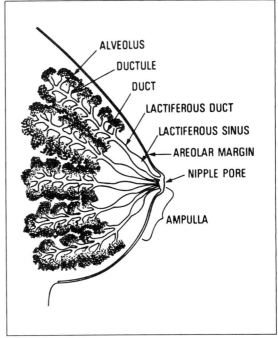

Fig. 57-8. Structure of the mature gland. Milk is synthesized in the epithelial cells lining the alveoli and secreted into the lumina of the alveoli, where most milk is stored. Milk is ejected from the alveoli through the lactiferous ducts into the lactiferous sinuses, in response to suckling. (From Rebar, R. W. Maternal serum hormone concentrations during pregnancy. In: Creasy, R. K., and Resnik, R. eds. *Maternal-Fetal Medicine: Principles and Practice.* Philadelphia: W. B. Saunders, 1989, P. 154.)

usually occurs within thirty seconds to a minute after a baby begins to suckle. Although milk ejection primarily responds to suckling, it can be conditioned. For example, sight of the infant, crying by the infant, or breast preparation can cause milk ejection; embarrassment or pain can inhibit it.

POSTPARTUM CHANGES IN THE MOTHER

The four-week to five-week period between delivery and resumption of ovulation and menstruation is referred to as the **puerperium.** It begins

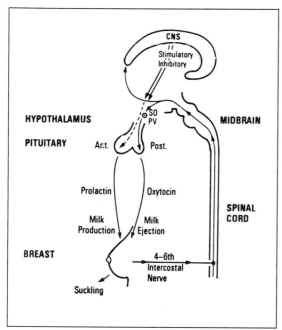

Fig. 57-9. Suckling reflex. Mechanical stimulation of the nipple generates nerve impulses that are transmitted to the spinal cord through the dorsal roots of the spinal nerves and then to the paraventricular and supraoptic nuclei of the hypothalamus. Oxytocin is synthesized in neurosecretory cells in these nuclei and secreted from the posterior pituitary. Oxytocin stimulates myoepithelial cells surrounding the alveoli of the mammary glands to contract, thus expelling milk. Milk ejection is often a conditioned response to psychogenic stimuli, such as sight of the infant. (From Rebar, R. W. Maternal serum hormone concentrations during pregnancy. In: Creasy, R. K., and Resnik, R., eds. *Maternal-Fetal Medicine: Principles and Practice.* Philadelphia: W. B. Saunders, 1989, P. 160.)

immediately after delivery of the placenta. Frequent strong myometrial contractions cause the uterus to rapidly involute and decrease in size. Within a week, it has decreased by 50 percent and by six weeks it has almost returned to its nonpregnant size. By the sixteenth day, the **endometrium** resembles that of the nonpregnant uterus in the proliferative phase of the menstrual cycle. The placental site of the endometrium requires several more weeks to recover. In gross appearance, the cervix resembles that in the non-

pregnant state within one week postpartum, although involution continues for at least six weeks. The vagina returns to its normal nonpregnant state in six to ten weeks.

Ovulatory function returns only gradually after parturition. The mechanism for **postpartum amenorrhea** is unknown. However, gonadotropin secretion, which is necessary for ovarian function, is minimal for two to three weeks after delivery, but the initial ovulation can occur as early as 27 days after delivery. In nonlactating women, the initial ovulation occurs at a mean of ten weeks; in nursing women, the initial ovulation takes place at a mean of 17 weeks. Women who breast-feed for less than 27 days ovulate at approximately the same time as women who do not breast-feed at all. In **nonlactating women,** the mean time to the first postpartum menses is seven to nine weeks, with 70 percent menstruating within 12 weeks. In lactating women, the time to the **first postpartum menses** increases as the duration of lactation increases. **Prolactin,** which is secreted in response to suckling, inhibits ovarian function, either directly or indirectly through inhibition of the hypothalamic–pituitary unit.

Maternal metabolism generally, but not always, returns to normal within six to eight weeks after parturition. **Cardiac output and blood volume,** which increase greatly during pregnancy, and **peripheral vascular resistance,** which decreases during pregnancy, return to baseline within six weeks. Most of the regression in the various components of **cardiovascular function,** such as heart rate, blood pressure, O_2 consumption, and total body water content, occurs during the first two weeks. **Renal function variables,** such as renal plasma flow, glomerular filtration rate, and serum creatinine clearance, increase during pregnancy and return to normal levels by six weeks postpartum. **Liver function,** which is altered by rising levels of estrogens during pregnancy, appears to return to normal within three weeks of delivery.

SUMMARY

Fusion of the male and female gametes and subsequent development of the offspring occur in the reproductive tract of the female. Spermatozoa are delivered from the penis of the male into the female reproductive tract through the vagina during coitus, the sexual act. If conception occurs, the resulting zygote begins to undergo a series of profound changes that enable it to develop into a full-term fetus. The conceptus, which develops in the uterus, regulates its environment by means of specific humoral signals. During the developmental period, which usually lasts approximately 39 weeks and is termed **gestation** or **pregnancy,** maternal metabolism is re-oriented toward providing nutrients to the conceptus. Pregnancy culminates in parturition, in which the fetus is expelled from the uterus. Further development and maturation of the infant, or neonate, occur outside the mother's body. For several months after birth, the neonate receives nutritional support from the mother in the form of breast milk. Development and maturation of the offspring are not actually completed until puberty.

NATIONAL BOARD TYPE QUESTIONS

Select the one most correct answer for each of the following

1. Propulsion of spermatozoa through the female reproductive tract to the fallopian tubes:
   A. Is due to rhythmic motion of the flagellum of the spermatozoon
   B. Is aided by contractions of the vagina and uterus
   C. Is retarded by the cervical mucus
   D. Is aided by the negative vaginal pressure generated during orgasm
   E. All of the above
2. The ovum:
   A. At the time of ovulation normally has undergone the second meiotic division, completed nuclear maturation, and extruded the second polar body
   B. Normally is not fertilized by a spermatozoon until after it passes through the fallopian tube and enters the uterus
   C. At the time of ovulation normally has completed the first meiotic division and extruded the first polar body, which lies between the zona pellucida and the vitelline membrane
   D. At the time of ovulation normally has the haploid number of chromosomes and the haploid quantity of DNA
   E. None of the above
3. Implantation of the conceptus:
   A. Normally occurs in the ampulla portion of the fallopian tube
   B. Is preceded by decidualization of the stromal cells of the endometrial epithelium
   C. Is inhibited by progesterone
   D. Occurs at the blastocyst stage when it consists of approximately 200 cells
   E. None of the above
4. Which of the following statements is *correct?*
   A. The maternal-facing and fetal-facing trophoblast cell membranes form the principal elements in the transplacental exchange of nutrients
   B. Prolactin is synthesized by amniotic and chorionic cells and secreted into the amniotic fluid
   C. Transfer of D-glucose from the mother to fetus occurs by active transport across the placenta
   D. Free fatty acids must be esterified to triglycerides prior to being transferred across the placenta to the fetus
   E. During pregnancy, maternal O_2-carrying capacity increases so that, at a given P_{O_2}, maternal hemoglobin can carry 20 to 30 percent more O_2 than fetal hemoglobin, thereby facilitating transfer of O_2 from the mother to the fetus
5. Biosynthesis of steriod hormones during pregnancy:
   A. Involves conversion of progestins to androgens by the placenta

B. Involves a shift of progesterone production from the corpus luteum of the ovary to the chorion during the eighth to ninth weeks of gestation

C. Is regulated by hCS

D. Involves rapid 16-hydroxylation of estradiol to estriol in syncytiotrophoblast cells of the placenta

E. Is inhibited by DHEA and DS of maternal and fetal origin

6. Which of the statements is *incorrect?*

A. The human fetus secretes hormones that initiate parturition when it has reached maturity

B. Placental synthesis and secretion of hCG reaches a maximum around the twelfth week of gestation and then declines

C. Estrogens promote parturition by increasing myometrial sensitivity to oxytocin and by enhancing production of PGE_2 and $PGF_{2\alpha}$

D. Relatively little PGE_2 is formed in the chorion

E. None of the above

7. Suckling causes:

A. Inhibition of oxytocin secretion

B. Increased estradiol secretion

C. Development of the alveoli and lactiferous ducts of the mammary gland

D. Secretion of prolactin

E. None of the above

ANNOTATED ANSWERS

1. E.
2. C.
3. D.
4. A.
5. B. This process is termed the **luteal—placental shift.**
6. A. Parturition is independent of the stage of development of the fetus itself.
7. D.

BIBLIOGRAPHY

Creasy, R. K., and Resnik, R., eds. *Maternal-Fetal Medicine: Principles and Practice.* Philadelphia: W. B. Saunders, 1989.

Knobil, E., et al, eds. *The Physiology of Reproduction.* New York: Raven Press, 1988.

Sadler, T. W. *Langman's Medical Embryology.* Baltimore: Williams & Wilkins, 1985.

Steven, D. H., ed. *Comparative Placentation: Essays in Structure and Function.* New York: Academic Press, 1975.

XII Integrative Physiology

58 Feedback Control and Integration

ERNEST C. FOULKES

Objectives

After reading this chapter, you should be able to:

Explain the differences between equilibrium and steady-state

Define homeostasis in terms of the maintenance of a preprogrammed steady-state

Describe the nature of the feedback control mechanisms that permit maintenance of homeostasis

Discuss the integration of feedback loops

Identify the gain or set-point of a control mechanism

Describe the pathophysiologic conditions that can interfere with the feedback loop

Distinguish between compensation and adaptation

As defined by the American Physiological Society, regulatory and integrative physiology "emphasizes relationships between organ systems and the control of physiological processes in the whole organism." Essentially all of physiology deals with integrative processes that serve, for example, to maintain constant body weight, core temperature, blood pressure, composition and volume of fluid compartments, and other properties, in the face of sometimes extreme variations in environmental influences or physiologic demands. Claude Bernard was one of the first to recognize this constancy of the internal environment, or to use Cannon's term, *homeostasis* (remaining the same). The science of **cybernetics,** as applied to physiology by Wiener and Rosenblueth some forty years ago, forms an essential component of the study of homeostasis.

The physiology of temperature regulation and exercise will be used primarily to illustrate the nature of integrative mechanisms. Choice of these topics is admittedly arbitrary. Temperature regulation, depending as it does on metabolism, exercise, circulation, respiration, the endocrine system, the autonomic nervous system, salt and water balance, and on other factors, is obviously an integrative function. *Exercise* is not only intimately related to temperature control, but also depends on the integration of many functions. Other aspects of integration, such as the control of blood pressure or the maintenance of acid–base balance, are discussed elsewhere in this book. An understanding of integration requires familiarity with both the autonomic nervous system and the various endocrine control mechanisms, which have already been covered in this book and to which the reader is referred.

STEADY-STATE

This chapter considers the nature of feedback loops, which are the basic unit of integrative mechanisms. Their existence explains how a number of variables can be integrated to establish a steady-state. The steady-state differs from the **equilibrium state,** which represents the lowest potential energy level of the system. For instance, after death a body will reach temperature equilibrium with its surroundings, as do all other energy-requiring gradients. In contrast, living animals expend considerable energy to maintain steady-states that are often far different from equilibrium. For example, mammals and birds can maintain a relatively constant body temperature despite large temperature variations in the environment.

In this discussion, it must be borne in mind that there is no one steady-state; instead, steady-state values can change when an individual adapts to new environmental factors, which can be either external or internal. Reproductive cycle and age represent internal factors, while agents such such drugs and pyrogens constitute external factors. For instance, high altitudes (low O_2 pressure), an external factor, require that the normal steady-state value of the hematocrit be increased to maintain tissue oxygenation. Such compensatory and adaptive changes are discussed later in this chapter. Fever, as described in Chapter 59, represents an instance where the steady-state value of the core body temperature is raised because of the influence of pyrogens, and the magnitude of the deviation from normal reflects the severity of the underlying disease (see the section "Pathophysiology of the Feedback Loop").

FEEDBACK LOOPS

Figure 58-1 illustrates a feedback loop. Normal feedback control is always negative, and tends to diminish any deviation from the desired steady-state. Factors that tend to influence the steady-state may be external or they can arise internally. A **sensor,** such as a baroreceptor involved in the maintenance of blood pressure or a chemoreceptor that detects the partial pressure of O_2, produces a **feedback signal** that relays this information to the **integrator**. Here information from various sensors about the actual state of the system is continuously integrated and compared with the desired (normal) state. The **feedback message** may consist of nervous impulses, as in the case of baroreceptors, or of chemical signals, as when circulating glucose levels modulate the central hunger drive.

Messages fed into the integrator generate an **error signal** that registers the deviation of the actual from the desired state. This signal then triggers an appropriate **compensatory response**. Examples of such compensatory mechanisms or feedback responses are **shivering,** which increases heat production when the body temperature falls below desired values, or increased **insulin secretion** during hyperglycemia. (The concept of compensation will be discussed further later in this chapter.) If the message remains constant for a given deviation from the normal steady-state, and the efficiency of the compensatory mechanisms is unaltered, the system will return to the programmed steady-state. If, however, the set-point of the integrator is changed to a new value, the steady-state will be changed, as illustrated by the effect of fever on temperature regulation (see Chapter 59).

Theoretically the same would be achieved by changing either the sensitivity of the integrator's response to an error signal or the efficiency of the compensating mechanisms, rather than the set-point of an integrator. Such changes refer to the **gain of the system** and are discussed in the section "Gain and Set-Point." The set-point of the integrator, or the gain of the whole system, may be influenced by internal factors such as age and reproductive cycles. External factors, such as diet, light–dark cycles, temperature, and drugs, may all critically affect the steady-state. *Arrow B* in Figure 58-1 might represent the indirect action of an external pyrogen on the integrator, while *arrow C* may represent dehydration that diminishes the sweating response to heat. (Chapter 59 discusses these processes in greater detail.)

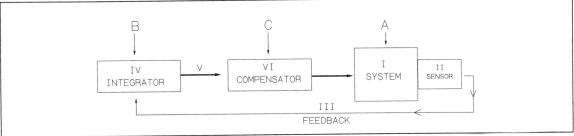

Fig. 58-1. Closed negative feedback loop. The loop consists of the system (*I*) whose steady-state, as detected by sensors (*II*), is maintained in face of a stress *A* at some predetermined value controlled in the integrator (*IV*). The integrator receives feedback messages (*III*) from many sensors. Any deviation from the desired steady-state in *I* generates an error message (*V*) that activates the compensator (*VI*). All sections of the feedback loop are subject to external influences, as indicated by arrows *B* and *C*.

Systems in the body are actually seldom governed by a single feedback loop, as illustrated by the multiple processes responsible for maintaining blood pressure. These may involve different sensors, including baroreceptors, chemoreceptors, or volume receptors. Different compensatory mechanisms, such as adjustments in cardiac output, or peripheral vascular resistance, or salt and water excretion, all contribute toward minimizing deviations from the desired steady-state pressure. The simultaneous and continuous functioning of multiple loops may be redundant under normal conditions, and may actually mask the malfunction of a specific loop. However, this redundancy furnishes both a backup mechanism and permits fine tuning and damping of the response of the system. Excessive oscillations from an accurately defined steady-state can thus be avoided.

INTEGRATION

The integration in Figure 58-1 takes the feedback information that describes the actual state and compares it to some **preset value,** then generates appropriate error signals to the compensators. Although there are important exceptions, integration is usually performed by the **central nervous system,** as Sherrington implied with the title of his classic volume on neurophysiology, *The Integrative Function of the Nervous System.*

This function involves **afferent and efferent pathways** and their integration, which operate to maintain the steady-state of the body as a whole; it may also involve cortical influences on the integrating centers. A classic example of such integration is **appetite control,** which is determined in part by higher cortical function and includes various psychologic and cultural influences. Ablation of the satiety center, however, leads to uncontrolled eating and gross obesity. The importance of balancing afferent and efferent activities is especially critical when many different sensors and compensating mechanisms are involved in the maintenance of the same steady-state. (Additional details on central control mechanisms are covered elsewhere in this book.)

Many **extrinsic or intrinsic factors** can influence the final steady-state values maintained in the body. For instance, optimum body temperature is presumably related to such intrinsic factors as the physicochemical properties of proteins. Extrinsic factors such as light–dark cycles, diet, and environmental temperature can prompt an animal to prepare for hibernation and permit its body temperature to drop to a new steady-state level. Another example of extrinsic factors are the fever-producing agents (pyrogens) that provoke increases in body temperature.

Peripheral sensors can also facilitate or inhibit the response of peripheral compensators to central feedback command. This probably explains why cooling an area of skin can reduce sweat production below that seen on control skin, although both areas are subject to the same central drive. In addition, some **integrating mechanisms** are clearly located outside the central nervous system. For instance, **insulin release** by the pancreas is triggered by elevated glucose levels in the pancreatic blood supply. Similarly, the **enteric nervous system,** along with its ganglia, participates in the integration of salt and water absorption. Feedback control of pancreatic enzyme secretion and the feedback from the renal tubule to the glomerulus (the tubuloglomerular reflex) are other examples of **local feedback loops**. Ultimately, however, the steady-state value for the system as a whole is determined by central integration.

Pancreatic control of the blood glucose levels represents an exception, in which sensing, integration, and compensation are all functions of the same organ. An example in which integration and sensing are partly carried out at the same site is the **hypothalamic heat integration,** which is controlled by the temperature of the hypothalamus.

GAIN AND SET-POINT

Two models of how the integrator evaluates the feedback and translates it into suitable responses by the compensating mechanisms are the **set-point model** and the **gain model.**

A common example of a set-point model is the thermostatically controlled air conditioner. When the temperature rises above the set-point, the apparatus is switched on and does not turn off until the temperature drops to the set-point. The response in this system is all-or-none. In the body the existence of multiple parallel loops fosters a graded response, which may lead to the recruitment of additional compensators. An example of this is the involvement of more skin areas in the sweating response with added heat stress.

Figure 58-2 illustrates the behavior of the set-point model in relation to body temperature. When core temperatures are below the set-point (37°C for the control, 39°C for the example of a patient with fever), the heat compensatory mechanisms are relatively inactive. Above the respective set-points, compensatory activity increases, and, for the sake of simplicity, this increase is assumed to be linear. The model shows that a one-degree rise above the set-point should produce a constant compensatory response, regardless of its actual value. Thus, if the body temperature is normally set at 37°C, a rise to 38°C would produce the same response as would a one-degree rise to 40°C in a feverish patient whose thermostat has been reset to 39°C. The nature of the compensatory responses to heat will be discussed later in this chapter and in Chapter 59.

In the **set-point model** of temperature control, an error signal is generated only when feedback data indicate that the desired steady-state temperature has been exceeded. In the **gain model** (Fig. 58-3), the compensator **continuously responds** to the feedback signal, such that there is some ongoing response even at quite low core temperatures. Thus, the higher the temperature rises, the greater the response. (A linear relationship is assumed here also.) The steady-state temperature is determined by the sensitivity with which the integrator responds to a small increment in input, that is, by the gain of the system. A decreased sensitivity, as in the hypothetical case of the patient with fever, will cause a smaller compensatory response to a given body temperature. As a result, the steady-state temperature will rise to a new level. Temperature control exhibits both set-point and gain characteristics, but it may be difficult to distinguish between these mechanisms in the context of some other complex biologic control systems.

AFFERENT AND EFFERENT PATHWAYS

Afferent and efferent pathways may be neural or humoral, or both, and are described in the chapters on the autonomic nervous system and endocrine physiology. Not all humoral pathways involve hormones, however. For example, the central hunger drive is largely stimulated by blood glucose concentrations, and secretion of

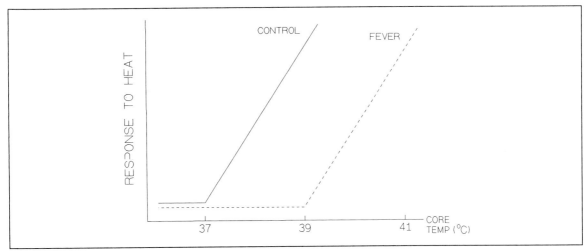

Fig. 58-2. Set-point model of temperature control. A constant response is seen to a one-degree (Celsius) rise of body temperature above the set-point in both a control (normal) subject and a patient with fever. (For details, see text.) (Based on: Mitchell, D., Snellen, J. W., and Atkins, A. R. Thermoregulation during fever: change of set-point or change of gain. *Pflugers Arch*. 321:293, 1970.)

Fig. 58-3. Functioning of a gain model of temperature control. The compensatory activity is minimal near the neutral ambient temperature of around 31°C. (see the section "Compensatory Mechanisms"). The normal subject responds more effectively to a rise in ambient temperature than does the patient with fever. In other words, fever is associated with a reduced gain of the system. (Based on: Mitchell, D., Snellen, J. W., and Atkins, A. R. Thermoregulation during fever: change of set-point or change of gain. *Pflugers Arch*. 321:293, 1970.)

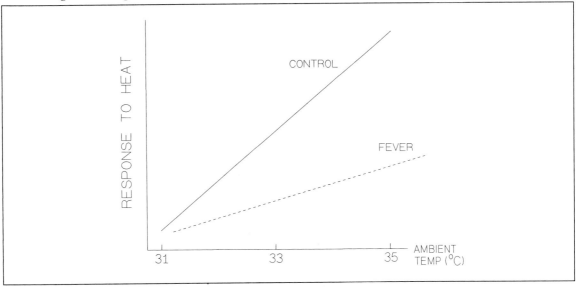

the antidiuretic hormone from the pituitary is triggered by diminished plasma osmolarity.

COMPENSATORY MECHANISMS

Compensators vary, of course, depending on the feedback loop involved and the steady-state to be maintained. Thus, for example, renal retention of salt and water will help compensate for **blood loss,** making the kidney the ultimate effector organ. Hematopoiesis is stimulated in the bone marrow. In the case of **zinc deficiency,** the effector organ is the intestine which responds by increasing zinc absorption. There are numerous other examples but only temperature regulation will be considered in detail here.

Temperature compensation responds to both high and low temperatures. When exposed to the cold, the body reacts in various ways to maximize heat production and conservation, and processes that promote heat loss are minimized. When the body is exposed to heat, it must maximize heat loss and minimize heat production. The **neutral temperature,** as illustrated in Figure 58-4, is that environmental temperature which requires minimal compensatory activity for maintaining the desired steady-state temperature of the body. For a resting and fasting human, this temperature (31°C) defines the basal condition of the body. The corresponding rate of metabolism and heat production is called the **basal metabolic rate** (BMR) and is discussed further in Chapter 59. Because no external work is done under resting conditions, all energy consumed must ultimately be liberated as heat. The temperature gradient from 37°C at the body core to 31°C in the neutral environment thus permits the basal heat production to be dissipated without the need for any compensatory activity. Heat in this case is lost mostly through conduction.

Compensatory mechanisms *1* to *5*, shown in Figure 58-4, represent the main physiologic processes that control body temperature in humans. Other mechanisms operate in different species, such as piloerection (fluffing of feathers or hair) in the cold or panting in the heat. Even though they may all depend on similar sensors and integrating mechanisms, they represent separate feedback loops, each with its own compensator. This is a further example of multiple feedback loops.

PATHOPHYSIOLOGY OF THE FEEDBACK LOOP

Because the maintenance of homeostasis depends on the normal function of a multitude of feedback loops, interference with any one loop will lead to loss of homeostasis only if not enough redundancy is afforded by parallel unaffected loops to control the steady-state. For example, vagotomy does not abolish control of blood pressure, but removal of the pituitary precipitates uncontrolled renal water loss (diabetes insipidus), which can only be counteracted by massive water intake. The etiology of many diseases can be readily explained by the inadequate performance of certain feedback loops.

If the loop shown in Figure 58-1 is opened or partially suppressed, the ability of the system to maintain control is destroyed. There are many ways in which this can happen. **Alterations in, or neutralization of, the sensors** could eliminate control. For instance, occlusion of the common carotid artery interferes with the ability of the carotid sinus to monitor blood pressure. Or the **afferent feedback message** may be altered, as occurs when the carotid sinus nerves are severed. This jeopardizes blood pressure control. The **integrator may also be suppressed;** an illustration of this is the abolition of hypothalamic activity during deep anesthesia which eliminates temperature control. For instance, when the body temperature of a patient is to be depressed in preparation for heart surgery, to reduce the metabolism rate and consequently the required cardiac output, the patient must be anesthetized. Without anesthesia, cooling would lead to uncontrolled shivering as the body attempts to compensate for the heat loss. This would place additional stress on the heart. The reverse is seen in cases of malignant hyperthermia, in which the

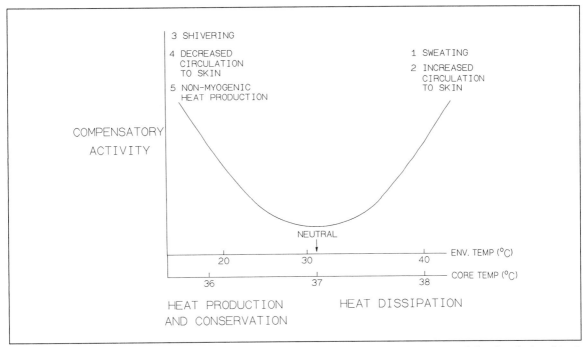

Fig. 58-4. Temperature compensation and neutral temperature. The neutral temperature is defined as that ambient temperature at which total heat compensatory activity is minimal. The temperature gradient from the body to the environment permits the basal heat production to be dissipated by conduction.

body temperature rises out of control; no increased steady-state temperature is established as in fever.

Other pathologic changes in the feedback loop can affect the **error signal** and the **compensating mechanisms**. The **efferent error signal** may be altered, as, for instance, in control of blood pressure by damage to the vagus nerves. The **compensator** may be destroyed, as occurs in diabetes mellitus when the absence of islet function in the pancreas prevents insulin secretion in response to hyperglycemia. The **compensating response** may be weakened. A well-documented instance of this is insulin-resistant **diabetes,** in which antibodies to insulin may interfere with the normal function of the compensatory response. Another example is **Addison's disease** (adrenocortical insufficiency), in which inadequate adrenal output of mineralocorticoids leads to excessive renal salt losses. The response of the compensating mechanism may also be reduced by **tachyphylaxis,** which is the progressively decreasing response of a system to prolonged drug therapy. This is illustrated by the ultimate reduction in renal sodium retention after prolonged steroid therapy.

As already mentioned, the action of the feedback loop, and therefore the steady-state, can also be affected by **external factors** such as drugs and pyrogens. **Antipyretic drugs** like aspirin lower fever (an increased steady-state body temperature) by direct effects on the integrator; **exogenous pyrogens** indirectly influence the temperature control center to raise body temperature. Systemic toxic or pathologic effects can also overwhelm the ability of normal control loops to maintain steady-state. For instance, renal glomerular damage could increase urinary protein loss, or hepatic intoxication could decrease synthesis of plasma protein. Both could ultimately precipitate a fall in plasma oncotic pressure, leading to edema and salt and water retention.

Even with normally functioning, intact feedback loops, the body may not be able to compensate for excessive stress. In such cases of an extreme imbalance between challenge and response, homeostasis cannot be maintained and reversible or irreversible changes will eventuate. For instance, the body has only a limited ability to compensate for excessive heat or cold, and hyperthermia or hypothermia are the consequence.

COMPENSATION AND ADAPTATION

The discussion so far has focused on the physiologic mechanisms that can react acutely to changes, leaving the steady-state undisturbed. When the stimulus to change ceases, the compensatory activity returns to baseline. When exposure to outside stimuli, such as heat, cold, or high altitudes, is prolonged, the system often **adapts,** possibly by attaining new steady-state levels.

Adaptation must be distinguished from compensation. The body will immediately compensate for heating by increasing sweat production; once heating stops, the system returns to baseline. In contrast, repeated and prolonged exposure to heat elicits **adaptive changes** that augment the efficiency of sweating and thus improve functioning in a hot environment. An adapted human can maintain a lower body temperature than can an unadapted one, largely because of increased sweating capacity and improved salt retention. Adaptation is a much slower process than is compensation, and does not decay as soon as the stimulus is removed. For example, a man who is used to working at high altitudes and low oxygen pressures will not lose his adaptive increase in blood hematocrit immediately upon return to sea level.

The characteristic properties of adaptation are well illustrated by the astronauts who spend varying periods in the gravity-free skylab. A de-

scription published in *New Yorker* magazine (September 6, 1976) best summarizes their medical condition: "You throw someone in a new environment and he's apt to have a tough time at first. But if he survives, he will tolerate it, and then begin to improve. It could be that the crews that were up the longest had the longest time to recover. It might take between three and six weeks for a man really to adapt to space."

Just as adaptation is a relatively slow process, so does de-adaptation require time. This is dramatically illustrated in Table 58-1, which documents that it took six years after the introduction of protective diving suits before Korean sponge divers had finally lost their ability to enter cold water without shivering.

All these cases ultimately depend on either the resetting of a set-point or on a change in the gain of some feedback loop. To maintain the overall steady-state of the body in the presence of continuous stress, the steady-state level of some specific systems may be altered. Thus, at high altitudes, the steady-state hematocrit of around 40 percent may be raised to 50 percent to ensure an adequate oxygen supply to the tissues. In the cold-adapted individual, the increased thyroid activity permits increased metabolic heat production. Note that these adaptive mechanisms apply

Table 58-1. Loss of Cold Adaptation in Korean Diving Women

Years	Basal Metabolic Rate (% of control)	Critical Water Temperature (°C)
1960–1977	130	25.9
1977—Introduction of Wet Diving Suits		
1980	100	27.9
1981	100	28.9
1982	100	29.7
1983	100	29.9

Critical water temperature is the temperature at which shivering could first be observed in absence of diving suit—that for non-divers was 29.9°C.
From: S. K. Hong, *News in Physiol. Sci.* 2:79, 1987.

to physiologic adaptation, and not to behavioral, cultural, and genetic adaptation.

Purely physiologic adaptation to temperature extremes can permit a human to survive for significant periods only over a range of about 10 to 45°C, depending on humidity, wind velocity, and other factors. With suitable protective clothing and a high caloric intake, on the other hand, humans can adapt to much colder temperatures, down to as low as −70°C, without an external source of heat. Heat adaptation does not significantly extend the temperature range of survival, but greatly enhances work efficiency in a hot environment. This is discussed further in Chapter 59. Humans also possess the short-term ability to compensate for much greater stresses: for instance, they can survive for 30 minutes or longer at a dry sauna temperature of well above 100°C without major changes in the core body temperature. The massive compensatory activity required, such as sweat loss at a rate as high as several liters per hour, can only be sustained for a limited period; dangerous hyperthermia would unavoidably result from prolonged exposure.

SUMMARY

This chapter re-emphasizes the importance of homeostasis, which is maintained by the efficient functioning of a multitude of negative-feedback loops. The nature of these loops and their component parts are discussed. Integration of loop activity is primarily, though not exclusively, a role of the central nervous system; the activity of the integrator is governed either by its set-point or its gain. Changes in integrating activity (resetting of the set-point) or from interference with the normal feedback loop can cause pathologic states. Normal feedback compensation is short-term and differs from adaptation, which is the actual resetting of new steady-states in response to prolonged alterations in the environment.

NATIONAL BOARD TYPE QUESTIONS

Select the one most correct answer for each of the following

1. Which of the following statements about a negative feedback loop in the body is *incorrect?*
   A. The feedback loop maintains steady-state by generating error signals that depend on the sensed deviation from the steady-state
   B. The steady-state is determined by the setting of the integrator
   C. The feedback loop may use central and/or peripheral sensors
   D. The efferent error signal is fed back directly to the system that is to be maintained at steady-state
   E. Resetting of the integrator may produce a new steady-state in the system
2. Oscillations in the steady-state values in various body compartments are minimized by which mechanism?
   A. Maximizing the speed of response of a feedback loop
   B. Increasing the sensitivity of the feedback loop
   C. Using only central sensors
   D. Relying on more than one feedback loop
   E. Increasing the power of the compensator
3. Indicate the one answer that correctly describes the relationship between physiologic adaptation and compensation.
   A. Adaptation is faster than compensation
   B. Adaptation is restricted to the same feedback loops as compensation
   C. Eskimos survive in their native environment because of adaptation and not compensation
   D. Unlike compensation, adaptation may lead to changes in the physiologic steady-state levels
   E. The process of adaptation is reversed, just as compensation stops, as soon as the inducing stress is removed
4. Which of the following disease states may be

associated with normally closed feedback loops?

A. Diabetes insipidus
B. Insulin-resistant diabetes mellitus
C. Addison's disease
D. Hypothermia
E. Neoplasia

5. The values of which of the following represent equilibrium rather than steady-state?

A. Blood volume
B. Blood pressure
C. Plasma osmolarity
D. Plasma pH
E. Skin temperature in cold water

ANNOTATED ANSWERS

1. D. The error signal is transmitted to the compensators, not directly into the system itself.

2. D. Increasing the speed and sensitivity of the loop or output of the compensator would not necessarily prevent overshoot, and the oscillation should be little affected by the location of the sensors or integrator.

3. D. Adaptation is slower to develop than compensation and may involve different mechanisms, as, for instance, with the thyroid in cold environments. Different loops may therefore be involved in the two processes. The normally high skin temperature of the Eskimoes shows that, to a significant extent, their adaptation to the cold is cultural and behavioral, rather than physiologic. The maintenance of normal oxygen supplies to tissue at high altitudes requires a raised steady-state value of the hematocrit. Adaptation is often noted long after the initial stimulus has been removed.

4. D. In hypothermia, the controlling feedback loops may normally be closed, but the capacity of the compensatory mechanisms to respond to the stress is exceeded. In diabetes insipidus, the secretion of antidiuretic hormone is depressed; in insulin-resistant diabetes mellitus, insulin antibodies may interfere with the normal function of the hormone; in Addison's disease, the salt-conserving steroids are no longer available. Neoplasia implies uncontrolled growth.

5. E. At maximum vasoconstriction, and when in contact with a good conductor of heat like water, the skin temperature will tend to equilibrate with that of water.

BIBLIOGRAPHY

Mitchell, D., Snellen, J. W., Atkins, A. R. Thermoregulation during fever: change of set-point or change of gain. *Pflugers Arch.* 321:293-302, 1970.

Samueloff, S., and Yousef, M. K. *Adaptive Physiology to Stressful Environments.* Boca Raton, FL: CRC Press, 1987.

59 Heat Control and Temperature Regulation

ERNEST C. FOULKES

Objectives

After reading this chapter, you should be able to:

Apply the concept of feedback control to the field of temperature regulation

Discuss temperature regulation in terms of energy balance, resulting from the input of food energy and output of external work, in addition to heat exchange

Define body temperature

Describe heat gain and loss by the body through conduction, convection, and radiation

Explain the importance of evaporative heat loss

Define the basal metabolic rate

Explain the principle of indirect calorimetry

Describe the compensatory mechanisms by which the body responds to heat and cold

Explain the importance of circulation in temperature control

Define fever and distinguish it from physiologic hyperthermia

---

The preceding chapter dealt with the properties of physiologic feedback loops and used temperature regulation to illustrate this process. This chapter goes into greater detail concerning temperature regulation. Because temperature reflects the heat content of a system, and heat is a form of energy, maintenance of a constant body temperature therefore represents a special aspect of the wider topic of energy balance.

The commonly used unit of heat energy is the **calorie,** which is the energy required to raise one gram of water 1°C. (Strictly speaking, the 1°C increment refers to a rise of from 14.5 to 15.5°C, which is approximately the point of highest density of water.) For physiologic purposes the large calorie, or **kilocalorie** (Calorie), where 1 Calorie equals 1,000 calories, is usually employed. Because units of energy and work are freely interchangeable, the energy content of food and work output may also be expressed in Calories. Work can also be measured in units that are based on a standard force acting over a standard distance. The equivalence of work and heat was first appreciated by Rumford 200 years ago, when he realized that the heat produced during the boring of gun barrels is proportional to the number of revolutions of the drill. Subsequently the **work equivalent of heat** (Joule's equivalent) was determined to be 4.18 joules per calorie, where the joule or newton·meter is the work done when a **force** of one newton acts over a **distance** of one meter. Finally, **power** is calculated as work/time, and measures the rate at which work can be performed; for instance, one watt (power) equals one joule (work) per second.

HEAT TRANSFER

Direct transfer of heat energy involves three processes: **conduction, convection,** and **radiation.** Heat can also be transferred indirectly, for instance as a result of physicochemical changes such as evaporation of water. Evaporation of water at body temperature requires 0.58 Cal per milliliter, as will be further discussed in relation to insensible heat loss from the body and sweating.

Heat transfer through a medium depends on its **conductivity.** Metals are better heat conductors than is water, which in turn conducts heat more efficiently than does air. Some representative values for the heat conductivity of different materials are listed in Table 59-1.

The **conductivity of air** varies with its water content and with density and pressure. Thus, severe heat loss through conduction may be encountered in the humid and high-pressure atmosphere of a diving bell or an underwater habitation. Because of its high air content, snow is a poorer heat conductor than water. The relatively high conductivity of water also accounts for the fact that an unprotected person can survive for only a very brief time in water that is near the freezing point.

The inverse of heat conductance is **insulation or resistance to heat flow.** The insulating properties of clothing, for instance, are expressed as 1/conductivity, in units of degrees centigrade per square meter per minute per Cal. In other words,

Table 59-1. Heat Conductivities

Substance	Cal/m²/1°C/min
Silver	600
Water	0.90
Snow (compacted)	0.30
Wood	0.18
Dry air at sea level	0.03

this measures the temperature gradient that the material can maintain for a given rate of heat flow.

Convection in the physiologic context is defined as the transfer of heat to moving air or water. Thus, heat loss from the body to the air is increased when air warmed at the skin is replaced by cold air. The chilling effects of cold air are accentuated by wind, creating the wind chill factor. Conversely, a wet diving suit restricts the movement of water in contact with the skin, and in this manner helps conserve heat.

Heat transfer by **radiation** describes an electromagnetic process that is independent of the intervening medium. When the skin temperature is less than the temperature of a distant source, such as a radiator or the sun, heat is gained by radiation. Inversely, heat may be lost to a distant sink, such as a cold window or the night sky.

Heat transfer by conduction, convection, or radiation can be modified by protective clothing that insulates (reduces conductance), minimizes convection, or reflects radiation. In the desert, heat gain by radiation from sun and sand can best be minimized by covering the skin with thin white clothing. At the other temperature extreme, the Eskimo in the Arctic wears highly insulating clothing to reduce conductive heat loss. As a result, a skin temperature is maintained that is little different from that seen in inhabitants of temperate regions, but this relatively high skin temperature is due to behavioral and not physiologic adaptation to the cold. The amount of added clothing that can be worn to resist extreme cold is limited, however, by its weight and bulk. In addition, the insensible perspiration from the skin, if trapped in the clothing, will increase heat conductance. Together with the respiratory heat loss (which will be discussed), this fixes at about −70°C, the maximum cold exposure humans can survive without help from outside sources of heat.

Another illustration of the role of convection and conduction in heat loss is the use of diving suits for protecting the diver against cold water. The dry suit excludes water from contact with the skin and traps low-conductance air; additional protective clothing may be worn to further lower the heat conduction between skin and water. In contrast, the wet suit traps water next to the skin but prevents its circulation; the water is warmed through contact with the skin. The low conductance of the diving suit, with its air pockets, minimizes heat loss from the trapped to the surrounding water. High pressures, however, compress these air pockets and thus reduce the insulating properties of wet suits. Additional insulating clothing can be worn under a dry suit to counteract the effects of high pressure.

BODY TEMPERATURE

The **heat content** of the body is reflected by its temperature. Although the mean body temperature is normally considered to be 37°C, there are actually significant variations, not only among different parts of the body, but also in any given region at different times. Because the skin is the main heat exchange organ of the human body, its temperature is determined by both the temperature and humidity of the environment, and by the needs of the body to conserve or dissipate heat. The temperature of the hands and feet may be several degrees lower than that in the core. Venous blood draining actively metabolizing tissues is likely to be warmer than pulmonary venous blood after it has undergone evaporative cooling in the lungs. The right heart is therefore likely to be warmer than the left heart. Similarly, rectal temperature exceeds oral temperature and, in fact, provides the most convenient approximation to mean core temperature. More accurate values, if required, can be obtained with a probe in the esophagus, or by measurement of urine temperature during voiding.

The **core temperature** in warm-blooded animals (homeotherms) is also influenced by intrinsic factors such as the menstrual cycle, or by diurnal and seasonal cycles. For instance, metabolic rate, and therefore body temperature, tend to fall somewhat during sleep and then rise again during the day. Major physiologic temperature shifts occur in some mammals during hibernation. Even reptiles and other normally cold-blooded animals (poikilotherms) can raise and maintain their body temperature above ambient levels in response to physiologic demands; certain snakes, for instance, incubate their eggs. The differences between homeotherms and poikilotherms are therefore only relative.

ENERGY BALANCE

Figure 59-1 is a modification of Figure 58-1, and illustrates the energy balance of the body with specific reference to heat energy and its control. The body gains or loses energy primarily by the processes indicated by *Arrows 1* to *4*. Heat transfer from the environment by conduction or radiation is described by *Arrow 1*. *Arrow 2* represents heat loss from the body to the environment by conduction and radiation, as well as through evaporation across the skin or in respiratory passages. Even under resting conditions, some water is lost through so-called **insensible perspiration.** Of particular importance is the heat lost from the respiratory passages. This is a necessary process because relatively cool and dry inspired air must be warmed and saturated with water at 37°C. Although some of the respiratory heat loss may be recovered upon exhalation by heat exchange at the external nares, the net respiratory heat loss may be very significant. This is especially true at high altitudes where low atmospheric oxygen pressure leads to the hyperventilation of cold and dry air.

Evaporative heat loss is also increased when the body responds to a heat load by secreting sweat. The ability to evaporate sweat is of course reduced at high humidities, so that relatively little heat can be dissipated from the body in a tropical jungle. While dry sauna temperatures of 100°C or higher may therefore be tolerated for lengthy periods, a wet sauna temperature above 37°C unavoidably leads to hyperthermia and possible heat stroke. The nature of the sweating response will be considered later in this chapter.

Arrow 3 in Figure 59-1 represents the major pathway of energy gain, with food the obvious fuel of the body. Table 59-2 summarizes the heat content of different classes of food. For several reasons, the total heat content that is physiologically

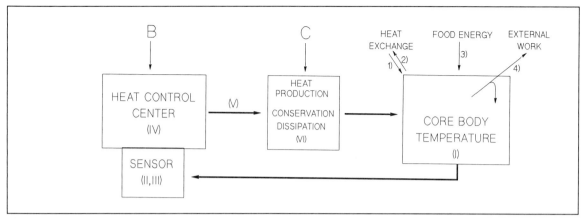

Fig. 59-1. Feedback loop of heat balance. The heat content of the body, as measured by its core temperature, is to be maintained at steady-state. No separate feedback message (*III*) is required because the main sensor (*II*) seems to be in the integrator or heat control center (*IV*). The error message (*V*) stimulates the appropriate compensatory mechanisms (*VI*). Arrows *1* to *4* represent energy input and output, as indicated. Arrows *B* and *C* represent external influences. (Compare with Figure 58-1.)

Table 59-2. Heat Content of Various Types of Food

Food	Physiologically available (Cal/gm)	O₂ Consumed (liters/gm)	CO₂ Produced (liters/gm)	RQ	Cal/liter O₂
Fat	9.0	1.96	1.39	0.71	4.6
Protein	4.0	0.94	0.75	0.80	4.2
Carbohydrate	4.0	0.81	0.81	1.00	4.9
Standard average				0.82	4.8

RQ = respiratory quotient, i.e., CO_2 produced/O_2 consumed.

available is less than that liberated by the combustion of these foods in a calorimeter. For instance, one of the end-products of protein metabolism is urea, whose chemical energy is lost to the body. The net energy gained by food intake may also be reduced if the food stimulates metabolism; the classic instance is the so-called **specific dynamic action** of proteins or amino acids.

The **minimal energy input** required to maintain steady-state is the **basal level,** which is described further in the section "Basal Metabolic Rate." To maintain this basal level in a patient on parenteral feeding may therefore require infusion of around 1,600 Cal per day. This corresponds to the heat content of 400 gm of glucose or of a fifth of 80-proof alcohol, neither of which can be infused without overloading the kidneys and other organs. Procedures have been worked out, however, that not only maintain patients at steady-state but also permit the normal growth of children by infusing fat suspensions in concentrated glucose and amino acid solutions into a central vein.

Energy is also lost from the body as **external work.** The quantitative relationship between food intake and work output can be illustrated by the calculation that walking for 1 hour at a rate of 3 miles per hour consumes 140 Cal. Given a heat content of alcohol of 7 Cal/gm, the one-hour exercise corresponds to the energy content of a glass of whiskey. Of course, increased muscle tone will raise the basal metabolic rate (BMR), thus providing an additional source of energy loss. In any event, prolonged exercise is required to compensate for excess caloric intake, and it is easy to see why the intake of more than 2,500 Cal per day in an adult with a sedentary life-style may lead to obesity.

The performance of, for example, 100 Cal of external work is associated with the production of approximately 400 Cal of heat, as indicated by the split *arrow 4* in Figure 59-1. In other words, the mechanical efficiency of the body only amounts to around 20 percent; the remaining 80 percent of the metabolic energy consumed is liberated as body heat. **Heat dissipation** during exercise can therefore limit the maximum work output. This is especially true of humans who, unlike the camel and some other species, cannot tolerate much of a rise in body temperature and thus cannot store appreciable amounts of heat. The camel, in contrast, accumulates heat during the day, readily tolerating a rise in its core temperature of 5°C. This heat is then dissipated by conduction during the cool of the night, without necessitating loss of water for evaporative cooling.

HEAT PRODUCTION AND INDIRECT CALORIMETRY

Zero external work is done under basal conditions, so that all the basic caloric expenditure is ultimately liberated in the body as heat. The same is essentially true of shivering, which may be regarded as a practically 100 percent efficient heat producer in response to cold stress. Additional details on shivering (the myogenic cold response) and nonmyogenic heat production are given in the section "Response to Cold."

The major source of heat is the **metabolic breakdown of food** (see Fig. 59-1; *process 3*); this also furnishes the energy needed for doing work. The physiologic combustion of one gram of each type of food consumes a characteristic number of liters of O_2, as shown in Table 59-2. At the same time, it liberates the stated volume of CO_2, and the ratio of CO_2 produced to O_2 consumed defines the respiratory quotient. A diabetic patient, because of the preferential utilization of fatty acids rather than glucose, is likely to exhibit a lower respiratory quotient than normal.

As shown in Table 59-2, a subject primarily using glucose as a source of energy will have a respiratory quotient of close to 1.00, and will accordingly liberate 4.8 Cal for each liter of O_2 consumed. This observation provides the basis for **indirect calorimetry.** The total heat output of the body is most accurately measured in a whole-body calorimeter, but this is not a convenient procedure and is not easily applicable in a work or exercise setting. In indirect calorimetry, expired air is collected in a bag for subsequent analysis, and the estimated number of liters of O_2 consumed is multiplied by 4.8. If the respiratory quotient is also determined, and is well below 1.00, then a value somewhat lower than 4.8 must be used in order to calculate the caloric yield per liter of O_2.

Indirect calorimetry can also be used to determine the heat output of a single organ. For instance, calories produced by the brain can be calculated, based on the fact that the metabolic substrate is primarily glucose. If O_2 consumption is determined as the product of blood flow and the arteriovenous difference in O_2 concentration, caloric consumption is yielded by the formula: $4.8 \times$ blood flow $\times$ O_2 concentration difference.

One disadvantage of indirect calorimetry is that it ignores the **O_2 debt,** which is the transient accumulation of lactic acid and other incomplete combustion products during exercise when glucose breakdown exceeds its oxidation to CO_2 and water. During hard and prolonged exercise, the O_2 debt will reach a steady-state value, at which time measurement of O_2 consumption will yield the correct value of caloric output. Indirect calorimetry performed on subjects engaged in short and violent exercise, such as a 100-yard dash, must be adjusted for the O_2 debt by continuing gas collection until respiration returns to normal.

However, return to normal may be delayed if metabolism is stimulated postexercise, independently of O_2 debt.

THE BASAL METABOLIC RATE

Even at rest, the body requires a significant expenditure of energy for the maintenance of homeostasis: blood pressure and muscle tone must be maintained, respiration continued, and ionic gradients restored. In addition, the body temperature must not fall much below 37°C. This basal requirement defines the BMR.

To assure basal conditions for the **measurement of BMR**, the subject must be awake, fasting, and resting horizontally, so that the need to pump blood against gravity is minimized. The ambient temperature must be neutral, as this is the temperature at which compensatory activity is minimal (see Fig. 58-4).

Under all these conditions, the BMR of a 70-kg male is around 80 Cal per hour, corresponding approximately to the heat output of a 100-watt bulb. The surface area of this average male is 1.73 m². Because heat exchange primarily involves the skin, the BMR is often expressed as 50 Cal/m² of body surface/hour, referred to as *1 MET*. During sleep, caloric requirements fall below basal levels, so that the daily basal requirement is 1,600 Cal, or somewhat less than that yielded strictly by multiplying 80 Cal per hour by 24 hours. Homeostasis, even under basal conditions, cannot be maintained on less than 1,600 Cal/day, as is made dramatically obvious in pictures of famine victims.

A variety of factors determine the BMR. For instance, the value for females generally is somewhat less than that of males, because females have a smaller muscle mass relative to their body surface area, and because of better insulation. Reduced muscle tone, as occurs in the absence of gravity or in advanced age, also lowers the BMR. Finally, changes in hormonal activity, especially of the thyroid gland, alter the BMR.

HEAT SENSORS AND INTEGRATION

Sensors that detect changes in body temperature are located in both peripheral areas and in the central nervous system. The chief sensory activity, however, and the integration of all the various feedback loops involved in the maintenance of a constant body temperature, takes place in the **preoptic hypothalamus.** This conclusion is based on the type of evidence illustrated in Figure 59-2. Among other findings, the suggested role of the hypothalamus in temperature control is supported by the existence of both heat-sensitive and cold-sensitive neurons in this area.

The data depicted in Figure 59-2 were obtained in a dog. The animal's initial rectal temperature was 37°C and, while resting in a room at 22°C, its metabolic rate as measured by O_2 consumption was close to basal. At *point A*, the temperature of the hypothalamus was raised, which can be achieved by perfusing the frontal sinuses or ears with warm water. The rectal temperature began to fall even though the room temperature was not lowered, and body temperature fell further when the ambient temperature was allowed to drop at *point B*. The continued warming of the hypothalamus in this situation is the equivalent of holding a candle under the thermostat in an otherwise cold room: no compensatory heat production or adequate heat conservation takes place. It was only at *point C*, when the hypothalamic or ear perfusion was stopped, that the dog's hypothalamus could sense the core body temperature and the signal went out to the compensating mechanisms. Metabolic heat production rose sharply, and body temperature began to return to normal. Clearly, it was the hypothalamic cooling that triggered this response. Even though central heat control resides in the hypothalamus, this does not exclude cortical influences from contributing to temperature homeostasis. For instance, the subjective feeling of cold may provide the motivation for warm-up exercises.

Spinal and peripheral sensors also participate in temperature control. A good example of the

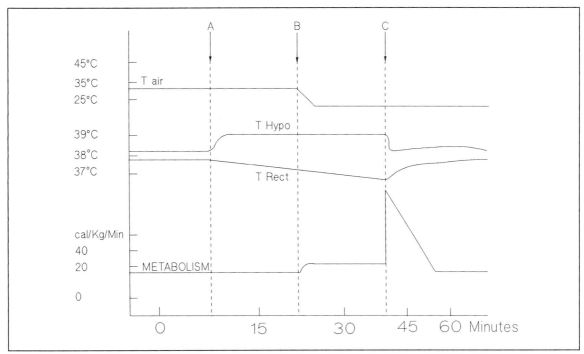

Fig. 59-2. Body temperature and heat production in a resting dog. The hypothalamic temperature was raised between points *A* and *C*; at *B* the room temperature was lowered. (Temperatures are: T-air = ambient; T-hypo = hypothalamic; *T-rect* = rectal, as a measure of core temperature.) (Based on: Hensel, H. *Fed. Proc.* 22:1156, 1963).

action of a peripheral mechanism is the reduced sweating rate observed when the forearm of a hot individual is cooled, even though the central heat drive from the hypothalamus has not been altered.

RESPONSE TO HEAT

Sweating

Secretion of sweat to reduce body temperature is the function of the two to three million eccrine sweat glands located over most of the body surface. Under extreme heat stress, they can produce sweat at a rate as high as 2 liters per hour. In composition sweat resembles a dilute ultrafiltrate of plasma. Unless salt and water loss are therefore replenished, high sweating rates may lead to cir-

culatory failure. The evaporation of one milliliter of sweat requires the expenditure of 0.58 Cal (the heat of evaporation of water). When this evaporation is diminished in a highly humid environment, a major source of heat dissipation is lost, so that humans cannot maintain a normal body temperature at an environmental temperature of 37°C at 100 percent relative humidity.

Sweat is first secreted as an isosmotic fluid into the proximal portion of the sweat duct, and salt reabsorption in the diluting segment of the duct produces the final form of sweat. The systemic administration of **cholinergic drugs** provokes sweating, and thermoregulatory sweating is abolished by **atropine,** showing that the postganglionic fibers innervating the sweat glands are cholinergic. However, sweating can also be stimulated by adrenergic agonists.

Vasodilatation and Cardiac Output

The evaporation of sweat provides only local cooling, and it is the **peripheral blood flow** that determines how much heat can be carried from the core to be dissipated on the body surface. If sweating has cooled the skin to 32°C, each liter of blood from a core temperature of 37°C can lose 5 Cal upon equilibration with skin temperature. In other words, the amount of heat lost depends on the evaporative cooling of the skin and the blood flow to the skin. A piece of steak, which has the same composition as human tissue, can evaporate as much water from its surface as can human skin, but, in the absence of circulation, it will be cooked at a temperature that humans can survive for some time.

An early response to heat exposure is therefore **decreased peripheral** (skin) **resistance** to blood flow. This **vasodilatation** permits increased peripheral blood flow, especially as cardiac output also begins to rise. This is illustrated in Table 59-3 for heat production at rest, as well as during strenuous or maximal exercise. For the purposes here, it will be assumed that the core temperature is 37°C and the skin temperature is cooled by heat conduction and sweat evaporation down to 32°C. The core-to-skin temperature gradient would therefore equal 5°C, such that each liter of blood from the core loses 5 Cal at the periphery. During strenuous exercise, peripheral blood flow can

thus carry 1.9 × 5 × 60, or 570 Cal per hour, to the skin for dissipation. This more than compensates for the heat production of 360 Cal per hour. If, however, work output is raised to maximal levels, with a caloric output of 600 Cal per hour, the blood flow to skeletal (and cardiac) muscle must take precedence over that to some other areas of the body, including the skin. The blood flow to the brain remains constant, but that to mesentery and kidneys is greatly decreased. These conditions initiate peripheral vasoconstriction, such that the peripheral resistance to blood flow increases. The total peripheral blood flow in this example can now dissipate only 0.6 × 5 × 60, or 180 Cal per hour, in the face of a caloric output of 600 Cal, and body temperature will rise rapidly to the point of exhaustion. These calculations ignore the contribution of respiratory heat loss to total heat dissipation, which is an important factor, especially at high rates of ventilation. Nevertheless, this example does illustrate how the ability to dissipate heat may limit the maximal work output.

Adaptation

As discussed in Chapter 58, humans are capable of some adaptation to extreme environments. This is further illustrated here for heat adaptation, as summarized in Table 59-4. The table shows that the skin temperature after adaptation

Table 59-3. Role of Circulation in Heat Control during Exercise

Variables	Rest	Strenuous Exercise	Maximal Exercise
O$_2$ consumption (liters/min)	0.14	1.20	2.00
Heat production (Cal/min)	0.7	4.8	8.0
Cardiac output (liters/min)	5.8	17.5	25.0
Skeletal muscle			
Blood flow (liters/min)	1.2	12.5	22.0
Vascular resistance (% of normal)	100	12.0	8.0
Skin			
Blood flow (liters/min)	0.5	1.9	0.6
Vascular resistance (% of normal)	100	40.0	115

Table 59-4. Adaptation to a Standard Workload in the Heat*

Variables	Before Adaptation	After Adaptation
Metabolism, from Q (Cal/hr)	+565	+520
Work output (Cal/hr)	− 50	− 50
Radiation gain (Cal/hr)	+ 20	+ 30
Conduction gain (Cal/hr)	+ 30	+ 40
Required net heat dissipation (Cal/hr)	565	540
Skin temperature (°C)	37.0	35.6
Core temperature (°C)	39.5	38.5
Gradient (°C)		
Environment → skin	3.0	4.4
Core → skin	2.5	2.9
Blood flow required to carry heat to skin (liters/min)	$565/(60 \times 2.5) = 3.7$	$540/(60 \times 2.9) = 3.1$

*Healthy young men performed 50 Cal of external work per hour at an ambient temperature of 40°C. Results are based on: Belding, H.S., In Yousef, M.K., ed. *Physiological Adaptation.* Academic Press, New York: 1972.

was lower than that before adaptation; this reflects the comparatively efficient sweating response to heat stress. Because of sweating, the temperature gradient from the environment to the skin rises, creating additional heat gain by radiation and conduction. At the same time, the gradient from the core to the skin also increases, so that each liter of blood conveys more heat from the core for dissipation at the periphery. Before adaptation, blood flow could only discharge all the heat accumulated in the core when the core-to-skin gradient reached 2.5°C. This corresponds to a core temperature of 39.5°C, a level associated with rapidly developing exhaustion. Under the same conditions, but after adaptation, the core temperature rises only to 38.5°C. This, together with the smaller cardiac output needed, explains how adaptation increases the efficiency of work in a hot environment.

RESPONSE TO COLD

Humans possess three main physiologic mechanisms for responding to cold: **shivering, nonmyogenic heat production,** and **vasoconstriction.** If the core body temperature in the cold declines by more than about 2°C, despite the maximal activity of these three compensators, hypothermia eventuates, with loss of coordination, inappropriate responses such as vasodilation, and inability to prevent further heat loss.

Shivering

The muscular (or myogenic) response to cold in humans is the shivering reflex. It is a noncoordinated activity of skeletal muscle that achieves no outside work, and therefore represents a 100 percent efficient source of metabolic heat. However, when violent, it can interfere with normal motor movement and it may actually increase heat loss by convection because it agitates the surrounding medium.

Nonmyogenic Heat Production

In response to cold the body can also reflexly increase heat production in nonmuscular tissues, particularly in adipose tissue. This is triggered by catecholamine secretion; this response is perhaps less important in humans than, for example, in rodents, which maintain relatively large brown fat deposits into adult life.

Vasoconstriction

The third major reflex response to cold in humans is peripheral vasoconstriction. This lowers skin temperature, and therefore the conductive heat loss that is determined by the temperature gradient from the skin to environment. The heat conductance of fully vasoconstricted skin approximates 0.12 Cal/m²/1°C/min, similar to that of cork. Such extreme vasoconstriction is not uniform over the entire body. For instance, circulation to the hands and feet is little depressed in the cold, which helps minimize the pain of excessive cooling of the extremities, but at the price of greater heat loss.

Obviously vasodilating drugs such as alcohol also elevate heat loss. While the monks at the St. Bernard Pass may have transiently restored the morale of travelers lost in the snow by providing brandy in flasks tied around the necks of their dogs, they may also have caused the victims to lose heat at a faster rate. Alcohol consumption in someone who is hypothermic is absolutely contraindicated until the victim has been returned to warm surroundings.

FEVER AND HYPERTHERMIA

Fever occurs when the steady-state core temperature of the body is raised above normal levels, either because the set-point is reset or the gain of the system is altered (see Chapter 58). Fever implies hyperthermia, but not all cases of hyperthermia constitute fever. This important distinction is predicated on the fact that the heat-producing or conserving mechanisms in fever are promoting an increased body temperature; whereas during exercise-induced hyperthermia, for instance, the cooling mechanisms are striving to return the body temperature to its normal steady-state. This is illustrated in Table 59-5, which compares data from two hyperthermic individuals, one who has just completed a 440-yard sprint and the other who suffers from fever. Both

Table 59-5. Thermal Response during Fever and Physiologic Hyperthermia*

Variables	Exercise Hyperthermia	Fever
Body temperature		
Actual	39	39
Steady-state	37	39
Environmental temperature		
Actual	31	31
Neutral	31	33
Responses		
Sweating	+	0
Vasoconstriction	0	+
Reflex heat production	0	+

*The two individuals were resting at 31°C. All temperatures are given in degrees Celsius. + = increased; 0 = no response.

are now resting at a normal neutral environment of 31°C. This is cooler than the neutral temperature for the fever patient, so this person reflexly produces heat by shivering. This reduces heat loss by vasoconstriction and the sweat response is also stopped. In the other person with exercise-induced hyperthermia, heat loss is increased through evaporation and vasodilatation, and extra heat production is minimized. Only when the fever "breaks" does the patient begin to sweat in order to restore the body temperature to normal.

The fever response can be observed in quite primitive animals, but its pathologic significance and survival value remain unclear. Fever usually results from the action of **endogenous pyrogens** on the hypothalamic heat control center. Different pyrogens are produced by a variety of cells in the body, under the influence of bacterial products and other compounds (the exogenous pyrogens). Antipyretic drugs such as aspirin interfere with the response to endogenous pyrogens by inhibiting cyclooxygenase activity. This enzyme catalyzes the synthesis of prostaglandins, which are involved in the hypothalamic response to pyrogens.

SUMMARY

The specific negative-feedback loops controlling temperature are described. They regulate the exchange of heat energy between the body and the environment in such a way that a stable core temperature can be maintained with the aid of well-defined compensatory mechanisms. The discussion therefore centers on heat gain yielded by metabolism or by energy uptake from the environment as well as on heat loss to the environment. The mechanisms that permit the body to increase net heat production in the cold and minimize it in the heat are also considered. Increased efficiency of the compensatory processes may elicit adaptation to a given stress. Neutral temperature is defined as the ambient temperature at which compensatory activity is minimal, and the resting metabolism at that temperature defines the BMR. The steady-state value of the deep body (core) temperature is controlled primarily by a heat loss center in the hypothalamus. When the set-point of this center (or the gain of the system) is raised, a higher steady-state body temperature is achieved, as in fever. Fever must be clearly distinguished from physiologic hyperthermia, which is an increase in body temperature resulting from an imbalance between heat loss and heat gain, not a change in the set-point or gain of the system.

NATIONAL BOARD TYPE QUESTIONS

Select the one most correct answer for each of the following

1. The neutral temperature defines:
   A. The point of minimal BMR
   B. An invariable temperature characteristic for each individual
   C. The optimal temperature for doing heavy work
   D. The point where resting O_2 consumption is minimal
   E. The point where heat production equals heat loss

2. Temperature homeostasis:
   A. Depends on a balance between caloric input and output
   B. Implies that deviations from normal are always corrected
   C. Does not require continuous regulatory activity
   D. Leads to a constant mean core temperature
   E. Does not require expenditure of energy

3. Which of the following statements about heat transfer is correct?
   A. Heat exchange by radiation between two points depends on the temperature of the intervening space
   B. Shivering does not affect convective heat loss
   C. The heat conductance of the skin limits the heat loss from the body during moderately severe exercise
   D. More heat is lost by conduction through air inside a diving bell deep under water than at the surface
   E. The heat conductance of snow exceeds that of water

4. Why can humans survive at a temperature that would cook a piece of steak?
   A. The water evaporation from outside surfaces in humans is greater than that of steak
   B. Core heat load cannot be dissipated by the steak
   C. The heat conductivity of the steak is greater than that of humans
   D. The water content of steak is less than that in humans
   E. There is more vasodilation in humans

5. Which of the following statements about BMR is correct?
   A. It is lower during sleep than wakefulness
   B. It is lower before than after a meal
   C. It is lower in the heat than in the cold
   D. It varies with gender, when expressed as Cal/kg body weight/minute
   E. For a 70-kg man it approximates 50 Cal per hour

ANNOTATED ANSWERS

1. D. There is a minimum metabolic rate, and therefore O_2 consumption, but the definition of BMR excludes the existence of maximal or minimal values. During work, the extra heat produced can be lost by passive conduction more efficiently at lower than at higher temperatures, without the need for evaporative heat loss. The neutral temperature is not invariable; for example, it is raised under the influence of pyrogens. At any steady-state, heat production equals heat loss.

2. A. By definition, homeostasis may involve an abnormal steady-state, as in fever. The mean core temperature changes during sleep–wake cycles, menstrual cycle, and other such circumstances. The maintenance of any steady-state requires continuous regulation, and therefore energy.

3. D. Heat conductance of air at high pressure exceeds that at low pressure. Radiation is a function only of the temperature difference between the source and sink, not the temperature of the intervening space. During moderately severe exercise, peripheral vessels are fully dilated and skin conductance is therefore high and not limiting. Shivering causes air movement. Air spaces in snow reduce its ability to conduct heat.

4. B. The composition of steak is the same as that of human muscle tissue, so that both water evaporation and heat conduction should be the same on the surface. Absence of circulation, however, limits core heat loss by steak. As steak cannot vasoconstrict, both humans and steak will be fully vasodilated.

5. D. Muscle, with its resting tone, has a relatively high resting metabolic rate, and muscle constitutes a greater proportion of total body mass in males than in females. The first three statements are meaningless, as the definition of BMR mandates a fasting condition, neutral temperature, and wakefulness. The BMR for a 70-kg man is approximately 80 Cal per hour (or 50 Cal/m²/hr).

BIBLIOGRAPHY

Benzinger, T. H. The human thermostat. *Sci. Am.* 204:134–147, 1961.

Dinarello, C. A., and Wolff, S. M. Pathogenesis of fever in man. *N. Engl. J. Med.* 298:607–612, 1978.

Dudrick, S. J., and Rhoads, J. E. New horizons for intravenous feeding. *J.A.M.A.* 215:939–949, 1971.

Quinton, P. M. Physiology of sweat secretion. *Kid. Int.* 32: (suppl.21);S102–S108, 1987.

60 Exercise Physiology

BARBARA N. CAMPAIGNE

Objectives

After reading this chapter, you should be able to:

Name the primary physiologic systems involved in exercise

Identify the three energy systems brought into use during exercise

Describe the determinants of O_2 use by the exercising muscle

List the components and discuss the function of the respiratory response to exercise

Name the three primary forms of input to the control of respiration during exercise

Identify the major component of the circulatory system that limits maximal O_2 consumption

Describe the neuroendocrine response to exercise

Outline and discuss the adaptation of the systems attributed to physical training

---

The coordinated physiological processes which maintain most of the steady states in the organism are so complex and so peculiar to living beings—involving, as they may, the brain and nerves, the heart, lungs, kidney and spleen, all working cooperatively—that I have suggested a special designation for these states, homeostasis.

Walter B. Cannon, 1932

While homeostasis is the maintenance of the internal environment at a constant level, exercise physiology can be considered the integration of all physiologic systems and processes involved in preserving a constant internal environment during exercise. This involves maintaining an equilibrium or balance in the internal environment. **Equilibrium** is usually maintained at rest and under steady-state conditions, and physiologic **steady-state** is reached when supply meets demand. An example of this is the delivery and use of O_2 and metabolic substrates to meet the de-

mand of working muscle during exercise. In addition, steady-state involves the maintenance of a relatively stable body temperature, pulse rate, blood pressure, and respiratory rate for a given period (see Chapter 58). This represents a very complex series of interactions because of the many systems and processes involved. However, if one views the systems individually, the total picture becomes increasingly clear. This chapter will examine independently several of the systems involved in the ultimate goal of deciphering the overall response to acute exercise and the adaptation to physical training. This chapter is not all inclusive, but will cover such aspects as the metabolic basis of the exercise response, involvement of the respiratory and circulatory systems, and a short section on the neuroendocrine response to exercise. There are many potential and unclear effects of exercise that are not described in this brief chapter. A few of these include sleep, the immune response, setting or resetting of appetite signals, gastrointestinal changes, and the psychoneuroendocrine response.

SYSTEMS INVOLVED

The **skeletomuscular systems** are the primary ones involved in **movement**. However, the respiratory, circulatory, and neuroendocrine systems also directly participate.

The **respiratory response** involves the lungs, as well as gas transport and exchange, and its regulation is a complex process. The **circulatory response** to exercise involves the heart and its function as a muscle pump, supplying the O_2 to the skeletal muscle, distributing other nutrients such as glucose, and dissipating heat. The hemodynamics of blood flow and vascular response are of prime importance during muscular activity. The **neural response** to exercise, including the organization of this response, will be described in this chapter as well. The **endocrine system** is intimately involved in fuel homeostasis, which is of major importance during exercise. The hormones involved in exercise also will be considered, along with their action. Finally, the **acute response** and the **long-term adaptation** to exercise will be compared. Short-term single bouts of muscular activity (acute effects) require immediate responses from the systems involved. When exercise is performed repeatedly over time at regular intervals, certain aspects of the physiologic systems adapt. This is the process of adaptation (chronic effects), and it will be discussed in the concluding portion of each section.

THE METABOLIC BASIS OF EXERCISE

The supply of biologic energy for physiologic activity, known as **bioenergetics,** is the basis for exercise. According to the **first law of thermodynamics,** energy can be neither created nor destroyed but only transferred from one form to another, or from one place to another. Accordingly, energy from the system, **cellular energy,**

is needed for exercise to take place, and, with the exhaustion of this energy, new energy has to be made available. Because the mechanical efficiency of muscle performing external work averages only around 20 percent (see Chapter 59), most of the energy used is released as heat that must be eliminated from the body.

During exercise, three different energy systems come into play. The **high-energy phosphate compounds** supply energy for the systems. Exercise that lasts only a few seconds uses several sources of immediate energy. During forceful exercise lasting from a few seconds to a minute, energy is primarily obtained from **glycolytic or nonoxidative sources.** The third energy system is **aerobic or oxidative,** and is used for muscle contraction that lasts two minutes or longer, referred to as **endurance.** Figure 60-1 illustrates the use of the three energy systems as a function of duration of activity (see also Fig. 13-11). As

Fig. 60-1. Logarithmic plot of average speed maintained versus time of event for men's world records in running. The presence of three curve components suggests the use of three energy systems. (Modified from McGilvery, R. W. *Biomedical Concepts.* Philadelphia: Saunders, 1975).

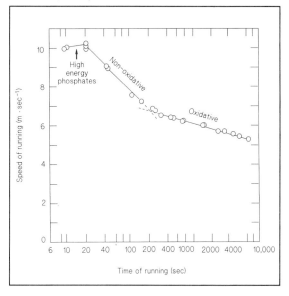

Figure 60-1 shows, no activity relies solely on one system; however, one will predominate during activity of a particular intensity and duration.

A review of the metabolic response in terms of substrate utilization is relevant to an understanding of exercise physiology. The **oxidation of fuel** is the primary pathway for energy generation and subsequent heat production during exercise. During sprinting or brief high-intensity exercise, glucose stored in the form of **muscle glycogen** is the major fuel source. Continuing exercise makes use of substrates carried in the blood in the form of glucose and free fatty acids. Prolonged exercise that lasts longer than two hours depletes glycogen stores, and free fatty acids become the dominant fuel source. The greatest source of potential energy in the body is fat. Fat reserves account for about 90,000 to 110,000 Cal of energy. In contrast, energy available in the form of carbohydrates constitutes less than 2,000 Cal. About 1,500 Cal (375 gm) of glycogen is stored intramuscularly; 400 Cal (100 gm) exists as liver glycogen, and 80 Cal (20 gm) of carbohydrate is available in the form of glucose in the extracellular fluids. **Protein** can serve as an important source of energy during prolonged exercise, such as marathons or ultraendurance events. However, it must be broken down to specific amino acids for energy release, and is thus a less efficient form of fuel.

Physical fitness is most often defined as an individual's ability to use O_2 during maximal exercise, given in liters per minute ($\dot{V}O_2$max). $\dot{V}O_2$max is determined by cardiac output (heart rate [HR] × stroke volume [SV]) and the arteriovenous O_2 difference [(a–v)O_2], expressed as milliliters per 100 ml of blood, as represented in the *Fick equation:* $\dot{V}O_2 = (HR \times SV) \times (a–v)O_2$. Often the intensity of exercise performed is defined as a percentage of $\dot{V}O_2$max. The effect of exercise intensity on fuel utilization has been studied, and it has been shown that at 50 percent of $\dot{V}O_2$max or less, glycogen accounts for less than 50 percent of the muscle substrate and only small amounts of blood glucose are used. When exercise is performed above 50 percent $\dot{V}O_2$max, carbohydrate use increases, leading to depletion of glycogen stores. At 70 to 80 percent $\dot{V}O_2$max, the muscle glycogen store is depleted at exhaustion, which occurs approximately 1½ to 2 hours after initiation. During exercise at extremely high intensity (90 to 100 percent $\dot{V}O_2$max), glycogen use is highest but depletion does not occur with exhaustion. At these high intensities, the intracellular pH and the buildup of metabolites, rather than fuel availability, appear to limit performance.

THE RESPIRATORY SYSTEM

Ventilation

Pulmonary ventilation requires work to expand the lung and overcome the resistance to movement of the lung tissue and the gas in the lung airways. The **work of breathing** can be obtained by calculating the area enclosed by the curve of the pressure–volume relationship. During exercise, both the respiratory rate and tidal volume increase, thus elevating the minute volume. There is an initial rise in tidal volume that may reduce the work performed by the lungs because of decreased flow resistance. At very high workloads with tidal volumes exceeding 50 percent of the vital capacity, the pulmonary work against elastic forces is heightened. Thus, during greatly increased workloads, breathing frequency continues to increase with no further change in the tidal volume. This markedly increases the work of breathing. During exercise, the anatomic dead space ventilation increases slightly, and, because tidal volume rises markedly, the ratio of the two decreases. This lower ratio represents a more efficient gas exchange brought about by an increase in the volume of air available to the alveoli for ventilation.

During the first few minutes of **submaximal exercise,** ventilation increases exponentially and remains nearly constant during the remainder of the exercise session. It has been demonstrated that the minute volume during a 2½-hour marathon is almost 70 percent of the maximal exercise ventilation.

Diffusion

Gas exchange between the alveoli and the pulmonary blood is known as the **diffusing capacity,** which is the volume of gas diffusing through the respiratory membrane at a pressure difference of 1 mm Hg each minute. During each breath, O_2 diffuses into the pulmonary capillaries through the alveolar membrane. Diffusion of O_2 through the alveolar membrane is directly proportional to the partial pressure difference of O_2 (Po_2) between the alveolus and capillary, and the cross-sectional area of the surface for gas exchange (see Chapter 31). With exercise, both the cross-sectional area and Po_2 gradient increase, permitting increased diffusion of O_2 across the alveolar–capillary membrane. The increased alveolar ventilation also produces a lower alveolar CO_2 and an increased diffusion pressure gradient for CO_2 from the capillaries into the alveolus and, therefore, into the atmosphere. An increase in the effective alveolar–capillary membrane area results from the opening of additional capillaries in the lung, as well as dilation of capillaries and alveoli (Fig. 60-2). Consequently, the capacity for diffusion increases almost threefold with exercise.

The pulmonary circulation (cardiac output) during exercise increases to match the increased ventilation. As blood flow exceeds the vascular resistance (dictated by the total pulmonary vascular cross-sectional area of arterioles), pulmonary arterial pressure rises. This augmented pulmonary blood flow decreases the amount of time the red blood cell (RBC) spends in the pulmonary capillaries. Under resting conditions, it takes the RBC about 0.75 second to travel through the pulmonary capillary; during heavy exercise the transit time is only 0.3 to 0.4 second. However, even though the actual time in the alveolar capillary is reduced, the equilibration of gases between the capillary blood and the alveolus can still be accomplished as proved by the fact that, during exercise, the arterial blood of healthy individuals remains fully oxygenated. The blood normally (at rest) stays in the lungs about three times longer than required for it to be fully oxy-

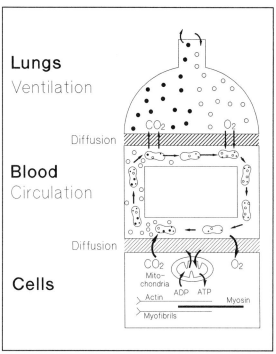

Fig. 60-2. The stages of oxygen uptake. (Modified from Frontera and Adams, Endurance exercise: normal physiology and limitations imposed by pathological processes [Part 1]. *The Physician and Sports Medicine* 14(8): 96, 1986).

genated. Figure 60-3 illustrates the ability of the lungs to perform, even under the increased demands of exercise.

Control

There are a minimum of three documented **primary afferent inputs** that control the output of the medullary inspiratory neurons. The first of these primary afferents are the **descending influences** from the central nervous system relative to locomotor activity. In a feed-forward manner, the motor cortex sends signals to the respiratory center to increase the rate and depth of breathing in a cognitive response to exercise. The second primary afferent consists of the **ascending influences** from exercising skeletal muscle. At the muscle level, afferents include the muscle spin-

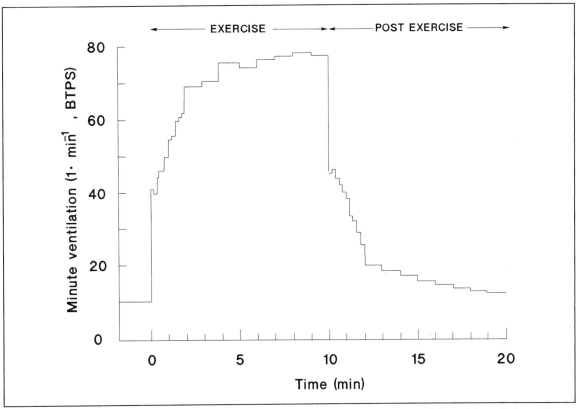

Fig. 60-3. The ventilatory response during and in recovery from steady-state exercise. (Adapted from: Dejours, P. Neurogenic factors in the control of ventilation during exercise. *Circ. Res.* XX, XXI(suppl 1):146–153, 1967.)

dle, Golgi tendon organs, and skeletal joint receptors. These send signals to the sensory cortex, which relays information to the respiratory center. The third primary afferent is the **humoral input** relative to the increasing production of CO_2. The central and peripheral chemoreceptors sense a change in the partial pressure of CO_2 (P_{CO_2}) and pH, and signal the respiratory center to increase respiration. The amount of neural input to inspiratory muscles is proportional to the rising metabolic demand. The mechanisms that control respiration during exercise are very effective. During all but very heavy exercise, the arterial P_{O_2}, P_{CO_2}, and pH remain almost the same. **Peripheral chemoreceptors** may also play a role in the ventilatory response to exercise. During exercise, arterial pH does decrease, and

arterial P_{O_2} is maintained or decreases slightly. It has been suggested that these factors stimulate the carotid bodies during exercise, which then send signals to the medullary respiratory center. The variation in arterial P_{O_2} occurring with exercise may increase the CO_2 and H^+ sensitivity of the carotid bodies.

During exercise, an important function of the respiratory control system is to minimize the effort of respiratory muscle. Although respiratory control mechanisms are poorly understood, there is evidence that, during exercise, the respiratory control system is "aware" of and responds to the mechanical needs of the chest wall. Examples of this mechanical proprioceptive feedback mechanism include: (1) **breathing frequency and tidal volume,** which are controlled by meeting a ventilatory demand sufficient to maintain tidal volume close to the linear portion of the pressure–volume relationship; that is, the tidal volume almost never exceeds 60 to 65 percent of vital capacity; (2) **activation of accessory respiratory muscles** (abductor muscles of the larynx, the parasternal intercostals, thoracic muscle, and expiratory abdominal muscle); (3) **normal impedance to air flow** in the lung, which occurs during exercise, must be accompanied by an inspiratory motor output that increases beyond the normal drive to breathe; this compensating inspiratory drive effort increases with increasing workload; and (4) **mechanical proprioceptive feedback** is far more effective than voluntary control of ventilation, which increases ventilatory work because of the extremely high pressures generated during voluntary expiration.

Adaptation to Physical Training

In contrast to the anatomic changes in the lung that accompany chronic stressful situations, such as hypoxia, the lung shows only very minor adaptation as the result of physical training. However, maximum ventilatory performance may be improved, such that a trained athlete can achieve and sustain a higher percent (75 percent) of the **maximal ventilatory volume** than can a nonathlete (68 percent). Athletes have also been shown to maintain a maximal ventilatory volume following exhaustive exercise, while nonathletes show reduced volumes under similar circumstances. These enhanced maximal ventilatory responses may result from specific respiratory muscle training. The major non-pulmonary-related effect of physical training on the pulmonary system is the reduction in ventilatory response to heavy, short-term, submaximal exercise. This is most likely induced by a decreased ventilatory drive, primarily attributable to decreased lactic acidosis, as well as to decreased levels of circulating catecholamines. During prolonged exercise, trained individuals also show enhanced **heat dissipation,** resulting in lower core body temperatures and this reduces the ventilatory re-

sponse. This is not seen in untrained individuals, though both trained and untrained individuals have elevated temperatures during exercise. It has not been established which humoral stimulus, if any, is responsible for increased ventilation during exercise. The decreased respiratory rate that occurs in trained athletes during submaximal exercise associated with physical training may reflect a decreased neural-induced recruitment of locomotor and respiratory muscles brought about by delayed muscle fatigue.

Pathophysiologic Limitations

Under pathologic conditions, such as chronic obstructive pulmonary disease, interstitial lung disease, and asthma, exercise capacity may be limited as a consequence of the pulmonary dysfunction. In such individuals exercise may be limited because of abnormal pulmonary mechanics, impaired gas exchange, ventilatory muscle fatigue, or a combination of these factors. These problems are manifested as abnormal exercise responses, including: dyspnea, arterial O_2 desaturation, and the development of respiratory acidosis during work. In fact, exercise can be used as a diagnostic procedure in the early stages of certain diseases. Pulmonary patients who have a low cardiovascular work capacity may be good candidates for exercise training programs that improve the physical work capacity.

THE CIRCULATORY RESPONSE TO EXERCISE

Muscle demand for metabolic fuel increases during exercise, and this demand is met by an increased supply of blood carrying necessary O_2 and metabolic substrates to the muscle. The changes in cardiovascular function that take place during exercise depend on both the type and intensity of exercise. **Dynamic-type activity** that involves the large muscle groups, such as running, swimming, and cycling, places the greatest demand on the circulatory system. Large increases occur in both the cardiac output (heart rate and stroke volume) and systolic blood pressure. The diastolic pressure tends to fall, so that the mean arterial pressure remains fairly constant. This indicates a match of the increase in cardiac output to the decrease in vascular resistance. **Static-type exercise,** such as weight lifting, which uses little muscle mass, causes moderate increases in cardiac output and heart rate, with accompanying increases in systolic, diastolic, and mean blood pressure caused by increases in sympathetic activity (norepinephrine release).

Cardiac Output and Heart Rate

The increase in cardiac output during exercise is partially mediated by a **decrease in the vascular resistance** combined with an **increase in the venous-filling pressure gradient.** The **sympathetic nervous system** also plays an important role in determining cardiac output during exercise. During **dynamic exercise,** the cardiac output increase is accomplished by increases in both heart rate and stroke volume. Heart rate rises in conjunction with O_2 consumption and workload. As $\dot{V}O_2$max is reached, cardiac output attains a plateau value in order to match lung perfusion to ventilation. The increase in cardiac output and heart rate depends on the type of exercise as well as on the individual's fitness, age, and sex. When **work output** is submaximal, cardiac output and heart rate rise and then plateau as O_2 transport requirements are met. **Exercise heart rate** is influenced by a host of **environmental factors,** such as anxiety (which also affects resting heart rate), dehydration, ambient temperature, and altitude. These environmental factors may not always be linked to changes in cardiac output, due to the inverse relationship between heart rate and stroke volume when "afterload" and "preload" factors are constant (non-exercise settings).

As previously mentioned, heart rate is not only affected by the intensity and duration of exercise, but also by the type of exercise performed. At the same percentage of $\dot{V}O_2$max, cycling elicits a higher heart rate than does treadmill running. This may be attributable to the workload placed on the muscles in relation to the size of the muscle group used. Another example of such a response is exhibited by work that involves only the arms, which causes a 10 percent greater heart rate response than does cycling at the same workload. Smaller muscle groups, such as the arms, seem to be used at a higher percentage of their maximum capacity during activity, than are the larger muscle groups, such as the legs during walking or running. In general, the heart rate is lower during exercises such as weight lifting. Thus, heart rate rises in response to the percentage of maximal contraction of the muscle mass involved as well as to the amount of muscle used.

The heart rate responds to the exercise changes during the **adaptive process** of physical training. In general, heart rate after physical training is lower than that in untrained subjects at any absolute submaximal workload. However, the heart rate response at a submaximal workload relative to the newly attained heart rate remains unchanged after training. During maximal exercise, heart rate is usually unchanged or somewhat lower in individuals who have undergone physical training than in those who have not.

Cardiac Output and Stroke Volume

Stroke volume appears to be the variable that responds most to the increased cardiac output demand during exercise. For example, the cardiac output of a trained athlete during maximal exercise may be as high as 38 liters per minute as compared to a sedentary individual whose maximal cardiac output is 23 liters per minute. Both groups may respond to a similar workload with a heart rate of 195 beats per minute but with different stroke volumes. The trained athlete will have a stroke volume on the order of 195 ml and the sedentary individual's will be 118 ml. One physiologic mechanism responsible for the greater stroke volume, both during exercise and at rest in trained athletes, is an **increased ventricular filling** during the diastolic phase of the cardiac cycle (due to a higher venous pressure gradient from the body periphery to the heart).

The enhanced end-diastolic volume stretches the fibers of the myocardium and this strengthens contraction, with a resulting more powerful ejection. This response is described as **Starling's law of the heart.** Another mechanism that elicits a greater stroke volume during exercise is the **increased systolic emptying** of the heart. This is due to the actions of epinephrine and norepinephrine, whose blood concentration increases during exercise and stimulates myocardial contractility with a resultant increased ejection of blood during the systolic phase of the heart. **Decreased systemic vascular impedance** also contributes to greater emptying of the left ventricle, as does direct **myocardial autonomic nervous system** stimulation.

Blood Flow and the Distribution of Cardiac Output

The distribution of cardiac output to different organs is in general directly related to and regulated by the **metabolic activity** of the tissue (Table 60-1). Under resting conditions in a comfortable environment, about 1 of the 5 liters of cardiac output perfuses skeletal muscle, while most of the remaining 4 liters is distributed to the digestive tract, spleen, brain, kidneys, and liver. During brief, intense exercise, the blood flow to the working muscles increases progressively with increasing intensity of work, while the fractional blood flow to the organs and skin decreases (see Table 60-1). Implications of this flow redistribution for heat dissipation have been discussed in Chapter 59.

The liver and kidneys utilize about 20 to 25 percent of their O_2 supply at rest, so that a reduced blood flow can be tolerated before function is compromised. The resulting relative decrease in blood supply (as a fraction of cardiac output) to the organs during high-intensity exercise can be tolerated for well over an hour. However, after prolonged submaximal exercise, reduced blood flow to the liver and kidneys compromises their function, and may account, in part, for the fatigue that eventuates.

O_2 Extraction Reserve

The heart uses about 75 percent of its O_2 supply at rest. Because of the greater demand on the heart during exercise, both the coronary blood flow and O_2 supply must increase. The four- to fivefold increase in cardiac output that occurs during exercise is accompanied by a similar increase in coronary circulation.

Table 60-1. Distribution of Cardiac Output Expressed as Blood Flow to Various Tissues, at Rest and during Light, Moderate, and Maximum Exercise

Tissue	Resting Blood Flow, ml-min (% of total cardiac output)	Exercise Blood Flow, ml-min (% of total cardiac output)		
		Light	Moderate	Maximum
Splanchnic	1,350 (27%)	1,100 (12%)	600 (3%)	300 (1%)
Renal	1,100 (22%)	900 (10%)	600 (3%)	250 (1%)
Cerebral	700 (14%)	750 (8%)	750 (4%)	750 (3%)
Coronary	150 (3%)	350 (4%)	750 (4%)	1,000 (4%)
Muscle	750 (15%)	4,500 (47%)	12,500 (71%)	22,000 (88%)
Skin (cool environment)	300 (6%)	1,500 (15%)	1,900 (12%)	600 (2%)
Other (lungs, bone, etc.)	650 (13%)	400 (4%)	400 (3%)	100 (1%)
Total		9,500	17,500	25,000

Modified from: Anderson, K. L. The cardiovascular system in exercise. In: Falls, H. B. *Exercise Physiology.* New York: Academic Press, 1968.

Oxygen Transport

In general, one liter of blood contains about 200 ml of O_2 when saturated. O_2 consumption at rest is about 250 ml per minute in both trained and untrained individuals, but about 1,000 ml per minute is potentially available if all the O_2 in the cardiac output is consumed. Thus, at rest, O_2 extraction is 25 percent and reserve is 75 percent.

O_2 is extracted more effectively during exercise in the trained individual than it is in an untrained subject. Sedentary individuals have a higher cardiac output during submaximal exercise than do trained athletes, whereas it is less in trained individuals under the same conditions. The greater O_2 requirement of exercising muscle is thus primarily supplied by an increased O_2 extraction during submaximal exercise. At rest, skeletal muscle consumes about 5 ml of O_2 per minute of the total 20 ml of O_2 that is available from the 100 ml of capillary blood supplied: $(a-v)O_2 = 5$ vol%. During submaximal exercise, the $(a-v)O_2$ increases to 15 vol%, both in sedentary individuals and in trained athletes. After physical training, the $(a-v)O_2$ increases by about 2 to 3 percent, causing about 80 percent of the O_2 to be extracted during submaximal exercise (Table 60-2). Because the capacity to increase $(a-v)O_2$ is limited, an increase in the stroke volume during maximal exercise, as well as increased density of muscle capillaries and mitochondria, contribute to the increased $\dot{V}O_2$max that occurs following physical training.

Table 60-2. Arteriovenous Oxygen Difference under Various Conditions

Condition	Arterial O_2 (ml/100 ml blood)	Venous O_2 (ml/100 ml blood)	Arteriovenous Oxygen Difference (%)
Rest	20	15	25
Submaximal exercise	20	5	75
Submaximal exercise (after physical training)	20	4	80

Pathophysiologic Limitations and Application

Exercise training has been found to reduce ischemia and improve ventricular function and aerobic capacity in men with stable coronary heart disease (CHD), though stroke volume and cardiac output are only slightly affected. However, the ability of skeletal muscle to receive and use O_2 can be improved, resulting in a greater $(a-v)O_2$ after training. Such patients can thus exercise at a given submaximal level with a lower cardiac output or perform higher workloads at a cardiac output similar to the pretraining one. Regular exercise thus reduces symptoms such as angina during submaximal efforts. Changes in heart function are modest in most cases, however. Marked increases in blood pressure occurring with heavy weight lifting should be avoided by those with CHD, due to the large oxygen demands placed on the myocardium.

Microcirculatory Changes

The microcirculation in muscle is also improved with physical training. Specifically, in endurance-trained muscle there is an increase in capillary density (number of capillaries per muscle fiber). Such an adaptation improves muscle fiber contact with the blood supply, which enhances the exchange of substrates and metabolic products. The $(a-v)O_2$ is also increased following training due to an increase in the metabolic capacity of muscle cells. For instance, mitochondrial size and number increase and there is increased oxidative enzyme activity. These changes improve aerobic ATP production without increasing lactate formation.

NEUROENDOCRINE RESPONSE DURING EXERCISE

The physiologic response necessary to maintain blood glucose homeostasis (approximately 90 mg/dl) during exercise is coordinated by two systems: the **autonomic nervous system** and the **endo-**

crine (hormonal) **system.** By means of chemical mediators, these systems coordinate the responses that maintain blood glucose concentration within the normal range during exercise. One example of the endocrine response to exercise is the decrease in insulin secretion as glucose levels fall. The declining plasma insulin concentration causes a decrease in the rate of glucose uptake while promoting fat utilization; these events contribute to the maintenance of normal blood glucose levels during exercise. The action of the autonomic nervous system includes sympathetic stimulation of cardiac contractility and frequency of contraction as well as the mobilization of fuels such as free fatty acids and glucose. At rest, the parasympathetic component of the autonomic nervous system permits fuel storage and reduces heart rate.

Control

Both the intensity of exercise and physical training affect the autonomic nervous system response to exercise. **Moderate exercise** has minimal effects on circulating blood catecholamine levels. As the intensity of exercise increases to 50 to 70 percent of $\dot{V}o_2$max, blood catecholamine levels rise markedly (Fig. 60-4). The release of catecholamines into the bloodstream during exercise is diminished following physical training because the exercise workloads impose less stress and cause diminished catecholamine release (Fig. 60-5).

Glucose and carbohydrate metabolism during exercise are affected greatly by epinephrine. By activating the beta receptors in muscle, **epinephrine** increases adenylate cyclase activity, intracellular free calcium levels, and, as a result, stimulates **glycogenolysis,** which is the breakdown of stored glycogen to form glucose. Glycogenolysis in the liver is also stimulated by catecholamines and to a greater extent by **glucagon,** and also increases with exercise and in response to decreasing blood glucose levels. However, during high-intensity exercise, the large increase in blood catecholamine levels suffices to stimulate hepatic glycogenolysis and elevate

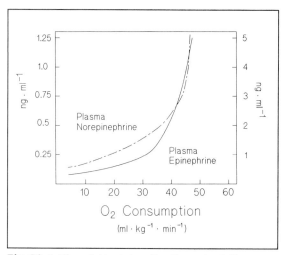

Fig. 60-4. The relative intensity of exercise influences the circulating levels of epinephrine and norepinephrine. Moderate exercise intensities produce essentially no increase in the catecholamine levels. Beyond 50 to 70 percent of maximum O_2 consumption, the catecholamine levels rise disproportionately. Maximum O_2 consumption occurs around 50 to 60 ml/kg/min. (Modified from Keul, V. J., Lehmann, L., and Wybitul, K. Zur wiung von burnitrolol auf hertz frequenz, metabolishe Grössen bei Korperarbeit und Leistungsverhalten. *Arzneim.-Gorsch/Drug Res.* 31:1, 1981.)

blood glucose concentrations. In addition, epinephrine indirectly affects glucose and glycogen metabolism by mobilizing free fatty acids from adipose tissue. At the onset of exercise, epinephrine released into the blood from the adrenal medulla quickly initiates **lipolysis** by stimulating hormone-sensitive lipase; this is followed by a slower, prolonged effect of growth hormone, which in turn maintains lipolysis. Other hormones that are important in fuel homeostasis during exercise are: **cortisol,** thyroid hormone, and antidiuretic hormone. Figure 60-6 illustrates how hormones mobilized during exercise influence the cellular energy systems such as glycolysis, lipolysis, and the Krebs' cycle. For example, as shown, a decrease in insulin and increase in glucagon bring about increased availability of glucose to the working muscle by means of glycogenolysis. This same combination of hormonal

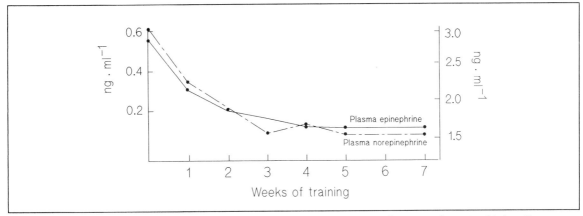

Fig. 60-5. Circulating catecholamine levels decrease in response to exercise of a given submaximal intensity, as a result of endurance training (Modified from: Winder, W. W. et al. Time course of sympathoadrenal adaptation to endurance exercise training in man. *J. Appl. Physiol.* 45:370, 1978.)

changes that occurs with exercise enhances lipolysis, thus making free fatty acids available as a fuel source for the working muscle.

Adaptation to Physical Training

Just as the neuroendocrine response is used to maintain glucose homeostasis during exercise, after physical training it also ensures "near normal" glucose levels. In general, following physical training, exercise of a given absolute or relative intensity imposes less metabolic stress and elicits a lower neuroendocrine response. Thus resting insulin levels are lower after physical training. It has been found that exercise enhances insulin sensitivity and may result in lower insulin requirements in insulin-dependent diabetics. An increase in cellular glucose transport following exercise has also been documented.

Pathologic Limitations

In disease states that affect the autonomic or central nervous systems the normal neuroendocrine response to exercise is limited. The autonomic response to exercise may be impaired in **autonomic neuropathy,** one of the complications associated with diabetes. Diabetic individuals with autonomic neuropathy may have an altered nor-

epinephrine and epinephrine response. This may be attributed to a reduced catecholamine release and possibly to lower clearance. Although the catecholamine response is blunted, in patients with diabetes who are in poor glucose control, the adrenergic response to exercise may be elevated. Whether physical training alters these abnormal responses to exercise has not been studied thoroughly.

SUMMARY

The physiologic response to exercise consists of a more or less efficient attempt to adjust to a new, high steady-state, even during heavy exercise. Thus, there are circulatory and respiratory adjustments, with neural and hormonal imputs acting as stimulating mechanisms to meet the added fuel demand of working muscle. While providing for the increased metabolic needs of the working muscle, the systems exquisitely sustain the basal requirements of the nonexercising tissues and organs for as long as possible. The efficiency of the homeostatic response to exercise is generally improved by physical training, in that the circulatory, neuroendocrine, and, to a lesser extent, respiratory systems adapt during training. Pathophysiologic limitations may alter the

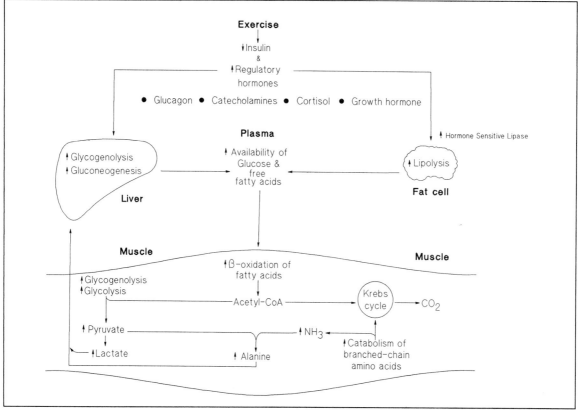

Fig. 60-6. The hormones mobilized during exercise and their influence on fuel supply to the muscles by means of several metabolic pathways within the muscle tissue, including glycolysis, lipolysis, and the Krebs' cycle. (*Acetyl-CoA* = acetyl coenzyme A.) (Modified from: Frontera and Adams, Endurance exercise: normal physiology and limitations imposed by pathological processes [Part 2]. *The Physician and Sports Medicine.* 14(8): 111, 1986.)

response to exercise and need to be carefully considered when exercise is recommended.

NATIONAL BOARD TYPE QUESTIONS

Select the one most correct answer for each of the following

1. Which physiologic system, or systems, show very little adaptation to physical training?
   A. Circulatory
   B. Respiratory
   C. Neuroendocrine
   D. Respiratory and neuroendocrine
   E. Circulatory and respiratory

2. After physical training, the (a–v)O$_2$ at submaximal exercise:
   A. Increases from pretraining values
   B. Is the same as at rest
   C. Is unchanged
   D. Decreases from pretraining values
   E. Is the same as at maximal exercise

3. The increase in O$_2$ consumption by muscle during maximal exercise after physical training is primarily supplied by:

A. Cardiac output and oxygen extraction
B. Oxygen extraction
C. Stroke volume
D. Cardiac output
E. None of the above

4. During weight lifting, heart rate increases with:
A. The percentage of $\dot{V}O_2$max
B. The amount of muscle mass involved
C. The duration of the lift
D. Breath holding
E. Strength development

5. The distribution of cardiac output during exercise goes mainly to the:
A. Brain
B. Kidneys
C. Skeletal muscle
D. Skin
E. Cardiac muscle

ANNOTATED ANSWERS

1. B. All three systems show adaptations; however, the respiratory system changes very little after physical training.

2. A. As shown in Table 60-2, the $(a-v)O_2$ rises after physical training. Because of greater O_2 extraction by the muscle, the venous O_2 level is lower after physical training.

3. A. The capacity to increase $(a-v)O_2$ is limited; therefore, cardiac output during maximal exercise is the primary contributing factor to the increased $\dot{V}O_2$max following physical training.

4. B. Heart rate during weight lifting increases with the amount of muscle mass involved and the percentage of maximal voluntary contraction.

5. C. During light, moderate, and maximum exercise, the cardiac output to the skeletal muscle is 47 percent, 71 percent, and 88 percent, respectively, of the total cardiac output. Any single tissue or organ receives no more than 15 percent, depending on the intensity of exercise.

BIBLIOGRAPHY

Brooks, G. A., and Fahey, T. D. *Exercise Physiology: Human Bioenergetics and Its Applications.* New York: John Wiley and Sons, 1984.

Dempsey, J. A., Aaron, E., and Martin, B. J. Pulmonary function and prolonged exercise. In: Lamb, D. R., and Murray, R., eds. *Perspectives in Exercise Science and Sports Medicine, Volume 1, Prolonged Exercise.* Indianapolis: Benchmark Press, 1988, Pp. 75–119.

Guyton, A. C. *Textbook of Medical Physiology,* 5th ed. Philadelphia: W. B. Saunders, 1976.

McArdle, W. D., Katch, F. L., and Katch, V. L. *Exercise Physiology: Energy, Nutrition and Human Performance,* 2nd ed. Philadelphia: Lea & Febiger, 1986.

Saltin, B, and Karlsson, J. Muscle glycogen utilization during work of different intensities. In: Pernow, B. and Saltin, B., eds: *Muscle Metabolism during Exercise.* New York: Plenum Publishing, 1971, Pp. 289–300.

Appendixes

Appendix I

Normal Values of Electrolytes, Metabolic Variables, and Hormones in Whole Blood, Plasma, or Serum

Constituent	Traditional Units	SI Units
Acetoacetate plus acetone (S)	0.3–2.0 mg/dl	3–20 mg/liter
Aldosterone (S)	3–20 ng/dl	83–277 pmol/liter
Ammonia (B)	12–55 μmol/liter	12–55 μmol/liter
Amylase (S)	4–25 units/ml	
Bilirubin (S)	Conjugated (direct) up to 0.4 mg/dl	up to 7 μmol/liter
Calcium (S)	8.5–10.5 mg/dl; 4.3–5.3 mEq/liter	2.1–2.6 mmol/liter
Carbon dioxide content (S)	24–30 mEq/liter	24–30 mmol/liter
Carotenoids (S)	0.8–4.0 μg/ml	1.5–7.4 μmol/liter
Chloride (S)	100–106 mEq/liter	100–106 mmol/liter
Cholesterol (S)	120–220 mg/dl	3.1–5.7 mmol/liter
Cortisol (S)	5–25 μg/dl	0.14–0.69 μmol/liter
Creatinine (S)	0.6–1.5 mg/dl	53–133 μmol/liter
Estradiol (P)	Women: basal, 20–600 pg/ml Ovulatory surge: >200 pg/ml Men: <50 pg/ml	74–221 pmol/liter >735 pmol/liter <184 pmol/liter
Glucose (B)	70–110 mg/dl	3.9–5.6 mmol/liter
Insulin (fasting) (P)	5–15 μU/ml	22–67 pmol/liter
Iron (S)	50–150 μg/dl	9.0–26.0 μmol/liter
Lactic acid (B)	0.6–1.8 mEq/liter	0.6–1.8 mEq/liter
Lipase (S)	Up to 2 U/ml	4.5–10 gm/liter

Constituent	Traditional Units	SI Units
Magnesium (S)	1.5–2.0 mEq/liter	0.8–1.3 mmol/liter
Osmolality (S)	280–296 mOsm/kg H_2O	
P_{CO_2} (arterial) (B)	35–45 mm Hg	
pH (arterial)	7.35–7.45	45–35 nmol/liter
Phosphatase, alkaline (S)	13–39 IU/liter (adults)	2.9–5.2 mmol/liter
Phosphorus, inorganic (S)	3.0–4.5 mg/dl	
P_{O_2} (arterial) (B)	75–100 mm Hg	
Potassium (P)	3.5–5.5 mEq/liter	
Progesterone (P)	Men and preovulatory and post-menopausal women: <2ng/ml	<6 nmol/liter
	Women, luteal peak: >5 ng/ml	>16 nmol/liter
Protein		
Total (S)	6.0–8.4 gm/dl	60–80 gm/liter
Albumin (S)	3.5–5.0 gm/dl	35–50 gm/liter
Globulin (S)	2.3–3.5 gm/dl	23–35 gm/liter
Sodium (S)	135–145 mEq/liter	135–145 mmol/liter
Testosterone (P)	Men: 300–1000 ng/dl	10–35 nmol/liter
	Women: <80 ng/dl	<2.8 nmol/liter
Thyroid-stimulating hormone (P)	0.5–4 μU/ml	
Thyroxine (P)	5–12 μg/dl	64–154 nmol/liter
Triiodothyronine (P)	70–190 ng/dl	1.1–2.9 nmol/liter
Transaminase (SGOT) (S)	7–24 U/liter	0.12–0.45 μmol/sec/liter
Urea nitrogen (BUN) (B)	8–25 mg/dl	2.9–8.9 mmol/liter
Uric acid (S)	3.0–7.0 mg/dl	0.18–0.42 mmol/liter

B = whole blood; P = plasma; S = serum.
Modified from: Scully, R. E. Case records of the Massachusetts General Hospital. *N. Engl. J. Med.* 314:39–49, 1986.

Appendix II

MEASUREMENT OF CARDIAC OUTPUT

Cardiac output is a clinically important measure that integrates ventricular performance with the metabolic needs of tissues. Although there are a number of techniques for measuring cardiac output, methods based on the Fick principle and indicator-dilution techniques are generally employed in the clinical setting.

The **Fick principle** is based on the law of conservation of mass, and states that the quantity of a substance produced or consumed by an organ is equal to the product of blood flow to that organ and the difference in arterial and venous concentration of that substance. Therefore, flow is equal to the amount of substance produced or consumed divided by the arteriovenous difference of the substance.

In clinical practice, oxygen is used as the substance consumed (VO_2); arterial and mixed venous (pulmonary arterial) blood is sampled and the arteriovenous oxygen difference ($A - VO_2$) is computed. Thus, pulmonary blood flow (Q) is calculated, which, in the absence of an intracardiac shunt, equals systemic blood flow. Lung oxygen uptake is assumed to equal tissue oxygen consumption. Thus, $Q = VO_2/A - VO_2$. This method requires right-heart catheterization and careful spirometry, as well as steady-state conditions during the measurement process.

An **indicator-dilution curve,** which plots the concentration of indicator against time, can also be used to determine cardiac output. The indicator-dilution method is actually an application of the Fick principle. The indicator substance can be introduced either by continuous infusion or by a single injection. A given volume of indicator (for example, indocyanine green) is injected into a peripheral or central vein and arterial blood is withdrawn at a constant rate for analysis. A mixing (cardiac) chamber is necessary, and any recirculation of the indicator needs to be corrected for. Before the indicator recirculates, the decline of indicator over time is exponential (Fig. A-1, top). The curve is corrected for recirculation by extrapolation of the exponential portion of the curve (Fig. A-1, bottom). The area under the corrected indicator-dilution curve divided by the duration of the dye curve represents the mean concentration of the indicator. After correcting for a one-minute interval, one gets the cardiac output:

$$\text{cardiac output} = \frac{60\,I}{c\,t}$$

Where I is the amount of indicator added; c is the mean concentration of indicator; and t stands for time. Indicator-dilution methods have the advantage of only minimally interfering with the circulation. However, they are inconvenient and their use in measuring cardiac output in humans has largely been replaced by the thermodilution method.

Thermodilution cardiac output methods use temperature as an indicator. Cold saline solution is injected into the central veins; it then mixes in the right ventricle, and the change in blood temperature is detected by a sensor located in the pulmonary artery. The cardiac output is computed from the time-dependent temperature change. Cardiac output may be overestimated in low-flow states, because of loss of indicator ability as the injectate warms during slow passage from the injection to the measuring sites. Although blood withdrawal and arterial puncture are unnecessary for this method, right-heart catheterization is required.

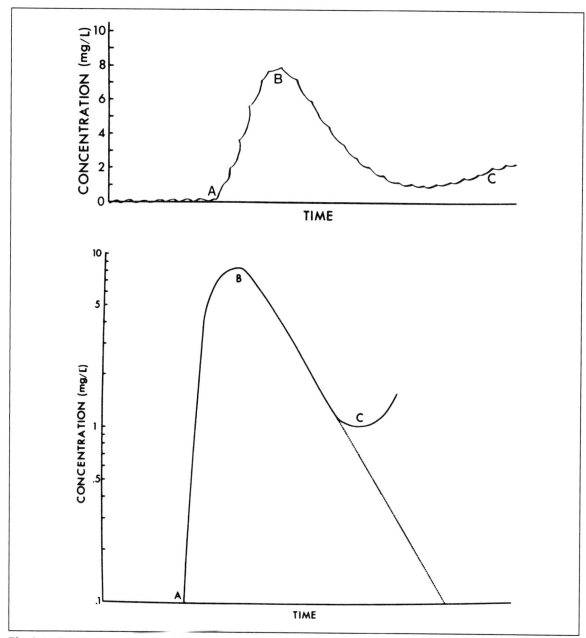

Fig. A-1. (Top) Indicator-dilution (time-concentration) curve. One milliliter of indocyanine green is injected into the pulmonary artery at time zero. Blood is withdrawn continuously through a densitometric curvette to derive an instantaneous concentration of dye. The dye first appears at *A*, peaks at *B*, and recirculates from *C*. (Bottom) The time-concentration curve is replotted on semilogarithmic paper. The exponential decline (*B–C*) is a straight line. Extrapolation of this line to baseline corrects for recirculation. (From: Grossman W., ed. *Cardiac Catheterization and Angiography*. Philadelphia: Lea & Febiger, 1986.)

Index

Index

The abbreviations *f* and *t* stand for figure and table, respectively.